Human Growth

A COMPREHENSIVE TREATISE

Second Edition

Volume 2

Postnatal Growth
Neurobiology

Human Growth

A COMPREHENSIVE TREATISE

Second Edition

Edited by Frank Falkner and J. M. Tanner

Volume 1	Developmental Biology Prenatal Growth
Volume 2	Postnatal Growth Neurobiology
Volume 3	Methodology Ecological, Genetic, and Nutritional Effects on Growth

Human Growth

A COMPREHENSIVE TREATISE

Second Edition

Volume 2
Postnatal Growth
Neurobiology

Edited by

FRANK FALKNER

School of Public Health
University of California at Berkeley
Berkeley, California
and School of Medicine
University of California at San Francisco
San Francisco, California

and

J. M. TANNER

Institute of Child Health
University of London
London, England

SPRINGER SCIENCE+BUSINESS MEDIA, LLC

Library of Congress Cataloging in Publication Data

Main entry under title:

Human growth.

Includes bibliographies and index.
Contents: v. 1. Developmental biology. Prenatal growth—v. 2. Postnatal growth. Neurobiology—v. 3. Methodology. Ecological, genetic, and nutritional effects on growth.
1. Human growth—Collected works. I. Falkner, Frank T., 1918- . II. Tanner, J. M. (James Mourilyan) [DNLM: 1. Growth. WS 103 H918]
QP84.H76 1985 612.6 85-19397

ISBN 978-1-4899-0524-6 ISBN 978-1-4899-0522-2 (eBook)
DOI 10.1007/978-1-4899-0522-2

Originally published by Plenum Press, New York in 1986
Softcover reprint of the hardcover 2nd edition 1986

Contributors

DONALD A. BAILEY • College of Physical Education, University of Saskatchewan, Saskatoon, Saskatchewan S7N 0W0 Canada

INGEBORG BRANDT • University Children's Hospital, University of Bonn, 5300 Bonn 1, Federal Republic of Germany

T. BERRY BRAZELTON • Child Development Unit, Children's Hospital Medical Center, Harvard Medical School, Boston, Massachusetts 02115

WILLIAM F. CROWLEY, JR. • Reproductive Endocrine Unit, Massachusetts General Hospital and Harvard Medical School, Boston, Massachusetts 02114

ARTO DEMIRJIAN • Université de Montréal, Montréal, Quebec, Canada H3C 3J7

IRVING M. FAUST • The Rockefeller University, New York, New York 10021

GILBERT B. FORBES • School of Medicine and Dentistry, The University of Rochester, Rochester, New York 14642

CHRISTINE GALL • Department of Anatomy, California College of Medicine, University of California, Irvine, California 92717

BEN GREENSTEIN • Department of Pharmacology, St. Thomas's Hospital Medical School, University of London, London SE1 7EH, England

MALCOLM A. HOLLIDAY • Department of Pediatrics, University of California, San Francisco Medical Center, San Francisco, California 94143

GWEN IVY • Center for the Neurobiology of Learning and Memory, University of California, Irvine, California 92717

FRANCIS E. JOHNSTON • Department of Anthropology, University Museum F1, University of Pennsylvania, Philadelphia, Pennsylvania 19104

GARY LYNCH • Center for the Neurobiology of Learning and Memory, University of California, Irvine, California 92717

PAMELA C. B. MACKINNON • Department of Human Anatomy, University of Oxford, Oxford, England

ROBERT M. MALINA • Department of Anthropology, The University of Texas, Austin, Texas 78712

WILLIAM A. MARSHALL • Department of Human Sciences, University of Technology, Loughborough, Leicestershire LE11 3TU, England

ROBERT L. MIRWALD • College of Physical Education, University of Saskatchewan, Saskatoon, Saskatchewan S7N 0W0 Canada

MICHAEL A. PREECE • Department of Growth and Development, Institute of Child Health, University of London, London WC1N 1EH, England

T. RABINOWICZ • Division of Neuropathology, University Medical Center, Geneva 1205, Switzerland

ALEX F. ROCHE • Division of Human Biology, Department of Pediatrics, Wright State University School of Medicine, Yellow Springs, Ohio 45387

PATRICK G. SULLIVAN • Department of Child Dental Health, The University of Sheffield, Sheffield S10 2SZ, England

JAMES M. TANNER • Department of Growth and Development, Institute of Child Health, University of London, London WC1N 1EH, England

COLWYN B. TREVARTHEN • Department of Psychology, University of Edinburgh, Edinburgh, Scotland EH8 9JZ

MARGARET E. WIERMAN • Reproductive Endocrinology Unit, Massachusetts General Hospital and Harvard Medical School, Boston, Massachusetts 02114

Preface to the Second Edition

It is a source of great satisfaction to us that a Second Edition of this treatise should be called for, especially because it has given us the opportunity to produce, we believe, a better book. Eighteen chapters, amounting to one-third of the whole, are new, and of these, 13 deal with subjects not covered at all in the First Edition. We have paid more attention to embryonic and fetal growth, with chapters on cell differentiation (Lehtonen and Saxén), embryonic growth (O'Rahilly and Müller), control of fetal size (Snow), regulation of fetal growth (D'Ercole and Underwood), and ultrasonic studies of fetal growth (Meire). At last the data are available for a chapter on the evolution of the human growth curve, by Elizabeth Watts. Large parts of the endocrine section have been rewritten (by Michael Preece, and by William Crowley and Margaret Wierman), and the genetics section has been largely recast, with new contributions by William Mueller and Ronald Wilson. Reynaldo Martorell has contributed a new chapter on growth in developing countries, and Tanner discusses growth surveys and standards as well as catch-up growth. Finally, there are two new chapters dealing with growth as a monitor of the health of populations—one by Tadeusz Bielicki, considering the contemporary scene, and the other by Robert Fogel, on the contribution that such studies are making to the economic history of the eighteenth and nineteenth centuries.

The appearance of Tanner's *A History of the Study of Human Growth* (Cambridge University Press, 1981) meant that we could drop the long historical chapter in Volume 3 of the First Edition, and that has provided space enough so that these new departures could all be fitted in without (or nearly without) increasing the total size of the book.

Our First Edition contributors have been meticulous in updating their chapters. As in the First Edition, our only regret is that one promised new Second Edition chapter—on the important subject of neuronal migration in brain development—failed to arrive.

To our considerable wonderment, our secretaries proved capable of working even harder than before, for this time all three volumes have been produced simultaneously. We, and more particularly our readers, owe the greatest debt to Karen Phelps at Berkeley, and to Jan Baines Preece and Susan Barrett in London. Without their gentle aid, neither we nor scarcely any of our contributors would have met the publisher's deadline. That they did all meet it is a marvel for which we give them our heartfelt thanks.

Frank Falkner
James Tanner

Berkeley and London

Preface to the First Edition

Growth, as we conceive it, is the study of change in an organism not yet mature. Differential growth creates form: external form through growth rates which vary from one part of the body to another and one tissue to another; and internal form through the series of time-entrained events which build up in each cell the specialized complexity of its particular function. We make no distinction, then, between growth and development, and if we have not included accounts of differentiation it is simply because we had to draw a quite arbitrary line somewhere.

It is only rather recently that those involved in pediatrics and child health have come to realize that growth is the basic science peculiar to their art. It is a science which uses and incorporates the traditional disciplines of anatomy, physiology, biophysics, biochemistry, and biology. It is indeed a part of biology, and the study of human growth is a part of the curriculum of the rejuvenated science of Human Biology. What growth is not is a series of charts of height and weight. Growth standards are useful and necessary, and their construction is by no means void of intellectual challenge. They are a basic instrument in pediatric epidemiology. But they do not appear in this book, any more than clinical accounts of growth disorders.

This appears to be the first large handbook—in three volumes—devoted to Human Growth. Smaller textbooks on the subject began to appear in the late nineteenth century, some written by pediatricians and some by anthropologists. There have been magnificent mavericks like D'Arcy Thompson's *Growth and Form.* In the last five years, indeed, more texts on growth and its disorders have appeared than in all the preceeding fifty (or five hundred). But our treatise sets out to cover the subject with greater breadth than earlier works.

We have refrained from dictating too closely the form of the contributions; some contributors have discussed important general issues in relatively short chapters (for example, Richard Goss, our opener, and Michael Healy); others have provided comprehensive and authoritative surveys of the current state of their fields of work (for exmaple, Robert Balázs and his co-authors). Most contributions deal with the human, but where important advances are being made although data from the human are still lacking, we have included some basic experimental work on animals.

Inevitably, there are gaps in our coverage, reflecting our private scotomata, doubtless, and sometimes our judgment that no suitable contributor in a particular field existed, or could be persuaded to write for us (the latter only in a couple of instances, however). Two chapters died on the hoof, as it were. Every reader will notice the lack of a chapter on ultrasonic studies of the growth of the fetus; the manuscript, repeatedly promised, simply failed to arrive. We had hoped, also, to include a chapter on the very rapidly evolving field of the development of the visual processes, but here also events conspired against us. We hope to repair these omissions in a second edition if one should be called for; and we solicit correspondence, too, on suggestions for other subjects.

We hope the book will be useful to pediatricians, human biologists, and all concerned with child health, and to biometrists, physiologists, and biochemists working in the field of growth. We thank heartily the contributors for their labors and their collective, and remarkable, good temper in the face of often bluntish editorial comment. No words of praise suffice for our secretaries, on whom very much of the burden has fallen. Karen Phelps, at Fels, handled all the administrative arrangements regarding what increasingly seemed like innumerable manuscripts and rumors of manuscripts, retyped huge chunks of text, and maintained an unruffled and humorous calm through the whole three years. Jan Baines, at the Institute of Child Health, somehow found time to keep track of the interactions of editors and manuscripts, and applied a gentle but insistent persuasion when any pair seemed inclined to go their separate ways. We wish to thank also the publishers for being so uniformly helpful, and above all the contributors for the time and care they have given to making this book.

Frank Falkner
James Tanner

Yellow Springs and London

Contents

I Postnatal Growth

Contents of Volumes 1 and 3

Volume 1

I Developmental Biology

Volume 3

I Methodology

II Ecological, Genetic, and Nutritional Effects on Growth

I

Postnatal Growth

1

Somatic Growth of the Infant and Preschool Child

FRANCIS E. JOHNSTON

1. Infant and Preschool Years

The infant and preschool-age years, from birth through 4 years of age, are crucial in the life of the individual. It is during this period that the neonate makes the adjustment from a sharply limited intrauterine environment, in which all stimuli are mediated through the maternal biological system, to the rich, varied, and often hostile, extrauterine world, in which the regulalory responses are direct functions of interactions between this new environment and the infant's or child's biological system.

The first 5 postnatal years are therefore years of change, compensation, and initiation, in which there is: (1) Adjustment to the striking environmental changes experienced; (2) Compensation for the new levels of stress which accompany the changes; and (3) Initiation of developmental and physiological regulation based upon internal mechanisms. It is no wonder that the preschool years are those which, apart from senescence, display the highest mortality rates; as many as 50% of all babies born will die by 5 years of age (Jelliffe, 1968). When viewed in the light of the very high growth rates that characterize this period, one accepts as logical the findings of many investigators that a significant percentage of the biological and behavioral variability encountered in later years has its genesis in the preschool period (see, e.g., Heald and Hollander, 1965; Huenemann, 1974; Mack and Johnston, 1976).

FRANCIS E. JOHNSTON • Department of Anthropology, University Museum F1, University of Pennsylvania, Philadelphia, Pennsylvania 19104.

1.1. Transition from Prenatal to Postnatal Life

Timiras (1972) summarized the major contrasts between the prenatal and postnatal environments (Table I). The contrasts involve the physical environment, physiological process, nutrient sources, and communication of sensory stimuli. During the prenatal period, when the transition from one environment to the other is effected, the infant undergoes systematic physiological changes that are partly due to the stresses of labor and delivery. The umbilical artery O_2 pressure may drop to zero, associated with profound acidosis (James, 1959) of pH 6.8 or lower. Changes in other systems and organs are also apparent.

The newborn infant must successfully make the transition from a regulatory system that was largely dependent on characteristics of the maternal organism to one based on his or her own genetic and homeostatic mechanisms. The continuing process of development, which has gone on since the initial cell divisions, becomes, during the preschool years, canalized, i.e., regularized and predictive of future events. Body size during the first year of life is poorly correlated with the adult size of the individual, but by 2 years of age self-correlations of the size of an individual with his or her adult height have stabilized at a value of about 0.8 in well-nourished populations (Figure 1).

Despite the appearance of self-regulatory systems, infants and young children are still heavily

Table I. Contrast between Prenatal and Postnatal Environments[a]

	Prenatal	Postnatal
Physical environment	Fluid	Gaseous
External temperature	Generally constant	Fluctuating
Sensory stimulation	Primarily kinesthetic or vibratory	Variety of stimuli
Nutrition	Dependent on nutrients in mother's blood	Dependent on food availability and digestive sufficiency
Oxygen supply	Passed from mother to fetus at placenta	Passed from lung surfaces to pulmonary vessels
Elimination of metabolic products	Discharged into mother's bloodstream	Elimination by lungs, skin, kidneys, and gastrointestinal tract

[a]Adapted from Timiras (1972).

dependent on older members of society. At birth, the human infant is relatively immature; the first few years of life are a time of intense maturation and socialization, characterized by rapid growth of biological structure, especially the brain (LaBarre, 1954; Meisami, 1975; McKenna, 1979).

During this period, the process of growth as an intrinsic property of the organism continues. Changes in the body's cell mass and its supporting components are evident from the earliest embryonic stages and constitute the broader set of dimensional changes known as *growth.* During the postnatal years, rates of growth are most rapid during infancy and childhood, at which time the child is especially sensitive to the surrounding environment. The growth status of an individual during these years is a widely used measure of the quality of this environment and, by extrapolation, of the environment of the total population.

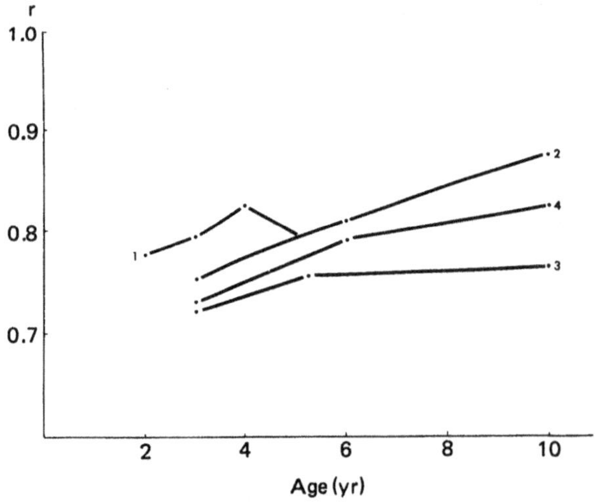

Fig. 1. Self-correlations of stature at 2–10 years of age with mature stature in four samples: 1, Aberdeen, sexes combined (Tanner *et al.,* 1956); 2, Berkeley boys; 3, Berkeley girls; 4, Berkeley, sexes combined (Tuddenham and Synder, 1954).

2. Description of Somatic Growth

Although there is no single definition of growth, virtually everyone defines it in quantitative terms. That is, rather than an elaboration of function, differentiation of tissues, or laying down of metabolic pathways, growth is defined as the increase or decrease of some measurable quantity of tissue. To the cellular biologist, growth may be thought of as either increase in the size of cells (hypertrophy) or increase in their number (hyperplasia) and is a part of an ongoing process throughout the life of the organism (Goss, 1964). To the auxologist, whose interests are focused on the quantification of these changes, and especially at the organismic level, growth is visualized as an increase in body mass in measurable dimensions. The cellular and the auxological are, of course, aspects of a single process. Since cellular dynamics involve not only hypertrophy and hyperplasia, but cellular destruction as well, growth may be negative when the decrease exceeds the increase, or it may be at equilibrium, when the decrease and increase in growth are equal within a population of cells (Goss, 1964).

Somatic growth is "described," i.e., measured, by the technique of anthropometry, a methodology originally developed by physical anthropologists, but now used and refined as well, by a number of disciplines (see Cameron, Volume 3, Chapter 1). The number of measurements that can be made on the body are almost limitless, and the selection of measurements must be dictated by the purpose of the study. The design engineer, interested in an adequate workspace, a safe piece of equipment, or clothing that fits, will be interested in one class of dimensions (e.g., Hertzberg *et al.*, 1963; Clauser *et al.*, 1972). The human biologist, interested in cause and effect and in relating physiological and ecological factors, will be interested in another class (Weiner and Lourie, 1969), and the health professional, evaluating health as a function of growth, yet another (Falkner, 1966; Jelliffe and Jelliffe, 1979). Some measurements will be common to virtually any study, others unique to only one; however, each investigator is interested in a set of dimensions that change with age and that interact with the individual's environment.

As a convenience, somatic growth, measured anthropometrically, may be visualized in terms of a few "clusters" of related measurements. These clusters relate biologically similar aspects of growth and permit a more rational selection of dimensions.

Various measurements are included within this category, all of which measure growth along some linear axis. Most common are dimensions along the longitudinal axis of the body, i.e., along the trunk and lower extremity, along the shafts of the long bones, and so forth. These dimensions are statistically related (Lindegaard, 1953) and, for practical purposes, are often reduced to relatively few. In general, these dimensions reflect skeletal growth at the metaphyses of endochondrally formed bones. The most commonly taken, of course, is stature. However, a number of other dimensions may be recorded as well (e.g., the sitting height), which, when subtracted from stature, yields an estimate of leg length. Other longitudinal measures include the length of individual bones, measured directly or from radiographs.

Size may also be measured in terms of body breadth, i.e., general measurements made in a plane perpendicular to its long axis. Most frequent are the biacromial breadth of the shoulder and the bicristal breadth of the hip, both of which reflect complex internal changes in the architecture of the shoulder and hip regions, including associated musculature and adipose tissue. Included are measurements of the breadths of the extremities of certain bones, the biepicondylar breadth of the humerus and the bicondylar breadth of the femur, as well as the breadths of the wrist and ankle joints, taken across the distal radius/ulna and tibia/fibula. These measurements are used in estimating the breadth factor of the skeleton or of the specific bones involved.

A somewhat different set of size measurements deals with the head and face. These measurements are usually taken for specific purposes and may involve detailed measurements of the face, the cranial base, or the brain case. Head circumference is one such measurement widely used in the evaluation of the health and nutritional status of infants and young children.

2.2. Measures of Mass and Composition

The basic measurement of body mass is weight. Together with stature, this composite dimension comprises the most widely taken pair of growth measurements. However, since it is such a composite, body weight is frequently divided into its components or into smaller units. This may be done in a laboratory, where the measurement of the weights of body segments (e.g., the arm) can be accomplished with greater precision and accuracy. The measurement of body composition; e.g., the determination of fat or lean body mass, may also be accomplished by the techniques of densitometry, whole-body counting (minerals), or fluid-space measurement.

However, in many cases, it is both necessary and desirable to use somatic measurements in an attempt to estimate body composition. This is true in studying large samples away from a laboratory, as well as when working with young age groups not suited for many of these techniques.

In addition to the above dimensions, somatic measurements of body composition include the taking of skinfolds and body circumferences. Most investigators use the data directly, and, as necessary, with appropriate statistical transformations to produce a gaussian distribution. Other investigators prefer to use regression formulas to predict whole-body composition from combinations of variables, but there are many sources of error in such a procedure (Johnston, 1982).

2.3. Measures of Shape and Proportion

By and large, humans grow proportionately, and the interrelationships between various measurements provide additional information of considerable value. The ratio of sitting height to stature can assist in diagnosing particular types of growth failure, as can the ratio of fat to lean body mass, arm span to stature, biacromial to bicristal breadth, and the like.

Measurements may be related to each other by means of an index or by the use of appropriate regression techniques:

1. Weight-for-height indices, of particular value in the study of obesity:
   a. Weight/height^2
   b. Weight/height$^{1.75}$
   c. $\sqrt[3]{\text{weight}}$/height
2. Calculation of relative weight:
   100 × (weight/standard weight)
3. Referring to various percentile tables.

These alternatives are discussed by Newens and Goldstein (1972) and by Billewicz *et al.* (1962).

The measurements that have been discussed constitute the main approaches to the evaluation of somatic growth in infants and preschool-age children. Although other measurements, involving the body as a whole, in terms of "constitution" (Carter, 1972) or using sophisticated noncartesian methods (Herron, 1972) are available, they are almost always applied to older age groups.

Table II. Coefficients of Variation of Body Length and Weight at Selected Ages among Denver Children[a]

Length (%)		Age	Weight (%)	
M	F	(years)	M	F
4.5	3.5	Birth	15.3	13.7
3.0	3.8	5	9.1	10.7
3.3	4.6	10	13.7	17.5
4.0	3.6	15	15.2	14.4

[a]Calculated from means and standard deviations presented by Hansman (1970).

3. Growth during Preschool-age Years

3.1. Size at Birth

Body size and composition at birth reflect fetal growth and need not concern us here. Even at this time, however, there is significant variability both within and between populations. For example, birth weights vary widely and systematically; on a worldwide basis (Meredith, 1970), there is an impressive range, and one may identify between-population factors (Adams and Niswander, 1973) that probably reflect both hereditary and ecological mechanisms.

The amount of within-sample variation at birth is also impressive. Table II presents the coefficients of variation (100 $\bar{x}/s$) for weight and length at selected ages, from the mixed longitudinal data of the Denver Child Research Council (Hansman, 1970). The relative variability at birth is similar to that seen at 5, 10, and 15 years of age.

Body composition at birth differs by sample as well. Anthropometric measurements of newborn infants of black, white, and Puerto Rican ancestry (Johnston and Beller, 1976) indicate differences in skinfold thickness and estimated arm-muscle mass by sex and socioeconomic status.

Thus, postnatal growth begins with significant variation among individuals and groups in virtually any measurement we may make. Since growth is cumulative over time, any attempt at explaining differences in the first 5 years of life must begin with the realization of the considerable variation already present at the beginning of this span.

3.2. Infancy

The first year of life is characterized by extremely rapid growth. Table III presents the mean increments of length and weight from birth to 1 year of a longitudinal sample of middle-class Finnish infants. In this example, the result is an increase of 50–55% in length and 180–200% in weight.

Postnatal growth velocities are highest, by far, immediately after birth, falling off sharply by 12 months. In the Finnish study, the length velocities fell from 45.8 cm/year between birth and 3 months of age to 14.4 cm/year in the 9 to 12-month interval

Table III. Mean Increments of Length and Weight in Finnish Children from Birth to 1 Year, Mixed Longitudinal Data[a]

	Length (cm)	Weight (kg)
Boys	28.3	7.22
Girls	25.9	6.34

[a]From Kantero and Tiisala (1971*a*).

in boys, with weight velocities dropping from 10.3 to 3.4 kg/year. The velocities in girls were 14% and 13% less for length and weight from birth to 3 months, but in the final 9 to 12-month interval, no sex difference was apparent. These are similar to the data presented by Tanner *et al.* (1966*a,b*), who reported instantaneous velocities at 0.16 year of 40.0 cm/year in boys and 36.0 cm/year in girls, dropping to 14.5 cm/year and 15.9 cm/year respectively, at 0.87 year.

Information on the growth of other body dimensions, studied longitudinally between birth and 1 year, is much less common. Some of the most complete data come from the Denver Child Research Council (Hansman, 1970). From the means at birth and 12 months of their mixed longitudinal sample, we may calculate the percentage increase of various anthropometric dimensions over the first year of life. The increase in length, hip (biiliac) breadth, and shoulder (biacromial) breadth averaged about 55% of the birth value in boys and about 54% in girls. Crown–rump lengths (the equivalent of sitting height) showed an increase of 45% in boys and 42% in girls, while for head circumference the values for boys and girls were 39% and 35%, respectively.

Additional data on other measurements during infancy are available (although cross-sectional), from the measurements of 4027 infants and children; in this study conducted by Snyder *et al.* (1975), the sample numbers vary by measurement, but 344 of the subjects were under 12 months of age. The sampling was conducted so as to provide a representative cross section, ethnically and socioeconomically, from 76 locations in eight states of the United States. These investigators present descriptive statistics for 30 linear dimensions and nine circumferences grouped, during infancy, into four 3-month periods. Many of the dimensions are of design-engineering significance, and this study is one of the most comprehensive for infancy in terms of both the range of measurements and the attention given to measurement techniques and equipment design.

There is but little information on skinfold thickness at birth, and certainly not enough to furnish comprehensive data on percentile distributions. Table IV presents the mean thicknesses of the triceps and subscapular folds in neonates, as reported in various studies; virtually all infants were measured within the first 3 days after birth. The range of variation among the means is quite striking: the differences between the low and high values range from 20 to 50% of the low. The roles of prenatal growth, gestation, and other factors are clearly of great significance.

After birth the skinfold thicknesses at the triceps and subscapular sites rise markedly to a peak, before starting to decline during the subsequent years. The revised standards suggested by Tanner and Whitehouse (1975) place this peak at 6 months for the subscapular, at 6 months for the triceps in boys, and at 1–1½ years for the triceps in girls. These peaks for the triceps are valid only for the 50th-percentile child, since they occur at other ages, earlier and later, for children on other percentiles. On the other hand, among mildly to moderately malnourished Guatemalan infants (Malina *et al.*, 1974), the peaks occurred in boys and girls, for both sites, at 3 months of age.

Finally, the difficulty in characterizing infant growth curves of skinfolds is indicated by the data of Fry *et al.* (1975). In contrast to the British or Guatemalan children, who have a clear peak in skinfold thickness, there is only a slight dip in the means (sexes combined) of triceps and subscapular folds of black, white, and Mexican-American infants. This occurred in the 4 to 6-month group among blacks and in the 6 to 8-month group among

Table IV. Skinfold Thickness in Newborn Infants

Country	Number	Triceps $\bar{x}$	Triceps SD	Subscapular $\bar{x}$	Subscapular SD
United States					
Whites[a]					
Males	32	4.0	1.1	3.8	1.2
Females	32	4.2	1.1	3.8	1.2
Blacks[a]					
Males	33	3.9	1.1	3.6	1.0
Females	34	4.3	1.1	4.0	1.1
Puerto Rican[a]					
Males	31	3.4	1.1	3.3	1.2
Females	34	3.9	1.3	3.5	1.3
England[b]					
Males	187	4.7	0.9	4.9	1.1
Females	144	4.7	1.0	5.2	1.2
Guatemala[c]					
Males	29	4.5	1.0	4.4	1.2
Females	16	4.7	0.9	4.3	0.8

[a]Johnston and Beller (1976).
[b]Gampel (1965).
[c]Malina *et al.* (1974).

the other two samples. While a rapid rise to a peak, followed by a subsequent decline, seems to be the case in most studies, the age at which the peak occurs and the duration and the intensity of the decline are not at all clear.

Small sample size may contribute significantly to this variability in pattern, and its role cannot be adequately assessed at present. Needed is the distribution of skinfolds at specific ages during infancy in large enough samples to reduce sampling error to a minimum. For example, Huenemann (1974) reported the means of four skinfolds (and their sum) from 228 male and 220 female 6-month-old infants from Berkeley, California. These and other means are presented in Table V. For the skinfolds, samples of this size have reduced the standard error of the mean (SEM) to about 1–2%.

Despite the probable decline in mean skinfold thickness after the infant peak (whenever it occurs), the arm continues to increase in size. The 3-month means of Snyder *et al.* (1975) for the circumference of the upper arm show a steady increase throughout this period, even at the 5th percentile. This suggests two possibilities: (1) A decrease in fatness in the upper arm concomitant with an increase in the underlying lean tissue; or (2) A thinning of the fat "ring" around the arm as it is carried outward by the expansion of the underlying tissue. In the latter case, the actual amount of fat, as estimated by the cross-sectional area, may increase despite a decrease in skinfold thickness (Tanner, 1962*a;* Johnston *et al.*, 1974).

Table V. Means of Anthropometric Dimensions of 228 Male and 220 Female Infants, 6 Months of Age, from Berkeley, California[a]

	Male		Female	
	Mean	SD	Mean	SD
Length (cm)	68.2	2.3	66.3	2.1
Weight (kg)	8.23	0.83	7.60	0.75
Skinfold (mm)				
Triceps	8.0	1.6	8.0	1.8
Subscapular	6.2	1.4	6.5	1.6
Suprailiac	4.9	1.7	5.4	1.9
Chest	3.7	0.9	3.8	1.2
Sum (of the mean)	22.8	4.5	23.7	5.2
Circumference (cm)				
Head	44.4	1.1	43.2	1.1
Upper arm (biceps)	15.4	0.9	15.0	0.9

[a]From Huenemann (1974).

There are innumerable studies on the growth of length and weight during the first year of life. The time-honored data of Stuart's Boston study (Vaughan, 1964), although based on a small sample, have been used over and over throughout the world to describe and compare, and they need not be reprinted here. In addition, for developed nations, there are other data sets from a variety of samples: American whites (Hansman, 1970), American blacks and Mexican-Americans (Fry *et al.*, 1975), British (Falkner, 1958; Tanner *et al.*, 1966*a,b*), Finnish (Bäckström and Kantero, 1971), Japanese (Terada and Hoshi, 1965*a*), and Polish (Chrzastek-Spruch and Wolanski, 1969), to name but a few. The most reliable of the available data, focusing on North American samples, have been reproduced by Roche and Malina (1983).

For the developing countries of the world, there are also many reports, for example, of infants from Guatemala (Yarbrough *et al.*, 1975), Mexico (Faulhaber, 1961; Scholl, 1975), Iran (Amirhakimi, 1974), Nigeria (Rea, 1971), and among technologically simple societies such as the Bushnegroes of Guyana (Doornbos and Jonxis, 1968), again to indicate but a few.

Despite the wide disparity in ecologies, ethnic backgrounds, and socioeconomic conditions, the variation in patterns of growth in infancy is not as large as might be expected. On the basis of several studies, Habicht *et al.* (1974) demonstrated that up until 6 months of age, children from a wide range of groups "generally grow rather uniformly in length and weight." After 6 months, they note that infants from developed countries and from high socioeconomic strata in developing countries cluster together. Those from lower strata in poorer countries are arrayed below this cluster.

Regardless of the group analyzed, somatic growth during infancy (excluding skinfolds) follows a simple curvilinear path. Various regression models have been used successfully to represent growth during this period, each involving a linear plus curvilinear function of age (Vandenberg and Falkner, 1965; Bialik *et al.*, 1973; Manwani and Agarwal, 1973; Scholl, 1975). The fit has been quite good and has permitted the estimation of the various curve parameters as formal tools for analyzing growth. Deviations from this mathematical regularity among individuals (Manwani and Agarwal,

1973; Scholl, 1975) are associated with nutritional and disease-related disturbances.

3.3. Early Childhood

The remaining preschool-age period, 1–4 years of age, may be called the years of early childhood. From the standpoint of somatic growth, they are years of developmental canalization, the achievement of regularity in the rate of increase of most dimensions. Many of the measurements that were previously increasing at a rapid and changing rate now show a striking decline in growth rate and achieve a steady, almost linear, increase until the onset of the adolescent spurt. The reduction in growth rate is particularly marked in those structures that display Scammon's "neural" (or brain and head) type of growth (see Tanner, 1962*a*). For example, the circumference of the head, which has increased by about 35% from birth to 1 year of age, increases only by about 10% through the 5th year, with most of that increase occurring between the ages of 1 and 2 years.

Table VI presents the means of head circumference as reported in four mixed longitudinal growth studies. The values are very close at similar ages and clearly demonstrate the deceleration in growth rate mentioned above. This deceleration may be seen in Figure 2, which presents the annual velocities of growth of head circumference, in centimeters per year, for the Finnish children (Kantero and Tiisala, 1971*b*) taken at 6-month measurements. Data from other investigations are not presented, since they cannot be converted to strictly comparable growth rates (i.e., based on longitudinal data and calculated semiannually). However, the shape of the curves in Figure 2 is characteristic of that in other studies; i.e., by 2 years of age, the increases are only in the range of 2 cm/year.

These data suggest a general similarity in the pattern of growth in head circumference across populations during the preschool-age years. This finding is supported by research on other human samples as well. Thus among urban Jamaican infants (Grantham-McGregor and Desai, 1973), longitudinal measurements indicate that the head circumference at 1 year is 33% greater than at birth. A similar difference is seen in Guatemalan Ladinos (Malina *et al.*, 1975); again, between 1 and 5 years, the means of these children had increased by 10%, consistent with the European data cited above.

The analysis of somatic growth during early

Table VI. Head Circumference (cm) in Four Mixed Longitudinal Samples

Age (months)	Finland[a]		Japan[b]		England[c]		United States[d]	
	Mean	SD	Mean	SD	Mean	SD	Mean	SD
				Boys				
3	41.0	1.2	41.1	0.9	45.7[e]	1.3	40.3	1.2
12	47.4	1.5	47.0	1.0	46.8	1.3	46.6	1.2
24	49.8	1.5	49.0	1.1	49.3	1.4	49.2	1.2
36	50.7	1.3	50.2	1.1	50.4	1.3	50.3	1.2
48	51.6	1.3	—	—	—	—	51.0	1.2
60	52.3	1.3	—	—	—	—	51.6	1.2
				Girls				
3	39.9	1.2	39.8	1.0	44.6[e]	1.2	39.5	1.1
12	46.3	1.3	45.6	1.1	45.7	1.2	45.6	1.3
24	48.8	1.4	47.9	1.3	47.9	1.1	48.1	1.4
36	50.0	1.4	49.0	1.4	49.5	1.1	49.2	1.4
48	50.3	1.4	—	—	—	—	50.2	1.5
60	51.3	1.5	—	—	—	—	50.7	1.5

[a]Kantero and Tiisala (1971*b*).
[b]Terada and Hoshi (1965*b*).
[c]Falkner (1958).
[d]Hansman (1970).
[e]Age: 39 weeks.

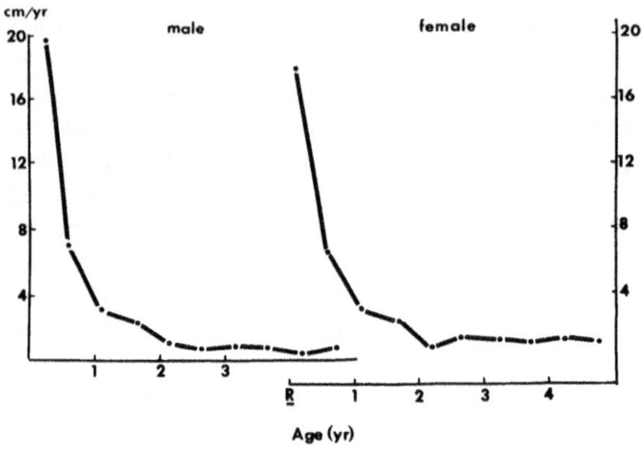

Fig. 2. Annual velocities of head circumference in Finnish children, measured at 6-month intervals. (Data from Kantero and Tiisala, 1971*b*.)

childhood is rendered more difficult by problems of measurement at these ages. In particular, it is important to determine whether length was measured in the erect or supine position, since the two techniques yield different results. Roche and Davila (1974) found that, between 2 and 5 years of age, recumbent length exceeds stature by median values ranging from 2.3 to 1.3 cm, decreasing sharply over the period. The approaches used differ among investigative groups. Kantero and Tiisala (1971*a*) measured stature after 21 months; Cruise (1973), in studying the growth of low-birth-weight children, measured supine length under 3 years, as did Faulhaber (1961) in her study of Mexican children. By contrast, Falkner (1958) and Terada and Hoshi (1965*a*) measured all lengths in the supine position for the 3 years of their studies, while in the study of Guatemalan Ladino children sponsored by the Nutrition Institute of Central America and Panama (INCAP), all measurements are made supine, even through age 7 (Yarbrough *et al.*, 1975). Differences in technique create systematic measuring error and spuriously inflate the variance between samples.

3.4. Length and Weight

From 1 through 4 years of age, growth in length and weight approach a nearly linear rate of increase, at least as indicated by distance curves of size against age. Figure 3 illustrates the curves of length and weight against age in males for two disparate samples of children—the first taken from the

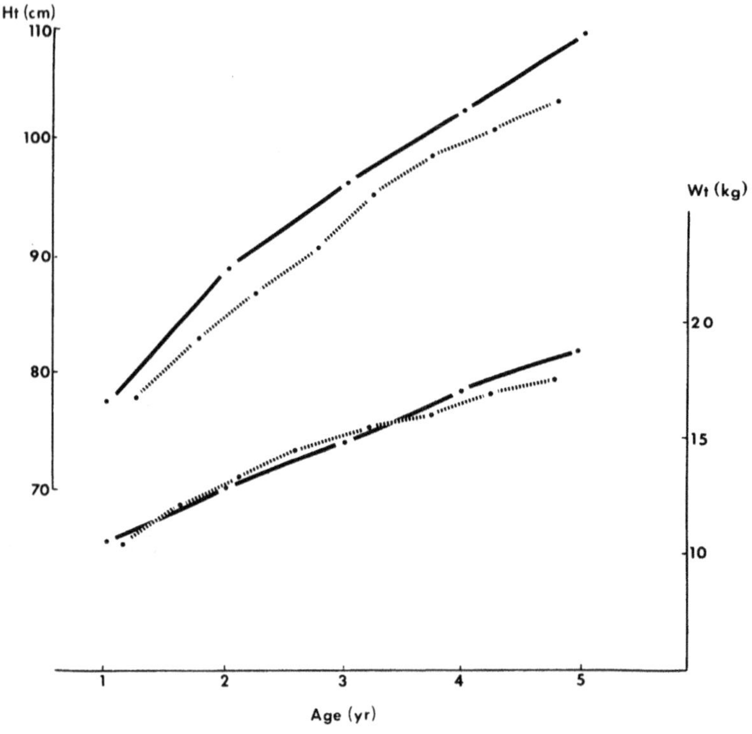

Fig. 3. Distance curves of length and weight against age in children from two samples: (—)Yugoslavia (Pogačnik, 1970). (---) England (Gore and Palmer, 1949).

survey of 3199 Slovenian children from Yugoslavia (Pogačnik, 1970) from birth through 6 years of age, and the second from a study of 5684 London children from 2 weeks through 4 years of age (Gore and Palmer, 1949). Both samples are cross-sectional, and the curves indicate a similarity of patterns (no smoothing has been done). While a slight deceleration is indicated, these means indicate a regular difference from age group to age group.

Pure longitudinal data over this age range are simply not available for samples of any size. However, Table VII presents incremental data from three studies in which a high level of sample stability is to be found from year to year. Growth increments, over 12-month periods, have been calculated, as indicated in Table VII, for weight and length among girls as an example of the change observed in individuals. A decreasing growth rate seems indicated but, after 2 years of age, this decrease is not consistent in all samples. An increase in the weight increment from 4 to 5 years, over that between 3 and 4 years, is indicated by all three studies.

The estimated percentage increase between 1 and 4 years of age may be seen, as calculated for 14 samples, in Table VIII. The data represent the differences between means, when adjusted to 1½ or 4½ years. There is a general consistency within each of the four columns, although the variability is greater for weight than for length. Unusual values are for Finnish children, with increases of less than 40%, and children from the Netherlands, with the greatest increases for all four values. The percentage increase in body weight is on the order of twice that for length, although it may be as little as 1.5 (Finnish girls) or as great as 2.3 (Guatemalan girls). In general, the increase is greater in females than males, although exceptions occur.

Table VII. Annual Increments of Length and Weight in Girls Aged 1–5 Years from Three Studies

Age (years)	United States[a]	United Kingdom[b]	Finland[c]
	Length (cm)		
1–2	12.8	11.4	11.2
2–3	8.7	8.4	8.5
3–4	8.0	7.5	8.4
4–5	7.1	6.7	8.5
	Weight (kg)		
1–2	2.5	2.7	2.8
2–3	2.0	2.0	1.9
3–4	2.1	2.0	2.1
4–5	2.2	2.2	2.3

[a]Differences between means of mixed longitudinal sample (Hansman, 1970).
[b]Average smoothed velocities, longitudinal data (Tanner *et al.*, 1966*b*).
[c]Summed increments (3-monthly between 1 and 2 years, semiannual thereafter) (Kantero and Tiisala, 1971*a*).

Table VIII. Percentage Differences between Length and Weight at Age 1 Year and Height and Weight at Age 4 Years

Study	Length M	Length F	Weight M	Weight F
Cross-sectional				
United States				
White[a]	26.6	31.4	47.5	42.4
Black[a]	29.5	29.5	61.6	53.1
Black (D.C.)[b]	31.9	30.1	63.5	63.1
Mixed[c]	28.4	30.4	54.6	62.2
United Kingdom (mixed)[d]	27.7	30.3	49.0	57.2
Canadian (national)[e]	30.1	29.3	60.2	58.0
Finland[f]	25.5	24.2	39.7	36.8
Netherlands (national)[g]	36.4	37.8	67.6	69.0
Czechoslovakia[h]	29.0	30.1	50.4	56.4
Yugoslavia[i]	29.3	30.2	52.9	56.0
Hong Kong[j]	28.9	29.3	52.4	53.7
Indian[k]	29.9	30.3	60.7	65.4
Mixed Longitudinal				
United States				
Boston[l]	30.4	32.0	52.4	57.2
Denver[m]	29.2	31.2	54.4	61.3
United Kingdom[n]	27.9	28.9	50.9	55.0
Guatemala[o]	30.9	32.8	65.8	72.2

[a]Abraham *et al.* (1975).
[b]Verghese *et al.* (1969).
[c]Snyder *et al.* (1975).
[d]Gore and Palmer (1949).
[e]Demirjian (1980).
[f]Bäckström and Kantero (1971).
[g]van Wieringen *et al.* (1971).
[h]Fetter and Suchy (1967).
[i]Pogačnik (1970).
[j]Chang *et al.* (1965).
[k]Indian Council of Medical Research (1972).
[l]Stuart and Meredith (1946).
[m]Hansman (1970).
[n]Tanner *et al.* (1966*b*).
[o]Yarbrough *et al.* (1975) (supine length at age 4).

3.5. Skinfold Thickness

Figures 4–7 present skinfold thicknesses from five samples of children of various nationalities and ethnic groups. The sample sizes are all quite adequate and some, such as the Yugoslavian sample (Pogačnik, 1970) are based on a total size of

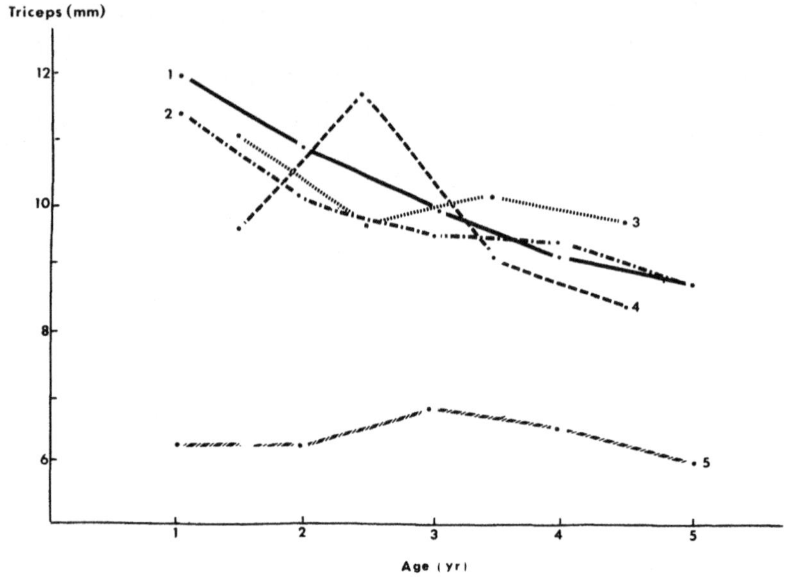

Fig. 4. Mean thickness of triceps skinfolds for boys aged 1–5 years, in five samples: 1, United Kingdom (Tanner and Whitehouse, 1975); 2, Yugoslavia (Pogačnik, 1970); 3, United States, white (Abraham *et al.*, 1975); 4, United States, black (Abraham *et al.*, 1975); 5, Guatemalan Ladino (Malina *et al.*, 1974).

over 3000. This study reported means plotted on the graphs shown in Figures 4–7. For the others, medians are used as the statistic of preference, although skewness at this age is small. Values for the British children are only estimates, taken from the graphs of Tanner and Whitehouse (1975).

In each case the age trends are the same; there is a gradual decrease in the thickness of the adipose layer, but with a number of exceptions. Between-sample variation is greatest at the younger ages, and the decrease in the medians seems to be greater where the skinfold thickness at 1 year is greater. In fact, among the Ladino sample from Guatemala (Malina *et al.*, 1975), where the medians are lowest,

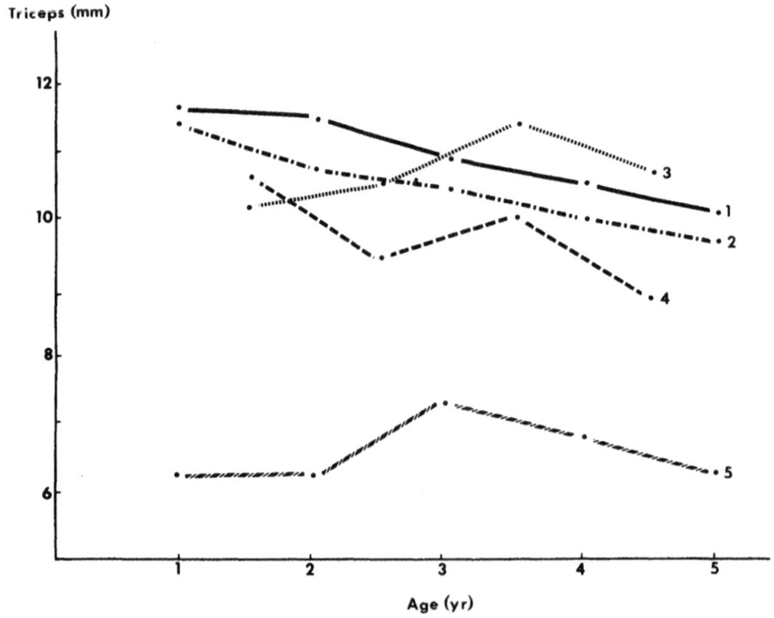

Fig. 5. Mean thickness of triceps skinfolds for girls, aged 1–5 years, in five samples: 1, United Kingdom (Tanner and Whitehouse, 1975); 2, Yugoslavia (Pogačnik, 1970); 3, United States, white (Abraham *et al.*, 1975); 4, United States, black (Abraham *et al.*, 1975); 5, Guatemalan Ladino (Malina *et al.*, 1974).

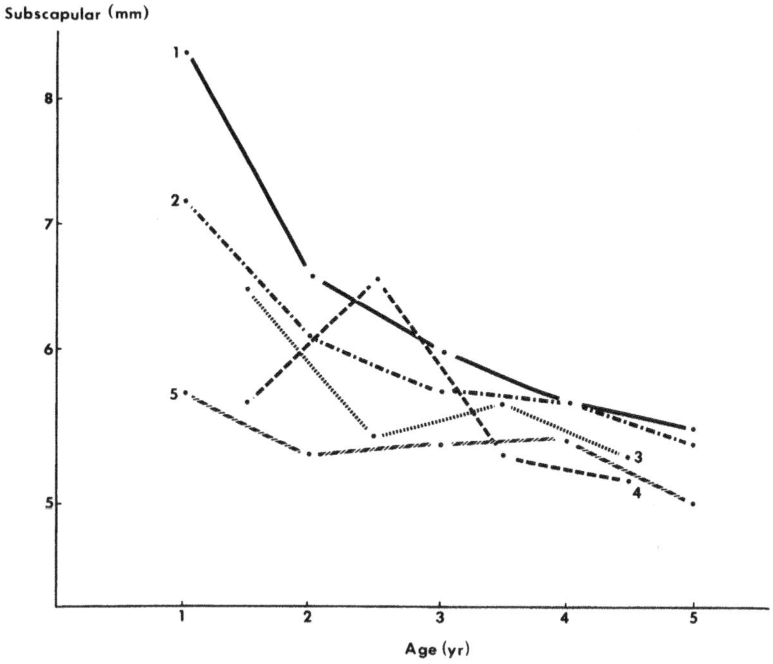

Fig. 6. Mean thickness of subscapular skinfolds for boys, aged 1–5 years, in five samples: 1, United Kingdom (Tanner and Whitehouse, 1975); 2, Yugoslavia (Pogačnik, 1970); 3, United States, white (Abraham *et al.*, 1975); 4, United States, black (Abraham *et al.*, 1975); 5, Guatemalan Ladino (Malina *et al.*, 1974).

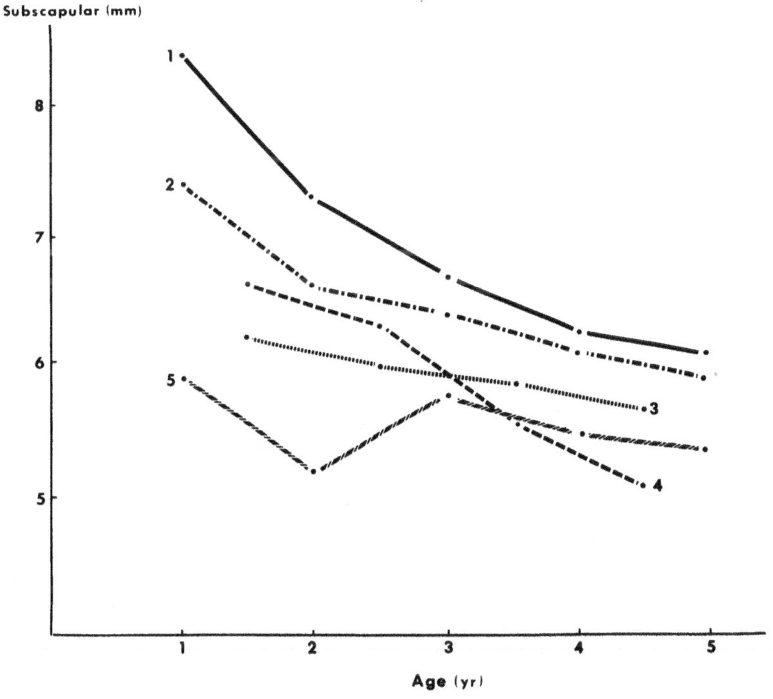

Fig. 7. Mean thickness of subscapular skinfolds for girls, aged 1–5 years, in five samples: 1, United Kingdom (Tanner and Whitehouse, 1975); 2, Yugoslavia (Pognačnik, 1970); 3, United States, white (Abraham *et al.*, 1975); 4, United States, black (Abraham *et al.*, 1975); 5, Guatemalan Ladino (Malina *et al.*, 1974).

markedly so for the triceps, there is no real decrease with age. In the literature there are other reports of skinfold measures among children of this age range (see, e.g., Schell and Johnston, 1986, for a review of native Americans). However, sample sizes are generally small, and any patterns that emerge confirm the conclusion of a continuation of the loss of fat, which began in infancy, throughout the preschool-age years. The response, however, to nutritional factors causing alterations in body fat may result in considerable individual and group variation, perhaps causing reversals in this trend.

3.6. Other Dimensions

Other dimensions grow in a more or less linear fashion, similar to height and weight. These include crown–rump length (or sitting height) (Falkner, 1958; Pogačnik, 1970; Abraham *et al.*, 1975; Snyder *et al.*, 1975; Demirjian, 1980), extremity lengths (Terada and Hoshi, 1966; Anderson *et al.*, 1956, 1964; Snyder *et al.*, 1975), and trunk and extremity circumferences (Wolanski, 1961; Fetter and Suchy, 1967; Pogačnik, 1970; Indian Council of Medical Research, 1972; Malina *et al.*, 1975).

These dimensions all tend to grow similarly to the general (or body) curve of Scammon (see Tanner, 1962*a*): during the preschool-age years there is, as with stature, a decreasing rate of growth until a linear rate is established. Growth in height, in fact, serves as a general model for most of the other somatic dimensions, save those of the head and those that reflect adipose tissue.

Even though most dimensions grow in the manner of stature, this does not necessarily mean that their rates of change are the same. Rather they are related in some exponential function, such that $y = x^a$, where y and x are dimensions and a is the exponent that reflects the systematic relationship between them.

This constant relationship between dimensions of growth is called *allometry* (Huxley, 1932) and is characteristic of many living forms. Exponent a is called the coefficient of allometry. Figure 8 shows the means of stature (crown–heel length in infants) and sitting height (crown–rump length) for two samples of boys aged 3 months to 4 years, one from the mixed longitudinal American sample from Denver (Hansman, 1970) and one from a large cross-sectional survey of Indian children and youth (Indian Council of Medical Research, 1972). The relationship between the two dimensions is constant in either sample, but the increase in sitting height is less than that for stature.

The steady change in various proportions of the body is a result of allometric growth. For example, the legs become longer relative to the trunk, while the trunk lengthens relative to its breadth as well as to the size of the head. In other words, the preschool-age years are those years in which the chunky physique of the newborn infant becomes transformed into the more elongated physique of the child. This trend continues throughout the years of childhood until the circumpubertal period.

Alongside this series of changes in shape is a considerable degree of independence in the growth of individuals from year to year. Increments of growth show relatively little correlation within individuals between various age periods. For example, Falkner (1958) computed correlation coefficients between the recumbent length increment during the period 4–13 weeks and length increment during the period 12–18 months. In 54 boys the correlation was 0.07, and in 50 girls it was 0.23; among all 104 the value was 0.01. Increments over adjacent age periods will frequently show a negative correlation if the periods are shorter than 1 year. This may reflect canalization of development, seasonal effects, or simply measuring error (Tanner, 1962*b*; Billewicz, 1967).

Increments of different dimensions do not seem to be highly correlated. Meredith (1962) calculated 240 correlations between pairs of increments from 5 to 10 years of 16 measurements in 67 white boys.

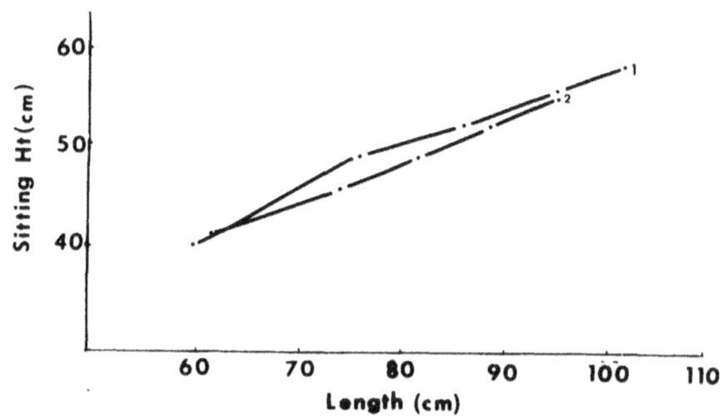

Fig. 8. Mean sitting height plotted against mean body length in two samples of boys, aged 3 months to 4 years of age: 1, United States, white (Hansman, 1970); 2, Indian (Indian Council of Medical Research, 1972).

Table IX. Coefficients of Correlation of Size at Various Ages from Three Samples of Children from Great Britain

	Aberdeen[a]		Newcastle-upon-Tyne[b]	London[c]	
Age	M	F	M and F	M	F
			Weight		
Birth to 3 years	0.49	0.17	0.36	—	—
Birth to 5 years	0.37	0.16	0.30	—	—
Birth to Adult	0.38	0.42	0.25	—	—
			Length		
Birth to 3 years	0.52	0.36	—	—	—
Birth to 5 years	0.42	0.34	—	—	—
Birth to adult	0.25	0.29	—	—	—
1 month to 1 year	—	—	—	0.19	0.54
1 to 2 years	0.85	0.76	—	0.84	0.80

[a]42 males, 38 females (Tanner *et al.*, 1956).
[b]201 males, 241 females (Miller *et al.*, 1972).
[c]60 males, 69 females (maximum) (Falkner, 1958).

Of the total of 240, 103 correlations were not significant, and another 71 showed only a "low" relationship (r = 0.31–0.45).

Table IX presents the correlations between either length or weight with itself at various ages, as determined in three samples of children from the British Isles. The values are almost all significant but, except for the 1 to 2-year value, all are low to moderate. Of interest is the general similarity of the r coefficients for birth and either 3 or 5 years to those for birth and adult weight. The progressive canalization is seen to be emerging and to have reached a more or less stable level by this time, although considerable independence still exists. Reed and Stuart (1959) commented on the dependence as well as the independence in their description of individual variation in growth patterns.

Even with the independence indicated by the above, a basic regularity can be seen to exist in growth between 1 and 4 years of age. This regularity permits the utilization of regression models for formal analysis, usually combining this age period with the subsequent years of childhood and, in some cases, using the first year of life as well. These models may be a function of a simple parabola of the form $y = a + bt + c \log t$, where t is age (Tanner *et al.*; 1956; Scholl, 1975) or a logistic curve that, when combined with a second logistic, permits analysis from the age of 1 year to maturity (Bock *et al.*, 1973). Other models include those discussed by Wingerd (1970), and the Jenss curve (Jenss and Bayley, 1937; Deming and Washburn, 1963), which approaches as an asymptote a straight line.

These models provide specific analytic frameworks for analyzing components of growth, e.g., the linear component and the intensity of deceleration. In particular they are valuable in studying individual variation across growth time, but they may be used in comparing samples as well.

4. Determinants of Growth and Its Variability in Infants and Preschool-age Children

Examined over short periods of time, growth occurs sporadically, with periods of no measurable increments interspersed among others in which the rate of change is quite high. Lampl (1983) studied the physical and behavioral development of a sample of Denver infants, measuring them on a weekly basis. Contrary to the widely-held view that physical growth occurs steadily, the infants in her study displayed significant incremental within-subject variability, even after she had corrected for the effects of intraobserver error.

On the other hand, when these short observational periods are averaged over longer times, a month or more, a basic underlying constancy emerges. This constancy is so striking that it permits, as noted above, the fitting of mathematical models with little error and leads to a conceptual model of internal self-regulation (Tanner, 1963).

At the same time, the amount of growth variability that may be catalogued is quite striking. This variation may be separated into significant between-subject and between-sample components. Any interpretation of growth up to 5 years of age must deal with the operation of those factors that promote growth variation. While a detailed coverage of the subject is not the major purpose of this chapter, it is necessary to view physical growth during these years within the context of factors that create, enhance, and maintain the variation that we may so easily observe.

4.1. Factors of Heredity

Basic to any discussion of growth variation is the influence of inheritance on variability. In this respect there are factors associated with family lines that promote variability among individuals and factors associated with populations that promote variability among groups.

The contrast between monozygotic and dizygotic twins provides a classic design for determining heritability, even though the degree to which data from twins may be generalized is still unclear. Vandenberg and Falkner (1965) used regression analysis to estimate the growth of 29 monozygotic (MZ) and 31 dizygotic (DZ) twin pairs from birth to 2 years of age. Using F ratios of within-pair variances, these workers found a significant hereditary effect on the growth rate and, particularly, on its deceleration. On the other hand, they found that the initial status (i.e., at birth), as estimated from the regression analysis, showed no genetic component. Similar findings have been reported by Wilson (1976, 1979), who analyzed twins from the same study. MZ twins showed a steadily increasing concordance of size, relative to DZ twins, until 8 years of age. This finding is in agreement with data on the genetics of birth weight (see the review by Hunt, 1966), which show only a minor contribution of the fetus's genotype to status at birth.

Starting from the position at birth, in which the effect of genetic factors on size have played only a small part, one may examine the expression of the infant's and child's genotype as a function of age. This has been done most commonly by means of the correlation of parental size with the size of their offspring at specific ages; a related strategy is to contrast the size of children grouped on the basis of parental size. There have been many such analyses; Mueller (1976) presented the results of 20 such studies. However, his work did not include any data on children under 5 years of age.

Table X presents r coefficients between parental height and, as an example, the height of their daughters from birth to 4 years of age taken from several studies. The correlation coefficients are seen to vary considerably among the studies by as much as 10-fold. By 4 years, when some stabilization has been reached, even with the same-sex parent the values (e.g., in fathers) range from 0.11 to 0.35. This may reflect the influence of between-family and within-family components of environmental variance and demonstrates once again that heritability, far from being a fixed notion, is sample specific and will vary significantly with the nature of the environment. On the other hand, the above sample sizes are small, and the confidence limits of the coefficients are quite wide; sampling error cannot be overlooked.

Parental size has also been shown to affect the height of children (Garn, 1961, 1966; Tanner *et al.*, 1970). In this earlier paper, reviewing genetic factors in growth, Garn showed that the statures of children differ from 1 year of age onward by parental mating type: tall × tall matings produce taller preschool-age children than do short × short parents. The other two cited publications have presented reference standards for height that take into account the mid-parental height. Garn indicates that a difference of 12 cm in mid-parental height accounts for a difference in the child at 1.0 years of age of 4 cm in boys and 1.6 cm in girls. At 4.0 years of age, this difference had increased to 6.8 cm and 7.0 cm, respectively. Tanner *et al.* (1970) found regressions of child height on mid-parental height ranging from 0.27 cm/year to 0.38 between 2 and 4 years of age; their values fall within those indicated by Garn.

Such correlations indicate the operation of hereditary factors. However, because of environmental similarities within families, the correlation coefficients may be substantially inflated or decreased. Mueller (1975) summarized the problem and pointed to the role of environmental covariance. Among older children and youths, he found a combined r value (parent/offspring) of 0.38 for children of European origin and of 0.21 for those of non-European origin. He suggests that the differences in

Table X. Correlations between Parental Height and Height of Daughters from Several Studies

Study	Age of daughter (years)					
	Birth	1	2	3	4	F/M[a]
Tanner and Israelsohn (1963)						0.11
Fathers	—	0.14	0.17	0.17	0.11	
Mothers	—	0.50	0.58	0.56	0.61	
Bouchalová and Gerylová (1970)[b]						0.43
Fathers	0.10	0.23	0.33	0.35	0.33	
Mothers	0.20	0.30	0.31	0.34	0.32	
Chrzastek-Spruch and Wolanski (1969)[c]						0.30
Fathers	−0.02	—	—	—	—	
Mothers	0.25	—	—	—	—	
Livson *et al.* (1962)[d]						—
Fathers	0.26	0.36	0.36	0.36	0.35	
Mothers	0.22	0.46	0.42	0.43	0.44	
Scholl (1975)						—
Fathers	0.11	0.32	0.42	0.46	—	
Mothers	0.21	0.43	0.28	0.30	—	
M. Russell (1976)						—
Fathers	—	—	—	0.26	—	
Mothers	—	—	—	0.22	—	

[a]Correlation between stature of father and mother.
[b]Partial *r*, corrected for father/mother correlation.
[c]Firstborn.
[d]Pooled data from studies in California, Ohio, and Oxford; *r* values read from graphs.

correlation indicate the harsher environment of most of the latter group.

The only controlled study of hereditary factors in skinfold thickness among preschool-age children is that by Brook *et al.* (1975), who calculated heritability estimates in MZ and DZ twins aged 3–15 years. Among those below age 10 years, the following heritabilities were obtained on the basis of intraclass correlations: for triceps, -0.25 ± 0.43 in boys and 0.56 ± 0.41 in girls; for subscapular, 1.23 ± 0.35 in boys and 0.47 ± 0.37 in girls (SEM in all cases). The large standard errors obviously account for the aberrant values in males, preventing firm conclusions. Brook and coworkers did suggest that environmental effects are strong below 10 years of age but less important thereafter.

4.2. Nutritional Factors

Nutritional factors are important sources of growth variation during the preschool-age years. The nutrient content of the diet interacts with other variables, such as physical activity and health status, to regulate, to a significant extent, the rate of growth. The precise nature of the nutritional ecosystem differs markedly from one society and one socioeconomic level to another. The relative abundance of calories as well as their source varies, as does the amount and quality of available protein. Specific nutrients, such as vitamins and trace elements, complete the range of dietary considerations.

Among populations not characterized by either undernutrition or overnutrition, the relationship of diet to growth is less clear. Roche and Cahn (1962) could not explain differences in skinfold thickness between British and Australian children on the basis of caloric intakes of the two groups. In another study, Sims and Morris (1974) found virtually no relationship between triceps skinfold thickness and biochemical indicators of nutritional status in 163 well-nourished preschool-age children. Although body weight showed more significant correlations, the results were still not completely clear. Significant clusters emerged only when certain socioeconomic and behavioral characteristics of the family were considered.

Within-sample relationships between growth and nutrient intake are likewise generally lacking for infants and preschool-age children. In one study, Mack and Kleinhenz (1974) followed a small group ($N = 5$) of infants aged 8–56 days, with total milk consumption as well as activity levels measured. The correlations were numerically high but were frequently not significant because of sample size. In general, the least active infants consumed more calories and gained weight more rapidly.

Early nutritional excess has been suggested as contributing significantly to later obesity, as indicated by a relationship between relative weight in infancy up to age 7 and by measures of relative weight up to 22 years of age (Miller *et al.,* 1972; Mack and Ipsen, 1974; Fisch *et al.,* 1975; Sveger *et al.,* 1975). The further statistical relationships among relative weight at 1 year, relative weight during adolescence, and the pregravid weight of the mother (Mack and Johnston, 1976) suggest a significant relationship among maternal characteristics, feeding practices, infant nutrition, and growth in childhood. However, the longer-term implications of this relationship are not nearly as clear as was once believed; most overweight 1-year-olds show no indication of obesity during the adolescent years (Johnston and Mack, 1978; Roche *et al.,* 1982).

In children suffering from undernutrition, the situation is much clearer; the relationship between protein-energy malnutrition (PEM) and growth is, by now, firmly established. In fact, growth status during childhood is accepted almost universally as the best indicator, in field studies, of nutritional status, at the levels of the individuals and the community (see, e.g., Gomez *et al.,* 1956; *Lancet,* 1970; Jelliffe, 1966; Tanner, 1976; Nowak and Munro, 1977; Waterlow *et al.,* 1977; Jelliffe and Jelliffe, 1979).

Descriptions of the growth of infants and preschool-age children living in environments characterized by chronic undernutrition (as well as related factors such as disease) come, to cite but a few, from Latin America (Cravioto, 1968; Cravioto and de Licardie, 1971; Yarbrough *et al.,* 1975; Graham *et al.,* 1980), North Africa (Wishaky and Khattab, 1962; Bouterline *et al.,* 1973), Turkey (Rharon, 1962), sub-Saharan Africa (Dean, 1960; McFie and Welbrown, 1962), Asia (Brink *et al.,* 1976; Vijayaraghavan and Gowrinath Sastry, 1976), and Australia (Kirke, 1969).

The response of the previously malnourished infant and preschool-age child to nutritional supplementation and therapy is still unclear. The general potential for catch-up growth has been described (Prader *et al.,* 1963), and some investigators have presented data to show striking catch-up, with no permanent physical deficit, after nutritional rehabilitation (Čabak and Najdanvic, 1965; Garrow and Pike, 1967). On the other hand, some workers (e.g., Graham, 1968) have suggested that the effects of early infant malnutrition are long term. Subsequent to this report, Graham and Adrianzen (1972) noted the need to consider the complexities of the nutritional ecosystem. In a prospective study of the outcome of clinically severe PEM among Mexican preschool-age children, Scholl *et al.* (1980) demonstrated that "any growth effects following the episode of severe PEM were accounted for by growth status preceding its onset."

4.3. Illness and Health Status

This rather heterogeneous collection of factors involves the effects on growth of disease and, as already included in this section, children whose birth weights were abnormally low. Low birth weight may be the result of a shortened gestation, i.e., prematurity, or of intrauterine growth retardation (IUGR) (i.e., small-for-gestational age). The etiology of these two causes of low birth weight are clearly quite different and must be considered separately.

The role of illness in reducing growth among infants and preschool-age children is not entirely clear. More often than not, groups characterized by a high incidence of illness also display other growth-deviating mechanisms, such as poor nutrition. In many cases it is difficult to separate the causes and, even then, the effects of the other factors may be found to outweigh those of illness.

In an early study, Evans (1944) contrasted increments of growth in several dimensions among 93 middle-class Iowa children, 2–5 years of age, grouped by frequency of respiratory and digestive tract illness. No relationship was found between illness, as measured by nursery school absences, and growth.

A decade later, Hewitt *et al.* (1955) analyzed annual height increments by illness severity, using five categories of illness, in the 650 children of the

Oxford Child Health Survey, studied at ages 1–5. These workers found a "small but definite diminution" in growth related to severity of illness, especially in boys.

These two studies failed to agree in their findings; possibly associated socioeconomic differences interacted with illness and biased the results. Both samples were from English-speaking populations, although the Iowa children were drawn from a higher social class.

One would expect clearer results among the developing nations, where episodes of illness are more frequent and acute. Among Guatemalan Ladinos, diarrheal disease has been associated with reduced increments in both length and weight, amounting to about 10% deficit (Martorell *et al.*, 1975). However, among rural children from southern Mexico (Condon-Paoloni *et al.*, 1977), the effects were not so severe. In the latter study, annual increments of length and weight from birth through 2 years were contrasted in children with high and low frequencies of illness in three categories: upper respiratory, lower respiratory, and diarrhea. Significant reductions were found only for weight increment and diarrhea, where there was a reduction of some 6% from the mean weight by 3 years of age.

4.4. Socioeconomic and Microenvironmental Factors

Interacting with the above are various factors—socioeconomic, sociocultural, and, in general, a part of the microenvironmental surroundings of the child. Some of these factors affect the individual directly; others exert their influence through intermediate mechanisms, such as nutrition. In sum, socioeconomic and microenvironmental factors form the complex developmental ecosystem that determines the development of the child during this crucial period. Many investigators have discussed the variety of factors that impinge on growth, but the ecosystem has been represented most elegantly by Cravioto and associates, who indicated the complexity of the situation in a developing country (Cravioto *et al.*, 1967; Cravioto, 1970).

In their study of Lebanese children, Kanawati and McLaren (1973) identified 20 factors that differentiate but that do not necessarily cause growth differences between thriving and nonthriving children. These factors ranged from disease prevalence, presence or absence of vaccination, and caloric intake, to the existence of an indoor toilet, father's education, and the size of the house. Kanawati and McLaren (1973) concluded that, in low socioeconomic conditions, "factors other than food and health are involved in adequate growth of the preschool child." A similar conclusion was drawn by Scholl (1975) and by Johnston *et al.*, (1980*a,b*), who identified demographic and parental biological characteristics associated with growth failure in Mexican children, and by Christiansen *et al.* (1975), who identified social factors in Bogota.

Socioeconomic conditions are significantly associated with infant and child growth; as examples, we may cite the data of Bouchalová and Gerylová (1970), Rea (1971), and Amirhakimi (1974) as typical for most locales. Such factors may differentiate rural and urban settings; Wolanski and Lasota (1964) reported quite striking differences between the two in Poland.

Even in the more affluent countries, socioeconomic effects are detectable in this age range. Garn and Clark (1975), reporting on the U.S. Ten-State Nutrition Survey, compared children of low with higher income groups. From ages 2–4, the difference in height reaches a maximum of 2%; weight, 5%; and triceps skinfold, 12%.

Birth weight is influenced by these factors as well, and Tanner (1974) has reviewed their role. However, the magnitude of their effects may vary in different settings (Bjerre and Värendh, 1975). That is, the results of the interaction of the various factors may be relatively specific for the ecosystem. Other factors identified include smoking (C. S. Russell *et al.*, 1968; Hardy and Mellits, 1972) and changing educational levels (de Licardie *et al.*, 1972).

5. Summary and Conclusions

The preschool-age years, from birth through age 4, are crucial years in determining future developmental events and health status even into adulthood. The transition from prenatal to postnatal existence, with an exponential increase in stimuli and stress, is accentuated by the dependence of the infant and the small child on the surrounding world.

The high growth rates and sharp inflections in various curves are characteristic of change from infancy to the regularity of childhood, associated with canalization, directionality, and self-regulation. The sensitivity, however, to environmental circumstances remains such that the preschool-age years provide us with an excellent age range in which to evaluate developmental adequacy and, by extension, the general health and nutritional status of the population to which these children belong.

ACKNOWLEDGMENTS. The assistance of Virginia Lathbury and Jean MacFadyen in preparing this manuscript is gratefullly acknowledged.

6. References

Abraham, S., Lowenstein, F. W., and O'Connell, D. E., 1975, Preliminary Findings of the First Health and Nutrition Examination Survey, United States, 1971–72, Anthropometric and Clinical Findings, Department of Health, Education and Welfare Publ. (HRA) 75-1229, U.S. Government Printing Office, Washington, D.C.

Adams, M. S., and Niswander, J. D., 1973, Birth weight of North American Indians: A correction and amplification, *Hum. Biol.* **15**:351.

Amirhakimi, G. H., 1974, Growth from birth to two years of rich urban and poor rural Iranian children compared with Western norms, *Ann. Hum. Biol.* **1**:427.

Anderson, M., Blais, M., and Green, W. T., 1956, Growth of the normal foot during childhood and adolescence, Length of the foot and interrelations of foot, stature, and lower extremity as seen in serial records of children between 1 and 18 years of age, *Am. J. Phys. Anthropol.* **14**:287.

Anderson, M., Messner, M. B., and Green, W. T., 1964, Distributions of lengths of the normal femur and tibia in children from 1 to 18 years of age, *J. Bone Joint Surg.* **46A**:1197.

Bäckström, L., and Kantero, R.-L, 1971, II. Cross-sectional studies of height and weight in Finnish children aged from birth to 20 years, *Acta Paediatr. Scand. (Suppl.)* **220**:9.

Bialik, O., Perirutz, E., and Arnon, A., 1973, Weight growth in infancy: A regression analysis, *Hum. Biol.* **45**:81.

Billewicz, W. Z., 1967, A note on body weight measurements and seasonal variation, *Hum. Biol.* **39**:241.

Billewicz, W. Z., Kemsley, W. W. W., and Thomson, A. M., 1962, Indices of adiposity, *Br. J. Prev. Soc. Med.* **16**:183.

Bjerre, I., and Värendh, G., 1975, A study of biological and socioeconomic factors in low birthweight, *Acta Paediatr. Scand.* **64**:605.

Bock, R. D., Wainer, H., Petersen, A., Thissen, D., Murray, J., and Roche, A., 1973, A parameterization for individual human growth curves, *Hum. Biol.* **45**:63.

Bouchalová, M., and Gerylová, A., 1970, The influence of some social and biological factors upon the growth of children from birth to 5 years in Brno, *Eleventh Reunion Coordinée des Recherches sur la Developpement de l'Enfant Normal,* Vol. 2, p. 43, Centre Internationale de l'Enfance, Paris.

Bouterline, E., Tesi, G., Kerr, G. R., Stare, F. J., Kallal, Z., Turki, M., and Hemaidan, N., 1973, Nutritional correlates of child development in Southern Tunisia. II. Mass measurements, *Growth* **37**:91.

Brink, E. W., Khan, I. H., Splitter, J. L., Staehling, N. W., Lane, J. M., and Nichaman, M. Z., 1976, Nutritional status of children in Nepal, 1975, *Bull. WHO* **54**:311.

Brook, C. G. D., Huntley, R. M. C., and Slack, J., 1975, Influence of heredity and environment in determination of skinfold thickness in children, *Br. Med. J.* **2**:719.

Čabak, V., and Najdanvic, R., 1965, Effect of undernutrition in early life on physical and mental development, *Arch. Dis. Child.* **40**:532.

Carter, J. E. L., 1972, *The Heath–Carter Somatotype Method,* Department of Physical Education, San Diego State College, San Diego, California.

Chang, K. S. F., Lee, M. M. C., Low, W. D., Chui, S., and Chow, M., 1965, Standards of height and weight of southern Chinese children, *Far East Med. J.* **1**:101.

Christiansen, N., Mora, J. O., and Herrera, M. G., 1975, Family social characteristics related to physical growth of young children. *Br. J. Prev. Soc. Med.* **29**:121.

Chrzastek-Spruch, H., and Wolanski, N., 1969, Body length and weight in newborns, and infant growth connected with parents' stature and age, *Genet. Pol.* **10**:257.

Clauser, C. E., Tucker, P. E., McConville, J. T., Churchill, E., and Laubach, L. L., 1972, Anthropometry of Air Force Women, AMRL-TR-70-5, Aerospace Medical Research Laboratory, Wright-Patterson Air Force Base, Ohio.

Condon-Paoloni, D., Johnston, F. E., Cravioto, J., and Scholl, T. O., 1977, Morbidity and growth of infants and young children in a rural Mexican village, *Am. J. Publ. Health* **67**:651.

Cravioto, J., 1968, Modificacion postnatal de fenotipo causada por la desnutricion, *Sobret. Gaceta Med. Mex.* **98**:523.

Cravioto, J., 1970, Complexity of factors involved in protein-calorie malnutrition, *Bibl. Nutr. Dieta* **14**:7.

Cravioto, J., and de Licardie, E. R., 1971, The long-term consequences of protein-calorie malnutrition, *Nutr. Rev.* **29**:107.

Cravioto, J., Birch, H. G., de Licardie, E. R., and Rosales, L., 1967, The ecology of weight gain in a preindustrial society, *Acta Pediatr. Scand.* **56**:71.

Cruise, M. O., 1973, A longitudinal study of the growth of low birth weight infants: I. Velocity and distance growth, birth to 3 years, *Pediatrics* **51**:620.

Dean, R. F. A., 1960, The effects of malnutrition on the growth of young children, in: *Modern Problems in Pediatrics* (F. Falkner, ed.), Vol. 7, pp. 111–122, S. Karger, Basel.

de Licárdie, E., Cravioto, J., and Zaldivar, S., 1972, Cambio educativo intergeneracional en la madre y cremiento fisico del niño rural en el primer año de la vida, *Bol. Med. Hosp. Infant. Mex. (Span. Ed.)* **29**:575.

Deming, J., and Washburn, A. H., 1963, Application of the Jenss curve to the observed pattern of growth during the first 8 years of life in 40 boys and 40 girls, *Hum. Biol.* **35**:484.

Demirjian, A., 1980, Nutrition Canada, Anthropometry Report, Growth Research Center, University of Montreal.

Doornbos, L., and Jonxis, J. H. P., 1968, Growth of the Bushnegro children on the Tapanahony River in Dutch Guyana, *Hum. Biol.* **40**:396.

Evans, M. E., 1944, Illness history and physical growth, *Am. J. Dis. Child.* **58**:390.

Falkner, F., 1958, Some physical measurements in the first three years of life, *Arch. Dis. Child.* **33**:1.

Falkner, F. (ed.), 1966, *Human Development,* W. B. Saunders, Philadelphia.

Faulhaber, J., 1961, El crecimiento en un grupo de niños Mexicanos, Instituto Nacional de Antropologia e Historia, Mexico City, Mexico.

Fetter, V., and Suchy, J., 1967, Anthropologic parameters of Czechoslovak young generation, current state, *Cesk. Pediatr.* **22**:97.

Fisch, R. D., Bilek, M. K., and Ulstrom, R., 1975, Obesity and leanness at birth and their relationship to body habitus in later childhood, *Pediatrics* **45**:521.

Fry, P. C., Howard, J. E., and Logan, B. C., 1975, Body weight and skinfold thickness in black, Mexican-American, and white infants, *Nutr. Rep. Int.* **11**:155.

Gampel, B., 1965, The relation of skinfold thickness in the neonate to sex, length of gestation, size at birth and maternal skinfold, *Hum. Biol.* **37**:29.

Garn, S. M., 1961, The genetics of normal human growth, *Genet. Med.* **2**:413.

Garn, S. M., 1966, Body size and its implications in: *Review of Child Development Research* (L. W. Hoffman and M. L. Hoffman, eds.), pp. 529–561, Russell Sage Foundation, New York.

Garn, S. M., and Clark, D. C., 1975, Nutrition, growth, development, and maturation: Findings from the Ten-State Nutrition Survey of 1968–1970, *Pediatrics* **56**:306.

Garrow, J. S., and Pike, M. D., 1967, The long term prognosis of severe infantile malnutrition, *Lancet* **1**:4.

Gomez, F., Ramos-Galvan, R., Cravioto, J., Chavez, R., and Vasquez, J., 1956, Mortality in second and third degree malnutrition, *J. Trop. Pediatr.* **2**:77.

Gore, A. T., and Palmer, W. T., 1949, Growth of the preschool child in London, *Lancet* **1**:385.

Goss, R. J., 1964, *Adaptive Growth,* Academic Press, New York.

Graham, G. G., 1968, The later growth of malnourished infants; effects of age, severity, and subsequent diet, in: *Calorie Deficiencies and Protein Deficiencies* (R. A. McCance and E. M. Widdowson, eds.), pp. 301–314, Little, Brown, Boston.

Graham, G. G., and Adrianzen, B., 1972, Late "catch-up" growth after severe infantile malnutrition, *Johns Hopkins Med. J.* **131**:204.

Graham, G. G., MacLean, W. C., Jr., Kallman, C. H., Rabold, J., and Mellits, E. D., 1980, Urban–rural differences in the growth of Peruvian children, *Am. J. Clin. Nutr.* **33**:338.

Grantham-McGregor, S. M., and Desai, P., 1973, Head circumference of Jamaican infants, *Dev. Med. Child. Neurol.* **15**:441.

Habicht, J. P., Martorell, R., Yarbrough, C., Malina, R. M., and Klein, R. E., 1974, Height and Weight standards for preschool children. How relevant are ethnic differences in growth potential?, *Lancet* **1**:611.

Hansman, C., 1970, Anthropometry and related data, anthropometry skinfold thickness measurements, in: *Human Growth and Development* (R. W. McCammon, ed.), pp. 103–154, Charles C Thomas, Springfield, Illinois.

Hardy, J. B., and Mellits, E. D., 1972, Does maternal smoking during pregnancy have a long-term effect on the child?, *Lancet* **2**:1332.

Heald, F. P., and Hollander, R. J., 1965, The relationship between obesity in adolescence and early growth, *Pediatrics* **67**:35.

Herron, R. E., 1972, Biostereometric measurement of body form, *Yearb. Phys. Anthropol.* **16**:80.

Hertzberg, H. T. E., Churchill, E., Dupertuis, C. W., White, R. M., and Damon, A., 1963, *Anthropometric Survey of Turkey, Greece, and Italy,* Macmillan, New York.

Hewitt, D., Westropp, C. K., and Acheson, R. M., 1955, Oxford Child Health Survey, effect of childish ailments on skeletal development, *Br. J. Prev. Soc. Med.* **9**:179.

Huenemann, R. L., 1974, Environmental factors associated with preschool obesity. I. Obesity in six-month old children, *J. Am. Diet. Assoc.* **64**:480.

Hunt, E. E., 1966, The developmental genetics of man, in: *Human Development* (F. Falkner, ed.), pp. 76–122, W. B. Saunders, Philadelphia.

Huxley, J., 1932, *Problems of Relative Growth,* Methuen, London.

Indian Council of Medical Research, 1972, *Growth and Physical Development of Indian Infants and Children,* Medical Enclave, New Delhi, India.

James. L. S., 1959, Biochemical aspects of asphyxia at birth, in: *Adaptation to Extrauterine Life* (T. K. Oliver, ed.), pp. 66–71, Thirty-first Ross Conference on Pediatric Research, Ross Laboratories, Columbus, Ohio.

Jelliffe, D. B., 1966, *The Assessment of the Nutritional Status of the Community,* World Health Organization, Geneva.

Jelliffe, D. B., 1968, Child Nutrition in Developing Countries, PHS Publication 1822, U.S. Government Printing Office, Washington, D.C.

Jelliffe, D. B., and Jelliffe, E. F. B. (eds.), 1979, *Human Nutrition, A Comprehensive Treatise,* Vol. 2: *Nutrition and Growth,* Plenum Press, New York.

Jenss, R. M., and Bayley, N., 1937, A mathematical method for studying the growth of a child, *Hum. Biol.* **9**:556.

Johnston, F. E., 1981, Physical growth and development

and nutritional status: Epidemiological considerations, *Fed. Proc.* **40**:2583.

Johnston, F. E., 1982, Relationships between body composition and anthropometry, *Hum. Biol.* **54**:221–245.

Johnston, F. E., and Beller, A., 1976, Anthropometric evaluation of the body composition of black, white, and Puerto Rican newborns, *Am. J. Clin. Nutr.* **29**:61.

Johnston, F. E., and Mack, R. W., 1978, Obesity in urban black adolescents of high and low relative weight at 1 year of age, *Am. J. Dis. Child.* **132**:862.

Johnston, F. E., Hamill, P. V. V., and Lemeshow, S., 1974, Skinfold thickness of youths 12–17 years, United States. National Center for Health Statistics, Ser. 11, No. 132, U.S. Government Printing Office, Washington, D.C.

Johnston, F. E., Newman, B. C., Cravioto, J. C., Le Licardie, E. R., and Scholl, T., 1980*a*, A factor analysis of correlates of nutritional status in Mexican children, birth to 3 years, in: *Social and Biological Predictors of Nutritional Status, Physical Growth, and Neurological Development* (L. S. Greene and F. E. Johnston, eds.), pp. 291–310, Academic Press, New York.

Johnston, F. E., Scholl, T. O., Newman, B. C., Cravioto, J., and de Licardie, E. R., 1980*b*, An analysis of environmental variables and factors associated with growth failure in a Mexican village, *Hum. Biol.* **52**:627.

Kanawati, A. A., and McLaren, D. S., 1973, Failure to Thrive, in Lebanon. II. An investigation of the causes, *Acta Paediatr. Scand.* **62**:571.

Kantero, R.-L., and Tiisala, R., 1971*a*, IV. Height, weight, and sitting height increments for children from birth to 10 years, *Acta. Paediatr. Scand. (Suppl.)* **220**:18.

Kantero, R.-L., and Tiisala, R., 1971*b*, V. Growth of head circumference from birth to 10 years, *Acta. Paediatr. Scand. (Suppl.)* **220**:27.

Kirke, D. K., 1969, Growth rates of aboriginal children in central Australia, *Med. J. Aust.* **2**:1005.

LaBarre, W., 1954, *The Human Animal,* University of Chicago Press, Chicago.

Lampl, M., 1983, *Episodic Growth Patterns in the First Year of Life,* Ph.D. thesis, University of Pennsylvania, Philadelphia.

Lancet, 1970, Classification of infantile malnutrition (editorial), *Lancet* **2**:302.

Lindegaard, B., 1953, Variations in human body build. A somatometric and X-ray cephalometric investigation on Scandinavian adults, *Acta Psychiatr. Neurol. Scand. Suppl.* **86**:1.

Livson, N., McNeill, D., and Thomas, K., 1962, Pooled estimates of parent–child correlations in stature from birth to maturity, *Science* **138**:818.

Mack, R. W., and Ipsen, J., 1974, The height–weight relationship in early childhood. Birth to 48 month correlations in an urban low-income Negro population, *Hum. Biol.* **46**:21.

Mack, R. W., and Johnston, F. E., 1976, The relationship between growth in infancy and growth in adolescence, *Hum. Biol.* **48**:693.

Mack, R. W., and Kleinhenz, M. E., 1974, Growth, Caloric intake, and activity levels in early infancy: A preliminary report, *Hum. Biol.* **46**:354.

Malina, R. M., Habicht, J.-P., Yarbrough, C., Martorell, R., and Klein, R. E., 1974, Skinfold thickness at seven sites in rural Guatemalan Ladino children birth through 7 years of age, *Hum. Biol.* **46**:453.

Malina, R. M., Habicht, J.-P., Martorell, R., Lechtig, A., Yarbrough, C., and Klein, R. E., 1975, Head and chest circumferences in rural Guatemalan Ladino children, birth to 7 years of age, *Am. J. Clin. Nutr.* **28**:1061.

Manwani, A. H., and Agarwal, K. N., 1973, The growth patterns of Indian infants during the first year of life, *Hum. Biol.* **45**:341.

Martorell, R., Yarbrough, C., Lechtig, A., Habicht, J.-P., and Klein, R. E., 1975, Diarrheal diseases and growth retardation in preschool Guatemalan children, *Am. J. Phys. Anthropol.* **43**:341.

McFie, J., and Welbrown, H. F., 1962, Effect of malnutrition in infancy on the development of bone, muscle, and fat, *J. Nutr.* **76**:97.

McKenna, J. J., 1979, Aspects of infant socialization, attachment, and maternal caregiving patterns among primates: A cross disciplinary review, *Yearb. Phys. Anthropol.* **22**:250.

Meisami, E., 1975, Early sensory influence on regional activity of brain ATPases in developing rats, in: *Growth and Development of the Brain* (M. A. B. Brazier, ed.), pp. 51–74, Raven Press, New York.

Meredith, H. V., 1962, Childhood interrelations of anatomic growth rates, *Growth* **26**:23.

Meredith, H. V., 1970, Body weight at birth of viable human infants: A world wide comparative treatise, *Hum. Biol.* **42**:217.

Miller, F. J. W., Billewicz, W. Z., and Thomson, A. M., 1972, Growth from birth to adult life of 443 Newcastle Upon Tyne schoolchildren, *Br. J. Prev. Soc. Med.* **26**:224.

Mueller, W. H., 1975, *Parent–Child and Sibling Correlations and Heritability of Body Measurements in a Rural Colombian Population,* Ph.D. thesis, University of Texas, Austin.

Mueller, W. H., 1976, Parent–child correlations for stature and weight among school aged children: A review of 24 studied, *Hum. Biol.* **48**:379.

Newens, E. M., and Goldstein, H., 1972, Height, weight and the assessment of obesity in children, *Br. J. Prev. Soc. Med.* **26**:33.

Nowak, T. S., Jr., and Munro, H. N., 1977, Effects of protein-calorie malnutrition on biochemical aspects of brain development, in: *Nutrition and the Brain,* Vol. 2: *Control of Feeding Behavior and Biology of the Brain in Protein-Calorie Malnutrition* (R. J. Wurtman and J. J. Wurtman, eds.), pp. 193–260, Raven Press, New York.

Pogačnik, T., 1970, Antropometrijski standardi otrok v. Sloveniji, *Glas. Antropol. Drus. Jogosl.* **7**:91.

Prader, A., Tanner, J. M., and von Harnack, G. A., 1963,

Catch-up growth following illness or starvation, *J. Pediatr.* **62**:646.

Rea, J. N., 1971, Social and economic influences on the growth of pre-school children in Lagos, *Hum. Biol.* **43**:46.

Reed, R. B., and Stuart, H. C., 1959, Patterns of growth in height and weight from birth to 18 years of age, *Pediatrics* **24**:904.

Rharon, H., 1962, Some observations on head circumferences in children with nutritional deficiencies, *Turk. J. Pediatr.* **4**:169.

Roche, A. F., and Cahn, A., 1962, Subcutaneous fat thickness and caloric intake in Melbourne children, *Med. J. Aust.* **1**:595.

Roche, A. F., and Davila, G. H., 1974, Differences between recumbent length and stature within individuals, *Growth* **38**:313.

Roche, A. F., Sierrogel, R. M., Chumlea, W. C., Reed, R. B., Valadian, I., Eichorn, D., and McCammon, R. W., 1982, *Serial Changes in Subcutaneous Fat Thickness of Children and Adults,* S. Karger, Basel.

Roche, A. F., and Malina, R. M., 1983, *Manual of Physical Status and Peformance in Childhood,* Vols. 1A and 1B, Plenum Press, New York.

Russell, C. S., Taylor, R., and Low, C. E., 1968, Smoking in pregnancy, maternal blood pressure, pregnancy outcome, baby weight and growth, and other related factors, *Br. J. Prev. Soc. Med.* **22**:119.

Russell, M., 1976, Parent–child and sibling–sibling correlations of height and weight in a rural Guatemalan population of preschool children, *Hum. Biol.* **48**:501.

Schell, L. M., and Johnston, F. E., 1986, Physical growth and development of North American Indian and Eskimo children and youth, in: *Handbook of North American Indians* (F. S. Hulse, ed.), U.S. Government Printing Office, Washington, D.C. (in press).

Scholl, T. O., 1975, *Body Size in Developing Nations: Is Bigger Really Better?,* Ph.D. thesis, Temple University, Philadelphia.

Scholl, T. O., Johnston, F. E., Cravioto, J., and de Licardie, E. R., 1980, A prospective study of the effects of clinically severe protein-energy malnutrition on growth, *Acta Paediatr. Scand.* **69**:331.

Sims, L. S., and Morris, P. M., 1974, Nutritional status of preschoolers, *J. Am. Diet. Assoc.* **64**:492.

Snyder, R. G., Spencer, M. L., Owings, C. L., and Schneider, L. W., 1975, Anthropometry of U.S. infants and children, SP-394, Society of Automotive Engineers, Warrendale, Pennsylvania.

Stuart, H. C., and Meredith, H. V., 1946, Use of body measurements in the school health program, *Am. J. Publ. Health* **36**:1365.

Sveger, T., Lindberg, T., Weibul, B., and Olsson, U. L., 1975, Nutrition, overnutrition, and obesity in the first year of life in Malmö, Sweden, *Acta Paediatr. Scand.* **64**:635.

Tanner, J. M., 1962*a*, *Growth at Adolescence,* 2nd ed., Blackwell, Oxford.

Tanner, J. M., 1962*b*, The evaluation of growth and maturity in children, in: *Protein Metabolism* (F. Gross, ed.), pp. 361–382, Springer-Verlag, Berlin.

Tanner, J. M., 1963, The regulation of human growth, *Child Dev.* **34**:817.

Tanner, J. M., 1974, Variability of growth and maturity in newborn infants, in: *The Effect of the Infant on its Caregiver* (M. Lewis and L. A. Rosenblum, eds.), pp. 77–103, Wiley, New York.

Tanner, J. M., 1976, Growth as a monitor of nutritional status, *Proc. Nutr. Soc.* **35**:315.

Tanner, J. M., and Israelsohn, W. J., 1963, Parent–child correlations for body measurements of children between the ages of 1 month and 7 years, *Ann. Hum. Genet.* **26**:245.

Tanner, J. M., and Whitehouse, R. H., 1975, Revised standards for triceps and subscapular skinfolds in British children, *Arch. Dis. Child.* **50**:142.

Tanner, J. M., Healy, M. J. R., Lockhart, R. D., MacKenzie, J. D., and Whitehouse, R. H., 1956, Aberdeen growth study, I. The prediction of adult body measurements from measurements taken each year from birth to 5 years, *Arch. Dis. Child.* **31**:372.

Tanner, J. M., Whitehouse, R. H., and Takaishi, M., 1966*a*, Standards from birth to maturity for height, weight, height velocity, and weight velocity: British children, 1965—I, *Arch. Dis. Child.* **41**:454.

Tanner, J. M., Whitehouse, R. H., and Takaishi, M., 1966*b*, Standards from birth to maturity for height, weight, height velocity, and weight velocity: British children, 1965—II, *Arch. Dis. Child.* **41**:613.

Tanner, J. M., Goldstein, H., and Whitehouse, R. H., 1970, Standards for children's height at ages 2–9 years allowing for height of parents, *Arch. Dis. Child.* **45**:755.

Terada, H., and Hoshi, H., 1965*a*, Longitudinal study on the physical growth in Japanese. (2) Growth in stature and body weight during the first three years of life, *Acta Anat. Nippon* **40**:166.

Terada, H., and Hoshi, H., 1965*b*, Longitudinal study on the physical growth in Japanese. (3) Growth in chest and head circumferences during the first three years of life, *Acta Anat. Nippon* **40**:368.

Terada, H., and Hoshi, H., 1966, Longitudinal study on the physical growth in Japanese. (4) Growth in length of extremities during the first three years of life, *Acta Anat. Nippon* **41**:313.

Timiras, P. S., 1972, *Developmental Physiology and Aging,* Macmillan, New York.

Tuddenham, R. D., and Snyder, M. M., 1954, Physical growth of California boys and girls from birth to eighteen years, *Univ. Calif. Publ. Child Dev.* **1**:183.

Vandenberg, S. G., and Falkner, F., 1965, Hereditary factors in human growth, *Hum. Biol.* **37**:357.

Van Wieringen, J. C., Wafelbakker, F., Verbrugge, H. P., and De Haas, J. H., 1971, *Growth Diagrams, 1965, Netherlands,* Wolters-Noordhoff, Groningen, Holland.

Vaughan, V. C., 1964, Growth and development, in:

Textbook of Pediatrics (W. E. Nelson, ed.), pp. 15–71, W. B. Saunders, Philadelphia.

Verghese, K. P., Scott, R. B., Teixeira, G., and Ferguson, A. D., 1969, Studies in growth and development, XII. Physical growth of North American Negro children, *Pediatrics* **44**:243.

Vijayaraghavan, K., and Gowrinath Sastry, K., 1976, The efficacy of arm circumference as a substitute for weight in the assessment of protein-calorie malnutrition, *Ann. Hum. Biol.* **3**:229.

Waterlow, J. C., Buzina, R., Keller, W., Lane, J. M., Nichaman, M. Z., and Tanner, J. M., 1977, The presentation and use of height and weight data for comparing the nutritional status of groups of children under the age of 10 years, *Bull. WHO* **55**:489.

Weiner, J. S., and Lourie, J. A., 1969, *Human Biology, A Guide to Field Methods,* IBP Handbook No. 9, F.A. Davis, Philadelphia.

Wilson, R. S., 1976, Concordance in physical growth for monozygotic and dizygotic twins, *Ann. Hum. Biol.* **3**:1.

Wilson, R. S., 1979, Twin growth: Initial deficit, recovery, and trends in concordance from birth to nine years, *Ann. Hum. Biol.* **6**:205.

Wingerd, J., 1970, The relation of growth from birth to 2 years to sex, parental size and other factors, using Rao's method of transformed time scale, *Hum. Biol.* **42**:105.

Wishaky, A., and Khattab, A., 1962, The effect of failure of growth due to undernutrition on the different parts of the body in infancy, *J. Egypt. Med. Assoc.* **45**:1029.

Wolanski, N., 1961, The new graphic method of the evaluation of the tempo and harmony of child physical growth, in: *Proceedings of the Seventh Meeting Deutsch Gesellschaft für Anthropologie* (W. Gieseler and I. Tillner, eds.), pp. 131–139, Musterschmidt-Verlag, Gottingen.

Wolanski, N., and Lasota, A., 1964, Physical development of countryside children and youth aged 2 to 20 years compared with the development of town youth of the same age, *Z. Morphol. Anthropol.* **54**:272.

Yarbrough, C., Habicht, J.-P., Malina, R. M., Lechtig, A., and Klein, R. E., 1975, Length and weight in rural Guatemalan Ladino children: Birth to seven years of age, *Am. J. Phys. Anthropol.* **42**:439.

2

Bone Growth and Maturation

ALEX F. ROCHE

1. Introduction

Growth is commonly defined as an increase in size, but more specificity is needed. An increase in size of the whole body or of an organ such as a bone may be caused by one or a combination of three processes: (1) Hyperplasia, or an increase in cell number, which involves duplication of DNA and cellular division; (2) Hypertrophy, or an increase in cell size, with true hypertrophy implying an increase in the size of the active functional elements of a cell, such as occurs in skeletal muscle with exercise; and (3) Storage of organic or nonorganic materials within or among cells. Each of these three processes occurs during bone growth, but the extent to which any one of them dominates depends on age or maturity and the part of the bone considered. Furthermore, these processes are often reversed in localized areas. While the overall change is an increase in size, some parts of a bone may become smaller or may be removed completely. In such areas, there is a reduction in cell number or cell size or in the amounts of organic or nonorganic material that are stored. Commonly, all these processes are combined.

Maturation is a more difficult concept. A widely used American dictionary defines maturation as: "the process of becoming mature." Unfortunately that source defines mature as "having undergone maturation." In regard to bone, maturation refers to the sequential alterations in this tissue as it changes from an embryonic anlage of a bone until each of its parts attains adult form and functional level. The characteristics of adult maturity are common to all normal young adults and do not show individual variation.

This chapter considers the growth and maturation of a typical long bone and an irregularly shaped bone from the early prenatal period to early adulthood. The changes during the growth and maturation of these typical bones occur in most other parts of the skeleton, with modifications that involve the extent and timing of the changes rather than their type or sequence. Separate descriptions are necessary for two complex regions: the craniofacial area and the pelvis. The former area is considered elsewhere in this treatise (Chapter 11), while the latter has been described in detail by several workers, including Reynolds (1945, 1947), Harrison (1958*a,b,* 1961), Coleman (1969), Moerman (1981, 1982). Attention is not given to factors that regulate bone growth—these have received adequate attention elsewhere, e.g., Bourne (1971), Ritzen *et al.* (1981), and Krabbe *et al.* (1982).

2. Prenatal Growth and Maturation of a Long Bone

The following description is appropriate for most long bones, although the timing of changes differs among these bones. Timing before birth is expressed as postconceptional age. Some alternative schemes are based on either crown–rump length or the maturity horizons of Streeter (1951); all these classification criteria are closely correlated. The

ALEX F. ROCHE • Division of Human Biology, Department of Pediatrics, Wright State University School of Medicine, Yellow Springs, Ohio 45387

term "gestational age" is common in the obstetrical literature. In some reports, gestational age equates with postconceptional age, but it is usually calculated without adjustment from the date of either the last menstrual period or the last normal menstrual period. The latter ages are systematically less than postconceptional ages.

Where a long bone is destined to develop, the first visible change is a condensation of mesenchyme. This packing of loosely arranged cells occurs at about 7 weeks of fetal life; soon afterward, the cells secrete compounds characteristic of cartilage. This occurs first in the central part of the condensed area; it is associated with a separation of the central cells and compression of the marginal cells into a membrane called the perichondrium. The cells of the inner layer of this membrane remain undifferentiated and retain the potential to form cartilage. The cells in the outer, more obvious, layer change to fibroblasts and sharply define the shape and size of the model of the future bone (Oliver, 1962). Even when first formed, this model resembles the adult bone in shape (Lewis, 1902; Hesser, 1926), although its ends are relatively wider and there are differences in structural details (Oliver and Pineau, 1959; Gardner and Gray, 1970). For example, the linea aspera of the femur is not present until the third prenatal month (Gardner and Gray, 1970). With chondrification, there is an indication of future joints in areas in which the mesenchyme persists as homogeneous interzones.

The model of a long bone enlarges by apposition from the deep layer of the perichondrium that surrounds it completely. This apposition involves the formation of chondrocytes (cartilage cells). The bone grows also by the division of chondrocytes within the model and by an increase in the intercellular material. Newly formed chondrocytes are concentrated at each end of the model, where the cells are densely packed. Rapid growth in these areas results in elongation of the model and increase in its width. The increase in width is more rapid at the ends than in the central part of the model, partly because the central cells do not divide. These central cells grow old, enlarge, develop vesicles and later vacuoles in their cytoplasm, and finally disintegrate. While these changes occur, the older chondrocytes secrete alkaline phosphatase, and the intercellular substance around them becomes calcified (Niven and Robison, 1934; Fell and Robison, 1934) (Figure 1A). This calcified area is usually too small to be visible radiographically, but exceptions occur in the vertebral bodies (Hadley, 1956).

Ossification begins around the central part of the model after this area of cartilage is calcified (Figure 1B). At first, osteoid tissue (uncalcified bone matrix) forms deep to the perichondrium as a collar. This osteoid tissue calcifies quickly to form periosteal bone in contact with the cartilage. The membrane external to this bone is called the periosteum; the deep cells in the periosteum are osteoblasts (bone-forming cells) that can become osteocytes (bone cells).

The calcified cartilage in the central part of the model is now vascularized by masses of cells that extend quickly from the periosteum through the periosteal bone. These masses contain osteoclasts (bone-removing cells) and undifferentiated cells. The cartilage cells are destroyed and blood vessels

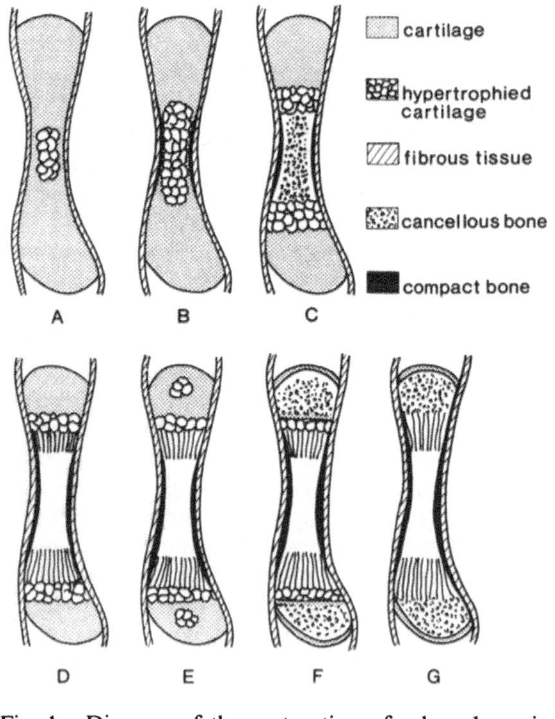

Fig. 1. Diagram of the maturation of a long bone in which the length of the bone has been kept constant. (A–G) Approximate age scale: (A) sixth prenatal week, (B) seventh prenatal week, (C) twelfth prenatal week, (D) sixteenth prenatal week to 2 years, (E) 2–6 years, (F) 6–16 years, and (G) adulthood. The clear area in (D–G) represents the marrow cavity. (Based on Roche, 1967.)

form over an area that elongates in the major axis of the bone. Some of the cells differentiate to become osteoblasts; others become hemopoietic (blood-forming) cells. The osteoblasts form bone around the remains of the calcified cartilage (Rambaud and Renault, 1864; Gray et al., 1957, Brunk and Sköld, 1962; Gardner, 1971) (see Figure 1B,C). Some workers, e.g., Ham (1974), illogically restrict the term "center of ossification" to this early endochondral ossification; the term should be applied to each discrete site of ossification. For example, in the femur, there are subperiosteal and endochondral centers near the mid-shaft, and other endochondral centers from much later near the ends of the bone. It is important that these centers of ossification not be called "growth centers" because the latter term connotes control over subsequent growth (Koski, 1968).

The ossified area extends toward the ends of the model both within the model (endochondral) and on its surface (subperiosteal) (see Figure 1C,D). The endochondral bone consists of trabeculae (spicules) with cores of calcified cartilage. These trabeculae are separated from each other by vascular tissue. The subperiosteal bone is more dense and is called "cortical bone" or "compact bone," although it is penetrated by numerous channels that contain blood vessels (Haversian systems). Near each end of the model, just beyond the ossified area, there is a narrow zone of calcified cartilage and next to it, still nearer the end, a zone in which the chondrocytes are hypertrophied.

Periosteal ossification extends more rapidly than endochondral ossification until it reaches the level of the zone of hypertrophied chondrocytes. While the ossified area is extending to occupy relatively more of the cartilaginous model, the model itself is elongating and becoming wider. Elongation is mainly due to the division of chondrocytes near the ends of the model and an increase in the intercellular substance. In the humerus, for example, the length of the ossified part of the model increases from 19% of the total length of the model at 10 weeks to 46% at 12 weeks, 71% at 17 weeks, and 79% at term (Gray and Gardner, 1969).

Soon the central trabeculae are absorbed, with the formation of a marrow cavity that elongates toward the ends of the bone and that enlarges laterally as the bone becomes wider (Figure 1D). The trabeculae around this marrow cavity widen as bone is deposited on their surfaces. By about the third month of fetal life, the bone has reached a level of maturation that is relatively fixed until birth or later (Figure 1D). This is in agreement with the generalization that organogenesis occurs mainly during the first three prenatal months.

The diaphysis (shaft) now has a dense cortex surrounding trabeculae and a marrow cavity. The ends of the bone are cartilaginous and, at the junctions between cartilage and bone (the future epiphyseal discs), there are transverse zones of hypertrophied chondrocytes covering both the dense outer cortex and the loose cancellous bone (trabeculae) present over the ends of the marrow cavity. These hypertrophic chondrocytes are in columns that have their long axes approximately parallel to that of the bone. The columns are separated by calcified cartilage strips around which bone is deposited. The importance of this area in bone elongation will be considered later. The radiographic appearance of the hand–wrist area in the early prenatal period is illustrated in Figure 2.

Bone growth is dependent not only on apposition but on resorption (Trueta, 1968). Without the latter, a normal adult shape would not be attained. One obvious aspect of remodeling concerns the width of the shaft. Early in development, the shaft flares near the cartilaginous end of the bone, and its width at this level is reduced as the bone elongates and the wider area is incorporated into a more central, narrower part of the shaft. This reduction in width involves the resorption of bone from the external surface of the cortex and concomitant apposition of bone on its deep surface. As a result of these processes, trabeculae formed in cartilage are incorporated into the cortex, which is of periosteal origin. In some areas, this change is so marked that the whole periosteal cortex is replaced by endochondral bone.

Remodeling, which occurs by a combination of apposition and resorption, is necessary also to retain the relative positions of muscular prominences and other features on the external surfaces of bones (Amprino and Cattaneo, 1937; Lacroix, 1951; Enlow, 1963; Gardner and Gray, 1970). Without this remodeling, features nearer the midpoint of the diaphysis than the epiphyseal zone would become relatively closer to this midpoint. There is also considerable internal remodeling throughout life that involves a rearrangement of the Haversian systems in the cortex and of the trabeculae. Some investigators believe that this remodeling occurs in anatomically discrete foci that can be called bone-remodeling units (Frost, 1964; Rasmussen and

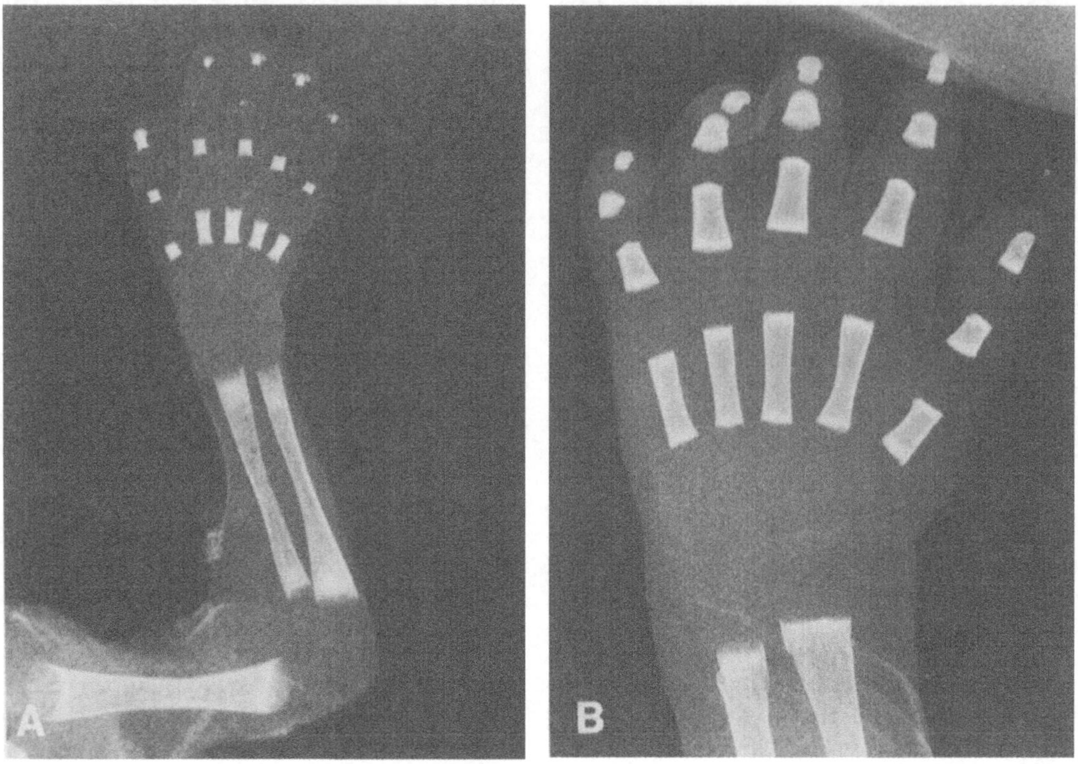

Fig. 2. Radiographs of silver-impregnated hand–wrist bones in fetal life. (A) At 12 weeks, all diaphyses have ossified except middle phalanx V, but proportionately large areas are not yet radiopaque. (B) At 16 weeks, note the considerable increase in the proportion of the hand–wrist area that is radiopaque and the progression toward adult proportions in the diaphyses.

Bordier, 1974; Parfitt, 1979). Further discussion of this important aspect is not included here, but the reader is referred to the early work of Amprino and Bairati (1936) and to a review by Lacroix (1971).

3. Postnatal Growth and Maturation of a Long Bone

The major early change from the prenatal state is the onset of ossification in the cartilaginous ends of the bone. These endochondral centers develop separately from the ossified shaft and are called "epiphyses" or "epiphyseal centers of ossification" (Figure 1F). In some bones, several centers develop at one end. For example, in the cartilage at the proximal end of the humerus there are separate centers for the head, greater tuberosity, and lesser tuberosity; these enlarge and fuse, forming a compound epiphysis.

At the site at which ossification will occur, the chondrocytes enlarge and become vesicular. The intercellular substance in this area calcifies and is invaded by vascular buds from adjacent cartilage canals (Figure 1E). Most of the calcified cartilage is removed; bone forms around the remnants (Figure 1F). Histologically, these changes are the same as those that occur when an endochondral center of ossification forms in the shaft. The ossified epiphyseal area enlarges rapidly within the cartilage, after the cartilage around the ossified area has passed through the same sequential changes that occurred in the first part of the cartilage to be ossified. Ossification does not extend throughout the cartilage at the end of the bone. The extreme end remains cartilaginous throughout life as articular cartilage and, for a long time, a disc of cartilage remains between the shaft and the ossified epiphysis. The disc is essential for further elongation; it persists until the bone reaches its adult length (Figure 1G).

At first, the ossified epiphysis is spherical and therefore appears circular in radiographs. Later its shape changes to approximate that of the cartilaginous end of the bone (Todd, 1930*a;* Greulich and Pyle, 1959; Pyle *et al.*, 1961; Gardner, 1971), when it is curved on all its surfaces except for the approximately flat aspect facing the shaft. This flat surface is covered by a thin layer of bone called the terminal plate, which was noted in radiographs as early as 1910 (Hasselwander, 1910). Only for a brief period after the onset of ossification does the ossified area enlarge, to any marked extent, on the surface adjacent to the epiphyseal zone (Payton, 1933; Siegling, 1941; Haines, 1975).

Soon after ossification begins, the epiphysis widens much more rapidly than does the shaft. Consequently, the ratio between epiphyseal width and shaft width (the latter being measured at the end of the shaft, i.e., the metaphysis) is very useful in assessing skeletal maturity (Mossberg, 1949; Scheller, 1960; Murray *et al.*, 1971; Tanner *et al.*, 1975; Roche *et al.*, 1975*b*), although it may be misleading in some pathological conditions, especially rickets (Acheson, 1966). While the ossified epiphysis widens, the epiphyseal disc also increases in width by the proliferation of chondrocytes deep in the cartilaginous epiphysis and in the margin of the epiphyseal disc and by the hypertrophy of chondrocytes in the disc (Heřt, 1972).

3.1. The Epiphyseal Disc

The layer of the disc near the ossified epiphysis consists of resting cartilage. A little nearer the diaphysis, the chondrocytes are arranged in columns that have their long axes parallel to the long axis of the bone. The chondrocytes in the columns divide and later mature before they hypertrophy, thus expanding the disc lengthwise. Soon after their maturation is complete, the nearby cartilage is calcified and the chondrocytes die. Most of the calcified cartilage is resorbed, and ossification occurs on the surface of the remnant. These latter changes occur on the diaphyseal aspect of the epiphyseal disc. After the ossified epiphysis of a long bone is well developed, elongation of the shaft occurs only by the formation of bone on the surface of the cartilage of the epiphyseal disc that faces the shaft. The disc remains approximately constant in thickness despite the marked growth changes in this area. The rate at which chondrocytes divide and enlarge is approximately balanced by the rate at which they die and are replaced by bone.

A short bone, e.g., proximal phalanx II, has an epiphysis at one end only. Surprisingly however, a popular textbook of histology states that the phalanges "do not develop epiphyseal centers of ossification" (Ham and Cormack, 1979). The nonepiphyseal end of the shaft is covered by articular cartilage and chondrocytes divide at the junction between this cartilage and the bone. Consequently, some elongation of the shaft occurs at the nonepiphyseal end. In short bones, between 2 and 11 years, about 20% of total elongation occurs at the nonepiphyseal end (Roche, 1965; Lee, 1968).

3.2. Increase in Shaft Width

Increase in shaft width occurs by the apposition of bone on the external surface due to the activity of osteoblasts in the deep layer of the periosteum. At the same time, bone is resorbed from the internal surface of the periosteum. As a result, the external diameter of the shaft and the diameter of the marrow cavity enlarge, but the cortex does not necessarily increase in thickness. In general, however, cortical thickness does increase with growth (Garn, 1970; Kimura, 1978; Low and Kung, 1982). In most bones, the processes of external apposition and internal resorption are unevenly distributed on the various aspects of the shaft (medial, lateral, anterior, posterior) with resultant changes in the position and shape of the cortical bone (Hattori *et al.*, 1979). Such a change is shown diagrammatically in Figure 3A, which represents the earlier cross section of a bone in which apposition and resorption

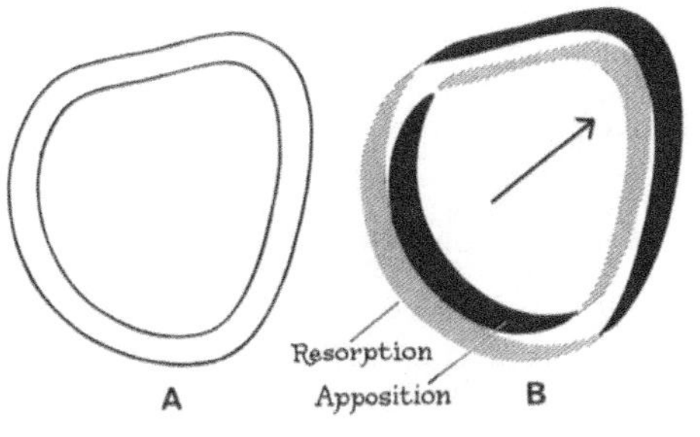

Fig. 3. Diagrammatic representation of the shaft of a long bone (A). As a result of resorption and apposition, cortical drift can occur in the direction of the arrow (B).

occur on different aspects of the external surface of the shaft with the opposite changes occurring on the corresponding parts of the internal surface of the cortex. As a result, the shaft drifts in the direction of the arrow while retaining the same shape. In general, the areas of apposition and resorption can be recognized from histological sections of dried bones (Enlow, 1963); limited inferences can be made from serial radiographs if bone scars (growth arrest lines) are present (Garn *et al.*, 1968).

3.3. Epiphyseal Fusion

At early ages, the end of the shaft consists of the surrounding cortex and the ends of trabeculae (Figure 4). Later, the end of the shaft is covered by a thin transverse layer of bone separating it from the cartilage of the epiphyseal disc (Hasselwander, 1910; Hellman, 1928; Todd, 1930*b;* Moss and Noback, 1958; Silberberg and Silberberg, 1961). This layer of bone, together with the calcified cartilage on its epiphyseal aspect, is visible radiographically, and its thickness is related to the rate of growth in stature (Park, 1954; Edlin *et al.*, 1976).

The division of chondrocytes in the disc becomes slower and finally ceases. During this phase, the epiphyseal disc becomes thinner and some elongation of the shaft occurs at the expense of the epiphyseal cartilage (Noback *et al.*, 1960). Finally, all the cartilage of the epiphyseal disc is replaced by bone. This occurs first peripherally or eccentrically—not necessarily in the central part of the disc (Stevenson, 1924; Todd, 1930*b;* McKern and Stewart, 1957; Haines and Mohiuddin, 1959; Haines, 1975). The last part of the epiphyseal disc to be replaced by bone is the circumferential margin. Consequently, in dried bones at the stage of nearly complete fusion, the epiphysis is fused to the shaft over a large area, but a circumferential groove remains at the level of fusion where the cartilage has not been replaced by bone.

In the area in which fusion occurs, a transverse layer of bone forms that is a combination of the termanal plate of the epiphysis and the plate of bone that covered the end of the shaft. This compound plate is visible radiographically. Usually it is resorbed soon after the completion of fusion, but it may persist for years later (Paterson, 1929; Sahay, 1941). After fusion is complete, the bone is adult in maturity, although there may be a radiopaque line at the level of fusion. The articular cartilage remains as the only remnant of the original model.

4. Growth and Maturation of an Irregularly Shaped Bone

In an irregularly shaped bone such as a carpal, ossification usually begins after birth. The early stages of chondrification and ossification are histo-

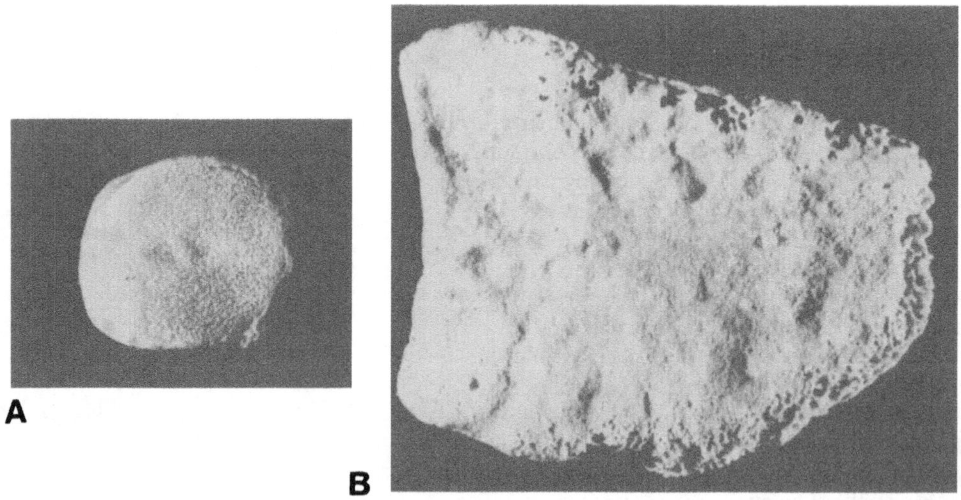

Fig. 4. The distal end of the diaphysis of the radius, viewed from the epiphyseal aspect, at 14 weeks of intrauterine life (A) and at 10 years (B). Before birth, the trabeculae of endochondral bone are not covered by a transverse layer of bone.

logically the same as those that occur within the models of long bones. Each carpal bone develops first as a condensation of embryonic connective tissue (Figure 5A). Subsequently, cavitation occurs in this connective tissue at the sites of future joints between the carpal and neighboring bones. After this occurs, the cartilaginous model, now resembling the future bone in shape, articulates with its neighbors on adjacent surfaces and is covered by a well-defined perichondrium on other surfaces (Figure 5B). It is invaded by blood vessels within cartilage canals, but further changes do not occur until ossification begins. Ossification starts in this cartilaginous model (Figure 5C) with the same histological processes as those described for endochondral ossification of long bones (Siffert, 1966). Occasionally, some carpals and tarsals have two primary centers, and the calcaneum always develops an ossified epiphysis. With the exception of the latter bone, the carpals and tarsals lack epiphyseal zones. At first, the ossified area expands rapidly in all directions (Gardner, 1971); later growth is more rapid in some directions than in others. The changes as the ossified area gradually matures to match the shape of an adult carpal (Figure 5D–F) are used to assess the skeletal maturity levels of children from radiographs. In adulthood, some surfaces of these bones are covered by periosteum and the others by articular cartilage, which is the only remnant of the cartilaginous model.

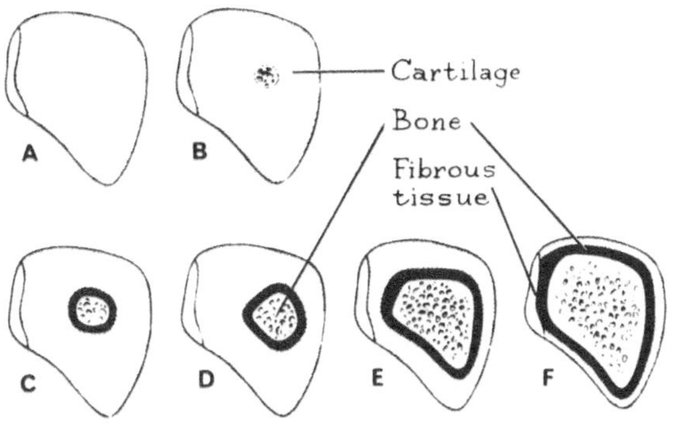

Fig. 5. Diagram of the maturation of a carpal bone in which size has been kept constant. (A–F) Approximate age scale: (A) prenatal, (B) 1–3 years, (C) 4–5 years, (D) 6 years, (E) 10 years, and (F) adulthood. (From Roche, 1967.)

5. Radiographic Changes

Calcified cartilage and bone are radiopaque; thus very small radiopaque areas may be calcified cartilage or bone surrounded by calcified cartilage. The only reliable radiographic evidence that bone is present is the recognition of trabeculae. As the ossified area enlarges, the changes in size and shape are visible radiographically in a two-dimensional view, but only those relating to shape are useful for the assessment of maturity. Measurements of size can be made with accuracy from radiographs, although the circumstances under which the radiographs were taken must be known so that the appropriate corrections for enlargement can be made (Tanner, 1962). If ratios are to be measured, variations in enlargement are of little importance (Poznanski *et al.*, 1978). The changes that occur during remodeling are not directly visible radiographically, although their occurrence can be inferred. Except insofar as they are reflected in changes of shape, they are not useful in the assessment of skeletal maturity.

6. Reference Data for Size and Shape

Bone lengths can be measured in various ways. Measurements of dried bones are inappropriate as a source of reference data because of obvious problems resulting from sample selection. Radiographic measurements are more reliable than those made on the living using calipers or tape measures (classic anthropometry) because of the difficulty of identifying landmarks and because measurements on the living have to made through overlying soft tissues. This difference in reliability was demonstrated convincingly by Maresh and Deming (1939) and by Day and Silverman (1952). One exception may concern the ulna and the tibia; Valk and coworkers reported that the lengths of these bones can be measured very accurately without radiographs using special apparatus (Valk, 1971, 1972; Valk *et al.*, 1983). These claims have not been validated independently.

The generalization that measurements on radiographs are preferable to those in the living patient is true only when the radiographs are taken under standardized conditions in which orientation is fixed and enlargement is known. Descriptions of appropriate radiographic techniques and of the principles involved have been given (Goff, 1960; Tanner, 1962; Shapiro, 1982).

The recommended reference values for the lengths of the major long bones are those suggested

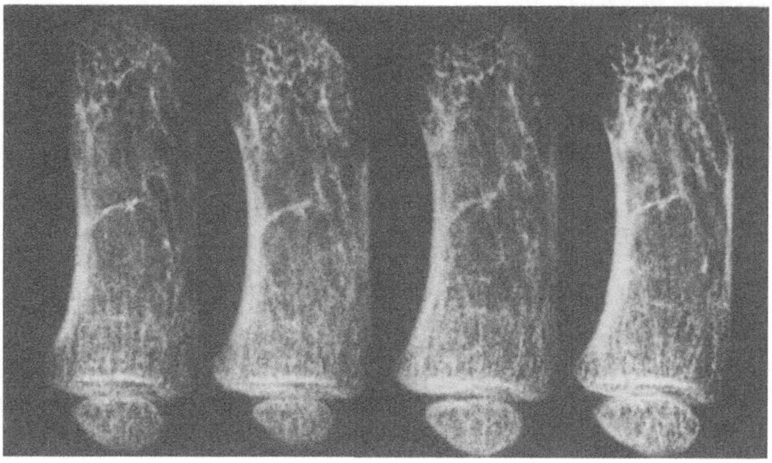

Fig. 6. Serial radiographs at 6-monthly intervals of a first metacarpal with the photographic enlargement adjusted to keep the total length constant. Note the fixed trabecular patterns that become relatively more distal with growth.

by Maresh (1970). Ratios between these values have been reported (Robinow and Chumlea, 1982). Dial calipers with sharp points can be used to measure radiographs and the distances recorded to the nearest 0.1 mm.* This level of accuracy is necessary for many research purposes. In other circumstances, useful data can be obtained using a transparent rule graduated to 1.0 mm; such a rule is adequate for the measurements required in the TW2 and RWT methods of assessing skeletal maturity (Tanner *et al.*, 1975; Roche *et al.*, 1975*b*). Serial standardized radiographs can be used also to determine the amounts of elongation at particular sites. For this purpose, natural bone markers, e.g., growth arrest lines, notches, or fixed trabecular patterns, are used as landmarks (Roche, 1965; Lee, 1968; Kimura and Takeuchi, 1982) (Figure 6).

Reference values for the lengths of the bones of the hand from birth to 15 months have been reported by Gefferth (1972) and from 2 years to adulthood by Garn *et al.* (1972*a*). Gefferth included the epiphyses whenever they were ossified; consequently, his reported lengths are dependent on levels of maturity. Garn and associates included the epiphyses in the distances measured, which were the maximum lengths in the long axis of each diaphysis. The data of Garn *et al.* (1972*a*) have been used to construct profiles of lengths within individual hands, and the deviations from these reference profiles in several diseases, e.g., hand–foot–uterus syndrome, Holt-Oran syndrome, and Turner syndrome, have been reported (Poznanski *et al.*, 1972*a,b*).

Some proportions involving the lengths of the short bones of the hand are also important in the clinical diagnosis of Turner syndrome because the fourth metacarpal is short in many of these XO individuals. This can be detected without measurement either by placing a ruler across the medial knuckles of the clenched fist, as described by Archibald *et al.* (1959), or by drawing a line on a radiograph corresponding to the plane of the ruler. As illustrated in Figure 7, in a normal individual, a

*One suitable instrument is Helios caliper, No. 4: $277.00 Mager Scientific Inc., 209 East Washington, Ann Arbor, Michigan 48108. More accurate data can be obtained with an electronic digitizer.

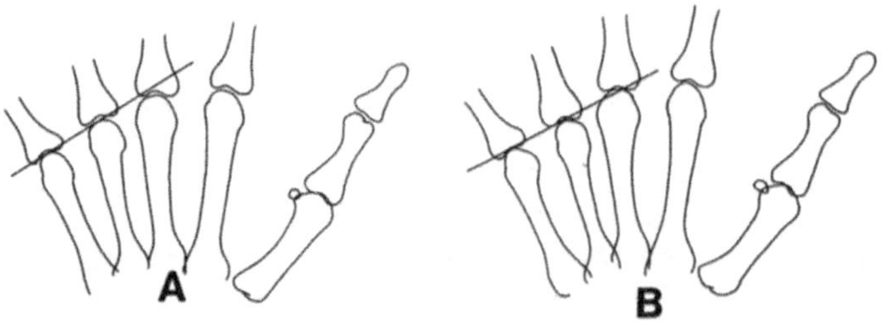

Fig. 7. Diagrams of the hand–wrist bones. (A) A tangent to the heads of the fifth and fourth metacarpals usually passes distal to the head of the third metacarpal (negative metacarpal sign). (B) A similarly constructed tangent passes across the head of the third metacarpal because the fourth metacarpal is short (positive metacarpal sign).

tangent to the most distal points on the heads of the fifth and fourth metacarpals passes distal to the head of the third metacarpal. In those with Turner syndrome and in some normal individuals, the tangent passes across the head of the third metacarpal. The shortness of the fourth metacarpal in Turner syndrome can be identified also by comparing the length of the fourth metacarpal with the combined lengths of the fourth proximal and distal phalanges in the same hand (Kosowicz, 1965).

Comparisons between lengths are used to determine whether brachymesophalangia (shortness of middle phalanx) of the fifth finger is present (Figure 8). Usually this can be recognized subjectively, but it is better to use measurements. The best choice is the ratio between the lengths of the fifth and fourth middle phalanges (Hewitt, 1963; Wetherington and Haurberg, 1974). This is preferable to criteria based on comparisons between the lengths of the fifth middle phalanx and other bones of the fifth ray because the correlation coefficients between the lengths of the bones of the hand are higher among bones in the same row, e.g., middle phalanges, than among bones in the same ray, e.g., metacarpal II, proximal phalanx II, middle phalanx II, and distal phalanx II (Roche and Hermann, 1970). Reference values are given in Table I.

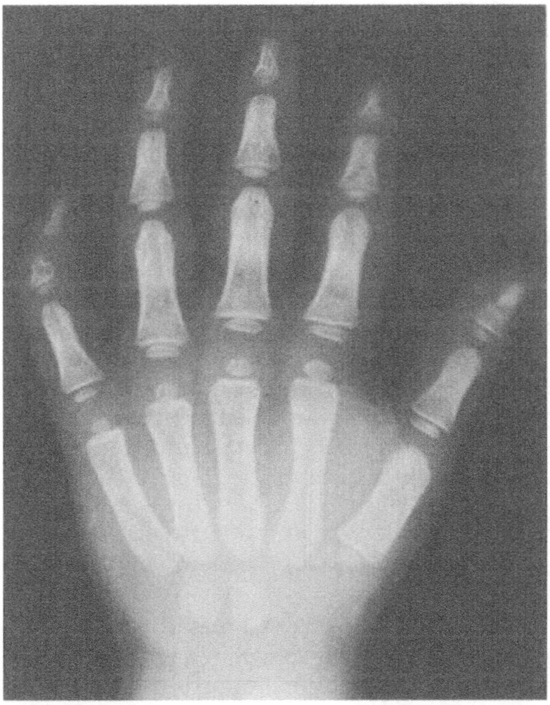

Fig. 8. Example of brachymesophalangia of the fifth digit.

Table I. Values for the Ratio (Middle Phalanx V/ Middle Phalanx IV) Lengths in the Hands of U.S. White Children[a,b]

Age (years)	Males		Females	
	Mean	SD	Mean	SD
1	0.66	0.06	0.65	0.06
4	0.67	0.05	0.66	0.04
9	0.69	0.04	0.68	0.04
Adult	0.73	0.03	0.71	0.03

[a]Lengths include epiphyses except at 1 year.
[b]Data from Garn *et al.* quoted by Poznanski (1974).

Reference values for bone widths have been reported for metacarpal II (external diameter, internal diameter, cortical thickness, and calculated cortical area) in white American children by Garn *et al.* (1971). The measurements were made at the midpoint of the bones, including the epiphyses; cortical areas were calculated on the assumption that the bones and the medullary cavities were circular. Corresponding data for Swiss children from 3 months to 11 years for metacarpal II and for metacarpals II–IV combined have been reported by Bonnard (1968). The influence of age and nutrition on cortical thickness and other measurements of the second metacarpal has been described by Garn (1970).

Cortical thickness and cortical area are used as indices of skeletal mass and therefore of calcium reserves (Dequeker, 1976; Meema and Meema, 1981). These measurements can be important not only in elderly women but in children with various diseases, e.g., renal disease, osteogenesis imperfecta. Measures of cortical thickness and cortical area are gradually being replaced by radiographic or photon absorptiometry, quantitative computerized tomography (CT), and neutron activation (Colbert and Bachtell, 1981; Mazess, 1981; Baziere, 1981; Cohn, 1981; Chesney and Shore, 1982; Tyson *et al.*, 1983).

In general, the reference data for shapes of bones are unsatisfactory. In part, this is attributable to the use of inappropriate methods. These complex shapes cannot be described accurately using a single ratio; e.g., length/width. Bone shapes, judged from radiographic outlines, have been summarized by curve-fitting methods. For example, Fourier analyses have been made that provide efficient summations of shape for both ectocranial and endocranial lateral silhouettes and for the outline of the distal end of the femur using parameters that can be interpreted readily (Lestrel, 1974, 1975; Lestrel *et al.*, 1977; Lestrel and Roche, 1976). Such approaches are suitable for research purposes only—it would be unreasonable to suggest that the reported parameters be used clinically as reference values.

Despite their limitations in regard to the description of skeletal shape, simple ratios between length and width can be useful clinically. If length/width ratios are calculated using data from metacarpals II–V combined, width being measured at the midpoint of each bone, the values increase from infancy to adulthood, being slightly lower in males (Sinclair *et al.*, 1960; Rand *et al.*, 1980). As expected, the ratio is high in Marfan syndrome. Kosowicz (1965) reported the ratio between the width of the subungual expansions and that of the shafts of the distal phalanges of the hand. This ratio is low in apical dystrophy, craniocleidodysostosis, and cretinism and is high in Turner syndrome (Macarthur and McCulloch, 1932; Brailsford, 1943, 1953; Kosowicz, 1965).

The shape of the carpal area is abnormal in Turner syndrome, if shape is determined from the angle between tangents to the medial and lateral parts of the proximal margin of the carpus. The normal range for this carpal angle is often given as 124.3° to 138.7° (Kosowicz, 1962, 1965) (Figure 9).

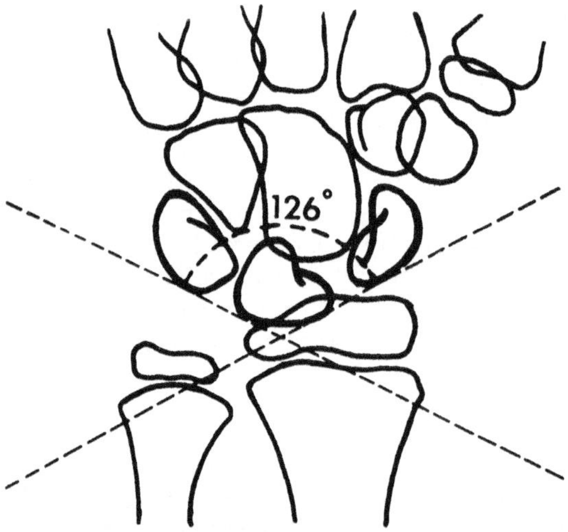

Fig. 9. Diagram of the carpal area to indicate the carpal angle. This is measured between a tangent to the proximal margins of the triquetral and lunate and a tangent to the proximal margins of the scaphoid and lunate. (From Kosowicz, 1962, 1965.)

It is important to note that it is about 10° higher in American blacks than whites, that it is 5° higher in the right hand than the left, and that it increases 5–10° between 5 and 14 years of age (Harper *et al.*, 1974). This angle is small in Turner syndrome. Also, the carpus is short in multiple epiphyseal dysplasia, the oropalatodigital syndromes, Turner syndrome, arthrogryposis and juvenile rheumatoid arthritis but long in achondroplasia (Poznanski *et al.*, 1978).

Many other skeletal dimensions are measured by radiologists. Some of the conditions in which they are useful are listed in Table II. Further details are available in the text of Lusted and Keats (1972).

7. Skeletal Weight

Data on skeletal weight have been reported for fetuses, infants, and children (Garrow and Fletcher, 1964; Trotter and Peterson, 1968, 1969*a,b,* 1970). Both before and after birth, the relationship between skeletal weight and body length approximates a sigmoid function (Figures 10 and 11), with larger increases in weight per centimeter as body length increases. This increase becomes very rapid after 170 cm of body length in boys. This spurt may be attributable to vagaries of sampling, but it could reflect late transverse growth of the trunk. The increase in skeletal weight at larger body lengths is much less marked in girls and in boys. Before birth, the weight of the femur in relationship to body length also has an approximately sigmoid function, following a curve that is almost parallel to that for total skeletal weight. This indicates that during fetal life the weight of the femur is a fixed proportion of the total skeletal weight. The relationship between the length of the femur and body length is essentially rectilinear for body lengths ranging from 23 to 176 cm. The means graphically represented in Figures 10 and 11 should be accepted as showing general trends only—the sample sizes are very small for all variables in the fetal period and for skeletal weight after birth.

8. Skeletal Maturation

The assessment of skeletal maturity from radiographs is based on the recognition of "maturity indicators." A maturity indicator is a radiographically visible feature that assists the determination of the maturity level of a bone (Roche, 1980). In the

Table II. Selected Skeletal Dimensions Measured on Radiographs with Their Clinical Relevance

Dimension	Use	Reference
Atlas–odontoid distance	Subluxation	Hinck and Hopkins (1960); Locke *et al.* (1966)
Spinal canal–AP diameter	Trauma, tumors, pathology of disc	Naik (1970); Hinck *et al.* (1962, 1965)
Vertebral column		
Interpedicular width	Achondroplasia	Hinck *et al.* (1966)
Disc and body height	Pathology of disc	Brandner (1970)
Shoulder, width of joint space	Dislocation	Arndt and Sears (1965)
Acetabular angle	Congenital dislocation of hip	Caffey *et al.* (1956)
Iliac angle and index	Down syndrome	Caffey and Ross (1958)
Axial relations within the foot	Congenital abnormalities	Templeton *et al.* (1965)

atlases, they are described in the text accompanying each standard; in the method of Tanner *et al.* (1975, 1983*a*) they are the criteria for stages; in the method of Roche *et al.* (1975*b*) they are called indicators and are divided into grades. After they have been recognized, a skeletal age is assigned to the radiograph, using a subjective or objective method of scoring. To be useful, an indicator must occur during the maturation of every child (Pyle and Hoerr, 1969); i.e., it must be universal. Because of limitations in the schedules of serial radiographic examinations, however, some indicator grades appear to be skipped in some children. Some of the appearances used as maturity indicators are shown in Figures 12 and 13. These illustrations of the distal end of the radius in boys have been enlarged so that the width of the end of the shaft is constant, thus removing overall size differences. Even to the inexperienced observer, there are clear differences in maturity between epiphyses with widely separated skeletal ages, e.g., 2.5 versus 3.5 years or 11.0 versus 14.0 years. It is equally evident that the recognition of the differences between, e.g., 10.0 and 11.0 years requires training and care and

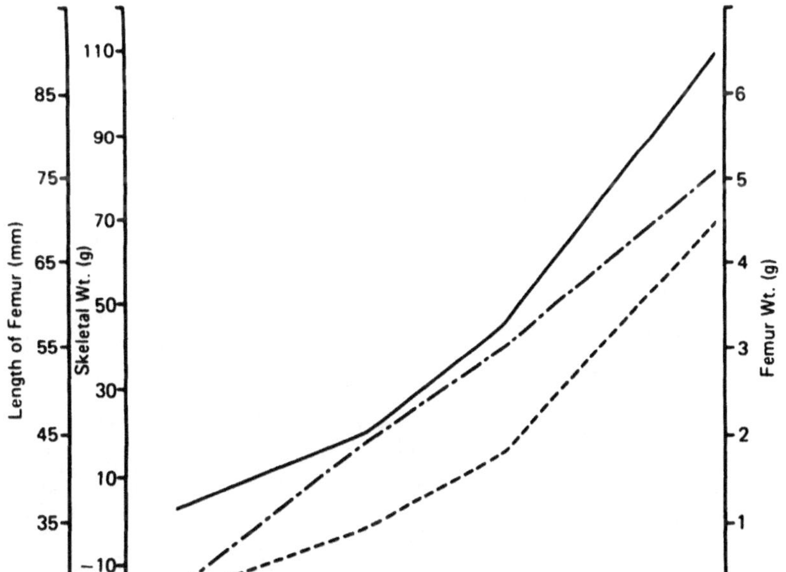

Fig. 10. Relationships of length of femur (—·—), skeletal weight (————), and weight of femur (- - - -) with body length during the prenatal period in U.S. white fetuses. (Based on Trotter and Peterson, 1969*a,b*.)

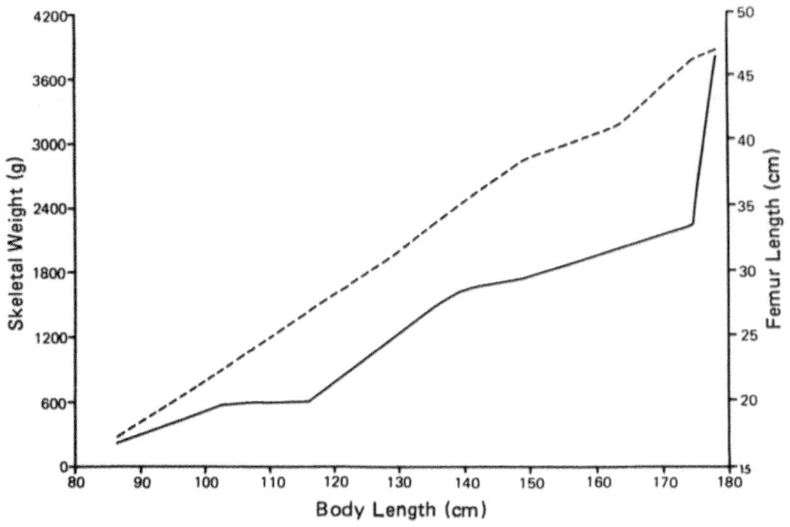

Fig. 11. Relationships of skeletal weight (—) and femur length (--) with body length after birth in U.S. white males. (Based on Maresh, 1970; Trotter and Peterson, 1970.)

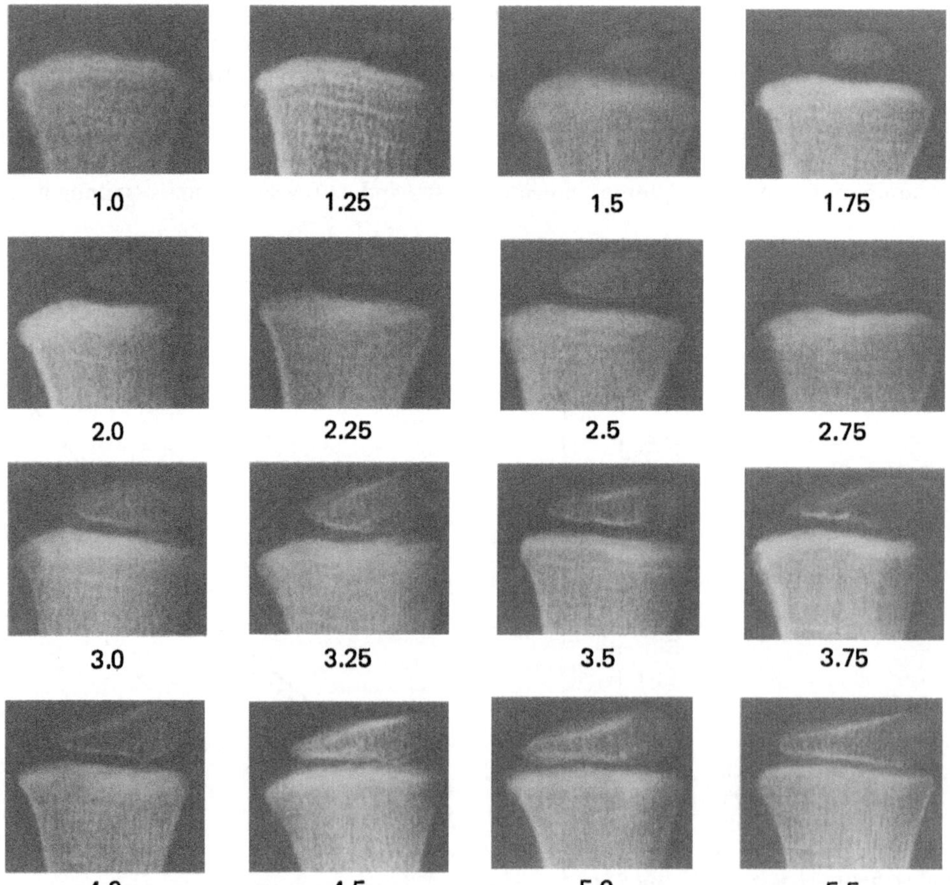

Fig. 12. The distal end of the left radius in boys aged 1.0–5.5 years, with diaphyseal width kept constant. The figure below each insert is the Greulich–Pyle skeletal age (years).

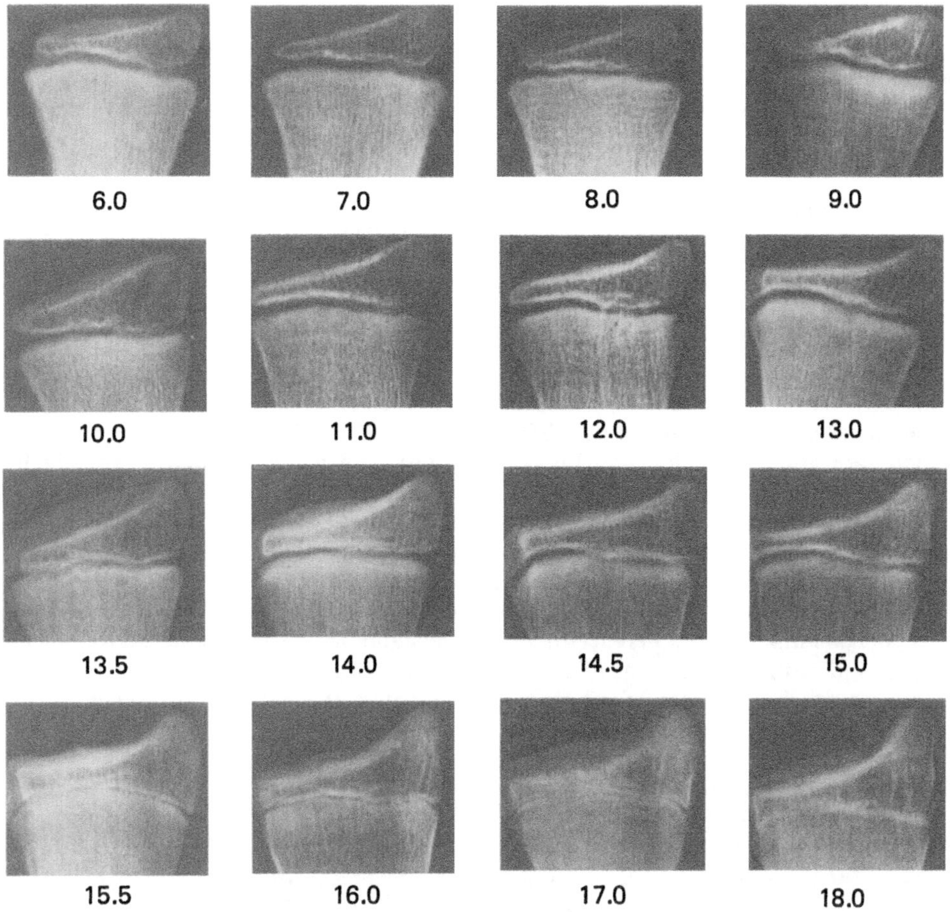

Fig. 13. The distal end of the left radius in boys aged 6.0–18.0 years, with diaphyseal width kept constant. The figure below each insert is the Greulich–Pyle skeletal age (years).

that some small differences, e.g., 7.0 versus 8.0 years, have to be carefully distinguished from individual variations in shape that may be present at the same maturity level.

Maturity indicators reflect the three-dimensional shapes of calcified or ossified areas. As bones become more mature, their external contours change because of differences in the rates of bone apposition and resorption at various surfaces. Parts of these surfaces that are apporximately parallel to the central axis of the radiographic beam produce dense white zones on a radiograph that may be maturity indicators. If these zones are long and narrow, they are called radiopaque lines. Radiographs used to assess skeletal age must be take under standardized conditions because the radiographic outline of a bone or the appearance of the shadow cast by a radiopaque zone or line depends on the inclination of the radiographic beam.

Maturity indicators provide information about the level of maturity in a particular bone at the time the radiograph was taken. Serial radiographs for an individual can show the duration of each indicator (or grade of an indicator), i.e., the period the bone remained at the same maturity level. Information concerning the duration of indicators (or grades) is useful in developing a scoring system because indicators of long duration are less informative than those of brief duration (Healy and Goldstein, 1976; Roche *et al.,* 1975*b*). Because the schedule of radiography in longitudinal growth studies is related to chronological age, not to the inception of maturity indicators, and because some scheduled visits are missed in all long-term series, this property of maturity indicators has been used only indirectly (Roche *et al.*, 1975*b*).

8.1. Number of Centers Method of Assessment

Some early methods for the assessment of skeletal maturity were based on the presence of ossification centers (Sontag *et al.*, 1939; Elgenmark, 1946). These methods have considerable appeal because radiographic positioning does not have to be controlled rigidly, the assessors need only minimal training, and observer errors are uncommon. While these methods are appropriate for preschool-age children and for areas that include many bones, e.g., hand–wrist and foot–ankle, if both the irregularly shaped bones and epiphyses are included (Yarbrough *et al.*, 1973), they are useless at older ages. One difficulty is that this method gives all centers the same weight despite their differences in representativeness and in the distributions of chronological ages over which the onset of ossification occurs. The problem is exemplified in Figure 14, which shows two hand–wrist bones with equal numbers of centers ossified that differ by 0.8 years in Greulich–Pyle skeletal age and by 35 points (0.7 years) on the TW2 scale.

8.2. Atlas Method of Assessment

The recognition and subjective grading of maturity indicators are the basis of the atlas method. This method was developed by the Cleveland school, first by Todd (1937) and then by others who have provided atlases of standards for the hand–wrist, foot–ankle, and knee (Greulich and Pyle, 1959; Hoerr *et al.*, 1962; Pyle and Hoerr, 1969; Pyle *et al.*, 1971). The atlas standards were derived from a group of white children of very high socioeconomic status living in Cleveland, Ohio, who were enrolled in the Brush Foundation Study. These classic atlases differ both in the area of the skeleton considered and in the manner in which the standards are presented. The Greulich–Pyle atlas for the hand–wrist has separate series of standards for boys and girls, each with their own set of corresponding skeletal ages expressed in years and

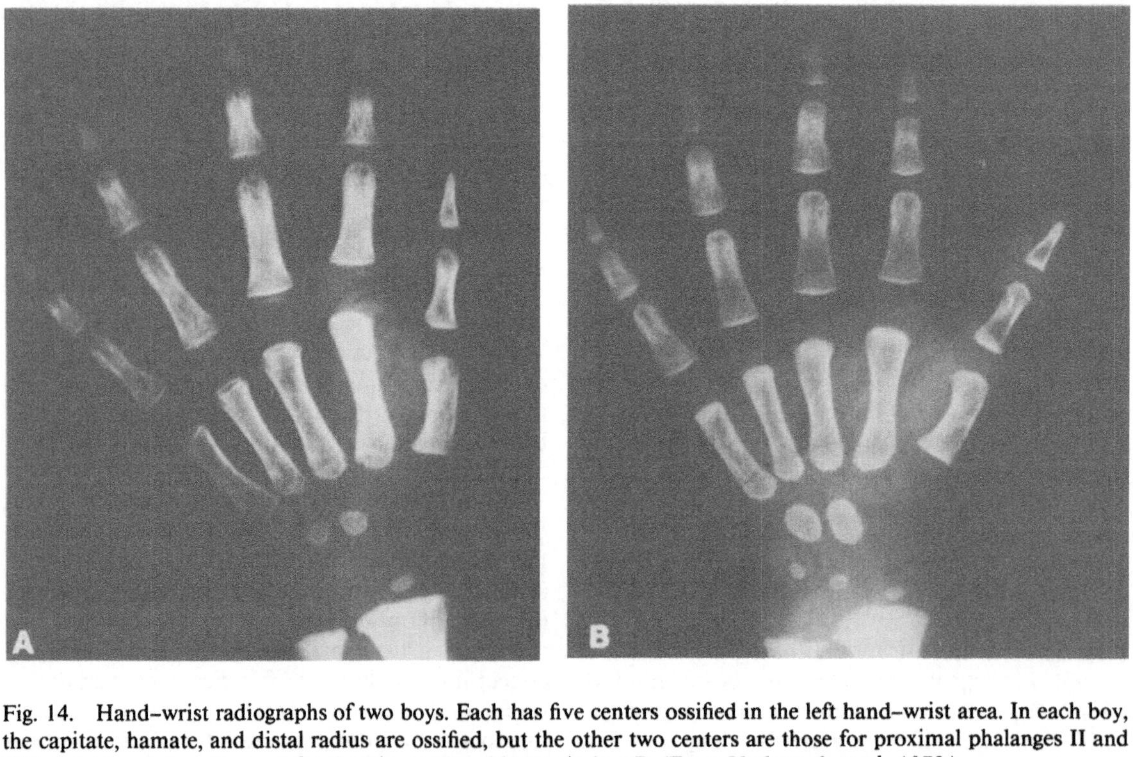

Fig. 14. Hand–wrist radiographs of two boys. Each has five centers ossified in the left hand–wrist area. In each boy, the capitate, hamate, and distal radius are ossified, but the other two centers are those for proximal phalanges II and III in boy A, whereas they are for the triquetral and lunate in boy B. (From Yarbrough *et al.*, 1973.)

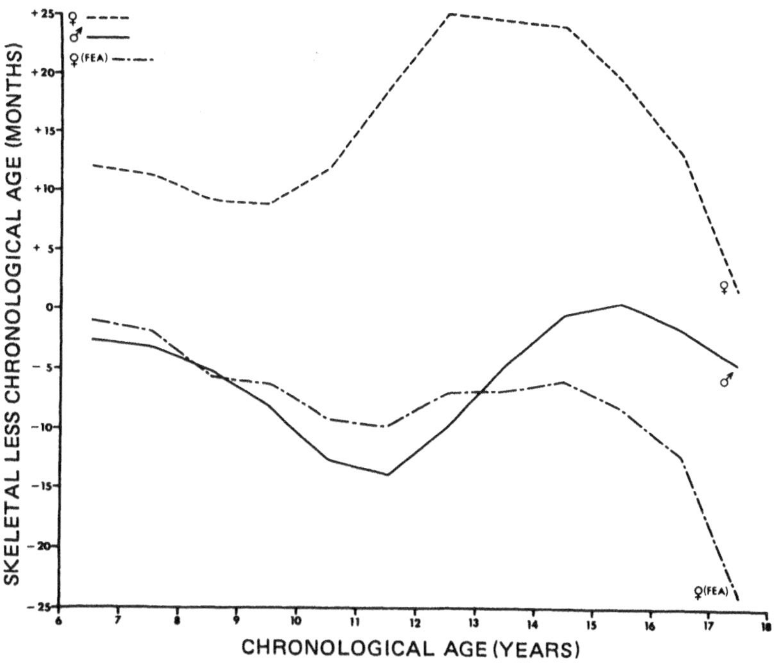

Fig. 15. Mean difference between skeletal age (Greulich–Pyle) and chronological age in a national probability sample of U.S. children and youth. (---) girls; (—) boys; (— - —) girls FEA. F EA, female equivalent ages (see text for details). (From Roche *et al.*, 1975*c*, 1976.)

months. The atlas of Pyle *et al.* (1971) for the hand–wrist and the atlases for the knee and foot–ankle (Hoerr *et al.*, 1962; Pyle and Hoerr, 1969) have only one set of standards for both sexes, but sex-specific skeletal age equivalents are provided for each standard. The standards approximate the median levels of maturity for Brush Foundation children with chronological ages matching the skeletal ages assigned to the standards. The children from whom the Cleveland atlases were derived were born between 1917 and 1942, and secular accelerations since then might be expected. In fact, the skeletal maturity standards for the hand–wrist in the Greulich–Pyle atlas are markedly in advance of the mean levels of maturity in national probability samples of U.S. children between 9 and 13 years of age (Roche *et al.*, 1974, 1976). This advancement is most marked at 11.5 years, when it is 13.8 months in boys and 10.7 months in girls (Figure 15). These data require explanation.

All radiographs taken in these national surveys were assessed, without the chronological age or the sex being known, against a set of male standards (Pyle *et al.*, 1971) essentially the same as those in the Greulich and Pyle atlas. Consequently, the means of the skeletal ages assigned to the girls (Figure 15) are in advance of the chronological ages, but those assigned to the boys (Figure 15) are less than the chronological ages. The skeletal ages assigned to the girls were transformed to what these workers believed they would have been had they been assessed against female standards. These female equivalent ages (FEA) were obtained using the bone-specific sex differences published by Pyle *et al.* (1971). The mean FEA (Figure 15) are less than the chronological ages, but the differences are smaller than those for the boys. These findings for U.S. children are of great importance to all who use the Greulich–Pyle and other Cleveland atlases. Children in Finland and Denmark also are about 0.7 years retarded in comparison with the Greulich–Pyle standards (Koski *et al.*, 1961; Andersen, 1968; Mathiasen, 1973). Particularly from 9 to 13 years, population groups should not be expected to be equivalent in maturity to the atlas standards, unless membership in the population groups studied is restricted to privileged children.

The unusual protocol followed in these national surveys has advantages and disadvantages. A totally unbiased estimate of skeletal maturity levels was obtained for boys, but the FEA are dependent on the accuracy of the sex differences in skeletal maturity reported by Pyle and associates (1971). The procedure applied obtained highly accurate estimates of sex differences (Figure 16). These estimates are close to those reported by Pyle *et al.* (1971), except after the age of 12 years when the actual sex differences are much smaller. In fact, at

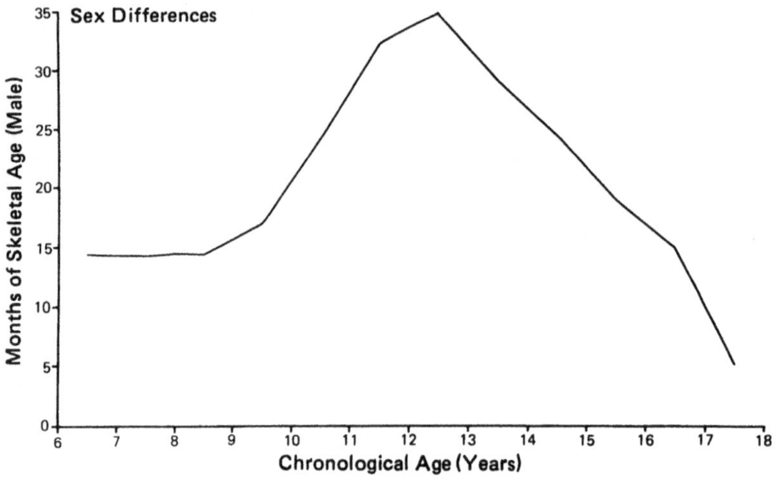

Fig. 16. Sex differences in skeletal maturity (Greulich–Pyle) from national probability samples of U.S. children and youths. (From Roche *et al.*, 1975*c*, 1976.)

17.5 years, the sex difference in skeletal age is only 3.9 months (Roche *et al.*, 1976), in at marked variance with previous reports (e.g., Roche, 1968; Pyle *et al.*, 1971; Tanner *et al.*, 1975). To repeat, when radiographs were assessed without knowledge of chronological age or sex, against a single set of standards, the mean difference between the ages assigned to boys and girls aged 17.5 years was, in fact, only 3.9 months. The reasons for these variations among studies have been considered in full elsewhere (Roche *et al.*, 1976).

When the atlas method is applied, the radiograph to be assessed is compared with the standards until one is found that is at the same maturity level as the radiograph. The skeletal age recorded for the radiograph is that of the standard it matches. Often it is desirable to interpolate between standards, and the procedure should be applied to individual bones. In the latter case, bone-specific skeletal ages are recorded. The way in which these bone-specific ages are combined affects the overall skeletal ages assigned to an area (Roche and Johnson, 1969). Few investigators other than some research workers, follow a definite scheme of combination—usually impressions are combined subjectively to a single-area skeletal age. Because the bones in the standards differ in maturity levels, there are differences among the bone-specific skeletal ages assigned to a single standard. This makes the atlas matching process less accurate if bone-specific skeletal ages are not recorded, but it is not known how much information is lost.

The most appropriate combination depends on two factors: The completeness with which the separate bones can be assessed and the purpose for which the area skeletal age is required. Skeletal ages cannot be assigned to bones until the appropriate parts are radiopaque, nor can skeletal ages, expressed in years be assigned to bones that have become adult. Consequently, a mean cannot be obtained for bone-specific hand–wrist skeletal ages except between about 5 and 13 years, but a median can be calculated with some reasonable assumptions, if skeletal ages have been assigned to more than half the bones. In theory, one combination could be significantly better than another when the skeletal age assessment is needed to estimate an outcome variable. In regard to the prediction of adult stature, reported findings are not in agreement. One group of workers found the median of all Greulich–Pyle bone-specific skeletal ages more useful than various weighted means of these ages (Roche *et al.*, 1975*a*). Others have reported that a combination of the maturity levels of the radius, ulna, and short bones was more useful in prediction than a combination with corresponding data from the carpals (Tanner *et al.*, 1975, 1983*b*).

Some have claimed that the range of bone-specific skeletal ages within a hand–wrist is related to the illness experience of a child (Pryor and Carter, 1938; Francis, 1939; Francis and Werle, 1939; Sontag and Lipford, 1943; Greulich, 1954; Greulich and Pyle, 1959; Hansman and Maresh, 1961); others have reported the contrary (Tiisala *et al.*, 1966; Acheson, 1966). Only recently have reference values for the range of bone-specific skeletal ages been made available (Roche *et al.*, 1978). These ranges, for national probability samples of U.S. children and youth, decrease in an approximately linear fashion with age except for a slight increase in girls from 8 to 11 years. The median ranges are about 15 months at 6.5 years, decreasing to 6 months at 17.5 years.

The accuracy of the atlas method, whether applied to an area or a bone, is unknown. Considerable data have been reported, however, concerning replicability and comparability among research workers (Mainland, 1953, 1954; Koski *et al.*, 1961; Moed *et al.*, 1962; Johnston, 1964; Acheson *et al.*, 1966; Andersen, 1968; Roche *et al.*, 1970, 1974, 1975*b,c;* Mazess and Cameron, 1971; Sproul and Peritz, 1971; Tanner *et al.*, 1983*b*) and pediatric radiologists (Johnson *et al.*, 1973), and during the training of assessors (Roche *et al.*, 1970). In general, repeated ratings (intraassessor) by experienced assessors of the hand–wrist using the atlas method differ by less than 0.5 years in 95% of cases; the difference between assessors (interassessor) are about 0.8 years in 95% of cases.

It has been claimed that maturity indicators appear in a fixed sequence for each bone (Pyle *et al.*, 1948; Greulich, 1954; Pyle and Hoerr, 1955; Greulich and Pyle, 1959; Tanner, 1962; Acheson, 1966) "no matter what the stock, parentage, economic standing, stature, weight or health of the child" (Todd, 1937). However, differences in sequence occur between parts of a bone, e.g., the distal end of the femur (Roche and French, 1970). If Todd's statement were true, Tanner and coworkers (1962, 1972, 1975) would not have needed to describe multiple criteria for maturity stages.

The sequence of indicators is far from fixed in a given area (Pryor, 1907; Todd, 1937; Robinow, 1942; Kelly and Reynolds, 1947; Pyle *et al.*, 1948; Abbott *et al.*, 1950; Greulich and Pyle, 1959; Garn and Rohmann, 1960; Christ, 1961; Hewitt and Acheson, 1961; Acheson, 1966; Garn *et al.*, 1966; Poznanski *et al.*, 1971, 1977; Yarbrough *et al.*, 1973; Roche *et al.*, 1978). It is clear that not all these differences in sequence are associated with ill health (Sontag and Lipford, 1943; Garn *et al.*, 1961*b;* Acheson, 1966), although this has been suggested (Francis, 1940; Pyle *et al.*, 1948). There is convincing evidence that unusual sequences tend to be aggregated in families (Garn *et al.*, 1966); while this does not prove that genetic factors are involved, such an explanation is plausible and is supported by estimates of the hereditability of maturation rate (Kimura, 1981).

8.3. TW2 Method of Assessment

The Tanner *et al.* (1975, 1983*b*) alternative to the atlas method is based on assigning numerical scores to bones depending on their levels of maturity. This concept was introduced by Acheson (1954), but the original method was unsatisfactory because the bone scores were ordinal, and only limited inferences can be made from summing such values (Lord, 1953). The Acheson ratings were recorded as 1, 2, 3, and so on, but they should not be added as ordinary numbers. Partly to avoid the temptation to do this, the TW2 stages are given letters; A, B, C, etc. Tanner and colleagues (1962, 1972, 1975, 1983*b*) improved and extended Acheson's method. Their TW2 method makes use of approximately nine maturity indicators (called "stages") for each of 20 hand–wrist bones. The bones of the second and fourth "rays"* are omitted because their skeletal maturity levels are correlated highly with those of corresponding bones in the other rays (Hewitt, 1963; Roche, 1970; Sproul and Peritz, 1971). However, if an underlying trait, such as skeletal maturity, is to be rated indirectly, some workers consider it preferable to use highly correlated indicators (Wilks, 1938; Wainer, 1976), as implied by Garn in his work on communality indices (Garn and Rohmann, 1959; Garn *et al.*, 1964). Not all agree. Tanner and co-workers adopted the principle of dropping bones with high correlations because they provide redundant information. They consider the variations among bones in levels to demonstrate differences in the rates of maturation, which is what they are trying to measure. Certainly, an assessment method restricted to highly correlated indicators would omit potentially important information from bones that tend to be divergent. The latter may be sensitive guides to environmental or genetic effects.

The scores assigned to the stages of each selected bone in the TW2 system were derived by assuming that all the observed stages of each bone of the individual reflect the same underlying quantity, the level of skeletal maturity for the hand–wrist, although the bones may be at different stages. Therefore the scores corresponding to stages were selected to minimize the disparity among the bone-specific scores for the individual, summed over all individuals in the standardizing group. This scoring system implies that the hand–wrist bones of an individual are normally at the same level of maturity, which is in agreement with Todd's (1937) concept of the evenly maturing skeleton. Although, as mentioned earlier, variations among bones are

*A ray of the hand consists of a metacarpal and its associated phalanges; e.g., metacarpal II, proximal phalanx II, middle phalanx II, and distal phalanx II.

common, this concept receives strong support from an analysis of the structure of bone-specific skeletal ages in the hand–wrist, foot–ankle, and knee assessed by the atlas method (Roche *et al.*, 1975*a*).

The TW2 mathematical method required at least two points of scale to be fixed. This was achieved by requiring the average of all the initial stages to be zero and the average of all the final stages to be 100, thus defining a scale from 0 to 100. Although the scores applied to each stage of each bone are objectively derived, as described above, an element of choice or subjectivity was introduced into some aspects of the system through these investigators' belief that the presence of many metacarpal and phalangeal bones in the hand–wrist would "swamp" the contribution of the carpals, radius, and ulna. Consequently, they promulgated two separate scores for maturity, one concerning the carpals only and the other the radius, ulna, and short finger bones (RUS). In the RUS score, furthermore, they dropped the second and fourth rays to diminish the number of finger bones and because of their high correlations in maturity with corresponding bones in other rays. Then they weighted the scores of the remaining finger bones so that together they only equaled in importance the joint scores of radius and ulna. This "biological" weighting is wholly separate from the mathematical scoring technique, although for convenience of the user the final scores incorporate both. In practice, the RUS system seems more informative than the carpals.

The TW2 system assumes a fixed sequence of maturity indicators for each bone, but some stages have multiple criteria. If only one criterion is given, this criterion must be satisfied in order for the stage to be assigned. However, if two criteria are given, it is sufficient if one is met, and if three criteria are given, two must be met for the stage to be assigned.

The maturity indicators in the TW2 method are satisfactory, except those based on relationships between the diameters of bones or the distances between bones. With such indicators, a pair of bones is being graded. In addition, some raters have been confused by the instruction "If at a particular stage a feature described in brackets is not present, this does not affect the rating: if it is, this confirms, but does not diagnose, the stage in question" (Tanner *et al.*, 1962, 1975). The descriptions in brackets could have been omitted without loss and have indeed been dropped in the latest revision (Tanner *et al.*, 1983*b*). In addition, there are problems at the upper end of the Tanner–Whitehouse scale, where a difference of one grade for one bone can alter the skeletal age for the entire hand–wrist by as much as 0.8 years (Tanner *et al.*, 1975). There is convincing evidence that the TW2 system is at least slightly more reliable than the Greulich–Pyle method (Acheson *et al.*, 1963, 1966; Johnston, 1963; Johnston and Jahina, 1965; Roche *et al.*, 1970; Johnson *et al.*, 1973; Tanner *et al.*, 1983*b*).

In the TW2 system, the bones are rated individually in a fixed order. Owing to the objectivity of this system, this presents no problem, whereas difficulties occur with the more subjective atlas method because the skeletal ages assigned to the bones assessed first can influence the ages assigned to the remaining bones. This problem is not overcome by assessing in a random order, as done by Sproul (Sproul and Peritz, 1971; Peritz and Sproul, 1971). The subject has been examined (Roche and Davila, 1976) using the Greulich–Pyle atlas with radiographs masked so that only one bone was visible at a time. In comparison with the findings when the whole of each radiograph was visible, both comparability and reliability of bone-specific skeletal ages were reduced; this reduction was particularly marked for the hamate and triquetral. Furthermore, when only one bone was visible at a time during assessment, the ranges of bone-specific skeletal ages within individual hand–wrist were increased.

8.4. RWT Method of Assessment

The Roche–Wainer–Thissen method is for the assessment of the knee (Roche *et al.*, 1975*b*); a corresponding system for the hand–wrist is being elaborated. The RWT method was developed using anteroposterior knee radiographs of "normal" southwestern Ohio children enrolled in the Fels longitudinal study. These radiographs had been taken between 1932 and 1972, but significant secular effects are absent. This was shown by correlating skeletal age (both RWT and Greulich–Pyle) with year of birth and by comparing the skeletal maturity levels of parents and children of the same sex at the same chronological age.

The RWT method, like all others, is based on maturity indicators. Roche *et al.* (1975*b*) listed all the maturity indicators reported, made the descriptions anatomically correct, and developed grading methods that were as objective as possible. Some indicators are graded by comparison with photographs and drawings, others by fitting standard curves to margins, and the remainder by measuring distances within bones to obtain ratios. The indicators that were retained were shown to be reliable, i.e., inter- and intraobserver differences were small. These indicators discriminated, i.e., they were present in some but not all children during a particular age range. This feature is age limited; an indicator will discriminate at some ages but not at others.

By definition, each grade of every indicator must occur during the maturation of an individual. Consequently, in an indicator with multiple grades, there must be ages at which either the least mature or the most mature grade is universal; however, because children mature at different rates, it is possible that there is no age at which a particular intermediate grade is present in all the children. When this occurred, serial radiographs were reviewed to be sure that the intermediate grade did indeed occur in the sequence of changes in each child. The criterion of universality was applied to the ratios by determining whether they became approximately constant at older ages. Although serial radiographs were used for this and some other purposes, the raw data were recorded without reference to other radiographs of the same child. The aim was to develop a method that would be applicable when only one radiograph was available.

The final indicators were valid. In the present context, validity refers to the quality by which a radiographic feature indicates skeletal maturity. A priori, the bones of a child must become more mature with increasing chronological age. Therefore, as pointed out by Takahashi (1956), the prevalence of grades of valid maturity indicators must change systematically with age until the most mature grade is universal. Prevalence data can provide only a general guide to validity because some children miss examinations. A better measure is obtained by reviewing serial data for individuals. "Reversals" were sought in serially organized data, i.e., changes in indicators with increasing age that were in the reverse direction from those expected from the group trends. This procedure is similar to that adopted by Hughes and Tanner (1970) when developing an assessment method for the rat and the method used in constructing the TW2 stages.

Useful maturity indicators must have the quality of completeness. This refers to the extent to which the indicator can be graded in a series of radiographs. As with some other criteria, an indicator may have this quality at some ages but not at others.

The values used to construct the RWT scale were the chronological age at which each indicator was present in 50% of children and the rate of change in each indicator's prevalence with age. These values were combined to one continuous index using latent trait analysis (Birnbaum, 1968; Semajima, 1969, 1972). This index was scaled so that the mean and variance of skeletal age in the standardizing group were equal to the mean and variance of chronological age.

This statistical method separates the within-age variance into two components: one attributable to real variation in maturity levels and one to sampling error, which is inherent in any scheme. Only the RWT method estimates this error component. The standard error tends to be small when many highly informative indicators are assessed that have grades with the age of 50% prevalence close to the chronological age considered. The standard error is large in the contrary circumstances and also when unusual combinations of indicator grades are observed.

The method is based on 34 maturity indicators for the femur, tibia, and fibula. At any particular age, only about one-half of these must be graded because there are limits to the age ranges during which each is useful. After the grades have been recorded, they are transferred to a computer. Published programs, that are also available on tape, provide the skeletal age appropriate for the radiograph together with the standard error of the particular estimate. The errors tend to be larger at some ages than at others, as shown by the total information curve (Figure 17). The information available from a knee radiograph is relatively great at 2 years, decreases until 6 years, and then remains approximately constant.

The RWT system has advantages and disadvantages. It is very helpful to know the accuracy of each assessment; the model permits the addition of corresponding values for other areas, e.g., hand–wrist, thereby providing a combined estimate for the two areas that uses the information from each. Reliability and comparability are about the same as with the TW2 method and are generally better than with the atlas method. Although an assessment does not require much time, access to a computer is essential. Others have applied computers to individual assessments for convenience, but not to obtain weighted estimates (Sempé and Capron, 1979).

8.5. Scale of Maturity

The scale applied in each method for the assessment of skeletal maturity is based on the assump-

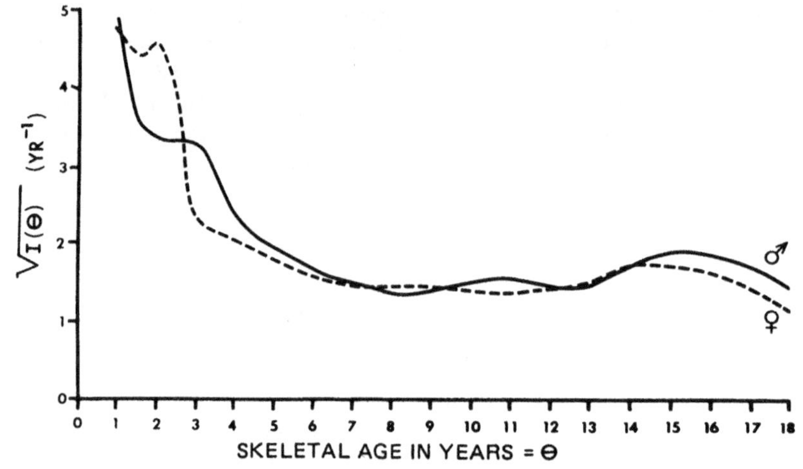

Fig. 17. The total information system curves for the RWT method (knee) in each sex. I, inverse of the error variance at Θ; Θ, skeletal age. (From Roche *et al.*, 1975*b*.)

tion that skeletal maturity is zero at conception and that all individuals reach the same level of complete maturity in young adulthood. The end point of the scale is the completion of epiphyseodiaphyseal fusion in bones with epiphyses and the attainment of adult shape in bones that do not develop epiphyses. Thus, each individual achieves the same amount of maturation between conception and adulthood, although their rates of maturation differ. In practice, the scale begins with the onset of ossification in epiphyseal centers or in the primary centers of irregular bones.

All maturity scales are divided into years of skeletal age, except in the TW2 method. However, even the TW2 scores are usually transformed to skeletal ages using published tables or graphs. Skeletal-age "years" are not known to be equivalent to each other even when the same system of assessment is used for the same area of the skeleton. These ordinal scales do not provide a measure of skeletal maturity, but they allow a maturation level to be assigned to a radiograph relative to pictorial standards for a whole area, individual bones, or individual indicators. Typically, girls achieve adult levels of skeletal maturity at younger chronological ages than do boys. Therefore, more maturation occurs during a typical female skeletal-age year than during a male skeletal-age year. Furthermore, bones differ in the chronological ages at which they reach adult maturity levels (Roche *et al.*, 1974); therefore, the amount of maturity achieved per skeletal-age year differs among bones.

The sex differences in skeletal maturation rates differ among bones (Garn *et al.*, 1966; Roche, 1968; Thompson *et al.*, 1973; Tanner *et al.*, 1983*a*). The measurement of skeletal maturation is also complicated by the fact that a given radiographic appearance does not necessarily indicate the same level of maturity in both sexes. Consequently, in the TW2 method, the sex differences between the scores for corresponding stages vary among bones. Similar variations occur with the RWT method in which each grade of each indicator has sex-specific values for rate of change in prevalence with increasing age and the age at which the threshold (50% prevalence) is crossed.

Population differences in skeletal maturity have been comprehensively discussed by Tanner *et al.* (1983*a*).

8.6. Influence of Variants on Assessments

Some problems arise if variants are present in the area to be assessed. When there are multiple centers of ossification in a single epiphyseal area, the width of the epiphysis should be regarded as the sum of the widths of its parts measured in the same straight line. Tanner *et al.* (1983*b*) give rules for assessing when some other hand–wrist variants are present. When the hook of the hamate varies markedly in form, including complete absence, ratings of this bone are to be based on other criteria. If the lunate develops from two centers of ossification, the general maturity of the bone is considered. In brachymesophalangia of the fifth finger, the epiphysis of the fifth middle phalanx fuses early; such a bone is rated at the maturity level of the third middle phalanx in the same hand–wrist.

8.7. The Area to Assess

The accuracy of any estimate of the maturity of the whole skeleton obtained from the assessment of only one area is limited because skeletal areas vary in maturity levels within normal individuals (Crampton, 1908; Reynolds and Asakawa, 1951; Garn *et al.*, 1964; Roche and French, 1970). Ideally, either the whole skeleton or the hemiskeleton would be assessed. There are several obvious difficulties, including the time needed for the assessment of so many bones, the expense of radiographic film, and most importantly, the risk of excessive radiation. An area should be chosen that can be radiographed with minimal radiation, as few radiographic views as possible should be used, and all useful relevant information should be obtained from each radiograph. Bilateral assessments are unnecessary. Although lateral asymmetry occurs in all phases of skeletal maturation, the differences are small and not of practical importance (Roche, 1962*a*).

Only two areas are commonly assessed. The area assessed most commonly is the hand–wrist, partly because it was the first region for which atlases became available (Howard, 1928; Flory, 1936; Todd, 1937; Greulich and Pyle, 1950). Furthermore, the standards in the well-known Greulich and Pyle atlas are much clearer than those in the Cleveland atlases for the knee or the foot–ankle. The hand–wrist has other advantages: it can be positioned easily, the subject receives very little irradiation, and many bones are present in a relatively small area. However, there are disadvantages: long age ranges occur during pubescence when the hand–wrist provides little information, and assessments of this area are not considered useful before 1 year (Sauvegrain *et al.*, 1962; Tanner *et al.*, 1975). During infancy, the knee can provide more information. Problems occur in assessments of the hand–wrist toward the end of maturation when some of the bones, e.g., distal phalanges, reach adult levels earlier than others (Todd, 1930*b;* Pyle *et al.*, 1961; Garn *et al.*, 1961*a;* Hansman, 1962). The designation "adult" can be applied to such bones, but skeletal ages cannot be assigned.

The area to be assessed should be determined partly by the reason for the assessment. For example, if a measure of skeletal maturity is required to estimate the potential for elongation at each knee joint of an anisomelic child, the knee should be assessed. When an assessment of skeletal maturation is needed for a child whose stature deviates from the mean, the knee is preferable to the hand–wrist because it is an important site of growth in stature and can provide a slightly more accurate estimate of the potential for growth in stature than can the hand–wrist (Roche *et al.*, 1975*a*), even when assessed using the atlas of Pyle and Hoerr (1969). The knee area includes only a few bones, but these are unusually informative (Roche *et al.*, 1975*b*). However, the knee is more difficult to position and, although the irradiation is very slight, it is double that required for the hand–wrist.

8.8. Noninvasive Assessment of Maturity

There has been considerable recent interest in the self-assessment of secondary sex characteristics. The reported studies, although somewhat deficient in design, show that youths can assign the standard grades with reasonable accuracy and with less embarrassment than if they are examined by others (Duke *et al.*, 1980; Morris and Udry, 1981; Neinstein, 1982).

This has led to a suggestion that another index of maturity could be obtained noninvasively without the use of radiography (Roche *et al.*, 1983). The underlying concept is that maturation is the part of development in which all normal individuals reach the same level as young adults. Present stature, expressed as a percentage of adult stature (relative stature), is such a measure because all individuals necessarily have a score of 100 in young adulthood. Since adult stature can be predicted with reasonable accuracy at some ages without the use of skeletal age, and therefore without radiography (Wainer *et al.*, 1978; see also Tanner *et al.*, 1983*b*), it is possible to estimate relative stature during childhood. These estimates are correlated significantly with recognized measures of skeletal and sexual maturity from 5 to 15 years in boys and from 3 to 13 years in girls. Reference data from the Fels longitudinal study are provided in Figure 18. These correlations are only slightly lower than those between generally accepted indices of maturity.

9. Minor Skeletal Variants

Major variants have been omitted from the present description; detailed accounts are available in O'Rahilly (1951), Schmorl and Junghanns (1971), and Caffey (1978). The minor variants considered

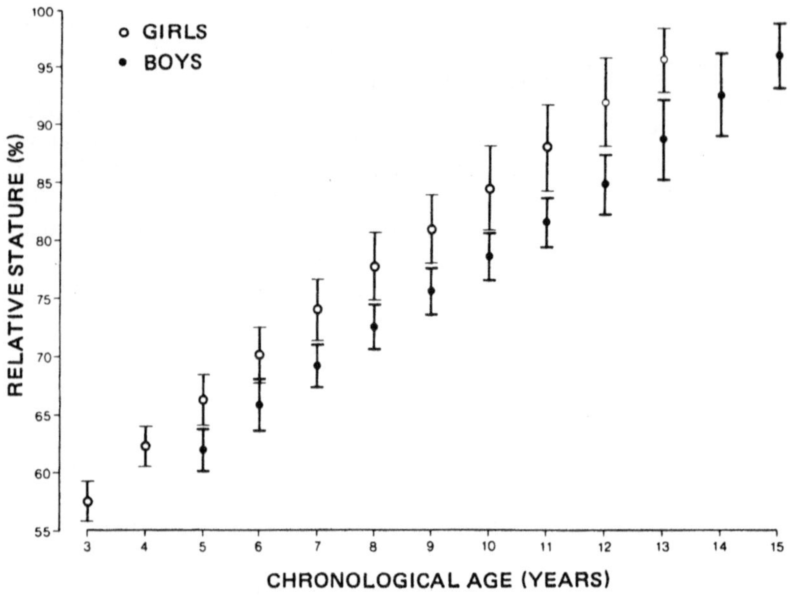

Fig. 18. Means ± 1 SD for relative statures of boys (•) and girls (○) in the Fels longitudinal study. Relative stature is present stature expressed as a percentage of predicted adult stature, when predictions are made without using skeletal age.

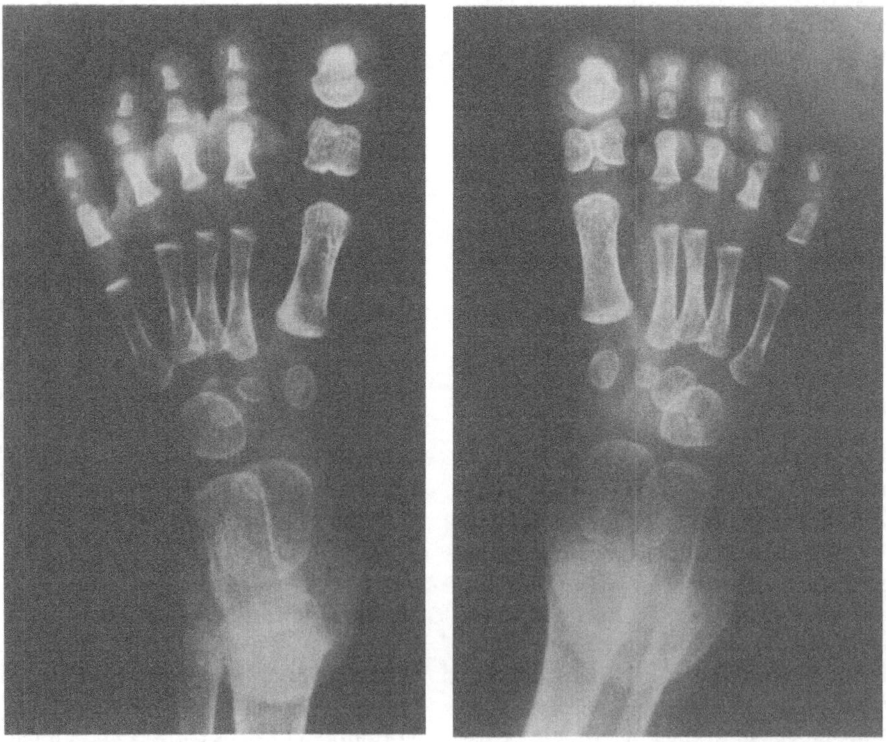

Fig. 19. Incompletely bifid first proximal phalanges in the feet of a child with Rubinstein–Taybi syndrome. (From Rubinstein and Taybi, 1963.)

have been described more fully by Köhler and Zimmer (1953) and Poznanski (1974).

Most duplications must be considered major variants, but minor expressions of this tendency occur as incomplete division of the subungual expansion of a terminal phalanx or complete or incomplete duplication of a bone (Figure 19). Although unusual, some bones may be incomplete. For example, the subungual expansion of the second distal phalanx of the foot is incomplete in about 30% of children aged 2–8 years (Roche, 1962*b*) (Figure 20). Some bones in the hand–wrist or the foot may be "incomplete" even when fully mature. For example, an adult ulna may lack a styloid process or a hamate may lack its hook (Figure 21).

In about 25% of children the diaphysis of the fifth middle phalanx of the foot has two centers of ossification (Figure 22). This will be seen only in radiographs taken between 2 and 3 years. This bone may be short (brachymesophalangia), a condition often associated with clinodactyly and with an increase in the width of the fifth middle phalanx (Roche, 1961; Garn *et al.*, 1972*b*) (Figure 8). Brachymesophalangia is unusually common in Down's syndrome (Roche, 1961; Chumlea *et al.*, 1981).

Variations occur at the proximal ends of the metacarpals, particularly the second, where cortical notches, pseudoepiphyses, and supernumerary epiphyses are relatively common (Kimura and Takeuchi, 1982). A notch in the cortex commonly represents the last stage of fusion between the diaphysis and either an epiphysis or a pseudoepiphysis (Figure 23A). A pseudoepiphysis differs from a epiphysis in that a separate center of ossification does not form in the cartilage at the end of diaphysis. Instead, diaphyseal ossification extends in continuity into this cartilage over a small central area, mushrooming within the cartilage (Figure 23B). The incidence of pseudoepiphyses on the second metacarpal has been reported as 14% in boys and 7% in girls, but supernumerary epiphyses are much less common (Iturriza and Tanner, 1969). Additional centers are common at the distal end of the first metacarpal also (Figure 24). It is not clear whether this flakelike epiphysis should be classified as a pseudoepiphysis or as a supernumerary epiphysis. Because the ossified area is concave on its diaphyseal aspect in all planes, it is impossible to determine from radiographs whether this flake ossifies independently or by extension from the diaphysis.

Radiopaque lines are common near the ends of the diaphyses of long bones (Figure 25). These are called "bone scars" or "lines of arrested growth." Commonly, but not always, these are associated

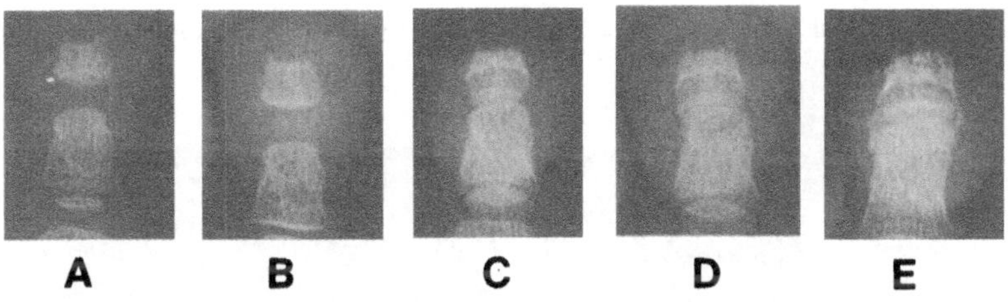

Fig. 20. Serial radiographs of an incomplete second distal phalanx in a girl at (A) 2.75 years, (B) 3 years, (C) 4.5 years, (D) 5.5 years, and (E) 8 years. (From Roche, 1962*b*.)

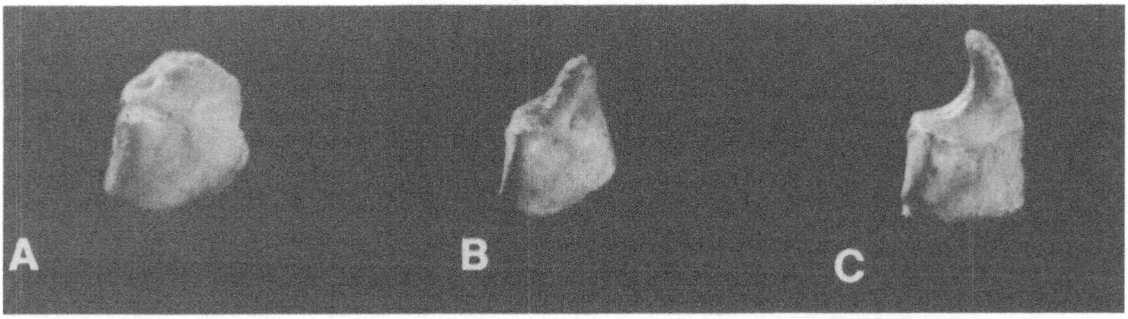

Fig. 21. Three left adult hamate bones viewed from the proximal aspect: (A) no hook, (B) small process in the possition of the hook, and (C) typical adult hook.

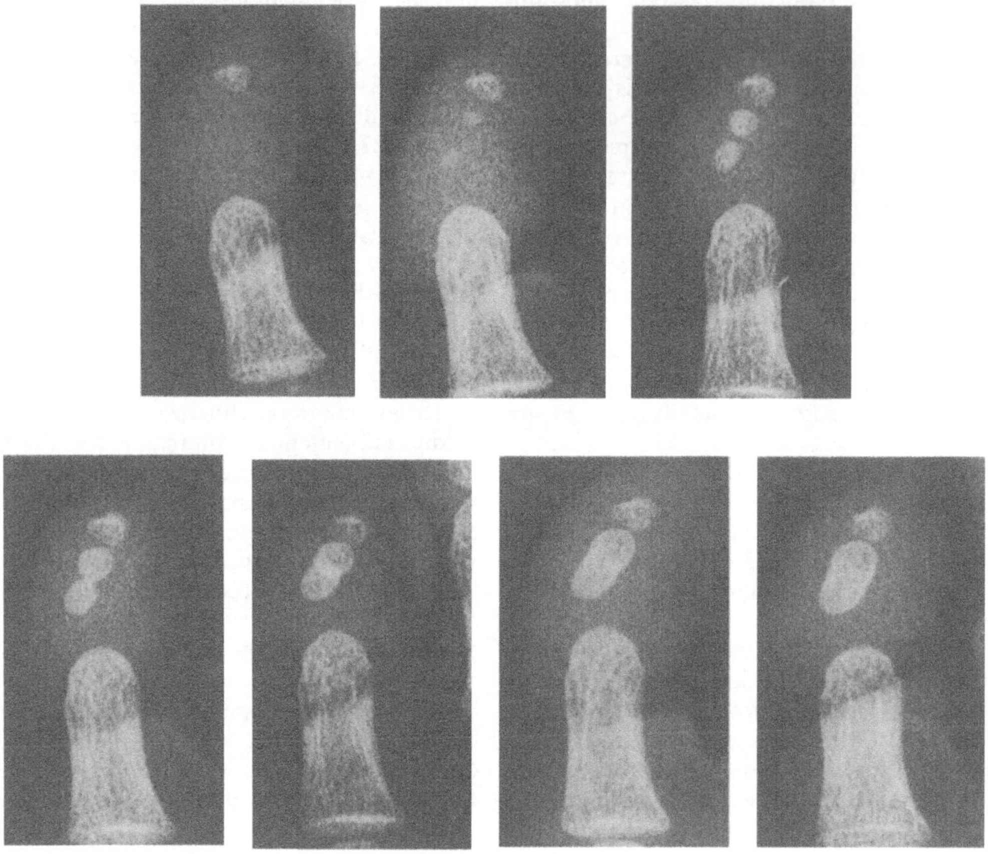

Fig. 22. Serial radiographs of a boy aged 2.5–4 years showing ossification of the shaft of the fifth middle phalanx of the foot from two centers. (From Roche and Sunderland, 1960.)

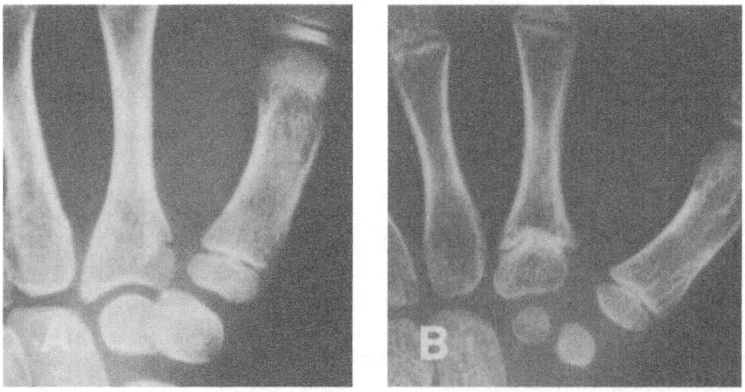

Fig. 23. Variations at the proximal end of the diaphysis of metacarpal II. (A) A notch in the cortex in a girl aged 7.5 years. (B) A pseudoepiphysis in a girl aged 10 years.

with diseases during growth (Hewitt *et al.*, 1955; Marshall, 1966). They are useful as signs of past ill health and as natural bone markers from which amounts of diaphyseal elongation can be measured.

Some variants occur commonly in epiphyses also. There may be more than one center of ossification in a single epiphyseal cartilage (Figures 26 and 27). When this occurs, the separate centers enlarge until, about 2 years after the onset of ossification, they coalesce to form a single center within the cartilage that matches the shape of epiphyses formed from single centers.

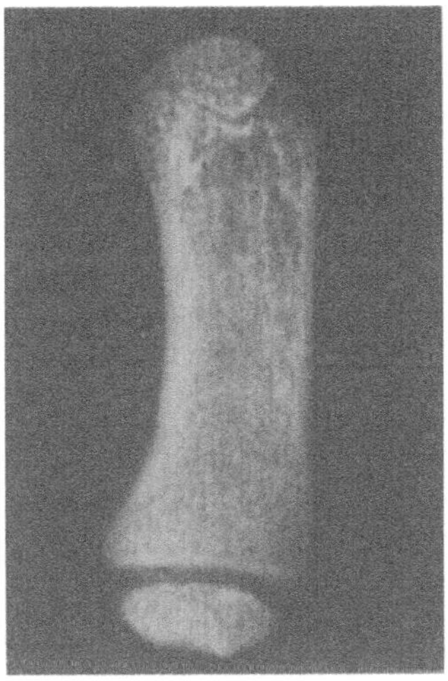

Fig. 24. An additional epiphysis at the distal end of the first metacarpal in a boy aged 7 years.

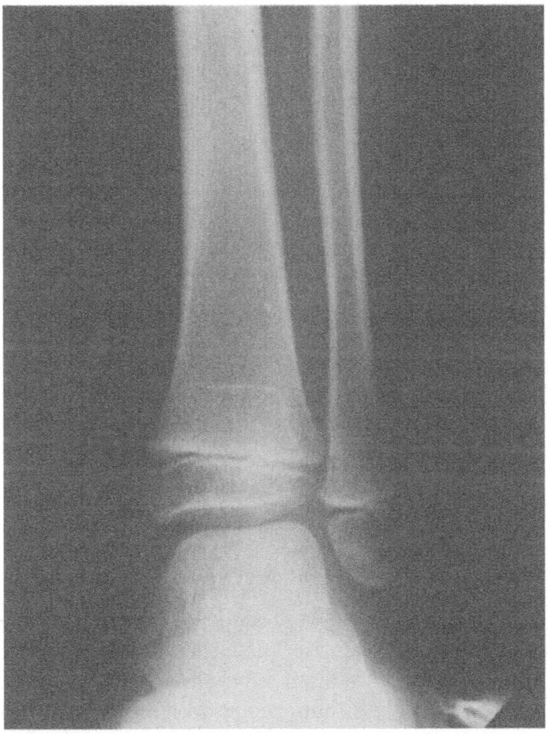

Fig. 25. Lines of arrested growth near the distal end of the tibia in a girl aged 3.5 years.

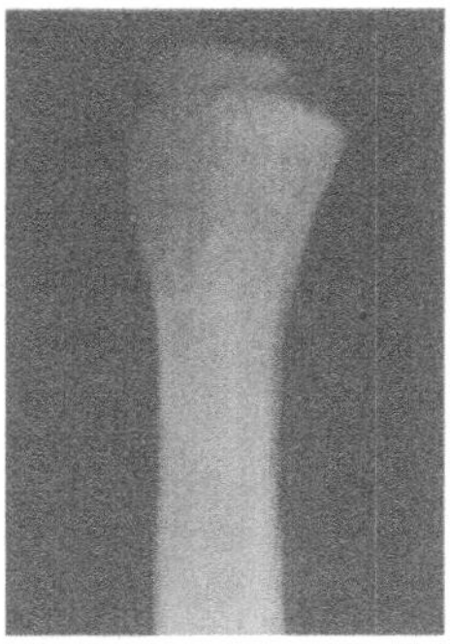

Fig. 26. Two centers of ossification in the epiphysis at the distal end of the ulna in a boy aged 7 years.

The ossified epiphysis may be conical with its diaphyseal aspect extending into a depression on the end of the diaphysis (Figure 28). Conical epiphyses are characteristic of several clinical conditions, e.g., achondroplasia, cleidocranial dysostosis (Caffey, 1978; Gellis and Feingold, 1968), but they occur in normal children as well.

10. Prediction of Adult Stature

Methods for the prediction of adult stature from childhood data are not free of errors. These errors are referred to as residuals, even if the prediction method considered is not based on regressions. Generally, stature at 18 years is accepted as adult, although there are small increments in stature after this (Roche and Davila, 1972). The common prediction methods are referred to by the following abbreviations: BP for the Bayley and Pinneau method (1952), RWT for the Roche-Wainer-Thissen method (1975*a*), and TW for the method of Tanner and others (1975, 1983*a,b*).

10.1. The Bayley–Pinneau Method

Methods to predict adult stature must be derived from serial data, but it is desirable that the method be applicable to data from a single examination. In the BP method, present stature and Greulich–Pyle skeletal age are used as predictors (Bayley and Pinneau, 1952). The details of how this skeletal age is obtained are not specified although these ages vary with the procedures used to obtain them (Roche and Johnson, 1969). The BP tables provide the percentages of adult stature achieved at particular skeletal ages, for groups of children who are accelerated, average or retarded in skeletal maturation. This tripartite division of the continuum of skeletal age–chronological age difference omits some information, but is simple to apply. The BP method can be used from 7 to 18.5 years of age in boys and from 6 to 18 years of age in girls. These age ranges

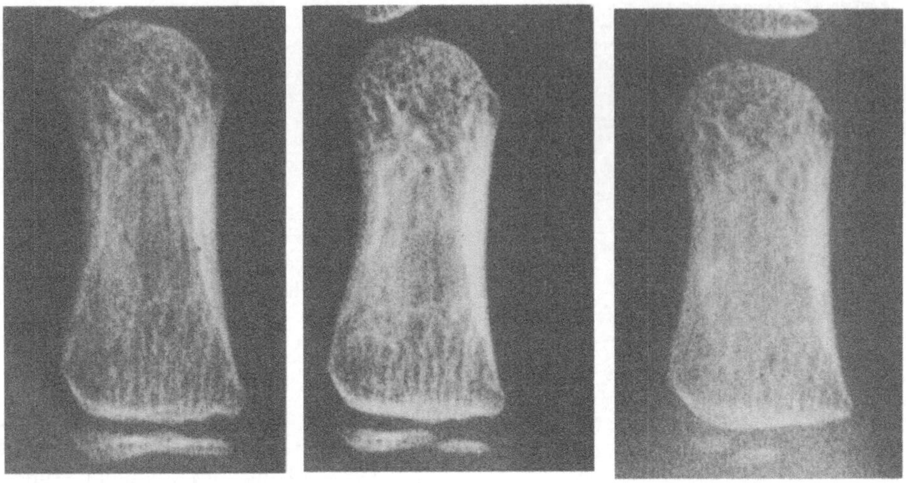

Fig. 27. Multiple centers of ossification at the proximal end of the first proximal phalanx in a girl at 3.5, 4.0, and 4.5 years.

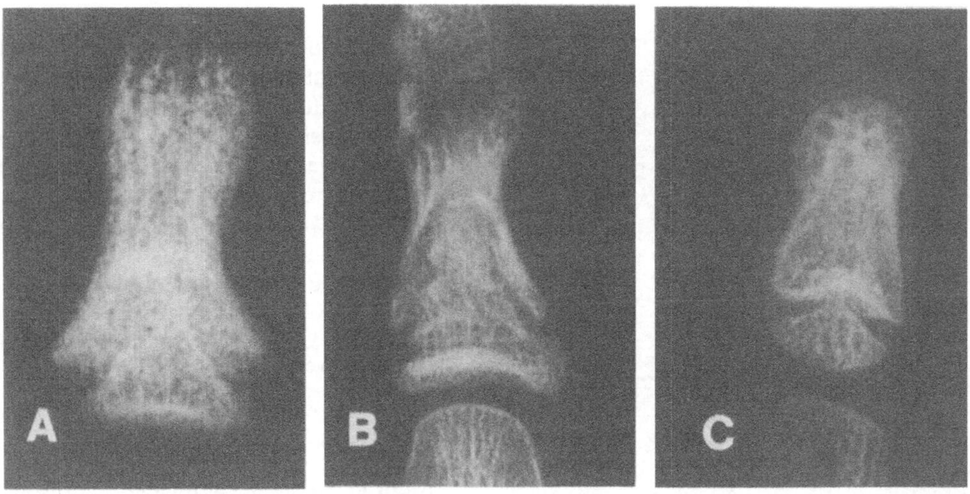

Fig. 28. Examples of conical epiphyses. (A) Second proximal phalanx (foot), (B) third proximal phalanx (hand), and (C) first distal phalanx (hand).

are appropriate for most clinical purposes but, near the upper ends, there is little or no further growth in stature.

10.2. The Roche–Wainer–Thissen Method

The RWT method uses a multiple regression equation in which adult height is estimated from recumbent length, weight, and skeletal age (median of Greulich–Pyle bone-specific skeletal ages) of the child together with mid-parent stature. Coefficients are given at all ages, although there are ages at which some of these variables contribute little (Roche *et al.*, 1975*a*). The method is applicable from 1 year to 16 years in boys and 14 years in girls, provided that less than one-half of the hand–wrist are adult. (Predictions for children with more than one-half the bones adult are unimportant because of such children's small growth potential.) Recumbent length was chosen in preference to stature because it can be measured at all ages, and its serial changes are more regular in some longitudinal studies. After age 2, stature measurements can be adjusted to approximate recumbent lengths (Roche and Davila, 1974). The median absolute errors, in replication samples, are about 3–4 cm (Roche *et al.*, 1975*a*). If skeletal age is unavailable, chronological age, adjusted for known population differences between chronological and skeletal ages, can be substituted in the RWT prediction equations. This slightly increases the residuals at 13 to 15 years in boys and 12 to 14 years in girls, but not at other ages (Wainer *et al.*, 1978; Harris *et al.*, 1980).

10.3. The Tanner–Whitehouse Method

The TW Mark I method uses stature, chronological age, and skeletal age as the independent variables in a multiple regression equation, with separate equations for pre- and postmenarcheal girls (Tanner *et al.*, 1975). The skeletal age is the RUS value from the Tanner–Whitehouse method. In 1983 Tanner and associates (1983*a*,*b*) published new estimation equations, called TW Mark 2. These differed from the old ones in that the standardizing sample had been extended beyond the normal children attending growth studies (on which GP, RWT, and TW Mark I are exclusively based) to include children attending a Growth Disorder Clinic with complaints of excessive shortness or tallness, although not gross pathological conditions. It has long been a criticism of height prediction systems that applying them to very tall, very short, or very delayed children represents an extrapolation of the standardizing estimation equations into areas in which they may not apply, and the new system seeks to answer this. For most of the clinic children, adult stature (often attained later than age 18) was measured at home by family members following written instructions. The new equations use the additional variates "increase in height during previous year," and "increase in RUS bone age during previous year," when available, at ages at which their inclusion increases the accuracy of prediction (11–15 in boys and 8–14 in girls). When applied to the standardizing groups, the accuracy of the TW Mark 2 equations is only slightly better than that of the TW Mark I, but the range of children covered is greatly increased. Thus Tanner and associates consider the Mark 2 method superior, especially for clinical use.

10.4. Other Methods

Onat (1983) recently published equations for estimating adult stature in girls, incorporating, for the first time, information on whether the puberty stages breast 2, pubic hair 2, and axillary hair 2 have been reached, as well as whether menarche has occurred. The standardizing sample consists of Turkish girls, mostly of high socioeconomic status; ages 9 upward are covered. The bone age used was median value bone-specific Greulich–Pyle, as in the RWT system. Weight was incorporated in the multiple regression analysis but was found to be noncontributory and dropped. Mid-parent height was also found to contribute little above age 12.0 years.

A new dimension was recently introduced into height prediction by the use of Bayesian methods of estimation (Bock and Thissen, 1980). In this method, the mean growth curve for the population is taken as the prior estimate, and successive modifications are made following one or more determinations of stature for a given child. There is no reason why bone ages should not be incorporated into this system, although this has not yet been accomplished.

10.5. Accuracy of Height Prediction

Comparisons of the accuracy of prediction between the commonly used methods can assist recommendations for use and could indicate ways in which accuracy might be improved.

There are two aspects: First, the size of random errors as represented by the residual standard deviations (or standard errors of estimation) obtained in the standardizing and in other groups; and Second, systematic errors as represented by a mean difference obtained when the equations are applied to groups other than the standardizing group. The first of these errors is by far the more important, since the second can be readily compensated once the degree of bias is known.

Figure 29 shows the residual SDs (above) and the correlation coefficients (below) reported for the TW Mark 2 equations applied to the standardizing group. Postmenarcheal girls are much more accurately predicted than are premenarcheal girls of the same age, as would be expected. The boys' residuals are greater than the girls', but this is linked to the fact that the adult stature SD is greater in men than in women; the correlations between predicted and actual height are about the same in the two sexes and rise smoothly from 0.8 at age 6 to a little over 0.9 at ages 14, 15, and 16 in boys and at ages 11, 12, and 13 in premenarcheal girls.

The RWT equation, also applied to the standardizing group, produces correlations (Roche *et al.*, 1975) of about 0.85 in boys at age 6, rising to 0.88 at age 10, but thereafter dropping at ages 11–14. In

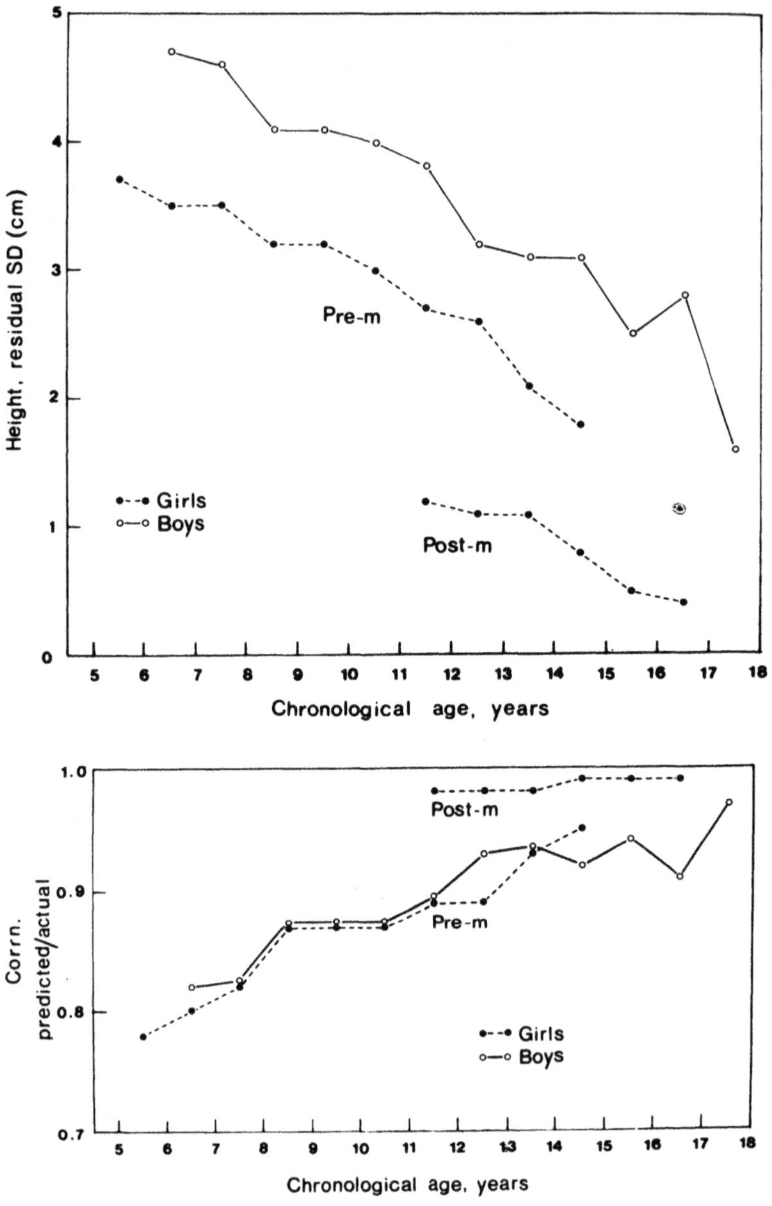

Fig. 29. Residual standard deviations of predicted height (above) and correlations between actual and predicted height (below) according to age: 3-, 4-, or 5-variate equations as appropriate. (From Tanner *et al.* 1983*b*.)

girls the correlation at age 6 is about 0.8, and it remains at about this level up to age 14. Comparing the systems on this criterion, Tanner *et al.* (1983*a*) wrote:

> Judged by residual standard deviations and correlation coefficients, TW Mark 2 predictions are more precise than those of Roche, Wainer and Thissen (1975*a*) at age 12 upwards, even when using the 3-variate equation, and more so using the 4-variate (this is with each judged by the fit of its own standardizing group). At ages 10 and 11 there seems little to choose between the systems, and below age 10 the RWT is better, presumably because of the inclusion of midparent height. In girls, the advantage of the TW Mark 2 system begins around age 10, and becomes substantial for the 4- and 5-variate equations thereafter. At ages 6 to 9 there is little difference between the systems. Judged on the same criteria, Onat's equa-

tions for girls, taking into account appearance or not of the secondary sex characteristics, approach the TW Mark 2 5-variable equation in precision, although the residuals remain slightly larger.

As regards systematic errors when applying equations to other than the standardizing groups, nothing is yet known about TW Mark 2. In normal children, the BP method is said to overpredict in girls by about 2 cm and to underpredict in boys by about 3 cm (Bayley and Pinneau, 1952); others report it tends to underpredict in girls and to overpredict in boys (Roche *et al.,* 1975*a*). The RWT method tends to overpredict in boys by about 1.0–1.5 cm; for girls, the RWT method tends to overpredict by about 2 cm in one group but not in another group (Roche *et al.,* 1975*a*). Lenko (1979) compared the BP, RWT, and TW Mark I methods but did not follow all the procedures for these methods, reducing the importance of his findings. The systematic errors tended to be smaller for the RWT and the TW Mark I methods, particularly in boys, except at older ages, when they were smaller for the BP method.

Schreiber *et al.* (1976) also compared these three methods, but some of the children included may not have reached adult stature, and not all the instructions of the original authors were followed. At 8 years, the systematic errors were smaller with the RWT and TW methods than with the BP method. Predictions at 14 years in boys and 12 years in girls were reasonably accurate except that the TW method underpredicted by a mean of 3.5 cm. The systematic errors tended to be smallest for the RWT method.

Harris and coworkers (1980) also reported that the BP, RWT, and TW Mark I errors were small except that the TW method tended to underestimate in each sex, as did the BP method in girls at ages 11 and 12 years. Similar findings were reported by Zachmann and colleagues (1978). Earlier, the latter workers reported small systematic errors for the TW Mark I method but overestimates of about 2.0 cm with the BP method (Zachmann *et al.,* 1976).

Only a few studies of the random errors are encountered when applying equations to groups that are substantially different from the standardizing ones. In one study, these errors were smaller with the BP method (Roche and Wettenhall, 1977; Lenko *et al.,* 1982), but contrary findings have been reported (Zachmann *et al.,* 1978). The BP method is said to overestimate for tall children without pathological conditions (Colle *et al.,* 1977; Hinkel and Schambach, 1980), although others have reported the BP, RWT, and TW Mark I methods to have low systematic errors in these children (Zachmann *et al.,* 1975, 1978). The random errors are smaller with the RWT method than with the others (Zachmann *et al.,* 1975, 1978). In children with endocrinopathies or idiopathic precocious puberty, the BP method is less inaccurate than the RWT or the TW method (Zachmann *et al.,* 1978), but there are large errors with all these methods.

It is necessary to make recommendations, yet these must be tentative pending further experience of the TW Mark 2 equations. A practical method should be based on data available from a single examination. The RWT and TW regression methods are appropriate, although they require more calculation than the BP method. This should not be a problem with the widespread use of desk calculators and microcomputers. It is emphasized that the instructions of the authors must be followed closely when prediction methods are applied and that the predicted value should be reported with its confidence limits.

Prediction methods should be applied only to children with nonpathological conditions. It should not be concluded without further study that the BP, RWT, and TW (or TW Mark 2) methods can be applied satisfactorily to children outside the United Kingdom and the United States (see discussion in Tanner *et al.,* 1983*b*). The errors of prediction made in routine clinics are likely to exceed those made in research circumstances due to decreases in the accuracy of the independent variables, particularly skeletal age. It is known that the RWT and TW methods work reasonably well for short or tall children without organic diseases and that, while the BP method may be best for children with pathological conditions, it provides only approximate estimates.

11. Conclusions

The potential value of radiographs in the study of child growth and development was realized very soon after their discovery, leading to the publication of a hand–wrist atlas by Poland in 1898. Much was achieved during the intervening 80 years by the development of appropriate methods for the measurement of size and shape and for the estimation of levels of maturity of the skeleton. These advances required large collections of suitable radiographs; further advances will be similarly dependent on the maintenance of radiographic libraries, some of which contain irreplaceable material.

As in almost every research field, much work remains to be done. For example, further information concerning three-dimensional trabecular patterns may soon be available from radiographs. In planning for the future, collections of serial standardized radiographs must be maintained and extended. The study of trivia should be omitted, and future research should be limited to features that meaningfully discriminate among children or increase our understanding of our own frame.

ACKNOWLEDGMENTS. This work was supported by grant HDAM-12252 from the National Institutes of Health. Gratitude is expressed to Anna Gregor for her help with the illustrations, to Joan Hunter and Lois Croutwater for secretarial assistance, and to Adel K. Abdel-Malek for advice.

12. References

Abbott, O. D., Townsend, R. O., French, R. B., and Ahmann, C. F., 1950, Carpal and epiphyseal development. Another index of nutritional status of rural school children, *Am. J. Dis. Child.* **79**:69.

Acheson, R. M., 1954, A method of assessing skeletal maturity from radiographs. A report from the Oxford Child Health Survey, *J. Anat.* **88**:498.

Acheson, R. M., 1966, Maturation of the skeleton, in: *Human Development* (F. Falkner, ed.), pp. 465–502, W. B. Saunders, Philadelphia.

Acheson, R. M., Fowler, G., Fry, E. I., Janes, M., Koski, K., Urbano, P., and Van der Werff Ten Bosch, J. J., 1963, Studies in the reliability of assessing skeletal maturity from x-rays, Part I, Greulich–Pyle Atlas, *Hum. Biol.* **35**:317.

Acheson, R. M., Vicinus, J. H., and Fowler, G. B., 1966, Studies in the reliability of assessing skeletal maturity from x-rays. Part III, Greulich–Pyle Atlas and Tanner–Whitehouse method contrasted, *Hum. Biol.* **38**:204.

Amprino, R., and Bairati, A., 1936, Processi di recostruzione e di riassorbimento nella sostanze compatta delle ossa dell'uomo. Ricerche su cento soggetti dalla nascita sino a tarda etá, *Z. Zellforsch. Mikrosk. Anat.* **24**:439.

Amprino, R., and Cattaneo, R., 1937, II. Substrato istologico delle varie modalitá di inserzioni tendinee alle ossa nell'uomo, Ricerche su individui di varia etá, *Z. Anat. Entwicklungsgesch.* **107**:680.

Andersen, E., 1968, Skeletal maturation of Danish school children in relation to height, sexual development, and social conditions, *Acta Paediatr. Scand. (Suppl.)* **185**:133.

Archibald, R. M., Finby, N., and de Vito, F., 1959, Endocrine significance of short metacarpals, *J. Clin. Endocrinol.* **19**:1312.

Arndt, J. H., and Sears, A. D., 1965, Posterior dislocation of the shoulder, *Am. J. Roentgenol.* **94**:639.

Bayley, N., and Pinneau, S. R., 1952, Tables for predicting adult height from skeletal age: Revised for use with Greulich-Pyle hand standards, *J. Pediat.* **40**:423.

Baziere, B., 1981, Partial body neutron activation—hand, in: *Non-Invasive Measurements of Bone Mass and Their Clinical Application* (S. H. Cohn, ed.), pp. 151–164, CRC Press, Boca Raton, Florida.

Birnbaum, A., 1968, Some latent trait models and their use in inferring an examinee's ability, in: *Statistical Theories of Mental Test Scores* (F. M. Lord and M. R. Novick, eds.) pp. 397–497, Addison-Wesley, Reading, Massachusetts.

Bock, R. D., and Thissen, D., 1980, Statistical problems of fitting individual growth curves, in: *Human Physical Growth and Maturation, Methodologies and Factors* (F. E. Johnston, A. F. Roche, and C. Susanne, eds.), Plenum Press, New York.

Bonnard, G. D., 1968, Cortical thickness and diaphysial diameter of the metacarpal bones from the age of three months to eleven years, *Helv. Paediatr. Acta* **23**:445.

Bourne, G. H., 1971, *The Biochemistry and Physiology of Bone,* Vol. III: *Development and Growth,* 2nd ed., Academic Press, New York.

Brailsford, J. F., 1943, Variations in the ossification of the bones of the hand, *J. Anat.* **77**:170.

Brailsford, J. F., 1953, *The Radiology of Bones and Joints,* 5th ed., J. and A. Churchill, London.

Brandner, M. E., 1970, Normal values of the vertebral body and intervertebral disc index during growth, *Am. J. Roentgenol.* **110**:618.

Brunk, U., and Sköld, G., 1962, Length of foetus and ossification of the toe and finger phalanges, *Acta Histochem.* **14**:59.

Caffey, J., 1978, *Pediatric X-Ray Diagnosis,* Vol. 2, 7th ed., Year Book Medical Publishers, Chicago.

Caffey, J., and Ross, S., 1958, Pelvic bones in infantile mongoloidism. 3. Roentgenographic features, *Am. J. Roentgenol.* **80**:458.

Caffey, J., Ames, R., Silverman, W. A., Ryder, C. T., and Hough, G., 1956, Contradiction of the congenital dysplasia-predislocation hypothesis of congenital dislocation of the hip through a study of the normal variation in acetabular angles at successive periods in infancy, *Pediatrics* **17**:632.

Chesney, R. W., and Shore, R. M., 1982, The noninvasive determination of bone mineral content by photon absorptiometry, *Am. J. Dis. Child.* **136**:578.

Chumlea, W. C., Malina, M., and Rarick, G. L., 1981, Brachymesophalangia of the fifth finger, stature, and weight in children with Down syndrome, *J. Ment. Defic. Res.* **25**:7.

Christ, H. H., 1961, A discussion of causes of error in the determination of chronological age in children by means of x-ray studies of carpal-bone development, *South Afr. Med. J.* **35**:854.

Cohn, S. H., 1981, Total body neutron activation, in: *Non-Invasive Measurements of Bone Mass and Their Clinical Application* (S. H. Cohn, ed.), pp. 191–214, CRC Press, Boca Raton, Florida.

Colbert, C., and Bachtell, R. S., 1981, Radiographic absorptiometry (Photodensitometry), in: *Non-Invasive Measurements of Bone Mass and Their Clinical Application* (S. H. Cohn, ed.), pp. 51–84, CRC Press, Boca Raton, Florida.

Coleman, W. H., 1969, Sex differences in the growth of the human bony pelvis, *Am. J. Phys. Anthropol.* **31:**125.

Colle, M. L., Alperin, H., and Greenblatt, R. B., 1977, The tall girl. Prediction of mature height and management, *Arch. Dis. Child.* **52:**118.

Crampton, C. W., 1908, Physiological age—A fundamental principle, *Am. Phys. Educ. Rev.* **13:**141–154, 214–227, 268–283, 345–358; reprinted in *Child Dev.* 1944, **15:**3.

Day, R., and Silverman, W. A., 1952, Growth of the fibula of premature infants as estimated in roentgen films: A method for assessing factors promoting or inhibiting growth, *Neo-Natal Stud.* **1:**111.

Dequeker, J., 1976, Quantitative radiology: Radiogrammetry of cortical bone, *Br. J. Radiol.* **49:**912.

Duke, P. M., Litt, I. F., and Gross, R. T., 1980, Adolescent's self assessment of sexual maturation, *Pediatrics.* **66:**918.

Edlin, J. C., Whitehouse, R. H., and Tanner, J. M., 1976, Relationship of radial metaphyseal band width to stature velocity, *Am. J. Dis. Child.* **130:**160.

Elgenmark, O., 1946, The normal development of the ossific centres during infancy and childhood. A clinical, roentgenologic and statistical study, *Acta Paediatr.* (Suppl. 1) **33:**79.

Enlow, D. H., 1963, *Principles of Bone remodeling. An Account of Post-Natal Growth and Remodeling Processes in Long Bones and the Mandible,* Charles C Thomas, Springfield, Illinois.

Fell, H. B., and Robison, R., 1934, The development of the calcifying mechanism in avian cartilage and osteoid tissue, *Biochem. J.* **28:**2243.

Flory, C. D., 1936, Osseous development in the hand as an index of skeletal development, *Monogr. Soc. Res. Child Dev.* **1:**139.

Francis, C. C., 1939, Factors influencing appearance of centers of ossification during early childhood, *Am. J. Dis. Child.* **57:**817.

Francis, C. C., 1940, The appearance of centers of ossification from 6–15 years, *Am. J. Phys. Anthropol.* **27:**127.

Francis, C. C., and Werle, P. P., 1939, The appearance of centers of ossification from birth to 5 years, *Am. J. Phys. Anthropol.* **24:**273.

Frost, H. M., 1964, Dynamics of bone remodeling, in: *Bone Biodynamics* (H. M. Frost, ed.), pp. 22–35, Little, Brown, Boston.

Gardner, E., 1971, Osteogenesis in the human embryo and fetus, in: *The Biochemistry and Physiology of Bone,* Vol. III (G. H. Bourne, ed.), pp. 77–118, Academic Press, New York.

Gardner, E., and Gray, D. J., 1970, The prenatal development of the human femur, *Am. J. Anat.* **129:**121.

Garn, S. M., 1970, *The Earlier Gain and the Later Loss of Cortical Bone in Nutritional Perspective,* Charles C Thomas, Springfield, Illinois.

Garn, S. M., and Rohmann, C. G., 1959, Commonalities of the ossification centers of the hand and wrist, *Am. J. Phys. Anthropol.* **17:**319.

Garn, S. M., and Rohmann, C. G., 1960, Variability in the order of ossification of the bony centers of the hand and wrist, *Am. J. Phys. Anthropol.* **18:**219.

Garn, S. M., Rohmann, C. G., and Apfelbaum, B., 1961*a,* Complete epiphyseal union of the hand, *Am. J. Phys. Anthropol.* **19:**365.

Garn, S. M., Rohmann, C. G., and Robinow, M., 1961*b,* Increments in hand–wrist ossification, *Am. J. Phys. Anthropol.* **19:**45.

Garn, S. M., Silverman, F. N., and Rohmann, C. G., 1964, A rational approach to the assessment of skeletal maturation, *Ann. Radiol.* **7:**297.

Garn, S. M., Rohmann, C. G., and Blumenthal, T., 1966, Ossification sequency polymorphism and sexual dimorphism in skeletal development, *Am. J. Phys. Anthropol.* **24:**101.

Garn, S. M., Silverman, F. N., Hertzog, K. P., and Rohmann, C. G., 1968, Lines and bands of increased density, *Med. Radiogr. Photogr.* **44:**58.

Garn, S. M., Poznanski, A. K., and Nagy, J. M., 1971, Bone measurement in the differential diagnosis of osteopenia and osteoporosis, *Radiology* **100:**509.

Garn, S. M., Hertzog, K. P., Poznanski, A. K., and Nagy, J. M., 1972*a,* Metacarpophalangeal length in the evaluation of skeletal malformation, *Radiology* **105:**375.

Garn, S. M., Poznanski, A. K., Nagy J. M., and McCann, M. B., 1972*b,* Independence of brachymesophalangia-5 from brachymesophalangia-5 with cone mid-5, *Am. J. Phys. Anthropol.* **36:**295.

Garrow, J. S., and Fletcher, K., 1964, The total weight of mineral in the human infant, *Br. J. Nutr.* **18:**409.

Gefferth, K., 1972, Metrische Auswertung der kurzen Röhrenknochen der Hand von der Geburt bis zum Ende der Pubertät; Längenmasse, *Acta Paediatr. Acad. Sci. Hung.* **13:**117.

Gellis, S. S., and Feingold, M., 1968, *Atlas of Mental Retardation Syndromes. Visual Diagnosis of Facies and Physical Findings,* U. S. Department of Health, Education, and Welfare, Social and Rehabilitation Service, Rehabilitation Services Administration, Division of Mental Retardation, U. S. Government Printing Office, Washington, D. C.

Goff, C. W., 1960, *Surgical Treatment of Unequal Extremities,* Charles C Thomas, Springfield, Illinois.

Gray, D. J., and Gardner, E., 1969, The prenatal development of the human humerus, *Am. J. Anat.* **124:**431.

Gray, D. J., Gardner, E., and O'Rahilly, R., 1957, The prenatal development of the skeleton and joints of the human hand, *Am. J. Anat.* **101:**169.

Greulich, W. W., 1954, The relationship of skeletal status to the physical growth and development of children, In:

Dynamics of Growth Processes (E. J. Boell, ed.), pp. 212–223, Princeton University Press, Princeton, New Jersey.

Greulich, W. W., and Pyle, S. I., 1950, *Radiographic Atlas of Skeletal Development of the Hand and Wrist,* 1st ed., Stanford University Press, Stanford, California.

Greulich, W. W., and Pyle, S. I., 1959, *Radiographic Atlas of Skeletal Development of the Hand and Wrist,* 2nd ed., Stanford University Press, Stanford, California.

Hadley, L. A., 1956, *The Spine. Anatomico-Radiographic Studies: Development and the Cervical Region,* Charles C Thomas, Springfield, Illinois.

Haines, R. W., 1975, The histology of epiphyseal union in mammals, *J. Anat.* **120:**1.

Haines, R. W., and Mohiuddin, A., 1959, A preliminary note on the process of epiphysial union, *J. Fac. Med. Baghdad* **1:**141.

Ham. A. W., 1974, *Histology,* 7th ed., J. B. Lippincott, Philadelphia.

Ham, A. W., and Cormack, D. H., 1979, *Histology,* 8th ed., J. B. Lippincott, Philadelphia.

Hansman, C. F., 1962, Appearance and fusion of ossification centers in the human skeleton, *Am. J. Roentgenol.* **88:**476.

Hansman, C. F., and Maresh, M. M., 1961, A longitudinal study of skeletal maturation, *Am. J. Dis. Child.* **101:**305.

Harper, H. A. S., Poznanski, A. K., and Garn, S. M., 1974, The carpal angle in American populations, *Invest. Radiol.* **9:**217.

Harris, E. F., Weinstein, S., Weinstein, L., and Poole, A. E., 1980, Predicting adult stature: A comparison of methodologies, *Ann. Hum. Biol.* **7:**225.

Harrison, T. J., 1958*a,* The growth of the pelvis in the rat—A mensural and morphological study, *J. Anat.* **92:**236.

Harrison, T. J., 1958*b,* An experimental study of pelvic growth in the rat, *J. Anat.* **92:**483.

Harrison, T. J., 1961, The influence of the femoral head on pelvic growth and acetabular form in the rat, *J. Anat.* **95:**12.

Hattori, K., Shindo, Y., Terazawa, T., and Morita, S., 1979, Age change of tibial cross section in fetal period, *Jikeikai Med. J.* **26:**195.

Hasselwander, A., 1910, Ossification des menschlichen Fusskeletts, *Z. Morphol. Anthropol.* **12:**1.

Healy, M. J. R., and Goldstein, H., 1976, An approach to the scaling of categorised attributes, *Biometrika* **63:**219.

Hellman, M., 1928, Ossification of epiphyseal cartilages in the hand, *Am. J. Phys. Anthropol.* **11:**223.

Heřt, J., 1972, Growth of the epiphyseal plate in circumference, *Acta Anat.* **82:**420.

Hesser, C., 1926, Beitrag zur Kenntnis der Gelenkentwicklung beim Menschen, *Morphol. Jahrb.* **55:**489.

Hewitt, D., 1963, Pattern of correlations in the skeleton of the growing hand, *Ann. Hum. Genet.* **27:**157.

Hewitt, D., and Acheson, R. M., 1961, Some aspects of skeletal development through adolescence. I. Variations in the rate and pattern of skeletal maturation at puberty, *Am. J. Phys. Anthropol.* **19:**321.

Hewitt, D., Westropp, C. K., and Acheson, R. M., 1955, Oxford Child Health Survey: Effect of childish ailments on skeletal development, *Br. J. Prev. Soc. Med.* **9:**179.

Hinck, V. C., and Hopkins, C. E., 1960, Measurement of the atlanto-dental interval in the adult, *Am. J. Roentgenol.* **84:**945.

Hinck, V. C., Hopkins, C. E., and Savara, B. S., 1962, Sagittal diameter of the cervical spinal canal in children, *Radiology* **79:**97.

Hinck, V. C., Hopkins, C. E., and Clark, W. M., 1965, Sagittal diameter of the lumbar spinal canal in children and adults, *Radiology* **85:**929.

Hinck, V. C., Clark, W. M., Jr., and Hopkins, C. E., 1966, Normal interpediculate distances (minimum and maximum) in children and adults, *Am. J. Roentgenol.* **97:**141.

Hinkel, G. K., and Schambach, H., 1980, Prediction of final height of very tall children, *Dtsch. Gesundheit.* **35:**1670.

Hoerr, N. L., Pyle, S. I., and Francis, C. C., 1962, *Radiographic Atlas of Skeletal Development of the Foot and Ankle, A Standard of Reference,* Charles C Thomas, Springfield, Illinois.

Howard, C. C., 1928, The physiologic process of the bone centers of the hands of normal children between the ages of five and sixteen inclusive; also a comparative study of both retarded and accelerated hand growth in children whose general skeletal growth is similarly affected, *J. Orthod. Oral Surg. Radiogr.* **14:**948, 1041.

Hughes, P. C. R., and Tanner, J. M., 1970, The assessment of skeletal maturity in the growing rat, *J. Anat.* **106:**371.

de Iturriza, J. R., and Tanner, J. M., 1969, Cone-shaped epiphyses and other minor anomalies in the hands of normal British children, *J. Pediatr.* **75:**265.

Johnson, G. F., Dorst, J. P., Kuhn, J. P., Roche, A. F., and Davila, G. H., 1973, Reliability of skeletal age assessments, *Am. J. Roentgenol.* **18:**320.

Johnston, F. E., 1963, Skeletal age and its prediction in Philadelphia children, *Am. J. Phys. Anthropol.* **21:**406.

Johnston, F. E., 1964, The relationship of certain growth variables to chronological and skeletal age, *Hum. Biol.* **36:**16.

Johnston, F. E., and Jahina, S. B., 1965, The contribution of the carpal bones to the assessment of skeletal age, *Am. J. Phys. Anthropol.* **23:**349.

Kelly, H. J., and Reynolds, L., 1947, Appearance and growth of ossification centers and increases in the body dimensions of white and Negro infants, *Am. J. Roentgenol.* **57:**477.

Kimura, K., 1978, Growth of cortical thickness of the second metacarpal based on chronological age and skeletal maturity, *Z. Morphol. Anthropol.* **69:**183.

Kimura, K., 1981, Skeletal maturity in twins, *J. Anthropol. Soc. Nippon* **89:**457.

Kimura, K., and Takeuchi, K., 1982, Notches at the nonepiphyseal end of the second metacarpal in Japanese-American hybrids, *Z. Morphol. Anthropol.* **73**:1.

Köhler, A., and Zimmer, E. A., 1953, *Grenzen des Normalen und Anfänge des Pathologischen im Röntgenbilde des Skelettes,* Georg Thieme Verlag, Stuttgart.

Koski, K., 1968, Cranial growth centers: Facts or fallacies?, *Am. J. Orthod.* **54**:566.

Koski, K., Haataja, J., and Lappalainen, M., 1961, Skeletal development of hand and wrist in Finnish children, *Am. J. Phys. Anthropol.* **19**:379.

Kosowicz, J., 1962, The carpal sign in gonadal dysgenesis, *J. Clin. Endocrinol. Metab.* **22**:949.

Kosowicz, J., 1965, The roentgen appearance of the hand and wrist in gonadal dysgenesis, *Am. J. Roentgenol.* **93**:354.

Krabbe, S., Transbøl, I., and Christiansen, C., 1982, Bone mineral homeostasis, bone growth, and mineralisation during years of pubertal growth: A unifying concept, *Arch. Dis. Child.* **57**:359.

Lacroix, P., 1951, *The Organization of Bones,* McGraw-Hill (Blakiston), New York.

Lacroix, P., 1971, The internal remodeling of bones, in: *The Biochemistry and Physiology of Bone* (G. H. Bourne, ed.) Vol. III, pp. 119–144, Academic Press, New York.

Lee, M. M. C., 1968, Natural markers in bone growth, *Am. J. Phys. Anthropol.* **29**:295.

Lenko, H. L., 1979, Prediction of adult height with various methods in Finnish children, *Acta Paediatr. Scand.* **68**:85.

Lenko, H. L., Mäenpää, J., and Perheentupa, J., 1982, Acceleration of delayed growth with fluoxymesterone, *Acta Paediatr. Scand.* **71**:929.

Lestrel, P. E., 1974, Some problems in the assessment of morphological size and shape differences, *Yearb, Phys. Anthropol.* **18**:140.

Lestrel, P. E., 1975, Fourier analysis of size and shape of the human cranium: A longitudinal study from four to eighteen years of age. Unpublished Ph.D. dissertation, University of California, Los Angeles.

Lestrel, P. E., and Roche, A. F., 1976, Fourier analysis of the cranium in trisomy 21, *Growth* **40**:385.

Lestrel, P. E., Kimbel, W. H., Prior, W. F., and Fleischmann, M. L., 1977, Size and shape of the hominoid distal femur: Fourier analysis, *Am. J. Phys. Anthropol.* **46**:281.

Lewis, W. H., 1902, The development of the arm in man, *Am. J. Anat.* **1**:145.

Locke, G. R., Gardner, J. I., and Van Epps, E. F., 1966, Atlas–Dens Interval (ADI) in children. A survey based on 200 normal cervical spines, *Am. J. Roentgenol.* **97**:135.

Lord, F. M., 1953, On the statistical treatment of football numbers, *Am. Psychol.* **8**:750.

Low, W. D., and Kung, L. S., 1982, Rate of growth of the second metacarpal bone in Chinese children, *Z. Morphol. Anthropol.* **73**:15.

Lusted, L. B., and Keats, T. E., 1972, *Atlas of Roentgenographic Measurement,* 4th ed., Year Book Medical Publishers, Chicago.

Macarthur, J. W., and McCulloch, E., 1932, Apical dystrophy. An inherited defect of hands and feet, *Hum. Biol.* **4**:179.

McKern, T. W., and Stewart, T. D., 1957, *Skeletal Age Changes in Young American Males Analyzed from the Standpoint of Age Identification,* Technical Report EP-45, Headquarters Quartermaster Research & Development Command, Environmental Protection Research Division, U.S. Army, Natick, Massachusetts.

Mainland, D., 1953, Evaluation of the skeletal age method of estimating children's development. I. Systematic errors in the assessment of roentgenograms, *Pediatrics* **12**:114.

Mainland, D., 1954, Evaluation of the skeletal age method of estimating children's development. II. Variable errors in the assessment of roentgenograms, *Pediatrics* **13**:165.

Maresh, M. M., 1970, Measurements from roentgenograms, heart size, long bone lengths, bone, muscle and fat widths, skeletal maturation, in: *Human Growth and Development* (R. W. McCammon, ed.), pp. 157–200, Charles C Thomas, Springfield, Illinois.

Maresh, M. M., and Deming, J., 1939, The growth of the long bones in 80 infants: Roentgenograms versus anthropometry, *Child Dev.* **10**:91.

Marshall, W. A., 1966, Problems in relating radiopaque transverse lines in the radius to the occurrence of disease, *Symp. Soc. Hum. Biol.* **8**:245.

Mathiasen, M. S., 1973, Determination of bone age and recording of wrist or skeletal hand anomalies in normal children, *Dan. Med. Bull.* **20**:80.

Mazess, R. B., 1981, Photon absorptiometry, in: *Non-Invasive Measurements of Bone Mass and Their Clinical Application* (S. H. Cohn, ed.), pp. 85–100, CRC Press, Boca Raton, Florida.

Mazess, R. B., and Cameron, J. R., 1971, Skeletal growth in school children: Maturation and bone mass, *Am. J. Phys. Anthropol.* **35**:399.

Meema, H. E., and Meema, S., 1981, Radiogrammetry, in: *Non-Invasive Measurements of Bone Mass and Their Clinical Application* (S. H. Cohn, ed.) pp. 5–50, CRC Press, Boca Raton, Florida.

Moed, G., Wight, B. W., and Vandergrift, H. N., 1962, Studies of physical disability: Reliability of measurement of skeletal age from hand films, *Child Dev.* **33**:37.

Moerman, M. L., 1981, A Longitudinal Study of Growth in Relation to Body Size and Sexual Dimorphism in the Human Pelvis, Unpublished Ph.D. dissertation, University of Michigan, Ann Arbor.

Moerman, M. L., 1982, Growth of the birth canal in adolescent girls, *Am. J. Obstet. Gynecol.* **143**:528.

Morris, N. M., and Udry, J. R., 1981, Validation of a self-administered instrument to assess stage of adolescent development, *J. Youth Adoles.* **9**:271.

Moss, M. L., and Noback, C. R., 1958, A longitudinal study of digital epiphyseal fusion in adolescence, *Anat. Rec.* **131**:19.

Mossberg, H. O., 1949, The x-ray appearance of the knee joint in obese over-grown children, *Acta Paediatr. Scand.* **38**:509.

Murray, J. R., Bock, R. D., and Roche, A. F., 1971, The measurement of skeletal maturity, *Am. J. Phys. Anthropol.* **35**:327.

Naik, D. R., 1970, Cervical spinal canal in normal infants, *Clin. Radiol.* **21**:323.

Neinstein, L. S., 1982, Adolescent self-assessment of sexual maturation, *Clin. Pediatr.* **21**:482.

Niven, J. S. F., and Robison, R., 1934, The development of the calcifying mechanism in the long bones of the rabbit, *Biochem. J.* **28**:2237.

Noback, C. R., Moss, M. L., and Leszczynska, E., 1960, Digital epiphyseal fusion of the hand in adolescence. A longitudinal study, *Am. J. Phys. Anthropol.* **18**:13.

O'Rahilly, R., 1951, Morphological patterns in limb deficiencies and duplications, *Am. J. Anat.* **89**:135.

Oliver, G., 1962, *Formation du Squelettes des Membres chez l'Homme,* Vigot Fréres, Paris.

Oliver, G., and Pineau, H. A., 1959, Embryologie de l'Humerus, *Arch. Anat. Pathol.* **7**:121.

Onat, T., 1983, Multifactorial prediction of adult height of girls during early adolescence allowing for genetic potential, skeletal and sexual maturity, *Hum. Biol.* **55**:443.

Parfitt, A. M., 1979, Quantum concept of bone remodeling and turnover: Implications for the pathogenesis of osteoporosis, *Calcif. Tissue Int.* **28**:1.

Park, E. A., 1954, Bone growth in health and disease, *Arch. Dis. Child.* **29**:269.

Paterson, R. S., 1929, A radiological investigation of the epiphyses of the long bones, *J. Anat.* **64**:28.

Payton, C. G., 1933, The growth of the epiphyses of the long bones in the madder-fed pig, *J. Anat.* **67**:371.

Peritz, E., and Sproul, A., 1971, Some aspects of the analysis of hand–wrist bone-age readings, *Am. J. Phys. Anthropol.* **35**:441.

Poland, J., 1898, *Skiagraphic Atlas. The Development of the Bones of the Wrist and Hand,* Smith, Elder, and Co., London.

Poznanski, A. K., 1974, *The Hand in Radiologic Diagnosis,* W. B. Saunders, Philadelphia.

Poznanski, A. K., Garn, S. M., Kuhns, L. R., and Sandusky, S. T., 1971, Dysharmonic maturation of the hand in the congenital malformation syndromes, *Am. J. Phys. Anthropol.* **35**:417.

Poznanski, A. K., Garn, S. M., Nagy, J. M., and Gall, J. C., Jr., 1972*a,* Metacarpophalangeal pattern profiles in the evaluation of skeletal malformation, *Radiology* **104**:1.

Poznanski, A. K., Garn, S. M., Gall, J. C., Jr., and Stern, A. M., 1972*b,* Objective evaluation of the hand in the Holt–Oram syndrome, *Birth Defects Orig. Art. Ser.* **8**:125.

Poznanski, A. K., Garn, S. M., Kuhns, L. R., and Shaw, H. A., 1977, Disharmonic skeletal maturation in the congenital malformation syndromes, *Birth Defects* **13**:45.

Poznanski, A. K., Hernandez, R. J., Guire, K. E., Bereza, U. L., and Garn, S. M., 1978, Carpal length in children—A useful measurement in the diagnosis of rheumatoid arthritis and some congenital malformation syndromes, *Radiology* **129**:661.

Pryor, J. W., 1907, The hereditary nature of variation in the ossification of bones, *Anat. Rec.* **1**:84.

Pryor, H. B., and Carter, H. D., 1938, Phases of adolescent development in girls, *Calif. West. Med.* **48**:89.

Pyle, S. I., and Hoerr, N. L., 1955, *Radiographic Atlas of Skeletal Development of the Knee. A Standard of Reference,* Charles C Thomas, Springfield, Illinois.

Pyle, S. I., and Hoerr, N. L., 1969, *A Radiographic Standard of Reference for the Growing Knee,* Charles C Thomas, Springfield, Illinois.

Pyle, S. I., Mann, A. W., Dreizen, S., Kelly, H. J., Macy, I. G., and Spies, T. D., 1948, A substitute for skeletal age (Todd) for clinical use: The red graph method, *J. Pediatr.* **32**:125.

Pyle, S. I., Stuart, H. C., Cornoni, J., and Reed, R. B., 1961, Onsets, completions, and spans of the osseous stage of development in representative bone growth centers of the extremities, *Monogr. Soc. Res. Child Dev.* **26(1)**:7.

Pyle, S. I., Waterhouse, A. M., and Greulich, W. W., 1971, *A Radiographic Standard of Reference for the Growing Hand and Wrist,* Prepared for the United States National Health Examination Survey, Case Western Reserve University, Cleveland, Ohio.

Rambaud, A., and Renault, C., 1864, *Origine et Dévelоppement des Os,* F. Chamerot, Paris.

Rand, T. C., Edwards, D. K., Bay, C. A., and Jones, K. L., 1980, The metacarpal index in normal children, *Pediatr. Radiol.* **9**:31.

Rasmussen, H., and Bordier, Ph.J., 1974, *The Physiological and Cellular Basis of Metabolic Bone Disease,* Williams & Wilkins, Baltimore.

Reynolds, E. L., 1945, Bony pelvic girdle in early infancy, *Am. J. Phys. Anthropol.* **3**:321.

Reynolds, E. L., 1947, The bony pelvis in prepubertal childhood, *Am. J. Phys. Anthropol.* **5**:165.

Reynolds, E. L., and Asakawa, T., 1951, Skeletal development in infancy; standards for clinical use, *Am. J. Roentgenol.* **65**:403.

Ritzen, M., Aperia, A., Hall, K., Larsson, A., Zetterberg, A., and Zetterstrom, R. (eds.), 1981, *The Biology of Normal Human Growth,* Raven Press, New York.

Robinow, M., 1942, Appearance of ossification centres: Grouping obtained from factor analysis, *Am. J. Dis. Child.* **64**:229.

Robinow, M., and Chumlea, W. C., 1982, Standards for limb bone length ratios in children, *Radiology* **143**:333.

Roche, A. F., 1961, Clinodactyly and brachymesophalangia of the fifth finger, *Acta Paediatr.* **50**:387.

Roche, A. F., 1962*a,* Lateral comparisons of the skeletal maturity of the human hand and wrist, *Am. J. Roentgenol.* **89**:1272.

Roche, A. F., 1962*b,* Incomplete distal phalanges in the foot during childhood, *Acta Anat.* **51**:369.

Roche, A. F., 1965, The sites of elongation of human metacarpals and metatarsals, *Acta Anat.* **61**:193.

Roche, A. F., 1967, The elongation of the mandible, *Am. J. Orthod.* **53**:79.

Roche, A. F., 1968, Sex-associated differences in skeletal maturity, *Acta Anat.* **71**:321.

Roche, A. F., 1970, Associations between the rates of maturation of the bones of the hand–wrist, *Am. J. Phys. Anthropol.* **33**:341.

Roche, A. F., 1980, The measurement of skeletal maturation, in: *Human Physical Growth and Maturation: Methodologies and Factors* (F. E. Johnston, A. F. Roche, and C. Susanne, eds.), pp. 61–82, Plenum Press, New York.

Roche, A. F., and Davila, G. H., 1972, Late adolescent growth in stature, *Pediatrics* **50**:874.

Roche, A. F., and Davila, G. H., 1974, Differences between recumbent length and stature within individuals, *Growth* **38**:313.

Roche, A. F., and Davila, G. H., 1976, The reliability of assessments of the maturity of individual hand–wrist bones, *Hum. Biol.* **48**:585.

Roche, A. F., and French, N. Y., 1970, Differences in skeletal maturity levels between the knee and hand, *Am. J. Roentgenol.* **109**:307.

Roche, A. F., and Hermann, R., 1970, Associations between the rates of elongation of the short bones of the hand, *Am. J. Phys. Anthropol.* **32**:83.

Roche, A. F., and Johnson, J. M., 1969, A comparison between methods of calculating skeletal age (Greulich–Pyle), *Am. J. Phys. Anthropol.* **30**:221.

Roche, A. F., and Sunderland, S., 1960, Observations on the developing fifth toe in normal children, *Acta Anat.* **41**:261.

Roche, A. F., and Wettenhall, H. N. B., 1977, Stature prediction in short boys, *Aust. Pediatr. J.* **13**:261.

Roche, A. F., Rohmann, C. G., French, N. Y., and Davila, G. H., 1970, Effect of training of replicability of assessments of skeletal maturity, *Am. J. Roentgenol.* **108**:511.

Roche, A. F., Roberts, J., and Hamill, P. V. V., 1974, *Skeletal Maturity of Children 6–11 Years,* United States Vital and Health Statistics Series 11, No. 140, U.S. Department of Health, Education and Welfare, Washington, D. C., pp. 1–62.

Roche, A. F., Wainer, H., and Thissen, D., 1975*a,* Predicting adult stature for individuals, *Monogr. Paediatr.* **3**:41–96.

Roche, A. F., Wainer, H., and Thissen, D., 1975*b, Skeletal Maturity. The Knee Joint as a Biological Indicator,* Plenum Press, New York.

Roche, A. F., Roberts, J., and Hamill, P. V. V., 1975*c, Skeletal Maturity of Children 6–11 Years: Racial, Geographic Area of Residence, and Socioeconomic Differentials,* United States Vital and Health Statistics Series 11, No. 149, U.S. Department of Health, Education and Welfare, Washington, D. C., 81 pp.

Roche, A. F., Roberts, J., and Hamill, P. V. V., 1976, *Skeletal Maturity of Youths 12–17 Years,* United States Vital and Health Statistics Series 11, No. 160, U.S. Department of Health, Education and Welfare, Washington, D. C., pp. 1–81.

Roche, A. F., Roberts, J., and Hamill, P. V. V., 1978, *Skeletal Maturity of Youths 12–17 Years: Racial, Geographic Area, and Socioeconomic Differentials,* United States Vital and Health Statistics Series 11, No. 167, U.S. Department of Health, Education and Welfare, Washington, D. C., pp. 1–98.

Roche, A. F., Tyleshevski, F., and Rogers, E., 1983, Noninvasive measurements of physical maturity in children, *Res. Q. Exer. Sport* **54**:364.

Rubinstein, J. H., and Taybi, H., 1963, Broad thumbs and toes and facial abnormalities. A possible mental retardation syndrome, *Am. J. Dis. Child.* **105**:588.

Sahay, G. B., 1941, Determination of age by x-ray examination (with reference to epiphyseal union), *Indian Med. J.* **35**:37.

Samejima, F., 1969, Estimation of latent ability using a response pattern of graded scores, *Psychometrika Monogr. (Suppl).* **34**:1.

Samejima, F., 1972, General model for free response data, *Psychometrika Monogr. Suppl.* **37**:18.

Sauvegrain, J., Nahum, H., and Bronstein, H., 1962, Etude de la Maturation Osseuse du Coude, *Ann. Radiol.* **5**:542.

Scheller, S., 1960, Roentgenographic studies on epiphysial growth and ossification in the knee, *Acta Radiol. (Suppl.)* **195**:35–37, 190–194.

Schmorl, G., and Junghanns, H., 1971, *The Human Spine in Health and Disease,* 2nd ed., Grune & Stratton, New York.

Schreiber, A., Patois, E., and Roy, M. P., 1976, Etude comparative de quatre méthodes de prédiction de la taille adulte, *Compte-Rendu de la XIIIe Réunion des Equipes Chargées des Etude sur la Croissance et le Développement de l'Enfant Normal, Rennes, 1976,* pp. 131–136.

Sempé, M., and Capron, J. P., 1979, Chronos: Analyse de la Maturation Squelettique par une Méthode Numerique Automatisée, *Pediatrie* **8**:833.

Shapiro, F., 1982, Developmental patterns in lower-extremity length discrepancies, *J. Bone Jt. Surg.* **64A**: 639.

Siegling, J. A., 1941, Growth of the epiphyses, *J. Bone Jt. Surg.* **23**:23.

Siffert, R. S., 1966, The growth plate and its affections, *J. Bone Jt. Surg.* **48A**:546.

Silberberg, M., and Silberberg R., 1961, Ageing changes in cartilage and bone, in: *Structural Aspects of Ageing* (G. H. Bourne, ed.), pp. 85–108, Pitman Medical, London.

Sinclair, R. J. G., Kitchin, A. H., and Turner, R. W. D., 1960, The Marfan syndrome, *Q. J. Med.* **53**:19.

Sontag, L. W., and Lipford, J., 1943, The effect of illness and other factors on the appearance pattern of skeletal epiphyses, *J. Pediatr.* **23**:391.

Sontag, L. W., Snell, D., and Anderson, M., 1939, Rate of appearance of ossification centers from birth to the age of five years, *Am. J. Dis. Child.* **58**:949.

Sproul, A., and Peritz, E., 1971, Assessment of skeletal age in short and tall children, *Am. J. Phys. Anthropol.* **35**:433.

Stevenson, P. H., 1924, Age order of epiphyseal union in man, *Am. J. Phys. Anthropol.* **7**:53.

Streeter, G. L., 1951, *Developmental Horizons in Human Embryos, Age Groups 11 to 23, Embryology Reprint,* Vol. 2, Carnegie Institute, Washington, D. C.

Takahashi, Y., 1956, *Studies on Skeletal Development in Japanese Children. Roentgenographic Indicators of Maturity in Bones of the Foot, Ankle, Knee and Elbow Areas,* U.S. Atomic Energy Commission, Oak Ridge, Tennessee.

Tanner, J. M., 1962, *Growth at Adolescence; with a General Consideration of the Effects of Hereditary and Environmental Factors upon Growth and Maturation from Birth to Maturity,* 2nd ed., Blackwell, Oxford.

Tanner, J. M., Whitehouse, R. H., and Healy, M. J. R., 1962, *A New System for Estimating Skeletal Maturity from the Hand and Wrist, with Standards Derived from a Study of 2,600 Healthy British Children,* International Children's Centre, Paris.

Tanner, J. M. Whitehouse, R. H., and Goldstein, H., 1972, *A Revised System for Estimating Skeletal Maturity from Hand and Wrist Radiographs, with Separate Standards for Carpals and Other Bones (TW II System),* International Children's Centre, Paris.

Tanner, J. M., Whitehouse, R. H., Marshall, W. A., Healy, M. J. R., and Goldstein, H., 1975, *Assessment of Skeletal Maturity and Prediction of Adult Height (TW2 Method),* Academic Press, London.

Tanner, J. M., Landt, K. W., Cameron, N., Carter, B. S., and Patel, J., 1983*a,* Prediction of adult height from height and bone age in childhood, *Arch. Dis. Child.* **58**:767.

Tanner, J. M., Whitehouse, R. H., Cameron, N., Marshall, W. A., Healy, M. J. R., and Goldstein, H., 1983*b, Assessment of Skeletal Maturity and Prediction of Adult Height,* 2nd ed., Academic Press, London.

Templeton, A. W., McAlister, W. H., and Zim, I. D., 1965, Standardization of terminology and evaluation of osseous relationships in congenitally abnormal feet, *Am. J. Roentgenol.* **93**:374.

Thompson, G. W., Popovich, F., and Luks, E., 1973, Sexual dimorphism in hand and wrist ossification, *Growth* **37**:1.

Tiisala, R., Kantero, R.-L., and Bäckström, L., 1966, The appearance pattern of skeletal epiphyses among normal Finnish children during the first five years of life, *Ann. Paediatr. Fenn.* **12**:64.

Todd, T. W., 1930*a,* The roentgenographic appraisement of skeletal differentiation, *Child Dev.* **1**:298.

Todd, T. W., 1930*b,* The anatomical features of epiphyseal union, *Child Dev.* **1**:186.

Todd, T. W., 1937, *Atlas of Skeletal Maturation,* C. V. Mosby, St. Louis.

Trotter, M., and Peterson, R. R., 1968, Weight of bone in the fetus. A preliminary report, *Growth* **32**:83.

Trotter, M., and Peterson, R. R., 1969*a,* Weight of bone during the fetal period, *Growth* **33**:167.

Trotter, M., and Peterson, R. R., 1969*b,* Weight of bone in the fetus during the last half of pregnancy, *Clin. Orthop.* **65**:46.

Trotter, M., and Peterson, R. R., 1970, Weight of the skeleton during postnatal development, *Am. J. Phys. Anthropol.* **33**:313.

Trueta, J., 1968, *Studies of the Development and Decay of the Human Frame,* W. B. Saunders, Philadelphia.

Tyson, J. E., Maravilla, A., Lasky, R. E., Cope, F. A., and Mize, C. E., 1983, Measurement of bone mineral content of preterm neonates, *Am. J. Dis. Child.* **137**: 735.

Valk, I. M., 1971, Accurate measurement of the length of the ulna and its application in growth measurements, *Growth* **35**:297.

Valk, I. M., 1972, Ulnar length and growth in twins with a simplified technique for ulnar measurement using a condylograph, *Growth* **36**:291.

Valk, I. M., Langhout Chabloz, A. M. E., Smals, A. G. H., Kloppenborg, P. W. C., Cassorla, F. G., and Schutte, E. A. S. T., 1983, Accurate measurements of the lower leg length and the ulnar length and its application in short term growth measurement, *Growth* **47**:53.

Wainer, H., 1976, Estimating coefficients in linear models: It don't make no nevermind, *Psychol. Bull.* **82**:213.

Wainer, H., Roche, A. F., and Bell, S., 1978, Predicting adult stature without skeletal age and without paternal data, *Pediatrics* **61**:569.

Wetherington, R. K., and Haurberg, S., 1974, Study of brachymesophalangia in a Hong Kong sample, *Am. J. Phys. Anthropol.* **41**:509.

Wilks, S. S., 1938, Weighting systems for linear functions of correlated variables when there is no dependent variable, *Psychometrica* **3**:23.

Yarbrough, C., Habicht, J.-P., Klein, R. E., and Roche A. F., 1973, Determining the biological age of the preschool child from a hand–wrist radiograph, *Invest. Radiol.* **8**:233.

Zachmann, M., Ferrandez, A., Murset, G., and Prader, A., 1975, Estrogen treatment of excessively tall girls, *Helv. Paediatr. Acta* **30**:11.

Zachmann, M., Ferrandez, A., Murset, G., Gnehm, H. E., and Prader, A., 1976, Testosterone treatment of excessively tall boys, *J. Pediatr.* **88**:116.

Zachmann, M., Sobradillo, B., Frank, M., Frisch, H., and Prader, A., 1978, Bayley-Pinneau, Roche-Wainer-Thissen, and Tanner height predictions in normal children and in patients with various pathologic conditions, *J. Pediatr.* **93**:749.

3

Adipose Tissue Growth and Obesity

Irving M. Faust

1. Introduction

Since the publication of the first edition of this treatise, there has been steady progress in the understanding of adipose tissue growth and development, particularly with regard to the problem of obesity. The results of a wide variety of experiments have begun to answer important questions at all levels from the cellular to the human/organismic. Extensive review articles written during this time chart the progress of these studies and suggest the directions to be taken by future research (e.g., Hirsch *et al.*, 1979; Kirtland and Gurr, 1979; Hausman *et al.*, 1980; Leibel *et al.*, 1983; Harris and Martin, 1984; Faust and Kral, 1985; Leibel, 1985b). Three particularly noteworthy books have also appeared during the past few years (Bonnet, 1981; Angel *et al.*, 1983; Van and Cryer, 1985), which are devoted almost entirely to detailed reviews of major subtopics in the areas of adipose tissue growth, adipose tissue metabolism and obesity. The present chapter focuses on a selection of recent findings pertaining to the growth of adipose tissue, the regulation of adipose mass, and aspects of these processes that bear on the problem of obesity.

2. Features of Brown and White Adipose Tissue

Adipose tissue makes up 15–30% of the total body weight of most adult humans. In male athletes, body fat may be less than 15% of body weight, while in individuals who are severely obese, it is substantially more than 30%. In the adult, virtually all the body fat is in the form of white adipose tissue. However, in the newborn child, there is also some brown adipose tissue distributed widely around the body. In the newborn, this tissue apparently functions as a source of heat. In many small animals of all ages, the thermogenic function of brown fat is extremely important. It elevates the body temperature of hibernators during arousal, and it is a major souce of body heat in small animals exposed to the cold. Brown adipose tissue has substantial capacity to metabolize substrate and generate heat because it contains numerous mitochondria in which oxidation is largely uncoupled from the production of ATP. It can also transfer rapidly the heat it produces to the rest of the body because it has an extremely rich vasculature (Himms-Hagen, 1983; Leibel *et al.*, 1983). After prolonged exposure of a small animal to the cold, the mass of the brown adipose tissue increases, thereby providing additional capacity for the generation of heat.

In some small animals, including the laboratory rat, brown fat appears to have an additional function, notably thermogenic dissipation of calories ingested in excess of needs for storage and metabolism. Such excess occurs when rats are fed diets with a high fat or high sugar content. The rats overeat, but do not use all the extra calories for increased lipid storage (Rothwell and Stock, 1979). The degree to which the energy dissipation function of brown fat contributes to long-term energy balance and to regulation of the size of lipid stores in

IRVING M. FAUST • The Rockefeller University, New York, New York 10021

rats and mice has not been determined. However, a defect in this function appears to be responsible, at least in part, for the obesity seen in *ob/ob* and *db/db* mice. (Evidence pertaining to this conclusion is reviewed by Himms-Hagen, 1983.) It is not clear that such a defect is basic to the etiology of the obesity in other obese rodents such as the Zucker rat and the hypothalamically lesioned rat. These animals do show a thermogenic response to diet that differs from normal, but there is no evidence that this abnormality is antecedent to the other disorders in these animals such as their high levels of food intake. It is very likely that many of the "abnormalities" exhibited by these animals (including the diminished thermogenic response to diet) will prove to be appropriate responses to some other disorder that is as yet undefined. The possibility that a defect in the thermogenic activity of brown fat is responsible for obesity in humans has been suggested (James and Trayhurn, 1981), but studies of brown adipose tissue thermogenesis in humans indicate that this is not likely (Astrup *et al.*, 1984). The most convincing studies of humans and of animals suggest that the obese enjoy a remarkable degree of normality in both food intake and energy expenditure (Ravussin *et al.*, 1982; Faust, 1984; Hirsch and Leibel, 1984). The key to these latter studies has been their focus on organisms that are in an unrestricted obese state rather than on organisms that are in the process of losing or gaining weight or maintaining normal weight (following weight reduction) by means of chronic calorie intake restriction.

White adipose tissue serves as the main energy store of the body. In the ordinary 70-kg adult, it consists of 10–30 × 10^9 fat cells, each of which contains, on average, about 0.5 μg of lipid. Since adipose tissue triglyceride has a caloric value of about 9 kcal/g (7 kcal/g tissue), the adipose tissue of the 70-kg adult contains a metabolizable energy store of about 100,000 kcal. Since the energy requirement of a relatively inactive individual undergoing total starvation is about 1100 kcal/day, white adipose tissue contains sufficient energy stores to sustain life for about 3 months. The mass of the adipose tissue in an adult often remains fairly constant over long periods of time, but the tissue is not inactive. Lipolysis and re-esterification occur at rapid rates at all times, regardless of the person's weight or nutritional status. The usual balance between rates of lipolysis and re-esterification, as well as the degree of imbalance that occurs when lipid stores are being increased or decreased, appears to be controlled by means of stimulation and inhibition of β- and α-adrenergic receptors on the fat cells (Ostman *et al.*, 1979). However, the relative preponderance of various adrenoreceptors and the functional response of these receptors to relevant agonists varies among different fat depots (LaFontan *et al.*, 1979; Smith *et al.*, 1979; Smith, 1980). This variation is probably the reason that these depots also vary in the degree to which they lose lipid during weight reduction (Krotkiewski *et al.*, 1975). Leibel (1985a) and Leibel *et al.* (1984) have reported the development of techniques for the study of fat cells from small samples of human adipose tissue that can provide detailed information on the relative activities of these receptors. Such information is potentially of great importance, since the incidence of certain diseases in the obsese has been linked to differences in accumulation and depletion of lipid in the various fat depots (see below).

2.1. Growth of White Adipose Tissue in Obese and Nonobese Organisms

The earliest indication of the production of mature adipocytes in humans is at about the 15th week of gestation. Shortly thereafter, the rate of adipocyte production becomes rapid and remains as such until about the 23rd week of gestation (Poissonnet *et al.*, 1983). Adipose tissue continues to grow during the remainder of the gestational period, particularly in subcutaneous sites (Knittle, 1978), but the rate of fat cell production is decreased (Poissonnet *et al.*, 1983). During the first 2 postnatal years, there are substantial increases in both the size and number of fat cells (Hager *et al.*, 1977; Knittle *et al.*, 1979). However, for the first 6 months of life, most adipose tissue growth appears to be caused by increases in fat cell size. Fat cell enlargement continues during the first year of life, but sometime between 6 and 12 months of age, the number of fat cells begins to increase as well. Between the second year of life and the onset of puberty in children of normal weight, the average size of fat cells either increases moderately or changes very little. The findings of the reported studies are not fully in agreement on this point. [It is also not clear whether fat cell size attains the adult level prior to, or after, the time of puberty (Hager *et al.*, 1977; Brook, 1978; Bonnet and Rocour-Brumioul, 1981).] However, it is generally agreed that the number of fat cells continues to increase slowly during this time of life in normal children (Bonnet and Rocour-Brumioul, 1981), while it increases rapidly in children who are clearly becoming severely obese (Knittle, 1978). By the time they reach adulthood, these obese children may have several times more fat cells than normal (Hirsch and Knittle, 1970; Salans *et al.*, 1973). In nonobese children, the rate of increase in fat cell number reaccelerates at about the time of puberty (Knittle, 1978; Bonnet and Rocour-Brumioul, 1981).

The normal growth of adipose mass that occurs during adulthood clearly involves enlargement of fat cells, but it is not clear whether the number of fat cells increases as well. The techniques that are used to determine total fat cell number in living people may not be adequate to detect increases that are limited to a few small sites or even one large site (Gurr and Kirtland, 1978). It is very difficult to obtain accurate values of the total number of fat cells in specific fat depots *ante mortem*. Thus, while it is possible that adipocyte hyperplasia contributes to the localized increases in adipose mass that often occur during adulthood (e.g., the accumulation of abdominal area fat in middle-aged men), there are no data that address this issue satisfactorily. At the moment, we can only be reasonably confident that the number of fat cells increases in at least some adults in whom severe obesity develops during adulthood (Hirsch and Batchelor, 1977).

In prenatal humans (Poissonnet *et al.*, 1983) and in rats (Hausman and Richardson, 1983), patterns of proliferation and enlargement of fat cells are inextricably associated with distinct patterns of vascularization. These patterns may reflect the role of the enzyme lipoprotein lipase (LPL) in the differentiation or lipid filling of adipocytes (Hietanen and Greenwood, 1977; Hausman and Richardson, 1983). LPL is produced in adipocytes but, apparently, to be maximally effective, it requires a close proximity between fat cells and large capillaries. This may be the reason that newly formed, rapidly enlarging fat cells are seen near large capillaries, while mature adipocytes of stable size are closest to capillaries that are relatively small (Hausman and Richardson, 1983). Whether or not this reasoning is correct, anatomical information of this type should prove of value in future studies that aim to assess more fully the nature of adipose tissue growth in adult humans.

In the male rat, which shows continued gradual growth throughout life, a slow rate of adipocyte production appears to continue indefinitely (Bertrand *et al.*, 1978, 1980; Faust and Miller, 1981). In male and female adult rats of certain strains, the feeding of a diet with a high fat or sugar content causes production of adipocytes to accelerate (Lemonnier, 1972; Faust *et al.*, 1978). In retroperitoneal fat depots of Osborne Mendel rats fed such a diet, adipocyte number has been seen to increase 100% in just 2 months (Faust *et al.*, 1978). Genetically obese rats show large increases in fat cell number that continue well into adulthood (Johnson *et al.*, 1978). Adult rats made obese by means of hypothalamic lesions also show accelerated production of adipocytes (Stern and Keesey, 1981), but this acceleration is greatly delayed if the rats are fed only chow (Faust *et al.*, 1984*b*). If the lesioned rats are fed a high-fat diet, proliferation of adipocytes accelerates soon after the time of lesioning and continues to occur at a high rate indefinitely. Thus, there is reason to believe that rats can make new fat cells at almost any point in the postnatal life span. Generalization from rat to human, in conjunction with the findings of Hirsch and Batchelor (1977), suggests that production of new fat cells may be possible at any time in the postnatal life of humans.

Histological examination of adipose tissue reveals the presence of very small adipocytes in adult animals that are not detected by nonhistological cell-counting techniques unless, or until, they enlarge (Kirtland *et al.*, 1975; Gurr and Kirtland, 1978; De Martinis and Francendese, 1982). It is thus possible that increases in fat cell number seen in adults are only apparent, reflecting a limitation of cell counting techniques rather than true hyperplastic growth (Jung *et al.*, 1978). If this were indeed the case, increases in the number of fat cells that appear to occur in adults would be limited to the number of very small fat cells. However, it has now been clearly shown indirectly, *via* biochemical assessment of rates of DNA synthesis (Klyde and Hirsch, 1979; Miller *et al.*, 1984) and directly by means of autoradiographic assessment of cell synthesis (Miller *et al.*, 1984) that *de novo* production of fat cells does occur in adult rats. When these animals are fed a high-fat diet, the synthesis of new fat cells is accelerated, and the degree of acceleration is consistent with known rates of increase in the number of mature fat cells. Thus, while very small fat cells clearly exist in adult animals, they are not the primary source of the large numbers of newly apparent large fat cells that are added during the development of obesity.

Knowledge as to which people are undergoing, or will undergo, excessive adipocyte hyperplasia would surely target candidates for intervention and might help further the development of safe and effective techniques for treatment. Current techniques for identifying children at risk for the development of severe hyperplastic obesity are inadequate. One might expect that obesity (or leanness) in childhood portends obesity (or leanness) later in life, but this is often not the case. Many normal-weight children become obese adults and many obese children become normal-weight adults. Even obese children of obese parents are not always obese as adults (Garn *et al.*, 1984).

A recently reported technique for predicting which children will become obese adults may be superior to others now in use. Rolland-Cachera *et al.* (1984) have shown that the age at which the weight/height index makes its second clear advance (generally between ages 5½ and 8 years) is a particularly good predictor of degree of obesity at age 16 years. These authors refer to the second advance as the "adiposity rebound." Children in whom the rebound occurs at an early age (5–6 years) are more likely to become obese adults than are children in whom the rebound occurs several years later. However, since obese children grow faster in all respects than their lean counterparts (Forbes, 1977), adiposity rebound may be just another measure of accelerated growth. Thus, while the observation of adiposity rebound is intriguing, whether it is indeed superior to other measures of growth as a predictor of obesity in adulthood is yet to be established. A technique for assessing rates of adipose tissue hyperplasia in small fragments of adipose tissue has also recently been reported (Miller *et al.*, 1984). Such fragments can be readily obtained from people by means of needle biopsy of the various subcutaneous fat depots. Perhaps this technique will prove useful, either alone or in conjunction with anatomical and anthropometric measures, in the assessment of levels and time patterns of adipocyte hyperplasia in those depots.

3. Effects of Undernutrition in the Very Young and of Severe Food Restriction in Adults

In the rat, nutritional manipulation during the first few postnatal weeks has a clear long-term impact on the number of fat cells produced. This is understandable, since most fat cells in the rat are normally produced during this time. Depending on the degree of deprivation imposed during this time and the strain of rat being studied, the total number of fat cells produced in some depots can be reduced by as much as 40% (Knittle and Hirsch, 1968; Johnson *et al.*, 1973; Faust *et al.*, 1980; Harris, 1980*a,b*). However, there is probably no long-term effect on the amount of lipid stored by each cell. About 6 months of *ad libitum* feeding appears to be sufficient to fully counter the effect of early underfeeding on fat cell size. In humans, growth of the adipose mass can be affected by the level of food intake of mothers during gestation (Ravelli *et al.*, 1976) as well as by the level of food intake during infancy and childhood (Ravelli *et al.*, 1976; Hager *et al.*, 1978; Sjostrom, 1981). It is thus of great interest to know whether manipulation of the level of caloric intake during certain periods of development can be an effective therapy for limiting the number of fat cells (and thus the amount of adipose tissue) produced by a child at risk for the development of hyperplastic obesity.

The experimental evidence to date, reviewed in detail by Leibel (1985*b*), argues that calorie-intake restriction early in life is not a simple or risk-free solution to the problem of obesity. In ordinary laboratory rats, the degree to which growth of adipose tissue is restrained by early underfeeding is not appreciably greater than the restraint that occurs in the growth of the rest of the body. When genetically obese rats are subjected to underfeeding for the first 8 months of life, the outcome is even more disheartening. Growth of nonadipose tissues (e.g., brain, muscle, and liver) is restrained to an even greater degree than is the growth of adipose tissue (Cleary *et al.*, 1980). Since adipose tissue growth occurs at the same time that other tissues are growing, the results of these studies should not be surprising. Figures 1 and 2 (from Pugliese *et al.*, 1983) demonstrate the deleterious effect that food intake restriction (self-imposed for the purpose of limiting the growth of adipose tissue) can have on the stature of a child. Corrective nutritional therapy begun early enough reverses most or all of the effect, but it is of little value once closure of the epiphyses has occurred. If there is ever to be clinically acceptable food intake restriction in people experiencing excessive adipocyte hyperplasia, it is essential that its effects be highly selective with regard to affected cell types. Discovery of pharmacological agents that would have such cell or organ-specific effect may result from a thorough understanding of the origin of fat cells and the factors that stimulate and limit the proliferation, differentiation, and maturation of their progenitors. There seems to be little hope that any pharmacological or dietary treatment will be devised that will safely eliminate fat cells that have already been formed (see below). Surgical treatment is limited to the removal of small amounts of adipose tissue, usually for the purpose of cosmetically altering certain body contours (Faust and Kral, 1985).

Whereas fat cells of the severely obese may be twice as large as those of normal-weight subjects, there is no clear upper limit to the number of fat cells. In retroperitoneal depots of adult rats, fat cell number can be induced to increase 10-fold in just 6 months (Faust *et al.,* 1984*b*). It has thus been of interest to ask whether diet could also be used to reduce the number of fat cells in the adult rat. The earliest studies to address this issue found no evidence to support such a possibility (Hollenberg and Vost, 1968; Hirsch and Han, 1969), but several subsequent studies reopened the question (Kasabuchi *et al.,* 1979; Sjostrom, 1981). The results of the most recent experiment to address this question seem to be definitive (Miller *et al.,* 1983). By means of severe, long-term food intake restriction, rats were made to lose half their body weight, which included virtually all adipose tissue triglyceride stores and significant amounts of adipose tissue DNA. Nevertheless, careful analysis of the adipose tissue

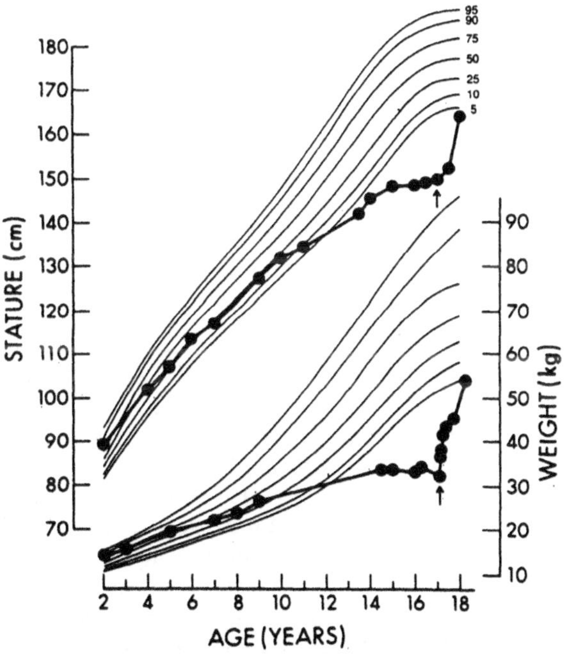

Fig. 1. Growth data for the most severely affected patient (case 2) in Pugliese *et al.* (1983). Note that inadequate weight gain preceded diminished linear growth and that adequate weight gain preceded an acceleration of linear growth. The arrows indicate initiation of nutritional therapy. (From Pugliese *et al.,* 1983).

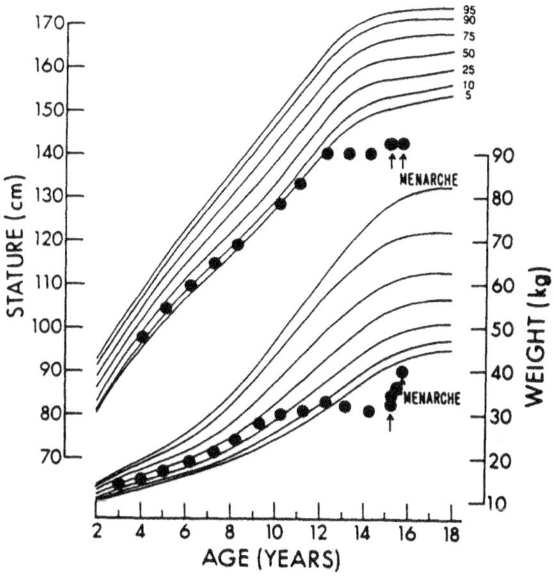

Fig. 2. Growth data for a patient whose growth was stunted by inadequate nutrition (case 6 in Pugliese *et al.,* 1983). Note that inadequate weight gain preceded a reduction in linear growth. The patient had a bone age of 14 years and was at Tanner's stage 4 of puberty when first seen. Menarche occurred soon after adequate weight gain had been reestablished. However, height changed minimally, and the patient probably had a permanent deficit of height potential. The arrows on the left indicate initiation of nutritional therapy. (From Pugliese *et al.,* 1983.)

showed that no adipocytes were lost during the course of the experiment (see Figure 3). The reduced levels of DNA in adipose tissue reflected exclusively the loss of endothelial, interstitial, and other nonadipocyte cells. The remarkable persistence of fat cells demonstrated by this experiment and by clinical observations argues strongly against the likelihood that any type of diet, with or without exercise, will ever be sufficient to reverse the hyperplastic component of severe obesity in rats or humans. It adds further strength to the argument that the solution to the problem of excessive production of fat cells is most likely to emerge from knowledge of the origin of fat cells and of the processes that guide the proliferation and differentiation of their precursors.

4. The Origin of Fat Cells

Regenerating adipose tissue in the lipectomized animal has proved a particularly useful model in

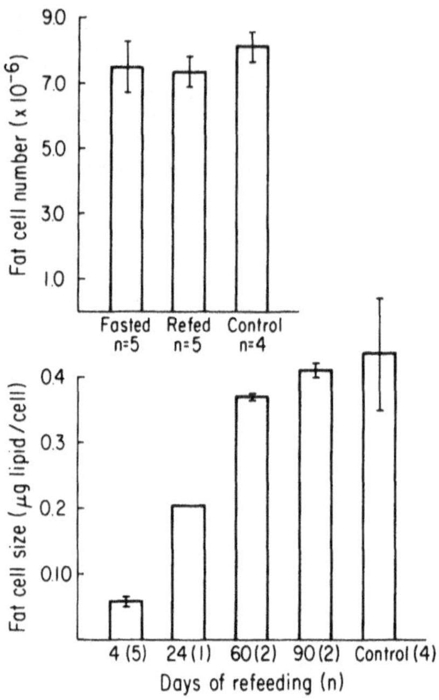

Fig. 3. Fat cell size and number of epididymal pads during 90 days of refeeding after extreme food deprivation. Values are means ±SE. (From Miller *et al.*, 1983.)

which to study the origin of the fat cell. Growth of adipose tissue is rapid, whereas there is little or no growth of most other tissues. The regenerating subcutaneous–inguinal fat depot of the rat has attracted the most recent interest because it is relatively easy to define anatomically and the degree to which it regrows is often substantial. Within a few days following surgical removal of the depot, a thin fascial sheath grows across the space between skin and muscle created by the removal. By 2 weeks postsurgery, the sheath is already heavily vascularized. As in the adipose tissue of rapidly growing young rats, differentiation and maturation of adipocytes on the sheath appear to take 6–9 days (Roth *et al.*, 1981). On the side of the sheath closest to the skin, mature adipocytes are evident adjacent to newly formed vasculature. By 4 weeks postsurgery, a thin layer of mature adipocytes is seen over a major portion of the sheath (Faust *et al.*, 1977*a*).

In the study by Roth *et al.* (1981), small, nonmembrane-bound lipid droplets were seen in several cell types. However, it is not clear in which cell type the droplets coalesce to form a large unilocular droplet—the major feature of the fat cell. Thus, it is not clear which cell type seen in this study is the progenitor of the adipocyte. During the earliest stages of regeneration, the most frequently seen cells with small lipid droplets appear to be macrophages. A less frequently seen cell type with lipid droplets looks like a typical spindle-shaped fibroblast. However, in virtually all studies of adipose tissue stromal–vascular cells in culture, the preadipocyte has an appearance similar to that of a fibroblast (e.g., Poznanski *et al.*, 1973; Van and Roncari, 1977; Bjorntorp *et al.*, 1978, 1980). In electron microscopic study of adipose tissue in preweanling rats, the adipocyte progenitors also seem to be fibroblastlike cells (Napolitano, 1963).

Various investigators have suggested that the site of origin of the preadipocyte is the capillary endothelium (Clark and Clark, 1940; Vodovar *et al.*, 1971; Desnoyers, 1977; Hausman *et al.*, 1980). A recent anatomical study of regenerating adipose tissue suggests that the population of pericytes lying adjacent to the capillary endothelium may be a major source of preadipocytes (Richardson *et al.*, 1982). In this study, a small portion of the inguinal depot of young rats was thermally ablated. A sequence of cellular events followed that culminated in the appearance of newly formed mature adipocytes. As in earlier studies of normal adipose tissue growth (Hollenberg and Vost, 1968; Greenwood and Hirsch, 1974; Roth *et al.*, 1981), new adipocytes did not appear until about the seventh day following ablation. Macrophages were abundant 24 hr following the time of ablation. A day later, the pericytes were enlarged and multiplying. After yet another day, there were many fibroblasts of two different types. One of these cell types appears to arise from differentiation of the activated pericytes and to become a mature adipocyte, while the other remains an ordinary fibroblast.

It is clear from the results of the studies cited above that regardless of whether a subcutaneous fat depot is removed completely or only partially, restoration of fat mass involves at least some growth of new tissue at the lipectomy site. However, regrowth is not limited to the lipectomy site. Some restoration of fat mass appears to occur as the result of increased (or compensatory) growth of subcutaneous adipose tissue not removed during surgery (Faust *et al.*, 1984*a*). The net result of the two growth responses to lipectomy is often perfect restoration of the mass and cellularity of the subcutaneous adipose tissue. When regeneration is complete, the total subcutaneous fat mass contains a normal number of fat cells of normal size (Faust *et al.*, 1977*a*, 1979, 1984*a*; Larson and Anderson, 1978). This suggests that there is some form of regulation of the entire subcutaneous adipose tissue mass. Regrowth of adipose tissue following lipectomy never occurs to any measurable excess. Original limitations of adipose tissue growth persist during regrowth. Growth and regrowth of subcutaneous adipose tissue thus appear to be regulated phenomena, in many ways similar to growth and regrowth of other regenerating organs such as the liver. However, a major dissimilarity between adipose tissue and most other organs makes comparisons of regrowth responses among them highly tenuous. To a large degree, the mass of a fat depot, and the number of parenchymal cells it contains, is a function of an animal's nutritional and environmental experience throughout life. It is hard to imagine how the sum of these experiences could possibly be structured into a biological parameter that guides the extent of adipose tissue regrowth. The crucial question that has not yet been addressed is whether otherwise identical rats that differ greatly in numbers of fat cells prior to lipectomy (because of nutritional experience) show correspondingly great differences in the absolute degree of adipose tissue regrowth that occurs following lipectomy.

5. Relationship of Patterns of Excessive Adipose Tissue Growth to Disease In Humans: Some Practical Implications

The general topic of adipose tissue growth is clearly a matter of central interest to many investigators concerned with the problem of obesity and its relationship to certain diseases in humans. There is no question that the severely obese suffer from increased medical morbidity and mortality. However, there is current debate as to whether there are major health risks associated with moderate obesity (Hubert *et al.*, 1983; Simopoulos and Van Itallie, 1984). Indeed, it has been suggested that moderate obesity may pose no health risk at all (Andres, 1980). In an attempt to resolve this issue, recent studies (Kissebah *et al.*, 1982; Kalkhoff *et al.*, 1983; Evans *et al.*, 1983; Krotkiewski *et al.*, 1983) have readdressed a suggestion made earlier by Vague (1956) that there are two anatomical types of obesity that can be distinguished both by the pattern in which the excess adipose tissue is distributed around the body and by the incidence of diseases generally associated with obesity. The results of these studies strongly suggest that the pattern of fat distribution in the obese may indeed be predictive of (and may even be responsible for) certain diseases of metabolism and of the cardiovascular system. The typically female pattern of excess adipose tissue accumulation (on hips, buttocks, and thighs) may carry little or no increased risk to health, but the more typically male pattern (enlargement of intraabdominal and abdominal–subcutaneous fat depots) is related to an increased incidence of disordered glucose metabolism, hypertension, heart attack, and stroke. Indeed, people who have large amounts of abdominal fat, but who would not otherwise be considered obese, may be at the greatest risk for heart attack and early death (Larsson *et al.*, 1984) (see Figure 4).

There are interesting and obvious implications of the above findings regarding a common practice of the life insurance industry. People who are above a certain ideal body weight (determined by age, sex, height, and sometimes frame) are subject to increased costs for life insurance, depending on the degree of overweight. As a reward for losing weight, these high-risk people may have their life insurance premiums lowered. People with relatively large amounts of abdominal fat, but who nevertheless have body weights within the range that is acceptable to the insurance industry, are not subject to increased life insurance costs. In light of the mounting evidence that adipose distribution, perhaps even more than degree of overweight, is related to degree of risk for disease and early death, assumptions regarding the actuarial risks of obese people should be reconsidered.

Consideration should also be given to the possibility that rewards for losing weight may cause an increase, rather than a decrease, in disease and mortality in some people. Those few overweight people who have benign female-type obesity but who diet successfully to maintain body weight at a normal level may prove to be less healthy than some people who do not diet and thus remain overweight. Basal metabolism is clearly subnormal in the weight-reduced obese (Leibel and Hirsch, 1984). Other abnormalities specifically attributable to the weight-reduced state (e.g., lowered thyroid hormone levels and leukopenia) are likely (Hirsch and Leibel, 1984). The extreme degree of calorie-intake restriction that weight-reduced obese people must continuously impose on themselves in order to avoid a return to their previous level of overweight may itself be deleterious to health. An individual who is consuming only a fraction of his or her usual daily caloric intake, in spite of an intense desire for food, may fail to consume all nutrients in the amounts needed for optimum well-being. Vitamin and mineral supplements are obviously very helpful and of great importance to the dieter, but they may not always counter all the deficiencies

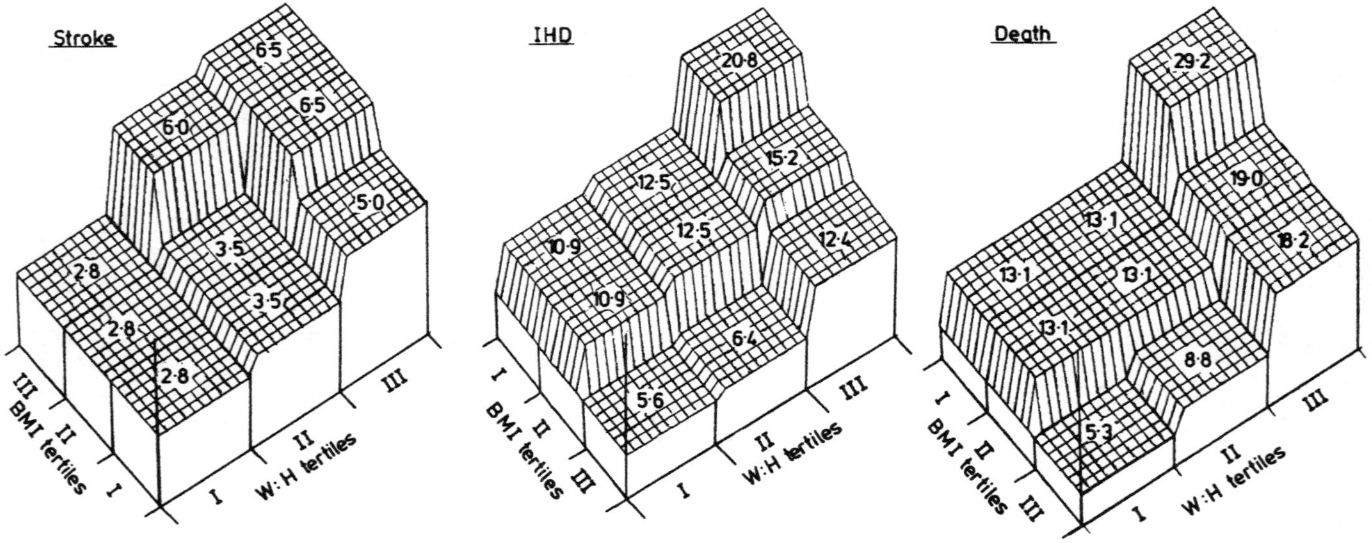

Fig. 4. Percentage probabilities of stroke, ischemic heart disease (IHD), and death from all causes in relation to tertiles of body mass index and waist to hip circumference (W:H) ratio. (BMI axes reversed for death and IHD.) (From Larsson *et al.*, 1984.)

and imbalances caused by severe food-intake restriction.

Since metabolic abnormalities often associated with obesity (e.g., hyperinsulinemia and hypertriglyceridemia) can be ameliorated by means of weight reduction (Olefsky *et al.*, 1974), it is certainly possible that effective treatment of adult male-type obesity would have a significant impact on the incidence of heart attack and stroke in this country. Calorie-restricted diets and exercise programs can be effective, but only as long as they are continued. Thus, they have been notoriously unsuccessful with regard to long-term weight control, particularly in the treatment of the severely obese. The various drugs that have been used in the treatment of obesity have been only moderately effective and often produce undesirable side effects. Perhaps safe and effective therapy for the treatment of obesity will emerge from the study of the biological systems underlying the production and distribution of fat cells. Alternatively, such therapy may emerge from an understanding of the processes that control the degree to which fat cells fill with lipid (see Section 5.3). For the immediate future, it would surely be worthwhile to develop a program of physician and patient education concerning the apparent relationship between adipose tissue distribution and health. Further clinical evaluation of the effect of moderate weight loss in the male-type obese on risk factors for cardiovascular and metabolic disease and on incidence of stroke, heart attack, and diabetes is essential. Also essential is an evaluation of the success of different weight-reduction therapies in this population. There is, as yet, no study of long-term maintenance of moderately reduced body weight in a population of male-type obese in which high risk factors for disease, other than overweight, have been established. The combination of a clear threat to health, clear and immediate benefits of weight loss, and the likelihood that risk factors may be significantly reduced with relatively moderate weight loss may prove adequate to provide a degree of therapeutic success not previously seen.

5.1. Hormonal Basis for Different Patterns of Adipose Tissue Growth

Overall growth of adipose tissue *in vivo* and proliferation and conversion of preadipocytes in culture vary as a function of the levels of certain hormones (Murakawa and Raben, 1968; Lemonnier and Alexiu, 1974; Kral and Tisell, 1976; Roncari and Van, 1978; Morikawa *et al.*, 1984). Site-to-site variations in adipose tissue growth also occur in response to some hormones (Krotkiewski, 1976; Krotkiewski and Bjorntorp, 1976; Krotkiewski *et al.*, 1980; Hendrikx *et al.*, 1980; Steingrimsdottir *et al.*, 1980). The sex hormones clearly direct the differential patterns of adipose tissue deposition that distinguish women from men. Extra amounts of adipose tissue may also accumulate in some depots more than in others in response to the hormones of pregnancy. This has been seen in the rat, in which the subscapular depot (a major subcutaneous fat depot in the rat) shows the greatest enlargement in response to pregnancy (Steingrimsdottir *et al.*, 1980*b*). Curiously, injections of progesterone have greatest impact on the parametrial (gonadal) depot of the rat (Krotkiewsky and Bjorntorp, 1976; Steingrimsdottir *et al.*, 1980*a*). Pregnancy-induced enlargement of fat depots occurs primarily as a result of hypertrophy of existing adipocytes. Short-term or widely spaced injections of progesterone promote enlargement of fat cells without causing any increase in their number. The reason for this is not known. Perhaps the relatively short durations of the reported experiments were inadequate to allow more than minimal hyperplasia. Perhaps a rapid succession of pregnancies or long-term intensive treatment with progesterone would support more substantial increases in the number of fat cells.

Hyperplastic growth of fat depots (or the lack of such growth) in the rat in response to either pregnancy or progesterone treatment is of particular interest because of the possible parallel condition of hyperplastic adipose tissue growth that may occur in women in response to pregnancy. There are many anecdotal reports of permanent weight gain resulting from pregnancy, and the results of a recent study suggest that this may indeed happen in some obese women (Garn and Pilkington, 1984). Perhaps a sufficient degree of enlargement of fat cells in certain depots during pregnancy promotes the proliferation of fat cells in those depots, as it does in dietary obesity in the rat (Faust *et al.*, 1978). Alternatively, the state of pregnancy may normally involve the presence of humoral factors that specifically inhibit the production of fat cells. Such factors may well be present in animals that periodically gain and lose large amounts of weight as they cycle through the stages of prehibernation and hibernation (Young *et al.*, 1982). If some women do indeed suffer a pregnancy-induced increase in body fat that does not reverse following pregnancy (presumably as a result of adipocyte hyperplasia), it would be of clear benefit to know which of the above two conditions normally prevails. If pregnancy does usually provide protection against adipocyte hyperplasia, it is important to know why some women may not be adequately protected, as well as whether inhibition of adipocyte proliferation continues during lactation.

It is commonly observed that high levels of cortisol, such as occur in Cushing's syndrome, and deficiency in the secretion of growth hormone both promote centripetal fat distribution in humans. Insulin administration promotes differential adipose tissue growth in the rat (Krotkiewski and Bjorntorp, 1976), and prolonged exposure to cold promotes hyperplastic growth of epidiymal fat pads while inhibiting the growth of retroperitoneal and subcutaneous fat depots (Miller and Faust, 1982). Further studies may reveal yet other situations and other humoral factors that favor the growth of one fat depot over another. It is of particular interest to learn whether particular features of the diet have differential effects on adipose tissue growth. Diets clearly vary with respect to the general type of adipose tissue growth that they promote (i.e., hyperplastic versus hypertrophic) (Lemonnier *et al.*, 1973; Borgeois *et al.*, 1983). As to the mechanisms underlying differential growth of adipose tissue, at least some understanding may be forthcoming from studies of the preadipocyte.

5.2. *Possible Cellular Basis for Different Patterns of Adipose Tissue Growth*

Much remains to be learned about the nature of the preadipocyte and its origins. Techniques for harvesting these cells and growing them in primary culture are still being refined and progress in culturing human preadipocytes has been slow. Studies of preadipocyte cell lines (Mackall *et al.*, 1976; Grimaldi *et al.*, 1978; Ailhaud, 1982; Green, 1983) are providing important insights regarding some likely activities of preadipocytes *in vivo,* but they are not well suited for the purpose of addressing certain key questions. In particular, they have provided no information as to whether there are intrinsic differences in the characteristics of preadipocytes from different anatomical regions that may be responsible for observed variations in the growth and development of adipose tissue in those areas. Such intrinsic cellular differences would point to embryogenesis and early developmental processes, rather than to current features of neural or vascular anatomy, as the basis for major differences in growth and metabolism among the various fat depots.

To address this issue, Nechad *et al.* (1983) compared cultures of preadipocytes from epididymal fat pads with those from interscapular and cervical brown adipose tissue. Using standard techniques of harvesting and plating large numbers of stromal–vascular cells from these depots, these workers obtained cultures rich in preadipocytes that differed appreciably from each other in terms of both growth and respiratory activity. This finding suggests that preadipocytes from the two examined areas are dissimilar, but the evidence is not conclusive. When experiments compare heterogeneous mixtures of cells from different sources it cannot be concluded with certainty that observed differences among cell colonies are caused by differing intrinsic characteristics of any particular cell type. Rather than being due to differences in cell features, variation among colonies may result from different combinations of cells in the original platings.

In another recently reported study of preadipocytes in culture, Djian *et al.* (1983) avoided the uncertainty inherent in studies of heterogeneous mixtures of cells, producing stronger support for the hypothesis that fat cells from different depots have differing intrinsic characteristics. Colonies of rat preadipocytes derived clonally (as well as from relatively large groups of stromal–vascular cells) were seen to vary in terms of both replication and differentiation as a function of the white adipose tissue depot (epididymal or retroperitoneal) from which they originated. The rate of production of mature fat cells in colonies derived from retroperitoneal depots was greater than in colonies derived from epididymal pads. This is in agreement with the findings of *in vivo* studies of effects of dietary obesity on adipocyte hyperplasia (Faust *et al.* 1978). This study demonstrates that primary cell culture can be used to make clear and meaningful distinctions among preadipocytes from different sources. Perhaps of greater importance, it introduces a technique that permits comparisions of preadipocytes that are unconfounded by differences in concentration or distribution of cell types in the depots from which the cells are taken. The close correspondence of the findings obtained with the two culture techniques used by Djian and associates suggests that numerous cell replications (necessary in cell cloning experiments) can occur without altering important distinguishing characteristics of different preadipocytes. Key differences in cell features (i.e., potential for replication, differentiation, and lipid filling) may persist in experiments with colonies derived from clones and thus may permit a more precise degree of characterization than is possible with cultures derived from heterogeneous populations of stromal–vascular cells. Such studies may show conclusively that the hormonal sensitivities of fat cells varies as a function of their depot of origin. Such differences, in turn, may prove the basis for site-to-site differences in growth and lipid storage and for the varied effects of the different depots on metabolism.

5.3. Control of the Size of Fat Cells

Although mature adipocytes cannot be eliminated by dietary means, obese people can reduce the size of their fat cells by various combinations of restricted calorie intake and increased calorie expenditure. Unfortunately, for most people, continuous dieting and regular exercise are the only safe and effective ways to achieve as well as maintain a reduction in the size of fat cells. For the great majority of people, the required discipline and self-control are burdens that are not endured indefinitely. Sooner or later, most people regain the elevated level of body weight they had maintained prior to weight loss. Perhaps if we had a greater understanding of the mechanisms that control the size of the fat cell lipid droplet, we would know how to intervene in a safe and effective manner to keep the fat cells small. A brief summary of studies that have begun to probe the mechanism of fat cell size control is presented in this section.

Obesity can be induced in a variety of ways, and it takes several forms. Neurological and endocrine disorders, genes, diets, viruses, and the normal rhythm of hibernation cycles can all cause obesity (Schemmel *et al.* 1970; Bray and York, 1971, 1979; Mrosovsky, 1976; Sclafani and Springer, 1976; Lyons *et al.*, 1982). In some cases, the excess adipose tissue is evenly distributed; in others, it is not. In some cases, there is an excess of fat cells; in other cases, there is no such excess. The one characteristic common to virtually all cases of obesity is that fat cells are abnormally large. In obese humans, normal (or subnormal) fat cell size is seen only as a response to restricted food intake. In animals, hyperplastic nonhypertrophic obesity can be produced experimentally, but only through the use of certain unusual diets or sequences of diets (Lemonnier *et al.*, 1973; Faust, 1980; Borgeois *et al.*, 1983).

There is also good reason to believe that fat cell size is not arbitrary. Rather, it seems to be the product of an equilibrium achieved, at least in part, through the action of a physiological system that counteracts increases or decreases in fat cell size by modifying energy intake and expenditure (Faust, 1984). Thus, when adipose tissue regenerates following lipectomy, newly formed fat cells adopt essentially the same size as the cells they replaced (Faust *et al.*, 1977*a*, 1979, 1984*a*). Early undernutrition affects total growth and fat cell number but has no permanent effect on fat cell size (Faust *et al.*, 1980). Certain diets can cause dramatic permanent increases in the number of fat cells in the rat, but there is no permanent effect on fat cell size (Faust *et al.*, 1978). The most dramatic evidence regarding the physiological control of fat cell size is the effect of altered adipose tissue morphology on spontaneous food intake. Lipectomized rats overeat less than do control rats when both are fed the same fattening diet (Faust *et al.*, 1977*b*). Likewise, rats with an overabundance of normal-size fat cells overeat more of a fattening diet than do their controls (Faust, 1980). The net result in both cases is that comparable fat cell enlargement can occur in experimental and control animals in spite of different levels of food intake.

Knowledge of the forces that cause the system regulating fat cell size to malfunction may well be a key to the understanding of obesity. Even the hyperplastic component of severe obesity may simply be a normal compensatory response to overenlargement of fat cells (Gale *et al.*, 1981; Faust and Miller, 1983). It is thus crucial for future studies to determine how information about the size of fat cells is communicated to other cells. Perhaps there are special chemical messengers from the fat cells to the CNS. Alternatively, the CNS may somehow assess levels of substrates that are byproducts of adipocyte metabolism (Leibel, 1977, 1985*a*). The attraction of the latter possibility is that it offers a readily understandable explanation for the fact that the level at which fat cell size stabilizes can vary markedly as a function of a wide variety of influences such as the composition of the diet. As our knowledge advances in the areas of regulation of fat cell size and adipose tissue histogenesis, so should the likelihood that human obesity will be ameliorated or prevented. It would be extremely gratifying to see advances in these areas during the next few years that match or exceed the recent advances discussed in this chapter.

ACKNOWLEDGMENTS. My own research in the areas of adipose tissue growth and adipose tissue mass regulation has been generously supported by the Irma T. Hirschl Trust, U.S. Department of Agriculture, the Nutrition Foundation, the National Institutes of Health (AM 20508), and the American Heart Association (84 1207).

6. References

Ailhaud, G., 1982, Adipose cell differentiation in culture, *Mol. Cell Biochem.* **49**:17–31.

Andres, R., 1980, Effect of obesity on total mortality, *Int. J. Obesity* **4**:381–386.

Angel, A., Hollenberg, C. H., and Roncari, D. A. K., 1983, *The Adipocyte and Obesity: Cellular and Molecular Mechanisms,* Raven Press, New York.

Astrup, A., Bulow, J., Christensen, N. J., and Madsen, J., 1984, Ephedrine-induced thermogenesis in man: No role for interscapular brown adipose tissue, *Clin. Sci.* **66**:179–186.

Bertrand, H. A., Masoro, E. J., and Yu, B. P., 1978, Increasing adipocyte number as the basis for perirenal depot growth in adult rats, *Science* **201**:1234–1235.

Bertrand, H. A., Lynd, F. T., Masoro, E. J., and Yu, B. P., 1980, Changes in adipose mass and cellularity through the adult life of rats fed ad libitum or a life-prolonging restricted diet, *J. Gerontol.* **35**:827–835.

Bjorntorp, P., Karlsson, M., Pertoft, H., Petterson, P., Sjostrom, L., and Smith, U., 1978, Isolation and characterization of cells from rat adipose tissue developing into adipocytes, *J. Lipid Res.* **19**:316–324.

Bjorntorp, P., Karlsson, M., Pettersson, P., and Sypniewska, G., 1980, Differentiation and function of rat adipocyte precursor cells in primary culture, *J. Lipid Res.* **21**:714–723.

Bonnet, F. P., 1981, *Adipose Tissue in Childhood,* CRC Press, Boca Raton, Florida.

Bonnet, F. P., and Rocour-Brumioul, D., 1981, Normal growth of human adipose tissue, in: *Adipose Tissue in Childhood* (F. P. Bonnet, ed.), pp. 81–107, CRC Press, Boca Raton, Florida.

Borgeois, F., Alexiu, A., and Lemonnier, D., 1983, Diet-induced obesity: Effect of dietary fats on adipose tissue cellularity in mice, *Br. J. Nutr.* **49**:17–26.

Bray, G. A., and York, D. A., 1971, Genetically transmitted obesity in rodents, *Physiol. Rev* **51**:598–646.

Bray, G. A., and York, D. A., 1979, Hypothalamic and genetic obesity in experimental animals: An autonomic and endocrine hypothesis, *Physiol. Rev.* **59**:719–809.

Brook, C. G. D., 1978, Cellular growth: Adipose tissue, in: *Human Growth,* Vol. 2: *Postnatal Growth* (F. Falkner and J. M. Tanner, eds.), pp. 21–33, Plenum Press, New York.

Clark, E. L., and Clark, E. L., 1940, Microscopic studies of new formation of fat in living adult rabbits, *Am. J. Anat.* **67**:255–285.

Cleary, M. P., Vasselli, J. R., and Greenwood, M. R. C., 1980, Development of obesity in the Zucker obese (fafa) rat in the absence of hyperphagia, *Am. J. Physiol.* **238**:E284–E292.

De Martinis, F. D., and Francendese, A., 1982, Very small fat cell populations: Mammalian occurrence and effect of age, *J. Lipid Res.* **23**:1107–1120.

Desnoyers, F., 1977, Etude morphologique chez le rat, du tissue adipeaux perirenal au stade de sa formation, *Ann. Biol. Anim. Biochem. Biophys.* **17**:787–798.

Djian, P., Roncari, D. A. K., and Hollenberg, C. H., 1983, Influence of anatomic site and age on replication and differentiation of rat adipocyte precursors in culture, *J. Clin. Invest.* **72**:1200–1208.

Evans, D. J., Hoffmann, R. G., Kalkhoff, R. K., and Kissebah, A. H., 1983, Relationship of androgenic activity to body fat topography, fat cell morphology, and metabolic aberrations in premenopausal women, *J. Clin. Endocrinol. Metab.* **57**:304–310.

Faust, I. M., 1980, Nutrition and the fat cell, *Int. J. Obesity* **4**:314–321.

Faust, I. M., 1981, Signals from adipose tissue, in: *The Body Weight Regulatory System: Normal and Disturbed Mechanisms* (L. A. Cioffi, W. P. T. James, and T. B. Van Itallie, eds.), pp. 39–43, Raven Press, New York.

Faust, I. M., 1984, The role of the fat cell in energy balance physiology, in: *Eating and Its Disorders (Association for Research in Nervous and Mental Diseases,* Vol. 62) (A. J. Stunkard, and E. Steller, eds.), pp. 97–107, Raven Press, New York.

Faust, I. M., and Miller, W. H., Jr., 1981, Effects of diet and environment on adipocyte development, *Int. J. Obesity* **5**:593–596.

Faust, I. M., and Miller, W. H., Jr., 1983, Hyperplastic growth of adipose tissue in obesity, in: *The Adipocyte and Obesity: Cellular and Molecular Mechanisms* (A. Angel, C. H. Hollenberg, and D. A. K. Roncari, eds.), pp. 41–51, Raven Press, New York.

Faust, I. M., and Kral, J. G., 1985, Growth of adipose tissue following lipectomy, in: *New Perspectives in Adipose Tissue Structure, Function and Development* (R. L. R. Van, and A. Cryer, eds.), Butterworths, London.

Faust, I. M., Johnson, P. R., and Hirsch, J., 1977*a,* Adipose tissue regeneration following lipectomy, *Science* **197**:391–393.

Faust, I. M., Johnson, P. R., and Hirsch, J., 1977*b,* Surgical removal of adipose tissue alters feeding behavior and the development of obesity in rats, *Science* **197**:393–396.

Faust, I. M., Johnson, P. R., Stern, J. S., and Hirsch, J., 1978, Diet-induced adipocyte number increase in adult rats: A new model of obesity, *Am. J. Physiol.* **235**:E279–E286.

Faust, I. M., Johnson, P. R., and Hirsch, J., 1979, Adipose tissue regeneration in adult rats, *Proc. Soc. Exp. Biol. Med.* **161**:111–114.

Faust, I. M., Johnson, P. R., and Hirsch, J., 1980, Long-term effects of early nutritional experience on the development of obesity in the rat, *J. Nutr.* **110**:2027–2034.

Faust, I. M., Johnson, P. R., and Kral, J. G., 1984*a,* Effects of castration on adipose tissue growth and regrowth in the male rat, *Metabolism* **33**:596–601.

Faust, I. M., Miller, W. H., Jr., Sclafani, A., Aravich, P. F., Triscari, J., and Sullivan, A. C., 1984*b,* Diet-dependent hyperplastic growth of adipose tissue in hypothalamic obese rats *Am. J. Physiol.* **247:**R1038–R1046.

Forbes, G. B., 1977, Nutrition and growth, *J. Pediatr.* **91:**40–42.

Gale, S. K., Van Itallie, T. B., and Faust, I. M., 1981, Effects of palatable diets on body weight and adipose cellularity in the adult obese female Zucker rat (fa/fa), *Metabolism* **30:**105–110.

Garn, S. M., and Pilkington, J. J., 1984, Comparison of three-year weight and fat change distributions of lean and obese individuals, *Ecol. Food Nutr.* **15:**7–12.

Garn, S. M., LaVelle, M., and Pilkington, J. J., 1984, Obesity and living together, *Marriage and Family Review* **7:**33–47.

Green, H., 1983, Adipogenic factors and the formation of adipocytes, in: *The Adipocyte and obesity: Cellular and Molecular Mechanisms* (A. Angel, C. H. Hollenberg, and D. A. K. Roncari, eds.), pp. 29–31, Raven Press, New York.

Greenwood, M. R. C., and Hirsch, J., 1974, Postnatal development of adipocyte cellularity in the normal rat, *J. Lipid Res.* **15:**474–483.

Grimaldi, P., Negrel, R., and Ailhaud, G., 1978, Induction of the triglyceride pathway enzymes and of lipolytic enzymes during differentiation in a "preadipocyte" cell line, *Eur. J. Biochem.* **84:**369–376.

Gurr, M. I., and Kirtland, J., 1978, Adipose tissue cellularity: A review. 1. Techniques for studying cellularity, *Int. J. Obesity* **2:**401–427.

Hager, A., Sjostrom, L., Arvidsson, B., Bjorntorp, P., and Smith, U., 1977, Body fat and adipose tissue cellularity in infants: A longitudinal study, *Metabolism* **26:**607–614.

Hager, A., Sjostrom, L., Arvidsson, B., Bjorntorp, P., and Smith, U., 1978, Adipose tissue cellularity in obese school girls before and after dietary treatment, *Am. J. Clin. Nutr.* **31:**68–75.

Harris, P. M., 1980*a,* Changes in adipose tissue of the rat due to early undernutrition followed by rehabilitation. 1. Body composition and adipose tissue cellularity, *Br. J. Nutr.* **43:**15–26.

Harris, P. M., 1980*b,* Changes in adipose tissue of the rat due to early undernutrition followed by rehabilitation. 2. Strain differences and adipose tissue cellularity, *Br. J. Nutr.* **43:**27–31.

Harris, R. B. S., and Martin, R. J., 1984, Lipostatic theory of energy balance: Concepts and signals, *Nutr. Behav.* **1:**253–275.

Hausman, G. J., and Richardson, R. L., 1983, Cellular and vascular development in immature rat adipose tissue, *J. Lipid Res.* **24:**522–532.

Hausman, G. J., Campion, D. R., and Martin, R. J., 1980, Search for the adipocyte precursor cell and factors that promote its differentiation, *J. Lipid Res.* **21:**657–670.

Hendrikx, A., Boni, L., and Kieckens, L., 1980, Influence of gonadal hormones on adipose-tissue cellularity in rats, *Int. J. Obes.* **4:**145–151.

Hietanen, E., and Greenwood, M. R. C., 1977, A comparison of lipoprotein lipase activity and adipocyte differentiaiton in growing male rats, *J. Lipid Res.* **18:**480–490.

Himms-Hagen, J., 1983, Brown adipose tissue thermogenesis in obese animals, *Nutr. Rev.* **41:**261–267.

Hirsch, J., and Batchelor, B. R., 1977, Adipose tissue cellularity and human obesity, *Clin. Endocrinol. Metab.* **26:**607–614.

Hirsch, J., and Han, P. W., 1969, Cellularity of rat adipose tissue: Effects of growth, starvation and obesity, *J. Lipid Res.* **10:**77–82.

Hirsch, J., and Knittle, J. L., 1970, The cellularity of obese and nonobese human adipose tissue, *Fed. Proc.* **29:**1516–1521.

Hirsch, J., and Leibel, R. L., 1984, What constitutes a sufficient psychobiologic explanation for obesity?, in: *Eating and its Disorders* (A. J. Stunkard and E. Stellar, eds.), pp. 121–130, Raven Press, New York.

Hirsch, J., Faust, I. M., and Johnson, P. R., 1979, What's new in obesity: Current understanding of adipose tissue morphology, in: *Contemporary Metabolism* (N. Freinkel, ed.) Vol. 1, pp. 385–399, Plenum Press, New York.

Hollenberg, C. H., and Vost, A., 1968, Regulation of DNA synthesis in fat cells and stromal elements from rat adipose tissue, *J. Clin. Invest.* **47:**2485–2498.

Hubert, H. B., Feinleib, M., McNamara, P. M., and Castelli, W. P., 1983, Obesity as an independent risk factor for cardiovascular disease: A 26-year followup of participants in the Framingham Heart Study, *Circulation* **67:**968–977.

James, W. P. T., and Trayhurn, P., 1981, Thermogenesis and obesity, *Br. Med. Bull.* **37:**43–48.

Johnson, P. R., Stern, J. S., Greenwood, M. R. C., Zucker, L. M., and Hirsch, J., 1973, Effect of early nutrition on adipose cellularity and pancreatic insulin release in the Zucker rat, *J. Nutr.* **103:**738–743.

Johnson, P. R., Stern, J. S., Greenwood, M. R. C., and Hirsch, J., 1978, Adipose tissue hyperplasia and hyperinsulinemia in Zucker obese female rats: A developmental study, *Metabolism* **27:**1941–1954.

Jung, R. T., Gurr, M. I., Robinson, M. P., and James, W. P. T., 1978, Does adipocyte hypercellularity in obesity exist?, *Br. Med. J.* **2:**319–321.

Kalkhoff, R. K., Hartz, A. H., Rupley, D., Kissebah, A. H., and Kelber, S., 1983, Relationship of body fat distribution to blood pressure, carbohydrate tolerance, and plasma lipids in healthy obese women, *J. Lab. Clin. Med.* **102:**621–627.

Kasabuchi, Y., Mino, M., Yoshioka, H., and Kusunoki, T., 1979, An autoradiographic study of new fat cell formation in adipose tissue in adult mice during malnutrition and refeeding, *J. Nutr. Sci. Vitaminol.* **25:**419–426.

Kirtland, J., and Gurr, M. I., 1979, Adipose tissue cellularity: a review. 2. The relationship between cellularity and obesity, *Int. J. Obesity* **3:**15–55.

Kirtland, J., Gurr, M. I., Saville, G., and Widdowson, E. M., 1975, Occurrence of "pockets" of very small cells in adipose tissue of the guinea pig, *Nature (London)* **256**:723–724.

Kissebah, A. H., Vydelingum, N., Murray, R., Evans, D. J., Hartz, A. J., Kalkhoff, R. K., and Adams, P. W., 1982, Relation of body fat distribution to metabolic complications of obesity, *J. Clin. Endocrin. Metab.* **54**:254–260.

Klyde, B. J., and Hirsch, J., 1979, Increased cellular proliferation in adipose tissue of adult rats fed a high-fat diet, *J. Lipid Res.* **20**:705–715.

Knittle, J. L., 1978, Adipose tissue development in man, in: *Human Growth,* Vol. 2: *Postnatal Growth* (F. Falkner and J. M. Tanner, eds.), pp. 295–315, Plenum Press, New York.

Knittle, J. L., and Hirsch, J., 1968, Effect of early nutrition on the development of rat epididymal fat pads: Cellularity and metabolism, *J. Clin. Invest.* **47**:2091–2098.

Knittle, J. L., Timmers, K., Ginsberg-Fellner, F., Brown, R. E., and Katz, D. P., 1979, The growth of adipose tissue in children and adolescents, *J. Clin. Invest.* **63**:239–246.

Krotkiewski, M., 1976, The effect of estrogens on regional adipose tissue cellularity in the rat, *Acta Physiol. Scand.* **96**:128–133.

Krotkiewski, M., and Bjorntorp, P., 1976, The effect of progesterone and of insulin on regional adipose tissue cellularity in the rat, *Acts Physiol. Scand.* **96**:122–127.

Krotkiewski, M., Sjostrum, L., Bjorntorp, P., and Smith, U., 1975, Reigonal adipose tissue cellularity in relation to metabolism in young and middle aged women, *Metabolism* **24**:703–710.

Krotkiewski, M., Kral, J. G., and Karlsson, J., 1980, Effects of castration and testosterone substitution on body composition and muscle metabolism in rats, *Acta Physiol. Scand.* **109**:233–237.

Krotkiewski, M., Bjorntorp, P., Sjostrom, L., and Smith, U., 1983, Impact of obesity on metabolism in men and women—Importance of regional adipose tissue distribution, *J. Clin. Invest.* **72**:1150–1162.

LaFontan, M., Dang-Tran, L., and Berlan, M., 1979, Alpha-adrenergic antilipolytic effect of adrenaline responsiveness to different fat depots, *Eur. J. Clin. Invest.* **9**:261–266.

Larson, K. A., and Anderson, D. B., 1978, The effects of lipectomy on remaining adipose tissue depots in the Sprague Dawley rat, *Growth* **42**:469–477.

Larsson, B., Svardsudd, K., Welin, L., Wilhelmsen, L., Bjorntorp, P., and Tibblin, G., 1984, Abdominal adipose tissue distribution, obesity and risk of cardiovascular disease and death: A 13-year follow up of participants in the study of men born in 1913, *Br. Med. J.* **288**:1401–1404.

Leibel, R. L., 1977, A biological radar system for the assessment of body mass, *J. Theoret. Biol.* **66**:297–306.

Leibel, R. L., 1985*a*, A radioisotopic method for the measurement of free fatty acid turnover and adrenoreceptor response in small fragments of human adipose tissue, *Int. J. Obes.* **9**.

Leibel, R. L., 1985*b*, Obesity, in: *Clinical Nutrition of the Young Child* (O. Brunser, F. R. Carrazza, M. Gracey, B. L. Nichols, and J. Senterre, eds.), Raven Press, New York (in press).

Leibel, R. L., and Hirsch, J., 1984, Diminished energy requirements in reduced obese patients, *Metabolism* **33**:164–170.

Leibel, R. L., Berry, E. M., and Hirsch, J., 1983, Biochemistry and development of adipose tissue in man, in: *Health and Obesity* (H. L. Conn, Jr., E. A. DeFelice, and P. Kuo, eds.), pp. 21–48, Raven Press, New York.

Leibel, R. L., Hirsch, J., Berry, E. M., and Gruen, R. K., 1984, Radioisotopic method for the measurement of lipolysis in small samples of human adipose tissue, *J. Lipid Res.* **25**:49–57.

Lemonnier, D., 1972, Effect of age, sex and site on the cellularity of the adipose tissue in mice and rats rendered obese by a high fat diet, *J. Clin. Invest.* **51**:2907–2915.

Lemonnier, D., and Alexiu, A., 1974, Nutritional, genetic and hormonal aspects of adipose tissue cellularity, in: *The Regulation of Adipose Tissue Mass* (J. Vague and J. Boyer, eds.), pp. 158–173, Excerpta Medica, Amsterdam.

Lemonnier, D., Alexiu, A., and Lanteaume, M. T., 1973, Effect of two dietary lipids on the cellularity of the rat adipose tissue, *J. Physiol. (Paris)* **66**:729–733.

Lyons, M. J., Faust, I. M., Hemmes, R. B., Buskirk, D. R., Hirsch, J., and Zabriskie, J. B., 1982, An obestiy syndrome in mice following virus infection, *Science* **216**:82–85.

Mackall, J. C., Student, A. K., Polakis, S. E., and Lane, M. D., 1976, Induction of lipogenesis during differentiation in a "preadipocyte" cell line, *J. Biol. Chem.* **251**:6462–6464.

Miller, W. H., Jr., and Faust, I. M., 1982, Alterations in rat adipose tissue morphology induced by a low temperature environment, *Am. J. Physiol.* **242**:E93–E96.

Miller, W. H., Jr., Faust, I. M., Goldberger, A. C., and Hirsch, J., 1983, Effects of severe long-term food deprivation and refeeding on adipose tissue cells in the rat, *Am. J. Physiol.* **245**:E74–E80.

Miller, W. H., Jr., Faust, I. M., and Hirsch, J., 1984, Demonstration of de novo production of adipocytes in adult rats by biochemical and autoradiographic techniques, *J. Lipid Res.* **25**:336–347.

Morikawa, M., Green, H., and Lewis, U. J., 1984, Activity of human growth hormone and related polypeptides on the adipose conversion of 3T3 cells, *Mol. Cell. Biol.* **4**:228–231.

Mrosovsky, N., 1976, Lipid programmes and life strategies in hibernators, *Am. Zool.* **16**:685–697.

Murikawa, S., and Raben, M. S., 1968, Effect of growth hormone and placental lactogen on DNA synthesis in rat costal cartilage and adipose tissue, *Endocrinology* **83**:645–650.

Napolitano, L., 1963, The differentiation of white adipose cells: an electron microscope study, *J. Cell Biol.* **18**:663–679.

Nechad, M., Kuusela, P., Carneheim, C., Bjorntorp, P., Nedergaard, J., and Cannon, B., 1983, Development of brown fat cells in monolayer culture. 1. Morphological and biochemical distinction from white fat cells in culture, *Exp. Cell Res.* **149**:105–118.

Olefsky, J., Reaven, G. M., and Farquhar, J. W., 1974, Effects of weight reduction on obesity. Studies of lipid and carbohydrate metabolism in normal and hyperlipoproteinemic subjects, *J. Clin. Invest.* **53**:64–76.

Ostman, J., Arner, P., Engfeldt, P., and Karger, L., 1979, Regional differences in the control of lipolysis in human adipose tissue, *Metabolism* **12**:1198–1205.

Poissonnet, C. M., Burdi, A. R., and Bookstein, F. L., 1983, Growth and development of human adipose tissue during early gestation, *Early Hum. Dev.* **8**:1–11.

Poznanski, W. J., Waheed, I., and Van, R., 1973, Human fat cell precursors. Morphologic and metabolic differentiation in culture, *Lab. Invest.* **29**:570–576.

Pugliese, M. T., Lifshitz, F., Grad, G., Fort, P., and Marks-Katz, M., 1983, Fear of obesity. A cause of short stature and delayed puberty, *N. Engl. J. Med.* **309**:513–518.

Ravelli, G. P., Stein, Z., and Susser, M. W., 1976, Obesity in young men after famine exposure in utero and early infancy, *N. Engl. J. Med.* **295**:349–353.

Ravussin, E., Burnand, B., Schutz, Y., and Jequier, E., 1982, 24 hour energy expenditure and resting metabolic rate in obese, moderately obese, and control subjects, *Am. J. Clin. Nutr.* **35**:566–573.

Richardson, R. L., Hausman, G. J., and Campion, D. R., 1982, Response of pericytes to thermal lesion in the inguinal fat pad of 10-day-old rats, *Acta Anat.* **114**:41–57.

Rolland-Cachera, M. F., Deheeger, M., Bellisle, F., Sempe, M., Guilloud-Bataille, M., and Patois, E., 1984, Adiposity rebound in children: A simple indicator for predicting obesity, *Am. J. Clin. Nutr.* **39**:129–135.

Roncari, D. A. K., and Van, R. L. R., 1978, Promotion of human adipocyte precursor replication by 17-beta-estradiol in culture, *J. Clin. Invest.* **62**:503–508.

Roth, J., Greenwood, M. R. C., and Johnson, P. R., 1981, The regenerating fascial sheath in lipectomized Osborne Mendel rats: Morphological and biochemical indices of adipocyte differentiation and proliferation, *Int. J. Obesity* **5**:131–143.

Rothwell, N. J., and Stock, M. J., 1979, A role for brown adipose tissue in diet-induced thermogenesis, *Nature (London)* **281**:31–35.

Salans, L. B., Cushman, S., and Weismann, R. E., 1973, Studies of human adipose tissue, adipose cell size and number in nonobese and obese patients, *J. Clin. Invest.* **52**:929–941.

Schemmel, R., Mickelson, O., and Gill, J. L., 1970, Dietary obesity in rats: Body weight and body fat accretion in seven strains of rats, *J. Nutr.* **100**:1041–1048.

Sclafani, A., and Springer, D., 1976, Dietary obesity in adult rats: Similarities to hypothalamic and human obesity syndromes, *Physiol. Behav.* **17**:461–471.

Simopoulos, A. P., and Van Itallie, T. B., 1984, Body weight, health, and longevity, *Ann. Intern. Med.* **100**:285–295.

Sjostrom, L., 1981, Can the relapsing patient be identified?, in: *Recent Advances in Obesity Research. III* (P. Bjorntorp, M. Cairella, and A. N. Howard, eds.), pp. 85–93, John Libbey, London.

Smith, U., 1980, Adrenergic control of human adipose tissue lipolysis, *Eur. J. Clin. Invest.* **10**:343–344.

Smith, U., Hammersten, J., Bjorntorp, P., and Kral, J. G., 1979, Regional differences and effect of weight reduction on human fat cell metabolism, *Eur. J. Clin. Invest.* **9**:327–332.

Steingrimsdottir, L., Brasel, J., and Greenwood, M. R. C., 1980*a*, Hormonal modulation of adipose tissue lipoprotein lipase may alter food intake in rats, *Am. J. Physiol.* **239**:E162–E167.

Steingrimsdottir, L., Greenwood, M.R.C., and Brasel, J., 1980*b*, Effect of pregnancy, lactation and a high-fat diet on adipose tissue in Osborne–Mendel rats, *J. Nutr.* **110**:600–609.

Stern, J. S., and Keesey, R. E., 1981, The effect of ventromedial hypothalamic lesions on adipose cell number in the rat, *Nutr. Rep. Int.* **23**:295–301.

Vague, J., 1956, The degree of masculine differentiation of obesities: A factor determining predisposition to diabetes, atherosclerosis, gout and uric calculus disease, *Am. J. Clin. Nutr.* **4**:20–34.

Van, R. L. R., and Cryer, A., 1985, *New Perspectives in Adipose Tissue: Structure, Function and Development*, Butterworths, London.

Van, R. L. R., and Roncari, D. A. K., 1977, Isolation of fat cell precursors from adult rat adipose tissue, *Cell Tissue Res.* **181**:197–203.

Vodovar, N., Desnoyers, F., and Francois, A. C., 1971, Origine et évolution des adipocytes mésénteriques de porcelet avant la naissance. Aspect ultrastructural, *J. Microsc.* **11**:265–284.

Young, R. A., Salans, L. B., and Sims, E. A. H., 1982, Adipose tissue cellularity in woodchucks: Effects of season and captivity at an early age, *J. Lipid Res.* **23**:887–892.

4

Growth of Muscle Tissue and Muscle Mass

ROBERT M. MALINA

1. Growth of Muscle Tissue

1.1. Histological Considerations

Muscle tissue consists of cylindrical, multinucleated cells called muscle fibers. A skeletal muscle contains many long fibers that have the ability to contract. This contractile property resides in specialized and interacting proteins localized in myofibrils making up the muscle fiber. Myofibrils have a transverse banding pattern that gives skeletal muscle its striated appearance. This banded pattern is produced by the alignment of a sequence of sarcomeres, the contractile units of the muscle fiber.

Muscle fibers are multinucleate, the number of cylindrical nuclei varying with the length of a fiber and with age. Nuclei are located at the periphery of a muscle fiber, just beneath the plasma or cell membrane, which immediately encloses the fiber's contents. There is some evidence (Landing *et al.*, 1974) to suggest that nuclei of a muscle fiber are arranged in an hexagonal array, with each nucleus covering a fiber area that is uniform in size and shape.

Closely related to muscle fibers are satellite cells, located at the periphery of muscle fibers, between the plasma and basement membranes. The latter is an outer coating of the fiber. Satellite cells appear to be important in muscle growth.

Muscle fibers have their origin in the middle germinal layer of the embryo, the mesenchyme. They arise from mononucleated myoblasts, which fuse to form myotubes. Myotubes are multinucleated syncytia that can be viewed as immature muscle fibers. Their myofibrils are distributed circumferentially with the nuclei at the central area of a syncytium. The nuclei then migrate to the periphery beneath the plasma membrane, and the myofibrils become more evenly distributed throughout the cytoplasm so that the cell now has the appearance of a proper muscle fiber (Fischman, 1972).

Goldspink (1972) suggests that the postembryonic growth of muscle tissue can be divided into two stages: (1) An early postembryonic stage during which myotubes develop into proper muscle fibers and increase in girth and length; and (2) A subsequent period of further growth in girth and length. The latter stage continues postnatally, and early during the postnatal period mature biochemical and physiological characteristics of the muscle develop.

In humans, it is thought that the number of muscle fibers increases prenatally and for a short period postnatally. Montgomery (1962) noted that the number of muscle fibers approximately doubles between the 32nd week of gestation and 4 months of age. However, the degree of postnatal increase in muscle fibers seems dependent on the organism's maturity status at birth, and the postnatal fiber increase shortly after birth should be considered an extension of the embryonic differentiation of muscle tissue (Goldspink, 1972).

Muscle fibers grow in length by an increase in the number of sarcomeres in series along myofibrils. Growth in length occurs primarily at the musculo-

ROBERT M. MALINA • Department of Anthropology, University of Texas, Austin, Texas 78712.

tendinous junction, and the number of sarcomeres along a muscle fiber is apparently adjusted to the functional length of a muscle (Goldspink, 1972). Needless to say, the differential growth of limb segments postnatally requires similar length increases in individual muscles.

Growth of a muscle in length also entails an increase in muscle fiber nuclei, an increase that continues after growth in fiber length has ceased (Goldspink, 1972; Burleigh, 1974; Cheek *et al.*, 1971). Since larger-diameter fibers generally have more nuclei, it can be concluded that growth of a muscle fiber in both girth and length is associated with an increase in the number of nuclei. Thus, the number of nuclei observed in fiber cross sections or along a fiber's length increases with age. The additional nuclei are apparently derived from satellite cells (Montgomery, 1962; Goldspink, 1972; Burleigh, 1974; Cheek *et al.*, 1971). The incorporation of these nuclei into the muscle fiber seemingly assures sufficient nuclear control for the new fiber components added through growth in length and girth. Note, however, that alternative explanations for the increase in nuclei with growth have been proposed (Burleigh, 1974).

1.2. Fiber Types

Mature muscle has two primary histochemical fiber types: (1) Type I (ST), slow-twitch fibers, characterized by high mitochondrial oxidative enzyme activity and by low phosphorylase and ATPase reactions; and (2) Type II (FT), fast-twitch fibers, characterized by high phosphorylase and ATPase reactions and by low activity of most dehydrogenases (Jennekens *et al.*, 1970; Brooke and Kaiser, 1970; Saltin *et al.*, 1977; Saltin and Gollnick, 1983). Type II fibers are sometimes subdivided into type IIa (FTa) and type IIb (FTb). The difference between the subpopulations of type II fibers apparently relates to differences in myosin molecules (Billeter *et al.*, 1981). Type IIa fibers are often viewed as oxidative–glycolytic, while type IIb are viewed as high glycolytic. However, Reichmann and Pette (1982) reported that in the case of human muscle, there is an almost complete overlap of the two populations of type II fibers when they are characterized by their aerobic oxidative capacity. Hence, it cannot be concluded that type IIa fibers are oxidative whereas type IIb fibers are not. A third subpopulation of type II fibers, type IIc (FTc), most likely represents undifferentiated fiber types that are not normally observed in adult muscle (Henriksson, 1979).

Prenatal differentiation of muscle fiber types is illustrated in Figure 1. The trends are based on samples of several muscles, including the quadriceps, rectus abdominis, deltoid, biceps brachii, soleus, and gastrocnemius, of fetuses and young children (Colling-Saltin, 1978*a,b*). Prior to 30 weeks gestation, most fibers are undifferentiated. Type I fibers of large size, probably the B fibers of Wohlfart (1937), appear between 20 and 30 weeks gestation, while type I fibers of normal size appear around 30 weeks. The relative distribution of the large type I fibers decreases during gestation and their presence at birth is rare (Colling-Saltin, 1978*a*). The relative distribution of normal-size type I fibers increases considerably from 30 weeks to term, comprising about 37% of the muscle fibers at term. Type II fibers also appear at 30 weeks, comprising about 25% of the muscle fibers between 31 and 37 weeks and 45% at term (Colling-Saltin, 1978*a*). Thus, at term, the relative fiber type frequencies are 38% type I (including about 1% of the large size), 45% type II (38% type IIa, 7% type IIb), and 17% undifferentiated. There is a gradual increase in type I and type II fibers during the first postnatal year, while the relative distribution of undifferentiated fibers falls. There is little difference in the relative fiber distribution in muscle tissue of 1-year-old children and adults. The major portion of the postnatal increase occurs in type I fibers at the expense of the undifferentiated fibers. Most of the increase occurs during the first month postnatally (Colling-Saltin, 1978*a*) and probably is an extension of embryonic differentiation.

The percentage distribution of type I fibers in children, adolescents, and adults is given in Table I. The data are derived only from nonathletic untrained individuals, although the degree of their activity may vary. Some highly trained athletes have different fiber-type distributions, e.g., endurance runners, sprinters, and swimmers (Costill *et al*, 1976*a*; Gollnick *et al.*, 1972; Nygaard and Nielsen, 1978; Saltin *et al.*, 1977). The vastus lateralis muscle is the most commonly sampled muscle, and there does not appear to be much variation among fiber-type distributions in the vastus lateralis, gastrocnemius, and deltoid. There are no sex differences in the percentage of type I fibers, but males have more variable distributions than do females, i.e., greater ranges (Hedberg and Jansson, 1976). There is no clear age-associated variation in the

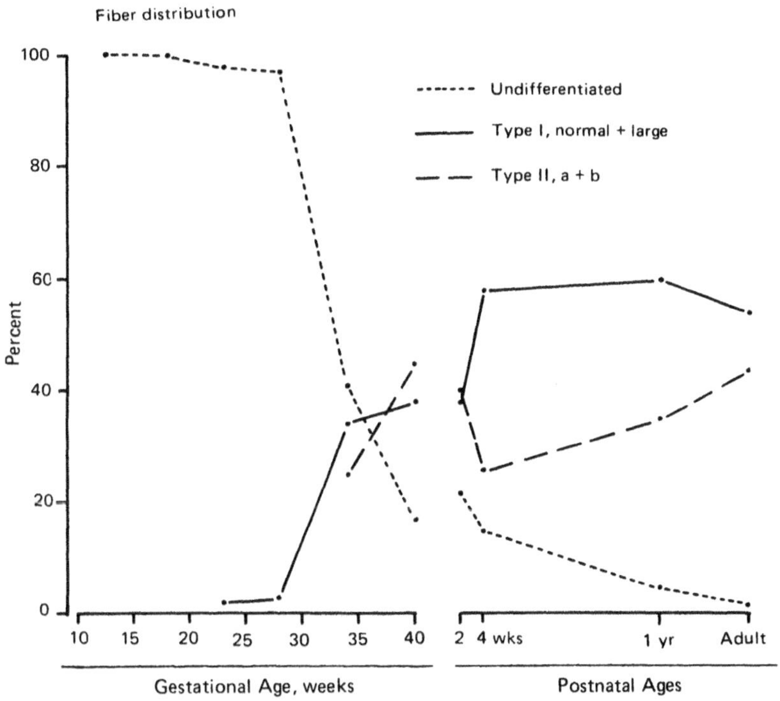

Fig. 1. Relative distribution of muscle fibers prenatally and postnatally. (Drawn from the data of Colling-Saltin, 1978*a*.)

percentage of type I fibers, with the exception of the increase noted by Larsson *et al.* (1978) with advancing age in small samples of adults. Note, however, that the percentage of type I fibers observed in preadolescent children is similar to that reported for endurance-trained youths and adults (Bell *et al.*, 1980).

Since elite endurance athletes have a high proportion of type I fibers in the muscles specifically involved in the sport activity, a question that merits concern is whether the fiber type distribution pattern is genetically determined or is a consequence of training. Komi *et al.* (1977) noted that monozygotic (MZ) twins (N = 13 pairs) were virtually identical in type I fiber distribution, while dizygotic (DZ) twins (N = 16 pairs) were quite variable. The estimated heritability (Holzinger) was 0.96. More recently, however, Lortie *et al.* (1985) noted only moderate genotype dependency in the percentage of type I fibers in MZ twins (N = 23 pairs, intraclass correlation of 0.66) compared with DZ twins (N = 10 pairs, intraclass correlation of 0.34) and brothers (N = 25 pairs, intraclass correlation of 0.14). In contrast, the percentages of type IIa and type IIb fibers showed little similarity in genetically related individuals, which would suggest nongenetic influences, e.g., training. These results support the possible transformation of type I to type II fibers (Jansson *et al.*, 1978) and fit well with the reversible transformation between type IIa and type IIb fibers observed by Andersen and Henriksson (1977).

The relative distribution of type IIa and type IIb fast-twitch fibers are shown in Table II. In the single sample of young children, the distribution of the two subpopulations of type II fibers is approximately equal. However, in the older samples, the percentage of type IIa fibers is about 1.5–2 times greater than the type IIb fibers.

1.3. Chemical Composition

Changes in the chemical composition of muscle during growth and development reflect histological changes. In the fetus, fibers are small, few in number, and widely separated by extracellular material. At term, fibers are still small but are greater in number and are more closely packed in the muscle. In adults, muscle fibers are larger in diameter with little space between them (Widdowson, 1969). These histological changes are illustrated in the decreasing concentrations of extracellular ions (sodium and chloride) and increasing concentrations of intracellular constituents (potassium and phosphorus) with

Table I. Percentage Distribution of Type I Slow-Twitch Fibers in Untrained Children, Adolescents, and Adults

Mean age or age range	N	Sex	Muscle sampled[a]	Percentage type I fibers: Mean	SD	SE	Range	Reference
2 months to 11 years	25	M + F	VL	58.9	12.0	—	—	Lundberg *et al.* (1979*a*)
6.4	13	M + F	VL	58.8	11.4	—	41–75	Bell *et al.* (1980)
	7	F	VL	55.6	8.5	—	41–67	Bell *et al.* (1980)
	6	M	VL	62.1	14.2	—	43–75	Bell *et al.* (1980)
11.2	5	M	VL	54.8[b]	—	3.4	45–60	Eriksson *et al.* (1973)
				48.9	—	3.4	40–55	Eriksson *et al.* (1973)
16.1	47	F	VL	51.8	9.1	—	—	Hedberg and Jansson (1976)
16.2	69	M	VL	53.9	12.2	—	—	Hedberg and Jansson (1976)
16.3	6	M	VL	43.6[c]	15.0	—	—	Fournier *et al.* (1982)
16.7	6	M	VL	47.2	8.3	—	—	Fournier *et al.* (1982)
16.8	6	M	VL	47.2[c]	15.7	—	—	Fournier *et al.* (1982)
17.1	6	M	VL	47.0	13.2	—	—	Fournier *et al.* (1982)
15–24	40	F	VL	49.1[d]	7.7	—	—	Komi and Karlsson (1978)
15–24	40	M	VL	55.9[d]	11.9	—	—	Komi and Karlsson (1978)
19–26	12	M	VL	43.0	—	—	26–59	Tesch *et al.* (1982)
	12	M	D	50.0	—	—	33–64	Tesch *et al.* (1982)
19–26	4	F	VL	36.4	—	—	27–42	Prince *et al.* (1977)
Adult	4	M	VL	35.5	—	—	20–48	Prince *et al.* (1976)
20–30	10	F	G	51.0	—	—	27–72	Costill *et al.* (1976*a*)
17–42	11	M	G	52.6	—	—	38–73	Costill *et al.* (1976*a*)
24.3	24	F	VL	50.7	—	—	28–71	Campbell *et al.* (1979)
21–29	6	F	VL	39.2	—	4.5	—	Ingjer (1979)
22–29	10	M	VL	44.0	—	—	—	Thorstensson *et al.* (1977)
23.1	9	M	VL	43.8	—	3.5	26–60	Green *et al.* (1979)
23.8	13	M	VL	53.3	18.6	—	25–81	Ivy *et al.* (1980)
27.4	19	M	G	57.7	—	2.5	—	Costill *et al.* (1976*b*)
24–30	12	M	D	46.0	—	6.8	14–60	Gollnick *et al.* (1972)
	12	M	VL	36.1	—	5.0	13–51	Gollnick *et al.* (1972)
31–52	14	M	D	45.2	—	2.7	33-59	Gollnick *et al.* (1972)
	14	M	VL	43.9	—	4.8	24–73	Gollnick *et al.* (1972)
20–29	11	M	VL	40.5	—	3.9	14–59	Larsson *et al.* (1978)
30–39	12	M	VL	36.5	—	1.6	27–44	Larsson *et al.* (1978)
40–49	10	M	VL	48.1	—	4.2	29–67	Larsson *et al.* (1978)
50–59	12	M	VL	51.7	—	3.0	28–68	Larsson *et al.* (1978)
60–65	10	M	VL	55.0	—	4.5	38–79	Larsson *et al.* (1978)

[a]Muscles sampled: VL, vastus lateralis; G, gastrocnemius; D, deltoid.
[b]Samples taken at 2 and 6 weeks of endurance training; differences not significant.
[c]Samples taken before and after a 3-month sprint running and endurance running (training) program, respectively; differences not significant.
[d]Monozygotic and dizygotic twins.

growth and development (Table III). In addition, there is a fall in the percentage of water. The size of the extracellular component of muscle with growth and development is first reduced by the increase in fiber number prenatally, and then by the increase in fiber size postnatally.

The decrease in the relative water content of muscle tissue is accompanied by an increase in total nitrogen (Table IV). Both nonprotein nitrogen and cellular protein nitrogen (sarcoplasmic and fibrillar) increase early in prenatal life to adulthood. The sarcoplasmic protein decreases during fetal development and then increases postnatally. The relative contribution of fibrillar protein, on the other hand, does not change much prenatally, but increases postnatally. Dickerson and Widdowson

Table II. Percentage Distribution of Type IIa and Type IIb, Fast-Twitch Fibers in Untrained Children, Adolescents, and Adults[a]

Mean age or age range	N	Sex	Percentage type II fibers: Type IIa Mean	SD	SE	Type IIb Mean	SD	SE	Reference
6.4	13	M + F	19.7	9.5	—	21.5	9.1	—	Bell *et al.* (1980)
	7	F	22.1	8.8	—	22.3	8.4	—	Bell *et al.* (1980)
	6	M	17.3	10.3	—	20.6	9.8	—	Bell *et al.* (1980)
16.1	47	F	32.8	9.3	—	14.8	6.8	—	Hedberg and Jansson (1976)
16.2	68	M	32.2	9.1	—	13.0	7.6	—	Hedberg and Jansson (1976)
16.3	6	M	36.3[b]	12.9	—	18.5	3.4	—	Fournier *et al.* (1982)
16.7	6	M	33.4	10.3	—	18.5	2.2	—	Fournier *et al.* (1982)
16.8	6	M	29.5[b]	4.3	—	21.9	14.1	—	Fournier *et al.* (1982)
17.1	6	M	36.9	14.2	—	15.1	6.5	—	Fournier *et al.* (1982)
23.1	9	M	36.9	—	4.4	18.0	—	7.8	Green *et al.* (1979)
21–29	6	F	39.6	—	3.8	12.8	—	6.0	Ingjer (1979)
20–29	7	M	34.0	—	2.8	24.3	—	4.1	Larsson *et al.* (1978)
30–39	11	M	38.6	—	2.4	22.1	—	2.1	Larsson *et al.* (1978)
40–49	7	M	33.6	—	3.1	15.0	—	4.5	Larsson *et al.* (1978)
50–59	9	M	28.7	—	2.7	19.4	—	3.9	Larsson *et al.* (1978)
60–65	7	M	19.6	—	2.1	12.9	—	4.3	Larsson *et al.* (1978)

[a]All studies cited sampled the vastus lateralis muscle.
[b]Samples taken before and after a 3-month sprint running and endurance running (training) program, respectively; differences not significant.

(1960) suggest that the rate at which fibrillar protein develops postnatally may be influenced by the functional activity of the muscle.

The absolute extracellular protein nitrogen increases through infancy and then decreases to adulthood (Table IV). As a percentage of total nitrogen, the extracellular fraction, however, increases to a maximum at about the time of birth and then decreases postnatally to a lower level in the adult. In a general way, the increase in the relative contribution of extracellula protein nitrogen appears to parallel the increase in muscle fiber numbers prenatally, and its decreasing relative contribution apparently occurs when muscle fibers are increasing in size (Dickerson and Widdowson, 1960; Widdowson, 1969).

Table III. Water and Electrolyte Composition of Human Skeletal Muscle Pre- and Postnatally[a]

	Fetus: 13–14 weeks	20–22 weeks	Infant: Full-term newborn	4–7 months	Adult
Water (g/100 g)[b]	91	89	80	79	79
Na (meq/kg)	101	91	60	50	36
Cl (meq/kg)	76	66	43	35	22
K (meq/kg)	56	58	58	89	92
P (mmoles/kg)	37	40	47	65	59
Na space (g/100 g)	80	71	43	35	26
Cl space (g/100 g)	67	58	35	29	18

[a]Adapted from Dickerson and Widdowson (1960).
[b]Results are expressed per unit of fresh muscle.

Table IV. Concentration of Nitrogen in Various Fractions of Human Skeletal Muscle (g/100 g Fresh Muscle) Pre- and Postnatally[a]

	Fetus		Infant		
	14 weeks	20–22 weeks	Full-term newborn	4–7 months	Adult
Total N	1.09	1.52	2.09	2.90	3.05
Nonprotein N	0.12	0.17	0.24	0.32	0.30
Sarcoplasmic protein N	0.36	0.37	0.39	0.50	0.67
Fibrillar protein N	0.57	0.87	1.09	1.70	1.99
Extracellular protein N	0.06	0.18	0.38	0.46	0.14
Sarcoplasmic protein (%)	33.0	24.3	18.7	17.2	22.0
Fibrillar protein (%)	52.3	57.2	52.2	58.6	65.2
Extracellular protein (%)	5.5	11.8	18.2	15.9	4.6

[a]Adapted from Dickerson and Widdowson (1960).

The lack of information for children and adolescents is especially apparent in the chemical composition data. Dickerson and Widdowson (1960), however, note that figures for the composition of muscle in an 11 and a 16 year old boy were similar to those for adults, and thus combined them with adult values. This would seem to suggest that muscle tissue attains chemical maturity during childhood, at least in terms of the parameters studied.

The oxidative potential of skeletal muscle as measured by succinate dehydrogenase (SDH) activity changes gradually prenatally. SDH activity is about 0.5 mmoles/kg wet weight per min from 12 to 25 weeks gestation and then increases gradually to reach 2–3 mmoles/kg wet weight per min at term. SDH activity doubles early postnatally, reaching about 5 mmoles/kg wet weight per min by 1 month of age (Colling-Saltin, 1978*b*). Values of SDH activity in untrained adults vary between 2.7 and 14.9 mmoles/kg wet weight per min (Gollnick *et al.*, 1972; Costill *et al.*, 1976*a*).

Glycolytic potential as measured by phosphofructokinase (PFK) activity shows a similar pattern of change, but higher levels are attained. PFK activity is less than 1 mmole/kg wet weight per min during the first half of gestation, after which it rises significantly to 3–4 mmoles/kg wet weight per min by 37 weeks, to 7 mmoles/kg wet weight per min in the first 2 weeks postnatally, and to 9–10 mmoles/kg wet weight per min by 4 weeks of age. Values of PFK activity in untrained adults vary between 17.7 and 38.0 mmoles/kg wet weight per min (Gollnick *et al.*, 1972).

Both oxidative and glycolytic enzymes change with training. Eriksson (1972) noted an increase of 30% and 8% in SDH and PFK activities, respectively, in five 11-year-old boys after 6 weeks of endurance training. Nevertheless, the PFK activity in the boys was low compared with adult values, which may suggest a difference in the magnitude of training responses in youngsters and adults. Gollnick *et al.* (1973) noted an increase of 95 and 11% in SDH and PFK activities, respectively, in six adult males after 5 months of endurance training. Comparing endurance and sprint training, Fournier *et al.* (1982) reported a 42% increase in SDH activity and no change in PFK activity after 3 months of endurance training, and a 21% increase in PFK activity and no increase in SDH activity after 3 months of sprint training in adolescent boys (N = 12, 16–17 years of age). The results of these studies illustrate age-associated variation in response to training, as well as the specificity of training responses. The direction of training responses observed in youths is similar to that observed in adults, but the magnitude of the responses varies.

1.4. Fiber Size and Area

Changes in muscle fiber diameter prenatally and postnatally are shown in Figures 2 and 3. Muscle fiber diameters increase gradually during gestation, and there is little difference among the various fiber types, with the exception of the type I fibers of large size (Figure 2). The diameters of normal type I and type II fibers at 1 year of age are smaller than type I fibers observed late in gestation (Colling-Saltin, 1978*a*).

The marked postnatal increase in muscle girth is attributable almost entirely to hypertrophy or continued growth of existing muscle fibers, and not to hyperplasia (Saltin and Gollnick, 1983). Muscle fibers increase in diameter linearly with age (Figure 3) and with body size postnatally (Bowden and Goyer, 1960; Aherne *et al.*, 1971). Increase in fiber diameter varies with the muscle studied and is apparently related to function or intensity of workload. During infancy and childhood, muscle fibers of boys and girls do not consistently differ in diameter, and adult diameters are apparently attained during adolescence (Brook and Engel, 1969). There is, however, a lack of muscle fiber data for middle childhood and adolescence.

Mean fiber areas in untrained individuals, derived primarily from the vastus lateralis muscle, are summarized in Table V. Sex differences are clearly apparent by 16 years of age. Fiber areas increase into the mid-20s in males, consistent with *in vivo* estimates of muscle mass (see Section 2.2 to 2.4). Type IIa fibers tend to be larger than type IIb, but differences in cross-sectional areas of type I and type II fibers are not consistent across studies, as summarized in Table V.

Regular training significantly influences fiber size and area. High resistance weight training in children, adolescents, and adults is accompanied by muscular hypertrophy and alterations in the ratio of type I to type II fiber areas (Malina, 1983). Strength training apparently results in a specific hypertrophy of type II fibers (Thorstensson, 1976; Krotkiewski *et al.*, 1979). Some experimental evidence with adult animals indicates that muscle fibers will divide by longitudinal fission ("fiber splitting") under the stress of an intensive weight-lifting regimen (Gonyea, 1980), although other data are not consistent with these observations (Gollnick *et al.*, 1981, 1983). In contrast, only minor changes in muscle size and significant increases in the relative area of muscle composed of type I fibers are characteristic of endurance-trained individuals (Gollnick *et al.*, 1973).

1.5. Muscle "Cell Size" and "Nuclear Number"

Growth of muscle tissue after the first few months of postnatal life is characterized by constancy in number of muscle fibers, an increase in fiber size, and a considerable increase in number of muscle nuclei (Montgomery, 1962). Changes in the number of nuclei during growth have recently received much attention in both humans and experimental animals. The number of nuclei in developing muscles has been estimated from measurements of DNA. Since DNA per nucleus is relatively constant at 6.2 pg per diploid nucleus, the total number of nuclei in a muscle tissue sample can be determined by dividing the total DNA in the tissue by 6.2. Cheek and colleagues have applied this method to assess muscular growth in a small sample of normal children (Cheek, 1968, 1975; Cheek and Hill, 1970; Cheek *et al.*, 1971). Using DNA content of a sample of gluteal muscle tissue as an index of nuclear or "cell" number and the ratio of protein to DNA as an index of "muscle cell size," Cheek and colleagues showed an increase in nuclear number or in DNA and an increase in amounts of protein or cell mass per nucleus or per unit DNA during growth, i.e., hypertrophy of normal-growing nuscle. Assuming that the gluteal biopsy sample is representative of body musculature

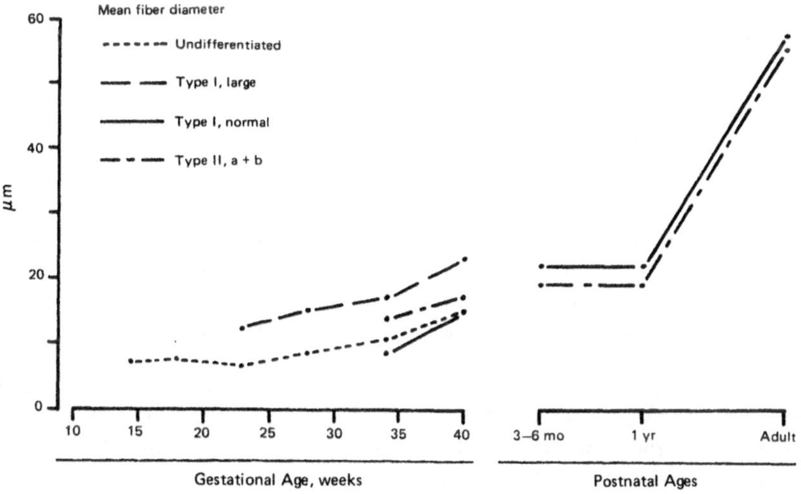

Fig. 2. Mean prenatal and postnatal changes in diameters of muscle fibers of different types. (Drawn from the data of Colling-Saltin, 1978*a*, 1980.)

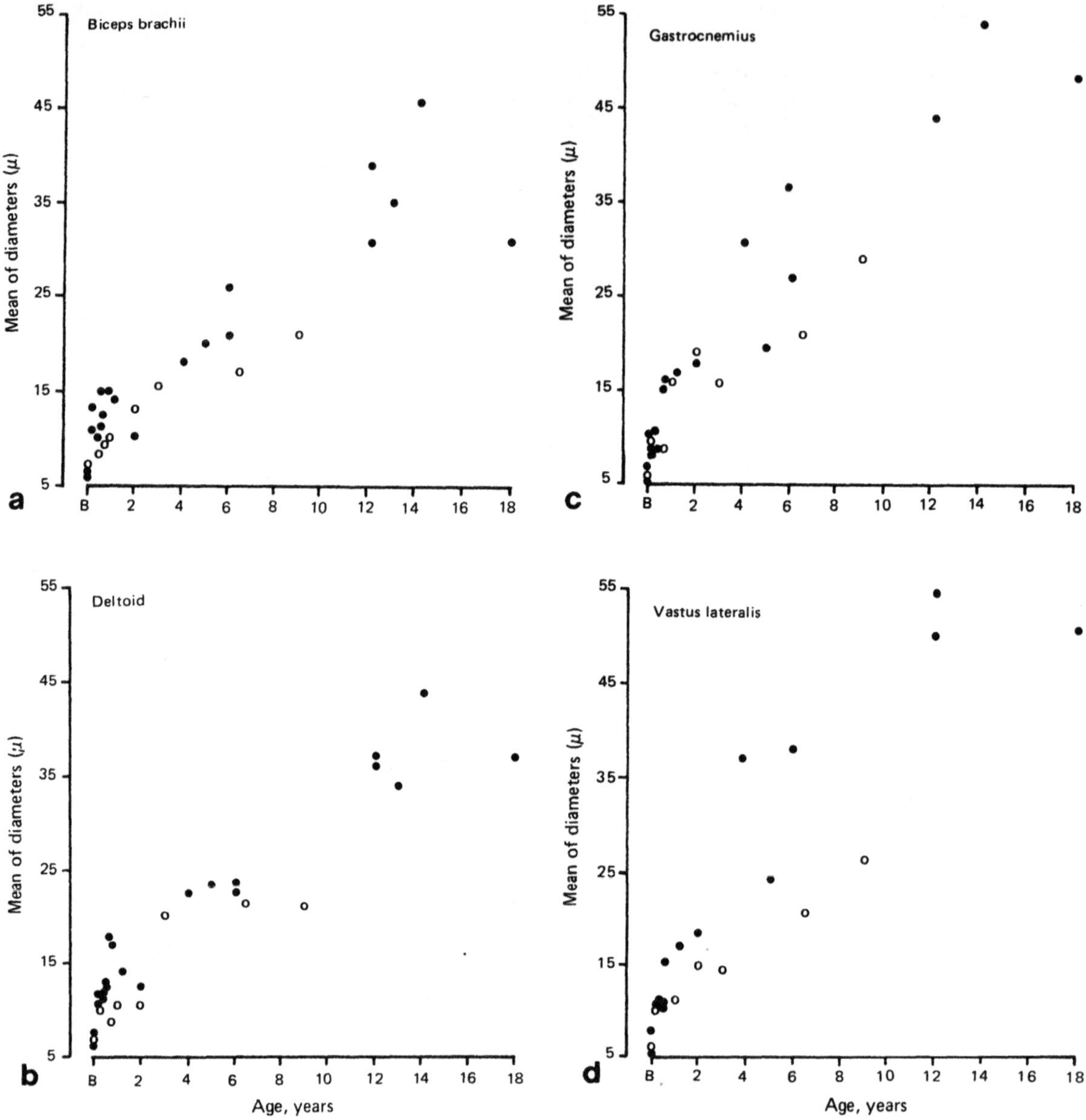

Fig. 3. Postnatal increase in fiber diameters of selected muscles. (a) biceps brachii; (b) deltoid; (c) gastrocnemius; (d) vastus lateralis. (●) Sexes combined (from Aherne *et al.*, 1971); (○) sexes combined (from Bowden and Goyer, 1960).

in general and using creatinine excretion per day (see Section 2.2) as an estimate of total muscle mass in the body, Cheek and colleagues generalized nuclear number and cell size estimates to total muscle mass in growing children.

On the basis of the preceding studies, boys have been found to undergo a 14-fold increase in muscle nuclear number from infancy through adolescence, while girls show a 10-fold increase. In terms of limits for muscle nuclear number, Cheek's (1975) estimates suggest a limit of 3.8×10^{12} nuclei for boys and of 2.0×10^{12} nuclei for girls. Relative to age, males show an increase in muscle nuclear number especially prior to 2 years of age and again after 9 years of age. The latter comprises the adolescent spurt in muscle mass. Prior to the adolescent spurt, sex differences in estimated nuclear numbers are not obvious. On the other hand, the increase in es-

Table V. Mean Fiber Areas of Different Types of Fibers in Youth and Adults

Mean age or age range	*N*	Sex	Muscle sampled[a]	Mean fiber areas (μm^2) I	II	IIa	IIb	Reference
16	45	F	VL	4310	—	4310	3920	Hedberg and Jansson (1976)[b]
20–30	25	F	VL	3948	—	3637	2235	Saltin *et al.* (1977)
16	70	M	VL	4880	—	5500	4900	Hedberg and Jansson (1976)[b]
20–30	10	M	VL	5310	—	6110	5600	Saltin *et al.* (1977)
20–30	10	F	G	3875	4193	—	—	Costill *et al.* (1976*a*)
17–42	11	M	G	5699	4965	—	—	Costill *et al.* (1976*a*)
27.4	19	M	G	5460	4947	—	—	Costill *et al.* (1976*b*)
23.1	9	M	VL	5285	—	5565	5146	Green *et al.* (1979)
19–26	4	F	VL	2784	—	3392	2425	Prince *et al.* (1977)
Adult	4	M	VL	3303	—	4105	3418	Prince *et al.* (1976)
21–29	6	F	VL	3798	—	5113	3758	Ingjer (1979)
20–29	11	M	VL	5666	6953	—	—	Larsson *et al.* (1978)
30–39	12	M	VL	6344	6975	—	—	Larsson *et al.* (1978)
40–49	10	M	VL	6754	6627	—	—	Larsson *et al.* (1978)
50–59	12	M	VL	5941	5954	—	—	Larsson *et al.* (1978)
60–65	10	M	VL	5591	5243	—	—	Larsson *et al.* (1978)

[a]Muscles sampled: VL, vastus lateralis; G, gastrocnemius.
[b]These values are based on the work of Hedberg and Jansson, as reported by Saltin *et al.* (1977). Values in the original report (Hedberg and Jansson, 1976) are approximately 10% too large owing to calibration error (B. Saltin, personal communication).

timated cell size or protein or cell mass per nucleus or per unit of DNA is only twofold for boys and girls. Cheek's data indicate larger protein:DNA ratios in girls at an earlier age, implying accelerated growth of "muscle cell size" in females. Males eventually display catch-up growth in this muscle tissue estimate and may eventually surpass females. Cheek's data, however, do not extend beyond 17 years of age.

The preceding estimates of nuclear number and cell size, although interesting, should be viewed with caution. The sample from which they are derived is small, 33 boys and 19 girls. Of the former, 17 boys are clustered between 0.18 and 1.37 years, while 16 range from 5.27 to 16.05 years. Only five of the boys are over 12 years of age. One can inquire, therefore, as to the accuracy of projections through male adolescence on the basis of such a limited data base. The 19 girls range from 3.50 to 17.08 years of age, only eight are over 11 years of age, and only two are sexually mature. In addition, generalizations from a sample of gluteal muscle tissue to the body as a whole need clarification. Methodological concerns relative to estimates of cell or nuclear number are also apparent. Muscle cells or fibers are multinucleate, and because of this, the amount of DNA in a muscle gives no information about the number of fibers in it. In addition, the ratio of protein or cell mass per nucleus or per unit DNA demonstrates nothing about the size of the muscle cell (Widdowson, 1970). What it shows, however, is the amount of cytoplasm associated with each nucleus; Widdowson (1970) emphasizes that this ratio increases in human muscle tissue, as in other organs, during postnatal growth until the proper functional relationship is attained. The amount of protein in the nucleus per unit DNA of skeletal muscle in the human newborn infant is 175 mg/mg and rises to 300 mg/mg in adult muscle (Widdowson, 1970).

Cheek and colleagues have referred to the increase in nuclear number as an increase in "muscle cell" number, choosing a working definition that assumes that each nucleus has jurisdiction over a finite volume of cytoplasm (see Cheek, 1968; Cheek *et al.*, 1971). Although this concept can be criticized and has added some confusion to the literature, it should be noted that the term "hyperplasia" does not necessarily have to be applied to whole muscle fibers (Goss, 1966).

The source of additional nuclei during growth is not clear. Satellite cells are commonly viewed as responsible for the increase in nuclei or DNA of muscle during growth (Cheek *et al.*, 1971; Goldspink,

1972; Burleigh, 1974; Saltin and Gollnick, 1983). However, alternative possibilities must be considered. Burleigh (1974), for example, indicates that increments in total DNA can also be the result of replication of nuclei in the connective tissue between muscle fibers, in the capillaries of the blood supply, and perhaps in fat cells. This would consequently imply other explanations for the increase in total muscle DNA or nuclear number. Interpretations of gross DNA estimates and generalizations from such estimates to the total muscle mass can also be obscured by other factors. Burleigh (1974) indicates two such considerations: (1) Muscle fibers differ in the extent to which their fibers elongate with age; and (2) Fibers may vary in the number of nuclei within them and in the number of dividing satellite cells that occur along their length. Thus, despite problems evident in the use and interpretation of the DNA unit and associated ratios, data derived from such estimates have provided insights into muscle tissue growth.

Studies of DNA content and thus number of muscle nuclei in growing animals undergoing regular training indicate a significant increase in DNA above that expected with normal growth (Buchanan and Pritchard, 1970; Bailey *et al.*, 1973; Hubbard *et al.*, 1974). The normal pattern of change in DNA content of skeletal muscle with growth is one of steady increase until puberty, with a rather constant level thereafter. The increased DNA in trained animals would seem to suggest that training is a significant factor influencing muscle nuclear number during growth. Training-induced increments in the DNA of skeletal muscle have also been observed in adult animals (Christensen and Crampton, 1965).

1.6. Limitations of Muscle Fiber and DNA Content Studies

The preceding provides an overview of changes in muscle tissue during growth. Sample sizes, however, are characteristically small and the samples are often limited to adults. Nevertheless, the data for children and youth are increasing.

A significant portion of the information is derived from single muscles, especially the vastus lateralis. Variation within (e.g., number of fibers examined and the depth at which a biopsy is taken) and between muscles must be recognized, while generalizations from a single muscle biopsy to the total body should be tempered with caution.

There is an approximately equal distribution of type I and type II fibers in the vastus lateralis, rectus femoris, and gastrocnemius muscles of the lower extremity and deltoid and biceps muscles of the arm. On the other hand, the soleus muscle has 25–40% more type I fibers than other lower extremity muscles, and the triceps muscle has 10–30% more type II fibers than other upper extremity muscles (Saltin *et al.*, 1977). Some data suggest that predominantly phasic muscles have a high percentage of type II fibers, while predominantly tonic muscles have a high percentage of type I fibers. However, many muscles perform both functions and show no predominance of either fiber type (Johnson *et al.*, 1973).

There is a need to assess comprehensively the relationship between muscle fibers and function. The percentage of type I fibers is moderately related to the time required to reach a specific force level in an isometric bilateral leg extension movement, $r = +0.48$ (Viitasalo and Komi, 1978), and to the isometric force of the quadriceps muscle group in males, $r = -0.55$ (Komi and Karlsson, 1978). Children with idiopathic late walking (mean age, 19 months) show reduced muscle fiber size, especially of type II fibers (Lundberg *et al.*, 1979*a*), whereas those with celiac disease (mean age, 14 months) have a reduced percentage of type I fibers and poor gross motor development (Lundberg *et al.*, 1979*b*). After dietary treatment in the latter group, the percentage of type I fibers increased, as did fiber diameter and gross motor development. The possible role of physical inactivity in such clinical conditions must be recognized. Similar changes occur with atrophy caused by inactivity. On the other hand, the influence of regular physical activity on muscle tissue morphometry and function is significant and must be considered in evaluating age and individual differences.

2. Growth of Muscle Mass

2.1. Dissection Studies

There is no proven method whereby total muscle mass can be measured *in vivo* (Garn, 1963; Durnin, 1969; Kreisberg *et al.*, 1970). Estimates of muscle mass derived from dissection studies (Table VI) indicate that muscle comprises about 23–25% of body weight at birth and about 40% of body weight in adults, although the range of reported values for adults varies. Furthermore, the adult dissection specimens are generally older adults so that one can perhaps assume a slightly higher percentage contribution of muscle mass to body weight in young adulthood, probably about 45% or more. Dissection data for the growing years are lacking. Relative to fat-free body weight, i.e., body weight minus total fat, muscle mass comprises approximately 45–50% of the fat-free body mass (Durnin, 1969).

2.2. Creatinine Excretion

Urinary creatinine excretion is frequently used as an index of total muscle mass in the body. Although creatinine is not solely a by-product of muscle metabolism, the volume of creatinine excreted is a function of the muscle mass. Creatinine excretion can be expressed simply in grams per day or relative to body weight (the creatinine coefficient). The former serves as an index of muscle mass, while the latter estimates the relative amount of muscle in the body (Novak, 1963; Cheek, 1968).

Muscle tissue development is affected by a variety of factors, including age, sex, degree of physical training, hormonal states, and metabolic states. Creatinine excretion is influenced by diet, exercise, emotional stress, the menstrual cycle, and certain disease states, in addition to normal day-to-day variation (Heymsfield *et al.*, 1983). Hence, creatinine is difficult to control and may be of limited utility in muscle mass estimates. Nevertheless, within the limitations imposed by the complexities of creatinine metabolism and excretion, urinary levels of creatinine, when collected under properly controlled conditions, offer the advantage of ready availability and thus provide a useful estimate of muscle mass.

Table VI. Muscle Mass as a Percentage of Body Weight[a]

Age	Sex	Body weight	Muscle mass (%)
Fetus, 6 months	—	401 g	22.75
Fetus, 6 months	—	491 g	22.71
Fetus, 6 months	—	—	24.60
Newborn	—	2.36 kg	23.30
Newborn	—	2.91 kg	24.03
Newborn	—	3.10 kg	25.05
Newborn	—	—	24.80
Adult	—	66.20 kg	43.40
Adult	—	—	43.07
Adult	M	69.67 kg	41.77
Adult	M	55.75 kg	41.36
Adult	M	76.51 kg	42.07
Adult, 35 years	M	70.55 kg	31.56
Adult, 46 years	M	53.80 kg	39.76
Adult, 60 years	M	73.50 kg	40.22
Adult, 48 years	M	62.00 kg	42.53
Adult	F	55.40 kg	35.83
Adult, 67 years	F	43.40 kg	23.13

[a]Based on dissection data collated from various sources (after Malina, 1969). (Unless noted sex not indicated in original data.)

The amount of creatinine excreted in the urine over a 24-hr period is used as an estimate of muscle mass. Mean urinary creatinine excretion increases with age during growth, especially in adolescence, and is greater in boys than in girls (Reynolds and Clark, 1947; Clark *et al.*, 1951; Novak, 1963; Zorab, 1969). From 10 to 18 years of age, the amount of creatinine excreted more than doubles in boys and almost doubles in girls.

A constant relationship between creatinine excreted and muscle mass has been suggested, although a definitive creatinine equivalence has not been established for humans (Heymsfield *et al.*, 1983). Talbot (1938) estimated that 1 g of creatinine excreted per 24 hr is derived from 17.9 kg of muscle tissue. By contrast, Cheek (1968) recommends a conversion factor of 20 as being more appropriate; i.e., each gram of creatinine excreted in the 24-hr urine sample is derived from 20 kg of muscle tissue. Heymsfield *et al.* (1983) report a creatinine equivalence of 17–20 kg of muscle tissue per gram of creatinine.

Although the creatinine equivalence varies, absolute muscle mass from childhood to young adulthood has been estimated from published creatinine excretion and body weight data using a creatinine equivalence of 20 (Figure 4). Absolute muscle mass increases with age, sex differences being small prior to the adolescent spurt but marked during adolescence. The estimates reported by Clark *et al.* (1951), mixed logintudinal; Novak (1963), cross-sectional; and Young *et al.* (1968), cross-sectional agree rather closely. The muscle mass estimates from Cheek's (1968, 1975) small sample are consistently lower, possibly the result of sampling and methodological variation, as Cheek's data are derived over a 3-day period in the presence of a low creatinine diet. However, the small sample of subjects above 3 years of age for whom creatinine excretion data are available (N = 19 boys and 19 girls) is somewhat unusual. Among the boys, 16 of the 17 for whom skeletal age (SA) data were available had skeletal ages less than their respective chronological ages (CA). The average SA–CA difference was −1.45 years. In contrast, 14 of the 19 girls had skeletal ages in advance of their chrono-

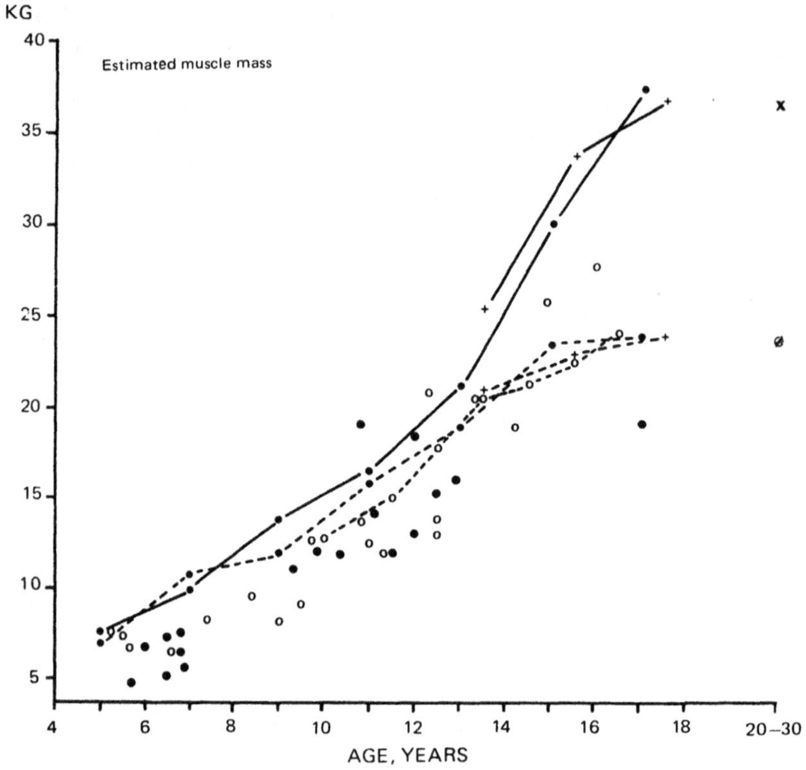

Fig. 4. Age changes and sex differences in muscle mass estimated from creatinine excretion and body weight. (●—●) Boys; (●---●) girls (from Clark *et al.*, 1951); (+—+) boys; (+---+) girls (from Novak, 1963); (o--o) girls (from Young *et al.*, 1968); (o) boys; (●) girls (from Cheek, 1968); (x) adult males (from Norris *et al.*, 1963); (ø) adult females, from Young *et al.*, 1963).

logical ages, an average SA–CA difference of +0.46 years. Note that muscular development and skeletal and/or sexual maturation are positively related (see Section 4), so that the lesser estimated muscle mass values might reflect maturity differences, especially in the boys.

Estimated muscle mass expressed as a percentage of body weight increases from 5 to 17 years of age in boys, from about 42 to 54% (Table VII). In girls, estimated muscle mass as a percentage of body weight increases from 5 through 13 years of age, from about 40 to 45%, decreasing somewhat after 13 years of age (Malina, 1969). The decrease in relative contribution of muscle mass to body weight is probably associated with the accumulation of fat during female adolescence. At all ages during childhood and adolescence, muscle mass contributes a greater percentage to body weight in boys than in girls. In young adults (20–29 years), the relative contribution of estimated muscle mass to body weight decreases somewhat relative to the late adolescent values, about 51% in males and 40% in females. This probably reflects a relative increase in fatness in both sexes during the third decade of life.

2.3. Potassium Concentration

Muscle tissue is rich in potassium, and the muscle mass of the body accounts for 50–70% of body potassium (Pierson *et al.*, 1974). As such, measures of potassium offer a marker for muscle tissue. Quite commonly, though, measures of potassium are used to derive estimates of lean body mass (LBM) and/or body cell mass (BCM) (Moore *et al.*, 1963; Malina, 1969, 1980; Forbes, 1972; Pierson *et al.*, 1974). The radioisotope ^{40}K concentration provides the basis from which LBM and BCM estimates of body composition are derived.

Potassium concentration of the body increases from infancy throughout childhood, the slope of increase being the same for both sexes until the adolescent spurt. Boys have consistently higher concentrations of potassium in the body than do girls in infancy and chilhood, with the sex difference becoming especially marked during adolescence (Burmeister, 1965; Flynn *et al.*, 1970; Novak, 1973). Boys, however, do not have a greater concentration of potassium in muscle tissue (Cheek, 1968).

Although postassium concentration or LBM estimates derived from them are frequently expressed

Table VII. Estimated Muscle Mass as a Percentage of Body Weight[a]

Age (years)	Percentage muscle	
	Males	Females
5	42.0	40.2
7	42.5	46.6
9	45.9	42.2
11	45.9	44.2
13	46.2	43.1
13.5	50.2	45.5
15	50.3	43.2
15.5	50.6	44.2
17	52.6	42.0
17.5	53.6	42.5
20–29	51.5	39.9

[a]Data are derived from average values for body weight and creatinine excretion (grams per 24 hr) collated from several sources (Malina, 1969).

relative to body weight, several analyses suggest stature as the standard for expressing potassium concentration or LBM, as body weight is too variable (Forbes, 1972, 1974; Flynn *et al.*, 1975). Potassium concentration per unit of stature relative to age is shown in Figure 5. There appears to be a rather marked increase during infancy, followed by a slower increase through childhood. During adolescence, boys show a sharp increase in potassium per unit stature, while girls show only a slight gain. This increase in potassium reflects the male spurt in muscle mass during adolescence. Estimates of LBM derived from other *in vivo* methods, e.g., total body water and densitometry, show general correspondence among themselves, although differing in quantitative estimates of lean tissue, and show a general correspondence to age- and sex-associated variation in muscle tissue (Malina, 1969, 1980; Forbes, 1974).

More recently, the ratio of total body potassium to total body nitrogen has been used to estimate muscle mass and nonmuscle lean tissue. Since muscle is potassium rich, a high ratio of potassium to nitrogen in the body indicates a high proportion of muscle (Burkinshaw *et al.*, 1979; Cohn *et al.*, 1980, 1983). Total body nitrogen is measured *via* neutron activation techniques, while total body potassium is measured with a whole-body counter. The procedure involves radiation exposure and two assumptions: (1) The concentrations of all potassium per unit nitrogen in muscle and nonmuscle lean tissue are the same in all individuals; and (2) The concentrations do not change when tissue is gained or lost. Furthermore, the error of estimation in the method is such that it cannot be applied to individuals; however, the error is sufficiently small to permit comparisons of group means (Burkinshaw *et al.*, 1979).

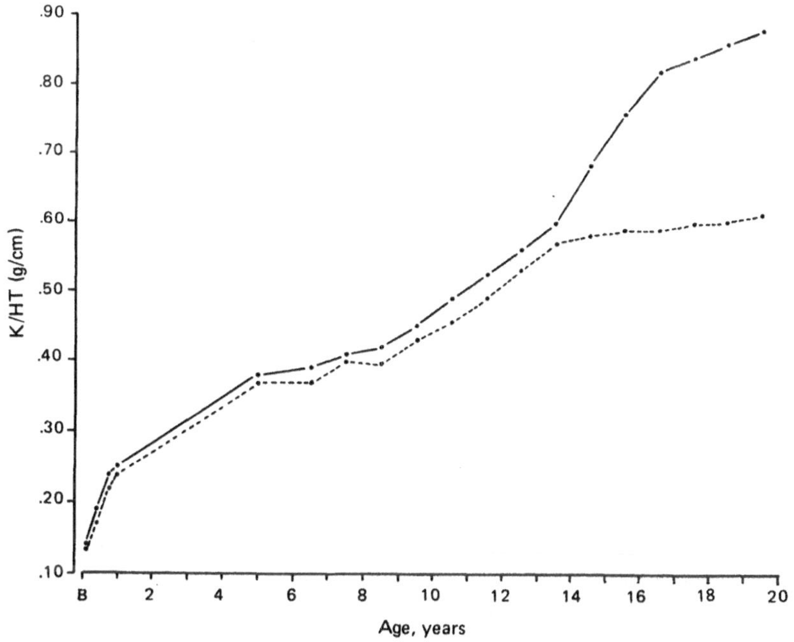

Fig. 5. Age changes and sex differences in potassium concentration per unit stature. (—) Boys (- - -) girls. Data for infancy are from Novak (1973), for 5 years from Flynn *et al.*, (1970), and for 6–20 years from Burmeister (1965).

To date, the method has been applied only to adults and yields estimates of absolute and relative muscle mass that are considerably lower than those obtained by dissection and creatinine excretion. For example, the estimated muscle mass of males ($N = 24$) and females ($N = 10$), 20–29 years of age, is 24.2 and 9.9 kg, respectively. After normalization for body weight, these estimates yield skeletal muscle masses that are, respectively, 30.4 and 16.5% of body weight in males and females (Cohn *et al.*, 1980). Given the underlying assumptions in the method and the radiation exposure, application of neutron activation analysis to estimates of muscle mass in children and youth seems limited at present.

2.4. Radiographic Studies

Muscle can be measured radiographically in a manner similar to that used to assess fat and bone. Arm, forearm, thigh, and calf are the most commonly measured areas, and the data are reported as summated measurements for "total limb muscle width," or as limb specific curves (see Malina, 1969; Tanner *et al.*, 1981). Qualitatively similar age and sex profiles during childhood and adolescence are provided by both the composite muscle-width curves and the limb-specific curves. Correlations between muscle widths measured on different limb segments indicate at best only moderate relationships that are slightly higher in males (Table VIII), but some evidence suggests a slight rise in the correlations during adolescence (Tanner *et al.*, 1981).

Muscles of the extremities increase in size with age from infancy through adolescence (Figure 6), and the growth curve generally resembles that for other estimates of muscle mass. Sex differences, although apparent, are small during childhood, boys having slightly wider muscles. By about 11 years of age, girls begin their adolescent spurt and have a temporary size advantage in muscle-width measures of the calf. There is, however, no temporary size advantage for girls in arm musculature. Boys then have their adolescent spurt in muscle mass and have considerably wider extremity muscles than do girls. Sex differences in muscle widths, established during adolescence, persist into adulthood and are more apparent for musculature of the upper extremities.

Velocities of growth in muscle widths are similar in boys and girls prior to adolescence (Figure 7). Males show well-defined spurts in both the arm and the calf musculature, while females show no clear evidence of an adolescent spurt. Rather, girls show a slight increase in the rate of muscle growth followed by a plateau that persists for 4 or 5 years.

When longitudinal data for radiographic muscle widths are aligned on peak height velocity (Tanner *et al.*, 1981), sex differences in the magnitude and perhaps timing of the adolescent muscle tissue spurt are apparent. Boys have a spurt in arm muscle width that is approximately twice the magnitude of that in females. In contrast, the peak in calf muscle width is only slightly greater in males. Peak velocities of arm and calf musculature occur after peak height velocity. For the arm, the peak occurs 3–4 months after peak height velocity in boys and about 6 months after peak height velocity in girls. For the calf, the gain in muscle width appears to be more or less constant from 1 year or so before to about 1.5 years after peak height velocity. When muscle widths are plotted relative to peak muscle

Table VIII. Reported Correlations between Muscle Widths Measured in the Arm, Thigh, and Calf

Areas compared	Age	Correlations: Males	Correlations: Females	Reference
Arm/calf	6.0–15.5 years	0.53	0.42	Malina and Johnston (1967*a*)
Arm/calf	Preadolescent	0.35	0.46	Tanner *et al.* (1981)
Arm/calf	Young adult	0.42	0.36	Tanner (1965)
Arm/thigh	Young adult	0.54	0.49	Tanner (1965)
Thigh/calf	Young adult	0.51	0.43	Tanner (1965)
Arm/calf	Adult, athletes	0.42	—	Tanner (1964)
Arm/thigh	Adult, athletes	0.30	—	Tanner (1964)
Thigh/calf	Adult, athletes	0.45	—	Tanner (1964)

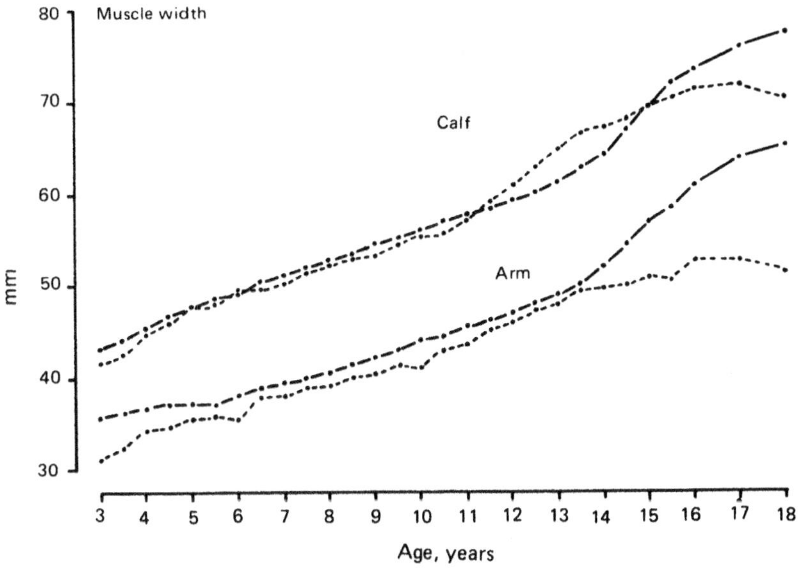

Fig. 6. Age changes and sex differences in muscle widths of the calf and arm. (—) Boys, (- - -) girls. (Drawn from the data of Tanner *et al.*, 1981.)

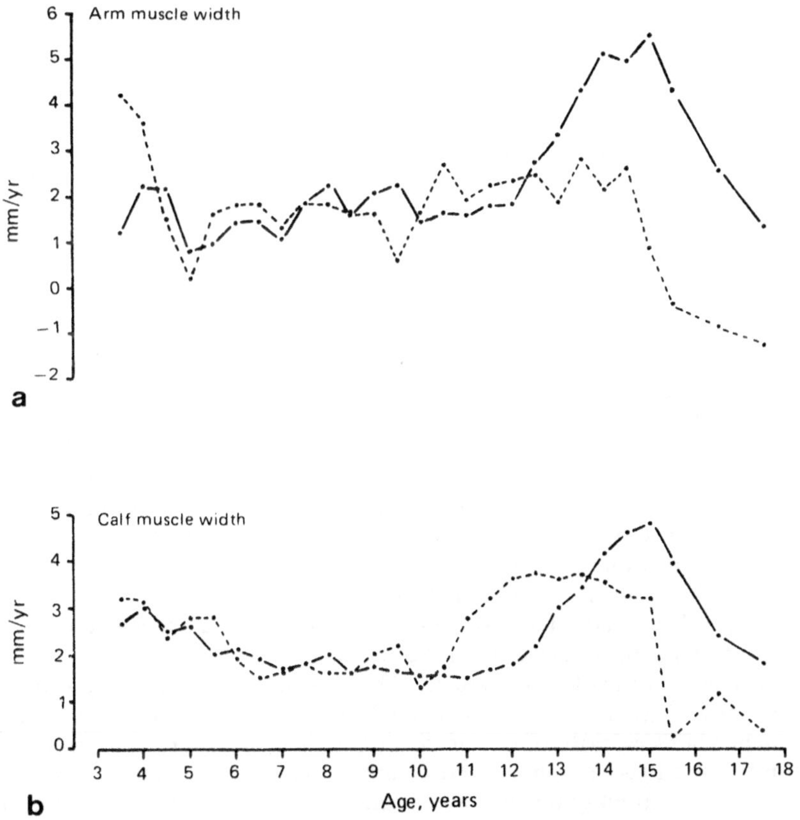

Fig. 7. Age changes and sex differences in whole-year velocities of muscle widths of (a) arm and (b) calf. (—) Boys, (- - -) girls. (Drawn from the data of Tanner *et al.*, 1981.)

velocities, it is apparent that the adolescent spurt in arm and calf muscle occurs over a 2-year period, which is about the same duration as the adolescent height spurt (Tanner *et al.*, 1981).

2.5. Estimated Arm Musculature

Muscle is the body's main protein reserve and, given the extent of protein-energy malnutrition in the developing world, estimates of muscle mass are important to field studies of nutritional status. Arm circumference and triceps skinfold, both measured at a level midway between the acromial and olecranon processes, are commonly used to estimate the muscle mass of the arm. The arm circumference is corrected for the outer perimeter of subcutaneous fat to yield an estimate of mid-arm muscle circumference or area. The procedure has several limitations, however. It assumes that the arm is a cylinder and that the subcutaneous fat is evenly distributed. Also, the size of the humerus is not considered. Variation is skinfold compressibility and errors in measurement are other important factors. Formulas for converting limb circumferences to estimates of muscle mass are presented by Malina (1980) and Heymsfield *et al.* (1982*b*), (see also Chapter 6, this volume).

Nomograms for the estimation of arm muscle circumference and area from arm circumference and the triceps skinfold in children, adolescents, and adults are provided by Gurney and Jelliffe (1973), while age- and sex-specific percentiles from early childhood through adulthood are given by Frisancho (1981). The pattern of age- and sex-associated variation in estimated arm musculature is similar to that for radiographic measures of arm muscle widths.

2.6. Interrelationships and Limitations

Dissection studies, creatinine excretion, potassium concentration, and neutron activation analysis provide only estimates of total muscle mass in the body. They do not indicate either regional variation in distribution of muscle tissue or possible age-, sex-, or race-associated variation in muscularity. The head and trunk, for example, constitute about 40% of the total weight of the musculature at birth, but only 25–30% at maturity. The muscles of the lower extremity show an increase from approximately 40% at birth to 55% of the total weight of the muscle mass at maturity, while the muscles of the upper extremity remain rather constant, comprising about 18–20% of the musculature during life (Scammon, 1923). Thus, body regions contribute differentially to the total muscle mass of the body.

Radiographic measures of muscle widths provide estimates of muscular development in the extremities, and the problem of muscle patterning, analogous to fat patterning, needs to be explored. This is significant in assessing racial differences in muscularity. For example, there are differences in calf muscle mass among races (Garn, 1963). Blacks have substantially smaller calf muscles relative to muscle development in the arm and thigh (Tanner, 1964; Malina 1973), while Japanese have a mesomorphic physique characterized by a dominant calf musculature (Kraus, 1951). Such observations would seem to suggest differences in muscle patterning.

Correlations between creatinine excretion and radiographic measures of muscle widths are moderately high ($r \sim 0.7$) (Reynolds and Clark, 1947; Garn, 1961), while estimated arm muscle area is highly correlated with creatinine excretion ($r = 0.94$) in a variety of normal and undernourished subjects (Heymsfield *et al.*, 1982*a*). Creatinine excretion and radiographic muscle widths are significantly related to total body potassium ($r \sim 0.9$), whereas correlations between creatinine excretion and other estimates of lean body mass are moderate to high ($r = 0.56$ to 0.95) (Malina, 1969; Boileau *et al.*, 1972; Forbes and Bruining, 1976). Correlations between estimated mid-arm muscle circumference and densitometrically estimated lean body mass are moderately high in young males ($r = 0.75$ to 0.85) (Johnston, 1981; Hugg and Malina, n.d.), but are somewhat lower in young females ($r \sim 0.60$) (Johnston, 1979). Lean body mass, however, is heterogeneous, and therefore cannot be regarded as an absolute indicator of muscle mass. Furthermore, the relationship between creatinine excretion and lean body mass is altered by physical training, with the correlation decreasing from 0.73 before training to 0.57 after training (Boileau *et al.*, 1972).

3. Development of Muscular Strength

The composition and size of muscle are functionally manifest in muscular strength, the capacity to exert force against a resistance. Strength of a muscle is related to its cross-sectional area.

Muscular strength increases linearly with chronological age from early childhood to approximately 13–14 years of age in boys, when there is a marked acceleration in development of strength through the late teens. This is shown in Figure 8 for grip strength, and the age trend and sex differences are generally similar for other strength tests. Strength, however, continues to increase into the third decade of life. In girls, strength improves linearly with age through about 15 years of age, with no clear evidence of an adolescent spurt. Boys demonstrate, on the average, greater strength than girls at all ages. Sex-related differences throughout childhood are consistent, although generally small. The marked acceleration of strength development during male adolescence magnifies the preadolescent sex difference (Metheny, 1941; Jones, 1949; Asmussen, 1962; Bouchard, 1966; Malina, 1974). With increasing age during adolescence, the percentage of girls whose performance on strength tests equals or exceeds that of boys drops considerably, so that after 16 years of age, few girls perform as high as the boys' average, and, conversely, practically no boys perform as low as the girls' average (Jones, 1949).

The relationship between strength development and general growth and maturation during male adolescence is such that the strength spurt is frequently considered as a maturity indicator. Maximum strength development occurs after peak velocity of growth in height and weight, the relationship being better with body weight (Stolz and Stolz, 1951; Carron and Bailey, 1974; Beunen *et al.*, n.d.). The adolescent peak in explosive and dynamic strength occurs more coincidentally with the peak in static strength (Beunen *et al.*, n.d.). The pattern of maximum strength development in girls is not so clear. The apex of strength development occurs more often after peak height in girls, but there is considerable variation. Furthermore, in more than one-half of the girls, the peak dynamometric strength gain precedes peak weight gain (Faust, 1977). Thus, the timing of peak strength development in adolescent girls is not as meaningful an indicator of maturity as in adolescent boys.

These data should be related to those for growth of muscle tissue during male adolescence. Peak growth in muscle widths occurs slightly after peak height velocity, while peak weight gain occurs about 6 months after peak height velocity. Thus, peak growth in muscle tissue occurs most likely during peak weight gain and probably after it, emphasizing the role of muscle mass in the adolescent weight gain. It also appears that muscle tissue increases first in mass and then in strength. This would seem to suggest a qualitative change in muscle tissue as adolescence progresses, or perhaps a neuromuscular maturation affecting the volitional demonstration of strength.

Strength is related to body size and lean body

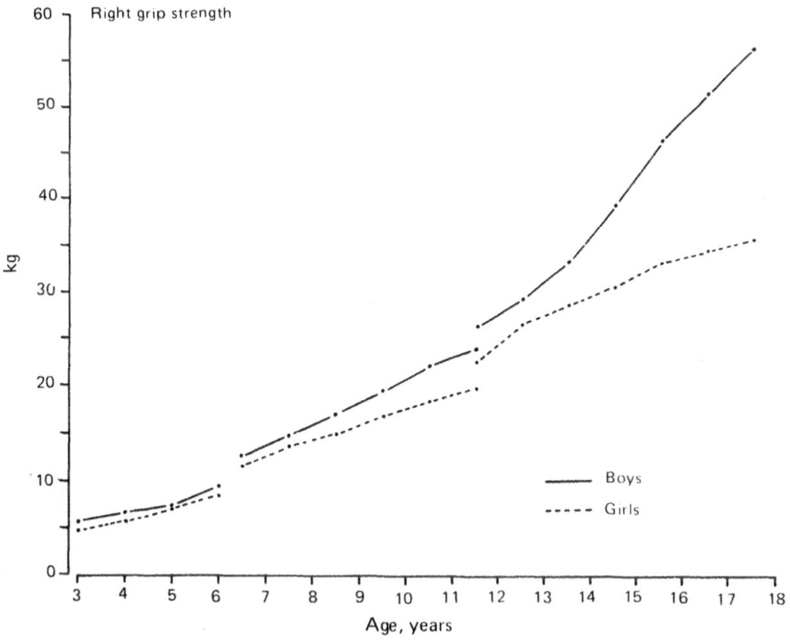

Fig. 8. Age changes and sex differences in muscular strength of the right grip. (—) Boys, (- - -) girls. Data for 3–6 years are from Krogman (1971), for 6.5–11.5 years from Malina (unpublished; see also Malina and Roche, 1983), and for 11.5–17.5 years from Jones (1949).

mass, so that sex differences in strength might relate to a size advantage in boys. This is true only for lower extremity strength. After adjusting for stature variation, sex differences in strength are not apparent in the lower extremity from 7 to 17 years of age. However, from 7 years of age on, boys are significantly stronger in upper extremity and trunk strength even after adjusting for sex differences in height (Asmussen, 1962).

During growth, strength of both boys and girls increases more than predicted on the basis of height alone (Asmussen and Heebøll-Nielsen, 1955). In boys, for example, strength improves out of proportion to gains in height or age, especially after 14 years of age or during adolescence. The predicted average yearly increase in strength in a sample of boys followed longitudinally from 10 to 16 years was approximately 12%, while the actual average yearly strength increase was about 23%, or about twice the predicted value (Carron and Bailey, 1974). These observations thus emphasize the magnitude of the male adolescent strength spurt. It might be plausible to hypothesize a relationship between the observed strength increments and the doubling of muscle nuclear number between 9 and 14 years of age in boys. The disproportionate increase in strength in male adolescence is more apparent in strength of the upper extremities than in the trunk or lower extremities (Asmussen, 1962; Carron and Bailey, 1974). It should be noted, in this regard, that the brachial muscle mass practically doubles during male adolescence (Baker *et al.*, 1958; Hunt and Heald, 1963; Malina and Johnston, 1967*b*).

Table IX. Correlations between Skeletal Maturity and Muscle Widths in the Arm and Calf[a]

Age	Area	Males	Females
6–60 months	Calf[b]	0.19	0.07
7–16 years	Calf[c]	0.35	0.36
7–16 years	Arm[d]	0.43	0.37
11–16 years	Calf[c]	0.51	—
11–16 years	Arm[d]	0.61	—
9–14 years	Calf[c]	—	0.39
9–14 years	Arm[d]	—	0.45

[a]Average correlations derived from *z*-transformed age-specific correlations weighted for sample size.
[b]Adapted from Hewitt (1958).
[c]R. M. Malina and F. E. Johnston (unpublished data)
[d]Adapted from Malina and Johnston (1967*b*).

4. Muscularity and Maturity Status

Maturity-associated variation in size and physique (Tanner, 1962; Malina and Rarick, 1973) is also evident for measures of muscle mass and strength. Correlations between skeletal age and radiographic muscle width measurements are low during early childhood and only slightly higher at older ages (Table IX). However, when correlations are limited to the circumpubertal years, ages 9–14 in girls and 11–16 years in boys, correlations between skeletal maturity and muscle widths of the arm and calf are moderately high, especially in boys.

The inference to be gained from the correlations in Table IX is that the more mature child in any age group has larger muscles, especially during male adolescence. This finding is also evident in comparisons of children grouped as early, average, and late maturers (Reynolds, 1946; Johnston and Malina, 1966). The early maturers have larger muscle measurements of the arm and calf, reflecting their larger body size.

Data on maturity differences in other measures of muscle mass are limited. The effects of sexual maturation of girls on creatinine excretion and musclarity estimates are shown in Table X. There are increasing creatinine excretion values and thus absolute muscle mass with age and advancing degrees of sexual maturity from premenarcheal to postmenarcheal stages. Sample size within specific age groups in the data of Young *et al.* (1968) are too small to illustrate maturity-associated variation within a single age category. When estimated muscle mass is viewed relative to body weight, there is no gradient with increasing maturity, although girls in the more advanced maturity categories (IV and V, postmenarcheal) have slightly greater relative muscle mass. Similar analysis of sexual maturity relative to muscle mass development of boys is lacking. Nevertheless, increased muscularity with advanced maturity status can be inferred from the elevated circulating levels of testosterone, which is specific for muscle tissue anabolism, with advancing stages of sexual maturity (Frasier *et al.*, 1969; August *et al.*, 1972).

Correlations between strength and skeletal maturity are moderate (0.35–0.63) in primary-grade boys (Rarick and Oyster, 1964), suggesting that the more mature boys are generally stronger. However, when the effects of body size are statistically controlled, correlations between skeletal maturity and strength are considerably reduced. Among boys di-

Table X. Creatinine Excretion and Estimated Muscle Mass in Girls Grouped According to Maturity Status[a]

Maturity group[b]	N	Age (years) Mean	Range	Creatinine excretion (g/24 hr)	Estimated muscle mass (kg)	Relative muscle mass (%)
I	18	10.2	9.3–11.9	0.635	12.70	39.3
II	25	11.3	9.5–13.1	0.793	15.86	40.8
III	12	12.9	11.7–15.6	0.928	18.56	39.5
IV	14	13.6	10.7–15.3	1.087	21.74	44.1
V	32	14.6	11.9–17.0	1.150	23.00	42.1

[a]Adapted from Young *et al.* (1968). Absolute and relative muscle mass are estimated from reported mean values using the conversion factor, 1 g creatinine = 20 kg muscle tissue.
[b]The maturity groups are defined by Young *et al.* (1968) as follows: I, premenarche, no secondary sex development; II, premenarche, some secondary sex development; III, premenarche at time of study, but reaching menarche within 6 months; IV, postmenarche, some secondary sex development; V, postmenarche, complete secondary sex development.

vided into the skeletally mature and immature groups, the former were found to be stronger, taller, and heavier (Rarick and Oyster, 1964). Thus, strength differences among young boys reflect primarily size differences.

During adolescence, maturity relationships with strength are more apparent for boys than they are for girls. Correlations between strength and skeletal maturity range from 0.43 to 0.64 in boys and from 0.35 to 0.12 in girls 12 to 16 years of age (Beunen *et al.*, 1976, 1978). The correlations for boys are greatest at 14 and 15 years of age (0.65 and 0.63, respectively), while those for girls are highest at 12 years of age (0.35) and decrease with age to 16 years (0.12).

With boys grouped into contrasting maturity categories, early-maturing boys are stronger age for age than their average- and late-maturing peers from preadolescence through adolescence (Jones, 1949; Clarke, 1971; Carron and Bailey, 1974; Beunen *et al.*, 1980). Strength differences between early and late maturers are especially apparent in boys aged 13–16 years, and the strength advantage for the early-maturing boys reflects their larger body size and muscle mass. When the effects of body weight are removed in comparing early- and late-maturing boys, strength differences between the contrasting maturity groups are reduced but not entirely eliminated (Carron and Bailey, 1974; Beunen *et al.*, 1980). Boys advanced in maturity status maintain their strength advantage.

Early-maturing girls are also stronger than their late-maturing peers during adolescence (Jones, 1949; Beunen *et al.*, 1976; Carron *et al.*, 1977). The differences are most apparent in girls aged 11–15 years and are reduced somewhat by 16 and 17 years of age. The differences between contrasting maturity groups of girls, however, are not as great as those between contrasting maturity groups of boys.

5. Overview

The growth of muscle tissue, muscle mass, and muscular strength has been described, and the limitations of the available data indicated. The effects of regular physical activity on muscle tissue have also been noted. However, other factors that regulate and influence the growth of muscle tissue have not been considered, such as the role of nerve supply to muscle, hormones, and nutrients. Detailed review of these factors is beyond the scope of this chapter.

ACKNOWLEDGMENTS. The constructive comments of Claude Bouchard, Marie-Christine Thibault, and Jean-Aime Simoneau of Laval University and of Eddie Coyle and John Ivy of the University of Texas at Austin are greatly appreciated.

6. References

Aherne, W., Ayyar, D. R., Clarke, P. A., and Walton, J. N., 1971, Muscle fibre size in normal infants, children and adolescents: An autopsy study, *J. Neurol. Sci.* **14**:171.

Andersen, P., and Henriksson, J., 1977, Capillary supply of the quadriceps femoris muscle of man: Adaptive response to exercise, *J. Physiol. (London)* **270**:677.

Asmussen, E., 1962, Muscular performance, in: *Muscle as a Tissue* (K. Rodahl and S. M. Horvath, eds.), pp. 161–175, McGraw-Hill, New York.

Asmussen, E., and Heebøll-Nielsen, Kr., 1955, A dimensional analysis of physical performance and growth in boys, *J. Appl. Physiol.* **7**:593.

August, G. P., Grumbach, M. M., and Kaplan, S. L., 1972, Hormonal changes in puberty. III. Correlations of plasma testosterone, LH, FSH, testicular size and bone age with male pubertal development, *J. Clin. Endocrinol. Metab.* **34**:319.

Bailey, D. A., Bell, R. D., and Howarth, R. E., 1973, The effect of exercise on DNA and protein synthesis in skeletal muscle of growing rats, *Growth* **37**:323.

Baker, P. T., Hunt, E. E., Jr., and Sen, T., 1958, The growth and interrelations of skinfolds and brachial tissues in man, *Am. J. Phys. Anthropol.* **16**:39.

Bell, R. D., Macdougall, J. D., Billeter, R., and Howald, H., 1980, Muscle fiber types and morphometric analysis of skeletal muscle in six-year-old children, *Med. Sci. Sports Ex.* **12**:28.

Beunen, G., Ostyn, M., Renson, R., Simons, J., and Van Gerven, D., 1976, Skeletal maturation and physical fitness of girls aged 12 through 16, *Hermes (Leuven)* **10**:445.

Beunen, G., Ostyn, M., Simons, J., Van Gerven, D., Swalus, P., and De Beul, G., 1978, A correlational analysis on skeletal maturity, anthropometric measures and motor fitness of boys 12 through 16, in: *Biomechanics of Sports and Kinanthropometry* (F. Landry and W. A. R. Orban, eds.), pp. 343–349, Symposia Specialists, Miami, Florida.

Beunen, G., Ostyn, M., Simons, J., Renson, R., and Van Gerven, D., 1980, Motorische vaardigheid somatische ontwikkeling en biologische maturiteit, *Genees. Sport* **13**:36.

Beunen, G., Malina, R. M., Van't Hof, M. A., Simons, J., Ostyn, M., Renson, R., and Van Gerven, D., Physical growth and motor performance of Belgian boys followed longitudinally between 12 and 19 years of age (submitted for publication).

Billeter, R., Heizmann, C. W., Howald, H., and Jenny, E., 1981, Analysis of myosin light and heavy chain types in single human skeletal muscle fibers, *Eur. J. Biochem.* **116**:389.

Boileau, R. A., Horstman, D. H., Buskirk, E. R., and Mendez, J., 1972, The usefulness of urinary creatinine excretion in estimating body composition, *Med. Sci. Sports* **4**:85.

Bouchard, C., 1966, Les différences individuelles en force musculaire statique, *Mouvement* **1**:49.

Bowden, D. H., and Goyer, R. A., 1960, The size of muscle fibers in infants and children, *Arch. Pathol.* **69**:188.

Brooke, M. H., and Engel, W. K., 1969, The histographic analysis of human muscle biopsies with regard to fiber types. 4. Children's biopsies, *Neurology (New York)* **19**:591.

Brooke, M. H., and Kaiser, K. K., 1970, Muscle fiber types: How many and what kind?, *Arch. Neurol.* **23**:369.

Buchanan, T. A. S., and Pritchard, J. J., 1970, DNA content of tibialis anterior of male and female white rats measured from birth to 50 weeks, *J. Anat.* **107**:185.

Burkinshaw, L., Hill, G. L., and Morgan, D. B., 1979, Assessment of the distribution of protein in the human body by in-vivo neutron activation analysis, in: *Nuclear Activation Techniques in the Life Sciences 1978*, pp. 787–797, International Atomic Energy Agency, Vienna.

Burleigh, I. G., 1974, On the cellular regulation of growth and development in skeletal muscle, *Biol. Rev.* **49**:267.

Burmeister, W., 1965, Body cell mass as the basis of allometric growth functions, *Ann. Paediatr.* **204**:65.

Campbell, C. J., Bonen, A., Kirby, R. L., and Belcastro, A. N., 1979, Muscle fiber composition and performance capacities of women, *Med. Sci. Sports* **11**:260.

Carron, A. V., and Bailey, D. A., 1974, Strength development in boys from 10 through 16 years, *Monogr. Soc. Res. Child Dev.* **39**(4).

Carron, A. V., Aitken, E. J., and Bailey, D. A., 1977, The relationship of menarche to the growth and development of strength, in: *Frontiers of Activity and Child Health* (H. Lavallée and R. J. Shephard, eds.), pp. 139–143, Editions du Pélican, Quebec.

Cheek, D. B., 1968, *Human Growth,* Lea and Febiger, Philadelphia.

Cheek, D. B., 1975, Growth and body composition, in: *Fetal and Postnatal Cellular Growth: Hormones and Nutrition* (D. B. Cheek, ed.), pp. 389–408, Wiley, New York.

Cheek, D. B., and Hill, D. E., 1970, Muscle and liver cell growth: Role of hormones and nutritional factors, *Fed. Proc.* **29**:1503.

Cheek, D. B., Holt, A. B., Hill, D. E., and Talbert, J. L., 1971, Skeletal muscle cell mass and growth: The concept of the deoxyribonucleic acid unit, *Pediatr. Res.* **5**:312.

Christensen, D. A., and Crampton, E. W., 1965, Effects of exercise and diet on nitrogenous constituents in several tissues of adult rats, *J. Nutr.* **86**:369.

Clark, L. C., Thompson, H. L., Beck, E. I., and Jacobson, W., 1951, Excretion of creatine and creatinine by children, *Am. J. Dis. Child.* **81**:774.

Clarke, H. H., 1971, *Physical and Motor Tests in the Medford Boys' Growth Study,* Prentice-Hall, Englewood Cliffs, New Jersey.

Cohn, S. H., Vartsky, D., Yasumura, S., Sawitsky, A., Zanzi, I., Vaswani, A., and Ellis, K. J., 1980, Compartmental body composition based on total-body nitrogen, potassium, and calcium, *Am. J. Physiol.* **239**:E524.

Cohn, S. H., Vartsky, D., Yasumura, S., Vaswani, A. N., and Ellis, K. J., 1983, Indexes of body cell mass: Nitrogen versus potassium, *Am. J. Physiol.* **244**:E305.

Colling-Saltin, A.-S., 1978*a,* Enzyme histochemistry on skeletal muscle of human foetus, *J. Neurol. Sci.* **39**:169.

Colling-Saltin, A.-S., 1978*b,* Some quantitative biochemical evaluations of developing skeletal muscles in the human foetus, *J. Neurol. Sci.* **39**:187.

Colling-Saltin, A.-S. 1980, Skeletal muscle development

in the human fetus and during childhood, in: *Children and Exercise.* IX (K. Berg and B. O. Eriksson, eds.), pp. 193–207, University Park Press, Baltimore.

Costill, D. L., Daniels, J., Evans, W., Fink, W., Krahenbuhl, G., and Saltin, B., 1976*a,* Skeletal muscle enzymes and fiber composition in male and female track athletes, *J. Appl. Physiol.* **40:**149.

Costill, D. L., Fink, W. J., and Pollock, M. L., 1976*b,* Muscle fiber composition and enzyme activities of elite distance runners, *Med. Sci. Sports* **8:**96.

Dickerson, J. W. T., and Widdowson, E. M., 1960, Chemical changes in skeletal muscle during development, *Biochem. J.* **74:**247.

Durnin, J. V. G. A., 1969, Muscular and adipose tissue and the significance of increase in body weight with age, in: *Physiopathology of Adipose Tissue* (J. Vague, ed.), pp. 387–389, Excerpta Medica, Amsterdam.

Eriksson, B. O., 1972, Physical training, oxygen supply and muscle metabolism in 11–13 year old boys, *Acta Physiol. Scand. (Suppl.)* 384.

Eriksson, B. O., Gollnick, P. D., and Saltin, B., 1973, Muscle metabolism and enzyme activities after training in boys 11–13 years old, *Acta Physiol. Scand.* **87:**485.

Faust, M. S., 1977, Somatic development of adolescent girls, *Monogr. Soc. Res. Child Dev.* **42**(1).

Fischman, D. A., 1972, Development of striated muscle, in: *The Structure and Function of Muscle,* Vol. I: *Structure,* Part 1 (G. H. Bourne, ed.), pp. 75–148, Academic Press, New York.

Flynn, M. A., Clark, J. Reid, J. C., and Chase, G., 1975, A longitudinal study of total body potassium in normal children, *Pediatr. Res.* **9:**834.

Flynn, M. A., Murthy, Y., Clark, J., Comfort, G., Chase, G., and Bentley, A. E. T., 1970, Body composition of Negro and white children, *Arch. Environ. Health* **20:**604.

Forbes, G. B., 1972, Relation of lean body mass to height in children and adolescents, *Pediatr. Res.* **6:**32.

Forbes, G. B., 1974, Stature and lean body mass, *Am. J. Clin. Nutr.* **27:**595.

Forbes, G. B., and Bruining, G. J., 1976, Urinary creatinine excretion and lean body mass, *Am. J. Clin. Nutr.* **29:**1359.

Fournier, M., Ricci, J., Taylor, A. W., Ferguson, R. J., Montpetit, R. R., and Chaitman, B. R., 1982, Skeletal muscle adaptation in adolescent boys: Sprint and endurance training and detraining, *Med. Sci. Sports Ex.* **14:**453.

Frasier, S. D., Gafford, F., and Horton, R., 1969, Plasma androgens in childhood and adolescence, *J. Clin. Endocrinol. Metab.* **29:**1404.

Frisancho, S. R., 1981, New norms of upper limb fat and muscle areas for assessment of nutritional status, *Am. J. Clin. Nutr.* **34:**2540.

Garn, S. M., 1961, Radiographic analysis of body composition, in: *Techniques for Measuring Body Composition* (J. Brožek and A. Henschel, eds.), pp. 36–58, National Academy of Sciences—National Research Council, Washington, D.C.

Garn, S. M., 1963, Human biology and research in body composition, *Ann. N.Y. Acad. Sci.* **110:**429.

Goldspink, G., 1972, Postembryonic growth and differentiation of striated muscle, in: *The Structure and Function of Muscle.* Vol. I: *Structure,* Part 1 (G. H. Bourne, ed.), pp. 179–236, Academic Press, New York.

Gollnick, P. D., Armstrong, R. B., Saubert, C. W. IV, Piehl, K., and Saltin, B., 1972, Enzyme activity and fiber composition in skeletal muscle of untrained and trained men, *J. Appl. Physiol.* **33:**312.

Gollnick, P. D., Armstrong, R. B., Saltin, B., Saubert, C. W. IV, Sembrowich, W. L., and Shepherd, R. E., 1973, Effect of training on enzyme activity and fiber composition of human skeletal muscle, *J. Appl. Physiol.* **34:**107.

Gollnick, P. D., Timson, B. F., Moore, R. L., and Riedy, M., 1981, Muscular enlargement and number of fibers in skeletal muscles of rats, *J. Appl. Physiol.* **50:**936.

Gollnick, P. D., Parsons, D., Riedy, M., and Moore, R. L., 1983, Fiber number and size in overloaded chicken anterior latissimus dorsi muscle, *J. Appl. Physiol.* **54:**1292.

Gonyea, W. J., 1980, Muscle fiber splitting in trained and untrained animals, *Ex. Sport Sci. Rev.* **8:**19.

Goss, R. J., 1966, Hypertrophy versus hyperplasia, *Science* **153:**1615.

Green, H. J., Thomson, J. A., Daub, W. D., Houston, M. E., and Ranney, D. A., 1979, Fiber composition, fiber size and enzyme activities in vastus lateralis of elite athletes involved in high intensity exercise, *Eur. J. Appl. Physiol.* **41:**109.

Gurney, J. M., and Jelliffe, D. B., 1973, Arm anthropometry in nutritional assessments: Nomogram for rapid calculation of muscle circumference and cross-sectional muscle and fat areas, *Am. J. Clin. Nutr.* **26:**912.

Hedberg, G., and Jansson, E., 1976, Skelettmuskelfiberkomposition. Kapacitet och intresse för olika fysiska aktiviteter bland elever i gymnasieskolan. Rapport 54, Pedagogiska Institute, Umeå.

Henriksson, K. G., 1979, Muscle histochemistry and muscle function, *Acta Paediatr. Scand. (Suppl.)* **283:**15.

Hewitt, D., 1958, Sib resemblance in bone, muscle and fat measurements of the human calf, *Ann. Human Genet.* **22:**213.

Heymsfield, S. B., Arteago, C., McManus, C., Smith, J., and Moffitt, S., 1983, Measurement of muscle mass in humans: Validity of the 24-hour urinary creatinine method, *Am. J. Clin. Nutr.* **37:**478.

Heymsfield, S. B., McManus, C., Smith, J., Stevens, V., and Nixon, D. W., 1982*a,* Anthropometric measurement of muscle mass: Revised equations for calculating bone-free arm muscle area, *Am. J. Clin. Nutr.* **36:**680.

Heymsfield, S. B., McManus, C., Stevens, V., and Smith, J., 1982*b,* Muscle mass: Reliable indicator of protein-energy malnutrition severity and outcome, *Am. J. Clin. Nutr.* **35:**1192.

Hubbard, R. W., Smoake, J. A., Matther, W. T., Linduska, J. D., and Bowers, W. S., 1974, The effects of growth and endurance training on the protein and DNA content of rat soleus, plantaris, and gastrocnemius muscles, *Growth* **38:**171.

Hugg, J. E., and Malina, R. M., n.d., Estimation of body composition from arm anthropometry (in preparation).

Hunt, E. E., Jr., and Heald, F. P., 1963, Physique, body composition, and sexual maturation in adolescent boys, *Ann. N.Y. Acad. Sci.* **110:**532.

Ingjer, F., 1979, Capillary supply and mitochondrial content of different skeletal muscle fiber types in untrained and endurance-trained men. A histochemical and ultrastructural study, *Eur. J. Appl. Physiol.* **40:**197.

Ivy, J. L., Withers, R. T., Van Handel, P. J., Elger, D. H., and Costill, D. L., 1980, Muscle respiratory capacity and fiber type as determinants of the lactate threshold, *J. Appl. Physiol.* **48:**523.

Jansson, E., Sjödin, B., and Tesch, P., 1978, Changes in muscle fibre type distribution in man after physical training: A sign of fibre type transformation? *Acta Physiol. Scand.* **104:**235.

Jennekens, F. G. I., Tomlinson, B. E., and Walton, J. N., 1970, The sizes of the two main histochemical fibre types in five limb muscles in man: An autopsy study, *J. Neurol. Sci.* **13:**281.

Johnson, M. A., Polgar, J., Weightman, D., and Appleton, D., 1973, Data on the distribution of fibre types in thirty-six human muscles: An autopsy study, *J. Neurol. Sci.* **18:**111.

Johnston, F. E., 1981, Anthropometry and nutritional status, in: *Assessing Changing Food Comsumption Patterns,* pp. 252–264, National Research Council, National Academy Press, Washington, D.C.

Johnston, F. E., and Malina, R. M., 1966, Age changes in the composition of the upper arm in Philadelphia children, *Human Biol.* **38:**1.

Jones, H. E., 1949, *Motor Performance and Growth,* University of California Press, Berkeley.

Komi, P. V., and Karlsson, J., 1978, Skeletal muscle fibre types, enzyme activities and physical performance in young males and females, *Acta Physiol. Scand.* **103:**210.

Komi, P. V., Viitasalo, J. H. T., Havu, M., Thorstensson, A., Sjödin, B., and Karlsson, J., 1977, Skeletal muscle fibres and muscle enzyme activities in monozygous and dizygous twins of both sexes, *Acta Physiol. Scand.* **100:**385.

Kraus, B. S., 1951, Male somatotypes among Japanese of northern Honshu, *Am. J. Phys. Anthropol.* **9:**347.

Kreisberg, R. A., Bowdoin, B., and Meador, C. K., 1970, Measurement of muscle mass in humans by isotopic dilution of creatinine-^{14}C, *J. Appl. Physiol.* **28:**264.

Krogman, W. M., 1971, *The Manual and Oral Strengths of American White and Negro Children Ages 3–6 Years,* Philadelphia Center for Research in Child Growth (see Malina and Roche, 1983).

Krotkiewski, M., Aniansson, A., Grimby, G., Björntorp, P., and Sjöström, L., 1979, The effect of unilateral isokinetic strength training on local adipose and muscle tissue morphology, thickness and enzymes, *Eur. J. Appl. Physiol.* **42:**271.

Landing, B. H., Dixon, L. G., and Wells, T. R., 1974, Studies on isolated skeletal muscle fibers, *Human Pathol.* **5:**441.

Larsson, L., Sjödin, B., and Karlsson, J., 1978, Histochemical and biochemical changes in human skeletal muscle with age in sedentary males, age 22–65 years, *Acta Physiol. Scand.* **103:**31.

Lortie, G., Simoneau, J.-A., and Bouchard, C., 1985, Muscle fiber type distribution and enzyme activities in biological sibs, dizygotic and monozygotic twins, in: *Human Genetics and Sport* (R. M. Malina and C. Bouchard, eds.) Human Kinetics Publishers, Champaign, Illinois (in press).

Lundberg, A., Eriksson, B. O., and Mellgren, G., 1979*a,* Metabolic substrates, muscle fibre composition and fibre size in late walking and normal children, *Eur. J. Pediatr.* **130:**79.

Lundberg, A., Eriksson, B. O., and Jansson, G., 1979*b,* Muscle abnormalities in coeliac disease: Studies on gross motor development and muscle fibre composition, size and metabolic substrates, *Eur. J. Pediatr.* **130:**93.

Malina, R. M., 1969, Quantification of fat, muscle and bone in man, *Clin. Orthop. Rel. Res.* **65:**9.

Malina, R. M., 1973, Biological substrata, in: *Comparative Studies of Blacks and Whites in the United States* (K. S. Miller and R. M. Dreger, eds.), pp. 53–123, Seminar Press, New York.

Malina, R. M., 1974, Adolescent changes in size, build, composition and performance, *Human Biol.* **46:**117.

Malina, R. M., 1979, The effects of exercise on specific tissues, dimensions, and functions during growth, *Stud. Phys. Anthrop. (Wrocław)* **5:**21.

Malina, R. M., 1980, The measurement of body composition, in: *Human Physical Growth and Maturation: Methodologies and Factors* (F. E. Johnston, A. F. Roche, and C. Susanne, eds.), pp. 35–59, Plenum Press, New York.

Malina, R. M., 1983, Human growth, maturation, and regular physical activity, *Acta Med. Auxol.* **15:**5.

Malina, R. M., and Johnston, F. E., 1967*a,* Relations between bone, muscle and fat widths in the upper arms and calves of boys and girls studied cross-sectionally at ages 6 to 16 years, *Human Biol.* **39:**211.

Malina, R. M., and Johnston, F. E., 1967*b,* Significance of age, sex and maturity differences in upper arm composition, *Res. Q.* **38:**219.

Malina, R. M., and Rarick, G. L., 1973, Growth, physique, and motor performance, in: *Physical Activity: Human Growth and Development* (G. L. Rarick, ed.), pp. 125–153, Academic Press, New York.

Malina, R. M., and Roche, A. F., 1983, *Manual of Physical Status and Performance in Childhood,* Vol. 2: *Physical Performance,* Plenum Press, New York.

Metheny, E., 1941, The present status of strength testing for children of elementary school and preschool age, *Res. Q.* **12:**115.

Montgomery, R. D., 1962, Growth of human striated muscle, *Nature (London)* **195:**194.

Moore, F. D., Olesen, K. H., McMurrey, J. D., Parker, H. V., Ball, M. R., and Boyden, C. M., 1963, *The Body Cell Mass and Its Supporting Environment*, W. B. Saunders, Philadelphia.

Norris, A. H., Lundy, T., and Shock, N. W., 1963, Trends in selected indices of body composition in men between the ages of 30 and 80 years, *Ann. N.Y. Acad. Sci.* **110**:623.

Novak, L. P., 1963, Age and sex differences in body density and creatinine excretion of high school children, *Ann. N.Y. Acad. Sci.* **110**:545.

Novak, L. P., 1973, Total body potassium during the first year of life determined by whole-body counting of ^{40}K, *J. Nucl. Med.* **14**:550.

Nygaard, E., and Nielsen, E., 1978, Skeletal muscle fiber capillarization with extreme endurance training in man, in: *Swimming Medicine IV* (B. Eriksson and B. Furberg, eds.), pp. 282–293, University Park Press, Baltimore.

Pierson, R. N., Jr., Lin, D. H. Y., and Phillips, R. A., 1974, Total-body potassium in health: Effects of age, sex, height, and fat, *Am. J. Physiol.* **226**:206.

Prince, F. P., Hikida, R. S., and Hagerman, F. C., 1976, Human muscle fiber types in power lifters, distance runners and untrained subjects, *Pflügers Arch.* **363**:19.

Prince, F. P., Hikida, R. S., and Hagerman, F. C., 1977, Muscle fiber types in women athletes and non-athletes, *Pflügers Arch.* **371**:161.

Rarick, G. L., and Oyster, N., 1964, Physical maturity, muscular strength, and motor performance of young school-age boys, *Res. Q.* **35**:523.

Reichmann, H., and Pette, D., 1982, A comparative microphotometric study of succinate dehydrogenase activity levels in Type I, IIa and IIb fibres of mammalian and human muscles, *Histochemistry* **74**:27.

Reynolds, E. L., 1946, Sexual maturation and the growth of fat, muscle and bone in girls, *Child Dev.* **17**:121.

Reynolds, E. L., and Clark, L. C., 1947, Creatinine excretion, growth progress and body structure in normal children, *Child Dev.* **18**:155.

Saltin, B., and Gollnick, P. D., 1983, Skeletal muscle adaptability: Significance for metabolism and performance, in: *Handbook of Physiology, Section 10, Skeletal muscle* (L. D. Peachey, sect. ed.), pp. 555–631, American Physiological Society, Bethesda, Maryland.

Saltin, B., Henriksson, J., Nygaard, E., and Andersen, P., 1977, Fiber types and metabolic potentials of skeletal muscles in sedentary man and endurance runners, *Ann. N.Y. Acad. Sci.* **301**:3.

Scammon, R. E., 1923, A summary of the anatomy of the infant and child, in: *Pediatrics* (I. A. Abt, ed.), pp. 257–444, W. B. Saunders, Philadelphia.

Stolz, H. R., and Stolz, L. M., 1951, *Somatic Development of Adolescent Boys*, Macmillan, New York.

Talbot, N. B., 1938, Measurement of obesity by the creatinine coefficient, *Am. J. Dis. Child.* **55**:42.

Tanner, J. M., 1962, *Growth at Adolescence*, 2nd ed., Blackwell Scientific Publications, Oxford.

Tanner, J. M., 1964, *The Physique of the Olympic Athlete*, Allen & Unwin, London.

Tanner, J. M., 1965, Radiographic studies of body composition in children and adults, *Symp. Soc. Study Human Biol.* **7**:211.

Tanner, J. M., Hughes, P. C. R., and Whitehouse, R. H., 1981, Radiographically determined widths of bone, muscle and fat in the upper arm and calf from 3–18 years, *Ann. Human Biol.* **8**:495.

Tesch, P., Karlsson, J., and Sjödin, B., 1982, Muscle fiber type distribution in trained and untrained muscles of athletes, in: *Exercise and Sport Biology* (P. Komi, ed.), pp. 79–83, Human Kinetics Publishers, Champaign, Illinois.

Thorstensson, A., 1976, Muscle strength, fibre types and enzyme activities in man, *Acta Physiol. Scand. (Suppl.)* 443.

Thorstensson, A., Larsson, L., Tesch, P., and Karlsson, J., 1977, Muscle strength and fiber composition in athletes and sedentary men, *Med. Sci. Sports* **9**:26.

Viitasalo, J. T., and Komi, P. V., 1978, Force–time characteristics and fiber composition in human leg extensor muscles, *Eur. J. Appl Physiol.* **40**:7.

Widdowson, E. M., 1969, Changes in the extracellular compartment of muscle and skin during normal and retarded development, *Bibl. Nutr. Dieta* **13**:60.

Widdowson, E. M., 1970, Harmony of growth, *Lancet* **1**:901.

Wohlfart, G., 1937, Über das Vorkommen verschiedener Arten von Muskelfasern in der Skelettmuskulatur des Menschen und einiger Säugetiere, *Acta Psych. Neurol. Scand. (Suppl.)* 12.

Young, C. M., Blondin, J., Tensuan, R., and Fryer, J. H., 1963, Body composition studies of "older" women, thirty to seventy years of age, *Ann. N.Y. Acad. Sci.* **110**:589.

Young, C. M., Bogan, A. D., Roe, D. A., and Lutwak, L., 1968, Body composition of pre-adolescent and adolescent girls. IV. Total body water and creatinine excretion, *J. Am. Diet. Assoc.* **53**:579.

Zorab, P. A., 1969, Normal creatinine and hydroxyproline excretion in young persons, *Lancet* **2**:1164.

5

Body Composition and Energy Needs during Growth

MALCOLM A. HOLLIDAY

1. Introduction

Energy requirement throughout growth relates to body composition in several ways. The components of energy requirement are basal metabolic rate (BMR), energy spent in physical activity and in response to cold, and energy used in growth. The most characteristic feature of this energy requirement is a decline in requirement per kilogram body weight from more than 100 to less than 45 kcal/kg per day (Figure 1). Much of this decline is caused by a decrease in BMR per kilogram. The decline in BMR per kilogram is secondary to a differential in the rate of growth of organs that have a high resting metabolic rate—brain, liver, heart, and kidney relative to the rate of growth of muscle mass, which has a low resting metabolic rate (Holliday, 1971).

In western society, physical activity throughout childhood (i.e., after 1 year of age) is also high relative to body weight and declines as we mature owing to the adoption of a sedentary life-style over the last several generations. The NRC/FNB (1980) recommendations reflect this. WHO/FAO recommendations (1973) are for a higher-calorie intake.

The high energy requirement per kilogram in infancy is attributable to a high requirement for energy to sustain growth. During the early months of life as much as 40% of total energy requirement is used for this purpose. By 1 year of age—the end of infancy—this has decreased to less than 5% and does not exceed 3% after 2 years of age even through the adolescent growth spurt.

During the past few years, information has developed in two areas: the energy requirements of low-birth-weight (LBW) infants and the nature of energy metabolism in individuals who are obese.

This chapter describes the relationship of energy requirement to body composition in adults and in children over the normal span of growth; it reviews the information on energy requirement of LBW infants as well as energy balance in malnutrition and obesity.

2. Body Composition and Energy Metabolism in Adults

Energy requirement during growth is characteristically expressed in relationship to weight. The decline in BMR per kilogram that is observed could represent a decline in metabolic activity of tissues or a decline in the relative mass of those systems that have a high requirement for energy. The latter appears to be the case. The analysis of the changes in body composition that occur with normal growth is important in supporting this conclusion (Holliday, 1971). This analysis begins with a description of body composition in the adult and the relative metabolism of the different organs that have been studied in adults. Basal metabolic rate is metabolism of cells when the individual is at rest. Cell mass for the purpose of analyzing BMR has two components—brain, liver, heart, and kidney on the one hand and muscle on the other (Widdow-

MALCOLM A. HOLLIDAY • Department of Pediatrics, University of California, San Francisco Medical Center, San Francisco, California 94143.

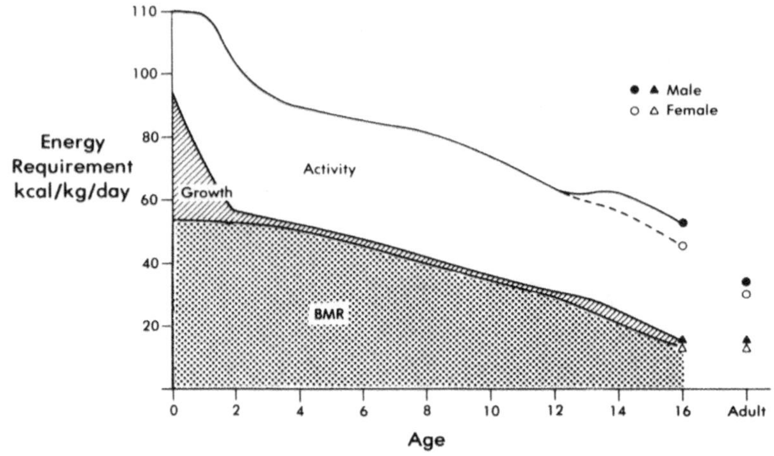

Fig. 1. Change in energy requirement per kilogram body weight during growth. Note the contribution of each component (BMR, growth, and activity).

son and Dickerson, 1964). Defining components of body weight that have little or no metabolic activity, e.g., adipose tissue and the extracellular phase, is also important because body weight is the reference measure for metabolic activity and body composition. Lean body mass (LBM) is body weight minus body fat (Figure 2).

2.1. Brain, Liver, Heart, and Kidney

These organs are considered together because they have high metabolic rates of the same order of magnitude. The composite is described as organ metabolic rate (OMR), which accounts for most of the BMR in the adult (Brozek and Grande, 1955) (Table I). Measurements of the mass of each of these organs in living subjects is currently impractical. Fortunately, individual differences when corrected for body şize and gender become relatively small, and average values for the sum of organ weights (ΣOW) are practical. These have been used in deriving the data in Table I. The data on liver metabolism are more accurately described as splanchnic metabolism and include the gastrointes-

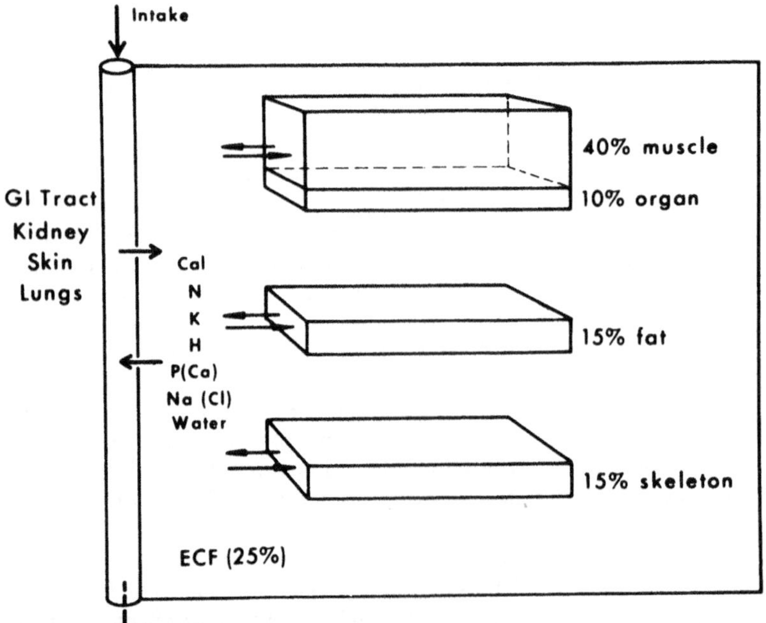

Fig. 2. Model of body composition of an adult male, in which size of specific components is indicated. Major nutrients exchanged between environment and body system through extracellular fluid (ECF) are indicated. (From Holliday 1976.)

Table I. Organ Metabolic Rate and Organ Size in Adults: Relationship to Basal Metabolic Rate

	Weight (kg)	VO_2 (ml/100 g per m)	OMR (hcal/da)	OMR[a]/BMR[b] (%)
Brain	1.40	4.2	414	23.3 (18.3)[c]
Liver	1.60	4.1	464	26.1 (26.4)
Heart	0.30	8.2	182	10.2 (9.2)
Kidney	0.30	5.5	116	7.1 (7.2)
Total 1–4	3.6	—	1177	66.7 —
Skeletal muscle	28.3	0.26	500	28.1 (25.6)
Total 1–5	31.9		1677	94.2 (86.7)

[a]Adapted from Holliday (1971).
[b]BMR for 70-kg man from Talbot (1938).
[c]Values reported by Brozek and Grande (1955).

tinal tract as well. To this extent, the metabolic rate of the liver is overestimated.

Of the four organs, brain, except with prolonged fasting, has a specific requirement for glucose as substrate for its energy requirement. The relationship of brain mass to total cell mass or body weight determines the minimum requirement for glucose, which, if given, will avoid the need for lipolysis and for gluconeogenesis using amino acids and glycerol as substrate (Cahill *et al.*, 1966; Owen *et al.*, 1967).

2.2. Muscle Mass

Muscle mass is considered separately from brain, liver, heart, and kidney because resting muscle metabolic rate (MMR) is approximately one-twentieth that of OMR. While it is true that BMR correlates closely with LBM mass and, for practical purposes, muscle mass, the nature of the relationship is not proportional and does not distinguish the different roles played by other organs relative to muscle mass. When an adjustment is made, the relationships are clearer. Among adults, the size of muscle is related to the size of the other organs, particularly in considering differences between men and women or between large and small persons of either gender.

Muscle mass can be estimated by any of several different methods, although the measurement is not precise. It may be estimated by measuring total cell mass and subtracting a constant for the combined weight of brain, liver, heart, and kidney or the sum of organ weight (ΣOW). Because ΣOW is small, variations in cell mass are due almost entirely to variations in muscle mass (Cheek, 1968).

Cell mass is estimated indirectly.* One method derives cell mass from intracellular fluid (ICF), which is the difference between total body water (TBW) and extracellular fluid (ECF) volume (Cheek, 1968), that is:

$$\text{Cell mass} = \frac{\text{TBW} - \text{ECF}}{0.75}$$

TBW is equal to the volume of distribution of deuterium or tritium oxide, or of antipyrene or urea. It is measured with an accuracy of 2% and a precision of $\pm 4\%$. ECF volume is not so accurately measured, however, (see Section 2.4). Hence, the estimate of ICF is relatively imprecise.

Cell mass also has been estimated from the measurement of total body potassium (TBK) (Flynn *et al.*, 1972) or from exchangeable ^{42}K (Corsa *et al.*, 1950). TBK can be measured with a precision of $\pm 5\%$ using liquid scintillation whole-body counting, but the method is not generally available and is tedious for the patient. Measuring exchangeable ^{42}K entails technical problems owing to the short half-life of ^{42}K and because radiation exposure is too high for use in children. Another problem that arises in using TBK as an index of cell mass is that the calculation of muscle mass from TBK assumes K concentration in cells to be constant (Patrick, 1977). While this is a safe assumption for healthy subjects, this is not the case in malnutrition (Metcoff *et al.*, 1960).

*Cell mass differs from lean body mass (LBM). Lean body mass is normal fat-free body weight plus a standard minimum of body fat. Because variation in body fat has little effect on cell mass, ECF volume, or supporting structures, LBM is useful as a reference standard in describing variations in body fat among normal individuals (see Section 2.4).

Muscle mass has been estimated from the quantity of creatinine excreted during a 24-hr period—the creatinine index. This method is simple and noninvasive when accurately timed, and complete urine samples can be collected. Several studies (Cheek, 1968; Forbes and Bruining, 1976) have concluded from available evidence that the creatinine index is a satisfactory predictor of muscle mass in most children and adults; patients with chronic renal failure may be an exception (Jones and Burnett, 1974).

Muscle mass estimated from measuring arm circumference (Jelliffe and Jelliffe, 1979) is useful for detecting gross deficiencies related to malnutrition or other conditions associated with severe muscle wasting.

The measurement of muscle protein and protein:DNA (Cheek, 1968), protein:K, and protein:P ratios (Delaporte *et al.,* 1976) also are useful in detecting major deviations of muscle composition from normal—principally those associated with muscle wasting.

2.3. Adipose Tissue Mass and Body Fat

Adipose tissue in adults contributes little to total metabolism. It is the energy bank account or stored energy of the individual and as such can vary from small to large.

The average male has 12% of his body weight as fat or more than 1000 kcal/kg of reserve energy. The average female has 25% or more than 2000 kcal/kg of reserve energy. Adipose tissue is made up of adipocytes and their supporting structures (Hirsch and Knittle, 1970) (see also Chapter 3, this volume) and is mostly located in skin or subcutaneous tissues. Most body fat* is in adipose tissue. In normal persons 65–70% of adipose tissue is fat. Adipose tissue either stores or releases fat as fatty acids and glycerol depending on whether dietary energy exceeds expenditure or the converse.

Individuals may have less than normal body fat—as low as 5%—and be healthy as long as they are not faced with sustained starvation or stress, e.g., marathon runners. Those with severe calorie deficiency have virtually no body fat (Garrow *et al.,* 1965). Calorie malnutrition is associated with reduced BMR, body fat, and muscle mass (Keys *et al.,* 1950). Excess body fat or overnutrition may range from slight overweight to gross obesity. Because obesity correlates significantly with heart disease, hypertension, and diabetes, its cause has become a major medical concern.

Malnutrition and overnutrition are characterized by changes in body fat, which are reflected in changes in body weight. The commonly used method for estimating nutritional status relative to body fat is the weight for height index, in which the weight of an individual is expressed as a percentage of the average weight of individuals having the same height. Undernutrition is suspected in cases in which the index is less than 80% (Waterlow and Alleyne, 1971). Among a population of severely undernourished children, the range was 50–90% and the average was 70% (Kerr *et al.,* 1973). Persons are classified as overweight when their index is 110–120% and as obese when it exceeds 120% (Davidson and Passmore, 1969). While a good correlation exists between body fat and weight/height index, there is obvious potential for errors, e.g., changes in muscle mass or ECF volume also affects weight and therefore height index independent of body fat. Variations in ECF volume are common in undernutrition, with dehydration occurring with acute gastrointeritis and edema a common feature of chronic calorie undernutrition, both clinical (Waterlow and Alleyne, 1971) and experimental (Keys *et al.,* 1950). Obese individuals may not simply be fat, but may be big as well; obese young people are taller and have a larger muscle mass (Forbes, 1964).

Direct measurements of body fat are difficult to make in humans (Keys and Brozek, 1953). Body fat is derived from measurement of body water; the derivation is based on the assumption that LBM or fat-free body mass has a water content of 72%. While this assumption is appropriate in normal and obese people, it is very uncertain in undernourished or edematous persons.

Body fat also can be derived from measurements of specific gravity or relative density obtained by weighing persons under water (Durnin and Taylor, 1960; Buskirk, 1961). This derivation depends on the fact that fat-free body mass has a fixed and high specific gravity and adipose tissue has a fixed and low specific gravity. The limitations of this method are similar to those deriving body fat from measurements of TBW. Persons in various health maintenance clubs have body fat measured by densitometry. This is the most practical means for following loss of adipose tissue when that is the goal.

Body fat in clinical practice and in the field is estimated from measurement of skin-fold thickness when the latter is measured in a systematic and

*Adipose tissue in normal individuals is 65% fat, 30% water, and 5% protein.

consistent manner (Tanner and Whitehouse, 1962). This is the preferred method in most instances. Body fat estimated by skinfold thickness agrees well with simultaneous estimates by densitometry in adults (Durnin and Rahaman, 1967) or body water in children (Brook, 1971), except in obese patients.

2.4. Extracellular Phase

Extracellular phase has fluid and structural components. The structural phase includes collagen, elastic tissue, and the organic and mineral phase of bone (see footnote in Section 2.2). ECF volume is a major component of body weight. It consists of plasma, interstitial fluid, connective tissue fluid, and fluids such as gastrointestinal and cerebrospinal fluid or so-called transcellular fluids (Edelman and Leibman, 1959). Extracellular fluid is the environment for cells (Gamble, 1947) and also is the interface between cells and the outside environment. Oxygen and nutrients from the environment reach cells through plasma and interstitial fluid, and products of cell metabolism leave cells via ECF.

However, ECF is neither an energy-using nor -generating system; ECF volume is important in relating energy requirement to body weight because variations in its size affect the use of body weight as a reference standard for energy requirement.

Changes in either concentration or volume of plasma and interstitial fluid are reflected in like changes in the other major component, connective tissue fluid. Interstitial fluid normally exists as a matrix or gel through which solutes readily pass. With interstitial edema, the gel becomes more fluid.

Measurement of ECF is imprecise. Its estimation is derived from the volume of distribution of substances distributed between plasma and interstitial fluid but excluded from cells. One set of compounds, the saccharides—insulin, mannitol, and sucrose—penetrate connective tissue fluids slowly and underestimate ECF volume (Nichols *et al.*, 1953). They are also rapidly cleared, so that conditions for measuring volume of distribution are poor. Isotopes of sodium, chloride, or bromide or stable bromide readily penetrate connective tissue fluid but are also distributed to a degree in cell fluid, leading to an overestimate of ECF volume. In an extensive review, Cheek (1961) compared the various volumes of distributions measured by different compounds in relationship to the various phases of ECF (Figure 3). He concluded that a corrected bromide space was the preferred method for estimating ECF in humans. Barratt and Walser

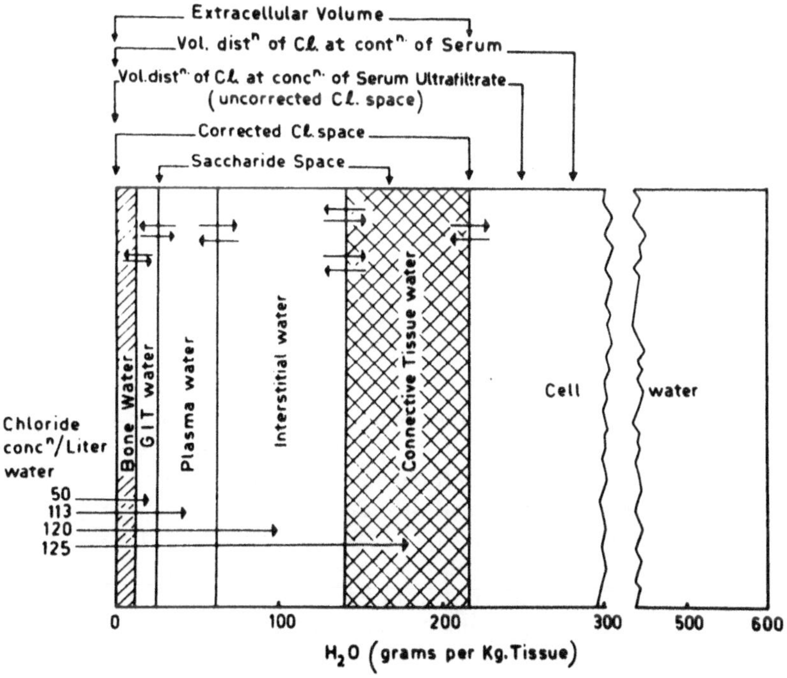

Fig. 3. Diagram of the anatomy of ECF and how various compounds used in measuring ECF volume are distributed. (From Cheek, 1961.)

(1968, 1969) subsequently compared sulfate spaces both *in vivo* and in tissue with bromide spaces and concluded that sulfate rapidly penetrates connective tissue and did not enter cells. Kaufman and Wilson (1973) have described a method for measuring bromide using fluorescent excitation analysis that makes measurement of bromide space using stable bromide safe, rapid and precise. Friis-Hansen (1957), who has reported ECF volumes of children, used thiosulfate, which appears to measure a space similar to sulfate space.

When the technique for fluorescent excitation analysis is available measurement of bromide space is most suited for estimating ECF volume in most patients, patients with uremia are an exception (Cotton *et al.*, 1979). Sulfate space both *in vivo* and *in vitro* analyses is probably the most accurate measure of ECF volume.

2.5. Gender Differences

Small differences in body composition exist between boys and girls at birth. Average length, weight, and muscle mass are greater in boys. Girls have more body fat. These differences are small and probably reflect the small influences exerted by male and female sex hormones operating in the prenatal period. The greater differences appear with sexual maturation. Puberty begins earlier for girls and, in a compositional sense, leads to a gain in body fat. For boys, puberty begins some 2 years later; body fat decreases and muscle mass and stature increase greatly.

No differences in energy metabolism are found between males and females when data are adjusted to LBM (James, 1983). Current results from calorimetry studies show good reproducibility of results and a relatively narrow coefficient of variations among normals. The fact that men have a larger lean body mass (both ΣOW and muscle mass) explains the larger BMR of men.

2.6. Summary

The size of the brain and other major organs is the major factor determining BMR. Except in unusual circumstances, the size of these organs can be estimated from normal data, and their demand for energy (OMR) is a function of their mass. Muscle mass is a large and variable cell mass among normal individuals. It has a low resting or BMR. Muscle mass is very sensitive to nutritional deficiency, i.e., it is the protein reserve of the body, and its size reflects the state of the reserve. Although it would be a useful means for evaluating nutritional status, measurement of cell mass, or muscle mass, is impractical in clinical practice. Fat mass is the retention of excess energy input over output. A normal reserve of fat is 12% of body weight in men and 25% in women. Body fat content that is less than 10% of body weight is suggestive of calorie deficiency, and body weight more than 120% of normal for age is regarded as obesity. Because fat is the major variable in body composition in normal individuals, variations in weight to height ratio are most often due to changes in fat. However, changes in muscle mass and ECF volume affect the accuracy of deriving fat from the weight for height index. Skinfold thickness on the other hand is a good method for estimating body fat. Extracellular fluid volume in health is a derivative of plasma, interstitial fluid, lymph, and connective tissue. Normal variations in body size affect ECF volume proportionally. Connective tissue, including bone, is relatively consistently 15% of lean body mass among healthy individuals. Interstitial fluid and plasma also are functions of lean body weight. Extracellular fluid volume may be decreased (dehydration) or increased (edema) with malnutrition; under these conditions, its estimation can increase the reliability of interpreting body weight.

3. The Changing Pattern of Body Composition as Growth Proceeds

Growth from birth at term to maturity encompasses an increase in weight from 3.5 to 50–70 kg. If LBW infants are included, the lower range is 1 kg (see below). Weight is the common reference base for describing changes in body composition during growth. This is usually done by expressing the weight or volume of a particular component as a percentage of body weight (Table II). This method is descriptive of body composition at the point of measurement, but it does not convey the dynamics of change with growth. A plot of data comparing weights of any component to total body weight over time expresses the relationship between the two more accurately than using percentage body weights (Miller and Weil, 1963). Growth rates of a particular component relative to body weight are derived from an allometric plot, i.e., log–log plot, of component weight to body weight over time (Kleiber, 1947).

Table II. Body Composition at Different Stages of Growth

Age (years)	Height (cm)	Body weight (kg)	% body weight: Organ weight[a]	Muscle mass[b]	Body fat[c]	ECF volume[d]
Low birth weight		1.1	21	<10	3	50
Newborn	50	3.5	18	20	12	40
0.25	60	5.5	15	22	11	32
1.5	80	11.0	14	23	20	26
5	110	19.0	10	35	15	24
10	140	31.0	8.4	37	15	25
Male aged 14	160	50	5.7	42	12	21
Male adult	180	70	5.2	40	11	19
Female aged 13	160	45	4.8	39	18	19
Female adult	162	55	4.4	35	20	16

[a]Sum of brain, liver, heart, and kidney (Holliday, 1971).
[b]Derived from creatinine excretion (Holliday, 1971).
[c]Interpolated values from literature (Widdowson and Dickerson, 1964; Friis-Hansen, 1971).
[e]Average of bromide space and thiosulfate space (Cheek, 1961; Friis-Hansen, 1957).

3.1. Brain, Liver, Heart, and Kidney

Brain, the principal user of glucose under normal conditions (Kennedy and Sokoloff, 1957), is a large fraction of an infant's weight, i.e., 17% of body weight and 75% of the weight of the four major organs. It grows rapidly during the first year of life, after which its growth rate decreases. Growth is nearly complete by the fifth year, when body weight is less than one-third of its mature size (Table III). The weight of brain, liver, heart, and kidney combined (ΣOW) increases, just as body weight increases in the first year. From 3 to 18 months of age, the ΣOW as a percentage of body weight is relatively stable, around 15%. Thereafter its growth rate is slower and its percentage body weight decreases to 5.1% at maturity (Table I) (Holliday, 1971).

3.2. Muscle Mass and Cell Mass

By contrast, muscle mass increases from a value of 22% at 3 months of age to 35% at 5 years of age and in males to more than 40% at maturity

Table III. Brain Size and Energy Requirement Relative to Body Weight and BMR at Different Stages of Growth

Body weight (kg)	Brain weight (g)	Brain wt/body wt (%)	Brain MR[a] (kcal/day)	BMR[b] (kcal/day)	BrMR/BMR (%)	Glucose needed[c] (g/day)	Glucose needed[c] (g/kg per day)
1.1	190	17	—	41			
3.5	475	14	(140)	161	(87)	35	10
5.5	650	12	(192)	300	(64)	48	8.5
11	1045	10	(311)	590	(53)	78	7
19 (5 years)	1235	6.5	367	830	44	92	5
31	1350	4.4	400	1160	34	100	3
50	1360	2.7	403	1480	27	101	2
70	1400	2.0	414	1800	23	103	1.5

[a]Derived from data of Kennedy and Sokoloff (1957), who presented data in children from 5 years of age and up. BrMR per 100 g brain weight was constant. BrMR for children under 5 years is assumed to be proportional to brain weight.
[b]Derived from data presented in Holliday (1971).
[c]The amount of glucose needed to support brain metabolic rate (BrMR) is calculated by dividing BrMR by 4 kcal/g.

(Table I)(Cheek, 1968). Females show a similar increase until puberty when the percentage tends to decrease.

The sum of organ mass and muscle mass—cell mass—averages 37% during infancy. Cell mass averages 45% of body weight from 5 years of age on. Cell mass as a fraction of body weight decreases in women after puberty.

3.3. Adipose Tissue Mass and Body Fat

Fat content of LBW infants is small—less than 3%. Normally during the last trimester of gestation, fat is added, so that infants born at term have 12% of body weight as fat. Fat continues to increase during the first year (Foman, 1966). During the second year of life, as infants turn their attention from eating and growing to walking and playing, the percentage of fat decreases. Thereafter, the child's fat content remains a stable fraction of body weight or, in some instances, increases. Often fat content of boys decreases during the pubertal growth spurt. The converse is true in the case of girls (Frisch *et al.,* 1973). The hypothesis relating the increases in body fat associated with puberty in girls to the endocrine changes that affect sexual maturation has been challenged (Kulin *et al.,* 1982).

There is much greater variation in body fat at any age or stage of growth than in other components, reflecting variations in calorie balance. When appetite is trimmed to activity, body fat will remain a relatively constant fraction of body weight. The controls regulating body fat are those that sense activity and set appetite.

3.4. Extracellular Phase

Extracellular fluid constitutes a large fraction of body weight at birth and an even larger fraction in infants of low birth weight. In the first postnatal weeks there is a contraction of ECF volume (Cheek, 1961). During infancy, ECF volume does not increase as rapidly as body weight, so that it decreases as a percentage of body weight (Friis-Hansen, 1971). This trend continues, but to a lesser degree after 1 year of age. The changes that occur during growth from infancy are influenced by the method of measurement. Cheek (1961) reported values for bromide space in normal infants in the first week of life to be 37% and in normal adults to be 22%. Friis-Hansen (1957), using thiosulfate, reported analogous values of 44% and 19%, respectively. The reason for the greater decline in values where thiosulfate was used is not clear. One possibility is that connective tissue in newborns is more readily penetrated by thiosulfate and becomes less so as growth proceeds (Table II).

3.5. Summary

Body compositional changes during growth reflect differential growth rates of the components. Organs grow proportionally with body weight in the first year, but they grow slower thereafter. Muscle mass, by contrast, grows at a more rapid rate than weight after the first year; at puberty muscle mass increases in males and decreases in females. In women, body fat increases at puberty. Expressing these changes as fractions of body weight at different ages provides a statement of composition, but it does not accurately portray the dynamics of a changing relationship.

Organ mass cannot be measured during growth, but its mass can be predicted from height and gender using normal data. Muscle mass can be estimated from creatinine excretion and body fat from measure of skinfold thickness. Extracellular fluid volume cannot be estimated accurately except by direct measurements of bromide or sulfate space. Changes in ECF volume following hydration (Cheek, 1956) or diuresis can be estimated from changes in body weight.

4. Body Composition and Energy Balance during Growth

Changes in body composition account for the change in BMR per kilogram that occurs with growth. The energy required as intake to balance energy expended is termed maintenance energy requirement. Some have restricted the term to mean a minimum requirement for an individual who is sedentary and who must exert little physical effort beyond feeding (Payne and Waterlow, 1971; FAO/WHO, 1973). Using this definition, maintenance energy requirement is approximately 1.5 × BMR. The more inclusive definition encompasses energy needed to balance all expenditures and varies between 1.6 and 2.1 × BMR.

The energy needed for growth is that which is stored as body fat and protein and the energy cost of processing it, the thermic effect of food, and the energy cost of synthesizing body fat and protein from substrate derived from diet (Garrow, 1974).

The energy cost of growth is appreciable only in the first year of life and especially in LBW infants. However, whereas the energy cost of growth is small, growth rate either in weight or stature is sensitive to food intake. Lowering caloric intake by as little as 10% usually reduces weight gain; lowering it more affects both weight gain and statural growth. Most of the cost of weight gain is the cost of adding muscle protein and adipose tissue fat.

4.1. Substrates Used

The nutrients needed to provide substrate for energy are carbohydrate, fat, and protein. The substrates derived from these nutrients are glucose, pyruvate, fatty acids, glycerol, ketones, and amino acids. Where nutrients are sufficient to provide substrate, energy balance exists. In most cases, substrates can be used interchangeably, and the mix of dietary carbohydrate, fat, and protein as precursors for substrates is not commonly a problem. Two exceptions are important. Glucose is the required substrate for brain metabolism. Glucose can only be derived from carbohydrate, glycerol, lactate, pyruvate, and some but not all amino acids. Amino acids, especially essential amino acids, are required to sustain body protein turnover and to support addition of new protein.

Under normal conditions, the requirements for glucose are set by the energy expenditure of brain (Kennedy and Sokoloff, 1957). A small additional quantity is needed to provide substrate to red cells. For an adult man, 25% of substrate for total BMR (1600 kcal/day) should be glucose, i.e., 400 kcal, or 100 g glucose/day. Where brain metabolism is a larger fraction of BMR, as it is in childhood, the percentage of substrate as glucose should be greater (Holliday, 1971; Cornblath and Schwartz, 1976). Stress incurs an increase in total energy expended, and the preferred substrate is glucose (Cuthbertson, 1964). When carbohydrate needs are not met from the diet, body protein will be catabolized to provide substrate for glucose synthesis. The normal male deprived of food but given 100 g glucose reduces protein losses to one-half (Gamble, 1946). For children the glucose requirement is higher (Table III).

4.2. BMR and the Sum of Organ Weight (ΣOW)

The changes in body composition that accompany growth have important effects on energy requirements. Both BMR and total energy intake increase with growth; during the first year they increase in proportion to weight gain, but thereafter they increase at a slower rate. Consequently, energy metabolism and requirements per kilogram decline after the first year of life (Fig. 1). The major share of basal energy is used by the organs, i.e., brain, liver, heart, and kidney (Table I). These organs are relatively larger at birth (Table II). The pattern of growth of these organs in relationship to change in body weight parallels the pattern of increase in BMR (Figure 4). The parallelism suggests an important relationship between organ mass, as reflected by the sum of brain, liver, heart, and kidneys (ΣOW), and BMR during growth.

This relationship is illustrated in Figure 5, where BMR is plotted against ΣOW. The slope of increase of BMR on organ weight is 555 kcal/kg, and the intercept is −232 (Figure 5, curve A). However, if BMR is corrected for the fraction from muscle metabolism, the balance is a composite of organ metabolism, or OMR. The slope of OMR on ΣOW 381 kcal/kg is close to the weighted average OMR of adult organs, 328 kcal/kg (Table I), and the intercept, −53, is near zero (Figure 5, curve B). This analysis indicates that OMR is proportional to organ weight during growth. The slope of regression of log OMR against log ΣOW is 1.08 ± 0.4, (Figure 6) and defines the relative rate of increase in OMR to ΣOW. Organ metabolic rate increases at the same rate as ΣOW. This relation exists because the OMR of the major organs are similar and the sum of their weights represents a relatively homogeneous mass with respect to energy metabolism. The growth rates of individual organs vary considerably from the growth rate of their sum (Figure 7). In the first year of life, brain metabolic rate accounts for more than one-half of BMR (Figure 8). Brain metabolic rate does not change under many conditions, and among the major organs, brain growth is least affected by rate of somatic growth rate and nutrition. The magnitude of brain metabolic rate determines the carbohydrate needed as substrate (Kennedy and Sokoloff, 1957; Owen *et al.*, 1967); e.g., an 11-kg child may need 78 g glucose (7 g/kg per day) to prevent gluconeogenesis, whereas an adult needs only 100 g (1.5 g/kg per day) (Table III) (Gamble, 1946). This relationship has been confirmed by studies relating glucose turnover rate to brain size during growth. Glucose turnover is proportional with brain size (Bier *et al.*, 1977).

The size of visceral organs is closely related to height (Coppoletta and Wolbach, 1933). The energy requirement of organs can be predicted in healthy persons when height is known.

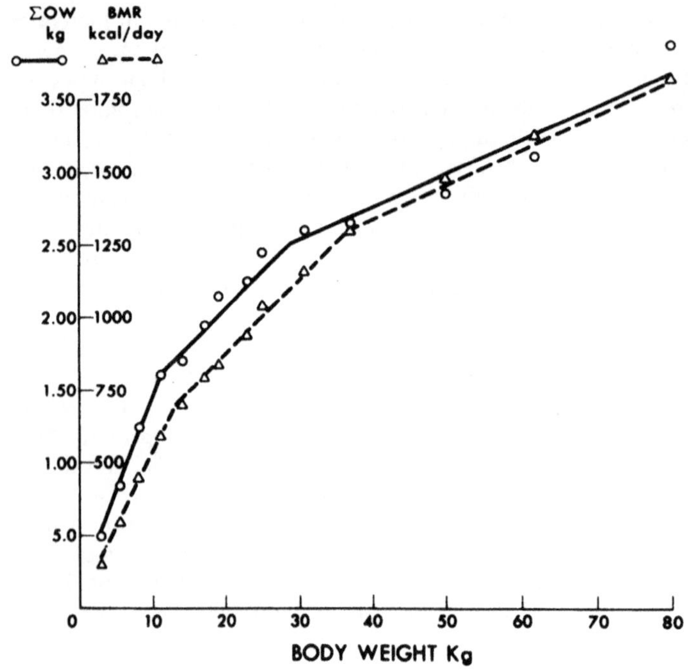

Fig. 4. Parallel relationship between BMR versus body weight and ΣOW versus body weight from infancy to maturity (males). (From Holliday, 1971.)

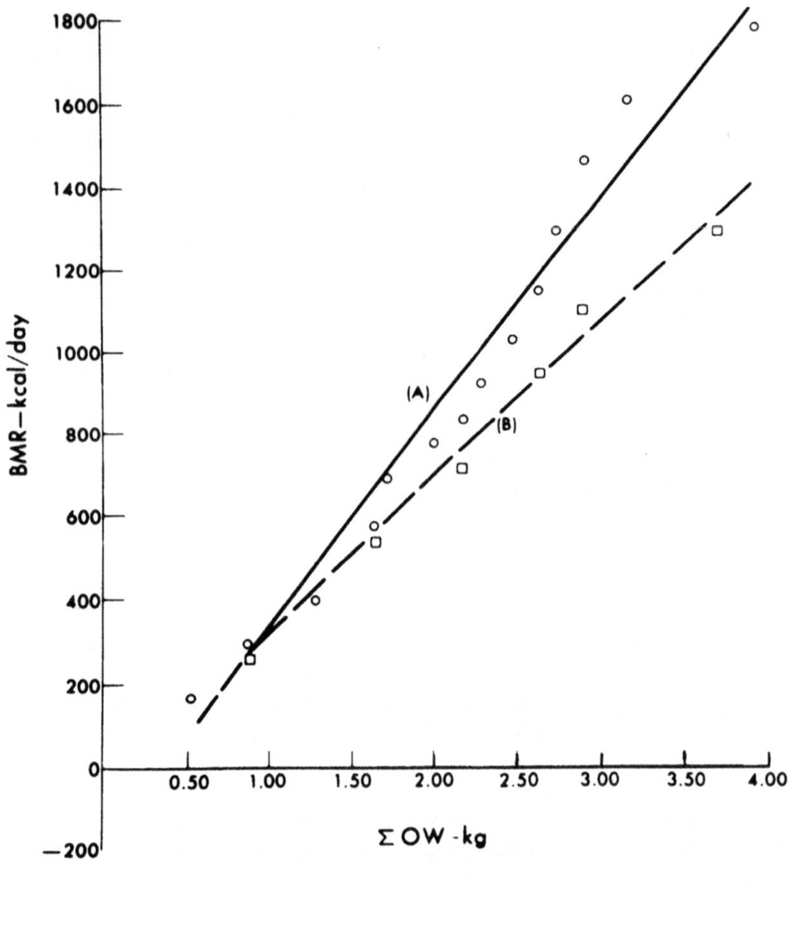

Fig. 5. Relationship between (A) BMR and sum of organ weight (ΣOW) and (B) BMR minus muscle metabolic rate versus ΣOW from infancy to maturity (males). (From Holliday, 1971.)

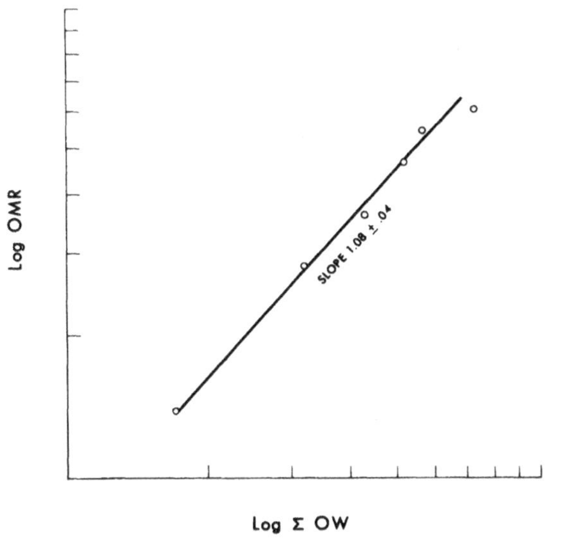

Fig. 6. Log organ metabolic rate (OMR) versus log ΣOW. Slope of 1.08 ± 0.04 is not significantly different from 1.0. Increases in OMR and ΣOW occur at the same rate. (Adapted from Holliday, 1971.)

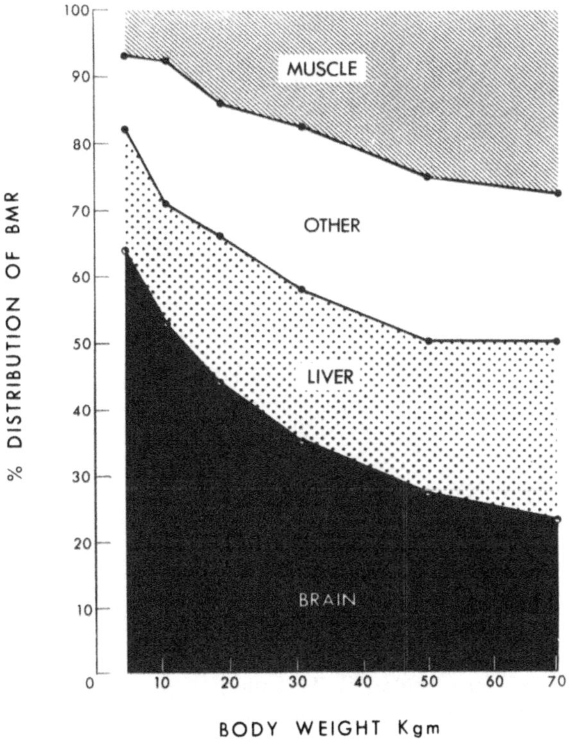

Fig. 8. Distribution of brain, liver, and muscle metabolic rates as percentage of total BMR at different body weights. (From Holliday, 1971.)

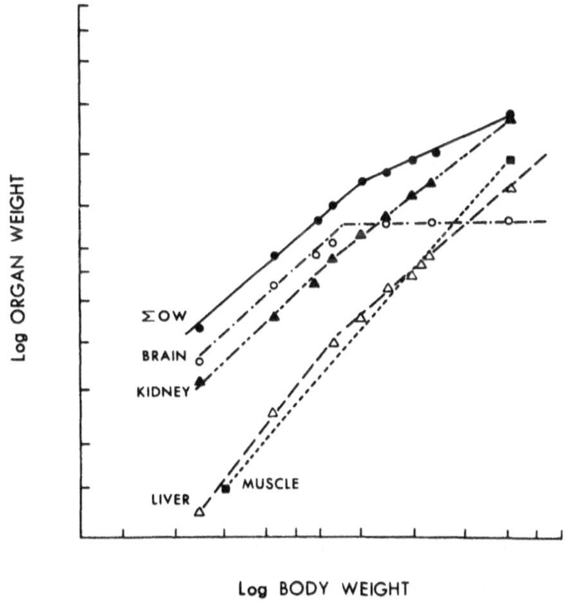

Fig. 7. Log of weight of brain, liver, kidney, muscle and ΣOW at different stages of growth. Log of individual weights adjusted so that all data might fit on a single log scale. The log of ΣOW is also provided.

4.3. Energy Requirement and Body Size

Because BMR per kilogram declines with growth, it is desirable to find a function of body size that is proportional to BMR so that a normal BMR independent of body size can be defined.

Among different species, BMR varies as weight$^{3/4}$ (Kleiber, 1947). In humans over a range of sizes, BMR is proportional to estimates of surface area after 1 year of age. Over the whole span of growth from term infants to adults, there are significant deviations of BMR/1.73 m^2 (Talbot, 1938). An alternative standard has been sought; weight$^{3/4}$ is not suitable (Figure 9).

Both BMR and recommended energy intake exhibit a changing relationship to body weight during growth that is empirical. Recommended energy intake has an easily remembered relationship to body weight that is useful in estimating energy requirements for clinical purposes and particularly for prescribing water requirements (Holliday and Segar,

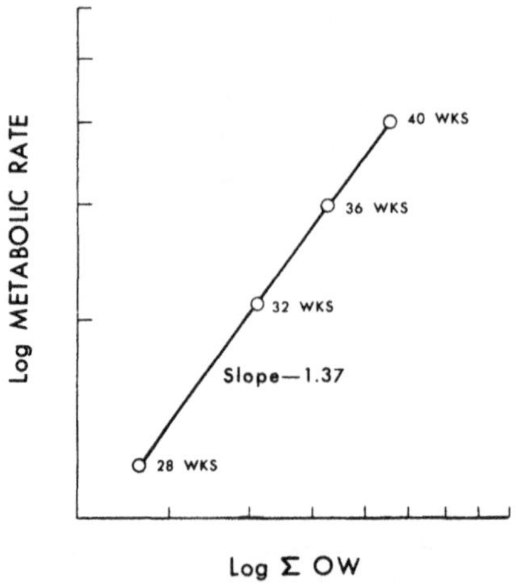

Fig. 9. Comparing observed BMR versus weight of children weighing 3.5–70 kg with theoretical relationship predicted from surface area and from weight$^{3/4}$. The largest discrepencies are noted in the weight range from 10 to 30 kg.

1957) (see Table IV). This method predicts results that fit the observed data with less variance than does surface area. Empirically BMR and energy requirement per kilogram of body weight are constant in the first year of life.

Muscle mass increases relative to weight with growth (Table II). Muscle mass per kilogram body weight is found to be increasing, while BMR per kilogram is decreasing. Basal metabolic rate related to LBM is the same among adults of both genders, but differences are significant during growth. Talbot (1938) used the relationship of BMR to height to define standards for BMR. At any height this method reduced variance among normals, but a table is required to describe the changing relationship of BMR to stature as growth proceeded. Height as a reference standard for BMR provides a means of examining BMR in patients when differences in body composition alter the result, e.g., in malnutrition.

Table IV. Relationship of Average Energy Requirement to Body Weight[a]

Average energy needed		Body wt	
(kg)	(kcal)	(kg)	(kcal)
3–10[b]	300–1000	25	1600
12	1100	35	1800
14	1200	45	2000
16	1300	55	2200
18	1400	70	2500

[a]Derived from Holliday Segar (1957).
[b]100 kcal/kg

4.4. Energy Intake and Renal Function during Growth

There is a proportional relationship between kidney function as measured by glomerular filtration rate (GFR) and both energy requirement and BMR during growth after 1 year of age (Table V). Since kidney size and GFR increase at the same rate during growth, BMR parallels kidney function and weight from 1 year of age on (Holliday *et al.,* 1967). This association may represent a selection process whereby kidney function that parallels energy need after weaning confers a biological advantage. The capacity to survive and grow in primitive societies is dependent on the capacity of the individual to meet energy needs. The foods used in meeting these needs are varied with respect to composition, and the timing of feeding is irregular, so that a capacity for tolerating a varied diet is needed. This tolerance is provided by the kidney. From this perspective, the parallelism among energy requirement, food intake, and renal function (food tolerance) during growth is an advantage. The modern counterpoint for this relationship is that patients who have impaired renal function have diminished tolerance to dietary variation (Holliday, 1976).

The lower GFR in the first year of life is appropriate to the diet characteristic of that period, i.e., human milk. It is low in protein, electrolytes, and mineral content; solute excretion is consequently low. Human milk is also relatively constant in composition, so that controls to adapt to variations are less in demand. The lower level of function reduces the rate of sodium reabsorption and energy expended, reducing energy required for BMR, i.e., the lower level of renal function in infancy is well adapted to the infant's diet and to economizing on energy expenditure.

4.5. Energy Requirement and Growth

The rate of growth in weight gain among normal infants is influenced by the spontaneous level of calorie intake. Whether hungry infants drive growth or more rapidly growing infants need and

Table V. Relationship of Energy Intake and Basal Metabolic Rate to Glomerular Filtration Rate at Different Stages of Growth

Age	Weight (kg)	Total energy intake (kcal/day)[a]	BMR[b]	GFR (ml/min)[c]	GFR (ml/min per 100 kcal)	
					Total intake	BMR
0.25	5.5	600	300	(11)	1.8	3.7
1.5	11	1100	590	29	2.6	4.9
5.5	19	1790	830	60	3.4	7.2
9.5	31	2380	1160	83	3.5	7.1
14.0	50	3020	1480	112	3.7	7.6
Adult	70	3100	1600	130	4.2	8.1

[a]From FAO/WHO (1973).
[b]From Holliday (1971).
[c]From Holliday (1972).

acquire more energy (food) is unclear. When energy density in formulas is changed, infants adjust volume intake to sustain a nearly constant calorie intake. However, according to a study by Foman *et al.* (1971), dilution of calorie density to 50 kcal/dl led to slightly less intake and weight gain. There was no effect on growth in length.

The energy cost of growth is related to the nature of weight gain that occurs with growth. During the rapid period of growth in infancy, weight gain is 40% fat, 10% protein, and 50% water (Foman, 1966). During other periods of growth, fat may be less than 20% of new weight, protein will be more, but water will be much more. The energy cost of adding 1 g of fat is 11 kcal, 9 kcal of which is stored as potential energy. Similarly, the cost of adding 1 g of protein is 9–11 kcal, but only 4 kcal is stored as potential energy. The energy cost of growth in infancy is 5 kcal/g. Where fat content is less than 30% of new weight, the energy cost of weight gain will be less (Spady *et al.*, 1976).

Breast-fed infants after 4 months of age gain weight at a slower rate than do *ad lib* formula-fed infants. The level of energy intake measured in these infants demonstrates that they receive fewer calories. The significance of this finding is in dispute (Waterlow and Thomson, 1979).

4.6. The Low-Birth-Weight Infant

The LBW infant has a low BMR related to body weight, which increases with growth from 1 to 3.5 kg (Sinclair *et al.*, 1967). The low value is a product both of the ΣOW, being smaller at 1 kg, and the level of metabolism in these organs, being lower in immature tissues. The slope relating log ΣOW to body over this period is 1.22 and of OMR to body weight is 1.37 (Holliday, 1971) (Figure 10).

The low BMR or resting metabolism of LBW in-

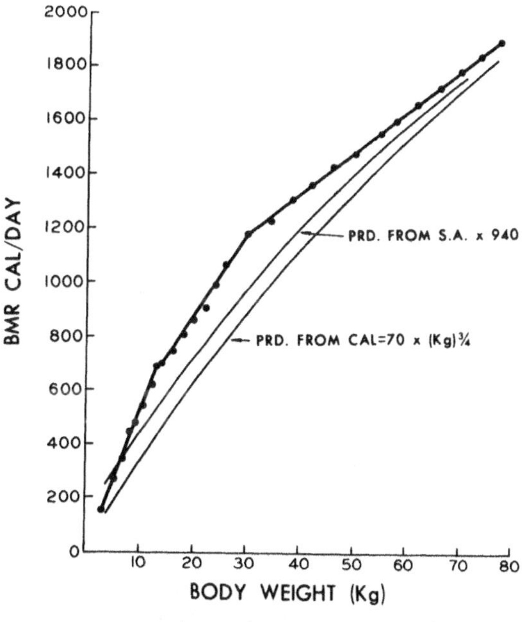

Fig. 10. Log metabolic rate versus ΣOW of low birth weight infants. Metabolic rate derived from VO_2 measured in 92 infants appropriate for gestational age under standard conditions (Sinclair, 1967), who weighed between 730 and 3840 g. Points interpolated at 28–40 weeks gestational age based on data divided into weight groups appropriate to that gestational age. (Adapted from Holliday, 1971.)

fants is associated with a faster rate of weight gain. The energy required for growth is much higher. The rate of growth of LBW infants in the range of 1.5 kg is 10–14 g/kg per day, and the cost is 5 kcal/g, i.e., 50–70 kcal/kg of energy intake is used to support growth in weights (Reichman *et al.,* 1982; Whyte *et al.,* 1983). When milk of lower density is fed, weight gain is less (Tyson *et al.,* 1983). The patterns of energy requirement and weight gain per kilogram are similar among LBW infants of varying size, increasing in the first few weeks of life and stabilizing thereafter. Rate of weight gain correlates well with energy intake and metabolic rate (Chessex *et al.,* 1981).

The growth of LBW infants is regulated by energy intake. Growth has a much higher requirement for other nutrients, e.g., protein and minerals, as well as electrolytes, and consequently administration of formulas with greater calorie density must be accompanied by additions of other nutrients used for growth.

5. Energy Balance and Weight Change

Restricted energy intake leads to changes in body composition characterized by malnutrition and retarded growth. Recovery entails an increase in intake of calories that restores body composition to normal. By contrast, obesity defines a state in which energy intake exceeds need and energy stored as fat becomes excessive. The relationship between energy intake and obesity is elusive, but current evidence points away from a metabolic abnormality.

5.1. Malnutrition and Recovery

The adaptation to malnutrition in previously normal adults is well documented. Activity declines dramatically when food intake is reduced to half-normal, or that equivalent to BMR, and fed at that level for 6 months. Basal metabolic rate decreases to two-thirds of control values, and calorie balance is not achieved. Both body fat and muscle mass decrease considerably (Keys *et al.,* 1950). The pattern resembles the effects of starvation (Cuthbertson, 1964; Forbes, 1970). Chronic undernutrition secondary to limited food availability can sustain a steady state when intake is approximately two-thirds of normal. Activity is less. Liver, heart, and kidney are smaller, and muscle mass is smaller, so that BMR is less, but BMR related to LBM is normal, and activity is reduced (Ashworth, 1968).

Recovery from malnutrition in adults is related to the rate of calorie intake. Given free access to food, Key's subjects ate more than 5500 kcal/day (85 kcal/kg per day) and gained 6 g/kg per day. The energy cost of their weight gain was 5 kcal/g (Keys *et al.,* 1950; Holliday, 1972).

Malnutrition in children affects growth as well as activity and BMR. Basal metabolic rate related to height is low in children with severe malnutrition. With recovery, BMR increases to values that are normal or above (Brooke and Ashworth, 1972). Of particular interest is the striking positive correlation between the postfeeding rise in oxygen consumption of infants fed a standard meal and their rate of weight gain during recovery from malnutrition (Ashworth, 1969). The energy cost of weight gain in infants recovering from malnutrition is 5 kcal/g weight gain, and the rate of weight gain may be 15 g/kg per day. The rate of weight gain correlates with energy intake (Spady *et al.,* 1976).

Children who experience moderate undernutrition and are growth retarded in height, weight, and skinfold thickness experience catch up growth during puberty and attain normal stature, albeit at a later age (Kulin *et al.,* 1982). More severe malnutrition in children appears to leave the children permanently short and behind developmentally. Distinguishing the effect of malnutrition *per se* from other adverse environmental factors is difficult (Malcolm, 1979) (Cravioto and Delicardie, 1979).

5.2. Obesity

Obesity is difficult to define but easy to recognize. Its definition noted earlier—weight for height in excess of 120% of normal—is empirical but serviceable (Davidson and Passmore, 1969). Obesity with onset after puberty usually entails a simple increase in body fat. Fat cell mass is increased, while fat cell numbers are less affected. Recent studies affirm that BMR in these persons when adjusted to LBM is normal. The thermic response to food is normal as well (James, 1983; Felig *et al.,* 1983). The concept that obesity, apart from unusual types, is a consequence of increased efficiency in foot utilization or of decreased activity of thyroid-related functions, e.g., activity of the sodium pump or sodium–potassium adenosine triphosphatase (Bray *et al.,* 1981) is not supported by these findings. The common type of obesity appears to be a disorder in appetite control. James (1983) reviews the evidence that normal individuals may increase BMR with overfeeding, whereas obese individuals do not, giving some basis for considering an inherent metabolic defect as a cause.

Obesity that has its onset in childhood differs significantly in that it is often associated with an increase in LBM and height (Forbes, 1964); such obesity is likely to be more intractible and severe. Fat cell number and size are increased. The evidence favors the development of abnormalities in appetite that may be developed early in life and may have genetic or largely learned causes (Bruch, 1973). The treatment of obesity remains an enigma (Dietz, 1983).

6. Summary

Energy expenditure from term birth to maturity is related to the size of those organs that have a high rate of energy expenditure and upon physical activity of muscle mass. In low-birth-weight infants, energy expenditure in relationship to cell mass is low; as growth proceeds, energy requirement increases relative to cell mass.

Malnourished individuals develop methods for conserving energy—BMR is low and the energy cost of activity is spread out over time as activity slows. During recovery, a state of increased metabolism develops that is largely but not completely due to the energy cost of weight gain. Obese individuals appear to have a normal metabolism in relationship to cell mass; their obesity is more likely to result from a defect in appetite regulation.

7. References

Ashworth, A., 1968, An investigation of very low calorie intakes reported in Jamaica, *Br. J. Nutr.* **22**:341.

Ashworth, A., 1969, Metabolic rates during recovery from protein-calorie malnutrition: The need for a new concept of specific dynamic action, *Nature (London.)* **223**:407.

Barratt, T. M., and Walser, M., 1968, Extracellular volume in skeletal muscle: A comparison of radiosulfate and radiobromide spaces, *Clin. Sci.* **35**:525.

Barratt, M., and Walser, M., 1969, Extracellular fluid in individual tissues in whole animals: The distribution of radiosulfate and radiobromide. *J. Clin Invest.* **48**:56.

Bier, D. M., and Leak, R. D., Haymond, M. W., Arnold, K. J., Greenke, L. D., Sperling, M. A. M., and Kipner, D. M., 1977, Measurement of "true" glucose production rates in infancy and children with 6,6-dideuteroglucose, *Diabetes* **26**:1016.

Bray, G. A., Kral, J. G., and Per Bjontorp, P., 1981, Hepatic sodium-potassium dependent ATPase in obesity, *N. Engl. J. Med.* **304**:1580.

Brook, C. G. D., 1971, Determination of body composition of children from skin fold measurements, *Arch. Dis. Child.* **46**:182.

Brooke, O. G., and Ashworth, A., 1972, Influence of malnutrition on the postprandial metabolic rate and respiratory quotient, *Br. J. Nutr.* **27**:407.

Brozek, J., and Grande, F. 1955, Body composition and basal metabolism in man: Correlation analysis versus physiological approach, *Hum. Biol.* **27**:22.

Bruch, H., 1973, *Eating Disorders: Obesity, Anorexia Nervosa and the Person Within,* Basic Books, New York.

Buskirk, E. R., 1961, Underwater weighing and body density: A view of procedures in techniques for measuring body composition: A review of Procedures, in: *Techniques for Measuring Body Composition* (J. Brozek and A. Henschel, eds.), p. 99, National Academy of Sciences, Washington, D.C.

Cahill, G. F., Jr., Herrera, M. G., Morgan, A. P., Soeldner, J. S., Steinke, J., Levy, P. L., Reichard, G. A., and Kipnis, D. M., 1966, Hormones—Fuel interrelationships during fasting, *J. Clin. Invest.* **45**:1751.

Cheek, D. B., 1956, Change in total chloride and acid base balance in gastroenteritis following treatment with large and small loads of sodium, *Pediatrics* **17**:839.

Cheek, D. B., 1961, ECF volume: Its structure and measurement and the influence of age and disease, *J. Pediatr.* **58**:103.

Cheek, D. B., 1968, *Human Growth: Body composition, Cell Growth, Energy and Intelligence,* p. 191, Lea & Febiger, Philadelphia.

Chessex, P., Reichman, B. L., Verellen, G. J. E., Putet, G., Smith, J. M., Heims, T., and Swyer, P. R., 1981, Influence of postnatal age, energy intake and weight gain on energy metabolism in the very low birth weight infant, *J. Pediatr.* **99**:761.

Coppoletta, J. M., and Wolbach, S. B., 1933, Body length and organ weights of infants and children, *Am. J. Pathol.* **9**:55.

Cornblath, M., and Schwartz, R., 1976, *Disorders of Carbohydrate Metabolism in Infancy,* 2nd ed., W. B. Saunders, Philadelphia.

Corsa, L., Olney, J. M., Steenberg, R. W., Ball, M. R., and Moore, F. D., 1950, The measurement of exchangeable potassium in man by isotope dilution, *J. Clin. Invest.* **29**:1280.

Cotton, J., Woodward, T. A., Carter, N., and Knochel, J. P., 1979, Resting skeletal muscle membrane potential as an index of uremic toxicity, *J. Clin. Invest.* **29**:1280.

Cravioto, J., and Delicardie, E. R., 1979, Nutrition, mental development and learning, in: *Human Growth* (F.

Falkner and J. M. Tanner, eds.), 1st ed., Vol. 3, p. 481, Plenum Press, New York.

Cuthbertson, D. P., 1964, Physical injury and the effect on protein metabolism, in: *Mammalian Protein Metabolism* (H. N. Munro and J. B. Allison, eds.), Vol. II, p. 19, Academic Press, New York.

Davidson, S., and Passmore, R., 1969, *Human Nutrition and Dietitics,* 4th ed., p. 385, Churchill Livingstone, Edinburgh.

Delaporte, C., Bergstrom, J., and Broyer, M., 1976, Variations in muscle cell protein of severely uremic children, *Kidney Int.* **10**:239.

Dietz, W. H., 1983, Childhood obesity: Susceptibility, cause and management, *J. Pediatr.* **103**:676.

Durnin, J. V. G. A., and Rahaman, M. M., 1967, The assessment of the amount of fat in the human body from measurement of skin fold thickness, *Br. J. Nutr.* **21**:631.

Durnin, J. V. G. A., and Taylor, A., 1960, Replicability of measurement of density of the human body as determined by underwater weighing, *J. Appl. Physiol.* **15**:142.

Edelman, I. S., and Leibman, J., 1959, Anatomy of the body fluids, *Am. J. Med.* **27**:256.

FAO/WHO Joint Ad Hoc Expert Committee, 1973, Energy and protein requirements, *WHO Tech. Rep. Ser.* 522.

Felig, P., Cunningham, J., Levitt, M., Hendler, R., and Nadel, E., 1983, Energy expenditure in obesity in fasting a postprandial state, *Am. J. Physiol.* **244**:E45.

Flynn, M. A., Woodruff, C., Clark, J., and Chase, G., 1972, Total body potassium in normal children, *Pediatr. Res.* **6**:239.

Foman, S. J., 1966, Body composition of the male reference infant during the first year of life, *Pediatrics* **40**:863.

Foman, S. J., Thomas, L. N., Filer, L. J., Ziegler, E. E., and Leonard, M. T., 1971, Food composition and growth of normal infants fed milk-based formulas, *Acta Pediatr. Scand. (Suppl.)* **223**:1.

Forbes, G. B., 1964, Lean body mass and fat in obese children, *Pediatrics* **34**:308.

Forbes, G. B., 1970, Weight loss during fasting: Implications for the obese, *Am. J. Clin. Nutr.* **23**:1212.

Forbes, G. B., and Bruining, G. J., 1976, Urinary creatinine excretion and lean body mass, *Am. J. Clin. Nutr.* **29**:1359.

Friis-Hansen, B., 1957, Changes in body water compartments during growth, *Acta Pediatr. Scand.* (Suppl. 110) **46**:1.

Friis-Hansen, B., 1971, Body composition during growth, *Pediatrics* **47**:264.

Frisch, R. E., Revelle, R., and Cook, S., 1973, Components of the critical weight at menarch and at initiation of the adolescent spurt: Estimated total water, lean body mass and fat, *Hum. Biol.* **45**:469.

Gamble, J. L., 1946, Physiological information gained from studies on the life raft ration, *Harvey Lect.* **42**:247.

Gamble, J. L., 1947, *Syllabus, Chemical Anatomy, Physiology and Pathology of ECF,* Harvard University Press, Cambridge, Massachusetts.

Garrow, J. S., 1974, *Energy Balance and Obesity in Man,* pp. 5–29, American Elsevier, New York.

Garrow, J. S., Fletcher, K., and Halliday, D., 1965, Body composition in severe infantile malnutrition, *J. Clin. Invest.* **44**:417.

Hirsch, J., and Knittle, J. L., 1970, Cellularity of obese and nonobese human adipose tissue, *Fed. Proc.* **29**:1516.

Holliday, M. A., 1971, Metabolic rate and organ size during growth from infancy to maturity, *Pediatrics* **47**:169.

Holliday, M. A., 1972, Calorie deficiency in children with uremia: Effect upon growth, *Pediatrics* **50**:590.

Holliday, M. A., 1976, Management of the child with renal insufficiency, in: *Clinical Pediatric Nephrology* (E. Lieberman, ed.), Part 1, p. 397, J. B. Lippincott, Philadelphia.

Holliday, M. A., and Segar, W. M., 1957, Maintenance need for water in parenteral fluid therapy, *Pediatrics* **19**:823.

Holliday, M. A., Pooter, D., Jarrah, A., and Bearg, S., 1967, Relation of metabolic rate to body weight and organ size, a review, *Pediatr. Res.* **1**:185.

James, W. P. T., 1983, Energy requirements and obesity, *Lancet* **2**:386.

Jelliffe, D. B., and Jelliffe, E. F. P., 1979, Epidermological considerations, in: *Human Growth* (F. Falkner and J. M. Tanner, eds.), 1st ed., Vol. III, p. 395, Plenum Press, New York.

Jones, J. D., and Burnett, P. D., 1974, Creatinine metabolism in individuals with decreased renal function, *Clin. Chem.* **30**:1204.

Kaufman, L., and Wilson, C. J., 1973, Determination of extracellular fluid volume by flourescent excitation analysis of bromide, *J. Nucl. Med.* **14**:812.

Kerr, D., Ashworth, A., Picou, D., Poulter, N., Seakin, A., Spady, D., and Wheeler, E., 1973, Accelerated recovery from infant malnutrition with high calorie feeding, in: *Endocrine Aspects of Malnutrition* (L. L. Gardner and P. Amacher, eds.), pp. 467–479, Kroc Foundation, Santa Ynez, California.

Kennedy, C., and Sokoloff, L., 1957, An adaptation of the nitrous oxide method to the study of cerebral circulation in children, *J. Clin. Invest.* **36**:1130.

Keys, A., and Brozek, J., 1953, Body fat in adult man, *Physiol. Rev.* **33**:245.

Keys, A., Brozek, J., Henschel, A., Michelson, O., and Taylor, H. L., 1950, *The Biology of Human Starvation,* University of Minnesota Press, Minneapolis.

Kleiber, M., 1947, Body size and metabolic rate, *Physiol. Rev.* **33**:245.

Kulin, H. E., Burbo, N., Mutie, D., and Sautner, 1982, The effect of chronic childhood malnutrition in pubertal growth and development, *Am. J. Clin. Nutr.* **36**: 527.

Malcolm, L., 1979, Protein-energy malnutrition and growth, in: *Human Growth* (F. Falkner and J. M. Tanner, eds.), 1st. ed., Vol. III, p. 361, Plenum Press, New York.

Metcoff, J., Silvestre, F., Antonowicz, I., Gordillo, G., and Lopez, E., 1960, Relations of intracellular ions to metabolic sequences in muscle in kwashiorkor, *Pediatrics* **26**:960.

Miller, I., and Weil, W. B., Jr., 1963, Some problems in expressing and comparing body composition determined by direct analysis, in: Body Composition, *Ann. N.Y. Acad. Sci.* **110**:153.

National Research Council–Food and Nutrition Board, 1980, *Recommended Dietary Allowances,* National Academy of Sciences, Washington, D.C.

Nichols, G., Nichols, N., Weil, W., and Wallace, W. M., 1953, The direct measurement of the extracellular phase of tissues, *J. Clin. Invest.* **32**:1299.

Owen, O. E., Morgan, A. P., Kemp, H. G., Sullivan, J. M., Herrera, M. G., and Cahill, G. F., 1967, Brain metabolism during fasting, *J. Clin. Invest.* **46**:1589.

Patrick, J., 1977, Assessment of body potassium stores, *Kidney Int.* **11**:476.

Payne, P. R., and Waterlow, J. C., 1971, Relative energy requirements for maintenance growth and physical activity, *Lancet* **2**:210.

Reichman, B. L., Chessex, P., Putet, G., Verellen, G. J. E., Smith, J. M., Heim, T., and Swyer, P. R., 1982, Partition of energy metabolism and energy cost of growth in the very low-birth-weight infant, *Pediatrics* **69**:446.

Sinclair, J. C., Scopes, J. W., and Silverman, W. A., 1967, Metabolic reference standards for the neonate, *Pediatrics* **39**:724.

Spady, D. W., Payne, P. R., Picou, D., and Waterlow, J. C., 1976, Energy balance during recovery from malnutrition, *Am. J. Clin. Nutr.* **29**:1073.

Talbot, F. B., 1938, Basal metabolism standards for children, *Am. J. Dis. Child.* **55**:455.

Tanner, J. M., and Whitehouse, R. H., 1962, Standards for subcutaneous fat in British children, *Br. Med. J.* **1**:446.

Tyson, J. E., Lasky, R. E., Mize, C. E., Richards, C. J., Blair-Smith, N., Whyte, R., and Beer, A. E., 1983, Growth metabolic response and development in very low-birth-weight infants fed banked human milk, or enriched formula. I. Noenatal findings, *J. Pediatr.* **103**:95.

Waterlow, J. C., and Alleyne, G. A. O., 1971, Protein malnutrition in children: Advances in knowledge in the last ten years, *Adv. Protein Chem.* **35**:117.

Waterlow, J. C., and Thomson, A. M., 1979, Observations on the adequacy of breast feeding, *Lancet* **2**:238.

Whyte, R. K., Haslam, R., Vlainic, C., Shannon, S., Samulski, K., Campbell, D., Bayley, H. S., and Sinclair, J. C., 1983, Energy balance and nitrogen balance in growing low-birth-weight infants fed human milk or formula, *Pediatr. Res.* **17**:891.

Widdowson, E. M., and Dickerson, J. W. T., 1964, Chemical composition of the body in: *Mineral Metabolism* (C. L. Comer and L. Bronner, eds.), Vol. II, p. 1, Academic Press, New York.

6

Body Composition in Adolescence

GILBERT B. FORBES

1. Introduction

Anatomists showed many years ago that some organs grow at rates that differ from that of the body as a whole. In recent decades we have witnessed the development of techniques that permit the assessment of certain features of body composition in the living subject. Humans rank among the fattest of mammals, and it is this fact, among others, which has stimulated the development of methods for estimating the relative proportions of lean and fat in our species; assays for total body Ca, P, and N have also been developed. These methods make it possible to define the contribution of each component to the adolescent growth process. Body composition data can also help in estimating certain nutritional requirements for growth. Although long a favorite of nutritionists, the metabolic balance method is not equal to this task (Wallace, 1959; Forbes, 1973, 1983), and one can look forward to more realistic estimates based on these newer techniques.

1.1. The Concept of Chemical Maturity

For more than a century biologists and chemists have known that young tissues differ from old in chemical composition (Bezold, 1857). The young body has a higher proportion of water and a lower proportion of ash; young bones contain less calcium, young muscle less potassium, and the ratio of extracellular fluid (ECF) volume to intracellular fluid (ICF) volume declines during growth (Friis-Hansen, 1957; Nichols *et al.*, 1968); indeed the average adult (of today) is fatter than the child.

A half-century ago Moulton (1923) put forth the concept of chemical maturity, namely that at some point during growth body composition approaches that of the adult, and hence is considered "mature." (Incidentally, aging reverses this process: for example, relative ECF volume increases, bone loses calcium.) Moulton concluded that mammals, including man, reach chemical maturity at about 4% of their life span. The age for man is not known precisely, and it is likely it varies for different tissues. Roentgenograms show that the skeleton is not completely ossified until after adolescence; on the other hand, skeletal muscle composition achieves values in the adult range for potassium, chloride, and water by about age six years (Nichols *et al.*, 1968). Cureton *et al.* (1975) suggest that "chemical maturity for potassium and the density of the fat-free body" has been achieved by the age of eight years. Data on composition of the newborn and adult, as determined by chemical analysis, are summarized in a paper by Forbes (1962).

1.2. The Fat-Free Body

Early workers noted that tissue samples were not always of uniform composition. These discrepancies were finally explained in large part by the work of such individuals as Pfeiffer (1887) and Magnus-Levy (1906), and later Hastings and Eichelberger (1937), who pointed out the necessity of accounting for the fat content of the tissue sample. Neutral fat does not bind water, nitrogen, or electrolyte, and it is now common practice to express the results of tissue analysis on a fat-free basis. It is this relative

GILBERT B. FORBES • School of Medicine and Dentistry, The University of Rochester, Rochester, New York 14642.

constancy of the composition of fat-free tissues, and of the fat-free body as a whole (Pace and Rathbun, 1945; Forbes and Lewis, 1956) that forms the basis for some of the methods now in use for estimating body composition in living man. Figure 1 shows values for chemically determined content of water, potassium, and calcium in man when these are calculated on a fat-free weight basis. It is obvious that major changes take place during growth: water content declines, as does Cl; K and Ca contents increase, as do N, P, and Mg. The fact that body Ca content increases more rapidly than K means that the ossified skeleton grows relatively more rapidly than soft tissues. It is also evident that there is reasonable compositional consistency for each age group when values are calculated on a fat-free basis.

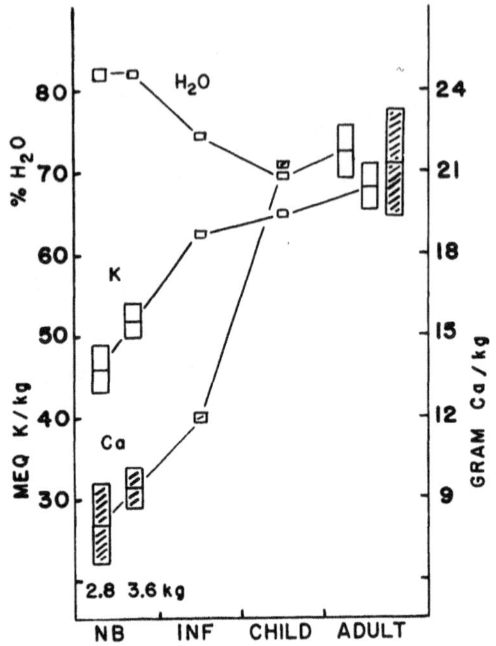

Fig. 1. Composition of fat-free human body by chemical analysis. Means ±SD for H_2O and K (clear boxes, scale at left) and Ca (filled boxes, scale at right). Data from Widdowson *et al.* (1951), Garrow *et al.* (1965), Forbes (1962).

1.3. Terminology

The terms fat-free weight (FFW) and lean body mass (LBM) are both in use. The author interprets these terms as synonymous, and prefers the latter as the more delicate of the two. As used in this chapter LBM means body mass minus ether-extractable fat and hence includes the stroma of adipose tissue.

Behnke and Wilmore (1974) have characterized the LBM as possessing 3–5% of its weight as "essential lipid," i.e., structural as opposed to storage lipid. Garrow (1974) defines LBM as total weight minus adipose tissue, i.e., fat together with its framework of adipocytes and connective tissue. Brozek *et al.* (1963) prefer to use an entity called "the reference man," designated to contain a certain amount of fat, to which can be added variable amounts of "obesity tissue." Womersley *et al.* (1976) have produced a sophisticated analysis of the contribution made by various techniques to the field of body composition.

Moore *et al.* (1963) have introduced the term body cell mass (BCM) to designate an entity whose mass (in grams) is calculated by multiplying total exchangeable K (mEq) by 8.33. Burmeister and Bingert (1967) also make use of this concept (although they assume BCM contains 92.5 mEq K/kg), which in essence defines that portion of the body which contains K at a given concentration. In discussing this concept Moore *et al.* (1963) admit to a degree of uncertainty about their chosen value, and state that it could be as high as 10.0 (or 100 mEq K/kg BCM), which brings it closer to the value chosen by Burmeister and Bingert (1967).

2. Techniques for Assessment of Body Composition

2.1. Dilution Procedures

These make use of a relationship long used for the analysis of static systems, namely $C_1V_1 = C_2V_2$, where C is concentration and V is volume. Since quantity (Q) of solute = CV, the volume of a solution (V_2) can be determined by adding a known quantity of solute and then, after mixing has occurred, measuring the final concentration (C_2):

$$V_2 = \frac{Q}{C_2} \tag{1}$$

This general principle has been used to estimate the volume of various body fluid compartments. However, the body is not a static system, and several assumptions have to be made: (1) The administered material enters the compartment in question quickly and becomes thoroughly mixed within a reasonable period of time; (2) The apparent volume of distribution coincides with that of the compartment in question; (3) If a steady-state concentration is not achieved, by virtue of metabolism and/or excretion of the administered material, the change in plasma concentration should be some simple function of time so that the concentration at zero time can be calculated; (4) The compartment in question is not undergoing changes in volume during the mixing period; and (5) The material administered does not significantly alter normal physiologic processes or homeostatic mechanisms. Obviously, allowance must be made for excretion of the administered material; if this is significant, equation (1) must be modified to read: $V_2 = (Q_{\text{administered}} - Q_{\text{excreted}})/C_2$. It is evident, therefore, that the prerequisites for such fluid volume assays are rigorous.

In the last analysis, it is difficult to conceive of a material which would completely satisfy assumption (2) above. Plasma and ECF represent the only avenues of communication of body cells with the outside world, so materials of all sorts are continuously being transported in and out of these compartments and at a very rapid rate. Such transfer rates are generally assumed to represent first-order reactions, and the mathematical technique known as kinetic analysis has been developed to a high art.

The materials used for estimation of body fluid volumes are listed in Figure 2 and Table I. The list is fairly long. Some materials can be given orally; some must be given intravenously. Not all yield the same result. In using ionic tracers (SO_4, Br, Cl, Na), appropriate corrections for the Donnan equilibrium must be applied.

Intracellular fluid (ICF) volume is estimated by difference (ICF = total H_2O − ECF), and it is clear that the calculated ICF volume will depend on the materials used to estimate ECF volume.

2.1.1. Isotopic Dilution

The advantage of radioisotopes lies in the fact that extremely small quantities can be measured with reasonable accuracy; hence the mass of administered material is so small as not to add appreciably to the *net* amount already present in the body, or to alter physiological mechanisms. As an example, 50 μCi of ^{24}Na, an amount sufficient to measure total exchangeable Na in an adult, weighs only 6×10^{-9} mg. There is, of course, a penalty for this tremendous advantage in the form of the accompanying radioactive emissions.

The dilution principle is the same, except that one attempts to measure a quantity rather than a volume. A known amount of the radioisotope is administered, and a period allowed for mixing to occur with its sister isotope in the body. At the end of the equilibration period, during which urine has been collected, a sample of urine or plasma is obtained for analysis. Then

$$\text{Total body } H_2O \text{ (ml)} = \frac{^3H_{\text{adm}} - {}^3H_{\text{exc}}}{(^3\text{H/ml serum } H_2O)}$$

$$\text{Total exchangeable Na (meq)} = \frac{^{24}\text{Na}_{\text{admin}} - {}^{24}\text{Na}_{\text{exc}}}{(^{24}\text{Na/meq Na in serum})} \qquad (2)$$

In like manner, total exchangeable K and Cl can be determined (Br is used for the latter since there is no suitable radioisotope of Cl).

The assumptions here are (1) that the radioisotope is handled by the body in the same way as its sister stable isotope; and (2) that mixing is complete, i.e., specific activity (X^*/X) is the same in all tissues. In the case of tritium and deuterium, mixing is complete in about 3 hr; in point of fact, these isotopes also exchange with the hydrogens of amino and carboxyl groups and so lead to an overestimation of body water content by a few percent (Foy and Schnieden, 1960; Bradbury, 1961; Tisavipat *et al.*, 1974). Radiosodium dilution underestimates total body Na because of incomplete exchange with Na of bone (Forbes and Perley, 1951). Recently total body neutron activation analysis has been compared with isotopic dilution (Rudd *et al.*, 1972; Ellis *et al.*, 1976), to yield an underestimate of about 25%.

Administered ^{42}K mixes rather slowly and not completely with body K. Comparisons of total body K by ^{40}K counting and by isotopic dilution indicate that the latter method underestimates body K content by 3–10% (Rundo and Sagild, 1955; Surveyor and Hughes, 1968; Talso *et al.*, 1960; Jasani and Edmonds, 1971; Davies and Robertson, 1973; Boddy *et al.*, 1974; Lye *et al.*, 1976). Administered radiocholoride mixes with body Cl in most tissues (brain is a clear exception) within a few minutes (Manery and Haege, 1941), but neither ^{36}Cl or ^{38}Cl is particularly suitable for human use. Bromide, either stable or radioactive, is therefore used; its volume of distribution is slightly larger than that of isotopic Cl and continues to increase for at least 21 hr (Gamble *et al.*, 1953). In the rat, however, total exchangeable Cl estimated by Br dilution has been found to equal chemically determined carcass Cl (Cheek and West, 1955).

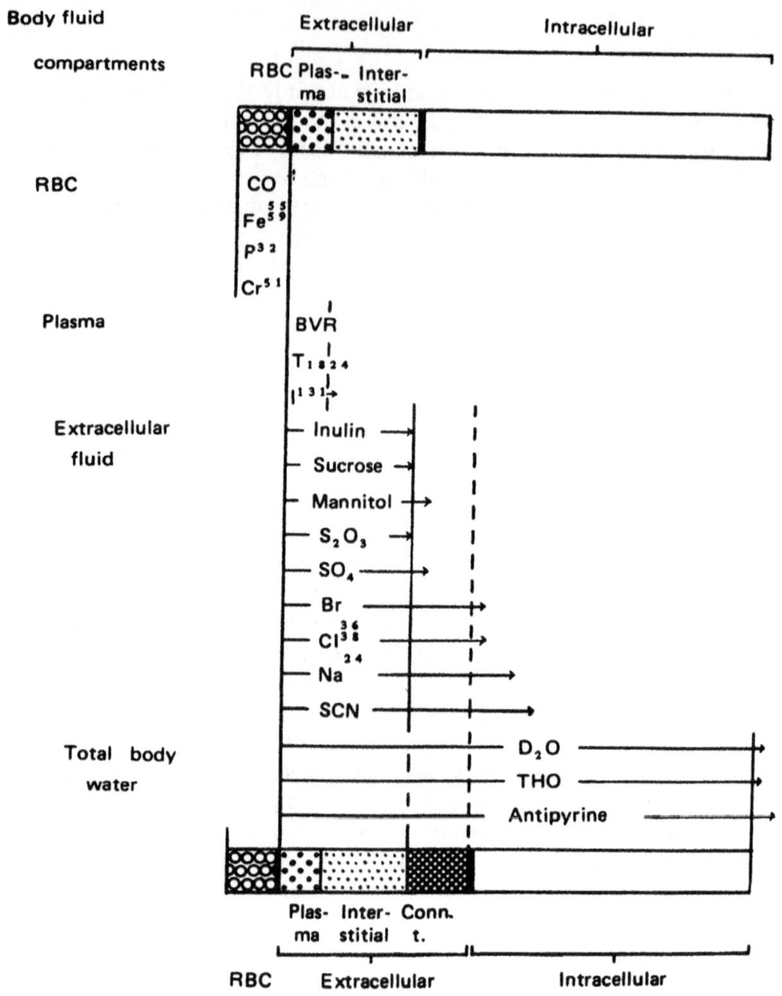

Fig. 2. Materials used to estimate body fluid compartments and their apparent volumes of distribution. T-1824 = Evans blue dye; ^{131}I = labeled serum albumin. (From Elkinton and Danowski, 1955.)

Table I. Additional Methods for Estimation of Body Fluid Volumes

Compartment	Material	Remarks
Extracellular fluid volume	^{82}Br ($t½$ 35 hr), Br	Stable Br analysis by fluorescent excitation (Kaufman and Wilson, 1973); ion-specific electrode
	$^{35}SO_4^{2-}$ ($t½$ 87 days)	Can be given p.o. (Bauer, 1976)
Total body water	*N*-Acetyl-4-amino antipyrine	Not bound to plasma proteins (Brodie *et al.*, 1951)
	Urea	McCance and Widdowson, 1951; Bradbury, 1961
	Alcohol	Give p.o., breath analysis (Grüner and Salmen, 1961), Loeppky *et al.* (1977)
	$H_2{}^{18}O$ dilution	Assay saliva (Schoeller *et al.* (1982); Dietz *et al.* (1982)
Total exchangeable K	^{42}K ($t½$ 12.4 hr)	
	^{43}K ($t½$ 22 hr)	Davies and Robertson (1973)
Total exchangeable Na	^{24}Na ($t½$ 15 hr)	^{22}Na($t½$ 2.6 years) has long residence time in bone
Total exchangeable Cl	Br, ^{82}Br	

It is customary to speak of the results of isotopic dilution assays as total exchangeable Na, K, and Cl. The radiation dose delivered to the body from the tracer amounts used in such studies is about 2 mrad/μCi.

2.2. Potassium-40

Nature has been kind to those studying body composition in devising an isotope of potassium that is long-lived ($t\frac{1}{2} = 1.3 \times 10^9$ year), emits a strong gamma ray (1.46 MeV), and is sufficiently abundant (0.012% of K) to permit its detection and quantitation in human subjects. The human adult body contains about 0.1 μCi (or 17 mg) of ^{40}K, which produces about 200,000 β-rays and 25,000 gamma rays each minute. The resultant radiation dose is about 0.016 rem/year, which is 16% of the total radiation background to man from all natural sources.

The body content of ^{40}K can be assayed in specially constructed counters, which essentially consist of a large shielded room (to reduce background radiation from cosmic and terrestrial sources) containing a γ-ray detector connected to a suitable recording apparatus. The detectors are of two types: Large thallium-activated NaI crystals, one or more of which are positioned near the subject; and Large hollow cylinders, or half-cylinders, the walls of which contain liquid or plastic scintillation material and into which the subject is placed so as to be completely or partially surrounded by the detector. These are referred to as having 4π or 2π geometry. The crystal type has a very good energy resolution, a low background rate, and low efficiency, while the plastic or liquid systems have very poor resolution, high background rate, and high efficiency.

Figure 3 illustrates the net γ-ray spectra for ^{137}Cs and ^{40}K obtained from a 4π and a crystal counter. The International Atomic Energy Agency (1970) has published a directory describing the features of the whole-body monitors now in use throughout the world. Rather elaborate instrumentation is needed to properly generate and interpret the γ-ray spectra obtained from human subjects. Most importantly, each instrument must be properly calibrated so that the recorded ^{40}K activity can be accurately translated into terms of body K content. The calibration must take into account subjects of varying size, the position of the subject relative to the detector, and the extent to which ^{40}K γ-rays are absorbed by the subject's body. The whole-body counter provides a noninvasive means for estimating body composition; the only hazard is claustrophobia.

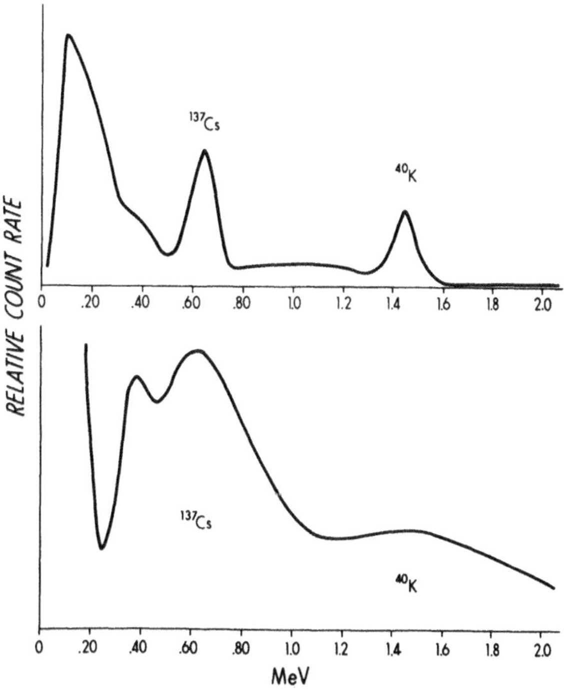

Fig. 3. Net γ-ray spectra from a NaI-Tl crystal counter (upper), and a 4π plastic scintillator (bottom) for adult subject containing 10 nCi ^{137}Cs (fallout level *circa* 1965) and 130 gK.

2.3. Body Density

Archimedes is credited with discovering that one can estimate the relative proportions of a two-component mixture, each component being of known density, by measuring the density of the whole system. Now it is known that fat has a density of 0.900 g/cm^3, and it has been estimated that lean has one of about 1.100 g/cm^3; thus, if the density of the body is determined, the relative proportions of lean and fat can be calculated. The formulation proceeds as follows: let Wf represent weight of fat, Wl weight of lean, Wb weight of body, Vf, Vl, and Vb their respective volumes. Since density (D) = W/V, the relationship $Vb = Vf + Vl$ can be written as

$$\frac{Wb}{Db} = \frac{Wf}{Df} + \frac{Wl}{Dl} \tag{3}$$

and since $Wl = Wb - Wf$, it follows that

$$\frac{Wb}{Db} = \frac{Wf}{Df} + \frac{Wb - Wf}{Dl}$$

Rearranging, one has

$$\frac{Wb}{Db} - \frac{Wb}{Dl} = \frac{Wf}{Df} - \frac{Wf}{Dl}$$

whence

$$Wb\left(\frac{1}{Db} - \frac{1}{Dl}\right) = Wf\left(\frac{1}{Df} - \frac{1}{Dl}\right)$$

which can be written as

$$\frac{Wf}{Wb} = \frac{\left(\frac{1}{Db} - \frac{1}{Dl}\right)}{\left(\frac{1}{Df} - \frac{1}{Dl}\right)}$$

A final rearrangement yields the following:

$$\frac{Wf}{Wb} = \left[\frac{1}{Db} \times \overbrace{\left(\frac{1}{\frac{1}{Df} - \frac{1}{Dl}}\right)}^{a}\right] - \underbrace{\left[\frac{1}{Dl} \times \left(\frac{1}{\frac{1}{Df} - \frac{1}{Dl}}\right)\right]}_{a'} \qquad (4)$$

Now if Df and Dl are known, the constants a and a' can be calculated, and equation (4) becomes

$$\text{Fraction fat}\left(\frac{Wf}{Wb}\right) = \frac{a}{Db} - a' \qquad (5)$$

and fraction lean is

$$1 - \frac{Wf}{Wb}$$

Body density can be measured by underwater weighing

$$Db = \frac{W(\text{air})}{W(\text{air}) - W(\text{water})} \qquad (6)$$

Appropriate corrections have to be made for the density of water and air and for the volume of air in the lungs, this last being estimated by nitrogen washout or helium dilution. For technical reasons gastrointestinal air is neglected; however, variations in the amounts of such air (0–500 ml in adults (Bedell *et al.*, 1956) are a source of error. This method requires a fair degree of subject cooperation and is obviously unsuited to young children.

Body volume can also be estimated by water displacement, by helium dilution, by air displacement, and even by special photographic techniques, whence density is simple W/V. Serious technical problems have limited the use of the latter two methods, and the presence of variable amounts of air in the gastrointestinal tract constitutes a real problem in using the helium dilution method for infants. The underwater weighing technique is the only one in common use. Details of these various techniques can be found in Whipple and Brozek (1963) and in Brozek and Henschel (1961).

2.4. Neutron Activation

Many elements respond to neutron bombardment by capturing the neutron, whereupon the atomic nucleus becomes unstable and emits particulate or electromagnetic radiation of characteristic energy. It is this property together with the decay rate that permits identification of the particular isotope produced by the neutron bombardment. This technique was first used some years ago for the analysis of tissue samples. The sample is exposed to a neutron beam of known flux, and the γ-rays emanating from the induced isotopes are detected by means of a scintillation counter connected to a multichannel pulse-height analyzer; hence the isotopes can be identified by their characteristic γ-ray spectra and quantitated. This method is extremely sensitive; for example, as little as 10^{-9} mEq Na can be assayed.

During the past decade, this technique has been applied to whole animals, including man. The subject is irradiated with neutrons, and then placed in a whole-body scintillation counter for detection of the induced radioactivity (Anderson *et al.*, 1964; Harvey *et al.*, 1973; Nelp *et al.*, 1970; Cohn and Dombrowski, 1971; Dombrowski *et al.*, 1973).

The reactions that have been employed are listed in Table II.

A survey of the technique, including *in vivo* estimation of thyroid iodine content, liver cadmium, and body hydrogen content can be found in Cohn (1981*a,b,*).

Thus the total body content of Ca, P, Na, Cl, and N can be determined in the living subject and with reasonable accuracy.* But there are formidable problems associated with this truly elegant technique: The equipment is very expensive and intricate (indeed there are only a few laboratories in the world capable of assaying human subjects); the radiation dose from neutrons and the neutron energy spectra must be precisely known; and the subject must be exposed to radiation. Although the radiation dose is very small (~30 mrad), there are only two reports on this technique for children (Archibald *et al.*, 1983, Harrison and McNeill, 1982).

2.5. Uptake of Fat-Soluble Gases

Cyclopropane, krypton, and xenon are much more soluble in fat than in lean tissues. A subject placed in an environment containing a known amount of such a gas will progressively take up the gas into the body, and the rate of uptake will depend almost entirely on the amount of fat contained therein. This method has been used successfully in estimating body fat content in adults and in laboratory animals (Lesser *et al.*, 1960, 1971; Hytten *et al.*, 1966). One attempt to use it for infants failed (Halliday, 1971); another was successful (Mettau *et al.*, 1977). It has the advantage of being the only technique available for direct estimation of body fat, lean weight being determined by difference. The disadvantages are the elaborate apparatus needed, the very long equilibration time (several hours of uninterrupted exposure), the multitude of corrections (e.g., temperature, pressure) necessary, and the sophisticated mathematical treatment that must be employed.

*Boddy *et al.* (1976) have used this technique to monitor changes in body composition of animals given a low Ca diet, and Cohn *et al.* (1974) have documented the reduced body Ca content in patients with osteoporosis.

Table II. Nuclear Reactions

Substance	Activation product half-life
^{48}Ca (0.18%)[a](η, γ)[b] ^{49}Ca	8.8 min
^{23}Na (100%) (η, γ) ^{24}Na	15 hr
^{37}Cl (25%) (η, γ) ^{38}Cl	37 min
^{31}P (100%) (η, α) ^{28}Al	2.3 min
^{14}N (99.6%) (η, 2η) ^{13}N[c]	10 min
^{26}Mg (11%) (η, γ) ^{27}Mg	9.5 min

[a]isotopic abundance.
[b]Nuclear reaction.
[c]Also n,γ (prompt gamma).

2.6. Urinary Creatinine Excretion

This compound is formed in muscle from phosphocreatine and, provided the diet does not contain excessive amounts of preformed creatinine, the quantity that appears in the urine is generally considered an index of muscle mass (Cheek, 1968). Indeed, Schutte *et al.* (1981) have demonstrated that total plasma creatinine (i.e., plasma Cr × plasma volume) is highly correlated with dissectable muscle mass in dogs, and that the former is correlated with urinary Cr excretion in humans.

A recent study of subjects of both sexes and of widely varying body size revealed an excellent correlation (r = 0.988) between LBM and creatinine excretion when two ^{40}K counts and three consecutive 24-hr collections of urine were made (Forbes and Bruining, 1976). The equation describing this relationship is LBM (kg) = 0.0291 Cr (mg/day) + 7.38. It is important to recognize that this equation, as well as those developed by others, has a positive intercept on the y axis; hence the ratio of LBM to urinary Cr excretion declines hyperbolically as Cr excretion increases, so that one cannot speak of a constant ratio. Using data on the fraction of LBM contributed by muscle (0.49) in humans (Forbes and Bruining, 1976), and solving the above equation for muscle: urinary Cr ratio, this turns out to be 21.7 kg/g for urinary Cr of 500 mg/day and 16.1 kg/g for one of 2000 mg/day.

Certain technical considerations must be kept in mind. Despite Shaffer's (1908) early dictum, creatinine excretion does vary somewhat from day to day, even in face of a constant diet or during fasting. The urine collection period must be accurately timed, since an error as small as 15 min represents 1% of a 24-hr period. It is advisable to make three consecutive 24-hr collections.

Picou *et al.* (1976) estimated muscle mass in infants and children by a complicated technique involving the administration of [^{15}N]creatine (^{15}N is a stable isotope). The time course of the urinary excretion of labeled creatine is determined, together with urinary creatinine and the creatine concentration in a sample of muscle. Kreisberg *et al.* (1970) used this same procedure in adults by administering [^{14}C]creatine. While this technique is of theoretical interest, it is very complicated and requires a muscle biopsy.

3-Methylhistidine, a nonmetabolizable amino acid, is known to be formed in muscle and to be linked to actin and myosin. Attempts are now being made to use the urinary excretion of this compound as an index of muscle protein turnover and of muscle mass (Long *et al.*, 1975). Studies in children (Ward *et al.*, 1981) and adults (Lukaski *et al.*, 1981) show reasonable correlations with LBM. One problem is the necessity that subjects be on a meat-free diet for at least 3 days prior to urine collection; and as is true for creatinine, there is some day-to-day variation in excretion.

2.7. Electrical Techniques

Hoffer *et al.* (1969) measured the impedance of a weak electrical current passed between the left ankle and the right wrist. When a correction is made for body height (squared), these workers found a good correlation ($r = 0.92$) with total body water. Of concern is the absence of reported studies on this technique in the intervening years.

A new technique is based on the change in electrical conductivity when the subject is placed in an electromagnetic field. This change is proportional to the electrolyte content of the body, and hence should theoretically reflect the amount of lean tissue present. The instrument is variously known as EMME or TOBEC. In a model designed for infants, Klish *et al.* (1983) found a good correlation with chemically determined LBM in rabbits, and in one designed for adults Presta *et al.* (1983*a,b*) found good correlations with both LBM (by density) and total body water; however the standard errors of estimate were rather high.

Neither technique is hazardous, and neither requires much cooperation of the part of the subject.

2.8. External Radiation Techniques

2.8.1. Bone

Roentgenograms of the hand are taken, in a manner to minimize parallax, and measurements are made of the cortex of the second metacarpal, either with a special caliper* or with a magnifying glass fitted with a reticule. Garn (1970) and Gryfe *et al.* (1971) prefer to measure cortex thickness at a point equidistant from the two ends of the bone, while Bonnard (1968) takes the reading where the cortex is thickest and includes the 3rd and 4th metacarpals as well as the 2nd. This measurement yields a value for cortex thickness [periosteal diameter (OD) minus endosteal diameter (ID)], from which one can calculate cross-sectional cortical bone area [$\pi/4\ (OD^2 - ID^2)$], assuming that the cortex is a regular cylinder at the point of measurement. Garn (1970) has postulated that metacarpal cortex thickness serves as an index of total skeletal mass.

Maresh (1966) has published data on bone, muscle, and subcutaneous fat widths measured radiographically at various points on the extremities of children and adolescents.

The technique of photon densitometry has been used by several investigators (Cohn, 1981*b;* Dequeker and Johnston, 1982). The bone, usually distal radius and ulna, is scanned traversely by a low-energy photon beam (^{125}I, 27 keV; ^{241}Am, 60 keV; or ^{153}Gd, 44 and 100 keV), and the transmission is monitored by a scintillation detector. The change in transmission as the beam is moved across the extremity is a function of the bone density in that region, and there is evidence that this is related to total skeletal mass, at least in adults (Manzke *et al.*, 1975; Christiansen and Rödbro, 1975*a;* Cohn and Ellis, 1975). Ringe *et al.* (1977) have published normal values for ages 5–90 years.

2.8.2. Muscle

Garn (1961) and Tanner (1962) discuss the technique involved in making satisfactory measurements of muscle widths from roentgenograms of the extremities (see Chapter 10). Since the entire width is measured, this procedure overlooks the possibility that there may be variable amounts of fat between the muscle layers.

2.8.3. Subcutaneous Fat Layer

With appropriate X-ray exposures, the subcutaneous fat layer can be identified and measured. The favored sites are the lateral chest wall and the region over the greater trochanter of the femur (Stuart *et al.*, 1940; Comstock and Livesay, 1963; Garn, 1961).

Mention should be made here of the use of ultrasound for estimation of subcutaneous fat thickness (Stouffer, 1963; Booth *et al.*, 1966).

*Helios Caliper fitted with needle points and a dial which can be read to the nearest 0.05 mm.

2.8.4. Muscle, Bone, and Fat

Computerized tomography offers the opportunity of defining these three components in an extremity, in a manner which permits one to calculate the cross-sectional area of each (Heymsfield *et al.*, 1979). Figure 4 compares the arms of a thin and an obese individual who happened to have the same LBM. The difference in subcutaneous fat thickness and cross-sectional area is striking indeed. Borkan *et al.* (1982) have used this technique to determine the relative amounts of intraabdominal and subcutaneous fat in obese individuals. It turns out that there is considerable variation in the relative size of these two fat depots.

In applying this technqiue to children one has to consider the radiation dose.

2.9. Anthropometry

In contrast to the techniques described above, anthropometry requires little in the way of apparatus. Calipers of various sorts and a tape measure are all one needs.

2.9.1. Thickness of Skin Plus Subcutaneous Tissue

This is variously called skinfold thickness and fatfold thickness. Since a double layer of skin at the usual sites is only 1.8 mm thick, the subcutaneous fat layer contributes the bulk of the measured value (Durnin and Womersley, 1974). The two most commonly used sites are over the triceps midway between elbow and shoulder and at the lower tip of the scapula. Others are the mid-biceps region, at the lower rib margin in the anterior axillary line, the periumbilical region, and over the iliac crest.

The measurement is made by grasping the subcutaneous tissue between thumb and forefinger, shaking it gently to (hopefully) exclude underlying muscle, and stretching it just far enough to permit the jaws of the spring-actuated caliper* to impinge on the tissue. Since the jaws of the caliper compress the tissue, the caliper reading diminishes for a few seconds and then the dial is read. In subjects with moderately firm subcutaneous tissue the measurement is easy to make, but those with flabby, easily compressible tissue and those with very firm tissue not easily deformable, present somewhat of a problem. In these latter two situations it may be difficult to achieve consistent readings.

The assumptions underlying the use of this method for estimating body fat content are two in number: first, that the thickness of the subcutaneous fat mantle reflects the total amount of fat in the body, and second, that the sites chosen for the measurement, either singly or in combination, represent the average thickness of the entire mantle. Neither assumption has been proved true. Despite the oft-repeated statement that subcutaneous fat makes up about half of the total body fat, there are no data to support such a contention. There are only two analyses of humans which bear on this point: a neonate in whom subcutaneous fat represented 42% of total body fat (Forbes, 1962) and an adult woman for whom the value was 32% (Moore *et al.*, 1968). Pitts and Bullard (1968) analyzed the bodies of a number of mammalian species and found a wide variation, from 4 to 43%.

A number of investigators have correlated skinfold thickness at one or more sites with body density or with body fat content estimated by total body water, ^{40}K counting, or densitometry (Parizková, 1977; Heald *et al.*, 1963; Nagamine and Suzuki, 1964; Sloan, 1967; Young *et al.*, 1968; Burmeister and Fromberg, 1970; Forbes and Amirhakimi, 1970; Hermansen and Döbeln, 1971; Brook, 1971; Durnin and Womersley, 1974; Ward *et al.*, 1975; Lohman, 1981). For the most part, the correlations are not very high, although some investigators claim that body fat estimates are reasonably accurate; other workers, such as Shepherd *et al.* (1983) and Haschke (1983), were very disappointed, and in a recent review Johnston (1982) states that "accurate estimates of whole body composition from anthropometry are not possible . . . investigators should utilize anthropometries directly instead of using them to estimate whole body composition."

2.9.2. Arm Circumference

By combining skinfold thickness (SF) with arm circumference at the same level, one can calculate a "corrected" arm circumference which represents the muscle and bone (M + B) component of the extremity. Novak *et al.* (1973) and others have used only the triceps skinfold for this measurement, yet

*Three are recommended: the Harpenden Caliper, H. E. Morse Co., Holland, Michigan: the Holtain-Harpenden, Holtain Ltd., Brynberian, Crymmych, Pembrokeshire, Wales; and the Lange caliper, Cambridge Scientific Industries, Inc., Cambridge, Maryland.

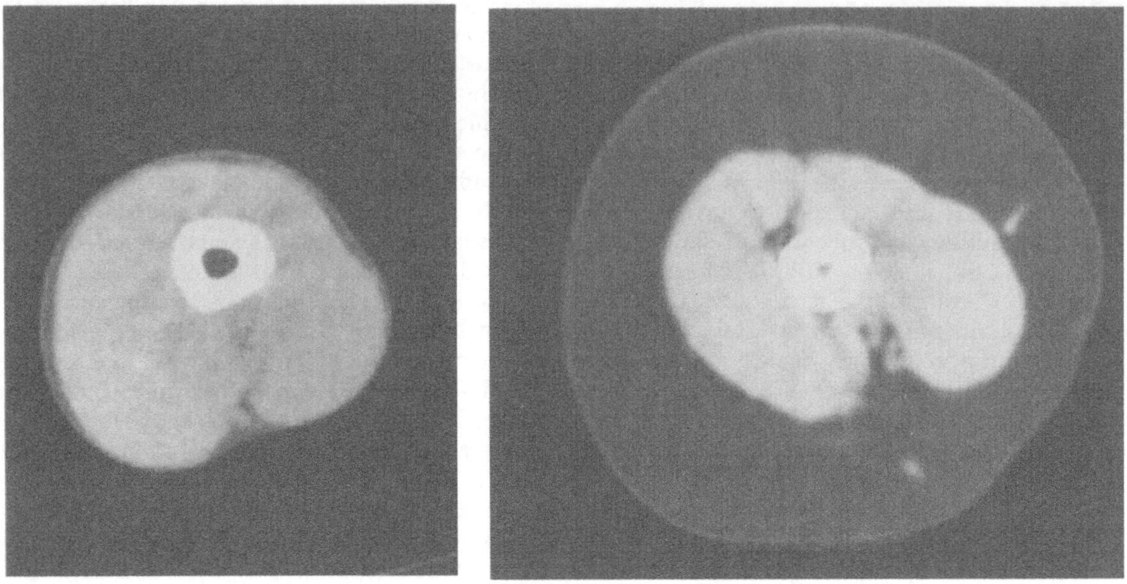

Fig. 4. Mid-arm CAT scan of a thin male and an obese female. Body weights are 62 and 102 kg, LBM 58 and 58 kg, fat 4 and 44 kg, arm circumference 29.5 and 43.0 cm, and triceps skinfold 3.2 and 39.5 mm, respectively.

for many subjects the subcutaneous tissue layer of the arm is not of uniform thickness, so a more appropriate formula would be (all measurements in centimeters)

$$\text{M} + \text{B circumference} = \text{arm circ.} - \frac{\pi}{2}(\text{triceps} + \text{biceps SF}) \tag{7}$$

$$\text{M} + \text{B area} = \frac{1}{4\pi}\left[\text{circ.} - \frac{\pi}{2}(T + \text{B SF})\right]^2 \tag{8}$$

Data on cross-sectional area of arm M + B and arm fat for 6–17-year-old American children have been published by the National Center for Health Statistics (1974). It is unfortunate that these values were based on triceps skinfold only.

2.9.3. Other Anthropometric Measurements

Behnke and Wilmore (1974) present a complicated formula for estimating lean weight. Body diameters are measured at six locations (biacromial biiliac, upper chest, bitrochanteric, wrist, and ankle), and these together with height are entered in the following formula:

$$\text{LBM} = \frac{(\text{sum of diameters})^2}{a} \times \text{height}^b \tag{9}$$

where a and b are constants, which differ for the sexes. Huenemann *et al.* (1974) used this method on a group of 16-year-old school children and compared the results with those obtained by the densitometric and ^{40}K technqiues. The correspondence for boys was very good: average of 55.2 kg lean weight by anthropometry, 55.7 by ^{40}K counting, and 55.9 kg by densitometry. The values for girls were, respectively, 47.3, 40.9, and 41.1 kg; hence for them the anthropometric estimates were clearly out of line.

Steinkamp *et al.* (1965) were able to predict body fat content in adults (by the H_2O-density method) rather well from multiple regression formulas involving various combinations of body circumferences, diameters, and skinfold thicknesses. Edwards and Whyte (1962) estimated body fat from skinfolds and height. Pollock *et al.* (1976) used equations involving skinfold thicknesses, body circumferences, and diameters to predict body density in young men, and Young *et al.* (1968) have used this approach in adolescent girls. Others include Michael and Katch (1968), Katch and Michael

(1968), Kjellberg and Reizenstein (1970), Döbeln (1959), Delwaide and Crenier (1973), and Lohman *et al.* (1975). Crook *et al.* (1966) combined skinfold thickness with percentage overweight.

2.9.4. Functions of Weight and Height

Physiologists commonly use weight$^{0.75}$ as an index of "active metabolic" weight. Moore *et al.* (1963) claim that body fat is a linear function of weight in adults; certainly the two should be well correlated in series where the weight range is large enough to include obese and underweight subjects. Since fat is a component of weight, they must perforce be related. While it is obvious that a man of average height weighing 150 kg is obese, one who weighs 100 kg at a height of 195 cm may well not be. It was this simple observation that led Behnke *et al.* (1942) to develop the densitometric method some three decades ago and to emphasize the importance of considering body weight as a two-compartment system—lean and fat.

The body mass index (BMI), which is W/ht^2, has been frequently used as an index of obesity and of undernutrition, since it is claimed that this index is independent of height, at least in adults. However, it is by no means certain that individuals of the same weight and height have the same body composition. Indeed, in adolescent males, Haschke (1983) found a poor correlation between BMI and fatness. Rolland-Cachera *et al.* (1982) and van Wieringen (1972) present data to show that this index varies with age in normal infants and children.

Cheek (1968) presented a formula for predicting LBM from height and weight ($aW + bH + c$) for children, and Hume (1966) and Hume and Weyers (1971) have done the same for the adult. Of more than passing interest here are the facts that: (1) The equations for children differ from those for the adult; (2) Those for males differ from those for females; and (3) Hume's 1966 equations differ appreciably, and inexplicably, from the ones published by him and his associate in 1971. Burmeister and Bingert (1967) attempted to predict body cell mass from the formula: $A^aW^bH^c$ (A is age; $a,b,$ and c are constants); four such equations were needed to encompass the entire age range for each sex. Wilmore and Behnke's (1970) admonition that anthropometric equations are valid only for the subject sample from which they were derived seems to have been borne out.

Flynn *et al.* (1972) and Forbes (1974) have related body K and LBM, respectively, to stature. However, the regression slopes and intercepts vary with age and sex, and although the correlation coefficients are statistically significant, they are not high enough to provide satisfactory predictability. Nevertheless, the effect of stature is great enough to make control of this variable mandatory in any study involving comparisons of LBM among groups of subjects.

Also of interest is the finding that LBM appears to be related to the cube of height (Forbes, 1974) and that the same is true of skeletal mass (Harrison and McNeill, 1982).

2.10. Cell Number and Cell Size

2.10.1. Muscle

A biopsy is done and the tissue analyzed for protein, intracellular H_2O, and DNA; an estimate of total muscle mass is derived from urinary creatinine excretion. Under the assumptions that each muscle cell has one nucleus, and that each nucleus contains 6.2 pg DNA, one can calculate total muscle cell number and relative muscle cell size.

Cheek (1968) has investigated the growth changes in muscle cell number and size in a small series of children and adolescents. Cell number increases from 0.2×10^{12} in the newborn to about 1.1×10^{12} at age 9 years, then rises to about 3.0×10^{12} at age 16 in the boy and 2.0×10^{12} in the girl. Cell size increases from about 150 mg protein/mg DNA in the baby to about 300 mg/mg in the adolescent.

2.10.2. Adipose Tissue

A biopsy is done, the tissue is analyzed for lipid, and the cells are counted. Total body fat is estimated. From these, adipocyte size (μg lipid per cell) and total adipocyte number can be calculated. Knittle's data (1976) show that cell size increases from about 0.35 μg lipid/cell in young children to about 0.50 in late adolescence, and cell number from 8×10^9 to about 32×10^9. Häger (1977) reports comparable values.

3. Calculation of Lean Body Mass and Fat

The three most widely used methods—total body H_2O, ^{40}K counting, and densitometry—all involve the assumption that the LBM is constant in composition. Water and K contents have actually been determined, but density cannot be measured and hence must be calculated from the densities and relative weights of the various tissues.

Data obtained by chemical analysis were shown in Figure 1 (also see Forbes, 1962). The adult body contains less H_2O but more K and Ca per unit fat-free weight, and also has a higher Ca/K ratio than the infant. Questions have been raised about the absolute constancy of fat-free body composition quite apart from the known effect of age (Wedgewood, 1963). In studies of pigs in whom body fat varied from 16 to 38% of body weight, Fuller *et al.* (1971) found that the K content of the fat-free body declined from 69.1 mEq/kg to 64.8 as H_2O content increased from 73.5 to 75.5%. The data of Ellis and Cohn (1975, and personal communication) show that the K/Ca ratio (g/g) of adult humans varies with stature, from about 0.10 in a 153-cm woman to 0.13 in a 188-cm man. Since bone has a lower H_2O and K content and a greater density than do soft tissues, this means that neither H_2O nor K form a constant fraction of the LBM, nor is its density absolutely constant.

Whole-body K/N ratios are reasonably constant in normal adult subjects (Morgan and Burkinshaw, 1983; Lukaski *et al.* (1981), with values of 1.8–1.9 mEq K/g N, which are close to that found by carcass analysis (Forbes, 1962). Archibald *et al.* (1983) find that this K/N ratio holds true in obese adolescents.

A few sex differences have been noted. Data cited above on the influence of stature on body K/Ca ratio imply that this ratio will on average be lower in women than in men. Compilation of data from several sources (see Forbes *et al.*, 1968) shows that the ratio of total exchangeable K to total body water is 5.8% less in adult females than in males. No such studies have been done on children and adolescents, so the scanty available data on ICF/total H_2O ratios were reviewed (see Forbes and Amirhakimi, 1970). These show that the female:male ratio for ICF/total H_2O is about 1.0 at age 9–12, about 0.98 at age 14, and about 0.95 at 17–18 years. Thus, a graduated correction factor was used for adolescent girls (see footnotes to equations 10 and 11). Incidentally, it has been shown that the ratio of ECF to ICF volume tends to increase in old age (Norris *et al.*, 1963).

Despite these diversions the weight of evidence is that LBM composition does not vary greatly among subjects of specified age (Cheek and West, 1955; Talso *et al.*, 1960; compilation by Forbes and Hursh, 1963; review by Sheng and Huggins, 1979).

The formulas used for the older child, adolescent, and adult are based on the results of chemical analysis:

$$\text{LBM(kg)} = \text{total body } H_2O \text{ (liter)}/0.73 \quad (10)$$

$$\text{LBM(kg)} = \text{total body K (meq)}/68.1 \text{ (males, and girls} < 13 \text{ year)*} \quad (11)$$

$$\text{LBM(kg)} = \text{total body K (meq)}/64.2 \text{ (females 18+ year)*} \quad (12)$$

whence

$$\text{Body fat} = \text{weight} - \text{LBM} \quad (13)$$

The equations used to calculate LBM and fat from body density are derived from measurements of the density of human depot fat and the major tissue components of the LBM: at 37°C fat has a density of 0.900 g/cm^3, water 0.993, protein 1.340, and mineral 3.000 (Brozek *et al.*, 1963). As noted in Section 2.3, the formula for calculating fraction fat is $Wf/Wb = (a/D) - a'$. Using values for fat and lean of 0.900 and 1.10,† respectively, the formula is

$$\frac{Wf}{Wb} = \frac{4.950}{D} - 4.500 \quad (14)$$

whence

$$Wl/Wb = 1 - Wf/Wb.$$

The constants a and a' have been assigned various values by various investigators, and there is a bewildering array of formulas from which to choose (Pearson *et al.*, 1968). The calculated fat

*Based on data referred to above. For adolescent girls the author has used a graduated value for the K content of the LBM: 68.1 meq/kg at age 12, 67.4 at 13, 66.7 at 14, 66.1 at 15, 65.4 at 16, 64.7 at 17, and 64.2 at 18 and above. If total exchangeable K is determined rather than ^{40}K content, the numerator of equations (11) and (12) should be multiplied by 1.05 or 1.10 depending on the extent to which one believes the ^{42}K dilution method underestimates body K content.

†Haschke (1983) has made some calculations to show that the density of the LBM is only 1.086 in 10–14-year-old males; hence equation (14) becomes $Wf/Wb = 5.255/D - 4.84$.

content will obviously depend on the particular formula selected by the investigator.

Siri (1961) has devised a formula utilizing data from both total body H_2O and density, which offers the advantage of being unaffected by the state of hydration, although it does require the assumption of a fixed mineral/protein ration of the body:

$$\frac{Wf}{Wb} = \frac{2.1366}{D} - \left(0.780 \times \frac{H_2O}{W}\right) - 1.374 \quad (15)$$

3.1. Precision of Various Techniques

The ultimate precision is limited by the fact that the human body is not a static system: body weight varies by a percent or so from day to day; a long drink of water can easily increase body H_2O content by 1%; an overnight thirst depletes body H_2O by about 0.5%.

Table III gives the variations found by investigators who have assayed their subjects on repeated occasions. It is evident that LBM cannot be estimated with a precision of much better than 3%.

4. Growth of Lean Body Mass

There are a number of reports on body K content of children and adolescents (Allen *et al.*, 1960; Meneely *et al.*, 1963; Oberhausen *et al.*, 1965; Burmeister and Bingert, 1967; Forbes, 1972*b;* Novak *et al.*, 1973; Flynn *et al.*, 1972), on total-body H_2O (Young *et al.*, 1968; Heald *et al.*, 1963; Schutte, 1980) and on density (Parizkovâ, 1977). Estimates of LBM can be made from these data.

Using various data sources, Haschke (1983) has compiled a listing of body components—fat, fat-free body mass, body protein, carbohydrate, osseous and nonosseous minerals, total-body water, and extracellular water—for the "male reference adolescent."

All are in agreement that the adolescent spurt in LBM is more intense in the adolescent boy than in the girl and that he achieves a mature value which is considerably larger. A collation of data obtained by various techniques showed a remarkable degree of agreement when correction was made for stature differences among the various groups (Forbes, 1972*b*).

Table III. Reproducibility Data

Technique	Coefficient of variation	Reference
^{40}K counting	1.9–2.9%	Shukla *et al.* (1973)
	2–3%	Forbes *et al.* (1968)
	1.4–4.1%	Johny *et al.* (1970)
	1.2–4.8%	Pierson *et al.* (1974)
Total exchangeable K	6.1%	Price *et al.* (1969)
	2.9%	Haxhe (1963)
	2.5%	Davies and Robertson (1973)
THO, D_2O, antipyrine dilution	2.5%	Price *et al.* (1969)
	1.8%	Haxhe (1963)
	9%	Greenway *et al.* (1965)
Density	0.0023 g/ml (SE)[a]	Durnin and Taylor (1960)
	0.0004–0.0043[a]	Buskirk (1961)
	0.0063[a]	Lohman (1981)
Cyclopropane, ^{85}Kr uptake	7–8% "uncertainty"	Lesser *et al.* (1971)
Skinfold thickness	16%	Greenway *et al.* (1965)
	6–24%	National Center for Health Statistics (1974)
Creatinine excretion	2–19%	Forbes and Bruining (1976)

[a]These values should be viewed within the context of the absolute range of density possibilities, namely 0.90 to 1.10 g/ml. One subject who assayed her body density on 10 occasions over a 4-month period, during which her body weight ranged from 52.5 to 53.6 kg, found a coefficient of variation of 0.6% (N. Butte, personal communication). The resultant coefficients of variation for LBM and fat will of course be dependent on the relative proportions of each; in this instance (body fat 9.4 kg), that for LBM was 2.3% and that for fat was 15.5%.

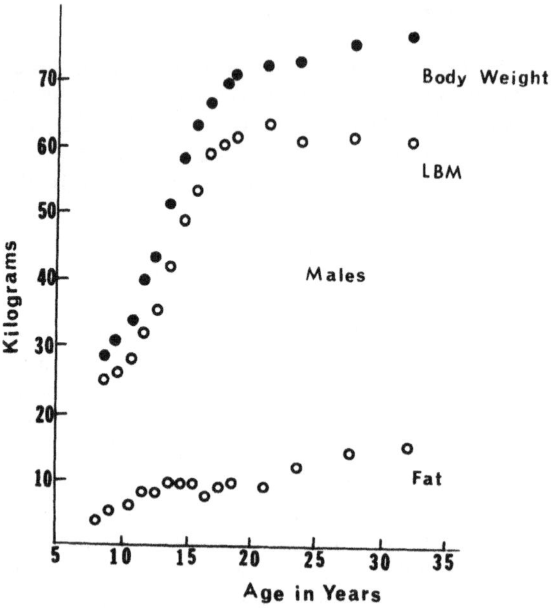

Fig. 5. Mean weight, LBM, and fat for 604 white males aged 8–35 years (author's data).

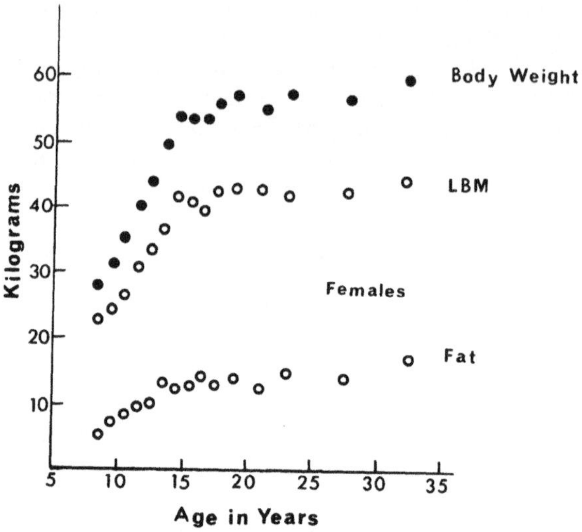

Fig. 6. Mean weight, LBM, and fat for 467 white females aged 8–35 years (author's data).

Figures 5 and 6 present the author's data on normal white subjects by ^{40}K counting* and compare the growth of the LBM with that of body weight. The average preadolescent boy has a slightly larger LBM than the girl, but it is only with the onset of puberty that the difference achieves real significance. The upswing in the boy is rapid and sustained, and maximum values are achieved by about age 20. [The word "maximum" is appropriate here, since there is a slow decline during the adult years (Forbes, 1976*a*).] The LBM spurt in girls is not as intense, and values close to the maximum are attained by age 18.

Figure 7 depicts changes in the LBM/height ratio during adolescence. Boys finally achieve a much higher value for this ratio than do girls. This means that the spurt in LBM is realtively greater than the spurt in height, a result that follows naturally from the known effect of androgens.

The variability in LBM is shown in Figure 8, where the hatched areas enclose the 95% confidence limits for our subjects. A number of studies have shown that LBM is a function of stature (Forbes, 1974), and the inserts in Figure 8 show the calculated regression lines for 22 to 25-year-old males and females. The former has a slope of 0.69 kg/cm, the latter one of 0.29 kg/cm. On average, then, tall individuals have a larger LBM than short individuals, an attribute that undoubtedly confers an advantage in athletics (Khosla, 1968). Data are also available to show that the regression slopes of LBM on height are rather gentle in childhood, and gradually steepen as adolescence progresses (Forbes, 1974).

Tanner (1974) has published curves depicting increments in radiographically determined muscle widths during adolescence. The peak velocity for males coincides with peak height velocity, as do those for biacromial and biiliac diameters; there is a lag of about a year for females. Parizkovå (1977) found that peak velocity in LBM growth coincides with peak height velocity in adolescent boys.

Malina (1969) has summarized the literature on growth of muscle (and bone and subcutaneous fat) as estimated by radiography for children and adolescents.

Figure 9 illustrates the sex ratios for body weight,

*Meneely *et al.* (1963) found the black females in their group to have a larger LBM, while the reverse was true for adolescent males. Cohn *et al.* (1977) found higher values for total K and Ca in black adults of both sexes. However, Schutte's (1980) findings in black adolescent males show that LBM is comparable to that of white adolescents. As might be expected, Japanese and Filipino young adults have a lower LBM than Caucasians (Nagamine and Suzuki, 1964; Novak, (1970). Eskimo women, on the other hand, have a slightly larger LBM than that of Caucasians, while the reverse is true for men (Shephard *et al.*, 1973). Some of these racial differences could be attributable to stature.

height, and LBM as functions of age in our subjects. The latter clearly exhibits the greatest degree of sexual dimorphism; on average the 15-year-old girl has an LBM 81% as large as the boy's, and by age 20 this has dropped to 68%. Sinc the LBM comprises the active metabolic tissues of the body, it is likely that certain nutritional needs and doses of certain medications possess greater sex differentials than would be anticipated from differences in total body weight.

5. Growth of Body Fat

Girls tend to accumulate more fat than boys. Slight sex differences in estimated body fat content have been detected in children (Novak, 1966*a;* Flynn *et al.*, 1970) and even in infants (Owen *et al.*, 1962), but it is not until adolescence that the difference becomes striking. Figures 5 and 6 show the age curve for body fat content, and Figure 7 depicts fat as a percent of body weight. Boys and girls both gain body fat early in adolescence; later the gain stops, even reverses temporarily, in boys, while girls continue to put on fat as adolescence proceeds. The adult years see a further accumulation in both sexes.

Body fat content tends to be more variable than lean. Burmeister and Bingert (1967) assayed ^{40}K content in several thousand German children. The range between the 10th and 90th percentiles of body fat is wide indeed, and the skewness in the data is evident from the fact that the distance from the 50th and the 90th percentile exceeds that from the 10th to the 50th percentile. These authors rightfully conclude that "The nongaussian distribution of weight (which is seen in most population studies) is thought to be due to the skewness in distribution of fat."

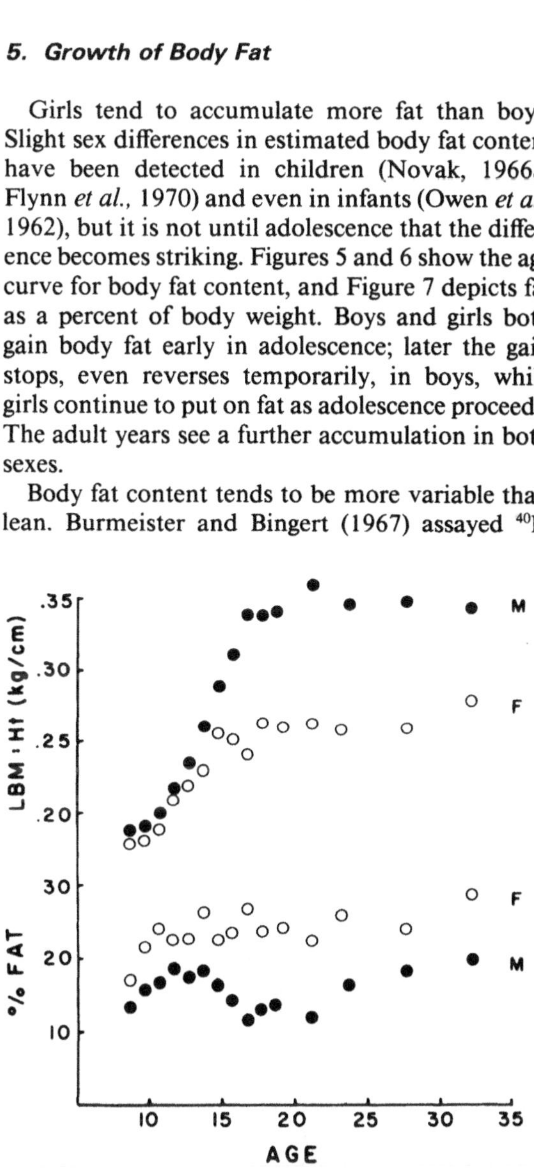

Fig. 7. Mean LBM/height ratio (kg/cm) (upper portion), and percent body fat (lower portion) (author's data).

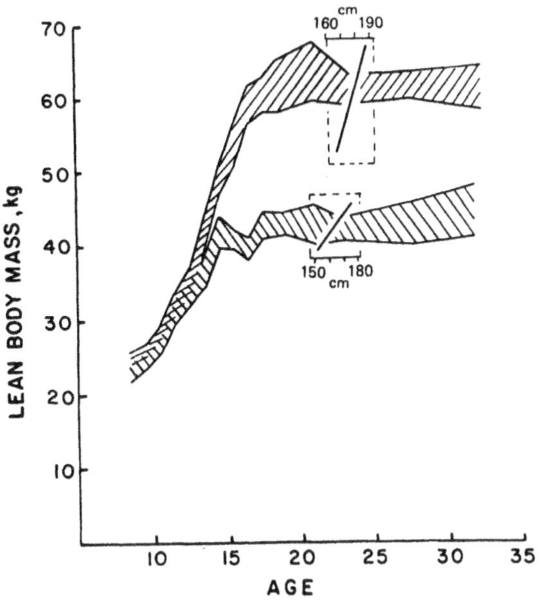

Fig. 8. Ninety-five percent confidence limits for LBM. Inserts show calculated regression lines for LBM against height in subjects 22–25 aged years: slope for males (N = 41) is much steeper (0.69 kg/cm) than for females (N = 29) (0.29 kg/cm) (author's data).

These findings are mirrored by data on skinfold thickness, for here one sees similar age and sex trends and a distinct tendency for skewness (Figure 10). Tanner and Whitehouse (1975) attempted to eliminate the skewness by plotting the logarithm of skinfold thickness, but achieved only partial success. Data from American adolescents do not differ greatly (National Center for Health Statistics, 1974).

Of considerable interest are the data on Shephard *et al.* (1973) on young Eskimo women, who contrary to popular belief proved to be no fatter than Caucasians.

A publication from the National Center for Health Statistics (1974) contains an interesting graph relating the half-yearly increments in triceps

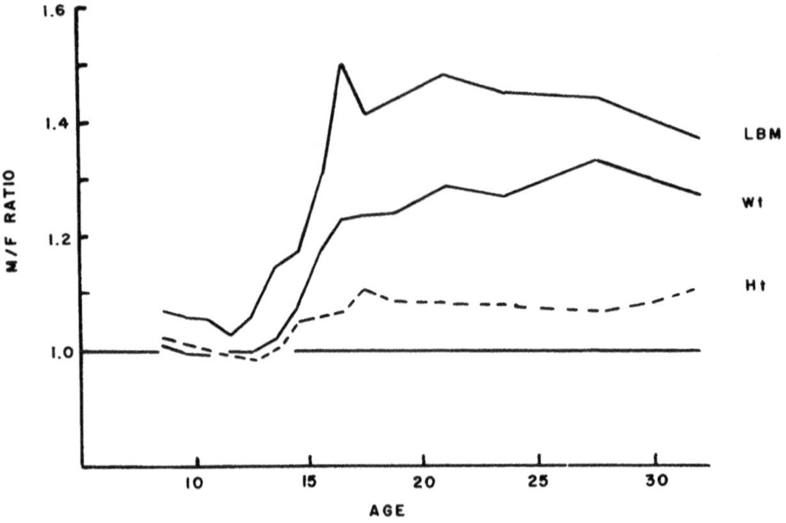

Fig. 9. Male:female ratios for LBM, weight, and height (author's data).

skinfold (SF) thickness to height velocity for children and adolescents (Figure 11). For boys, velocity for triceps SF is positive in the earlier years, with a peak at about age 10; then velocity declines and actually becomes negative, with a nadir that nicely coincides with peak height velocity at about age 13½ years. The velocity curve for girls is quite similar in shape, the entire curve being shifted upward, so that the nadir of the velocity dip just touches zero; but here, too, the nadir coincides in time with

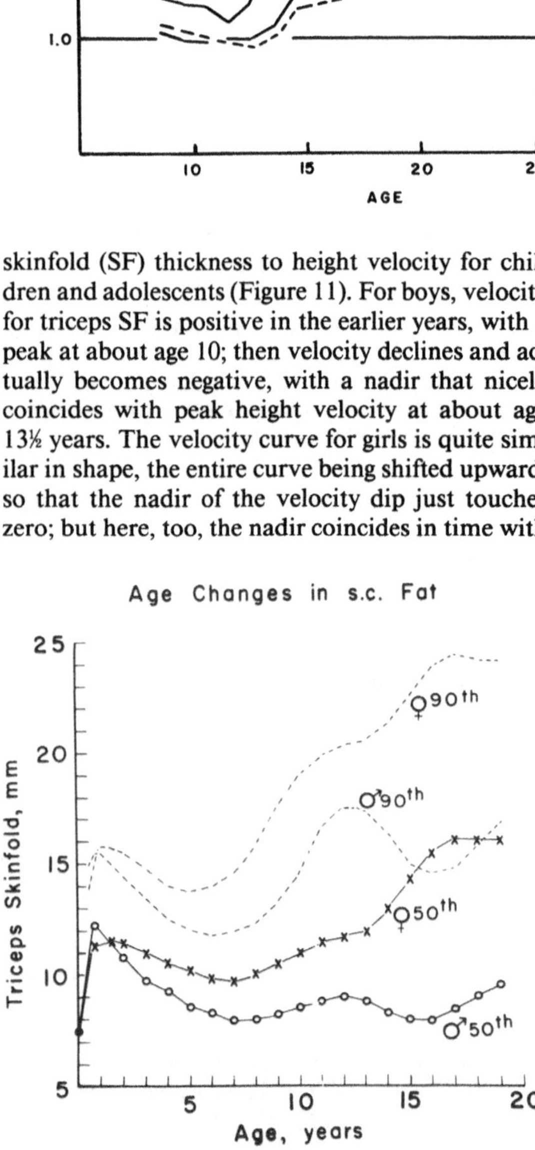

Fig. 10. Skinfold thickness at the mid-triceps region for British children. (Redrawn from Tanner and Whitehouse, 1975).

peak height velocity at about 12 years of age. The later years of adolescence witness a positive triceps SF velocity once again. Tanner's (1974) published curves for radiographically determined subcutaneous fat thicknesses at various sites are similarly related to peak height velocity.

This interesting coincidence, in which it appears that triceps SF and height velocity curves resemble mirror images of each other in both sexes, raises the possibility that a single hormonal influence is responsible for both phenomena.

In a series of publications, Frisch and associates (1970, 1974) have put forth the hypothesis that menarche is somehow triggered by the attainment of a certain body weight (48 kg) and by the acquisition of a certain percent of body fat (17%). What is often overlooked in considering her hypothesis is that she calculated body fat from height and weight rather than measuring it by one or more of the techniques described above.

Many years ago in a longitudinal study Shock (1943) found changes in BMR, blood pressure, and pulse rate in association with menarche, and more recently the author has shown that the perimenarchal years are associated with a number of changes that occur *en echelon:* LBM, pelvic breadth, metacarpal cortex thickness, bone age, and urinary hydroxyproline excretion, as well as body fat (Forbes, 1981, unpublished). Hence this event is accompanied by a wide variety of physiological, anatomical, and of course hormonal changes, so one cannot credit body fat with being the primal influence. Scott and Johnston (1982) have offered substantial criticism of Frisch's critical weight (fat) hypothesis.

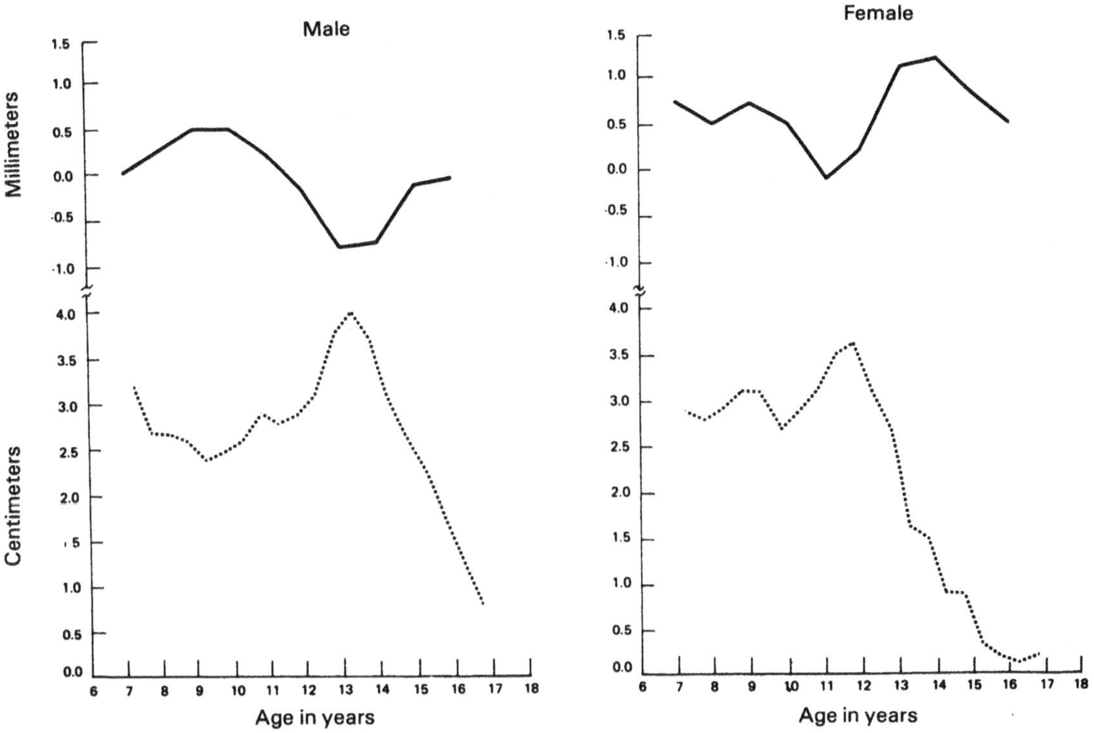

Fig. 11. Pseudo-velocity curves for triceps skinfold thickness and stature for U.S. males and females. (—) Median triceps skinfold differences (smoothed by a two-period moving average). (· · ·) Mean height differences (smoothed by a three-period moving average). (Reproduced from the National Center for Health Statistics, 1974, p. 13.)

6. Changes in Total-Body Calcium

An earlier illustration (Figure 1) depicted the Ca enrichment of the LBM which takes place during growth, as determined by chemical analysis. Trotter and Hixon's (1974) data on skeletal weights show a progressive increase in skeletal weights during childhood and adolescence, to a maximum of about 4400 g (dry fat-free weight) in males and 3100 g in females (once again the word "maximum" is appropriate, for skeletal weight declines during the adult years).

Figure 12 shows changes in estimated total body Ca during childhood and adolescence. One set of curves is from the data of Garn (1970), who estimated skeletal weight from measurements of the thickness of the mid-cortex of the second metacarpal bone, whence skeletal weight $\times$ 0.25 = body Ca (more than 99% of body Ca is in bone). The second is based on data derived from photon densitometry of the distal radius and ulna (Christiansen and Rödbro, 1975*a,b*; Christiansen *et al.*, 1975). The two sets of estimates agree rather well for preadolescent children and for adults, but there is a marked difference for the adolescent years. Garn's subjects were American, Christiansen's were Danish. The basis for Garn's (1970) extrapolation of metacarpal cortex thickness to total skeletal weight is not clear—indeed he has been criticized (Mazess and Cameron, 1972)—while Christiansen and Rödbro (1975*a*) checked their photon measurements against the weight and Ca content of various human bones. On the other hand, the latter investigators' subjects showed some irregularities in height growth.

This discrepancy between the two sets of estimates is unfortunate, for now any estimate of the rapidity of Ca accretion during adolescence is open to doubt, and this is just the time in life when such an estimate would be most helpful from a nutritional point of view. Schuster (1970) did photon measurements on large numbers of children, but

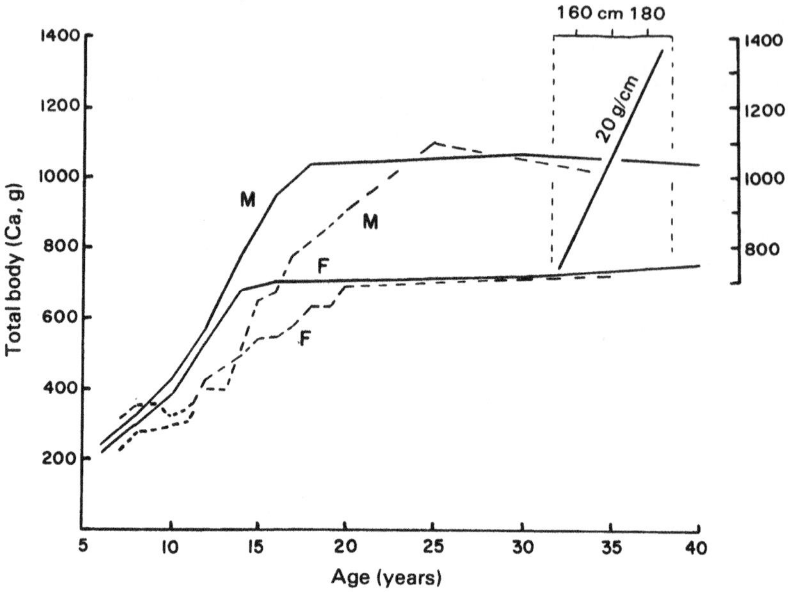

Fig. 12. Total-body calcium as a function of age. (—) from Garn's (1970) estimates of skeletal weight derived from metacarpal cortex thickness; (---) data from Christiansen and Rödbro (1975*b*) and Christiansen *et al.* (1975) based on photon scans of the forearm; (insert) data from Ellis and Cohn (1975, and personal communication) by neutron activation: regression of body Ca on stature for 6 women, 9 men, aged 30–39 years.

made no attempt to extrapolate these values to skeletal weight.

It is evident that the sex difference in total body Ca is appreciable and of the same order of magnitude as the sex difference in LBM (Figure 8). At age 20 years the male–female ratio is 1.47/1 for Garn's data and 1.31:1 for Christiansen's; the ratio for Trotter and Hixon's (1974) adult skeletons (dry weight) is 1.4:1, and for a series of adult Russian skeletons (wet weight) it is about 1.2:1 (Borisov and Marei, 1974); by neutron activation the ratio for body Ca is 1.4:1. These are consistent with the sex ratio for LBM (Figure 9).

As is the situation for LBM, stature exerts an influence on body Ca content, and indeed this is one plausible reason for the larger body Ca content of men. The insert in Figure 12 shows that, on average, body Ca in young adults increases by 20 g for each cm increment in height. Hence a 186-cm male would be expected to contain 640 g more Ca than a 154-cm woman (1370 − 730 g Ca), a remarkable difference and one that certainly is of nutritional importance. The series of skeleton weights reported by Borisov and Marei (1974) also demonstrate the effect of stature: the tallest men had skeletons weighing about 12 kg, the shortest women ones weighing only about 8 kg (wet weight).

Cohn *et al.* (1974) also determined total body phosphorus in adults by neutron activation. The ratio of body P:Ca (g/g) averaged 0.48 in males and 0.50 in females. No assays are available for children and adolescents.

7. Effect of Exercise and Physical Training

It is well known that the athletically inclined are apt to have a larger LBM than sedentary individuals (Parizková, 1963; Novak, 1966*b;* Behnke and Wilmore, 1974). Yet the former may be naturally better endowed.

Figure 13 depicts changes in weight and LBM for adolescent and young adult subjects who have participated in exercise programs, weight training, or competitive sports. Although strength and endurance can be enhanced by such training, and although some body fat can be lost, the recorded increases in LBM are modest indeed. However when androgens are added clear-cut increases in LBM and weight do occur, along with a diminution of body fat.* These drugs are known to promote nitrogen retention, but they can also interfere with gonadotrophin production and they can cause liver abnormalities. Exercise and androgen effects have been reviewed by Forbes (1982, 1984) and by Wilmore (1983).

*Although there is some debate as to whether androgens can promote muscle strength and athletic performance, the recent publicity (summer 1983) about their widespread use indicates that athletes believe that they do.

8. Effect of Nutrition

8.1. Obesity

It is plainly obvious that obese individuals carry a lot of excess fat, and assays reveal that in some the fat burden exceeds 50%. What is less well known is that they also tend to have a modest increase in LBM as compared with nonobese individuals of similar height and age. A compilation by Forbes and Welle (1983) showed that on average the obese have 10–25% more LBM than their age and height peers. A small number, however, possess a normal LBM. Knittle (1976) and Häger (1977) find that obese children have an excessive fat cell size and an even greater increase in total fat cell number.

Both Forbes (1964) and Cheek *et al.* (1970) report that children who become obese early in life tend to have a larger increment in LBM together with an advance in bone age and stature as compared with those whose obesity is of more recent onset.

The usual form of human obesity thus differs from certain types of experimental obesity in animals, for hypothalamic lesions in the rat are accompanied by a reduction in carcass nitrogen and in body length as well as an increase in body fat.

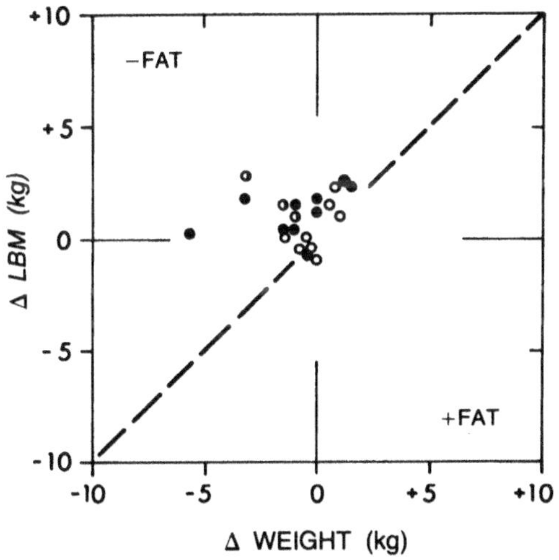

Fig. 13. Change in LBM plotted against change in body weight as a consequence of exercise and/or training. Group averages: (○) moderate exercise; (●) vigorous exercise; (◑) obese subjects. Line of identity; values to the left indicate decrease in body fat, to the right an increase in fat. (From Forbes, 1983*b*, p. 83, with permission Am. Med. Assoc.)

8.2. Deliberate Overfeeding

In assessing the results of overfeeding experiments, Brozek *et al.* (1963) estimated that "obesity tissue" was only two-thirds fat, the remainder being ECF and "cell residue." Of the weight gained by the subjects of Goldman *et al.* (1975), 69% was fat and 31% was lean; furthermore, calculations from their data show that the increment in LBM, as well as total weight, was directly proportional to the total excess calories consumed. The author has observed three obese adolescents who after a period of weight reduction abandoned their diet and gained weight. In each instance there was also a gain in LBM.

8.3. Anorexia Nervosa

These adolescents, most of whom are girls, have a reduced LBM as well as less body fat (Russell *et al.*, 1983; Ljunggren *et al.*, 1957), and during recovery LBM and fat both increase. Despite severe weight loss, plasma protein and blood hemoglobin concentrations are usually normal.

8.4. Deliberate Underfeeding

Low-energy diets, even though adequate in protein, lead to negative nitrogen balance and to some loss of LBM in obese adolescents. In the study reported by Merritt *et al.* (1980), the ratio of N loss to weight loss was 5.0 g N/kg, and in that of Dietz and Schoeller (1982) it was 3.6 g N/kg. This difference may well be due to the fact that the former group were given less food (680 kcal/day) than the latter (960 kcal/day). Two studies of longer duration have been reported in which body composition techniques were used. Brown *et al.* (1983) fed a group of obese adolescents 600 kcal/day for 5 months; average weight loss was 30 kg, of which 7.8 kg represented LBM. Archibald *et al.* (1983) fed their subjects 880 kcal/day for about 3 months; average weight loss was 12.2 kg, of which 6 kg (by ^{40}K counting) or 2.6 kg (body N by neutron activation) represented LBM.

None of the above subjects achieved normal weight, so it is impossible to predict whether these LBM losses would have continued to the point of detriment to their well-being. However, experience with adults given very-low-energy diets (300 kcal/

day) shows that serious consequences can occur (Lantigua *et al.*, 1980).

Sufficient data are now at hand to permit a general statement on the effects of overfeeding and underfeeding in man, namely that appreciable gains or losses of body weight almost always involve alterations in LBM (Forbes, 1984). In this sense body fat and LBM are companions, and it has yet to be demonstrated that mammals can lose significant amounts of body fat without some erosion of the LBM, the only exception being the hibernating bear.

9. Body Composition in Abnormal States

9.1. Muscular Dystrophy (Duchenne Type)

Children with this disorder have a markedly reduced body K content, which progressively departs from normal as the disease progresses (Griggs *et al.*, 1983). However, Borgstedt *et al.* (1970) found no alterations in body K content of unaffected relatives.

9.2. Cystic Fibrosis

Shepherd *et al.* (1983) and Forbes *et al.* (1979) have shown that these children have a reduced muscle mass as well as less body fat. The former investigators were able to bring about an improvement in both body components by dietary supplements.

9.3. Chromosomal Disorders

East *et al.* (1976) reported that males with a 47,XXY chromosome complement tend to have a moderately reduced LBM as determined by ^{40}K counting, to the point where the values were about midway between the usual male and female values. The situation for the 47,XYY male is less clear; the values were slightly reduced but the reduction was of questionable statistical significance.

9.4. Growth-Hormone Deficiency

Tanner and Whitehouse (1967) and Collipp *et al.* (1973) have documented a decrease in subcutaneous fat thickness in hypopituitary dwarfs given growth hormone, an effect that would be anticipated from the known metabolic actions of this hormone. The latter workers also noted an increase in body K content. Cheek's (1968) data show an increase in muscle mass with treatment.

10. Pregnancy

Hytten and Leitch (1971) have reviewed the changes in body composition that occur during normal pregnancy: blood volume, ECF volume, and total body water all show an increase. Seitchik (1967) studied 99 women, some of whom were teenagers: their average weight gain was 16 kg, of which 8 kg was water; fat-free body mass calculated by the combined H_2O–density method increased by 9 kg. The 40 women studied by Godfrey and Wadsworth (1970) gained an average of 19 kg weight and 6.7 g K. MacGillivray and Buchanan (1958) found a mean increase of 770 meq in total exchangeable Na and one of 560 meq in total exchangeable K in eight pregnant women. King *et al.* (1973) studied 10 teenage girls during the last trimester of their pregnancy: average weight gain was 77 g/day and average K accretion was 3.4 meq/day. In 50 women studied both early and late in pregnancy, the average gain in weight was 0.46 kg/week, of which 0.30 kg was LBM (Forbes, 1982). Pregnancy thus results in a gain of both lean and fat, but the relative proportions of the two are difficult to determine with exactitude for in such studies one is measuring two subjects simultaneously, one of which is known to differ in composition from the other.

11. Heredity

Brook *et al.* (1975) in a study of mono- and dizygotic twins found a high coefficient of heredibility for skinfold thickness, and metacarpal cortex thickness also appears to be strongly influenced by heredity (Smith *et al.*, 1973). It remains to be determined whether other aspects of body composition have a genetic basis.

12. Nutritional Implications

The LBM comprises the bulk of the active metabolic tissue, so it is no accident that BMR is strongly correlated with the size of the lean body (Döbeln, 1956), as is blood volume (Muldowney, 1961). In sedentary individuals, total energy expenditure also shows a strong correlation with LBM

(Webb, 1981); thus, for these two features of energy metabolism, the sex difference that is manifest on a body-weight basis has disappeared.

The marked sex difference in the intensity and duration of the adolescent spurt in LBM dictates that caloric and certain nutrient needs are higher for adolescent boys than for girls. For example, Hawkins (1964) has made some calculations to show that the boy requires as much iron at this time of life as does the menstruating girl. Increments in body Ca, P, N, Mg, and Zn should exhibit a similar degree of sexual dimorphism. Note should be taken of the documented LBM–stature and body Ca–stature relationships, which suggest that nutrient needs for growth are considerably higher in tall adolescents.

Many adolescent boys carry rather small amounts of fat, so one must decry the efforts of some high school wrestling coaches at encouraging them to lose weight in preparation for contests. Strenuous exercise demands energy, so the young athlete must have a sufficient energy intake.

ACKNOWLEDGMENTS. The author's work has been supported by grants from the National Institute of Child Health and Human Development and under a contract with the U.S. Energy Research and Development Administration at the University of Rochester Biomedical and Environmental Research Project, and has been assigned Report No. UR-3490-1121.

13. References

Allen, T. H., Anderson, E. C., and Langham, W. H., 1960, Total body potassium and gross body composition in relation to age, *J. Gerontol.* **15**:348.

Anderson, J., Osborn, S. B., Tomlinson, R. W. S., Newton, D., Rundo, J., Salmon, L., and Smith, J. W., 1964, Neutron-activation analysis in man *in vivo:* A new technique in medical investigation, *Lancet* **2**:1201.

Archibald, E. A., Harrison, J. E., and Pencharz, P. B., 1983, Effect of a weight-reducing high-protein diet on the body composition of obese adolescents, *Am J. Dis. Child.* **137**:658.

Bauer, J. G., 1976, Oral administration of radioactive sulfate to measure extracellular fluid space in man, *J. Appl. Physiol.* **40**:648.

Bedell, G. N., Marshall, R., Dubois, A. B., and Harris, J. H., 1956, Measurement of the volume of gas in the gastrointestinal tract. Values in normal subjects and ambulatory patients, *J. Clin. Invest.* **34**:336.

Behnke, A. R., and Wilmore, J. H., 1974, *Evaluation and Regulation of Body Build and Composition,* Prentice-Hall, Englewood Cliffs, New Jersey.

Behnke, A. R., Feen, B. G., and Welham, W. C., 1942, The specific gravity of healthy men, *J. Am. Med. Assoc.* **118**:495.

Bezold, A. von, 1857, Untersuchungen über die Vertheilung von Wasser, organischer Materie und anorganische Verbindungen im Thierreiche, *Z. Wiss. Zool.* **8**:486.

Boddy, K., Lindsay, R., Holloway, I., Smith, D. A. S., Elliott, A., Robertson, I., and Glaros, D., 1976, A study of changes in whole-body calcium, phosphorus, sodium and nitrogen by neutron activation analysis *in vivo* in rats on a calcium-deficient diet, *Clin. Sci. Mol. Med.* **51**:399.

Bonnard, G. D., 1968, Cortical thickness and diaphyseal diameter of the metacarpal bones from the age of three months to eleven years, *Helv. Paediatr. Acta* **30**:445.

Booth, R. A. D., Goddard, B. A., and Paton, A., 1966, Measurement of fat thickness in man: A comparison of ultrasound, Harpenden calipers, and electrical conductivity, *Br. J. Nutr.* **20**:719.

Borgstedt, A., Forbes, G. B., and Reina, J. C., 1970, Total body potassium and lean body mass in patients with Duchenne dystrophy and their female relations, *Neuropadiatrie* **1**:447.

Borisov, B. K., and Marei, A. N., 1974, Weight parameters of adult human skeleton, *Health Phys.* **27**:224.

Borkan, G. A., Gerzof, S. G., Robbins, A. H., Hults, D. E., Silbert, C. K., and Silbert, J. E., 1982, Assessment of abdominal fat content by computed tomography, *Am. J. Clin. Nutr.* **36**:172.

Bradbury, M. W. B., 1961, Urea and deuterium-oxide spaces in man, *Br. J. Nutr.* **15**:177.

Brodie, B. B., Berger, E. Y., Axelrod, J., Dunning, M. F., Porosowska, Y., and Steele, J. M., 1951, Use of N-acetyl-4-aminoantipyrine (NAAP) in measurement of total body water, *Proc. Soc. Exp. Biol. Med.* **77**:794.

Brook, C. G. D., 1971, Determination of body composition of children from skinfold measurements, *Arch. Dis. Child.* **46**:182.

Brook, C. G. D., Huntley, R. M. C., and Slack, J., 1975, Influence of heredity and environment in determination of skinfold thickness in children, *Br. Med. J.* **2**:719.

Brown, M. R., Klish, W. J., Hollander, J., Campbell, M. A., and Forbes, G. B., 1983, A high protein, low calorie liquid diet in the treatment of very obese adolescents: Long term effect on lean body mass, *Am. J. Clin. Nutr.* **38**:20.

Brozek, J., Grande, F., Anderson, J. T., and Keys, A., 1963, Densitometric analysis of body composition: Revision of some quantitative assumptions, *Ann. N.Y. Acad. Sci.* **110**:113.

Burmeister, V. W., and Fromberg, G., 1970, Depotfet, bestimmt nach der Kalium-40 Methode, und seine Beziehung zur Hautfaltendicke, *Arc. Kinderh.* **180**:228.

Burmeister, W., and Bingert, A., 1967, Die quantitativen

Veränderungen der menschlichen Zellmasse zwischen dem 8 und 90 Lebensjahr, *Klin. Wochenschr.* **45**:409.

Buskirk, E. R., 1961, Underwater weighing and body density: A review of procedures, in: *Techniques for Measuring Body Composition* (J. Brozek and A. Henschel, eds.), pp. 90–107, National Academy of Science, Washington, D.C.

Cheek, D. B., 1968, *Human Growth*, Lea and Febiger, Philadelphia.

Cheek, D. B., and West, C. D., 1955, An appraisal of methods of tissue chloride analysis: The total carcass chloride, exchangeable chloride, potassium and water of the rat, *J. Clin. Invest.* **34**:1744.

Cheek, D. B., Schultz, R. B., Parra, A., and Reba, R. C., 1970, Overgrowth of lean and adipose tissue in adolescent obesity, *Pediatr. Res.* **4**:268.

Christiansen, C., and Rödbro, P., 1975*a*, Estimation of total body calcium from the bone mineral content of the forearm, *Scand. J. Clin. Lab. Invest.* **35**:425.

Christiansen, C., and Rödbro, P., 1975*b*, Bone mineral content and estimated total body calcium in normal adults, *Scand. J. Clin. Lab. Invest.* **35**:433.

Christiansen, C., Rödbro, P., and Nielsen, C. T., 1975, Bone mineral content and estimated total body calcium in normal children and adolescents, *Scand. J. Clin. Lab. Invest.* **34**:507.

Cohn, S. H., Abesamis, C., Zanzi, I., Aloia, J. F., Yasumura, S., and Ellis, K. J., 1977, Body elemental compositions: Comparison between black and white adults, *Am. J. Physiol.* **232**:E419.

Cohn, S. H., and Dombrowski, C. S., 1971, Measurement of total-body calcium, sodium, chlorine, nitrogen and phosphorus in man, *J. Nucl. Med.* **12**:499.

Cohn, S. H., and Ellis, K. J., 1975, Predicting radial bone mineral content in normal subjects, *Int. J. Nucl. Med. Biol.* **2**:53.

Cohn, S. H., Ellis, K. J., Wallach, S., Zanzi, I., Atkins, H. L., and Aloia, J. R., 1974, Absolute and relative deficit in total-skeletal calcium and radial bone mineral in osteoporosis, *J. Nucl. Med.* **15**:428.

Collipp, P. J., Curti, V., Thomas, J., Sharma, R. K., Maddaiah, V. T., and Cohn, S. H., 1973, Body composition changes in children receiving human growth hormone, *Metabolism* **22**:589.

Comstock, G. W., and Livesay, V. T., 1963, Subcutaneous fat determinations from a community-wide chest X-Ray survey in Muscogee County, Georgia, *Ann. N.Y. Acad. Sci.* **110**:475.

Crook, G. H., Bennett, C. A., Norwood, W. D., and Mahaffey, J. A., 1966, Evaluation of skinfold measurements and weight chart to measure body fat, *J. Am. Med. Assoc.* **198**:39.

Cureton, K. J., Boileau, R. A., and Lohman, T. G., 1975, A comparison of densitometric, potassium-40 and skinfold estimates of body composition in prepubescent boys, *Hum. Biol.* **47**:321.

Davies, D. L., and Robertson, J. W. K., 1973, Simultaneous measurement of total exchangeable potassium and sodium using ^{43}K and ^{24}Na, *Metabolism* **22**:133.

Delwaide, P. A., and Crenier, E. J., 1973, Body potassium as related to lean body mass measured by total water determination and by anthropometric method, *Hum. Biol.* **45**:509.

Dequeker, J., and Johnston, C. C., Jr. (eds.), 1982, *Non-invasive Bone Measurements: Methodological Problems*, IRL Press, Oxford.

Dietz, W. H., Jr., and Schoeller, D. A. 1982, Optimal dietary therapy for obese adolescents: Comparison of protein plus glucose and protein plus fat, *J. Pediatr.* **100**:638.

Döbeln, W. von, 1956, Human standard and maximal metabolic rate in relation to fat-free body mass, *Acta Physiol. Scand. (Suppl.)* **37**:*126*.

Döbeln, W. von, 1959, Anthropometric determination of fat-free body weight, *Acta Med. Scand.* **165**:37.

Dombrowski, C. S., Wallach, S., Shukla, K. K., and Cohn, S. H., 1973, Determination of whole body magnesium by *in vivo* neutron activation, *Int. J. Nucl. Med. Biol.* **1**:15.

Durnin, J. V. G. A., and Taylor, A., 1960, Replicability of measurements of density of the human body as determined by underwater weighing, *J. Appl. Physiol.* **15**:142.

Durnin, J. V. G. A., and Womersley, J., 1974, Body fat assessed from total body density and its estimation from skinfold thickness: Measurements on 481 men and women aged from 16 to 72 years, *Br. J. Nutr.* **32**:77.

East, B. W., Boddy, K., and Price, W. H., 1976, Total body potassium content in males with X and Y chromosome abnormalities, *Clin. Endocrinol.* **5**:43.

Edwards, K. D. G., and Whyte, H. M., 1962, The simple measurement of obesity, *Clin. Sci.* **22**:347.

Ellis, K. J., and Cohn, S. H., 1975, Correlation between skeletal calcium mass and muscle mass in man, *J. Appl. Physiol.* **38**:455.

Ellis, K. J., Shukla, K. K., Cohn, S. H., and Pierson, R. N., Jr., 1974, A predictor for total body potassium in man based on height, weight, sex and age: Application in metabolic disorders, *J. Lab. Clin. Med.* **83**:716.

Ellis, K. J., Vaswani, A., Zanzi, I., and Cohn, S. H., 1976, Total body sodium and chlorine in normal adults, *Metabolism* **25**:645.

Flynn, M. A., Murthy, Y., Clark, J., Comfort, G., Chase, G., and Bentley, A. E. T., 1970, Body composition of Negro and white children, *Arch. Environ. Health* **20**:604.

Flynn, M. A., Woodruff, C., and Chase, G., 1972, Total body potassium in normal children, *Pediatr. Res.* **6**:239.

Forbes, G. B., 1962, Methods for determining composition of the human body, *Pediatrics* **29**:477.

Forbes, G. B., 1964, Lean body mass and fat in obese children, *Pediatrics* **34**:308.

Forbes, G. B., 1972, Growth of the lean body mass in man, *Growth* **36**:325.

Forbes, G. B., 1973, Another source of error in the metabolic balance method, *Nutr. Rev.* **31**:297.

Forbes, G. B., 1974, Stature and lean body mass, *Am. J. Clin. Nutr.* **27**:595.

Forbes, G. B., 1976*a*, The adult decline of lean body mass, *Hum. Biol.* **48**:161.

Forbes, G. B., 1976*b*, Biological implications of the adolescent growth process: Body composition, in: *Nutrient Requirements in Adolescence* (J. I. McKigney and H. N. Munro, eds.), pp. 57–66, M. I. T. Press, Cambridge, Massachusetts.

Forbes, G. B., 1981, Pregnancy in the teenager: Biologic aspects, in: *Pregnancy and Childbearing during Adolescence* (E. R. McAnarney and G. Stickle, eds.), pp. 85–90, Alan R. Liss, New York.

Forbes, G. B., 1983*a*, Unmeasured losses of potassium in balance studies, *Am. J. Clin. Nutr.* **38**:347.

Forbes, G. B., 1983*b*, Some influences on lean body mass: Exercise, androgens, pregnancy, and food, in: *Diet and Exercise: Synergism in Health Maintenance* (P. L. White and T. Mondeika, eds.), pp. 75–91, American Medical Association, Chicago.

Forbes, G. B., 1984, Body composition as affected by physical activity and nutrition, *Fed. Proc.* **44**:343.

Forbes, G. B., and Amirhakimi, G. H., 1970, Skinfold thickness and body fat in children, *Hum. Biol.* **42**:401.

Forbes, G. B., and Bruining, G. J., 1976, Urinary creatinine excretion and lean body mass, *Am. J. Clin. Nutr.* **29**:1359.

Forbes, G. B., and Hursh, J. B., 1963, Age and sex trends in lean body mass calculated from K^{40} measurements: With a note on the theoretical basis for the procedure, *Ann. N.Y. Acad. Sci.* **110**:255.

Forbes, G. B., and Lewis, A., 1956, Total sodium, potassium, and chloride in adult man, *J. Clin. Invest.* **35**:596.

Forbes, G. B., and Perley, A. M., 1951, Estimation of total body sodium by isotopic dilution, *J. Clin. Invest.* **30**:558.

Forbes, G. B., and Welle, S. L., 1983, Lean body mass in obesity, *Init. J. Obesity* **7**:99.

Forbes, G. B., Schultz, F., Cafarelli, C., and Amirhakimi, G. H., 1968, Effects of body size on potassium-40 measurement in the whole body counter (tilt-chair technique), *Health Phys.* **15**:435.

Forbes, G. B., Schwartz, R. H., and Nelson, L. A., 1979, Body composition studies in patients with cystic fibrosis, Cystic Fibrosis Foundation, Atlanta, Georgia (abstract).

Forbes, G. B., Lantigua, R. A., Amatruda, J. M., and Lockwood, D. H., 1981, Errors in potassium balance, *Am. J. Clin. Nutr.* **34**:105.

Foy, J. M., and Schnieden, H., 1960, Estimation of total body water (virtual tritium space) in the rat, cat, rabbit, guinea pig and man, and of the biological half-life of tritium in man. *J. Physiol. (London)* **154**:169.

Friis-Hansen, B., 1957, Changes in body water compartments during growth, *Acta Paediatr. (Suppl.)* **110**:46.

Frisch, R. E., 1974, Critical weight at menarche, initiation of the adolescent growth spurt, and control of puberty, in: *Control of Onset of Puberty* (M. Grumbach, G. Grave, and F. Mayer, eds.), pp. 403–423, Wiley, New York.

Frisch, R. E., and Revelle, R., 1970, Height and weight at menarche and a hypothesis of critical body weight and adolescent events, *Science* **169**:397.

Fuller, M. F., Houseman, R. A., and Cadenhead, A., 1971, The measurement of exchangeable potassium in living pigs and its relation to body composition, *Br. J. Nutr.* **26**:203.

Gamble, J. L., Jr., Robertson, J. S., Hannigan, C. A., Foster, C. G., and Farr, L. F., 1953, Chloride, bromide, sodium, and sucrose spaces in man, *J. Clin. Invest.* **32**:483.

Garn, S. M., 1961, Radiographic analysis of body composition, in: *Techniques for Measuring Body Composition* (J. Brozek and A. Henschel, eds.), pp. 36–58, National Academy of Science, Washington, D.C.

Garn, S. M., 1970, *The Earlier Gain and the Later Loss of Cortical Bone in Nutritional Perspective*, Charles C Thomas, Springfield, Illinois.

Garrow, J. S., Fletcher, K., and Halliday, D., 1965, Body composition in severe infantile malnutrition, *J. Clin. Invest.* **44**:417.

Godfrey, B. E., and Wadsworth, G. R., 1970, Total body potassium in pregnant women, *J. Obstet. Gynaecol. Br. Commonw.* **77**:244.

Goldman, R. F., Haisman, M. F., Bynum, G., Horton, E. S., and Sims, E. A. H., 1975, Experimental obesity in man, in: *Obesity in Perspective* (G. A. Bray, ed.), pp. 165–186, U.S. Department of Health Education and Welfare, Publ. No. (NIH) 75-708, Washington D.C.

Greenway, R. M., Littell, A. S., Houser, H. B., Lindan, O., and Weir, D. R., 1965, An evaluation of the variability in measurement of some body composition parameters, *Proc. Soc. Exp. Biol. Med.* **120**:487.

Griggs, R. C., Forbes, G. B., Moxley, R. T., and Herr, B. E., 1983, The assessment of muscle mass in progressive neuromuscular disease, *Neurology* (New York) **33**:158.

Grüner, O., and Salmen, A., 1961, Vergleichende Körperwasserbestimmungen mit Hilfe von N-Acetyl-4-amino Antipyrin und Alkohol, *Klin. Wochenschr.* **39**:92.

Gryfe, C. I., Exton-Smith, A. N., Payne, P. R., and Wheeler, E. F., 1971, Pattern of development of bone in childhood and adolescence, *Lancet* **1**:523.

Häger, A., 1977, Adipose cell size and number in relation to obesity, *Postgrad. Med. J.* **53**(2):101.

Halliday, D., 1971, An attempt to estimate total body fat and protein in malnourished children, *Br. J. Nutr.* **26**:147.

Harrison, J. E., and McNeill, K. G., 1982, Normalization for bone mass measurements by neutron activation analysis, in: *Non-Invasive Bone Measurements: Methodological Problems* (J. V. Dequeker and C. C. Johnston, Jr., eds.), pp. 129–135, IRL Press, Oxford.

Harvey, T. C., Dykes, P. W., Chen, N. S., Ettinger, K. V., Jain, S., James, H., Chettle, D. R., Fremlin, J. H., and Thomas, B. J., 1973, Measurement of whole-body nitrogen by neutron-activation analysis, *Lancet* **2**:395.

Haschke, F., 1983, Body composition of adolescent males, *Acta Paediatr.(Suppl.)* **307**:1.

Hastings, A. B., and Eichelberger, L., 1937, The exchange of salt and water between muscle and blood, *J. Biol. Chem.* **117**:73.

Hawkins, W. W., 1964, Iron, copper, and cobalt, in: *Nutrition: A Comprehensive Treatise* (G. H. Beaton and E. W. McHenry, eds.), Vol. 1, pp. 309–372, Academic Press, New York.

Haxhe, J. J., 1963, *La Composition Corporelle Normale,* Librairie Maloine S. A., Paris.

Heald, F. P., Hunt, E. E., Jr., Schwartz, R., Cook, C. D., Elliot, O., and Vajda, B., 1963, Measures of body fat and hydration in adolescent boys, *Pediatrics* **31**:226.

Hermansen, L., and Döbeln, W., von, 1971, Body fat and skinfold measurements, *Scand. J. Clin. Lab. Invest.* **27**:315.

Heymsfield, S. B., Olafson, R. P., Kutner, M. H., and Nixon, D. W., 1979, A radiographic method of quantifying protein-calorie undernutrition, *Am. J. Clin. Nutr.* **32**:693.

Hoffer, E. C., Meador, C. K., and Simpson, D. C., 1969, Correlation of whole-body impedance with total body water volume, *J. Appl. Physiol.* **27**:531.

Huenemann, R. L., Hampton, M. C., Behnke, A. R., Shapiro, L. R., and Mitchell, B. W., 1974, *Teenage Nutrition and Physique,* Charles C Thomas, Springfield, Illinois.

Hume, R., 1966, Prediction of lean body mass from height and weight, *J. Clin. Pathol.* **19**:389.

Hume, R., and Weyers, E., 1971, Relationship between total body water and surface area in normal and obese subjects, *J. Clin. Pathol.* **24**:234.

Hytten, F. E., and Leitch, I., 1971, *The Physiology of Human Pregnancy,* 2nd ed., Blackwell, Oxford.

Hytten, F. E., Taylor, K., and Taggart, N., 1966, Measurement of total body fat in many by absorption of ^{85}Kr, *Clin. Sci.* **31**:111.

International Atomic Energy Agency, 1970, *Directory of Whole Body Radioactivity Monitors,* Vienna.

Jasani, B. M., and Edmonds, C. J., 1971, Kinetics of potassium distribution in man using isotope dilution and whole-body counting, *Metabolism* **20**:1099.

Johnston, F. E., 1982, Relationships between body composition and anthropometry, *Hum. Biol.* **54**:221.

Johny, K. V., Worthey, B. W., Lawrence, J. R., and O'Halloran, M. W., 1970, A whole body counter for serial studies of total body potassium, *Clin. Sci.* **39**:319.

Katch, F. I., and Michael, E. D., Jr., 1968, Prediction of body density from skinfold and girth measurements of college females, *J. Appl. Physiol.* **25**:92.

Kaufman, L., and Wilson, C. J., 1973, Determination of extracellular fluid volume by fluorescent excitation analysis of bromine, *J. Nucl. Med.* **14**:812.

Khosla, T., 1968, Unfairness of certain events in the Olympic Games, *Br. Med. J.* **4**:111.

King, J. C., Calloway, D. H., and Morgen, S., 1973, Nitrogen retention, total body ^{40}K and weight gain in teenage pregnant girls, *J. Nutr.* **103**:772.

Kjellberg, J., and Reizenstein, P., 1970, Body composition in obesity, *Acta Med. Scand.* **188**:161.

Klish, W. J., Forbes, G. B., Gordon, A., and Cochran, W. J., 1984, New method for the estimation of lean body mass in infants (EMME instrument): Validation in nonhuman models, *J. Ped. Gastrointest. Nutr.* **3**:199.

Knittle, J. L., 1976, Discussion, in: *Nutrient Requirements in Adolescence* (J. I. McKigney and H. N. Munro, eds.), pp. 76–83, M.I.T. Press, Cambridge, Massachusetts.

Kreisberg, R. A., Bowdoin, B., and Meador, C. K., 1970, Measurement of muscle mass in humans by isotopic dilution of creatine-^{14}C, *J. Appl. Physiol.* **28**:264.

Lantigua, R. A., Amatruda, J. M., Biddle, T. L., Forbes, G. B., and Lockwood, D. H., 1980, Cardiac arrhythmias associated with a liquid protein diet for the treatment of obesity, *N. Engl. J. Med.* **303**:735.

Lesser, G. T., Perl, W., and Steele, J. M., 1960, Determination of total body fat by absorption of an inert gas: Measurements and results in normal human subjects, *J. Clin. Invest.* **39**:1791.

Lesser, G. T., Deutsch, S., and Markofsky, J., 1971, Use of independent measurement of body fat to evaluate overweight and underweight, *Metabolism* **20**:792.

Linquette, M., Fossati, P., Lefebvre, J., and Checkau, C., 1969, The measurement of total water, exchangeable sodium and potassium in obese persons, in: *Pathophysiology of Adipose Tissue* (J. Vague, ed.), pp. 302–316, Excerpta Medica Foundation, Amsterdam.

Ljunggren, H., Ikkos, D., and Luft, R., 1957, Studies on body composition, *Acta Endocrinol.* **25**:187.

Loeppky, J. A., Myhre, L. G., Venters, M. D., and Luft, U. C., 1977, Total body water and lean body mass estimated by ethanol dilution, *J. Appl. Physiol.* **42**:803.

Lohman, T. G., 1981, Skinfolds and body density and their relation to body fatness, *Hum. Biol.* **53**:181.

Long, C. L., Haverberg, L. N., Young, V. R., Kinney, J. M., Munro, H. N., and Geiger, J. W., 1975, Metabolism of 3-methyl histidine in man, *Metabolism* **24**:929.

Lukaski, H., and Mendez, J., 1980, Relationship between fat-free weight and urinary 3-methylhistidine excretion in man, *Metabolism* **29**:758.

Lukaski, H. C., Mendez, J., Buskirk, E. R., and Cohn, S. H., 1981*a,* A comparison of methods of assessment of body composition including neutron activation analysis of total body nitrogen, *Metabolism* **30**:777.

Lukaski, H. C., Mendez, J., Buskirk, E. R., and Cohn, S. H., 1981*b,* Relationship between endogenous 3-methylhistidine excretion and body composition, *Am. J. Physiol.* **240**:E302.

Lye, M., May, T., Hammick, J., and Ackery, D., 1976,

Whole body and exchangeable potassium measurements in normal elderly subjects, *Eur. J. Nucl. Med.* **1**:167.

MacGillivray, I., and Buchanan, T. J., 1958, Total exchangeable sodium and potassium in non-pregnant women and in normal and pre-eclamptic pregnancy, *Lancet* **2**:1090.

Magnus-Levy, A., 1906, Physiologie des Stoffwechsels, in: *Handbuch der Pathologie des Stoffwechsels* (C. von Noorden, ed.), p. 446, Hirschwald, Berlin.

Manery, J. F., and Haege, L. F., 1941, The extent to which radioactive chloride penetrates tissues, and its significance, *Am. J. Physiol.* **134**:83.

Manzke, E., Chestnut, C. H. III, Wergedal, J. E., Baylick, D. J., and Nelp, W. B., 1975, Relationship between local and total bone mass, *Metabolism* **24**:605.

Maresh, M. M., 1966, Changes in tissue widths during growth, *Am. J. Dis. Child.* **111**:142.

Mazess, R. B., and Cameron, J. R., 1972, Growth of bone in school children: Comparison of radiographic morphometry and photon absorptiometry, *Growth* **36**:77.

McCance, R. A., and Widdowson, E. M., 1951, A method of breaking down the body weights of living persons into terms of extracellular fluid, cell mass and fat, and some applications of it to physiology and medicine, *Proc. R. Soc. Ser. B* **138**:115.

Meneely, G. R., Heyssel, R. M., Ball, C. O. T., Weiland, R. L., Lorimer, A. R., Constantinides, C., and Meneely, E. U., 1963, Analysis of factors affecting body composition determined from potassium content in 915 normal subjects, *Ann. N.Y. Acad. Sci.* **110**:271.

Merritt, R. J., Bistrian, B. R., Blackburn, G. L., and Suskind, R. M., 1980, consequences of modified fasting in obese pediatric and adolescent patients. I. Protein-sparing modified fast, *J. Pediatr.* **96**:13.

Mettau, J. W., Degenhart, J. J., Visser, H. K. A., and Holland, W. P. S., 1977, Measurement of total body fat in newborns and infants by absorption and desorption of nonradioactive xenon, *Pediatr. Res.* **11**:1097.

Michael, E. D., Jr., and Katch, F. I., 1968, Prediction of body density from skinfold and girth measurements of 17-year-old boys, *J. Appl. Physiol.* **25**:747.

Moore, F. D., Lister, J., Boyden, C. M., Ball, M. R., Sullivan, N., and Dagher, F. J., 1968, The skeleton as a feature of body composition: Values predicted by isotope dilution and observed by cadaver dissection in adult female, *Hum. Biol.* **40**:135.

Morgan, D. B., and Burkinshaw, L., 1983, Estimation of non-fat body tissues from measurements of skinfold thickness, total body potassium and total body nitrogen, *Clin. Sci.* **65**:407.

Moulton, C. R., 1923, Age and chemical development in mammals, *J. Biol. Chem.* **57**:79.

Muldowney, F. P., 1961, Lean body mass as a metabolic reference standard, in: *Techniques for Measuring Body Composition* (J. Brozek and A. Henschel, eds.), pp. 212–222, National academy of Science, Washington, D.C.

Nagamine, S., and Suzuki, S., 1964, Anthropometry and body composition of Japanese young men and women, *Hum. Biol.* **36**:8.

National Center for Health Statistics, 1974, Skinfold thickness of youths 12–17 years, U.S. Department of Health Education and Welfare, Publ. No. (HRA) 74-614, Series 11, No. 132, Rockville, Maryland.

Nelp, W. B., Palmer, H. E., Murano, R., Pailthorp, K., Hinn, G. M., Rich, C., Williams, J. L., Rudd, T. G., and Denney, J. D., 1970, Measurement of total body calcium (bone mass) *in vivo* with the use of total body neutron activation, *J. Lab. Clin. Med.* **76**:151.

Nichols, B. L., Hazlewood, C. F., and Barnes, D. J., 1968, Percutaneous needle biopsy of quadriceps muscle: Potassium analysis in normal children, *J. Pediatr.* **72**:840.

Novak, L. P., 1966*a*, Total body water and solids in six- to seven-year-old children: Differences between the sexes, *Pediatrics* **38**:483.

Novak, L. P., 1966*b*, Physical activity and body composition of adolescent boys, *J. Am. Med. Assoc.* **197**:891.

Novak, L. P., 1970, Comparative study of body composition of American and Filipino women, *Hum. Biol.* **42**:206.

Novak, L. P., Tauxe, W. N., and Orvis, A. L., 1973, Estimation of total body potassium in normal adolescents by whole body counting: Age and sex differences, *Med. Sci. Sports* **5**:147.

Oberhausen, E., Burmeister, W., and Huycke, E. J., 1965, Das Wachstum des Kaliumbestandes im Menschen gemessen mit dem Gänzkörperzähler, *Ann. Paediatr.* **205**:381.

Owen, G. M., Jensen, R. L., and Fomon, S. J., 1962, Sex-related difference in total body water and exchangeable chloride during infancy, *J. Pediatr.* **60**:858.

Pace, N., and Rathben, E. N., 1945, Studies on body composition, *J. Biol. Chem.* **158**:685.

Pearson, A. M., Purchas, R. W., and Reineke, E. P., 1968, Theory and potential usefulness of body density as a predictor of body composition, in: *Body Composition in Animals and Man* (J. T. Reid, ed.), Publ. 1598, pp. 153–185, National Academy of Science, Washington, D.C.

Pfeiffer, L., 1887, Über den Fettgehalt des Körpers, *Z. Biol.* **23**:340.

Picou, D., Reeds, P. J., Jackson, A., and Poulter, N., 1976, The measurement of muscle mass in children using [^{15}N]creatine, *Pediatr. Res.* **10**:184.

Pitts, G. C., and Bullard, T. R., 1968, Some interspecific aspects of body composition in mammals, in: *Body Composition in Animals and Man* (J. T. Reid, ed.), Publ. 1598, pp. 45–70, National Academy of Science, Washington, D.C.

Pollock, M. L., Hickman, T., Kendrick, A., Jackson, A., Linnerud, A., and Dawson, G., 1976, Prediction of body density in young and middle-aged men, *J. Appl. Physiol.* **40**:300.

Presta, E., Wang, J., Harrison, G. G., Björntorp, P., Harker, W. H., and van Itallie, T. B., 1983, Measurement of total body electrical conductivity: A new method for estimation of body composition, *Am. J. Clin. Nutr.* **34**:735.

Presta, E., Segal, K. R., Gutin, B., Harrison, G. G., and van Itallie, T. B., 1983, Comparison in man of total body electrical conductivity and lean body mass derived from body density: Validation of a new body composition method, *Metabolism* **32**:524.

Price, W. F., Hazelrig, J. B., Kreisberg, R. A., and Meador, C. K., 1969, Reproducibility of body composition measurements in a single individual, *J. Lab. Clin. Med.* **74**:557.

Ringe, J. D., Rehpenning, W., and Kuhlencordt, F., 1977, Physiologie Änderung des Mineral gehalts von Radius und Ulna in Abhängigkeit von Lebensalter und Geschlecht, *Fortschr. Röntgenstr.* **126**:376.

Rolland-Cachera, M. F., Sempé, M., Guilloud-Bataille, M., Patois, E., Pequignot-Guggenbuhl, F., and Fautrad, V., 1982, Adiposity indices in children, *Am. J. Clin. Nutr.* **36**:178.

Rudd, T. G., Pailthorp, K. G., and Nelp, W. B., 1972, Measurement of non-exchangeable sodium in normal man, *J. Lab. Clin. Med.* **80**:442.

Rundo, J., and Sagild, U., 1955, Total and "exchangeable" potassium in humans, *Nature (London)* **175**:774.

Russell, D. McR., Prendergast, P. J., Darby, P. L., Garfinkel, P. E., Whitwell, J., and Jeejeebhoy, K. N., 1983, A comparison between muscle function and body composition in anorexia nervosa: the effect of refeeding, *Am. J. Clin. Nutr.* **38**:229.

Schoeller, D. A., Dietz, W., van Santen, E., and Klein, P. D., 1982, Validation of saliva sampling for total body water determination by $H_2{}^{18}O$ dilution, *Am. J. Clin. Nutr.* **34**:591.

Schuster, W., 1970, Über Methoden und Ergebnisse quantitativer Mineralsalzbestimmungen am kindlichen Skelett, *Arch. Kinderh.* **180**:256.

Schutte, J. E., 1980, Growth differences between lower and middle income black male adolescents, *Hum. Biol.* **52**:193.

Schutte, J. E., Longhurst, J. C., Gaffney, F. A., Bastian, B. C., and Blomqvist, C. G., 1981, Total plasma creatinine: An accurate measure of total striated muscle, *J. Appl. Physiol.* **51**:762.

Scott, E. C., and Johnston, F. E., 1982, Critical fat, menarche, and the maintenance of menstrual cycles: A critical review, *J. Adolescent Health Care* **2**:249.

Seitchik, J., 1967, Total body water and total body density of pregnant women, *Obstet. Gynecol.* **29**:155.

Shaffer, P., 1908, The excretion of kreatinin and kreatin in health and disease, *Am. J. Physiol.* **23**:1.

Shephard, R. J., Hatcher, J., and Rode, A., 1973, On the body composition of the Eskimo, *Eur. J. Appl. Physiol.* **32**:3.

Shepherd, R. W., Thomas, B. J., Bennett, D., Cooksley, W. G. E., and Ward, L. C., 1983, Changes in body composition and muscle protein degradation during nutritional supplementation in nutritionally growth-retarded children with cystic fibrosis, *J. Ped. Gastroenterol. Nutr.* **2**:439.

Shock, N. W., 1943, The effect of menarche on basal physiological functions in girls, *Am. J. Physiol.* **139**:288.

Shukla, K. K., Ellis, K. J., Dombrowski, C. S., and Cohn, S. H., 1973, Physiological variation of total-body potassium in man, *Am. J. Physiol.* **224**:271.

Siri, W. E., 1961, Body composition from fluid spaces and density: Analysis of methods, in: *Techniques for Measuring Body Composition* (J. Brozek and A. Henschel, eds.), pp. 223–244, National Academy of Science, Washington, D.C.

Sloan, A. W., 1967, Estimation of body fat in young men, *J. Appl. Physiol.* **23**:311.

Smith, D. M., Nance, W. E., Kang, K. W., Christian, J. C., and Johnston, C. C., Jr., 1973, Genetic factors in determining bone mass, *J. Clin. Invest.* **52**:2800.

Steinkamp, R., Cohen, N. L., Gaffey, W. R., McKey, T., Bron, G., Siri, W. E., Sargent, T. W., and Isaacs, E., 1965, Measures of body fat and related factors in normal adults—II: A simple clinical method to estimate body fat and lean body mass, *J. Chron. Dis.* **18**:1291.

Stouffer, J. R., 1963, Relationship of ultrasonic measurements and X-rays to body composition, *Ann. N.Y. Acad. Sci.* **110**:31.

Stuart, H. C., Hill, P., and Shaw, C., 1940, The growth of bone, muscle and overlying tissue as revealed by studies of roentgenograms of the leg area, *Monogr. Soc. Res. Child Dev.* V: (No. 3, serial No. 26) 1–190.

Surveyor, I., and Hughes, D., 1968, Discrepancies between whole-body potassium content and exchangeable potassium, *J. Lab. Clin. Med.* **71**:464.

Talso, P. J. Miller, C. E., Carballo, A. J., and Vasquez, I., 1960, Exchangeable potassium as a parameter of body composition, *Metabolism* **9**:456.

Tanner, J. M., 1974, Sequence and tempo in the somatic changes in puberty, in: *Control of the Onset of Puberty* (M. M. Grumbach, G. D. Grave, and F. E. Mayer, eds.), pp. 448–470, Wiley, New York.

Tanner, J. M., and Whitehouse, R. H., 1967, The effect of human growth hormone on subcutaneous fat thickness in hyposomatotrophic and panhypopituitary dwarfs, *J. Endocrinol.* **39**:263.

Tanner, J. M., and Whitehouse, R. H., 1975, Revised standards for triceps and subscapular skinfolds in British children, *Arch. Dis. Child.* **50**:142.

Tisavipat, A., Vibulsreth, S., Sheng, H.-P., and Huggins, R. A., 1974, Total body water measured by desiccation and by tritiated water in adult rats, *J. Appl. Physiol.* **37**:699.

Trotter, M., and Hixon, B. B., 1974, Sequential changes in weight, density, and percentage ash weight of human skeletons from an early fetal period through old age, *Anat. Rec.* **179**:1.

Wallace, W. M., 1959, Nitrogen content of the body and its relation to retention and loss of nitrogen, *Fed. Proc.* **18**:1125.

Ward, G. M., Krzywicki, H. J., Rahman, D. P., Quaas, R. L., Nelson, R. A., and Consolazio, C. F., 1975, Relationship of anthropometric measurements to body fat as determined by densitometry, potassium-40, and body water, *Am. J. Clin. Nutr.* **28**:162.

Ward, L. C., Miller, M., Thomas, B. J., Cooksley, W. G., and Shepherd, R., 1981, 3-Methylhistidine excretion in children: Relationship with creatinine, body weight and fat-free mass, *IRCS Med. Sci.* **9**:725.

Webb, P., 1981, Energy expenditure and fat-free mass in men and women, *Am. J. Clin. Nutr.* **34**:1816.

Wedgewood, R. J., 1963, Inconstancy of the lean body mass, *Ann. N.Y. Acad. Sci* **110**:141.

Widdowson, E.M. McCance, R.A., and Spray, C.M., 1951, The chemical composition of the human body, *Clin. Sci.* **10**:113.

Wieringen, J. C. van, 1971, *Secular Changes in Growth,* p. 193, Netherlands Institute for Preventive Medicine, Leiden.

Wilmore, J. H., 1983, Body composition in sport and exercise: Directions for future research, *Med. Sci. Sports Exercise* **15**:21.

Wilmore, J. H., and Behnke, A. R., 1970, An anthropometric estimation of body density and lean body weight in young women, *Am. J. Clin. Nutr.* **23**:267.

Wilmore, J. H., Girandola, R. N., and Moody, D. L., 1970, Validity of skinfold and girth assessment for predicting alterations in body composition, *J. Appl. Physiol.* **29**:313.

Womersley, J., Durnin, J. V. G. A., Boddy, K., and Mahaffy, M., 1976, Influence of muscular development, obesity, and age on the fat-free mass of adults, *J. Appl. Physiol.* **41**:223.

Young, C., Sipin, S. S., and Roe, D. A., 1968, Body composition of pre-adolescent and adolescent girls, *J. Am. Diet. Assoc.* **53**:25, 357.

14. Suggested Readings

Brozek, J., 1961, Body composition, *Science* **134**:920.

Brozek, J. (ed.), 1965, *Human Body Composition: Approaches and Applications,* Pergamon Press, Oxford.

Brozek, J., and Henschel, A. (eds.), 1961, *Techniques for Measuring Body Composition,* National Academy of Science-National Research Council, Washington, D.C.

Cohn, S. H., 1981, *In vivo* neutron activation analysis: State of the art and future prospects, *Med. Phys.* **8**:145.

Cohn, S. H. (ed.), 1981, *Non-invasive Measurements of Bone Mass and Their Clinical Application,* CRC Press, Boca Raton, Florida.

Elkinton, J. R., and Danowski, T. S., 1955, *The Body Fluids,* Williams & Wilkins, Baltimore.

Garrow, J. S., 1974, *Energy Balance and Obesity in Man,* American Elsevier, New York.

Garrow, J. S., 1981, *Treat Obesity Seriously,* Churchill Livingstone, London.

Malina, R. M., 1969, Quantification of fat, muscle, and bone in man, *Clin. Orthop.* **65**:9.

Moore, F. D., Olesen, K. H., McMurray, J. D., Parker, H. V., Ball, M. R., and Boyden, C. M., 1963, *The Body Cell Mass and Its Supporting Environment,* W. B. Saunders, Philadelphia.

National Academy of Sciences, 1968, Body composition in Animals and Man, Publ. 1598, Washington, D.C.

Parizková, J., 1977, *Body Fat and Physical Fitness,* Martinus Nijhoff, The Hague.

Sheng, H.-P., and Huggins, R. A., 1979, A review of body composition studies with emphasis on total body water and fat, *Am. J. Clin. Nutr.* **32**:630.

Tanner, J. M., 1962, *Growth at Adolescence,* 2nd ed., Blackwell, Oxford.

Whipple, H. E., and Brozek, J. (eds.), 1963, Body composition, *Ann. N.Y. Acad. Sci.* **110**:1.

7

Physical Activity and Growth of the Child

DONALD A. BAILEY, ROBERT M. MALINA, and ROBERT L. MIRWALD

1. Introduction

> [a] condition of strain, the result of stress, is a direct stimulus to growth itself. This indeed is no less than one of the cardinal facts of theoretical biology. The soles of our boots wear thin, but the soles of our feet grow thick, the more we walk upon them: therefore it would seem that the living cells are "stimulated" by pressure, or what we call "exercise", to increase and multiply.
>
> D'Arcy Thompson
> *On Growth and Form* (1917)

Whereas knowledge about the effect of exercise on child growth has been expanding rapidly in recent years, it is still quite amazing how many gaps exist in our understanding of this subject since D'Arcy Wentworth Thompson wrote his classic treatise *On Growth and Form* almost 70 years ago. A survey of the literature reveals that while much is known, much remains to be learned.

DONALD A. BAILEY and ROBERT L. MIRWALD • College of Physical Education, University of Saskatchewan, Saskatoon, Saskatchewan S7N 0W0 Canada. ***ROBERT M. MALINA*** • Department of Anthropology, The University of Texas, Austin, Texas 78712.

Reviews dealing with the topic of exercise and growth (Espenschade, 1960; Rarick, 1974; Bailey *et al.*, 1978*a;* Malina, 1979, 1980, 1983*a*) have identified a major problem faced by most investigators working with growing children. This has to do with the fact that physical training may induce changes in the same direction and of approximately the same magnitude as expected growth changs. Thus, it becomes difficult, if not impossible, to separate out exercise change from normally expected growth change when studying children. For this reason, much of the early literature was incomplete, arriving at the general conclusion that a certain minimum physical activity is necessary to support normal growth, but failing to identify what this minimum is, or should be, and what effect more intensive activity may have.

It should be emphasized that physical activity is only one of the many factors that may affect the growing child. Thus, the precise role of properly graded training programs in influencing growth and development is difficult to define and not completely understood.

The general aim of this chapter is to consolidate what is known in order that problems may be viewed in perspective. Specifically, the sections that follow select, describe, and evaluate the pertinent literature on physical activity, focusing on the differences in response to exercise between children and adults.

2. General Considerations

Before proceeding to an examination of the differences in the response to exercise between children and adults and the effects of physical activity and training on growth, there is a need for a brief discussion dealing with study design and techniques for assessing training and growth changes. Also, there is a need to clarify certain exercise-related terms and concepts.

2.1. Physical Activity and Training

Physical activity is not necessarily the same as regular physical training. Physical activities are obviously a part of training programs, but not all physical activities qualify as training. Physical training refers to the regular, systematic practice of specific physical activities; e.g., calisthenics, lifting weights, isometric exercises, running, and games or sport activities, performed at specific intensities for a defined duration. Training programs vary in kind or type; i.e., endurance running or swimming, strength training, sprint training, or skill training. The effects of such programs are generally specific to the type of training stimulus (Edgerton, 1976; Fox, 1977; McCafferty and Horvath, 1977; Saltin and Rowell, 1980; Bouchard *et al.*, 1981), although training effects induced by one kind of program may be more general. For example, running apparently has more general effects than cycling. Thus, training is not a single entity, but varies in kind, intensity and duration. It can be viewed as a continuum, ranging from relatively mild work to severely stressing activity.

In studies of training during growth, programs vary in type, intensity, and duration and are often described as mild, moderate, or severe without more specific definition. At times, youngsters are simply defined as "active" or "inactive." These labels are often based on teacher and/or coach assessment of frequency and duration of sports participation, as well as self-reported activity levels. There is thus a need to qualify and quantify training programs; e.g., the number of sessions per week, the duration of workouts, and the distance covered during a workout and at what intensity. For example, age-group swimmers in some programs may swim about 4500 m per session at varying time intervals 6 days per week, while in other programs they may swim only 2500 m per session 3 days per week.

2.2. Study Designs

Data from a variety of studies have been used to make inferences about the effects of regular physical activity on human growth and maturation:

1. Experimental studies compare trained (treatment) and untrained (control) groups. The training stimulus usually varies in type, intensity, and duration, so that problems are generally encountered in defining and quantifying the training stimulus within and between studies. Selection of subjects, motivation to train, and control of outside activity are also critical factors. In addtion, variable and composite age groups of children and youth are used, as are small samples. Attrition rates tend to be high, and most studies are short term. There are few experimental studies in which the training factor has been regularly applied and changes monitored over a sufficiently long period during growth. Difficulties inherent in conducting longitudinal programs with children and youth are obvious. Hence, a significant amount of experimental data are derived by extrapolation from studies of animals. Although reasonable generalizations can be made for human growth, the concept of species specificity must be recognized. A limitation of studies monitoring changes associated with regular training during youth is that the focus is not the growth and maturation of children. Many studies have as their focus physiological changes associated with training, e.g., maximal aerobic power and metabolic substrates. Growth observations are usually made in passing or indirectly, whereas maturity status is generally not considered.
2. Comparisons of athletes and nonathletes during childhood and adolescence are commonly used to make inferences about the effects of physical training on growth and maturation. It is assumed that the athletes had been training regularly, and differences in growth and maturation relative to nonathletes are attributed to the training programs required for the specific sports. Problems with such an approach are the definition of an athlete at young ages and subject selection. Youngsters proficient in sports are undoubtedly selected for skill and in some sports for size. Size, physique, strength, and motor skill proficiency are related, and an individual's strength and motor ability may

in turn influence his or her level of habitual activity. Maturity differences also characterize youngsters who excel in sports, and these differences are especially apparent over the adolescent years. Males who are successful in sports competition are more often than not somewhat advanced in biological maturity status. This probably reflects the size, strength, and performance advantages associated with earlier maturation. By contrast, females who excel in sports tend to be average or late in biological maturity status. Swimmers tend toward the average, while female athletes in other sports tend to be late maturing. Late-maturing girls tend to be more linear in physique and leaner, both factors that may be more suitable for sports performance. Another factor that must be considered in comparing young athletes and nonathletes is the role of social circumstances, i.e., socialization into or away from sports. Social factors may interact or vary with a youngster's growth and maturity progress and in turn influence the child's sporting or activity pursuits.

3. Comparisons of adult athletes with nonathletes or the general population are also used to make inferences on the effects of regular physical activity during growth. It is assumed that the adult athletes began training during their youth, and the differences relative to nonathletes reflect training effects on growth and maturation processes. Most recently, such an approach was used to make inferences on the effects of early training on the sexual maturation of young girls (Frisch *et al.*, 1981). Problems associated with subject selection, age of onset of training, motivation to persist in training programs, and so on are similar to those already mentioned.

4. Some activities require extreme levels of unilateral effort; i.e., tennis, baseball pitching, and specific manual occupations. Such specialized activity is occasionally used to illustrate training effects. The individual is his or her own control, as the dominant limb (trained) is compared with the nondominant limb (untrained). Observations from such studies are ordinarily limited to skeletal and muscular observations of the upper extremity.

5. The cotwin control method involves comparisons of monozygotic twins discordant for regular physical activity. One member of the twin pair is regularly trained, while the other member is not and/or follows his or her usual pattern of activity. Data from such studies primarily consider physiological variables (Klissouras *et al.*, 1972; Bouchard and Malina, 1983).

6. Clinical and experimental observations of prolonged bed rest, immobilization with casts, and muscular inactivation as in nerve injuries also provide insights into the role of physical activity in developing and maintaining the integrity of skeletal and muscular tissues. These are perhaps more apparent in the muscular atrophy and loss of skeletal mineral content associated with prolonged inactivity or disuse.

2.3. Techniques for Assessing Change

Traditionally, the evaluation of physical growth has relied on the application of anthropometric techniques. Comparison of these physical measures with known standards permits an appraisal of the size and shape status of the individual relative to age and sex peers. Similarly, the measurement of body composition and function has added to the assessment of a child's growth status. Indices of developmental or biological age define an individual's progress toward maturity as related to a standard population. These more commonly used techniques have proved invaluable to the study of normal and abnormal growth and will continue to be so. However, researchers are continually in search of more efficient approaches. Obviously, the complicated problems associated with exercise and growth require sophisticated research solutions that must look beyond the limits of traditional methodology.

3. Adaptation to Exercise

Physical activity subjects the developing organism to a variety of stresses that may give rise to any number of significant responses. Whether adapting to repeated exercise exposure or to a single bout of exercise, the growing child undergoes changes. The magnitude of these changes varies with the timing, duration, and intensity of the exercise stimulus. While our knowledge of children's adaptation to exercise is still deficient, the following section endeavors to amplify what is known about the role of exercise on the growing organism.

3.1. Exercise and Stature

Although some early studies suggested an increase in stature with regular training (Beyer, 1896; Schwartz *et al.*, 1928; Adams, 1938), the observed changes were small, subject selection was not controlled, and no allowance was made for maturity status. A study by Astrand *et al.* (1963) on the functional capacity of elite female swimmers in Sweden is often cited suggesting that training stimulates growth in stature. A close examination of the data for the 30 swimmers, however, indicates that the girls were in fact taller than average at 7 years of age, and apparently entered adolescence at an earlier age than the Swedish reference data for the 1950s. The apparent acceleration in statural growth was not related to the intensity of training. Rather, it was probably related to the swimmer's somewhat earlier maturation. Other studies of elite young female swimmers suggest the same thing (Malina *et al.,* 1982; Meleski *et al.,* 1982).

In a study of young boys (11–13 years of age) selected for swimming training, Milicer and Denisiuk (1964) noted slightly greater average increments in stature over 2 years of training compared with a control sample. The apparent rate difference in linear growth most likely reflects maturity variation. The young swimmers included a greater percentage of early maturers (25%) compared with the control sample (17%) and did not include any late maturers.

Three studies that assessed the effects of endurance running training on boys aged 10–15 did not include measures of maturity status, although the boys' statures at the start of training corresponded to accepted reference data. In Ekblom's (1969) study, five 11-year-old boys trained for 32 months and showed a somewhat accelerated growth rate in stature compared with Swedish reference data. By contrast, the control group of four boys did not. Since maturity status was not controlled, the boys could have experienced all or part of their adolescent spurts, given the normal variation in timing and intensity of the male adolescent growth spurt (Tanner, 1962).

Eriksson's (1972) observations are similar. A sample of 12 boys, 11–13 years of age, increased, on the average, 3.5 cm in stature after 16 weeks' endurance training. An average gain of 3.5 cm over 4 months, however, would correspond to an annual gain of about 10 cm, which would seem to suggest the adolescent growth spurt. Since maturity status was not controlled, it is difficult to ascertain whether the observed changes are the result of training or simply of normal adolescent growth. The six boys, 10–15 years of age, followed by Daniels and Oldridge (1971), trained for 22 months in endurance running. Their statures equaled the reference data at the start of the study but, in contrast to the above studies, were slightly below the standard at the termination of the study.

While there is no strong evidence to suggest that regular training stimulates growth in stature, it is likewise apparent that the experience of training and competition in sports does not appear to have a negative effect on height. This is relevant to the frequently misquoted early study of Rowe (1933), who compared the growth in stature and weight of male athletes and nonathletes, aged 13.7–15.7, over a 2-year period. Rowe noted that the athletes were taller but that they grew at a slower rate over the 2 years. This observation has been misinterpreted by some as evidence of a negative growth due to athletic participation, without consideration of Rowe's qualifying observation that the differences between groups probably reflected differential timing of the adolescent spurt.

In a study by Mirwald *et al.* (1981) of highly active and inactive boys studied over a 10-year period from 7 to 16 years, no difference in stature between groups was observed at any time during the study when developmental age was taken into consideration.

When the literature is critically analyzed, taking into account the importance of controlling for maturity status, it is clear that regular physical activity has no apparent effect on stature in growing individuals.

3.2. Exercise and Maturation

3.2.1. Skeletal Development

Although regular activity functions to enhance skeletal mineralization and density, it does not accelerate or delay skeletal maturity as assessed in growth studies, using hand–wrist X-ray films. Cerny (1969) monitored the skeletal maturity of boys engaged in three different activity levels of training. There were no skeletal maturity differences among the three groups at the start, during, and at the completion of the study. Rather, variation in skeletal maturity within the different activity groups was greater than bewteen. Kotulan *et al.* (1980) followed the skeletal maturity of young male athletes training regularly for cycling, rowing, and ice hockey from 12 to 15 years of age. Over the three year period, the gains in skeletal maturity varied between 3.0 and 3.6 years in the athletes and did not differ from control subjects and youngsters who started the training program but dropped out. In a sample of elite female athletes including gymnasts, figure skaters, tennis players, volleyball players, and football players, Novotny (1981) assessed their skeletal maturity initially and after 3 or 4 years of regular training. The young athletes were rated as advanced, normal, or retarded in skeletal maturity. After the period of regular training, only 19 (21%) of the 89 girls changed categories, while 70 (79%) remained in the same skeletal maturity category. Of the small number who changed categories, 11 shifted from advanced to normal or from normal to delayed, while 8 shifted from delayed to normal or from normal to advanced. In addition, at the beginning and end of the study, mean chronological and skeletal ages of the young athletes did not differ significantly.

The results of these three studies from Czechoslovakia would thus seem to indicate that skeletal maturity as assessed in growth studies is not affected by regular physical training in young adolescent boys and girls.

3.2.2. Age of Menarche

Intensive physical training has been suggested as a factor that may delay menarche in young girls (see Malina, 1983*b*). The data that deal with intensive training and menarche are quite limited and not well controlled for other factors that influence the time of menarche. Inferences on the role of training are largely based on the observation that menarche occurs, on the average, later in athletes than in nonathletes, and later in those who began training prior to the maturational event than in those who began training after menarche.

The relationship between training and menarche is currently of considerable interest, given the conclusion of Frisch *et al.* (1981) that intense physical activity does in fact delay menarche. This conclusion is based on a small sample of university athletes consisting of 12 swimmers and 6 runners who began training before menarche. The correlation between age at menarche and years of training before menarche was 0.53, a moderate relationship. Correlation, however, indicates only a relationship between two variables and does not imply a cause–effect sequence. The conclusions of this study need to be more thoroughly studied. Data from a small sample of Olympic volleyball players ($N = 18$) show no correlation ($r = -0.05$) between age at menarche and duration of training before menarche (Malina, 1983*b*).

A corollary of the suggestion that training delays menarche is the postulate that changes in weight or body composition associated with training may function to delay menarche. This is related to the critical weight or critical fatness hypothesis put forward by Frisch (1976). This hypothesis has been discussed at length by many investigators (Johnston *et al.*, 1975; Malina, 1978; Trussell, 1980; Scott and Johnston, 1982), with the conclusion that the data do not support the specificity of weight or fatness as the critical variable for the onset of menarche. There is a moderately high correlation between age at menarche and skeletal maturity and a reduced variance in skeletal ages at the time of menarche (Tanner, 1962; Malina, 1978). Furthermore, the process of skeletal maturation is influenced by gonadal and other hormones. Thus, if the hormonal responses to regular training were to influence sexual maturation, one might expect them to influence skeletal maturation as well, especially during the pubertal years. This is clearly not the case in the three studies of training and skeletal maturity already discussed.

3.3. Exercise, Body Weight, and Composition

Regular physical activity is an important factor in the regulation and maintenance of body weight. The composition of body weight is frequently thought of as a two-compartment system with known densities for the fat and nonfat (lean body) components. The assumptions underlying this model have been questioned (Lohman, 1982; Ross and Marfell-Jones, 1982; nevertheless, the literature on changes in body composition with exercise relies heavily on this system. Regular training generally results in an increase in lean body mass and in a corresponding decrease in body fat in children and youth. Training produces similar effects in adults, often without any appreciable change in body weight (Parizkova, 1977), but results are not consistent across studies (Forbes, 1978). The magnitude of change in body composition with regular activity varies with the intensity and duration of the program, and the changes are dependent on continued activity. Parizkova (1977), for example, reported fluctuations in body fatness and density that were proportional to the intensity of training in 10 young gymnasts aged 13–18 followed longitudinally over 5 years. Fat levels decreased as the girls engaged in training for the competitive season and increased once again as training tapered.

Body composition changes considerably during normal growth and maturation so that it is difficult to separate effects of training from those associated with normal growth. Furthermore, the continuity of fatness levels from childhood through adolescence is rather weak (Cronk *et al.*, 1983), emphasizing the variation in fatness associated with growth and maturation.

In one of the more comprehensive studies of training and body composition, Parizkova (1977) followed boys engaged in different levels of sports participation and training over a 7-year period from 11 to 18 years. Three levels of training were compared: (1) regularly trained (intensive, 6 hr/week); (2) trained but not on a regular basis (in sport schools, about 4 hr of organized exercise per week); and (3) untrained (about 2.5 hr/week, including school physical education). Sample sizes for the three groups at the conclusion of the study, however, were small: 8, 18, and 13, respectively. The groups did not differ either in anthropometric characteristics at the beginning of the study or in body composition. During the course of the study and at its end, the most active boys had significantly more lean body mass and less fat than did the least and moderately active boys, who differed among themselves only slightly.

Von Dobeln and Eriksson (1972) reported significant changes in body composition in nine boys aged 11–13 after an endurance training program. Using potassium measurements as an index of muscle mass, the increase in potassium following training was about 6% greater than expected, while the gain in body weight was 5% less than expected relative to linear growth. It should be noted that the boys gained an average of 3.5 cm in stature over the 16-week program (Eriksson, 1972), indicating that the adolescent growth spurt had occurred during the program. A significant increase in muscle mass is known to accompany male adolescence; therefore, the reported changes probably reflect growth as well as training. The results of the above two studies summarize quite well the information on training and body composition. Youngsters regularly engaged in physical activity programs, be they formal training for sport or recreational activities, are generally leaner, i.e., have more lean body mass and less fat than do those who are not regularly active (Ruffer, 1965; Malina *et al.*, 1982). Two questions remain: Are the changes in body composition associated with regular activity greater than those associated with normal growth and maturation? How persistent are the training-associated changes? The increase in lean body mass observed in youth regularly trained over a several-year period would seem to suggest an increase greater than expected with normal growth. It should be noted, however, that muscle mass, and therefore lean body mass, continues to increase into the mid-20s. On the other hand, most of the variation in body composition with activity or inactivity is associated with fatness, which fluctuates inversely with the training stimulus. Changes in response to short-term training programs are most likely related to fluctuating levels of fatness, with only minimal changes in lean body mass.

Methods of estimating body composition *in vivo* are indirect, and most are based on models derived from adults (Novak, 1963). Hence, it must be recognized that the estimates reported in any study include a certain degree of measurement variability. Such error is especially important in longitudinal studies, in which it may be compounded in repeated measurement sessions. In addition, prediction equations are sometimes used to monitor training-mediated changes in body composition. However, evidence indicates that equations based on skinfolds may not be sufficiently sensitive to detect lean body mass changes with training (Wilmore *et al.*, 1970; Lohman, 1982).

3.4. Exercise and Specific Tissues

A reasonably extensive body of literature considers the effects of physical activity programs on specific tissues and functions. The subsequent discussion considers some of the available information dealing with the effects of regular training on bone, muscle, and fat tissue. Some of this material, along with other relevant information, is discussed elsewhere in this volume in the chapters dealing with growth of the specific tissues (see Chapters 2–4, this volume).

3.4.1. Bone

Bone formation is regulated by complex genetic and biochemical mechanisms. Physical activity, muscle strength, and weight bearing have been shown to be major factors involved in the expression of these mechanisms and, as such, are important determinants of bone development. Trueta (1968) reported that to keep childrens' bones growing at the required rate, intermittent compression of the entire growth cartilage through gravity, weight bearing, and muscle contraction is essential. According to Evans (1973), the real stimulus for bone formation and growth comes from tensile or compressive force aided by muscular contraction and weight bearing. Houston (1978) documented this in a case study of an infant born with a missing tibia in the left leg. Orthopedic surgeons moved the fibula centrally in a child aged 2 years to carry the weight of the child in the absence of the missing tibia. After 16 months of weight bearing, the fibula, having done the work of the tibia, assumed essentially the size, shape, and strength of the tibia.

The basic shape of a long bone is genetically determined and established embryonically in cartilage. The development of this cartilaginous model into adult bone can be affected by such influences as nutrition, exercise, and disease. Bone growth entails changes in length and width, changes in density, and the maintenance of shape and integrity. When one speaks about growth of bone or about the effects of exercise on bone growth, care must be taken to distinguish among the above processes.

Experimental studies of developing animals (Saville and Smith, 1966; Saville and Whyte, 1969; King and Pengelly, 1973; Kiiskinen, 1977) indicate greater skeletal mineralization and density and wider more robust bones with prolonged physical training. Observations on adult humans engaged in prolonged unilateral activity as in tennis (Buskirk *et al.*, 1956; Jones *et al.*, 1977; Huddleston *et al.*, 1980) or baseball pitching (King *et al.*, 1969; Tullos and King, 1972) indicate similar results, i.e., wider, more robust bones, and increased bone mineral in the preferred arm compared with the nonpreferred arm. Since most adults began formal training during childhood, the evidence would suggest a training-mediated response. Activity-related bone mineralization data for children are limited. In a study of bone mineralization of the dominant and nondominant arms of amateur baseball players aged 8–19, Watson (1973) reported significant mineralization and width differences between the dominant and nondominant humeri, but not for the radii and ulnae. The differences in mineral content betwen the dominant and nondominant humeri increased with age, which would suggest a training effect, assuming that the older boys participated in the specialized throwing activity longer than the younger boys in the sample.

It should be noted that these studies of unilateral activities indicate rather localized increases in bone mineralization. Comparing the densities of the distal femora of young adult athletes and nonathletes, Nilsson and Westlin (1971) noted greater femoral densities in the athletes and a clear gradient of higher densities with increasing activity levels, from nonexercising through exercising controls and ordinary athletes to the top athletes.

Dalen and Olsson (1974) observed a 20% increase in the trabecular bone of the extremities among cross-country runners, while Aloia *et al.* (1978) found an increased total bone mass in marathon runners. It is suggested that the differences relative to nonrunners may represent beneficial effects of regular physical activity on bone mineralization during growth.

One area of concern that needs further study has to do with extremely active young women. Recent reports based on small, uncontrolled studies have suggested that high levels of training may be accompanied by luteal phase irregularities in some young women, suggesting an anovulatory pattern (Shangold *et al.* 1979; Prior *et al.*, 1982). This has led to a concern that demineralization of bones in some active young women, secondary to an estrogen deficit, may be a potential danger (Jacobs, 1982; Gonzalez, 1982; Caan, 1983). These findings are in apparent disagreement with other studies that have shown bone density increases in young people who were physically active (Dalen and Olsson, 1974; Jones *et al.*, 1977; D. C. Cumming *et al.*, 1981). There is a need for further study in this area; as in all things, it is in the extreme that there is a need for concern. Just as extremes in dieting may lead to anorexia nervosa with associated skeletal problems, extremes in exercise may carry with it some inherent osteoporotic problems secondary to exercise-induced amenorrhea.

Experimental literature on training and specific bone lengths indicates reduced bone lengths in rats (Lamb *et al.*, 1969; Tipton *et al.*, 1972) and mice (Kiiskinen, 1977) exposed to voluntary forced swimming and to moderate or intensive running. Corresponding data for humans are not available. The observations made by Kato and Ishiko (1966) suggest that excessive compressive forces on the epiphyses of the knee may obstruct growth and thus reduce stature. The sample on which this suggestion is based, however, came from an economically poor and nutritionally substandard background. By contrast, the experimental and clinical data of Viteri and Torun (1981) suggest that regular activity during rehabilitation from protein-energy malnutrition facilitates recovery, including linear growth. Buskirk *et al.* (1956) reported longer bones of the dominant compared with the nondominant arm of elite tennis players and suggested that the difference is attributable to the effects of vigorous activity on bone growth during the adolescent years.

Given the evidence on activity and growth of a bone in length, it seems that the conclusion reached by Steinhaus (1933) more than 50 years ago may still be plausible, i.e., the pressure effects of physical activity may stimulate epiphyseal growth to an optimal length, but excessive pressure can retard linear growth. There is an obvious need to evaluate the effects of activity on the epiphyseal growth plate. When dealing with youngsters, it is difficult to define excessive pressure. As noted earlier, elite young athletes apparently grow as well in stature as nonathletes even after controlling for maturity differences and recognizing the possible role of selection for small body size in some sports.

3.4.2. Muscle

Increases in muscle size, strength, and endurance are probably among the best recognized effects of muscular extertion. "The fact that persistent use of muscles causes enlargement and a correlated increase in their strength has been known ever since there were boys..." (Steinhaus, 1955). Since growth of muscle tissue as well as the effects of physical training on muscle tissue are considered in detail in Chapter 10 of this volume, only a summary is presented here.

On the basis of early muscle physiology studies on animals (Morpurgo, 1897; Siebert, 1928; Holmes and Rasch, 1958; Hettinger, 1961), it has generally been held that muscle growth due to exercise is the result of hypertrophy or of simple enlargement of existing fibers. Recently, however, a greater number of muscle fibers have been reported to occur as a result of weight-lifting exercises in cats (Gonyea and Bonde-Petersen, 1978; Gonyea, 1980). This finding has been disputed by Gollnick *et al.* (1981), who report that in rats the number of fibers in skeletal muscle is fixed and that increase in size as a result of growth or training is the result of hypertrophy. Most of these investigations have used adult animals as subjects. The effect of exercise on young animals during growth is not as well understood. Owing to ethical limitations, longitudinal histological data on growing human subjects are not available.

Early biochemical studies on growing animals undergoing regular training indicated a significant rise in skeletal muscle DNA concentration above that expected for normal growth (Buchanan and Pritchard, 1970; Bailey *et al.*, 1973; Hubbard *et al.*, 1974). The increased DNA in trained animals was interpreted as suggesting that training may have been a significant factor influencing muscle nuclei number in growing animals. Problems in interpretation of the biochemical results on muscle tissue have clouded these early suggestions (Edgerton, 1973, 1976).

There is no question that morphological and biochemical changes take place in skeletal muscle in reponse to exercise. Regular physical training commonly results in hypertrophy of skeletal muscle. The degree of hypertrophy varies with the intensity of the training stimulus. Hypertrophy is accompanied by an increase in contractile substances (Helander, 1961), myofibrils (Goldspink, 1964), enzyme activity (Holloszy, 1967), and strength. The concept of the specificity of training must be emphasized. Muscular hypertrophy is associated primarily with high-resistance training activities, such as weight training.

Most of the available human data on muscle hypertrophy are derived from adults, and data on muscle tissue responses to training in developing subjects are not as extensive as for adults. In one of the few studies on adolescent boys (Fournier *et al.*, 1982), muscle hypertrophy was observed to have occurred following endurance training, but not following sprint training. Cross-sectional area of slow-twitch fibers and of some fast-twitch fibers increased 10–30% following 3 months of endurance training. Although there seems to be no strong evidence to suggest that fiber type distribution in children can be changed as a result of training (Eriksson *et al.*, 1974; Jacobs *et al.*, 1982), the relative area of a muscle composed of slow- or fast-twitch fibers may change in response to exercise. The direction of the change depends on the type of training stimulus. While muscle data for children and adolescents are scarce, those studies that have been done indicate that the pattern of muscle adaptation to training is similar to that of adults (Bar-Or, 1983). Persistence of the changes with the cessation of training ordinarily have not been considered.

Changes in oxidative enzyme activity in adolescents following endurance training have been reported in addition to changes in rate-limiting enzymes of anaerobic glycolosis (Eriksson *et al.*, 1974; Fournier *et al.*, 1982). While the direction of the response is similar to that observed in adults, the magnitude differs. Thus, the evidence is strong that physical training is an important modifying component in terms of muscle metabolic capacity in growing children.

In conclusion, it would seem that there is still much to be learned relative to the mechanics of muscle growth in response to exercise in humans. Morpurgo's (1897) statement might still apply today, especially if it were applied to growing children:

> As certain as is the fact that the mass of voluntary muscle increases in response to greater work, so uncertain is our knowledge concerning the mechanism that underlies the enlargement. There is no lack of assertations in the literature that deal with this subject in more or less decisive fashion . . . but exhaustive proof is everywhere lacking.

3.4.3. Adipose Tissue

Given the current interest in adipose tissue cellularity during growth (Knittle, 1978; Roche, 1981; Chumlea *et al.,* 1981, 1982), and the generally favorable influence of training on body fatness, one can inquire into the possible effect of regular training during growth on adipose tissue cellularity. Bjorntorp *et al.* (1972), for example, reported a training-associated reduction in fat cell size in young adult soccer athletes and middle-aged endurance athletes. Krotkiewski *et al.* (1979), on the other hand, observed a reduction in subcutaneous fat thickness with strength training in adult women but no significant changes in estimated fat cell size. Rather, the evidence suggested that the decreased thickness of subcutaneous tissue was a function of altered muscle thickness, i.e., "the same fat now surrounds an increased muscle volume" (Krotkiewski *et al.,* 1979).

Some experimental evidence suggests that training initiated early in the life of rats (at preweaning ages) effectively reduced the rate of fat cell accumulation and thus resulted in a significant reduction in the number of fat cells and body fatness later in life (Oscai *et al.,* 1972, 1974). On the other hand, endurance running begun after 7 weeks of age in rats did not affect adipose cell number, but significantly reduced adipose cell size (Booth *et al.,* 1974; Askew and Hecker, 1976). These results thus indicate an important role for regular activity in regulating fat cell size; however, in order for training to influence fat cell number, the program must be initiated very early in the life of rats. By about 7 weeks of age in rats, the pattern of fat cell proliferation is apparently established, so that adipose cells continue to increase in number, even though a training stimulus may be present.

It is not certain whether the preceding experimental observations can be applied to developing children. If so, it may require that training programs be initiated early in life, perhaps as early as early childhood, to have an influence on adipose tissue cellularity. However, the present information on developmental age trends and sex differences, as well as regional variability in estimated adipose cell number and size, is variable across studies (Kirtland and Gurr, 1979). For a complete discussion on adipose tissue cellularity in the growing child, the reader is referred to Chapter 2 and 11 of this volume.

3.5. Exercise and the Oxygen Transport System

In studying the physiological response of the oxygen transport system to exercise, another dimension is added beyond the simple morphological or quantitative observed change. Qualitative changes may occur in children during growth and or training with or without a quantitative alteration. Documentation of changes with age at the morphological level are easily observed and measured, but evidence of qualitative change at the biochemical level is limited. Methodological and ethical constraints prevent the necessary intrusive investigations that would provide more definitive evidence of cellular and subcellular mechanisms operating at different stages of child development. There is a shortage of data available at the cellular level in humans. Therefore, as in other areas of child growth, major gaps exist in understanding, defining, and explaining the processes involved in "normal" physiological response to exercise.

3.5.1. Aerobic Changes with Growth

One parameter that has been studied extensively in children is maximal aerobic power. This index represents the maximum volume of oxygen taken up per unit of time under maximal exertion conditions and is generally accepted as the best available measure of the efficiency of the oxygen transport system. When results are expressed in absolute units (liters per minute) values for both boys and girls increase with age, reaching a peak at maturity (Astrand, 1952; Shephard, 1966, 1982). Before puberty, absolute values for aerobic power are approximately the same for boys and girls (Shephard, 1982), although Bar-Or (1983), in summarizing the result of 19 separate investigations, reports a slight advantage for boys. For both sexes, values grow at the same rate until the age of 12. Boys show a spurt of aerobic power at puberty that is closely aligned with the adolescent growth spurt (Mirwald *et al.*, 1981). Development in girls after puberty is slower. Part of this sex-related difference can be explained by differences in lean body mass development between sexes during adolescence. This was demonstrated by Davies *et al.* (1972), who plotted absolute aerobic power against muscle mass used to perform an exercise test, when regression lines virtually identical for both sexes resulted.

To compare maximal aerobic power in children who differ in body size and mass, a relative rather than absolute expression must be used. The most common way of expressing aerobic power to dissociate size has been to use relative units (milliliters per kilogram body weight per minute). Relative maximal aerobic power expressed in these terms remains relatively constant for boys in the 50-ml/kg per min range from 6 to 18 years (Shephard, 1966, 1971, 1982; G. R. Cumming, 1967), although some studies show a slight decline during adolescence (Bailey *et al.*, 1978*b;* Bar-Or, 1983). In young girls, relative values are closely comparable to those of boys until the age of 10, at which time they show a progressive decline (Shephard, 1982). Bar-Or (1983). The results of a number of studies suggest that sex differences may be apparent earlier than the age of 10 and that the age-related decline in relative maximal aerobic power in girls also starts at an earlier age. Factors contributing to these age–sex differences in relative aerobic power are increased adiposity in girls with a corresponding relative decrease in lean body mass during adolescence and a possibly reduced hemoglobin concentration in adolescent girls (Astrand, 1952). Diminished voluntary participation in vigorous physical activity occurs in girls at adolescence as well. This cultural component, while difficult to document, probably represents a substantial reason for the sexual discrepancy. Whatever the reason, relative maximal aerobic power in girls is some 15–20% less than in boys following puberty (Shephard, 1982).

It should be noted that a number of investigators have questioned the use of body weight as the basis for expressing relative maximal aerobic power to control for size, especially in growing children (Asmussen and Heeboll-Nielson, 1955; Von Dobeln and Eriksson, 1972; Bailey *et al.*, 1978*b;* Ross and Marfell-Jones, 1982). The use of a linear dimension such as stature to control for growth change is conceptually preferable to the physical scientist and also has the advantage of being largely independent of extrinsic environmental factors. This is an important consideration when trying to partition out growth effects from the effects of physical training.

3.5.2. Aerobic Changes with Physical Training

The traditional approach to studying the influence of exercise, physical activity, or athletic training on functional growth of the oxygen transport system has been to compare athletes with nonathletes or trained subjects with untrained subjects or to compare subjects classified according to various levels of physical activity. Most of these studies have been of short duration involving pre- and post-test measurements and are subject to some or all of the constraints outlined in Sections 2.1 and 2.2.

A number of studies have examined specialized populations. For example, swimmers exposed to a regular training stimulus of high intensity have been studied by several investigators (Astrand *et al.*, 1963; Engstrom *et al.*, 1971; Andrew *et al.*, 1972; Gibbins *et al.*, 1972; Cunningham and Eynon, 1973; Vaccaro and Clarke, 1978). The results of these studies have been equivocal in terms of absolute values, some showing positive effects beyond growth, others not so sure. The same situation holds true with studies using different types of running training. A number of studies have shown positive effects beyond growth (Ekblom, 1969, 1971; Eriksson, 1972, 1978; Eriksson and Koch, 1973; Brown *et al.*, 1972). Other studies using different running regimens have found no change in physiological parameters over and above normally expected growth changes (G. R. Cumming *et al.*, 1967, 1972; Daniels and Oldridge, 1971; Daniels *et al.*, 1978; Mocellin and Wasmund, 1973, Stewart and Gutin, 1976). It is difficult to generalize on the basis of these studies because of the considerable variability in the intensity and duration of the training stimulus and the nonrandom selection of subjects.

Another approach taken to evaluate the influence of physical activity on functional growth has been the application of an enriched physical education program to a subsample of students in a school setting (G. R. Cumming *et al.*, 1969; Kemper, 1973; Kemper *et al.*, 1978; Bar-Or and Zwiren, 1973; Shephard, 1982). There are a number of unique limitations to studies of this type (Shephard, 1982), and the results are difficult to interpret. In spite of the enrichment program, the differences in physical activity patterns during childhood are probably not very large, and outside school activity is a confounding factor.

Several longitudinal studies, each using a different approach, have attempted to separate out exercise-induced changes in function from normally occurring growth changes. In a 10-year study, Bailey (1973) with Mirwald *et al.* (1981) classified normal children into activity groups on the basis of their habitual physical activity patterns as determined by questionnaire and interview. Parizkova (1973) with Sprynarova (1974) classified children on the basis of sport club participation and training. In a training study, Ekblom (1969) used dimensional theory to interpret the results of physiological testing, and Andersen *et al.* (1976) used a similar approach assuming normal activity patterns for children. Jequier *et al.* (1977) with Shephard (1982) increased school physical education time by 5 hr/week for an experimental group in an effort to evaluate effects of activity. Longitudinal studies have the advantage of permitting investigators to define the adolescent growth period more precisely (Tanner, 1962), and a number of the above studies, plus others, have suggested that during the adolescent period the child may be especially susceptible to training effects (Astrand, 1967; Ekblom, 1969; Schmucker and Hollman, 1974; Sprynarova, 1974; Kobayashi *et al.*, 1978; Mirwald *et al.*, 1981). Other workers (Berg and Bjure, 1974; Lammert *et al.*, 1982) are not as certain.

A major consideration in studies looking at training and growth differences is the role of heredity. Early data from twin studies suggest that the principal determinant of variability in maximal aerobic power among individuals who have lived under similar environmental conditions is genetic (Klissouras, 1971, 1972; Klissouras *et al.*, 1972; Holmer and Astrand, 1972; Komi *et al.*, 1973; Howald, 1976; Weber *et al.*, 1976). Weber *et al.* (1976) suggest that the "old hypothesis that more might be gained by introducing extra exercise at a time when the growth impulse is the strongest is no longer tenable." However, caution is warranted in accepting this conclusion. Sample size in this study was extremely small. Bouchard (1978) calculated that 16% of the variance in aerobic power was attributable to an environment–heredity interaction. This figure is more than double the 7% interaction estimated by Weber *et al.* (1976). Clearly, although genetic factors contribute significantly to an individual's aerobic capacity, training is an important environmental component. A detailed discussion of the genetics of physiological fitness and motor performance has been provided elsewhere by Bouchard and Malina (1983).

Taking into consideration all the constraints and limitations inherent in studying training-mediated responses during growth, what conclusions can be made? A consistent finding is that physical activity or training has a small or limited effect on maximal aerobic power prior to adolescence (see Section 4.1.2). Since most activity tasks proceed at submaximal work rates, the use of maximal aerobic power as a measure for evaluating the efficiency of the oxygen transport system may be misleading in prepubescent children (Stewart and Gutin, 1976). Physical training or high levels of physical activity may lead to improvements in submaximal efficiency that are independent of changes at maximal effort (Stewart and Gutin, 1976; Lussier and Buskirk, 1977; Mirwald and Bailey, 1984). This is an area that needs further investigation.

At adolescence, the effect of training or high levels of physical activity is less clear, and it is difficult to draw any firm conclusions. It has been suggested that the effectiveness of aerobic training is greatest at or around the time of peak height velocity in boys (Kobayashi *et al.*, 1978; Mirwald *et al.*, 1981). Biologically, this would seem reasonable, in view of the marked changes taking place in endocrine function at this state of development.

To summarize, in spite of a surge of interest in the child, in growth, and in exercise among scientists in a number of disciplines, our understanding of the child's physiological response to exercise is still fragmentary. This is primarily attributable to an inherent methodological constraint that has and continues to be a major challenge for investigators working in this area. As Bar-Or (1983) states:

> In adults, changes in function between pre- and postintervention can be attributed with fair certainty to the conditioning program. Not so with children or adolescents. Here, changes due to growth, development, and maturation often outweigh and mask those induced by the intervention. It is intriguing that many of the physiologic changes that result from conditioning and training also take place in the natural process of growth and maturation.

4. *Child–Adult Differences in the Physiological Response to Exercise*

The purpose of this section is to highlight a number of child–adult differences in physiological response to exercise. Consideration of these differences in adaptation is important in determining whether observed changes in children over time are the result of training, growth, or both. More complete and extensive treatment of the topic may be found in a specialized pediatric work physiology reference by Bar-Or (1983).

4.1. The Aerobic System

The ability to move and perform work depends on muscular contraction that results from of the continuous breakdown and reconstruction of adenosine triphosphate (ATP). But the available stores of ATP can activate a muscle for only a fraction of a second. As quickly as ATP is split into free energy, it must be rebuilt, a process that also requires energy. Of the three methods employed by the muscles to build ATP, two processes work anaerobically (without oxygen). Adult–child differences in these processes are discussed in Sections 4.2.1 and 4.2.2. The third and principal process is the oxygen-dependent aerobic pathway described by the Krebs (TCA) cycle. This most important pathway is dependent on the oxygen-transfer system of the body; some components of this system show a differential response to exercise between the child and the adult.

4.1.1. Components of the Oxygen Transport System

The adult response to exercise is often used as the benchmark for comparing the response of children. The assumption is that the child's oxygen-delivery system will function and perform in the same manner as the adult's system. This assumption may not be tenable, and a number of studies have provided initial evidence suggesting that not only are the morphological capacities of children and adults different, but the functional capacities may be different as well.

During both submaximal and maximal effort, children have a higher ventilation per unit of oxygen uptake and a correspondingly poorer ventilatory efficiency than that of adults (Robinson, 1938; Astrand, 1952). By contrast, the alveolar ventilation accounts for a slightly larger proportion of total external ventilation in children than in adults (Shephard, 1971, 1977). Children have a higher respiratory frequency than that of adults performing the same task (Rutenfranz *et al.*, 1981). Lung volumes are of a size that would be expected on the basis of body dimensions (Astrand, 1952; Bailey *et al.*, 1979), and the diffusing capacity of the child's lungs is at least as great as the adult's whether related to body weight or to oxygen uptake (Shephard *et al.*, 1969; Shephard, 1971). During near-maximal exercise, young adolescents do appear to have a lower diffusing capacity at a given level of functional residual capacity (FRC) than that of adults (Koch and Eriksson, 1973). Nevertheless, as in adults there is no evidence to indicate that the aerobic power of healthy children is limited by pulmonary factors. While there are quantitative differences, a child's pulmonary response to exercise appears to be quite similar to that of an adult.

Children do, however, have shorter oxygen uptake transients than that of adults (Robinson, 1938; Macek and Vavra, 1980, Freedson *et al.*, 1983). In a study comparing 10- and 11-year-old boys with young adults, Macek and Vavra (1980) found lower oxygen deficits in the children and that boys could increase their oxygen uptake at a markedly higher rate than could the adult group. The young boys reached 55% of their final oxygen uptake within 30 sec and a steady state within 2 min. The adults reached only 33% of their final oxygen uptake within the first 30 sec and required some 3–4 min to reach a steady state. The results of these studies lead to a number of speculations: (1) Because of their shorter oxygen transients, children do not need to resort as much to anaerobic pathways (hence the smaller oxygen deficit and lactate production); (2) Shorter transients are compensatory for low glycolytic capacity in children (Bar-Or, 1983); and (3) Shorter transients in children are a reflection of a smaller body and the resulting shorter circulation time (G. R. Cumming, 1978).

From an analysis of the oxygen conductance equation, Shephard (1971, 1977) established that blood transport is the dominant factor limiting the overall transport of oxygen in both children and adults. The number of red blood cells and the concentration of hemoglobin in children are lower relative to adults, which means that the oxygen-binding capacity is accordingly lower. Moreover, the child's total body hemoglobin is less than expected on the basis of body stature (Astrand and Rodahl, 1977).

Submaximal heart rate in children declines with age (Robinson, 1938; Astrand, 1952; Ulbrich, 1971;

Andersen *et al.*, 1974; Bouchard *et al.*, 1977; Wirth *et al.*, 1978; Yamaji *et al.*, 1978). In studying boys aged 8–18 performing a standard cycle-ergometer task at an absolute workload, Bouchard *et al.* (1977) found that the submaximal heart rate can be as much as 30–40 beats higher in an 8-year-old than an 18-year-old. Such a difference is partially attributable to the greater relative exercise intensity performed by the younger children, but it is also found at equal relative metabolic loads (Wirth *et al.*, 1978). The higher rate among young children is biologically sound, as it compensates for a lower stroke volume.

A number of studies report maximal heart rates for children and adolescents in the 195–215-beat/min range (Robinson, 1938; Astrand, 1952; Bar-Or *et al.*, 1971; Andersen *et al.*, 1974). These cross-sectional studies report a decline with age only after maturity has been reached. In a longitudinal study of boys aged 7–16, Mirwald *et al.* (1984) make the same observation. However, if one takes the difference betwen submaximal and maximal heart rates as reflecting a specific cardiac reserve, an adolescent has a distinctly greater reserve than does a prepubescent child (Bar-Or, 1983; Mirwald and Bailey, 1984).

Cardiac output is the product of stroke volume and heart rate. Children have a markedly lower stroke volume than that of adults at all levels of exercise. This lower stroke volume is compensated for in part by a higher heart rate, but the end result is a lower cardiac output (Bar-Or *et al.*, 1971; Eriksson, 1971; Eriksson *et al.*, 1971*b;* Eriksson and Koch, 1973; Drinkwater *et al.*, 1977). It is not clear whether the somewhat lower cardiac output of children is of any biological significance. Quite possibly the concomitant higher arteriovenous oxygen difference in children is sufficient to compensate the oxygen transport system during submaximal exercise (Bar-Or, 1983). A potential handicap due to low cardiac output may exist during maximal exercise when peripheral oxygen extraction can no longer rise (Eriksson and Koch, 1973) or when the child is exposed to the combined stresses of exercise and extreme heat. At all levels of exercise, the stroke volume of boys is somewhat higher than in girls (Bar-Or *et al.*, 1971; Godfrey *et al.*, 1971; G. R. Cumming, 1977). In line with the lower cardiac output and stroke volume, a child has a lower exercise blood pressure than that of an adult, and boys have a higher peak systolic pressure than that of girls (Riopel *et al.*, 1979).

What effect do these child–adult differences in the components of the oxygen transport system have on the most widely accepted index used to measure the efficiency of the system, namely, maximal aerobic power?

4.1.2. Aerobic Power Differences

Maximal aerobic power represents the greatest volume of oxygen that can be consumed per unit of time under conditions of maximal exertion. It therefore represents both the maximal energy output of the aerobic energy process and the functional and anatomical interrelationships of heart, blood, lungs, and skeletal muscle.

Compared with adults, values for relative maximal aerobic power are high in children. However, if one looks at metabolic reserve, i.e., the difference between maximal oxygen uptake and oxygen uptake needed for a given task, children are shown to be at a disadvantage. The higher oxygen cost in young children in performing a given task is probably the result of mechanical inefficiency (Daniels *et al.*, 1978). The relative oxygen cost of walking or running is higher among children (Astrand, 1952; Skinner *et al.*, 1971; MacDougall *et al.*, 1979; Krahenbuhl *et al.*, 1979). MacDougall *et al.*, (1979) found that an 8-year-old child running at a 180-m/min pace is operating at 90% of maximal aerobic power, while a 16-year-old running at the same speed is operating at ony 75% of maximum. Thus, compared with an adult, the child is not as aerobically efficient as might be expected from looking at the high relative maximal aerobic power values.

Curiously, in spite of the above consideration, children grade physical effort lower than adolescents, and adolescents perceive the same effort as being less strenuous than adults. Bar-Or (1977) conducted a study on more than 1000 male subjects ranging in age from 7 to 68 years, who performed an identical cycle ergometry test. The subjective perceived effort was correspondingly lower, the younger the individual, although the relative intensity of effort demonstrated by heart and circulatory reactions was equally great. These data suggest that certain types of physiological strain are perceived to be less stressful by children than by older persons.

The trainability of the aerobic system in children as compared with adults has also been an area of study. Saltin (1969) suggested that young adults undergoing a training program are more trainable in terms of maximal aerobic power than are older persons. With regard to children, the evidence is not nearly so clear cut. In fact, the preponderance of evidence suggests that the trainability of aerobic power in young preadolescents is lower than expected in spite of improved athletic or motor performance (Mocellin and Wasmund, 1973; Schmucker and Hollmann, 1974; Bar-Or *et al.*, 1974; Stewart and Gutin, 1976; Gilliam and Freedson, 1980; Yoshida *et al.*, 1980). Studies with adolescents are equivocal, with some reporting expected improvements following training (Ekblom, 1969; Weber *et al.*, 1976; Sprynarova *et al.*, 1978) and others not (G. R. Cumming *et al.*, 1967; Daniels *et al.*, 1978; Hamilton and Andrew, 1976).

What explanation can be offered for the lower-than-expected improvement in aerobic power following training in young children? Bar-Or (1983) offers four possibilities: (1) With growth the child becomes mechanically more efficient; (2) Anaerobic capacity may be improving as a result of training, (3) Measurement techniques may not be sensitive enough to measure aerobic changes in children; and (4) Free time activity is so high in young children that differences between control subjects who do not participate in the training program and those children who do are beyond detection.

In sum, the question of whether children are less sensitive than adults to aerobic training needs further investigation. This is also the question of whether children have an inferior maximal aerobic power compared with adults. Until definitive studies are carried out, conclusions will continue to depend on the point of reference and on the basis for comparison.

4.2. The Anaerobic System

In addition to the aerobic system, the anaerobic system can be employed by the working muscles to build ATP. The anaerobic process can function in the absence of oxygen and involves two pathways: (1) The anaerobic alactic pathway, i.e., ATP and creatine phosphate (CP) dependent; and (2) The anaerobic lactic pathway, i.e., glycogen dependent. Muscle contractions that result from anaerobic reactions can only be sustained for short periods of time, in contrast to aerobic work, which can be carried on for many minutes or hours.

4.2.1. Anaerobic Alactic Pathway

Concentrations of energy-rich phosphagens, ATP and CP, in children are similar to adult values (Eriksson, 1972, 1978; Eriksson *et al.*, 1973). Moreover, when the rates of depletion of these compounds during exercise are expressed as a percentage of maximal aerobic power, children and adults are not found to differ. Therefore, children should be capable of intense brief activities of less than 20 sec, similar to adults.

4.2.2. Anaerobic Lactic Pathway

In physical activities lasting less than 2–3 min, the glycolytic pathway constitutes the predominant source of muscle energy, with the resulting by-product of lactic acid. In young subjects the concentrations of lactic acid both after exhaustive effort and during submaximal work are lower relative to adult values (Astrand, 1952; Shephard *et al.*, 1969; Eriksson *et al.*, 1971*a*, 1973; Eriksson, 1972; Shephard, 1977). The lower lactate concentration in children may reflect a slower rate of glycolysis. A number of studies support this contention that children have a markedly lower anaerobic lactate capacity, reflecting a qualitative deficiency in the muscle (Karlsson, 1971; Eriksson and Saltin, 1974; Eriksson, 1980).

The resting concentration of glycogen and the rate of its anaerobic utilization are lower in children, and therefore children are at a functional disadvantage when performing strenuous activities lasting 20–60 sec. It has been suggested that the ability of boys to produce lactate during maximal exercise depends on sexual maturity (Eriksson *et al.*, 1971); in animal studies it has been linked to the level of circulating testosterone (Krotkiewski *et al.*, 1980). Further support for a hormonal link is provided in comparing mature females with mature males. Females demonstrate lower anaerobic performance than males, and there is a smaller age-related difference among females in anaerobic capacity (Davies *et al.*, 1972). Further studies are required to delineate the rate of glycolysis controlling for age, sex, and sexual maturity.

Phosphofructokinase, the glycolytic rate-limiting enzyme, has been found to be less active in the muscle cells of 11- to 13-year-old boys (Eriksson *et al.*, 1973, 1974) or 16- to 17-year-old boys (Fournier *et al.*, 1982) than in young adults. This finding offers a reason for the slower rate of glycolysis in children.

An additional indicator of anaerobic capacity is the contractibility of the muscle at higher degrees of acidosis. Children do not reach the level of acidosis reached by adolescents or young adults (Gaisl and Buchberger, 1977). Irrespective of absolute pH values, there is an age-related increase with maximal acidosis.

The accumulating evidence supports the contention that preadolescent children are not mature in their capability to derive energy from the anaerobic lactate pathway. Glycogen concentration and its rate of utilization, glycolytic capacity, lactic acid concentration, concentration of PFK enzyme, and tolerance of acidosis are lower in the preadolescent child than in the adult. With sexual maturation, the age-related differences disappear.

In summary, anaerobic capacity in the child is lower both in absolute and in relative terms. This lower capacity as compared with adults is largely due to a qualitative deficiency in the muscles of the child.

4.3. Temperature Regulation

Because of morphological and functional characteristics, heat regulation in children is insufficient, particularly when exposed to very hot or very cold climatic conditions (Bar-Or, 1980). The child has a larger surface area:body mass ratio than that of the adult, hence a greater heat transfer to or from the body (Bar-Or, 1983). In mild climatic conditions, the geometric difference between children and adults is advantageous. But in extreme temperatures, when heat transfer should be minimized, children are at a disadvantage. Although children produce and transfer heat by conduction, convection, and radiation as effectively as adults, they have a lower sweating rate and evaporative capacity (Wagner *et al.*, 1972; Haymes, *et al.*, 1975; Drinkwater *et al.*, 1977; Davies, 1981; Inbar *et al.*, 1981). The lower sweating rate in children is apparent both in absolute terms and when normalized per unit of surface area (Bar-Or, 1983). The population density of sweat glands is relatively greater in children than in adolescents or adults. The output per gland rather than the number of glands limits the sweating rate of children (Bar-Or, 1980). Although low sweat production helps children conserve water, it hampers their ability to sustain high metabolic production in a hot climate.

Another age-related difference in adaptation to exercise in heat is the rate of acclimatization. The process of adaptation takes considerably longer in children than in adults (Bar-Or and Inbar, 1977; Inbar *et al.*, 1981).

5. Muscular Strength and Motor Performance

5.1. Muscular Strength

The relationship between strength and muscle mass with age-, sex-, and maturity-associated variations in the ability to perform muscular work has been considered in detail in Chapter 4 of this volume. What follows is a brief summary of strength changes associated with growth and training.

Strength increases with chronological age; there are sex-related differences in the development of strength, but these differences are not great until approximately 13 years of age (Jones, 1947; Asmussen, 1973). The development of strength is characterized by a growth curve very similar to the curves in height and weight. The apex of the male adolescent strength spurt occurs after both the adolescent height and weight spurts (Stolz and Stolz, 1951; Carron and Bailey, 1974). The pattern of maximum strength development in girls is not so clear. While the apex of strength development occurs more often after peak height velocity in girls, there is a wide variation in the timing of the peak strength gains (Faust, 1977).

In the case of young women, menarche provides an accurate focal point of sexual maturity to which the growth and development of strength can be related. Jones (1947) found that regardless of the age at which menarche fell, the peak rate of growth in strength occurred shortly before this reference point. Since menarche occurs after peak height velocity (Marshall and Tanner, 1969), the sequence in terms of peak strength in girls would appear to be peak height velocity followed by the apex of strength development, followed by menarche.

Individual differences in strength tend to be highly stable over successive years but decrease in stability as the time between measurements is increased (Rarick and Smoll, 1967; Carron and Bailey, 1974). Strength measurements also show a moderately high degree of heritability (Bouchard and Malina, 1983). The issue of trainability of muscular strength is relevant, and some data suggest that children and adolescents may be more sensitive to a training stimulus than are adults.

Rohmert (1968) studied the increase in isometric strength in several muscle groups in 8-year-old boys and girls who were given a series of standard training programs. He also conducted a parallel series of studies on adults for comparative purposes. The children were found to be more trainable than the adults who were given the same relative training stimulus. These findings have been interpreted as showing that children are naturally adapted to a lower degree of strength utilization in daily life and are therefore more sensitive to strength training. This interpretation was supported by the experiments by Ikai (1967), who found that another parameter, muscular endurance, could be increased considerably more in children, especially in the age bracket 12–15 years, than in adults. Similarly, Nielsen *et al.* (1980) reported a greater increase in isometric strength after 5 weeks of training for young girls (7–13 years of age) than for older girls (14–19 years of age).

In contrast to these studies, Vrijens (1978) found greater strength improvement in postpubescent boys than in prepubescent boys in response to a strength training program. Hettinger (1961) reported less improvement in isometric strength following training for adolescents of both sexes as compared with young adults. Lammert *et al.* (1982), in a study of muscular endurance of boys and girls, reported that the period in which the pubertal growth spurt occurs is that period in which children are less sensitive to training as compared with later or earlier in their development.

Clearly, the data on trainability of strength are far from conclusive. There is a need for more carefully controlled studies before definitive statements relative to optimal ages and sensitive periods can be made.

5.2. Motor Performance

Proficiency in motor activities is an important component of the child's behavioral repertoire and is essential for participation in active games and athletics. This section briefly considers the effects of age trends and sex differences on performance in selected fundamental motor tasks requiring power, speed, agility, and coordination. Although motor skills are many, those commonly used to assess a youngster's motor ability are running, jumping, and throwing tasks, as well as variants of these skills (Espenschade and Eckert, 1974).

Motor performance improves with age in boys aged 17–18. Note that at these ages the boys are generally in the last year of high school, after which the available data are not as extensive. Some data for college students indicate continued improvement in motor performance into the early 20s. Jumping and throwing ability show some improvement during male adolescence, suggestive of a performance spurt in these power events. Performance of girls improves through childhood but reaches a plateau at approximately 14 years of age and shows little improvement afterward (Espenschade and Eckert, 1974).

Sex-related differences are slight during childhood when biological age is considered; boys, on the average, perform better than girls in running, jumping, and throwing tasks. The sex-related differences in performance become progressively more marked during adolescence. The slopes of the performance curves are steep for boys, while those for girls are rather flat at this time. In a longitudinal series of children (Espenschade, 1940), average performances of girls in all motor tasks except throwing fall within 1 standard deviation of the means for boys during early adolescence. From 14 years of age on, however, the average performance of girls is consistently outside the limit desired by 1 standard deviation below the boys' averages. Few girls approximate the throwing performance of boys as adolescence progresses (Espenschade, 1940; Malina, 1974).

Some studies suggest that the stability of motor performance is high. Willimczik (1980) studied motor development in children aged 6–10. Students with high coordination at the beginning of the study maintained or improved their standing as compared with other students at the end of the study.

A child's level of motor performance is in part related to age, size, physique, body composition, and biological maturity status. These biological correlates of performance have been treated elsewhere in depth (Malina, 1975). Evidence suggests negative effects of excess body weight, fatness, and endomorphy on motor performance items involving movement of the entire body. The magnitude of correlations of size, physique, composition, and maturity status with motor performance, however, is generally low to moderate, and as such has limited predictability. Furthermore, the effects of size, physique, and maturity on performance are generally more apparent at the extremes of comparisons, i.e., extreme endomorphs relative to extreme ectomorphs, or early-compared to late-maturing children.

Maturity-associated variation in motor performance is marked in boys, with early-maturing boys generally superior to their late-maturing peers. On the other hand, motor performance of adolescent girls is poorly related to maturity status, with some suggestion of better performance levels in later-maturing girls (Espenschade, 1940; Clarke, 1971; Malina and Rarick, 1973).

Data on the genetics of motor performance are not extensive, and heritability estimates vary from task to task (Bouchard and Malina, 1983). Some skills show a substantial genetic component, while others do not. This is a reflection of the complex interrelationships among biological characteristics and motor performance and would seem to suggest a possible role for other factors in determining a youngster's performance. These include social class, ethnicity, birth order, rearing practices, ordinal position, and motivation, to mention a few (Malina, 1983*c*).

6. Childhood and Adolescent Athletics

Participation in organized athletic competition is a reasonably well-established feature of childhood and youth throughout the world. The extent of participation by young people in sports is partly highlighted by the decreasing age of participants at the international level in many sports and by the greater opportunitites for sports participation by young girls. Hence, a consideration of the growth and maturity characteristics of young athletes is warranted, especially since many are engaged in rigorous training programs. However, several problems must be recognized and addressed.

The first is a matter of definition. What is an athlete? Young athletes are usually defined by success on school teams (especially football, ice hockey, baseball, and track), in national and international competitions (especially swimming, track, and gymnastics), and in selected club and age-group competitions.

A second problem deals with selection. Successful young athletes are a highly selected group, usually because of skill, but sometimes because of size and physique in some sports or in positions within a sport. Selection may be by oneself, by parents and coaches, or by both. Self-selection is indeed a critical factor, e.g., it may be determined by, or be dependent on, a youngster's motivation to train and compete. Selection also occurs to some extent by default, i.e., there are dropouts. Some youngsters choose not to participate for reasons of ability; others may drop out for reasons unrelated to sport, such as changing social interests. Many participants are at the novice level. However, as the level of competition increases, the number of successful participants becomes progressively smaller, and studies of young athletes generally focus on the successful participants.

A third problem deals with the variation of biological maturity in anthropometric and body composition characteristics. Children advanced in skeletal and sexual maturation differ in size, physique, and body composition from children who are average or delayed in these indices of biological maturation. The differences are especially pronounced during the adolescent phase of growth, ages 9–16. A youngster's maturity is also related to physical performance. The relationship, however, differs for boys and girls. Among boys, early maturation is positively related to strength and motor performance. Among girls, the relationship of strength and motor performance to biological maturity is inconsistent. Better performances are generally reported for girls who mature later, and the differences between contrasting maturity groups are more apparent in later adolescence (Malina and Rarick, 1973; Malina, 1975).

A fourth problem relates to the determinants of successful athletic performance. A youngster's anthropometric, compositional, and maturity characteristics do not, in themselves, determine successful performance. They are rather a part of a complex matrix of biological, psychological, and sociocultural characteristics related to performance (see Malina, 1983*c*).

A comprehensive review of the growth, maturity, and body composition characteristics of young athletes is beyond the scope of this chapter but has appeared elsewhere (Malina, 1982; Malina *et al.*, 1982). In general it can be said that young athletes of both sexes grow as well as nonathletes. The experience of athletic training and competition does not harm the physical growth and development of the youngster. The young trained athlete generally has a lesser percentage of body weight as fat and a greater lean body mass. Maturity relationships are not consistent across sports. More often than not male athletes tend to be maturationally advanced compared with nonathletes. On the other hand, female athletes tend to be delayed in maturity, with the possible exception of swimmers. Maturity-associated variation in size and body composition is therefore a significant factor in comparing athletes and nonathletes, especially during the circumpubertal years.

Young athletes who train extensively for competition reach much the same level of physiological condition attained by adult athletes who undergo an equivalent training program. There is, however, a need for follow-up studies of young athletes, centering on the question of the persistence of training-associated changes during childhood.

7. Concluding Remarks

Only selected aspects of physical growth and maturation relative to the stress of regular physical training have been considered in this chapter. Other areas of research are undoubtedly relevant. For instance, there is a need for more detailed examination of hormonal responses to regular training as well as their possible relationships to growth and maturation. While physical activity is obviously essential for normal growth, just how much activity is necessary during the years of active growth is unknown. The role of regular activity in the biological maturation process is not clearly established. There is a need for the development of sensitive methods of monitoring and quantifying the amount and intensity of normal physical activity in children and youth. Further studies, using different training programs, are obviously necessary, especially longitudinal observations of children and youth for whom the training stimulus is carefully defined and quantified. Only with continued adherence to appropriate research design and acceptable methodology will our knowledge of the underlying physiological processes involved in training and growth be extended.

8. References

Adams, E. H., 1938, A comparative anthropometric study of hard labor during youth as a stimulator of physical growth of young colored women, *Res. Q. Am. Assoc. Health Phys. Ed.* **9**:102–108.

Aloia, J. F., Cohn, S. H., Babu, T., Abesamis, C., Kalici, N., and Ellis, K., 1978, Skeletal mass and body composition in marathon runners, *Metabolism* **27**:1793–1796.

Andersen, K. L., Seliger, V., Rutenfranz, J., and Berndt, I., 1974, Physical performance capacity of children in Norway. Part II. Heart rate and oxygen pulse in submaximal and maximal exercises—Population parameters in a rural community, *Eur. J. Appl. Physiol.* **33**:197–206.

Andersen, K. L., Seliger, V., Rutenfranz, J., and Skrobak-Kaczynski, J., 1976, Physical performance capacity of children in Norway, *Eur. J. Appl. Physiol.* **35**:49–59.

Andrew, G. M., Becklake, M. R., Guleria, J. S., and Bates, D. V., 1972, Heart and lung function in swimmers and nonathletes during growth, *J. Appl. Physiol.* **32**:245–251.

Askew, E. W., and Hecker, A. L., 1976, Adipose tissue cell size and lipolysis in the rat: Response to exercise intensity and food restriction, *J. Nutr.* **106**:1351–1360.

Asmussen, E., 1973, Growth in muscular strength and power, in: *Physical Activity, Human Growth and Development* (G. L. Rarick, ed.), pp. 60–79, Academic Press, New York.

Asmussen, E., and Heeboll-Nielsen, K. R., 1955, A dimensional analysis of physical performance and growth in boys, *J. Appl. Physiol.* **7**:593–603.

Astrand, P. O., 1952, *Experimental Studies of Physical Working Capacity in Relation to Sex and Age,* Munksgaard, Copenhagen.

Astrand, P. O., 1967, Commentary—Symposium on physical activity and cardiovascular health, *Can. Med. Assoc. J.* **96**:760.

Astrand, P. O., and Rodahl, K., 1977, *Textbook of Work Physiology,* 2nd ed., McGraw-Hill, New York.

Astrand, P. O., Engstrom, L., Eriksson, B., Karlberg, P., Nylander, I., Saltin, B., and Thoren, C., 1963, Girl swimmers. With special reference to respiratory and circulatory adaptation and gynecological and psychiatric aspects, *Acta Paediatr. Scand. (Suppl.)* **147**:1–75.

Bailey, D. A., 1973, Exercise, fitness and physical education for the growing child—A concern, *Can. J. Publ. Health* **64**:421–430.

Bailey, D. A., 1976, The growing child and the need for physical activity, in: *Child in Sport and Physical Activity* (J. G. Albinson and G. M. Andrew, eds.), pp. 81–93, University Park Press, Baltimore.

Bailey, D. A., Bell, R. D., and Howarth, R. E., 1973, The effect of exercise on DNA and protein synthesis in skeletal muscle of growing rats, *Growth* **37**:323–331.

Bailey, D. A., Malina, R. M., and Rasmussen, R. L., 1978*a,* The influence of exercise, physical activity, and athletic performance on the dynamics of human growth, in: *Human Growth,* Vol. 2: *Postnatal Growth* (F. Falkner and J. M. Tanner, eds.), pp. 475–505, Plenum Press, New York.

Bailey, D. A., Ross, W. D., Mirwald, R. L., and Weese, C., 1978*b,* Size dissociation of maximal aerobic power during growth in boys, in: *Pediatric Work Physiology* (J. Borms and M. Hebbelinck, eds.) pp. 140–151, Karger, Basel.

Bailey, D. A., Mirwald, R. L., Weese, C. H., Rasmussen, R. L., and Stirling, D., 1979, Data analysis techniques in a longitudinal growth study using vital capacity as an example variable, in: *Motorische Entwicklung. Proceedings from International Motoric Symposium* (K. Willimczik and M. Grosser, eds.), pp. 225–238, Verlag Karl Hofmann, Schorndorf.

Bar-Or, O., 1977, Age-related changes in exercise perception, in: *Physical Work and Effort* (G. Berg, ed.), pp. 255–266, Pergamon Press, Oxford.

Bar-Or, O., 1980, Climate and the exercising chld—A review, *Int. J. Sports Med.* **1**:53–65.

Bar-Or, O., 1983, *Pediatric Sports Medicine for the Practitioner,* Springer-Verlag, New York.

Bar-Or, O., Inbar, O., 1977, Relationship between perceptual and physiological changes during heat acclimatization in 8- to 10-year-old boys, in: *Frontiers of Activity and Child Health* (H. Lavallee and R. J. Shephard, eds.), pp. 205–214, Editions du Pelican, Quebec.

Bar-Or, O., and Zwiren, L. D., 1973, Physiological effects of increased frequency of physical education classes and of endurance conditioning on 4- to 10-year-old girls and boys, in: *Pediatric Work Physiology* (O. Bar-Or, ed.), pp. 183–198, Wingate Institute, Natanya.

Bar-Or, O., Shephard, R. J., and Allen, C. L., 1971, Cardiac output of 10- to 13-year old boys and girls during submaximal exercise, *J. Appl. Physiol.* **30**:219–223.

Bar-Or, O., Zwiren, L. D., and Ruskin, H., 1974, Anthropometric and developmental measurements of 11- to 12-year-old boys, as predictors of performance 2 years later, *Acta Paediatr. Belg. (Suppl.)* **28**:214–220.

Berg, K., and Bjure, J., 1974, Preliminary results of long-term physical training of adolescent boys with respect to body composition, maximal oxygen uptake, and lung volume, *Acta Paediatr. Belg. (Suppl.)* **28**:183–190.

Beyer, H. G., 1896, The influence of exercise on growth, *J. Exp. Med.* **1**:546–558.

Bjorntorp, P., Grimby, C., Sanne, H., Sjostrom, L., Tibblin, G., and Wilhelmsen, L., 1972, Adipose tissue fat cell size in relation to metabolism in weight-stabile, physically active men, *Horm. Metab. Res.* **4**:178–182.

Booth, M. A., Booth, M. J., and Taylor, A. W., 1974, Rat fat cell size and number with exercise training, detraining, and weight loss, *Fed,. Proc.* **33**:1959–1963.

Bouchard, C., 1978, Genetics, growth and physical activity, in: *Physical Activity and Human Well-Being* (F. Landry and W. A. R. Organ, eds.), pp. 29–45, Symposia Specialists, Miami.

Bouchard, C., and Malina, R. M., 1983, Genetics of physiological fitness and motor performance, *Ex. Sport Sci. Rev.* **11**:306–339.

Bouchard, C., Malina, R. M., Hollmann, W., and Leblanc, C., 1977, Submaximal working capacity, heart size and body size in boys 8 to 18 years, *Eur. J. Appl. Physiol.* **36**:115–126.

Bouchard, C., Thibault, M. C., and Jobin, J., 1981, Advances in selected areas of human work physiology, *Yrbk. Phys. Anthropol.* **24**:1–36.

Brown, C. H., Harrower, J. R., and Deeter, M. F., 1972, The effects of cross-country running on pre-adolescent girls, *Med. Sci. Sports* **4**:1–5.

Buchanan, T. A. S., and Pritchard, J. J., 1970, DNA content of tibialis anterior of male and female white rats measured from birth to 50 weeks, *J. Anat.* **107**:185.

Buskirk, E. R., Andersen, K. L., and Brozek, J., 1956, Unilateral activity and bone and muscle development in the forearm, *Res. Q. Am. Assoc. Health Phys. Ed.* **27**:127–131.

Cann, C. E., 1983, Hypothalamic amenorrhea in premenopausal women associated with bone loss, *Calcif. Tissue Int.* **35**:A49.

Carron, A. V., and Bailey, D. A., 1974, Strength development in boys from 10 through 16 years, *Monogr. Soc. Res. Child Dev.* **39(4)**:1–37.

Cerny, L., 1969, The results of an evaluation of skeletal age of boys 11–15 years old with different regime of physical activity, in: *Physical Fitness Assessment* (V. Seliger, ed.), pp. 56–59, Charles University, Prague.

Chumlea, W. C., Knittle, J. L., Roche, A. F., Siervogel, R. M., and Webb, P., 1981, Size and number of adipocytes and measures of body fat in boys and girls 10 to 18 years of age, *Am. J. Clin. Nutr.* **34**:1791–1797.

Chumlea, W. C., Siervogel, R. M., Roche, A. F., Mukherjee, D., and Webb, P., 1982, Changes in adipocyte celluarity in children ten to 18 years of age, *Int. J. Obesity* **6**:383–389.

Clarke, H. H., 1971, *Physical and Motor Tests in the Medford Boys' Growth Study,* Prentice-Hall, Englewood Cliffs, New Jersey.

Cronk, C. E., Roche, A. F., Kent, R., Eichorn, D., and McCammon, R. W., 1983, Longitudinal trends in subcutaneous fat thickness during adolescence, *Am. J. Phys. Anthropol.* **61**:197–204.

Cumming, D. C., Strich, G., Brunsting, L. A., Greenberg, L., Ries, A., Yen, S., and Rebar, R., 1981, Acute exercise related endocrine changes in women runners and nonrunners, *Fertil. Steril.* **36**:421–422.

Cumming, G. R., 1967, Current levels of fitness, *Can. Med. Ass. J.* **96**:868–877.

Cumming, G. R., 1977, Hemodynamics of supine bicycle exercise in "normal" children, *Am. Heart J.* **93**:617–622.

Cumming, G. R., 1978, Recirculation times in exercising children, *J. Appl. Physiol. Respir. Environ. Exercise Physiol.* **45**:1005–1008.

Cumming, G. R., Goodwin, A., Baggley, G., and Antel, J., 1967, Repeated measurements of aerobic capacity during a week of intensive training at a youth's track camp, *Can. J. Physiol. Pharmacol.* **45**:805–811.

Cumming, G. R., Goulding, D., and Baggley, G., 1969, Failure of school physical education to improve cardiorespiratory fitness, *Can. Med. Assoc. J.* **101**:69–73.

Cumming, G. R., Garand, T., and Borysyk, L., 1972, Correlation of performance in track and field events with bone age, *J. Pediatr.* **80**:970–973.

Cunningham, D. A., and Eynon, R. B., 1973, The working capacity of young competitive swimmers, 10–16 years of age, *Med. Sci. Sports* **5**:227–231.

Dalen, N., and Olsson, K. E., 1974, Bone mineral content and physical activity, *Acta Orthop. Scand.* **45**:170–174.

Daniels, J., and Oldridge, N., 1971, Changes in oxygen consumption of young boys during growth and running training, *Med. Sci. Sports* **3**:161–165.

Daniels, J., Oldridge, N., Nagle, F., and White, B., 1978, Differences and changes in VO_2 among young runners 10 to 18 years of age, *Med. Sci. Sports* **10**:200–203.

Davies, C. T. M., 1981, Thermal responses to exercise in children, *Ergonomics* **24**:55–61.

Davies, C. T. M., Barnes, G., and Godfrey, S., 1972, Body composition and maximal exercise performance in children, *Hum. Biol.* **44**:195–214.

Drinkwater, B. L., Kupprat, I. C., and Denton, J. E., 1977, Response of prepubertal girls and college women to work in the heat, *J. Appl. Physiol. Respir. Environ. Exercise Physiol.* **43**:1046–1053.

Edgerton, V. R., 1973, Exercise and the growth and development of muscle tissue, in: *Physical Activity, Human Growth and Development* (G. L. Rarick, ed.), pp. 1–31, Academic Press, New York.

Edgerton, V. R., 1976, Neuromuscular adaptation to power and endurance work, *Can. J. Appl. Sport Sci.* **1**:49–58.

Ekblom, B., 1969, Effect of physical training in adolescent boys, *J. Appl. Physiol.* **27**:350–355.

Ekblom, B., 1971, Physical training in normal boys in adolescence, *Acta Paediatr. Scand. (Suppl.)* **217**:60–62.

Engstrom, I., Eriksson, B. O., Karlberg, P., *et al.*, 1971, Preliminary report on the development of lung volumes in young girl swimmers, *Acta Paediatr. Scand. (Suppl.)* **217**:73–76.

Eriksson, B. O., 1971, Cardiac output during exercise in pubertal boys, *Acta Paediatr. Scand. (Suppl.)* **217**:53–55.

Eriksson, B. O., 1972, Physical training, oxygen supply and muscle metabolism in 11- to 15-year old boys, *Acta Physiol. Scand. (Suppl.)* **384**:1–48.

Eriksson, B. O., 1978, Physical activity from childhood to maturity: Medical and pediatric considerations, in: *Physical Activity and Human Well-Being* (F. Landry and W. A. R. Orban, eds.), pp. 47–55, Symposia Specialists, Miami.

Eriksson, B. O., 1980, Muscle metabolism in children—A review, *Acta Paediatr. Scand. (Suppl.)* **283**:20–27.

Eriksson, B., and Koch, G., 1973, Effect of physical training on hemodynamic response during submaximal exercise in 11–13 year old boys, *Acta Physiol. Scand.* **87**:27–39.

Eriksson, B. O., and Saltin, B., 1974, Muscle metabolism during exercise in boys aged 11 to 16 years compared to adults, *Acta. Paedatr. Belg. (Suppl.)* **28**:257–265.

Eriksson, B. O., Karlsson, J., and Saltin, B., 1971*a*, Muscle metabolites during exercise in pubertal boys, *Acta Paediatr. Scand. (Suppl.)* **217**:154–157.

Eriksson, B. O., Grimby, G., Saltin, B., 1971*b*, Cardiac output and arterial blood gases during exercise in pubertal boys, *J. Appl. Physiol.* **31**:348–352.

Eriksson, B. O., Gollnick, P. D., and Saltin, B., 1973, Muscle metabolism and enzyme activities after training in boys 11 to 13 years old, *Acta Physiol. Scand.* **87**:485–487.

Eriksson, B. O., Gollnick, P. D., and Saltin, B., 1974, The effect of physical training on muscle enzyme activities and fiber composition in 11-year-old boys, *Acta Paediatr. Belg. (Suppl.)* **28**:245–252.

Espenschade, A., 1940, Motor performance in adolescence, *Monogr. Soc. Res. Child Dev.* **5(24)**:1–126.

Espenschade, A., 1960, The contributions of physical activity in growth, *Res. Q. Am. Assoc. Health Phys. Ed.* **31**:351–364.

Espenschade, A., and Eckert, H. M., 1974, Motor development, in: *Science and Medicine of Exercise and Sport,* 2nd ed. (W. R. Johnson and E. R. Buskirk, eds.), pp. 322–328, Harper & Row, New York.

Evans, F. G., 1973, *Mechanical Properties of Bone,* Charles C Thomas, Springfield, Illnois.

Faust, M. S., 1977, Somatic development of adolescent girls, *Monogr. Soc. Res. Child Dev.* **42(1)**:1–45.

Forbes, G. B., 1978, Body composition in adolescence, in: *Human Growth,* Vol. 2: *Postnatal Growth* (F. Falkner and J. M. Tanner, eds.), pp. 239–272, Plenum Press, New York.

Fournier, M., Ricci, J., Taylor, A. W., Ferguson, R. J., Montpetit, R. R., and Chairman, B. R., 1982, Skeletal muscle adaptation in adolescent boys: sprint and endurance training and detraining, *Med. Sci. Sports Exercise* **14**:453–456.

Fox, E. L., 1977, Physical training: methods and effects, *Orthop. Clin. North Am.* **8**:533–548.

Freedson, P., Gilliam, T. B., Sady, S., and Katch, V. L., 1983, Transient VO_2 characteristics in children at the onset of exercise, *Eur. J. Appl. Physiol.* **4**:167–173.

Frisch, R. E., 1976, Fatness of girls from menarche to age 18 years, with a nomogram, *Hum. Biol.* **48**:353–359.

Frisch, R. E., Gotz-Welbergen, A. V., McArthur, J. W., Albright, T., Witschi, J., Bullen, B., Birnholz, J., Reed, R. B., and Hermann, H., 1981, Delayed menarche and amenorrhea of college athletes in relation to age of onset of training, *J. Am. Med. Assoc.* **246**:1559–1563.

Gaisl, G., and Buchberger, J., 1977, The significance of stress acidosis in judging the physical working capacity of boys aged 11 to 15, in: *Frontiers of Activity and Child Health* (H. Lavallee, and R. J. Shephard, eds.), pp. 161–168, Editions du Pelican, Quebec.

Gibbins, J. A., Cunningham, D. A., Shaw, D. B., and Eynon, R. B., 1972, The effect of swimming training on selected aspects of the pulmonary function of young girls—A preliminary report, in: *Training: Scientific Basis and Application* (A. W. Taylor ed.), pp. 139–143. Charles C Thomas, Springfield, Illinois.

Gilliam, T. B., and Freedson, P. S., 1980, Effects of a 12-week school physical fitness program on peak VO_2 body composition and blood lipids in 7 to 9 year old children, *Int. J. Sports Med.* **1**:73–75.

Godfrey, S., Davies, C. T. M., Wozniak, E., and Barnes, C. A., 1971, Cardio-respiratory response to exercise in normal children, *Clin. Sci* **40**:419–431.

Goldspink, G., 1964, The combined effects of exercise and reduced food intake on skeletal muscle fibers, *J. Cell Comp. Physiol.* **63**:209–219.

Gollnick, P. D., Timson, B. F., Moore, R. L., and Riedy, M., 1981, Muscular enlargement and number of fibers in skeletal muscles of rats, *J. Appl. Physiol. Resp.* **50**:936–944.

Gonyea, W. J., 1980, Role of exercise in inducing increases in skeletal muscle fiber number, *Am. Physiol. Soc.* **48**:421–426.

Gonyea, W. J., and Bonde-Petersen, F., 1978, Alterations in muscle contractile properties and fiber composition after weight-lifting exercise in cats, *Exp. Neurol.* **59**:75–84.

Gonzalez, E. R., 1982, Premature bone loss found in some nonmenstruating sportswomen, *J. Am. Med. Assoc.* **248**:513–514.

Hamilton, P., and Andrew, G. M., 1976, Influence of growth and athletic training on heart and lung functions, *Eur. J. Appl. Physiol.* **36**:27–38.

Haymes, E. M., McCormick, R. J., and Buskirk, E. R., 1975, Heat tolerance of exercising lean and obese prepubertal boys, *J. Appl. Physiol.* **39**:457–461.

Helander, E. A., 1961, Influence of exercise and restricted activity on the protein composition of skeletal muscle, *Biochem. J.* **78**:478–482.

Hettinger, T. L., 1961, *Physiology of Strength,* Charles C Thomas, Springfield, Illinois.

Hettinger, T. L., 1976, Influence of growth and athletic training on heart and lung functions, *Eur. J. Appl. Physiol.* **36**:27–38.

Holloszy, J. O., 1967, Biochemical adaptations in muscle: Effects of exercise on mitochondrial oxygen uptake and respiratory enzyme activity in skeletal muscle, *J. Biol. Chem.* **242**:2278–2282.

Holmer, I., and Astrand, P. O., 1972, Swimming training and maximal oxygen uptake, *J. Appl. Physiol.* **32**:510–513.

Holmes, R., and Rasch, P. J., 1958, Effect of exercise on number of myofibrils per fiber in sartorius muscle of the rat, *Am. J. Physiol.* **195**:50–55.

Houston, C. S., 1978, The radiologist's opportunity to teach bone dynamics, *J. Can. Assoc. Radiol.* **29**:232–239.

Howald, H., 1976, Ultrastructure and biochemical function of skeletal muscle in twins, *Ann. Hum. Biol.* **3**:455–462.

Hubbard, R. W., Smoake, J. A., Matther, W. T., Linduska, J. D., and Bowers, W. S., 1974, The effects of growth and endurance training on the protein and DNA content of rat soleus, plantarius, and gastrocnemius muscles, *Growth* **38**:171–185.

Huddleston, A. L., Rockwell, D., Kulund, D. N., and Harrison, R. B., 1980, Bone mass in lifetime tennis athletes, *J. Am. Med. Assoc.* **244**:1107–1109.

Ikai, M., 1967, Trainability of muscular endurance as related to age, in: *Proceedings of ICHPER Tenth International Congress, Vancouver, B. C., Canada.*

Inbar, O., Bar-Or, O., Dotan, R., and Gutin, B., 1981, Conditioning versus exercise in heat as methods for acclimatizing 8- to 10-year-old boys to dry heat, *J. Appl. Physiol. Respir. Environ. Exercise Physiol.* **50**:406–411.

Jacobs, H. S., 1982, Amenorrhea in athletes, *Br. J. Obstet. Gynecol.* **89**:498–499.

Jacobs, I., Sjodin, B., and Svane, B., 1982, Muscle fiber type, cross-sectional area and strength in boys after 4 years' endurance training, *Med. Sci. Sports Exercise* **14**:123 (abst).

Jequier, J. C., Lavallee, H., Rajic, M., Beaucage, C., Shephard, R. J., and Labarre, R., 1977, The longitudinal examination of growth and development: History and protocol of the trois rivieres regional study, in: *Frontiers of Activity and Child Health* (H. Lavallee and R. J. Shephard, eds.), pp. 49–54, Editions du Pelican, Quebec City, Canada.

Johnston, F. E., Roche, A. F., Schell, L. M., and Wettenhall, N. B., 1975, Critical weight at menarche: Critique of a hypothesis, *Am. J. Dis. Child.* **128**:19–23.

Jones, H. E., 1947, Sex differences in physical abilities, *Hum. Biol.* **19**:12–25.

Jones, H. H., Priest, J. D., Hayes, W. C., Tichenor, C. C., and Nagel, D. A., 1977, Humeral hypertrophy in response to exercise, *J. Bone Jt. Surg.* **59A**:204–208.

Karlsson, J., 1971, Muscle ATP, CP and lactate in submaximal and maximal exercise, in: *Muscle Metabolism During Exercise* (B. Pernow and B. Saltin, eds.), pp. 383–393, Plenum Press, New York.

Kato, S., and Ishiko, T., 1966, Obstructed growth of children's bones due to excessive labor in remote corners, in: *Proceedings of International Congress of Sport Sciences* (K. Kato, ed.), p. 479, Japanese Union of Sports Sciences, Tokyo.

Kemper, H. C. G., 1973, The influence of extra lessons of physical education on physical and mental development of 12–13 year old boys, in: *Physical Fitness* (B. Seliger, ed.), Charles University Press, Prague.

Kemper, H. C. G., Verschuur, R., Ras, K. G. A., Snel, J., Splinter, P. G., and Tavecchio, L., 1978, Investigation into the effects of two extra physical education lessons per week during one school year upon the physical development of 12- and 13-year old boys, in: *Pediatric Work Physiology* (J. Borms and M. Hebbilinck, eds.), pp. 159–166, Karger, Basel.

Kiiskinen, A., 1977, Physical training and connective tissue in young mice—Physical properties of achilles tendons and long bones, *Growth* **41**:123–137.

King, D. W., and Pengelly, R. G., 1973, Effect of running exercise on the density of rat tibias, *Med. Sci. Sports Exercise* **5**:68–69.

King, J. W., Brelsford, H. J., and Tullos, H. S., 1969, Analysis of the pitching arm of the professional baseball pitcher, *Clin. Orthop.* **67**:116–123.

Kirtland, J., and Gurr, M. I., 1979, Adipose tissue cellularity: A review. 2. The relationship between cellularity and obesity, *Int. J. Obes.* **3**:15–55.

Klissouras, V., 1971, Heritability of adaptive variation, *J. Appl. Physiol.* **31**:338–344.

Klissouras, V., 1972, Genetic limit of functional adaptability, *Int. Z. Angew. Physiol. Einschl. Arbeits-Physiol.* **30**:85–94.

Klissouras, V., Pirnay, F., and Petit, J. M., 1972, Adaptation to maximal effort: Genetics and age, *J. Appl. Physiol.* **35**:288–293.

Knittle, J. L., 1978, Adipose tissue development in man, in: *Human Growth,* Vol. 2: *Postnatal Growth* (F. Falkner and J. M. Tanner, eds.), pp. 295–315, Plenum Press, New York.

Kobayashi, K., Kitamura, K., Miura, M., Sodeyama, H., Murase, Y., Miyashita, M., and Matsui, H., 1978, Aerobic power as related to body growth and training in Japanese boys: A longitudinal study, *J. Appl. Physiol. Respir. Environ. Exercise Physiol.* **44**:666–672.

Koch, G., and Eriksson, B. O., 1973, Effect of physical training on anatomical R-L shunt at rest and pulmonary diffusing capacity during near-maximal exercise in boys 11 to 13 years old, *Scand. J. Clin. Lab. Invest.* **31**:95–105.

Komi, P. V., Klissouras, V., and Karvinen, E., 1973, Genetic variation in neuromuscular performance, *Int. Z. Angew. Physiol.* **31**:289–304.

Kotulan, J., Reznickova, M., and Placheta, Z., 1980, Exercise and growth, in: *Youth and Physical Activity* (Z. Placheta, ed.), pp. 61–117. J. E. Purkyne University Medical Faculty, Brno.

Krahenbuhl, G. S., Pangrazi, R. P., and Chomokos, E. A., 1979, Aerobic responses of young boys to submaximal running, *Res. Q. Am. Assoc. Health Phys. Ed.* **50**:413–421.

Krotkiewski, M., Aniansson, A., Grimby, G., Bjorntorp, P., and Sjostrom, L., 1979, The effect of unilateral isokinetic strength training on local adipose and muscle tissue morphology thickness and enzymes, *Eur. J. Appl. Physiol.* **42**:271–281.

Krotkiewski, M., Kral, J. G., and Karlsson, J., 1980, Effects of castration and testosterone substitution on body composition and muscle metabolism in rats, *Acta Physiol. Scand.* **109**:233–237.

Lamb, D. R., Van Huss, W. D., Carrow, R. D., Heusner, W. W., Weber, J. C., and Kertzer, R., 1969, Effects of prepubertal physical training on growth, voluntary exercise, cholesterol and basal metabolism in rats, *Res. Q. Am. Assoc. Health Phys. Ed.* **40**:123–133.

Lammert, O., Anderson, B., and Froberg, K., 1982, The effect of training in relation to chronological age and development in children 9 to 17 years, in: *Human Adaption. A Workshop on Growth and Physical Activity* (P. Russo, ed.), pp. 17–29, Cumberland College of Health Sciences, Sydney.

Lohman, T. G., 1982, Body composition methodology in sports medicine, *Phys. Sportsmed.* **10(12)**:47–58.

Lussier, L., and Buskirk, E. R., 1977, Effects of an endurance training regimen on assessment of work capacity in pre-pubertal children, *Ann. N.Y. Acad. Sci.* **301**:734–747.

MacDougall, J. D., Roche, P. D., Bar-Or, O., and Moroz, J. R., 1979, Oxygen cost of running in children of different ages; maximal aerobic power of Canadian school-children, *Can. J. Appl. Sports Sci.* **4**:237 (abstract).

Macek, M., and Vavra, J., 1980, The adjustment of oxygen uptake at the onset of exercise: A comparison between prepubertal boys and young adults, *Int. J. Sports Med.* **1**:75–77.

Malina, R. M., 1974, Adolescent changes in size, build, composition and performance, *Hum. Biol.* **46**:117–131.

Malina, R. M., 1975, Anthropometric correlates of strength and motor performance, *Exercise Sport Sci. Rev.* **3**:249–274.

Malina, R. M., 1978, Adolescent growth and maturation: Selected aspects of current research, *Yrbk. Phys. Anthropol.* **21**:63–94.

Malina, R. M., 1979, The effects of exercise on specific tissues, dimensions, and functions during growth, *Studies Phys. Anthropol. (Wroclaw)* **5**:21–52.

Malina, R. M., 1980, Physical activity, growth, and functional capacity, in: *Human Physical Growth and Maturation* (F. E. Johnston, A. F. Roche, and C. Susanne, eds.), pp. 303–327. Plenum Press, New York.

Malina, R. M., 1982, Physical growth and maturity characteristics of young athletes, in: *Children in Sport,* rev. ed. (R. A. Magill, M. J. Ash, and F. L. Smoll, eds.,), pp. 73–96, Human Kinetics Publishers, Champaign, Illinois.

Malina, R. M., 1983*a,* Human growth, maturation, and regular physical activity, *Acta Med. Auxol.* **15**:5–27.

Malina, R. M., 1983*b,* Menarche in athletes: A synthesis and hypothesis, *Ann. Hum. Biol.* **10**:1–24.

Malina, R. M., 1983*c,* Socio-cultural influences on physical activity and performance, *Bull. Soc. Roy. Belg. Anthropol. Prehist.* **94**:155–176.

Malina, R. M., Meleski, B. W., and Shoup, R. F., 1982, Anthropometric, body composition and maturity characteristics of selected school-aged athletes. Pediatr. Clin. North. Am. 29:1305–1323.

Malina, R. M., and Rarick, G. L., 1973, Growth, physique and motor performance, in: *Physical Activity: Human Growth and Development* (G. L. Rarick, ed.), pp. 125–153, Academic Press, New York.

Marshall, W. A., and Tanner, J. M., 1969, Variations in patterns of pubertal changes in girls, *Arch. Dis. Child.* **44**:291–303.

Marshall, G. R., Bailey, D. A., Leahy, R. M., and Ross, W. D., 1980, Allometric growth in boys of ages seven to sixteen years studied longitudinally, in: *Kinanthropometry* (M. Ostyn, G. Bennen, and J. Simons, eds.), Vol. II, pp. 371–380, University Park Press, Baltimore.

McCafferty, W. B., and Horvath, S. M., 1977, Specificity of exercise and specificity of training: A subcellular review, *Res. Q. Am. Assoc. Health Phys. Ed.* **48**:358–371.

Meleski, B. W., Shoup, R. F., and Malina, R. M., 1982, Size, physique, and body composition of competitive female swimmers 11 through 20 years of age, *Hum. Biol.* **54**:609–625.

Milicer, H., and Denisiuk, L., 1964, The physical development of youth, I. in: *International Research in Sport and Physical Education* (E. Jokl and E. Simon, eds.), pp. 262–285, Charles C Thomas, Springfield, Illinois.

Mirwald, R. L., and Bailey, D. A., 1984, Longitudinal comparison of aerobic power and heart rate responses at submaximal and maximal workloads in active and inactive boys aged 8 to 16 years, *Third International Congress of Auxology,* in: *Human Growth and Development* (J. Borms, R. Hauspie, A. Sand, C. Susanne, and M. Hibbelink, eds.), pp. 561–570, Plenum Press, New York.

Mirwald, R. L., Bailey, D. A., Cameron, N., and Rasmussen, R. L., 1981, Longitudinal comparison of aerobic power on active and inactive boys aged 7.0 to 17.0 years, *Ann. Hum. Biol.* **8(5)**:405–414.

Mocellin, R., and Wasmund, U., 1973, Investigations on the influence of a running-training programme on the cardiovascular and motor performance capacity in 53 boys and girls of a second and third primary school class, in: *Pediatric Work Physiology* (O. Bar-Or, ed.), pp. 279–285, Wingate Institute, Natanya.

Morpurgo, B., 1897, Veber activats hypergrophic den wilkurlichen muskeln, *Virchows Arch* **150**:552.

Nielsen, B., Nielsen, K., Behrendt Hansen, M., and Asmussen, E., 1980, Training of "functional muscular strength" in girls 7–19 years old, in: *Paediatric Work Physiology* (K. Berg and B. Eriksson, eds.), Vol. IX, pp. 69–78, University Park Press, Baltimore.

Nilsson, B. E., and Westlin, N. E., 1971, Bone density in athletes, *Clin. Orthop.* **77**:179–182.

Novak, L. P., 1963, Age and sex difference in body density and creatinine excretion of high school children, *Ann. N.Y. Acad. Sci.* **110(2)**:545–577.

Novotny, V., 1981, Veranderungen des Knochenalters im Verlauf einer mehrjahrigen sportlichen Belastung, *Med. Sport* **21**:44–47.

Oscai, L. B., Spirakis, C. N., Wolff, C. A., and Beck, R. J., 1972, Effects of exercise and of food restriction on adipose tissue cellularity, *J. Lipid Res.* **13**:588–592.

Oscai, L. B., Babirak, S.P., McGarr, J. A., and Spirakis, C. N., 1974, Effect of exercise on adipose tissue cellularity, *Fed. Proc.* **33**:1956–1958.

Parizkova, J., 1973, Body composition and exercise during growth and development, in: *Physical Activity: Human Growth and Development* (G. L. Rarick, ed.), pp. 97–124, Academic Press, New York.

Parizkova, J., 1977, *Body Fat and Physical Fitness,* Martinus Nijhoff, The Hague.

Polgar, G., and Promadhat, V., 1971, *Pulmonary Function Testing in Children: Techniques and Standards,* W. B. Saunders, Philadelphia.

Prior, J. C., Cameron, K., Yuen, B. G., and Thomas, J., 1982, Menstrual cycle changes with marathon training: Anovulation and short luteal phase, *Can. J. Appl. Sports Sci.* **7**:173–177.

Rarick, G. L., 1974, Exercise and growth, in: *Science of Medicine of Exercise and Sports,* 2nd ed. (W. R. Johnson, ed.), pp. 306–321, Harper & Row, New York.

Rarick, G. L., and Smoll, F. L., 1967, Stability of growth in strength and motor performance from childhood to adolescence, *Hum. Biol.* **B9**:295–306.

Riopel, D. A., Taylor, A. B., and Hohn, A. R., 1979, Blood pressure, heart rate, pressure rate product and electrocardiographic changes in healthy children during treadmill exercise, *Am. J. Cardiol.* **44**:697–704.

Robinson, S., 1938, Experimental studies of physical fitness in relation to age, *Int. Z. Angew. Physiol. Einschl. Arbeitsphysiol.* **10**:251–323.

Roche, A. F., 1981, The adipocyte-number hypothesis, *Child. Dev.* **52**:31–43.

Rohmert, W., 1968, Rechts-Links- Verlich bei isometrischem Armmuskeltraining mit verschiedenem Trainingereiz bei achtzahrigen Kinder, *Int. Z. Angew. Physiol. Einschl. Arbeitsphysiol.* **26**:363.

Ross, W. D., and Marfell-Jones, M. J., 1982, Kinanthropometry, in: *Physiological Testing of the Elite Athlete* (J. D. MacDougall, H. Wenger, and H. Green, eds.), pp. 75–115, Canadian Association of Sport Sciences, Ottawa.

Ross, W. D., Marshall, G. R., Vajda, A. S., and Roth, K., 1978, Dimensionalitat und proportionalitat von kraftleistungen, *Leisstungssport* **8**:195–205.

Rowe, F. A., 1933, Growth comparisons of athletes and non-athletes, *Res. Q. Am. Assoc. Health Phys. Ed.* **4**:108–116.

Ruffer, W. A., 1965, A study of extreme physical activity groups of young men, *Res. Q. Am. Assoc. Health Phys. Ed.* **36**:183–196.

Rutenfranz, J., Andersen, K. L., Seliger, V., Klimmer, F., Berndt, I., and Ruppel, M., 1981, Exercise ventilation during the growth spurt period: comparison between two European countries, *Eur. J. Pediatr.* **136**:135–142.

Saltin, B., 1969, Physiological effects of physical conditioning, *Med. Sci. Sports* **1**:50–56.

Saltin, B., and Rowell, L. B., 1980, Functional adaptations to physical activity and inactivity, *Fed. Proc.* **39**:1506–1513.

Saville, P. D., and Smith, R., 1966, Bone density, breaking force and leg muscle mass as functions of weight in bipedal rats, *Am. J. Phys. Anthropol.* **25**:35–39.

Saville, P. D., and Whyte, M. P., 1969, Muscle and bone hypertrophy: Positive effect of running exercise in the rat, *Clin. Orthop.* **65**:81–88.

Schmucker, B., and Hollmann, W., 1974, The aerobic capacity of trained athletes from 6 to 7 years of age on, *Acta. Paediatr. Belg. (Suppl.)* **28**:92–101.

Schwartz, L., Britten, R. H., and Thompson, L. R., 1928, Studies in physical development and posture. 1. The effect of exercise in the physical condition and development of adolescent boys, *Publ. Health Bull.* **179**:1–38.

Scott, E. C., and Johnston, F. E., 1982, Critical fat, menarche, and the maintenance of menstrual cycles: A critical review, *J. Adolescent Health Care* **2**:249–260.

Shangold, M., Freeman, R., Thysen, B., and Gatz, M., 1979, The relationship between long distance running, plasma progesterone, and luteal phase length, *Fertil. Steril.* **31**:130–133.

Shephard, R. J., 1966, World standards of cardiorespiratory performance, *Arch. Environ. Health* **13**:664–672.

Shephard, R. J., 1971, The working capacity of school children, in: *Frontiers of Fitness* (R. J. Shephard, ed.), pp. 319–344, Charles C Thomas, Springfield, Illnois.

Shephard, R. J., 1977, *Endurance Fitness*, University of Toronto Press, Toronto.

Shephard, R. J., 1982, *Physical Activity and Growth*, Yearbook Medical Publishers, Chicago.

Shephard, R. J., Allen, C., Bar-Or, O., Davies, C. T. M., Degre, S., Hedman, R., Ishii, K., Kaneko, M., LaCour, J., di Prampero, P., and Seliger, V., 1969, The working capacity of Toronto school children, *Can. Med. Assoc. J.* **100**:560–566, 705–714.

Siebert, W. W., 1928, Investigations on hypertrophy of the skeletal muscle, reprinted from *Z. Clin. Med.* **109**:350, in: *Classical Studies on Physical Activity*, 1968 (R. C. Brown and G. S. Kenyon, eds.) pp. 229–236, Prentice-Hall, Englewood Cliffs, N.J.

Skinner, J. S., Bar-Or, O., Bergsteinova, V., Bell, C. W., Royer, D., and Buskirk, E. R., 1971, Comparison of continuous and intermittent tests for determining maximal oxygen intake in children, *Acta Paediatr. Scand. (Suppl.)* **217**:24–28.

Sprynarova, S., 1974, Longitudinal study of the influence of different physical activity programs on functional capacity of the boys from 11 to 18 years, *Acta Paediatr. Belg. (Suppl.)* **28**:204–213.

Sprynarova, S., and Reisenauer, R., 1978, Body dimensions and physiological indicators of physical fitness during adolescence, in: *Physical Fitness Assessment* (R. J. Shephard and H. Lavallee, eds.), pp. 32–37, Charles C Thomas, Springfield, Illinois.

Sprynarova, S., Parizkova, J., and Irinova, I., 1978, Development of the functional capacity and body composition of boy and girl swimmers aged 12–15 years, in: *Pediatric Work Physiology* (J. Borms and M. Hebbelinck, eds.), pp. 32–38, Karger, Basel.

Steinhaus, A. H., 1933, Chronic effects of exercise, *Physiol. Rev.* **13**:103–147.

Steinhaus, A. H., 1955, Strength from Morpurgo to Muller—A half century of research, *J. Assoc. Phys. Ment. Rehab.* **9**:147–150.

Stewart, K. J., and Gutin, B., 1976, Effects of physical training on cardiorespiratory fitness in children, *Res. Q. Am. Assoc. Health Phys. Ed.* **47**:110–120.

Stolz, H. R., and Stolz, L. M., 1951, *Somatic Development of Adolescent Boys*, Macmillan, New York.

Tanner, J. M., 1962, *Growth at Adolescence*, 2nd ed., Blackwell, Oxford.

Thompson, D. W., 1966, *On Growth and Form* (J. T. Bonner, ed.), abr. ed., p. 238, Cambridge University Press, Cambridge.

Tipton, C. M., Matthes, R. D., and Maynard, J. A., 1972, Influence of chronic exercise on rat bones, *Med. Sci. Sport* **4**:55.

Trueta, J., 1968, *Studies of the Development and Decay of the Human Frame*, W. B. Saunders, Philadelphia.

Trussell, J., 1980, Statistical flaws in evidence for the Frisch hypothesis that fatness triggers menarche, *Hum. Biol.* **52**:711–720.

Tullos, H. S., and King, J. W., 1972, Lesions of the pitching arm in adolescents, *J. Am. Med. Assoc.* **220**:264–271.

Ulbrich, J., 1971, Individual variants of physical fitness in boys from the age of 11 up to maturity and their selection for sports activities, *Med. Sports* **24**:118–136.

Vaccaro, P., and Clarke, D. H., 1978, Cardiorespiratory alterations in 9–11 year old children following a season of competitive swimming, *Med. Sci. Sports* **10**:204–207.

Viteri, F. E., and Torun, B., 1981, Nutrition, physical activity, and growth, in: *The Biology of Normal Human Growth* (M. Ritzen, A. Aperia, K. Hall, A. Larsson, A. Zetterberg, and R. Zetterstrom, eds.), pp. 265–273, Raven Press, New York.

Von Dobeln, W., and Eriksson, B. O., 1972, Physical Training, maximal oxygen uptake and dimensions of the oxygen transporting and metabolizing organs in boys 11 to 13 years of age. *Acta Paediatr. Scand.* **61**:653–660.

Vrijens, J., 1978, Muscle strength development in the pre- and post-pubescent age, *Med. Sport (Basel)* **11**:152–158.

Wagner, J. A., Robinson, S., Tzankoff, S. P., and Marino, R. P., 1972, Heat tolerance and acclimatization to work in the heat in relation to age, *J. Appl. Physiol.* **33**:616–622.

Watson, R. C., 1973, *Bone Growth and Physical Activity in Young Males*, Unpublished doctoral dissertation, University of Wisconsin, Madison.

Weber, G., Kartodihardjo, W., and Klissouras, V., 1976, Growth and physical training with reference to heredity, *J. Appl. Physiol.* **40**:211–215.

Willimczik, K., 1980, Development of motor control capability (body coordinatio) of 6 to 10 year old children—Results of a longitudinal study, in: Kinanthropmetry 11 (M. Ostyn, G. Beunen and J. Simons, eds.), pp. 328–346, University Park Press, Baltimore.

Wilmore, J. H., Girandola, R. N., and Moody, D. L., 1970, Validity of skinfold and girth assessment for predicting alterations in body composition, *J. Appl. Physiol.* **29**:313–317.

Wirth, A., Trager, E., Scheele, K., *et al.*, 1978, Cardiopulmonary adjustment and metabolic response to maximal and submaximal physical exercise of boys and girls at different stages of maturity, *Eur. J. Appl. Physiol.* **29**:229–240.

Yamaji, K., Miyashita, M., and Shephard, R. J., 1978, Relationship between heart rate and relative oxygen intake in male subjects aged 10 to 27 years, *J. Hum. Ergol.* **7**:29–39.

Yoshida, T., Ishiko, I., and Muraoka, I., 1980, Effect of endurance training on cardiorespiratory functions of 5-year-old children, *Int. J. Sports Med.* **1**:91–94.

8

Puberty

WILLIAM A. MARSHALL and JAMES M. TANNER

1. Definition

In this chapter the word "puberty" refers collectively to the morphological and physiological changes that occur in the growing boy or girl as the gonads change from the infantile to the adult state. These changes involve nearly all the organs and structures of the body but they do not begin at the same age nor take the same length of time to reach completion in all individuals. Puberty is not complete until the individual has the physical capacity to conceive and successfully rear children.

In the past the word "adolescence" was used synonymously with "puberty." More recently it has become common practice to use "adolescence" to refer to the psychological changes associated with puberty. However, the acceleration of somatic growth which is part of the physical change of puberty is still usually referred to as the adolescent spurt. As this chapter is not concerned with psychology, the synonymous use of the two words need not be ambiguous.

The principal manifestations of puberty are as follows:

1. The adolescent growth spurt; i.e., an acceleration followed by a deceleration of growth in most skeletal dimensions and in many internal organs.
2. The development of the gonads.
3. The development of the secondary reproductive organs and the secondary sex characters.
4. Changes in body composition, i.e., in the quantity and distribution of fat in association with growth of the skeleton and musculature.
5. Development of the circulatory and respiratory systems leading, particularly in boys, to an increase in strength and endurance.

These and other minor changes that accompany them will be the subject of this chapter. Puberty is, of course, the result of developmental processes in the neuroendocrine system, but these are discussed elsewhere (see Chapter 10, this volume).

2. The Adolescent Growth Spurt

If serial measurements of the statures of a typical West European boy and girl are plotted against age, curves similar to those shown in Figure 1 are obtained. The girl's curve begins to rise more steeply at about the age of 10.5 and the boy's at about 12.5. This inflection represents the adolescent spurt in stature which occurs, on average, about 2 years earlier in girls than in boys. As we shall see, the spurt occurs at different ages among children in any population, whereas the mean age may vary considerably from one population to another.

WILLIAM A. MARSHALL • Department of Human Sciences, University of Technology, Loughborough, Leicestershire LE11 3TU, England. **JAMES M. TANNER** • Department of Growth and Development, Institute of Child Health, University of London, London WC1N 1EH, England.

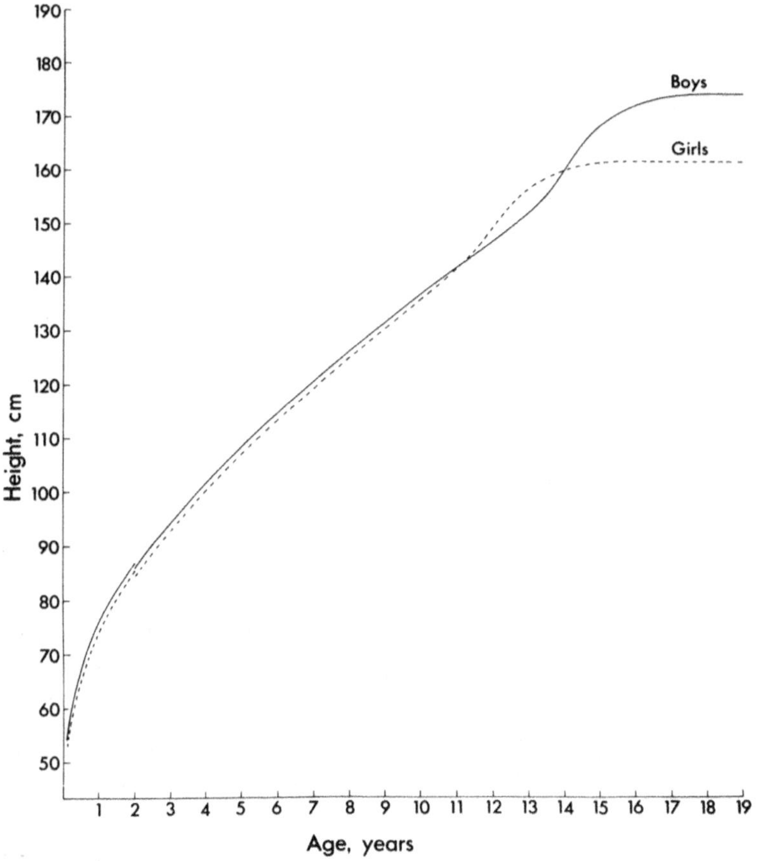

Fig. 1. Height at different ages of a hypothetical boy and girl of mean birth length, who grew at the mean rate and experienced the adolescent growth spurt at the mean age for their sex. Each finally reached the mean adult stature. (Reproduced from Tanner *et al.*, 1966.)

Velocity curves, in which the speed of growth (in centimeters per year) is plotted against age, show the adolescent spurt more clearly. It appears as a sharp increase in velocity, which rises to a maximum and then immediately begins to decrease again (Figure 2). The maximum velocity is referred to as peak height velocity (PHV).

The absolute value of peak height velocity varies from one child to another. Marshall and Tanner (1969, 1970) found a mean value of 10.3 cm/year with a standard deviation of 1.54 cm/year in 49 healthy boys who were measured every 3 months by a single skilled observer, R. H. Whitehouse. The average peak velocity for 41 girls in the same study was 9.0 cm/year with a standard deviation of 1.03 cm/year. The value for each subject was estimated by drawing a smooth curve through a plot of velocities. It is only by fitting a curve in some way that the moment of peak height velocity can be identified with reasonable confidence. The velocity measured over 3 months, 6 months, or a year does not represent the actual peak velocity because the growth rate passes through the final stages of its acceleration and begins to decelerate within a very short period of time. When the velocity is calculated over a whole year centered on the peak, i.e., including 6 months before and 6 months after it, the average is in the region of 9.5 cm/year for boys and 8.4 cm/year for girls. Thus, for about 1 year during adolescence the velocity of growth is nearly twice the velocity in either sex just before the adolescent spurt begins (about 5 cm/year). During the year in which a boy attains his PHV he usually gains between 7 and 12 cm in stature, while a girl in the corresponding year gains between 6 and 11 cm.

During the past few years, the longitudinal studies of growth initiated in a number of countries under the auspices of the International Children's Centre (see description in Tanner, 1981) have been completed on subjects followed from birth to maturity. Major studies of growth at adolescence have

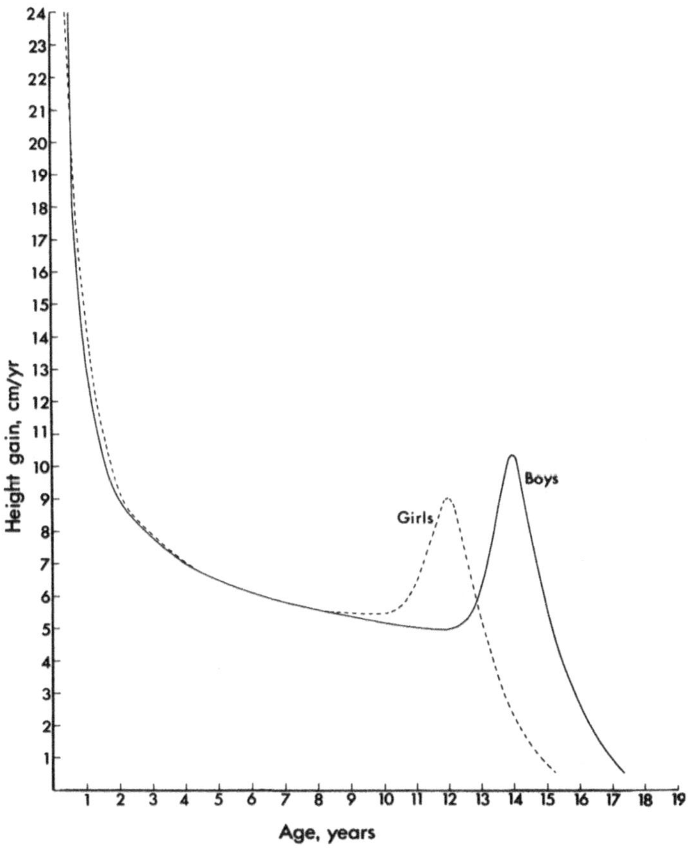

Fig. 2. Growth velocity, in centimeters per year, at different ages of the boy and girl whose statures are shown in Figure 1. (Reproduced from Tanner *et al.*, 1966.)

been published on the Zurich (Largo *et al.*, 1978) and Stockholm (Karlberg *et al.*, 1976; Taranger *et al.*, 1976) cohorts as well as further work on the London subjects (Tanner *et al.*, 1976). There have also been four longitudinal studies covering the years of adolescence only—one in Newcastle-upon-Tyne, England (Billewicz *et al.*, 1981*a,b*, 1983*b*) involving 564 boys and 669 girls measured every 6 months from age 9 to 17, a second on 338 middle-class girls of Newton, Massachusetts, measured, although only by parents, two or three times a year from 6 to 18 (Zacharias and Rand, 1983); a third on a sample of nearly 600 twins and 750 singleton controls in urban areas throughout Sweden, measured by school nurses twice a year from 10 to 16 (girls) or 18 (boys) (Lindgren, 1976, 1978; Ljung *et al.*, 1977); and a fourth study on some 200 boys and 250 girls in Wroclaw (Bielicki, 1975, 1976; Welon and Bielicki, 1979). In the Stockholm, Newcastle, Wroclaw, and Swedish twin studies, PHV and age at PHV were estimated graphically; in the Zurich study cubic splines were fitted—a formalized version of the same procedure. In the Newton data, however, Preece–Baines (PB) curves were fitted. Brown and Townsend (1982) also used PB curves for fitting longitudinal data on 62 Australian aboriginals, as did Hauspie *et al.* (1980*a*) in fitting longitudinal data on some 100 middle-class Indians in West Bengal, and Billewicz and McGregor (1982) in a very sophisticated birth to maturity study of the children in two West African villages, in Gambia.

The European and American series all produce values quite similar to the British ones quoted above. Zacharias and Rand (1983) make the point that the intensity of the adolescent spurt varies so much that in a few girls it is hardly noticeable. The Zurich data give the mean PHV for girls as 7.1 cm/year with a SD of 1.0 cm/year, so that one might occasionally expect a peak of only 5.1 cm/year. The mean value for the minimum velocity just before the spurt begins ("takeoff velocity") was 4.8 cm/year (SD 0.7), so that in this group also there could well have been a few girls in whom the spurt could only be made visible by frequently repeated measurements and sophisticated computer programs. The mean PHV for the Stockholm girls was nearer to that of the British, being 8.6 cm/year, the urban Swedish girls 8.2 cm/year, Wroclaw girls 8.3 cm/year (SD 1.1), Newcastle girls 8.0 cm/year, and the PB-estimated PHV for the Newton girls themselves 7.8 cm/year (SD 1.6).

The boys' mean PHV was estimated as 9.9 cm/year in the Stockholm study and 9.7 cm/year in the urban Swedish sample, 9.6 cm/year in the Newcastle study, and again rather low, 9.0 cm/year, in the Zurich spline-fitted subjects.

The height of the spurt is inversely related to the age at which it occurs, earlier-maturing children having higher PHVs. The correlations are of the order of −0.3 to −0.6. Largo *et al.* (1978) attributed the relationship solely to the diminution of prepubertal velocity with age; late maturers take off on their adolescent spurt from a lower prepubertal platform, as it were, and, in the Zurich data, the actual rise above the platform was independent of age. The shape of the velocity curve at the point immediately before takeoff is also in question. It has often been thought that there is a dip below the level indicated in the smoothed curves of Figure 2, especially, perhaps, in late-maturing boys (cf. Stuetzle *et al.,* 1980). It has also been said that the shape of the curve of the adolescent spurt is not actually symmetrical but that it has a slower rise and faster fall (Largo *et al.,* 1978).

One thing is very clear: There is no relationship in normal boys or in girls between age at PHV and final height. On average, early maturers and late maturers end up at exactly the same height (although, at least in girls, they do not have exactly the same body build, the early maturers being, as young adults, more fatty and muscular, the late maturers more linear). The magnitude of the adolescent spurt is also unrelated to final height.

The relationship between age at PHV and age at maximal rate of gain in weight apparently varies with the general rate of maturation. Lindgren (1978) grouped the 700 girls and boys of the Swedish urban sample into early, average, and late maturers, according to age at PHV. The mean difference in time between peak weight velocity and peak height velocity was 0.9, 0.5 and 0.2 years, respectively, for the girls and 0.6, 0.3, and 0.1 years, respectively, for the boys.

There is a close relationship between age at PHV and age at menarche; indeed, it is perhaps the closest relationship between any two events occurring in different systems during puberty (see also Ellison, 1982). The correlation is typically between 0.8 and 0.9. Menarche occurs on average about 1.3 years after PHV, with a range of 0–2.5 years. Only some 1% of girls, if that, have menarche before PHV.

Tanner *et al.* (1976) examined the correlations between different measurements (height, sitting height, leg length, and biacromial and bi-iliac diameters) for the variables, age of takeoff, age of peak velocity, peak velocity, and total adolescent gain. For the ages, the correlations were high (peak velocity 0.72–0.93), but for the peak velocity and total amount gained, they were low (amount gained 0.30–0.64). These workers concluded that at puberty there must be considerable differences between individuals in the shape changes that occur.

Nearly all skeletal dimensions accelerate at adolescence, but the effect is not uniform throughout the skeleton, and there is a greater increase in the length of the trunk than in the length of the legs, so that the proportion of the total stature which is due to the trunk rises during adolescence (Eveleth, 1978). Also, the spurt begins at different times in different parts of the body. Leg length reaches its peak growth velocity about 0.6 years before trunk length. The foot has its peak velocity earliest of all and attains its adult size before any other region except possibly the head. Some adolescents have disproportionately large feet, and girls are occasionally worried by this. However, the situation is usually self-limiting, as the earlier cessation of growth in the feet results in their not being disproportionately big by the time the child is fully grown.

In the arm, there appears to be a similar gradient to that in the leg, so that the forearm reaches its peak velocity about six months before the upper arm but some time after the hand (Maresh, 1955; Welon and Bielicki, 1979; Cameron *et al.,* 1982; see also Low, 1978). The peak velocities of shoulder width and hip width occur at about the same time as the peak velocity of sitting height (Tanner *et al.,* 1976; Welon and Bielicki, 1979).

The cranium grows very slowly after early childhood, but the growth of both head length and head breadth is accelerated slightly at adolescence (Shuttleworth, 1939). There is some disagreement as to whether this growth is due mainly to expansion of the cranial cavity, to an increase in the thickness of bone, or to an increase in thickness of the soft tissue. One radiographic longitudinal study showed that the lengths of frontal and parietal arcs increased only by 1–2% after age 12, but the thickness of the bones and scalp tissue increased by some

15% at adolescence (Roche, 1953; Young, 1957, 1959). However, Singh *et al.* (1967), who carried out a radiographic longitudinal study of head width from age 9 to 14 years, found that growth in both sexes was due mainly to increase in the thickness of soft tissue and in the width of the cranial cavity, while thickness of bone increased very little. The width of the cranial cavity increased more in boys than in girls, and they had a greater increase in soft tissue width.

The forward growth of the forehead at adolescence is mainly due to the development of the brow ridges and frontal sinuses, according to Björk (1955), who carried out a radiographic study of 243 boys at age 12 and again at age 20. However, the middle and posterior cranial fossae do enlarge, and the cranial base posterior to the sella turcica becomes lower (Zuckerman, 1955). The cranial base also increases in length (Ford, 1958).

In the face, the growth of most dimensions accelerates to reach maximum velocity a few months after peak velocity in stature (Nanda, 1955; Thompson *et al.*, 1976; Hulanicka, 1978; Baughan *et al.*, 1979; Bishara *et al.*, 1981; Ekstrom, 1982; Mitani, 1977). The greatest change is in the mandible, where 25% of the total growth in height of the ramus occurs between the ages of 12 and 20 years, in contrast to the cranial base, where only 6–7% of the total growth occurs during this period. According to Savara and Tracey (1967), the increase during adolescence is greatest in mandibular length and least for bigonial width. However, the data of Buschang *et al.* (1983) suggest very little adolescent spurt in mandibular length and a much greater spurt in the height of the ramus. The mandible as a whole becomes longer and thicker, so that it projects further forward in relationship to the face. These changes are much more marked in boys than in girls and may be accentuated by further apposition of bone at the mandibular symphysis between 16 and 23 years of age.

The literature on maxillary growth is somewhat unsatisfactory and difficult to interpret, as different workers have employed different techniques. Forward growth is apparently accelerated, particularly above the upper incisor teeth, so that prognathism is increased, although less so than in the lower jaw. Savara and Singh (1968) found that some measures of height and width showed greater acceleration than length. Burke (1979) used serial stereophotogrammetry to study mid-facial growth in twins aged 9–16 years. Boys had a considerably greater spurt than girls, as well as being always larger.

In boys, the nose usually grows forward and downward relative to the rest of the face, but this change is not detectable in all children.

Growth in the axial diameters of the eye results in an increasing tendency to myopia as adolescence progresses (Tanner, 1962; Largo, 1978). Largo found in the Zurich study that at age 10, 13% of boys and 19% of girls wore glasses; at age 16, 20% of both sexes wore glasses.

3. *Mathematical Description of Adolescent Growth*

The adolescent spurt may be described mathematically by fitting curves to series of measurements taken on the same children at different ages, i.e., to longitudinal data. The Gompertz and logistic functions, both of which give sigmoid curves, have been widely used in the past. They are discussed by Marubini and Milani in Chapter 4, Vol. 3.

Deming (1957) was the first to fit the Gompertz curve to longitudinal growth data. Marubini *et al.* (1972) fitted both Gompertz and logistic functions to measurements of height, sitting height, leg length and biacromial diameter taken at 3-monthly intervals in 23 boys and 18 girls. The fit of the curves was equally good for both sexes and was slightly better for sitting height, leg length, and biacromial diameter than for stature. The residuals, i.e., the discrepancies between the actual measurements and the fitted curves, were not significantly greater than the errors of measurement.

Bock and associates described a double logistic function, i.e., a sum of the two logistics, for fitting growth data from childhood to adulthood (Bock *et al.*, 1973; Thissen *et al.*, 1976). Use of this function requires that the child's final mature size be known, although it may sometimes be possible to estimate it, within satisfactory limits of error, from another function. According to Preece and Heinrich (1981), a triple summation of logistic functions, suggested by Bock and Thissen (1976, 1980), offers a much better fit to the data than does the double logistic, but at the expense of a great number of parameters of increasing complexity to analyze. Preece and Baines (1978) developed a much simpler model that has been used by a number of investigators. Not only does it fit the whole of growth from childhood to maturity, but it has the advantage that final height need not be known because it is estimated by the function. The equations of the summed logistics and the Preece–Baines curve are also discussed by Marubini and Milani (Chapter 4, Vol. 3).

Hauspie *et al.* (1980*b*) compared the results of fitting different curves with data on the growth in height of 35 Belgian girls followed from birth to 18.0 years. Over adolescence, the logistic fitted the data rather better than the Gompertz, with significantly lower pooled residual mean squares. Over the age range 2–18 years, the Preece–Baines model 1 showed significantly lower residual mean squares than did the double logistic. A mean-constant curve obtained by the double logistic function estimated mean PHV as being slightly greater and occurring earlier than the corresponding curve obtained from Preece–Baines Model 1. The latter model, using all available data for each child, from the second birthday onward, gave a mean peak height velocity of 7.3 cm/year (SD 1.0) at a mean age of 11.5 years (SD 1.0). As compared with graphical estimates, the Preece–Baines fit gave a significantly lower estimate of PHV, while the double logistic function estimated the mean age at PHV significantly too early. Another most interesting model has been proposed by Stuetzle *et al.* (1980) in which the preadolescent and adolescent components interact, the advent of the second component acting to switch off the action of the first (see discussion in Section 3).

4. Growth of the Heart, Lungs, and Viscera

Several investigators have described an adolescent spurt in growth of the transverse diameter of the heart, as measured radiographically (Bliss and Young, 1950; Lincoln and Spillman, 1928; Maresh, 1948). According to Simon *et al.* (1972), who carried out a longitudinal study of chest radiographs, this spurt is of about equal magnitude in girls and boys, and the age at peak velocity for heart and lung diameter coincides with the age at peak height velocity. Growth in length of the lung apparently occurs somewhat later and does not reach peak velocity until about 6 months after the peak for lung width. The magnitude of the spurt in the lungs is similar in both sexes, but the mean width of the lung in girls does not exceed that of boys at any stage. There is probably an adolescent spurt in all the abdominal viscera, including the liver and kidneys and the nonlymphatic portion of the spleen. Evidence on this point, however, is limited.

The growth in length of the pharynx is accelerated with the result that the hyoid bone becomes much lower in relation to the mandible. Before puberty, the two bones are at approximately the same level. There is a less marked adolescent growth in the anteroposterior diameter of the pharynx (King, 1952). Kahane (1978) studied sex differences in growth of the larynx in postmortem material. Although there is little sex difference prepubertally, after puberty the angle of the junction of the thyroid laminae becomes about 90° in the male, in contrast to 120° in the female, and the male vocal folds reach an average length of 11 mm, as contrasted to the female's 4 mm.

5. Sex Differences in Size and Shape Arising from Adolescence

Apart from the development of the secondary sex characters, the most noticeable sex differences that arise at puberty are the greater stature of the male and his broader shoulders as compared with the wider hips of the female. Until girls enter their adolescent growth spurt there is only a slight difference in stature between the two sexes. At this point, the girls tend to become taller than boys; however, the difference is not very great.

Figure 3 shows the growth curves of three boys and three girls, each of whom experienced the adolescent spurt at the average age for their sex. One child of each sex is of average stature, another is tall, at the ninety-seventh percentile, and the third short at the third percentile. The sex difference in stature arising from the earlier adolescent spurt in girls is of sufficient magnitude to make the short girl taller than the short boy, not as tall as the average boy. The average girl becómes taller than the average boy, but only the tall girl is taller than the ninety-seventh percentile boy. A calculation based on the data of Tanner *et al.* (1966) shows that, at 13.9 years, 7% of girls who reach PHV at the average age are taller than 13-year-old boys who are at the ninety-seventh percentile and also reach PHV at the average age for their sex. In practical terms, the difference is most important to the tall girl and the small boy. The former may realize one day that she is taller than all her male peers and be concerned lest she should become an abnormally tall woman. Similarly, the small boy finds that he is shorter than all the girls of his age, and this too may cause concern. In both cases, the situation is rectified when the boys experience their adolescent spurts and the growth of the girls begins to slow down (unless there is some actual pathology responsible for the girl's tallness or the boy's shortness).

Boys have, on average, about 2 more years of

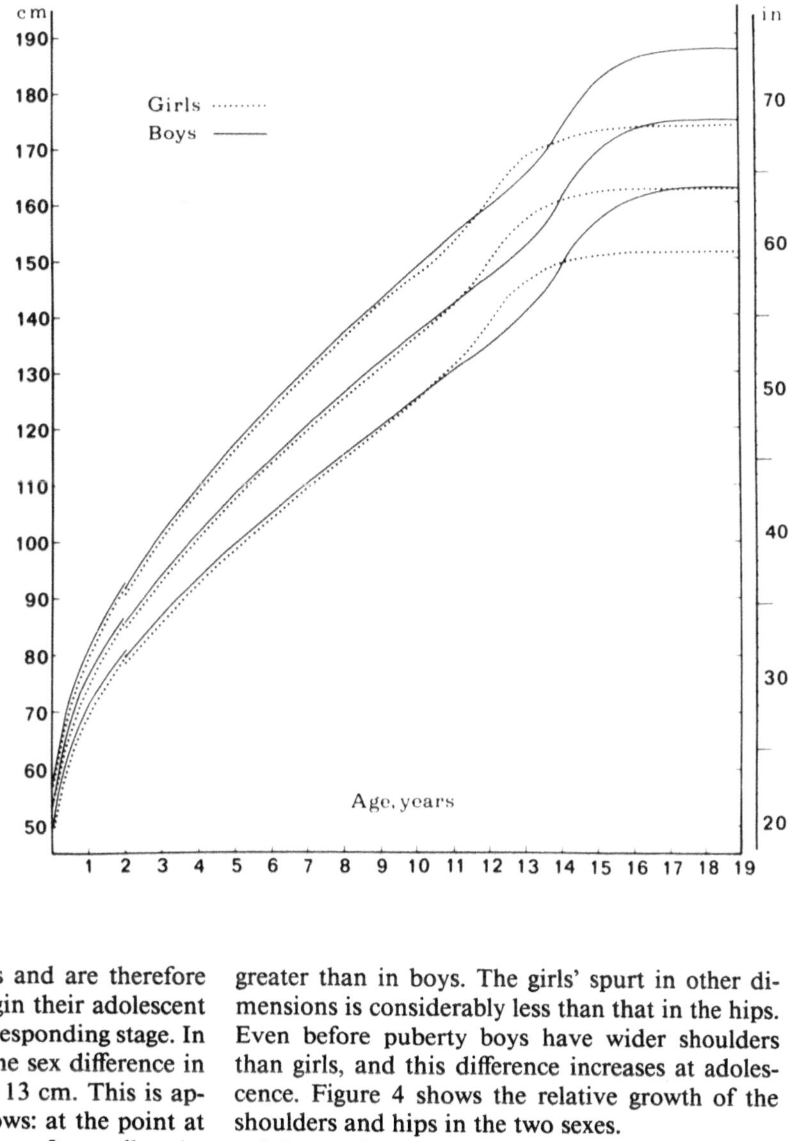

Fig. 3. Growth curves of three girls and three boys in whom the adolescent spurt occurred at the average age for their sex and produced the average increase in height. The statures of the tallest boy and girl are consistently near the 97th percentile for their sex. The lower pairs are near the 50th and 3rd percentiles, respectively. (Reproduced from Marshall, 1970.)

preadolescent growth than girls and are therefore some 9 cm taller when they begin their adolescent spurts than girls were at the corresponding stage. In European and American data the sex difference in adult stature amounts to about 13 cm. This is approximately partitioned as follows: at the point at which girls start their spurt, boys are 2 cm taller; the delay in the boys' spurt accounts for a further 7 cm, and the boys' spurt itself is some 4 cm greater than the girls'. The longer period of preadolescent growth in boys is also largely responsible for the fact that men's legs are longer than women's in relationship to the length of the trunk because the legs grow relatively faster than the trunk immediately before adolescence.

The sex difference in the relative widths of shoulders and hips is attributable to the fact that girls have a greater adolescent spurt in hip width than that of boys when hip width is related to stature, although, in absolute terms, the increase is not greater than in boys. The girls' spurt in other dimensions is considerably less than that in the hips. Even before puberty boys have wider shoulders than girls, and this difference increases at adolescence. Figure 4 shows the relative growth of the shoulders and hips in the two sexes.

It has to be remembered that not all sex differences in the skeleton arise at puberty. The forearm, for example, has greater sex dimorphism than the upper arm, and this is due to a greater rate of growth before puberty, notably at ages 6–8 (Tanner, 1962; Low, 1978).

6. Lymphatic Tissues

Apart from the subcutaneous fat, the lymphatic tissues and thymus are the only major structures that do not show an adolescent spurt in growth. The weight of the thymus increases from birth to a

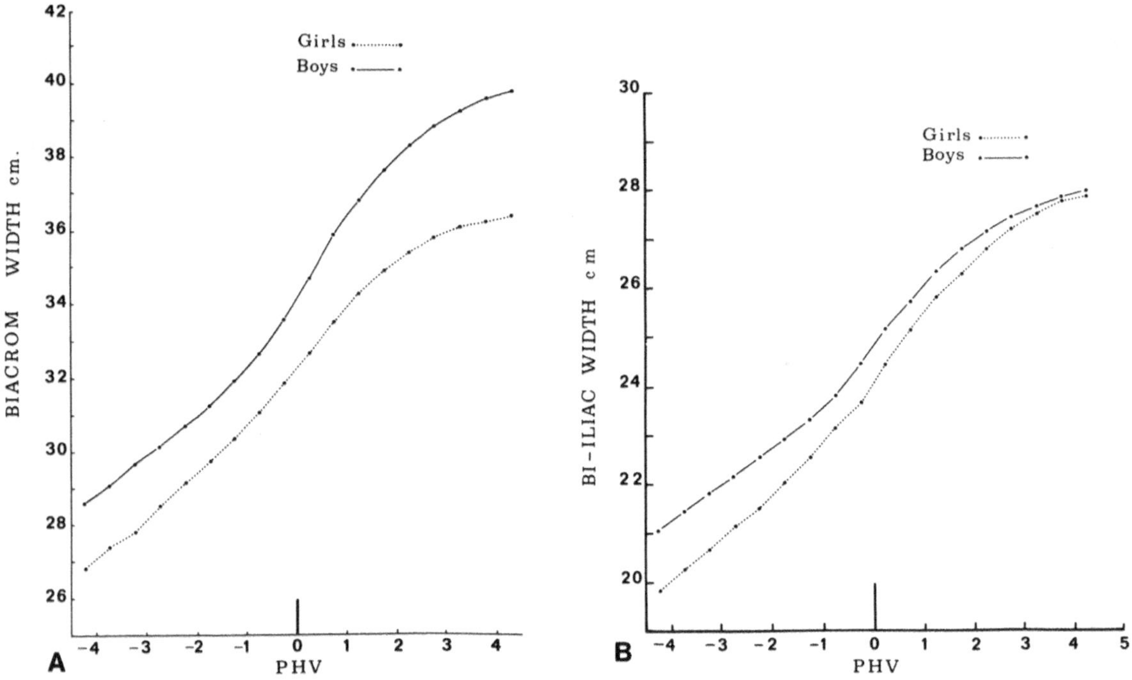

Fig. 4. Growth in width of (A) shoulders and (B) hips in girls and boys. In order to eliminate the difference in age at which the adolescent spurt occurs in the two sexes, the measurements have been plotted against a scale of years before and after peak height velocity (PHV). (Reproduced from Marshall, 1970.)

maximum within the age range 11–15 and then decreases (Boyd, 1936). This observation has been confirmed by other workers but is based only on cross-sectional data from material obtained at autopsy. Turpin *et al.* (1939) carried out a study of the thymus shadow in tomographic X-ray films that indicated that the involution took place at the time of the adolescent spurt. The thymus was found to be at its maximum size at about age 12 in girls and age 14 in boys, with a decrease occurring the following year. However, these data too were cross-sectional, and the number of subjects at each age was relatively small. The lymphatic parenchyma of the thymus may reach its maximum size slightly before the gland as a whole, as a small proportion of the organ is composed of connective tissue the absolute weight of which continues to increase throughout puberty. The lymphatic tissue in other organs, such as the spleen, intestine, appendix, and mesenteric lymph nodes, regresses at about the same time as the thymus (Scammon, 1930). In a series of 300 cases of death within 10 hr from an injury, Hwang and Krumbhaar (1940) showed that the amount of lymphatic tissue in the appendix was less in the 11 to 20-year age group than it was in the 1 to 10-year age group, although the remainder of the appendix showed an increase with age and some indication of an adolescent spurt. In the spleen, the absolute amount of lymphatic tissue increased slightly between these 10-year periods, but this increase was not nearly as great as in the nonlymphatic tissue, so that the percentage of total weight accounted for by lymphatic structures dropped from 10.8 to 7.7 (Hwang *et al.*, 1938). Scammon's data on the spleen showed an actual decrease in lymphatic tissue at adolescence.

7. Development of the Reproductive Organs and Secondary Sex Characters: Male

7.1. The Testes

For many years the study of testis growth in the living subject was hampered by lack of satisfactory and acceptable methods of measuring testis size. Nowadays, however, measurements that are sufficiently accurate for clinical purposes can be taken with the Prader orchidometer (Figure 5), which is a series of models of known volume with the shape of ellipsoids (Zachmann *et al.*, 1974). These are mounted on a string in order of increasing size, and each model is numbered according to its volume in milliliters. The instrument is held in one of the observer's hands and the testis is palpated gently with the other hand. The models are similarly palpated until the one that appears to be most nearly the same size as the testis is identified. The volume of this model is then read.

Testes with volumes of 1, 2, and occasionally 3 ml on this scale are found in prepubertal boys, but a greater volume almost always indicates that puberty has begun. The volume of the adult testis is found to vary between 12 and 25 ml when measured by this technique. There is very little increase in size of the testes before puberty. At puberty rapid growth occurs. Figure 6 shows the 10th, 50th, and 90th percentiles of testis size in boys at different ages (from Tanner, 1978, based on Dutch, Swiss, and Swedish males). At peak height velocity, the mean testicular volume is usually 12 ml (Taranger *et al.*, 1976).

Growth in size of the testes at puberty is associated with the development of their reproductive and endocrine functions. In the prepubertal testis, the interstitial tissue has a loose appearance and contains no Leydig cells. The seminiferous tubules are cordlike structures ranging between 50 and 80 μm in diameter and having no lumen. The lumen does not become distinct until puberty, although

Fig. 5. The Prader orchidometer.

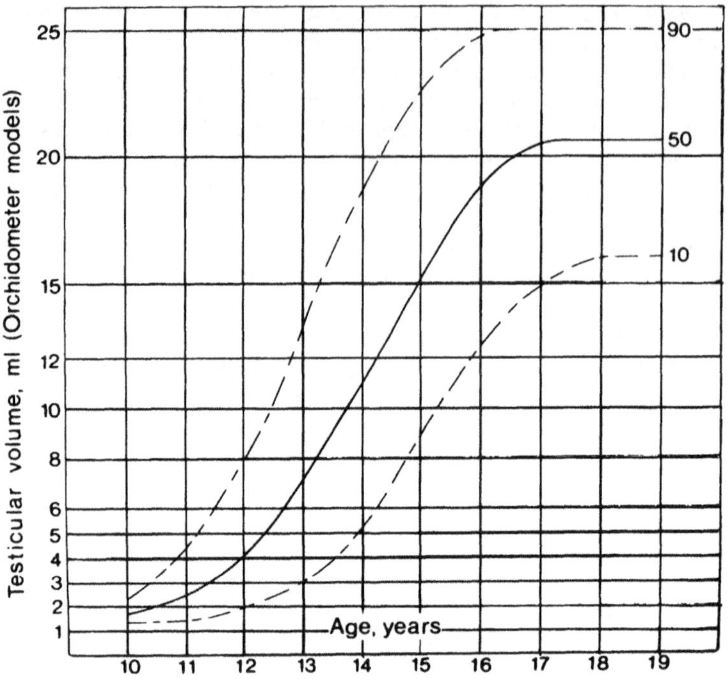

Fig. 6. Growth curve of testis size. Estimated 10th, 50th, and 90th percentiles, based on Dutch, Swiss, and Swedish males. (Data from van Wieringen *et al.*, 1971; Zachmann *et al.*, 1974; Taranger *et al.*, 1976.)

the first signs of it usually appear around the age of 6 years. During childhood, the number of spermatogonia in the tubules increases, although the Sertoli cells remain undifferentiated. At puberty, there is a considerable increase in the size and tortuosity of the tubules. A thin tunica propria, containing elastic fibers, develops.

Differentiation of Sertoli cells and division of the basally situated spermatogonia represent the beginning of the changes in the germinal epithelium, which lead gradually to spermatogenesis. The precise chronological relationship between the beginning of maturation in the seminiferous tubules and the differentiation of the Leydig cells is uncertain, but it is probable that the Leydig cells appear at the same time that mitotic activity begins in the germinal epithelium. By the time the Leydig cells are fully differentiated histologically, the meiotic divisions that lead eventually to the development of spermatids and spermatozoa have already begun.

7.2. Accessory Sex Organs

The epididymis, seminal vesicles, and prostate increase very little in weight before puberty and then show a rapid increase beginning at about the same time as enlargement of the testes (Figure 7).

7.3. Penis and Scrotum

There is no precise scale for measuring the development of the external genitalia, but for descriptive purposes the process may be divided into five stages, as shown in Figure 8. The criteria for each stage (modified from Tanner, 1962) are as follows:

Stage 1: The infantile state which persists from birth until puberty begins. During this time the genitalia increase slightly in overall size, but there is little change in general appearance.

Stage 2: The scrotum has begun to enlarge, and there is some reddening and change in texture of the scrotal skin.

Stage 3: The penis has increased in length, and there is a smaller increase in breadth. There has been further growth of the scrotum.

Stage 4: The length and breadth of the penis have increased further and the glans has developed. The scrotum is further enlarged, and the scrotal skin has become darker.

Stage 5: The genitalia are adult in size and shape.

The appearance of the genitalia may satisfy the criteria for one of these stages for a considerable time before the penis and scrotum are sufficiently developed to be classified as belonging to the next stage. It is sometimes important to distinguish between the moment at which a child's genitalia first fulfill the conditions of a particular stage, e.g., stage 3, and the length of time he remains in that stage, before passing into stage 4. The moment of arrival at a given stage may be indicated by the abbreviation G2, G3, and so on, while at other times the genitalia are described, for example, as being in stage 3, or as G3+.

The variation in ages at which different boys arrive at each stage of genital development is shown in Figure 9, which is based on data reported by Marshall and Tanner (1970), van Wieringen *et al.* (1971), and Taranger *et al.* (1976) on boys of British, Dutch, and Swedish origin, respectively. The Swiss boys reported by Largo and Prader (1983*a*) are slightly earlier. The range indicated in the figure is 1 SD on either side of the mean, i.e., it includes about 66% of the population. The genitalia may attain stage 2 at any time after the 9th birthday but need not do so before the 14th. Thus a 14-year-old, or occasionally even a 15-year-old, boy whose genitalia remain infantile may not be abnormal. Complete maturity of the genitalia (stage 5) may be reached before the 13th birthday, but some boys are not fully developed until they are 18 years of age, or occasionally even older.

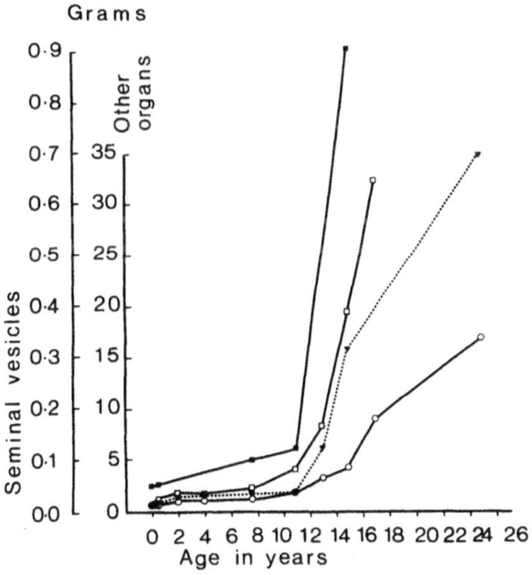

Fig. 7. Mean weights of male reproductive organs at various ages. (■——■) seminal vesicles; (▼····▼) testes; (□——□) testes plus epididymides; (O——O) prostate. (Reproduced from Marshall and Tanner, 1974.)

Standards for ages at which boys are encountered within a particular stage are naturally somewhat different and are given by Tanner (Chapter 5, Vol. 3). An alternative way of recording a child's progress through the stages of puberty is given by Taranger and Karlberg (1976). This is based on probit lines fitted to successive stages (see Section 10.3). Such lines give centiles for ages at entry and the position of the observed individual is regarded as somewhere between two successive lines, most probably equidistant from each.

Duke *et al.* (1980) suggested that some adolescents can be relied on to make correct self-ratings when given copies of the samples shown in Figures 8 and 10. However, larger-scale studies need to be done on this before self-rating is introduced as a general method of monitoring.

7.4. Pubic Hair

The increase in distribution of the pubic hair, like the development of the genitalia, may be described in five stages (Tanner, 1962). After the first recognizable pubic hair has appeared (stage 2), the stages are based entirely on the distribution of hair, and not its density. The observer must not classify sparse hair as being in a lower stage than its overall distribution allows.

Stage 1: There is no true pubic hair, although there may be a fine velus over the pubes similar to that over other parts of the abdomen.

Stage 2: Sparse growth of lightly pigmented hair is usually straight or only slightly curled. This usually begins at either side of the base of the penis.

Stage 3: The hair spreads over the pubic symphysis and is considerably darker and coarser and usually more curled.

Stage 4: The hair is now adult in character but covers an area considerably smaller than in most adults. There is no spread to the medial surface of the thighs.

Stage 5: The hair is distributed in an inverse triangle as in the female. It has spread to the medial surface of the thighs but not up the linea alba or elsewhere above the base of the triangle.

Stage 5 may be taken conveniently as the end point of adolescent growth in the pubic hair, although in early manhood the hair usually spreads beyond this triangular pattern. Some workers have

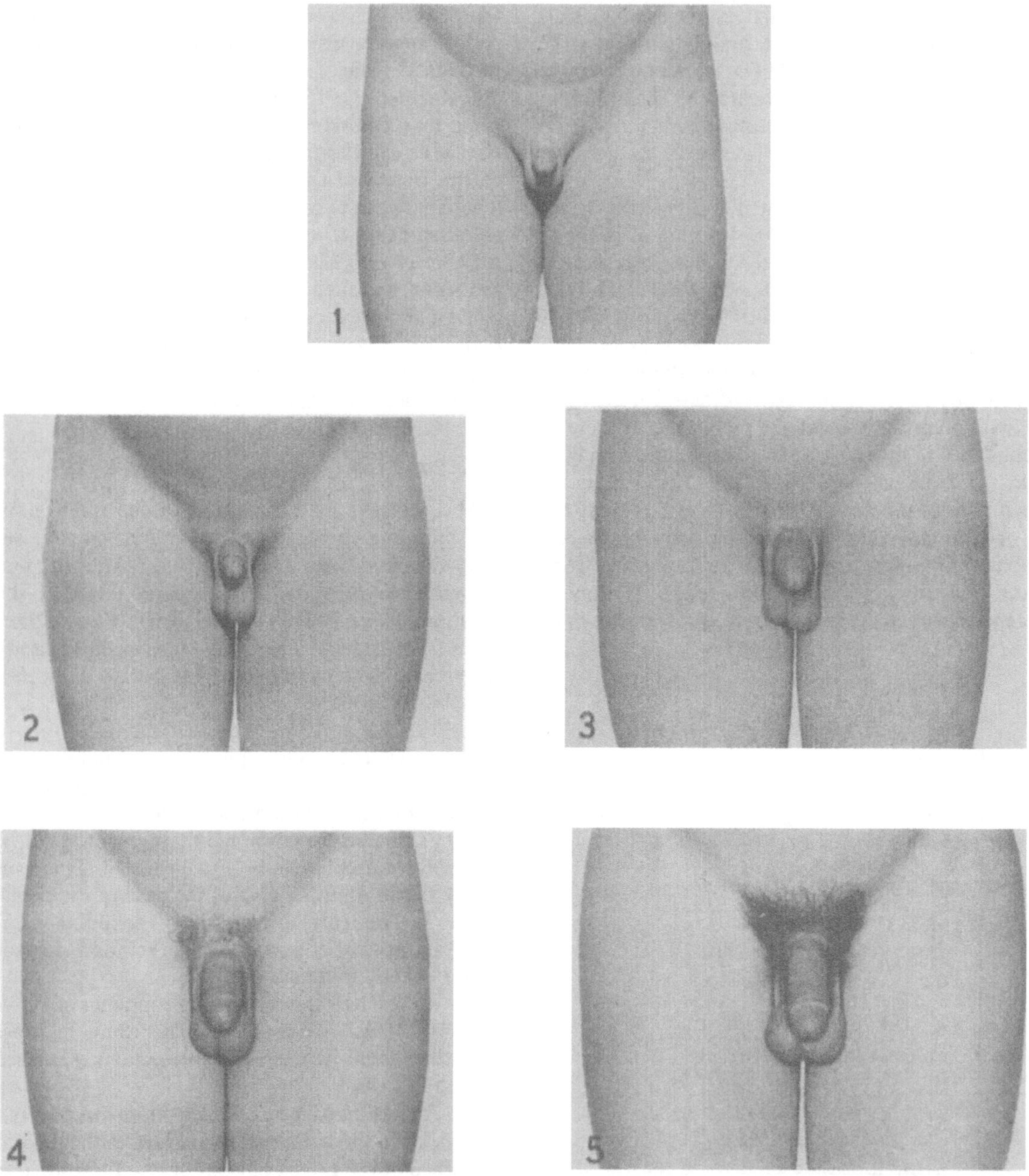

Fig. 8. Stages of development of the penis and scrotum. (Reproduced from Tanner, 1962.)

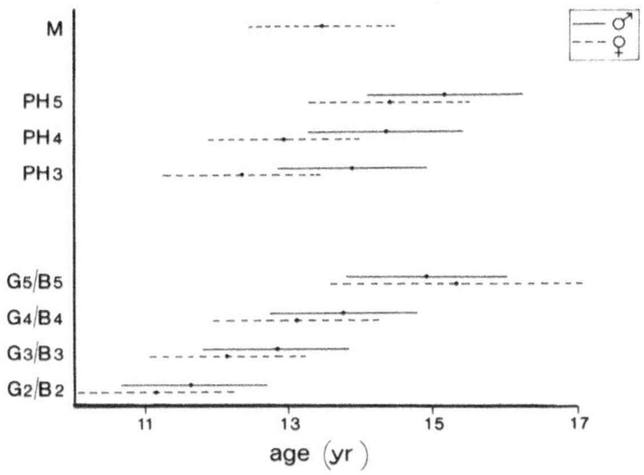

Fig. 9. Variation in age on reaching stages of puberty in boys and girls. PH2, PH3, etc., = pubic hair stages; G2, G3, etc., = stages for penis and scrotum; B2, B3, etc., = female breast stages; M = menarche. Each horizontal line extends for one standard deviation on either side of the mean.

used stage 6 to indicate the spread of hair onto the abdominal wall. However, the full adult hair distribution is seldom reached before the mid-20s.

As in the case of the genitalia, we must distinguish the moment at which a child's pubic hair reaches a given stage, e.g., PH3, and the period of time for which he is in stage 3 or PH3+. Figure 9 shows the variation of age at which the various stages of pubic hair growth are reached in European boys (Marshall and Tanner, 1970; van Wieringen *et al.*, 1971; Taranger *et al.*, 1976). Differences among populations are discussed in Section 10.7.

7.5. Axillary and Facial Hair

Neyzi *et al.* (1975*b*) observed the first appearance of axillary hair at an average of 13.5 in Turkish boys, i.e., about 1.3 years later than the onset of pubic hair growth, and in Swedish boys the interval was 1.5 years (Taranger *et al.*, 1976). The beginning of facial hair growth at the corners of the upper lip comes about 1 year later. The next stage of facial hair growth, i.e., with hair on the whole of the upper lip and some on the cheeks and chin, is not seen for about 2 more years. In general, it appears that axillary hair is not usually seen until the development of the genitalia is well advanced and 1 or 2 years after the first pubic hair is developed. However, this interval varies greatly from one child to another, and occasionally the axillary hair may appear before the pubic hair. The relationship between facial and axillary hair is also variable, and hair may appear at the two sites almost simultaneously. The manner of spread of facial hair is, however, fairly constant. Usually the first sign is an increase in length and pigmentation of the hairs at the corner of the upper lip; this hair then spreads medially until the whole of the upper lip is covered. The upper part of the cheek and the region just below the lower lip in the midline are usually the next sites at which hair appears, later followed by the sides and lower part of the chin. Hair seldom grows on the chin before the development of both genitalia and pubic hair is complete (Tanner, 1962).

7.6. The Breast

During puberty, the horizontal diameter of the male areola increases on average from about 14 mm to about 20 mm. The vertical diameter increases from approximately 11.3 to 17.0 mm (Roche *et al.*, 1971). In many boys, there is also enlargement of the underlying breast tissue, sometimes to an extent that may cause discomfort or embarrassment. This, however, is usually a temporary situation, and the enlargement regresses in a year or so, although occasionally it may persist or even increase. Rarely, the enlargement of the breasts may be sufficient to cause real psychological or social difficulties and may not regress. Although this condition may warrant surgical treatment on cosmetic grounds, it is probably an extreme variation of normal rather than a pathological state (Roche *et al.*, 1971).

7.7. Variation in Rate of Progress through Pubertal Changes

Boys vary greatly, not only in the age at which they reach any given stage of development of the genitalia or the pubic hair, but also in the time they take to pass from one stage to another or through the whole sequence of changes leading to sexual maturity. If we know at what age a boy's genitalia reached stage 2, this does not give us reliable information about when he will reach stage 5 or even stage 3.

Marshall and Tanner (1970) studied the variation in the time which normal boys take to pass through the various stages of genital and pubic hair development. The duration of any one stage can be expressed as the difference between the age at which the boy reached it (e.g., G2) and the age at which he reached the next one (G3). The duration

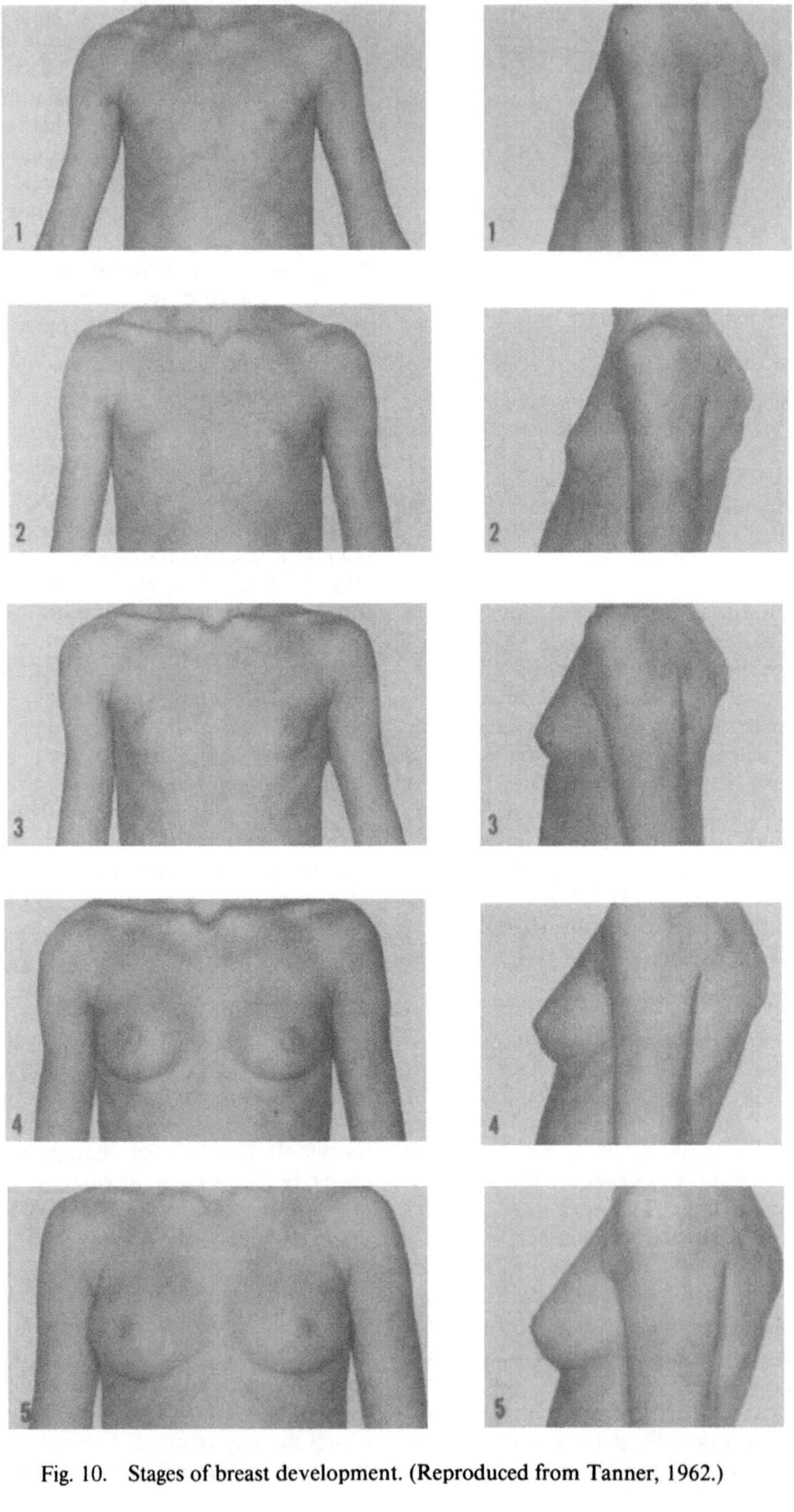

Fig. 10. Stages of breast development. (Reproduced from Tanner, 1962.)

of the whole process of genital development is the time interval between G2 and G5. The duration of pubic hair stages is similarly defined. Table I gives percentiles for the variations in duration of the different stages in the genitalia and pubic hair. The duration of pubic hair stage 2 (PH2–PH3) and the total duration of pubic hair growth (PH2–PH5) is probably underestimated because the observations were made on serial photographs that probably did not reveal the first appearance of pubic hair. It should be noted that the 97.5 percentile for G2–G3 is less than the 2.5 percentile for G2–G5. This implies that some boys pass through the whole process of their genital development in less time than others take to progress from G2 to G3. Some boys will complete their genital development in less than 2 years, while others may take 5 or more years to complete the process. There is no clear relationship between the rate of progress through the stages of genital or pubic hair development and the age at which they begin. Similar results were obtained by Largo and Prader (1983*a*).

7.8. Breaking of the Voice

The voice change in boys is a gradual event that is difficult to evaluate in a clinical examination. Taranger *et al.* (1976) give a mean age of "beginning to break" of 13.9 years, which is the same as entry to G4 and attainment of PHV, and a mean age of "having broken" just one year later.

7.9. Seminal Emission

Laron *et al.* (1980) reported the age at which boys had achieved their first ejaculation of semen during wakefulness, whether by masturbation or spontaneously. In Israel, the age range was 12.5–16.5 years, and testis size between 5 and 20 ml. The average age was about 14.5. Hirsch *et al.* (1979) reached similar conclusions by studying the presence of sperm in urine samples. Richardson and Short (1978) also studied this but failed to use probits on their data, giving instead, and erroneously, the mean age of just the proportion of the boys who produced sperm. It is clear, all the same, that some boys do produce sperm at an early stage in their external genital development.

7.10. Interrelationships of the Changes at Puberty in Boys

At any one time, a boy's genitalia and pubic hair are not necessarily at the same stage of development. Also boys reach peak height velocity when they are in different stages of either pubic hair or genital development. Table II shows the percentage of boys studied by Marshall and Tanner (1970), who were in each stage of genital development when they were observed for the first time (in 3-monthly examinations) to be in each successive pubic hair stage. The great majority of subjects were in genital stages 3+ or 4+ when the first growth of pubic hair was seen, and only 1% had any growth of pubic hair before there were visible changes in the external genitalia. However, Taranger *et al.* (1976) and Largo and Prader (1983*a*), examining subjects rather than photographs, found only 51 and 63%, respectively, of their boys in G2+ still in PH1+. van Wieringen *et al.* (1971) showed that testicular volume also varied greatly in boys who were in any given stage of either pubic hair growth or development of the penis or scrotum. Daniel *et al.* (1982) have also given testis volumes (determined by a somewhat different method) at various G and PH stages.

Table I. Percentiles of Variation in Length of Different Stages[a]

	Percentiles		
Interval	2.5	50	97.5
G2–G3	0.41	1.12	2.18
G3–G4	0.24	0.81	1.64
G4–G5	0.38	1.01	1.92
G2–G5	1.86	3.05	4.72
PH2–PH3	0.11	0.44	0.87
PH3–PH4	0.31	0.42	0.54
PH4–PH5	0.20	0.72	1.45
PH2–PH5	0.82	1.59	2.67

[a]Data from Marshall and Tanner (1970).

Table II. Percentage of Boys in Each Stage of Genital Development When First Seen in Each Pubic Hair Stage[a]

	Percentage in genital stage				
Pubic hair stage	1	2	3	4	5
2	1	13	45	41	0
3	0	4	17	75	4
4	0	0	6	65	29
5	0	0	0	10	90

[a]Data from Marshall and Tanner (1970).

8. The Reproductive Organs and Secondary Sex Characters: Female

8.1. The Ovaries

At birth, the cortex of the ovary is made up of groups of primordial follicles separated by connective tissue. Each primordial follicle contains one primordial oocyte ringed by a layer of small undifferentiated cells; those follicles that are farthest from the surface epithelium are the largest. At this time, the medulla consists of loose fibrous connective tissue, blood vessels, and nerves, while the epithelium consists mainly of cuboidal cells. Throughout childhood, follicles develop and regress, apparently at various stages of development (Peters *et al.,* 1976). As puberty approaches, the number of large follicles increases; some may grow to a considerable size, but most will still regress, and there are usually a number of anovulatory menstrual cycles before the first ovulation occurs.

As the pelvic cavity enlarges, the ovaries, uterine tubes, and uterus come to lie relatively lower in the pelvis and, by the time of menarche, they are in their adult position (the ovaries having been at the pelvic rim at birth). The uterine tubes increase in diameter, and more complex folds develop in the tubal mucosa, where the epithelium becomes ciliated.

During the 2 years preceding menarche, the ovaries increase considerably in size and are said to weigh between 3 and 4 g 1 year before menarche as compared with about 6 g at the time of the first period.

8.2. The Vagina

The vagina, which is approximately 4 cm long at birth, increases only 0.5–1 cm during childhood. Further growth usually begins before the secondary sex characters appear and continues until menarche or a little later. By the time the effects of estrogen on the vaginal mucosa can be detected, the length is in the region of 7–8.5 cm, and at menarche it is usually 10.5–11.5 cm. In childhood, the lower third of the vagina is fixed and inelastic owing to the rigidity of the musculofascial components of the pelvic floor.

At birth, the mucosa is hypertrophied and folded into high ridges as a result of the action of maternal hormones. The thickness is due mainly to the superficial layer, which consists of 30–40 layers of cells that are large and irregularly shaped. There is little cornification, and large quantities of glycogen are present. The mucosal hypertrophy in the newborn is more marked than that seen at any other time of life except during pregnancy. Smears taken from the vagina of newborn infants are cytologically very similar to those of postmenarcheal girls. Later, when the effects of maternal estrogen have disappeared, the immature vaginal epithelium is similar to that over the external cervix, with a well-defined basal layer and intermediate zone but little or no superficial layer. There is no cornification. Smears taken at this stage are made up of small round cells with dark-staining cytoplasm and large nuclei. Cytological changes in the vaginal epithelium begin before there is any development of the breast or the pubic hair and are usually the first clear indication that puberty is starting. In late childhood, vaginal smears show fewer of the small cells typical of the basal layer than are found in smears taken at an earlier age. The intermediate cells are more frequent and larger, with medium-size nuclei, and their cytoplasm stains less darkly. As the amount of estrogen in the circulation rises, the number of superficial cells in the smear becomes greater until an adult smear is obtained. Usually 5–15% of the cells are from the superficial layer before there are other signs of sexual development and before the mucosa of the vulva or distal half of the vagina shows any visible sign of estrogen action.

Immediately after birth, the pH of the vaginal contents is in the range of 5.5–7.0. Within the first day or so, lactic acid, produced by lactobacilli appearing in the vaginal flow, lowers the pH to between 5 and 4 and then, as estrogen stimulation is withdrawn over the next few days, the pH becomes neutral and then alkaline. In early childhood, the pH is about the same as it is at birth, although little vaginal fluid is present. A year or so before menarche, the amount of fluid begins to increase and its reaction again becomes acid.

8.3. The Vulva

At birth, the genital or labioscrotal swellings that develop in the fetus are still present. They are large at first, but flatten out after a few weeks. These folds are the precursors of the labia majora, which do not develop as distinct structures until late childhood.

The labia minora and the clitoris are also much larger at birth in relationship to the other vulval structures than they will be later, and the hymen is an inverted cone protruding outward into the vestibule. It is quite thick but becomes thinner very soon as the stimulus of maternal estrogen is withdrawn. Its central opening at this time is usually only about 0.5 cm in diameter.

About 1 week after birth, the effects of maternal hormone wear off, and the vulva attains the appearance which persists for about seven years. The genital swelling and labia minora are flatter, while the clitoris does not grow as much as other structures and therefore appears smaller. During early childhood, the vulval mucosa is thin, and the hymen is also thin and less protruding. Gradually, deposition of fat thickens the mons pubis and enlarges the labia majora, the surfaces of which begin to develop fine wrinkles, which become more marked during the immediate premenarcheal period. The clitoris increases slightly in size while the urethral hillock becomes more prominent. The hymen thickens, and its central orifice enlarges to about 1 cm. The vestibular (Bartholin's) glands become active.

8.4. The Uterus

The uterus of the newborn is usually between 2.5 and 3.5 cm long. The cervix comprises about two-thirds of the whole organ, but shrinks rapidly during the postnatal period. The external os is not completely formed.

The myometrium is thick, but the endometrium measures only 0.2–0.4 mm in depth and consists of sparse stromal cells with a surface layer of low cuboidal epithelium. The uterine glands of some infants are well developed. According to Ober and Bernstein (1955), 68% of 169 infants had uteri with endometria in the proliferative phase, while 27% showed some degree of secretory activity and progestational changes were found in 5%. However, this activity subsides very quickly, and the endometrium remains in a quiescent state from a short time after birth until shortly before menarche.

At the age of 6 months, the uterus is only about 80% of its size at birth, most of this regression having occurred at the cervix while the size of the corpus is reduced only slightly. By the fifth year, the uterus has again returned to approximately its neonatal size and growth continues slowly after this, but it is not until the premenarcheal period that the uterus gains a size and shape similar to that of the adult organ. During childhood, the axis of the uterus is in the craniocaudal plane, and there is no uterine flexion. There is very little growth of the corpus in early childhood, and it is not until about the tenth year that its length is approximately equal to that of the cervix. At this stage, the growth of the corpus is due to myometrial proliferation, and there is little development of the endometrium until shortly before menarche. Before this, the uterine cavity is lined by a single layer of low cuboidal cells, and there is no evidence of secretory activity.

Shortly before menarche, the cervix develops its adult shape and increases in size. The cervical canal becomes larger as the cervical glands become active, although the cervix is rather long as compared with the corpus. The adolescent growth of the uterus is greater in the corpus than in the cervix, so that by the time of menarche the two portions are of approximately equal length.

Shortly before menarche, the cervical epithelium produces copious clear secretions that tend to form threads and deposit fernlike crystals when they are dried. This type of secretion is an index of estrogen stimulation and is found in the mid-portion of the ovarian cycle in postmenarcheal girls and women.

Little is known of the motility or secretions of the uterine tubes in premenarcheal children. Tubal movements increase in amplitude and frequency during the preovulatory phase of the ovarian cycle. After ovulation, the movements decrease but secretory activity is greatest in the ovulatory and post-ovulatory phases of the cycle.

8.5. The Breasts

The breasts are frequently engorged at birth in both sexes and sometimes produce a whitish secretion known in folklore as witch's milk. The engorgement and secretion disappear very quickly, and the further development of the breasts may conveniently be divided for descriptive purposes into five stages, formalizing prior descriptions (Tanner, 1962). The stages are illustrated in Figure 10:

Stage 1: The infantile stage persists from the immediate postnatal period until the onset of puberty.

Stage 2: This is the bud stage, during which the breast and papilla are elevated as a small mound, and the diameter of the areola is increased. The development of this appearance is the first indication of pubertal development of the breast.

Stage 3: The breast and areola are further enlarged and present an appearance rather like that of a small adult mammary gland with a continuous rounded contour.

Stage 4: The areola and papilla are further enlarged and form a secondary mound projecting above the corpus of the breast.

Stage 5: This is the typical adult stage with a

smooth rounded contour, the secondary mound present in stage 4 having disappeared.

Some girls apparently never attain stage 5, and stage 4 persists until the first pregnancy or even beyond, while a few girls never exhibit stage 4 and pass directly from stage 3 to stage 5. The range of ages at which each stage of breast development is reached in British girls (Marshall and Tanner, 1969) is shown in Figure 9. As in the case of the boys' genitalia, we must distinguish between the moment of attaining a given breast stage (e.g., B2, B3) and the time which is spent in that stage.

The breasts may begin to develop at any age from about eight years onwards, and they do so in most normal girls before the 13th birthday. In more advanced girls, the breasts may be fully mature by the 12th birthday, but in others the process of development is not completed before the age of 19 and occasionally even later. These general statements are compatible with the data available from Europe and North America.

8.6. Pubic Hair

The five stages used to describe pubic hair growth in girls are essentially the same as those for boys, but the site at which hair is first seen is usually the labia, although sometimes it is the mons pubis. Figure 9 shows the variation of age at reaching each pubic hair stage in the girls studied by Marshall and Tanner (1969), but their values for PH2 are probably too high, as their photographic technique does not reveal minimal hair growth on the labia. van Wieringen *et al.* (1968) gave a 50th percentile age of 11.3 for PH2 with the 10th and 90th percentiles at 9.5 and 13.0 years, and Taranger *et al.* (1976) a 50th percentile of 11.5 years. Largo and Prader (1983*b*) found a value of 10.4 years. These results were based on direct inspection of the children and are probably more accurate than those given by Marshall and Tanner. For the later stages of pubic hair, there is no significant discrepancy among the four studies. For other populations, see Section 10.6. The adult distribution of hair is usually attained between the ages of 12 and 17.

8.7. Cutaneous Glands

The apocrine glands of the axilla and vulva usually begin to function early in puberty and, at about the same time, the sebaceous glands and sweat glands of the general body skin also become more active.

8.8. Interrelationship between Events of Puberty in Girls

The breasts and pubic hair do not necessarily begin to develop at the same time. Most girls' breasts have begun to develop, and some have even reached stage 4, before there is any growth of pubic hair. However, the reverse situation with pubic hair appearing before the breast buds, although less usual, is entirely normal. Axillary hair usually appears when the breasts are in stage 3 or 4, but in a few normal girls there may be growth of axillary hair even before there is any breast development.

Menarche is usually a late event in puberty and occurs most commonly when the breasts are in stage 4, although about 25% of girls experience their first menses while they are at breast stage 3, and about 10% do not do so until their breasts are fully mature. In the Newcastle study, 0.5% of girls had menarche, apparently normally, while in stage B1+, and 2.5% were in B2+ (Billewicz *et al.*, 1981*a*).

In girls the adolescent growth spurt begins early in relationship to the other changes of puberty, and it is extremely rare for a girl to menstruate before she has passed peak height velocity. Some 30–40% of girls attain PHV while their breasts are in stage 2, so their growth is slowing down throughout the remainder of puberty. It is therefore not justifiable to assume that a girl whose breasts are just beginning to develop has still the greater part of the adolescent spurt before her. She may have already experienced it. Most girls reach peak height velocity when their breasts are in stage 3, although about 10% do not do so until their breasts are in stage 4.

The fact that annual radiographs of the pelvis were taken in the Fels longitudinal study before the dangers of this technique were realized permitted an analysis of the adolescent growth of the birth canal (Moerman, 1982). Although it would seem that the peak velocity of measurements of both inlet and outlet occurred at about the same time as peak height velocity, growth continued longer after menarche than did growth in height. Thus some 13% of total growth occured after menarche (compared with 4% for height and 10% for biiliac diameter) and 3% continued after menarche plus 2 years. This was for inlet and outlet dimensions; internal bispinous diameter went on growing for longer still. The results could be of importance in relationship to teenage pregnancies: Moerman found that girls with early menarche had smaller pelves at menarche than did girls with late menarche, a difference greatly diminished by the time of menarche plus 2 years. Whether early-menarche girls have on average a longer anovulatory period than late-menarche girls is unknown.

9. Sex Differences in Timing of Events

Boys differ from girls in that the growth spurt occurs relatively late in relationship to the development of the genitalia; when the genitalia are beginning to develop, one can assume that the whole of the adolescent spurt is yet to come. As we have seen, the adolescent growth spurt occurs on average about 2 years earlier in girls than in boys, so that the typical 12-year-old girl is near her maximum growth rate (PHV), whereas most boys at this age are still growing at a preadolescent rate or are in the early stages of the growth spurt. There are, of course, many early-maturing boys who will experience the adolescent growth spurt before late-maturing girls.

Despite the marked difference between girls and boys in the timing of the adolescent growth spurt, there is a smaller difference in the age at which the first secondary sex characteristics appear. On average, boys' genitalia began to develop only a year later than the girls' breasts in the Stockholm study and only 0.3 year later in the Swiss study. The end stages B5/G5 and PH5 were reached at identical ages in the two sexes in Stockholm, some 0.8 year apart in Zurich. From the social point of view, girls appear to have an earlier puberty because their growth spurt and breast development can be easily observed even when they are fully clothed, while corresponding changes in boys, i.e., the development of the genitalia, are not obvious when they are clothed. Those changes that do affect the external appearance of boys, i.e., the growth spurt, the development of facial hair, and the breaking of the voice, do not occur until the genitalia are approaching maturity and thus at a considerably later age than the visible changes in girls.

10. Variation among Population and Social Groups

There is now a vast amount of information about the age of menarche in different populations. Information has also been gained about other aspects of sexual development.

The importance of genetic factors in determining the age at menarche was demonstrated by Tisserand-Perrier (1953), who found that the mean difference in menarcheal age between 46 pairs of identical twins, in a French sample, was 2.2 months, while that between 39 pairs of nonidentical twins was 8.2 months. In the Swedish longitudinal study of twins born in 1954–1955, the difference in age at menarche in 28 monozygous (MZ) pairs averaged 3.5 months, and between 48 dizygous (DZ) pairs 8.5 months; the intrapair correlations were 0.93 and 0.62 (Fischbein, 1977*b;* see also Fischbein 1977*a* and Hauspie *et al.*, 1982, for similar studies on age at PHV). In a study of Punjabi MZ and DZ twins, Sharma (1983) also found a strong genetic component in the regulation of pubertal change. In clinical practice, children experiencing late puberty often have a history of similarly late maturity in one or both parents but, at the population level, comparison between children and their parents does not provide a satisfactory basis for studying the inheritance of age at menarche, as children are often brought up in environmental conditions different from those experienced by their parents during childhood. Also, it is often impossible to obtain reliable data on the mothers' ages at menarche. However, when the environment is reasonably uniform and provides the basic requirements of good nutrition and health, the variability of menarcheal age within the population is largely attributable to genetic differences.

Differences between the mean age of menarche of different populations may be magnified or obscured by the use of unreliable methods to estimate the mean age of menarche. It is therefore important to recognize the advantages and disadvantages of the different methods.

10.1. Age at Menarche: Cross-sectional Retrospective Method

In the cross-sectional retrospective method, each subject is asked the age at which she began to menstruate. This is unsatisfactory because the subject's recollection may be inaccurate, while some girls may deliberately give false answers. In the Swedish longitudinal study (Bergsten-Brucefors, 1976), the known ages at menarche in 339 girls were compared with the ages recalled some 4 years after the event. The correlation between the actual and recalled age was 0.81 ± 0.05. Only 63% of girls recalled the date correctly to within 3 months. Damon *et al.* (1969) found a correlation of 0.78 between the actual date in 60 women and the date recalled 19 years later.

A second problem with this method is that the result will be biased if the sample of girls asked their recollected ages at menarche includes girls who have not yet begun to menstruate. Wilson and Sutherland (1949), for example, showed that the difference in the mean age between two samples was attributable to the inclusion of more premenarcheal girls in the sample having the lower mean.

Bias also results from the common practice of stating one's age as that at the preceding birthday. Thus, for example, a girl who experienced menarche at the age of 12.75 years may say that the event occured when she was 12. In a large sample, this kind of error would lead to the mean being underestimated by just 0.5 years but, in a smaller sample, the bias is less consistent.

Thus the method of data collection by recollection is clearly unsatisfactory, but it was used in nearly all the older surveys.

10.2. Age at Menarche: Longitudinal Method

In the longitudinal method, a group of subjects is seen repeatedly, and each girl is asked on every occasion if she has yet begun to menstruate. If the interval between visits is short, the precise date can usually be determined with reasonable confidence. This is the most accurate method of determining the age at menarche in individuals, but a very large sample would be required to give a reliable estimate of the population mean. In view of the financial and administrative difficulties involved, no very large samples have been studied. Zacharias *et al.* (1976) succeeded in following prospectively 633 middle-class, European-descended girls born around 1960 and living in Newton, Massasschuetts; the mean menarcheal age was 12.8 ± 0.05 years, SD 1.2 years, and range 9.1–17.7 years. The Newcastle study also followed a large number of girls prospectively (see Section 10.4).

10.3. Age at Menarche: Status Quo Method

In the status quo method, each subject in a representative sample of the population need only be asked her precise age at the time of the questioning and whether she has yet begun to menstruate. This leads to a record of the percentage of affirmative answers at each successive age. To this percentage distribution, a probit or logit transformation is applied. This method does not reveal the age of menarche in any individual, but it is the best method for estimating the median and variance for the population. Figure 11 illustrates the method. The percentage of postmenarcheal girls at each age is shown, and the probits corresponding to those percentages are plotted. A line is fitted to these probits, and the age corresponding to 50% as indicated by this line is the median.

It should be noted that the plotted percentages are estimates from samples and do not represent exact single observations. It is therefore not appropriate to use the simple least-squares method for fitting the line. The reader is referred to standard statistical tests dealing with probit analysis for further discussion of this point.

10.4. Age at Menarche in Different Populations

Table III lists the ages at menarche that had been reported up to 1983 in various populations. All mean values except those specifically marked are based on the status quo method. The variation between populations is considerable.

In some countries the age at menarche differs widely between social groups. For example, in northeast Slovenia, mean menarcheal ages of 13.3, 13.7, and 14.2 were reported for girls of good, medium, and poor social standing (Kralj-Cercek, 1956). Similar differences, i.e., 13.4, 13.8, and 14.1, were reported for rich, middle-class, and poor girls, respectively, who were attending schools in, and in the rural areas around, Cairo (Attalah *et al.,* 1983). In Turkish girls, Neyzi *et al.* (1975*a*) observed that not only menarche but all stages of breast development and pubic and axillary hair growth occurred later in the lower socioeconomic classes as

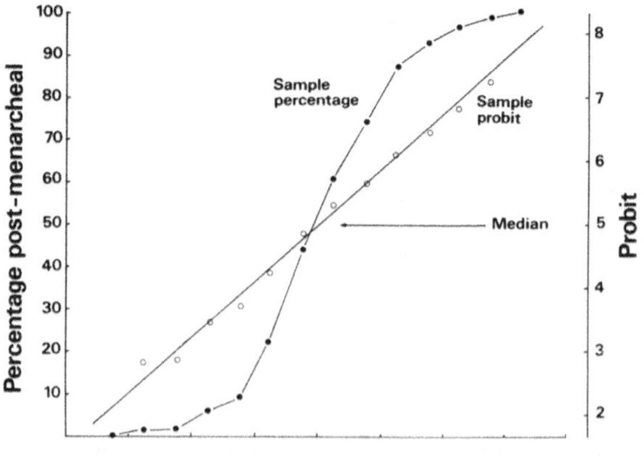

Fig. 11. Determination of age at menarche by the status quo method. The sigmoid curve indicates the percentage of girls in the sample (made at London, 1966) who had experienced menarche by each age. From this, the probits were determined and a straight line fitted through them.

Table III. Age at Menarche in Various Populations

Country	Year	Mean ±SE	Source
Europe			
Denmark: Copenhagen	1964	13.2 ± 0.20	Anderson (1968)
England			
London	1966	13.0 ± 0.03	Tanner (1973)
Northeast	1967	13.3 ± 0.03	Roberts *et al.* (1975)
Newcastle	ca. 1975	13.4 ± 0.04	Billewicz *et al.* (1981*b*), prospective
Finland	1969	13.2 ± 0.02	Kantero and Widholm (1971)
France			
Paris	1966	13.0 ± 0.11	Crognier *et al.* (1979)
Rural	1976	13.1 ± 0.18	Crognier *et al.* (1979)
Greece: Athens, middle-class	1979	12.6 ± 0.07	Dacou-Voutetakis *et al.* (1983)
Holland			
All	1965	13.4 ± 0.03	Van Wieringen *et al.* (1971)
Utrecht	1972	13.4 ± 0.09	Venrooij-Ijsselmuiden *et al.* (1976)
Hungary			
West	1965	13.1 ± 0.01	Eiben (1972)
North	1976	12.8 ± 0.19	Panto (1980)
Komarom	1980	12.8 ± 0.09	Farkas *et al.* (1981)
Szeged	1981	12.7 ± 0.03	Farkas *et al.* (1982)
Iceland: Reykjavik	1973	13.1 ± 0.10	Magnusson (1978)
Italy			
Cararra	1968	12.6 ± 0.04	Marubini and Barghini (1969)
Naples, rural	1969	12.5 ± 0.02	Carfagna *et al.* (1972)
Veneto	1976	12.8 ± 0.04	Gallo (1977)
Jugo-Slavia: Zagreb	1973	12.6 ± 0.05	Prebeg (1978)
Norway: Oslo	1970	13.2 ± 0.01	Bruntland and Walløe (1973)
Poland			
Warsaw	1976	12.7 ± 0.03	Laska-Mierzejewska *et al.* (1982)
Rural	1976	13.4 ± 0.02	Laska-Mierzejewska *et al.* (1982)
Romania			
Towns	ca. 1960	13.5 ± 0.06	Cristescu *et al.* (1964), recalculated
Villages	ca. 1960	14.6 ± 0.07	Cristescu *et al.* (1964), recalculated
Sweden: Stockholm	1967	13.1 ± 0.08	Furu (1976)
Switzerland: Zurich	ca. 1970	13.4 ± 0.10	D. Flug, R. H. Largo, and A. Prader, unpublished, prospective
USSR: Moscow	1970	13.0 ± 0.08	Miklashevskaya *et al.* (1972)
North and South America			
Argentina: La Plata, middle-class	1978	12.5 ± 0.05	Lejarraga *et al.* (1980)
Canada: Montreal, French	1969	13.1 ± 0.04	Jenicek and Demirjian (1974)
Chile: Santiago, middle-class	1970	12.6 ± 0.12	Rona and Pereira (1974)
Cuba			
Havana	1973	12.8	Jordan *et al.* (1974)
Rural	1973	13.3	Jordan *et al.* (1974)
Mexico			
Xochimilco	1966	12.8 ± 0.18	Diaz de Mathman *et al.* (1968), recalculated
Rural Oaxaca	1976	14.3 ± 0.20	Malina *et al.* (1977)
United States			
European origin	1968	12.8 ± 0.04	MacMahon (1973)
African origin	1968	12.5 ± 0.11	MacMahon (1973)
Venezuela			
Caracas, middle-class	1976	12.0 ± 0.04	Limongi (1977)
Carabobo, nonmanual	1978	12.3 ± 0.10	Farid-Coupal *et al.* (1981)

(continued)

Table III. (continued)

Country	Year	Mean ±SE	Source
Middle East			
Egypt			
Cairo, well off	1976	12.6 ± 0.29	Attallah (1978)
Cairo, middle-class	1976	13.1 ± 0.17	Attallah (1978)
Cairo, rural villages	1976	13.9 ± 0.18	Attallah (1978)
Nubian	1966	15.2 ± 0.30	Valšik *et al.* (1970)
India			
Madras, well off	1975	12.9 ± 0.10	Roberts *et al.* (1977)
Warangel, poor	1975	14.1 ± 0.20	Roberts *et al.* (1977)
Lucknow, school	1967	14.5 ± 0.17	Koshi *et al.* (1971), recalculated
Iraq			
Baghdad, well off	1969	13.6 ± 0.06	Shakir (1971)
Baghdad, poor	1969	14.0 ± 0.05	Shakir (1971)
Israel: Jerusalem	1977	13.3 ± 0.45	Belmaker (1982)
Nepal: Chumik	1981	16.2 ± 0.13	Beall (1983)
Turkey: Istanbul, well off	ca.1970	12.4 ± 0.10	Neyzi *et al.* (1975*a*)
Far East			
Australia: Sydney	1970	13.0	Jones *et al.* (1973)
Hong Kong			
Better off	1978	12.4 ± 0.18	Low *et al.* (1982)
Worse off	1978	12.7 ± 0.18	Low *et al.* (1982)
Japan[a]	1966–1967	12.9 ± 0.01	Yanagisawa and Kondo (1973)
New Guinea			
Bundi, highlands	1967	18.0 ± 0.19	Malcolm (1970)
Kaipit, lowlands	1967	15.6 ± 0.25	Malcolm (1969)
New Zealand			
Maori	1969	12.7 ± 0.07	New Zealand Department of Health (1971)
Non-Maori	1969	13.0 ± 0.02	New Zealand Department of Health (1971)
Singapore			
Rich	1968	12.4 ± 0.09	Aw and Tye (1970)
Average	1968	12.7 ± 0.09	Aw and Tye (1970)
Poor	1968	13.0 ± 0.04	Aw and Tye (1970)
Africa			
Nigeria			
Ibadan, better off	1974	13.3 ± 0.06	Oduntan *et al.* (1976)
Enugu, better off	1978	13.2 ± 0.13	Ucha *et al.* (1979)
Senegal: Dakar	1970	14.6 ± 0.08	Bouthreuil *et al.* (1972), recalculated.
Somalia: Mogadish, better off	1978	13.1 ± 0.18	Gallo *et al.* (1980)
Sudan: Khartoum, better off	1980	13.4 ± 0.14	Attallah *et al.* (1983)
Uganda: Kampala, better off	1960	13.4 ± 0.16	Burgess *et al.* (1964)

[a]More recent data, but not by probits, indicates ~12.5 in 1978 (see Nagu *et al.*, 1980; Hoshi *et al.*, 1981).

compared with girls in the higher socioeconomic classes. A similar social class difference was observed in Jewish girls in Jerusalem (Belmaker, 1982). The difference is less marked as conditions improve; for example, in Hong Kong the upper-lower class difference was 0.62 years in 1962 and 0.44 in 1978 (Low *et al.*, 1982).

In the Netherlands, de Wijn (1966) found a mean age at menarche of 13.8 years among girls whose fathers were in the lower and middle socioeconomic groups, while the daughters of those in the higher classes gave a value of 13.5. Mother's education may be at least as important as father's occupation (Farkas, 1980; Oduntan *et al.*, 1976).

Brundtland *et al.* (1980) reported on differences in menarcheal age between social strata in Oslo (Norway) in 1928, 1952, 1970, and 1975. Menarcheal age was lowest in the higher strata until the 1950s, but this trend reversed, and in 1975 the lower strata had the lowest menarcheal age. This was associated with a change to greater weight for age in the lower strata, a reversal of the former situation.

In the Stockholm survey of 1967, however, there was no difference at all between social classes; the values being 13.05, 13.14, and 13.04 in three groups classified by father's and mother's occupations (Furu, 1976). The same was true among middle-class families in Newton, Mass. (Zacharias *et al.*, 1976).

Much or all of the social class difference is associated with differences in the number of children in the family (Roberts *et al.*, 1971). In the survey done in Northeast England by Roberts *et al.* (1975), remarkable for its statistical sophistication, families with five or more children gave a mean menarcheal age of 13.62 years, those with three or four children 13.42, and those with one or two children 13.10. No effect of father's occupation could be demonstrated when allowance was made for family size. However, the families surveyed lived in "better class dormitory suburb areas," and the equally precise analysis by Billewicz *et al.* (1981*b*) of 699 families drawn at random in the city of Newcastle-upon-Tyne showed a clear nonlinear interaction between the effects of occupation and family size. In nonmanual class families, age at menarche was unrelated to family size, but it was found to be related among the manual classes, and especially in the unskilled manual class. The total social class difference in menarcheal age (as might perhaps be expected in England around 1975) was almost entirely due to a delayed age of menarche among girls from large families with fathers in manual occupations (Billewicz *et al.*, 1981*b*). In the middle-class families in Newton, Massachusetts, Zacharias *et al.* (1976) found no relationship of menarcheal age to number of siblings. The effect of social class and family size is probably principally linked to nutrition, but the effects of chronic gastrointestinal and respiratory disease and the conditions under which pregnancy occurs cannot be ignored either. In Poland, there are rather large differences in menarcheal age between urban and rural populations; these differences diminish, but do not entirely disappear, when allowance is made both for father's level of education and for family size (Laska-Mierzejewska *et al.*, 1982).

Modern European populations differ only slightly in average age at menarche, as shown in Table III. In the western part of the continent, most population means lie between 13.0 years (London) and 13.4 years (Holland). Mediterranean girls, however, reach menarche appreciably earlier. Mean values of 12.5 years have been reported in a poor rural population near Naples (Carfagna *et al.*, 1972) and 12.6 years in an urban population (Carrara) in northern Italy (Marubini and Barghini, 1969). A similarly low value (12.6 years) was recently reported for middle-class Greek girls in Athens (Dacou-Voutetakis *et al.*, 1983), for girls in Zagreb (Prebeg, 1978), and for rich girls in Cairo (Attallah, 1978). North American girls of European ancestry reach menarche slightly earlier than those in Northwest Europe. In a national sample of the United States, the mean value was 12.8 years (MacMahon, 1973). French-Canadians living in Montreal had the same value (13.1 years) as French girls in France (13.0 and 13.1 years), as shown in Table III. In Australia and New Zealand, the means are similar to those in London, i.e., 13.0 years (Jones *et al.*, 1973).

Some racial differences appear to override the effects of social conditions. Well-off Chinese girls living in Hong Kong have menarche at a median age of 12.4 years (Low *et al.*, 1982), and even poor Chinese girls in the same population begin to menstruate as early as Europeans in much better economic circumstances (Lee *et al.*, 1963). In Japan there has been a remarkable secular trend (see Section 11) in age at menarche; the most recent values give a mean of 12.5 years (Nagai *et al.*, 1980; Hoshi and Kouchi, 1981). Even Japanese living in a rural area in Brazil had a mean of 12.8 years (Eveleth and Souza Freitas, 1969). In New Zealand, the Maori at 12.7 years are earlier than New Zealand girls of European ancestry (New Zealand Department of Health, 1971).

Among Africans living in Africa, menarche is generally late, e.g., 13.4 years in Uganda and 14.1 years in the Nigerian Ibo (Burgess and Burgess, 1964; Tanner and O'Keefe, 1962). However, earlier means may yet be demonstrated in well-off Africans. Afro-Americans are usually early with an average age of 12.5 years (MacMahon, 1973). The Cuban population, largely of European–African mixture, averages 13.2 years over the whole country, but in Havana the mean is about 12.4 years as compared with about 13.3 years in rural areas (Jordan *et al.*, 1974). This implies that the late menarche among Africans in Africa may be due entirely to environmental factors. In favorable conditions, Africans may mature as early as Asiatics and Southern Europeans.

In most Middle Eastern populations, the mean age of menarche is rather late, but this is the result of economic circumstances, as well-off girls living in Istanbul have one of the earliest menarches so far recorded, 12.4 ± 0.1 (Neyzi *et al.,* 1975*a*). On the other hand, in Baghdad, even well-off girls are late at 13.6 years (Shakir, 1971). In India, there is a difference of 1.5 years or more between the rich, who average about 12.8 years (Madhavan, 1965), and the poor, who average 14.4. The latest recorded ages of menarche are in the Melanesians of New Guinea, where different studies have given values ranging from 15.5 to 18.4 years (Malcolm, 1970; Wark and Malcolm, 1969). There is a clear difference in age at menarche between town and country girls, the urban girls being earlier in every comparison so far reported (Eveleth and Tanner, 1976).

The effects of climate on age at menarche are difficult to distinguish from those of other associated factors such as race. Roberts (1969) approached this problem by using data from 39 samples studied by the status quo method. His samples included 21 from Europe, 5 from eastern Asia, 6 from India and neighboring territories in south Asia, and 7 from Africa. Covariance analysis showed that, within races, the regression of menarcheal age on mean temperature was significant ($r = -0.40$). The multiple correlation between menarcheal age and mean temperature and year of investigation combined was 0.52. After these two effects had been taken into account, differences between the adjusted means among racial groups remained highly significant.

Social conditions in infancy may be particularly important in determining the age at which menarche occurs. In Norway, for the period between 1900 and World War II, Liestøl (1982) found that the correspondence between the gross domestic product and the age of menarche was greatest when GNP was related to the year of birth of the women concerned. When values of the GNP corresponding to later childhood were used, the correlation was lower. Liestøl also found that in English longitudinal data of the National Survey of Health and Development done in the 1950s age at menarche was related to the interval between the birth of the girl being considered and her next-earlier sibling. It was found that the longer the spacing, the earlier the menarche. Ellison (1981) demonstrated among many countries a parallelism between declines in menarcheal age and infant mortality, with a lag of about 30 years.

Factors that are difficult to classify in socioeconomic, climatic, or genetic terms may also influence the age at menarche. Girls who are blind and also those who are deaf seem to have a significantly earlier menarche than controls (Zacharias and Wurtman, 1964, 1969; Malina and Chumlea, 1977; Buday, 1981). The reasons for this are not clear. Girls who take up ballet dancing or track athletics (but not, oddly enough, competitive swimming) on average have a delayed menarche (Warren, 1980; Frisch *et al.,* 1980; reviewed in Malina, 1983). Here also the causative factors are unclear; the extreme expenditure of energy during training may be one, but social pressure causing self-selection for participation in athletics is probably another.

10.5. Age at Menarche as Subject to Stabilizing Selection

There is preliminary evidence that age at menarche, like birth weight, is subject to stabilizing selection. Wyshak (1983*b*) showed that women with early menarche (before age 12) and those with late menarche (after age 14) had a higher proportion of unsuccessful pregnancies than did women with menarcheal ages closer to the mean. This effect was greater among Catholics, in whom the number of abortions (interfering with the selection process) was presumed to be lower.

10.6. Population Differences in Age of Pubertal Stages: Girls

There are a number of reports of mean ages, estimated by probits, of transition from one puberty stage to another. Estimates are made by the status quo method, exactly as for menarche. Taking the breast stages as an example, a sample of girls is examined and the observed stages recorded. Then the percentage stage 2 or beyond is plotted against age; the percentage stage 3 or beyond is plotted and so on. In middle-class girls in Athens, Dacou-Voutetakis *et al.* (1983) found mean transition (or entry-to-stage) ages as follows: B2 10.6, B3 11.8, B4 12.2, B5 14.2; PH2 10.5, PH3 11.6, PH4 12.6, PH5 14.3. The standard errors were about 0.07 years. Neyzi *et al.* (1975*a*) give similar data for Istanbul girls, New Zealand Department of Health (1971) for New Zealand girls, van Wieringen *et al.* (1971) for Dutch girls, Belmaker (1982) for Israeli girls, Andersen (1968) for Danish girls, and Lee *et al.* (1963) and Low *et al.* (1982) for Hong Kong girls.

Dutch girls reached B2 on average at age 11.0 years, some 6 months later than Greek and Israeli girls, and 1 year later than middle-class Turkish or well-off Chinese girls. The longitudinal series of data in Zurich, Stockholm, London, Newcastle, and Paris give figures for B2 similar to those in the Dutch survey (Largo and Prader, 1983*b*; Taranger *et al.*, 1976; Marshall and Tanner, 1969; Billewicz *et al.*, 1981*a*; Roy *et al.*, 1972). For PH2, Dutch girls (11.3 years) were nearly a year later than Greek, Turkish, and Israeli girls, but the Chinese girls (11.8) were relatively delayed, as were Maori girls compared with European-descended New Zealanders (New Zealand Department of Health, 1971). Stockholm girls averaged 11.5 years and Zurich girls, surprisingly, 10.4 years; the value for London girls is not comparable for PH2, as the ratings were done on the basis of photographs alone.

Harlan *et al.* (1980) reported on American girls examined in the National Health Examination Survey Cycle III, 1966–1970. They give tables of the numbers of girls in each cell of a cross-classification by breast and pubic hair stages at ages 12, 13, 14, and 15–17 for 2688 white and 500 black girls. The girls are too old to permit valid calculations of ages at entry to stages other than 4 and 5. Americans of African descent were significantly earlier-maturing in both breast and pubic hair stages than were Americans of European descent. This is in line with the findings for menarcheal age, but Harlan *et al.* (1980) state that the racial difference in secondary sex character ratings remains significant even when only premenarcheal and only postmenarcheal girls are compared.

10.7. Population Differences in Age of Pubertal Stages: Boys

Ages for entry to puberty stages have been reported for Dutch boys by van Wieringen *et al.* (1971), for Turkish boys by Neyzi *et al.* (1975*b*), for New Zealand boys by the New Zealand Department of Health (1971) for American boys by Harlan *et al.* (1979*b*), for Chinese boys by Chang *et al.* (1966), for Egyptian boys by Hafez *et al.* (1981), for Swedish boys by Taranger *et al.* (1976), for Swiss boys by Largo and Prader (1983*a*), and for British boys by Marshall and Tanner (1970).

PH2 was reached on average at 11.8 years in Dutch, at 12.2 years in Swiss, at 12.5 years in Swedish, and at 11.8 years in well-off Turkish boys. In upper-class Hong Kong boys, PH2 was later (13.0 years), as in the case of girls. In contrast to the situation in girls, however, no difference was observed between American boys of African and of European origins. Indian boys growing up in the Himalayas at altitudes of 2000–3000 m had PH2 at 13.5 years (Singh and Sidhu, 1981). Twin boys in Chandigarh, just south of the Himalayas, had an equally late age at PH2 (Sharma, 1983).

10.8. Population Differences: The Adolescent Spurt and Undernutrition

The adolescent height spurt seems to be resistant to moderate degrees of undernutrition. Middle-class Indians living in rural areas near Calcutta measured throughout their childhood by Das (Hauspie *et al.*, 1980*a*) were smaller than British subjects as adults, but nearly all of this was due to preadolescent growth; the fitted curves over adolescence were very similar (Hauspie *et al.*, 1980*b*). Similarly, in rural Hyderabad a group of boys who were small at age 5 due to undernutrition gained very nearly as much during puberty as those who were tall at age 5 (Satyanarayama *et al.*, 1980). Children in Gambia who were chronically undernourished during part of the year gained as much during adolescence as British standards, although much less from birth to age 5 (Billewicz and McGregor, 1982). Australian aboriginals actually gained more than British during puberty, but Brown and Townsend (1982) think there might have been an element of catchup growth in this, with nutrition becoming better at just about the time of adolescence, an explanation that may to some degree apply to the other groups as well.

11. The Trend toward Earlier Puberty

In the last century the age at menarche has become progressively earlier in Europe, North America, and several other parts of the world. The data collected over the past 100 years are naturally variable in their reliability. However, when the results from several studies are combined (as in Figure 12), they are remarkably consistent in indicating that menarche has been occurring earlier, at an average rate of some 3 or 4 months per decade. This is not

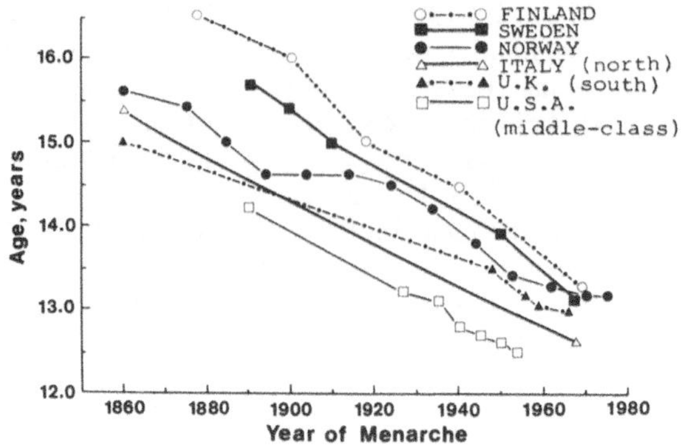

Fig. 12. Secular changes in age at menarche, 1860–1980.

to say that the trend has always been as linear as such an approximation might imply. The detailed study of Oslo Maternity Hospital records by Brudevoll *et al.* (1979) shows a rather rapid decline from 1860 to 1890 followed by a period of little or no decline until around 1920, when a steep decline began. This continued until around 1960, after which there has been little change (see Figure 2 in Chapter 15, this volume). It must be remembered also that the values plotted for the nineteenth century in Figure 12 refer to working-class women. Middle-class women in the period 1860–1900 had menarche on average 1 year (London and Manchester) to 18 months (Copenhagen and Berlin) earlier. In the United Kingdom, from 1910 to 1960, there was a drop of 4.4 months per decade in the working class and 2.8 months per decade in the middle class (Tanner, 1981; see also Manniche, 1983). Wyshak's (1983*a*) data give a value of 3.2 months per decade for middle-class Americans, 1925–1955 (year of menarche), while Goodman's (1983) data give a value of 2.3 months per decade for Europe-descended Hawaiians 1915–1955, with Hawaiians of Japanese or Chinese descent having the greater trend of about 6.1 months per decade.

Other notable studies of the trend in menarcheal age are those of Laska-Mierzejewska *et al.* (1982), who gave a trend from 1965–1975 of 3 months per decade in Polish cities and 7 months per decade in rural areas; Panto (1980), who gave a decline of 3 months per decade in Hungary 1955–1975; and Venrooij-Ijsselmuiden *et al.* (1976), who gave around 2.5 months per decade for Dutch girls, 1955–1972. Ducros and Pasquet (1978) reviewed the available French data, which point to little trend between 1850 and 1890 (cf. Norway) and a trend of some 3.5 months per decade from 1890 to 1970. Nagai *et al.* (1980) reviewed the figures for Japan; there seems to have been only a slight trend from 1900 to 1935; from then until 1950 the trend reversed, but that was followed by a steep decline in menarcheal age at the rate of some 11 months per decade until 1975, when the trend leveled out to nearly zero. Clearly, the trend cannot continue forever in any country, and there is evidence that in Norway and southern England it has all but stopped (Brundtland and Walløe, 1973; Tanner, 1973).

Data on children's heights and weights are in keeping with the notion that the adolescent spurt has also become progressively earlier (e.g., Ljung *et al.*, 1974; Tanner, 1981). In Oslo children, the age of maximum increment (the population equivalent to PHV) shifted by 1.5 years in boys and by 1.0 years in girls between 1920 and 1950; from 1950 to 1970 no further shift occurred (Brundtland *et al.*, 1980).

Tanner *et al.* (1982) fitted Preece–Baines model I curves to the annual mean values from ages 5 to 17 of Japanese school data collected in 1957, 1967, and 1977. Between 1957 and 1977, the maximal increments in height, sitting height, and leg length all became earlier by about 1 year in boys and a little less in girls. In urban Japanese girls, Hoshi and Kouchi (1981) found that secular changes in age at peak height velocity were similar to those in menarche.

12. Adolescent Sterility

Menarche does not necessarily imply fertility, nor does amenorrhea necessarily denote sterility. There are several recorded instances of conception occurring before menarche but, in most girls, ovulation does not occur until there have been a number of menstrual periods (see Chapter 9, Vol. 2). In a group of 209 girls living in their parental homes, Metcalf *et al.* (1983) found that only 25% of cycles were ovulatory (by pregnanediol excretion) during the first 2 years after menarche, 45% during the next 2 years, then successively 63, 72, and 83% in the subsequent years. Girls living away from home had fewer ovulatory cycles in the period 5–8 years after menarche than did girls still at home. In the Newcastle study, the duration of the first 36 cycles was followed in 298 girls. The length between periods decreased from an average of 50 days for the first cycle to 30 days for cycles 31–36. Individual subjects became gradually more regular: the within-person SD fell from 23 days for cycles 1–6 to 11 days for cycles 7–12, and finally to 5 days for cycles 31–36. The number of short cycles remained the same, but the proportion of long (> 57 days) cycles dropped from 27 to 1%. The individual's variability was unrelated to age at menarche (Billewicz *et al.*, 1980). Following up the girls of the Harvard School of Public Health longitudinal growth study, Gardner (1983) found a significant relationship between dysmenorrhea and menstrual irregularity in adolescence and poor gynecological health in middle age.

In populations in which early sexual intercourse, without contraception, is common, it is apparently unusual for pregnancy to occur until sometime after menarche. A striking example of this was reported by Gorer (1938). The Himalayan Lepchas believed that puberty in girls was the result of sexual intercourse. Girls were therefore betrothed and occasionally married from about the age of 8 years and boys from the age of 12. From this time onward intercourse might take place regularly, and yet pregnancy rarely occurred before the 22nd year. Menarche in this population probably occurred about the age of 13 and the first seminal emission in boys at the age of about 15. However, there appeared to be a high overall rate of adult sterility in this population, and it would be unwise to suggest that the results were typical of other groups. For further discussion, see the review by Ashley Montagu (1957).

13. Puberty and Skeletal Maturation

There is no clear relationship between skeletal maturation, represented as bone age, and the development of the secondary sex characters in normal children. Marshall (1974) showed that the chronological age of 74 girls, whose breasts had just begun to develop (B2), had a mean value of 11.0 years with a standard deviation of 1.07, whereas the RUS skeletal age measured by the TW2 method (Tanner *et al.*, 1975) had a mean value of 11.1 years with a standard deviation of 1.05. Thus, the skeletal age was just as variable as the chronological age (see Figure 13). This was also true of a sample of 97 boys, in which the mean ages at G2 were 11.5 years for chronological age and 11.7 years for RUS age, with standard deviations of 1.07 and 1.13, respectively. When maturity is reached, the situation is similar. In 35 girls at B5, the standard deviation of chronological age was 0.95 and that of the RUS age was 0.91. In 70 boys at G5 the standard deviation of chronological age was 0.99 and that of the RUS age 0.81. None of these small differences was statistically significant. At peak height velocity, the standard deviation of RUS age was slightly less than that of chronological age, but again the difference was not statistically significant. However, the point at which children of either sex attained 95% of their mature height was closely linked to bone age, the standard deviations of their RUS ages being 0.27 and 0.22 for girls and boys, respectively, compared with standard deviations of chronological age of 0.77 and 0.74 ($p < 0.001$). Also, as pointed out earlier, age at menarche is quite closely linked to bone age (see Figure 14), the standard deviation of RUS age at this point being 0.48, compared with the standard deviation of chronological age of 0.84 ($p < 0.001$).

Clearly, the view that puberty begins at a more or less constant bone age is untenable as far as children who reach puberty within the ±2 SD range of variation are concerned. However, when puberty is unusually advanced or delayed, some relationship to skeletal maturation is apparent. In pathological

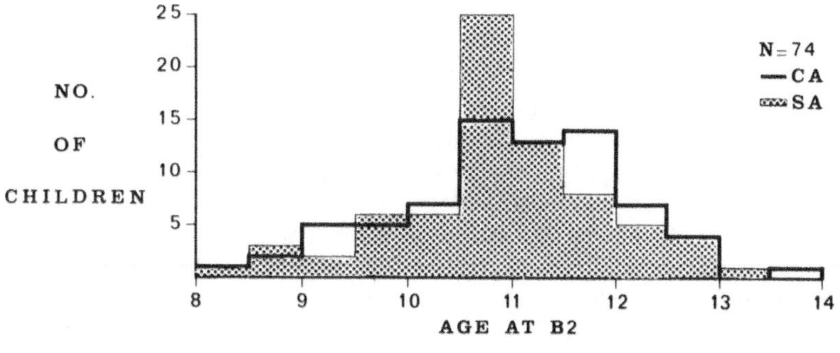

Fig. 13. Distribution of chronological age and of skeletal age in girls on reaching breast stage 2 (B2). (Reproduced from Marshall, 1974.)

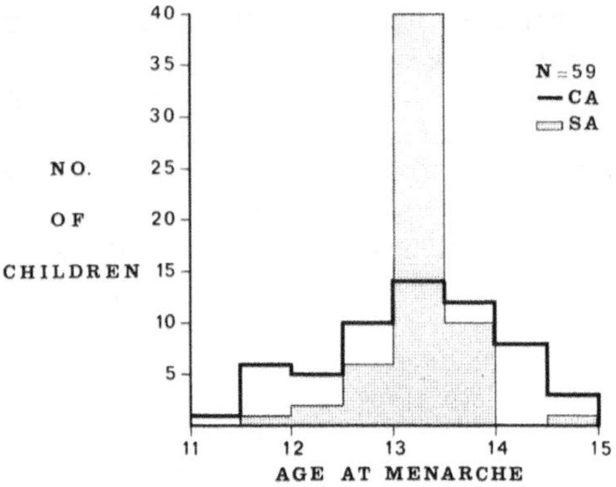

Fig. 14. Variation of chronological age and of skeletal age in girls at menarche. (Reproduced from Marshall, 1974.)

precocious puberty, the bone age is always advanced, although this might be interpreted as the inevitable result of the circulating sex hormones that caused the early puberty rather than a demonstration of a direct relationship between puberty and bone age. When puberty is unusually delayed, the secondary sex characteristics do not usually begin to develop until the bone age is in the range within which these changes take place in normal children.

The small variation of bone age, as compared with chronological age, at the attainment of 95% mature height no doubt reflects an overall relationship between the skeletal age and the proportion of the child's total growth that has been completed. Such a relationship is implied by the usefulness of bone age in predicting adult height (Tanner *et al.*, 1975). However, by the same argument, we should not expect skeletal age to be of any help in predicting when any of the changes of puberty other than menarche will occur.

Orthodontists, in particular, have an interest in predicting when the adolescent spurt will occur, because some of their treatments are most effective when applied at that time. Houston *et al.* (1979) examined prediction from ossification events (i.e., bone age stages) and came to the conclusion implied above—that bone age was not useful in predicting the beginning of the spurt except in subjects who were markedly advanced or delayed. Hägg and Taranger (1982) showed that dental emergence was not useful in predicting the spurt either (see also Filipsson and Hall, 1976).

14. Prediction of Age at Menarche

Bone age is useful in predicting the age at which menarche will occur. In a premenarcheal girl we can predict when menarche is most likely to occur if we know the mean and standard deviation of age at menarche in the population. We can also give an upper age limit before which the girl will almost certainly have begun to menstruate if she is normal. The range of error in these predictions can be reduced if we make proper use of additional information such as the bone age. It is, however, essential to recognize that the statistical basis of the prediction changes as the age of the premenarcheal girl increases.

14.1. Prediction on the Basis of Chronological Age Only

Let us consider a hypothetical population in which the mean age at menarche is 13.0 years and the standard deviation is 1 year (Figure 15A). A girl whose age is less than 10.5 years has a 95% probability that menarche will occur between the ages of 11.0 and 15.0. In the absence of any additional information, it must be concluded that she is most likely to menstruate at the modal age, i.e., 13.0 years.

When a girl's age is already within the distribution, i.e., over 10.5 years, the properties of a truncated normal distribution must be taken into account. The shading in Figure 15B indicates the distribution of age at menarche in girls who are premenarcheal at the age of 12.0 years. Such a girl might menstruate almost immediately, but she is most likely to do so at the modal age of this distribution, which is again 13.0 years. Girls who are still premenarcheal at an age beyond the population mode, e.g., at 14.0 years, have ages at menarche distributed according to the shaded area in Figure 15C. Statistically speaking, each of the these girls is more likely to menstruate immediately than at any other later time.

The method of calculating the 95% upper limit in each of the above distributions is as follows. In complete gaussian distributions such as that in Figure 15A we find the normal deviate, i.e., the number of standard deviations from the mean, corresponding to 95% of the population using a statistical table (e.g., Table IX of Fisher and Yates, 1963). This has the value of 1.64. Therefore, where the mean is 13.0 years and the SD 1.0 year, 95% of

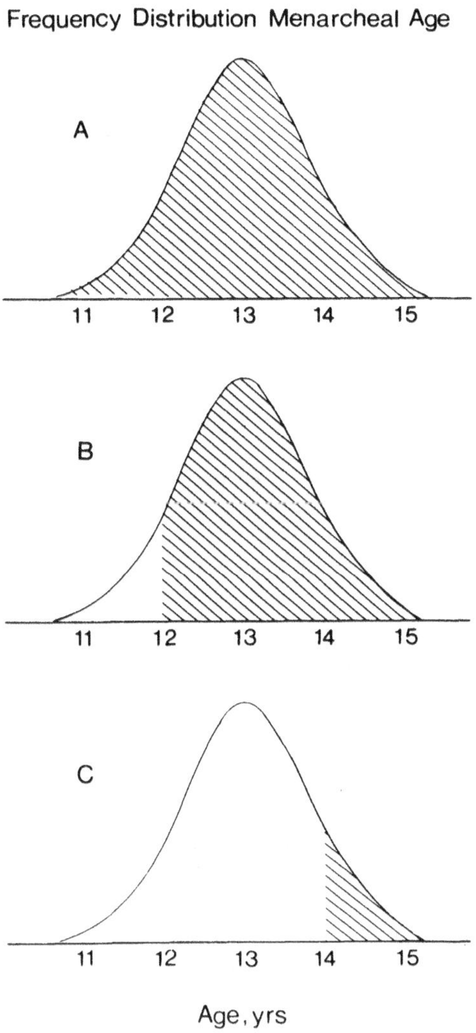

Fig. 15. Changes in distribution of possible age at menarche in premenarcheal girls of different chronological ages. (For discussion, see text.) (Reproduced from Marshall and de Limongi, 1976.)

girls aged 10 or less can be expected to menstruate by the age of 14.64. To find the upper limit in the case of premenarcheal 12-year-old girls, we have first to estimate the percentage of the total population who are still premenarcheal at this age: 12.0 years has a normal deviate of -1, which gives us a figure of 16% of the population. The remaining 84% are premenarcheal. Ninety-five percent of this remainder will have menstruated by an age at which the normal deviate corresponds to the 16% who have already menstruated plus 95% of the remainder. This deviate is found for

$$16 + (95 \times 84\%) = 95.8\% \text{ of the population}$$

The normal deviate is 1.73. Therefore, 95% of girls who are premenarcheal at age 12 will have menstruated by the age of $13.0 + 1.73 = 14.73$ years. This method can also be applied to premenarcheal girls above the modal age and up to 3 SD above the mean for the population.

Figure 16 shows the result of applying the above method to girls who are premenarcheal at given chronological ages. Also shown are the ages at which girls in the different age groups are most likely to menstruate (see Marshall and de Limongi, 1976).

14.2. Prediction Using Chronological and Skeletal Age

Bone age might be incorporated into the calculations by computing a multiple regression of age at menarche on bone age and chronological age. However, Marshall and de Limongi (1976) used a simple regression of age at menarche on the difference between bone age and chronological age (BA − CA). It was assumed that age at menarche in girls at any give value of (BA − CA) had a gaussian distribution about the regression line with a standard

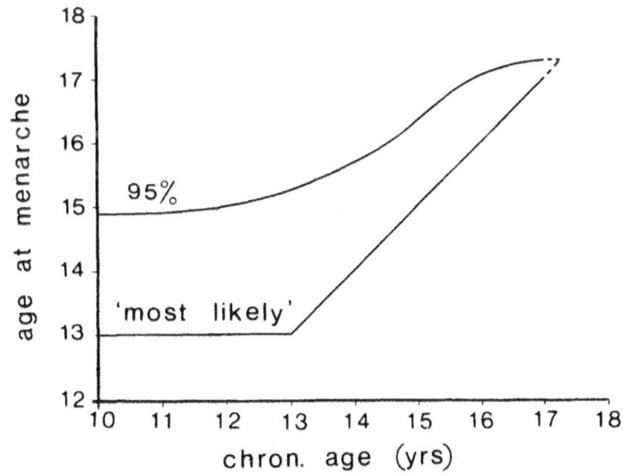

Fig. 16. (Lower curve) Most likely age at menarche in premenarcheal girls of given chronological ages. (Upper curve) Age by which menarche will have occurred in 95% of girls who are premenarcheal at given chronological ages. (Reproduced from Marshall and de Limongi, 1976.)

deviation equal to the residual standard deviation (RSD) about the regression.

At any given value of (BA − CA), a premenarcheal girl whose chronological age is more than 2 RSD below the regression line (curve A in Figure 17) may experience menarche at any age defined by the limits ±2 RSD but is most likely to do so at the age indicated by the regression line, curve A1. (Note that the same vertical axis in Figure 17 is used to represent both the chronological ages of the subjects under discussion and the predicted mean age at menarche.) If her age were less than 2 RSD below the line (curve B in Figure 17), she might menstruate immediately and would be most likely to do so at the age indicated by curve B1. A girl whose age is already above the regression line C is most likely to experience menarche immediately (C1).

The age at which 95% of girls in any of the above categories will have menstruated can be calculated at given values of BA − CA in exactly the same way as when the bone age was not considered (above). The values at each (BA − CA) may then be joined to form a line as illustrated in Figure 17, which also shows the regression line and RSD. Figure 18 shows the age at which menarche will have occurred in 95% of girls who are premenarcheal at given values of BA − CA at given chronological ages.

Marshall and de Limongi (1976) demonstrated that the advantage of including bone age in the prediction increases as bone age advances in relationship to chronological age. When bone age is advanced by 2 years, the error of prediction is reduced by 50% of more, and the prediction is improved in all cases except in 13-year-olds and 14-year-olds with very retarded bone ages.

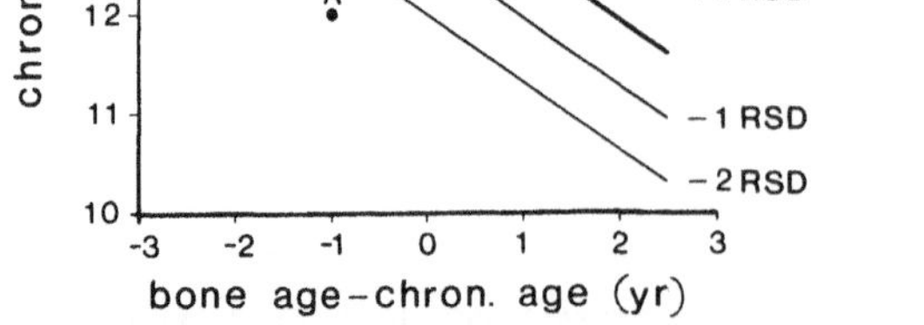

Fig. 17. Regression (with residual standard deviations) of age at menarche on (BA − CA). For explanation of examples A, B, C, and the curved line marked 95% at age 13, see text. (Reproduced from Marshall and de Limongi, 1976.)

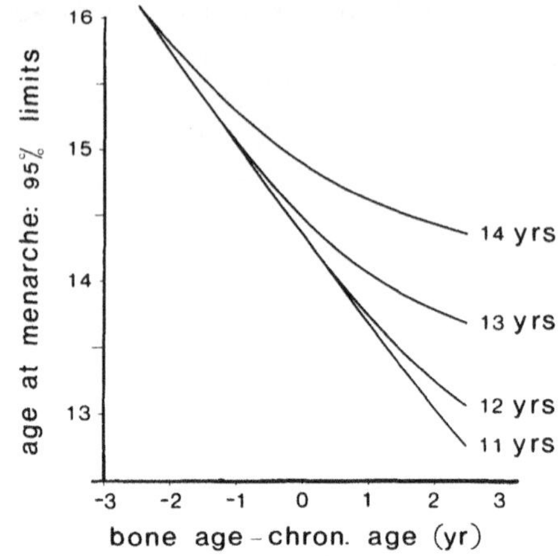

Fig. 18. Age at which menarche will have occurred in 95% of girls who are premenarcheal at given values of (BA − CA). Separate lines apply to girls of different chronological ages as indicated. (Reproduced from Marshall and de Limongi, 1976.)

In girls below the age of 10 years, bone age is not of much help in predicting age at menarche. In an unpublished study, Marshall found a correlation of 0.49 between bone age at age 8 and age at menarche. Simmons and Greulich (1943) found a similar correlation for girls aged 7 years. This level of correlation is not high enough to be useful in predicting age at menarche for an individual child.

Frisancho *et al.* (1969) used the radiographic appearance of the adductor sesamoid and the fusion of the epiphyses of the distal phalanx of the second finger to estimate age at menarche. However, as fusion does not occur in this phalanx until after menarche, this method cannot be used to predict when menarche will occur in an individual, although it may be of some value in retrospective studies of populations.

The secondary sex characteristics are of limited value in predicting age at menarche. As we have seen above, Marshall and Tanner (1969) demonstrated that the interval from the beginning of breast development to menarche varies in different girls from 6 months to more than 4 years. The observation that 26% of girls begin to menstruate in breast stage 3 and 62% do so in breast stage 4 does provide a basis for the claim that any girl in breast stage 4 is likely to experience menarche in the next year or so. But there remain some 10% of girls who do not menstruate until after their breasts have reached stage 5.

Frisch (1974) reported a technique for predicting age at menarche on the basis of height and weight, which she used to estimate the water content of the body. The range of error in these predictions appears to be similar to that found when chronological age alone is used, with due allowance for the properties of the truncated normal distribution. There is as yet no entirely satisfactory method for predicting age at menarche, but at most ages the most accurate method currently available is that which uses bone age, estimated by the TW2 method, in conjunction with chronological age (Marshall and de Limongi, 1976).

15. Body Composition at Puberty

The lean body mass increases quickly during adolescence to reach a maximum of about 63 kg at age 20 in males and about 42 kg at age 17 in females (Forbes, 1972). The sex difference in lean body mass is greater than that in either height or weight, and the spurt at adolescence is much more pronounced in males. Females have a higher fat content, which increases throughout adolescence, whereas values for total body fat in males decline between the ages of 15 and 17 and then increase slightly.

Tanner and co-workers measured the widths of bone, muscle, and fat in the limbs, as shown on radiographs taken by a special technique to eliminate parallax errors (Tanner, 1965; Tanner *et al.*, 1981). The widths of bone, muscle, and fat were measured midway down the upper arm and at the maximal width in the calf. The study was longitudinal, so an age of peak height velocity could be calculated for each subject. Figures 19 and 20 show the velocities of muscle and fat plotted against years before and after PHV, in order to avoid the difficulties created by the adolescent changes occurring at different ages in different subjects (see also Baughan *et al.*, 1980; Cronk *et al.*, 1983*a;* Billewicz *et al.*, 1983*b*).

In boys the rate of growth of muscle in the limbs becomes maximal at approximately the same time as peak velocity in stature. The rate of gain in fat changes in the opposite direction. Most children gain limb fat steadily from the age of about 8 years until adolescence, but the rate of increase in thickness of the subcutaneous fat layer becomes slower as the growth of the skeleton and muscle begins to accelerate. The minimal rate of fat gain is almost coincident with the maximum gain in bone and muscle. It can be seen in Figure 20 that the minimum rate of gain in fat in boys has a negative value, i.e., the fat is being lost. The typical boy becomes thinner in the limbs during adolescence, while the average girl continues to get fatter but at a slower rate during the adolescent spurt than either before or after. The decrease in rate of fat accumulation is more marked in the limbs than in the trunk. Venrooij-Ijsselmuiden (1978) reported longitudinal data from 628 Dutch girls and 504 boys in which the bone–muscle and fat cross-sectional areas of the left arm were calculated from skinfold measurements and the limb circumferences. The girls had more subcutaneous fat than the boys, and this difference increased after puberty. Muscle–bone area was about identical in the two sexes until

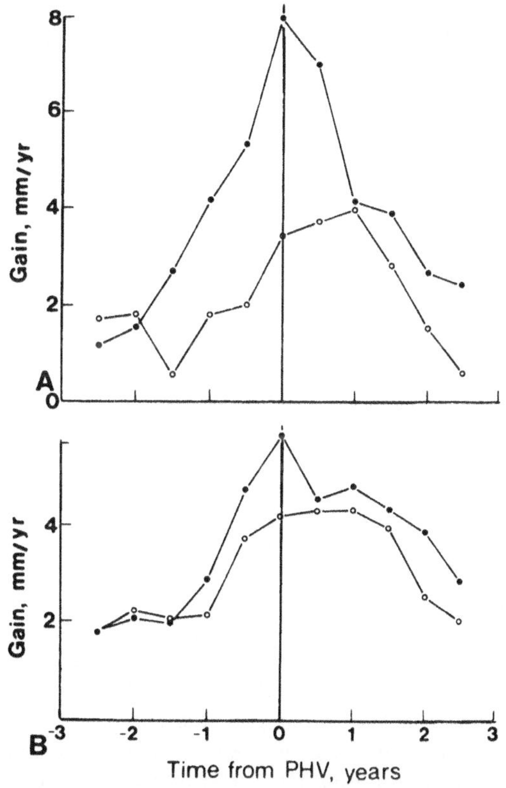

Fig. 19. Mean velocities of arm muscle widths (A) and calf muscle widths (B) in boys (●) and girls (○) at puberty. Individuals classified by years before or after their own peak height velocity (PHV). (From Tanner *et al.*, 1981.)

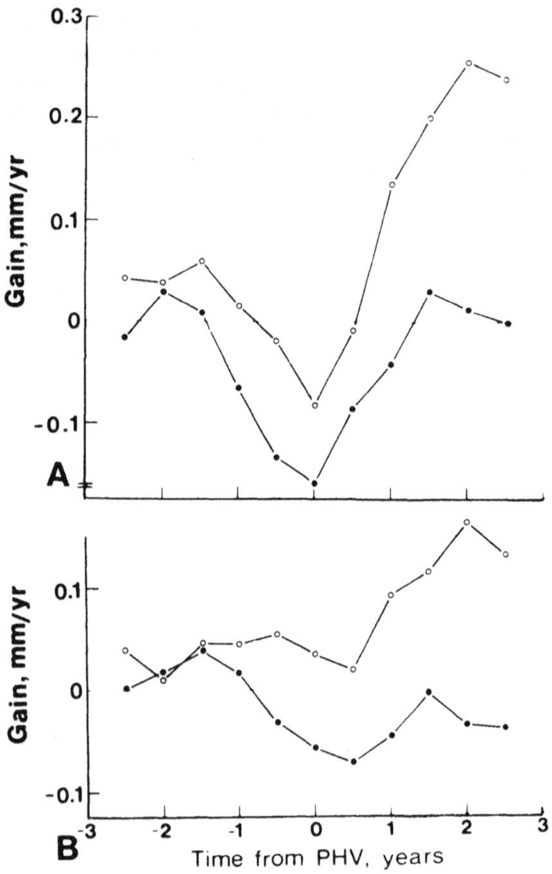

Fig. 20. Mean velocities of arm fat widths (A) and calf fat widths (B) in boys (●) and girls (○) at puberty. Individuals classified by years before or after their own peak height velocity (PHV). (From Tanner *et al.*, 1981.)

the age of 13. Thereafter, in association with the male adolescent growth spurt, the boys attained greater muscle–bone areas than did the girls.

In both boys and girls, at a given age or height, body weight varies with pubertal status, as does the change in weight accompanying a given change in height. However, the relationship between weight and height also varies with age, so that both chronological age and pubertal status should be taken into account in constructing weight for height standards in a well-nourished healthy population (Billewicz *et al.*, 1983*a*).

Changes in body density have been measured during puberty on the basis of Archimedes' principle (Heald *et al.*, 1963; Parizkova, 1961). The fact that boys gain more muscle and tend to lose fat is reflected by a greater increase in body density among boys than girls.

Cronk *et al.* (1983*b*) noted that there is only weak continuity between childhood fatness levels and adolescent patterns of fatness, while there is greater continuity between adolescent and adult fatness. Chumlea *et al.* (1981) could find no relationship of adipocyte size to age in girls during puberty and only a questionably significant negative relationship in boys.

16. Physiological Changes: Strength and Work Capacity

Puberty brings with it many physiological changes. At the same time as the muscles grow in size and strength, the heart and lungs become bigger, not only in absolute terms but in relationship to total body size. The systolic blood pressure rises, and the heart rate becomes slower, while a considerable testosterone-induced rise in hemoglobin concentration in the blood in boys gives it a greater capacity for carrying oxygen (Tanner, 1962; Cassimos *et al.*, 1977; Goldring *et al.*, 1977; Hoffmann *et al.*, 1976; Krabbe *et al.*, 1978).

The spurt in bone growth is accompanied by increased plasma inorganic phosphate and alkaline phosphatase levels. In a longitudinal study, Round *et al.* (1979) showed that the alkaline phosphatase levels rose and fell in parallel with the height velocity curve, peaking at PHV. The level of inorganic phosphate also rose and fell, but with its peak some four months earlier than PHV. Serum urate levels rose sharply in boys from stage G2 to G4 but showed little change in girls (Round, 1980; see also Harlan *et al.*, 1979*a;* Bennett *et al.*, 1976). Hydroxyproline excretion has also been studied longitudinally; its level also rose and fell, with a correlation between age at peak hydroxyproline level and age at PHV of 0.88 in boys and 0.78 in girls. Absolute level related both to body size and to amplitude of the height peak (Clark and Zorab, 1978). Vitamin D metabolites also rise during the adolescent spurt (Aksnes and Aarskog, 1982).

There have been a number of studies of blood lipids and lipoproteins during puberty (e.g., Morrison *et al.*, 1977, 1978, 1979; Ibsen *et al.*, 1980; Christensen *et al.*, 1980). Plasma cholesterol decreases in boys during early and mid-puberty but rises between G4 and G5; the fall seems to be due to high-density lipoprotein and the rise to low-density lipoprotein. In girls cholesterol also falls during mid-puberty and rises at late puberty. Triglyceride levels appear to increase throughout puberty in both sexes, although perhaps falling at the end of puberty in girls. None of these changes, however, is as clear cut as those relating to bone and muscle growth, and the physiological mechanisms remain to be clarified.

As a result of all the cardiorespiratory changes accompanying puberty (discussed in more detail in Chapter 4, Vol. 2), the individual becomes not only stronger but able to endure hard physical work over a longer period of time. There is no justification for the popular belief that boys outgrow their strength in puberty. They become progressively stronger, although their maximum strength may not be reached until their growth in height and sexual development are more or less complete. There may be a period during which a boy looks like a mature man but has not yet developed the strength and stamina suggested by his appearance. Nevertheless, he is considerably stronger than he was the previous year.

Those boys who experience the adolescent growth spurt earlier than their fellows have an advantage not only in size but also in strength and endurance. They have an advantage in sports and other activities that may influence their social status in the peer group, but these advantages will be lost as the others approach maturity in size and strength. It is sometimes advisable to warn the early-maturing athletic boy that his physical prowess may be only temporary and that he will not always be bigger and stronger than his peers. On the other hand, late-maturing boys sometimes need reassurance to the effect that they will not always be smaller and weaker than most of their peers.

17. References

Aksnes, L., and Aarskog, D., 1982, Plasma concentrations of vitamin-D metabolites in puberty—Effect of sexual maturation and implications for growth, *J. Endocrinol. Metab.* **55**:94–101.

Andersen, E., 1968, Skeletal maturation of Danish school children in relation to height, sexual development and social conditions, Universitäts Forlaget, Aarhus.

Attallah, N. L., 1978, Age at menarche of schoolgirls in Egypt, *Ann. Hum. Biol.* **5**:185–189.

Attallah, N. L., Matta, W. M., and El-Mankoushi, M., 1983, Age at menarche of schoolgirls in Khartoum, *Ann. Hum. Biol.* **10**:185–188.

Aw, E., and Tye, C. Y., 1970, Age at menarche of a group of Singapore girls, *Hum. Biol.* **42**:329–336.

Baughan, B., Demirjian, A., Levesque, G. Y., and Lapalme-Chaput, L., 1979, The pattern of facial growth before and during puberty, as shown by French-Canadian girls, *Ann. Hum. Biol.* **6**:59–76.

Baughan, B., Brault-Dubuc, M., Demirjian, A., and Gagnon, G., 1980, Sexual dimorphism in body composition changes during the pubertal period—shown by French-Canadian children, *Am. J. Phys. Anthrop.* **52**:85–94.

Beall, C. M., 1983, Age at menopause and menarche in a high-altitude Himalayan population, *Ann. Hum. Biol.* **10**:365–370.

Belmaker, E., 1982, Sexual maturation of Jerusalem schoolgirls and its association with socio-economic factors and ethnic group, *Ann. Hum. Biol.* **9**:321–328.

Bennett, D. L., Ward, M. S., and Daniel, W. A., 1976, Relationship of serum alkaline phosphatase concentrations to sex maturity ratings in adolescents, *J. Pediatr.* **88**:633–636.

Bergsten-Brucefors, A., 1976, A note on the accuracy of recalled age at menarche, *Ann. Hum. Biol.* **3**:71–73.

Biclicki, T., 1975, Inter-relationships between various measures of maturation rate in girls during adolescence, *Studies Phys. Anthropol., Polish Acad. Sci.* **1**:51–64.

Bielicki, T., 1976, On the relationships between maturation rate and maximum velocity of growth during adolescence, *Styd. Phys. Anthropol., Polish Acad. Sci.* **3**:79–84.

Billewicz, W. Z., and McGregor, I. A., 1982, A birth-to-maturity longitudinal study of heights and weights in two West African (Gambian) villages, 1951–1975, *Ann. Hum. Biol.* **9**:309–320.

Billewicz, W. Z., Fellowes, H. M., and Thomson, A. M., 1980, Postmenarcheal menstrual cycles in British (Newcastle-upon-Tyne) girls, *Ann. Hum. Biol.* **7**:177–180.

Billewicz, W. Z., Fellowes, H. M., and Thomson, A. M., 1981*a*, Pubertal changes in boys and girls in Newcastle-upon-Tyne, *Ann. Hum. Biol.* **8**:211–219.

Billewicz, W. Z., Fellowes, H. M., and Thomson, A. M., 1981*b*, Menarche in Newcastle-upon-Tyne girls, *Ann. Hum. Biol.* **8**:313–320.

Billewicz, W. Z., Thomson, A. M., and Fellowes, H. M., 1983*a*, Weight-for-height in adolescence, *Ann. Hum. Biol.* **10**:119–124.

Billewicz, W. Z., Thomson, A. M., and Fellowes, H. M., 1983*b*, A longitudinal study of growth in Newcastle-upon-Tyne adolescents, *Ann. Hum. Biol.* **10**:125–133.

Bishara, S. E., Jamison, J. E., Peterson, L. C., and Dekock, W., 1981, Longitudinal changes in standing height and mandibular parameters between the ages of 8 and 17 years, *Am. J. Orthod.* **80**:115–135.

Björk, A., 1955, Cranial base development. A follow-up X-ray study of the individual variation in growth occuring between the ages of 12 and 20 years and its relation to brain case and face development, *Am. J. Orthod.* **41**:198–225.

Bliss, C. I., and Young, M. S., 1950, An analysis of heart measurements of growing boys, *Hum. Biol.* **22**:271–280.

Bock, R. D., and Thissen, D., 1976, Fitting multi-component models for growth in stature, in: *Proceedings of the Ninth International Biometric Conference,* Vol. 1, pp. 431–442, The Biometric Society, Raleigh, North Carolina.

Bock. R. D., and Thissen, D., 1980, Statistical problems of fitting individual growth curves, in: *Human Physical Growth and Maturation, Methodologies and Factors* (F. E. Johnston, A. F. Roche, and C. Susanne, eds.), pp. 265–290, Plenum Press, New York.

Bock, R. D., Wainer, H., Peterson, A., Thissen, D., Murray, J., and Roche, A. F., 1973, A parameterization for individual growth curves, *Hum. Biol.* **45**:63–80.

Bouthreuil, E., Niang, I., Michaut, E., and Darr, V., 1972, Etude préliminaire de la puberté de la fille à Dakar, *Ann. Pédiatr.* **19**:685–690.

Boyd, E., 1936, Weight of the thymus and its component parts and number of Hassall corpuscles in health and disease, *Am. J. Child.* **51**:313–335.

Brown, T., and Townsend, G. C., 1982, Adolescent growth in height of Australian aboriginals analysed by the Preece–Baines function: A longitudinal study, *Ann. Hum. Biol.* **9**:495–506.

Brudevoll, J. E., Liestøl, K., and Walløe, L., 1979, Menarcheal age in Oslo during the last 140 years, *Ann. Hum. Biol.* **6**:407–416.

Brundtland, G. H., and Walløe, L., 1973, Menarcheal age in Norway: Halt in the trend toward earlier maturation, *Nature (London)* **241**:478–479.

Brundtland, G. H., Liestøl, K., and Walløe, L., 1980, Height, weight and menarcheal age of Oslo schoolchildren during the last 60 years, *Ann. Hum. Biol.* **7**:307–322.

Buday, J., 1981, Age at menarche in girls with sensory deprivation, *Acta Med. Auxol.* **13**:131–139.

Burgess, A. P., and Burgess, H. J. L., 1964, The growth pattern of East African schoolgirls, *Hum. Biol.* **36**:177–193.

Burke, P. H., 1979, Growth of the soft tissues of middle third of the face between nine and 16 years, *Eur. J. Orthod.* **1**:1–13.

Buschang, P. H., Baume, R. M., and Nass, G. G., 1983, Craniofacial growth maturity gradient for males and females between 4 and 16 years of age, *Am. J. Phys. Anthropol.* **61**:373–381.

Cameron, N., Tanner, J. M., and Whitehouse, R. H., 1982, A longitudinal analysis of the growth of limb segments in adolescence, *Ann. Hum. Biol.* **9**:211–220.

Carfagna, M., Figurelli, E., Matarese, G., and Matarese, S., 1972, Menarcheal age of schoolgirls in the district of Naples, Italy, in 1969–1970, *Hum. Biol.* **44**:117–125.

Cassimos, Ch., Varlamis, G., Karamperis, S., and Katsouyannopoulos, V., 1977, Blood pressure in children and adolescents, *Acta Paed. Scand.* **66**:439–443.

Chang, K. S. F., Ng, P. H., Chan, S. J., and Lee, M. M. C., 1966, Sexual maturation of Southern Chinese boys in Hong Kong, *Pediatrics* **37**:804–811.

Christensen, B., Glueck, C., Kwiterovich, P., Degroot, I., Chase, G., Heiss, G., Mowery, R., Tamir, I., and Rifkind, B., 1980, Plasma cholesterol and triglyceride distributions in 13,665 children and adolescents—Prevalence study of the Lipid Research Clinics Program, *Pediatr. Res.* **14**:194–202.

Chumlea, W. C., Knittle, J. L., Roche, A. F., Siervogel, R. M., and Webb, P., 1981, Size and number of adipocytes and measures of body fat in boys and girls 10 to 18 years of age, *Am. J. Clin. Nutr.* **34**:1791–1797.

Clark, S., and Zorab, P. A., 1978, Hydroxyproline centiles for normal adolescent boys and girls, *Clin. Orthop.* **137**:217–226.

Cristescu, M., Bulai-Stirbu, M., and Feodorovici, C., 1964, L'influence des facteurs géographiques et sociaux sur le développement des enfants, *Ann. Roumain. Anthropol.* **1**:65–80.

Crognier, E., and da Rocha, M.-A.T., 1979, Age at menarche in rural France, *Ann. Hum. Biol.* **6**:167–170.

Cronk, C. E., Mukherjee, D., and Roche, A. F., 1983*a*, Changes in triceps and subscapular skinfold thickness during adolescence, *Hum. Biol.* **55**:707–722.

Cronk, C. E., Roche, A. F., Kent, R., Eichorn, D. and McCammon, R. W., 1983*b*, Longitudinal trends in subcutaneous fat thickness during adolescence, *Am. J. Phys. Anthropol.* **61**:197–204.

Dacou-Voutetakis, C., Kontza, D., Lagos, P., Tzonou, A., Katsarou, E., Antoniadis, S., Papazisis, G., Papadopoulos, G., and Matsaniotis, N., 1983, Age of pubertal stages including menarche in Greek girls, *Ann. Hum. Biol.* **10**:557–564.

Damon, A., Damon, S. T., Reed, R. B., and Valadian, I., 1969, Age at menarche of mothers and daughters, with a note on accuracy of recall, *Hum. Biol.* **41**:161–175.

Daniel, W. A., Feinstein, R. A., Howardpeebles, P., and Baxley, W. D., 1982, Testicular volumes of adolescents, *J. Pediatr.* **101**:1010–1012.

Deming, J., 1957, Application of the Gompertz curve to the observed pattern of growth in length of 48 individual boys and girls during the adolescent cycle of growth, *Hum. Biol.* **29**:83–122.

de Wijn, J. F., 1966, Estimation of age at menarche in a population, in: *Somatic Growth of the Child* (J. J. van der Werff ten Bosch and A. Haak, eds.), pp. 16–24, Stenfert-Kroese, Leiden.

Diaz de Mathman, C., Rico, V. M. L., and Galvan, R. R., 1968, Crecimiento y desarrallo en adolescentes femeninos, 2. Edad de la menarquia, *Bol. Med. Hosp. Infant (Mexico)* **25**:787–794.

Ducros, A., and Pasquet, P., 1978, Evolution de l'âge d'apparition des premières régles (ménarche) en France, *Biomét. Hum.* **13**:35–43.

Duke, P. M., Litt, I. F., and Gross, R. T., 1980, Adolescents self-assessment of sexual maturation, *Pediatrics* **66**:918–920.

Eiben, O. G., 1972, Genetische und demographische Faktoren und Menarchealter, *Anthropol. Anz.* **33**:205–212.

Ekstrom, C., 1982, Facial growth rate and its relation to somatic maturation in healthy children, *Swed. Dental J. (Suppl.)* **11**:5–99.

Ellison, P. T., 1981, Morbidity, mortality and menarche, *Hum. Biol.* **53**:635–644.

Ellison, P. T., 1982, Skeletal growth, fatness and menarcheal age: a comparison of two hypotheses, *Hum. Biol.* **54**:269–282.

Eveleth, P. B., 1978, Differences between populations in body shape of children and adolescents, *Am. J. Phys. Anthropol.* **49**:373–382.

Eveleth, P. B., and Souza Freitas, J. A., 1969, Tooth eruption and menarche in Brazilian-born children of Japanese ancestry, *Hum. Biol.* **41**:176–184.

Eveleth, P. B., and Tanner, J. M., 1976, *Worldwide Variation in Human Growth,* Cambridge University Press, Cambridge.

Farid-Coupal, N., Contreras, M. L., and Castellano, H. M., 1981, The age at menarche in Carabobo, Venezuela, with a note on the secular trend, *Ann. Hum. Biol.* **8**:283–288.

Farkas, G., 1980, Veränderungen des Menarche—Medianwertes nach dem Beruf der Mutte, *Arztliche Jugendkd.* **71**:62–67.

Farkas, G., and Nagy, J., 1981, Uber das menarchealter von mädchen aus Komarom, *Acta Biol. Szeg.* **27**:223–227.

Farkas, G., and Szekeres, E., 1982, On the puberty of girls in Szeged, Hungary, *Acta Biol. Szeg.* **28**:1–4.

Filipsson, R., and Hall, K., Correlation between dental maturity, height development and sexual maturation in normal girls, *Ann. Hum. Biol.* **3**:205–210.

Fischbein, S., 1977*a,* Intra-pair similarity in physical growth of monozygotic and of dizygotic twins during puberty, *Ann. Hum. Biol.* **4**:417–430.

Fischbein, S., 1977*b,* Onset of puberty in MZ and DZ twins, *Acta Genet. Med. Gemellol.* **26**:151–158.

Fischer, R. A., and Yates, F., 1963, *Statistical Tables for Biological, Agricultural and Medical Research, 6th ed.,* Oliver and Boyd, Edinburgh.

Forbes, G. B., 1972, Growth of the lean body mass in man, *Growth* **36**:325–338.

Ford, E. H. R., 1958, Growth of the human cranial base, *Am. J. Orthod.* **44**:498–506.

Frisancho, A. R., Garn, S. M., and Rohmann, C. G., 1969, Age at menarche. A new method of predicting and retrospective assessment based on hand X-rays, *Hum. Biol.* **41**:42–450.

Frisch, R. E., 1974, A method of prediction of age of menarche from height and weight at ages 9 through 13 years, *Pediatrics* **53**:384–390.

Frisch, R. E., Wyshak, G., and Vincent, L., 1980, Delayed menarche and amenorrhea in ballet dancers, *N. Engl. J. Med.* **303**:17–18.

Furu, M., 1976, Menarcheal age in Stockholm girls, 1967, *Ann. Hum. Biol.* **3**:587–590.

Gallo, P. G., 1977, The age at menarche in some populations of the Veneto, North Italy, *Ann. Hum. Biol.* **4**:179–181.

Gallo, P. G., and Mastriner, M. F., 1980, Growth of children in Somalia, *Hum. Biol.* **52**:547–562.

Gardner, J., 1983, Adolescent menstrual characteristics as predictors of gynaecological health, *Ann. Hum. Biol.* **10**:31–40.

Goldring, D., Londe, S., Sivakoff, M., Hernandez, A., Britton, C., and Choi, S., 1977, Blood pressure in a high school population: I. Standards for blood pressure and relation of age, sex, weight, height and race to blood pressure in children 14 to 18 years of age, *J. Pediatr.* **91**:884–889.

Goodman, M. J., 1983, Secular changes in recalled age at menarche, *Ann. Hum. Biol.* **10**:585.

Gorer, G., 1938, *Himalayan Village,* Michael Joseph, New York.

Hägg, U., and Taranger, J., 1982, Maturation indicators and the pubertal growth spurt, *Am. J. Orthod.* **82**:299–309.

Hafez, A. S., Salem, S. I., Cole, T. J., Galal, O. M., and Massoud, A., 1981, Sexual maturation and growth pattern in Egyptian boys, *Ann. Hum. Biol.* **8**:461–468.

Harlan, W. R., Cornoni-Huntley, J., and Leaverton, P. E., 1979*a,* Physiologic determinants of serum urate levels in adolescence, *Pediatrics* **63**:569–575.

Harlan, W. R., Grillo, G. P., Cornoni-Huntley, J., and Leaverton, P. E., 1979*b,* Secondary sex characteristics of boys 12 to 17 years of age—United States Health Examination Survey, *J. Pediatr.* **95**:293–297.

Harlan, W. R., Harlan, E. A., and Grillo, G. P., 1980, Secondary sex characteristics of girls 12 to 17 years of age—The United States Health Examination Survey, *J. Pediatr.* **96**:1074–1078.

Hauspie, R. C., Das, S. A., Preece, M. A., and Tanner, J. M., 1980*a,* A longitudinal study of the growth in height of boys and girls of West Bengal (India) aged six months to 20 years, *Ann. Hum. Biol.* **7**:429–441.

Hauspie, R. C., Wachholder, A., Baron, G., Cantraine, F., Susanne, C., and Graffar, M., 1980*b,* A comparative study of the fit of four different functions to longitudinal data of growth in height of Belgian girls, *Ann. Hum. Biol.* **7**:347–358.

Hauspie, R. C., Das, S. R., Preece, M. A., and Tanner, J. M., 1982, Degree of resemblance of the pattern of growth among sibs in families of West Bengal (India), *Ann. Hum. Biol.* **9**:171–174.

Heald, S. P., Hunt, E. E., Schwartz, R., Cook, C. D., Elliott, E., and Vajdan, B., 1963, Measures of body fat and hydration in adolescent boys, *Pediatrics* **31**:226–239.

Hirsch, M., Shemesh, J., Modan, M., and Lunenfeld, B., 1979, Emission of spermatozoa. Age of onset, *Int. J. Androl.* **2**:289–298.

Hoffmann, A., Bruppacher, R., Gutzwiller, F., and De Roche, Ch., 1976, Blutdruck in der Adoleszenz. Messergebnisse und ihre Beziehung zür körperlichen Entwicklung bei 745 gesunden Basler Schulern, *Helv. Paediatr. Scand.* **31**:121–129.

Hoshi, H., and Kouchi, M., 1981, Secular trend of the age at menarche of Japanese girls with special regard to the secular acceleration of the age at peak height velocity, *Hum. Biol.* **53**:593–598.

Houston, W. J. B., Miller, J. C., and Tanner, J. M., 1979, Prediction of the timing of the adolescent growth spurt from ossification events in hand–wrist films, *Br. J. Orthodont.* **6**:145–152.

Hulanicka, B., 1978, Facial morphology during adolescence; a longitudinal study, *Stud. Phys. Anthropol. Polish Acad. Sci.* **4**:35–48.

Hwang, J. M. S., and Krumbhaar, E. B., 1940, The amount of lymphoid tissue of the human appendix and its weight at different age periods, *Am. J. Med. Sci.* **199**:75–83.

Hwang, J. M. S., Lipincott, S. W., and Krumbhaar, E. B., 1938, The amount of splenic lymphatic tissue at different ages, *Am. J. Pathol.* **14**:809–819.

Ibsen, K. K., Lous, P., and Andersen, G. E., 1980, Lipids and lipoproteins in 350 Danish schoolchildren ages 7 to 18 years, *Acta Paediatr. Scand.* **69**:231–234.

Jenicek, M., and Demirjian, A., 1974, Age at menarche in urban French-Canadian girls, *Ann. Hum. Biol.* **1**:339–346.

Jones, E. L., Hemphill, W., and Mayers, E. S. A., 1973, *Height, Weight, and Other Physical Characteristics of New South Wales Children:* Part I: *Children Aged 5 Years and Over,* Health Commission of New South Wales, Sidney.

Jordan, A., Bebelagua, A., Ruben, M., and Hernandez, J., 1974, Pubertal development and the age of menarche, Cuba 1973, in: *Paediatrica XIV: Proceedings of the Fourteenth International Conference of Pediatrics,* pp. 46–48, Editorial Medica Panamerica S. A., Buenos Aires.

Kahane, J. C., 1978, A morphological study of the human prepubertal and pubertal larynx, *Am. J. Anat.* **151**:11–19.

Kantero, R. L., and Widholm, O., 1971, The age of menarche in Finnish girls in 1969, *Acta Obstet. Gynecol. Scand. (Suppl.)* **14**:7–18.

Karlberg, P., Taranger, J., Engstrom, I., Karlberg, J., Landstrom, T., Lichtenstein, H., Lundstrom, B., and Svennberg-Redegren, I., 1976, The somatic development of children in a Swedish urban community. I. Physical growth from birth to 16 years and longitudinal outcome of the study during the same period, *Acta Paediatr. Scand. (Suppl.)* **258**:7–76.

King, E. W., 1952, A roentgenographic study of pharyngeal growth, *Angle Orthod.* **22**:23–37.

Koshi, E. P., Brasad, B. G., and Bhushan, V., 1971, A study of the menstrual pattern of school girls in an urban area, *Indian J. Med. Res.* **58**:1647–1652.

Krabbe, S., Christensen, T., Worm, J., Christiansen, C., and Transbol, I., 1978, Relationship between haemoglobin and serum testosterone in normal children and adolescents and in boys with delayed puberty, *Acta Paediatr. Scand.* **67**:655–658.

Kralj-Cercek, L., 1956, The influence of food, body build and social origin on the age at menarche, *Hum. Biol.* **28**:398–406.

Largo, M. A., 1978, Visusveränderungen im Verlaufe der Pubertät und Augendominanz im Alter von 10 Jahren, *Helv. Paediatr. Acta* **32**:59–70.

Largo, R. H., and Prader, A., 1983*a,* Pubertal development in Swiss boys, *Helv. Paediatr. Acta* **38**:211–228.

Largo, R. H., and Prader, A., 1983*b,* Pubertal development in Swiss girls, *Helv. Paediatr. Acta* **38**:229–243.

Largo, R. H., Gasser, Th., Prader, A., Stuetzle, W., and Huber, P. J., 1978, Analysis of the adolescent growth spurt using smoothing spline functions, *Ann. Hum. Biol.* **5**:421–434.

Laron, Z., Arad, J., Gurewitz, R., Grunebaum, M., and Dickerman, Z., 1980, Age at first conscious ejaculation—A milestone in male puberty, *Helv. Paediatr. Acta* **35**:13–20.

Laska-Mierzejewska, T., Milicer, H., and Piechaczek, H., 1982, Age at menarche and its secular trend in urban and rural girls in Poland, *Ann. Hum. Biol.* **9**:227–234.

Lee, M. C., Chang, K. S. F., and Chan, M. M. C., 1963, Sexual maturation of Chinese girls in Hong Kong, *Pediatrics* **32**:389–398.

Lejarraga, H., Sanchirico, F., and Cusminsky, M., 1980, Age at menarche in urban Argentinian girls, *Ann. Hum. Biol.* **7**:579–581.

Liestøl, K., 1982, Social conditions and menarcheal age: The importance of early years of life, *Ann. Hum. Biol.* **9**:521–538.

Limongi, Y. de, 1977, El desarollo puberal de los adolescentes venezolanos, *Acta Cient. Venez.* **28**:160–164.

Lincoln, E. M., and Spillman, R., 1928, Studies on the hearts of children. II: Roentgen-ray studies, *Am. J. Dis. Child.* **35**:791–810.

Lindgren, G., 1976, Height, weight and menarche in Swedish urban school children in relation to socio-economic and regional factors, *Ann. Hum. Biol.* **3**:501–528.

Lindgren, G., 1978, Growth of schoolchildren with early, average and late ages of peak height velocity, *Ann. Hum. Biol.* **5**:253–267.

Ljung, B.-O., Bergsten-Brucefors, A., and Lindgren, G., 1974, The secular trend in physical growth in Sweden, *Ann. Hum. Biol.* **1**:245–256.

Ljung, B.-O, Fischbein, S., and Lindgren, G., 1977, A comparison of growth in twins and singleton controls of matched age followed longitudinally from 10 to 18 years, *Ann. Hum. Biol.* **4**:405–416.

Low, W. D., 1978, Growth of the upper limb, arm and forearm in Chinese children, *Z. Morphol. Anthropol.* **69**:172–182.

Low, W. D., Kung, L. S., and Leong, J. C. Y., 1982, Secular trend in the sexual maturation of Chinese girls, *Hum. Biol.* **54**:539–550.

MacMahon, B., 1973, *Age at Menarche: United States,* DHEW Publication No. (HRA) 74-1615, NHS Series 11, No 133, National Center of Health Statistics, Rockville, Maryland.

Madhavan, S., 1965, Age at menarche of South Indian girls belonging to the states of Madras and Kerala, *Indian J. Med. Res.* **53**:669–673.

Magnusson, T. E., 1978, Age at menarche in Iceland, *Am. Am. J. Phys. Anthropol.* **48**:511–514.
Malcolm, L. A., 1969, Growth and development of the Kaipit children of the Markham Valley, New Guinea, *Am. J. Phys. Anthropol.* **31**:39–51.
Malcolm, L. A., 1970, Growth and development in the Bundi children of the New Guinea highlands, *Hum. Biol.* **42**:293–328.
Malina, R. M., 1983, Menarche in athletes: A synthesis and hypothesis, *Ann. Hum. Biol.* **10**:1–24.
Malina, R. M., and Chumlea, C., 1977, Age at menarche in deaf children, *Ann. Hum. Biol.* **4**:485–488.
Malina, R. M., Chumlea, C., Stepick, C. O., and Lopez, F. G., 1977, Age at menarche in Oaxaca, Mexico, schoolgirls, with comparative data for other areas of Mexico, *Ann. Hum. Biol.* **4**:551–558.
Manniche, E., 1983, Age at menarche: Nicolai Edvard Ravn's data on 3385 women in mid-19th century Denmark, *Ann. Hum. Biol.* **10**:79–82.
Maresh, M. M., 1948, Growth of the heart related to bodily growth during childhood and adolescence, *Pediatrics* **2**:382–404.
Maresh, M. M., 1955, Linear growth of long bones of extremities from infancy through adolescence, *Am. J. Dis. Child.* **89**:725–742.
Marshall, W. A., 1974, Inter-relationships of skeletal maturation, sexual development and somatic growth in man, *Ann. Hum. Biol.* **1**:29–40.
Marshall, W. A., and de Limongi, Y., 1976, Skeletal maturity and the prediction of age at menarche, *Ann. Hum. Biol.* **3**:235–243.
Marshall, W. A., and Tanner, J. M., 1969, Variation in the pattern of pubertal changes in girls, *Arch. Dis. Child.* **44**:291–303.
Marshall, W. A., and Tanner, J. M., 1970, Variation in the pattern of pubertal changes in boys, *Arch. Dis. Child.* **45**:13–23.
Marshall, W. A., and Tanner, J. M., 1974, Puberty, in: *Scientific Foundations of Paediatrics* (J. A. Davis and J. Dobbing, eds.), pp. 124–151, William Heinemann, London.
Marubini, E., and Barghini, G., 1969, Richerche sull'éta media di comparsa della puberta nella populazione scolare feminile di Cararra, *Minerva Pediatr.* **21**: 282–285.
Marubini, E., Resele, L. F., Tanner, J. M., and Whitehouse, R. H., 1972, The fit of the Gompertz and logistic curves to longitudinal data during adolescence on height, sitting height and biacromial diameter in boys and girls of the Harpenden Growth Study, *Hum. Biol.* **44**:511–524.
Metcalf, M. G., Skidmore, D. S., Lowry, G. F., and Mackenzie, J. A., 1983, Incidence of ovulation in the years after the menarche, *J. Endocrinol.* **97**:213–219.
Miklashevskaya, N., Solovyeva, J. S., Godina, E. Z., and Kondik, V. M., 1972, Growth processes in man under conditions of the high mountains, *Trans. Moscow Soc. Nat.* **43**:181–194.
Mitani, H., 1977, Occlusal and craniofacial growth changes during puberty, *Am. J. Orthod.* **72**:76–84.
Moerman, M. L., 1982, Growth of the birth canal in adolescent girls, *Am. J. Obstet. Gynecol.* **143**:528–532.
Montagu, M. F. A., 1957, *The Reproductive Development of the Female,* Julian Press, New York.
Morrison, J. A., deGroot, I., Edwards, B. K., Kelly, K. A., Rauh, J. L., Mellies, M., and Glueck, C. J., 1977, Plasma cholesterol and triglyceride levels in 6775 school children aged 6–17, *Metabolism* **26**:1199–1211.
Morrison, J. A., deGroot, I., Edwards, B. K., Kelly, K. A., Mellies, M. J., Khoury, P., and Glueck, C. J., 1978, Lipids and lipoproteins in 927 schoolchildren ages 6 to 17 years, *Pediatrics* **62**:990–995.
Morrison, J. A., Laskarzewski, P. M., Rauh, J. L., Brookman, R., Mellies, M., Frazer, M., Khoury, P., deGroot, I., Kelly, K., and Glueck, C. J., 1979, Lipids, lipoproteins and sexual maturation during adolescence: The Princeton Maturation Study, *Metabolism* **28**:641–649.
Nagai, N., Matsumoto, K., Mino, T., Takeuchi, H., and Takeda, S., 1980, The secular trends in the menarcheal age and the maximum growth age in height for Japanese schoolgirls, *Wakayama Med. Rep.* **23**:41–45.
Nanda, R. S., 1955, The rates of growth of several facial components measured from serial cephalometric roentgenograms, *Am. J. Orthod.* **14**:658–673.
New Zealand Department of Health, 1971, *Physical Development of New Zealand School Children, 1969,* Department of Health Special Report No. 38, Department of Health, Wellington, New Zealand.
Neyzi, O., Alp, H., and Orhon, A., 1975*a,* Sexual maturation in Turkish girls, *Ann. Hum. Biol.* **2**:49–60.
Neyzi, O., Alp, H., Yalcindag, A., Yakacikli, S., and Orhon, A., 1975*b,* Sexual maturation in Turkish boys, *Ann. Hum. Biol.* **2**:251–259.
Ober, W., and Bernstein, J., 1955, Observations on the endometrium and ovary in the newborn, *Pediatrics* **16**:445–460.
Oduntan, S. O., Ayeni, O., and Kale, O. O., 1976, The age of menarche in Nigerian girls, *Ann. Hum. Biol.* **3**:269–274.
Panto, E., 1980, Age at menarche and body development in girls based on a cross-sectional study in Eger (Northern Hungary), *Coll. Anthropol.* **4**:163–173.
Parizkova, J., 1961, Age trends in fat in normal and obese children, *J. Appl. Physiol.* **16**:173–174.
Peters, H., Himelstein-Braw, R., and Faber, M., 1976, The normal development of the ovary in childhood, *Acta Endocrinol.* **82**:617–630.
Prebeg, Z., 1978, Secular trend of menarche in Zagreb school girls, *Coll. Anthropol.* **2**:87–92.
Preece, M. A., and Baines, M. J., 1978, A new family of mathematical models describing the human growth curve, *Ann. Hum. Biol.* **5**:1–24.
Preece, M. A., and Heinrich, I., 1981, Mathematical modelling of individual growth curves, *Br. Med. Bull.* **37**:247–252.
Richardson, D. W., and Short, R. V., 1978, Time of onset

of sperm production in boys, *J. Biosoc. Sci. (Suppl.)* **5**:15–24.

Roberts, D. F., 1969, Race, genetics and growth, *J. Biosoc. Sci. (Suppl.)* **1**:43–67.

Roberts, D. F., Rozner, L. M., and Swan, A. V., 1971, Age at menarche, physique and environment in industrial north-east England, *Acta Paediatr. Scand.* **60**: 158–164.

Roberts, D. F., Danskin, M. J., and Chinn, S., 1975, Menarcheal age in Northumberland, *Acta Paediatr. Scand.* **64**:845–852.

Roberts, D. F., Chinn, S., Girija, B., and Singh, H. D., 1977, A study of menarcheal age in India, *Ann. Hum. Biol.* **4**:171–177.

Roche, A. F., 1953, Increase in cranial thickness during growth, *Hum. Biol.* **25**:81–92.

Roche, A. F., French, N. Y., and Davilla, D. H., 1971, Areolar size during pubescence, *Hum. Biol.* **43**:210–223.

Rona, R., and Pereira, G., 1974, Factors that influence age of menarche in girls in Santiago, Chile, *Hum. Biol.* **46**:33–42.

Round, J. M., 1980, Changes in plasma urate, creatinine, alkaline phosphatase and the 24-hour excretion of hydroxyproline during sexual maturation in adolescents, *Ann. Hum. Biol.* **7**:83–88.

Round, J. M., Butcher, S., and Steele, R., 1979, Changes in plasma inorganic phosphorus and alkaline phosphatase activity during the adolescent growth spurt, *Ann. Hum. Biol.* **6**:129–136.

Roy, M. P., Sempé, M., Orssaud, E., and Pedron, G., 1972, Evolution clinique de la puberté de la fille, *Arch. Fr. Pédiatr.* **29**:155–168.

Satyanarayana, K., Nadamuni Naidu, A., and Narasinga Rao, B. S., 1980, Adolescent growth spurt among rural Indian boys in relation to their nutritional status in early childhood, *Ann. Hum. Biol.* **7**:359–368.

Savara, B. S., and Singh, I. J., 1968, Norms of size and annual increments of seven anatomical measures of the maxillae in boys from 3 to 16 years of age, *Angle Orthod.* **38**:104–120.

Savara, B. S., and Tracey, W. E., 1967, Norms of size and annual increments for five anatomical measures of the mandible from 3 to 16 years of age, *Arch. Oral Biol.* **12**:469–486.

Scammon, R. E., 1930, The measurement of the body in childhood, in *The Measurement of Man* (J. A. Harris, C. M. Jackson, D. G. Patterson, and R. E. Scammon, eds.), pp. 173–215, University of Minnesota Press, Minneapolis.

Shakir, A., 1971, The age of menarche in girls attending school in Baghdad, *Hum. Biol.* **43**:265–270.

Sharma, J. C., 1983, The genetic contribution to pubertal growth and development studied by longitudinal growth data on twins, *Ann. Hum. Biol.* **10**:163–172.

Shuttleworth, F. K., 1939, The physical and mental growth of girls and boys aged 6 to 19 in relation to age at maximum growth, *Monogr. Soc. Res. Child. Dev.* **4** (3):291.

Simmons, K. E., and Greulich, W. W., 1943, Menarcheal age and the height, weight and skeletal age of girls aged 7 to 17 years, *J. Pediatr.* **22**:518–548.

Simon, G., Reid, L., Tanner, J. M., Goldstein, H., and Benjamin, B., 1972, Growth of radiologically determined heart diameter, lung width, lung length from 5 to 19 years with standards for clinical use, *Arch. Dis. Child.* **47**:373–381.

Singh, I. J., Savara, B. S., and Newman, M. T., 1967, Growth of the skeletal and non-skeletal components of head width from 9–14 years of age, *Hum. Biol.* **39**:182–191.

Singh, S. P., and Sidhu, L. S., 1981, Pubertal development of Gaddi Rajput boys of Dhaula Dhar range of the Himalayas, *Z. Morphol. Anthropol.* **72**:89–98.

Stuetzle, W., Gasser, Th., Molinari, L., Largo, R. H., Prader, A., and Huber, P. J., 1980, Shape-invariant modelling of human growth, *Ann. Hum. Biol.* **7**:507–528.

Tanner, J. M., 1962, *Growth at Adolescence,* 2nd ed., Blackwell Scientific Publications, Oxford.

Tanner, J. M., 1965, Radiographic studies of body composition, in: *Body Composition, Symposia of the Society for the Study of Human Biology* (J. Brozek, ed.), Vol. 7, pp. 211–236, Pergamon, Oxford.

Tanner, J. M., 1973, Trend towards earlier menarche in London, Oslo, Copenhagen, The Netherlands, and Hungary, *Nature (London)* **243**:95–96.

Tanner, J. M., 1975, Growth and endocrinology of the adolescent, in: *Endocrine and Genetic Diseases of Childhood and Adolescence,* 2nd ed. (L. I. Gardner, ed.), pp. 14–64, W.B. Saunders, Philadelphia.

Tanner, J. M., 1978, *Foetus into Man,* Open Books, London, and Harvard University Press, Cambridge, Massachusetts.

Tanner, J. M., 1981, *A History of the Study of Human Growth,* Cambridge University Press, Cambridge.

Tanner, J. M., and O'Keefe, B., 1962, Age at menarche in Nigerian schoolgirls with a note on their heights and weights from age 12 to 19, *Hum. Biol.* **34**:187–196.

Tanner, J. M., Whitehouse, R. H., and Takaishi, M., 1966, Standards from birth to maturity for height, weight, height velocity and weight velocity; British children, 1965, *Arch. Dis. Childh.* **41**:454–471; 613–635.

Tanner, J. M., Whitehouse, R. H., Marshall, W. A., Healy, M. J. R., and Goldstein, H., 1975, *Assessment of Skeletal Maturity and Prediction of Adult Height,* Academic Press, London.

Tanner, J. M., Whitehouse, R. H., Marubini, E., and Resele, F., 1976, The adolescent growth spurt of boys and girls of the Harpenden Growth Study, *Ann. Hum. Biol.* **3**:109–126.

Tanner, J. M., Hughes, P. C. R., and Whitehouse, R. H., 1981, Radiographically determined widths of bone, muscle and fat in the upper arm and calf from age 3–18 years, *Ann. Hum. Biol.* **8**:495–518.

Tanner, J. M., Hayashi, T., Preece, M. A., and Cameron, N., 1982, Increase in length of leg relative to trunk in Japanese children and adults from 1957 to 1977; comparison with British and with Japanese Americans, *Ann. Hum. Biol.* **9**:411–424.

Taranger, J., and Karlberg, P., 1976, The somatic development of children in a Swedish urban community: VII. Graphic analysis of biological maturation by means of maturograms, *Acta Paediatr. Scand. (Suppl.)* **258**:136–146.

Taranger, J., Lichtenstein, H., and Svennberg-Redegren, I., 1976, The somatic development of children in a Swedish urban community: VI. Somatic pubertal development, *Acta Paediatr. Scand. (Suppl.)* **258**:121–135.

Thissen, D., Bock, R. D., Wainer, H., and Roche, A. F., 1976, Individual growth in stature: A comparison of four growth studies in the USA, *Ann. Hum. Biol.* **3**:529–542.

Thompson, G. W., Popovich, F., and Anderson, D. L., 1976, Maximum growth changes in mandibular length, stature and weight, *Hum. Biol.* **48**:285–293.

Tisserand-Perrier, M., 1953, Etudes de certains processus de croissance chez les jumeaux, *J. Génét. Hum.* **2**:87–102.

Turpin, R., Chassagne, P., and Lefèbvre, J., 1939, La mégalothymie prépubertaire. Etude planigraphique du thymus au cours de la crossiance, *Ann. Endocrinol.* **1**:358–378.

Uche, G. O., and Okorafor, A. E., 1979, The age at menarche in Nigerian urban school girls, *Ann. Hum. Biol.* **6**:395–398.

Valsik, J. A., Strouhal, E., Hussein, F. H., and El-Nofely, A., 1970, Biology of man in Egyptian Nubia, *Mater. Pr. Antropol.* **78**:93–98.

Venrooij-Ijsselmuiden, M. E. van, 1978, Mixed longitudinal data on height, weight, limb circumferences and skinfold measurements of Dutch children, *Hum. Biol.* **50**:369–384.

Venrooij-Ijselmuiden, M. E. van, Smeets, H. J. L., and Werff Ten Bosch, J. J. Van der, 1976, The secular trend in age at menarche in the Netherlands, *Ann. Hum. Biol.* **3**:283–284.

Wark, M. L., and Malcolm, M. A., 1969, Growth and development of the Lumi children of the Sepik district of New Guinea, *Med. J. Aust.* **2**:129–136.

Warren, M. P., 1980, The effects of exercise on pubertal progression and reproductive function in girls, *J. Endocrinol. Metab.* **51**:1150–1157.

Welon, Z., and Bielicki, T., 1979, The timing of adolescent growth spurts of eight body dimensions in boys and girls of the Wroclaw Growth Study, *Stud. Phys. Anthropol., Polish Acad. Sci.* **5**:74–79.

Wieringen, J. C. van, Wafelbakker, F., Verbrugge, H. P., and Haas, J. H. de, 1971, *Growth Diagrams 1965 Netherlands: Second National Survey on 0–24 year-olds,* Wolfters–Noordhoff, Groningen, Netherlands.

Wilson, D. C., and Sutherland, I., 1949, The age at menarche, *Br. Med. J.* **2**:130–132.

Wyshak, G., 1983*a*, Secular changes in age at menarche in a sample of US women, *Ann. Hum. Biol.* **10**:75–78.

Wyshak, G., 1983*b*, Age at menarche and unsuccessful pregnancy outcome, *Ann. Hum. Biol.* **10**:69–74.

Yanagisawa, S., and Kondo, S., 1973, Modernization and physical features of the Japanese with special references to leg length and head form, *J. Hum. Ergol.* **2**:97–108.

Young, R. W., 1957, Post-natal growth of the frontal and parietal bones in white males, *Am. J. Phys. Anthropol.* **15**:367–386.

Young, R. W., 1959, Age changes in the thickness of the scalp in white males, *Hum. Biol.* **31**:74–79.

Zacharias, L., and Rand, W. M., 1983, Adolescent growth in height and its relation to menarche in contemporary American girls, *Ann. Hum. Biol.* **10**:209–222.

Zacharias, L., and Wurtman, R. J., 1964, Blindness: Its relation to age at menarche. *Science* **144**:1154–1155.

Zacharias, L., and Wurtman, R. J., 1969, Blindness and menarche, *Obstet. Gynecol.* **33**:603–608.

Zacharias, L., Rand, W. M., and Wurtman, R. J., 1976, A prospective study of sexual development and growth in American girls: The statistics of menarche, *Obstet. Gynecol. Surv.* **31**:325–337.

Zachmann, M., Prader, A., Kind, H. P., Haflinger, H., and Budliger, H., 1974, Testicular volume during adolescence, *Helv. Paediatr. Acta* **29**:61–72.

Zuckerman, S., 1955, Age change in the basicranial axis of the human skull, *Am. J. Phys. Anthropol.* **13**:521–539.

9

Prepubertal and Pubertal Endocrinology

MICHAEL A. PREECE

1. Introduction

The processes of growth and development imply increases in cell size and number with all the accompanying requirements for protein and nucleic acid synthesis. For orderly growth, the availability of appropriate nutrients as the building blocks of macromolecular synthesis is fundamental. It has become clear over recent years that there is also a requirement for a network of supervising processes to control many or all of these metabolic steps. Most of the supervisory processes are controlled by the endocrine system. This network of interrelating hormones links the activity of the neuroendocrine system with cellular operations both in controlling relatively unchanging processes, which are little affected by the aging process, and in entraining processes that subserve changes and occur at specific times of life such as the development of puberty.

Endocrinology has developed dramatically over the past 25 years largely because of the availability of improved methods of measuring the very small amounts of the various hormones in the bloodstream at any one time. We have also begun to realize that there are complex interrelationships between one hormone and another such that the system can often not be studied hormone by hormone but requires more of a synthetic approach if we are to understand its overall functions.

This chapter makes no attempt at presenting comprehensive review of the endocrine system in childhood. Instead, emphasis is placed on those hormones that, according to current knowledge, appear to have a direct effect on the regulation of the growth process. This function essentially revolves about the hypothalamopituitary axis and the hormones that depend on it. The notable addition to this group is insulin, which is also briefly discussed. Many other hormones, particularly of the gut and its related organs and also the catecholamines, have importance in the provision of energy substrates for growth, and in a similar way parathyroid hormone and the vitamin D metabolites are important for normal bone development. However, as these last two groups of hormones do not appear to be directly involved in the control of growth and pubertal development, they are not considered further here.

The chapter is divided into three parts: A review of the hypothalamopituitary endocrine system with no particular reference to maturational stage; A more detailed description of the changes seen with age and pubertal development in each of the hormone systems; and finally an attempt to describe some of the interrelationships insofar as they are at present known.

2. The Hypothalamopituitary Endocrine System

Figure 1 is a schematic representation of the set of five hormone systems that function through the agency of the hypothalamus and the anterior pitu-

MICHAEL A. PREECE • Department of Growth and Development, Institute of Child Health, University of London, London WC1 1EH, England.

itary gland. There is a striking consistency in the general pattern of this system, with interrelationship between the hypothalamus and the pituitary depending on a set of peptide releasing or release-inhibiting factors or hormones. The pituitary gland secretes a number of protein hormones which act on a variety of target endocrine glands. In turn, these glands produce a number of hormones of smaller molecular weight that appear to exert the tissue effects of each system. The hypothalamic releasing hormones are conventionally referred to as factors, such as prolactin (PRL)-release-inhibiting factor, when their biological activity has been demonstrated but before they have been isolated, characterized, and synthesized. Once these latter steps have been achieved, they are usually attributed the title of hormone, such as thyrotrophin-releasing hormone (TRH). Thus far, four substances have attained this status; i.e., TRH (Schally *et al.*, 1969) gonadotrophin-releasing hormone (GnRH) (Schally *et al.*, 1971), growth hormone-releasing hormone (GHRH) (Spiess *et al.*, 1982), and growth hormone release-inhibiting hormone, or somatostatin (GHRIH) (Brazeau *et al.*, 1973). The latter hormone is not included in Figure 1. Corticotrophin-releasing factor (CRF) has been isolated from ovine hypothalami, and its sequence in this species, but not yet in humans, is known. TRH and GnRH are small-molecular-weight peptides of 3- and 12-amino acid residues. GHRH and CRF are much larger, the former consisting of either 40- or 44-residues and the latter probably of 41, in sheep at least. (See Chapter 10 of this volume for a further discussion of the neuroendocrine system in relationship to pituitary–gonadal function.)

Each of the hypothalamic hormones acts on the anterior pituitary gland mostly stimulating, or in two cases inhibiting, release of a specific pituitary protein hormone. In the case of GnRH, its action on the pituitary is complex and is related to subtle episodic secretions, as discussed in Chapter 10. The net effect of this is secretion of the two gonadotrophins luteinizing hormone (LH) and follicle-stimulating hormone (FSH), both of which have molecular weights of about 30,000. The gonadotrophins consist of two subunits, α and β, in which the α-subunit is common to the two hormones, but the β-subunits are specific. Chorionic gonadotrophin, produced by the placenta during pregnancy and essential for the maintenance of the fetoplacental unit, also shares the common α-subunit but has a third β-subunit. These hormones are glycoproteins having important carbohydrate moieties covalently linked to the protein parts of the molecule.

The actions of the gonadotrophins are principally on the gonads, but each gonadotrophin affects a different part of the gonad in a highly specific way. In principle, LH is responsible for the endocrine function of both the ovary and the testis, stimulating the secretion of estradiol by the former and testosterone by the latter. In contrast, FSH acts on the follicular cells of the ovary influencing ovulation and on the testicular tubules influencing spermatogenesis. There is a negative feedback loop from the gonad to either the hypothalamus or pituitary, more likely the former. The sex steroids in

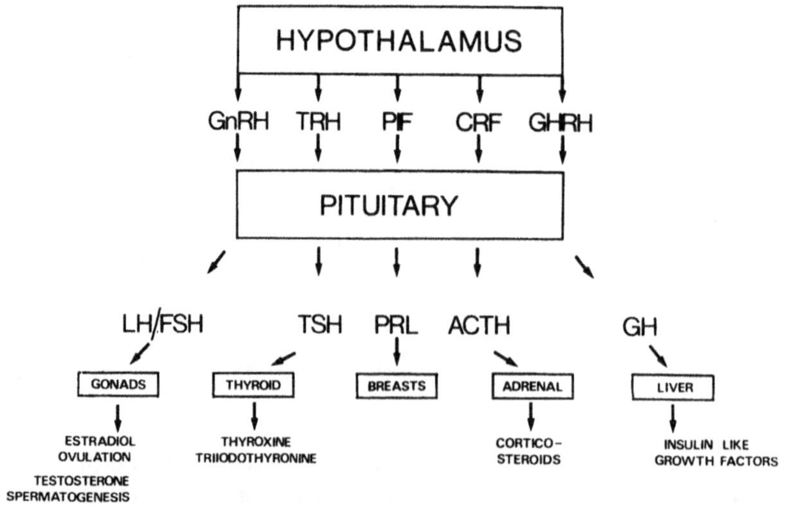

Fig. 1. Schematic diagram of the hypothalamopituitary system of hormones. ACTH, adrenocorticotrophic hormone; CRF, adrenocorticotrophin-releasing factor; FSH, follicle-stimulating hormone; GH, growth hormone; GHRH, growth hormone-releasing hormone; GnRH, gonadotrophin-releasing hormone; LH, luteinizing hormone; PIF, prolactin-release-inhibiting factor; PRL, prolactin; TRH, thyrotrophin-releasing hormone; TSH, thyroid-stimulating hormone.

the male have a straightforward negative feedback, but in the mature female the relationship is biphasic with positive feedback. The nature of the negative feedback loop for the FSH system is less well understood but involves the release of a nonsteroidal hormone, inhibin, from the Sertoli cells of the testis and granulosa cells of the ovary (Franchimont *et al.*, 1981).

The result of TRH stimulation of the anterior pituitary is the release of thyrotrophin or TSH. This is another glycoprotein, which has a common α-subunit with the gonadotrophins, but with its own highly specific β-subunit. It acts directly on the thyroid gland, stimulating the synthesis and release of the two principal thyroid hormones thyroxine (T4) and triiodothyronine (T3). We believe that the latter is the active hormone and that T4 only acts by being converted to T3 at the periphery by deiodinase enzymes (Robbins and Rall, 1983).

The interrelationships of prolactin-release-inhibiting factor (PIF) and PRL are still far from clear. The identity of PIF remains uncertain, but there is some evidence that it may be dopamine and/or α-aminobutyric acid (GABA) (Franks, 1983). Prolactin is a large protein hormone of molecular weight 22,000, the actions of which are very uncertain. It is involved in lactation but seems to have many other potential functions, at least as judged by the widespread effects of its excessive production in pathological states. In contrast to all the other pituitary protein hormones, there is no evidence of a further hormone dependent on PRL for its release in the same way as the thyroid hormones depend on TSH. No other specific target organ has been identified except for the breast tissue.

Corticotrophin-releasing factor, possibly in concert with vasopressin (Gillies and Lowry, 1982), modulates the secretion and release of adrenocorticotrophic hormone (ACTH) (Grossman *et al.*, 1982). This is only one of a complex family of related hormones synthesized in the pituitary gland. As the metabolism of ACTH does not seem to be significatly related to normal growth and development and yet forms such an intricate network of its own involving the opioid peptides and the melanocyte-stimulating hormones (MSH), it is not discussed further here; a detailed discussion can be found in Rees and Lowry (1983). Apparently independently of the other hormones of its family, ACTH stimulates the adrenal cortex to produce a whole group of corticosteroid hormones that have varying significance in protein synthesis and other fundamental cellular activities. One group of adrenal hormones has substantial androgenic activity, although it is much less potent in this respect per mole than testosterone or its derivatives.

Finally, GHRH is the most recently characterized of the hypothalamic releasing hormones. It stimulates the release of growth hormone (GH) from the anterior pituitary gland. This is another large protein hormone of molecular weight 21,500, which has many similarities to PRL. Growth hormone in primates also has PRL-like actions; so much so, indeed, that PRL was long thought not to exist in the higher primate as an independent hormone. The actions of growth hormone, as the name would suggest, have many and profound effects on growth of the animal, but it can by no means act alone and the integrity of the whole pituitary system is necessary for its optimal functioning. It appears to work through two mechanisms: a direct effect, mostly on muscle and fat tissue leading to the mobilization of energy stores (Kostyo and Isaksson, 1977), and an indirect effect through the insulin-like growth factors (IGFs) on skeletal growth and RNA, DNA, and protein synthesis (Preece and Holder, 1982). The IGFs are also referred to as the somatomedins, but the former nomenclature is used in the present discussion. The major organ for synthesis and release of the IGFs is thought to be the liver, although there is increasing evidence that many other tissues may be capable of this synthesis as well.

3. Hormonal Changes through Childhood and Adolescence

The changes in the hypothalamus are dealt with in Chapter 10 of this volume. This section covers the changes in production of those hormones that are synthesized and secreted by the pituitary gland or its target organs. In many cases our knowledge is incomplete. It is first important to observe that many, if not all, hormones go through periods of episodic release at certain times of life or, indeed, may always be released in an episodic manner. Unfortunately, most of our present knowledge is based on levels of serum hormones taken under basal conditions, as this is really the only practical way, for ethical reasons, in which data can be collected. Changes in these levels do give us some insight into the progression of events as the child grows older, but they must be viewed with the clear understanding that they represent only a shadow of the truth and that much more subtle changes occur that can only be seen when detailed, multisample 24-hr profiles are studied.

3.1. The Pituitary-Gonadal Axis

The pattern of secretion of the gonadotrophins and principal gonadal steroids related to age are illustrated in Figures 2 and 3. It has proved difficult to combine data from different studies for a number of reasons. As was observed by Tanner (1981), published studies differ in the way the data are reported, and few indeed have adopted appropriate longitudinal methods of analysis, even when the data were originally collected longitudinally. Furthermore, even after converting published results to the same units of measure (the SI system is used here), there are substantial differences in absolute values; this is especially true of the gonadotrophins. It has generally been held that earlier immunoassays tended to produce higher values of peptide hormone concentrations because of nonspecific interference in the assays, and this may provide at least a partial explanation.

Thus the curves shown in the figures should be considered qualitative. They were obtained from the data of Faiman and Winter (1974), Apter *et al.* (1978*a*), Sizonenko *et al.* (1976), and Preece *et al.* (1984) and from some additional unpublished data from M. A. Preece and associates. These studies were selected because the data were longitudinal and were analyzed in a longitudinal manner by methods similar to those of Tanner and Gupta (1968). It was clear that the relative changes in hormone levels from one stage of puberty to another were consistent in the varous studies although absolute values differed. The combined curves were obtained by making the assumption that mean values of a given hormone in a particular stage of puberty should be the same for all similar populations (this may not be valid, but it seemed a reasonable assumption). Thus, differences between, say, mean LH concentration in genitalia stage 3 in different studies, should be due to a combination of various methodological factors rather than biological ones. By comparing the whole set of such values for each hormone, it was possible to adjust the longitudinal means for the various chronological ages of each study minimizing differences attributable to methodological factors. The overall means were then calculated and graphically smoothed giving the curves in Figures 2 and 3; the direction of adjustment was such that the magnitudes of the hormone concentrations approximate most closely to those from Preece *et al.* (1984). They probably represent a reasonable consensus at the present time and the general patterns of hormone secretion are probably correct. Figure 4 shows longitudinal mean values for the same hormones according to breast or genitalia stage. Here only the data of Preece *et al.* (1984, and unpublished) have been used.

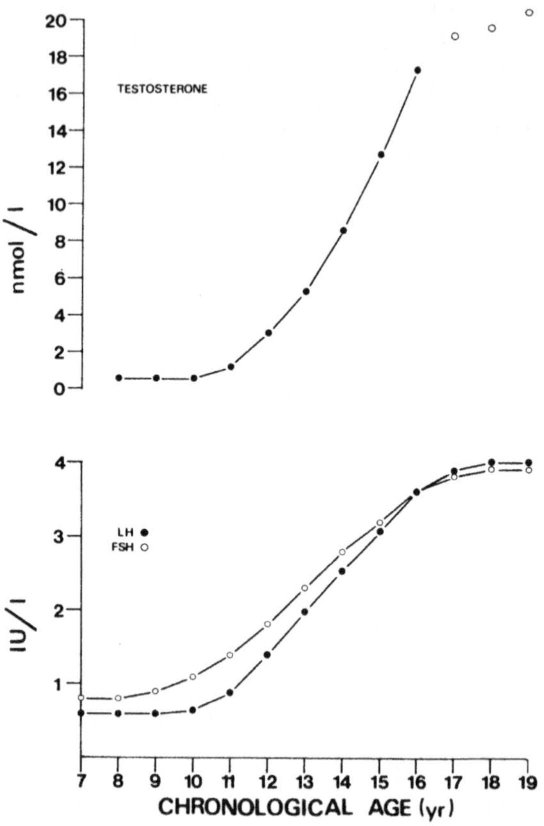

Fig. 2. The pattern of change, related to chronological age, of the gonadotrophins and testosterone in boys. See text for a description of the method of calculation of the means and the data sources. For testosterone at ages above 16 years, no longitudinal data existed; the three open circles represent cross-sectional data from Pakarinen *et al.* (1979).

Follicle-stimulating hormone begins to rise earlier than LH, particularly in girls, and reaches a peak around the time of menarche in girls and at a testis volume of about 20 ml in boys. Thereafter, the level either stays constant or may decline slightly. Certainly in studies of urinary FSH excretion, there is a marked decline about 1 year after the age of peak height velocity (Tanner *et al.*, 1978). The relative timing of the LH and FSH profiles varies somewhat, but in general the lag in LH rises relative to FSH is less marked in boys than girls. Compared with boys, girls have greater concentrations of both gonadotrophins than do boys, especially in the case of FSH.

Boyar *et al.* (1974, 1976) recorded plasma LH levels every 20 min throughout 24 hr in 14 adolescent children. When LH first began to rise, at a bone age of 8 "years," initially it did so in bursts released during sleep. Later in puberty, the secretory peaks occurred during wakefulness as well, but the quantities of LH released during sleep remained predominant until puberty was practically complete. This result was confirmed with slightly different methodology by Lee *et al.* (1976*a*), and the same group showed a similar phenomenon for FSH (Lee *et al.*, 1978). In boys, the LH surges are followed, about 30 min later, by high levels of testosterone (Parker *et al.*, 1975). In this way, adult sex hormone levels, in the male, are first reached at

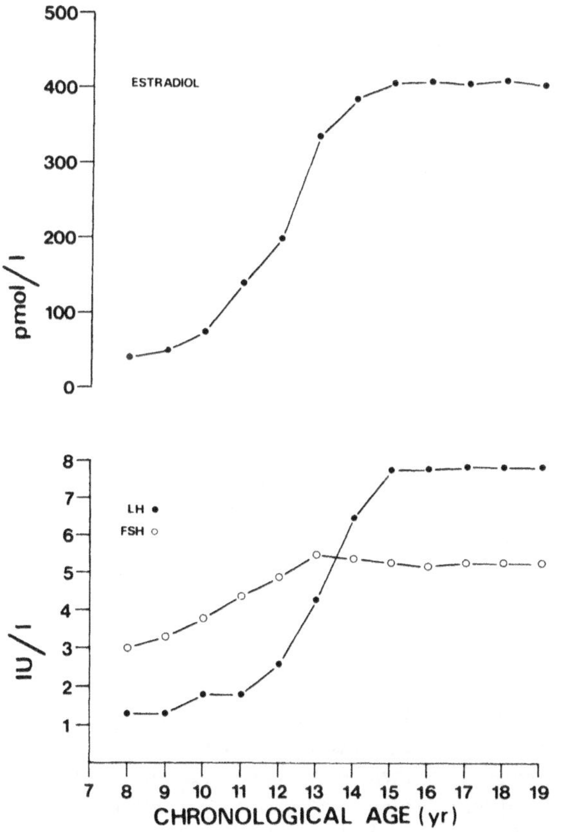

Fig. 3. The pattern of change, related to chronological age, of the gonadotrophins and oestradiol in girls. See text for details.

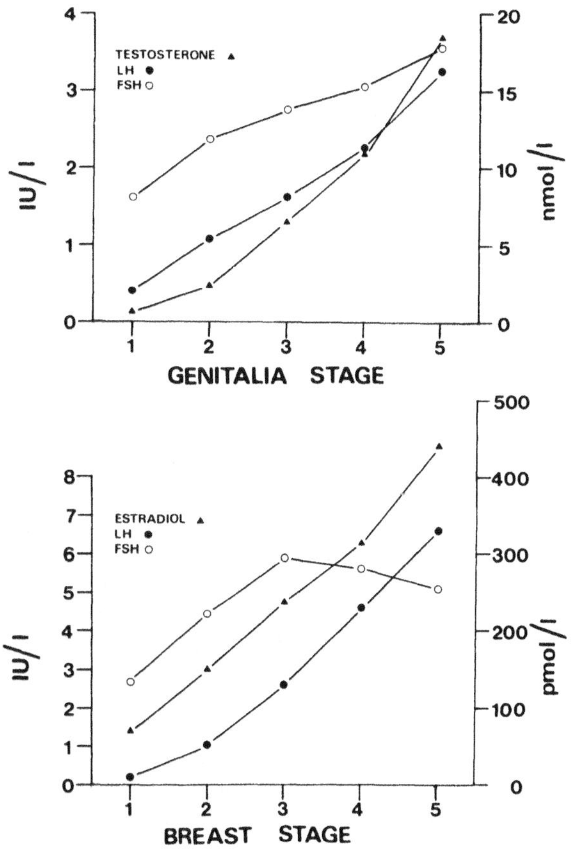

Fig. 4. The pattern of change in the principal sex hormones related to genitalia stages in boys (top) and breast changes in girls (bottom). (Data from Preece *et al.*, 1984; M. A. Preece, unpublished results.)

night and only later during the daytime. In girls, estradiol showed no such night rise but had a circadian rhythm with a peak at around 3 PM (Boyar *et al.*, 1976). The nighttime peaks of both LH and FSH occur before any other signs of puberty are manifest. Testosterone concentrations in boys (see Figure 2) follow a very similar course to that of LH. The steepest part of the rise seems slightly earlier than peak height velocity and therefore coincides with the peak velocity in sitting height. This aspect is considered a little further in Section 4. The earliest part of the increase, up to around 6 nmoles/liter will occur even in agonadal males (Forest *et al.*, 1979) and presumably represents testosterone derived from adrenal androgen metabolism. In girls, testosterone may rise to similar levels, also in early puberty (Apter and Vihko, 1977). Testosterone in girls is presumably derived from adrenal and perhaps ovarian androstenedione and remains at this level, although it may fluctuate somewhat in the menstrual cycle. In males, the testosterone levels go on rising to around 20 nmoles/liter by genitalia stage 5 and go still further in young adults. Clearly, testosterone levels continue to rise, even after the main morphological changes of puberty have been completed.

In early and middle puberty, there is a striking circadian rhythm with higher levels of testosterone appearing at night, following LH surges (Large and Anderson, 1979; Parker *et al.*, 1975). Estradiol levels in boys show no circadian rhythm, so that in the early stages of puberty boys tend to have ratios of estradiol to testosterone that may be higher than at any other time of life. It has been suggested (Large and Anderson, 1979) that this contribues to the development of gynaecomastia at this stage of puberty.

It seems most likely that some of the pubertal effects of testosterone are effected by conversion to dihydrotestosterone (DHT) by the enzyme 5α-reductase. Changes in blood dihydrotestosterone (DHT) levels during puberty closely parallel those of testosterone itself (Gupta *et al.*, 1972, 1975; Pakarinen *et al.*, 1969). However, conversion of testosterone to DHT takes place in the receptor tissues themselves such that circulating levels of DHT are always very low compared with those of testosterone.

There is a further sex difference in the metabolism of testosterone that arises at puberty (Zumoff *et al.*, 1976). When a dose of ^{14}C-labeled testosterone is injected intravenously, about 70% is recovered in the urine. There, nearly two-thirds is present as androsterone plus etiocholanolone. In prepubertal children of both sexes, androsterone is the predominant product and remains so in adult men. In young adult women, however, eticholanolone becomes predominant. Most of the metabolism of testosterone, with the notable exception of conversion to DHT, is hepatic (Ishimaru *et al.*, 1978), and this change in the handling of testosterone at puberty presumably reflects sex differences in liver enzymes that are known to exist in certain other animals.

In girls, estradiol concentrations more or less follow the levels of LH as testosterone does in boys (see Figure 2). The most dramatic rise occurs between breast stages 2 and 4 (Figure 4) thus between about 11 and 13 years of age. As menarche is approached, levels of estradiol have greater and greater variance as a result of the onset of cycling, which is developed to a variable degree by the time of menarche (Winter and Faiman, 1973; see also Section 3.2). The concentrations of estradiol in the years preceding the development of puberty are less clear (Reiter and Root, 1977). Using cross-sectional data, Radfarr *et al.* (1976) showed a difference between mean estradiol levels in girls less than 8 years of age and girls greater than 8 years of age but still in breast stage 1. However, neither Ducharme *et al.*, (1976) nor Lee *et al.*, (1976*b*) found any change before age 10.

In a manner somewhat analogous to the testosterone levels in girls, boys develop low levels of estradiol up to about 80 pmoles/liter, which is equivalent to the levels seen at breast stage 2 in girls. Male estradiol is derived partly from conversion of testosterone in the testes and, perhaps more importantly, by conversion of androstenedione and testosterone in peripheral tissue, particularly fat cells (Kelch *et al.*, 1972).

3.2. The Menstrual Cycle

Discussion of puberty in girls would be incomplete without reference to the menstrual cycle. Figure 5, redrawn from Apter *et al.* (1978*b*), illustrates the pattern of gonadotrophin, estradiol (E_2), and progesterone secretion throughout the menstrual cycle in 10 girls between 1.6 and 4.5 years after menarche. The hormone cycles have been centered about the LH peak. A dramatic rise in E_2 precedes the mid-cycle LH peak, coincident with a small peak in FSH. Following ovulation, progesterone levels show a prolonged rise, declining at the end of the luteal phase. There is a second, more sustained, E_2 peak at the same time.

This regular cyclical behavior of the hormones is not immediately established at menarche. Using the presence of a clear luteal progesterone peak to indicate an ovulatory cycle, Apter (1980) found that only 14% of cycles are ovulatory in the first year postmenarche, with 50% ovulatory by 3 years and 87% by 6 years.

3.3. The Pituitary-Thyroid Axis

That T4 and T3 are necessary for normal growth and development has been known for many years. Deficiency of these hormones results in general retardation of growth and organ maturation with reduced rates of cell division in all proliferating tissues (Greenberg *et al.*, 1974). As soon as a deficiency of T4 is made good by replacement therapy, there is enhanced protein synthesis and a re-

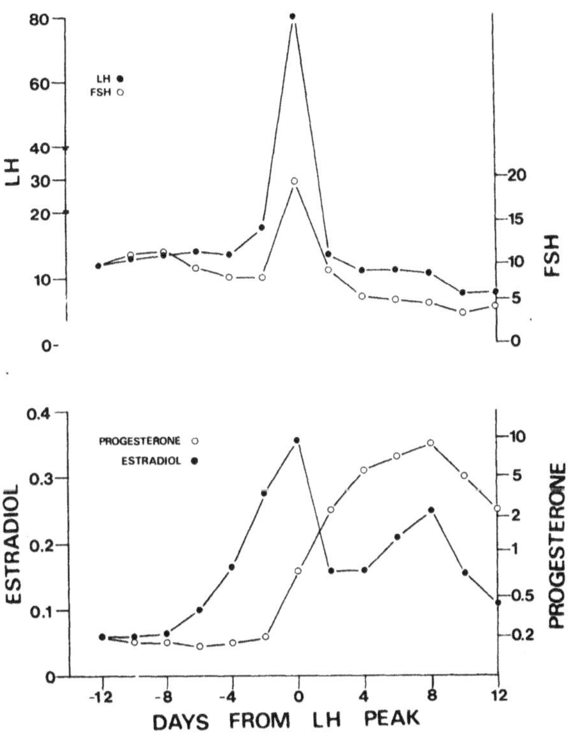

Fig. 5. The cyclical pattern of secretion of the gonadotrophins, estradiol, and progesterone for 10 girls from 1.6 to 4.5 years after menarche. The means are centered about the LH peak. (Data from Apter *et al.*, 1978*b*.)

covery of normal statural growth (Prader *et al.*, 1963). In an analogous way, but far less commonly, an excess of thyroid hormones may be associated with a mild acceleration of growth and skeletal maturation.

Because of the requirement for continuing secretion of thyroid hormones throughout the whole of growth, it is not surprising that there is little change in thyroid hormone levels during childhood. Once the rather high values of TSH, T3, and T4 of the perinatal period (see Chapter 18, Vol. 1) have declined, these hormones show little change throughout the rest of childhood until the onset of puberty. At this time, however, there is a tendency for a fall in T4 concentration (Fisher *et al.*, 1977; Sack *et al.*, 1982). The data for T3 concentrations are a little less certain. Avruskin *et al.* (1973) failed to demonstrate any changes, but Sack *et al.* (1982) showed a fall in T3, particularly in late puberty. As none of these studies was longitudinal, some of the discrepancies may be related to sampling.

Thyroid-stimulating hormone secretion shows no evidence of changes throughout childhood, although it should be said that most assays for this hormone are insufficiently sensitive to reach the lower end of the normal range and there may be subtle changes related to maturation. Thyroxine-binding globulin (TBG) falls strikingly throughout puberty in boys. This may be related to rising levels of testosterone, which do tend to lower serum TBG concentrations. It is notable that the fall in T4 and serum TBG are in parallel such that there do not seem to be changes in free T3 and T4 but only in their total concentrations (Parra *et al.*, 1980).

3.4. Prolactin

In girls, PRL shows a late rise in puberty, and in those studies in which it has been related to pubertal stages the maximum increments in PRL secretion seem to be related to the transition between breast stages 3 and 4 (Preece *et al.*, 1980). Once menarche has been achieved, there is a tendency in some studies for PRL levels to fall slowly to the normal adult level.

In boys PRL concentrations are much the same as girls at the start of puberty but then fall very gradually through the course of puberty to adult values by the age of 16–17 years (Preece *et al.*, 1984).

3.5. The Pituitary-Adrenal Axis

Normal secretion of glucocorticoid hormones, which are all C_{21} steroids, is essential for survival but does not appear to be a major controlling factor in human growth. When glucocorticoids are present in excess, either because of disease of the pituitary or adrenal glands or because of exogenous steriod therapy, severe stunting of growth may occur (Preece, 1976). There have been many proposals as to the site of this action of the glucocorticoids including suggestions that they inhibit growth hormone release in the pituitary, that they suppress IGF generation in the liver, and that they affect cartilage metabolism directly. It is not certain which of these actions is the most important, although the last is perhaps the favored candidate; glucocorticoids certainly have profound effects on chondrocyte metabolism at many levels (Silberman, 1983).

Secretion of ACTH and cortisol, the principal glucocorticoid, is episodic with a very strict diurnal rhythm. More frequent secretory episodes of ACTH occur in the early morning than at other times, and plasma cortisol levels are higher at that time (Gallagher *et al.*, 1973). During childhood and puberty, cortisol production increases in proportion to body size (Savage *et al.*, 1975), but the concentration of cortisol in serum remains relatively constant; mean levels may be slightly higher in mature males than in females (Zumoff *et al.*, 1974). In obese children, cortisol production may rise in proportion to body weight. By contrast, in severe malnutrition cortisol production rates are low but serum cortisol levels may be elevated because of an even greater reduction in cortisol metabolism (Smith *et al.*, 1975).

We have to pay much more attention to the second major group of adrenal steroids, the adrenal androgens. These are all C_{19} steriods variously related to testosterone, which is itself secreted by the adrenal in small quantities. Adrenal androgens are produced in great variety but the principal ones are dehydroepiandrosterone (DHA), its sulfate (DHAS), and androstenedione.

There are substantial data on the behaviour of DHA throughout childhood and adolescence. The patterns are not the same in both sexes, although in one important aspect there is similarity. In both boys and girls, DHA secretion increases markedly at about 7 years of age before any of the usual morphological criteria of puberty are visible. The preadolescent fat spurt may be the one exception to this statement, as it commences at about this time (Tanner, 1981). In girls, DHA concentrations rise steeply from about 7 years of age until mid-puberty at about 13–15 years of age. Some data show more of a plateau at this time, with a sharper increase later (Apter *et al.*, 1979). In boys, there is a generally slower rise in DHA levels until about the age of 16 years, with a rather steep rise in the latter half of puberty.

Androstenedione shows an essentially similar pattern, although the prepubertal rise is relatively delayed compared with DHA with no clear elevation until past 10 years of age in girls (Apter, 1980; Ilondo *et al.*, 1982) and even later in boys at 12–14 years of age (Ilondo *et al.*, 1982).

There has been much debate as to the mechanism of the onset of increased adrenal androgen secretion at 7–8 years of age. It has been suggested that there may be an unidentified pituitary or extrapituitary factor that specifically stimulates adrenal androgen secretion when acting on ACTH-primed adrenal glands (Parker and Odell, 1979). It has also been postulated that there may be an age-related shift of response to ACTH in adrenal steroid biosynthesis (Rich *et al.*, 1981) or that a gradient of intraadrenal cortisol secretion might stimulate the growth of the zona reticularis (Anderson, 1980). The first possibility has now been considerably strengthened. Parker *et al.* (1983) reported the isolated of a glycoprotein of molecular weight 60,000 from pituitary extracts that stimulated DHA production by isolated dog adrenal cells but did not affect cortisol production.

Kelner and Brook (1983) published longitudinal data of urinary adrenal androgen metabolite secretion in boys. Their data suggested that adrenarche was accompanied by a fall in 3β-hydroxysteroid dehydrogenase activity, a small rise in 21-hydroxylase activity and a fall in 11β-hydroxylation. There was a slight rise in 17α-hydroxylase activity and a marked rise in 17,20-lyase activity. Despite this, cortisol metabolite excretions changed little throughout childhood and adolescence. Their data are equally compatible with either a cortical androgen-stimulating hormone or the effects being solely due to ACTH accompanied by intraadrenal changes.

3.6. The Growth Hormone Axis

Probably more than any other pituitary hormone, growth hormone is secreted in an exceedingly pulsatile manner such that for many hours of the day the blood concentration is at undetectable levels in normal individuals (Finkelstein *et al.*, 1972). For this reason, single samples of blood taken for GH assay are of even less value than for most other hormones in assessing patterns of change with age. Studies of children and adolescents using frequent sampling techniques over 24-hr periods (Finkelstein *et al.*, 1972) have shown some discrepancies but with a consensus that the production of GH is age related and greatest during childhood and puberty. In a relatively limited number of individuals, there were clear changes both in numbers of GH peaks and in their intensity from childhood through adolescence (Finkelstein *et al.*, 1972). At all times, secretion of GH occurs predominantly at night and is associated particularly with deep sleep. The frequency and magnitude of the nocturnal pulses increases dramatically at the time of puberty and in adults returns to an overall pattern of GH secretion that is lower than in the prepubertal child.

These changes in GH secretion are mirrored in the changes in serum concentration of the IGFs. Of these two peptides, both closely related to insulin in structure, insulin-like growth factor I (IGF-I) is most closely dependent on GH in terms of its secretion and is probably the more direct mediator of its actions. Unlike GH, both IGFs show relatively constant levels throughout the day, possibly because of a specific circulating binding protein that tends to even out any pulsatility in synthesis of the IGF in response to the peaks of GH secretion (Furlanetto, 1980). Serum concentrations of IGF-I are relatively low at birth but rise steadily through childhood with a clear peak during adolescence (Hall *et al.*, 1981; Underwood *et al.*, 1984) (see Table I). At least superficially, it would appear that the surge in IFG-I secretion at puberty can be closely related to the changes in GH secretion at that time. The relationships of both changes to the pattern of skeletal growth during that period are discussed in Section 4. Although similar data are not available for GH, data for the IGFs suggests a close relationship between the adolescent peak of IGF-I secretion with the middle stages of puberty (Rosenfield *et al.*, 1983; Underwood *et al.*, 1984).

3.7. Pancreatic Insulin

Observation of the insulin-like actions of the IGFs has raised the possibility that insulin itself may be a hormone necessary for human growth. It has been known for a number of years that insulin stimulates protein synthesis (Manchester, 1972) and can stimulate the growth of cartilage or other cells *in vitro* (Schwartz and Amos, 1968). While insulin is not an essential mediator of the actions of GH, it does seem that the full expression of the latter's anabolic effect requires the presence of insulin (Milman *et al.*, 1951). In addition, GH stimulates production not only of IGF but of insulin itself (Curry and Bennett, 1973). It is possible that these hormones act synergistically in regulating cell growth or even that insulin itself may exhibit pure growth factor activity. In some children following surgical removal of craniopharyngioma, normal growth is maintained despite unmeasurable GH levels. In such children, immunoreactive insulin levels are higher than in other hypopituitary patients (Costin *et al.*, 1976). Supporting *in vitro* evidence has been reported by Nagarajan and Anderson (1982), who showed direct stimulation of growth of F9 embryonal carcinoma cells by insulin through its own receptor.

3.8. The Pineal Gland

The previous discussion of the serum profiles of various pituitary and other hormones during maturation has made no comment about the mechanisms responsible for initiating the pubertal changes. The maturation of the hypothalamus and the changes in GnRH secretion are obviously fundamental, but it is still unclear whether a further system initiates those events. There has been some evidence, mostly of a circumstantial nature, that the pineal gland may be involved in some way, perhaps playing an antigonadotrophin effect in prepuberty. Melatonin, an important pineal indole, has antigonadal effects in laboratory animals (Wurtman *et al.*, 1963); furthermore, pineal tumors are sometimes associated with precocious sexual maturation in boys (Waldhauser and Wurtman, 1983). There has been disagreement in the literature in attempts to show a direct relationship between serum melatonin and pubertal progression. On the one hand Silman *et al.*, (1979) demonstrated a rather dramatic fall in melatonin from early to mid-puberty in boys, although not in girls, in daytime blood samples. By contrast, Lenko *et*

Table I. Mean Concentrations of IGF-I (U/ml) in Boys and Girls According to Age Group or Genitalia/Breast Stage[a]

Age group (years)	Boys (N = 369)	Girls (N = 477)	Genitalia/breast stage	Boys (N = 114)	Girls (N = 277)
0–3	0.3	0.5	—	—	—
3–6	0.5	0.7	1	0.7	1.5
6–11	0.8	1.5	2	1.3	2.2
11–13	1.1	2.6	3	2.6	2.6
13–15	2.1	2.6	4	2.3	2.3
15–18	1.8	1.8	5	2.1	2.4

[a]Confidence limits may be found in Underwood *et al.* (1984).

al., (1981) found no such relationship. More recently, Waldhauser *et al.* (1984) demonstrated a significant fall in serum melatonin in nighttime samples but not in daytime samples. Clearly this debate will continue for some time. At this stage, it seems at least possible that the pineal gland has a role in the onset and progression of puberty, although it remains unclear as to whether this is a direct regulatory function or is just an associated epiphenomenon.

4. Interrelationships between Auxological and Endocrinological Events

Apart from relating different hormonal levels with the successive stages of puberty there has generally been little attempt to relate hormonal events with auxological ones; this is because of a lack of adequate data. Unfortunately, growth studies, by and large, have concentrated on auxological variables and hormonal studies on hormonal ones. Thus, within any given study, which by definition needs to be longitudinal, there is often a lack of the variables necessary for making detailed comments. Further, we now realize that studies of this type are often inadequate to the job because of the problems of pulsatile hormone secretion.

Recently, Preece *et al.* (1984) published data for boys in the Chard longitudinal study showing a more even distribution between auxological and endocrine variables. Only data for boys are available, but some interesting speculations can be made particularly relating to the adolescent growth spurt. Greater insight into the relationships between the skeletal events at puberty and various hormonal changes was seen when the hormone concentrations were related to peak height velocity age (i.e., years before and after peak height velocity for each individual) rather than to chronological age. This is shown in Figure 6. It can be seen that maximal changes in serum testosterone occurred during the year encompassing the age of attainment of peak height velocity.

When the relationship between serum testosterone and height velocity itself was investigated, it was possible to draw further conclusions. An investigation of the regression of height velocity on serum testosterone concentration in boys grouped according to the number of years before or after peak height velocity demonstrated a quite striking pattern. In boys from 2 to 0 years before peak height velocity, there was a significant positive

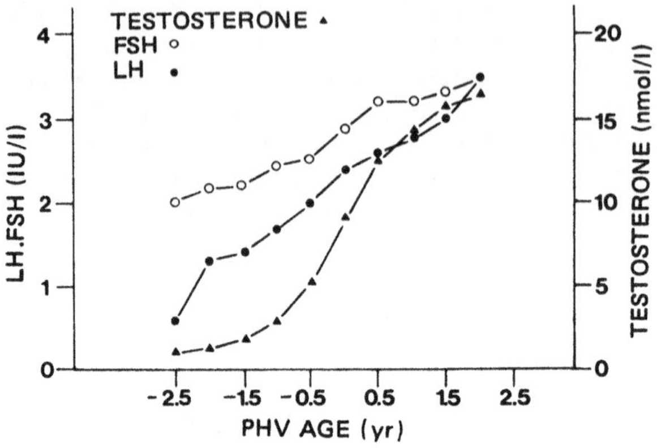

Fig. 6. Mean gonadotrophin and testosterone values for boys, plotted by years before or after the age of peak height velocity. (From Preece *et al.*, 1984.)

regression of velocity on testosterone concentration. By contrast, in the year following the attainment of peak height velocity, the regression became significantly negative and remained so, at least until 3 years after peak height velocity. Thus, there seems to be a gradual change from early puberty where there is a significant positive correlation between serum testosterone concentration and height velocity to a negative correlation in the later stages. This would be compatible with the principal effect of testosterone in early puberty being to stimulate cartilage growth but in later puberty to rapidly advance epiphyseal fusion. Although by no means conclusive, this information certainly adds credence to the belief that it is testosterone secretion that has a dominant effect on the adolescent growth spurt in boys.

With the observation that GH and IGF secretion increase remarkably at puberty, it does seem possible that these hormones may have an important role in the causation of the adolescent growth spurt. The maximum increase in IGF concentrations in boys occurs at a stage of puberty appropriate for the adolescent growth spurt, about G3 or G4. In girls, however, it does not—IGF values are at a peak also at stages B3 and B4 (Underwood *et al.*, 1984), which is rather late for the adolescent growth spurt in girls (Tanner, 1962).

Bioassayable IGFs studied in relationship to years before and after peak height velocity (Preece *et al.*, 1984) showed the peak of IGF activity to be at the age of peak height velocity in boys, but this was followed by substantial between-individual variation in the next year or so. In some boys, peak

IGF concentrations were not attained until well after the peak height velocity. Rosenfeld *et al.* (1983), using a radioimmunoassay for IGF-I, found that peak IGF values were often not seen until a year or so following peak height velocity age; this was true of both sexes. It would seem inappropriate to propose the increase in IGF concentrations as causal to the growth spurt, and in many ways it seems equally likely that the changes in IGF are secondary to the other hormonal changes, in parallel with the growth spurt to some extent.

There is striking between-individual variation in the rate of progression through puberty (see Chapter 8, this volume). The Chard data allow some exploration of this phenomenon. Although there were no strong correlations between the time taken for boys to progress through the genitalia stages and hormonal levels at the start, middle, or end of puberty, there was a weak, albeit statistically significant, negative correlation between testosterone concentration at the beginning of puberty and time of transit. Thus higher testosterone concentrations at change from G1 to G2 were associated with faster progression through puberty (Preece *et al.*, 1984).

5. Conclusions

There is still much that we do not know about the changes in the endocrine system through childhood and adolescence. There do seem to be two important groups of hormonal systems. On the one hand there are those like the mechanisms controlling glucocorticoid or thyroid hormone secretion, where there is little real change with maturation, although if either system is disturbed there may be profound effects on normal growth. The second group are those particularly involving the reproductive system, in which there are clear important causal relationships between changes in hormonal secretion and the whole maturation process. Without these hormones, normal growth and development will not occur, nor will the normal changes of the progression from childhood to adult develop.

Because of the pulsatile nature of the secretion of most hormones relevant to growth and development, data such as presented here have limited value but do give useful indications as to the interplay of some hormones during puberty. There is, however, another level of organization and control into which we have little insight—the target cells themselves. Whether this could function by alterations in receptor function or intracellular changes in response to hormonal stimulation is not clear. Perhaps a combination of the two mechanisms is operative. The changing response of the growth plate to testosterone may well be an example of this mechanism, but from the data available it is not possible to decide whether the change from accelerated growth to maturation results from receptor modulation of from quite distinct intracellular changes.

6. References

Anderson, D. C., 1980, The adrenal androgen-stimulating hormone does not exist, *Lancet* **2:** 454–456.

Apter, D., 1980, Serum steroids and pituitary hormones in female puberty: A partly longitudinal study, *Clin. Endocrinol.* **12:**107–120.

Apter, D., and Vihko, R., 1977, Serum pregnenolone, progesterone, 17-hydroxyprogesterone, testosterone and 5-alpha-dihydrotestosterone during female puberty, *J. Clin. Endocrinol. Metab.* **45:**1039–1048.

Apter, D., Pakarinen, A., and Vihko, R., 1978*a,* Serum prolactin, FSH and LH during puberty in girls and boys, *Acta Paediatr. Scand.* **67:**417–423.

Apter, D., Viinikka, L., and Vihko, R., 1978*b,* Hormonal pattern of adolescent menstrual cycles, *J. Clin. Endocrinol. Metab.* **47:**944–954.

Apter, D., Pakarinen, A., and Hammond, G. L., *et al.,* 1979, Adrenocortical function in puberty, *Acta Paediatr. Scand.* **68:**599–604.

Avruskin, T. W., Tang, S. C., Shenkman, L., Mitsuma, T., and Hollander, C. S., 1973, Serum triiodothyronine concentrations in infancy, childhood, adolescence and pediatric thyroid disorders, *J. Clin. Endocrinol. Metab.* **37:**235–237.

Boyer, R. M., Rosenfeld, R. S., Kapen, S., *et al.,* 1974, Simultaneous augmented secretion of luteinising hormone and testosterone during sleep, *J. Clin. Invest.* **54:**609–618.

Boyar, R. M., Wu, R. H. K., Roffwatg, H., Kapen, S., Weitzman, E. D., Hellman, L., and Finkelstein, J. W., 1976, Human puberty—24 hour estradiol patterns in prepubertal girls, *J. Clin. Endocrinol. Metab.* **43:**1418–1421.

Brazeau, P., Vale, W., Burgus, R., Ling, N., Butcher, M., Rivier, J., and Guillemin, R., 1973, Hypothalamic polypeptide that inhibits the secretion of immunocreactive pituitary growth hormone, *Science* **179:**77–79.

Costin, G., Kogut, M. D., Phillips, L. S., and Daughaday, W. H., 1976, Craniopharyngioma: The role of insulin in promoting post-operative growth, *J. Clin. Endocrinol. Metab.* **42:**370–379.

Curry, D. L., and Bennett, L. I., 1973, Dynamics of insulin release by perfused rat pancreases: Effect of hypophysectomy, growth hormone, adrenocorticotrophic hormone and hydrocortisone, *Endocrinology* **93**:602–609.

Ducharme, J. R., Forest, M. G., Peretti, E. de, Sempé, M., Collu, R., and Bertrand, J., 1976, Plasma adrenal and gonadal sex steroids in human pubertal development, *J. Clin. Endocrinol. Metabl.* **42**:468–476.

Faiman, C., and Winter, J. S. D., 1974, Gonadotrophins and sex hormone patterns in pubescence: Clinical data, in: *Control of the Onset of Puberty* (M. M. Grumbach, G. D. Grave, and F. E. Meyer, eds.), pp. 32–55, Wiley, New York.

Finkelstein, J. W., Roffwarg, H. P., Boyar, R. M., Kream, J., and Hellman, L., 1972, Age-related change in the 24-hour spontaneous secretion of growth hormone, *J. Clin. Endocrinol. Metab.* **35**:665–670.

Fisher, D. A., Sack, J., Oddie, T. H., Pekary, A. E., Hershman, J. M., Lam, R. W., and Parslow, M. E., 1977, Serum T4, TBG, T3-uptake, reverse T3 and TSH concentrations in children 1–15 years of age, *J. Clin. Endocrinol.* **45**:191–198.

Forest, M. G., Peretti, E., de, and Bertrand, J., 1979, Developmental patterns of the plasma levels of testosterone, Δ^4-androstenedione, 17-alphahydroxyprogesterone, dehydroepiandrosterone and its sulfate in normal infants and prepubertal children, in: *The Endocrine Function of the Human Adrenal Cortex* (M. Serio, ed.), pp. 561–582, Academic Press, New York.

Franchimont, P., Henderson, K., Verhoeven, G., Hazee-Hagelstein, M-T., Charlet-Renard, C., Demoulin, A., Bourguignon, J-P., and Lecomte-Yerna, M-J., 1981, Inhibin: Mechanisms of action and secretion, in: *Intragonadal Regulation of Reproduction* (P. Franchimont and C. P. Channing, eds.), pp. 167–191, Academic Press, New York.

Franks, S., 1983, Prolactin, in: *Hormones in Blood* 3rd ed. (C. H. Gray and V. H. T. James, eds.), Vol. IV, pp. 109–136, Academic Press, New York.

Furlanetto, R. W., 1980, The somatomedin-C binding protein—Evidence for a heterologous subunit structure, *Clin. Endocrinol. Metab.* **51**:12–19.

Gallagher, T. F., Yoshida, K., Roffwarg, H. D., Fukushima, D. K., Weitzman, E. D., and Hellman, L., 1973, ACTH and cortisol secretory patterns in man, *J. Clin. Endocrinol. Metab.* **36**:1058–1068.

Gillies, G. E., and Lowry, P. J., 1982, in: *Frontiers in Neuroendocrinology* (W. F. Ganong and L. Martinit, eds.), Vol. 7 pp. 45–74, Raven Press, New York.

Greenberg, A. H., Najjar, S., and Blizzard, R. M., 1974, Effects of thyroid hormone on growth, differentiation and development, in: *Handbook of Physiology,* Section 7: *Endocrinology,* Vol. 3: Thyroid (M. A. Greer and D. H. Solomon, eds.), pp. 377–389, Williams & Wilkins, Baltimore.

Grossman, A., Kruseman, A. C., Perry, L., Tomkin, S., Schally, A. V., Coy, D. H., Rees, L. H., Comaru-Schally, A-M., and Besser, G. M., 1982, New hypothalamic hormone, corticotropin-releasing factor, specifically stimulates the release of adrenocorticotropic hormone and cortisol in man, *Lancet* **1**:921–922.

Gupta, D., McCafferty, E. M., and Rager, K., 1972, Plasma 5-alpha-dihydrotestosterone in adolescent males at different stages of sexual maturation, *Steroids* **19**:411–431.

Gupta, D., Attanasio, A., and Raaf, S., 1975, Plasma estrogen and androgen concentrations in children during adolescence, *J. Clin. Endocrinol. Metab.* **40**:636–643.

Hall, K., Sara, V. R., Enberg, G., and Ritzen, E. M., 1981, Somatomedins and postnatal growth, in: *Biology of Normal Human Growth* (M. Ritzen *et al.,* eds.), pp. 275–283, Raven Press, New York.

Ilondo, M. M., Vanderschueren-Lodeweyckx, M., Vlietinck, R., Pizarro, M., Malvaux, P., Aggermont, E., and Eeckels, R., 1982, Plasma androgens in children and adolescents, *Horm. Res.* **16**:61–77.

Ishimaru, T., Edmiston, W. A., Pages, L., and Horton, R., 1978, Splanchnic extraction and conversion of testosterone and dihydrotestosterone in Man, *J. Clin. Endocrinol. Metab.* **46**:528–533.

Kelch, R. P., Jenner, M. R., Weinstein, R., Kaplan, S. L., and Grumbach, M. M., 1972, Estradiol and testosterone secretion by human, simian and canine testes, in males with hypogonadism and in male pseudohermaphrodites with feminizing testis syndrome, *J. Clin. Invest.* **51**:824–830.

Klenar, C. J. H., and Brook, C. G. D., 1983, A mixed longitudinal study of adrenal steroid excretion in childhood and the mechanism of adrenarche, *Clin. Endocrinol.* **19**:117–129.

Kostyo, J. L., and Isaksson, O., 1977, Growth hormone and the regulation of somatic growth, *Int. Rev. Physiol.* **13**:255–274.

Large, D. M., and Anderson, D. C., 1979, Twenty-four hour profiles of circulating androgens and oestrogens in male puberty with and without gynaecomastia, *Clin, Endocrinol.* **11**:505–521.

Lee, P. A., Plotnick, L. P., Steele, R. E., Thompson, R. G., and Blizzard, R. M., 1976*a,* Integrated concentrations of luteinising hormone and puberty, *J. Clin. Endocrinol. Metab.* **43**:168–172.

Lee, P. A., Xenakis, T., Winer, J., 1976*b,* Puberty in girls: Correlation of serum levels of gonadotropins, prolactin, androgens, estrogens and progestins with physical changes, *J. Clin. Endocrinol. Metab.* **43**:775–784.

Lee, P. A., Plotnick, L. P., Migeon, C. J., and Kowarski, A. A., 1978, Integrated concentrations of follicle stimulating hormone and puberty, *J. Clin. Endocrinol. Metab.* **46**:488–490.

Lenko, H. L., Laing, U., Aubert, M. L., Paunier, L., and Sizonenko, P. C., 1981, Hormonal changes in puberty. VII. Lack of variation of daytime melatonin, *J. Clin. Endocrinol. Metab.* **54**:1056–1058.

Manchester, K. L., 1972, The effect of insulin on protein synthesis in diabetes, *Diabetes* **21**:447–452.

Milman, A. E., de Moor, P., and Luckens, F. D. W., 1951, Relation of purified pituitary growth hormone and insulin in regulation of nitrogen balance, *Am. J. Physiol.* **166**:354–363.

Nagarajan, L., and Anderson, W. B., 1982, Insulin promotes the growth of F9 embryonal carcinoma cells apparently by acting through its own receptor, *Biochem. Biophys, Res. Commun.* **106**:974–980.

Pakarinen, A., Hammond, G. L., and Vihko, R., 1979, Serum pregnenolone, progesterone, 17 alpha-hydroxyprogesterone, androstenedione, testosterone, 5-alpha-dihydrotestosterone and androsterone during puberty in boys, *Clin. Endocrinol.* **11**:465–474.

Parker, L. N., and Odell, W. D., 1979, Evidence for existence of cortical-androgen stimulating hormone, *Am. J. Physiol* **236**:616–620.

Parker, D. C., Jadd, H. J., Rossman, L. G., and Yen, S. S. C., 1975, Pubertal sleep–wake patterns of episodic LH, FSH and testosterone release in twin bodys, *J. Clin. Endocrinol. Metab.* **40**:1099–1109.

Parker, L. N., Lifrak, E. T., and Odell, W. D., 1983, A 60,000 moleculare weight human pituitary glycopeptide stimulates adrenal androgen secretion, *Endocrinology* **113**:2092–2096.

Parra, A., Villalpando, S., Junco, E., Urquieta, B., Alatorre, S., and Garcia Bulnes, G., 1980, Thyroid gland function during childhood and adolescence, changes in serum TSH, T4, T3, thyroxine binding globulin, reverse T3 and free T4 and T3 concentrations, *Acta Endocrinol.* **93**:306–314.

Prader, A., Tanner, J. M., and von Harnack, G. A., 1963, Catch-up growth following illness or starvation, *J. Pediatr.* **62**:646–659.

Preece, M. A., 1976, The effect of administered corticosteroids on the growth of children, *Br. Postgrad. Med. J.* **52**:625–630.

Preece, M. A., and Holder, A. T., 1982, The somatomedins: A family of serum growth factors, in: *Recent Advances in Endocrinology and Metabolism* (J. L. H. O'-Riordan, ed.), Vol. 2, pp. 47–72, Churchill Livingstone, Edinburgh.

Preece, M. A., Cameron, N., Baines Preece, J. C., and Silman, R., 1980, Auxological and serum hormonal changes during puberty, in: *Pathophysiology of Puberty* (E. Cacciari and A. Prader, eds.), Vol. 36, pp. 79–88, Academic Press, London.

Preece, M. A., Cameron, N., Donmall, M. C., Dunger, D. B., Holder, A. T., Baines Preece, J. C., Seth, J., Sharp, G. M., and Taylor, A. M., 1984, The endocrinology of male puberty, in: *Proceedings of the IIIrd Congress of Auxology, Brussels, 1982* (J. Borms, R. Hauspie, A, Sand, C. Susanne, and M. Hebbelinck, eds.), pp. 23–37, Plenum, New York.

Radfar, N., Ansusingha K., and Kenny, F. M., 1976, Circulating bound and free estradiol and estrone during normal growth and development and in premature thelarche and isosexual precocity, *J. Pediatr.* **89**:719–723.

Rees, L. H., and Lowry, P. J., 1983, Adrenocorticotrophin and lipotrophin, in: *Hormones in Blood*, 3rd ed., (C. H. Gray and V. H. T. James, eds.), Vol. IV, pp. 279–284, Academic Press, New York.

Reiter, E. O., and Root, A. W., 1977, Effect of an infusion of Gn-RH upon levels of sex hormones in prepubertal and pubertal girls: Evidence for relative ovarian insensitivity, *Steroids* **30**:61–69.

Rich, B. H., Rosenfield, R. L., Lucky, A. W., Helke, J. C, and Otto, P., 1981, Adrenarche: changing adrenal response to adrenocorticotropin, *J. Clin. Endocrinol. Metab.* **52**:1129–1136.

Robbins, J., and Rall, J. E., 1983, The iodine containing hormones, in: *Hormones in Blood*, 3rd ed. (C. H. Gray and V. H. T. James, eds.), Vol. IV, pp. 219–265, Academic Press, New York.

Rosenfeld, R. I., Furlanetto, R., and Bock, D., 1983, Relationship of somatomedin-C concentrations to pubertal changes, *J. Pediatr.* **103**:723–728.

Sack, J., Bar-On, Z., Shemesh, J., and Becker, R., 1982, Serum T4, T3 and TBG concentrations during puberty in males, *Eur. J. Pediatr.* **138**:136–137.

Savage, D. C. L., Forsyth, C. C., McCafferty, E., and Cameron, J., 1975, The excretion of individual adrenocortical steroids during normal childhood and adolescence, *Acta Endocrinol. (Kbh.)* **79**:551–567.

Schally, A. V., Redding, T. W., Bowers, C. V., and Barrett, J. F., 1969, Isolation and properties of porcine thyrotropin-releasing hormone, *J. Biol. Chem.* **244**:4077–4088.

Schally, A. V., Arimura, A., Baba, Y., Nair, R., Matsuo, H., Redding, T. W., Debeljuk, L., and White, W. F., 1971, Isolation and properties of the FSH and LH-releasing hormone, *Biochem. Biophys. Res. Commun.* **43**:393–399.

Schwartz, A. G., and Amos, H., 1968, Insulin dependence of cells in primary culture: Influence on ribosome integrity, *Nature (London)* **219**:1366–1367.

Silberman, H., 1983, Hormone and cartilage, in: *Cartilage* (B. K. Hall, ed.), pp. 327–368. Academic Press, New York.

Silman, R. E., Leone, R. M., Hooper, R. J. L., and Preece, M. A., 1979, Melatonin, the pineal gland and human puberty, *Nature (London)* **282**:301–302.

Sizonenko, P. C., Paunier, L., and Carmignac, D., 1976, Hormonal changes during puberty. IV: Longitudinal study of adrenal androgen secretions, *Horm. Res.* **7**:288–302.

Smith, S. R., Bledsoe, T., and Chhetri, M. K., 1975, Cortisol metabolism and the pituitary-adrenal axis in adults with protein-calorie malnutrition, *J. Clin. Endocrinol. Metab.* **40**:43–52.

Spiess, J., Rivier, J., Thorner, M., and Vale, W., 1982, Sequence analysis of a growth hormone releasing factor from a human pancreatic islet tumor, *Biochemistry* **21**:6037–6040.

Tanner, J. M., 1962, *Growth at Adolescence*, Blackwell, Oxford.

Tanner, J. M., 1981, Endocrinology of puberty, in: *Clinical Paediatric Endocrinology* (C. G. D. Brook, ed.), pp. 207–223, Blackwell, Oxford.

Tanner, J. M., and Gupta, D., 1968, A longitudinal study of the excretion of individual steroids in children from 8 to 12 years, *J. Endocrinol.* **4**:139–156.

Tanner, J. M., Whitehouse, R. H., Jackson, D., and Gupta, D., 1978, A longitudinal study of 24-hour urinary excretion rates of LH, FSH, sex steroids, creatinine and hydroxyproline throughout puberty, *Proceedings of the 17th Annual Meeting of the European Society of Paediatric Endocrinology, Athens, 1978.*

Underwood, L. E., Smith, E. P., Van Wyk, J. J., Clemmons, D. R., D'Ercole, A. J., Pandian, M. R., Preece, M. A., and Moore, W. V., 1986, Somatomedin-C/insulin-like Growth factor I: Regulation and clinical applications, in: *Human Growth Hormone* (S. Raiti, ed.), Plenum Press, New York (in press).

Waldhauser, F., and Wurtman, R. J., 1983, The secretion and actions of melatonin, in: *Biochemical Actions of Hormones* (G. Litwack, ed.), pp. 187–225, Academic Press, New York.

Waldhauser, F., Weiszenbacher, G., Frisch, H., Zeitalhuber, U., Waldhauser, N., and Wurtman, R. J., 1984, Fall in nocturnal serum melatonin during prepuberty and pubescence, *Lancet* **1**:362–365.

Winter, J. S. D., and Faiman, C., 1973, The development of cyclic pituitary gonadal function in adolescent females, *J. Clin. Endocrinol. Metab.* **37**:714–718.

Wurtman, R. J., Axelrod, J., and Chu, E. W., 1963, Melatonin, a pineal substance: Effect on the rat ovary, *Science* **141**:277–278.

Zumoff, B., Fukushima, D. K., Weitzman, E. D., Kream, J., and Hellman, L., 1974, The sex difference in plasma cortisol concentration in man, *J. Clin. Endocrinol. Metab.* **39**:805–808.

Zumoff, B., Bradlow, H. L., Finkelstein, J., Boyar, R. M., and Hellamn, L., 1976, The influence of age and sex on the matabolism of testosterone, *J. Clin. Endocrinol. Metab.* **42**:703–706.

10

Neuroendocrine Control of the Onset of Puberty

MARGARET E. WIERMAN and WILLIAM F. CROWLEY, JR.

1. Introduction

Sexual maturation has been variously defined by biologists in the past, but all definitions have included the orderly sequence of maturational events culminating in reproductive competence. Despite the relatively uniform agreement as to the essential components of this process, so critical to the propagation of the species, there is little understanding of its regulation and even less of its control. What is clear is that this development stage, termed puberty in the human, is composed of at least three distinct beiological processes: gonadal maturation, adrenal maturation, and a somatic growth spurt.

Gonadal maturation is by far the most important of these components, since it orchestrates the appearance of secondary sexual characteristics *via* steroidogenesis and ensures the ability to reproduce *via* gametogenesis. Recent studies have demonstrated, however, that gonadarche, the beginning of gonadal maturation, represents the end point of the prior development of a highly integrated neuroendocrine system with a fascinating and complex ontogeny of its own. Adrenarche, as first described by Albright *et al.* (1942), refers to a comparable but generally antecedent development of adrenal androgen secretion. While it shares several features with gonadarche, even less is understood about its regulation. The somatic growth spurt accompanying these two processes appears to be a direct effect of the superimposition of gonadal and adrenal sex steroids on the individual's inherent growth program (see Chapter 9, this volume).

It is important to view sexual maturation not as an isolated event, but rather as a developmental continuum, the initiation of which appears to reside within the CNS. Puberty is thus the clinical translation of this highly integrated program of interactions among extrahypothalamic centers, the hypothalamus, the anterior pituitary gland, and the peripheral sex hormone-producing target organs, the gonads and adrenals.

This review accordingly focuses on the neuroendocrine aspects of these processes, as recent evidence suggests that the key to the initiation and regulation of puberty resides within the CNS. As few species other than the primate are characterized by a long period of latency followed by such a discrete period of development before achieving reproductive competence, the primary focus of this chapter is the pubertal process in the primate and, whenever possible, the human.

2. Models for the Study of Puberty

Much of the insight into the pubertal process has been derived from the study of sexual development in one of three circumstances. Chronicling the events of a normal puberty in the human or in the primate (Boyar *et al.*, 1972; Weitzman *et al.*, 1976; Kulin *et al.*, 1969; Faiman and Winter, 1974; Bercu *et al.*, 1983*a,b*) has been useful in determining the

MARGARET E. WIERMAN and WILLIAM F. CROWLEY, JR. • Reproductive Endocrine Unit, Massachusetts General Hospital and Harvard Medical School, Boston, Massachusetts 02114.

ontogeny of each of the components of the hypothalamic–pituitary–gonadal axis during adolescence. Several constraints on such studies, however, have limited the number of these observations in normals. On the other hand, in precocious (Boyar *et al.*, 1973; Crowley *et al.*, 1981; Matthews *et al.*, 1982) or delayed (Savage *et al.*, 1981) puberty, a more detailed evaluation is often warranted, and thus considerable information has been forthcoming from the study of these "experiments of nature." The most abundant and useful data regarding the control of puberty, however, have resulted from observations in which an experimental manipulation of the hypothalamic–pituitary–gonadal axis has been possible. The study of various circumstances in which a component of the reproductive system has been ablated, either experimentally or by the presence of a congenital defect, may be combined with replacement of the missing component to permit the testing of specific hypotheses relating to regulation of the intact system. Moreover, as the locus of such defects can occur either at the neuroendocrine level (as in various models of GnRH deficiency) or at the gonadal level (as in gonadal dysgenesis or castration), the importance of each component in the regulation of puberty may be examined.

Each of these models has contributed significantly to the concept that reproductive competence is the final product of a highly integrated system of controls initiated by the hypothalamic secretion of the gonadotrophin-releasing hormone (GnRH), which in turn effects synthesis in and release from the anterior pituitary of both gonadotrophins, luteinizing hormone (LH) and follicle-stimulating hormone (FSH) (cf. Figure 1). These two glycoprotein hormones then functionally bifurcate the gonad into a compartment responsible for steroidogenesis which is under the control of LH, and a portion responsible for maturation of the gamete, regulated by FSH. Since puberty is but the final step of this developing process of sexual development, a complete understanding of its integrity requires an examination of the ontogeny of the system from fetal life onward.

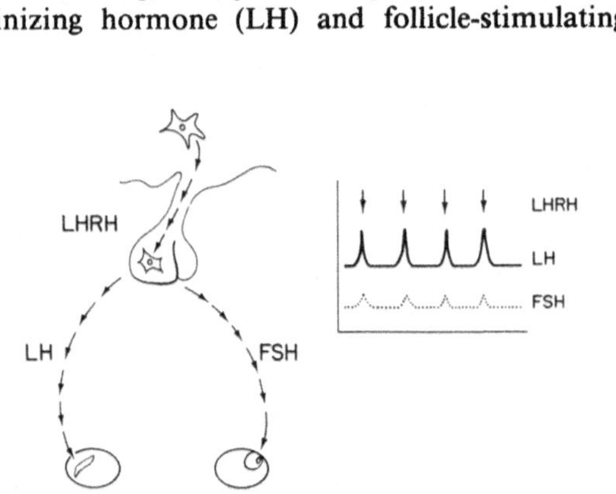

Fig. 1. Schematic diagram outlining the current understanding of the hypothalamic-pituitary-gonadal axis during reproductive maturation. FSH, follicle-stimulating hormone; LH, luteinizing hormone; LHRH, luteinizing hormone-release hormone.

3. Ontogeny of the Hypothalamic-Pituitary-Gonadal Axis

3.1. In Utero Development

Although radioimmunoassays have indicated the presence of GnRH in crude extracts of the brain as early as 4.5 weeks gestation (Winter *et al.*, 1974), its localization to neurons within the hypothalamus has been first noted by immunofluorescence by the 10–14th week (Kaplan *et al.*, 1976; Bugnon *et al.*, 1977). The hypothalamic portion of the neuroendocrine axis is partially functional by the fifth month of gestation as indicated by the ability of human hypothalamic fragments to release GnRH *in vitro* following administration of the opiate receptor antagonist, naloxone (Rasmussen *et al.*, 1983). This study not only demonstrates the functional integrity of the GnRH-secreting neurons within the fetal hypothalamus, but also suggests that at least some of the neuroendocrine modulators of GnRH secretion that are evident following reproductive maturation are in place as early as the 21st week of development.

The hypophyseal-portal system undergoes refinement from day 60 to 100 (Thliveris and Currie, 1980) coincident with the presence of immunocytochemical evidence of gonadotrophins within the anterior pituitary by 10 weeks gestation (Kaplan *et al.*, 1976). *In vitro* studies have demonstrated that these gonadotrophs also release LH and FSH following exposure to GnRH, once again illustrating the potential functional integrity of the neuroendocrine axis by mid-gestation (Takagi *et al.*, 1977; Clements *et al.*, 1976; Siler-Khodr *et al.*, 1974; Kaplan *et al.*, 1976). At this point an interesting and recurrent theme of sexual dimorphism of pituitary gonadotrophin physiology first emerges. Extensive evaluation of pituitary gonadotrophin content as well as direct measurements in fetal blood (cf. Figure 2) have documented that there is an earlier

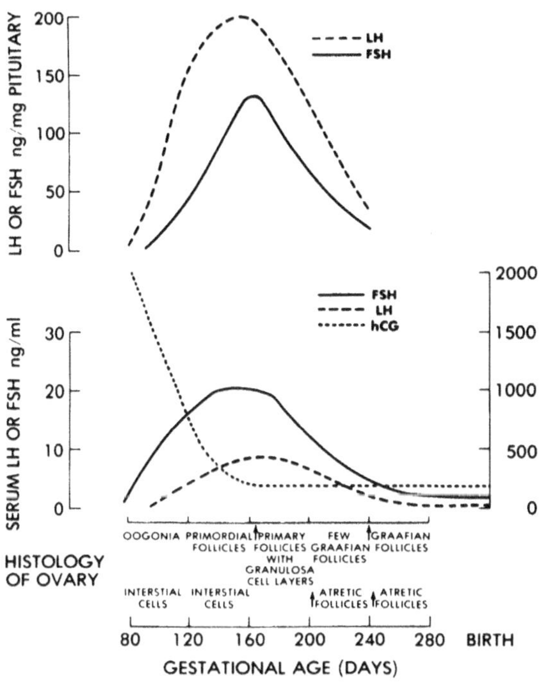

Fig. 2. Patterns of the pituitary content of LH and FSH (top) and serum levels of LH, FSH, and hCG (bottom) correlated with developmental histology of the fetal ovary, across gestation. (From Kaplan SL and Grumback, 1978.)

peak of both gonadotrophins (by 20 weeks) and a marked difference of FSH content and concentration in female fetuses as compared with males of the same age (Kaplan *et al.*, 1976, 1978; Siler-Khodr and Khodr, 1980). The pituitary LH:FSH ratio of 50:1 in males remains relatively constant throughout gestation, whereas this ratio decreases to 2:1 by 24 weeks in the female. Such a change is attributable to a relative increase in FSH content of the pituitary in the female and foreshadows a recurrent pattern of a more abundant FSH profile in females of all ages.

At the level of the gonad, sexual differentiation has been initiated by 10–12 weeks, and external genital development is nearly complete in the male by the end of the first trimester (DiZerega and Ross, 1980; Siiteri and Wilson, 1974). The fact that these morphological changes in the gonad begin prior to the appearance of significant amounts of pituitary gonadotrophins and that they can also occur in anencephalic fetuses lacking a hypothalamus and gonadotrophins (Kaplan *et al.*, 1976) has suggested that placental hCG, at least in the male, may well play a primary role in coordinating germ cell migration and Leydig cell differentiation. On the other hand, in the female, gonadal development, especially the progression of oocyte maturation *in utero,* appears to parallel the above-mentioned emergence of FSH secretion from the anterior pituitary during gestation (Kaplan *et al.*, 1976). These data suggest that both the placenta and the fetal pituitary are important for reproductive development *in utero.*

Thus, it appears that during the developmental period of gestation, the basic neuroendocrine-gonadal framework that will subsequently guide sexual development in firmly in place. Data concerning its integration is more scarce. However, the close temporal relationships of the appearance of certain components as well as their *in vitro* activities suggest that early communication within the hypothalamic-pituitary-gonadal axis may well occur.

3.2. The Neonatal Period

The immediate postpartum and neonatal period is one of intense activity of the reproductive system and one in which the critical nature of an episodic program of neuroendocrine communication first becomes manifest. GnRH is secreted in a pulsatile fashion from the hypothalamus, as attested by the episodic discharge of both LH and FSH during this period (Huhtaniemi *et al.*, 1979; Bercu *et al.*, 1983*b*; Tapanainen *et al.*, 1982; Waldhauser *et al.*, 1981). These gonadotrophins are biologically active, as demonstrated by the gonadally produced sex steriod secretion also observed during this period (Bercu *et al.*, 1983*a*). Increased levels of bioactive and immunoactive LH, increased radioimmunoassayable FSH levels, and increased responsiveness to exogenous GnRH testing are thus all characteristics of this period of reproductive development.

Waldhauser *et al.* (1981) demonstrated that each of six infants studied was observed to have an episodic pattern of gonadotrophin secretion quite similar in nature to that subsequently observed in the adult (cf. Figure 3). Of note was the fact that male infants secreted LH as the predominant gonadotrophin, whereas the females all displayed an FSH predominance. Studying male infants with cryptorchidism from 3 to 39 days, Tapanainen *et al.* (1982) demonstrated an increasing responsiveness of both gonadotrophins to GnRH testing, up to a peak response evident at 1–3 months postdelivery. There-

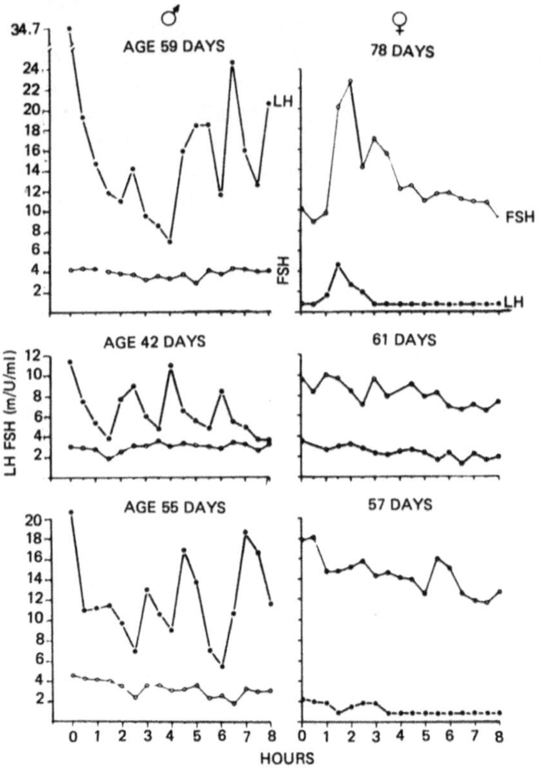

Fig. 3. Serum LH and FSH values sampled at 30-min intervals in three male and three female infants. (From Waldhauser *et al.*, 1981.)

after, this response declined to a minimum at 12 months postpartum. Similar studies in the female (Job *et al.*, 1977) have also noted increased responsiveness of FSH to GnRH until 1 year of age. Taken together, these studies document a high degree of neuroendocrine integration within the first year of life.

Gonadal steroid secretion parallels these changes in gonadotrophins during the neonatal period. In the newborn monkey, elevation of testosterone levels correlates well with the increased bioactive and immunoactive LH levels throughout this period during baseline sampling (Bercu *et al.*, 1982*a*) and following GnRH infusions (Huhtaniemi *et al.*, 1979) (cf. Figure 4). In fact, the episodic fluctuations of testosterone mirror LH pulses quite precisely (Steiner and Bremner, 1981), suggesting a one-to-one translation of each burst of neuroendocrine secretion at the gonadal level. Similar changes in the gonad have been noted in large cross-sectional studies of testosterone levels in male infants, who have low androgen levels on day 1 following delivery quickly followed by a rebound to nearly adult levels by 2 months of age. As in the monkey, the correlation of these gonadal changes with serum LH was quite significant (Forest and Cathiard, 1975; Winter *et al.*, 1975, 1976; Gendrel *et al.*, 1980; Bidlingmaier *et al.*, 1983).

These observations indicate a remarkable degree of functional integration of the CNS, anterior pituitary, and gonads during the neonatal period. In addition, many of the critical features of control of the reproductive system that become apparent again at puberty, such as intermittency of secretion, are initially manifest during this brief period of neonatal activity. The reason for this short but active stage of development of the reproductive axis is unclear. However, it may represent a preliminary period of anatomical and functional organization of the gonads, and other target organs for sex steroids, which may be critical for subsequent orderly development at puberty. The later appearance of CNS defects, such as problems in spatial ability noted in GnRH-deficient subjects who presumably lacked the transient elevation of androgens during this period (Hier and Crowley, 1982), may represent the necessity for this early period of activation of the reproductive axis.

3.3. *Prepubertal Period*

Following this brief period of activity of the reproductive system neonatally, a relatively quiescent stage of childhood ensues. The dampening of neuroendocrine secretion during this period is so complete that several studies have been unable to document any significant evidence of gonadotrophin secretion during this time (Hayes and Johanson, 1972; Boyar *et al.*, 1972; Chipman, 1980). More recent evidence employing urinary gonadotrophin determinations (Kulin *et al.*, 1972), GnRH testing in the prepubertal period (Roth *et al.*, 1973; Reiter *et al.*, 1976), and frequent sampling of serum gonadotrophin levels with more sensitive radioimmunoassays (Parker *et al.*, 1975; Judd *et al.*, 1977; Lee *et al.*, 1978; Jakacki *et al.*, 1982; Ross *et al.*, 1983) have all demonstrated a low but detectable level of activity of the neuroendocrine-gonadal axis during the prepubertal period. Waldhauser *et al.* (1981) sampled hormonal levels in two prepubertal children for 8 hr and observed that their LH levels were suppressed to a greater extent than was FSH in both sexes, implying a relative suppression of hypothalamic secretion of GnRH. Jakacki *et al.* (1982), using frequent sampling for 6–22 hr, documented the presence of discernible pulsatile release of LH in 50% of children with bone ages less than 10 years and in all four subjects with bone ages under 5 years (cf. Figure 5). These data imply a relative suppression of hypothalamic secretion of GnRH during childhood. The hypothalamic–pituitary axis is evidently intact during this period but functions at a considerably reduced level of activity.

The precise cause of this mysterious period of neuroendocrine latency in childhood has been the subject of much speculation. One theory, based on

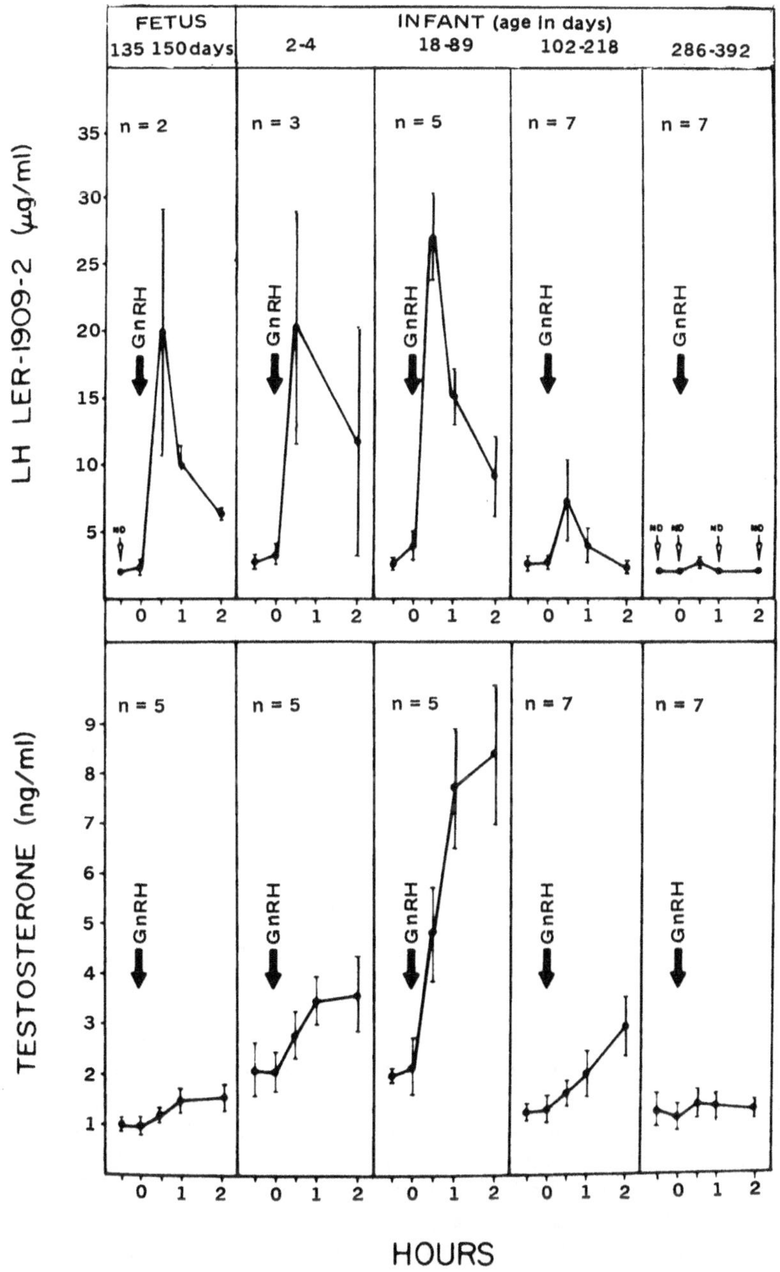

Fig. 4. Mean response (±SEM) of circulating LH and testosterone levels to GnRH infusions in male fetal and infant monkeys. (From Huhtaniemi *et al.*, 1979.)

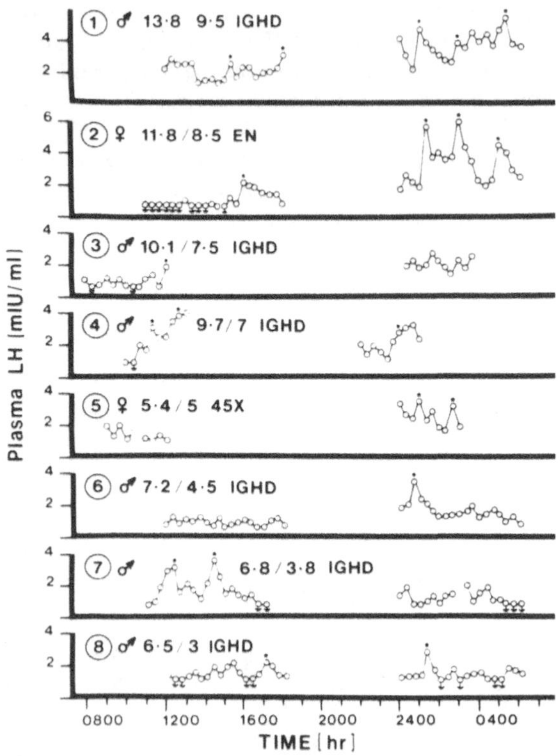

Fig. 5. Evidence of pulsatile LH release in eight prepubertal children. Chronologic age/bone age values are shown in upper left corners of the individual graphs. EN, endocrinologically normal; IGHD, isolated growth hormone deficiency; 45X = Turner syndrome. (From Jakacki *et al.*, 1982.)

experimental data demonstrating a marked suppression of gonadotrophin secretion during sex steroid infusion in prepubertal children, has postulated the existence of a putative "gonadostat" within the CNS. This gonadostat is postulated to be exquisitively sensitive to the low levels of sex steroids secreted during the prepubertal period, so that it is responsible for the suppression of gonadotrophin secretion during childhood (Kulin *et al.*, 1969, 1972). The acquisition of this sensitivity to negative feedback would then mark the transition from the neonatal period to the relative suppression of gonadotrophin secretion in childhood. Conversely, the loss of this feedback sensitivity would account for the reactivation seen at the onset of puberty.

A more recent synthesis of the phenomenology of this period has been derived from two of the previously mentioned models of pubertal development. Study of children whose gonads have either been removed surgically or who have gonadal dysgenesis has permitted the chronicling of the changes in the hypothalamic-pituitary axis throughout childhood and adolescence without interference from gonadal steriods (Conte *et al.*, 1980; Maruca *et al.*, 1983; Ross *et al.*, 1983). These direct observations of the neuroendocrine changes in this open loop system have revealed a qualitatively similar involution of gonadotrophin secretion in agonadal and normal children as they pass from the first 1–2 years of life into the quiescence of the prepubertal years (cf. Figure 6). There is also a rise in gonadotrophins at adolescence in the agonadal children similar to that observed in normal subjects. While the replacement of sex steroids in the agonadal patient may temporarily delay the subsequent pubertal rise of gonadotrophins, eventual emergence of elevated gonadotrophin secretion still occurs. These observations indicate that the key elements of the involution and reemergence of reproductive activity that mark the beginning and end of the period of latency are under the direct control of the CNS and ascribe a relatively minor role to sex steroid modulation.

Other evidence in support of this assertion, together with data of value in further localizing the site of prepubertal quiescence to the arcuate nucleus of the hypothalamus is supplied by studies in Rhesus monkeys. Knobil and colleagues (Wildt *et al.*, 1980) induced full reproductive maturation in arcuate-lesioned animals by instituting a physiologic replacement schedule of pulsatile GnRH administration in these hypogonadotrophic monkeys. When a similar schedule was employed in intact prepubertal female monkeys, gonadotrophin levels rose and there was a prompt institution of folliculogenesis (cf. Figure 7). Interestingly, despite the achievement of estradiol levels that could evoke an LH surge in adults, no mid-cycle gonadotrophin surge was usually observed in the first few cycles. This observation is reminiscent of the anovulatory cycles frequently noted in adolescent girls (Hansen *et al.*, 1975) and suggests a role of pituitary maturation in adolescent development.

These experiments substantiate the findings in agonadal subjects and confirm that the reproductive quiescence that bridges the gap between the neonatal period and the onset of puberty is due to a relative suppression of hypothalamic secretion of GnRH. They also suggest that further understanding of the control of the onset of puberty will be forthcoming only by a careful examination of the modulator(s) of GnRH secretion during the prepubertal years.

4. Adrenarche

It is during this period of relative reproductive silence in childhood that adrenal maturation is initiated. Termed adrenarche by Albright *et al.* (1942), this process refers to the regrowth of the zona reticularis of the adrenal cortex, the zone that was largely responsible for the large size of the fetal adrenal and that regresses markedly following delivery (Migeon, 1981; Jaffe *et al.*, 1981). Older studies followed this process by the appearance of pubic and axillary hair in girls and the increasing amounts of 17-ketosteroids in the urine of both sexes. More recently, serum levels of dehydroepiandrosterone sulfate (DHEAS) have been substituted as a more convenient marker of this process (Sizonenko *et al.*, 1976; Korth-Schutz *et al.*, 1976*b*; Reiter *et al.*, 1977; Ilondo *et al.*, 1982). This steroid is chiefly adrenal (95%) in origin, has a long serum half-life, and exhibits small differences between the two sexes—all features that enhance its clinical usefulness as a marker of adrenal androgen secretion (Korth-Schutz, 1976*a*).

Recent studies of the mechanisms of adrenarche have shown that this process largely reflects the development of the enzymes 17,20-lyase and 17α-hydroxylase within the zona reticularis (Scheibinger *et al.*, 1981; Rich *et al.*, 1981). It is interesting to note that glucocorticoid secretion, when expressed per square millimeter of body surface, remains virtually constant from the neonatal period of senescence, whereas adrenal androgen secretion undergoes a distinctly different ontogeny. Following the involution of the fetal adrenal postpartum, early

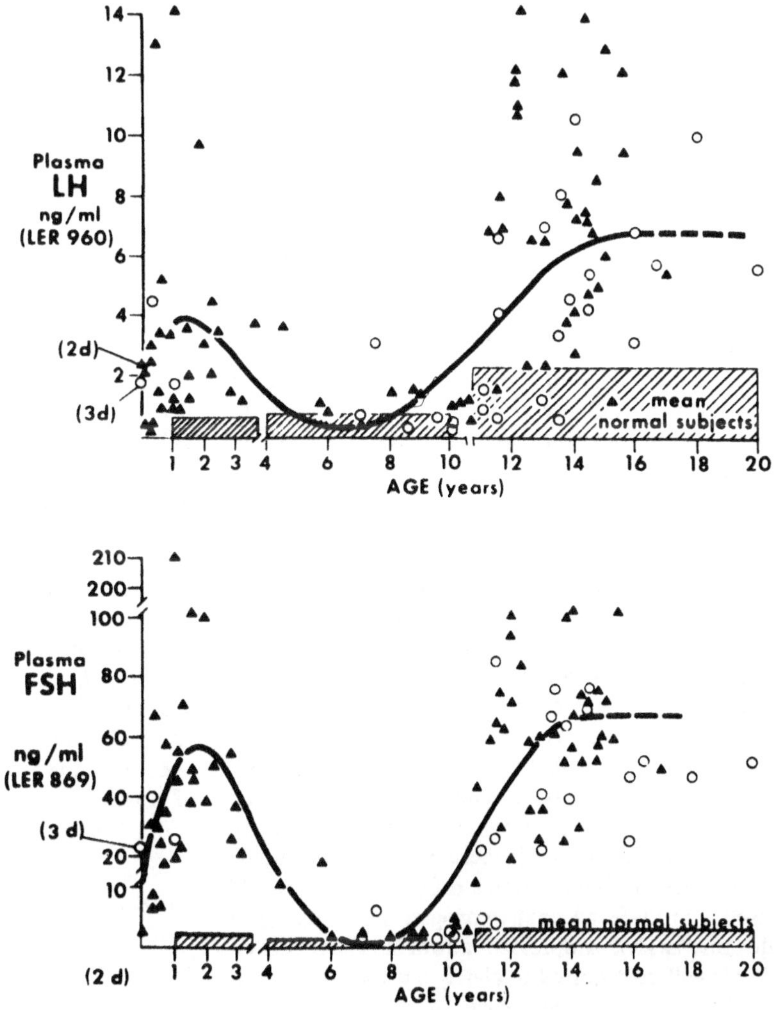

Fig. 6. LH and FSH levels during the neonatal childhood and pubertal years in agonadal children demonstrating the triphasic pattern of secretion of the hypothalamic-pituitary axis when unrestrained by gonadal steroids. (From Conte *et al.*, 1975.)

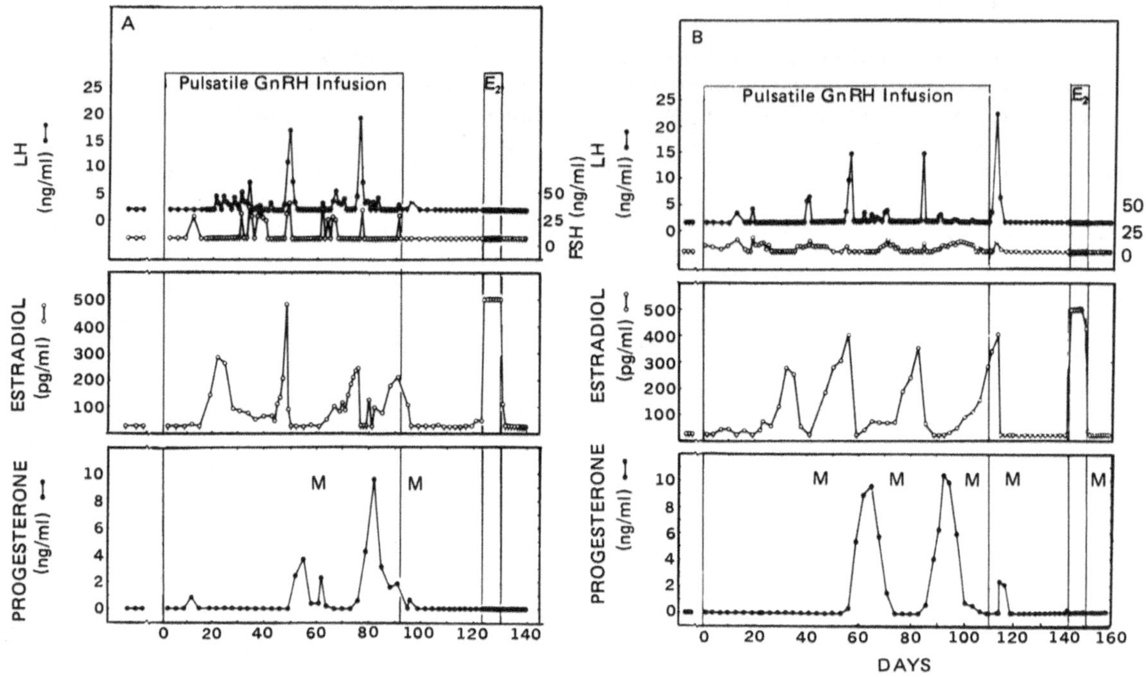

Fig. 7. Induction of ovulatory menstrual cycles in prepubertal monkeys by the administration of pulsatile GnRH. Note the first cycle in each case is anovulatory. (From Wildt *et al.*, 1980.)

childhood is characterized by low levels of adrenal androgens until ages 6–9, when their secretion increases steadily into the third decade. A plateau then occurs until late in the fourth decade, when a second period of suppression is encountered, which is permanent.

Two theories have been postulated to account for the regulatory control of this phenomenon. The first suggests that adrenarche is under the direct control of ACTH but requires years of adrenocortical stimulation to be expressed (Anderson, 1980). Proponents of this theory point to the ability of glucocorticoids to suppress and ACTH to stimulate adrenal androgen secretion following adrenarche. A second theory suggests that a separate adrenal androgen stimulating factor exists within the pituitary which remains to be isolated and characterized (Parker *et al.*, 1983). Proponents of this theory have obtained preliminary bioassay data which indicate that crude pituitary extracts differing from ACTH itself stimulate adrenal androgen secretion from dexamethasone-suppressed animals. Further evidence will be required to clarify such a distinction.

The temporal precedence of adrenarche over gonadarche as well as the observation that patients with uncontrolled congenital adrenal hyperplasia often enter puberty prematurely has stimulated speculation as to whether adrenal androgen secretion might represent the long-sought trigger for sexual maturation. However, several investigators have put forth a convincing case for the dissociation of these two processes (Sklar *et al.*, 1980; Cutler *et al.*, 1980; Smail *et al.*, 1982; Plant and Zorub, 1984). Sklar and colleagues (cf. Figure 8) noted that a normal gonadarche occurs in the absence of prior adrenal androgen exposure in subjects with juvenile Addison's disease and that a premature onset of adrenal androgen secretion does not generally accelerate the timing of puberty. Conversely, the presence, absence, or suppression of gonadarche does not appear to alter the timing or progression of adrenal maturation (Sklar *et al.*, 1980; Wierman *et al.*, 1984).

5. Puberty

5.1. Hypothalamus

The final development of reproductive maturation begins in early adolescence with an amplification of the pulsatile secretion of GnRH from the hypothalamus. There are several lines of evidence showing that this episodic secretion of GnRH is absolutely essential for full expression of reproductive maturation.

Knobil and colleagues (Belchetz *et al.*, 1978) using hypothalamic-lesioned and prepubertal Rhesus monkeys, were able to institute normal functioning of the pituitary gonadal axis only by administering GnRH in an intermittent fashion. Continuous GnRH was ineffective in stimulating pituitary gonadotrophin release (cf. Figure 9). Moreover, when pulsatile administration was changed to a continuous mode of GnRH delivery, suppression of both FSH and LH secretion occurred and could only be restored by reverting to an episodic pattern of stimulation (cf. Figure 10). Similar findings resulted from attempts to induce puberty in GnRH-deficient men, using long-acting agonists of GnRH, which produced a prolonged occupancy of the GnRH receptor on the gonadotroph (Crowley *et al.*, 1982). Suppression of gonadotrophin secretion occurred, which was dependent on the dosage and frequency of the GnRH analogue; it could only be reversed by pulsatile delivery of the shorter-acting natural sequence GnRH. Each of these experimental models, when subsequently exposed to a low dosage of GnRH administraiton at a physiological frequency, was capable of responding both at the pituitary and gonads, in a fashion identical to that seen in normal puberty (Knobil *et al.*, 1980; Hoffman and Crowley, 1982).

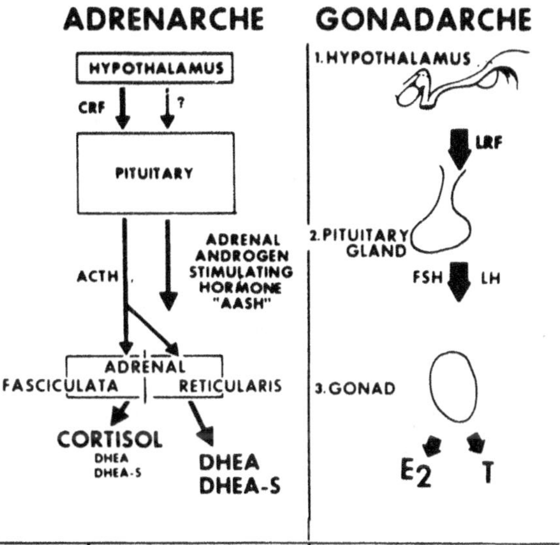

	ADRENARCHE	GONADARCHE
PREMATURE ADRENARCHE	+	−
PRIMARY ADRENAL INSUFFICIENCY	−	+
IDIOPATHIC PRECOCIOUS PUBERTY (ONSET < 6 YRS)	−	+
IDIOPATHIC PRECOCIOUS PUBERTY (ONSET > 6 YRS)	+	+
SYNDROME GONADAL DYSGENESIS	+	−
ISOLATED GONADOTROPIN DEFICIENCY	+	−
CONSTITUTIONAL DELAYED ADOLESCENCE	−	−

Fig. 8. Schematic representation of the interrelationships between adrenarche and gonadarche in various clinical disorders of sexual development. (From Sklar *et al.*, 1980.)

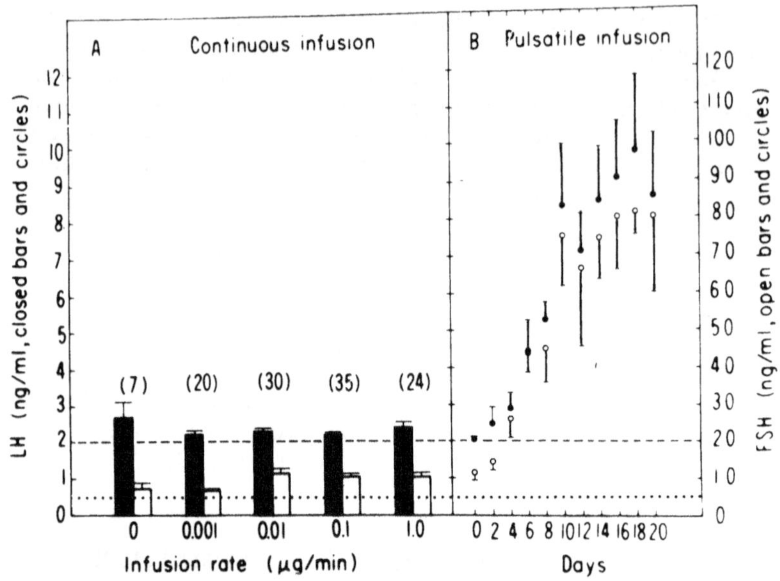

Fig. 9. The failure of continuous GnRH infusion to reestablish gonadotrophin secretion in hypothalamic-lesioned Rhesus monkeys, whereas a pulsatile mode of administration corrects their hypogonadotrophic state. (From Belchetz *et al.*, 1978.)

From these observations, a working model of the reproductive endocrine system has evolved as portrayed in Figure 1. Hypothalamic GnRH secretion, which is episodic in nature, results in the pulsatile discharge of gonadotrophins from the anterior pituitary. These gonadotrophins then initiate gonadarche and complete the process of sexual maturation. A corollary of this hypothesis is that each episode of gonadotrophin release is preceded by, and an accurate reflection of, an antecedent burst of hypothalamic secretion of GnRH. The onset of puberty is then characterized by an activation of the hypothalamic secretion of GnRH. The low amplitude of pulsatile secretion of GnRH characteristic of the prepubertal period gives way to a sleep-entrained activation of GnRH-induced gonadotrophin secretion in early puberty. The amplitude of each pulse increases as early puberty progresses. Whether a change in frequency also occurs is controversial. Several workers (Jakacki *et al.*, 1982; Penny *et al.*, 1977; Corley *et al.*, 1981) observed an increase in the frequency of gonadotrophin pulse during puberty, but Ross *et al.* (1983), studying agonadal subjects, observed an increase in the frequency of LH only, together with a somewhat paradoxial decrease in the frequency of FSH pulses. As puberty progresses, the striking day–night differences become less marked, gonadotrophin pulses

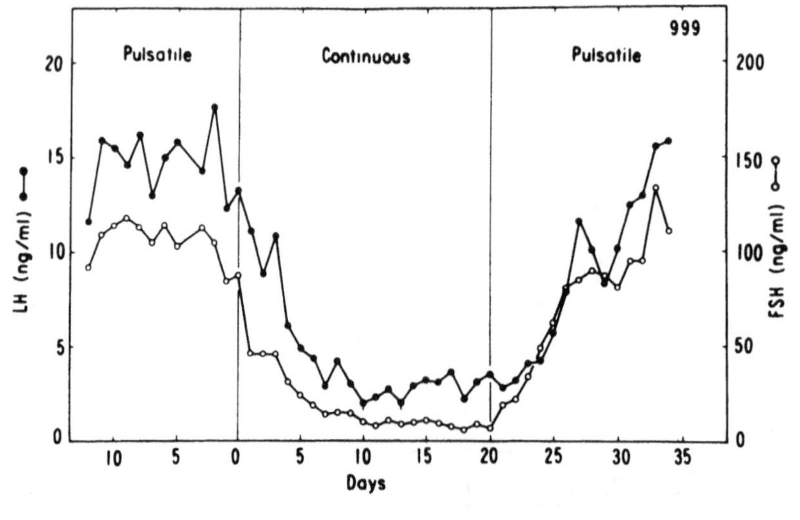

Figure 10. The critical dependency of pituitary gonadotrophin secretion of hypothalamic-lesioned Rhesus monkeys upon the mode of GnRH delivery. Whereas pulsatile GnRH has restored gonadotrophin secretion, continuous delivery ablates LH and FSH release only to have it restored by reversion to a pulsatile delivery pattern. (From Belchetz *et al.*, 1978.)

occur during both day and night, and eventually little difference in pattern of gonadotrophin secretion is noted during the 24-hr period (cf. Figure 11).

5.2. Pituitary

Despite the predominant influence of hypothalamic secretion of GnRH on the initiation of puberty, it is clear that pituitary and gonadal amplification of GnRH secretion have important roles in sexual maturation. While the amplitude and frequency changes in GnRH secretion produce sizable changes in the quantity of gonadotrophins presented to the gonads during puberty, it is clear that alterations in the qualitative nature of the gonadotrophins also occur in cross-sectional studies of puberty. Lucky *et al.* (1980), employing a sensitive *in vitro* Leydig cell bioassay, demonstrated an increase in the biopotency of the LH secreted. This rise in the ratios of biologically to immunologically active (B/I) secreted LH in puberty has been confirmed by others (Reiter *et al.*, 1982). It suggests an active role of the pituitary in controlling the maturational process of puberty.

Along related lines of investigation, changes in the molecular forms of FSH secreted have been noted to occur in several animal species. Several workers have described molecular heterogeneity of FSH that is dependent on both the species studied and the steroid hormone milieu (Chappel *et al.*, 1983 in the rat; Peckham *et al.*, 1973; Peckham and Knobil, 1976, in the monkey; Wide and Lundberg, 1981; Wide, 1983; Wide and Hobson, 1983; Wide and Wide, 1984, in the human). These changes have been ascribed to alterations in the glycosylation of the polypeptide core that may affect the binding and potency of the gonadotrophins toward their gonadal receptors. Further studies are required to characterize these changes more precisely and to correlate them with developmental markers, patterns of GnRH secretion, and gonadal maturation.

A final role for the pituitary is suggested in the development of the steroid hormone feedback mechanisms at puberty (Kulin, *et al.*, 1972; Dierschke *et al.*, 1974; Knobil, 1974, 1980). When estrogens are administered to fully matured females of several species, an initial suppression of gonadotrophin secretion is observed, followed by a subsequent positive surge of both LH and FSH. This biphasic effect of estrogen on the hypothalamic-pituitary axis in sexually mature females represents the highest degree of integration of steroid hormones with the hypothalamus and anterior pituitary. As such, it constitutes the end of a series of maturational events of the neuroendocrine regulation of puberty. In GnRH-deficient animals receiving a fixed regimen of hypothalamic replacement, the role of the pituitary in generating the LH surge can be examined in detail, since all changes observed in this model can be ascribed only to the pituitary secretion and feedback effects of sex steroid hormones. As can be seen in Figure 7, the development of an LH surge in response to a mid-cycle estradiol challenge in these animals did not occur in the initial GnRH-induced cycles. Despite the fact that sufficient FSH was secreted to ensure folliculogenesis with mid-cycle estrogen levels, no LH surge occurred until the second cycle. A similar event has been noted in the dysfunctional uterine bleeding observed secondary to the anovulatory menstrual cycles of pubertal girls (Fraser *et al.*, 1973; Hansen *et al.*, 1975). It appears that although the initiation of sexual maturation is attributable to a hypothalamic GnRH message, the pituitary has an important role in modifying the CNS initiative from above and in developing appropriate responses to modulation of gonadal products from below.

5.3. Gonads

The final end point of this complicated train of neuroendocrine events is gonadarche, which follows closely the activation of the CNS program. As can be seen in Figure 12 (Boyar *et al.*, 1974), during the period of sleep-entrained gonadotrophin pulsation, testosterone secretion is promptly stimulated from the Leydig cell in the male (Judd *et. al.*, 1977). Similar secretion of estrogen in response to longer intervals of gonadotrophin secretion in the female occurs before menarche (Fraser *et al.*, 1973; Hansen *et al.*, 1975). These alterations suggest that the gonad is readily responsive to gonadotrophins during the late prepubertal period and that an augmented secretion of gonadotrophins at puberty is relatively promptly translated into sex steroid secretion.

As puberty progresses, the gonad also acquires an ability to amplify the pituitary gonadotrophin message and to release even greater amounts of sex steroids for a given input of gonadotrophins. Human chorionic gonadotrophin testing during puberty demonstrates a continued enhancement of testosterone secretion to fixed doses of hCG, perhaps reflecting increased Leydig cell size, numbers and/or

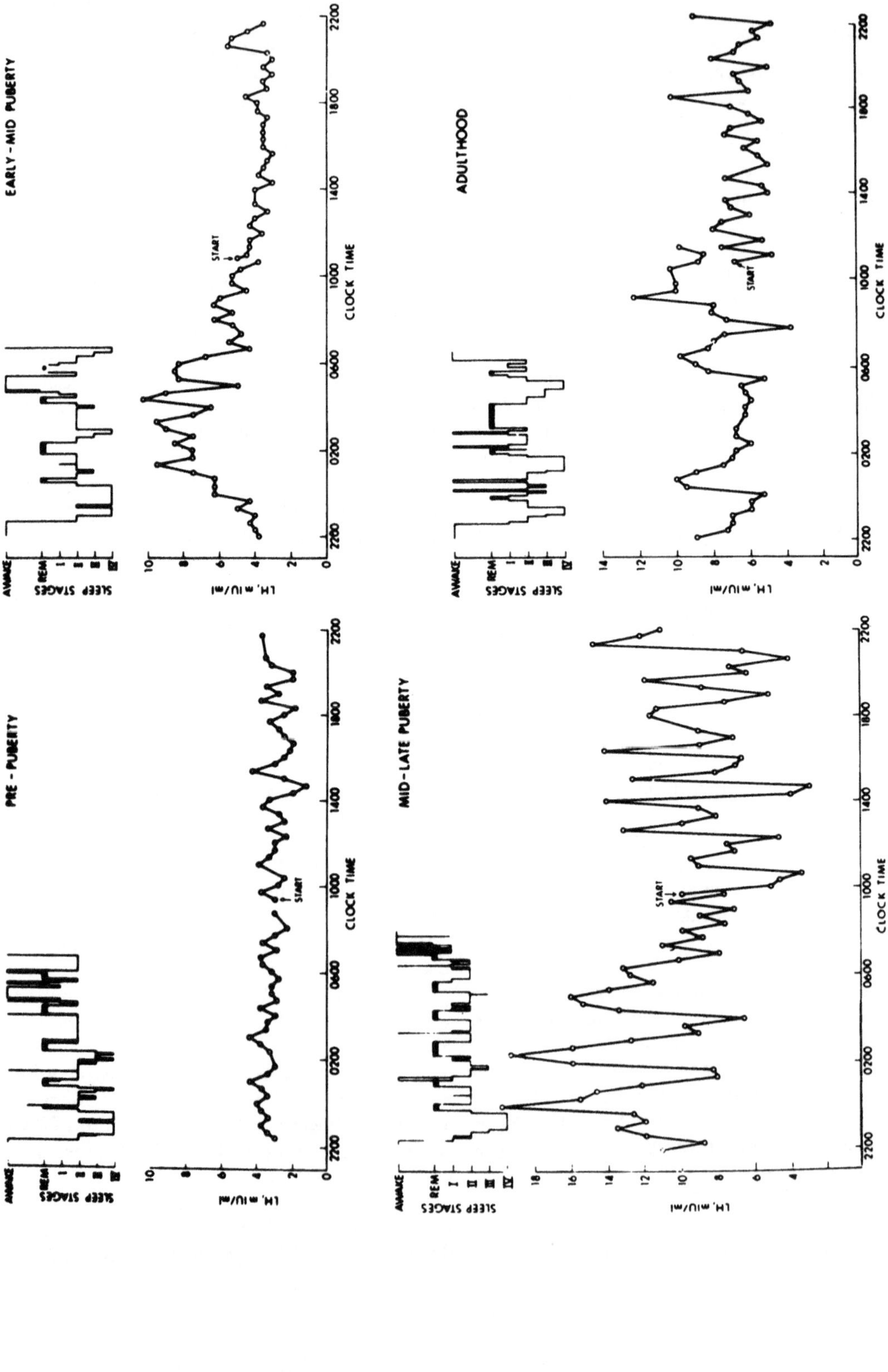

Fig. 11. The 24-hr pattern of LH secretion in a representative prepubertal girl (9 years), early pubertal (15 years), late pubertal (16 years), and young adult (23 years) males. Sleep stage is depicted for each nocturnal sleep period. (From Weitzman *et al.*, 1976.)

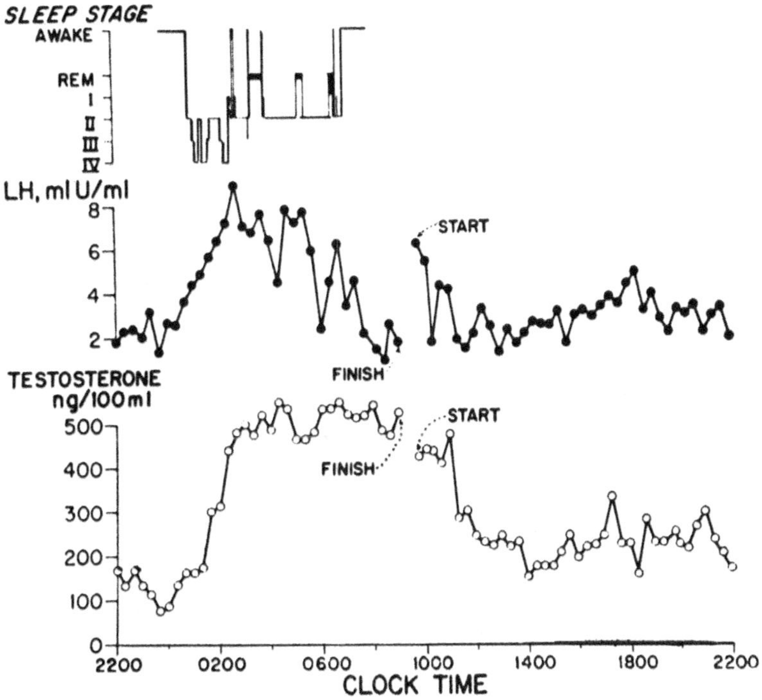

Fig. 12. The 24-hr pattern of LH and testosterone secretion observed in a mid-pubertal male. Note the prompt gonadal response accompanying the nocturnal pattern of gonadotrophin secretion. (From Boyar *et al.*, 1974.)

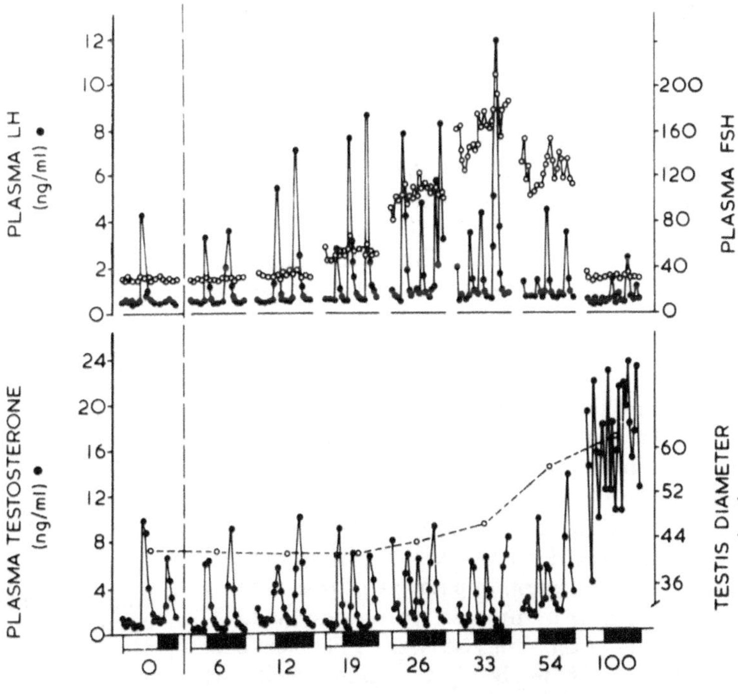

Fig. 13. Enhanced gonadal responsiveness to progressively smaller LH pulses across sexual maturation in the seasonal-breeding ram. (From Lincoln and Short, 1980.)

responsiveness (Winter *et al.*, 1972). Similarly, seasonal breeding rams require progressively less and less gonadotrophin input to achieve the same output of testosterone as sexual maturation progresses (Lincoln and Short, 1980) (cf. Figure 13). It appears that in the final stages of sexual maturation, gonadal amplification of the neuroendocrine message represents the finishing touches of gonadarche.

6. Summary

Although a considerable amount of information has been obtained concerning the chronology of events in puberty and the localization of its initiation within the hypothalamus, precious little data are available as to the control of its onset. Currently, three major hypotheses are under investigation. The role of melatonin as a negative modulator of GnRH secretion that declines in influence as puberty progresses has been put forth by several investigators employing measurements of varying degrees of specificity. Using the most sensitive and specific assays to date, Waldhauser *et al.* (1984) observed decreasing levels of melatonin secretion measurable in the peripheral blood across the various stages of sexual development. These investigators suggested that the well-described antigonadotrophic effect of melatonin, which clearly dampens GnRH secretion in lower animals, is responsible for the inactivity of the neuroendocrine system during the juvenile period. A second theory postulates a fall at the time of puberty in the activity of a hypothalamic enzyme that degrades GnRH (Advis *et al.*, 1983). Such a post-translational modification of GnRH activity within the cell and its fall at puberty would account for the rising levels of GnRH secretion observed at puberty. Finally, on the basis of several clinical conditions in which various metabolic derangements are associated with alterations in the timing or pase of puberty, several authors have focused on the critical effect of weight or body fat content as a metabolic index triggering the onset of sexual maturation (Frisch and McArthur, 1974).

No definite statement can be made at this time as to the precise trigger(s) of sexual maturation. Although the final common pathway of this effect is clearly mediated by modulation of hypothalamic secretion of GnRH, little else is clear. The next decade of research should be highly informative in this regard and should perhaps yield more specific information regarding the long-sought signal that first suppresses and then rekindles the neuroendocrine control of reproduction in the human.

7. References

Advis, J. P., Krause, J. E., and McKelvy, J. F., 1983, Evidence that endopeptidase-catalyzed luteinizing hormone releasing hormone cleavage contributes to the regulation of median eminence LHRH levels during positive steroid feedback, *Endocrinology* **112:**1147.

Albright, F., Smith, P. H., and Fraser, R., 1942, A syndrome characterized by primary ovarian insufficiency and decreased stature: Report of 11 cases with a digression on hormonal control of axillary and pubic hair, *Am. J. Med. Sci.* **204:**625.

Anderson, D. C., 1980, The adrenal androgen-stimulating hormone does not exist, *Lancet* **2:**454.

Belchetz, P. E., Plant, T. M., Nakai, Y., Keogh, E. J., and Knobil, E., 1978, Hypophyseal responses to continuous and intermittent delivery of hypothalamic gonadotropin-releasing hormone, *Science* **202:**631.

Bercu, B. B., Lee, B. C., Pineda, J. L., Spiliotis, B. E., Denman, D. W., Hoffman, H. J., Bioron, T. J., and Sachs, H. C., 1983*a*, Male sexual development in the monkey. I. Cross-sectional analysis of pulsatile hypothalamic-pituitary-testicular function, *J. Clin. Endocrinol. Metab.* **6:**1214.

Bercu, B. B., Lee, B. C., Spiliotis, B. E., Pineda, J. L., Denman, D. W., Hoffman, H. J., and Bioron, T. J., 1983*b*, Male sexual development in the monkey. II. Cross-sectional analysis of pulsatile hypothalamic-pituitary secretion in castrated males, *J. Clin. Endocrinol. Metab.* **6:**1227.

Bidlingmaier, F., Dorr, H. G., Eisenmenger, W., Kuhnle, U., and Knorr, D., 1983, Testosterone and androstenedione concentrations in human testis and epididymis during the first two years of life, *J. Clin. Endocrinol. Metab.* **57:**311.

Boyar, R. M., Finkelstein, J. W., David, R., Roffwarg, H., Kapen, S., Weitzman, E. D., adn Hellman, L., 1972, Synchronization of augmented luteinizing hormone secretion with sleep during puberty, *N. Engl. J. Med.* **287:**582.

Boyar, R. M., Finkelstein, J. W., David, R., Roffwarg, H., Kapen, S., Weitzman, E. D., and Hellman, L., 1973, Twenty-four hour patterns of plasma luteinizing hormone and follicle stimulating hormone in sexual precocity, *N. Engl. J. Med.* **289:**282.

Boyar, R. M., Rosenfeld, R. S., Kapen, S., Finkelstein, J. W., Roffwarg, H. P., Weitzman, E. D., and Hellman, L., 1974, Stimultaneous augmented secretion of luteinizing hormone and testosterone during sleep, *J. Clin. Invest.* **54:**609.

Bugnon, C., Bloch, B., and Fellman, D., 1977, Etude immuno-cytologique des neurones hypothalamiques à LHRH chez le foetus humain, *Brain Res.* **128:**249.

Chappel, S. C., Ulloa-Aguirre, A., and Coutifaris, C.,

1983, Biosynthesis and secretion of follicle-stimulating hormone, *Endocrine Rev.* **4**:179.

Chipman, J. J., 1980, Pubertal control mechanisms as revealed from human studies, *Fed. Proc.* **39**:2391.

Clements, J. A., Reyes, F. I., Winter, J. S. D., and Faiman, C., 1976, Studies on human sexual development. III. Fetal pituitary and serum, and amniotic fluid concentrations of LH, CG, and FSH, *J. Clin. Endocrinol. Metab.* **42**:9.

Conte, F. A., Grumbach, M. M., and Kaplan, S. L., 1975, A diphasic pattern of gonadotrophin secretion in patients with the syndrome of gonadal dysgenesis, *J. Clin. Endocrinol. Metab.* **40**:670.

Conte, F. A., Grumbach, M. M., Kaplan, S. L., and Reiter, E. O., 1980, Correlation of luteinizing hormone-releasing factor-induced luteinizing hormone and follicle-stimulating hormone release from infancy to 19 years with the changing pattern of gonadotrophin secretion in agonadal patients: Relation to the restraint of puberty, *J. Clin. Endocrinol. Metab.* **50**:163.

Corley, K. P., Valk, T. W., Kelch, R. P., and Marshall, J. C., 1981, Estimation of GnRH pulse amplitude during pubertal development, *Pediatr. Res.* **15**:157.

Crowley, W. F., Comite, F., Vale, W., Rivier, J., Loriaux, D. L., and Cutler, G. B., 1981, Therapeutic use of pituitary desensitization with a long-acting LHRH agonist: A potential new treatment for idiopathic precocious puberty, *J. Clin. Endocrinol. Metab.* **52**:370.

Crowley, W. F., Vale, W., Rivier, J., and McArthur, J. W., 1982, LHRH in hypogonadotropic hypogonadism, in: *LHRH Peptides as Male and Female Contraceptives* (G. I. Zatuchni, J. D. Shelton, and J. J. Sciarra, eds.), p. 321, J. B. Lippincott, Philadelphia.

Cutler, G. B., and Loriaux, D. L., 1980, Adrenarche and its relationship to the onset of puberty, *Fed Proc.* **39**:2384.

Dierschke, D. J., Weiss, G., and Knobil, E., 1974, Sexual maturation in the female Rhesus monkey and the develeopment of estrogen-induced gonadotropic hormone release, *Endocrinology* **94**:198.

DiZerega, G. S., and Ross, G. T., 1980, Clinical relevance of fetal gonadal structure and function. *Clin. Obstet. Gynecol.* **23(3)**:849.

Faiman, C., and Winter, J. S. D., 1974, Gonadotropins and sex hormone patterns in puberty: Clinical data, in: *Control of the Onset of Puberty,* (M. M. Grumbach, G. D. Grave, and F. E., Mayer, eds.), p. 32, Wiley, New York.

Forest, M. G., and Cathiard, A. M., 1975, Pattern of plasma testosterone and Δ^4-androstenedione in normal newborns: Evidence for testicular activity at birth, *J. Endocrinol. Metab.* **41**:977.

Fraser, I. S., Michie, E. A., Wide, L., and Baird, D. I., 1973, Pituitary gonadotropins and ovarian function in adolescent dysfunctional uterine bleeding, *J. Clin. Endocrinol. Metab.* **37**:407.

Frisch, R. E., and McArthur, J. W., 1974, Menstrual cycles: Fatness as a determinant of minimum weight for height necessary for their maintenance or onset, *Science* **185**:949.

Gendrel, D., Chaussain, J.-l., Roger, M., and Job, J.-C., 1980, Simultaneous postnatal rise of plasma LH and testosterone in male infants, *J. Pediatr.* **97**:600.

Hansen, J. W., Hoffman, H. J., and Ross, G. T., 1975, Monthly gonadotrophin cycles in pre-menarcheal girls, *Science* **190**:161.

Hayes, A., and Johanson, A., 1972, Excretion of follicle-stimulating hormone (FSH) and luteinizing hormone (LH) in urine by pubertal girls, *Pediatr. Res.* **6**:18.

Hier, D. B., and Crowley, W. F., 1982, Spatial ability in androgen-deficient men, *N. Engl. J. Med.* **306**:1202.

Hoffman, A. R., and Crowley, W. F., 1982, Induction of puberty in men by long-term pulsatile administration of low-dose gonadotropin-releasing hormone, *N. Engl. J. Med.* **307**:1237.

Huhtaniemi, I.,Koritnik, D. R., Korenbrot, C. C., Mennin, S., Foster, D. B., and Jaffee, R. B., 1979, Stimulation of pituitary-testicular function with gonadotropin-releasing hormone in fetal and infant monkeys, *Endocrinology* **1**:109.

Ilondo, M. M., Vanderschueren-Lodeweyckx, M., Vlietinck, R., Pizarro, M., Malvaux, P., Eggermont, E., and Eeckles, R., 1982, Plasma androgens in children and adolescents. Part I. Control subjects, *Horm. Res.* **16**:61.

Jaffe, R. B., Seron-Ferre, M., Crickard, K., Koritnik, D., Mitchell, B. F., and Huhtaniemi, I. T., 1981, Regulation and function of the primate fetal adrenal gland and gonad, *Recent Prog. Horm. Res.* **37**:41.

Jakacki, R. I., Kelch, R. P., Sauder, S. E., Lloyd, J. S., Hopwood, N. J., and Marshall, J. C., 1982, Pulsatile secretion of luteinizing hormone in children, *J. Clin. Endocrinol. Metab.* **55**:453.

Job., J.-C., Chaussain, J.-L., and Garnier, P. E., 1977, The use of luteinizing hormone-releasing hormone in pediatric patients, *Horm. Res.* **8**:171.

Judd, H. L., Parker, D. C., and Yen, S. S. C., 1977, Sleep-wake patterns of LH and testosterone release in prepubertal boys, *J. Clin. Endocrinol. Metab.* **44**:865.

Kaplan, S. L., and Grumbach, M. M., 1978, Pituitary and sex steroids in the human fetus and subhuman primate fetus, *J. Clin. Endocrinol.* **7**:489.

Kaplan, S. L., Grumbach, M. M., and Aubert, M. L., 1976, The ontogenesis of pituitary hormones and hypothalamic factors in the human fetus. Maturation of central nervous system regulation of anterior pituitary function, *Recent Prog. Horm. Res.* **34**:161.

Knobil, E., 1974, On the control of gonadotropin secretion in the Rhesus monkey, *Recent Prog. Horm. Res.* **30**:1.

Knobil, E., 1980, The neuroendocrine control of the menstrual cycle, *Recent Prog. Horm. Res.* **36**:53.

Knobil, E., Plant, T. M., Wildt, L., Belchetz, J., and Marshall, G., 1980, Control of the Rhesus monkey menstrual cycle: Permissive role of hyptohalamic gonadotropin-releasing hormone, *Science* **207**:1371.

Korth-Schutz, S., Levine, L. S., and New, M. I., 1976*a*,

Dehydroepinadrosterone sulfate (DS) levels, a rapid test for abnormal adrenal androgen secretion, *J. Clin. Endocrinol. Metab.* **42:**1005.

Korth-Schutz, S., Levine, L. S., and New, M. I., 1976*b,* Serum androgens in normal prepubertal and pubertal children and in children with precocious adrenarche, *J. Clin. Endocrinol. Metab.* **42:**117.

Kulin, H. E., and Reiter, E. O., 1973, Gonadotropins during childhood and adolescence: A review. *Pediatrics* **51:**260.

Kulin, H. E., Grumbach, M. M. and Kaplan, S. L., 1969, Changing sensitivity of the pubertal gonadal hypothalamic feedback mechanism in man, *Science* **165:**1012.

Kulin, H. E., Grumbach, M. M., and Kaplan, S. L., 1972, Gonadal-hypothalamic interaction in prepubertal and pubertal man: Effect of clomiphene citrate on urinary follicle stimulating hormone and luteinizing hormone and plasma testosterone, *Pediat. Res.* **6:**162.

Lee, P. A., Plotnick, L. P., Migeon, C. J., and Kowarski, A. A., 1978, Integrated concentrations of follicle stimulating hormone and puberty, *J. Clin. Endocrinol. Metab.* **46:**488.

Lincoln, G. A., and Short, R. V., 1980, Seasonal breeding: Nature's contraceptive, *Recent Prog. Horm. Res.* **36:**1.

Lucky, A. W., Rich, B. H., Rosenfield, R. L., Fang, V. S., and Roche-Bender, N., 1980, LH bioactivity increases more than immunoreactivity during puberty, *J. Pediatr.* **97:**205.

Maruca, J., Kulin, H. E., and Santner, S. J., 1983, Perturbations of negative feedback sensitivity in agonadal patients undergoing estrogen replacement therapy, *J. Clin. Endocrinol. Metab.* **56:**53.

Matthews, M. M., Parker, D. C., Rebar, R. W., Jones, K. L., Rossman, L., Carey, D. E., and Yen, S. S. C., 1982, Sleep-associated gonadotrophin and oestradiol patterns in girls with precocious sexual development, *Clin. Endocrinol.* **17:**601.

Migeon, C. J., 1981, Physiology and pathology of adrenocortical function in infancy and childhood, in: *Pediatric Endocrinology* (R. Collu, ed.), p. 465, Raven Press, New York.

Parker, D. C., Judd, H. L., Rossman, L. G., and Yen, S. S. C., 1975, Pubertal sleep–wake patterns of episodic LH, FSH and testosterone release in twin boys, *J. Clin. Endocrinol. Metab.* **40:**1099.

Parker, L. N., Lifrak, E, T., and Odell, W. D., 1983, A. 60,000 molecular weight human pituitary glycopeptide stimulates adrenal androgen secretion, *Endocrinology* **113:**2092.

Peckham, E. D., and Knobil, E., 1976, Qualitative changes in the pituitary gonadotropins of the male Rhesus monkey following castration, *Endocrinology* **98:**1061.

Peckham, W. D., Yamaji, T., Dierschke, D. J., and Knobil, E., 1973, Gonadal function and the biological and physiochemical properties of follicle-stimulating hormone, *Endocrinology* **92:**1660.

Penny, R., Olambiwonnu, N. O., and Frasier, S. D., 1977, Episodic fluctuations of serum gonadotropins in pre- and post-pubertal girls and boys, *J. Clin. Endocrinol. Metab.* **45:**307.

Plant, T. M., and Zorub, D. S., 1984, A study of the role of the adrenal gland in the initiation of the hiatus in gonadotropin secretion during prepubertal development in the male Rhesus monkey *(Macaca mulatta), Endocrinology* **114:**560.

Rasmussen, D. D., Liu, J. H., Wolf, P. L., and Yen, S. S. C., 1983, Endogenous opioid regulation of gonadotropin-releasing hormone release from the human fetal hypothalamus *in vitro, J. Clin. Endocrinol. Metab.* **57:**881.

Reiter, E. O., Root, A. W., and Duckett, G. E., 1976, The response of pituitary gonadotropes to a constant infusion of luteinizing hormone-releasing hormone (LHRH) in normal prepubertal and pubertal children and in children with abnormalities of sexual development, *J. Clin. Endocrinol. Metab.* **43:**400.

Reiter, E. O., Fuldauer, V. G., and Root, A. W., 1977, Secretion of the adrenal androgen dehydropiandrosterone sulfate, during normal infancy, childhood, and adolescence, in sick infants, and in children with endocrinologic abnormalities, *J. Pediatr.* **90:**766.

Reiter, E. O., Beitins, I. Z., Ostrea, T., and Gutai, J. P., 1982, Bioassayable luteinizing hormone during childhood and adolescence and in patients with delayed pubertal development, *J. Clin. Endocrinol. Metab.* **54:**155.

Rich, B. H., Rosenfield, R. L., Lucky, A. W., Helke, J. C. and Otto, P., 1981, Adrenarche: Changing adrenal response to adrenocorticotropin, *J. Clin. Endocrinol. Metab.* **52:**1129.

Ross, J. L., Loriaux, D. L., and Cutler, G. B., 1983, Developmental changes in neuroendocrine regulation of gonadotropin secretion in gonadal dysgenesis, *J. Clin. Endocrinol. Metab.* **57:**288.

Roth, J. C., Grumbach, M. M., and Kaplan, S. L., 1976, Effect of synthetic luteinizing hormone-releasing factor on serum testosterone and gonadotropins in prepubertal, pubertal and adult males, *J. Clin. Endocrinol. Metab.* **37:**680.

Savage, M. O., Preece, M. A., Cameron, N., Jones, J., Theintz, G., Penfold, J. L., and Tanner, J. M., 1981, Gonadotrophin response to LHRH in boys with delayed growth and adolescence, *Arch. Dis. Child.* **56:**552.

Scheibinger, R. J., Albertson, B. D., Cassorla, F. G., Bowyer, D. W., Gellhoed, G. W., Cutler, G. B., and Loriaux, D. L., 1981, The developmental changes in plasma adrenal androgens during infancy and adrenarche are associated with changing activities of adrenal microsomal 17-hydroxylase and 17,20-desmolase, *J. Clin. Invest.* **67:**1177.

Siiteri, P. K., and Wilson, J. D., 1974, Testosterone formation and metabolism during male sexual differentiation in the human embryo, *J. Clin. Endocrinol. Metab.* **38:**113.

Siler-Khodr, T. M., and Khodr, G. S., 1980, Studies in human fetal endocrinology. II. LH and FSH content and concentration in the pituitary, *Obstet. Gynecol.* **56**:176.

Siler-Khodr, T. M., Morgenstern, L. L., and Greenwood, F. C., 1974, Hormone synthesis and release from human fetal adenohyophyses *in vitro, J. Clin. Endocrinol. Metab.* **39**:891.

Sizonenko, P. C., Paunier, L., and Carmignac, D., 1976, Hormonal changes during puberty. IV. Longitudinal study of adrenal androgen secretions, *Horm. Res.* **7**:288.

Sklar, C. A., Kaplan, S. L., and Grumbach, M. M., 1980, Evidence for dissociation between adrenarche and gonadarche: Studies in patients with idiopathic precocious puberty, gonadal dysgenesis, isolated gonadotropin deficiency, and constitutionally delayed growth and adolescence, *J. Clin. Endocrinol. Metab.* **51**:548.

Smail, P. J., Faiman, C., Hobson, W. C., Fuller, G. B., and Winter, J. S. D., 1982, Further studies on adrenarche in nonhuman primates, *Endocrinology* **111**:844.

Steiner, R. A., and Bremner, W. J., 1981, Endocrine correlates of sexual development in the male monkey, *Macaca fascicularis, Endocrinology* **109**:914.

Takagi, S., Yoshida, T., Tsubata, K., Ozaki, H., Fujii, T. K., Nomura, Y., and Sawada, M., 1977, Sex differences in fetal gonadotropins and androgens, *J. Steroid Biochem.* **8**:609.

Tapanainen, K., Koivisto, M., Huhtaniemi, I., and Vihko, R., 1982, Effect of gonadotropin-releasing hormone on pituitary-gonadal function of male infants during the first year of life, *J. Clin. Endocrinol. Metab.* **55**:689.

Thliveris, J. A., and Currie, R. W., 1980, Observation on the hypothalamohypophyseal portal vasculature in the developing human fetus, *Am. J. Anat.* **157**:441.

Waldhauser, F., Weibenbacher, G., Frisch, H., and Pollak, A., 1981, Pulsatile secretion of gonadotropins in early infancy, *Eur. J. Pediatr.* **137**:71.

Waldhauser, F., Weiszenbacher, G., Frisch, H., Zeitlhuber, U., Waldhauser, M., and Wurtman, R. J., 1984, Fall in nocturnal serum melatonin during prepuberty and pubescence, *Lancet* **1**:362.

Weitzman, E. B., Boyar, R. M., Kapen, S., and Hellman, L., 1976, The relationship of sleep and sleep stages to neuroendocrine secretion and biological rhythms in man, *Recent Prog. Horm. Res.* **32**:309.

Wide, L., 1983, Male and female forms of human follicle-stimulating hormone in serum, *J. Clin. Endocrinol. Metab.* **55**:682.

Wide, L., and Hobson, B. M., 1983, Qualitative difference in follicle-stimulating hormone activity in the pituitaries of young women compared to that of men and elderly women, *J. Clin. Endocrinol. Metab.* **56**:371.

Wide, L., and Lundberg, P. O., 1981, Hypersecretion of an abnormal form of follicle-stimulating hormone associated with suppressed luteinizing hormone secretion in a woman with a pituitary adenoma, *J. Clin. Endocrinol. Metab.* **53**:923.

Wide, L,. and Wide, M., 1984, Higher plasma disappearance rate in the mouse for pituitary follicle-stimulating hormone of young women compared to that of men and elderly women, *J. Clin. Endocrinol. Metab.* **58**:426.

Wierman, M., Beardsworth, D., Crawford, J., Crigler, J., Bode, H., Kushner, D., and Crowley W., 1984, Adrenarche and growth during LHRH agonist ($LHRH_a$) administration, *Soc. Pediatr. Res.* **502**:179A.

Wildt, L., Marshall, G., and Knobil, E., 1980, Experimental induction of puberty in the infantile female Rhesus monkey, *Science* **207**:1373.

Winter, J. S. D., Taraska, S., and Faiman, C., 1972, The hormonal response to hCG stimulation in male children and adolescents, *J. Clin. Endocrinol.* **34**:348.

Winter, J. S. D., Eskay, R. L., and Porter, J. C., 1974, Concentration and distribution of TRH and LRH in the human fetal brain, *J. Clin. Endocrinol. Metab.* **39**:960.

Winter, J. S. D., Faiman, C., Hobson, W. C., Prasad, A. V., and Reyes, F. I., 1975, Pituitary gonadal relations in infancy. I. Patterns of serum gonadotropin concentration from birth to four years of age in man and chimpanzee, *J. Clin. Endocrinol. Metab.* **40**:545.

11

Skull, Jaw, and Teeth Growth Patterns

PATRICK G. SULLIVAN

1. Introduction

The growth of the human skull has been recorded in many ways over the centuries, ranging from the drawings found in the tombs of ancient Egypt to the computer-drawn plots of modern times. Although both the advent of radiography and the data-processing power of computer technology have permitted a much improved understanding of the overall patterns of skull growth, the stimuli and control mechanisms that govern the growth of an individual are still subject to debate and controversy.

Developmentally, the structure of the skull can be divided into two components: one, the neurocranium, concerned with the support and protection of the brain, and the other, the viscerocranium (nasofacial complex), concerned with the mechanisms for respiration, mastication, and speech. The relationship and evolution of these components has to be appreciated for a full understanding of skull growth. The configuration of the human skull, together with that of the other members of the suborder Anthropoidea, differs markedly from the rest of the animal kingdom. Perhaps the most obvious and certainly the most significant difference is in the relative proportions of the neurocranium and viscerocranium. Human evolutionary development has been accompanied by a marked increase in brain capacity, which has resulted in enlargement of the neurocranium. Furthermore, as the neural component has increased in volume, it has also become progressively repositioned relative to the remainder of the cranial architecture. During the same evolutionary time span, other changes have taken place. Thus with the adoption of an upright posture, the opening of the foramen magnum has been relocated from the posterior to the inferior surface of the skull. The summation of these changes is seen in the progressive movement of the viscerocranium from in front of the neural component to a position beneath the forebrain (Figure 1).

The growth pattern of the human skull in its passage from neonatal to adult form is a measure of the difference in growth characteristics between the neural and visceral components. The neurocranium ceases development at a comparatively earlier age, and the pattern of skull growth reflects the dynamic relationship between the later-developing facial complex and the base of the skull. The change in shape between the infant and the adult skull shown in Figure 2 is related to the relative increase in size of the nasofacial complex (viscerocranium) when compared with the calvaria (neurocranium). Between birth and adult life the volume of the calvaria increases four times, and the volume of the facial region about 12 times. Furthermore, as 80% of the postnatal growth of the calvaria occurs during the first 2 years, the major part of calvarial growth is completed long before facial growth is terminated.

Although the skull may be viewed developmentally as being capable of division into two com-

PATRICK G. SULLIVAN • Department of Child Dental Health, The University of Sheffield, Sheffield S10 2SZ, England.

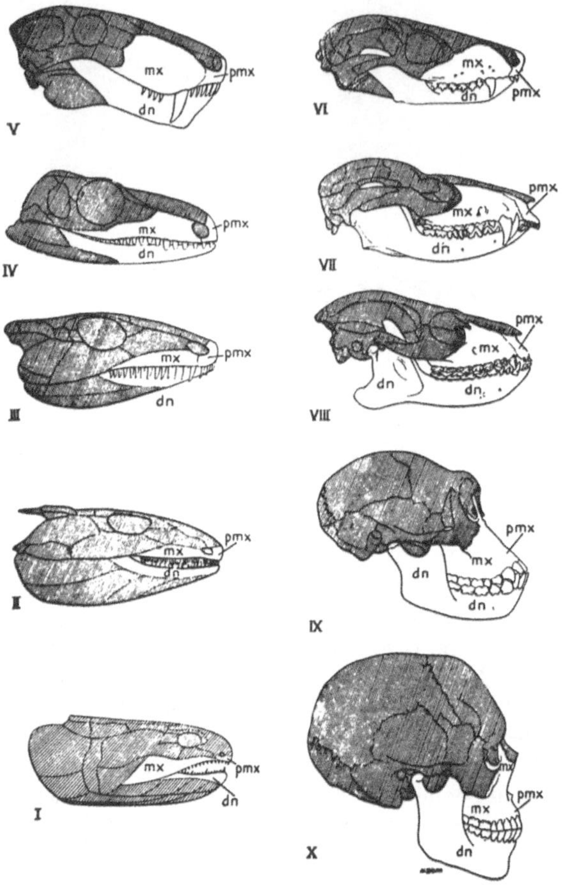

Fig. 1. Evolution of the maxillary bones from lobe-finned fish to man (pmx, premaxilla: mx, maxilla: and dn, dentary). In the ascending series, consisting of primitive amphibian, stem reptile, and three stages of mammal-like reptiles, the dentary and maxilla increase greatly in size. In the earlier adult mammals the dentary becomes the sole bone of each half of the lower jaw, and the premaxilla and maxilla together form the subocular surface of the face. The transition from this stage to that of man involves chiefly the deepening and anteroposterior shortening of the maxillary bones (including the premaxilla, maxilla, and dentary). (I) Lobe-finned fish, Devonian age (essentially Rhizodopsis). After Traquair, Watson, Bryant. (II) Primitive amphibian (Palaeogyrinus), lower Carboniferous. After Watson. (III) Primitive cotylosaurian reptile (Seymouria), Permocarboniferous. After Broili, Williston, Watson. (IV) Primitive theromorph reptile (Mycterosaurus), Permocarboniferous. After Williston. (V) Gorgonopsian reptile (Scymnognathus). After Broom. (VI) Primitive cynodont reptile (Ictidopsis), Triassic. After Broom, Haughton. (VII) Recent opossum. (VIII) Primitive primate (Notharctus), Eocene. After Gregory. (IX) Anthropoid (female chimpanzee), Recent. (X) Man. Recent. (From Gregory, 1927.)

ponent regions, this is not demonstrated by the bone architecture. The demands of evolution have resulted in the development of a complex structure made up from 22 individual bones. Thus change in size and shape is achieved by the integrated activity of many component bones, each bone having its own individual growth pattern, which may differ from the others in the complex in both direction and rate of growth.

2. Methods Employed to Record Skull Growth

Three mechanisms are available to bring about changes in shape of a bone: ossification of cartilage, new bone addition at a suture (Pritchard *et al.*, 1956), and remodeling by deposition and resorption at the periosteal and endosteal surfaces. All three mechanisms are involved in skull growth, each making a varying contribution to different stages in the achievement of adult skull form.

The growth pattern followed by a point on a given bone surface can only be appreciated by comparison with a fixed point, which may be taken as an unchanging fiducial marker, constant in all three spatial planes. One of the complicating factors in the appreciation of skull growth is that there are no absolute fixed points from which growth can be said to have taken place. Any selected anatomic or radiographic landmark is itself liable to movement by the growth activity of both the bone on which the landmark is situated and that of the adjacent bones.

A landmark on the surface of a bone of the skull may move in space relative to a fixed point by the following mechanisms: (1) Growth may take place by surface bone deposition; (2) Growth may occur at a suture between the bone bearing the landmark and an adjacent bone. In this case because of the new bone formation at the sutural interface between the bones, the landmark is moved away from the fixed point by the bodily movement of the bone on which the landmark is located; (3) The landmark-bearing bone may be carried away from the fixed point by the movement of an intervening bone nearer to the fixed point.

An ideal recording method should be able to register not only change in position of selected landmarks, but also the three-dimensional shape of a constituent skull bone. The method should also permit the measurement of small changes in the same individual to allow longitudinal growth analysis (Tanner, 1951). The nearest approach to these requirements is found in the use of intravital dyes in experimental animals (Hunter, 1798; Harris, 1960; Suzuki and Matthews, 1966), where the detailed relationship between the successive dye-marked contours and the surface of the bone may be established and measured by the use of a three-dimensional reconstruction method (Sullivan, 1972).

In the study of human skull growth, where the examination must obviously be limited to the use of noninjurious methods, these techniques are not applicable, and longitudinal growth recording is limited to the use of direct craniometry and radiography.

Several workers have employed methods of direct craniometry on the living subject; for instance, Goldstein (1936) examined 100 Jewish males at biannual intervals between 2 and 21 years of age. All measurement techniques employed in face and jaw growth studies are liable to error, the source of the error residing in the characteristics of the measurement method. In the case of direct craniometry, the error lies in the accurate location of surface landmarks through the overlying tissues.

Other investigators have made extensive use of radiography, and this approach confers the ability to measure the relative positions of anatomical structures within the skull as well as on the surface. There are two basic drawbacks to the use of radiographic method: (1) The necessity of achieving the same head orientation at successive recordings; and (2) The distortion encountered when a "shadow" of a complex three-dimensional object is thrown onto a two-dimensional film.

A variety of methods have been suggested to obtain the same head position on successive radiographs. Lynsholm (1931) employed an optical method whereby the head was positioned by means of images in a mirror system. However, Broadbent (1931) adapted an instrument previously employed for anthropological measurements for use as a radiographic craniostat (Figure 3), and instruments based on this design have become widely used for skull-growth recording.

The instrument employs two adjustable ear rods, inserted into the external auditory meati, and a pointer is adjusted to contact the left infraorbital foramen. The ear rods are positioned such that they are coincident in a beam of X-rays, and a lateral view of the skull is recorded.

A magnification error is inherent in recording different parts of the same skull in the lateral skull view. Those structures farthest from the film will be magnified to a greater degree than those nearest to the film. This error is minimized by setting the X-ray source at a distance from the subject to obtain as near a parallel beam of X-rays as possible. The remaining magnification error may be overcome by measurement of a frontal radiograph of the same skull. From this view, the distance of a structure from the X-ray film may be calculated and an appropriate correction factor applied to the measurement taken from the lateral film. A further means of calibrating these errors was described by Adams

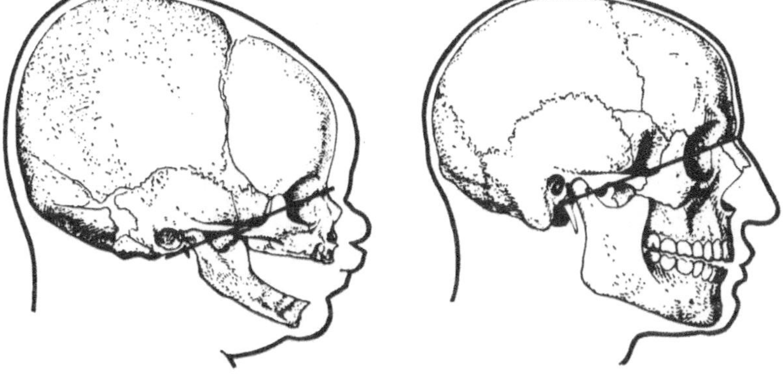

Fig. 2. The jaws at birth compared to the adult. (After Tulley and Campbell, 1975.)

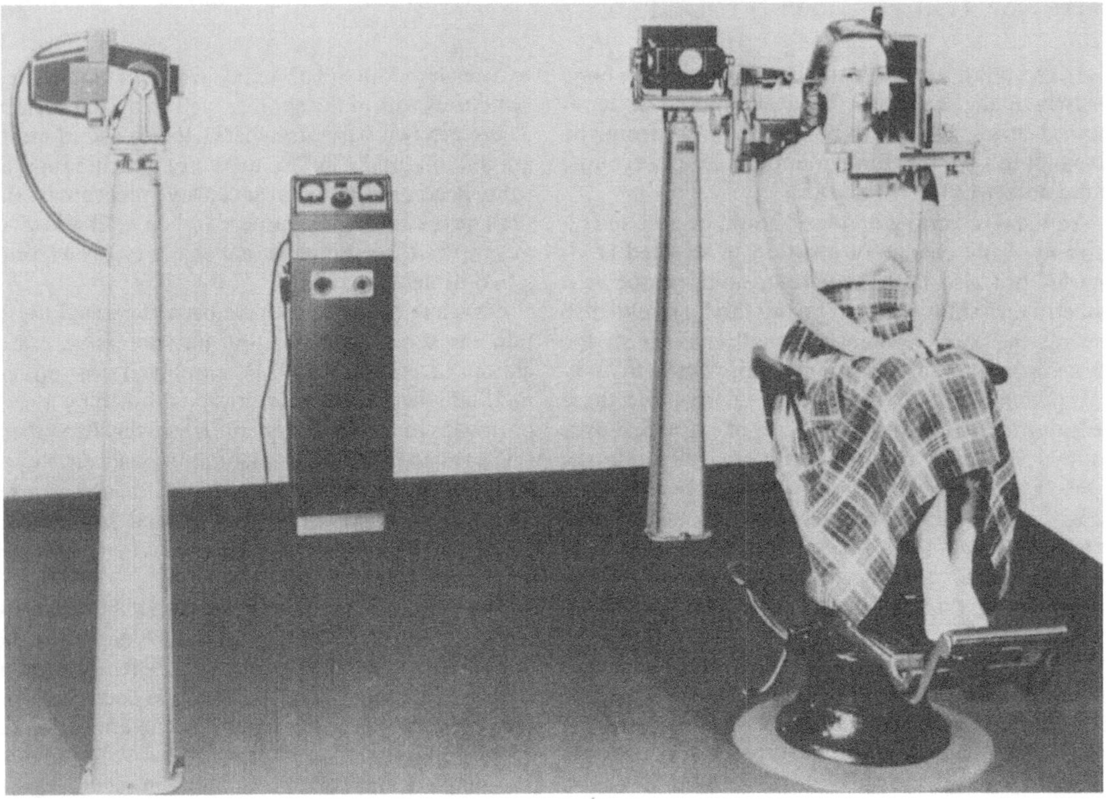

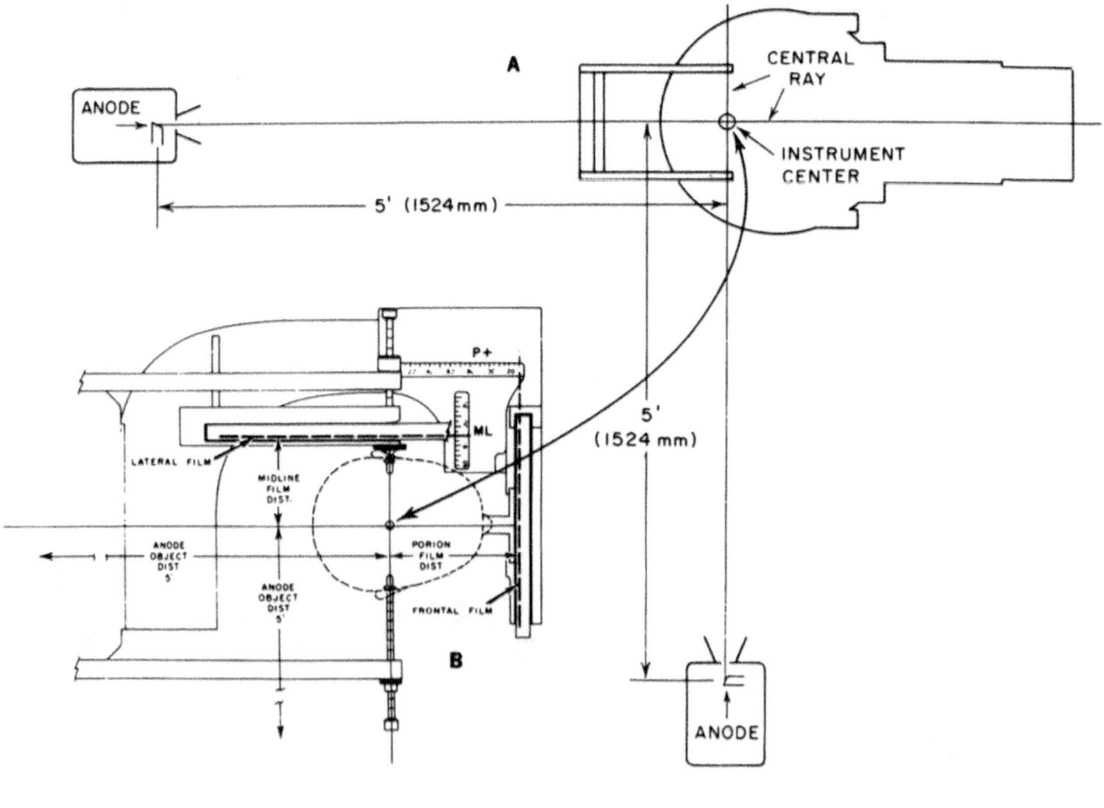

A
ANODE
CENTRAL RAY
INSTRUMENT CENTER
5' (1524mm)
5' (1524 mm)
P+
ML
LATERAL FILM
MIDLINE FILM DIST.
ANODE OBJECT DIST 5'
PORION FILM DIST
ANODE OBJECT DIST 5'
FRONTAL FILM
B
ANODE

(1940), who used a radiopaque scale projected onto the radiograph.

The relative movement of all parts of the craniofacial complex during growth has already been stressed. Thus, if cephalometric radiographs are used, either in the case of a cross-sectional survey on different individuals or in a longitudinal survey on the same individual, a uniform superimposition method is required to provide a true representation of the growth changes. Broadbent evolved a method based on the identification of certain anatomical landmarks associated with the skull base that could be identified with most certainty on a cephalometric radiograph. These landmarks were employed as the point of origin for the reference planes used for the superimposition of successive radiographs. He described a reference plane, the Bolton plane, drawn between the suture between the frontal and nasal bones and the highest point in the notch posterior to the occipital condyle. This plane was used to demonstrate the growth pattern of the facial complex (Figure 4). Since 1931, many other planes and landmarks have been used for fiducial purposes (Krogman and Sassouni, 1957). The selection of these landmarks was made on the assumption that as the base of the skull has completed most of its growth at a relatively early stage, it may be considered as a fixed base against which the growth of the facial skeleton takes place. The statistical reliabilities of some of these planes and points were examined by Bjork and Solow (1962).

Moss *et al.* (1983) reexamined the possibility of identifying a fixed reference point from which the growth of all or part of the skull could be measured. They used a computer-assisted statistical method in evaluating both animal and human data, but concluded that such a reference point could not be defined.

Bjork (1955) contributed a major advance in the use of radiography by developing a method for the insertion of small metallic pins into the bones of the jaws. The pins are inserted beneath the periostal surface and are unaffected by subsequent changes in bone shape. When radiographs are taken, the pins appear as small radiopaque points. If tracings taken from successive radiographs are superimposed on the radiopaque points, the true change in bone shape is demonstrated (Figure 5). An alternative superimposition method, based on the position of the auditory canals, was published by Delattre and Fenart (1958). They suggested that in view of the function of the semicircular canals in the appreciation of spatial position, they might be used for the superimposition of skull radiographs.

Enlow (1966) adopted a method whereby tracings taken from radiographs were superimposed according to growth patterns derived from histological studies of areas of bone deposition and resorption found in postmortem skulls. Although not a quantitative method, this approach has proved valuable in identifying the growth mechanisms contributing to regional skull growth. Several investigators have used the three-dimensional positions of points in the radiographic image (Broadbent 1931; Ricketts *et. al.*, 1972).

In 1972, Savara (1972) described a method whereby the spatial positions of selected landmarks were abstracted from radiographs taken simultaneously at right angles. By a process of triangulation the landmarks are positioned in space by means of three measurements. However, the difficulty in identifying the same point on the two radiographs is such as to render this approach of limited accuracy.

The ability of a computer to store and process large volumes of information has led to the establishment of data banks based on the recordings obtained from radiographic surveys. Walker (1972) stored tracings using a computer-based technique, and as part of his survey developed an atlas of the mean stages for progress of skull growth at given ages throughout the adolescent period (Figure 6).

The use of radiography, the only practical approach to the study of human jaw growth, has played an important part in understanding the trends of human skull growth. However, the study of the detailed movements of forming bone surfaces is limited both by the resolution of the clinical radiograph and by its inability to record the shape of a contour in three dimensions.

3. Regional Growth Patterns

The skull is a complex structure formed from many component bones which articulate along an

Fig. 3. Broadbent–Bolton radiographic cephalometer used in the Bolton Study Case Western Reserve University, Cleveland, Ohio. (From Broadbent *et al.*, 1975.) The Bolton cephalometric floor plan shows: (A) relation of the anodes, the central rays and the instrument centre; (B) the relationship of the instrument centre to the lateral and frontal films.

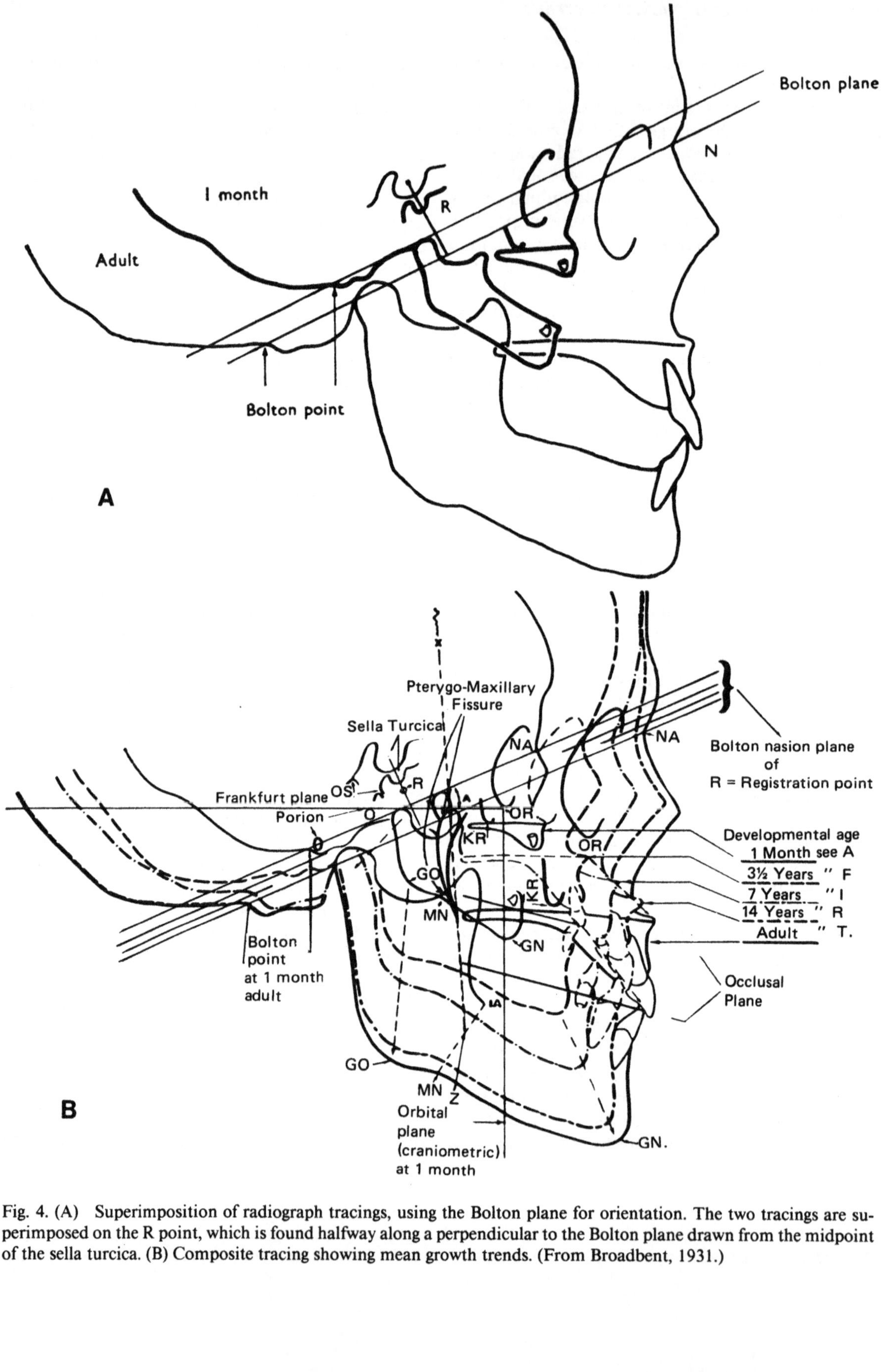

Fig. 4. (A) Superimposition of radiograph tracings, using the Bolton plane for orientation. The two tracings are superimposed on the R point, which is found halfway along a perpendicular to the Bolton plane drawn from the midpoint of the sella turcica. (B) Composite tracing showing mean growth trends. (From Broadbent, 1931.)

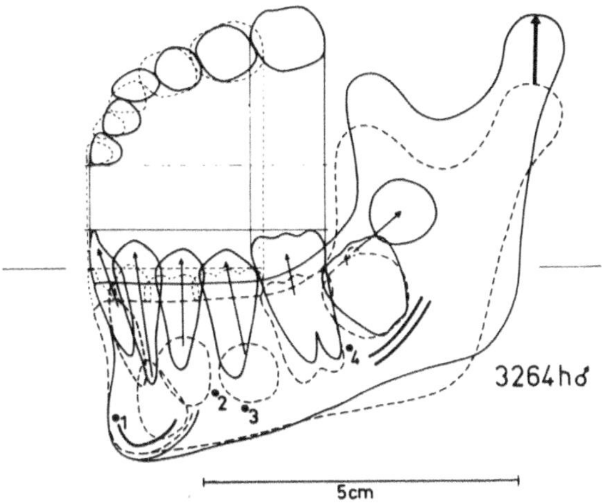

Fig. 5. Growth of the mandible analyzed by superimposition on implant markers. The tracings are superimposed on the images of four pins inserted into the bone. Broken line, age 5 years, 8 months; solid line, age 10 years, 8 months. (From Bjork, 1963.)

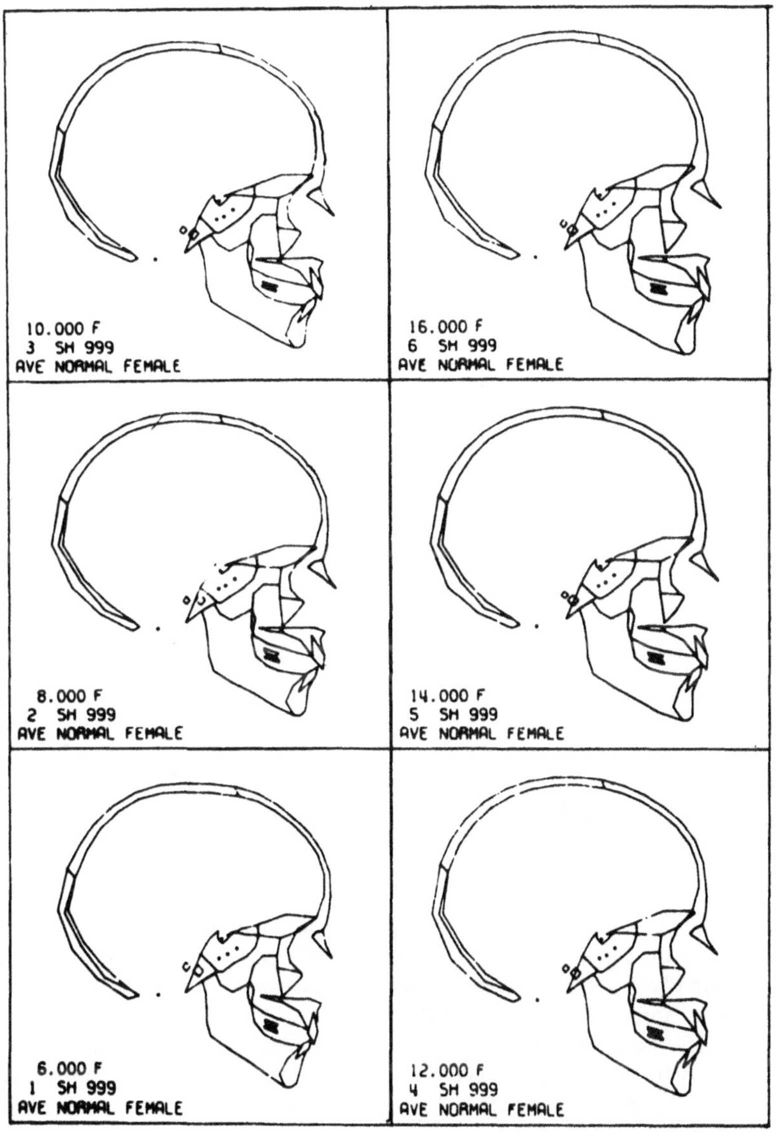

Fig. 6. Atlas of average skull dimensions for girls from 6 to 16 years of age. Sample size for each group is approximately 100. (From Walker, 1972.)

intricate pattern of junctions. Thus any individual change in size of a component bone must be accompanied by a balanced readjustment in the adjacent bones. Similarly, more distant bones must also adjust in size such that an integrated progressive change in shape is seen throughout the growth period. It is thus misleading to analyze skull growth on a regional basis, as if each region grows in isolation without regard to the over-all skull-growth pattern. However, the different regions of the skull do vary one from another, not only in the obvious morphological and functional characteristics, but also in the balance between the growth mechanisms employed to bring about the integrated change in size and shape.

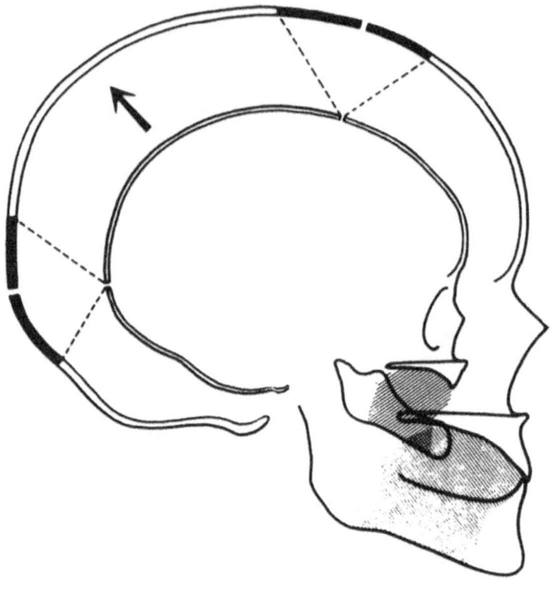

Fig. 7. Radial movement of the calvarial bones following expansion of the brain. Compensatory sutural growth shown in black. (After Salzmann, 1966.)

3.1. *Calvaria*

The cranial vault is constructed from the frontal and parietal bones and parts of the temporal, occipital, and sphenoid bones. These bones are platelike in structure and are formed by ossification in membrane.

At birth the bones are separated anteriorly, posteriorly, and at other locations in the calvarium by six fontanelles. Each fontanelle is bridged by a fibrous membrane, which becomes progressively reduced in size by bone formation around the periphery of the adjacent bones. The anterior fontanelle is the last to close, at about 18 months, leaving the bones of the cranial vault in contact along a complex pattern of suture lines.

The increase in size of the cranial vault necessary to accommodate the progressively enlarging brain is accomplished by bone deposition along the lines of the sutural system. Figure 7 diagrams the new bone formation that has taken place at the coronal and lambdoidal sutures in order to accomodate the increase in size of the cranial contents.

The curvature of the surface of a large sphere is less than that on a small sphere. The adult calvarium is larger than the infant's and shows a corresponding reduction in curvature. Thus, in addition to increase in circumference by sutural deposition, the bones of the cranial vault undergo a progressive flattening. The reduction in curvature is carried out by selective bone deposition and resorption at areas over both the inner and outer surfaces of the bones.

Originally the sutures were regarded as primary growth centers (Weinmann and Sicher, 1955), that is the new bone laid down at the suture interface was thought to be responsible for the movement apart of the adjacent bones. This has been questioned by several investigators (Moss, 1954; Scott, 1956), and it is now more generally accepted that sutural growth is a secondary phenomenon, involved with roofing over the separation between the bones created by the effect of increase in size of the enclosed soft tissues. Thus increase in brain volume will cause a potential separation at, for instance, the lambdoidal and coronal sutures, which is filled in by bone formation at the suture.

The progressive growth pattern of the calvaria has been demonstrated by Brodie (1953), using the superimposition of tracings taken from lateral skull radiographs (Figure 8). The spacing of the tracings over the period from 6 months to 4 years indicates a rapid phase of growth. After this age, the tracing lines become more coincident, indicative that only a relatively small increase in sagittal cranial growth occurs after 4 years, shown mainly by an increase in the anteroposterior dimension.

3.2. *Cranial Base*

The cranial base is largely composed of bones formed by the ossification of cartilage precursors. The base is formed from the basal part of the occipital, the sphenoid, the petrous part of the temporal, and the ethmoid bones. The main postnatal growth mechanism found in this region is provided by en-

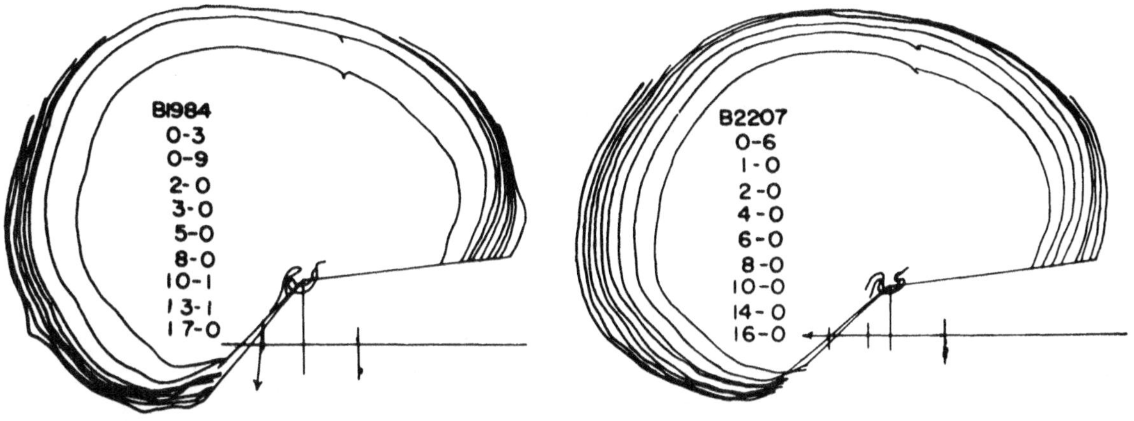

Fig. 8. Progressive growth of the calvaria. Tracings from radiographs at ages (years and months) shown inside the composite drawings. (From Brodie, 1953.)

dochondral ossification at the sphenoccipital synchondrosis (Figure 9) which is active until about 12–15 years and fuses at about 17–20 years of age. Cartilage replacement at this site is related to forward growth of the whole anterior cranial segment.

Anterior movement in this manner causes an increase in the anteroposterior dimension of the nasopharynx and carries the upper facial skeleton forward. At the same time, the glenoid fossae (the sites of articulation of the mandibular condyles) are carried posteriorly and inferiorly (Figure 13). Thus the anteroposterior growth of the mandible must be greater, when compared to the maxilla, if the bony bases carrying the dentition are to be related in a harmonious fashion. The direction of growth also has a vertical component which tends to open the space between the maxilla and mandible associated with the vertical development of the alveolar processes.

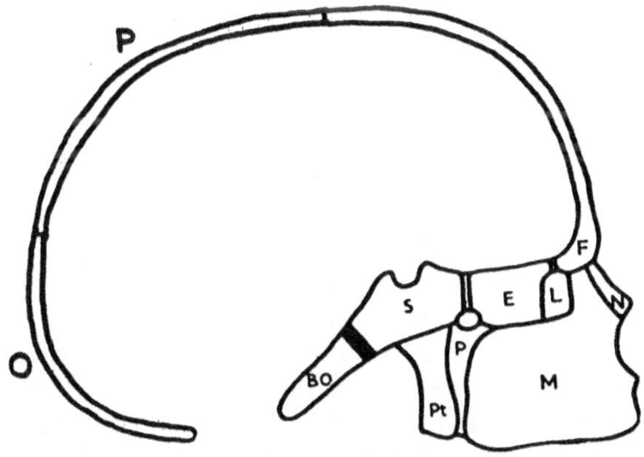

Fig. 9. The bones of the calvaria, cranial base, and upper face. Position of the sphenooccipital synchondrosis shown in black. M, maxilla; N, Nasal bone; F, frontal bone; L, lacrimal bone; E, ethmoid: P, vertical plate of palatine bone; S, body of the sphenoid; Pt, Pterygoid plate; BO, Basioccipital; Pa, Parietal.

The process of endochondral ossification in a synchondrosis appears to be similar to that found in the epiphyseal cartilage of a long bone, except that, unlike the epiphyseal cartilage, ossification occurs on two surfaces. In the case of the sphenooccipital synchondrosis ossification takes place on both the sphenoidal and the occipital faces of the cartilage.

The presence of an area of cartilage with interstitial growth potential, which has been shown to be active during the period of skull growth, has led to its implication as a primary growth center for the base of the skull. Evidence for this view is obtained by examination of the effects of developmental disorders that affect the formation and growth of cartilage. For instance, in achondroplasia, the base of the skull is shortened, with a flattened palate, sunken bridge to the nose, and a general reduction in the development and size of the facial region. Conversely, the growth of the vault of the skull, which does not rely on the ossification of cartilage for growth, is unhindered (Brash, 1956).

The growth characteristics of cartilage were studied experimentally by Koski and Ronning (1970), using an intracerebral transplantation technique in the rat. They found that the growth capacity of the transplanted cartilage depended on the origin of the cartilage, that from the epiphysis showing a continuing ability to proliferate while the growth of the synchondral cartilage was much reduced. Nevertheless, synchondral cartilage grows normally *in situ*. Thus, unlike an epiphyseal plate, it appears that the sphenooccipital synchondrosis requires an additional extrinsic factor for the stimulation of normal growth.

Although the deepening of the middle cranial fossae on either side of the midline is related to increase in the temporal lobes of the brain, Scott (1967) suggested that growth of the midline cranial base is largely independent of the growth of the brain. He drew his evidence from the finding that patients suffering from microcephaly show a normal range of cranial-base dimensions despite the marked reduction of brain volume.

Whether the synchondrosis acts as a primary growth center or not, it must not be regarded as the only growth mechanism participating in cranial base growth. If independent endochondral ossification were to occur in the sphenooccipital synchondrosis, the skull would be split into two parts, anterior and posterior to the synchondrosis. No splitting takes place because a synchronized process of sutural growth and surface remodeling takes place in the adjoining bones, with the effect of producing an integrated change in size and shape of the whole bone complex. In this instance all three mechanisms of endochondral ossification, sutural growth, and surface remodeling participate in the achievement of an harmonious change in skull form.

3.3. Upper Face

The bones of the upper face ossify in membrane and are developmentally associated with the capsules of the eyes, nose, and mouth cavity.

The upper facial region consists of the nasal, lacrimal, maxilla, zygomatic, vomer, palatine, and pterygoid bones. It will be seen that this regional complex is closely related to the anterior segment of the cranium formed by frontal, ethmoid, and sphenoid bones (Figure 9). Thus any relative forward growth of the anterior cranial base will carry the upper facial region with it into a more anterior position.

The growth of the nasomaxillary complex takes place in a downward and forward direction and is related to the growing eyeballs and nasal septum (Figure 10). The increase in height of this region contributes to the relative increase in height of the face compared with the neutral part of the cranium and is related to the enlargement of the maxillary sinus and the development and eruption of the dentition.

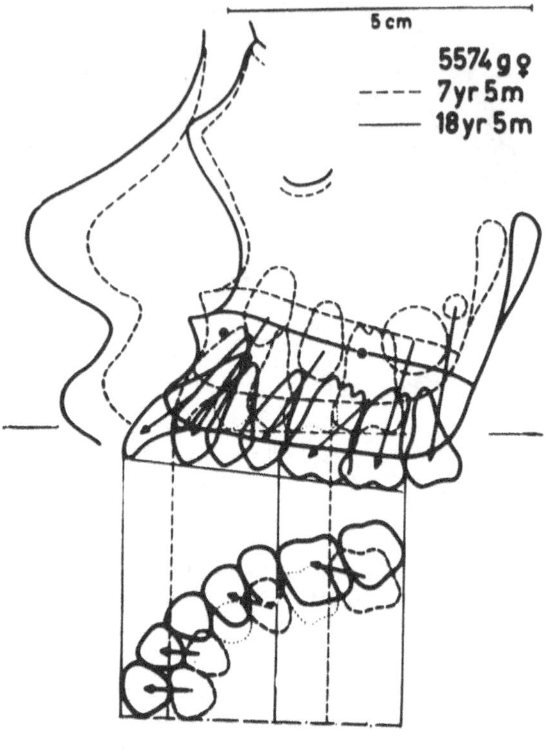

Fig. 10. Growth of the maxilla analyzed by superimposition on implant markers. (From Bjork, 1964.)

The nasofacial region is attached to the anterior skull base by a complex series of sutures, including the frontomaxillary, zygomaticotemporal, zygomaticosphenoidal, zygomaticomaxillary, ethmomaxillary, ethmofrontal, nasofrontal, frontolacrimal, palatine, and vomerine sutures. The orientation of this circumfacial system is such that sutural growth will result in the downward and forward movement of the facial complex when compared with the base of the skull (Scott, 1967).

It is suggested by Enlow (1975) that more vertical displacement takes place by sutural growth in the posterior part of the facial complex than in the anterior region. The balance is said to be maintained anteriorly by surface deposition and resorption. This involves the concept of sutural slippage, the relative movement of adjacent bones past one another by adjustment at the sutural interface, described by Latham (1968).

The marked increase in vertical height of the

maxilla, which is associated both with increase in the size of the maxillary sinus and with development of the alveolar bone, is brought about by surface deposition and internal resorption (Figure 11).

A measure of the increase in size of the maxilla is given by the change in size of the maxillary sinus. At birth the sinus occupies a small depression in the lateral wall of the nasal cavity, but by adult life the small depression has expanded to a large cavity.

The growth pattern shown by the maxilla, the major bone in the upper face, provides an example of the operation of the three mechanisms of bone movement described in Section 2. If the movement of a landmark on the anterior surface of the maxilla from soon after birth to adult life is examined (point A in Figure 4a), all three growth mechanisms are found to participate in the movement.

First, the landmark (point A) will move in a mainly downward direction by bone deposition on the inferior-facing surfaces of the maxilla (Figure 11). An equivalent, integrated degree of bone resorption is found on the superiorly facing surfaces. This remodeling is associated with the increase in maxillary height that occurs with the enlargement of the maxillary sinus and the development and eruption of the dentition.

Second, the landmark will move following sutural growth at the circumnasal suture system. The effect on point A of growth at these sutures is in a downward and forward direction, but in this instance the movement is achieved by the bodily movement of the whole maxilla.

Finally, the landmark is repositioned in a forward direction by endochondral bone formation at the sphenooccipital synchondrosis. In this instance the whole anterior cranial base has been repositioned forward by an enlargement of the middle cranial fossa.

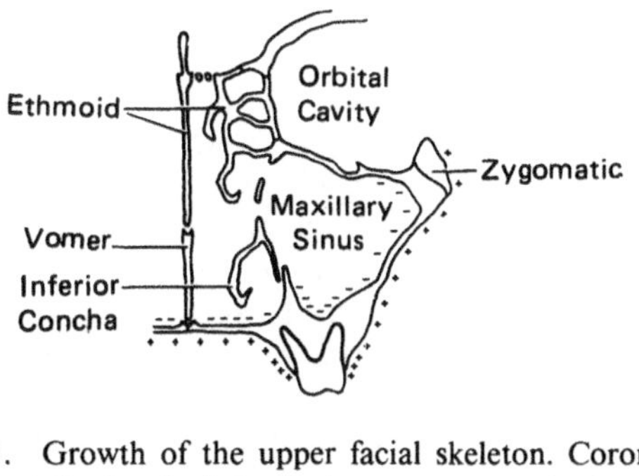

Fig. 11. Growth of the upper facial skeleton. Coronal section at the level of the first permanent molar tooth. Surface deposition, +; resorption, −. (After Enlow, 1975.)

Furthermore, it will be appreciated that the movement of one point on the surface of the maxilla, in a downward and forward direction, does not mean that all the points in the maxilla are growing in the same direction at the same rate. For instance, at stages during the development of the molar teeth, the posterior aspect of the maxilla is growing rapidly in a distal direction.

This increase in anteroposterior length is obtained by deposition at the posterior borders of the alveolar arch to form the maxillary tuberosities. The next molar tooth to be added to the dental arch is found developing in the tuberosity. Thus at 3 years, the tuberosity contains the developing first permanent molar, at 6 years the developing second permanent molar, and at 12 years the developing third permanent molar. As the current molar tooth erupts the posterior development of the tuberosity continues to accommodate the next tooth in the series.

The part played in maxillary growth by sutural deposition and surface remodeling is well recognized; the contribution made by ossification of cartilage is more controversial. In the midline of the nasomaxillary complex is found the septal cartilage, which articulates in infancy with the ethmoid sphenoid, vomer, and premaxillary bones (Figure 12). This structure is derived from the cartilage of the embryonic nasal capsule and is thus formed from primary cartilage. At the septoethmoidal junction the cartilage is replaced by bone in a zone of endochondral ossification (Baume, 1961). Furthermore the cartilage is attached anteriorly to the premaxilla by a septopremaxillary ligament (Latham, 1969). The growth potential of cartilage and the anatomy of this region have prompted several authors (e.g., Scott, 1967) to suggest that the nasal septum acts as a primary growth center for the normal development of the upper facial region. This view is supported by experiments involving the resection of all or part of the nasal septum in experimental animals, with a resulting distortion in the normal growth pattern of the facial region (Wexler and Sarnat, 1961; Sarnat and Wexler, 1966, 1967). A similar result was found by Grange and Johnston (1974), who severed the septopremaxillary attachment in rats.

These results are contradicted by the work of Moss *et al.* (1968) in the rat and by Stenstrom and Thilander (1972) in the guinea pig. These papers suggest that the nasal septum acts solely as a structural component and has no primary determining effect on facial growth. Moss *et al.* (1968) also claim

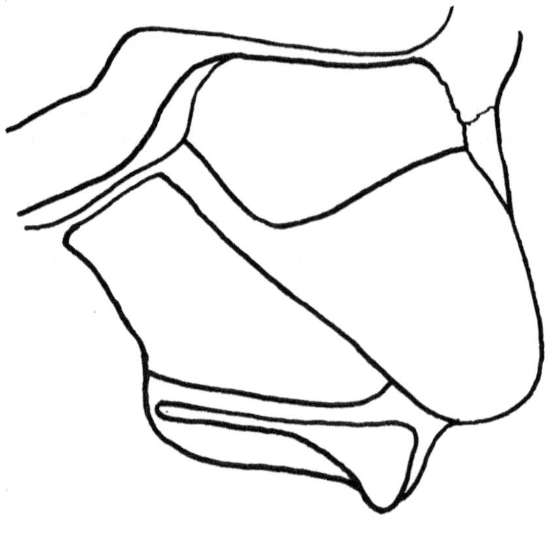

Fig. 12. The antral septum and adjacent bones in a young child.

that in humans, congenital absence of the cartilaginous nasal septum leads to little reduction in facial growth.

4. Mandible

The human mandible is a membrane bone which forms in close association with Meckel's cartilage, the first branchial arch cartilage. At birth the bone is in two parts, joined in the midline by the symphysis menti, which closes by the end of the first year of life.

No other bone of the skull has been more extensively studied, the basic growth pattern being first described by John Hunter in the 18th century. The mandible is unusual in having a secondary growth cartilage located posteriorly under the surface of the articular condyle. The secondary growth cartilage differs from a synchondrosis or epiphyseal cartilage in that it is derived from a membrane condensation and is not a remnant of a primary cartilage precursor. Not only does the cartilage differ in origin but also in structure. It is covered with a layer of fibrous tissue that forms the articular surface, unlike other examples of cartilage, which are bounded by bone on both sides of the area of interstitial cartilaginous growth.

The overall change in shape is shown in Figure 2. The most striking difference is found in the relative increase in height of the ramus when compared with the body of the mandible. This change in proportion is reflected by the decrease in the angle between the posterior and inferior borders of the bone.

The growth to adult dimensions is brought about by a combination of endochondral ossification of the condylar cartilage and by remodeling over the bone surface. The larger part of the increase in the height of the mandibular body is brought about by addition to the alveolar bone on the superior surface, and similarly most of the increase in length occurs through a process of progressive distal relocation of the ramus (Figure 13). The movement of the ramus is effected by deposition on the posterior border and resorption from the anterior border of the bone. This process was first described by Hunter (1798) and subsequently confirmed in a classical experiment by Humphrey (1863). He inserted wire rings into both the anterior and posterior borders of the mandibular ramus in the pig. After growth the anterior ring became extruded from the bone and was found in the soft tissues, while the posterior ring became more deeply embedded.

The remodeling changes seen in the development of an individual mandible, over a 5-year period are shown in Figure 5. In this case, the two mandibular tracings have been superimposed on the images of metal pins inserted into the bone, and the change in outline represents the true contour change experienced by the individual.

As with other sites of cartilage proliferation, the mandibular secondary growth cartilage has been suggested as a primary growth center for mandibular development. This view is supported by the observation that damage to the condyle at an early age may be followed by a reduction in mandibular growth (Rushton, 1944). Subsequent studies in experimental animals have provided results which suggest that following condylectomy only minor changes (Sarnat, 1963) or no change (Giannelly and Moorrees, 1965) in mandibular growth pattern occurs. It may well be that modification of the growth pattern following early damage may be brought about by contraction or fixation by scar tissue rather than by loss of the growth potential of the condylar cartilage. Moss and Rankow (1968) claim that careful surgical resection of the condyle in childhood produces little reduction in mandibular growth.

The matter is further complicated by the effects of pathological conditions on the secondary growth cartilage. In eosinophilic adenoma of the pituitary gland (which causes gigantism before and acromegaly after the fusion of the epiphyses) the mandibular growth cartilage reacts like epiphyseal cartilage, and a characteristic overgrowth of the mandible occurs. However, in the hereditary disease achondroplasia, unlike the long bones (epiphyseal cartilage), the mandible continues to grow with a normal growth pattern. Furthermore, transplantation studies have shown that, unlike epiphyseal cartilage, transplanted condylar cartilage has very little independent growth potential (Koski and Ronning, 1965). Thus it would seem that on balance the evidence is against the role of the condylar cartilage as a primary growth center.

This view is supported by a series of experiments carried out by Petrovic and coworkers, who studied the effect of appliances fitted to the teeth of rats (Petrovic *et al.*, 1981). The appliances were designed to hold the mandible in a protrusive position, with the condyle forward out of its correct articulation. Petrovic found that this posture led to an increased rate of growth in the secondary cartilage of the condyle. These experiments have been confirmed in primates (McNamara and Carlson, 1979) and in young adult animals (McNamara *et al.*, 1982).

5. Dentition

In humans two sets of teeth are formed. The deciduous dentition (20 teeth) starts to erupt at 6 months and is completed by about 2½ years. Eruption of the permanent dentition (32 teeth) commences at 6 years and the last of the deciduous teeth are shed by 12–14 years. The permanent dentition is usually completed by the end of the second decade.

The timing of both the development and eruption of the teeth has been studied by several investigators, for instance Moorees *et al.* (1963*a,b*). The increase in arch length necessary to accommodate the increase in number of teeth is achieved both by expansion of the dental arch and by the addition of bone at the ends of the arch. The shape of the dental arch is established with the eruption of the deciduous dentition, and the arch expands in size until about 8 years (Figure 14) (Sillman, 1964). After this age, only slight increase in transverse arch width is found, the major enhancement in length being provided by increase at the posterior ends of the arch. In the upper jaw arch length is increased by bone deposition in the tuberosity region of the maxilla. In the lower jaw, space for eruption of the posterior teeth is provided by the distal relocation of the mandibular ramus.

The positions of the teeth in the vertical dimension are influenced by: (1) Movements associated with the initial eruption of the teeth; (2) Movement associated with surface deposition and remodeling of the alveolar bone after tooth eruption [Confirmation of this form of vertical movement is

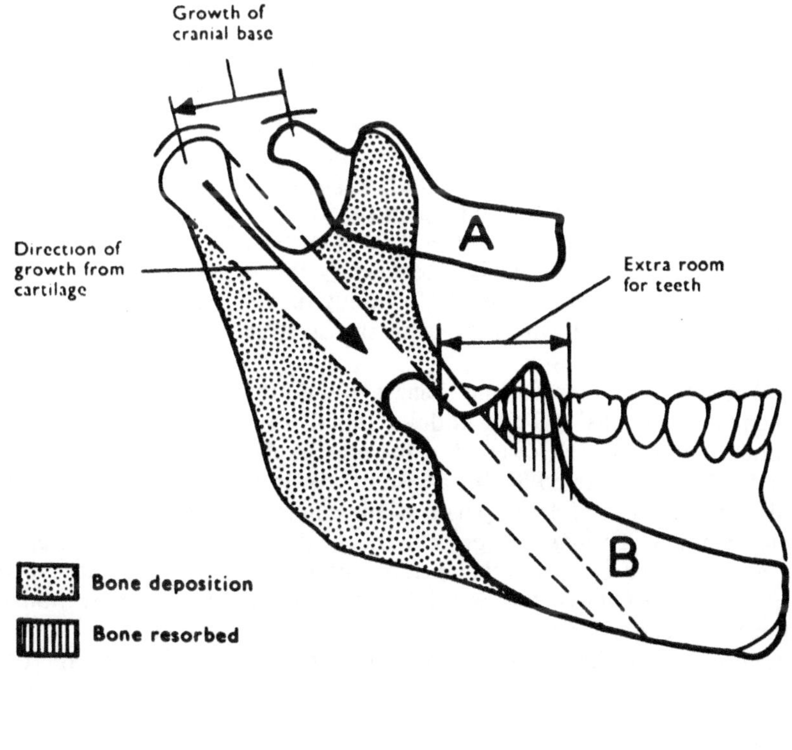

Fig. 13. Growth changes in the mandible. (After Weinmann and Sicher, 1955.)

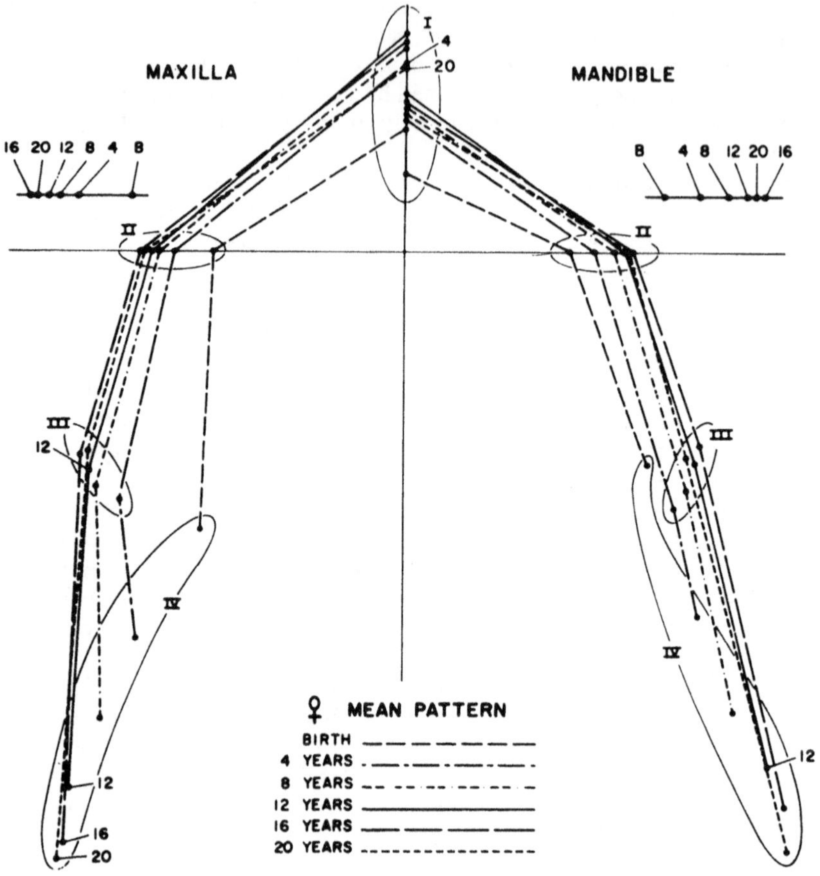

Fig. 14. Composite mean growth pattern of male maxilla and mandible from birth to 20 years of age. (From Sillman, 1964.)

found in the misnamed clinical condition of tooth "submergence" (Figure 15). In this case, a single tooth becomes ankylosed to the surrounding bone and does not experience the vertical development shown by the adjacent teeth. Although originally in occlusion with the teeth of the opposing arch, the tooth is overtaken by the vertical development of the neighboring teeth, and is subsequently found below the surface of the bone.]; (3) Bodily movement of the bone on which the teeth and alveolar bone are based.

The growth and remodeling changes of both the mandibular ramus and the middle cranial fossa produce a progressive lowering of the mandibular arch. The space created by this aspect of growth is filled by the vertical development of the craniofacial complex. During this process the teeth are maintained in occlusion by the operation of the factors mentioned above, but to a differing degree in the two arches. For instance, the vertical movement of the maxillary teeth is effected by surface remodeling to a greater degree than that of the mandibular teeth (Enlow, 1975).

6. Analysis of the Patterns of Skull Growth

The shape of the calvarium is established relatively early in life (Figure 8), and subsequent change is limited mainly to increase in size. The facial skeleton shows a more extensive postnatal change, which is not complete until the second decade and is capable of showing a much greater range of individual variation.

The overall growth of the facial complex in a downward and forward direction was demonstrated by Broadbent *et al.* (1975) using data taken from 5000 radiographs (Figure 4). The progressive change in facial form shown in the diagram is based on a cross-sectional study and represents the mean pattern.

A similar approach to skull growth, demonstrat-

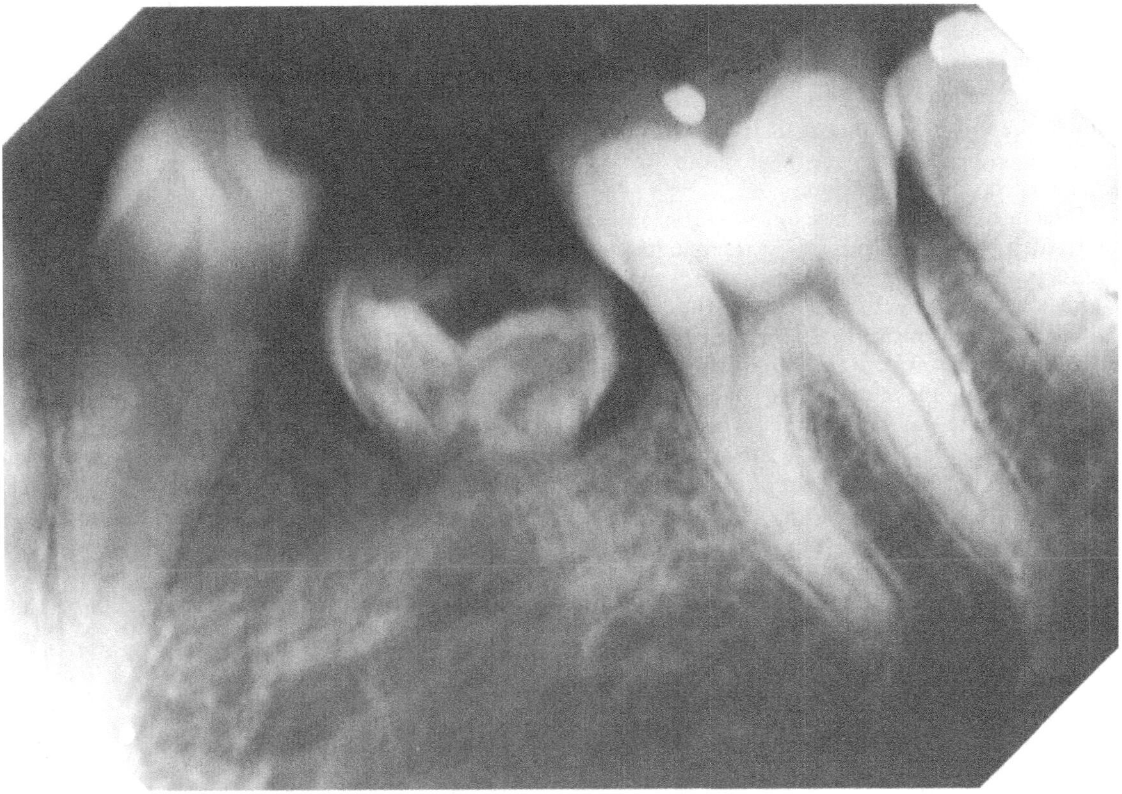

Fig. 15. Radiograph showing relative submergence of a previously fully erupted tooth. The tooth has become ankylosed to the supporting alveolar bone and has not experienced the occlusal movement shown by the adjacent teeth.

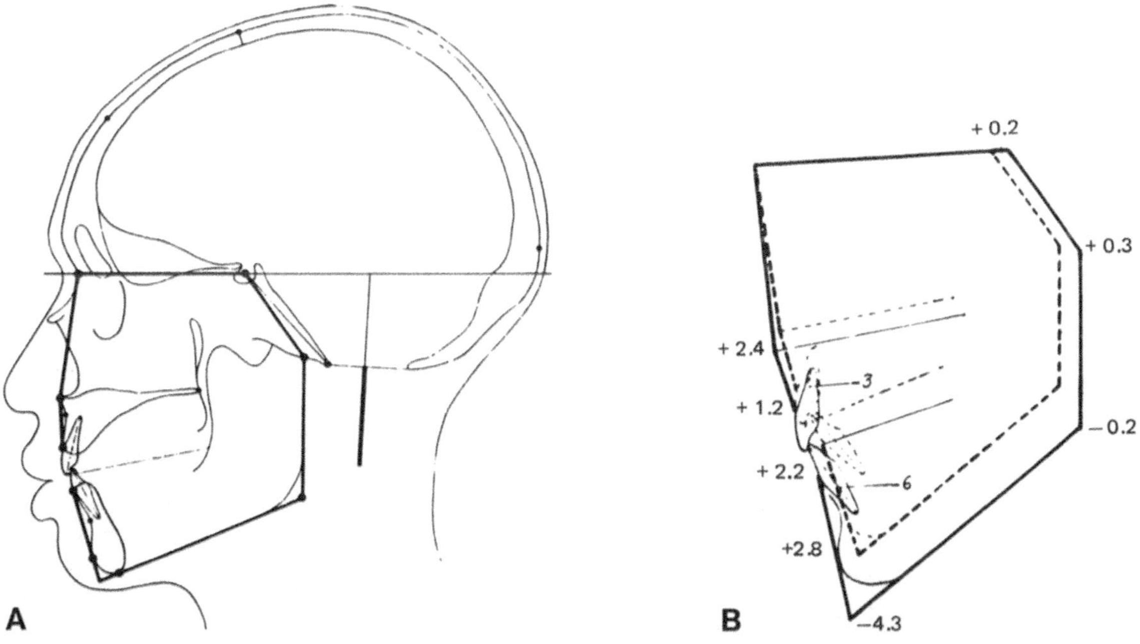

Fig. 16. Diagram to show the change in facial pattern between 12 and 20 to 21-year-old Swedish males. (From Bjork 1951.) Region of a radiograph tracing (A) from which (B) is taken. Measurements in millimeters.

ing the mean growth pattern, was adopted by Walker (1972; Figure 6). Broadbent *et al.* (1975) plotted the mean profile at annual intervals deriving his data from recordings taken from 100 subjects at each stage in development.

If the mean pattern of facial growth is examined in detail, a gradual increase in facial prognathism is found, with the bony bases growing forward relative to the anterior base of the skull (Bjork, 1951). The chin grows forward more than the base of the nose, and both grow forward more than the area of bone supporting the teeth. These changes are illustrated in Figure 16. Many other workers (Wylie, 1947; Bjork, 1947; Downs, 1948) developed analytical strategies for the assessment of facial shape and form that have been extensively used in planning orthodontic and surgical treatment. For instance, the Steiner analysis (Steiner, 1953, 1959, 1960) relates the positions of the teeth and jaws to the cranial base and permits comparison of these relationships with the population average.

Other authorities have developed methods that have sought to illustrate the discrepancy between the individual and the mean by the use of graphical representation. Voorhies and Adams (1951) drew out a polygon, based on the strategy of Downs (1948), representing the range of variation about the mean of 10 selected measurements taken from cephalometric radiograph tracings (Figure 17). The measurements record the configuration of both the skeletal and the dental elements in the skull tracing. The value for each measurement taken from any individual radiograph is entered on the chart, and a straight line is drawn between the points. The "shape" described by this line is characteristic of the overall skull configuration of the individual concerned.

Moorrees and Yem (1955) described a method to

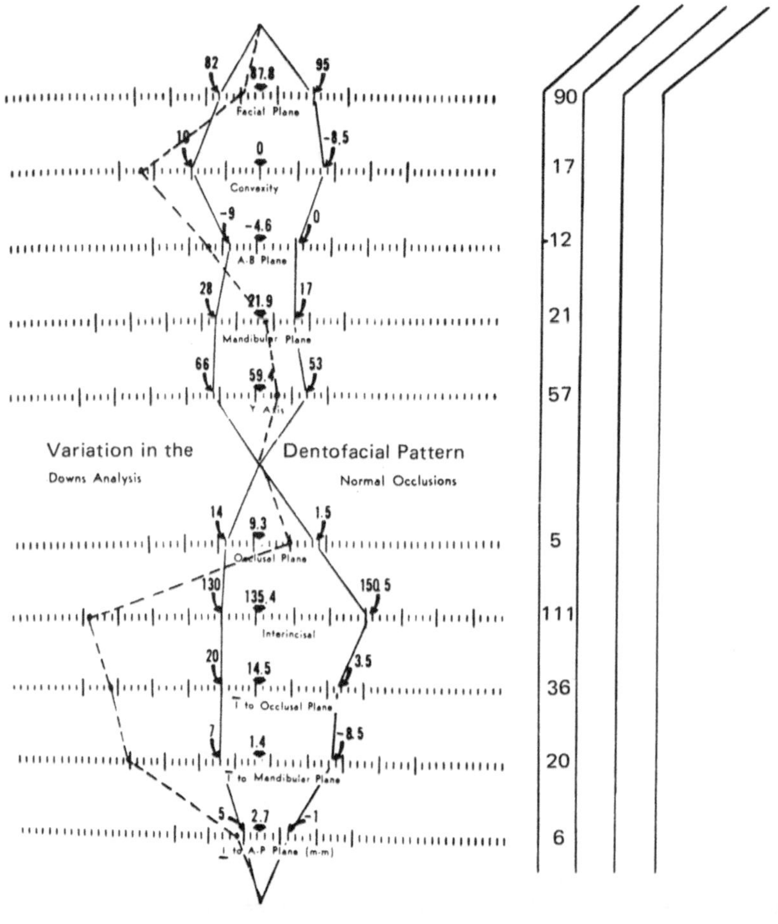

Fig. 17. Polygon demonstrating the relationship between the dentofacial pattern of an individual (broken line) and that of the population (solid line). The upper half of the polygon records the range of skeletal measurements, and the lower half refers to the range of dental measurements. (From Voorhies and Adams, 1951.)

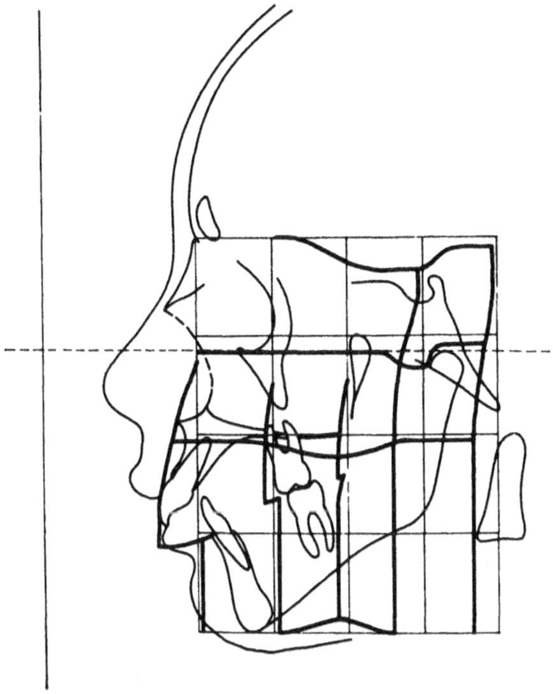

Fig. 18. Mesh analysis of craniofacial growth in an 11-year-old boy. See text for an outline of mesh construction procedure. (From Moorrees and Kean, 1958.)

illustrate the skull growth changes based on the coordinated analysis method described by D'Arcy Thompson (1917). The changes in shape and position are demonstrated by the derivation of a rectangular mesh superimposed on tracings of cephalometric radiographs taken before and after treatment (Figure 18). The relationships of certain radiographic landmarks to the horizontal and vertical mesh lines in the pretreatment tracing are changed in order to achieve a similar proportionate distance of landmarks to mesh lines as observed in the tracing made before treatment. The resulting distortion of the mesh lines calls attention to differences in the position of landmarks within various rectangles of the diagram.

This approach was extended by Moorrees and Kean (1958) to examine the relationship between head orientation at rest and certain of the intracranial reference lines. All their radiographs were taken using an optical method to ensure a "natural head position." A true vertical marking registered on the radiograph was employed as a reference line in the construction of the rectangular mesh. The relationship between the cranial reference lines and the mesh was used to orient the cranial base to the vertical and to study the relationship between anatomical structure and the posture and profile of the head. In addition they derived the average facial pattern of 50 female students. The ranges of individual variation about the mean are shown in Figure 19.

The early studies of Broadbent and of Brodie suggested that the pattern of facial growth occurred in an even manner in a downward and forward direction (Figure 4B). With the completion of longitudinal growth studies came the realization of the range of the pattern of growth. Individual patterns

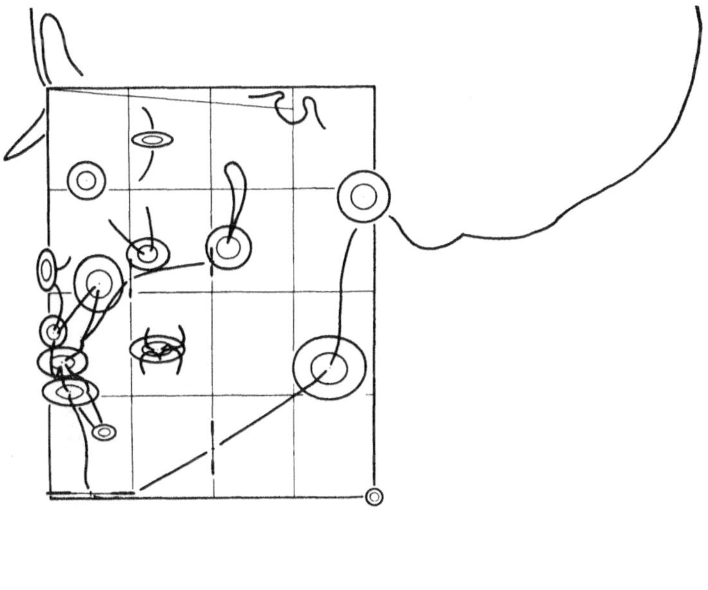

Fig. 19. The average facial pattern, based on 50 North American females, The concentric ovals show the ranges of individual variation around the mean at the one and two standard deviation limits. (From Moorrees and Kean, 1958.)

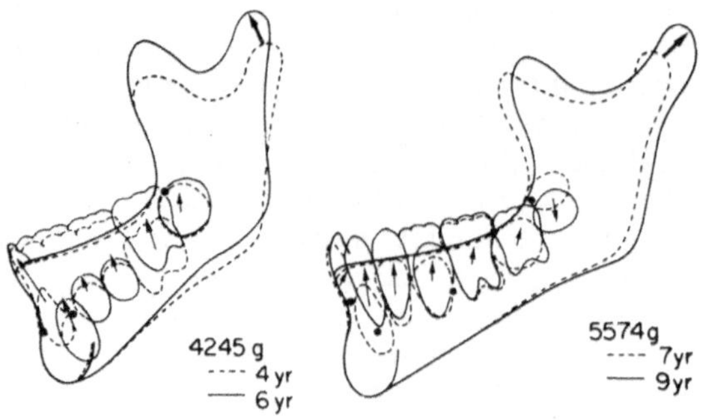

Fig. 20. Variation in mandibular growth pattern. Examples illustrated show extremes of difference in pattern. (From Bjork, 1955.)

can differ markedly from each other and from the mean. The possible range of individual variation was particularly well demonstrated by the use of the fiducial marker-pin technique developed by Bjork (Figure 20). In a study of mandibular growth in 45 males, Bjork found that mandibular growth could vary widely between a predominantly downward and backward to a predominantly downward and forward pattern (Figure 21). Bjork also found that as well as moving slightly forward the jaws can rotate when compared with the remainder of the skull (Bjork and Skieller, 1983) (Figure 22). The rotation is, on average, in an upward and forward direction, more marked in the mandible than in the maxilla.

Although a mildly prognathic growth pattern, with mandibular development similar to that seen in Figure 21B, is representative of the mean, some individuals diverge from this pattern. In Figures 21A and 21B the growth patterns vary markedly; in cases such as these at the extremes of the range of variation it is clinically important to predict the outcome of the growth pattern. A gross divergence from the mean will lead to a marked degree of facial disharmony when full adult size has been reached. Whereas it is possible to treat some of these cases by orthodontic techniques alone, others require a combined surgical–orthodontic approach. In either case, the appropriate treatment will depend on recognizing the abberrent growth pattern at an early stage in the development of the jaws.

Considerable efforts have been made to develop methods to predict facial growth. Johnston (1975) constructed and tested an onlay template designed to be placed over a tracing taken from a patient's radiograph. The template is inscribed with a series of scales recording the movement with growth of six radiographic landmarks. The intervals along the scales are derived from survey data and represent the mean displacement of the landmark over a one year period. To obtain the prediction the landmarks are "moved" along the scale by the appropriate number of divisions. From these new positions a predictive tracing is derived.

This approach was employed by Ricketts and associates using a computer-based technique to make predictions of individual growth (Ricketts, 1973; Ricketts *et al.* 1972). He collected information on skull growth both from his own and other published studies. The growth patterns found in these data were analyzed according to age, sex, size, and ethnic type and used to calculate the growth baselines for the appropriate populations. Suitable growth increments are abstracted from the stored

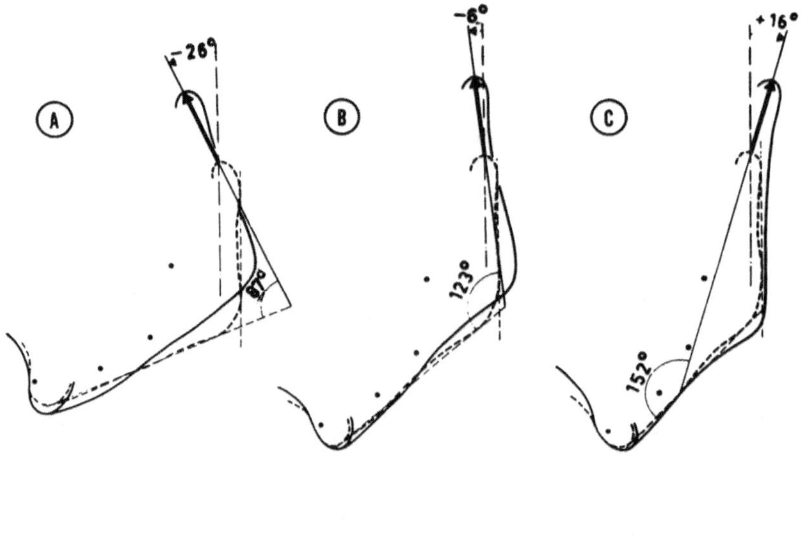

Fig. 21. Diagram illustrating range of mandibular growth pattern in 45 males. B illustrates the mean; and A and C show the extremes of condylar growth direction. (From Bjork, 1963.)

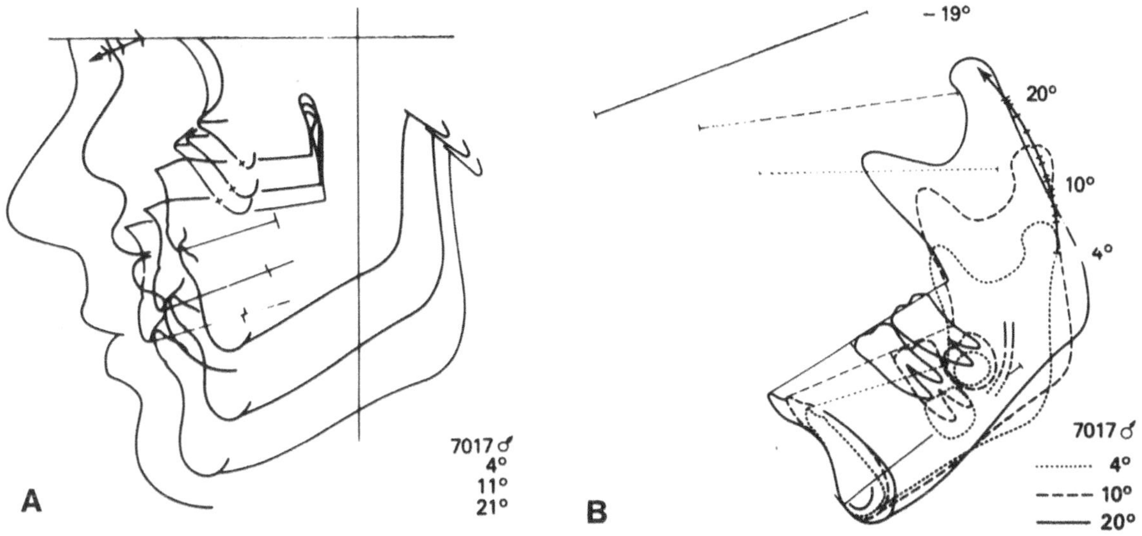

Fig. 22. Diagram to show the rotation of the mandible with growth. (From Bjork and Skieler, 1983.) (A) The relationship of the mandible to a plane recording the anterior base of the skull (sella–nasion plane) is shown at three stages during growth. (B) The three stages have been resuperimposed on fixed natural landmarks in the mandible, the change in orientation of the reference plane indicates the rotation brought about by differential growth.

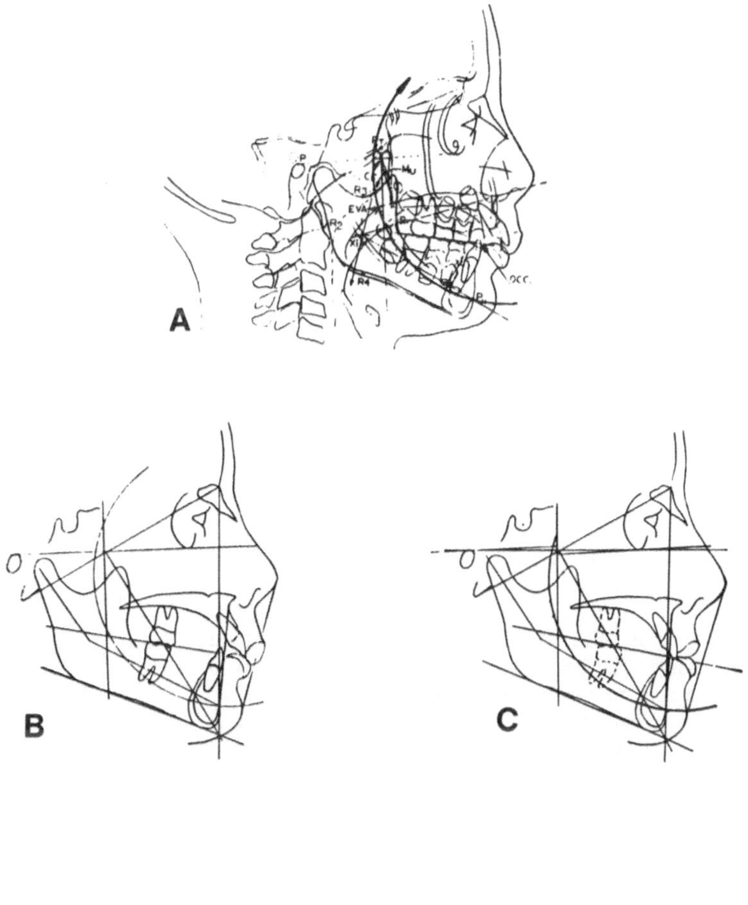

Fig. 23. Ricketts' arcial method for the prediction of facial growth. (A) Tracing of a girl at 8 years of age with an arc drawn through the protuberance menti or suprapogonion and up through a point called EVA at the center of the upper anterior quadrant of the ramus. A third point formed with the distance EVA to PM as its radius will produce the arc as demonstrated. It is then hypothesized that this is the direction in which the mandible grows. (B) The prediction without treatment at the age of 14½ years after principal growth has ceased. (C) The idealized positioning of the teeth after treatment. (From Ricketts *et al.*, 1972).

data by the computer program and are applied to a radiograph tracing to draw the skull form to be expected after a selected time interval. The forecast shape is demonstrated by a computer-drawn plot of the skull. The prediction may be further modified by the anticipated effect of orthodontic treatment on the positions of the teeth. On the basis of these data, Ricketts suggested that the basic growth pattern of the mandible follows a circular path, the particular curve for an individual being derived from a geometric construction based on tracings taken from the lateral skull radiograph (Figure 23).

The accuracy of this method of growth prediction was examined by Mitchell *et al.* (1975). These workers studied eight cases using the position of metal implants in the jaws to compare the predicted tracing and the tracing taken at the end of the growth period. They claimed an acceptable accuracy of prediction in six of the eight cases studied.

The growth pattern of the mandible was examined by Moss and Salentijn (1970) and by Moss *et al.* (1974), using an implanted metal pin superimposition method. These authors also suggested a curving path for mandibular growth, in this case based on a logarithmic spiral (Figure 24). These investigators claimed that the radiographic positions of three neurovascular foramina—the mandibular foramen, mental foramen, and foramen ovale—are found to lie, during growth, along the path of a logarithmic spiral.

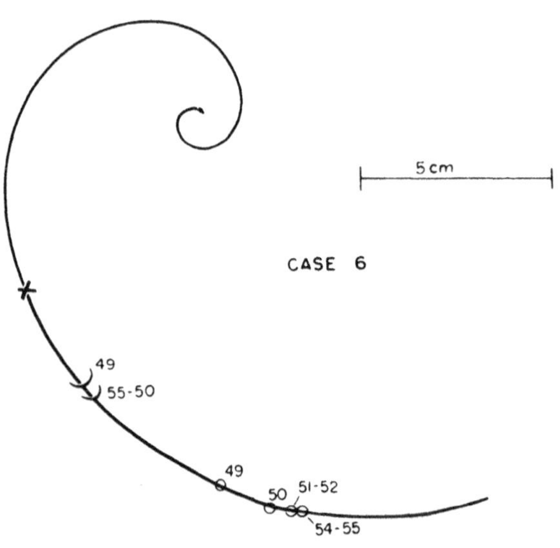

Fig. 24. Analysis of mandibular growth by the use of a unitary logarithmic spiral. With constant registration at foramen ovale (x) and with the spiral, held in an arbitrary position, the absolute "movement" of the sites of the mandibular (O) and mental (o) foramina are demonstrated at annual intervals over a 6-year period. The numerals indicate calendar years; e.g., 49 = 1949. (From Moss *et al.*, 1974.)

There is some support for the concept that the nerve foramina and canals might in some way act as orienting points or pathways during skull growth. Koski (1973) drew attention to the relatively unchanging orientation of the inferior dental and other nerve canals during growth. He found that if tracings of lateral skull radiographs were superimposed and scaled to overcome differences in size, the orientation of the inferior dental canal in the mandible was relatively unchanging during growth.

The "stability" of the inferior dental canal has also been mentioned by Bjork (1963). On the basis of his metal pin measurements in the mandible, he suggested that three radiographic landmarks are relatively unchanging during growth and may be used as fiducial markers. In addition to the outline of the inferior dental canal, the other reference contours made up the lower margin of a molar tooth germ prior to root formation and the inner cortical structure of the inferior border of the symphysis.

The contributions made by the different mechanisms for the increase in bone size, whether sutural growth, cartilage replacement, or surface remodeling, to the growth of the regions of the skull have been mentioned previously in this chapter. However, the control system by which the effects of these mechanisms are integrated into the overall growth of the skull are still little understood. Whereas transplantation studies suggest that epiphyseal cartilage (the growth mechanism for long bones) appears to possess an independent growth potential, which would perhaps explain the elongation of a long bone, both the sphenooccipital synchondrosis and the condylar cartilage appear to require an additional stimulus to achieve their full potential.

On the basis of studies in experimental animals (Moss, 1954; Watanabe *et al.*, 1957; Sarnat, 1963), it is now generally accepted that sutural growth at a bone interface is secondary to the increase in size of the tissues contained within the bone. For example the increase in size of the neural tissue associated with growth of the brain tends to separate the bones of the cranial vault, which provides the stimulus for bone formation along the suture systems of the calvaria. This concept of the adaptation of the bone elements to the requirements of the contained tissues was greatly extended by Moss (1968) in his functional matrix theory. Moss suggested that the head could be considered as an aggregation of functional units, each unit responsible for a separate activity. The functional unit is divided into a skeletal component and a functional matrix that includes all the other tissues, organs, or spaces necessary to carry out the function.

The functional matrices may be of two types depending on whether they contain muscles attached to the bone *via* a periosteum (periosteal matrix) or consist of functional cavities (capsular matrix). The capsular matrices may include tissue masses, such as are found in the orbital and intracranial cavities, or may contain functional spaces as in the nose, mouth, and pharynx. Moss suggested that the capsular matrix acts by producing displacement of the surrounding skeletal units, examples being the growth of calvaria or circumorbital bones. Evidence for this view was supplied by experiments in the rat carried out by Young (1959), who found that reduction in intracranial contents led to a decrease in the size of the calvaria, whereas an enlargement followed the induction of experimental hydrocephalus.

The periosteal matrix is taken to operate by the stimulation of bone deposition and resorption. This view is supported by the experiments of Washburn (1947) and Avis (1961), who found that the extirpation of certain of the muscles of mastication can lead to an almost complete absence of the associated area of bone into which the muscle would normally have been inserted; i.e., absence of the related skeletal component.

Thus the theory would appear to suggest that the stimulus to bone growth is transmitted by the mechanical forces applied by the functional matrix. The effect of pressure causing bone resorption, and of tension causing deposition, is well established (Reitan, 1951); were this not so, the orthodontic movement of teeth would not be possible. Furthermore, Frost (1964) put forward theories based on the tendency to deformation of the surface of a bone when placed under conditions of stress. He suggested that bone areas which tend to a more concave contour under load were associated with deposition, whereas areas tending to a more convex contour lead to resorption.

The investigation of Bassett (1972) may form the basis of an explanation of how the effects on bone structure are mediated. He found that bone exhibits piezoelectric properties, i.e., deformed bone can produce weak electric potentials. It has also been shown experimentally that small electric potentials are capable of stimulating bone formation (Bassett *et al.,* 1964). The sign of the potential is associated with the nature of the change; i.e., bone deposition accompanies a negative potential, and resorption is associated with a positive potential. Perhaps a feedback system relying on minute deformation-induced potentials is responsible for the change of bone contour.

Facial growth is controlled by the inherent genetic constitution and the environmental influences upon an individual. Similarly, the development of a malocclusion is subject to the effects of both hereditary and environmental factors. The correct relationship of the two dental arches depends on three basic characteristics. The first is whether there is a disproportion in the relative positions of the bone arches in which the teeth develop; if there is a disproportion, the teeth will not erupt into a harmonious interdigitated relationship.

Second, the molding effect of the oral musculature affects the position and shape of the dental arches. The morphology and activity of the tongue inside and the circumoral musculature of the lips and cheeks outside dictate a balance position for the teeth on eruption. If the skeletal bony base elements are malrelated, but the soft tissue environment is favorable, the teeth, on eruption, are moulded into angulations which tend to compensate for the skeletal disproportion. However, this compensatory effect is limited and capable of compensating satisfactorily for only minor degrees of skeletal malrelation.

Third, the degree of dental crowding affects the positions of the teeth. During the process of evolution there has been a reduction in the size of the bone arches supporting the teeth. Despite the human dentition consisting of less teeth than the usual mammalian dentition (32 compared with 44), the bone arches are frequently too short to accommodate the complete dentition in a smooth arch. Thus, on development and eruption of the teeth competition for space occurs, resulting in crowding and displacement of individual teeth. In some populations a degree of dental crowding is a majority finding, constituting the "normal" state of affairs.

In addition to these three basic factors, other influences of a more local nature may operate in individual cases. These influences are many in number and include the failure of teeth to develop, the loss of teeth due to decay, the development of supernumerary teeth, and thumb- or finger-sucking habits causing tooth displacement.

The previous discussion has been concerned with the morphology and development of the bones of the skull. Much less attention has been paid to the measurement and analysis of the soft tissues of the face, although the assessment of facial pattern has an important bearing both on the success of orthodontic treatment and on planning surgical treatment for severe discrepancies of the facial skeleton. The measurement of facial form is complex. As Darcy Thompson pointed out in 1917, it is almost impossible to arrive at a qualitative assessment of change in a shape. One approach to this problem is to use subjective rankings of aesthetic appearance, a technique used by Tedesco *et al.* (1983), but this method is not based on direct morphological measurement. Other workers have investigated the use of facial measurements. Burke and Beard (1967) recorded the three-dimensional shape of the face by the construction of contour

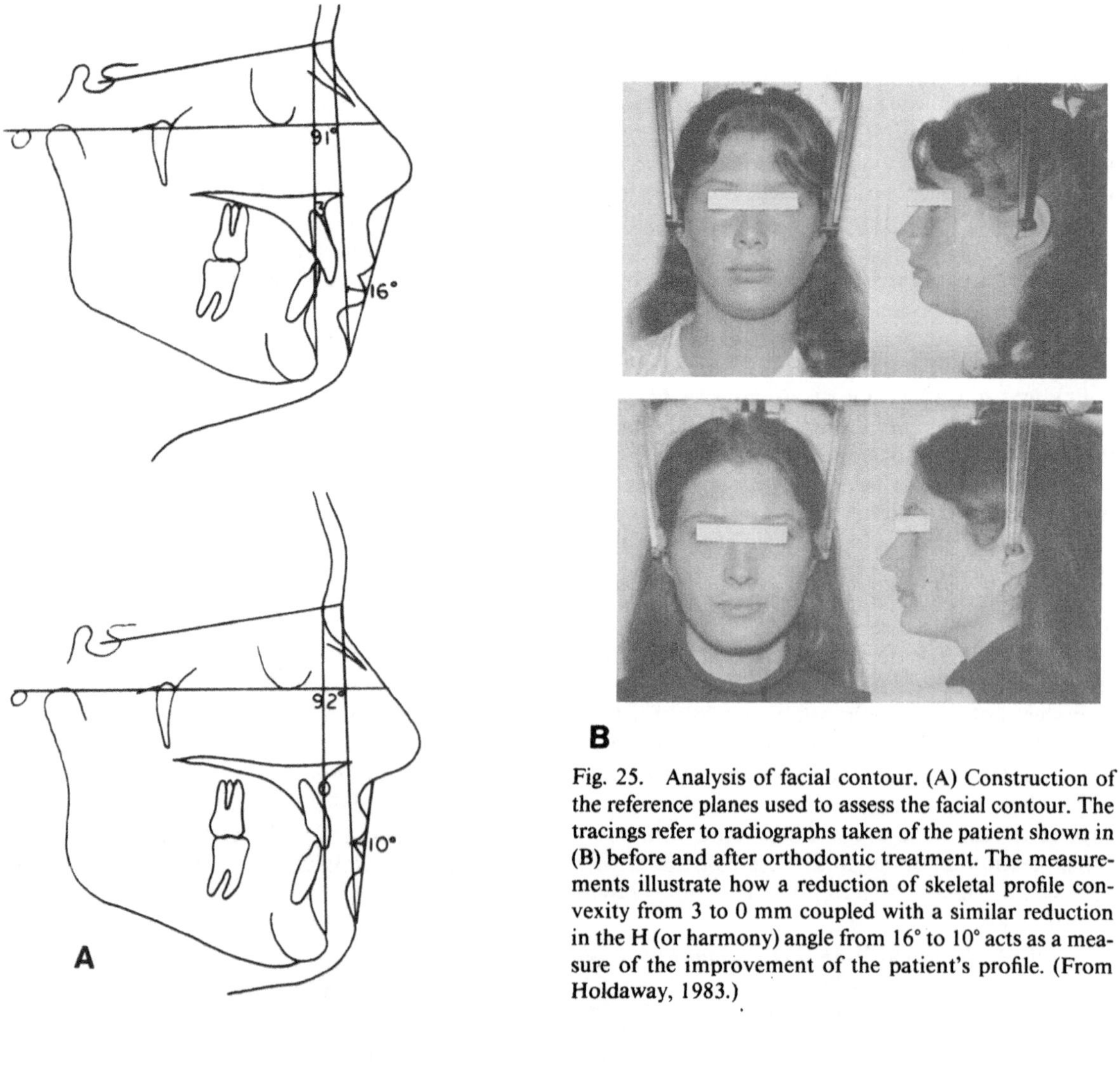

Fig. 25. Analysis of facial contour. (A) Construction of the reference planes used to assess the facial contour. The tracings refer to radiographs taken of the patient shown in (B) before and after orthodontic treatment. The measurements illustrate how a reduction of skeletal profile convexity from 3 to 0 mm coupled with a similar reduction in the H (or harmony) angle from 16° to 10° acts as a measure of the improvement of the patient's profile. (From Holdaway, 1983.)

maps using a stereo photogrammetric technique, and Burke (1983) suggested a series of measurements to characterise facial form.

Other workers have employed the same lateral skull radiographs as used to record the structure of the bones of the skull to study the soft tissue profile of the face. If an aluminum wedge filter is placed in an appropriate position, the soft tissue profile may be recorded on the same radiograph as the shape of the bones. The reliability of the soft tissue profile seen on skull radiographs was investigated by Hillesund *et al.* (1978).

Holdaway (1983) suggested an analysis that compares the soft tissue profile with the contour of the underlying bone. He studied radiograph tracings taken from patients with 'harmonious' facial profiles and suggested the use of two facial planes to act as baselines for measurement (Figure 25). The relationship between the baselines and the landmarks found in the facial bones are measured and compared with standard tables to estimate whether the measurements fall within the range of facial harmony.

The interplay between hereditary and environmental factors is complex and presents formidable problems for investigation. In this respect, several investigators have drawn attention to the effects that some forms of orthopedic or orthodontic treatment can have on the facial growth pattern. Logan (1962), Alexander (1966), Rock, and Baker (1972), and others have described the effect of the Milwaukee brace on the facial growth pattern. The Milwaukee brace is used in the treatment of scoliosis (abnormal lateral curvature of the spine) and consists of a metal braced caliper or corset. The appliance is designed to provide traction to the vertebral column, yet still permit the patient to stand and walk. The bottom end of the brace rests against the iliac crests of the pelvis, and the upper end against the underside of the mandible and occipital bones of the skull. The pressure exerted on the mandible is transmitted to the whole facial region and over a period of time distorts the normal growth pattern.

Similarly, changes in facial growth have also been reported due to orthodontic treatment techniques. These include the use of orthodontic appliances to cause rapid expansion of the mid-palatal suture (Haas, 1961) and the use of headgear harnesses to provide extraoral traction to orthodontic appliances (Klein, 1957; Graber *et al.,* 1967). Thus, it would appear that the normal growth pattern is capable of modification by the operation of external environmental factors.

A classic approach to the differentiation between hereditary and environmental effects lies in the study of pairs of twins, and this form of investigation has been applied to the patterns of skull growth (Lundstrom, 1954; Hunter, 1965). Differences in growth pattern found between the members of monozygous pairs must be due to environmental factors. Dudas and Sassouni (1963) studied the growth of 12 pairs of monozygotic (MZ) twins and 10 pairs of dizygotic (DZ) twins. These investigators compared skull growth between members of each pair of twins and between the two groups. They concluded that of the 15 skeletal measurements recorded in the investigation, only four appeared to be under strong hereditary (or weak environmental) influence.

The four measurements record the relative anteroposterior position of two points on the mandible compared with the base of the skull and also the total and lower part of the anterior facial height.

7. Conclusions

The present state of knowledge concerning the subject of skull growth presents a confusing picture. The mechanisms that have evolved to bring about change in bone shape have been explored in some detail, but the stimuli that initiate and control their operation are still subject to controversy. Whether the requirements of the essential activities of the skull provide the control or whether there exists a more direct genetic control mechanism has yet to be conclusively established. The studies on twins appear to implicate a mixture of hereditary and environmental factors, but more investigation is required into this and all other aspects of skull growth.

A major stimulus for future research arises from the clinical need to predict facial growth. Current methods based on population data have proved unreliable in cases showing an aberrant growth pattern—just those cases in which prediction is most important.

8. References

Adams, J. W., 1940, Correction of errors in cephalometric roentgenograms, *Angle Orthod.* **10**:3–13.

Alexander, R. B., The effects on both position and maxillo facial vertical growth during treatment of scoliosis with the Milwaukee brace, *Am. J. Orthod.* **52**:161–189.

Avis, V., 1961, The significance of the angle of the mandible: An experimental and comparative study, *Am. J. Phys. Anthropol.* **19**:55–61.

Bassett, C. A. L., 1972, Biophysical properties affecting bone structure, in: *The Biochemistry and Physiology of Bone,* (G. H. Bourne, ed.), Vol. III pp. 1–76, Academic Press, New York.

Bassett, C. A. L., Pawluk, R. J., and Becker, R. O., 1964, Effects of electric currents on bone in vivo, *Nature (London)* **202**:652–654.

Baume, L. J., 1961, The postnatal growth activity of the nasal cartilage septum, *Helv. Odontol. Acta.* **47**:881–901.

Bjork, A., 1947, The face in profile, *Sven. Tandlaek. Tidskr.* **40**:56.

Bjork, A., 1951, The significance of growth changes in facial pattern and their relationship to changes in occlusion, *Dent. Rec.* **7**:197–208.

Bjork, A., 1955, Facial growth in man, studies with the aid of metallic implants, *Acta. Odontol. Scand.* **13**:9–34.

Bjork, A., 1963, Variations in the growth pattern of the human mandible: Longitudinal radiographic study by the implant method, *J. Dent. Res.* **42**:400–411.

Bjork, A., 1964, Sutural growth of the upper face studied by the implant method, *Trans. Eur. Orthod. Soc.* **49**:49–65.

Bjork, A., and Skeiller, V., 1983, Normal and abnormal growth of the mandible. A synthesis of longitudinal cephalometric implant studies over a period of 25 years, *Eur. J. Orthod.* **5**:1–46.

Bjork, A., and Solow, B., 1962, Measurements on radiographs, *J. Dent. Res.* **41**:672–683.

Brash, J. C., 1956, *The Aetiology of Irregularity and Malocclusion of the Teeth,* Part 1, Dental Board of U.K.

Broadbent, B. H., Sr., Broadbent, B. H., Jr., and Golden, W. H., 1975, *Bolton Standards of Dentofacial Developmental Growth,* C. V. Mosby, St. Louis, Missouri.

Broadbent, H., 1931, A new X-ray technique and its application to orthodontics, *Angle Orthod.* **1**:45–66.

Brodie, A. G., 1953, Late growth changes in the human face, *Angle Orthod.* **23**:146–157.

Burke, P., 1983, Serial stereophotogrammetric measurement of the soft tissues of the face, *Br. Dent. J.* **155**:373–379.

Burke P., and Beard L. F. H., 1967, Stereophotogrammetry of the face, *Am. J. Orthod.* **53**:769–782.

D'Arcy Thompson, W., 1917, *On Growth and Form,* Cambridge University Press, Cambridge.

Delattre, A., and Fenart, R., 1958, La méthode vestibulaire appliqué à l'étude du crane. Son champ d'application, *Z. Morphol. Anat.* **1**:90–114.

Downs, W. B., 1948, Variations in facial relationships, their significance in treatment and prognosis, *Am. J. Orthod.* **34**:812–840.

Dudas, M., and Sassouni, V., 1963. The hereditary components of mandibular growth, a longitudinal twin study, *Angle Orthod.* **43**:314–323.

Enlow, D. H., 1966, A morphogenetic analysis of facial growth, *Am. J. Orthod.* **52**:283–299.

Enlow, D. H., 1975, *Handbook of Facial Growth,* W. B. Saunders, Philadelphia.

Frost, H. M., 1964, *The Laws of Bone Structure,* Charles C Thomas, Springfield, Illinois.

Giannelly, A. A., and Moorrees, C. F. A., 1965, Condylectomy in the rat, *Arch. Oral Biol.* **10**:101–106.

Goldstein, M. S., 1936, Changes in dimension and form of the face and head with age, *Am. J. Phys. Anthropol.* **22**:37–89.

Graber, T., Chung, D., and Aoba, J., 1967, Dentofacial orthopedics versus orthodontics, *J. Am. Dent. Assoc.* **75**:1145–1166.

Grange, R. J., and Johnston, L. E., 1974, The septo-premaxillary attachment and mid-facial growth, *Am. J. Orthod.* **66**:71–81.

Gregory, W. K., 1927, The palaeomorphology of the human head: Ten structural stages from fish to man: Part 1. The skull in norma lateralis, *Q. J. Biol.* **2**:267.

Haas, A. J., 1961, Rapid expansion of the maxillary dental arch and nasal cavity by opening the mid-palatal suture, *Angle Orthod.* **31**:73–90.

Harris, W. H., 1960, A microscopic method of determining rates of bone growth, *Nature (London)* **188**:1038–1039.

Hillesund, E., Fjeld, D., and Zachrisson, B. U., 1978, Reliability of soft tissue profile in cephalometrics, *Am. J. Orthod.* **74**:537–550.

Holdaway, R. A., 1983, A soft-tissue cephalometric analysis and its use in orthodontic treatment planning, *Am. J. Orthod.* **84**:1–28.

Humphrey, G. M., 1863, Results of experiments on the growth of the jaws, *Br. J. Dent. Sci.* **6**:548–550.

Hunter, J., 1798, Experiments and observations on the growth of bones, in: *Hunter's Works, 1837* (D. F. Palmers, Ed.), Longman, London.

Hunter, W. S., 1965, Study of inheritance of craniofacial characteristics as seen in lateral cephalograms of 72 like-sexed cases, *Trans. Eur. Orthod. Soc.* **41**:59–69.

Johnston, L. E., 1975, A simplified approach to prediction, *Am. J. Orthod.* **67**:253–257.

Klein, P. L., 1957, An evaluation of cervical traction on the maxilla and the upper first permanent molar, *Angle Orthod.* **27**:61.

Koski, K., 1973, Variability of the facial skeleton: An exercise in roentgen-cephalometry, *Am. J. Orthod.* **64**:188–196.

Koski, K., and Ronning, O., 1965, Growth potential of transplanted components of the mandibular ramus of the rat III, *Suom. Hammaslaak Seur Toim* **61**:292–297.

Koski, K., and Ronning, O., 1970, Growth potential in intracerebrally transplanted cranial base synchondrosis in the rat, *Arch. Oral Biol.* **15**:1107–1108.

Krogman, W. M., and Sassouni, V., 1957, *A Syllabus in Roentgenographic Cephalometry,* Philadelphia Center for Research in Child Growth, Philadelphia.

Latham, R. A., 1968, The sliding of cranial bones at sutural surfaces during growth, *J. Anat.* **102:**593.

Latham, R. A., 1969, The septopremaxillary ligament and maxillary development, *J. Anat.* **104:**584–586.

Logan, W. R., 1962, The effect of the Milwaukee brace on the developing dentition, *Dent. Pract.* **12:**447–452.

Lundstrum, A., 1954, The importance of genetic and nongenetic factors in the facial skeleton studied in 100 pairs of twins, *Trans. Eur. Orthod. Soc.* **30:**92–106.

Lynsholm, E., 1931, Apparatus and technique for roentgen examination of the skull, *Acta Radiol. Suppl.* **12.**

McNamara, J. A., and Carlson, D. S., 1979, Quantitative analysis of temporomandibular joint adaptation to protrusive function, *Am. J. Orthod.* **76:**593–611.

McNamara, J. A., Hinton, R. J., and Hoffman, D. L., 1982, Histologic analysis of temporomandibular joint adaptation to protrusive function in young adult rhesus monkeys *(Macaca mulatta), Am. J. Orthod.* **82:**288–298.

Mitchell, D. L., Jordan, J. F., and Ricketts, R. M., 1975, Facial growth with metallic implants in mandibular growth prediction, *Am. J. Orthod.* **68:**655–659.

Moorrees, C. F. A., and Yem, P. K.-J., 1955, An analysis of changes in the dentofacial skeleton following orthodontic treatment, *Am. J. Orthod.* **41:**526–538.

Moorrees, C. F. A., and Kean, M. R., 1958, Natural head position a basic consideration in the interpretation of cephalometric radiographs, *Am. J. Phys. Anthropol.* **16:**213–234.

Moorrees, C. F. A., Fanning, E. A., and Hunt, E. E., 1963*a,* Age variation of formation stages for ten permanent teeth, *J. Dent. Res.* **42:**1490–1502.

Moorrees, C. F. A., Fanning, E. A., and Hunt, E. E., 1963*b,* Formation and resorbtion of three deciduous teeth in children, *Am. J. Phys. Anthropol.* **21:**205.

Moss, M. L., 1954, The growth of the calvaria in the rat, *Am. J. Anat.* **94:**333.

Moss, M. L., 1968, The primacy of functional matrices in orofacial growth, *Dent. Pract. Dent. Rec.* **19:**65–73.

Moss, M. L., and Rankow, R. M., 1968, The role of the functional matrix in mandibular growth, *Angle Orthod.* **39:**95–103.

Moss, M. L., and Salentijn, L., 1970, The logarithmic growth of the human mandible, *Acta Anat.* **77:**341–360.

Moss, M. L., Bromberg, B. E., Song, I. C., and Eisenman, G., 1968, The passive role of nasal septal cartilage in mid-facial growth, *Plast. Reconstr. Surg.* **41:**536–542.

Moss, M. L., Salentijn, L., and Ostreicher, H. P., 1974, The logarithmic properties of active and passive mandibular growth, *Am. J. Orthod.* **66:**645–664.

Moss, M. L., Skalak, R., Shinoznka, M., Patel, H., Moss-Salentijn, L., Vilmann, H., and Melita, P., 1983, Statistical testing of an allometric centered model of craniofacial growth, *Am. J. Orthod.* **83:**5–18.

Petrovic, A., Stutzman, J., and Gasson, N., 1982, The final length of the mandible: Is it generally predetermined?, in: *Craniofacial Biology,* Monograph No. 10, *Craniofacial Growth Series,* Ann Arbor, 1981 (D. S. Carlson, ed.), pp. 105–126, Center for Human Growth and Development, University of Michigan.

Pritchard, J. J., Scott, J. H., and Girgis, F. G., 1956, The structure of cranial and facial sutures, *J. Anat.* **90:**73–86.

Reitan, K., 1951, The initial tissue reaction incident to orthodontic tooth movement as related to the influence of function, *Acta. Odontol. Scand.* Suppl. 6.

Ricketts, R. M., 1973, New findings and concepts emerging from the clinical use of the computer, *Trans. Eur. Orthod. Soc.* **1973:**507–515.

Ricketts, R. M., Bench, R. W., Hilgers, J. I., and Schulhof, A. B., 1972, An overview of computerized cephalometrics, *Am. J. Orthod.* **61:**1–28.

Rock, W. P., and Baker, R., 1972, The effect of the Milwaukee brace upon dentofacial growth, *Angle Orthod.* **42:**96–102.

Rushton, M. A., 1944, Growth at the mandibular condyle in relation to some deformities, *Br. Dent. J.* **76:**57–68.

Salzmann, J. A., 1966, *Practice of Orthodontics,* Vol. 1, J. B. Lippincott, Philadelphia.

Sarnat, B. G., 1963, Postnatal growth of the upper face: Some experimental considerations, *Angle Orthod.* **33:**139–161.

Sarnat, B. B., and Wexler, M. R., 1966, Growth of the face and jaws after resection of the septal cartilage in the rabbit, *Am. J. Anat.* **118:**755–760.

Sarnat, B. G., and Wexler, M. R., 1967, Rabbit snout growth after resection of central linear segments of nasal septal cartilage, *Acta. Oto-Laryngol.* **63:**467–478.

Savara, B. S., 1972, The role of computers in dentofacial research and the development of diagnostic aids, *Am. J. Orthod.* **61:**231–245.

Scott, J. H., 1954, The growth of the human face, *Proc. R. Soc. Med. (London)* **47:**91.

Scott, J. H., 1956, Growth at facial sutures, *Am. J. Orthod.* **42:**381.

Scott, J. H., 1967, *Dento-facial Development and Growth,* Pergamon, London.

Sillman, J. H., 1964, Dimensional changes of the dental arches: Longitudinal study from birth to 25 years, *Am. J. Orthod.* **50:**824–842.

Steiner, C. C., 1953, Cephalometrics for you and me, *Am. J. Orthod.* **39:**729–755.

Steiner, C. C., 1959, Cephalometrics in clinical practice, *Angle Orthod.* **29:**8–29.

Steiner, C. C., 1960, The use of cephalometrics as an aid to planning and assessing orthodontic treatment, *Am. J. Orthod.* **46:**721–735.

Stenstrom, S. J., and Thilander, B. L., 1972, Effects of nasal septal cartilage resections on young guinea pigs, *Plast. Reconstr. Surg.* **45:**160–170.

Sullivan, P. G., 1972, A method for the study of jaw growth using a computer-based three-dimensional recording technique, *J. Anat.* **112:**457–470.

Suzuki, H. K., and Matthews, A., 1966, Two colour fluorescent labelling of mineralising tissues with tetracycline and 2,4-bis(N,N′-di-(carboxymethyl)aminomethyl) fluorescein, *Stain Technol.* **41**:57–60.

Tanner, J. M., 1951, Some notes on the reporting of growth data, *Hum. Biol. 23:93–159.*

Tedesco, L. A., Albino, J. E., Cunat, J. J., Green, L. J., Lewis, E. A., and Slakter, M. J., 1983, A dental-facial attractiveness scale, *Am. J. Ortho.* **83**:38–43.

Tulley, W. J., and Campbell, A. C., 1975, *A Manual of Practical Orthodontics,* 3rd ed., Wright and Sons, Bristol, England.

Voorhies, J. M., and Adams, J. W., 1951, Polygonic interpretation of cephalometric findings, *Angle Orthod.* **21**:194–197.

Walker, G. F., 1972, A new approach to the analysis of craniofacial morphology, *Am. J. Orthod.* **61**:221–230.

Washburn, S. L., 1947, The relation of the temporal muscle to the form of the skull, *Anat. Rec.* **99**:239–248.

Watanabe, M., Laskin, D. M., and Brodie, A., 1957, The effect of antotransplantation on growth of the zygomaticomaxillary suture, *Am. J. Anat.* **10**:319.

Weinmann, J. P., and Sicher, H., 1955, *Bone and Bones: Fundamentals of Bone Biology,* 2nd ed., Kimpton, London.

Wexler, M. R., and Sarnat, B. G., 1961, Rabbit snout growth: Effect of injury to the spetovomeral region, *Arch. Oto-Laryngol.* **74**:305–313.

Wylie, W. L., 1947, Assessment of antero-posterior dysplasia, *Angle Orthod.* **17**:97.

Young, R. W., 1959, The influence of cranial contents on postural growth of the skull of the rat, *Am. J. Anat.* **103**:383–415.

12

Dentition

ARTO DEMIRJIAN

1. Introduction

The dental system is an integral part of the human body; its growth and development can, and should, be studied in parallel with other physiological maturity indicators, such as bone age, menarche, and height. Knowledge about the development of dentition can be applied not only in dentistry and orthodontics, but in such diverse fields as physical anthropology, endocrinology, nutrition, and forensic odontology as well.

There is considerable latitude in the study of dental maturation, since various criteria are available for evaluation. Dental maturity can be based on not one, but two, complete sets of dentition. Both sets can be observed from the beginning, or at least from early stages, of calcification (formation). Furthermore, the clinical emergence of both sets of teeth and the shedding of the deciduous teeth can serve as yet other maturity indicators. However, the study of dental development, unlike that of skeletal development, has unfortunately not progressed to the stage where norms for dental maturity have been wholly clarified, standardized, and universally accepted. This obstacle should soon be overcome as newer techniques for the assessment of dental maturation are applied.

The most commonly used parameter for assessing dental maturity is dental *eruption,* which is both rapid and convenient to record. The term eruption is used almost universally, albeit incorrectly, to indicate the clinical appearance of the cusps of a tooth through the gum. In fact, eruption is not an instantaneous phenomenon corresponding to the moment when a tooth pierces the gum; rather it is the continuous upward movement of the dental bud, from the depth to the edge of the alveolar bone, and further into the occlusal plane.

Historically, however, data on eruption refer to the clinical appearance of a tooth in the oral cavity. The first use of eruption as an indicator of physical maturity occurred in England (Saunders, 1837), when the Factory Act stipulated that a child without a second permanent molar would not be permitted to work in the factories. For about the first quarter of the twentieth century, Beik (1913), Bean (1914), and Cattell (1928) used dentition as a maturity indicator for school entrance purposes.

Up to the 1940s, data concerned with dental development were mostly descriptive and were based on histological sections from very small samples. Errors from this unreliable approach persist even to the present day, for most eruption is still compared with standards derived from histological studies. For example, Legros and Magitot (1893) first reported on the chronology of the formation of tooth buds; their study was based on histological sections. Regardless of the fact that "they based their conclusions about the onset of calcification on sections from a single foetus" (Kraus, 1959), the results reported by Legros and Magitot were used for many decades, even in textbooks of histology and embryology.

In their well-known article, Logan and Kronfeld (1933) criticized the Legros and Magitot material and its unchallenged use for 40 years. Yet, their new material was hardly better, having been derived from a heterogeneous sample composed of 3 newborns, 13 infants up to 12 months, and 9 chil-

ARTO DEMIRJIAN • Université de Montréal, Montréal, Quebec, Canada, H3C 3J7.

dren from 1 to 15 years of age. Moreover, Logan and Kronfeld claimed that some of their subjects had pathological conditions. Obviously, the results of their study, also based on histological sections, are too biased and unreliable to be of any use as a guide to the chronology of dental development.

It is not very edifying to trace the history of repetitive errors in this field. It is interesting to note, however, that as late as 1940, Schour and Massler published a two-part article featuring the chronology of growth of human teeth, a table modified from that of Logan and Kronfeld. The authors described the continuous development not only of the permanent teeth, but of the deciduous teeth as well. Yet the Logan and Kronfeld article makes no reference to the deciduous dentition, and it is difficult to find reference to any other sample in Schour and Massler's paper. We suspect that histological and roentgenographic material was used without distinction. However, Schour and Massler cautioned their readers that "since the X-ray film is a record of density, the stage of tooth development, as indicated in histological sections, will be slightly earlier than that indicated by X-rays." Although roentgenographic representations and anatomical findings are not identical, this distinction has sometimes been disregarded.

Today, studies of dental maturity use and correlate histological, radiological, and clinical data. The study of the prenatal phase of deciduous dental development is still based on the histological approach. However, in postnatal growth studies, radiological (formation) and clinical (emergence) data can be used, either independently or in parallel. It must be remembered that formation and eruption involve two essentially different processes that may be influenced differently by genetic, environmental, and hormonal factors. The study of dental development may be further complicated by such factors as the prenatal environment and its influence on the formation of the deciduous teeth, the possible physiological or pathological influences on one or both sets of teeth, the shedding of the deciduous dentition, and the effect of the deciduous teeth in the timing of the emergence of their permanent successors.

2. Clinical Emergence

As indicated in the preceding sections, *eruption* is erroneously used to denote the appearance of a tooth in the oral cavity. The correct term for the piercing of the gum is *clinical emergence,* or simply *emergence.* The piercing of the alveolar bone, seen in a radiographic image and referred to as bony or alveolar eruption, should similarly be called *alveolar emergence.* From this point on in this chapter, the use of the term eruption is confined to the dynamic movements of the tooth, both before and after actual emergence. This process resembles the pre- and postnatal growth of an infant, where the actual birth of the child corresponds to emergence and is an event of short duration during the development of the child.

The literature is replete with reports of research into the chronology and timing of dental emergence. These studies have been conducted in an effort to establish population standards for clinical, anthropological, medicolegal, and other purposes. Several such standards have also been used as criteria for the determination of biological (dental) age.

Emergence has been analyzed in many different ways. Individual teeth, specific groups of teeth, and the entire dentition (deciduous and permanent) have been studied in cross-sectional population samples, in single-age groups, and also longitudinally, with groups of subjects closely monitored throughout the crucial years of their growth and development. Groups of children from many parts of the world have been observed in an effort to trace any developmental differences that may be related to race, sex, nutrition, socioeconomic levels, hormonal conditions, and other intrinsic or extrinsic factors.

A thorough review of the subject was presented by Tanner (1962); one of our objectives is to bring Tanner's review up to date. As much as possible, the deciduous dentition is treated separately from the permanent dentition. However, in discussions of both, information gained about the various intrinsic and extrinsic influences mentioned is emphasized.

3. Emergence of Deciduous Dentition

The emergence of the deciduous dentition takes place between the 6th and 30th months of postnatal life. Although this is a relatively short period of time, the study of dental maturity in these age groups is useful in growth studies concerned with environmental and hereditary factors. For the evaluation of maturity, radiographic assessment of the deciduous dentition is a more meaningful approach, since the development of every stage of each tooth can be followed regularly between birth and the third year of life, rather than during the more limited period of actual emergence, from the 6th to the 30th months. However, the difficulty of developing adequate radiological techniques for infants, as well as the hazards of the radiation itself, represent two significant disadvantages to this method. Nevertheless, Moorrees *et al.* (1963*b*) studied the development of the mandibular deciduous canines and molars and presented normative data, using the Fels Institute's material.

3.1. Sex Differences

Many investigators have reported no difference between boys and girls in the timing of emergence (Falkner, 1957; Yun, 1957; Lysell *et al.*, 1962; Bailer, 1964; Roche *et al.*, 1964; McGregor *et al.*, 1968; Friedlaender and Bailit, 1969; Brook and Barker, 1972, 1973; Billewicz, 1973; Robinow, 1973; Trupkin, 1974; El Lozy *et al.*, 1975; Barrett and Brown, 1976; Nystrom, 1977). Other investigators, however, have found an earlier emergence pattern among boys (Doering and Allen 1942; Robinow *et al.*, 1942; Meredith, 1946; MacKay and Martin, 1952; Ferguson *et al.*, 1957; Leighton, 1968; Burdi *et al.*, 1970; Infante, 1974; Taranger *et al.*, 1976; Tanguay *et al.*, 1984).

In a recent study of French-Canadian children (Tanguay *et al.*, 1984), we found that the emergence of the deciduous dentition in boys was advanced by 1 month over that in girls. A multivariate analysis of variance yielded a significant overall sex-related difference. The only exception was the first deciduous molar for which no difference was detected. Using an identical procedure for data collection in a longitudinal sample of Finnish children, Nystrom (1977), obtained different results. He found no difference between boys and girls, except for the first molar, for which girls were ahead. We do not have an answer to the question: Why does the first deciduous molar have a different pattern of sexual dimorphism?

Only one group of investigators (Bambach *et al.*, 1973) reported a delayed emergence among boys as compared with girls, but the difference was not statistically significant.

3.2. Racial and Ethnic Differences

The emergence of dedicuous teeth has been investigated in different racial and ethnic groups. Ferguson *et al.*, (1957) reported that the emergence of deciduous teeth is delayed in white Americans, as compared with black Americans, during the first year of life. Yun (1957) noted that the emergence of deciduous teeth is slower in Korean children than in American children. McGregor *et al.* (1968) reported that American white, English, and French children were ahead of Gambian children with respect to the emergence of deciduous teeth. The Gambian children caught up, though, by 2 years of age. This latter trend was more firmly established

Table I. The Age of Emergence (in months) of the Deciduous Dentition: Boys[a,b]

Teeth	No.	Skewness	Kurtosis	Percentiles 3e	5e	10e	25e	50e	75e	90e	95e	97e	Interquartile range
							Maxilla						
I_1	201	0.09	−0.33	5.70	6.02	6.53	7.43	8.49	9.63	10.71	11.39	11.84	2.20
I_2	194	0.32	0.36	5.87	6.31	7.02	8.28	9.81	11.46	13.06	14.07	14.74	3.18
C	178	−0.15	0.19	12.44	13.03	13.97	15.62	17.56	19.61	21.55	22.76	23.56	3.99
M_1	189	0.13	0.19	11.26	11.72	12.45	13.72	15.20	16.76	18.23	19.13	19.74	3.04
M_2	140	−0.10	−0.21	19.95	20.78	22.09	24.38	27.04	29.84	32.48	34.11	35.20	5.46
							Mandible						
I_1	200	0.20	0.10	3.90	4.23	4.75	5.70	6.86	8.11	9.34	10.11	10.63	2.41
I_2	188	0.02	0.63	6.95	7.46	8.27	9.72	11.48	13.37	15.21	16.36	17.13	3.65
C	173	−0.30	0.38	12.98	13.54	14.42	15.97	17.77	19.66	21.45	22.56	23.30	3.70
M_1	186	0.10	−0.13	11.49	11.93	12.63	13.84	15.25	16.72	18.11	18.97	19.54	2.88
M_2	138	0.09	−0.41	19.19	20.00	21.29	23.52	26.13	28.88	31.48	33.08	34.14	5.36

[a]From Lauzier and Demirjian (1981).
[b]Transformation: square root. Square root transformation has been used to normalize variables.

through the work of Mukherjee (1973) and Billewicz *et al.* (1973). The former compared Indian children, and the latter compared Hong Kong children with American, British, and French samples. Like McGregor, they found a delay in both the Indian and Hong Kong children. Bambach *et al.* (1973) compared results from a sample of Tunisian children with those of McGregor *et al.* (1968) and also found a delay in the emergence of deciduous teeth of the African children as compared to the European and American groups. In every case, it was reported that the delay was eliminated by the age of 1½ to 2 years. Friedlaender and Bailit (1969) concluded that the emergence of deciduous dentition is complete by the end of the 19th month of life, regardless of ethnic origin or race, and is independent of the age at which the first tooth appeared.

However, one should keep in mind that there is a wide variation among individuals even within the same ethnic group. One of the most recent studies (Lauzier and Demirjian, 1981), in which the date of the emergence was recorded very carefully by the parents, shows that a difference of up to 15 months may exist between the third and 97th percentile values for the age of emergence of a given tooth (see Tables I and II). The median values for this population ranged from 6.9 months (mandible, M_1, boys) to 27.9 months (maxilla, M_2, girls). A chart showing the chronology of emergence for clinical use has been published by the same investigators (see Figure 1).

3.3. Socioeconomic Status

Most studies that attempt to correlate socioeconomic status and dental emergence have been concerned with the permanent dentition. Bambach and colleagues (1973) attempted to relate deciduous tooth emergence to socioeconomic conditions, but they found no significant correlations. The emergence of deciduous dentition seems to be influenced more by genetic factors than by environmental or socioeconomic conditions.

3.4. Nutrition

Nutrition also been investigated extensively (McGregor *et al.*, 1968; Cifuentes and Alvarado, 1973; Infante and Owen, 1973; Trupkin, 1974; Delgado *et al.*, 1975). In general terms, no correlation has been found between nutrition and the emergence of the deciduous teeth. Dental emergence might be delayed in cases of severe protein-calorie malnutrition (PCM). However, even in those cases, the effect of malnutrition is greater on other parameters, i.e., height, weight, and bone age. El Lozy *et al.* (1975) reported that less severe malnutrition led to a delayed emergence of deciduous teeth among Tunisian children. This fact may be true of developing countries, even though dental development is less affected than somatic and skeletal growth (Boutourline-Young, 1969; Habicht *et al.*, 1974).

Table II. The Age of Emergence (in months) of the Deciduous Dentition: Girls[a,b]

				Percentiles									Inter-
Teeth	No.	Skewness	Kurtosis	3e	5e	10e	25e	50e	75e	90e	95e	97e	quartile range
						Maxilla							
I_1	164	−0.25	−0.34	6.19	6.56	7.15	8.19	9.42	10.75	12.01	12.80	13.33	2.56
I_2	160	0.38	0.07	6.13	6.62	7.40	8.82	10.53	12.40	14.21	15.35	16.12	3.58
C	128	−0.21	−0.25	13.28	13.86	14.79	16.41	18.31	20.31	22.20	23.37	24.15	3.90
M_1	145	0.25	−0.03	11.51	11.93	12.59	13.73	15.06	16.45	17.75	18.56	19.09	2.71
M_2	109	0.08	−0.60	20.67	21.52	22.85	25.16	27.86	30.69	33.37	35.02	36.11	5.53
						Mandible							
I_1	165	0.26	−0.19	3.96	4.32	4.91	5.99	7.31	8.75	10.17	11.07	11.68	2.77
I_2	159	−0.07	−0.40	7.70	8.24	9.11	10.67	12.54	14.57	16.52	17.75	18.57	3.90
C	133	−0.16	−0.28	13.14	13.77	14.77	16.52	18.58	20.76	22.82	24.11	24.96	4.23
M_1	147	−0.02	−0.24	11.67	12.10	12.78	13.96	15.32	16.76	18.10	18.93	19.48	2.80
M_2	100	−0.04	−0.09	20.22	21.01	22.25	24.40	26.90	29.53	32.01	33.53	34.54	5.13

[a]From Lauzier and Demirjian (1981).
[b]Square root transformation has been used to normalize variables.

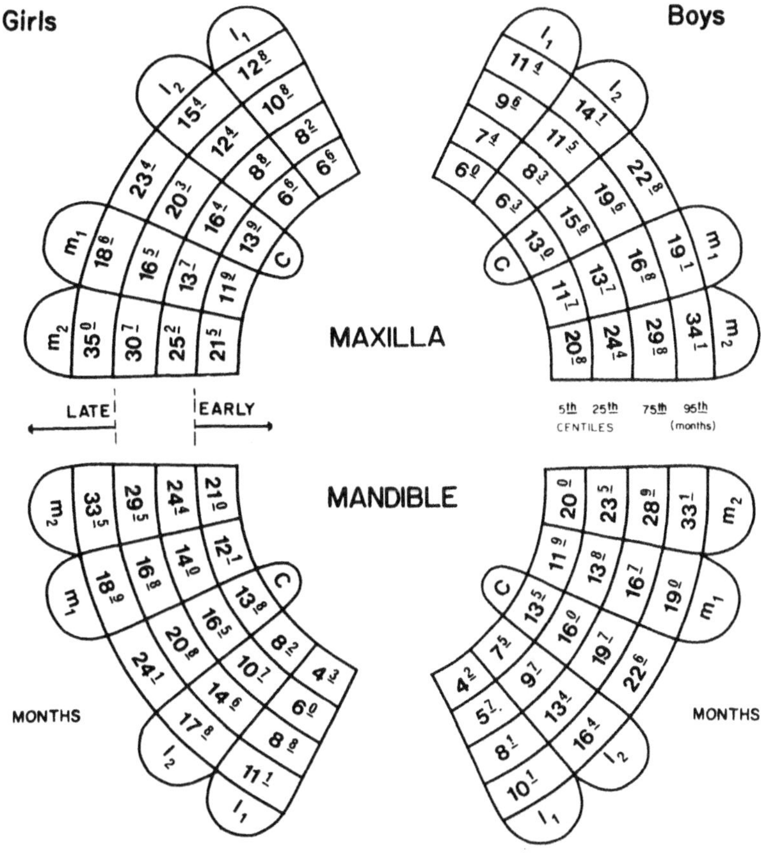

Fig. 1. Emergence of the deciduous dentition in French-Canadian children, (From Lauzier and Demirjian, 1981.)

The next logical step would be to investigate severe maternal malnutrition in poor communities in an effort to trace not only enamel hypoplasia, but developmental and eruptive patterns of the deciduous teeth as well. With the information currently available, emergence and formation patterns of the dediduous dentition seem to be programmed in fetal life (Jelliffe and Jelliffe, 1973).

3.5. Other Factors

Low birth weight and prematurity are two other factors that have been linked to delayed emergence of the deciduous teeth. Some correlations between birth weight and the number of teeth in the mouth have been reported (Billewicz *et al.*, 1973; Infante and Owen, 1973; Trupkin, 1974; Delgado *et al.*, 1975), children with higher birth weight tending to have more teeth in the mouth at any given age. Although birth weight seems to be closely related to the emergence of the deciduous teeth, there is no evidence this comes from variation in the length of gestation. Advanced emergence has also been correlated to body length in both sexes and to head circumference in boys (Infante and Owen, 1973). This relationship between the length of an infant's body and the emergence of deciduous dentition was corroborated by Falkner (1957).

No correlation has been found between deciduous dental emergence and skeletal maturity (Robinow, 1973; Billewicz *et al.*, 1973). McGregor *et al.* (1968) also came to the same conclusion that there is no connection between emergence and/or the number of teeth in the mouth and the general physiological growth of the child. Slight trends suggesting such a relationship have been observed, but none has proved statistically significant. Since there are so few definitive correlations between deciduous tooth emergence and other physiological parameters, such as skeletal maturation, size, and sex (Falkner, 1957), it appears that the deciduous dentition is remarkably independent of other morphological processes.

4. Emergence of Permanent Dentition

Cross-sectional surveys have been the major vehicle for studying the emergence of the permanent dentition. Means and standard deviations of the time of emergence of each tooth can be derived from these data. Cumulative incidence curves can be constructed from the cross-sectional data, which show the percentage of children, at a given age, in whom a certain tooth has emerged (Cattell, 1928; Boas, 1933; Klein *et al.,* 1937; Hellman, 1943; Hurme, 1948, 1949; C. Dahlberg and Maunsbach, 1948; Godeny, 1951; Leslie, 1951; Clements *et al.,* 1953*b;* Tanner, 1962). In more recent studies (Clements *et al.,* 1953*a;* Kihlberg and Koski, 1954; A. A. Dahlberg and Menegaz-Bock, 1958; Gates, 1964; Miller *et al.,* 1965; Perreault *et al.,* 1974; Mayhall *et al.,* 1977; Demirjian and Levesque, 1980; Koyoumdjisky-Kaye *et al.,* 1981), probit transformation has been adopted as the best method for analyzing these data.

The pattern and timing of the emergence of the permanent dentition have been studied both cross-sectionally and longitudinally. In longitudinal studies, if subjects are only seen at yearly intervals, it is likely that the exact timing of emergence will be missed. For this reason, A. A. Dahlberg and Menegaz-Bock (1958) suggested that the age of emergence be placed halfway between the ages immediately before, and immediatey after, this tooth appeared. If such an adjustment is not made, data resulting from longitudinal studies will always give values too high for the age of the emergence.

4.1. Symmetry

It is now widely accepted that there is symmetry in the emergence times between the teeth of the right and left sides (Clements *et al.,* 1957*b;* Gates, 1964; Lee *et al.,* 1965; Nanda and Chawla, 1966; Lysell *et al.,* 1969; Krumholt *et al.,* 1971; Perreault *et al.,* 1974; Helm and Seidler, 1974; Billewicz and McGregor, 1975). The emergence of the upper and lower teeth varies according to the individual tooth. The first to emerge is always the lower central incisor. The lower molars usually emerge earlier than their upper counterparts, while the upper premolars precede the lower premolars (A. A. Dahlberg and Menegaz-Bock, 1958; Nanda, 1960; Gates, 1964; Lee *et al.,* 1965; Nanda and Chawla, 1966; Houpt *et al.,* 1967; Lyseel *et al.,* 1969; Garn *et al.,* 1973; Helm and Seidler, 1974).

4.2. Sex

Several investigators have found significant sex-related differences in the emergence of the permanent dentition (Hurme, 1949; A. A. Dahlberg and Menegaz-Bock, 1958; Garn *et al.,* 1959; Fanning, 1961; Halikis, 1961; Lee *et al.,* 1965; Miller *et al.,* 1965; Haupt *et al.,* 1967; Krumholt *et al.,* 1971; Brook and Barker, 1972; Robinow, 1973, Helm and Seidler, 1974; Mayhall *et al.,* 1977). Girls are usually 1–6 months ahead of boys. For certain teeth, such as the canines, this difference can be as great as 11 months in some populations. The sexual dimorphism for the other teeth is somewhat less extensive.

For the third molar, there is no agreement among investigators concerning sexual dimorphism in its development and emergence. As early as 1936, Saito found an earlier emergence of this tooth among girls before the age of 13 years. Garn *et al.* (1962) studied its calcification and emergence and in spite of a female precocity in the emergence of all other permanent teeth, did not find any demonstrable sexual dimorphism for this tooth. The latest study on third molar development and emergence was done by Levesque *et al.* (1981) in an investigation of more than 4600 cases. At the median level, boys were ahead of girls by 6 months in both alveolar (from X-rays), and clinical emergence. The ages for clinical emergence were 18.5 years for boys and 19.0 years for girls. These ages are lower than the mean ages obtained by Helman in 1936 for Americans (20.5 years for boys and 20.8 years for girls). This sexual dimorphism is also true for the developmental stages of the root.

Figure 2 summarizes the combined developmental and emergence patterns among boys and girls from birth to maturity for the deciduous and permanent dentitions in French-Canadian children. Up to age 3, boys are ahead of girls by an average of 1 month, during the emergence of the deciduous teeth. From 3 to 6 years, the sexual differences are practially nonexistent, during the development period of the permanent dentition. But from 6 to 17 years of age, girls are clearly advanced over boys. By the end of adolescence, there is a second crossover in the developmental curves and, for the root formation stages of the third molar, the boys are again more advanced than girls.

4.3. Race and Ethnic Origin

Hereditary factors and ethnic origin exert considerable control over the emergence of the permanent

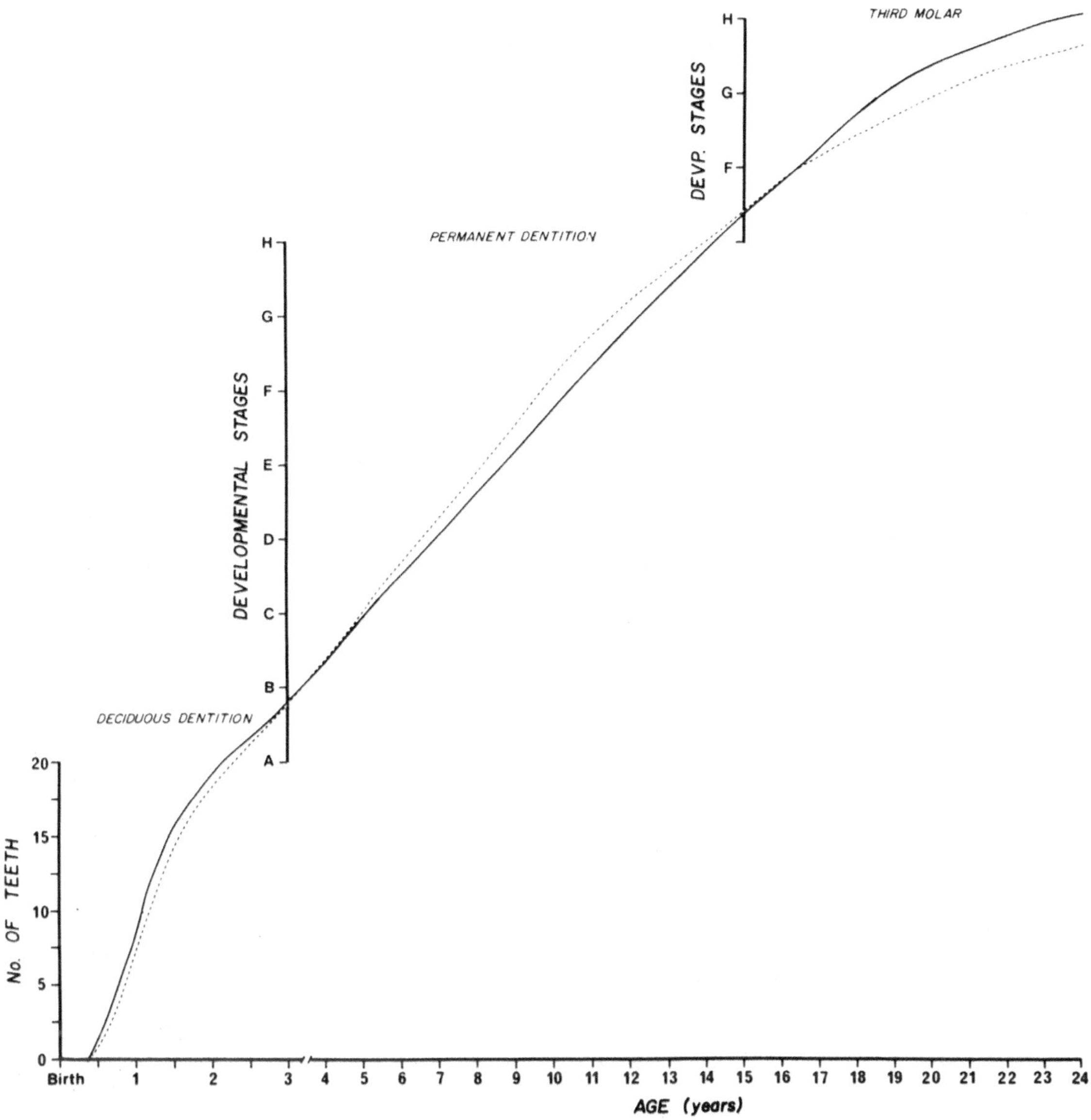

Fig. 2. Sexual dimorphism in the emergence and formation of the deciduous and the permanent dentition from birth to maturity; (—) boys; (· · ·) girls.

teeth. Dental emergence is earlier in American blacks than in American whites (Garn *et al.*, 1973). Likewise, it is earlier in American blacks than in Europeans, sometimes by as much as 1½ years (Houpt *et al.*, 1967; Krumholt *et al.*, 1971). New Guinea children are also ahead of American and British children (Voors and Metselaar, 1958; Brook and Barker, 1972). Lee *et al.* (1965) compared a group of Hong Kong children with samples of American (white), British, New Zealand, Hawaiian, and Chinese children; differences in the timing of the emergence were attributed to the ethnic origins of the various groups of children. Northwestern Ontario Indian children show an earlier emergence than do French-Canadian children, the largest difference being for maxillary second premolars (Mayhall *et al.*, 1977). Koyoumdjisky *et al.* (1977), using a modified Bjork *et al.* (1964), classification, found a delay in the emergence of canines and incisors in a group of Israeli children compared with other populations.

Not only is the timing of permanent tooth emergence under ethnic influences, but individual teeth emerge according to certain genetically controlled patterns. The timing of the emergence of the anterior and the posterior segments of dentition is different for Pima Indians, Hong Kong children, American, and British children (Leslie, 1951; A. A. Dahlberg and Menegaz-Bock, 1958; Lee *et al.*, 1965). Dahlberg and Menegaz-Bock (1958) offered a possible explanation for this phenomenon, i.e., that the "posterior teeth emerge earlier in a population subject to premature shedding of the deciduous molars." Leslie (1951) concluded that the later the deciduous molar is lost, the more accelerated is the emergence of the succeeding premolars. Racial differences have a more significant effect on the incisors and molars, and a less important one on the premolars and canine groups (Garn *et al.*, 1973).

4.4. Socioeconomic Conditions

Socioeconomic conditions have been variously defined by the occupation of the father, family income, parents' education, or the family housing and environment. These criteria have been used either simultaneously or individually to establish the social level of a family. Socioeconomic conditions are known to have a definite effect on general body growth, but several investigators (Lee *et al.*, 1965; Houpt *et al.*, 1967; Garn *et al.*, 1973; Mukherjee, 1973) have concluded that either there is no such impact on permanent tooth emergence, or at most, only an insignificant influence as compared with general somatic growth. Clements *et al.* (1953*a*, 1957*a,b*) found dental emergence to be earlier among English children whose fathers are in the upper occupational levels. Yet, the Ten-State Nutrition Survey (1972) conducted in the United States found that white children from high- and low-income states do not differ significantly in the age of emergence of the permanent teeth; among black children, the emergence was earlier in the higher-income states.

Studies comparing rural and urban populations (Adler, 1958) attributed the early emergence of teeth to urbanization. Others investigators (Clements *et al.*, 1975*a,b*) found a tendency toward early emergence in children from rural schools as compared with those from urban elementary schools. There is no agreement on this point, as Helm and Seidler (1974) found only a small and insignificant difference in the tooth emergence patterns of rural and urban Danish children. Consequently, we agree with Garn *et al.* (1973) that "population differences in tooth emergence exceed socioeconomic differences."

4.5. Nutrition

The study of nutrition in relationship to dental development has been of special interest to many investigators. Both the emergence and the formation of deciduous and permanent teeth have been studied in the developing and Western countries (Voors, 1957); Niswander, 1963; Garn *et al.*, 1973; Billewicz and McGregor, 1975). As nature–nurture studies in dentistry are quite variable from one population to another, the results have always been compared and correlated to such anthropometric parameters as height, weight, and arm circumference. The literature is full of such statements as "children with fewer teeth are, in general, shorter" or "long-term undernutrition may have some effect on eruption." Severe undernutrition affects the skeletal and dental systems but affects the latter to a lesser degree. Statistically significant correlations between dental emergence and nutrition always remain low.

4.6. Hormones

Few studies have investigated the role of hormones in the process of dental emergence or development because of the failure of clinicians to extend the systematic diagnosis of pathological conditions to the area of the dentition. For example, consistent dental follow-up in cases of endocrinopathies is clearly lacking. Such studies, when they are undertaken, should adopt radiographic techniques rather than emergence, to obtain a complete picture of the developmental history of the teeth.

In their longitudinal study of two children with pituitary insufficiency, Cohen and Wagner (1948) came to the conclusion that the bony structures (maxilla and mandible) and dentition showed independent growth patterns and were affected differently as a result of different development pattern and different origins (ectodermal and mesodermal). Because of the lack of synchronization between bone and dental tissues, positional arrangements, such as changes in the vertical dimension and overbite, become more complicated. Wagner *et al.*

(1963) conducted a longitudinal study of 11 children with idiopathic sexual precocity, androgenic virilism, or adrenocortical hyperplasia. Dental development was recorded radiologically. Odontogenesis was advanced in most children with either congenital adrenocortical hyperplasia or androgenic virilism. It is worthwhile to stress again that sexual and skeletal maturity on the one hand, and dental development on the other, are dissociated. Garn *et al.* (1965*a*) studied several children with various endocrinopathies, such as thyroid abnormalities, athyreosis, and hypopituitary dwarfism. They separated nonendocrine developmental delays from endocrine-related cases and found that in congenital hypothyroidism and hypopituitarism, tooth emergence is slightly delayed. By contrast, emergence is delayed by 10% and skeletal development by 60%, among athyrotic children. In constitutional and endocrine precocities, the degree of dental advancement remains low. Dental development also tends to be advanced in children with Turner's syndrome, but this phenomenon cannot be explained on endocrine grounds alone.

We can see that even serious endocrinopathies, which severely retard somatic growth and maturation, exert only a minor effect on dentition. Garn *et al.* (1962, 1965*a*) attributed the movement of the second premolar (PM_2) and the second molar (M_2) to steroid hormones; they found little or no correlation among dental, sexual, and bone maturation before puberty. Garn and coworkers proposed that sex-linked genes might be concerned in timing of tooth eruption, but Mather and Jinks (1963) pointed out that this did not follow from their data. In 1977 Edler investigated radiographically 14 cases of confirmed growth hormone (GH) deficiency and found an average delay of 28% in skeletal age compared with 9% only in dental age. Beris *et al.* (1977) reported a slight effect of GH therapy toward the reduction of this lag in dental age; Myllerniemi *et al.* (1978) found that of 24 patients treated with GH for 4 years most showed an acceleration in dental development.

It is hoped that future studies in this field will shed more light on the influence of hormones on dental development and on their use in the treatment of pathological cases.

5. Secular Trend in Dental Maturity

Several researchers have investigated a secular trend in puberty, height, weight, and general growth over a period of several decades. During recent decades, stature has increased more than other measurements, such as that of head or upper-arm circumference and weight (Tanner, 1962).

It has been assumed that the dental system, being an inherent part of the body, must also be influenced by this trend as well, although perhaps to a lesser degree. Yet the development of deciduous dentition is quite independent of postcranial morphological processes (Falkner, 1957), and no secular trends have been detected. It is well known that the timing of the shedding of deciduous teeth has a definite effect on the emergence of the permanent teeth (Clements *et al.*, 1957*a*). The early shedding of deciduous teeth advances the emergence of the permanent ones, or perhaps we could reverse this statement and say that an earlier development of the permanent teeth causes early shedding of the deciduous teeth. This latter remains to be shown. However, the loss of a deciduous tooth, when very premature, can retard the emergence of its permanent successor (A. A. Dahlberg and Menegaz-Bock, 1958; Brook, 1973). In disputing this apparent relationship, Butler (1962) argued that there is no evidence that the early shedding of the deciduous teeth is the major cause for the early emergence of permanent teeth; rather, he attributed the phenomenon to advanced human growth and development generally, as observed during this century.

When one of the earliest studies done in the United States on Hebrew infants (Boas, 1927) is compared with studies done in the early 1940s (Robinow *et al.*, 1942; Sandler, 1944; Doering and Allen, 1942), the earlier emergence of the deciduous teeth in these later studies is quite apparent (an average advance of 4 months for central incisors). In comparing the 1940 studied with more recent investigations (Lauzier and Demirjian, 1981), the trend is still apparent, but to a lesser degree (0.2–1.4 months for different teeth) in both sexes. The differences may represent a secular trend in the emergence of the deciduous dentition; or they may be only an expression of variability among ethnic groups.

As for the permanent dentition, Miller *et al.* (1965) compared a sample of 2000 children with the results found by Ainsworth (1925) and also concluded that emergence is earlier today. Brook (1973) also favors the theory that the earlier emergence of permanent teeth parallels early puberty and an increase in height and weight noted in the last two decades. The early formation and emergence pattern has also been noted for the third molar by Garn *et al.* (1962). Comparison of the results of a Canadian study (Levesque *et al.*, 1981) with those of Hellman (1936) indicates that there may be a tendency for the third molar to emerge earlier by 2 years for boys and by 1.8 year for girls. Is this an expression of a secular trend for the dentition? No deciduous tooth is shed prior to the emergence of the third molar, and racial differences are irrelevant in comparing Americans and Canadians. Is the third molar therefore the front runner among the permanent teeth in the establishment of a secular trend? These questions should be answered in the future as more studies complete the picture.

6. Concept of Dental Age

Several approaches have been used to assess dental age on the basis of the number of teeth present in the mouth at each chronological age. Although the use of dental emergence seems to be a simple approach for predictive or age-determination purposes, there are inherent limitations to the method: (1) While the application may be useful for population groups, it must be recognized that there is a high probability of inaccuracies for a given child; and (2) Little information is available during periods when no variation in the number of teeth occurs. If predictions are to be made for individuals, such factors as low birth weight, local pathological conditions, supernumerary teeth, the early or late loss of deciduous teeth, and some general pathological conditions should be taken into consideration. such factors as low birth weight, local pathological conditions, supernumerary teeth, the early or late loss of deciduous teeth, and some general pathological conditions should be taken into consideration.

Cattell (1938) was the first to conduct such a study and assign a dental maturity age according to the number of emerged teeth in the mouth. Later investigators (Voors and Metselaar, 1958; Moorrees *et al.*, 1963*a*; McGregor *et al.*, 1968) worked along the same line. It is understandable that teeth, being less susceptible to environmental changes, can be regarded as an ideal instrument for determination of chronological age. Bailey (1964) devised a linear regression formula:

$$Y = AX + B$$

where Y is the number of emerged teeth and X is the age in months, to predict age from dental emergence of New Guinean children. This approach was later criticized by several workers (Meredith, 1971; Brook and Barker, 1972; Billewicz *et al.*, 1973).

Using linear regression equations, Malcolm and Bue (1970) developed a system of predicting height, according to the number of teeth in the mouth. Following studies among six different New Guinea populations, they based their system on four different stages, wherein 4, 4–10, 12, or 14–26 teeth were present in the mouth. A similar approach was used by Filipsson (1975), for dental-age determination, by using the curve of the total number of emerged permanent teeth. In a later paper, Filipsson and Hall (1975) attempted to predict final body height from the age at which the first 12 permanent teeth emerged, as well as corresponding height for that age.

The concept of dental age although simple to visualize and compare with either chronological age or other developmental ages, should be replaced by a scale of maturity, ranging from immaturity to complete maturity, which can be expressed by figures, say between 0 and 100. This dental maturity scale could later be transformed and presented in terms of dental age, if needed.

7. Evaluation of Dental Maturity by Developmental Stages

Up to now, we have reviewed dental maturity according to the criterion of emergence. It was Gleiser and Hunt (1955) who first stated that "the calcification of a tooth may be a more meaningful indication of somatic maturation than is its clinical emergence." Since then, this concept has steadily gained acceptance among workers in human biology (Garn *et al.*, 1958; Nolla, 1960; Fanning, 1961; Moorrees *et al.*, 1973*a*; Demirjian *et al.*, 1973) for the following reasons:

1. Although the emergence of a tooth is a fleeting event and its precise time very difficult to determine, its calcification is a continuous process that can be assessed by permanent records, such as X-ray films. Although several workers agree that clinical emergence occurs when the first tip of the cusp pierces the gum, one still finds in the literature several different definitions of *eruption,* ranging all the way from the moment the alveolar bone is broken to the attainment of the occlusal plane.
2. The emergence of a tooth is disturbed by dif-

ferent exogenous factors, such as infection or the premature extraction of its deciduous predecessor. It is known that the early extraction of a deciduous tooth delays the emergence of its permanent successor and that late extraction favors early emergence (Dahlberg and Menegaz-Bock, 1958; Fanning, 1962; Brook, 1973). We can also cite the influence of crowding, as the emergence of the permanent teeth is subject to the available space in the arch.

3. If we take the emergence of a tooth as an indicator of maturity, we have to consider each tooth individually. We cannot assess the overall picture of the dentition unless we deal with the total count of the teeth, making the evaluation of maturity a very crude measurement.
4. The use of clinical emergence as a maturity indicator is limited to certain ages. The emergence of the deciduous teeth can only be used between the ages of 6 and 30 months, after which time there is no activity until the first permanent molar emerges, at about the age of 6 years. Excluding third molars, the last permanent tooth emerges at approximately 12 years of age. The third molar is not taken into consideration because of the inconsistency and great variation in the timing of its emergence. Besides, the third molar is the only tooth to emerge between the ages of 14 and 20 years, and it would be too imprecise to base estimates of dental maturity for such an age span on just one tooth. Thus, we see that it is impossible, with emergence data alone, to assess the dental maturity of a child between 2½ and 6 years, and after 12 years of age, as there is no clinical emergence of teeth during these intervals.

For all these reasons, dental formation or calcification, which is a continuous developmental process, should be considered a better measure of physiological maturity than dental emergence.

The process of assessing dental maturity from X-ray films differs from that of other techniques used in human biology, in which length measurements are the basis of the assessment. For example, both stature and bigonial diameter are measures of the distance between two well-defined points. On the other hand, the assessment of dental maturity, like skeletal maturity, should be based on biological criteria rather than on length or width measurements alone. The latter are really only estimates of growth, as distinct from maturity.

Like the bones of the hand–wrist area, teeth undergo different sequences of maturational stages. The first stage is the actual formation of the crypt, and the final stage is the fully mature tooth, defined as the closure of the apex. During this maturational process, one sees continuous changes in the size and shape of the teeth. Each tooth follows the same sequence; in order to study the entire process, arbitrary stages must be selected and fully described that trace the entire developmental process from beginning to end. These stages must (1) Describe the major developmental stages of the tooth, (2) Be clearly defined (not merely on the basis of length increases), and (3) Be objective enough to be reproducible. Unlike the skeletal system, the dental system has two overlapping developmental periods for two sets of teeth. The developmental period for the deciduous teeth extends from the third month of intrauterine life to the third year of postnatal life. For the permanent teeth, it is from the age of 6 months postnatal to 14–15 years, excluding the third molars.

Radiographic evaluation of the development of both dentitions, along with the resorption of the deciduous teeth, is effective in the assessment of dental maturity. Evaluation of a single tooth or group of teeth can be useful in certain clinical situations. However, when estimating maturity, it is preferable to evaluate the entire dentition, either deciduous or permanent, or both, since every tooth contributes to the dental maturity process (Prahl-Andersen and Van der Linden, 1972). However, to study every tooth is prohibitively time consuming and expensive and involves certain technical difficulties as well. Acceptable modifications have therefore been devised. For example, several workers (Nolla, 1960; Moorrees, *et al.*, 1963*a;* Liliequist and Lundberg, 1971; Demirjian *et al.*, 1973) demonstrated a high degree of correlation between the developmental stages of the right and left sides of the maxilla and mandible. As a result, it is possible to study only one side of the arch. This type of selection is also made in bone-age assessments, where the hand–wrist area is usually selected for evaluation.

Most investigators prefer to limit their observations to the mandibular teeth (Gleiser and Hunt, 1955; Demisch and Wartmann, 1956; Demirjian *et al.*, 1973) and to the maxillary incisors and canines (Fanning, 1961; Moorrees *et al.*, 1963*a*; Liliequist and Lundberg, 1971), because the radiographic view of most maxillary teeth is obstructed during the early developmental period of the permanent teeth (1–6 years); i.e., the tooth buds are still in the maxillary bone and the bony structures are superimposed in the premolar and molar area.

Because the radiographic view of the mandibular teeth is so clear, it has been proposed that, for purposes of standardization, all mandibular teeth (excluding the third molars) on the left side be selected for study. Once a group of teeth has been agreed on for assessment, a maturity scale can be constructed. The concept of a maturity scale is different from that of a scale of height, for example, in that all subjects pass through the same series of points on the scale, starting with a stage designated as 0 for complete immaturity and concluding with 100, corresponding to complete maturity. Scores between these two extremes are assigned to the degree of development of the specific teeth under consideration.

The first study to assess dental formation using radiographic techniques was conducted by Hess *et al.* (1932), who evaluated the physiological maturity of children. Gleiser and Hunt (1955) studied 25 boys and 25 girls longitudinally, from birth to age 10, and traced the development of the first permanent mandibular molar. They described 13 different stages of maturity, from the calcification of the top of the cusp to the closure of the apex. The description of the stages is based on length criteria, corresponding to one-half of the crown or three-fourths of the root completed, for example. They found that although "calcification and eruption are not identically synchronized in all children, on average clinical emergence occurs soon after one third of the root is completed."

In a longitudinal study of 255 boys and girls Garn *et al.* (1958) defined five stages of the development of PM_2 and M_2: three for calcification and two for eruption. All assessments of the calcification stages, as well as of the alveolar emergence and occlusal attainment, were made from radiographs. Clinical emergence was not assessed. These workers found that the sequences of calcification and emergence alternated in 55% of subjects. Thus, they concluded that a stage of calcification, for example, cannot be a prediction criterion for a stage of eruption. Once again, the lack of synchronization between different developmental stages and the variability among individual teeth was demonstrated.

Nolla (1960) described another technique for evaluating the development of the permanent dentition, based on 10 length stages, such as one-half of the crown or two-thirds of the root. A score of 1–10 was assigned to each stage, starting from the presence of the crypt without calcification to the completion of the apical end. These developmental stages were studied longitudinaly for both the maxillary and mandibular teeth in 25 boys and 25 girls.

Fanning (1961) applied Gleiser and Hunt's 13 stages of dental development to all mandibular teeth and maxillary incisors. For greater precision, she added a total of seven more stages: two initial stages; two for cleft formation in the molars; and three to describe the different degrees of apex closure. In fact, these added stages served only to detract from the precision of the method. It proved too difficult to differentiate between successive stages, and measurements became little more than subjective estimates.

Moorees *et al.* (1963*b*) used the longitudinal records from the Fels Institute sample to study the formation and resorption of the deciduous canines and molars. Weighted estimates of the mean attainment age for each stage were obtained using a standard deviation of 2.042 log age-since-conception units. A graphic representation of the chronology of tooth formation and root resorption was provided for each tooth. However, these investigators cautioned that "standard score ratings for the various teeth should be presented preferably as a range of maximal and minimal deviates. They should not be averaged, in recognition of the individuality of each tooth."

Moorrees *et al.* (1963*a*) also studied the developmental pattern of eight mandibular and two maxillary incisor permanent teeth. The incisors were assessed from intraoral radiographs of 134 Boston children, while uniradicular and multiradicular posterior teeth were assessed from lateral radiographs of 246 children from the Fels sample. The results of this study were presented in chart form. They are "designed specifically for determining the dental maturity of an individual child, for each tooth separately." The usefulness of this system can be very well appreciated in orthodontic clinics, where formative stages of individual teeth and their emergence are important for treatment-planning purposes. The developmental chart of the maxillary and mandibular incisors for the girls is exemplified by Figure 3.

Following a study of 287 children, Liliequist and Lundberg (1971) published a system of evaluation for dental development based on the radiographic assessment of the mandibular teeth (excluding the third molars) and maxillary incisors and canines. The upper premolars and molars were not exam-

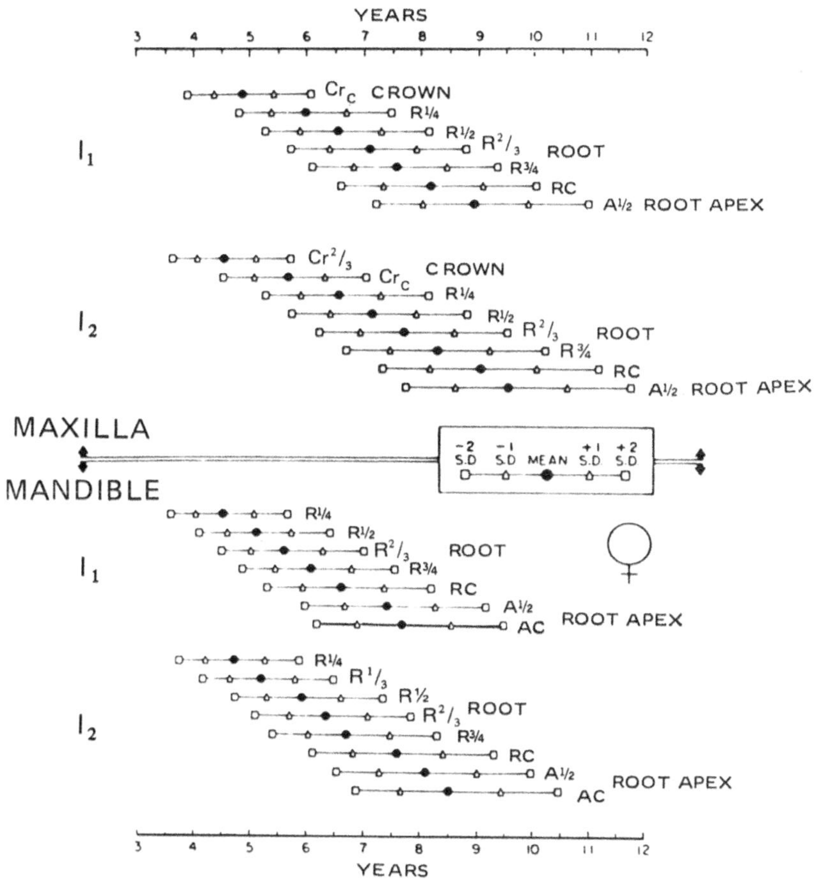

Fig. 3. Norms of the formation of permanent maxillary and mandibular incisor roots of girls, including terminal stages of crown formation of the maxillary incisors. (From Moorrees *et al.*, 1963*a*.)

ined because of the difficulty in obtaining a clear image of these teeth on X-ray film. Eight developmental stages were defined between the noncalcified crown and the completion of root development. In this system, the intervals between each successive stage were considered equal. In other words, scores of 1–8 were assigned to each stage, respectively.

Gustafson and Koch (1974) constructed a tooth development diagram based on previously published data between 1909 and 1970 in a system similar to that presented by Garn *et al.* (1958). Gustafson and Koch did not consider the differences between histological and radiographic findings. They described four stages—three for calcification and one for emergence—for deciduous teeth.

Using the last two methods, Crossner and Mansfeld (1983) attempted to estimate the unknown chronological age of adopted non-European children. Although it is true that a high correlation exists between dental and chronological ages, one should be very careful when applying developmental standards based on one population, to children from another population.

Some of the systems and stages of dental development described above are illustrated in Table III. It should be noted that these systems are based on absolute values, for the lengths of the teeth, crowns, or roots. If dental radiographs from longitudinal studies are not available, however, such a system is difficult to apply. Even an experienced investigator can have trouble distinguishing between such criteria as one-fourth or one-third of a root length. Moreover, absolute lengths of teeth can be highly variable.

An acceptable alternative for these systems of assessing dental maturity should be both convenient and specific and should contain just enough stages to avoid confusion and subjectivity. A new approach was developed by Demirjian *et al.* (1973) based on the same principle as the bone-age assessment suggested by Tanner *et al.* (1975). Using a sys-

Table III. Comparative Table of "Stages of Dental Formation" According to Different Investigators

	Fanning (1961)	Moorrees *et al.* (1963*a*)	Gleiser and Hunt (1955)	Nanda and Chawla (1966)	Nolla (1960)	Liliequist-Lundberg (1971)	Demirjian *et al.* (1973)	Gustafson and Koch (1974)	Garn *et al.* (1958)
Presence of crypt	1	—	—	—	1	1	—		—
Initial cusp formation	2	1	—	—	2	—	1 (A)	1	1
Coalescence of cusps	3	2	1	—	—	—	—	—	—
Occlusal surface completed	4	3	2	1	—	—	2 (B)	—	—
Crown ⅓	—	—	—	—	3	—	—	—	—
Crown ½	5	4	3	2	—	2	3 (C)	—	—
Crown ⅔	6	—	4	—	4	—	—	—	—
Crown ¾	—	5	—	3	5	—	—	—	—
Crown formation completed	7	6	5	4	6	3	4 (D)	2	—
Initial radicular formation	8	7	6	5(⅛)	—	—	—	—	2
Initial radicular bifurcation	8 A,B	8	—	—	—	—	—	—	—
Root ¼	9	9	7	6	—	4	5(E)	—	—
Root ⅓	10	—	8	7(⅜)	7	—	—	—	—
Root ½	11	10	9	8	—	—	—	—	—
Root ⅔	12	—	10	9(⅝)	8	5	6(F)	—	—
Root ¾	13	11	11	10	—	6	—	—	—
Root completed	14	12	12	11(⅞)	9	7	7(G)	3	—
Apex ½ closed	—	13	—	—	—	—	—	—	—
Apex closed[a]	15	14	13	12	10	8	8(H)	—	3

[a]Apex closure: ¼, ½, ¾.

tem that utilizes a new technique of panoramic radiographs (see Figure 4), weighted scores are assigned to each of the seven left mandibular teeth. Eight stages of development, from calcification of the tip of the cusp to the closure of the apex, were designated 0 (for no calcification) and by letters A–H, corresponding to the eight stages (see Figure and Table IV). Letters rather than numbers were selected so as not to leave the false impression that each stage is equidistant from the others.

7.1. Assigning the Ratings

The ratings for our new approach are based on the following criteria:

1. The mandibular permanent teeth are rated in the following order: second molar, (M_2); first molar (M_1); second premolar (PM_2), first premolar (PM_1); canine, lateral incisor, central incisor.
2. The rating is assigned by carefully following the written criteria for each stage and by comparing the tooth with the diagrams and X-ray films shown in Figure 5. The illustration should only be used as an aid, not as the sole source for comparison. For each stage, there are one, two, or three written criteria, marked a, b, and c (see Table IV). If only one criterion is given, this must be met for the stage to be recorded. If two criteria are given, it is sufficient for the first one to be met for the stage to be recorded as reached. If three criteria are given, the first two criteria must be met for the stage to be considered as having been reached. At each stage, in addition to the criteria for that stage, the criteria for the previous stage must be satisfied. In borderline cases, the earlier stage is always assigned.

3. There are no absolute measurements to be taken. A pair of dividers is sufficient to compare the relative length (crown to root). To determine apex closure stages, a magnifying glass is not necessary. The ratings should be made with the naked eye.
4. The crown height is defined as the maximum distance between the highest tip of the cusps and the cementoenamel junction. When the buccal and lingual cusps are not at the same level, the midpoint between them is considered the highest point.
5. If there is no sign of calcification, a rating of 0 is given. The crypt formation is not taken into consideration.

7.2. Dental Formation Stages

Once the ratings are assigned, each stage is then given a numerical score, according to the mathematical technique of Tanner *et al.* (1975) and Healy and Goldstein (1976). The sum of these scores provides an estimate of an individual's dental maturity on a scale from 0 to 100. The scores and percentile standards were calculated separately for boys and girls between the ages of 3–16 years, based on a standard sample of 1446 boys and 1482 girls, aged 2–20 years, of French-Canadian origin.

Although we have no absolute standard by which to judge the variability of a dental maturity system, it should have two general properties: (1) Maturity scores must reflect the continuous nature of biological development and succeed each other in a smooth logical progression; and (2) The variability in the maturity score, at each age, should be sufficient to reflect the natural variability of the general population.

The system just described has two main shortcomings. First, it is necessary to rate all seven teeth. However, in many children, one or more teeth may be missing, and it may not be possible to substitute the corresponding tooth from the right side of the mandible. Also, for practical reasons, it is often simpler, using a standard X-ray machine to take a radiograph of fewer than seven teeth. Thus, a system based on fewer teeth would be preferable. Naturally, some information and precision will be lost, with the result that slightly different components of dental maturity may be measured. Second, the standard sample had an insufficient number of children in the highest and lowest age groups. This meant that the earlier stages of some teeth could

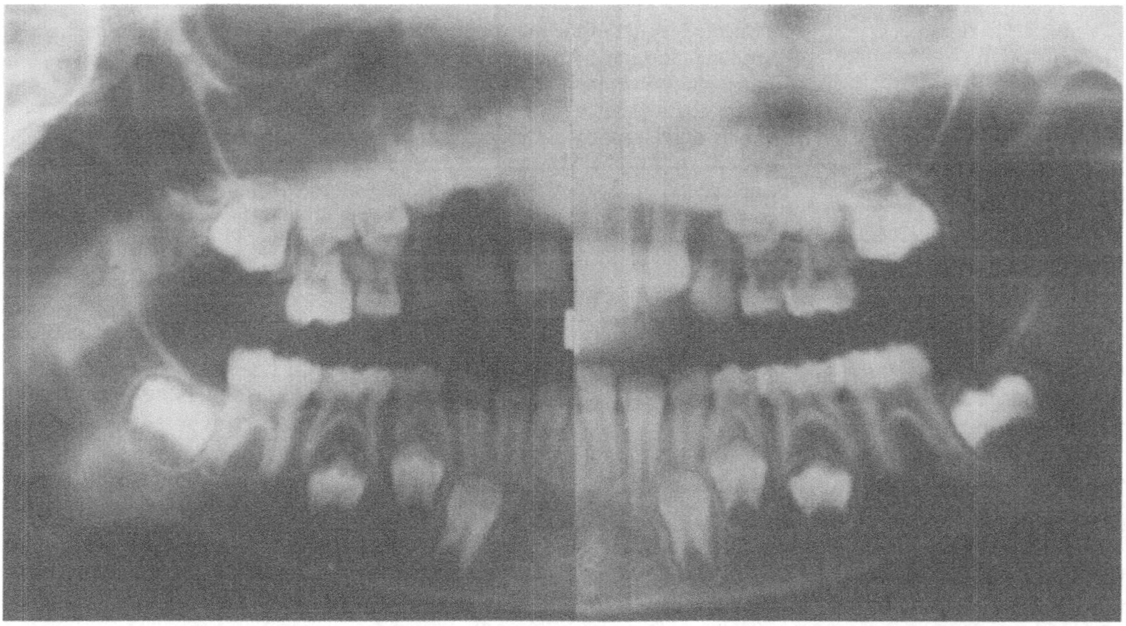

Fig. 4. Panoramic radiograph of the maxilla and the mandible gives a complete picture of the whole dentition. The posterior teeth of the upper jaw are superimposed with the palative and the zygomatic bones; the lower jaw shows a clearer view for a better assessment.

Table IV. Description of Dental Formation Stages[a]

Stage	Criteria	Description
A		In both uniradicular and multiradicular teeth, a beginning of calcification is seen at the superior level of the crypt, in the form of an inverted cone or cones. There is no fusion of these calcified points.
B		Fusion of the calcified points forms one or several cusps, which unite to give a regularly outlined occlusal surface.
C	a	Enamel formation is complete at the occlusal surface. Its extension and convergence toward the cervical region is seen.
	b	The beginning of a dentinal deposit is seen.
	c	The outline of the pulp chamber has a curved shape at the occlusal border.
D	a	The crown formation is completed down to the cementoenamel junction.
	b	The superior border of the pulp chamber in uniradicular teeth has a definite curved form, being concave toward the cervical region. The projection of the pulp horns, if present, gives an outline like an umbrella top. In molars, the pulp chamber has a trapezoidal form.
	c	Beginning of root formation is seen in the form of a spicule.
E		*Uniradicular teeth*
	a	The walls of the pulp chamber now form straight lines, whose continuity is broken by the presence of the pulp horn, which is larger than in the previous stage.
	b	The root length is less than the crown height.
		Molars
	a	Initial formation of the radicular bifurcation is seen in the form of either a calcified point or a semilunar shape.
	b	The root length is still less than the crown height.
F		*Uniradicular teeth*
	a	The walls of the pulp chamber now form a more or less isosceles triangle. The apex ends in a funnel shape.
	b	The root length is equal to or greater than the crown height.
		Molars
	a	The calcified region of the bifurcation has developed further down from its semilunar stage to give the roots a more definite and distinct outline, with funnel-shaped endings.
	b	The root length is equal to or greater than the crown height.
G	a	The walls of the root canals are now parallel (distal root in molars).
	b	The apical ends of the root canals are still partially open (distal root in molars).
H	a	The apical end of the root canal is completely closed (distal root in molars).
	b	The periodontal membrane has a uniform width around the root and the apex.

[a]From Demirjian *et al.* (1973).

not be included, since they were not adequately represented in the sample. Consequently, percentile standards could not be provided for the extreme age groups (3 and 14 years).

For these reasons, the standard sample was enlarged to include a total of 2047 boys and 2349 girls, aged 2–20 years (Demirjian and Goldstein, 1976). Two stages, which were excluded in the initial study, could now be included, i.e., stage A of the PM_1 and stage C for the I_1. The 3rd and 97th percentile estimates for the maturity standards could also now be included. The new scores and percentile curves for the seven teeth are provided in the Table V and Figures 6 and 7. The comparative curves of the maturity scores for boys and girls are given in Figure 8.

For practical purposes, when panoramic radiographs cannot be used, it is convenient to take two periapical radiographs of the molar and the premolar areas. Therefore, the use of two molars and a premolar (M_2, M_1, PM_2, and PM_1) has been considered as the basis of a second system for dental assessment, for which scores and standards have also been established (see Table VI and Figures 9 and 10).

Since the development of the mandibular incisor is chronologically almost the same as that of the first molar, the central incisor can be substituted for

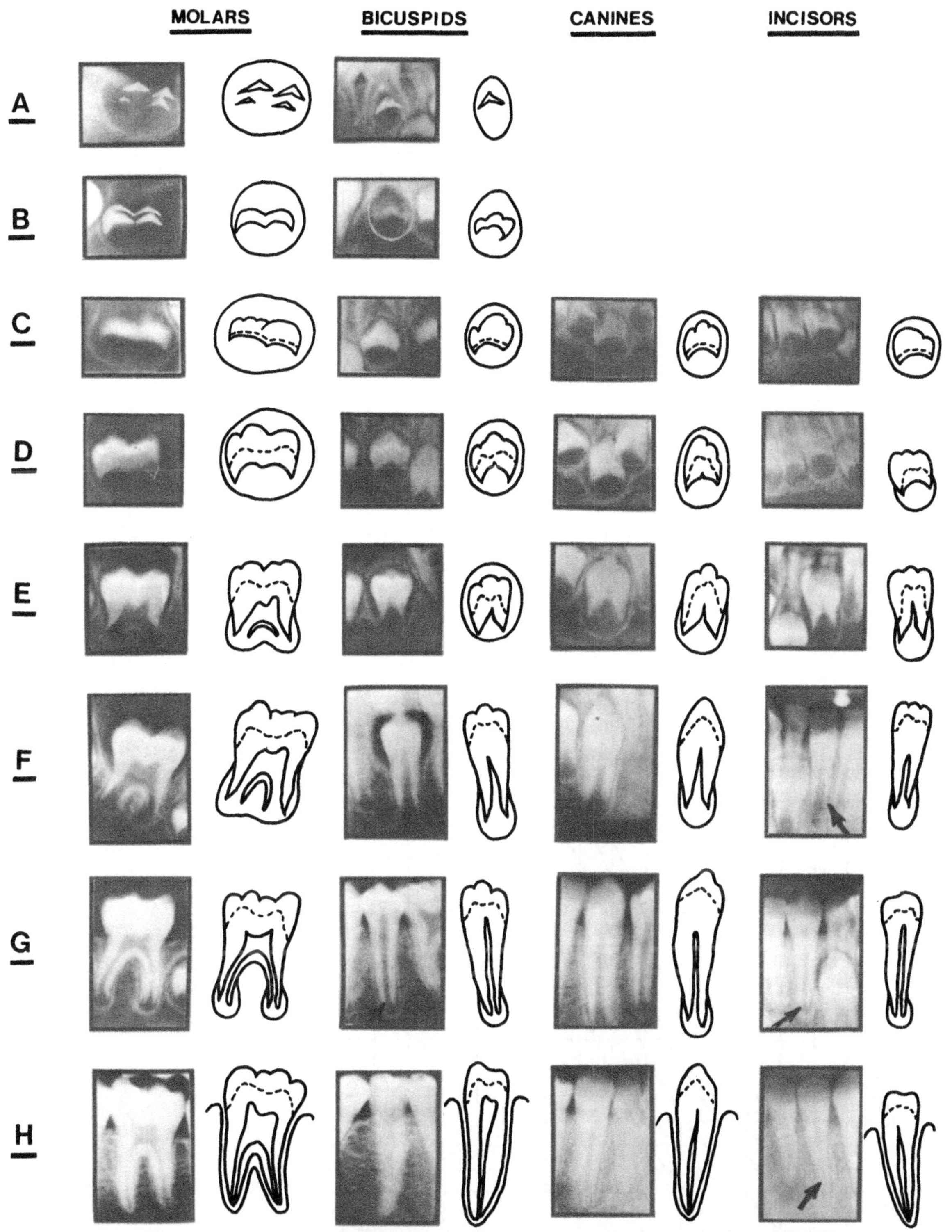

Fig. 5. The developmental status of each group of teeth (molars, bicuspids, canines, incisors) is defined for stages A–H. The definition of each stage of the permanent dentition is based on biological criteria (see text).

Table V. Self-Weighted Scores for Dental Stages, Seven Teeth (Mandibular Left Side)[a]

Tooth	Stage								
	0	A	B	C	D	E	F	G	H
					Boys				
M_2	0.0	1.7	3.1	5.4	8.6	11.4	12.4	12.8	13.6
M_1				0.0	5.3	7.5	10.3	13.9	16.8
PM_2	0.0	1.5	2.7	5.2	8.0	10.8	12.0	12.5	13.2
PM_1		0.0	4.0	6.3	9.4	13.2	14.9	15.5	16.1
C				0.0	4.0	7.8	10.1	11.4	12.0
I_2				0.0	2.8	5.4	7.7	10.5	13.2
I_1				0.0	4.3	6.3	8.2	11.2	15.1
					Girls				
M_2	0.0	1.8	3.1	5.4	9.0	11.7	12.8	13.2	13.8
M_1				0.0	3.5	5.6	8.4	12.5	15.4
PM_2	0.0	1.7	2.9	5.4	8.6	11.1	12.3	12.8	13.3
PM_1		0.0	3.1	5.2	8.8	12.6	14.3	14.9	15.5
C				0.0	3.7	7.3	10.0	11.8	12.5
I_2				0.0	2.8	5.3	8.1	11.2	13.8
I_1				0.0	4.4	6.3	8.5	12.0	15.8

[a]From Demirjian and Goldstein (1976).

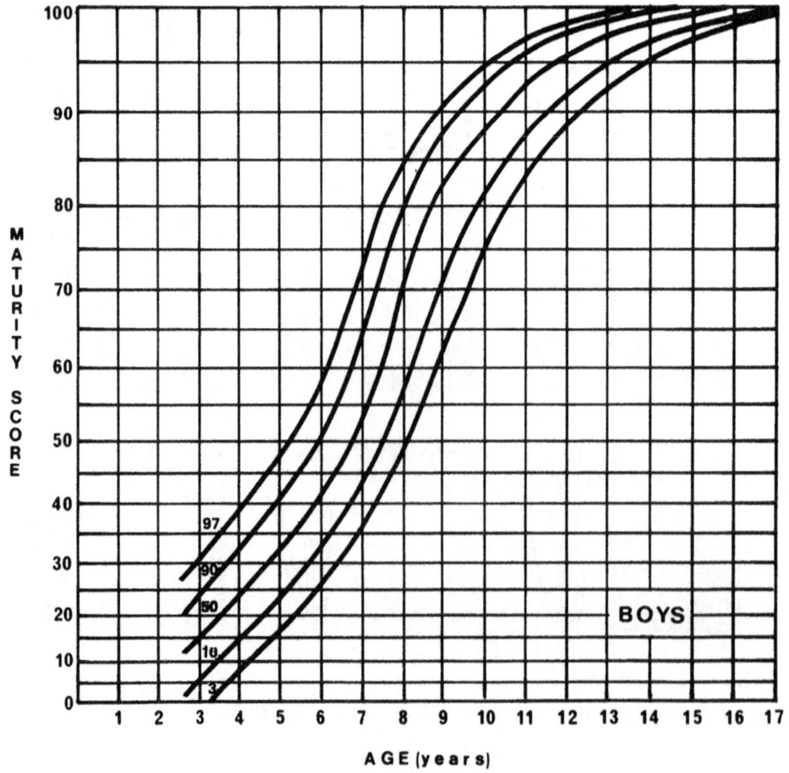

Fig. 6. Dental maturity percentiles scored for boys aged 3–17 years (based on seven left mandibular teeth).

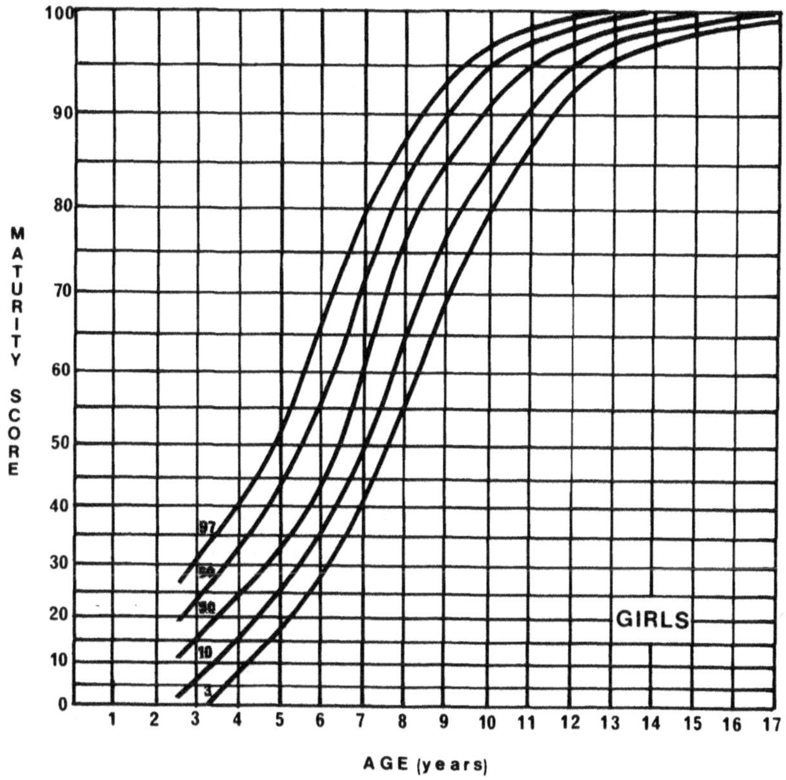

Fig. 7. Dental maturity percentiles scored for girls aged 3–17 years (based on seven left mandibular teeth).

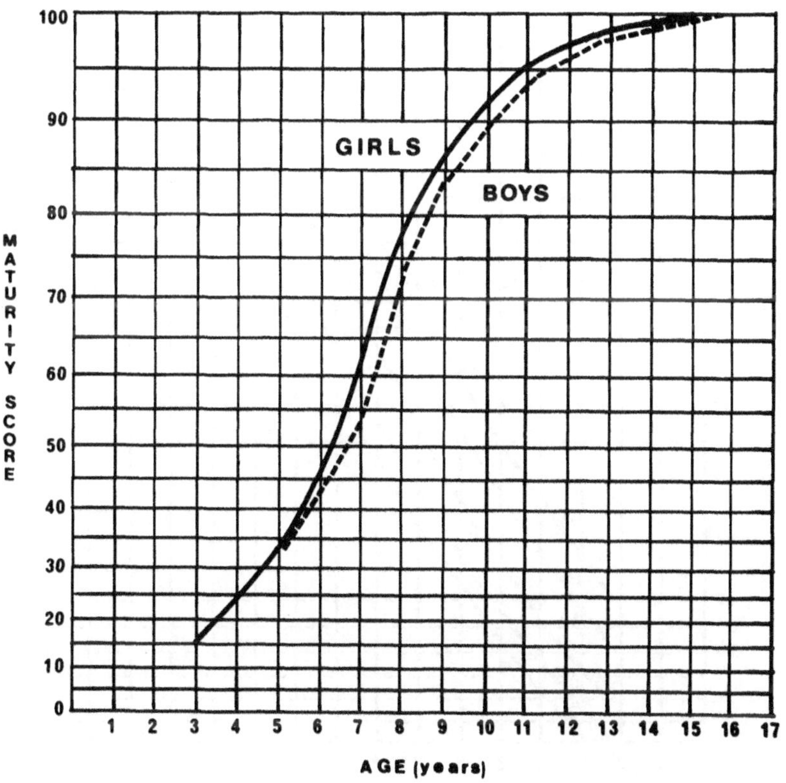

Fig. 8. Comparative curves (median) of dental maturity for boys and girls 3–17 years (based on seven left mandibular teeth).

Table VI. Self-Weighted Scores for Dental Stages, Four Teeth Mandibular Left Side)[a]

Tooth	Stages 0	A	B	C	D	E	F	G	H
					Boys				
M_2	0.0	3.2	6.2	9.9	14.4	18.4	20.7	21.9	23.3
M_1				0.0	8.0	12.6	16.9	21.8	27.4
PM_2	0.0	3.1	5.6	9.5	13.7	17.4	20.1	21.4	22.5
PM_1		0.0	5.9	10.7	15.7	20.7	23.8	25.4	26.8
					Girls				
M_2	0.0	3.6	6.1	9.9	15.3	19.2	21.7	23.0	24.2
M_1				0.0	5.4	9.8	14.3	20.1	25.9
PM_2	0.0	3.7	5.8	9.8	14.7	18.1	20.8	22.3	23.3
PM_1		0.0	4.6	9.2	15.1	20.2	23.3	25.1	26.6

[a]From Demirjian and Goldstein (1976).

the molar in older age groups, where the latter is frequently missing. This principle forms the basis of a third system of dental assessment, which assigns separate scores and standards for another group of four teeth (M_2, PM_2, PM_1, I_1). This third system requires an additional periapical film of the incisor area (see Table VII and Figures 11 and 12).

All three systems use equal "biological" weights for each tooth. The correlation among the three systems is quite high, the correlation coefficient being 0.7–0.9 for each age group between 6 and 16 years. This could be anticipated, since most of the same

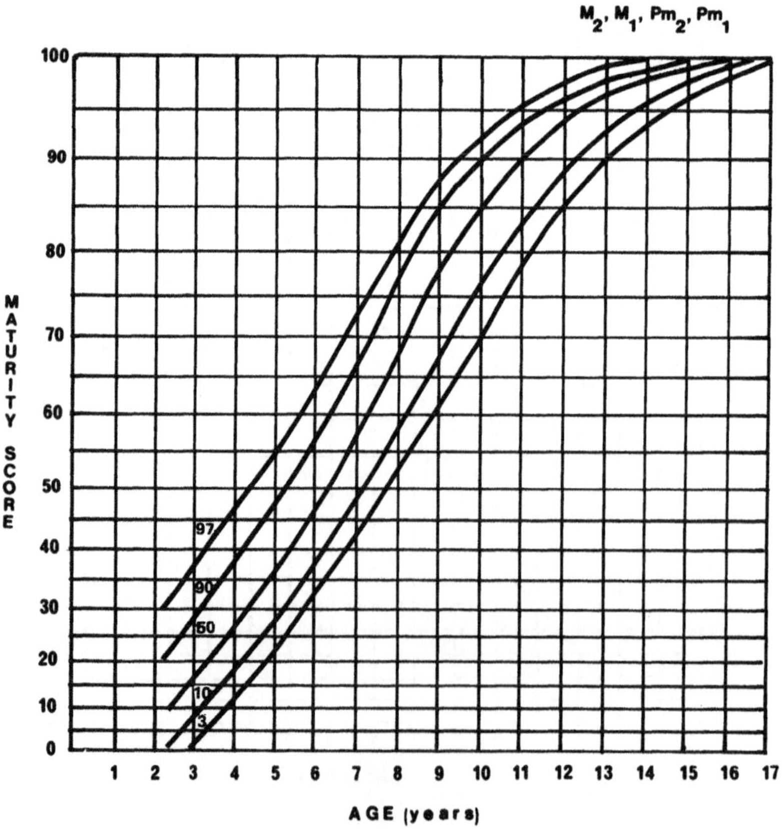

Fig. 9. Dental maturity percentiles scored for boys aged 3–17 years (4 teeth I) (based on M_2, M_1, PM_2, PM_1).

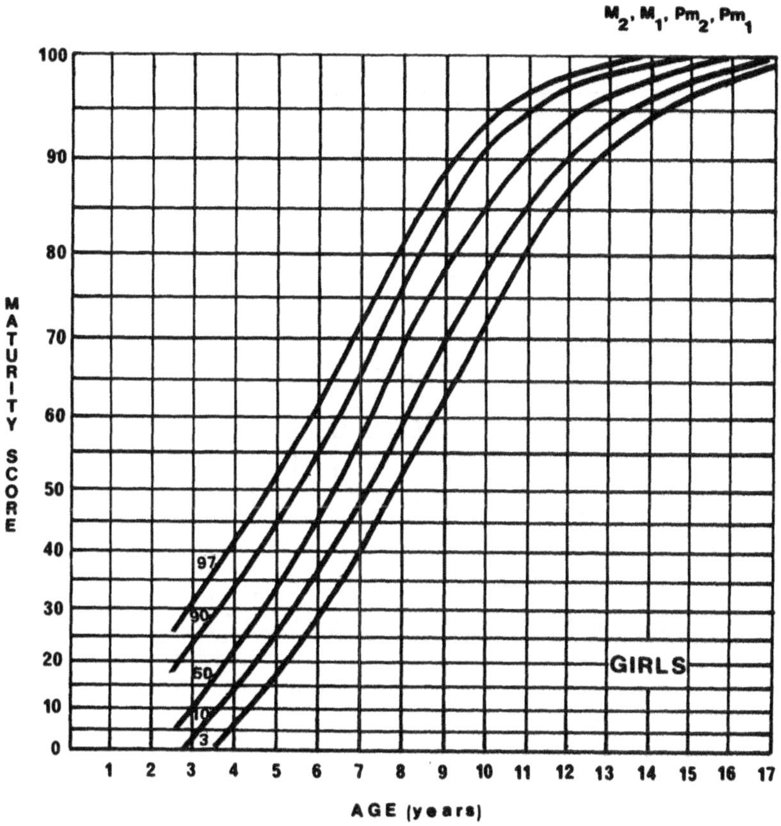

Fig. 10. Dental maturity percentiles scored for girls aged 3–17 years (4 teeth I) (based on M_2, M_1, PM_2, PM_1).

teeth are involved in all three scoring systems. What we have done is to devise two subsystems, each comprising convenient groups of teeth for rating purposes. The results of comparing these systems with the seven-tooth system and with each other raise the possibility that somewhat different aspects of maturity are being measured. In order to study whether this is the case, we need to compare children with longitudinal records, using the different system. If the systems prove comparable, one or the other may be selected, according to whether a full panoramic radiograph is available, whether a particular tooth is missing, and other factors. The reasons for selecting one system over the other

Table VII. Self-Weighted Scores for Dental Stages, Four Teeth (Mandibular Left Side)[a]

Tooth	Stages								
	0	A	B	C	D	E	F	G	H
				Boys					
M_2	0.0	3.3	6.1	9.9	15.0	19.7	21.3	22.1	23.5
PM_2	0.0	3.2	5.6	9.6	14.2	18.8	20.9	21.7	22.8
PM_1		0.0	7.1	11.6	16.9	22.8	25.8	26.8	27.9
I_1				0.0	7.4	11.5	14.6	18.9	25.7
				Girls					
M_2	0.0	3.4	6.3	10.2	15.7	20.0	21.5	22.3	23.5
PM_2	0.0	3.7	6.2	10.3	15.1	19.1	21.0	21.7	22.8
PM_1		0.0	5.9	10.2	16.2	21.9	24.6	25.6	26.8
I_1				0.0	8.1	12.2	15.6	20.7	27.0

[a]From Demirjian and Goldstein (1976)

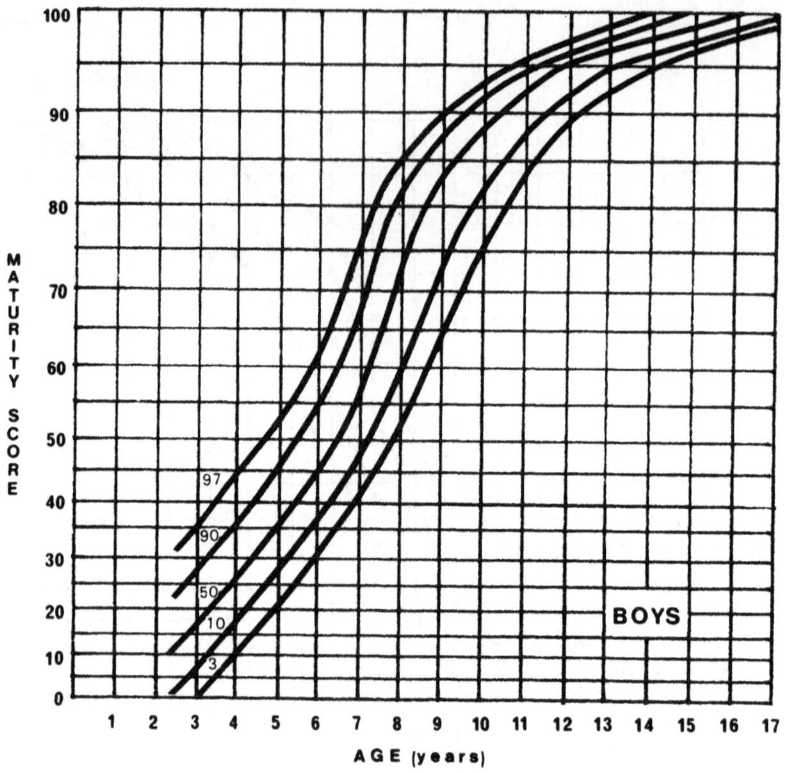

Fig. 11. Dental maturity percentiles scored for boys aged 3–17 years (4 teeth II) (based on M_2, PM_2, PM_1, I_1).

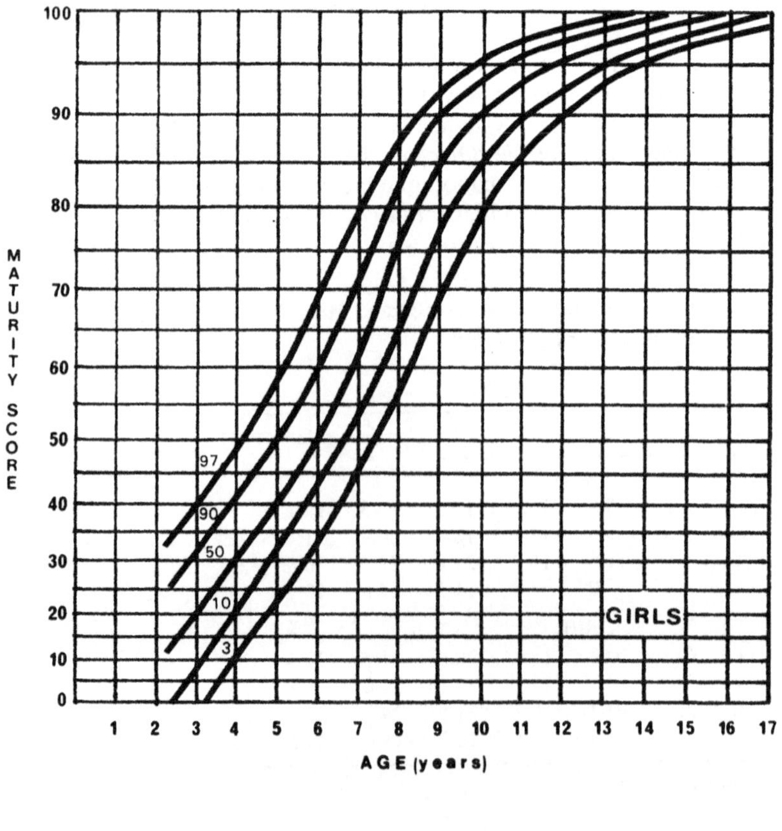

Fig. 12. Dental maturity percentiles scored for girls aged 3–17 years (4 teeth II) (based on M_2, PM_2, PM_1, I_1).

should be recorded when reporting a maturity score.

The percentile curves enable us to assess the centile position of an individual's maturity score. If required, the maturity score can be converted to dental age by finding the age at which the 50th percentile value equals the given maturity score. However, since dental development is a maturational process and the scores reflect this maturation, we consider it more appropriate to use maturity scores, rather than dental age, when comparing the development of individuals or tracing the progress of a particular child.

When using the scoring system and standards presented here, it should be remembered that the sample is entirely of French-Canadian origin. We do not know as yet how representative the results are for children generally. We would conjecture, however, that the scores for the eight stages will not vary much between populations, but that the maturity standards may change appreciably in different populations. It is possible to study differences in the average maturity of different populations using the present scoring systems with relatively small samples.

Demirjian and Lamarche (1979) made a preliminary study to assess the dental maturity of the children from the Harpenden growth study. A slight tendency toward early maturity in British children was found.

In the Nymegen study in Netherlands, Prahl-Andersen *et al.* (1979) reported a delay of dental maturity among Dutch children, compared with the standardizing sample of Montreal. The amplitute of the delay seems to be age related; in other words, the delay is more pronounced after 8 years of age.

Proy *et al.* (1981) found an advance in the dental maturity of French children in Lyon as compared with the Montreal sample. Even though their technique was standardized and checked with the Canadian group, on average an advancement of 9 months was observed in the dental development of French children. These workers also stress the importance of having national standards of dental maturation. Studies are under way in a Swedish group (U. Hagg, 1983, personal communication) for the creation of such standards.

8. Correlations between Dental Development and Other Maturity Indicators

Many cross-sectional and longitudinal studies have been conducted to establish possible correlations between the development of the dental system and other physiological indicators, such as bone age, height, menarche, and circumpubertal growth. Most studies have used dental emergence of both the deciduous and/or permanent teeth as the maturity indicator. However, some have investigated the formation or calcification pattern of the teeth in relationship to craniofacial data in an attempt to establish the etiology of malocclusions in orthodontic practice. Because of the different methodologies and approaches used in these studies, it is instructive to look at them in some detail.

8.1. Correlations Using Dental Emergence Criteria

As early as 1942, Robinow *et al.* used factor analysis to establish correlations between the emergence of the deciduous teeth and the appearance of ossification centers, attained height, and age at onset of walking. These workers found no significant correlation among any of these parameters and concluded that the development of the dentition is a relatively independent growth process. Two decades later, Lysell *et al.* (1962) came to similar conclusions. This is not too surprising, since bones and teeth have different origins. Teeth are at least partially of epithelial origin, while bone is derived from the mesoderm.

The relationship of the emergence of the deciduous dentition to birth weight has been reported in Section 3.5. Falkner (1957) reported that small babies have a faster rate of dental emergence than that of larger babies. However, he found no correlation between skeletal maturation and dental emergence. These results have been corroborated by Robinow (1973), Billewicz *et al.* (1973), and McGregor *et al.* (1968).

Doubtless, because of the greater availability of data from older children (school and medical records), the emergence of permanent dentition and/or the number of permanent teeth in the mouth have been more thoroughly investigated in relationship to somatic growth, skeletal development, and puberty. Earlier studies employed the Greulich and Pyle atlas (1950) as the reference base for skeletal development. The reference used most frequently for dental emergence was the series of dental charts compiled by Schour and Massler (1940).

Sutow *et al.* (1954) investigated the correlation between the number of permanent teeth in the mouth and skeletal development in Japanese children aged 6–14 years. These investigators found that the children who had the greatest number of teeth in each group were also the most skeletally advanced. Nevertheless, differences within any single age group were not found to be statistically significant. The observation illustrates the importance of distinguishing age-specific correlations between different maturity indicators from correlations using pooled ages, which are confounded by ongoing maturation in both systems.

In a longitudinal study, Lamons and Gray (1958) came to the same conclusion, i.e., that there is a larger correlation between dental emergence and chronological age than between dental emergence and bone age. In his longitudinal study of 43 girls and 29 boys, Meredith (1959) assessed dental development by the emergence of the canine and the first and second mandibular molars. A very low correlation was found between the emergence of these teeth and age of maximum circumpubertal height growth. Using the longitudinal records of the Denver study, Nanda (1960), came to the same conclusion and found a low and nonsignificant correlation (0.2 for boys and 0.3 for girls) between dental emergence and age of peak height velocity at puberty. The correlation was slightly higher (0.6) between the completion of the permanent dentition and age at menarche. Another interesting corollary to this finding was that no close relationship has been found between the growth of the face, which follows the general somatic pattern, and the emergence of the teeth. This knowledge is of importance in the diagnosis and treatment of malocclusions.

Eveleth (1960) compared dental emergence and age at menarche in American children living in Rio de Janeiro, Brazil, with that of their counterparts living in the United States. Again, only a very low correlation was found between these two indices in both populations. Bjork and Helm (1967) also failed to find any significant correlation between dental age and skeletal age and height or age at menarche.

Thus, the evidence indicates that the skeletal system, as well as height and the onset of puberty, develop largely independently of the dental system. For further information on this, we now turn to studies using the newer quantitative measures of dental formation. As there are few studies in this field, and these have varied in the precise measures used and in the way in which the results were presented, we shall review them individually in some detail.

In one of the first correlation studies using dental formation, Brauer and Bahador (1942) compared the calcification process with emergence for deciduous and permanent of teeth in 415 children. These workers found that calcification age and emergence age do not necessarily correspond to chronological age. They also stressed that without radiographic data, a diagnosis based solely on dental maturity using emergence could be inaccurate in at least 50% of patients. The conclusions of Brauer and Bahador were corroborated by Garn *et al.* (1960), who found that tooth calcification and eruption processes are independent of one other.

Since it appears well established that the two dental processes are autonomous, most investigators correlate tooth calcification with other physiological maturity indicators. In one early study, Demisch and Wartmann (1956) studied the calcification of the third molar in relationship to bone age among 151 children. Within any 1-year chronological age group, the correlation coefficient between bone age and dental calcification was about 0.45 for both boys and girls.

Hotz (1959) and Green (1961) concluded independently that there is a high correlation between dental and chronological age, but not between dental age and skeletal age. Both workers found large variations in skeletal age; as a result, Green suggested that dental age should be adopted as a more reliable criterion of somatic growth evaluation.

Lewis and Garn (1960) also found less variability, as assessed by the coefficient of variation in dental development than in skeletal development among equivalent age groups, when they applied their five stages of tooth development to 255 children. This finding suggests that dental development can be a better criterion of biological maturation than of ossification (Figure 13). They went a step further by suggesting that dental age could be used to determine an unknown chronological age. It is also interesting to note that in the same study, Lewis and Garn (1960) found a weaker correlation between dental calcification stages and dental emergence than among different stages of calcification.

Garn *et al.* (1962) studied the calcification and emergence patterns of the third molar in relationship to bone age and chronological age. Because of the third molar's inconsistent development, Garn and co-workers found a low correlation between the development of this tooth and menarche and skeletal development. A higher correlation was subsequently found between the development of the second molar and the other maturity indicators.

Gron (1962) compared dental emergence, calcification of the root, and skeletal development in

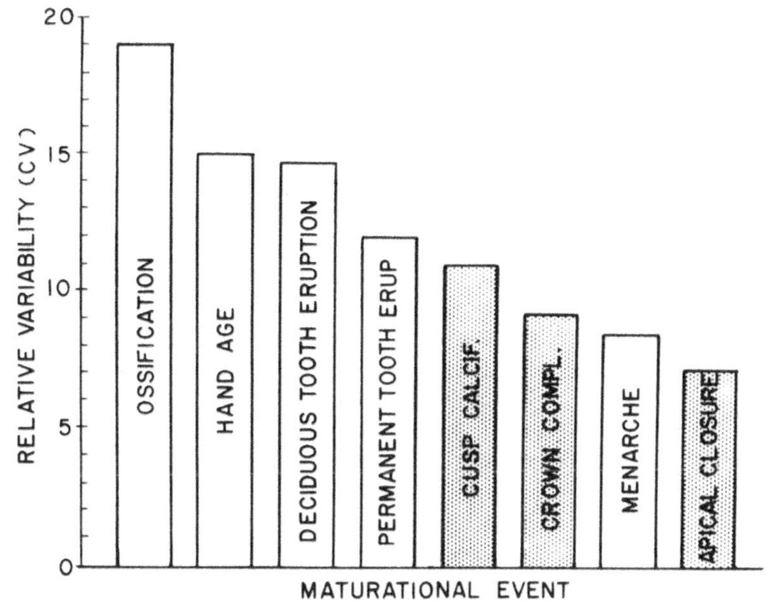

Fig. 13. Relative variability, (CV), of different maturational events. (From Lewis and Garn i.e., the coefficient of variability 1960.)

874 children. She found a higher correlation between dental emergence and root formation than between dental emergence and chronological or skeletal age. Thus, she suggested that emergence and root formation data be used for clinical and treatment purposes in orthodontics.

Steel (1965) investigated the relationships between dental and physiological maturation in a group of 12-year-old children. Each child was evaluated within 1 week of his twelfth birthday. No direct interdependence was detected between dental and skeletal maturation, and furthermore, the maturity stages of all three teeth investigated (mandibular second premolar and the second and third molars) varied independently. Steel concluded that this problem might be overcome through the creation of a system using the combined results of the maturity of all teeth.

Gyulavari (1966) confirmed Steel's findings concerning the variability of individual tooth development and attributed the contradictory results of different authors to this fact. Following his study of low-birth-weight Hungarian children, Gyulavari corroborated the view that dental and skeletal systems are independent of one other and stressed the necessity of establishing dental standards for each population.

Lacey (1973) investigated 40 children whose heights were below the third percentile. Skeletal age was assessed by Tanner's method (Tanner *et al.*, 1972) and dental development by the appearance of the root development of the second molar. Certain associations between dental and skeletal development were found, but Lacey concluded that, in normal children, the influences controlling dental development are different from those that determine bone age. More recently, Anderson *et al.* (1975) found that dental development, in both sexes, is more strongly related to morphological development than to skeletal age. These investigators also pointed out the sex specificity and individuality in the development of the teeth.

Using a longitudinal sample of 32 Canadian children from the Burlington study, Liebgott (1978) investigated the relationships of dental age to skeletal age as well as to the time of the peak of pubertal growth in mandibular length. His results also show little relationship between maturation of the dentition and general growth.

The developmental stages of the permanent dentition have also been used as a measure of prediction of the emergence of individual teeth (Demirjian and Levesque, 1980). This can be used for the longitudinal follow-up of patients in the orthodontic and pedodontic clinics, as well as for treatment planning purposes.

One of the most recent studies (Engstrom *et al.*, 1983) investigated the relationship of the lower third molar development and skeletal maturation in children between 10 and 18 years of age. Even with all ages pooled together, these workers came to the conclusion only that "the lower third molar development on the whole seems to be correlated with skeletal maturation." But the correlation was probably due to this agepooling; we should stress once more the importance of establishing age-specific correlations to find out the real value of these relationships.

At the Growth Centre of the University of Montreal, some 200 boys and girls have been followed longitudinally since 1967. Their dental development and emergence have been studied in detail, along with their anthropometric measurements, skeletal maturation, and nutritional habits. We analyzed the correlations among these different parameters within each age group; low correlation coefficients have been found, especially between dental age and skeletal age and/or stature (0.1–0.3) between the ages of 7 and 13 years (Table VIII). The correlations between stature and skeletal age are always higher. A decrease is noted in girls at age 13, attributable to the beginning of puberty.

9. Conclusion

During the first half of this century, the clinical emergence of teeth, referred to as eruption, was used as the sole criterion of dental development. Population standards have been established on the basis of very few and unsatisfactory samples obtained from aborted fetuses or living subjects. Deciduous and permanent dentitions have been studied separately for this purpose, as one might expect. However, their study is limited to the time of appearance of each tooth in the mouth; e.g., for the deciduous dentition, this includes the period between the 6th and the 30th months, while for the permanent dentition, it is between 6th and 12th years (omitting the third molar). On the other hand, using dental calcification to assess dental development extends the age range from birth to 15 years without interruption. Although an early emergence pattern has been established for girls for the permanent dentition, it is still unclear as far as the deciduous dentition is concerned. In both the deciduous and permanent dentitions, hereditary factors seem to be more influential than environmental factors such as nutrition or socioeconomic status.

In the middle of this century, biologists throughout the world began to use dental calcification, rather than clinical emergence, as a maturity indicator of dental development. This latter approach has many advantages over the previous one; the most obvious is that the dental system can be assessed as a whole, like the skeletal system, instead of assessing the emergence of individual teeth, as is done for clinical purposes. Different methods are described for the evaluation of dental calcification as a maturity indicator.

As with the studies using dental emergence, the evidence indicates the dental and skeletal maturation are substantially independent. As the two maturational processes have different embryological origins, with possible differences in genetic control, this is not too surprising. The same environmental factors, such as nutrition, affect both processes, but to differing degrees. With the possible exception of gross nutritional deprivation, not generally found

Table VIII. Correlation Coefficients for Dental Age with Skeletal Age and Stature[a–d]

	Girls				Boys			
Age	No.	DA/SA	DA/STAT	SA/STAT	No.	DA/SA	DA/STAT	SA/STAT
7	87	0.18	0.24	0.51	97	0.29	0.27	0.39
8	118	0.11	0.18	0.66	127	0.25	0.29	0.42
9	121	0.20	0.19	0.59	128	0.28	0.32	0.55
10	141	0.26	0.35	0.57	160	0.24	0.21	0.51
11	198	0.17	0.20	0.59	213	0.15	0.20	0.56
12	165	0.18	0.17	0.57	187	0.28	0.23	0.58
13	65	0.09	0.25	0.32	84	0.14	0.19	0.66

[a]Dental age (DA) calculated by unrevised seven-tooth weighting system and converted to ages, according to Montreal standards.
[b]Skeletal age (SA) according to the system devised by Tanner *et al.* (1972).
[c]As the data on stature (STAT) come from an unfinished longitudinal study, it has not been possible to express stature in maturity form, i.e., as a proportion of final adult height.
[d]Unpublished data from Montreal Growth Centre.

in Western countries, dental maturation has been found to be relatively impervious to environmental factors.

To the human biologist, the study of the development of the dentition adds another dimension to the growing body of knowledge of human growth and development. Dental maturity should be considered, along with skeletal development and stature, as a major component of the developing physiological unit. To the clinical orthodontist, the developmental life of individual teeth may be of prime importance, but he or she also needs a method of evaluating the overall developmental status of the mouth being treated—this is provided by a dental maturity measure. If this tool is to have scientific use, meticulous longitudinal studies are required to elucidate the physiological relationships involved and to produce the necessary measurement standards.

10. References

Adler, P., 1958, Studies on the eruption of the permanent teeth. IV. The effect upon the eruption of the permanent teeth of caries in the deciduous dentition, and of urbanisation, *Acta Genet. Stat. Med.* **8**:78.

Ainsworth, M. J., 1925, The incidence of dental disease in children, *Med. Res. Counc. (GB) Spec. Rep. Ser.* 97.

Anderson, D. L., Thompson, G. W., and Popovich, F., 1975, Interrelationships of dental maturity, skeletal maturity, height and weight from age 4 to 14 years, *Growth* **39**:453.

Bailey, K. V., 1964, Dental development in New Guinean infants, *J. Pediatr.* **64**:97.

Bambach, M., Saracci, R., and Young, H. B., 1973, Emergence of deciduous teeth in Tunisian children in relation to sex and social class, *Hum. Biol.* **45**:435.

Barrett, M. J., and Brown, T., 1976, Eruption of deciduous teeth in Australian aborigines, *Aust. Dentr. J.* **11**:43.

Bean, R. B., 1914, Eruption of teeth as physiological standard for testing development, *Pedagog. Sem.* **21**:596.

Beik, A. K., 1913, Physiological age and school entrance, *Pedagog, Sem.* **20**:283.

Beris, R. B., Hayles, A. B., Isaacson, R. J., and Sather, A. M., 1977, Facial growth response to human growth hormone in hypopituitary dwarfs, *Angle Orthod.* **47**:193.

Billewicz, W. Z., 1973, A note on estimation of calendar age on the basis of development of primary teeth (monogr. no. 28), *J. Trop. Pediatr. Environ.* **19**:243.

Billewicz, W. Z., and McGregor, I. A., 1975, Eruption of permanent teeth in West African (Gambian) children in relation to age, sex and physique, *Ann. Hum. Biol.* **2**:117.

Billewicz, W. Z., Thomson, A. M., Baber, F. M., and Field, C. E., 1973, The development of primary teeth in Chinese (Hong Kong) children, *Hum. Biol.* **45**:229.

Bjork, A., and Helm, S., 1967, Prediction of the age of maximum puberal growth in body height, *Angle Orthod.* **37**:134.

Bjork, A., Krebs, A., and Solow, B., 1964, A method for epidemiological registration of mal-occlusion, *Acta Odont. Scand.* **22**:27.

Boas, F., 1927, The eruption of deciduous teeth among Hebrew infants. *J. Dent. Res.* **7**:245.

Boas, F., 1933, Studies in growth—II, *Hum. Biol.* **5**:429.

Boutourline, E., and Tesi, G., 1972, Deciduous tooth eruption in a region of southern Tunisia, *Hum. Biol.* **44**:433.

Boutourline-Young, H., 1969, The arm circumference as a public health index of protein-calorie malnutrition of early childhood. XII. Arm measurements as indicators of body composition in Tunisian children, *J. Trop. Pediatr.* **15**:222.

Brauer, J. C., and Bahador, M. A., 1942, Variation in calcification and eruption of the deciduous and permanent teeth, *J. Am. Dent. Assoc.* **29**:1373.

Brook, A. H., 1973, The secular trend in permanent tooth eruption times (Monogr. no. 28), *J. Trop. Pediatr. Environ. Child Health* **19**:206.

Brook, A. H., and Barker, D. K., 1972, Eruption of teeth among the racial groups of Eastern New Guinea: A correlation of tooth eruption with calendar age, *Arch. Oral Biol.* **17**:751.

Brook, A. H., and Barker, D. K., 1973, The use of deciduous tooth eruption for the estimation of unknown chronological age (Monogr. no. 28). *J. Trop. Pediatr. Environ. Child Health* **19**:234.

Burdi, A. R., Garn, S. M., and Miller, R. L., 1970, Developmental advancement of the male dentition in the first trimester, *J. Dent. Res.* **49**:889.

Butler, D. J., 1962, The eruption of teeth and its association with early loss of the deciduous teeth, *Br. Dent. J.* **112**:443.

Cattell, P., 1928, Dentition as a measure of maturity, *Harvard Monographs in Education,* No. 9, 91 pp., Harvard University Press, Cambridge, Massachusetts.

Cifuentes, E., and Alvarado, J., 1973, Assessment of deciduous dentition in Guatemalan children (Monogr. no. 28), *J. Trop. Pediatr. Environ. Child Health* **19**:211.

Clements, E. M. B., Davies-Thomas, E., and Pickett, K. G., 1953*a,* Time of eruption of permanent teeth in Bristol children in 1947–48, *Br. Med. J.* **1**:1421.

Clements, E. M. B., Davies-Thomas, E., and Pickett, K. G., 1953*b,* Order of eruption of the permanent human dentition, *Br. Med. J.* **1**:1425.

Clements, E. M. B., Davies-Thomas, E., and Pickett, K. G., 1957*a,* Age at which the deciduous teeth are shed, *Br. Med. J.* **1**:1508,

Clements, E. M. B., Davies-Thomas, E., and Pickett, K. G., 1957*b,* Time of eruption of permanent teeth in British children at independent, rural and urban schools, *Br. Med. J.* **1**:1511.

Cohen, M. M., and Wagner, R., 1948, Dental development in pituitary dwarfism, *J. Dent. Res.* **27**:445.

Crossner, C. G., and Mansfeld, L., 1983, Determination of dental age in adopted non-European children, *Swed. Dent. J.* **7**:1.

Dahlberg, A. A., and Menagaz-Bock, R. M., 1958, Emergence of the permanent teeth in Pima Indian children, *J. Dent. Res.* **37**:1123.

Dahlberg. G., and Maunsbach, A. B., 1948, The eruption of the permanent teeth in the normal population of Sweden, *Acta Genet Stat. Med.* **1**:77.

Delgado, H., Habicht, J.-P., Yarbrough, C., Lechtig, A., Martorell, R., Malina, R. M., and Klein, R. E., 1975, Nutritional status and the timing of deciduous tooth eruption, *Am. J. Clin. Nutr.* **28**:216.

Demirjian, A., and Goldstein, H., 1976, New systems for dental maturity based on seven and four teeth, *Ann. Hum. Biol.* **3**:411.

Demirjian, A., and Lamarche C., 1979, Dental development of British and French-Canadian children, p. 227, in: *Proceeding of the First International Congress of Auxology, 12–16, April 1977*, Centro Auxologico Italiano, Rome.

Demirjian, A., and Levesque, G. Y., 1980, Sexual difference in dental development and prediction of emergence, *J. Dent. Res.* **59**:1110.

Demirjian, A., Goldstein, H., and Tanner, J. M., 1973, A new system of dental age assessment, *Hum. Biol.* **45**:211.

Demisch, A., and Wartmann, P., 1956, Calcification of the mandibular third molar and its relation to skeletal and chronological age in children, *Child Dev.* **27**:459.

Doering, C. R., and Allen, M. F., 1942, Data on eruption and caries of the deciduous teeth, *Child Dev.* **13**:113.

Edler, R. I. 1977, Dental and skeletal ages in hypopituitary patients, *J. Dent. Res.* **56**:1145.

El Lozy, M., Reed, R. B., Kerr, G. R., Boutourline, E., Tesi, G., Ghamry, M. T., Stare, F. J., Kallal, Z., Turki, M., and Hemaidan, N., 1975, Nutritional correlates of child development in southern Tunisia. IV. The relation of deciduous dental eruption to somatic development, *Growth* **39**:209.

Engstrom, C., Engstrom, H., and Sagne, S., 1983, Lower third molar development in relation to skeletal maturity and chronological age, *Angle Orthod.* **53**:97.

Eveleth, P. B., 1960, Eruption of permanent dentition and menarche of American children living in the tropics, *Hum. Biol.* **39**:60.

Falkner, F., 1957, Deciduous tooth eruption, *Arch, Dis. Child.* **32**:386.

Fanning, E. A., 1961, A longitudinal study of tooth formation and root resorption, *N. Z. Dent. J.* **57**:202.

Ferguson, A. D., Scott, R. B., and Bakwin, H., 1957, Growth and development of Negro infants, *J. Pediatr.* **50**:327.

Filipsson, R., 1975, A new method for assessment of dental maturity using the individual curve of number of erupted permanent teeth, *Ann. Hum. Biol.* **2**:13.

Filipsson, R., and Hall, K., 1975, Prediction of adult height of girls from height and dental maturity at ages 6–10 years, *Ann. Hum. Biol.* **2**:355.

Friedlaender, J. S., and Bailit, H. L., 1969, Eruption times of the deciduous and permanent teeth of natives of Bougainville Island, Territory of new Guinea: A study of racial variation, *Hum. Biol.* **41**:51.

Garn, S. M., and Rohmann, C. G., 1962, X-linked inheritance of developmental timing in man, *Nature (London)* **196**:695.

Garn, S. M., Lewis, A. B., Koski, P. K., and Polacheck, D. L., 1958, The sex difference in tooth calcification, *J. Dent. Res.* **37**:561.

Garn, S. M., Lewis, A. B., and Polacheck, D. L., 1959, Variability of tooth formation, *J. Dent. Res.* **38**:135.

Garn, S. M., Lewis, A. B., and Polacheck, D. L., 1960, Interrelations in dental development. 1. Interrelationships within the dentition, *J. Dent. Res.* **39**:1049.

Garn, S. M., Lewis, A. B., and Bonne, B., 1962, Third molar formation and its development course, *Angle Orthod.* **32**:270.

Garn, S. M., Lewis, A. B., and Blizzard, R. M., 1965*a*, Endocrine factors in dental development, *J. Dent. Res.* **44**:243.

Garn, S. M., Lewis, A. B., and Kerewsky, R. S., 1965*b*, Genetic, nutritional and maturational correlates of dental development, *J. Dent. Res.* **44**:228.

Garn, S. M., Sandusky, S. T., Nagy, J. M., and Trowbridge, F. L., 1973, Negro-Caucasoid differences in permanent tooth emergence at a constant income level, *Arch. Oral Biol.* **18**:609.

Gates, R. E., 1964, Eruption of permanent teeth of new South Wales school children, Part I. Ages of eruption, *Aust. Dent. J.* **9**:211.

Gleiser, I., and Hunt, E. E., Jr., 1955, The permanent mandibular first molar: Its calcification, eruption and decay, *Am. J. Phys. Anthropol.* **13**:253.

Godeny, E., 1951, Studies on the eruption of the permanent teeth. 1. The age at the eruption of the different teeth in the normal school population in Hungary, *Acta Genet. Stat. Med.* **2**:331.

Green, L. J., 1961, The interrelationships among height, weight, and chronological, dental and skeletal ages. *Angle Orthod.* **31**:189.

Greulich, W. W., and Pyle, S. I., 1950, *Radiographic Atlas of Skeletal Development of the Hand and Wrist*, 190 pp., Stanford University Press, Stanford, California.

Gron, A. M., 1962, Prediction of tooth emergence, *J. Dent. Res.* **41**:573.

Gustafson, G., and Koch, G., 1974, Age estimation up to 16 years of age based on dental development, *Odont. Rev.* **25**:297.

Gyulavari, O., 1966, Dental and skeletal development of children with low birth weight, *Acta Paiediatr. Acad. Sci. Hung.* **7**:301.

Habicht, J. P., Martorell, R., Yarbrough, C., Malina, R. M., and Klein, R. E., 1974, Height and weight standards for pre-school children. How relevant are ethnic differences in growth potential? *Lancet* **1**:611.

Halikis, S. E., 1961, Variability of eruption of permanent teeth and loss of deciduous teeth in western Australian children. 1. Times of eruption of permanent teeth, *Aust. Dent. J.* **6**:137.

Healy, M. J. R., and Goldstein, H., 1976, An approach to the scaling of categorized attributes, *Biometrika* **63**:219.

Hellman, M., 1943, the phase of development concerned with erupting the permanent teeth, *Am. J. Orthod.* **29**:507.

Helm, S., and Seidler, B., 1974, Timing of permanent tooth emergence in Danish children, *Community Dent. Oral Epidemiol.* **2**:122.

Hess, A. F., Lewis, J. M., and Roman, B., 1932, Radiographic study of calcification of teeth from birth to adolescence, *Dent. Cosmos* **74**:1053.

Hotz, R., 1959, The relation of dental calcification to chronological and skeletal age, *Eur. Orthod. Soc.* **1959**:140.

Houpt, M. I., Adu-Aryee, S., and Grainger, R. M., 1967, Eruption times of permanent teeth in the Brong Ahafo region of Ghana, *Am. J. Orthod.* **53**:95.

Hurme, V. O., 1948, Standards of variation in the eruption of the first six permanent teeth, *Child Dev.* **19**:211.

Hurme, V. O., 1949, Ranges of normalcy in the eruption of the permanent teeth, *J. Dent. Child.* **16**:11.

Infante, P. F., 1974, Sex difference in the chronology of deciduous tooth emergence in white and black children, *J. Dent. Res.* **53**:418.

Infante, P. F., and Owen, G. M., 1973, Relation of chronology of deciduous tooth emergence to height, weight, and head circumference in children, *Arch. Oral Biol.* **18**:1411.

Jelliffe, E. F. P., and Jelliffe, D. B., 1973, Deciduous dental eruption, nutrition and age assessment (Monogr. no. 28), *J. Trop. Pediatr. Environ. Child Health* **19**:194.

Kihlberg, J., and Koski, K., 1954, On the properties of the tooth eruption curve, *Fin. Tandlak Sallak. Forh.* **50**:6.

Klein, H., Palmer, C. E., and Kramer, M., 1937, Studies on dental caries. II. The use of the normal probability curve for expressing the age distribution of eruption of the permanent teeth, *Growth* **1**:385.

Koyoumdjisky-Kaye, E., Baras, M., and Grover, N. B., 1977, Stages in the emergence of the dentition; an improved classification to Israeli children, *Growth* **41**:285.

Koyoumdjisky-Kaye, E., Baras, M., and Grover, N. B., 1981, Emergence of the permanent dentition: Ethnic variability among Israeli children, *Z. Morphol. Anthropol.* **72**:267.

Kraus, B. S., 1959, Calcification of the human deciduous teeth, *J. Am. Dent. Assoc.* **59**:1128.

Krumholt, L., Roed-Petersen, B., and Pindborg, J. J., 1971, Eruption times of the permanent teeth in 622 Ugandan children, *Arch. Oral Biol.* **16**:1281.

Lacey, K. A., 1973, Relationship between bone age and dental development, *Lancet* **2**:736.

Lamons, F. F., and Gray, S. W., 1958, A study of the relationship between tooth eruption age, skeletal development age, and chronological age in sixty-one Atlanta children, *Am. J. Orthod.* **44**:687.

Lauzier, C., and Demirjian, A., 1981, L'émergence des dents primaires chez l'enfant canadien-français, *Union Med. Can.* **110**:12.

Lee, M. M. C., Chan, S. T. Low, W. D., and Chang, K. S. F., 1965, Eruption of the permanent dentition of southern Chinese children in Hong Kong, *Arch. Oral Biol.* **10**:849.

Legros, C. H., and Magitot, E., 1893, *Chronologie des follicles dentaires chez l'homme,* Congrès de Lyon, Lyon, France.

Leighton, B. C., 1968, Eruption of deciduous teeth, *Practitioner* **200**:836.

Leslie, G. H., 1951, *A Biometrical Study of the Eruption of the Permanent Dentition of New Zealand Children,* 72 pp., Government Printing Office, Wellington.

Levesque, G. Y., Demirjian, A., and Tanguay, R., 1981, Sexual dimorphism in the development, emergence, and agenesis of the mandibular third molar, *J. Dent. Res.* **60**:1735.

Lewis, A. B., and Garn, S. M., 1960, the relationship between tooth formation and other maturational factors, *Angle Orthod.* **30**:70.

Liebgott, B., 1978, Dental age: Its relation to skeletal age and the time of peak circumpuberal growth in length of the mandible, *J. Can. Dent. Assoc.* **5**:223.

Liliequist, B., and Lundberg, M., 1971, Skeletal and tooth development, *Acta Radiol.* **11**:97.

Logan, W. H. G., and Kronfeld, R., 1933, Development of the human jaws and surrounding structures from birth to the age of fifteen years, *J. Am. Dent. Assoc.* **20**:379.

Lysell, L., Magnusson, B., and Thilander, B., 1962, Time and order of eruption of the primary teeth, *Odont. Rev.* **13**:217.

Lysell, L., Magnusson, B., and Thilander, B., 1969, Relations between the times of eruption of primary and permanent teeth—A longitudinal study, *Acta Odont. Scand.* **27**:271.

Malcolm, L. A., and Bue, B., 1970, Eruption times of permanent teeth and the determination of age in New Guinean children, *Trop. Geogr. Med.* **22**:307.

Mather, K., and Jinks, J. L., 1963, correlation between relatives arising from sex-linked genes, *Nature (London)* **198**:314.

Mayhall, J. T., Belier, P., and Mayhall, M. F., 1977, Permanent tooth emergence timing of Northern Ontario Indians, *Ont. Dent.* **54**:8.

McGregor, I. A., Thomson, A. M., and Billewicz, W. Z., 1968, The development of primary teeth in children from a group of Gambian villages, and critical examination of its use for estimating age, *Br. J. Nutr.* **22**:307.

Meredith, H. V., 1946, Order and age of eruption for the deciduous dentition, *J. Dent. Res.* **25**:43.

Meredith, H. V., 1959, Relation between the eruption of selected mandibular permanent teeth and the circumpubertal acceleration in stature, *J. Dent. Child.* **26**:75.

Meredith, H. V., 1971, Growth in body size: A compendium of findings on contemporary children living in different parts of the world, *Adv. Child Dev. Behav.* **6**:153.

Miller, J., Hobson, P., and Gaskell, T. J., 1965, A serial study of the chronology of exfoliation of deciduous teeth and eruption of permanent teeth, *Arch. Oral Biol.* **10**:805.

Moorrees, C. F. A., Fanning, E. A., and Hunt, E. E., Jr., 1963*a*, Age variation of formation stages for ten permanent teeth, *J. Dent. Res.* **42**:1490.

Moorrees, C. F. A., Fanning, E. A., and Hunt, E. E., Jr., 1963*b*, Formation and resorption of three deciduous teeth in children, *Am. J. Phys. Anthropol.* **21**:205.

Mukherjee, D. K., 1973, Deciduous dental eruption in low income group Bengali Hindu children (Monogr. no. 28), *J. Trop. Pediat. Environ. Child Health* **19**:207.

Myllarniemi, S., Lenko, H. L., and Perheentupa, J., 1978, Dental maturity in hypopituitarism, and dental response to substitution treatment, *Scand. J. Dent. Res.* **86**:307.

Nanda, R. S., 1960, Eruption of human teeth, *Am. J. Orthod.* **46**:363.

Nanda, R. S., and Chawla, T. N., 1966, Growth and development of dentitions in Indian children. I. Development of permanent teeth, *Am. J. Orthod.* **52**:837.

Niswander, J. D., 1963, Effects of heredity and environment on development of dentition, *J. Dent. Res.* **42**:1288.

Nolla, C. M., 1960, The development of the permanent teeth, *J. Dent. Child.* **27**:254.

Nystrom, M., 1977, Clinical eruption of deciduous teeth in a series of Finnish children. *Proc. Finn. Dent. Soc.* **73**:155.

Perreault, J. G., Demirjian, A., and Jenicek, M., 1974, Emergence des dents permanentes chez les enfants canadiens-français, *J. Assoc. Dent. Can.* **40**:306.

Prahl-Andersen, B., and Van der linden, F. P. G. M., 1972, The estimation of dental age, *Eur. Orthod. Soc.* **1972**:535.

Prahl-Andersen, B., Kowalski, C. J, and Heydendael, 1979, *A Mixed-Longitudinal Interdisciplinary Study of Growth and Development,* Academic Press, New York.

Proy, E., Sempe, M., and Ajacques, J. C., 1981, Etude comparée des maturations dentaire et squelettique chez des enfants et adolescents français, *Rev. Orthop. Dentofac.* **15**:3.

Robinow, M., 1973, The eruption of the deciduous teeth (factors involved in timing) (monogr. no. 28), *J. Trop. Pediatr. Environ. Child Health* **19**:200.

Robinow, M., Richards, T. W., and Anderson, M., 1942, The eruption of deciduous teeth, *Growth* **6**:127.

Roche, A. F., Barkla, D. H., and Maritz, J. S., 1964, Deciduous eruption in Melbourne children, *Aust. Dent. J.* **9**:106.

Saito, H., 1936, Rontgenologische Untersuchungen uber die Ertwicklung des dritten Molaren, *Kakubuo-Gakka-Zasslis* **10**:156.

Sandler, H. C., 1944, The eruption of the deciduous teeth, *J. Pediatr.* **25(2)**:140.

Saunders, E., 1837, *The Teeth a Test of Age, Considered with Reference to the Factory Children,* H. Renshaw, London,

Schour, I., and Massler, M., 1940, Studies in tooth development: The growth pattern of human teeth, Parts I and II, *J. Am. Dent. Assoc.* **27**:1778; 1918.

Steel, G. H., 1965, The relation between dental maturation and physiological maturity, *Dent. Pract.* **16**:23.

Sutow, W. W., Terasaki, T., and Ohwada, K., 1954, Comparison of skeletal maturation with dental status in Japanese children, *Pediatrics* **14**:327.

Tanguay, R., Demirjian, A., and Thibault H. W., 1984, Sexual dimorphism in the emergence of the deciduous teeth. *J. Dent. Res.* **63**:65.

Tanner, J. M., 1962, *Growth at Adolescence,* 2nd ed., pp. 143–155, Blackwell, Oxford.

Tanner, J. M., Whitehouse, R. H., Healy, M. J. R., and Goldstein, H., 1972, *Standards for Skeletal Age,* Centre International de l'Enfance, Paris.

Tanner, J. M., Whitehouse, R. M., Marshall, W. A., Healy, M. J. R., and Goldstein, H., 1975, *Assessment of Skeletal Maturity and Prediction of Adult Height: TW2 Method,* Academic press, London.

Taranger, J., Lichtenstein, H. and Svennberg-Redegren, I., 1976, III. Dental development from birth to 16 years, *Acta Pediatr. Scand. (Suppl.)* **258**:83.

Ten-State Nutrition Survey 1968–1970, 1972, Part III, Centers for Disease Control, Atlanta, Georgia, U.S. Department of Health, Education, and Welfare, Publ. No. (HSM) 72-8131.

Trupkin, D. P., 1974, Eruption patterns of the first primary tooth in infants who were underweight at birth, *J. Dent. Child.* **41**:279.

Voors, A. W., 1957, The use of dental age in studies of nutrition in children, *Doc. Med. Geogr. Trop.* **9**:137.

Voors, A. W., 1973, Can dental development be used for assessing age in underdeveloped communities? (Monogr. no. 28), *J. Trop. Pediatr. Environ. Child Health* **19**:242.

Voors, A. W., and Metselaar, 1958, The reliability of dental age as a yardstick to assess the unknown calendar age, *Trop. Geogr. Med.* **10**:175.

Wagner, R., Cohen, M. M., and Hunt, E. E., Jr., 1963, Dental development in idiopathic sexual precocity, congenital adrenocortical hyperplasia and adrenogenic virilism, *J. Pediatr.* **63**:506.

Yun, D. J., 1957, Eruption of primary teeth in Korean rural children, *Am. J. Phys. Anthropol.* **15**:261.

II

Neurobiology

13

Neuroembryology and the Development of Perceptual Mechanisms

COLWYN B. TREVARTHEN

1. Introduction

A considerable part of developmental brain research is concerned with how cellular mechanisms of perception are built. This chapter attempts a review of the field, but does not claim to cover all complex issues brought to light. There is good reason to be selective, as there are difficult conceptual problems. First, we must define perception so that this psychological function—the uptake of information by sensory systems—can be related to the many developmental processes that contribute to it. Somehow we must compare selective and organizing processes in perception with pattern-making and pattern-using systems in the growth of the entire behavioral system: body and brain.

Perception is an effect of stimuli, but perceptual images arise in brain cell networks. No aspect of perception is independent of the structure of the cell system in which it arises. There can be no perception without accurate and complex patterning of nerve cell connections by growth. It would seem likely that integrative, selective, or constructive processes in perception are derived from the embryogenic processes that give form to neural systems. In particular, the necessary collaborative intermodal action of separate perceptual channels and the ease with which perception directs a variety of equivalent movements may be a consequence of the way in which neural systems originate within the embryo organism.

Understanding brain growth presents enormous problems. Neurogenesis is a vast sorting and selection process in which the most active and sensitive cellular elements of the growing body generate patterned interactions of unparalleled complexity. Nevertheless, at least in the early stages, the developing tissues of the central and peripheral nervous system are subordinated within the overall pattern formation and differentiation of the embryo. Later, the outgrowth of axons and selective formation of synaptic linkages between nerve cells, sometimes far removed from each other in the body, completely transform the processes of nerve cell differentiation. Spontaneous excitatory activity in neurons favors transmission of inductive effects with refined precision between distant assemblies of cells undergoing differentiation.

While still immature and undergoing large-scale regulatory changes, networks of brain cells may be transformed by excitation, some of which is stimulated by the environment. Because they are in close contact with the outside world, growing perceptual mechanisms might seem to be particularly susceptible to influence from stimuli. On the other hand, dependable rules by which the primary sensory zones of the brain categorize and sort input for control of behavior may be expected to be adap-

COLWYN B. TREVARTHEN • Department of Psychology, University of Edinburgh, Edinburgh, Scotland EH8 9JZ.

tively predetermined in the genetic material and to show effects of strong selective pressures in evolution. Are the self-regulatory processes of nerve net formation capable of specifying rules for perception, or does experience of the regular features of the outside world have to put the necessary structure into perceptual systems? This question becomes particularly pertinent when we consider the sensory zones of the mamalian cortex. These recently evolved mechanisms, when mature, make fine discriminations among patterns of sensory stimulation that override and extend the more ancient perceptual mechanisms of the brain stem.

With the great advances of recent decades in neurobiology, especially in techniques for visualizing and testing the structure and biochemistry of developing nerve networks, there is intense interest in the relationship between brain histogenesis and various kinds of learning. Refined physiological, immunological, and radiographic tracer techniques add to behavioral methods of assaying effects of experimental surgery on growing brains. These analytical techniques have produced reliable direct evidence of cell changes in discrete response to biased patterns of stimuli, especially in immature animals or in those undergoing metamorphosis. By surgical interference, or by artificially restricting or shaping stimuli, it is possible to test the anatomical plasticity and self-regulation of growing perceptual mechanisms. It must be kept in mind, however, that the drastic methods of this research carry a danger of distorting the natural growth process beyond recognition. They may not reveal the brain's normal control of its own development when conditions are not so unbalanced. Sensory deprivation methods invariably interfere with the motor output of the brain, which normally has the power to select and reject information for perception.

Natural perception is never simply a matter of neural processing of input from arrays of receptors, although it is often described in this oversimplified way. It is a self-regulating sensorimotor activity. In normally developing brains, the education that perception mechanisms receive from the environment is controlled by the patterned activity of motor organs, some of which, like the muscles of the lens of the eye, are adapted exclusively to the purpose of regulating sensory input. The neuroembryology of perception must include an account of the growth and development of movements that gate perception. In turn, movements are a property of motor organs directly dependent on body form. Therefore, from the start of what may be called perceptogenesis, we have to look into individuation of the body as a whole, its polarity and symmetry, and the positioning of secondary receptor and motor structures in relationship to the body axis.

The higher forms of intelligence elaborate greatly on whatever representation of significant objects in their world are present at birth. New programs for effective behavior become integrated by the growing and learning brain to increase the conceptual basis for perception. This cognitive development changes perception, as memories are formed of important events, and new instances reinforce and affirm past perceptions. Cognitive concepts qualify the ground rules for immediate perceptual control of movement, without usurping them. It is necessary to look for evidence of growth in brain mechanisms that mediate these later, higher-level developments in perception that must be analyzed in large systems of nerve cells. Evidence from the effects of brain surgery or trauma in humans reveals a widespread anatomy of consciousness in the cerebral hemispheres with surprisingly long development and surprising regularity of final form. Even these determinants of perception bear the stamp of genetic instruction.

It is important to emphasize that brain epigenetics is a life-long process. In a human embryo, the organs of the body are not yet either fully differentiated or functionally mature, but they have attained a fairly close approximation to adult form and adaptive tissue structure by the end of the embryonic period. The brain, however, is still undifferentiated. All the higher cerebral structures of mammals differentiate into recognizable rudiments of the adult organs postembryonically, during the fetal period. At birth the human brain, although not much smaller than the adult brain, is much less developed than the rest of the body, as it is destined to go through exceedingly complex morphogenetic changes long after birth with little gain in bulk. That is, neuroembryology of perception is not restricted to growth of perceptual machinery in the embryo.

Indeed, the evidence concerning brain growth makes it virtually impossible to place an upper bound to the generation of structure in brain tissue. We cannot place its end early in the postnatal period. We have to consider that significant new development may continue throughout life. Nevertheless, in the whole panoply of the brain's functional systems, those for perception do attain maturity most quickly. The most important period for the attainment of basic function in perception is early infancy.

Embryo and fetus are prefunctional stages. The nerve net becomes active, but the excitation is intrinsic or spontaneous. It carries almost no information about the outside world. In spite of evidence that fetuses have some powers of psychological response to stimuli, perception of the world is rudimentary before birth. Even if the conditions for receiving informative stimuli were adequate inside the womb, the fetus has exceedingly restricted powers of action. Although they might categorize auditory, vestibular, kinesthetic, mechanoreceptive, tactile, and taste stimuli, fetuses can do very little. They perceive only to a very limited degree. In the neonate and the infant, we see the most significant advances in perceptual growth, as the subject develops specific adaptations to a rich environment for action. From recent research on perceptuomotor coordination in this critical period of life has come the most exciting psychological evidence on innate processes for perception.

Our main focus is on human perceptual growth, but it is necessary to bring in information from animal experiments as well. Experiments on growth of fetal and postnatal perception systems have been done with birds and mammals. Knowledge of the embryo phase comes partly from invertebrates but mainly from fish and amphibia. The embryological approach to human perception is thus also phylogenetic in true darwinian spirit. Nevertheless, caution has to be exercised; data from forms with very different kinds of life and powers of perception require careful translation for comparison with the human condition. Human brain development, like human consciousness, is unique. The long postponement of free existence permits formulation of new adaptations and new kinds of perception.

Information on the growth of the human embryo and fetus may be found in Arey (1965), Barth (1953), and Hamilton *et al.* (1962). An excellent atlas containing illustrations of developmental stages by Böving (1965) is reproduced as an appendix to Falkner (1966), which also presents an account of fetal brain development (Larroche, 1966).

2. Perceptual Systems and Action Modes

A working hypothesis for this review involves the dependency of the development of perceptual awareness and of effective voluntary movements on differentiation of new refinements and new levels of integration within a scheme laid down before birth. In this process, the environment exercises an increasing influence, selecting (validating or eliminating) specific elements of sensory reaction from among alternatives that exist in the neural structures created *a priori* by growth, until a stable system has been created that is adapted to detect the defining features of natural events and objects affecting behavior.

Perceptual control of movements is obtained not so much by the growing together of initially separate sensory impressions or sensorimotor arcs, but by reinforcing part of an initial excess of patterns of connection in a hierarchically structured and unified control system, one that never loses unity while it generates large numbers of temporary states of action. Perception and movement control each other in innately specified servo loops. In more complex forms of action, conscious percepts, while highly responsive to what they find in the external world, are determined in part by prewired coordinative strategies and in part by acquired templates (feature assemblies in memory). Mentally synthesized images, models (Craik, 1943), schemata (Bartlett, 1932; Neisser, 1976), or implicit hypotheses about reality (Gregory, 1970) may also set in motion very elaborate patterns of action with, at the outset, little perceptual determination (i.e., little sensory feedback control) in the mature subject (Bernstein, 1967; Greene, 1972; Whiting, 1984). Thus perception becomes an intermittent but ever-ready regulator of action patterns set up by neural systems that accurately predict the environmental conditions for their success.

The mechanisms for perception constitute a set of systems for finding information or for detecting invariants in the environmental arrays of stimuli, both information and invariants being defined with respect to possible modes of action of the brain and body (J. J. Gibson, 1966; Sperry, 1952; Trevarthen, 1978*a*). Sifting out of meaningful or useful information from complex ensembles of sensory input may be treated as a problem in physics or computation (Marr, 1982; Ullman, 1979), but artificial perceiving mechanisms are profitably modelled on natural mechanisms capable of movement in a space–time frame (Reichardt and Poggio, 1981). The following complementary components of active perception apply to all modalities (see also Trevarthen, 1968*a*,*b*, 1978*a*, 1984*c*).

2.1. Body Guidance

This component of active perception is activated in locomotion, and it supports standing or sitting as a limit case. It informs about the layout of surroundings and gives advance information on the distances, separations, gradients, densities, and resistances of surfaces (J. J. Gibson, 1966). Displacement and steering of the body require widespread coordinated muscle activity jointly and simultaneously guided by dynamic input from the distance receptors of sight and hearing, assisted by olfaction, gustation, and pickup of radiant heat. Direct contact information about the "feel" of the relatively unresistant media of air and water, as well as the unyielding surfaces that stimulate touch, pressure, and temperature awareness, must then be integrated into the advance picture obtained by the distance receptors.

The picture of reality is obtained primitively and most unambiguously when the subject is moving about, but in mature consciousness layout and form can be seen in a single glance, without information from the subject's displacement. The world may also be seen in pictures. Some space-dividing rules for perceiving by the static geometry and configuration of stimulus information may derive from embryogenic processes, while others must be learned. Their relation to the perception processes that guide movement is probably close, but it is not in fact known.

Because one can see, hear, smell, and feel the layout of surroundings richly when not moving, or moving very little, it is often concluded that stimuli that arise only when the body is active (in the vestibular organs, joints, muscle sensors, and visceral mechanoreceptors) are in dominant, if largely unconscious, control of locomotion and shifts of posture. It is relatively simple to verify that this is not so (J. J. Gibson, 1952). Intracorporeal information is integrated closely with proprioceptive (body-sensing) information from the teleoceptive (sensing at a distance) organs of sight, hearing, and so forth (Lee, 1978; Lee and Lishman, 1975; Dichgans and Brandt, 1974). The definition of a proprioceptive organ should depend on the relationship of its receptor functions to the body as a generator of movement, not on the type of stimulus energy to which the receptor organ is sensitive. Both vision and audition function proprioceptively in regulating body posture, stationarity, and movement. Their perceptions also develop with locomotion.

To perceive movement of one coherent self in relationship to surroundings, the mechanism of central motor control for the whole body must receive equivalent and coincident images of space layout from the various modalities. A unified anatomical system of connections within the brain in which body movements are unambiguously and coherently represented is therefore essential to the primary perception of a space in which one behaves. Because the propulsive forces made by the body are directly related to its form and the geometry of skeletomuscular systems, the common field for perceptual control of movements must in principle be body-shaped or somatotopic in design. Guidance of locomotion (whole-body displacement) with respect to a reference outside the body requires localizing of that reference in a body-centered field representing possible movements. This field of control can be called the behavioral field (Trevarthen, 1968*b*, 1972, 1974*b*,*c*, 1978*a*,*b*).

Stimulus effects produced by self-movement in stable surroundings must be distinguished from motion of external origin. This again may be solved by somatotopic representation or an equivalent ordered set of loci. The image of body-produced acts in such a representation may function like a background for detecting locus and direction of motion generated outside. Such somatotopic representation would be involved in an efference-copy system (von Holst, 1954) and in a system designed to detect stimulus invariants from feedback to guide body displacement (Gibson, 1954, 1966).

2.2. Locating an Object

Within the above perceptual field for the whole body is a system for perception of objects by parts of the body that move separately or in combination against the objects. The class "object" is defined by reference to acts of use, and individual objects are distinguished by their kinds of use. Food is defined by the act of eating; manipulanda (tools) are defined by their handling properties, and so forth. J. J. Gibson (1979) calls the information for use "affordance." Correspondence must be established between advance detection of objects (e.g., by sight or odor) and their direct, final, or consummatory use. Obviously, the actual movement of "use" cannot construct object perceptions which are formed predictively (Metzger, 1974), but close equivalence of perception and act is necessary.

Since the movements for using objects are part of, or a subset of, the movements of locomotion, perception of objects must be a subset of perception of a space–time frame for locomotion (Lee and Young, 1984). Objects are located in the body-centered behavior space. Since the different object-locating or -orienting acts are to cooperate in apprehension of common physical causes (e.g., hands and eyes must together apprehend a thing as one object, not two, unless they are acting on separable parts of the thing), then not only space, but time as well, must be unified within the perception of objects. The different mechanisms for sensing objects must act in one frame of space and time together and be transitively equivalent.

Since objects might lie anywhere in the common behavior field of the subject and still have the same properties (identity), whatever mechanism apprehends an object as separated from the background must be able to function independently of locations in the primary field. This independence is the converse of the orienting ability, which enables acts of capture to be set up from many starting points in the primary field (Trevarthen, 1978*a*; Gross and Mishkin, 1977. See the two cortical mechanisms of vision described by Mishkin and colleagues, 1983).

2.3. Identification and Analysis

The task of distinguishing thousands of kinds of objects, and the elements of which they are composed, requires detection of distinctive features. This is more taxing on the neural mechanism if objects are to be perceived at long distance, or telescopically; the stimuli that reach the subject from far off are attenuated. Also, many relevant features are perceived only when they are analytically teased out from the ensemble of the object's stimulation and from temporary details contributed by the background; e.g., shadows, reflections, and overlapping surfaces.

The critical motor process in identifying is a serial focalization that involves deployment of discrete, high-resolution perception mechanisms that have narrowly restricted fields of detection but many neural elements (Adrian, 1946; Whitteridge, 1973). Scanning movements that regulate sampling in the general field of perceived space determine chains of focalized experience—fixations of the fovea of the eyes, setting of the frequency–response range of the ear, sniffing with the nose, touching with fingertips, and tasting with the tongue. The movements of focalizing need to be precise and rapid. They must be integrated within patterns of body displacement or body transformation (body part on body part articulation). They must also be governed by perception of layout and configuration which can operate without movements, as in seeing a landscape by lightning flash. Seeing what things are requires quick assimilation of many pieces of information. This process would fail unless focalizing to get the right details were highly efficient and obedient to knowledge held in the perceiving system (Yarbus, 1967).

2.4. Modeling Properties of Object Existence and Persistence in Time

All percepts must have, in addition to spatiotemporal equivalence in different modalities, properties of use and identifying features, the further attributes of continuity, permanence, or specified change (Michotte, 1963; Johansson *et al.,* 1980). The system requires images in nerve activity, or excitatory threshold changes, that persist beyond stimulation or that connect separate events in stimulation. The most general form of change to be perceived is that caused by motion of objects relative to the perceiver or to each other and the solid surfaces of surroundings. The quality and dimensions of independent motion must be correctly detected in perception, if it is to serve for guidance and placing of adaptive action (J. J. Gibson, 1954). Objects are detected by their own surface-bounded volumes and forms and by qualities of color, texture, hardness, resonance, odor, and taste, which remain invariant in perception while the objects are displaced about the body and receptors, so that object perception requires the operation of the principle of constancy. For vision, the rules of constancy are as follows: (1) The object holds the same position if only the subject is displacing; (2) An object remains the same size when perceived at different distances; (3) An object keeps the same form when rotated relative to the perceiver or when apprehended by perception piecemeal in different parts; and (4) An object remains the same color in spite of changes in spectral composition of the illumination it receives. Objects may also deform, dissolve, or swell. These changes and underlying processes must be perceived through representation of their defining attributes in the brain. Finally, objects must be perceived to change in perceptibility but not existence when they go into or behind masses that screen them. This requires perception of the forms of surfaces and masses in the space around the objects.

2.5. Perception of Movement

A wide variety of commanding factors of biological relationship between the perceiver and other animate beings makes discrimination of movements a basic factor in perception of most animals. It requires detection of the special invariants that define animation (e.g., self-regeneration, tempo, rhythm, grouped bursts, or purposeful aim). Attributes of the quality of animation are also important (e.g., strength/weakness, aggressiveness/friendliness, and stealth/openness). Control over the perception of other beings involves communication; i.e., intersubjectivity (Trevarthen, 1977). Acts of the subject may regulate the acts of others, and perception of others includes perception of their various expressions of communication in several modalities at once. Coordination between predator and prey or between animals coordinating purposeful behavior or motivations may require extremely fast reactions to perceived movements.

Mechanisms for subject perception and for intersubjectivity thus have unique requirements that may develop (phylogenetically and ontogenetically) out of self-regulatory perception requirements. Both perception of self and perception of others seem to refer back to an image of the body as a system that is coherent and polarized in design and activity. Research with infants indicates that inborn evaluations in perception are effective in detecting persons (e.g., from voice qualities, color, face configuration) and that a comparable set of rules is capable at birth of generating expressive movements that others will perceive as communications (Trevarthen, 1983*a*, 1984*d*).

There is reason to believe that different classes of perceptuomotor strategy require mediation by different innate (prefunctionally wired in) mechanisms in the brain, and this is the main justification for presenting the above summary here. They may have distinctive developmental histories. Some of the differing requirements for active information uptake for perception are reflected in anatomical specializations of the receptors; even at the point at which stimuli are first received, they are segregated to serve in different levels of the process. Thus, in vision, the eye shows one set of mechanisms for picking up information for perception of the layout of surroundings (the rod system of the periphery, serving ambient vision) and another for focal detection of objects and their identifying features (the cone system of the fovea, serving focal vision). Motor structures inside and surrounding the eye regulate the relative stimulation between these two receptor systems by varying the coherence, spatial density, and intensity of light patterns in the retinal images, and by varying eye displacement relative to the sources of these light patterns (Figure 1). Rotary movements of the eyes alternating with fixations favor focal vision; continuous translatory or linear displacements of the center of the eye, as in locomotion, favor ambient vision by causing high velocity stimulation in the far-extrafoveal parts of the retinal image. A similar analysis could be made of the perceptual appreciation of objects by the hands and arms; sense organs for mass and bulk are different from those for feeling surface features, texture, temperature, and hardness, and they require different motor patterns to control them. Indeed, this multiplicity of perceiving extends to all the classical modalities, each of which has its own anatomical organ for focusing on information after it is located in the space of behavior.

Not only may different perception systems in one modality grow at different rates, but they are likely to interact with those of other modalities during the course of development in quite different ways. It is reasonable to suppose that whereas the primary space-defining general field of body action is set up at an early phase of development for all modalities jointly, focal operations of different modalities may develop at different times early in postnatal life depending on the maturation of motor activities to which they are most closely allied. We may expect a series of differentiations of perceptual systems in infants in relation to their emerging patterns of action. For example, the growth of object prehension and manipulation in the first year may occasion particular development in touch and vision and their association, and developments from babbling to speech would be expected to be accompanied by changes in auditory and visual perception of mouth movement. Writing must involve the emergence of new visuomanual and audiomanual perception devices in the brain. There is no reason to assume that such changes in perceptual differentiation and reintegration (association) are just a passive consequence of learning, independent of brain development.

3. Early Embryo

3.1. Beginnings of the Nervous System: Origins of Brain–Body Mapping

At 2½ weeks postconception, when the human embryo is a two-layered disk less than 1 mm in di-

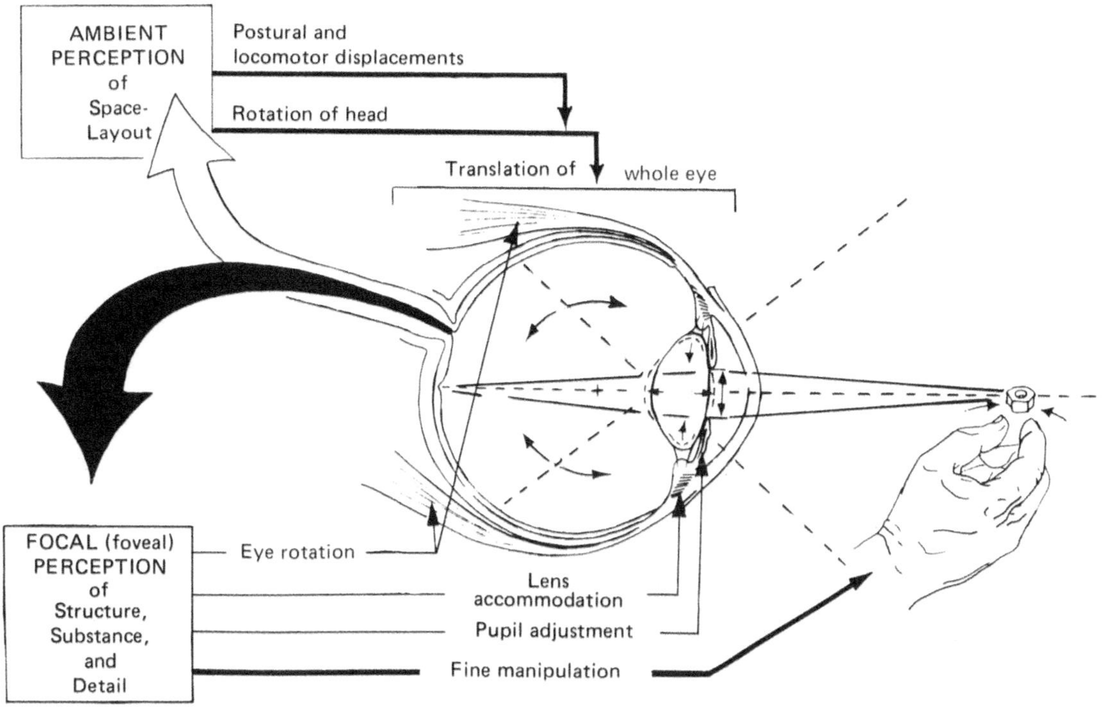

Fig. 1. Peripheral components of the sensorimotor control mechanisms of human vision. Heavy arrows represent movements that regulate input to several modalities. Thin arrows are movements that tune visual input exclusively. Translocations of the eye deform the whole retinal image (velocity pattern of stimulation). Focal adjustments shift the retinal center (fovea) relative to a stabilized retinal image, change the contrast and spatial frequency of the image, regulate binocular convergence and registration and, in the case of manipulation, displace, deform or rearrange visual features or objects. Not shown are movements of brows and eyelids, interceptive or throwing movements that change motion of objects, or gestural and expressive movements of face, hands, or whole body that are aimed to affect vision in another subject.

ameter, the nervous system is identifiable only by its place in the bilaterally symmetrical body, the cells having the same appearance as ectoderm (Figure 2A). Surgical rearrangement studies with amphibian embryos show that neuroectoderm is induced to differentiate from ectoderm by some factor from the notochord mesoderm beneath it (Jacobson, 1970; Needham, 1942; Saxén and Toivonen, 1962; Waddington, 1966). When pieces are removed from early embryos, the whole cell array can regulate or reorganize a normal body form by remodeling the remainder; i.e., by morphallaxis (T. H. Morgan, 1901).

The form of the embryo body is created by ordered cell divisions and cell migrations (Gustafson and Wolpert, 1967). In tissue culture, disordered embryo cell aggregates move to reorganize the body pattern exhibiting preferential tissue affinity for contact with their own kind to recreate the original layers or masses (Holtfreter, 1939). This seems to depend on a chemical communication system aided by close junctions at points between the walls of embryonic cells (Loewenstein, 1968). The nature of the communication is unknown (Barondes, 1970; Saxén, 1972; McMahon and West, 1976). Biochemical experiments with social amebae, the cells of which undergo patterned movements to congregate in bodies with definite form and polarity, suggest that transmission between cells in contact controls hormonelike cyclic nucleotide molecules that regulate basic intracellular biochemistry (McMahon, 1974). In turn the molecular or biochemical events become contained within patterns of cell mechanics and controlled intracellular communication (Wolpert, 1971). The position of cells in the array determines how they will differentiate into body parts.

Neural-plate transplant experiments with am-

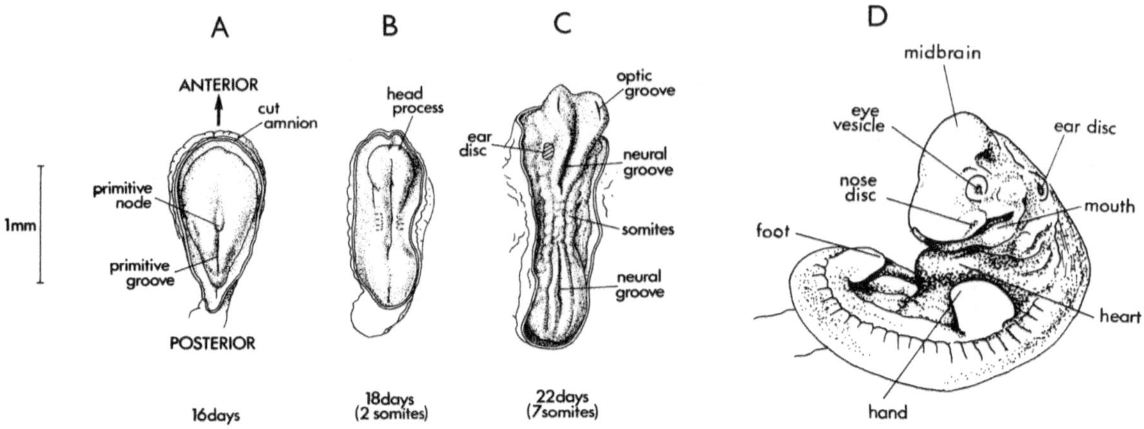

Fig. 2. (A–C) Human embryos in early stages of the formation of the nervous system. (D) By the end of the first month (late somite embryo) special receptor structures are clearly differentiated. (Based on scanning micrograph from Lennart Nilsson.)

phibian embryos shows that the anteroposterior polarity of the plate is determined before the mediolateral axis, at a stage equivalent to the 2½-week-old human embryo (Roach, 1945) (Figure 2A). In the early embryo of amphibia, the organizer of cell differentiation and source of polarity for the whole body is a group of cells at the dorsal lip of the blastopore (W. H. Lewis, 1907; Spemann and Mangold, 1924). These cells give rise to head mesoderm, including the anterior tip of the notochord. In the embryo brain, a small cluster of cells constituting the presumptive diencephalon may serve the same function in marking the polarity of adjacent brain fields (Chung and Cooke, 1975).

The embryo rapidly elongates when the nerve plate is formed. Midline neuroectoderm cells curl by contraction of their outer (dorsal) walls to fold in a gutterlike neural groove (Figure 2B,C). At the same time, the neural crest cells at the edges of the plate multiply and rise up to form walls that turn inwards and eventually fuse. In this manner, a tubular CNS is pinched off along the mid-dorsal line. The neural plate cells already form a mosaic of regions, each having a tendency to form a particular part of the brain (Jacobson, 1959).

Before 3 weeks, the embryo develops clearly demarcated head and trunk (Figure 2B). The head will consist principally of brain and ectoderm, with rather diminutive pockets of muscle-forming mesoderm, but in the trunk the neural tube will be flanked by large muscle plates. A gradient develops with increasing proportion of mesoderm to ectoderm toward the caudal end of the body. Experiments involving unsticking of embryo cells from each other, then mechanically mixing them, indicate that early neuroectoderm cells, capable of growing into either brain or spinal cord, respond to their surroundings. If artificially combined with a mixture rich in mesoderm, the neuroectoderm cells differentiate into spinal structures. If in a mixture low in mesoderm, the same neuroectoderm generates brain structures (Toivonen and Saxén, 1968).

While the neural tube is closing, cells of lateral plate mesoderm begin to aggregate into pairs of somites or muscle blocks (Figure 2B,C). The original polarized and bilaterally symmetric field of the body is cut by segmentation into a set of equivalent subfields (Figure 3). We do not know what form of communication controls the primary field of the body, but apparently segmentation permits it to be reiterated many times within an initially single field. Subsequent developments indicate that the segments of the nervous system form mutually compatible moduli, each impressed with a somatotopic layout, that can be grouped by formation of axonal links into equivalent subsystems, each capable of acting as an agent for the body or as one of a limited set of agents.

Cells from the dorsal ectoderm bordering the neural plate make the greater part of the input mechanism; from the viscera, the body surface, and the special visual, auditory, and vestibular sense organs to the brain. The olfactory sensory neurons grow from ectoderm over the anterior end of the brain. The sensory nerves from neural crest conform to the segmental pattern of the somites, making a chain of dorsal ganglia.

In the late-somite embryo, the main receptors

and effectors are already making their appearance (Figure 4A). The head, with nose, eyes, and ears (including vestibular organs), consists of a constellation of anterior segments. These segments have small muscle primordia, most of which become specialized to control uptake of sensory information. The mouth and tongue (formed of the appendages of anterior segments of prevertebrate ancestors) will grow into a complex organ for taking in food while tasting and masticating it *en route* and for grasping, exploring, or breaking up objects. In higher vertebrates the visceral arch musculature is elaborated into the muscles of the jaws and face that, in man, form a unique apparatus for expressive communication. The muscles of the neck, together with those that turn the eyes and ears with respect to the head, are important in controlling the aim of sensory experience. The forelimbs and hindlimbs appear in the embryo as paired lobes with distal thickenings. They develop their own segmentation and in later development coopt neural sensory and motor branches from several adjacent trunk segments. In humans, the hands function as organs of gestural and symbolic expression. Elaborate brain systems required to coordinate the special sensorimotor systems develops in postembryonic stages.

3.2. Morphogenesis of Sense Organs

Sense organs form from neural crest and ectoderm next to it at the border of the neuroectoderm plate. In the formation of each special receptor, for sight, olfaction, taste, or hearing, an infolding of ectoderm generates a pouch that lies against the brain. Sense cells develop in the pouch, in a cluster of neural crest cells, or in a lobe of the brain that meets it. The pituitary forms in comparable manner as an ectodermal pocket that grows upward to meet an outgrowth from the floor of the brain near the tip of the notochord. In primitive vertebrates, the portion that corresponds with the anterior lobe

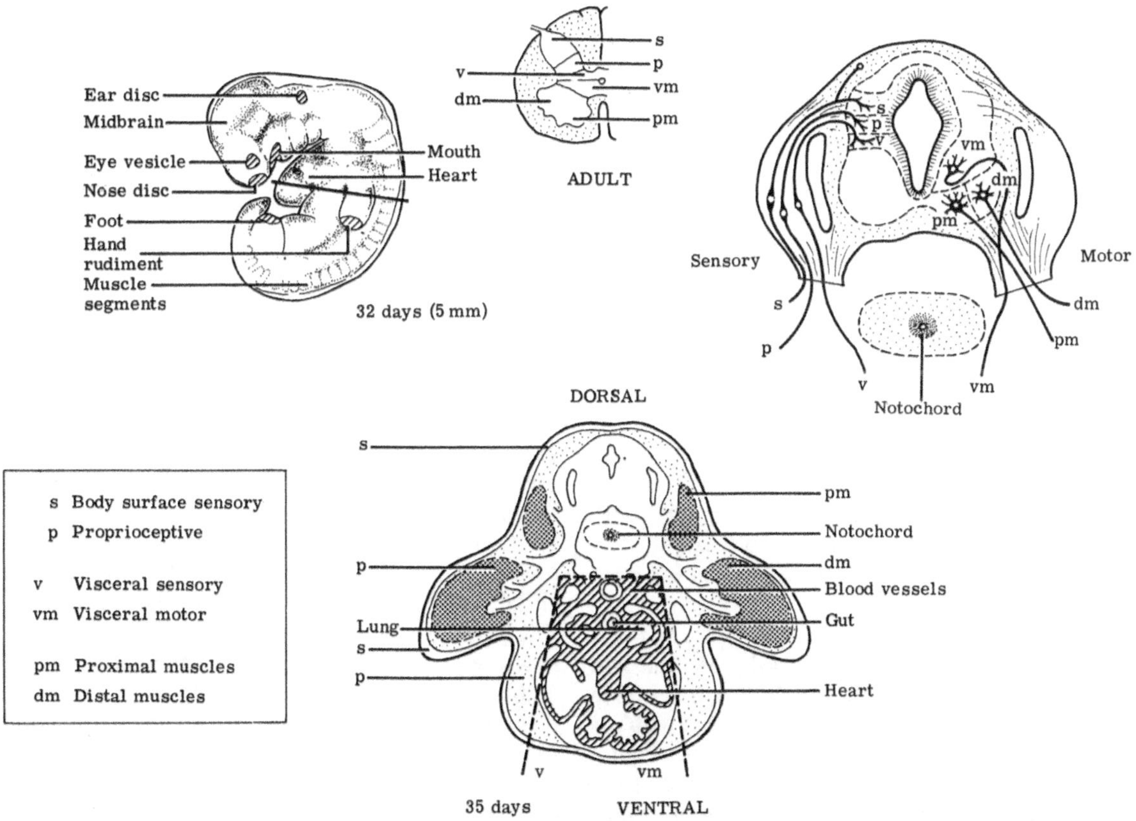

Fig. 3. Somatotopic arrangement of neurons, axon pathways, and connections in the trunk of a human embryo. Body territories are represented in a map in the spinal cord. Visceral motor neurons migrate from their place of origin.

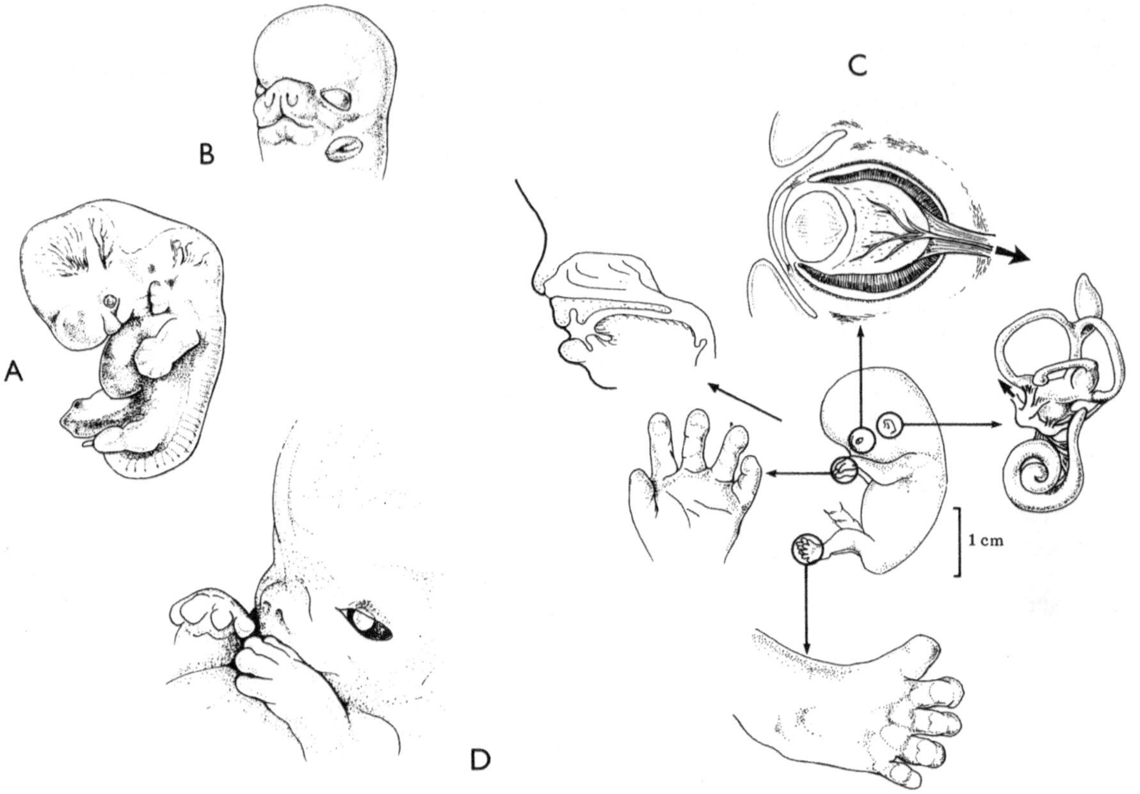

Fig. 4. Formation of human special receptors in the last 20 days of the embryo period. (A) At 40 days, hands, eyes, outer ear, and foot begin to form. (B) At 50 days, the face shows lips and nostrils. (C) At 60 days, eye, inner ear, hand, and foot are well formed, and neural tracts begin to form central connections. (D) At about 7 weeks, before neurogenic movements, head, eyes, and hands appear to be cooriented to a point a short distance in front of the face. (Based on photograph from Lennart Nilsson.)

of the hypophysis cerebri (pituitary gland) is a chemical sense organ (Sarnat and Netsky, 1974).

Receptors of the skin and of the muscles, joints, and internal organs are connected to the brain by axons that grow from spinal ganglia (Figure 3). Migration of neural crest cells to form these ganglia between the ectodermal structures and CNS, both of which they innervate, is comparable with the formation of the optic vesicle as an outpushing from the end of the brain.

At 30 days, eye cup and lens, octocyst, or inner ear and an adjacent mass of neural crest that will form the vestibulocochlear ganglion, nasal pouch and olfactory lobe of the brain, and mouth with rudimentary tongue are all clearly identifiable (Figure 4A). During the next 30 days, each of these structures attains a visible adaptive design with a rudimentary receptor epithelium placed in characteristic relationship to the accessory structures (lens and eye chamber, semicircular canal, sacculus and cochlea of the inner ear, olfactory pouch, muscular tongue, and lips).

The distance receptors, eye and inner ear, are particularly interesting because in their geometric form they already express how they will obtain information about the shape of the outside world and the nature of external events (Figure 4C). The geometry of eye and lens determines the optic image and the way in which movements will transform it. The geometry of the semicircular canals determines how the cupula receptors will be excited by acceleration of the body in given directions. The sacculus is a detector oriented to regulate the correct position of the head in the gravitational field. The form of receptor mechanisms of the cochlea, which, like the central foveal structures of the retina, mature late in the fetal period, determines how the hair cells will respond differentially to sounds of different frequency aided by the pinna of the ear (Klosovskii, 1963). The shape of the turbinal bones of the nose is critical for regulation of air streams caused by breathing and thus for space perception by olfaction (Von Békésy, 1967).

Parallel with the formation of these head organs, the perception organs of limbs undergo closely similar changes. By the end of the first 30 days, the primordia of hands and feet are clear, with thickened ectoderm forming placodes in which highly specialized receptor structures will grow (Figure 4A). The digits achieve essentially their adult appearance in the next 20 days. Thus, by the end of the second month, all the specialized organs of perception have reached a close approximation to their adult forms (Figure 4C). This remarkable preparation for complex function under cerebral control takes place before the nerve fibers connect receptor cells to the CNS or motor neurons of the CNS to the voluntary muscles.

3.3. Origin of Sensory Maps

Growth of receptor nerves into the brain involves an orderly series of inductive contacts between cell groups that have previously been systematically transformed by infolding or migration. When and how specification of neuroblasts for afferent nerve circuits is achieved has been the topic of many experimental studies (W. H. Lewis, 1906; Harrison, 1935, 1945; Spemann, 1938; Stone, 1960; Gaze, 1970; Jacobson, 1970; Hunt, 1975*a,b*; Gaze *et al,* 1979*a*; Hamburger, 1981; Hunt and Cowan, 1986). The first question concerns the origin of cell differences that are expressed in selective growth of the primary axonal connections in the late embryo. To tackle this, it is necessary to make operations on receptor primordia from the moment they can be identified as well as at later stages.

Most of this research has been performed on the eyes of amphibia (Figure 5). Basically, the experiment involves cutting out the eye rudiment at different stages, reimplanting it in different locations or different orientations in the embryo body, then tracing the formation of connections. Earlier eye-transplant experiments (Stone, 1944; Sperry, 1945; Székely, 1957) employed orienting behavior of the adult to measure the eye–brain connections; more recent studies use direct electrophysiological mapping of optic nerve presynaptic terminals on the optic tectum of the midbrain after metamorphosis (Jacobson, 1968*a*; Gaze, 1970) and histochemical or radiographic tracer techniques for locating nerve fibers, cell bodies, and terminals (e.g., Crossland *et al.,* 1974; Fawcett and Gaze, 1982).

Before the first mature neuroblasts appear at the center of the eye opposite the lens and cease to divide, each retina gains a program for specification of polarity as a unit, independent of surroundings, and, as long as pieces of retina are not removed from the growing eye, this program remains to determine the directions of behavioral movements under visual guidance. Recent transplant experiments indicate that the anlage of the eye receives orientational specification early, when it consists of a very few hundred cells in an indented protrusion from the embryo brain (Hunt, 1975*a,b*; Hunt and Berman, 1975). The small group of cells destined to become the retina may receive orientational specification—not from surrounding epidermis of the body surface as previously thought, but from the pigmented epithelium or outer wall of the two layered eye cup in which retinal cells form the inner wall (Gaze *et al.,* 1979*a*). As in all primordial receptor epithelia, the first group of cells is in tight communication with permeable intercellular junctions that later become broken (Dixon and Cronly-Dillon, 1972).

Surgical studies of determination in embryo amphibian eyes are complicated by the capacity of even small parts of the eye stalk or epithelium to replace or reorganize retinal cells; transplanted eye rudiments also sometimes undo an intended rotation by derotation. Such spontaneous responses may confuse interpretation of the determination of positional specificites in the retina.

Development of all vertebrate embryos follows essentially the same pattern, although the rate of development varies greatly. De Long and Coulombre (1965) and Crossland *et al.* (1974) have found that removal of retinal segments from eye cups of chick embryos results in the predicted gaps in the histologically verified projection to the optic tectum later in development. The latter study gives a more acurate picture of the process. If the operation was performed on the chick before 45 hr of incubation (stage 11), no defect occurred. The eyes repaired themselves and generated a complete map. After 48 hr, the destination of retinal quadrants on the rectum, where retinal axons will terminate, appears fully determined. This stage may be identified, with respect to eye development, with the human embryo at 4 weeks postconception (3.5 mm, Böving, 1965).

It is precisely at this stage that, in all vertebrates including man, a marked change is taking place in

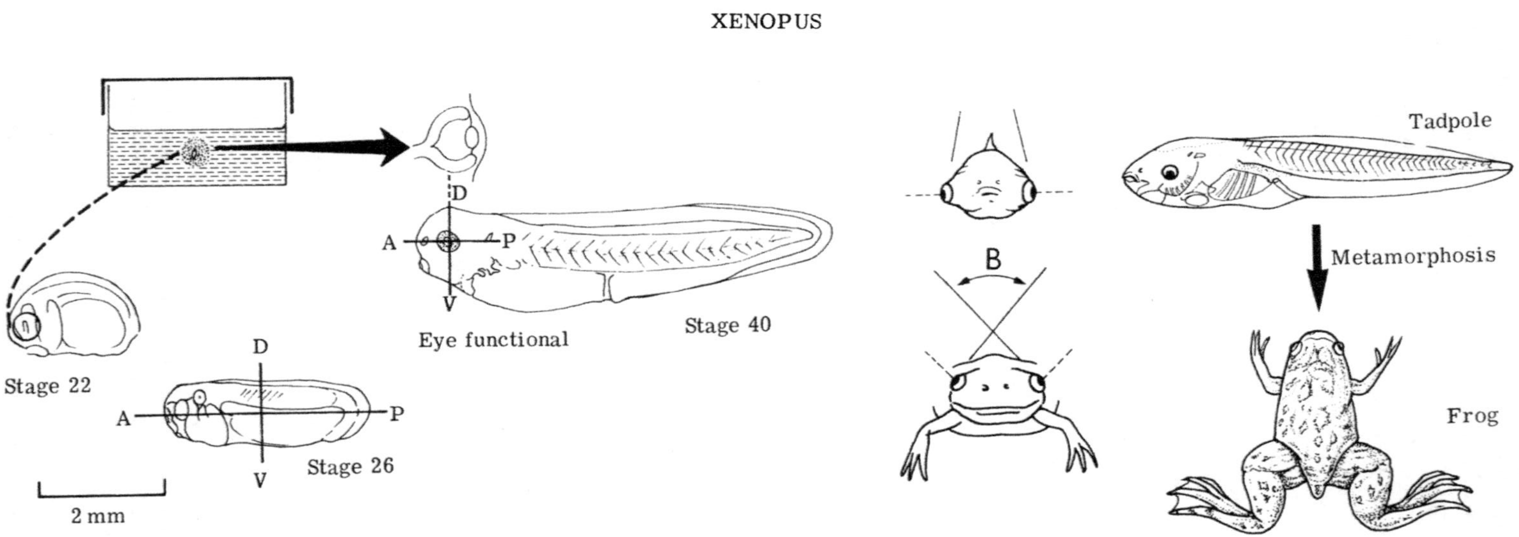

Fig. 5. Embryology of visual structures of *Xenopus*. Eye primordia incorporate polarity markers as early as embryo stage 22, at which stage an excised eye will undergo morphogenesis when isolated *in vitro*. The swimming tadpole at stage 40 has a functional eye. This visual projection is reworked in metamorphosis when the eyes migrate upwards and the dorsal binocular field (B) increases considerably in size.

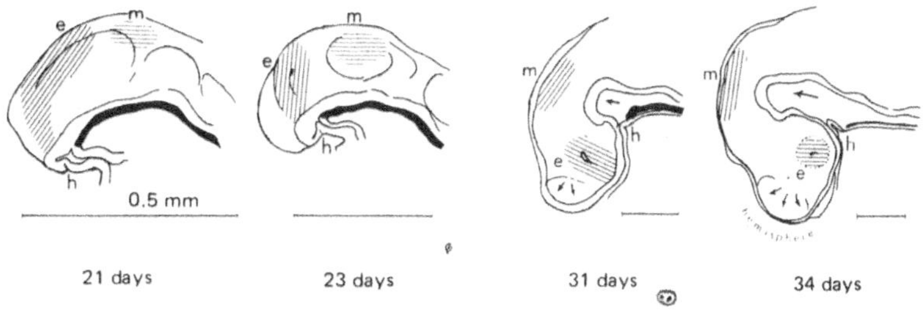

Fig. 6. Human brain in mid-sagittal section to show flexion. e, retinal primordium; m, midbrain optic tectum; h, Rathke's pouch (pars anterior of hypophysis); notochord, black. (Based on Bartelmez, 1922; Bartelmez and Blount, 1954; Bartelmez and Dekaban, 1962.)

the shape of the brain (Figure 6). Known as flexion, this bends the front end of the brain round the tip of the notochord. It results in an approximately 108° rotation of the eye rudiment relative to the body axis and to the notochord. At the same time, the eye changes from a rounded outpushing of the brain to a hollow spoon shape, the borders of which are closing to form the optic cup.

The existence of the inverted retinal image challenges any nativistic theory of the visual projection. If seeing things upright is based on brain circuitry formed under tight genetic control, it is necessary to introduce an inverting principle to explain how the retina may be specified for orientation according to a cytochemical field common to all body parts and yet anticipate the inversion of the retinal image by optical projection of light in the eye. Either an already specified eye map is anatomically rotated with respect to other somatotopic maps of the brain, or the polarities of its somatotopic code are switched chemically. In the light of recent experiments with *Xenopus,* it seems that the former explanation is the correct one.

Experiments with embryo eyes of amphibia and birds prove that correspondence between points in visual space around the body and loci in the space of movements of the body is determined in general layout by some differentiation of cells in retina, giving them a specified connective affinity to the tectum. Whatever additional anatomical refinements must be introduced to attain the full precision of visuomotor circuits by sorting of nerve cell connections prenatally or postnatally, this basic geometric correspondence for orienting is predetermined genetically and is set up early in the formation of the embryo. Presumably the same applies to humans. It remains to be determined how each retinal ganglion cell axon finds its way to the correct tectal location. According to one view, the postsynaptic tectal cells must be specified along cytochemical gradients like the ganglion cell array of the retina. Chung and Cooke (1975) claim evidence that the source of orientational polarity information for the tectal primordium of *Xenopus,* at least until stage 37, is in the diencephalic region near the tip of the notochord. This region regenerates if it is taken out. It may be the organizer of all the brain fields, including the eye primordia.

It is likely that all the special perceptual systems that later come to represent the subject in exploration of objects and surroundings are given basic coordination with the general somatotopic field of brain and body by similar early biochemical processes. Mapping of rudimentary peripheral fields to central fields begins before nerve cells conduct impulses, but embryo cells are probably transmitting information actively through the epithelial sheets of body and brain from earlier stages. Information from eyes, ears, nose, and mouth is subsequently carried in an impulse code and mapped into one behavior field in central nerve circuits because before the first axons grow out, and before the larval forms of lower vertebrates achieve active mobility, their primordia have been specified to relate to a common body map. The polarity or orientation of this map was triggered in the brain cell arrays in the early somite embryo before the neural plate closed to form the rudiment of brain and spinal cord. The formation of the first neuroblasts, immature nerve cells that divide no further, occurs in all parts of the primary nervous system at the stage after axial specificities sense organs are irreversible. In a widespread breakup of an intercellular federation of tightly connected embryonic cells, the loss of cell contacts may indeed be responsible for the observed fixation of polarities, or it may represent a final separation of independent clonal lines of multiplying cells that propagate the cytochemical differences established several cell generations previously.

4. Late Embryos and Lower Vertebrates: Growth of the Nerve Net

4.1. Patterning of Early Nerve Fiber Tracts by Growth Cone Activity

The neural tube of a 30-day human embryo has a mitotically active ependymal layer and a surface mantle of neuroblasts, cells beginning differentiation into neurons. In the second month, the mobile neuroblasts deploy in a complex pattern of layers and clusters that clearly express a latent differentiation of types of cell and selective affinities between them. The remarkable patterning of nerve axons, when they do suddenly begin to grow out from neuroblasts and thread intricately in the brain from the thirtieth day onward, expresses affinities among these cells for invisible structure in the cells they contact. The mantle (later gray matter) becomes covered by a mass of axons or marginal layer that will form the white matter in the mature CNS.

The directional growth of axons demonstrates that they can choose one path from among many possible ones. A first step in understanding this was made by Ramón y Cajal (1929), who observed a swelling at the tip of an immature nerve fiber that he called the "growth cone" (Figure 7). He described this structure as actively pushing and seeking to find the correct path and the correct ending among the brain cells. Later Harrison (1907, 1910) grew neuroblasts in tissue culture and observed the activity of the growth cone and outgrowth of the axon in isolation. Subsequent research has shown that the exceedingly fine membranous and filamentous extensions of the growth cone move actively over surfaces and may penetrate extensively between cell membranes in tissues. The process of filipodia extrusion is reversible, with threads being multiplied and then cut off or withdrawn to produce different net directions of advance. Growth of the axon tip thus involves a contractile or protrusive activity of the protoplasm and cell membrane similar to that seen in ameboid locomotion of protozoa (Jacobson, 1970; Allen and Haberey, 1973).

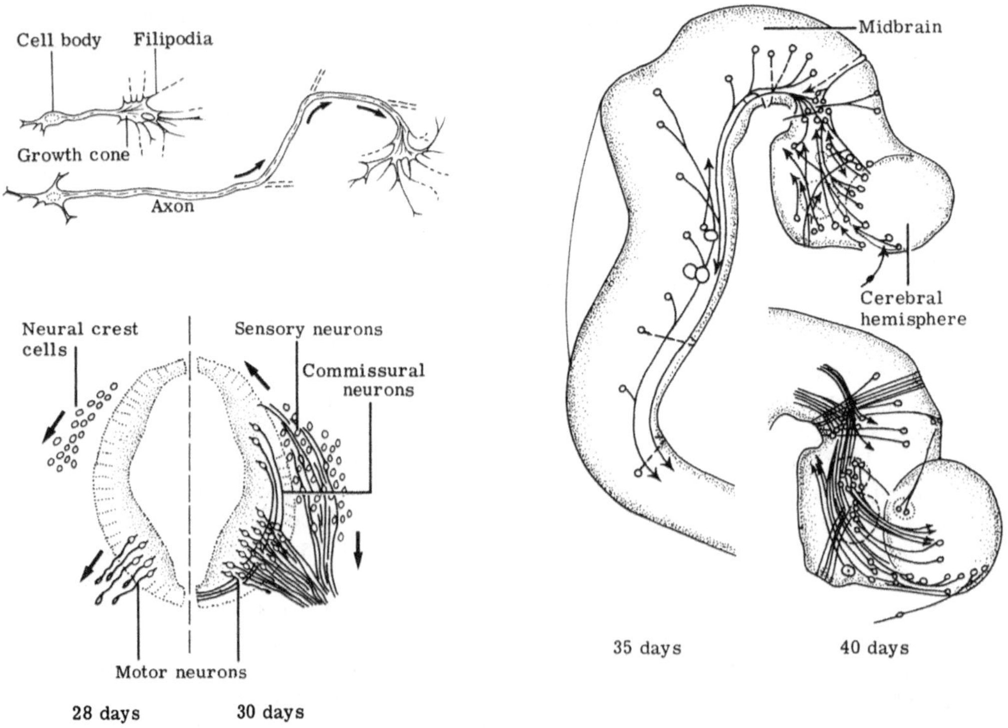

Fig. 7. Growth of axons in human embryos. Neurons produce axons guided by selective retention and extension of filipodia of the growth cone. Different kinds of neurons show different preformed paths of growth. Long tracts interconnect brain nuclei with spinal cord after the end of the first month of gestation. (B after Windle, 1970.)

Axoplasm flows to the axon from the cell body, but material circulates down and up the axon, offering the possibility of both intrinsic directives from nucleus and cell soma to the growth tip, as well as feedback from the growth tip and the media it contacts. The process offers a means of communication between growing nerve fibers and between each nerve cell and cell membrane of the medium. If the media are chemically or physically differentiated, the axon can choose a pathway and guide its growth. The problem of pattern formation in the embryo nerve net is like the problem of behavioral development in the organism. Behaviors adapt to patterns in experience, but may take forms that do not reflect the form of stimulation. Neuroblasts in different parts have different inherent capacities for reacting to the medium and to each other in for mation of stable nerve chains.

In recent years evidence has been accumulating that formation of selective connections may in many, if not most, parts of the CNS, involve elimination from an initial excess of axon branches and intercellular contacts. Growing nerve cells may first spread out randomly and then sort each other into arrays by comparison of some property related to their locus of origin. Final sorting may not involve a chemical code distinguishing each cell and specifying how pre- and postsynaptic cells should pair off. A number of different selective elimination mechanisms proposed to account for the formation of precise neural projections between groups of neurons that originated in remote parts of the brain, are discussed in Section 4.5. Selective retention of some sensory connections may involve differential activity in postsynaptic cells or in interneurons.

Baker and collaborators (1984, 1985) have examined the factors guiding selective termination of dorsal ganglion axons in the dorsal spinal cord. They maintained separated spinal cord segments of rat fetuses in tissue culture, each with its pair of dorsal ganglia loosely attached by connective tissue. The cord was removed and sliced into segmental discs at a stage before the ganglionic neroblasts have produced axons (approximately 15 days; equivalent to the 28 days human embryo in Figure 7). In a few days the dorsal ganglionic axons penetrated the isolated but grossly normal differentiating cord slice at various points on its circumference. In spite of often widely ectopic entry, the axons terminated selectively in the dorsal quadrant of the cord as in normal development. If, however, the activity of interneurons in the cord was suppressed by chronic administration of tetrodotoxin prior to recording, axons spread throughout the side of the cord that they had entered. Evidently in the normally active tissue the dorsal terminals were sustained by spontaneous activity of the interneurons while growing nerve cells may first spread out randomly, then sort each other into arrays by comparison of some property related to their locus of origin. Final sorting may not involve a chemical code distinguishing each cell and specifying how pre- and postsynaptic cells should pair off. A number of different selective elimination mechanisms proposed to account for the growth of precise neural projections between groups of neurons that originated in remote parts of the brain are discussed in Section 4.5.

That there is some selective recognition of the media in which growth occurs is proved by the criss-crossing of axon fascicles from different cell groups to make different directions of growth within one and the same medium. Fiber tracts are formed that cut across one another as they grow, headed for quite different sites (Figure 7).

Using Bjorklund's brain graft technique, it is possible to innervate adult rat caudate-putamen, previously denervated by a lesion to the substantia nigra, with dopaminergic cells taken from ventral mesencephalic tissue of late embryos—a stage at which neurons have finished multiplying but remain undifferentiated (Bjorklund and Stenevi, 1979; Dunnett *et al.*, 1983). Transplants to dorsal striatum relieve the spontaneous motor disorder (persistent circling) caused by the nigral lesion, and lateral transplants relieve sensorimotor (attentional neglect) symptoms. The embryo cells grow axons, as revealed by histofluorescence, selectively into caudate-putamen, bypassing other cortical and subcortical tissues. This technique, and particularly a refinement that injects suspensions of dissociated embryo neural tissue, offers the possibility of testing growth preferences of neurons from different parts of the embryo brain in mature brain.

4.2. Growth of Optic Connections

In lower vertebrates the principal visual projection is completely crossed. Nerve fibers grow across the retina in converging lines, down the optic nerve, through the chiasma to the other side of the brain, and thence in an orderly array over the surface of the optic tectum by way of two branches of each optic tract (brachia) (Scalia and Fite, 1974) (Figures 8 and 9). Within the tectum, each axon

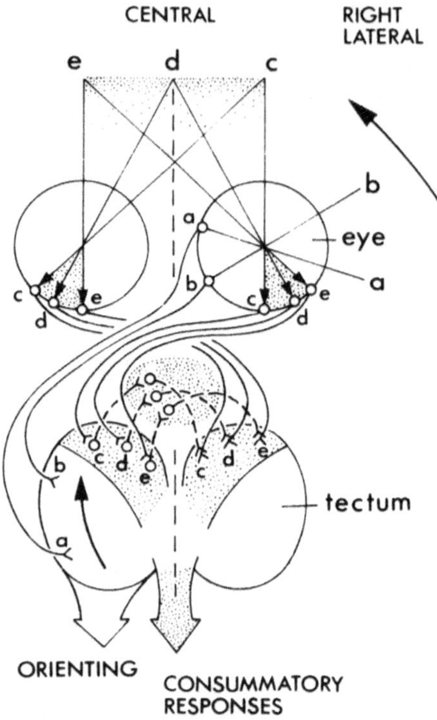

Fig. 8. Mapping of visual field in body coordinates on visual tectum of vertebrates. The binocular field involves commissural connections *via* nuclei in the ventral diencephalon.

must establish synaptic contacts with cells at a particular locus and depth. Discriminatory visual functions require further patterning of optic terminals among the dendrites of tectal cells. The first step in this orderly mapping is taken as the growth cones of visual fibers find their way in the optic nerves. Fibers also grow from each eye to nuclei in the contralateral diencephalon and mesencephalic tegmentum. Pioneer fibers grow from the first formed ganglion cells at the center of the eye. In the human eye, this occurs on about day 35 of gestation, when the embryo is 12 mm is length. Within hours, a fascicle of fibers enters the optic chiasm from each optic nerve (Mann, 1964).

The precise timing of events in development of the eye and tectum and the march of projections between them has been intensively studied in *Xenopus* (Nieuwkoop and Faber, 1956; Jacobson, 1968*a*,*b*; Gaze and Keating, 1972; Chung *et al.*, 1974; Gaze *et al.*, 1974; Hunt, 1975*a*; Fawcett and Gaze, 1982). When visual axons reach the anterior end of the tectum, the number of rods and cones in the retina approximates that in the adult eye but the cell laminae of the tectum are just starting to form. At stage 57, when the projection is nearly complete, the larvae escape from visual stimuli appearing above them by swimming forward, and an optokinetic bending of the tail may be clearly observed (Mark and Feldman, 1972). Comparable growth of retinotectal connections has been traced in the chick (Cowan *et al.*, 1968; De Long and Coulombre, 1965; Crossland *et al.*, 1975). Sperry and Hibbard (1968) noted the extraordinary similarity in size and cell content of the retinal and tectal layers in the *Rana* tadpole, in which the tectum is slightly smaller in area than is the retina. Presumably this reflects a common developmental plan.

Through the late embryo and larval stages of anurans and into the adult stage of fish and urodeles, both retina and optic tectum add new cells and new connections. Formation of retinotectal connections is delayed to fetal or even postnatal stages in mammals. In the mouse almost all the neurons of the tectum originate in the eleventh through thirteenth days of gestation, just before the optic nerve fibres reach the colliculus (De Long and Sidman, 1962). Schneider (1973) demonstrated extensive plastic reorganization of retinocentral connections in the neonatal hamster after surgery (see below). The hamster is born at a stage when retinotectal connections are still forming.

Embryo eyes of amphibia may be deprived of half their substance, then closed up to make single vesicles (Gaze, 1970; Berman and Hunt, 1975; Hunt and Berman, 1975). These undergo regulatory changes to produce different patterns of positional information, as described above. Alternatively, double nasal or double temporal eyes carrying duplicate front or back halves of retina only may be constructed (Gaze *et al.*, 1963). Such eyes reform apparently normal globes that grow like normal eyes, adding new radially specified ganglion cells to the retinal circumference (Straznicky and Tay, 1977; Gaze *et al.*, 1979*a*), then grow neural projections. Just how surgically halved eyes are regulated to make a whole retina is not known, but the same types of phenomena occur when similar surgical interference is made with the formation of the whole body at early stages of embryogenesis and in developing limb and sense-organ primordia of insects, amphibia, and birds (Spemann, 1938; Weiss, 1939; Wolpert, 1971; French *et al.*, 1976).

Growth of retinal axons along the optic pathway has been followed by histological methods in *Xenopus* after surgical modification of the eye vesicles. Axons from eyes with portions of the retina removed follow the appropriate path, choosing the correct brachium and entering the tectum at the right place in spite of losing interfiber ordering *en route* (Fawcett, 1981; Fawcett and Gaze, 1982) (see Figure 9). The evidence supports Attardi and Sperry's conclusion from regeneration experiments with fish that outgrowing axons from each part of the retina follow a chemoaffinity trace to enter the tectum with correct orientation to establish retinotectal mapping. The pathfinding does not seem to be a mechanical result of fibers keeping an arrangement related to a pioneering fiber, the ordering of ganglion cells in the retina, or a timetable of axon growth that follows the age at which ganglion cells differentiated. Optic tracts show fiber arrangements that suggest optic axons are progressively attracted to the correct tectal zone after taking rather independent paths at first (Scholes, 1979; Bunt, 1982; Scalia and Arango, 1983).

Guidance of optic axons in the embryo brain has been further elucidated by forcing them to enter at points far distant from the normal position ventral to the diencephalon, so they have to wander through tissues they would not normally encounter. Eyes replacing or additional to normal eyes have been grafted to forebrain, midbrain and hindbrain regions of amphibian embryos, and to genetically eyeless axolots. The latter can be given essentially normal vision and vestibulomotor responses by adding eyes (Hibbard and Omberg, 1975; Schwenk and Hibbard, 1977) and this is achieved by growth of optic connections that seek to follow the normal pathways of eyed animals (Harris, 1982). Misplaced (heterotopic) grafted eyes can form a visual projection to the tectum (Sharma, 1972), and the outgrowing axons show capacity to seek the optic chiasm and tract after growing long distances in the brain. When establishing connections in brains with fully occupied visual centers, abnormal contacts may be made from ectopic eyes to other sensory zones, but a preference is shown for the optic pathway and optic tectum (Hibbard, 1986). Optic fibers from supernumerary eyes can reach the tectum of the host and compete for terminal space,

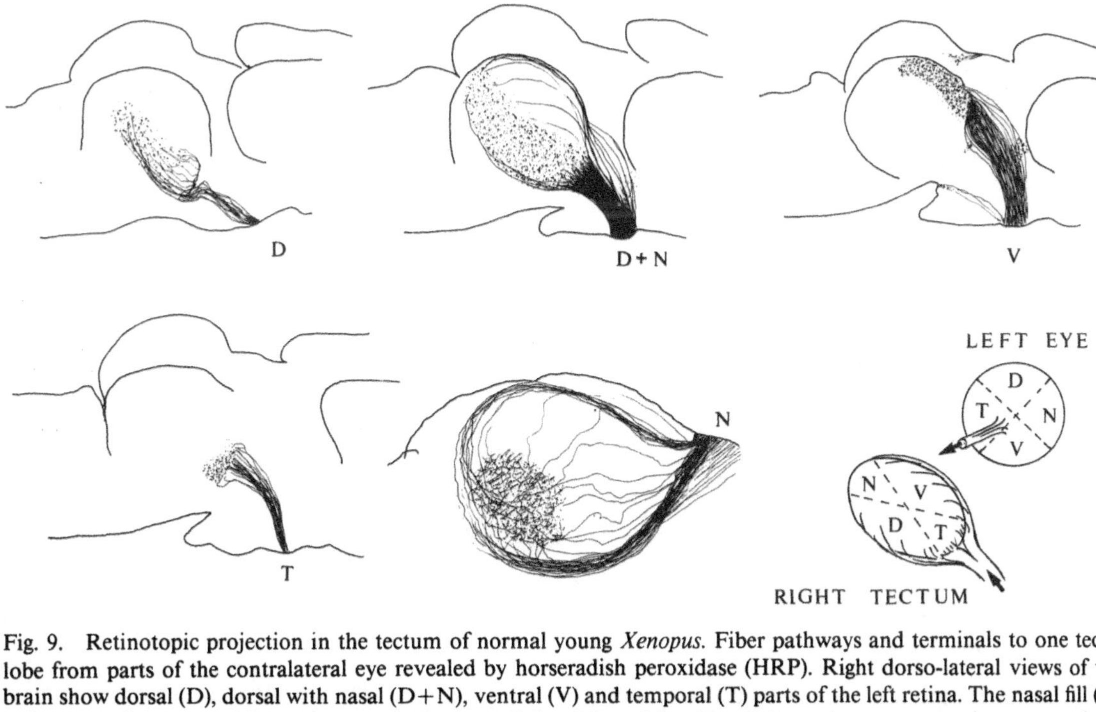

Fig. 9. Retinotopic projection in the tectum of normal young *Xenopus*. Fiber pathways and terminals to one tectal lobe from parts of the contralateral eye revealed by horseradish peroxidase (HRP). Right dorso-lateral views of the brain show dorsal (D), dorsal with nasal (D+N), ventral (V) and temporal (T) parts of the left retina. The nasal fill (N) is from the right eye and is seen to pass by both medial and lateral brachia of the optic tract to terminate posteriorly in the left tectum. Dorsal fibers take the lateral brachium, ventral fibers the medial brachium. Choice of route is apparently determined by chemoaffinity cues. (From Fawcett and Gaze, 1982.)

generating alternate bands of terminal endings (see Figure 10) (Constantine-Paton and Law, 1978; Constantine-Paton, 1983). These alternate eye-dominance columns have also been produced when optic fibers from both normal eyes innervate one tectal lobe after the other has been removed (Law and Constantine-Paton, 1980). The same patterning of eye-dominance stripes is formed when double nasal or double temporal eyes, made in the embryo stage, innervate an optic tectum of *Xenopus* (Fawcett and Willshaw, 1982).

Optic nerve regrowth studies give evidence for a chemoaffinity mechanism that is capable of guiding axons between alternative pathways and that specifies strongest affinity between the retinal cells and their normal pathways and terminations in optic tract and tectum. Optic nerves have been forced to regenerate through oculomotor or olfactory nerve stumps (Hibbard, 1986). Some fibers get lost, but many show guided growth to the optic chiasm or tectum. After reaching the tectum, a different fiber-sorting process produces bands when two retinae converge in one map.

Optic axons have to make progressively more specific local choices in growth to establish connections that will serve discrete visual discriminations and visuomotor coordinations. Research with fish and amphibia shows that the sorting process depends on interaction between an initial excess of nerve cell outgrowths and developing tectal tissue. The process has to accommodate a large increase in cell numbers and growth in size of both eye and tectum, even after the visual circuits are functioning. The direction of addition of tectal tissue may be a factor in the determination of the direction of alignment of ocular-dominance stripes when there is double innervation of the tectum (Fawcett and Willshaw, 1982).

Gaze (1970) pointed out that there is more than a 100-fold increase in the number of retinal cells in *Xenopus* and a far greater increase in *Rana* between the time the eye disc is provided with positional markers and the adult stage. Subsequently formed cells must have any retinotopic code handed on to them. Virtually all the adult amphibian retina is formed by mitosis at the ciliary margin of the eye (Glucksmann, 1940; Stone, 1959; Jacobson, 1968*b*; Straznicky and Gaze, 1971; Hollyfield, 1971), and chick and mouse eyes grow similarly (Crossland *et al.*, 1975; Sidman, 1961).

While the eye grows radially, with ganglion cells added in rings from the center outward throughout larval and early juvenile growth, the tectum grows asymmetrically along the posterior and medial borders or neck (Straznicky and Gaze, 1972; Crossland *et al.*, 1975). Observation of this discrepancy in *Xenopus* has led Gaze and Keating (1972) to question the notion of a chemical encoding of affinity between ganglion cells and tectal cells. Radiotracer techniques confirm that cells are added around the whole edge or rim of the optic cup, but only to the caudomedial boundary of the tectum (Straznicky and Gaze, 1971, 1972; Gaze *et al.*, 1974, 1979*b*). This noncongruent expansion requires the retinal projection to slide over the tectum by repositioning of synapses (Gaze *et al.*, 1979*b*).

Crossland *et al.* (1975) have shown that, in the chick, successively established optic terminals oc-

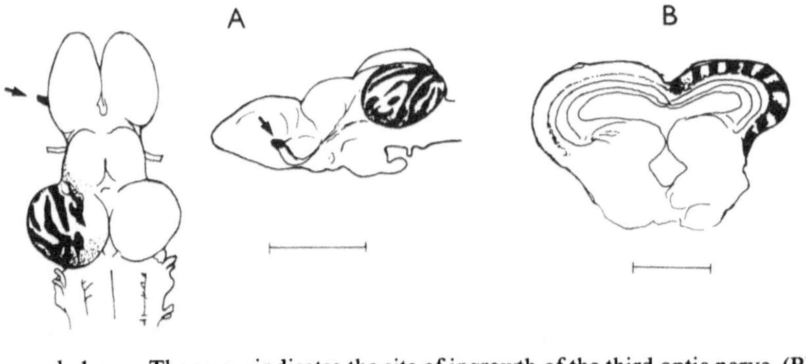

Fig. 10. Banded pattern observed in doubly innervated tectal lobes of *Rana pipiens* (common leopard frog). (A) Dorsal and lateral view of the brain of a stage XI tadpole whose third optic nerve was filled with horseradish peroxidase (HRP). The tectal innervation pattern of the supernumary eye was revealed through a diaminobenzidine (DAB) procedure on the whole brain. Bar equals 1 mm. The arrow indicates the site of ingrowth of the third optic nerve. (B) Autoradiographic demonstration of the striping pattern in a coronal section through the tectal lobes of a 2-month postmetamorphic frog whose normal, coinnervating eye was injected intraocularly with [^{3}H]proline. Bar equals 1 mm. (Drawn from photographs in Constantine-Paton, 1983.)

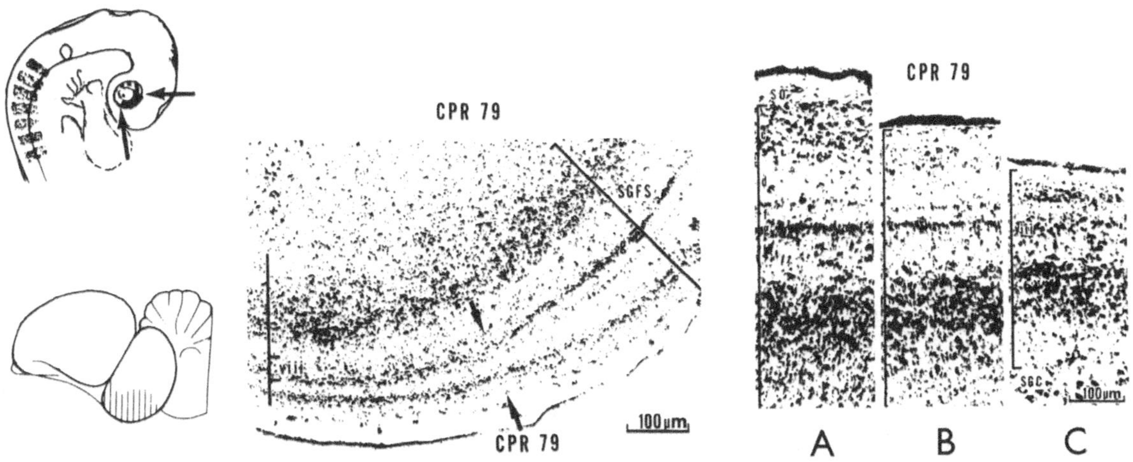

Fig. 11. Development of retinotectal projection in the chick after ablation of caudal half of the right optic cup at stage 17 (Hamburger and Hamilton, 1951), early on the third day of incubation. Chick sacrificed at the end of incubation (day 20). Retinal axons grow over the cross-hatched region (corresponding to the ablated retina and nasal visual field) and terminate on the dorsal and posterior tectum. (Center) Photomicrograph through junction of innervated and noninnervated left optic tectum. Laminae disappear from the noninnervated *stratum griseum et fibrosum superficiale* (SGFS), which is reduced in thickness (to left of arrows). (A) normal control of right tectum; (B) innervated part of left tectum; (C) noninnervated left tectum. The SGFS of B is reduced in thickness but has normal lamination. C lacks laminae a–f and has appearance of a tectum totally denervated by enucleation of the eye at day 3 of incubation. (From Crossland *et al.*, 1974.)

cupy bands around the map locus that corresponds to the center of the retina, where ganglion cells first develop. They consider their findings to give strong support to the hypothesis that retinotectal connections form according to a predetermined chemoaffinity mechanism. If pieces of retina are cut out of the eyes of chick embryos before the tectum is innervated, at days 4 and 5 of incubation, the tectal projection manifests corresponding holes, observed by histological methods, in the finished projection at 12 days (De Long and Coulombre, 1965; Crossland *et al.*, 1974) (Figures 11 and 12). The same result is described below with regeneration of connections in adult goldfish (Attardi and Sperry, 1963; Meyer and Sperry, 1976) and for original growth of retinal input to the tectum from depleted eyes in newborn hamsters (Frost and Schneider, 1979). In both cases, fibers made visible with silver stain or autoradiographic methods, either newly formed or newly regenerated, grew selectively to locate the appropriate tectal region. Expansion of regenerated inputs from a half retina over the tectum of a goldfish can be prevented or reversed in the competing presence of fibers from an intact eye (Schmidt, 1978; Sharma and Tung, 1979). Similarly, when a fraction of the retinal input from one eye is deflected into a denervated tectum on the ipsilateral side it is prevented from forming connections if normal input to the host region is regenerated at the same time (Meyer, 1979). These experiments demonstrate the competitive process establishing the retinotectal innervation.

The first retinotectal map of *Xenopus* is diffuse and crowded in the anterior half of the available surface (Gaze and Keating, 1972; Chung *et al.*, 1974; Gaze *et al.*, 1974). The map determined electrophysiologically grows progressively more detailed and accurate after terminals have spread over the whole tectum. It is likely that this process involves an intense selection among an excess of terminal branches and among synaptic endings on tectal cells (Gaze and Hope, 1976; Prestige and Willshaw, 1975). It may reflect progressive refinement in connective affinities of tectal cells. An alternative hypothesis is that the correct retinal axon endings are sorted within the tectum by selective reinforcement of presynaptic or postsynaptic neighbors by synchronized impulses that encode neighborhoods in the retinal array. This is the hypothesis suggested by Willshaw and von der Malsburgh (1976). It is compatible with what is known of the movements of nerve cell processes during pattern-

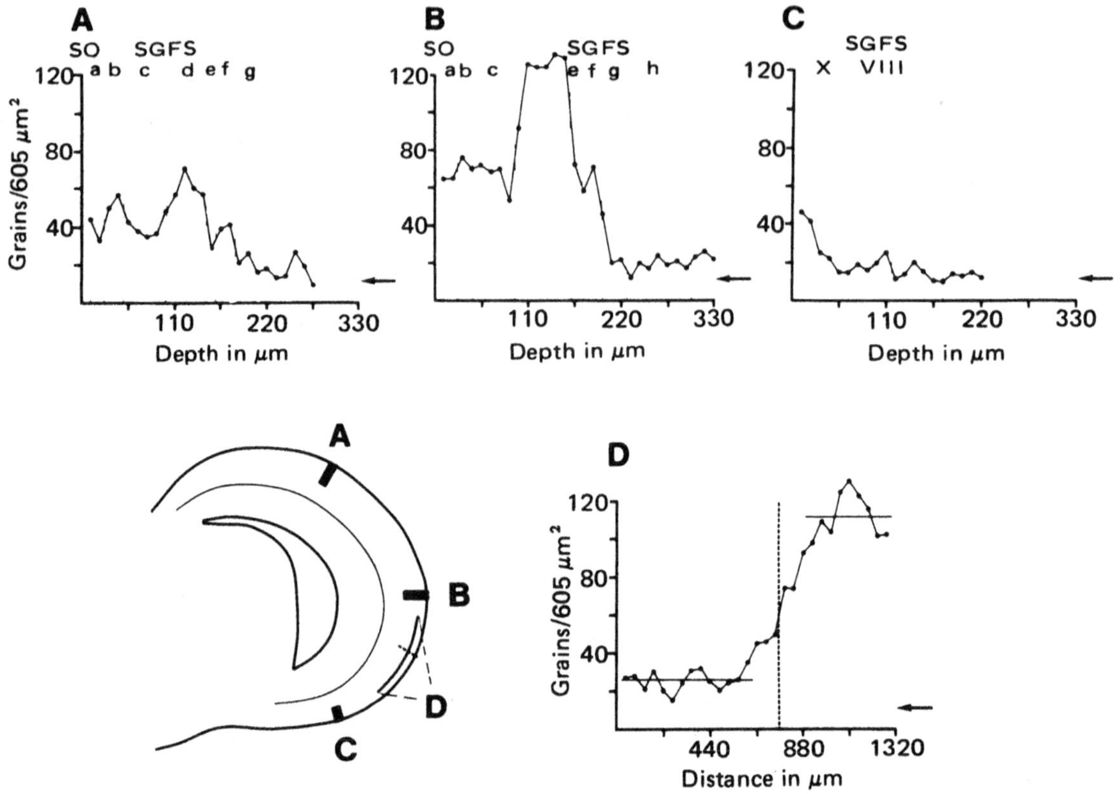

Fig. 12. Same chick as that in Figure 11. Radioactive [^{3}H]proline injected into the operated eye on day 20. Embryo sacrificed 6 hr later. Radioactive protein was transported down the optic nerve fibers to the tectum. Grain density plots from the autoradiographs show distribution of labeled terminals in the innervated (A and B) and noninnervated areas (C) as well as transversely across the junction region (D). (From Crossland *et al.*, 1974.)

ing of retinotectal links in *Xenopus* and in other forms. However, it cannot explain the orientation of the retinotopic map on the tectum. Death of cells in retina and tectum may also eliminate misfit elements (Glucksmann, 1940, 1965), but this seems to become more significant later when the projection is resculptured in metamorphosis (see Section 4.3).

Surgical transpositions and rotations of embryo eyes and deletions on duplications of retinal parts prove that the cells in the eye vesicle contain positional information that can guide formation of orderly connections and this information is propagated without confusion as the eye grows. As we have seen, retinal axons are guided along the path of the optic tract by some property that permits this positional differentiation to be retained, at least to retrieve overall orientation and topographical order before entry to the tectal rim. Surgical ablations and rearrangements on the developing tectum indicate that it too has markers that guide retinal axons to the approximate site of the cells with which they must retain functional connections if orderly visual function is to be produced (Chung and Cooke, 1978; Yoon, 1986).

Rotation of the tectum primordium causes development of rotated retinotectal maps. It is not yet certain that this is a consequence of position specification in the tectal primordium before it is innervated or the result of an induction of positional cues from an ordered array of ingrowing axons that might be guided into the tectum by diencephalic cells carried with the transplanted midbrain roof. One tectum was kept free of optic terminals in *Xenopus* by removal of one eye in the embryo; after metamorphosis, a piece of the tectum was rotated or translocated, and fibers from the remaining eye were directed into the virgin tectal tissue (Straznicky, 1978). Fibers formed an appropriate rotated or translocated projection, indicating that tectal tissue, although never previously innervated, possessed positional information that was capable of guiding visual axons that had entered in an abnormal and disordered way.

These surgical studies have shown that the process of sorting is an elastic one, with graded affinities permitting considerable variety in the form of maps. On the other hand, they also prove that the mechanism that first forms the visual projection to the midbrain is one that operates without correction from the relationship of stimuli to behavior. Primitive visuomotor coordination in vertebrates is an innate product; i.e., it is formed before sensorimotor function. The functional nerve circuits, though not simply the expression of preestablished one-to-one cell affinities, result from a negotiated differentiation in which preestablished cell differences carry essential positional information to create new intercellular communications. As a result of these new interactions, further refinement is produced in positional determination. The patterns of nerve growth are both the consequence of cell typing and the cause of further cell specification for function in the network.

Compound (double nasal and double temporal) eyes, made by operating on embryo vesicles, make abnormal projections to the optic tectum, exhibiting expansion to fill the whole tectal area in spite of lacking half the population of retinal positions. This has been interpreted to show that tectal cells do not have specific affinity for particular retinal cells (Gaze *et al.,* 1963; Gaze and Keating, 1972). More recent experiments indicate that the cells of compound eyes do retain their original positional information, that this may be expressed in choice of pathway to the tectum and in preferential affinity for half the tectum and that double temporal eyes are appropriately selective in their innervation of a "virgin" tectum (Straznicky and Gaze, 1982; Fawcett and Gaze, 1982) (Figure 13). Fawcett and Willshaw (1982) have shown that the terminals of double nasal and double temporal eyes, surgically constructed in the embryo, form themselves into stripes that are lined up in the tectum parallel with the body axis. This anatomical evidence of a double sorting process in the course of growth of the visual projection is discussed below.

Regenerating axons transport large amounts of protein and 4S RNA toward their endings and interference with this may prevent reestablishment of the retinotectal map (Murray and Grafstein, 1969; Murray and Forman, 1971; Ingoglia and Sharma, 1978). These events are probably associated with the activity of the growth cone, production of axon material and communication between cells permitting selective formation of intercellular contacts. Once contact is established, the transport of material down the axons continues for a time while collaterals multiply, but it is then turned off.

4.3. Metamorphosis of the Amphibian Visual Projection: Influences from Stimuli

In seeing tadpole stages of *Xenopus* (stages 50–57), the eyes face laterally; the most important visual response is forward flight triggered by a shadow entering the field from any direction. The highest sensitivity is in the posterior dorsal field. After metamorphosis (stages 60–65), the eyes face obliquely upward and, since the animal is an active swimmer that uses its limbs, the most significant part of the field is presumably anterior (Figures 5 and 14). Changes observed in the tectal map (Gaze and Keating, 1972) would appear to fit these morphological adjustments in the eye and the functional changes in visual response: change from filter-feeding and tail-swimming tadpole into legged adult must be associated with change in vision. A transformation in color vision of frogs at metamorphosis was found by Muntz (1962), who recorded corresponding change in neurophysiological responses of the dicephalon to color stimulation. Pomeranz (1972) correlated ganglion cell dendritic tree anatomy and visual physiology in tadpoles and adults of frogs *(Rana)* showing that there are adaptive changes at metamorphosis in both retinal visual detectors and retinotectal connections.

In the spinal sensorimotor system of the frog and in the distal parts of the limbs, a major contribution to remodeling at metamorphosis comes from programmed and selective cell death (Saunders and Fallon, 1966; Hughes, 1968; Prestige,1970). Waves of cell death have been observed in the retina of developing eyes of *Rana* (Glucksmann, 1940). Cell death is regulated by thyroid hormone (Kollros, 1968), and Levi-Montalcini (1966, 1984, 1986) discovered a powerful active protein, nerve growth factor, which, when released from an unknown source, determines the survival and growth of sympathetic ganglia in chick and mouse (see Jacobson, 1970, for a review of hormonal regulation of differentiation in the developing nervous system).

Metamorphic change in eye position of frogs enlarges the binocular visual field. During metamorphosis of *Xenopus* (stages 45–66), this change is correlated with development of a bridge of nerve fibers which projects information from the anteromedial part of each tectum *via* one or two synapses through the postoptic commissure to the other side of the brain (Beazley *et al.*, 1972; Keating and Gaze, 1970). This creates an ipsilateral projection (each eye projecting to the tectal lobe on the same side), which is added to the completely crossed primary projection. The intertectal pathway terminates in register with the preestablished map of the central (anterior) visual field on the other side. An accurate binocular convergence of visual information is created by which stimuli from one locus in the binocular visual field may product excitation from both eyes at two points, one in each tectum, and a duplicate representation of the central binocular field results (Gaze and Jacobson, 1962).

Electrophysiological mapping experiments combined with surgery and manipulation of visual experience in one eye show that convergence of crossed and ipsilateral (intertectal) projections is achieved in *Xenopus* by selective reinforcement of the latter intertectal terminals under the influence of visual stimulation (Gaze *et al.*, 1970; Beazley *et al.*, 1972; Keating, 1976).

Spatiotemporal patterning of excitation by stimuli brings about a further step in binocular registration between visual inputs roughly guided by chemoaffinity cues, to calibrate the maps (Beazley, 1975). In normal development the ipsilateral projection is at first retinotopic, but it is diffuse in its distribution on the tectum. Later it becomes more precise. If one eye is removed in the embryo the intertectal projection from the remaining eye is

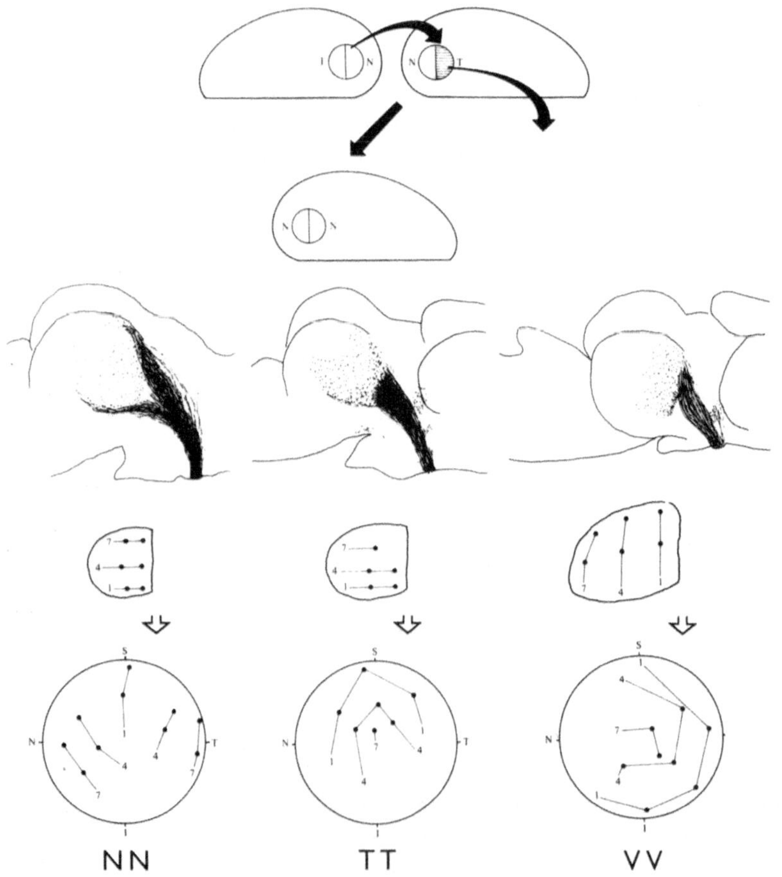

Fig. 13. Abnormal retinotectal projections in *Xenopus.* Compound eyes, made by doubling half eyes, mirror-image fashion, about one axis, are prepared by surgery at early embryo stages as shown for a double nasal (NN) eye above. Subsequent projections, although appropriately oriented, show expansion of the tectal map, terminals ending over a wider territory than normal for those retinal parts. Fibers from NN eyes form a wide tract and strong medial and lateral brachia. Fibers from VV eyes enter the tectum via the medial brachium and fibers from TT eyes form a narrow tract and enter the tectum directly from its rostral extremity. Thus fibers for each type of compound eye follow pathways appropriate to the embryonic origin of the retina forming the compound eye. Electrophysiological mapping after growth of functional connections, shows characteristic mirror symmetries for each type of compound eye. Sites in the visual field of a compound left eye are recorded in electrodes inserted in a grid pattern on the right tectum. Visual fields, marked with nasal, temporal, superior, and inferior poles, are shown below charts of the tectum with arrow indicating anterior. (From Fawcett and Gaze, 1982.)

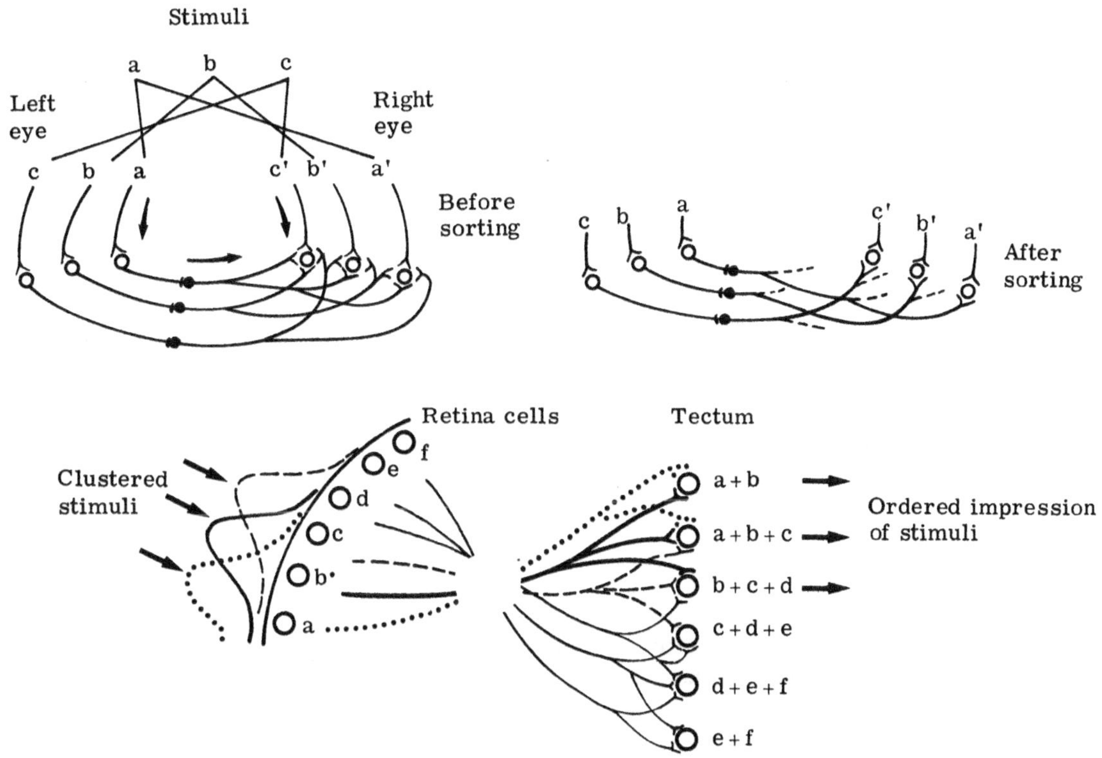

Fig. 14. Determination of neuronal interconnections decided by postsynaptic effects of excitation. In postmetamorphic *Xenopus* binocular convergence, involving transmissions from each tectum *via* the nucleus isthmi of the diencephalon to the other side, requires visual stimulation to attain accurate interocular registration. Initially, diffuse projections become more precisely wired by elimination of noncorresponding inputs (Keating and Gaze, 1970). Willshaw and Von der Malsburg (1976) proposed a model that sorts terminals to reproduce the order of the locations of cells of origin in a receptor array. Terminations from different retinal cells are retained or reinforced if they excite postsynaptic cells coincidentally. Incoherent inputs are eliminated.

again retinotopic but unrefined (Gaze *et al.*, 1970). The same results are obtained if the animals are reared through metamorphosis in the dark.

The critical results follow inversion of one eye before growth of the ipsilateral projection, but after permanent polarization of the retina (Gaze *et al.*, 1970; Beazley, 1975). This experiment confirms that exposure to light may cause formation of a correctly registered binocular map on the tectum ipsilateral to the inverted eye (contralaterally transplanted from one embryo to another) and an inverted binocular map on the other side. This would appear to prove that further sorting of connections may occur among the terminals of one intertectal pathway, even to the extent of inverting the natural tendencies for nerve connection. Possibly the neurons involved belong to a later generation of cells that are fully segregated only after they make functional contact with active neurons. They remain plastic, like the eye of *Xenopus* was at an earlier stage. However, there is no evidence to support a similar conclusion for *Rana,* which appears to develop the binocular projection according to intrinsic coding of ganglion cells, intertectal relay cells, and tectal cells and dendrites (Jacobson and Hirsch, 1973; Scarf and Jacobson, 1974). Udin (1983) obtained anatomical evidence confirming Keating's demonstration by electrophysiological techniques (Gaze *et al.*, 1970; Keating, 1976) that in *Xenopus* the visuotopic convergence between direct retinotectal axons and the intertectal input relayed by way of the nucleus isthmi results in redirection of isthmal axons. She used HRP to reveal abnormal changed trajectories of the isthmal axons in the tectum after one eye of the tadpole had been rotated 180°. Rerouting of the axons was presumably caused by selective increase in terminals coactivated by retinal stimulation, leading to transformation of the retinotopically directed initial input of intertectal fibers to the tectum ipsilateral to the rotated eye.

For functional interpretation of the binocular projections, it is important to remember that binocular stereopsis is not the only way in which visual circuits may measure distance. Monocular kinetic stereopsis, by motion parallax effects, is almost certainly more primitive. In swimming forms it is a reliable source of information for guidance of forward progression (Trevarthen, 1968*a*). Ingle (1976*b*) demonstrated that frogs may have depth detection with cues available only to one eye. Evidently this is done by accommodation of the lens to vary focus. In kinetic depth perception and perception of causal relations in events, a given retinal locus may participate in many different arrays of excitation that signify different space information or different types of event. Any morphogenetic specification of nerve circuits for this kind of perceptual function, comparable, say, with the innate motion and event detectors of the static, predatory frog (Lettvin *et al.*, 1959; Ingle, 1976*a*) would be ready for adapting in evolution to specify depth detection by binocular stereopsis.

4.4. Regenerated Afferent Connections in Adult Fish and Amphibia

Whatever the process by which selective nerve connections are built up in embryos, and whatever the influence of receptor excitation by patterned stimulation in metamorphosis, in the adults of lower vertebrates the primary projection from retina to optic tectum may end up with its many neural elements precisely specified. In regeneration, functional connections are reformed from eye to brain according to a fixed somatotopic arrangement that is unresponsive to environmental stimulation. This was shown by the classical experiments of Sperry in the 1940s with frogs, salamanders, and fish (Sperry, 1943, 1944, 1945). If the eyes of these animals are rotated in the orbit, maladaptive prey-catching or swimming movements persist for months, showing no signs of correction by learning (Figure 15). If the optic nerves from the inverted eyes are cut, new retinotectal connections develop and these reconstitute precisely mapped, but inverted and therefore totally maladaptive responses. Regrown optic axons make accurately selective connections to reinstate not only orienting responses, but also color discrimination and detection of distinctive invariants in size, form, and motion of stimuli. They do so even if forced to wander in disarray through an interval of nerve in which the orderly fabric of nerve sheaths is destroyed, or if routed into wrong branches of the optic nerve. To attain the place and type of cell for which they are specified, the axons will make long detours through inappropriate territories of the tectum that have been made vacant by denervation. The terminal branches of axons penetrate the tectum to end on the right dendrites in the optic layer and so rebuild specialized responses to different patterns of stimuli (Maturana *et al.*, 1959).

The results of a large number of studies that induced regeneration of afferent nerves, central interneuronal tracts, and motor nerves proved that both cells of origin and the cells on which synapses are formed must either be specified in a cytochemical or molecular code capable of sorting them out in large numbers (Sperry, 1951*a,b*) or be first roughly sorted by chemoaffinity, and then precisely ordered by a self-sorting mechanism. Anatomical studies of regenerated fibers suggest that the growing axon tips are influenced by attractive forces at points along their course (Attardi and Sperry, 1963; Sperry, 1963; Fawcett and Gaze, 1982) (Figures 9 and 13). However, it is not possible by these techniques alone to determine exactly how selection takes place or the limits of its precision. Electrophysiological studies suggest that fine branches of growing nerve fibers spread widely at first to create a diffuse projection (Gaze, 1970). They may sort out later to form the highly patterned pathways that are made visible by staining. On the other hand, selective Nauta staining of terminal aborizations (Roth, 1974) and autoradiographic studies (Meyer and Sperry, 1976) with goldfish and amphibia tend to confirm that the axon of a given ganglion cell may make route errors, but eventually sends terminals to only one small site in the tectum (Udin, 1977; Meyer, 1980; Cook, 1982; Fawcett and Gaze, 1982).

There are clearly important species differences; young adult goldfish and salamanders are still undergoing growth of eye and tectum. They may therefore respond to surgery by regulation that permits considerable plasticity of connections. Adult tree frogs show more refined specificity of connections that are less plastic. Attempts to resolve this

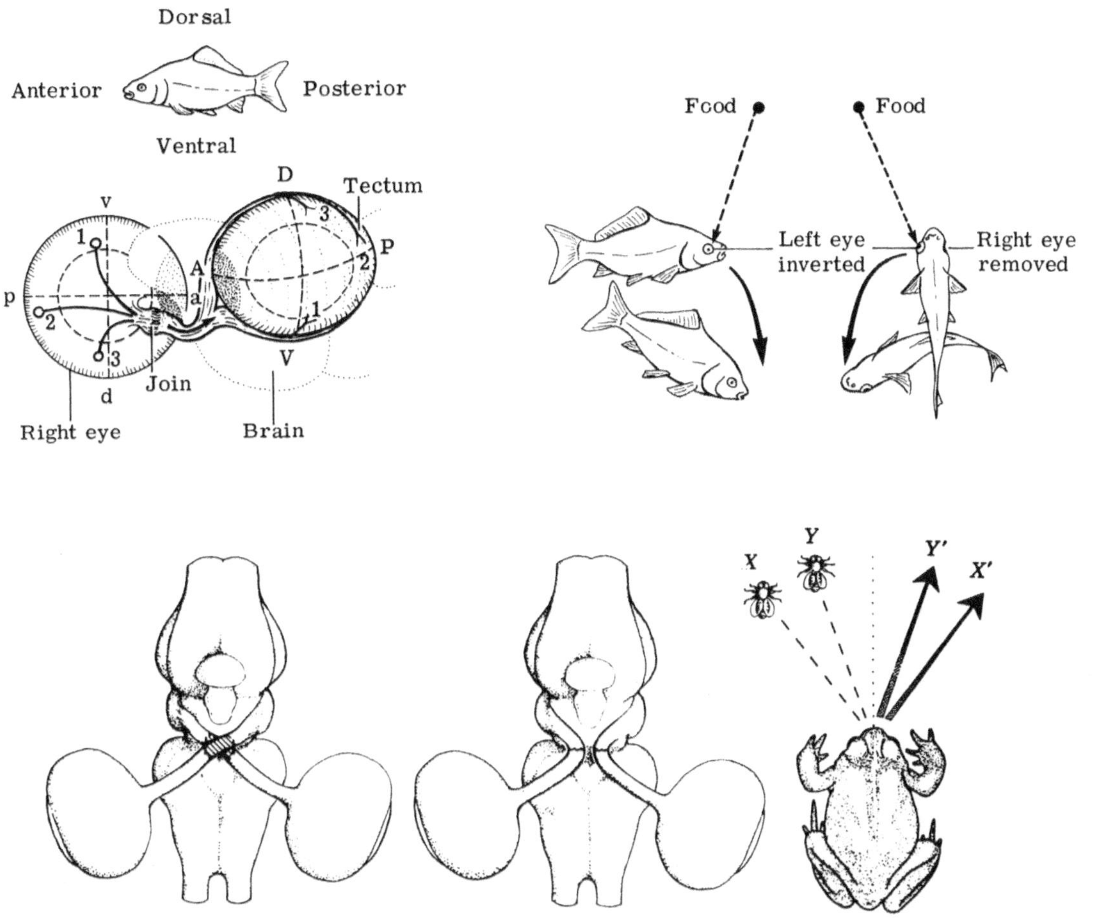

Fig. 15. Goldfish retinal cells reconnect through a join in which regenerating axons are disordered and establish a retinotopic map on the contralateral tectum, the axons passing by the lateral and medial branches of the optic tract. Connections regenerated from an inverted eye lead to permanent maladaptive orienting responses. (Bottom) Sperry's drawing of the behavioral consequences of uncrossing the optic nerves in the toad and regrowth of axons to the wrong sides of the brain. Again, maladaptive orienting responses prove mirror retinotopic segregation of regenerated terminals.

question have involved removal of parts of the retina or of the tectum before regeneration (Gaze and Sharma, 1970; Yoon, 1971; Sharma, 1972; Keating, 1976; Meyer, 1975; Meyer and Sperry, 1976; Ingle and Dudek, 1977; Sturmer, 1981). In some instances, axons of part of the retina appear to become gradually expanded, spreading out over vacated territory to form a map with correctly ordered loci, or the whole retina may regenerate a compressed projection to an available half tectum. These results clearly rule out a rigid one-to-one chemospecificity principle in favor of a polarized field of specificities, as in a magnet, and a competitive fiber-sorting process of some kind (Cook and Horder, 1974; Gaze, 1974; Gaze and Hope, 1976; Hope *et al.,* 1976; Willshaw and von der Malsburg, 1976; Keating, 1976). It has been claimed that an expanded projection over the tectum can change the connective affinities of tectal cells (Schmidt, 1978), but Meyer (1986) tested this hypothesis and showed that forced mislocation of retinal axons does not change preferential connective affinities of tectal cells in 2 years.

Compression of the whole retinal projection to a half tectum is reported under conditions in which it is not likely that the tectal cells have died and regenerated. Yoon (1972) made reversible compression of input to a half tectum by inserting a barrier of gelfilm across the path of regenerating axons. Later, when the gelfilm becomes absorbed, the projection expands to fill the normal whole territory. Apparently expansion and compression of the retinotectal projection may persist when plastic adjustment to reversal of eye orientation is not possible (Sharma and Gaze, 1971; Yoon, 1973, 1986; Meyer and Sperry, 1976).

Numerous experiments (reviewed by Gaze, 1974; Gaze *et al.*, 1979*b*; Meyer and Sperry, 1976; Meyer, 1986; Yoon, 1986) show that regeneration of abnormal connections to tecta of fish and amphibia obey basic somatotopic affinities but also show adjustments due to dynamic interaction between competing elements during growth. When a whole retina sends axons into a tectum reduced by about 50% by surgical ablation, the map of synapses that forms immediately is guided by somatotopic principles so that only the appropriate half of the retina connects to the reduced tectum. The fibers corresponding to the ablated tectal territory are left to crowd densely at the cut border where they probably do not form synapses. A subsequent adjustment process causes the first-formed map to be gradually pushed back, with the crowded fibers without proper home creating a competitive pressure that causes them to roll back the correctly located competitors progressively. The synaptic array is reorganized by a process which involves axon–axon or intersynaptic competition. After several months, a topographically ordered but compressed projection is achieved, and the fish recovers a visual field without scotoma. Scott (1975) showed that this field functions visually to permit acquisition of a conditioned respiratory response to color stimuli. The function of the field in directing orienting movements is not known. Yoon's decompression experiment (Yoon, 1972) shows that the compressed map was in an unstable condition.

Observation of the behavior of regenerating axons when negotiating enforced detours or bypassing obstacles in their path suggests that an inertial growth force is guided by information in the medium or among the population of fibers, with different axons reacting differently at a given site in the array (Sperry, 1965; Sperry and Hibbard, 1968; Hibbard, 1986). Fibers also show a strong growth preference to return to the correct laminae when they are deflected on traversing an incision scar in the tectum (Meyer and Sperry, 1976). Yoon (1984) showed that regenerating retinal axons seek the correct lamina when they enter blocks of tectal tissue that have been inserted upside-down, with surface and deep layers inverted with respect to the host tectum. Ingle and Dudek (1977) observed laterally displaced connections after incisions were made in the tectal afferent fibers of *Rana pipiens*.

Meyer and Sperry (1973) investigated regeneration of retinotectal connections in adult tree frogs with tectal lesions. Both behavioral and electrophysiological mapping indicated that in these animals the locus specifications are less flexible than in the goldfish. Tectal ablation causes appropriate scotomata in the visual field, and these last unchanged for at least 9 months of experience of moving about and catching prey visually. Evidently, this species develops more narrowly specified visuomotor circuits. Comparable limited compression with removal of part of the tectum was obtained with *Rana* (Udin, 1977) and newborn hamsters (Finlay *et al.*, 1979).

The plasticity of the retinotectal projection brought out by experiments with whole eyes growing into part tecta has also been explored by means of translocations, rotations, and inversions of pieces of tectum in goldfish (Yoon, 1973, 1975, 1986) and *Xenopus* (Jacobson and Levine, 1975). The electrophysiological maps show that regenerating retinal fibers frequently terminate according to the original polarity of the rearranged tectal pieces, in spite of the discrepancy this creates in the overall map. When small blocks of tectum are interchanged without alteration of their polarity, regenerated terminals are appropriate to the original location of the blocks, proving that topographic address is specified with affinities for retinal loci (Yoon, 1986), and not just polarity (as hypothesized by Gaze and Hope, 1976; Hope *et al.*, 1976).

Plasticity observed in formation of functional connections between the skin and spinal circuits of amphibia suggests that there is a flow of specification from skin to nervous system in the direction of nerve conduction (Miner, 1956). This in turn would favor the idea that the process of circuit selection involves transport of material down the axon by a mechanism linked to discharge of the action potential in the nerve membrane. Regenerating dorsal spinal roots of frogs appear to form connections widely in the skin without selection of locus or receptor type (Jacobson and Baker, 1969; Johnston *et al.*, 1975). Recovery of function after surgical disarrangement may result in this case from reorganization of the effective junctions the dorsal root cells make with spinal interneurons (Miner, 1956; Baker, 1972). According to this hypothesis, place in the skin is chemically coded and contact with the dendrites of the afferent neurons provides a means for transporting the coded label into the cord and influencing the axon of that cell to make the right connections centrally (see Mark, 1975), a kind of end-organ induction (cf. Hunt and Cowan, 1986). Van der Loos (1977) found that each whisker follicle on the muzzle of a mouse (follicles appear on the twelfth gestational day) induces formation of a mass of about 2000 neurons in layer IV of the somatosensory cortex—a whisker barrel. Neopallial matrix cells destined to be barrel neurons cease dividing on day 14 and barrels first appear on postnatal day 3. Extra follicles cause extra cortical barrels by cascading induction over three synapses. Van der Loos (1979) suggests that somatotopic maps of the integument in the somato sensory cortex are caused by genetically regulated structure of the integument.

4.5. *Theoretical Problems in Embryogenesis of Nerve Nets*

The above studies of epigenetic processes in visual and tactile projection pathways open up a new perspective on the pattern-forming process of developing nerve circuits, one in which a balance of preferences in the differentiated cytochemical and biophysical environment, created jointly by the target tissue and invading axons the way a magnetic field of force is created jointly by magnet and iron filings, determines how functional synapses will be ordered among cells and dendrites. In principle, the process may be compared with other morphogenetic processes that create complex adaptive machinery in cells and tissues by regulating interactions of elements. But it should be remembered that the special properties of neurons allow a new order of complexity and rapidity of adjustment between intercommunicating growth elements that may lie far apart in the body yet intimately affect one another. The refined circuits of the nerve net certainly benefit from nerve impulse conduction in their processes of differentiation and in the transmission of growth response to stimuli.

The capacity of the growing brain circuits to regenerate and adjust in the face of drastically changed conditions or mutilation is an aspect of their capacity to generate elaborate integrative systems prefunctionally when left to grow normally. Lower vertebrates, as well as mammals, show adjustment of the preestablished nerve circuits at critical times in their growth. This adjustment permits them to acquire more finely adaptive perception processes with the aid of stimulus regularities from the environment (Ingle, 1978). Further discussion of dual instruction models of the formation of retinotectal connections may be found in Gaze and Keating (1974), Willshaw and von der Malsburg (1976), Meyer and Sperry, (1976), Fraser and Hunt (1980), Whitelaw and Cowan (1981), Law and Constantine-Paton (1981), Fawcett and Willshaw (1982), Constantine-Patton (1983) and Meyer (1983, 1986).

The chemoaffinity hypothesis of specification in nerve circuits for which Sperry accumulated experimental evidence during the 1940s and 1950s and which was given its full formulation during the 1960s still stands during the 1980s (Hamburger, 1981; Levi-Montalcini, 1986; Meyer, 1986). It replaced concepts of relatively unselective mechanical guidance of nerve growth and functional selection of connections and has withstood presentation of alternative models as the primary explanation for nerve net formation. Although the code of positional information and the mechanism of intercellular recognition that determines selective retention of effective synaptic contacts are unknown, studies of developmental plasticity increasingly bring evidence for a chemoaffinity factor operating before synapses are formed and confirm Sperry's insight that neurogenesis is to be perceived as a magnificent elaboration of epigenetic principles that govern morphogenesis and cell differentiation in embryos (Hunt and Cowan, 1986).

Hebb (1949) suggested that associative linkages in the cortex could be acquired by synachronous presynaptic and postsynaptic activity. He thought perception processes of infants would be built up by formation of cortical cell assemblies, sensory analyzers being constructed by orderly scanning movements that would give regular pattern to stimuli. His hypothesis clearly underestimates the contribution of intrinsic pattern-forming processes in growth of cortical tissue. Contemporary neurobiologists still seek to explain nerve nets in terms of functional selection from disorderly arrays, but they admit that this complements embryogenic determination.

Willshaw and von der Malsburg (1976) have developed a model like Hebb's that, in theory, is capable of ordering the projections from one field of neurons to another with unlimited precision. It depends on an impulse synchronization according to neighborhoods in the afferent cell array. Adjacent cells are coupled electrically, and the synchrony of activity decreases with their distance apart. Limited zones of interaction between source cells are determined with the aid of inhibition. Terminals spreading and intermingling in the target region of the CNS are reinforced or eliminated according to the synchronization of impulses in a competition that results from either presynaptic or postsynaptic interaction. Creeping of jostling of projections in the target region permits a continuous improvement in the accuracy of mapping until equilibrium when all sites are filled. The impulse patterns would presumably be effective in determining cell contacts because they result in neurotransmitters or other neurosecretions being released locally in controlled concentrations.

Impulse coding of position according to neighborhoods is presumed to be effective in absence of environmental stimulation because nerve circuits show spontaneous electrical activity, and they are formed prenatally in great refinement. Willshaw and von der Malsburg (1976) call the polarity marking, magnification controls, and so forth 'boundary conditions', and they leave these fundamental morphogenetic factors unaccounted for.

Mark (1974) proposed that plasticity in formation of adaptive circuits after surgical alteration in afferent, interneuronal, and efferent links may be explained in terms of competitive selection among synapses in an initially overdispersed or overconnected network. He concludes that the genetically determined chemical labeling somehow makes possible strengthening of elements that match up functionally, and suppression or elimination of less-well-matched ones. However, the experiments cited above and by Schneider (1973) and Hubel *et al.* (1977) show that dynamic reorganizations, including expansions, compressions, and decompressions of projection systems, can be mediated by interaction of elements that have not yet formed synapses. Mark further proposes that an invisible deactivation of permanent synapses may be responsible for the formation of memory traces in mature brain circuits. The concept of competitive synaptic inactivation without loss of synapses has not received support from experiments with neuromuscular preparations of lower vertebrates that it was originally proposed to explain (Scott, 1975). On the other hand, Duffey *et al.* (1976) showed that suppression of vision in an unused eye of a kitten may, indeed, be due to inhibitory suppression rather than loss of connections. They reversed this inhibition to reevoke cortical responses for the unused eye by administering bicuculline with blocks inhibition by γ-aminobutyric acid (GABA).

Changeux *et al.* (1973) formulated a general theory of the development of functional neural nets that requires selective stabilization of synapses in an initially overconnected system with many redundant axon collaterals and terminals. Although admitting that genetic specification of affinities between neural elements defines the basic layout of the CNS and central-peripheral relationships, this theory also proposes to explain learning (Changeux *et al.*, 1973; Changeux and Danchin, 1976). Evidence supporting the theory comes from development of cerebellar and neocortical systems in fetal and postnatal mammals discussed below and its macromolecular basis has been sought in the acetylcholine receptors of neuromuscular and fish electric organ functions. Receptor sites for neuromuscular transmitters proliferate in developmental stages and are both labile and mobile. A subset of them is aggregated, stabilized, and made quasi-permanent beneath exploratory nerve terminals by synchronous excitation of both pre- and postsynaptic cell membranes (Changeux, 1979). On the presynaptic side, the elimination of redundant nerve endings, axon collaterals or whole motorneurons may be a consequence of the coincidence of activity in the terminals and postsynaptic cell. Such a selection would contribute to specification of neuronal networks. Evidence for selective elimination of nerve connections in mammals is discussed in the following section.

5. The Fetus of Reptiles, Birds, and Mammals

5.1. Lower and Higher Vertebrate Brains

The nervous system of adult lower vertebrates is matched in reptiles, birds, and mammals by the system present at the end of the embryo period. The first motor, sensory, and interneuronal connections in humans at this stage (Windle, 1970) may resemble the brain of the tiger salamander described by Herrick (1948). Formation of this basic nerve network in a human embryo in advance of any receptor excitation (Figure 7) suggests that specifications for a body-centered neural field for perceptual guidance of whole-body action may be retained as a scaffold for higher mental proceses and more complex patterns of action. The latter are prepared in the fetal period by growth of additional brain structures in an unstimulating, highly-controlled intrauterine environment.

Much research on brain growth makes the assumption that the cerebral hemispheres and other structures that came to dominate over the primitive reflex system of the brain and spinal cord are formed by basically the same processes as those that regulate nerve circuit growth in fish, salamanders, and frogs. This is supported by the universal mapping of central nuclei, tracts, and cortical fields in linked somatotopic arrays, like those of the primitive visual system. Evolutionary changes in vertebrate brains also support the concept of a basic common design that is elaborated without loss of its original layout (Sarnat and Netsky, 1974). It is clear, however, that new morphogenetic mechanisms and a much greater variety of somatotopic representations specialized for different sensorimotor relationships have been evolved that are adapted to integrate brain growth to fit a richer course of experience.

Coghill (1929) proposed that the total pattern of behavior that integrates swimming, orienting, and feeding reactions of the salamander becomes differentiated into the separate mechanisms of more advanced behavior. This would leave even the highest and most elaborate functions with the imprint of their origin from the basic system. However, in the evolution of fetal forms that develop in an egg or uterus, certain individuated sensorimotor structures or reflexes concerned with specialized parts of the body may achieve a functional isolation while the rest of the system is dormant or not yet developed (Anokhin, 1964; Gottlieb, 1973; Hamburger, 1973; Oppenheim, 1973). Movements of fetal maintenance, hatching, or birth clearly are specialized, and they are given a developmental boost relative to other behavioral structures. Adaptations to conditions arising before birth, and individuated subfunctions that emerge after birth tend to break up the inherent suprareflex unity of the system. This does not deny the importance of Coghill's notion of intrinsic differentiation of total pattern nerve networks. Nor does it justify belief in an extreme pavlovian associationist view of adaptive development in higher intelligence.

5.2. *Growth of the Fetal Brain*

Human brain growth, reviewed by Larroche (1966), is briefly summarized here. The hemispheres appear as thin-walled sacs from the unpaired end chamber of the brain while flexion is completing at 30 days (Bartelemez and Dekaban, 1962; Hamilton *et al.*, 1962) (Figure 6). They lack nerve connections when the first retinotectal projections and other sensorimotor circuits of brain stem and cord are established (Windle, 1970). Neuroblasts of the forebrain begin to appear just in front of the optic stalk. These later grow axons to the hypothalamus and diencephalon. Olfactory lobe neuroblasts appear at this time. The first neuroblasts within the thickened mantle layer of the hemisphere wall correspond to the corpus striatum, which is a conspicuous bipartite swelling by 40 days. Neurons in it produce fibers that join the caudally directed medial longitudinal tract. The thin dorsal wall (pallium) remains undifferentiated until the end of the second month, when neuroblasts begin to accumulate in the primordial hippocampus and pyriform cortex (paleopallium). The latter receives axons of secondary olfactory neurons of the olfactory bulb.

Neocortical neuroblasts are produced early in the fetus, beginning in the lateral cerebral wall and extending rapidly into the roof and medial wall above the hippocampus. Frontal and occipital poles are the last to reach this stage, which is complete in humans by 65 days (50 mm). The lateral and dorsal part of the cerebral vesicle grows with the thalamus, receiving from it an influx of nonolfactory afferent axons. Thus, neocortex is committed from its earliest fetal origins to receive sensory information from the body surface and special distance receptors.

The presumptive parietal area, which is multimodal in the adult, is the first to show neuroblast formation, the most rapid to develop, and the first area to receive afferent axons from the thalamic neurons. Visual and auditory cortices receive afferent connections later. Phylogenetic comparisons indicate that a multimodal cortical afferent field is primitive in mammals; the unimodal, primary areas evolving out of it in relation to increasing demands for high resolution in perception (Diamond and Hall, 1969). In the cat the first stage of form vision appears to be worked up on the parietotemporal extrastriate cortex; areas 19, 20, and 21 that receive input *via* the midbrain and pulvinar (Sprague *et al.*, 1981).

Electrophysical unit studies and refined anatomical and metabolic mapping bring out many and varied somatotopic maps throughout the parietal area of primates, some with clearly different stimulus-analyzing properties (Allman *et al.* 1981; Zeki, 1974; Van Essen and Maunsell, 1983; Mishkin *et al.*, 1983). The nonspecific parietal cortex is almost certainly not a random blend of the receptor modalities, even in primitive mammals. Similar parallel organization in the visual projections of birds indicates a common anatomical scheme for all land vertebrates (Benowitz and Karten, 1976). In the adult monkey, its cells are active prior to the execution of an act of orientation of hand or eyes to an object in extracorporeal space. They are proposed to execute a command function directing praxic or exploratory acts and readying the processes of perceptual analysis (Mountcastle *et al.,* 1975; Mesulam, 1983).

The cerebral hemispheres and thalamus enlarge greatly in the first fetal month, and the hemispheres develop their characteristic ram's-horn shape with distinct frontal, temporal, and occipital lobes, the latter covering the midbrain by the end of this month (Larroche, 1966).

Nerve cell formation in the hemispheres occupies the third to fifth month in humans (Dobbing and Smart, 1974) and begins with movement of the cell bodies of neuroepithelial cells from the region of active nuclear division near the internal limiting membrane (ventricular zone) out toward the pial surface, where the neuroblasts separate off to form a dense cortical cell plate (Angevine, 1970; Berry and Rogers, 1966; Berry, 1974; Rakic, 1972, 1974; Sidman and Rakic, 1973). The mature neurons at any given level are relegated inward by the migration past them of later formed cells. The result is a series of strata, but distinct laminae only appear when the nerve cells grow axons and dendrites in the sixth to eighth fetal months. Thus, cells of laminae VI and V in the visual cortex, which will project to subhemispheric sites, are produced before those of layers II and I, which project intracortically or by commissures to the other hemisphere. The order of cell maturation follows this order of production, except that small Golgi type II cells produced later in ontogeny may migrate between the laminae. These cells are thought to be inhibitory in function and may contribute to postnatal memory function (Altman, 1967; Jacobson, 1974). Thalamic afferents terminate mainly in cells of layer IV.

Patterns of cell generation and migration in the growing brain have been revealed in fetal mammals by injection, into the gravid female, of tritiated amino acids that label cells at the time of cell division, giving a picture of cell birthdays (Angevine, of this phase (Sidman and Rakic, 1973). The second spurt, from the twentieth week to 3 months after birth, and continuing at a slower rate for about 2 years after that, is correlated with glial multiplication, dendrite growth, synapse multiplication, selective death and removal of axon collaterals (Innocenti, 1983), and myelination of axons (Yakovlev and Lecours, 1967). Gross genetic defects of cerebral formation are related to the first growth spurt, which is sensitive to radiation insult and poi-1970; Rakic, 1974). The concept of a highly specific spatiotemporal pattern of neuroblast dispersal in fetal brain growth (Coghill, 1924; Hamburger, 1948) has been strongly confirmed. Movement of cortical cell bodies is radial, producing columns at right angles to the surface, by migration of neuroblasts among glial fibers between the two limiting membranes of the cortex (Rakic, 1976, 1977; Sidman and Rakic, 1973). The columns remain an anatomical unit in subsequent organization of intercellular connections (Jones, 1975; Rakic, 1976, 1977; Goldman and Nauta, 1977), and they relate directly to functional components in the adult (Hubel and Wiesel, 1962; Mountcastle, 1957; Szentagothai and Arbib, 1975). Angevine (1970) suggested that the cytochemical labeling of cortical cells that determines the formation of highly selective functional cell groups in the fetus and in postnatal maturation may follow the clonal production of cells from neuroepithelial cells, with molecular tags being handed on and elaborated in the process of cell division and migration.

Axons of thalamic and brain stem origin arrive in the cortex early in the fetal period (Marin-Padilla, 1970; Morest, 1970), and these may determine the depth at which migrating cortical cells stop creeping outwards (Sidman and Rakic, 1973). Synapses form on the dendrites at the inner and outer boundaries of the cortical plate soon after the plate begins to form (Cragg, 1972; Molliver *et al.,* 1973). Early tangential axons probably contribute important information to direct the collection of migrating cells into layers in several regions of the CNS. This exemplifies how formation of complex central neuron mechanisms involves inductive interactions between surfaces of cells brought together by migration and axon growth. Work on the somesthetic cortex of the mouse suggests that while the laminar organization is locally generated, tangential somatotopic maps may be induced from the integument *via* the thalamocortical projections (Van der Loos and Dörfl, 1978).

Rates of increase of DNA in the whole human brain show two spurts of cell multiplication (Dobbing, 1981; Dobbing and Smart, 1974) (Figure 16). The first relates to neuroblast production in the 10th to 18th weeks after conception, differentiation of nondividing neurons occurring toward the end

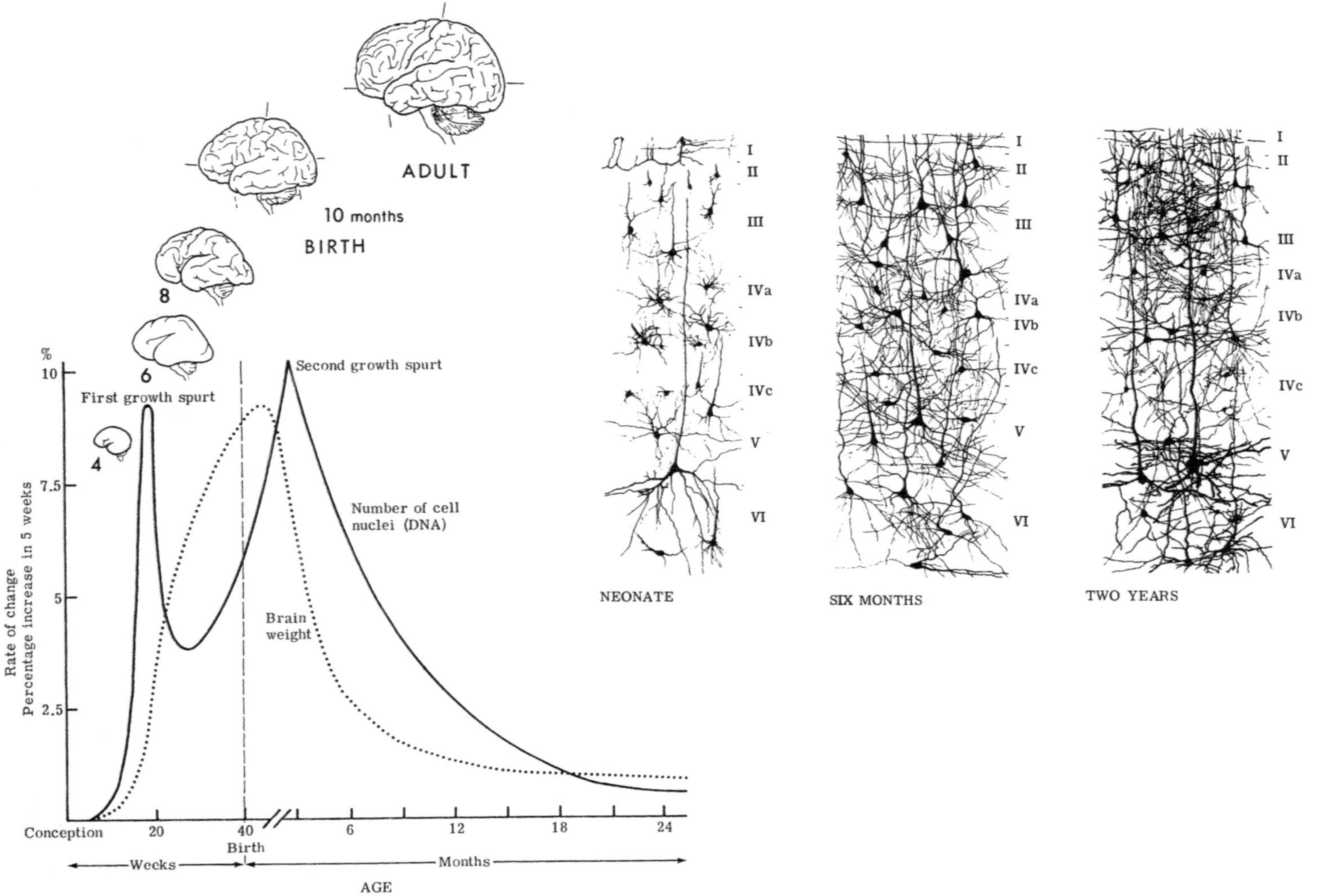

Fig. 16. (Left) Velocity curves showing incremental rates of DNA and fresh weight of whole human brain. The DNA curve has two peaks, one reflecting neuron multiplication in the mid-gestational period and one for glial multiplication associated with increase of brain weight, dendrite development, and synaptogenesis in the first postnatal months. (After Dobbing, 1971.) (Right) Dendrite growth in the visual cortex of an infant. (Conel, 1939—1963).

soning. Malnutrition of the mother during the period of glial cell multiplication, when the child is dependent on her body for nutrition, can cause permanent deficiencies in behavior (Dobbing, 1981). It is an important social fact that both prenatal and postnatal starvation can profoundly affect psychological functions.

The intricate structure of the adult cortex at the cell level, with many distinct forms of pyramidal and nonpyramidal cells in highly ordered arrays, is determined in essentials prenatally. Most of the classical description of brain histology by the Spanish anatomist Ramón y Cajal (1909–1911) was carried out by silver staining of fetal mouse brains and the immature brains of birds. He found the immature brain clearer in structure, and his small budget limited his work to small easily bred animals. To a remarkable degree, the main cell types and the tracts are evident in the human brain at birth (Conel, 1939–1963). However, postnatal developments, although adding little to brain bulk, greatly add to function of brain circuits for years after birth (Conel, 1939–1963; Shkol'nik-Yarros, 1971; Yakovlev and Lecours, 1967). Growth of dendrites and formation of synapses may go on until adulthood. The relative maturity of the brain at birth differs greatly in different species (Dobbing, 1981) (Figure 17). In humans, pyramidal cell dendrites studded with synapse-bearing spines grow throughout childhood years. With senility these cell extensions shrink as intellectual functions decline (Scheibel *et al.*, 1975).

The great commissural links between the hemispheres grow in relation to the development of the late-maturing cortical zones they connect. The first is the anterior commissure between pyriform cortices and olfactory bulbs on the two sides. Soon afterward, the relatively small hippocampal commissure joins the hippocampal areas. The enormous neopallial commissure (corpus callosum), now es-

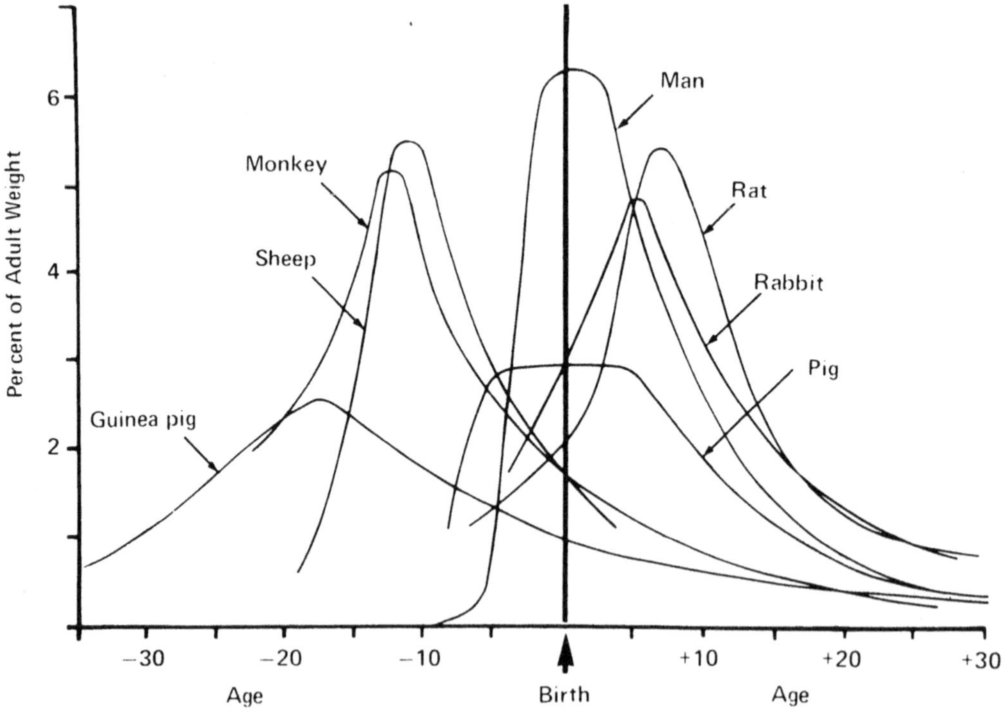

Fig. 17. The brain growth spurts of mammals expressed as first-order velocity curves of the increase in weight with age. Rates of weight gain are expressed as a percentage of adult weight for each time unit. The units of time for each species are as follows: guinea pig, days; rhesus monkey, 4 days; sheep, 5 days, pig, weeks; man, months; rabbit, 2 days; rat, days. (From Dobbing, 1981.)

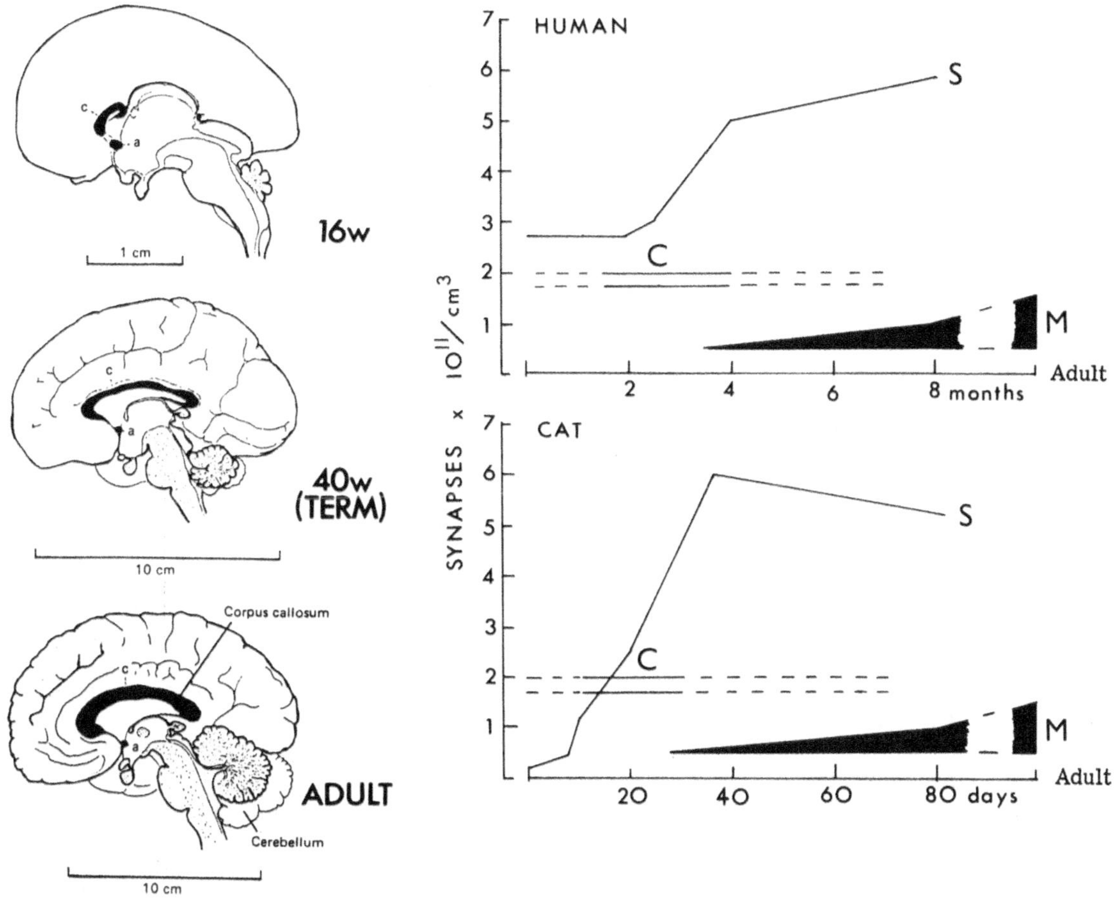

Fig. 18. (Left) Growth of the human corpus callosum. There is a marked increase in proportional bulk of the callosum postnatally, although this period is marked by the loss of many axons. Increase is associated with myelination. (Right) Time relationships between elimination of callosal axons (reshaping, C), development of synapses in the visual cortex (S) and myelination of the corpus callosum in cat and human (M). There is an intense period of loss of juvenile callosal connections in the cat, followed by longer-lasting reorganization of uncertain duration (interrupted lines). Time of callosal reshaping in humans is hypothetical. (After Innocenti, 1983.)

timated to have 800 million fibers in the adult (Koppel and Innocenti, 1983), first appears as a few axons passing over the hippocampal commissure at 10 weeks (60 mm) (Figure 18). It grows long after birth, doubling in cross-sectional area from 2 years after birth to adulthood (Hewitt, 1962; Rakic and Yakovlev, 1968). This increase of size is caused by enlargement of the axons and development of myelin sheaths (Luttenberg, 1966), and it conceals loss of a large number of axons accompanying cortical differentiation (Koppel and Innocenti, 1983). The posterior end (splenium), which communicates perceptual processes of parietal and occipital regions, shows relatively greater enlargement in infancy, but less development later. In some individuals part or all of the corpus callosum fails to develop (Loeser and Alvord, 1968; Probst, 1973). The cause of callosal agenesis is unknown. It is accompanied by abnormal development of a longitudinal tract in the medial wall of either or both hemispheres (Probst's bundle) and a radial pattern of sulci on the medial face of the hemispheres.

In the fifth lunar month of gestation (with reference to conception, as throughout this chapter) the occipital lobe enlarges and becomes pointed. At this stage, cell proliferation in the cortex becomes greatly reduced and differentiation begins. The primary sulci are formed, deepening to the seventh month. In the somatosensory cortex, trunk regions mature histologically before the limbs, and forelimb areas develop in advance of those for the hindlimbs. The characteristic splitting of layer IV in the striate cortex is evident from the eighth month. Presumably it reflects proliferation of dendrites from large stellate cells destined to receive synapses from the thalamic axons. The recently rediscovered anatomical asymmetries of the human cerebral cortex are detectable as early as the fifth month of gestation (Witelson and Pallie, 1973; Wada *et al.*, 1975; LeMay, 1976; Chi *et al.*, 1977).

The cerebellum undergoes a protracted and elaborate development wholly comparable with that of the cerebral hemispheres and certainly significant for perceptual development (Larroche, 1966; Sidman and Rakic, 1973). The phylogenetically old flocculonodular lobes linked to the vestibular system differentiate first. The cerebellar hemispheres, which mature later, receive input from both proprioceptive and exteroceptive modalities and the latter participate exproprioceptively in the direct regulation of body movement. There is an important visual input to the cerebellum and pons *via* the posterior parietal cerebral cortex (area 18) (Glickstein and Gibson, 1976; Glickstein, 1984; Magnin *et al.*, 1983). The complex layered histological structures of the neocerebellum, with highly specialized cell types, develop at different rates in different vertebrates, although the general order of events is the same in all (Jacobson, 1970). Precocial animals (e.g., chick, guinea pig, sheep) are born with the cerebellum at an advanced stage of maturation. It is relatively immature in helpless altricial newborns (e.g., rat, cat). In humans, Purkinje cells can be detected from the twenty-fourth week of gestation, reaching their final position by 36 weeks (Larroche, 1966). Structural differentiation begins during the last weeks of pregnancy, and large changes occur during the first 6 months after birth. Ramon y Cajal (1929) observed that at birth the Purkinje cells of an infant have 20–24 collaterals, but 2 months later these are reduced to one. Cerebellar development continues thereafter for several years. Early development of the rat cerebellum is highly sensitive to malnutrition or hormonal imbalance. This is used as a model system in studies of biochemical regulation of intrauterine cerebral growth (Balazs *et al.*, 1975).

5.3. Fetal Behavior and Perception

Spontaneous and reflex movements of fetal birds and mammals have been charted to trace the development of structures underlying movement and perception (Hamburger, 1973; Oppenheim, 1974). Human fetuses aborted at different stages have been subject to behavioral tests (Hooker, 1952; Humphrey, 1964, 1969), but spontaneous movements have been neglected while attention has been directed to charting reflex responses to artificial discrete stimuli. Observations of living fetuses *in utero* by X-radiography, ultrasonics, and physiological studies of the amniotic fluid (Liley, 1972) give evidence for control of adaptive movements (heartbeat, swallowing amniotic fluid, shifts of posture) before birth (Prechtl, 1984). Other movements (e.g., limb extensions, thumb-sucking, respiratory movements) may contribute to modeling of structures needed after birth. Intrauterine muscle activity plays a part in selective elimination of motoneurone in the chick (Oppenheim and Nunez, 1982).

On the whole, the most striking feature of the fetal condition is the limited behavior despite the exceedingly complex brain structure already outlined. Neurological examination of fetuses born prematurely suggests that they are capable only of weak and stereotyped movements until the last trimester, when marked changes occur (Saint-Anne Dargassies, 1966). Aborted fetuses are in inauspicious conditions, and their normal level of activity has probably been underestimated. There is physiological evidence that the levels of activity of the developing central nervous circuits is regulated by inhibition that, with a delay of a few weeks after the first reflex circuits are completed, may greatly qualify the patterns of action generated in fetal motor organizer systems and their sensory control (Humphrey, 1969; Oppenheim and Reitzel, 1975). Postnatally, a degree of relative inaction and comparatively simple attentional reaction to stimuli appears to pass away rapidly with the termination of the neonatal period at 4–6 weeks. This development coincides with a doubling of synaptic density in the visual cortex and massive changes in neocortical cell populations (Innocenti, 1983).

Developments either side of term indicate that this period is a clearly marked and specialized stage of brain growth. In ungulates, the time of birth is under control of the fetal brain and humoral system. It is therefore probable that, under optimal conditions, human birth is regulated by the fetal brain to occur when neural circuits have achieved a specific readiness for access to the external world. Postnatal developments in cats and monkeys, reviewed below, show that the developmental state of the brain at or soon after birth determines the influence of stimulation on the maturation and refinement of perceptual circuits.

The first spontaneous movements of humans occur at 5½–6 weeks gestational age (Prechtl, 1984). The precocious appearance of touch reflexes and primary vestibular circuits (Gottlieb, 1971*b*), as well as postural reflexes and compensatory eye movements, suggest that among the first organized efferent-afferent feedback loops are those of coordination of body parts into a mechanical system capable of controlled displacement of parts. Older fetuses adjust posture and react to positional changes caused by the mother moving her body. Finely individuated local face and hand movements appear as early as 12–16 weeks (Humphrey, 1964, 1969; Prechtl, 1984). They express neuromotor organization in brain stem and cord adapted to manual and oral prehension of objects and they may also be parts of innate motor patterns for communication by grimace or gesture with which humans are richly endowed. At 20 weeks, the human fetus actively regulates its nutrition and protein metabolism by swallowing amniotic fluid (Liley, 1972). There is evidence of taste preference for sugar and rejection of bitter substances.

Fetal behavior could influence the development of perceptual nerve circuits through feedback stimulation of receptors. Self-patterning of excitation is part of the environment to which growing nerve circuits are adapted to respond. However, all the experimental work done so far on effects of fetal stimulation, self-produced or from outside sources, seems to point the other way. Fetal mobility is shown to be patterned indpendently of reafferent effects that have a significant influence only postnatally (Hamburger, 1963; Oppenheim, 1974; 1982). Oppenheim raised a question regarding the accessibility of nerve systems, when they become responsive, to cumulative influence from stimulation. An early form of undiscriminating habituation to repeated stimuli appears to protect the fetus from learning sensorimotor links. Then fragmentary responses to repeated stimulation become more consistent.

Virtually nothing is known about the role of prenatal self-stimulation in man, the most remarkable studies having been done with birds. Gottlieb (1971*a*) showed that prehatching calls of ducklings may be one factor in the normal development of their ability to hear the mother on hatching. Precocial birds have to be born with a near-ready ability to follow the maternal assembly call that is species specific. Vestibular, tactile, and mechanoreceptor systems may all be stimulated before birth or hatching as a result of movements of the fetus or of the mother. This could aid normal differentiation of their structure. There is, however, no evidence that it does so.

Intrauterine recordings of light and sound in the womb show that human fetuses could perceive low-frequency vibrations and sharp knocking sounds from the air round the mother, and large changes in level of light falling on the mother's belly. The loudest sounds are those originating in the mother's stomach, heartbeat, respiration, and vocal organs. Behavioral reactions of fetuses to rhythmic or sudden external sounds are frequently reported. It seems likely that fetuses can perceive and learn to recognize rhythms and intonation contours of the mother's speech for which they show a preference almost immediately after birth (Busnel and Grenier-Deferre, 1983; Mehler and Bertoncini, 1979; De Casper and Fifer, 1980).

The earliest spontaneous movements of embryos show activity cycles and rhythmical or pulsating bursts of movement (Hamburger, 1969; Oppenheim, 1974; 1982). It is likely that pacemaker systems are an autogenous component of growing central networks or that they contribute in the early development of perception rather than the other way about (Hamburger, 1963). Experimental studies show that sensory feedback does not determine pacemaker origin (Carmichael, 1926). The importance of spontaneous rhythms in both normal and pathological behavior of infants is stressed by Wolff (1966, 1968). It is also a fundamental in Lashley's analysis of serial ordering processes in action and perception (Lashley, 1951) and in Bernstein's theory of human motor coordination (Bernstein, 1967; Whiting, 1984).

It is important, when considering fetal perception, to note that the special sense organs, having attained their basic form in the late embryo, are cut off from stimulation by morphological changes in the early fetal period of neuron production (Arey, 1965; Hamilton *et al.*, 1962). The eyelids grow over the cornea to fuse at 7½ weeks. They reopen at 6 months. The ear ossicles develop within a spongy mesoderm which remains to block transmission until the last fetal months, when a cavity forms around the ossicles. The tympanic cavity remains obliterated by endodermal thickening and swelling and is excavated shortly after birth in association with changes that accompany the onset of pulmonary respiration. Available evidence suggests that auditory discrimination becomes possible only after 6 months (Busnel and Granier-Deferre, 1983). From the second to sixth gestational months, the nostrils are closed by epithelial plugs. Fine individuated finger movemented occur from the end of the fourth month, at which time the sensory structures of the skin, including hair follicles, are clearly differentiated.

The development of premature infants in the last trimester of gestation, when dendrite expansion and glial multiplication are beginning in the hemispheres, indicates formation of circuits throughout the CNS that follows a tightly controlled schedule in considerable independence of body weight and in resistance to the radically changed environmental conditions of hospital care (Larroche, 1967; Saint-Anne Dargassies, 1966; Friedman and Sigman, 1981). At 25 weeks, the neurological and metabolic machinery of an infant is sufficiently advanced for it to be viable in an incubator. The change to this level of competence is a sudden one, the 6-month-old fetus having achieved a characteristic state of relative functional readiness (Saint-Anne Dargassies, 1966). At 7½ months, the fetus is more alert and better coordinated, waking easily to stimulation, and showing the first spontaneous eye movements with lids wide open. The doll's-eye reflex (rotating of the eyes to oppose lateral head rotation), quick eye closure to a dazzling light, a sluggish pupillary light reflex, all attest to rapid emergence of the fundamental motor controls for seeing. This is the age at which the infant may raise his hands to his face, place them in the mouth, and suck them actively and rhythmically. It is the period at which exploratory, unstimulated grasping first occurs. During the eighth and ninth lunar months, the infant develops muscle tone from lower to upper limbs and, by the end of the ninth lunar month, legs and arms lie in flexion at rest and the neck may briefly support the head, but mobility is less than at early stages of prematurity.

Electrical signs indicate a progressive activation of circuits from brain stem to cortex. The brain stem reticular formation, which will regulate sleep, spontaneous orienting, and consciousness after birth, exhibits patterned electrical activity first in the pons, at 10 weeks (Bergström, 1969). The activity then advances through the midbrain reticular formation and basal ganglia, spontaneous rhythmic brain waves appearing in scalp electrodes at 24 weeks (Dreyfus-Brisac, 1967). At the time the fetus becomes viable, the EEG becomes more regular with bursts of theta waves (4–6 cps) synchronized within each hemisphere. Within a week or two, this isosynchronism diminishes. At 30 weeks, higher-frequency bursts become more common (10–14 cps), this being the period of most rapid development of the EEG (between 24 and 36 weeks). There are still no electroencephalic reactions to stimuli and no sleep–wake differences. Activities at this time are probably of subcortical origin.

A distinction between sleep and waking tracings appears at 37 weeks. The waking EEG is similar to that of the full-term infant but the sleep record differs, characteristics of the much younger (30-week) EEG reappearing in light sleep. The development of a distinct wakeful state may reflect maturation of the cortex or reticular formation, or both of these. The stages of EEG development form reliable indicators of conceptional age. Compared with adults, newborns have an immature sleep cycle in which all phases are less clearly marked. The process of maturation continues for years after birth. An awake newborn shows clear reactions of alerting to sounds, touch, changes in light, vibration, and may orient to follow changes in position of sounds or visual stimuli (Peiper, 1963; Beintema, 1968). Exploratory behavior is rudimentary during the first weeks, although dishabituation tests demonstrate that perceptual impressions of considerable refinement may be formed (see below).

Electrographic responses evoked by flashes of light and sounds are readily obtained in neonates. They are large in amplitude, simplified in form, and long in latency. Thus, while demonstrating considerable complexity in cortical nerve networks, they also show that those networks are far from mature. Only slow photic driving is possible (<3/sec). Response to stimulation includes a flattening of the bioelectrical activity. Habituation has been observed before 37 weeks in respiration, body movements, and myographic activity.

5.4. *Plasticity in the Fetal Brain*

The Syrian hamster is born at a stage corresponding to about the first or second fetal month of man, when major cell migrations of the developing cortex are taking place. The circuits of midbrain and thalamus are still forming. Schneider (1969) found that adult hamsters show reciprocal defects with ablations of superior colliculus and visual cortex. Animals lacking cortex retain visual orienting, but

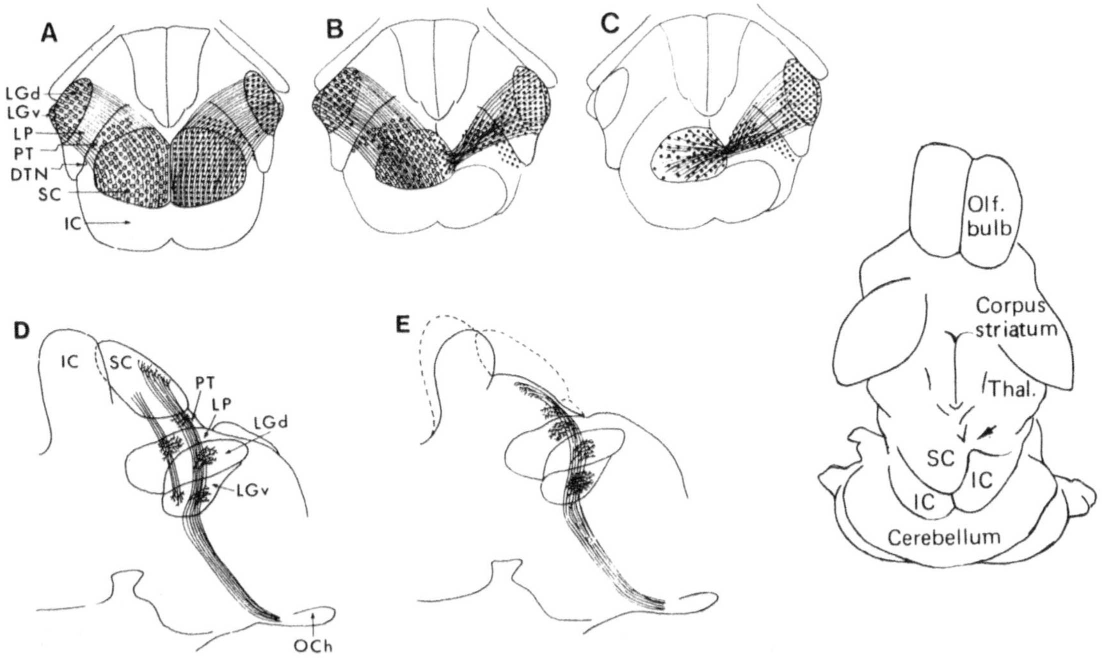

Fig. 19. (Right) Brain stem of Syrian hamster with an anomalous bundle of retinal axons crossing the midline to the anteromedial part of the left tectum (SC) after removal of surface layers of the right tectum at birth. Top left: Diagrams of visual connections show terminations of left (open circles) and right optic tracts (closed circles): (A) normal hamster; (B) superficial layers of right colliculus at day of birth, with anomalous connections formed on the left medial tectum and to other nuclei along the tracts; (C) similar to previous, but with removal of the right eye at birth, denervating the left tectum and permitting extensive invasion by anomalous crossover input from left eye. (Bottom left) Reconstructed lateral views of the hamster brain. D: Normal projections: (E) anomalous connections after removal of the surface layers of the superior colliculus in the neonate. LGd and LGv, dorsal and ventral nuclei of the lateral geniculate body; LP, nucleus lateralis posterior; PT, pretectal area; DTN, dorsal terminal nucleus of the accessory optic system; SC, superior colliculus; IC, inferior colliculus; OCh, optic chiasma. (From Schneider, 1973, 1976.)

lose pattern vision; those lacking colliculus are deficient in orienting, but disciminate patterns. If the same operations are performed on newborn hamsters, there is a high degree of functional recovery (Schneider, 1973; Schneider and Jhaveri, 1974). Anatomical investigations reveal a number of compensating changes in nerve connections in the growing brain (Figures 19 and 20). These rewirings indicate that the retina, colliculus, and geniculocortical cells share a common somatotopic cytochemical code which confers a limited power of substitution in circuits between them (Schneider, 1976). Response to the code depends on competitive interaction between terminals, and patterning is greatly disturbed if sites for termination or contingents of terminals are removed surgically. Visual terminals deprived of their normal target region by ablation of the superior colliculus may even invade deafferentiated areas of the inferior colliculus that normally receive only auditory afference (Kalil and Schneider, 1975).

In the young colliculus, the cells of deeper laminae are acceptable to retinal axons for formation of ordered connections when the superficial laminae have been removed (Figure 19). Moreover, a reduced colliculus, made small by ablation of the superficial layers near birth, may receive a compressed input of retinal axons (Schneider and Jhaveri, 1974; Finlay *et al.,* 1979) like that produced in the goldfish by the same type of surgery (Meyer and Sperry, 1976). Misdirected axons, terminating on inappropriate locations, were also observed. Finally, retinal axons may cross the midline to terminate in the inappropriate colliculus in an abnormal mirror array; this is facilitated if the invaded colliculus has been denervated by removal of the other eye (Finlay *et al.,* 1979). Behavioral tests (orienting to food) confirm the anatomical

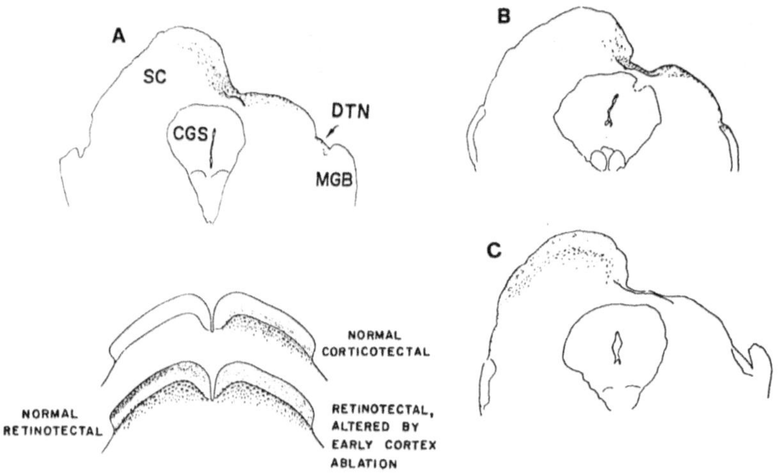

Fig. 20. Syrian golden hamster. Competitive interactions of retinal and corticofugal terminals in the tectum. Silver-stained coronal sections. Degenerated axons (dashed lines) and terminals (dots). (A) Surface layers of right tectum and right eye removed at birth. Extensive recrossing of fibers from the left eye to the left tectum (cf. Figure 18C). (B) Right tectal lesion at birth, left eye removed in adult 5 days before sacrifice. Projection to medial zone of left tectum only (cf. Figure 18B). (C) Degeneration from right eye removed in adult. Right tectal lesion at birth. Complementary to B. (From Schneider, 1973; Schneider and Jhaveri, 1974.)

findings in this study, which offers important evidence concerning the mechanisms for substitution, competition, and reformation of circuits after extensive damage to the still-developing brain. Functions intrinsic to the axon-producing cell and its terminal arborizations ("pruning" factors) are contrasted by Schneider (1973, 1976, 1981) with factors that operate in populations of axons seeking termination together in a potential end site (axon-ordering factors, or axon–axon interactions).

The experiments show that plasticity of the midbrain visual system extends into postembryonic stages of development in mammals. This may reflect evolution of a developmental strategy to make way for incorporation of the later-formed projections from forebrain visual fields. Developmental plasticity of cerebral circuits on a small scale extends far into postnatal stages even in humans, but only the early fetal stages may exhibit such drastic regenerated changes as seen in the hamster (Guth, 1975). Schneider (1981) proposes that operation of competitive axon ordering factors in brains developing after lesions or infection may account for some human behavior disorders and mental pathologies. Plastic responses in postnatal development of the human cerebral cortex are discussed below.

Development of monocular dominance zones in the visual system of the monkey is completed in the superior colliculus and lateral geniculate in the last half of gestation. The sorting of afferents from the two eyes into laminae of the LGN appears to involve competitive elimination of a temporary excess of axons (Rakic, 1979; Rakic and Riley, 1983). At first, retinal terminals from each eye spread through all laminae. Distinct laminae with terminals from the eyes separated develop between 90 and 140 days (gestation totals 165 days). This separation is prevented if one eye is removed on day 64. Axons from the LGN reach the cortex at about 80 days and the striate cortex develops uniocular territories shortly after birth (Rakic, 1977, 1979). Throughout the visual projection, segregation of terminals from the two eyes from originally overlapping populations is sensitive to removal of one eye or, in the case of the visual cortex, reduced stimulation of one eye. There is a competitive interaction between terminals of the two eyes, and this is regulated by the relative activation of the visual cells of the two retinae (Hubel *et al.*, 1977; Blakemore and Vital-Durand, 1981; Swindale, 1982).

6. Late Fetus, Birth, and Infancy

6.1. General Features of Postnatal Brain Growth

Birth occurs in most mammals around the time that cortical neurons are rapidly developing and synapses are forming over the dendrites (Figures 16 and 17). Cortical mechanisms for visual perception are more completely formed in preyed-upon species born aboveground. Primates show a short period of postnatal maturation, but nest-born altricial young have poorly developed cortices at birth. The visual cortex of monkeys and their visuomotor coordinations, appear to develop about four times as fast as in the human infant (Teller, 1981). Cortical development is exceptionally complex and prolonged in humans. Recent psychological studies indicate that the first half of infancy is the most significant in the formation of working perception systems, but cognitive developments transform perceptual skills long after that.

The surface topography of the cortex reflects expansion of dendrites and maturation of white matter. Most of the secondary folds of the human cortex are present at birth (Figure 16). The tertiary folds begin to be prominent in the latter part of the second year, but they go on developing, possibly into adulthood (Larroche, 1966; Yakovlev, 1962). In the primary projection areas, it is possible to trace a close inborn relationship between cortical gyri and the prominent afferent zones of the periphery (Adrian, 1946; Whitteridge, 1973). Indeed, expansion of somatotopic zones for particular receptors may be induced by the integument and afferent tracts that mature before the cortex (Van der Loos, 1979). In humans, a difference in the folding of the cortex in the left and right hemispheres correlates with the language comprehension; an asymmetry in the left temporal planum is detectable in the middle fetal stage (Galaburda *et al.*, 1978).

There is conspicuous proliferation of dendrites in early infancy (Conel, 1939–1963; Shkol'nik-Yarros, 1971) (Figure 16), and branches of nonpyramidal "stellate" cells that receive thalamic afferents (Jones, 1975) appear more spinous in infants than later in life (Marin-Padilla, 1970; LeVay, 1973). Many synaptic contacts are formed just prior to birth, and a large number are added in the early postnatal period. Nissl granules and neurofibrils develop postnatally along with the maturation of cortical dendrites, and glia multiply to fill expanding spaces between nerve cells (Dobbing and Smart, 1974; Dobbing, 1981) (Figure 16). The cortex is less mature at birth than the nuclei of the central gray and brain stem. The association cortices grow more slowly than the primary projection zones and may retain an embryonal plasticity throughout life (Yakovlev, 1962). Although both dendrites and synaptic arrays on them are immature at birth, anatomical studies with radioactive tracers show that major fiber systems of the newborn monkey cortex are organized at birth and that transcrotical and commissural projections are distributed in a uniform system of columnar modules (Goldman and Nauta, 1977). Extensive refinements in commissural and associational axonal projections in the hemispheres and in pyramidal tracts result from selective elimination of collaterals with little reduction in numbers of cell bodies (Innocenti, 1981; Stanfield, 1984) (Figure 18). These developments accompany an increase in the number of synapse and are shortly followed by myelination of the retained axons. They are responsible for patterning of the cortical mechanisms for separate sensory modalities, and for orderly associative relationships between different sensory analyzers and modalities.

A remnant germinal layer of cells is found near the head and body of the caudate in the cerebral hemispheres of mammals. This subependymal of subventricular zone continues to proliferate in the adult rodent, generating cells that differentiate into small Golgi type II neurons after migrating considerable distances. It has been proposed that most of these cells are inhibitory and that they assist in maturation of cortical response patterns, by removing unstimulated components from function, and they may contribute to learning (Altman, 1967).

6.2. Synaptogenesis and Differentiation of Cortical Networks

In all mammals studied there is a well-defined period of cortical cell branching followed by a rapid accumulation of synapses. In sheep, a precocial form, synapse proliferation in the cortex takes place about the middle of gestation (60 days) (Astrom, 1967). In cats, which are born at a much earlier state of development, it is postnatal, during a peroid after eye opening when visual performance is showing marked improvement. In monkeys, this phase is completed in 6 weeks after birth, while in humans it probably occupies about 6 months.

Cragg (1975) traced the development of synapses through the visual system of the cat for visual cortex corresponding to 60° in the periphery of the visual field (Figure 21). Optic axons reach the lateral geniculate nucleus at least 20 days before birth, but synapses first appear within the near central retina only nine days before birth, and ribbon-containing (inhibitory?) bipolar synapses do not appear until birth. Few retinal receptors possess outer segments with disks at birth, but five days later disk-bearing outer segments have developed. The lateral geniculate body shows a rapid growth of synapses from 5 days after birth to a maximum of about 4×10^{11} synapses/cm^3 at 25 days. This process precedes growth in size of geniculate neuron cell bodies by about 10 days.

Parallel with the development of retinogenicu-

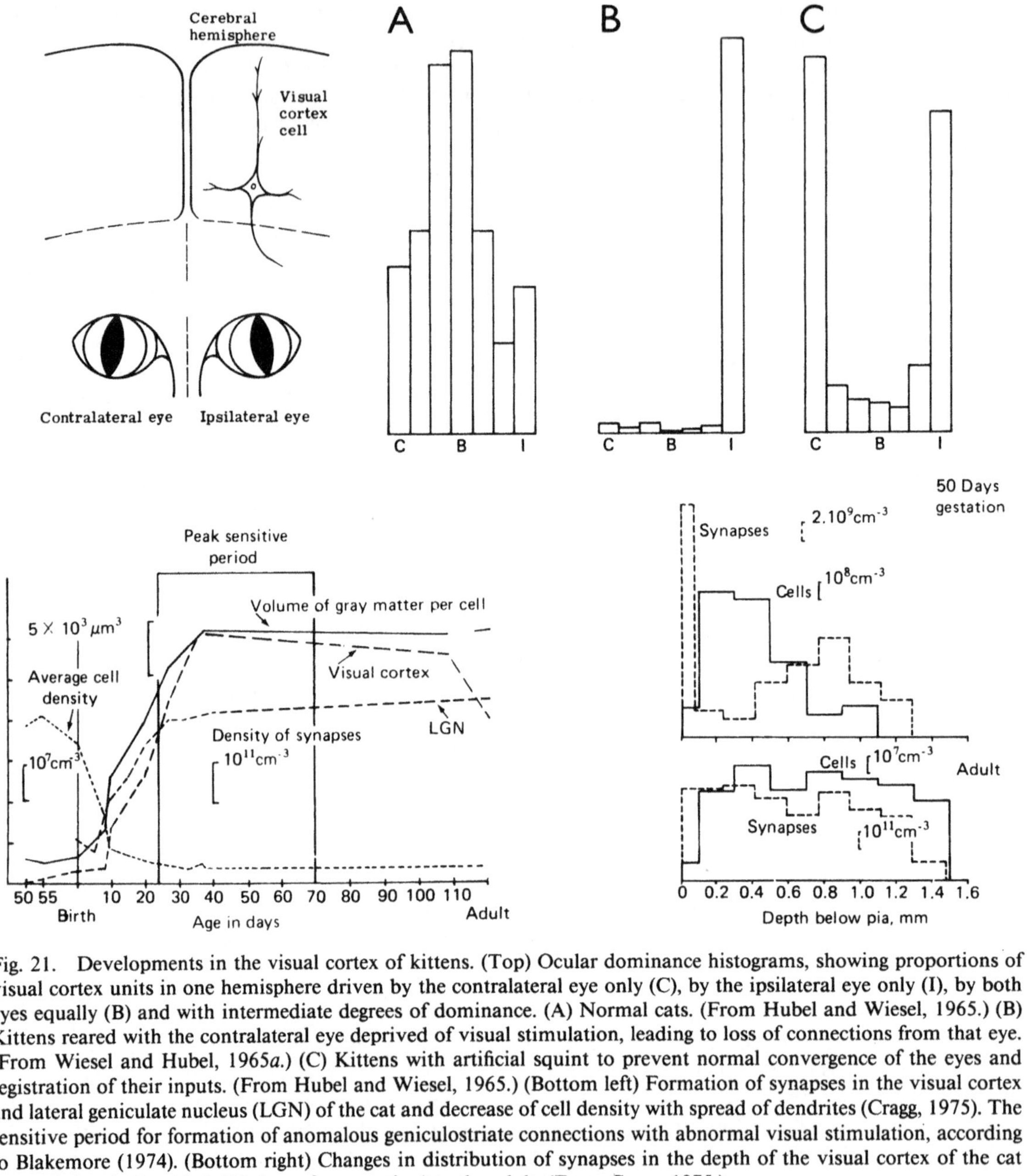

Fig. 21. Developments in the visual cortex of kittens. (Top) Ocular dominance histograms, showing proportions of visual cortex units in one hemisphere driven by the contralateral eye only (C), by the ipsilateral eye only (I), by both eyes equally (B) and with intermediate degrees of dominance. (A) Normal cats. (From Hubel and Wiesel, 1965.) (B) Kittens reared with the contralateral eye deprived of visual stimulation, leading to loss of connections from that eye. (From Wiesel and Hubel, 1965*a*.) (C) Kittens with artificial squint to prevent normal convergence of the eyes and registration of their inputs. (From Hubel and Wiesel, 1965.) (Bottom left) Formation of synapses in the visual cortex and lateral geniculate nucleus (LGN) of the cat and decrease of cell density with spread of dendrites (Cragg, 1975). The sensitive period for formation of anomalous geniculostriate connections with abnormal visual stimulation, according to Blakemore (1974). (Bottom right) Changes in distribution of synapses in the depth of the visual cortex of the cat from 15 days before birth (50 days after gestation) to the adult. (From Cragg, 1975.)

late synapses, but delayed relative to them by about two days, geniculostriate connections grow from day 7 to a maximum of $6.5 \times 10^{11}/cm^3$ about day 35. This fits the time of appearance of dendritic spines on stellate neurons of the striate cortex (Adinolfi, 1971). The first-formed synapses, probably not all from visual axons, appear just above and below the optical plate of cortical cells, 2 weeks before birth.

The volume of gray matter per cell of visual cortex, a measure of dendrite and axon growth between the cell bodies, increases about a week ahead of the synapse population (Anker and Cragg, 1975; Cragg, 1975). Myelin starts to appear three weeks after birth (Marty and Scherrer, 1964). The average number of synapses associated with one cortical neuron rises to a peak of about 13,000 at 7 weeks after birth. In the adult cat the synapses and neurons are slightly less dense in the cortex, apparently because glial cells develop. There is no evidence of loss of synapses.

Synaptic development in the cat visual pathway starts before patterned afferent input begins; the animal is not yet born and receptor cells are immature. Cragg suggests that it may be important in all sensory pathways that the first patterns of synaptic connectivity be established in absence of stimulation. But, the main increase of synapses in the lateral geniculate nucleus and cortex takes place when the visual system is being used, in the fourth and fifth weeks (Figure 21).

The cortex of the cat is responsive to electrical shocks applied to the optic nerve at birth before vision starts (Marty and Scherrer, 1964) and single-cell responses to flash stimulation are reported at 4–6 days after birth (Huttenlocker, 1967). Cortical units recorded by Hubel and Wiesel (1963) at a week after birth although sluggish, were selectively responsive to direction of target displacement. Eyelids open from 1 to 2 weeks after birth, but the optic media only become fully clear by the fourth week (Barlow and Pettigrew, 1971). Evidently selective responses of cortical cells for orientation of a visual target and binocular disparity develop during the fourth and fifth weeks after birth (Barlow and Pettigrew, 1971; Pettigrew, 1972) (Figure 21). Effect of manipulation of visual experience on responses of cortical cells are described below. Stimulus-dependent changes in cortical function occur after synapses are completely formed. They may involve selective cytochemical repression, but not necessarily anatomical loss of synapses (Mark, 1974). There is evidence that GABA-active inhibition by cells entering the cortex and developing connections postnatally may be responsible for suppression of inputs from one eye to binocular neurons after that eye has been deprived of stimulation from birth (Duffy *et al.*, 1976).

In the monkey retinogeniculate and retinocollicular pathways are fully differentiated before birth, but the striate area circuits are still developing in the first 1 or 2 months after birth (Rakic, 1976, 1977, 1979; Wiesel and Hubel, 1974). Monocular deprivation experiments confirm this (Hubel *et al.*, 1977). The fovea of the monkey appears to be clearly differentiated at birth (Ordy *et al.*, 1962), but it is not certain that foveal cones or their retinal connections are in a mature functional state.

In humans the foveal cones are generally described as immature at birth with end segments conspicuously elongated during the first months after birth (Mann, 1964). This conclusion, which may come from data on pathological eyes, was recently questioned both on the basis of evidence of vision in the central part of the field of newborns and because of doubts cast on the reliability of the anatomical data (Haith, 1978). Nevertheless, normal neonatal spatial resolution is less than one-twentieth that of an adult (Braddick and Atkinson, 1979; Gwiazda *et al.*, 1980). This development is probably due to anatomical differentiation of retina as well as central visual circuits postnatally, and the fovea almost certainly does change. Striate cells undergo considerable elaboration of dendrites in early infancy with the formation of dendritic spines (Conel, 1939–1963; Shkol'nik-Yarros, 1971); subsequently there is loss of spines and other surface irregularities on dendrites. These changes accompany a rapid development of perception; they correlate with improvements in acuity and with adaptive changes in amblyopia or astigmatism (Barlow, 1975; Braddick and Atkinson, 1979; Gwiazda *et al.*, 1980; Hickey, 1981).

Innocenti (1983) suggests that in the cerebral cortex of mammals, including humans, a large transitory excess of axon collaterals from cortical neurons is eliminated by phagocytosis at the same time as the number of synapses is rapidly increasing on cortical dendrites. This leads to selective patterning of interconnections between groups of cortical cells in tangential bands or patches of cortex with either other cortical areas (in the same hemisphere or the other hemisphere via the corpus callosum) or with subcortical nuclei (Figure 18). Exuberant overproduction of axons spreading widely then selective removal of certain cohorts at the time synapses are formed would appear to be a mechanism whereby cortical systems accommodate to developments in other parts of the brain and to each other. Cells that lose callosal collaterals retain associational connections within the same hemisphere, which may explain how Probst's bundle is formed in the fetus of human congenital accallosals (Probst, 1973). In the newborn kitten there are projections from primary and secondary auditory cortex to visual areas 17 and 18 that are eliminated by postnatal day 38 (Innocenti and Clarke, 1984). Formation of modality specific systems and regulation of intermodal associative processes would seem to depend on this large-scale selection among exploratory axon collaterals.

Siamese cats have a genetic deformity that leads to more retinal axons crossing the optic chiasm than normal, which is associated with abnormal retention of interhemispheric connections between visual areas (Shatz, 1977). This suggests that exuberant collaterals may be a mechanism for plastic adaptation of cortical circuits to genetic variation of the periphery (see Innocenti, 1983; Stanfield, 1984). Schneider's observations of anatomical responses to surgery in fetal rodent brains led to a similar concept of plasticity in cerebral axonal projections that may have significance for human pathology and recovery from brain trauma (Schneider, 1981; Schneider *et al.*, 1984).

Modification of the cortical mapping of sensory processing in the hemispheres of humans born deaf (Neville, 1980; Neville and Bellugi, 1978; Bellugi *et al.*, 1983) is an example of postnatal plasticity that may involve unusual retention of axon collaterals in childhood when hemispheric asymmetry of function and the callosal pathways that must mediate it are undergoing development (Trevarthen, 1984*b*).

6.3. Myelogenesis and Maturation of Cerebral Integrative Systems

Myelin may be differentially stained and its progressive accumulation followed in nerve tracts. It used to be thought that myelin was an indicator of the start of nerve conduction. This was disproved by Langworthy (1928), who studied the movements of embryo marsupials when they crawl up the belly of the mother to enter the pouch under their own steam. At this stage the marsupial brain has no myelin.

Nevertheless, myelin deposition changes the precision and strength of transmitted patterns of excitation. Its appearance on a given axon is probably closely correlated with maturation of synaptic relations and of telodendria and dendrites of that cell. In the development of cortical areas, the order of cytoarchitectonic maturation correlates well with myelogenesis in the axons on the same cells (Yakovlev, 1962). Its differential appearance in growth of the brain is undoubtedly a valuable indicator of developmental change after formation of axons and outgrowth of the main dendrite branches (Larroche, 1966; Yakovlev, 1962; Yakovlev and Lecours, 1967; Lecours, 1975, 1982). Innocenti (1983) relates myelination of callosal axons to synaptogenesis and to shaping of interhemispheric and transcortical connections by elimination of axon collaterals in rats, cats, and humans.

Myelination starts in the second half of uterine life in humans, but in certain parts of the brain it has a life-long course (Figure 22). The large afferent bundles in the brain stem myelinate well before the association fibers and reticular formation, and the hemispheres are almost devoid of myelin before birth. All the pathways of sensation and equilibration are myelinated in the mid-fetal stages, generally from below upward. In the cord, the motor roots are myelinated before the afferent roots. The vestibular root is in advance of other cranial nerves, and the white matter of the cerebellar lamellae in the flocculus is very precocious. Thus, the cycle of myelination of the vestibular–cerebellar system, which begins during the sixth fetal month, is brief and entirely prenatal.

Marty and Scherrer (1964) have followed myelinization of the optic nerves of the cat. In humans, myelination spreads from the chiasm toward the optic nerve and optic tract in the thirteenth week of gestation. The thalamic optic radiations begin their cycle of myelogenesis just before term and are completed in the fifth postnatal month. The cycle is thus brief and occurs postnatally, coinciding with foveal maturation and other important postnatal changes in vision summarized in Section 6.5.

The nonspecific thalamic radiations, cortical association fibers, corpus callosum, and reticular formation show the most protracted cycle of myelinogenesis, suggesting that the final process of maturation of the brain circuits, continuing long after birth, is concerned with refinement of the integrative and generative processes that derive in evolution and in embryogenesis from the core mechanisms of the brain. Although the special cortical sensory and motor organs have evolved into distinct cortical structures in the higher terrestrial vertebrates, and although in fetal stages they initially lag behind the parietal association cortex in maturity, late fetal and postnatal developments reverse this relationship. More than 80 years ago, Flechsig (1901) demonstrated that association cortices are later myelinating in infancy than the primary and secondary projection zones (Figure 23).

The nonspecific thalamic radiations begin mye-

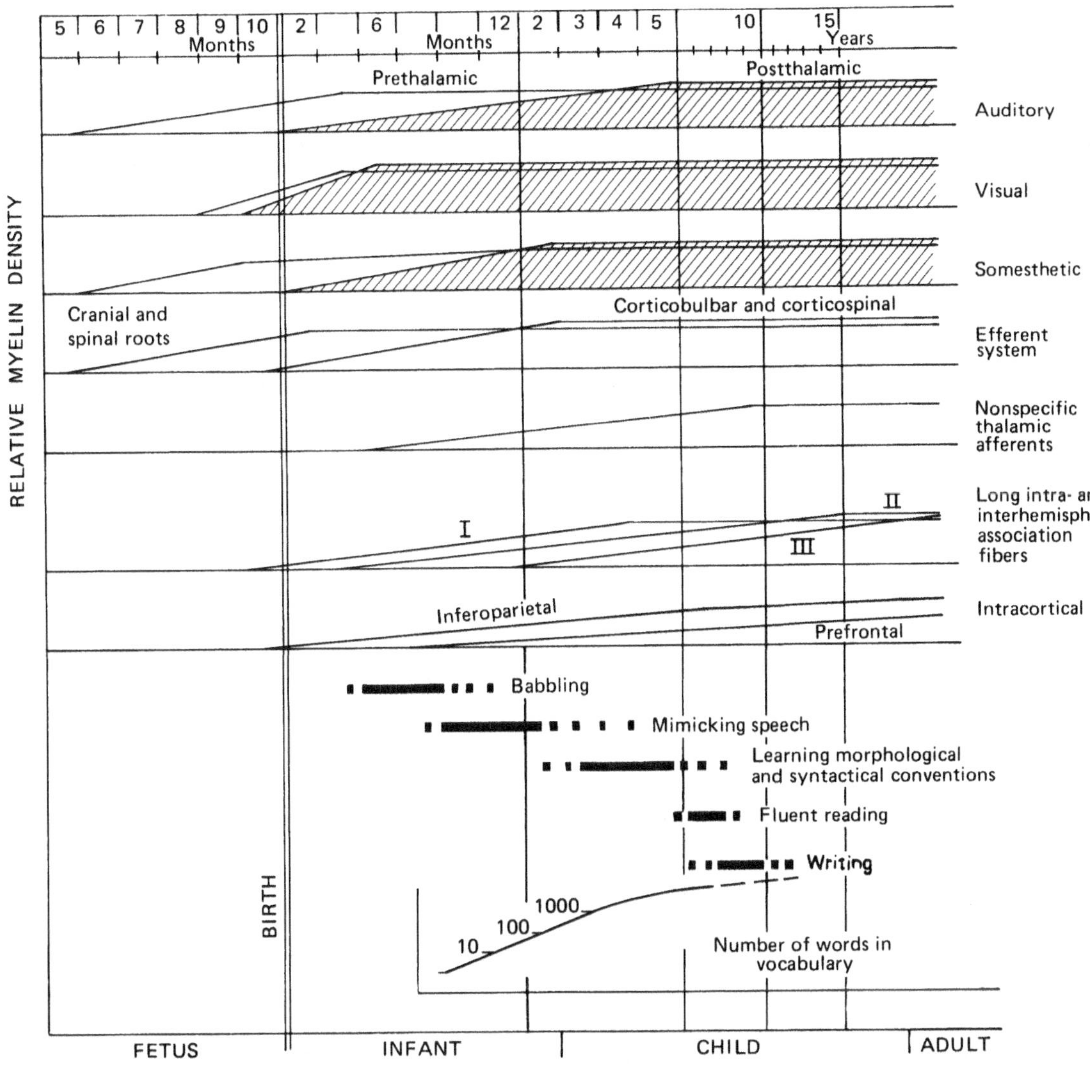

Fig. 22. Human brain development. Myelinization of cerebral systems, and correlations with the development of language functions. (After Lecours, 1975, 1982; Yakovlev and Lecours, 1967.)

linization at about 3 months after birth and do not achieve maturity until at least 5 years later. The short tangential fibers of the intracortical nerve net in nonspecific association areas of the hemispheres begin their myelin cycle about 1 or 2 years after birth and do not complete it until several decades later. Lecours (1975; 1982) emphasized the parallels between maturation of speech and language and the growth to maturity of postthalamic acoustic afferent pathways, intracortical neuropile in primary and secondary acoustic cortices, association pathways linked with these, and other sensory and motor structures involved in regulating sound production in speech (Figure 22). A similar analysis could be made of the development of the eye–hand system, which requires convergence of excitation within the parietal and frontal cortex from visual and somesthetic areas, which mature at an earlier stage. This system becomes the chief language mechanism of deaf signers (Bellugi and Poizner, 1984).

Yakovlev and Lecours (1967) present an over-

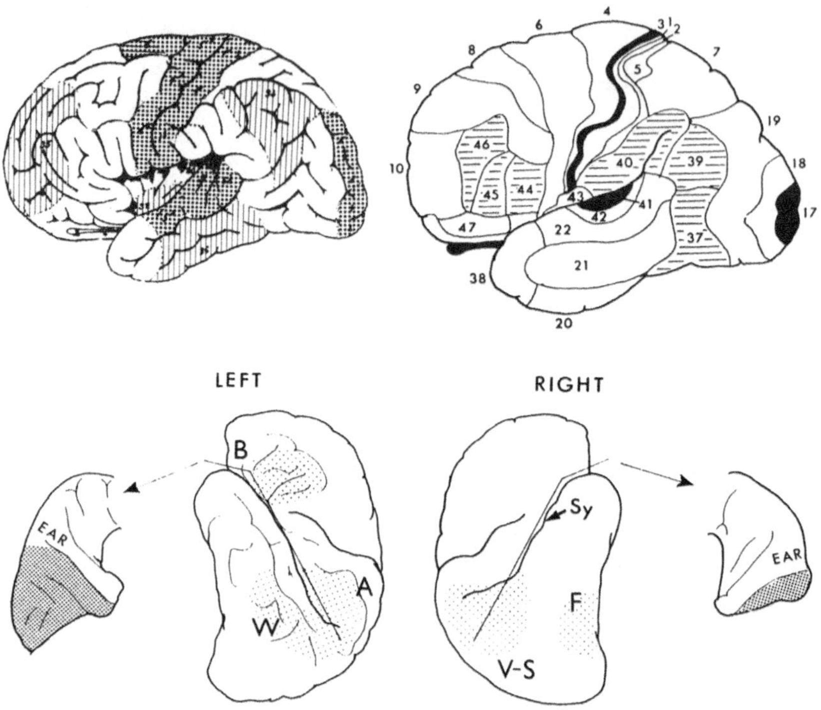

Fig. 23. (Top left) Flechsig's 1901 diagram of intracortical myeling zones, modified to show his three maturation groups: Group 1 (dots), myelin formation advanced at birth, visible to the naked eye; group 2 (white) start to show myelin before one month after birth; group 3 (vertical lines), start myelination in infancy or childhood. (Top right) Brodmann's map of tissue types in the cortex, showing primary receptor areas (black) and territories that are rudimentary or absent in apes (horizontal lines). (Bottom) Anatomical asymmetries of the hemispheres include enlarged left temporal planum posterior to auditory cortex and Sylvian Sulcus (Sy) straight on left side but bent dorsally on right side. Stippled areas contain lateralized psychological systems as demonstrated by neuropsychological testing. B, Broca's speech articulation area; W, Wernicke's speech perception area; A, andular gyrus, lesions produce defects in reading-writing or manual sign perception; V-S, visuospatial area (lesions produce visuoconstructive disorders); F, facial identification (lesions may produce prosopagnosia).

view of psychological maturation in relationship to the myelinogenetic cycles in different brain systems.

6.4. Adult Sensory Pathways: Evidence of Innate Patterning

The classical anatomy of receptors, neural projections, nerve cell bodies in different nuclear regions of the brain, and neurometer systems is being greatly enriched by modern techniques. The whole picture is one of a minutely designed system. This explains why most neuroanatomists are more nativists than empiricists in their ideas of brain ontogeny. However, an epigenetic approach, seeking to reconcile these two ways of explaining perceptual function, must accept that specific forms of elements in the perceptual mechanism may be determined by intrinsic developmental processes and yet be open to modification, selection, and so on, by environmental effects. Postnatal maturation of perception systems is indeed dependent on stimulus-induced excitation.

Antenatal morphogenesis of receptors and effectors in the embryo and fetus of mammals commits them in highly specific ways to the functions they perform later. The same may be said for details of their histology and especially for the patterns of neural cells in receptor epithelia and projection systems which we have already discussed. Here we shall draw attention to some details of the input mechanisms of the brain that are clearly preadaptive to perception.

The retina has elaborate antenatal determination of structure. The primate eye is nearly complete at birth, although the foveal system appears to complete its differentiation after birth. In monkeys LGN cells representing the fovea gain in spatial resolution in the first year and there is a parallel gain in acuity of cortical visual cells (Blakemore and Vital-Durand, 1981, 1982). Other mammals have less mature eyes at birth, but the visual system of the lamb is clearly more mature (Clark *et al.*, 1979; Kennedy *et al.*, 1983).

The anatomical basis of feature extraction in the central visual field is best known for the visual system of the cat and monkey, this research being pioneered by Hubel and Wiesel (1962, 1963, 1965, 1968, 1974*a,b*). Combination of anatomical studies and unit recording has provided a detailed description that discloses staggering detail and precision in nerve tissue. There are important correlations between psychophysiological phenomena and details of the central visual anatomy. Now much of the controversy about the degree to which environmentally patterned stimulation may select, stimulate, or otherwise influence the growth process that forms particular feature extractors has been settled or at least changed in its terms (Barlow, 1975; Swindale, 1982). Other receptor modalities appear to have similar design to vision, but what type of features are extracted in them is less clear than for vision (Evans, 1974; Webster and Aitkin, 1975; Mountcastle, 1974; Jones, 1975).

The cat and monkey retinal ganglion cells have circular receptor fields with center-surround organization, and these represent different solid angles of the visual field. They project to the lateral geniculate nucleus, superior colliculus, and a number of other diencephalic and mesencephalic sites. The lateral geniculate layers exhibit somatotopic mapping of the field (Malpeli and Baker, 1975), an orderly segregation of the input of the two eyes, and a segregation of color categories of input. The projection from geniculate neurons to the striate cortex forms an orderly map of visual field locations (Talbot and Marshall, 1941; Daniel and Whitteridge, 1961; Whitteridge, 1973) and a systematic segregation by which circular receptor fields are linked to form edge detectors of definite orientation. Each edge detector element has the property of responding to displacement in the field of a boundary between areas of differing brightness or wavelength, with the direction of displacement that produces the greatest amount of excitation at right angles to the angle of inclination of the boundary, which also has an optimal length or extent. The rate of displacement of the edge segment or line segment is also a factor in producing excitation of the striate cells. It is conceivable that stationary edge detectors evolve out of velocity detectors that are insensitive to stationary stimuli. Evidence from developments in cat, rabbit, and monkey suggests that oriented edge detectors may arise by anatomical modification of velocity detectors. A wide variety of velocity tunings are found in visual cortical cells (Orban *et al.*, 1978). Other properties such as disparity selectivity, contrast sensitivity, and spatial frequency tuning depend on refined structure of visual cortex. These are less well understood. Some visual cortical neurons show refined sensitivity to the ratio of velocities of stimulation in retinal areas of the two eyes to which they respond (Cynader and Regan, 1982; Regan and Cynader, 1982). This provides an important source of data to monitor both self-motion and the displacement of objects in the three-dimensional visual field (Beverley and Regan, 1980; Regan and Beverley, 1982).

In layer IV of the visual cortex, columns or lines of adjacent cortical cells in a penetration at right angles to the cortical surface display the same orientation sensitivity. In each orientation column one eye predominates, and columns concerned with one eye are lined up to form bands or slabs 0.4 mm wide that divide up the cortex into meandering ribbon-like territories that, in the monkey, after histological processing, look like the stripes of a zebra's back, alternate territories receiving afference predominantly from one eye (LeVay *et al.*, 1975). The ocular dominance slabs are laid out in a clear maplike pattern of lines that intersect the vertical meridian of the somatotopic visual field at right angles and line up parallel to the horizontal meridian (Figure 24).

Tangential electrode penetrations in the visual cortex of monkeys and cats show a uniform mapping of units that respond to serially ordered orientations of edges (Hubel and Wiesel, 1974*a*). Evidently, the pattern is complex. There are suggestions that isoorientation columns are the same width as monocular dominance columns, but laid out at right angles to them (Hubel *et al.*, 1978; Humphrey *et al.*, 1980; Swindale, 1982). In passing through about 0.8 mm of cortex there is approximately a 180° rotation in the direction of the preferred orientation of cortical cells. This hypercolumnar unit corresponds functionally and in width with two ocular dominance slabs, one for the left eye and one for the right.

Hypercolumn assemblies have remarkable anatomical uniformity over the striate cortex of the monkey, corresponding to a patch of 0.8–1 mm of cortex for all parts of the explored central 20° of the visual field, the size of the region in visual degrees

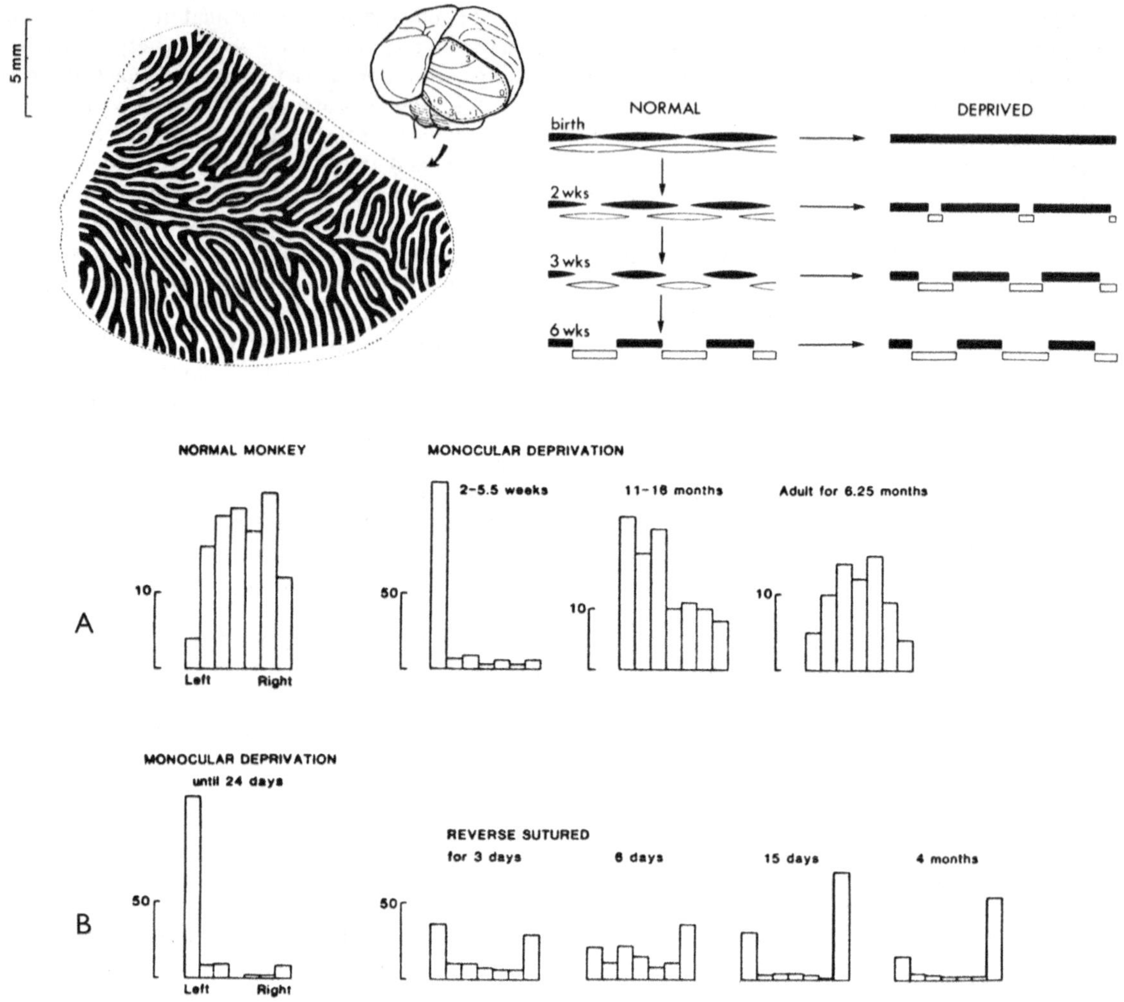

Fig. 24. (Top) Ocular dominance bands in the striate cortex of the monkey. Territories of one eye, black; the other, white. The visual field on the right occipital lobe is shown with horizontal lines 1°, 3°, and 6° above (+) and below (−) the horizontal meridian (0). (From LeVay *et al.*, 1975.) (Right) A hypothetical scheme for effects of eye closures in newborn monkeys on segregation of optic terminals from the left and right eyes after birth. The width of the black and white bars represents presumed density of terminals. It is supposed that competition normally occurs between the eyes and that weaker inputs retract to make way for stronger. Strength of terminals is enhanced by stimulation or weakened by deprivation (from Hubel *et al.*, 1977). (Bottom) Ocular dominance histograms for visually responsive neurons outside layer IVc in monkey visual cortex. The extreme left bin in each histogram contains cells monocularly driven by the left eye; the right bin contains cells monocularly equally responsive, although both eyes and the other four bins are for cells more or less strongly dominated by left or right eye. (A) One normal 8-month-old animal and 3 monkeys whose right eyes had been deprived by lid-suture for the periods of time shown. Monocular deprivation, especially in early weeks, causes pronounced reduction in cells dominated by the right eye. (B) A series of monkeys all deprived in the right eye until about 4 weeks of age. The first was recorded at that stage, but the others were then reverse-sutured (closing the left eye and opening the right) and given varying durations of forced use of the right eye. There is a rapid switch in the dominance of cortical cells. (From Blakemore and Vital-Durand, 1981.)

following the cortical magnification factor. Inside one of these unit territories, individual units represent a random scattering of receptor fields within this location. The two monocular territories of a hypercolumn represent corresponding locations in the two eyes.

Color-detection elements, innately wired to segregate four primary colors into opponent pairs, may also be distributed over the cortex in an orderly pattern (Hubel and Wiesel, 1968; Zeki, 1973). Probably the mapping of ocularity, orientation, and color intersect to permit optimal distribution of the different elements within the somatotopic map of the visual field. The functional advantages of overdispersion (orderly gridlike or alternating bandlike progression) have been discussed by Hubel and Wiesel (1974*a*).

In a vertical penetration down a line of cells through all striate cortical layers in one location in the map of the visual field, ocular dominance and orientation may remain constant. The vertical and transcortical connections seem to permit combination of the primary categories into an orderly hierarchy of increasingly specific and elaborate detection elements (Hubel and Wiesel, 1962, 1968). The simplest featured detectors are cells in layer IV that receive direct input from the thalamus. Transcortical projection, reciprocally connecting areas of cortex into functional systems (Van Essen and Maunsell, 1983; Mishkin *et al.*, 1983), permits integrative combination of receptor field types into higher-order detectors and streams of processing, by means of long intracortical axons. There are several parallel, differently organized maplike projections of visual functions on the cerebral cortex of rats, cats, and monkeys (Allman, 1975 *et al.*, 1981; Zeki, 1974; Van Essen and Maunsell, 1983). It is probably a general feature of brains of birds and mammals for different classes of perceptual processing to be represented in systems of centers that assemble percepts in a number of different ways, through ascending and descending links between midbrain cortex and nuclei, thalamic nuclei and cerebral areas, as well as by reciprocal transcortical connections. Motion analysis may be carried out in a specialized middle temporal cortical area in the monkey that receives input from several occipital and parietal visual areas (Van Essen and Maunsell, 1983). It is probable that a considerable part of the visual processing in extrastriate areas of both cat and monkey is informed by input *via* the pulvinar that bypasses area 17, and some of these processes may not require input from area 17.

Units in area 17 of the adult cat are apparently calibrated to respond to different binocular disparities, thereby encoding depth or distance from the subject for a single stimulus that excites corresponding regions of the two eyes (Barlow *et al.*, 1967). Stereopsis units are recorded in the prestriate cortex of the monkey. Indeed, most feature-detecting processes in monkey visual cortex appear more elaborate than their equivalents in the cat (Blakemore and Vital-Durand, 1981).

The theory of a prefunctionally wired in anatomical system of combinations that permit progressive categorization and recombination of stimulus features is derived from the above remarkable anatomical and physiological findings. Nevertheless, the details remain incomplete, and we possess only the outline of a theory of the fundamental morphogenetic principles by which the network is built. The bands or stripes and patches of approximately 0.3–0.5 mm in width in different arrangements for different features detecting cell types would seem to indicate that a very orderly pattern is formed by combining reiterated modular elements such as the one would expect from the operation of intersecting morphogenetic rules (cf. Hubel and Wiesel, 1974*a*). Models of systems that produce such a segregation require a competition between mixtures of contrasting inputs that tend to cluster, so that like elements are kept close together. This could be achieved by stabilizing adjacent synapses if they were from the same presynaptic population or from presynaptic cells that are excited at the same time (LeVay *et al.*, 1975; Constantine-Paton, 1983). There is increasing evidence for an active sprouting of terminals in development to produce a mixture at end sites, where axons from different cell groups converge, then a sorting into clumps or bands that maximize the homogeneity of type of endings in any one small locale. Selective stabilization of a subset of projections from one cortical area to another clearly has large-scale effects in postnatal development of cortical systems.

Comparison of findings in a number of mammals suggests that the geniculostriate projection and parallel extrageniculate projections have the same general principles of structure in the different forms but that the preponderance of oriented edge detection in the striate and circumstriate cortex of cat and monkey may be an adaptation associated with more refined oculomotor control and refined vision under conditions of fixation (Drager, 1975). There are also indications that the fundamental unit of detection for motion stimuli is a rate of change in visual angle, rather than in distance on the retina, which differs greatly in forms with different-sized eyes. This brings up the important possibility that the findamental unit is a morphogenetic building block of the egocentric somatotopic behavioral field (an element of behavior space). Daniels and Pettigrew (1976) propose that the extreme anatomical refinements of connections required for binocular stereopsis is a major factor in the evolution of the special characteristics of the visual cortex of cat and monkey as compared with the rabbit.

In all forms, there is a bias for central binocular representation in the geniculostriate cortex and a comparatively strong representation of the peripheral field in the superior colliculi. This accords with a functional distinction between colliculus and striate cortex that has been brought out by behavioral studies of the effects of surgery to brain stem and forebrain visual centers or split-brain surgery (Schneider, 1967, 1969; Trevarthen, 1968*b;* Trevarthen and Sperry, 1973). According to the two-visions theory, midbrain visual cortex and striate cortex are complementary components in a visual mechanism that is tightly integrated through vertical connections and by convergence on a common motor system. The latter is organized as a hierarchy of somatic elements such that axial and proximal musculature functions as a context for distal motor adjustments. The functional properties of collicular units depend on projections from the cortex to the colliculus that confer form-discrimination capacity (Wickelgren and Sterling, 1969). However, at least in cats and more primitive mammals, the midbrain-pulvinar projection to areas 19, 20, and 21 of the cortex has some form-discriminating capacity independent of the geniculostriate system (Sprague *et al.,* 1977, 1981). It is generally agreed that the midbrain cortex is inherently prewired for perception of spatial relationships among stimuli and for regulation of attentional localization related to oculomotor scan (Ingle and Sprague, 1975). It is also organized to coordinate vision and audition with touch where parts of the body, such as limbs or vibrissae, project into the visual field (Drager and Hubel, 1975; Finlay *et al.,* 1978), and in cats it mediates fast correction of aim for paw extensions (Alstermark *et al.,* 1984).

Psychological tests of visual awareness in monkeys with cortical lesions carefully defined by anatomical criteria led Mishkin and colleagues (1983) to propose that there are two cortical pathways—one subserving object vision and the other spatial vision. Anatomical investigations and metabolic mapping of neural activity confirm that there are two streams of projections from striate and prestriate cortex that lead separately to either inferior parietal cortex implicated in spatial deployment of visual exploration and landmark discrimination or to inferior temporal cortex that is essential for object discrimination.

6.5. Postnatal Development and Plasticity of the Visual Cortex

Psychological recordings from cortical cells of visually naive monkeys prevented from receiving any patterned visual stimuli led to the conclusion that the above-described anatomy of retinocortical projections and intracortical connections is determined without benefit of patterned stimulation at birth (Wiesel and Hubel, 1974). This conclusion has been revised in the light of anatomical observations that show cortical circuits to be only partly formed in the newborn animal (Hubel *et al.,* 1977) (Figure 24). The development of visual tissue is, for a brief period, largely indifferent to deprivation of patterned stimuli, however (LeVay *et al.,* 1980). Orientation columns and ocular dominance columns and their organization in unit hypercolumns are present after 3 or 4 weeks of visual deprivation consequent to suturing of both eyelids at birth. Precise orientation preference was observed in cortical cells within minutes of access to stimuli, and the orderly array of responses recorded subsequently was insensitive to order of previous stimuli during the electrophysiological mapping. The monkey visual system appears to be remarkably mature in these respects soon after birth. Simple complex and hypercomplex cells exist, and orientation responses are obtained to an "on and off" of stationary slits of light, dark bars, or edges (LeVay *et al.,* 1980).

Despite these signs that the visual cortical mechanisms of baby monkeys can undergo considerable segregation for visual function without instruction from the environment, there are now clear indications that their morphogenesis is influenced by stimulation (Swindale, 1982). It has been shown that the foveal contrast sensitivity of striate cortex neurones develops throughout the first year of life, evidently because of neural reorganizations in circuits of the retina or LGN as well as cortical changes (Blakemore and Vital-Durand, 1982) (Figure 25). Behavioral measures of contrast sensitivity in monkeys show parallel changes (Boothe *et al.,*

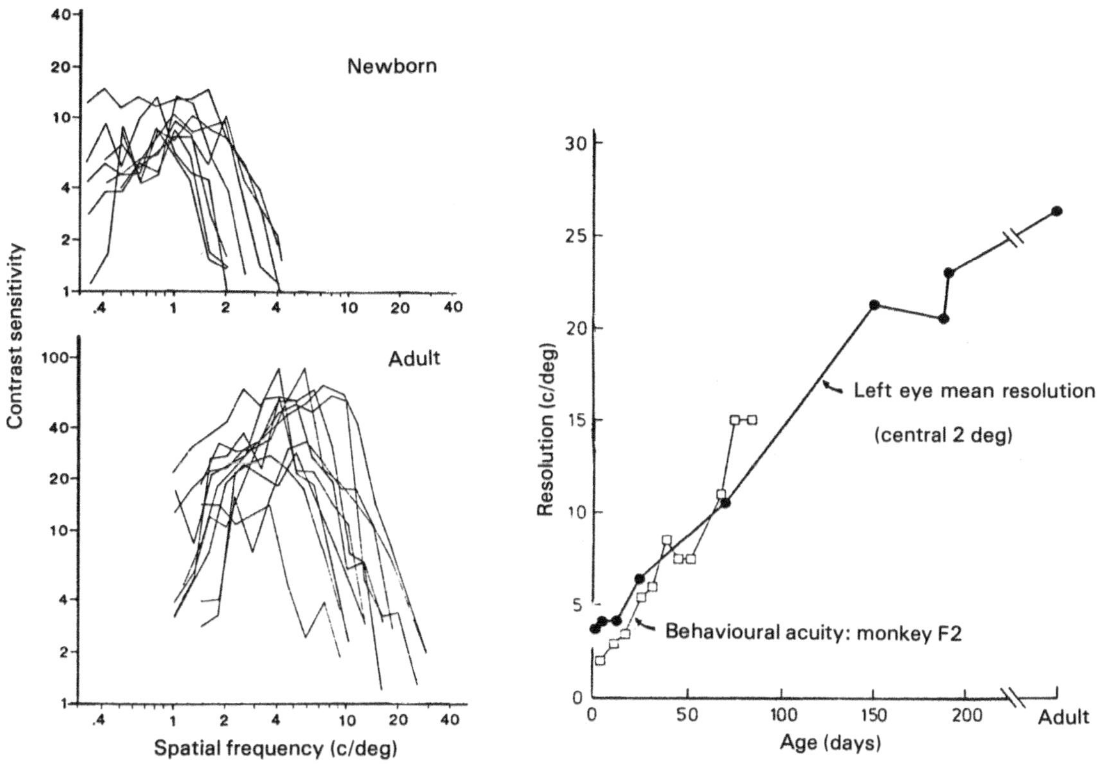

Fig. 25. Development of visual resolution in monkeys. (From Blakemore and Vital-Durand, 1981.) (Left) Contrast sensitivity–spatial frequency functions for neurons from the primary visual cortex of newborn and adult monkey. Grating patterns in optimal orientation for each cell were drifted across the receptive field. Threshold contrasts were measured for each grating. (Right) Postnatal improvement of acuity. Filled circles plot mean resolution for all cells in the LGN laminae connected to left eye cells for the central 2° of the visual field. c/deg, central degrees. Squares show behavioral determinations of visual acuity for one monkey. (From Teller *et al.*, 1978.)

1980). These and other improvements can be interfered with by abnormal stimulation.

Binocular convergence of inputs to cortical cells is highly sensitive to visual deprivation or interocular imbalance of input near birth in the monkey. Although an accurate pattern of binocular convergence on cells of deeper layers of the cortex for corresponding locations and corresponding forms of stimulation may be observed at 2 days with no chance for their formation by patterned visual input, deprivation of patterned vision for a few weeks after birth, even after permitting 3 weeks normal access to experience first, leads to destruction of most of the binocular convergence, most cells giving responses only to one eye (Figure 24). Hubel *et al.* (1977) found, by injecting one eye of a monkey with tritium-containing compounds the day after birth and following the radioactive terminals in layer IVc at 1 week, that the inputs from the two eyes, although already in the form of parallel bands, are only slightly segregated. Electrical recordings with tangential penetrations showed considerable overlap of inputs at 8 days and much more advanced segregation at 30 days. Huber and colleagues concluded that the ocular dominance columns are not fully developed until some weeks after birth. Both anatomical and electrophysical techniques confirm that the lateral geniculate body is highly differentiated at birth, unlike the cortex (cf. Rakic, 1976, 1977).

The terminals from the two eyes retract from overlapping regions in the visual cortex by some kind of competitive repulsion to produce regular sharply defined monocular bands with centers separated by 0.4 mm (Figure 24). The segregation of fibers is apparently due to a self-sorting mechanism like that proposed by Willshaw and von der Malsburg (1976). Blakemore and Vital-Durand (1981) conclude that deprivation of all visual stimulation arrests the formation of orientation-selective cells as well as maintaining the low immediate postnatal contrast sensitivity and cutoff spatial frequency. The process of segregation that takes place if the monkey is reared in the dark (Wiesel and Hubel, 1974) is highly sensitive to monocular deprivation. Precise segregation of monocular inputs to cells in layer IVc of the striate cortex is prevented by monocular deprivation during the first postnatal month of a monkey's life (Hubel *et al.*, 1977; Blakemore *et al.*, 1981; LeVay *et al.*, 1980). Cells outside layer IV, binocularly driven in the normal monkey, respond only to the experienced eye when one eye is occluded in the first few weeks after birth. Suturing one eye of a monkey for different periods after birth demonstrates that the critical period for monocular deprivation to prevent development of binocular neurones is in the first 6 weeks, but some effect is still seen after deprivation around the end of the first year (Figure 24). Reverse suturing experiments show that a remarkably rapid switching of terminals occurs, so that cells that were driven by one eye now covered are quickly taken over by the other newly opened eye (Blakemore and Vital-Durand, 1981). Anatomical evidence suggests that the original change is not caused, as originally proposed, by degeneration of inactive terminals, but that both reorganizations are due to an active sprouting of axons to invade sites weakened by deprivation of excitation (Blakemore *et al.*, 1981; Swindale, 1982).

The results reported with monkeys contrast with those for kittens in which orientation detectors and binocular disparity detectors have been said to be created by coincidence of stimulation from patterned light (Blakemore, 1974; Daniels and Pettigrew, 1976). The argument concerns the degree to which the different connections, on which visual distinction of many features depends, can be genetically determined. There is now no doubt that environmental input is necessary for normal circuits to develop or to be retained, but the evidence for creation of detector circuits out of less organized or unorganized and highly plastic growing circuits is less conclusive; a considerable intrinsic regulation by intercellular excitation that is not shaped by stimuli is indicated for both cat and monkey. Fluctuations in density of input from one eye to layer IV of the striate cortex that appear in tangential sections of the visual cortex dark-reared kittens are eliminated if activity of retinal ganglion cells is blocked by administering Tetrodotoxin to both eyes (Stryker, 1981). This shows that early stages of segregation of monocular dominance columns are driven by spontaneous retinal activity. The comparatively immature visual system of the newborn kitten becomes highly sensitive to deprivation or abnormal patterning of visual input a short time after birth when synaptic connections are forming in the visual cortex (Figure 21). Binocular disparity detection is destroyed by early postnatal deprivation of vision in one eye (Daniels and Pettigrew, 1976). It also adapts to a change in visual directions caused by placing a prism in front of one eye of a young kitten (Shlaer, 1971). Preferential orientations of cortical cells evolve separately for the two eyes in kittens that are prevented from having simultaneous binocular experience (Blakemore and Van Sluyters, 1974). Removal of one eye of a kitten at birth changes the effects of visual deprivation of visual experience. Significantly more orientation selective cells remain in the enucleated kittens than in the binocularly deprived kittens, and the deprivation resistant cells respond preferentially to horizontal and vertical orientations and are recorded mainly in the cortex contralateral to the remaining eye (Frégnac *et al.*, 1981). Apparently, the development of oblique orientation cells requires binocular afferents in the visual pathway. In the foveal projection area of monkeys, vertical and horizontal sensitive cells are represented more strongly than cells sensitive to other orientations (Blakemore *et al.*, 1981; Leventhal, 1983).

According to current information, reviewed by Swindale (1982), more refined patterns of connection within the visual system of the cat and monkey are built with the aid of patterned stimuli that link simultaneously activated cells into permanent assemblies with strong functional commitment. Stationary edge detectors, specific for orientation, and binocularly responsive cells, by which disparity may be detected to permit depth detection by binocular stereopsis, are the most dependent on coherent and sufficiently varied binocular stimulation. On the other hand, it is certain that all acquired powers of visual discrimination are set up within a highly elaborate innate "wiring" of nerve circuits that defines basic dimensions and categories or elements of vision and also the fundamentals of visuomotor coordination.

To understand how genetic instructions and environmental information combine to determine visual circuits, it is necessary to distinguish each anatomical level of the visual mechanism according to its time of development in the general plan of growth. The cat and monkey differ in the maturity of their basic visual machinery at birth, but both show a progressive addition of visual analyzing components postnatally. The different components will have differing and regulated sensitivity to experience, as well as to input from parts of the brain, such as the reticular formation, that are either spontaneously active or stimulated indirectly, perhaps by other modalities (Rauschecker and Singer, 1979). Later developed mechanisms, even basic receptor pathways, (e.g., the foveal detector system, which is essential for the highest resolution in central vision of the monkey or man, and the central neural circuits to which the fovea projects) may benefit from an education that is controlled by previously "wired-in" visuomotor circuits. Visual fixation and conjugate saccadic displacement of the direction of gaze are present at birth in monkeys and also in man (Haith, 1978). The "diet" of visual detail will therefore be subject to the control of previously formed mechanisms that choose sites for fixation. They depend on an accurate and reliable motor system that is capable of correcting errors in convergence. Proprioceptive input from the orbit, signaling eye position, plays a role in the final process of maturation of the visual pathway (Buisseret and Gary-Bobo, 1979). Active visuomotor experience is necessary to maintain and develop specific cells after the third week of postnatal life in kittens (Imbert, 1979) and eye movements to fixate visual targets, active from birth, are essential to their visuomotor development (Hein *et al.,* 1979).

Interocular transfer of perceptual fatigue effects that lead to illusions of motor or tilt have been used by Hohman and Creutzfeldt (1975) to obtain evidence that the binocular neurons of humans are sensitive or labile over the first 3 years of life. During this period, the experience of the two eyes is highly correlated in three-dimensional space by means of innate acts of selective visual attending. Banks *et al.* (1975) have also found that correction for congenital esotropia must be carried out before 2 years if it is to be permanently effective. Slater and Findlay (1975) claim evidence that binocular stereopsis is functional in the first days after birth. Other evidence for environmental effects on directionally sensitive elements in humans is reviewed by Barlow (1975). Leehey *et al.* (1975) found that the dominance of elements sensitive to horizontal and vertical contours as compared with diagonals is not formed by pattern of retinal stimulation.

Abnormal stimulation of immature perception systems may distort their growth without revealing how they normally develop. It is likely that unspecified or unrestrained exposure of neonatal circuits may impair their usefulness for perception. As yet, we do not have a description of the natural processes by which stimulation is regulated in early human development (selective information uptake) and their part in formation of higher levels of the perceptual apparatus of the brain is uncertain. At most we know a few basic looking strategies of neonates in artificial circumstances (Haith, 1978; Bronson, 1982). Until we have more complete data there is little profit in speculating about the adaptive advantage of leaving circuit formation to "instruction" from the environment.

Many elements of the visual mechanism of young mammals are sensitive at critical stages of their development to effects of drastically reduced or distorted stimulation (Blakemore, 1974; Tees, 1976). Retina, dorsolateral geniculate nucleus, superior colliculus, and visual cortex all show growth changes or degeneration of elements (cells, dendritic spines, or synapses) if the eyes are kept in the dark or prevented from receiving patterned stimuli by suturing the eyelids (Valverde, 1971; Blakemore, 1974; Ganz, 1978; Tees, 1976). Competitive interactions between components are revealed by monocular deprivation or by the effect of squint by which one eye is made to deviate from the normal convergence with the other eye (Hubel and Wiesel, 1965; Hubel *et al.,* 1977; Wiesel and Hubel, 1963, 1965). Destruction may be much greater with monocular than with binocular deprivation and the pattern of binocular convergence is destroyed by unbalanced or discordant stimulation of the two eyes (Hubel *et al.,* 1977; Guillery, 1972) (Figures 21 and 24). Some effects of monocular deprivation or other restrictive stimulation are reversible, indicating that the structures put out of action may recover when conditions change (Blakemore and Vital-Durand, 1981; Blakemore *et al.,* 1981; Hickey, 1981) (Figure 24).

The colliculi of the cat, which show postnatal maturation immediately before that of the visual cortex (Stein *et al.,* 1973; Cragg, 1975), develop some functions normally if the visual cortex is removed (Sherman, 1972), but it is claimed that binocularity and direction selectivity of collicular cells develops under the influence of corticocollicular projections (Wickelgren, 1972). Again, there appears to be functional competition for synaptic space within the colliculus tissue between retinal and striatal inputs. Anatomical evidence for this has been obtained for the neonate hamster (Schneider, 1973) (Figure 20). The monkey colliculus shows ocular dominance "clumps" that become clearer prenatally during the last half of gestation (Rakic, 1976, 1977, 1979).

Monocularly deprived kittens lose binocular cells from the visual cortex completely (Wiesel and Hubel, 1965; Ganz *et al.,* 1972). A period of reversed vision, exposing the eye previously deprived on its own, can partly cure the effects of monocular deprivation (Chow and Steward, 1972; Blakemore and Van Sluyters, 1974). There is a close correspondence between the period of maximum sensitivity to monocular deprivation and the rapid growth of synaptic terminals in the visual cortex (Cragg, 1975; Hubel and Wiesel, 1970; Blakemore, 1974; Blakemore and Van Sluyters, 1974).

In newborn monkeys in which one eye has been removed or kept from patterned stimuli by suturing the eyelids, the imbalance of stimulation changes development, and the terminals in the striate cortex for the deprived eye occupy abnormally narrow regions, while those from the other eye have larger than normal bands. The process takes between 4 and 8 weeks to complete, so it coincides with the observed maturation of ocular dominance columns and the separation of terminal arborizations (Hubel *et al.,* 1977) (Figure 24). The evidence suggests that the narrowing results from a takeover by competing terminals from the normal eye of regions normally in the territory of the terminals made weak by deprivation. Cortical dimensions do not change, and the spacing of eye dominance bands remains about 0.4 mm. The deprivation produces greater loss of territory in the hemisphere ipsilateral to the deprived eye, indicating that the crossed afference to the cortex is stronger, and recalling the situation observed in the developing binocular visual system of *Xenopus.*

This evidence for competition between the visual circuits for the two eyes during development fits with an earlier finding that binocular deprivation in cats produces a less severe effect in cortical cell responses than predicted from monocular closures (Wiesel and Hubel, 1965). There are other indications of competition between inputs from the eyes in the development of the lateral geniculate nucleus (LGN) and cortex of cats (Guillery, 1972; Sherman *et al.,* 1974). The monocular crescents of the projection are protected because they lack competing terminals from the other eye, and regions of the deprived eye corresponding to patches where retina has been removed in the other eye do not show weakened unit responses. The effect is clearly due to interaction between terminals from the two eyes, and it may depend on a sorting principle that uses impulse synchrony to encode place of origin (Willshaw and von der Malsburg, 1976).

The population of orientation detectors in the striate cortex of the cat can be modified to conform to experience in a field of stripes all going in one direction (Blakemore and Cooper, 1970; Hirsch and Spinelli, 1970). Van Sluyters and Blakemore (1973) and Pettigrew and Freeman (1973) produced spot detectors by raising kittens in a spotty environment. Detectors of this form are rare in the normal cat cortex; they are closer to an immature receptor field.

6.6. Behavior with Deprivations of Experience: Mechanisms of Visuomotor Coordination

Spalding (1873) showed that chickens are able to perform accurate pecks within 2 or 3 days of hatching and that this act would mature without experience of trial and error. Hess (1956) showed that maturation improved the coherence of aim even if this was displaced by prismatic bending of vision, proving that feedback or failure to hit did not guide the maturation process. More recent studies with nonmammals are reviewed by Ingle (1978).

Cats are born blind and weak but can orient to visual targets as soon as their eyes open; after 1 month of development, they show rapid attainment of full visual control over movements. Monkeys and humans have some orienting capacity at birth and this competence progressively improves. Deprivation experiments reveal a complex interaction of maturation, environmental facilitation, and environmental formation of structure in the building of visuomotor systems (Ganz, 1975; Tees, 1976).

Walking and steering around objects is profoundly deficient in cats reared in total darkness (Riesen and Aarons, 1959). Although they still possess efficient space perception by auditory and tactile cues, exposure to vision causes total disruption of their locomotion. Similar disorientation and confusion are reported for humans when their vision is restored by removal of cataracts (Von Senden, 1960; Gregory and Wallis, 1963). Cats may show good recovery from effects of rearing in the dark, and monkeys deprived for less than 10 weeks recover so quickly (in 0–4 days) that one must conclude that they possess a program for developing orientation in space that is destroyed by longer deprivation (Fantz, 1965; Wilson and Riesen, 1966).

Response to approach of an object (looming) is absent after rearing in the dark, but it recovers after exposure to the effects of locomotion in light arrays. Human infants show a complex defensive reaction to looming of a shadow from 2 weeks of age (Ball and Tronick, 1971; Bower, 1974). They possess limited abilities at birth to orient vision in space and coordinate sound with seeing to do so (Haith, 1978). They show a high sensitivity to human baby talk and orient their eyes towards a voice within the first few hours (Alegria and Noirot, 1978).

Aiming of the limbs toward surfaces for support, or toward objects of prehension, develops after birth, and it is sensitive to deprivation of feedback from vision of limb movement. Extending of the paws to an approaching surface develops in the first postnatal month in kittens but does not depend on experience with patterned light (Hein *et al.*, 1970*a,b*). Deprivation in darkness causes loss of depth detection, but recovery is often rapid in a patterned light environment (Riesen and Aarons, 1959; Walk and Gibson, 1961; Wiesel and Hubel, 1965; Baxter, 1966; Wilson and Riesen, 1966; Ganz and Fitch, 1968; Hein, 1970; Tees, 1976). The rapidity of recovery indicates that the underlying competence (specific anatomical organization of structure) may be present without experience of visual information.

The perception of structural subdivision of surfaces, and corresponding orienting of the paws to left or right for accurate placing on projections, has a different maturation in the cat from triggered extension in the general direction of the approaching surface (Hein and Held, 1967). This compares with the monkey in which anatomical and behavioral studies show two separate systems, one for general ballistic aiming involving the proximal musculature in adjustment to target position and the other concerned with local structure and aiming of the palm and fingers (distal musculature) under tactual and foveal control (Kuypers, 1973; Brinkman and Kuypers, 1973).

Reaching matures more rapidly in monkeys than in humans, but it appears to undergo the same succession of stages in a compressed schedule. Film and TV records prove that coordinated reaching toward objects in the visual field is present in outline in human neonates and that the complex sequencing of distal and proximal movements of reaching and grasping is innately programmed (Trevarthen, 1974*a,c*, 1984*a*; Trevarthen *et al.*, 1981, Hofsten, 1982, 1984). However, arm extension is stereotyped and inadequate in extent for the first 4 months (McGraw, 1943). Hofsten (1980, 1983) showed that from the onset of efficient reaching (16 weeks), the infant adjusts arm extension to the velocity of an object to obtain interception. Visual guidance of reaching matures in step with motor coordination. Observations of institutionalized infants who have much less than normal opportunity to reach and grasp objects show that deprivation can retard development of eye–hand coordination (White *et al.*, 1964).

Proprioceptively guided arm extension and sustained aim are achieved by infants before the development of manipulation with fine finger movement under central visual control. The hand is oriented to fit objects in different positions from 18 weeks, but hand shaping develops considerably over the following months (Hofsten and Fazel-Zandy, 1984). In the monkey correctly directed reaching matures in a few days after birth, but again, accurate grasping takes longer to mature. These developments correlate with anatomical changes in the cortical projection to spinal motor centers. The direct corticomotorneuronal projection which is concerned with refined motor control of the hand and dependent on tactile and visual guidance grows postnatally in primates (Lawrence and Kuypers, 1965). It shows a slower cycle of maturation in apes than in monkeys and is yet slower to mature in humans (Kuypers, 1973). It is probable that the efferent mechanisms controlling arm and hand movement mature in relationship to, and probably slightly in advance of, the mechanisms for visual perception of details of form and space. In human infants, accommodation, on which efficient fixation and visual resolution of detail depend, is detectable at birth and attains maturity in 6 months, as monocular acuity reaches about 75% of adult levels (Braddick and Atkinson, 1979; Gwiazda *et al.*, 1980; Trevarthen *et al.*, 1981).

Perception of depth in control of locomotion has been examined using the visual cliff in which information about a drop is restricted to the visual modality and dependent principally on motion parallax effects generated by eye displacement (Walk and Gibson, 1961). Visual deprivation, again, leads to temporary loss rather than destruction of this visuomotor coordination (Ganz, 1975; Tees, 1976). In the rat, maturational and experimental factors in development of depth perception have been demonstrated by Tees (1976).

Perception of objects in motion and of the stable world when the subject is in movement or being displaced is helped by tracking movements of the eyes. Tracking of a complete visual array (as in compensating for body displacement) is different from pursuit of an object moving relative to the general array, but both need periodic displacement of gaze to other orientations. Recentering or redirection of gaze against motion information has very different requirements from the following movements. Pursuit of objects going behind other objects or making complex hard-to-follow paths of displacement requires predictive capture with saccadic (ballistic) jumps that are aimed with respect to the probable future location of the target. All these different aspects of visual orienting need to be distinguished in studies of their maturation postnatally.

Since eye movements, and all other movements displacing receptors, cause change in stimulation like that which occurs when things displace in the world, the brain must have some means of distinguishing what is self-produced from what is of external origin. In most cases, there are distinctive features of self-produced receptor displacement in the array of stimulation; therefore, as Gibson (1966) pointed out, it is unnecessary to postulate a reafference mechanism (von Holst, 1954; Teuber, 1960) in all instances (see also Regan and Beverley, 1982).

Optokinetic tracking of moving stripes in a large surrounding field is innate in kittens, but the resolution of the system improves greatly over the first month. In monkeys and humans, optokinetic tests show that visual acuity for moving striations improves more slowly. Smooth tracking of objects against a stable background is absent in infants for several months after birth, during which only saccadic pursuit is observed (Haith, 1978), but smooth optokinetic following of a striped array may be obtained in neonates (McGinnis, 1930; Atkinson and Braddick, 1979) and is present in monkeys on the day of birth (Ordy *et al.*, 1962). From the outset, head and eye are coordinated in pursuit movements of infants (Trevarthen *et al.*, 1981), although the head rotations become stronger over the first month at the same time that the infant develops strategies for predicting the appearance of objects moving against complex backgrounds and behind surfaces (Mundy-Castle and Anglin, 1973; Bower, 1974; Haith and Campos, 1977). Changes in head mobility suggest that maturation of neck muscles and their activity appears to parallel that of the proximal forelimb muscles (Trevarthen, 1974*d*). Deprivation studies with animals confirm that pursuit movements are innately determined, but sensitivity of the different components to postnatal experience has not been closely studied. Vital-Durand and Jeannerod (1974) and Flandrin and Jeannerod (1975) distinguish two levels in optokinetic response in kittens, the more precise function, which adjusts to velocity of the stripe rotation, depending more on postnatal experience and deriving input from the striate cortex. Hein *et al.* (1979) show binocular interaction and eye mobility are necessary for development of visual orienting in kittens.

Jeannerod (1982) presented a review of evidence that different levels in an innate hierarchy of visuomotor control mature at different times and showed differing degrees of sensitivity to deprivation effects. Earlier developed ballistic orienting, used within a field of locations specified in the peripheral visual field to one level of precision, is less susceptible to deprivation effects than a later-developed guided orienting function that is capable of a much higher precision. Jeannerod separately invokes the foveal (geniculostriate) visual projection and the peripheral (colliculopulvinar-extrastriate) projection. Robinson and Fish (1974) use this distinction to explain effects of depriving one eye of a cat of visuomotor experience for the first 18 months of life. The disadvantaged eye lacked photopic vision in the central part of the visual field and exhibited strabismus or squint. Visual orienting in low light, however, remained (see also Hein and Diamond, 1971). Kittens kept in the dark or in stroboscopic light fail to develop directional sensitivity and normal ocular dominance in units of the superior colliculus that are partly dependent on input from the striate cortex for their differentiation (Flandrin and Jeannerod, 1975).

Loss of the striate cortex in humans leaves a very weak visuomotor function that has been overlooked until recently (Perenin and Jeannerod, 1975; Pöppel *et al.*, 1973; Weiskrantz *et al.*, 1974). In spite of a total deficiency in form vision due to occipital lesions, the deprived parts of the visual field still possess discrimination of direction and of the occurrence of events in light. Destriate monkeys recover vision of space and motion (ambient vision) (Humphrey, 1974; Pasik and Pasik, 1971). It is possible that the vision of space remaining after loss of the geniculostriate component is based on connections *via* the colliculi. Bronson's argument that human newborns have ambient (subcortical) vision and grow foveal (focal or cortical) vision later is, however, not well supported by developmental findings and is probably not a useful simplification (Bronson, 1974). The late-developing extrastriate areas of the visual cortex send fibers to thalamic nuclei and the superior colliculi and probably contribute to ambient vision. Undoubtedly both systems are present but immature at birth. Peripheral vision is limited in newborns as is central acuity (Haith, 1978).

Restriction in perception of the body parts and of their joint use as well as reduction in the active control of a subject in gaining experience produce severe defects of vision (Held and Hein, 1963; Hein and Held, 1967; Held and Bauer, 1967). This finding led Held to propose that visual adaptation and visual maturation require exercise of self-produced effects and convergence of feedback from acts with some internal record in the brain of the expected or likely consequences of each act generated (Held and Freedman, 1963). It is possible, however, that what is missing is not a correlation between stimuli and an internal signal of acts performed, but a motor control of access to stimuli for parts of the perceptual mechanism that discriminate inside contexts established by the innate general orienting abilities. There may be very different experiences of form in stimuli in the experimental and control groups of animals and differences in the consequences (reinforcements) of acts performed (Taub, 1968; Ganz, 1975).

Held and Hein (1963) showed, by elegant use of a machine by which one kitten transported a confined, near-immobile partner around a striped visual environment, that inactivity in early life can lead to pronounced, but temporary, deficiencies in visuomotor performance, in spite of non-self-produced visual stimulation. Hein *et al.* (1979) demonstrated a probable role of eye-movement proprioception in development of spatial vision. In the light of recent findings on postsynaptic factors in development of visual cortex projection assemblies by selective elimination of uncorrelated presynaptic terminals, Held's corollary discharge may be represented by a centrally generated excitation of the neurones that are targets for visual afference.

Kittens deprived of vision of their paws, by wearing an opaque paper collar when in light, developed paw extension to an approaching surface, but failed to orient the paws to projections from the edge. Visual guidance of the limbs was also deformed in attempts to hit at objects (Held and Hein, 1963; Hein and Held, 1967; Hein *et al.*, 1970*a,b*; Hein and Diamond, 1971). There is rapid recovery of good eye–paw coordination after the normal conditions of visual control of movement are permitted. It appears that the remarkable specificity of effects of depriving perceptual monitoring of acts during early development may be caused by suppression of parts of a mechanism that is innately structured in general design, but in which component parts mature in rivalry.

Experiments by Held and Bauer (1967, 1974) with infant monkeys gave results similar to those for the kittens. Deprived of vision of their arms and hands, monkeys lost normal eye–hand coordination and developed abnormal attention to their hands as objects of visual interest. But after 1 month or rearing without vision of arms and hands, recovery of normal function was rapid. It is important to consider the possibility that in this experiment an imbalance was caused by competition between the complementary proximal and distal mechanisms governing reach and grasp of objects. Compulsive hand regard looks like an exaggerated form of interest in manipulation under visual control, but without an object. This behavior is highly active in normal 1-month-old monkeys, but it is usually applied to an object. A different kind of deprivation procedure, allowing practice of aimed reaching under touch guidance during the period that the animals were prevented from seeing their hands, supports the conclusion that hand watching is reduced if nonvisual orienting of the arm is not allowed to degenerate (Walk and Bond, 1971). This study appears to invoke the distinction between ballistic reaching and visually guided fine grasping. Only the latter showed marked deficiency after deprivation of sight of arm and hand. However, Bauer and Held (1975) reported that the practice in guided reaching outside visual control has limited application over the visual field, being progressively less effective for directions further away from the trained direction.

The above studies do not give support to the notion that reaching movements are mapped by vision of the hands moving in visual space in early infancy. Their extension to infants reared in institutions fails to prove that human eye–hand coordination is acquired by learning (White *et al.*, 1964). They do, however, indicate that exercise plays a significant role in the maturation of visuomotor control and that full calibration of eye–hand coordination requires seeing the hand in action. Bower (1974) reports observations that human infants gain predictive adjustment of muscle contraction to the weight of a seen object in the second year. He claims that guidance of reaching by audition, by vision, and by mechanoproprioception mature at different times in a somewhat competitive interaction that is guided by experience obtained in the first 2 years. Fine manipulation shows developmental improvement for years after this.

Development of the mechanisms for space vision and visuomotor coordinations could conceivably be based largely on an innate program. However, the demonstrated influence of the environment appears to play a constructive role, selectively retaining afferent elements that provide accurate feedback information for movements. In the case of skills or discriminations that must become closely adapted in order to be modified by change in the features of the environment, as well as reformulations of perception to accommodate change in dimensions of body parts through growth, there must be a neural mechanism for modifying the processes of perception. Increasing evidence suggests that the mechanisms of neural maturation and of such adaptations to experience are closely related. Even newborns have motor functions that control their visual stimulation, and this probably regulates development of visual circuits in the brain. Haith (1978) suggested that the central generating systems for eye movements are adapted to generate maximum cortical neural excitation by seeking optimal patterns of stimulus change on the retina. For example, moving to scan across boundaries, or aiming to explore points where contours change direction. A similar thesis is put forward by Bronson (1982).

6.7. Learning and Modification of Perception: Relationship to Brain Growth

Visual stimuli leave highly detailed impressions that last for some short time in the perceptual mechanism (Neisser, 1967). These temporary images are constantly replaced, but in the process a selection of information is retained. Remarkably detailed, but incomplete, memories may remain to affect later behavior. Evidently there is a process, or a series of processes, that transfers temporary impressions into a more permanent store that may be ready for use in appropriate situations, in preparation for the right matching event. Categorization of impressions confers meaning and adaption to related but new impressions. In other words, retained effects of experience generalize to other effects and become part of the process of perceiving.

In infancy and childhood, there is an elaborate development of mental schemata that confer increasingly powerful understanding of objects and perception of their properties and uses. A major change occurs toward the end of the first year of life; many psychologists believe that a human being first possesses imagistic representation and thought at this age. But this is only the beginning of a process that continues throughout the years of childhood affecting all aspects of psychological function. The acknowledged master of this domain is Piaget (1953, 1970), whose studies have been taken up and elaborated by developmental psychologists prodigiously since the 1960s. Unfortunately there is almost no correlation between this developmental psychology and sciences of brain growth. We have indeed very little real knowledge of the neural mechanisms of the critical events in perception or the formation of images and memories. It helps to consider the advantages of coding complex arrays of stimuli according to a set of prewired features. But doing so leaves unclear how features in stimuli become associated together in perception of forms and things, and how they may be variously interpreted or ignored by the mind (Neisser, 1976). Presumably many rules for generalization and perceptual integration in terms of phenomenal reality are derived from built-in structures of the brain, even though they may only attain full maturity through use.

Goldman (1971, 1972, 1976; Goldman *et al.*, 1970) demonstrated changes in developing psychological functions of young monkeys following surgery to dorsolateral and orbital regions of the frontal cortex and to related subcortical nuclei, at different times after birth. In part, these changes are comparable with adjustments in the distribution of visual terminals between alternative central sites on a competitive basis. However, more recent results show that compensation is limited.

Goldman removed areas that in the adult are essential for certain psychological functions of perception and memory. In the 6 months between 12 and 18 months after birth, effects of removal of orbitofrontal cortex, head of the caudate nucleus, and dorsomedial nucleus of the thalamus are as selective and severe as those caused by the same ablations in young adults, but lesions of the dorsolateral cortex do not produce the same effects, apparently because this cortex is relatively immature and not performing its function as in the adult (Goldman, 1976). If the dorsolateral animals are retested at 2 years of age, when the brain is more mature, they show severe deficits on the critical delayed response tests. Orbitofrontal lesions fail to have the usual effect on monkeys much earlier, at 2½ months of age.

This area matures earlier than the dorsolateral area, and effects of its ablation appear by 1 year of age.

Differences between very young and older monkeys in the effects of cerebral lesions thus appear to be related to differing maturation rates of the removed areas. They are not easily interpreted in terms of compensatory changes and functional redistribution in the young brains. A powerful effect of an inflexible developmental program, continuing long after birth, is indicated.

At the same time that geniculostriate and local projection systems in the visual cortex of kittens are showing extensive selective elimination of an excess of axon collaterals and/or active terminals, to form precise patterns of eye dominance, orientation preference, and grating sensitivity, Innocenti (1980, 1983) described a much more widespread sculpting of associational fibers in the kitten brain. Overproduction of axon collaterals is followed by selective elimination of a considerable proportion of the long fibers to be found at birth. When observing the patterning of callosal communication between visual cortices of kittens, Innocenti found that at or soon after birth, callosal axons from one cortex spread to all areas of the other hemisphere though they do not penetrate far into the gray matter. Correspondingly, most areas of the cortex contained cells that projected through the callosum. In the adult cat, callosal cells are restricted to certain cortical laminae, and they are absent from primary sensory areas and some other regions. Those that do remain project in a highly selective pattern of bands to particular cell layers in the other hemisphere. This pattern emerges in kittens between 10 and 30 days after birth, when the number of synapses is increasing rapidly and discrete visual responses are appearing in cortical cells (Figure 21). It is achieved by many cells losing a collateral that passes across the corpus callosum, while projections are retained to unknown sites in the same hemisphere. The number of cortical cells does not show a parallel fall. About 70% of callosal axons are lost, leaving 23 million in the adult cat, and the process of loss is followed shortly by the start of myelination of the axons, which continues until the cat is an adult and more than compensates for loss in bulk of the callosum due to axon loss (Koppel and Innocenti, 1983). Within a hemisphere, there is a comparable elimination of projections from visual to auditory cortex. Binocularly deprived kittens show normal reduction of callosal connections, but kittens that are monocularly deprived, or that lose binocular coordination through an induced squint, retain callosal projections that would normally be lost (Innocenti and Frost, 1979).

Innocenti concludes that definite association systems, although partially guided by intrinsic pathfinding principles, are also constructed by elimination from an exuberance of collaterals to leave each region of the cortex a very specific set of projections to other cortical sites. The same principle applies to other projections to and from the cortex (Stanfield, 1984). One consequence of this transformation of corticocortical connections is the plastic response to a peripheral genetic abnormality of the optic tracts in Siamese cats already described. Innocenti believes that exuberancy of axon collaterals is a mechanism whereby the brain may adjust to varied genetic or environmental influences.

6.8. Growth of Perception in Infants and Children

As reviewed, for example, by E. J. Gibson (1969), theories of perceptual development cover a wide terrain of possibilities. The empiricists leave biological processes of development a small role to play. Those that emphasize cognition, imagination, inference, attentional readiness, or problem solving with insight place perception subordinate to the development of mental skills and strategies for coping with the world. Those that do allow innate determination in the growth of perceptuomotor structures *per se* (Koffka, 1931; Werner, 1961; Gesell *et al.*, 1934) are handicapped by the insufficiency of knowledge of the neural circuits specifically concerned with processing perceptual information. The Russian physiologically oriented school (Zaporozhets, 1965) attributes the process of development to pavlovian conditioning and practice rather than to growth and differentiation of generative or interpretive structures adapted to perception.

The most comprehensive theory of psychological development is that of Piaget (1953, 1970), who conceives the structures of intelligence, in the form of schemata for acts or reasoning, to grow by a self-regulating process of internal equilibration while keeping an active assimilatory relationship to external conditions. Perceptual uptake of information onto schemata that determine acts of exploration or praxis gives rise to more and more powerful mental representations. The process of perception thereby develops in increasingly dependency on thought or reasoning. Piaget's concept of early perceptual development from the reflexes of the sensorimotor period of infancy, while biological or embryogenic in form, does not offer precise directives for a neurobiological study. His great contribution lies in defining terms for a program of systematic growth of logical operations, and especially the development of concepts pertaining to the object as a psychological creation. He describes the mind as progressively achieving freedom from an innate tendency to perceive situations egocentrically and to become committed to one object of interest. This is overcome by a spontaneous process of decentration, which generates more complete representations of the conditions for versatile behavior. In other words, perceptions are part of immediate and direct self-regulation early in development, and then increasingly part of an extended model of reality in which the self may have multiple places or in which several selves (subjects) are interacting through communication.

In E. J. Gibson's view, which follows the perceptual theory of J. J. Gibson (1966), the problem of perceptual growth is one of the formation of increasingly efficient skills for selective uptake of information from an information-rich environment, and not one of associative learning. Children gain in skill at seeing what is in the world, taking longer to learn more subtle distinctions and combinations in experience. They gain in perceptual skills as they gain in cognitive mastery and are unable to carry out the distinctively human task of perceiving meaning in coded stimuli, built of pictures, in words, or in symbolic objects, without cultural training. Nevertheless, they are amazingly precocious in attempting acts on objects and in actually achieving communication with persons. How this comes about is a major biological mystery.

Bower (1974, 1978) presents an analysis of the development of intelligence in infants based on experiments which show attainment of steps in the development of the object concept earlier than proposed by Piaget. Marked changes in perceptual abilities in the first 2 years are explained by Bower as the consequence of an epigenetic process of mutual adjustment between rules for perceiving and due to differentiation of specific distinctions within abstract formulations of goals for interest. Direct perception by vision, hearing, or touch is portrayed as limited by growth of body parts that change certain metrical dimensions affecting information uptake, and by insufficiency of motor coordination. Each perceptual modality is dependent on exercise for development, and associative affinities between modalities, while prepared for by antenatal growth of a common orienting field, are developed postnatally and are highly sensitive to disuse or exercise.

The lack of information on growth changes in perception may be attributable to the fact that perception achieves efficiency rapidly; it may mature in a few months, as Bower proposes. There is physiological evidence that the human peripheral visual system undergoes rapid maturation in the first six months in correlation with the mastery of orienting and prehension (Pirchio *et al.*, 1978). Visual awareness of detail, ability to accommodate the lens and achieve accurate foveal fixation of both eyes by convergence adjustments, binocular stereopsis, and smooth tracking of a displacing object are all achieved by 6 months after birth (Gwiazda *et al*, 1980; Braddick and Atkinson, 1979; Aslin, 1981; Aslin and Jackson, 1979; Banks, 1980; Haith, 1980; Trevarthen *et al.*, 1981) (Figure 26). Held *et al.* (1984) have found that female infants mature in vernier acuity ahead of males in the period 4–6 months after birth and they suggest that the differences in perception reflect effects of differing hormonal environments on maturation of the visual cortex. In the past psychologists have tended to consider perception in terms that do not permit one to see growth in the information-selecting processes that might be identified with growth of particular brain parts.

Concepts of a monotonically ascending perceptual power, by learning from simple reactions to complex integrations, are seriously upset by recent demonstrations that young infants are capable of categorical judgments for many subtle and complex patterns of stimuli, and that they perceive within several modalities together in one preformed space of attending and acting (Haith and Campos, 1977; E. J. Gibson and Spelke, 1983). Perceptual discrimination begins earlier than it should if it is a practiced skill, and the findings argue strongly for an autonomous and largely prenatal formation of a general psychological process for perception of space and objects, independent of skillful action and, to a large degree, independent of cognitive achievements that require behavioral exploration of objects and their relationships.

Observations of the selectivity in orienting and alerting reactions of infants, and of their deliberate control of head-turning, sucking, or limb movements in conditioning tests, reveal that within the first few months infants are able to perceive sights, sounds, and so on over which they have no direct locomotor, manipulative, or articulatory control (e.g., see Fantz, 1963; Papousek, 1961, 1969; Sequeland and DeLucia, 1969; Lipsitt, 1969; Eisenberg, 1975; Bower, 1974; Walk and Gibson, 1961; McKenzie and Day, 1972; Eimas *et al.*, 1971; Trehub,

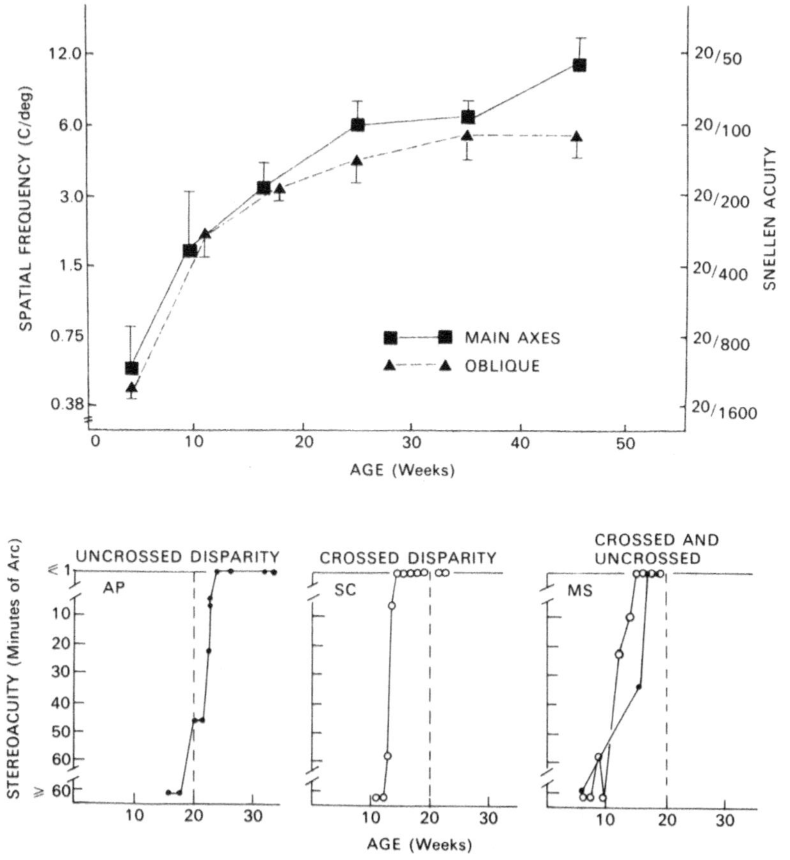

Fig. 26. Development of visual acuity in humans. (Top) Mean spatial frequency at which infants preferred either a main axis (vertical or horizontal) (■) or an oblique grating (▲) to a homogeneous gray field. Preferential looking technique. (From Gwiazda *et al.*, 1980.) (Bottom) Stereoacuity as a function of age in three infants. Subject AP was tested with uncrossed disparity in stimuli projected through polaroids at right angles, and viewed by the infant through goggles with polarizing axes at right angles. Subject SC was presented with crossed disparity. Subject MS was tested with both crossed (○) and uncrossed (●) disparities. Uncrossed disparities, leading to perception of an object behind the focal plane, are detected a few weeks later than are crossed disparities.

1973; Bornstein, 1975). Infants can select, integrate, and categorize stimuli from objects, perceive the distance, direction, size, displacement, orientation, and color of an object, and detect the supramodal or amodal connection between its separate indications in receptors of eye, ear, hand, and mouth. They are also able to follow the passage of an object behind a screen and anticipate its reappearance. Even more complex perceptions must underlie their special interests in the movements of persons and especially in their acts of communication. All these highly complex perceptual integrations are made in some degree while movements of orienting and focusing of receptors, although already well patterned, are definitely immature, and in absence of controlled locomotion or object prehension. As soon as they can crawl, infants show perception of the drop distances of horizontal surfaces from their eyes (Walk and Gibson, 1961; Tees, 1976).

Evidently the growth of the brain before birth permits differentiation of a neural field for specifying phenomenal perceptions. The characteristics of this field, when revealed in infancy, suggest that it is not composed of independent, unimodal image-forming devices. A coherent space for behavior is defined and in this stimulus distinctions are made in many equivalent perceptual subfields that process stimuli in distinct ways (Trevarthen *et al.*, 1981; Gibson and Spelke, 1983).

The reactions of infants 2–3 months old to signals brought under their motor control by artificial devices show that perception is linked to generation of movement at a highly elaborate level soon after birth (Papousek, 1969; Lipsitt, 1969; Kalnins and Bruner, 1973).

Bower (1974) emphasized certain developmental transformations, including some apparent regressions, in the abilities of infants to perceive by different sense modalities. Correlated with developments in action, such as mastery of reaching to grasp objects, which occurs at 4–5 months, or the development of standing and walking, these perceptual changes are not monotonically progressive and they cannot be explained as cumulative learning effects. They argue for competitive interaction between self-organizing cerebral components. These are prepared as functional alternatives or complements for the neural regulation of acts by pickup of perceptual information. Observations with blind, deaf, or armless infants show that processes directed to elicit or receive input through the missing organs grow for a time as if they were to be used. Then, at a stage at which they would normally mature with the benefit of exercise and stimulation, they decline (Drever, 1955; Axelrod, 1959; Lenneberg, 1967; Freedman, 1964; Fraiberg, 1974, 1976). The effects may reflect overgrowth by intact systems and may lead to deficiencies of other forms of mental activity that are not solely dependent on the defective channel.

All aspects of psychological maturation in humans are integrated with interpersonal processes that permit perception and action to be constantly extended and enriched, not only by the subject on his or her own, but also by joint activity with others who can act as agents, and also show and teach. It is therefore not surprising to find that there is a strong innate predisposition to cooperative, interpersonal experience, although this topic has been remarkably neglected in research. The process of sharing experience of surroundings and things becomes systematic in the second half of the first year of life, long before speech, and early enough to affect the basic maturation of object perception, although whether it actually does so is open to investigation. It is latent in the exceedingly complex person-perceiving functions and expressive actions of infants 2–3 months of age (Trevarthen, 1977, 1983, 1984*a*; Trevarthen *et al.,* 1981).

Of all the achievements of infants brought to light by recent observational research, those pertaining to recognition of and communication with persons are the most remarkable. That they permit visual identification of faces and hands as organs of expression and visual and auditory perception of expressive movements of many kinds during the first months after birth has been established by imitation tests (Maratos, 1982; Uzgiris, 1974; Meltzoff and More, 1977; Field *et al.,* 1981). Infants are selectively responsive to the sight of human faces and to human voice sounds (Frantz, 1963; Lewis, 1969; Wolff, 1969) and categorize speech sounds in basic phonological forms (Eimas *et al.,* 1971; Trehub, 1973; Eisenberg, 1975; Mehler *et al.,* 1978).

Evidently infants perceive people preferentially. To do this they must have an innate knowledge of them as different from other events existing in the environment. The rapid development of game playing, appropriate signaling of interest and orientation, and reciprocal "conversation"-like performance in exchanges with people regulated by both participants shows that the brain of an infant is adapted to receive and give experience in exchange with others (Brazelton *et al.,* 1975; Schaffer, 1977; Stern, 1985; Trevarthen, 1983, 1984*a*). This process undoubtedly expresses an organization of the human brain for cooperative intelligence, but the neural systems involved in perceiving persons and what they do are virtually unimaginable on the basis of present knowledge.

6.9. Development of Asymmetry of Mental Functions in the Human Brain

That the cerebral hemispheres of humans are inherently specialized to perform very different psychological functions was demonstrated more than 100 years ago with the discovery of relationships between hemispheric site of brain injury and particular neurological or psychological abnormalities (Vinken and Bruyn, 1969; Hecaen and Albert, 1978; Trevarthen, 1984*b*). Complex mental processes were implicated that bear no obvious relationship to the embryogenetic somatotopic mapping of each half of the body and each half of body-centered perceptual space in the opposite half of the brain. Although the neuroanatomical basis for the lateralized functions is just beginning to be revealed (Galaburda *et al.,* 1978) (Figure 23), these high-level psychological functions offer the best indirect evidence for protracted postnatal differentiation of adaptive cerebral mechanisms in humans.

In most cases, disorders of speaking, writing, or understanding of language, as well as inability to perform analytical thought or serial reasoning as in mathematical calculation, follow lesions in the left hemisphere. Accumulated evidence of many kinds leads to the conclusion that, in the adult population, nearly 90% of such disorders are genotypic right-handers; of these more than 99% have left-hemisphere dominance for language. Of the approximately 10% of people who are left-handed or with mixed handedness, about 56% have left-hemisphere language dominance (Levy, 1976).

Lesions of the right hemisphere may be diagnosed by demonstration of disorders in visuospatial perception, in body image, and in recognition of complex perceptual *gestalts,* including such diverse entities as faces and musical chords. Recent interpretations take the view that the hemispheres develop different kinds of perceptual processing in an orderly topography of functional regions, the right being specialized for parallel or simultaneous synthesis of images signifying layout or configuration, the left performing an analytic function dependent on brain processes that specify sequences or chains of relationship between identified elements.

Understanding of the brain mechanisms of these complementary psychological processes of the hemispheres has been advanced by detailed study of consciousness and reasoning in a few epileptic patients in whom the cerebral hemispheres have been surgically disconnected to control the spread of severe seizures that were unresponsive to drugs (Sperry *et al.*, 1969; Sperry, 1970, 1974; Trevarthen, 1984*b*). An elaborate series of tests with split-brain patients shows that certain perceptual processes, especially generation of images of the sounds of words on which comprehension and, above all, production or speech or writing depends, are confined to the left hemisphere (Gazzaniga and Sperry, 1967; Levy and Trevarthen, 1977; Zaidel, 1978). The ability visually to perceive whole forms from fragmentary stimuli, or the unanalyzed configuration of complex terms seen, heard, or felt as whole identities simultaneously, are much more easily performed by the right hemisphere (Nebes, 1974; Levy *et al.*, 1972; Levy and Trevarthen, 1977). Findings with patients with lateralized brain injury have thus been confirmed and clarified.

Comparable asymmetries are also brought out in normal subjects by psychophysical tests that measure the rate and efficiency of perception with different kinds of stimuli confined to the left or right visual field or hand (Bryden, 1982; Bradshaw and Nettleton, 1983; Corballis, 1983). The effects observed, although consistent, are weak and require statistical methods for their demonstration. To test lateralization of auditory perception, conflicting stimuli are presented to the two ears (dichotic stimulation); this procedure causes the bilateral attention processes that normally mask asymmetries at the periphery to be partially blocked. Hemispheric differences in perception of sounds become apparent as differences in hearing by the two ears.

There is no doubt that hemispheric lateralization of mental function is under some heritable control, and it has been suggested that a relatively simple one- or two-gene model may determine both handedness (or hemispheric hand control) and hemispheric dominance for language (Annett, 1964; Levy and Nagylaki, 1972; Levy, 1976). Corballis and Beale (1976) accept a theory of Morgan (1976) that lateralization of brain functions may be a result of an asymmetry of a cytoplasmic factor in the egg cell. Performance of identical twins on intelligence tests compared with the performance of nonidentical subjects matched for age indicates that visuospatial functions associated with the right hemisphere are also under genetic control. Reading disability (dyslexia) in schoolchildren correlates with indistinct manual dominance and confusions of left–right sense indicative of incomplete lateralization of functions in the cerebral hemispheres (Corballis and Beale, 1976). It has been suggested that performance differences of adolescents on tests of verbal and spatial abilities are related to maturation rates more strongly than to sex. Late-maturing children were also more lateralized for speech than those maturing early. Sex differences in mental abilities may therefore be due to differences in cerebral functional development correlated with differential rates of biological maturation (Waber, 1976). Geschwind and Behan (1982) have found correlations among reading difficulties, lefthandedness, and immune disorders. They suggest that hemispheric lateralization in the fetus may be regulated by testosterone slowing development of the left hemisphere while at the same time favoring later immune disorder.

The effects of one-sided brain injury at different ages show that the functions of the hemispheres become progressively more distinct over a period of years after birth (Lenneberg, 1967; Hecaen, 1976). Brain injury before or at birth rarely produced specific and isolated language disorders later. Growth of language may be retarded, but the degree of retardation has not seemed to correlate with the site of lesion, if the injury occurs before the age of 4 (Basser, 1962; Kraschen, 1973). Symptoms of language disorder are claimed to occur with right-hemisphere lesions only in children less than 5 years old, unless they are genetically predisposed to center language control bilaterally or in the right hemisphere (Kraschen, 1973, 1975; Hécaen, 1976).

Woods and Teuber (1978) suggest that the apparent involvement of the right hemisphere in language is, for most cases, an artifact of postoperative infection of the remaining left hemisphere, which was common in cases before the adequate antibiotic treatment was available. With injuries in early childhood there is a progressive clarification of the initial clinical effects with age of trauma, which correlates with mastery of language in normal development (Lenneberg, 1967). Injury to or removal of the left hemisphere of children in the first few years causes the child to become mute or to have disordered perception of speech. The effect of partial damage depends on the focus of lesion, whether toward frontal parts of the hemisphere (mutism), or exclusively in the temporal area (auditory perceptual defect) (Hécaen, 1976). Brain-injured young children do not exhibit logorrhea, and they rarely show paraphasia.

Usually a brain-injured child recovers an approximately normal level of language function completely inside a few weeks or a year, even after dominant (left) hemisphere lesions of a size that would produce permanent aphasia in an adult. Permanent effects of early trauma to the undeveloped speech control mechanisms of the young child can only be revealed by rigorous testing.

However, such testing proves that there is a difference in the language capacities of the hemispheres at birth. Dennis (1983) examined language functions in subjects with left or right hemisphere removed in early infancy and demonstrated characteristic left hemisphere and right hemisphere deficits in their language comprehension and particularly in reading and writing when the children were entering adolescence.

Tests with adults who have had early childhood brain lesions prove that the functional recovery may involve unusual development of language in the right hemisphere (Milner, 1974) and this may, in turn, protect their verbal functions at a later date against epileptic foci developing in the left hemisphere (Landsdell, 1962, 1969). Even so, the specialization of the left hemisphere for generative language and the right for visuospatial and constructional functions has definitely begun at birth and is reflected in more subtle psychological defects of patients who have had brain damage before or near birth (Woods and Teuber, 1973, 1978; Rudel and Denkla, 1974; Rudel *et al.*, 1974).

The anatomical foundations for asymmetrical functions of the hemisphere are now known to be established before birth. The hidden upper surface of the temporal lobe inside the Sylvian fissure of the left hemisphere (planum temporale), corresponding to Wernicke's area or the auditory association cortex, is enlarged in the majority of adult brains (Geschwind and Levitsky, 1968; Geschwind, 1974; Galaburda *et al.*, 1978) (Figure 23). This area is also enlarged in most fetuses and newborns (Teszner *et al.*, 1972; Witelson and Pallie, 1973; Wada *et al.*, 1975). Damage to it in an adult causes severe defects in perception of speech and writing. There are also enlargements of the parietal operculum on the left side above the Sylvian fissure (LeMay, 1976). Recent progress in understanding of the genetics, and epigenetics, of language disorders brings this subject within the scope of developmental brain research (Ludlow and Cooper, 1983). Galaburda (1983) has attempted a neuroembryological explanation of anatomical abnormalities in hereditary dyslexia that would link this late developing disorder in language processing with deviant migration of neuroblasts into particular zones of neocortex and thalamus in mid-fetal stages.

Psychological tests of young babies, mentioned above, show that perceptual categorization of speech sounds into entities that correspond to consonants and syllables in the production of mature speakers occurs very early in infancy, probably from birth (Eimas *et al.*, 1971; Trehub, 1973; Fodor *et al.*, 1975). Furthermore, EEG records have revealed distinctive evoked potentials for speech sounds in the left hemisphere of neonates as well as adults, as compared to the potentials evoked by melodies or noise (Molfese, 1983; Molfese *et al.*, 1975). These findings strongly support the view that complex perceptual mechanisms, specialized to regulate speech, are innate in man and functioning in infancy at some level (cf. Studdert-Kennedy, 1983).

The condition of language in the normal nondominant right hemisphere of right-handers has been brought out most clearly by studies of cases in which the left hemisphere has been removed after language is developed, and in split-brain subjects with hemispheres disconnected after the early teens (Smith, 1966; Gazzaniga and Sperry, 1967; Zaidel, 1978). The isolated or disconnected right hemisphere appears to have the vocabulary for comprehension of all grammatical classes of words of about a 15-year-old, but it is deficient in syntactic generative functions and is unable to formulate the acoustic image for a word to denote an object identified by nonlinguistic means. It may only speak a few automatic words or short phrases. Evidently there is accumulation of a lexicon throughout adulthood within the "minor" hemisphere, but by about 5 years of age the capacities to generate speech and to comprehend or recall context-free speech sequences more than one or two words in length are located entirely on the left side. Basic to this class of defect due to development of a strong lateralization of knowledge is loss by the right hemisphere of the function that may evoke the sound of a word for an object from the visual or tactile image of the object, or from its remembered identity (Levy and Trevarthen, 1977).

Recent clinical and experimental data suggest that perception of speech as an adjunct to the process of language generation is lateralized in the dominant hemisphere within the first 5 years after birth. After this, language development is funneled into the left hemisphere, probably by development of different neural systems with long cycles of differentiation. Receptive semantic function continues to develop in the minor hemisphere, possibly at a progressively reduced rate relative to the left hemisphere (Zaidel, 1978). Relocation of speech-synthesizing functions in the right hemisphere when the left is injured early in life may be supposed to involve a neural transformation within developing circuits analogous to that which permits optic terminals from a normally stimulated eye to invade the territory of a stimulus-deprived eye in layer IV of the striate cortex of the newborn monkey (Hubel *et al.*, 1977).

Any competitive interaction between territories in the neocortex of the hemispheres for representation of higher psychological processes and resultant asymmetrically patterned mechanisms must involve transmission of information through the corpus callosum, unlike the interactions of the striate cortex input that may occur directly between terminal arborizations of lateral geniculate cells representing the two eyes. If the conditions of interhemisphere communication are changed while the hemispheres are actively developing, then differentiation of speech-regulating and visuospatial mechanisms is imperfect, and the right hemisphere retains language functions it would normally lose. Conversely, injury or loss to the right hemisphere spoils the development of language receptive functions by forcing left hemisphere circuits to be occupied more than is normal with visuospatial operations. It may be presumed that learning of language, and practice of performative skills that depend largely on a strong inventory of visual identities and on resolution of visuospatial confusions inherent in the variety of layouts which the visual world presents, would together favor hemisphere differentiation toward an optimal dual-hemisphere system that is highly asymmetrical.

Presence of the commissures would appear to be necessary for hemispheric complementarity to be completed. This conclusion is supported by the results of tests on a patient with congenital absence of the corpus callosum—a malformation of the brain that comes about in the late embryoric period (Saul and Sperry, 1968). This subject showed none of the split-brain effects and no evidence of hemispheric specialization of function. It is possible that each hemisphere carried, besides the mechanisms necessary to generate speech, full bilateral representation of the visual field. If so, the development of mirror complementary striate area projections is a consequence of competition between left and right field projections that is mediated through the neocortical commissures. Less likely is the possibility of an abnormal linking of striate areas *via* an enlarged anterior commissure or through the thalamus or midbrain. A higher than average verbal IQ associated with a mild right hemisphere syndrome (Sperry, 1974) may indicate that invasion of the minor hemisphere by language functions had preempted circuits that would otherwise have mediated the visuospatial processes. This accords with Levy's view of the evolution and development of lateralized language functions in competition with visuospatial abilities (Levy, 1969).

The timetable of development for mechanisms of language revealed by the study of the effects of brain injury at different ages correlates well with our meager knowledge of maturation of cortical and subcortical circuits for audition and for motor control of speech (Lenneberg, 1967). Developments after the age of 5 years in posterior "association" cortex tally with the optimal period for acquisition of the higher complex intermodal skills of reading and writing. The main anatomical finding, principally from studies of myelinization of tracts and intracortical axons and their relationship to developments in language abilities is summarized in Figure 22, based on Lecours (1975).

It should be noted that this method of portraying language development fails to indicate the role played by innate structures that define the deep grammatical forms of human communication that are already evident in the behavior of infants, before words are used to give wealth and precision to messages of intent and understanding (Bruner, 1975; Trevarthen, 1974*c*, 1977; Halliday, 1975). Once again we are led to consider the contrasting roles in development of brain systems that either: (1) Process afferent information according to invariants in it that define layouts, events, objects, structures, and substances; (2) Coordinate movement; or (3) Relate present experience to an accumulated knowledge and understanding of what the world has to offer. Each of these three kinds of brain functions affect perception, but only the first is directly and exclusively perceptual (Trevarthen, 1983).

7. Conclusions

During the 1970s, advances in neurobiology intensified controversy about the causes of neural mappings that perform perceptual analysis of stimuli. New evidence for plasticity in the growth of retinotectal connections of amphibia obtained by Gaze and colleagues, and in the early postnatal development of visual analyzers in the cortex of cats and monkeys from Pettigrew, Blakemore, and others, led to the proposal that inherent chemical affinity mechanisms and guided nerve growth could provide no more than a rough outline of functional anatomy. This was a challenge to the view of the 1950s and 1960s that brain mechanisms were patterned first by an intrinsic mechanism, Sperry's chemoaffinity theory giving the best supported explanations. But regeneration experiments with fish and amphibia and developmental studies with embryo chicks and fetal hamsters continued to support Sperry's theory, and fetal and newborn monkeys were found by Hubel and Wiesel to develop elaborate segregation of ocular dominance territories and orientational specificity in visual cells, independent of stimulation. The application of new anatomical tracer techniques brought direct evidence for highly patterned segregations of axon terminals in the lateral geniculate nucleus and visual cortex and along corticocortical pathways of baby monkeys. A major new insight was achieved with the demonstration that both segregation of monocular dominance territories and development of higher-order binocular cells involve competitions in which the making of effective intercellular linkages depended on the balance of activity levels of presynaptic cells. The binocular projection on the optic lobes of *Xenopus,* involving mapping between a direct retinal input and an intertectal relay, was shown by Gaze and Keating to be formed by the same kind of process.

The 1980s seem to be leading us swiftly toward an entirely new conception of development in perceptual systems. Nature–nurture controversies seem increasingly pointless besides a flood of evidence for elaborate epigenetic processes that utilize both intrinsic prepatterning guided by chemoaffinity and selective response to both environmental and intrinsically generated excitation to carve out integrative mechanisms. Cell–cell interactions throughout the nervous system in embryo, fetus, and postnatal stages produce selective elimination of connections that arise in waves of excessive axon and synapse proliferation.

Mark, Changeux, Schneider, Innocenti, Willshaw and von der Malsburg, Meyer, Yoon, Constantine-Patton and others, deriving evidence from optic projections of fish and amphibia; neuromuscular connections of fish, salamanders, and rats; cerebellum of rats; and midbrain, thalamus or cortex of fetal hamsters and monkeys, and newborn kittens and monkeys, have proposed a two-stage structural segregation. Axons spread more widely and more prolifically than had been imagined, although not randomly, and they undergo complex mutual adjustments with loss of a large proportion of their collaterals through fetal and postnatal stages. It appears that embryogenic mechanisms that construct consistent anatomical arrangements in peripheral receptors and CNS intergrade with responsive cell growth. This, in the end, can retain highly specific impressions of stimulation in a memory that conditions future behavior. Recent work on monkey and cat visual cortex supports the same conception. There is no frontier between innate and acquired in brain growth.

Where anatomical developments of the primary visual cortex show modification with imbalance or incoordination of input from the two eyes in immediate postnatal stages, and refinement or tuning of primary feature detector systems is dependent on coincident retinal stimulations and systematic orienting of the two eyes to pick up features in the light array, perception of objects and events in all modalities at once must be the result of patterned growth in more extensive transcortical and corticosubcortical connections, such as those studied by Mishkin and colleagues, van Essen and Zeki in the monkey, and Sprague and Berlucchi in the cat.

Plasticity in the cortex appears to provide for adjustment to peripheral errors of development or trauma to receptor structures, and it is a means whereby accurate and orderly representation of the spatial and functional relations between retinal cells, and of the timing of their excitation by dynamic patterns of stimuli, can be conveyed into cortical cell arrays. The development of high-acuity temporospatial discrimination in kittens or baby monkeys and human infants in an early postnatal period results from this cortical growth response as Blakemore and Vital-Durand have discussed. But the process is certainly not a passive response of brain tissue. First, it is clear from the work of Hein and Imbert that brain-controlled oculomotor activity and proprioception from the orbit contribute to the development of visual representation of space for behaviour. And the stabilization of retinocortical projections and binocular integrations is also dependent on the level of intrinsic excitation of postsynaptic cells, for example from reticular formation input to the cortex, as Rauchecker and Singer demonstrated. In other words, the selective retentions of even primary visual couplings in formation of elementary cortical analyzers is a consequence of coincidence between internal "motivational" activity and certain frequently recurring patterns of input from receptor arrays.

There is increasing evidence that both primary circuit formation in embryo and fetus and postnatal refinement of perceptual systems with the help of experience are directed by intrinsic neuronal activity. The role of activity in interneurons of the reticular formation, limbic system, and so on, in guiding selective processes in sensory projections must now assume central theoretical importance. A model may be found in the embryo rat spinal cord, where primary dorsal root connections to the developing spinal cord nuclei are sorted by activity of interneurons (Baker *et al.*, 1984; Baker, 1985).

The recent psychophysical studies of infant vision by Held, Gwiazda and others, and Braddick and Atkinson give accurate data on improvements in spatial resolution, contrast sensitivity and oculomotor control of the retinal image. There is gratifying agreement with these psychological findings and the inferences that can be safely made from anatomical and physiological studies of the monkey visual system, with due allowance for the slower growth of the human visual apparatus. The findings have immediate application to control of damage to vision that can be caused by eye abnormalities, defective oculomotor coordination, or covering of one eye in infancy. But these developments seem only to concern the maturation of an efficient detector of visual detail. They do not relate to the development of intelligent perception of meaningful objects and their affordances.

Measurements of object recognition, problem solving and language in children show clearly that perception develops throughout childhood. Studies of evoked potentials and neuropsychological investigations support the idea that this development involves extensive differentiation in widespread cortical mechanisms that extend the processes of nerve-circuit formation through adolescence (Bradshaw and Nettleton, 1983; Corballis, 1983; Kirk, 1983; Trevarthen, 1983, 1984). Cerebral asymmetry of cognitive and perceptual functions develops slowly, and this appears to correlate with myelination of parts of the cortex remote from the primary areas. Innocenti's studies of the development of callosal and intrahemispheric association tracts in kittens offer a model for how human transcortical and intermodal mechanisms may be built as children acquire perceptual skills involved in speaking, recognizing individual persons by face, dress, and so on, reading and writing, using mathematical reasoning, playing music, finding their way about in a city, making functional mechanisms. All these activities, with many more, show evidence of cerebral functional asymmetry.

Hemispheric perceptual mechanisms show both extensive plasticity and strong intrinsic or innate organization. Both functional and anatomical changes result from hemispherectomy in infancy, deafness and learning of a visuomanual sign language, and social and educational deprivation. There are sex differences, and Geschwind and colleagues show that these may be linked to individual differences in cerebral growth patterns consequent on differences in the hormonal environment of the brain in development.

Recent studies of infants indicate that powerful intrinsic motives govern their access to experiences and that interpersonal communication enters into control of infant learning very early. Awareness of persons, manifest in the first two months, provides most impressive evidence for an integrated perception of phenomena by intermodal coordination. Experiments by E. J. Gibson and colleagues and by Alegria and Noirot, Mehler, Murray (Trevarthen, *et al.* 1981), Meltzoff, Field, and others, indicate that very young infants can detect persons from the dynamic patterns of visual and auditory stimulations and that the modalities integrate together immediately. Imitation studies show that even neonates relate sight and sound of another person to their own expressive impulses. They can identify their own body parts with the same parts of another person.

The reaching studies of von Hofsten strengthen earlier conclusions of Bower and Trevarthen that infants are capable of perceiving objects in space before they can manipulate them, and can project their arms in the right direction. But von Hofsten's proof that infants can perceive the velocity of an approaching object and make accurate interception from 4 months of age goes further. It shows that event perception is elaborately prepared for in the visual system and highly effective in detecting motions of objects at a distance as soon as accurate focalizing movements and a high degree of monocular acuity have been achieved.

Our ignorance of the growth of perceptual systems is most apparent in the area of intermodal association and the direct regulation of movements from dynamic sensory feedback. The intensive anatomical and physiological studies of primary and adjacent visual cortex can, presumably, be generalized to other modalities where fine discriminations depend on orderly, iterated, and complementary dispersion of afferent axons in an analyzer matrix. But the association cortices clearly have a vital role to play in object and event awareness, and they respond to higher-order stimulus information that is difficult to measure with psychophysical techniques. It would seem likely that, for a time at least, rather large-scale anatomical researches in growing brains with various experimental manipulations at critical stages will provide the most useful data on functions that undergo programmed development throughout childhood. We need to know how functional plasticity in these systems, related learning to perceive meanings, depends on: (1) Programmed generation of an exuberance of cells; axons, connections or potential connections; and (2) Selective retention or reinforcement of elements to provide new representations of experience that specify how intrinsic motives can be satisfied. This is a question of adaptive epigenesis, not an attempt to resolve a nature vs. nurture dispute. Increasing knowledge of the genetical control of processes of neurogenesis that shape cortical and subcortical perceptual systems in the fetus and after birth, along the lines of Galaburda's analysis of the "anatomical phenotype" of a dyslexic brain (Galaburda, 1983), will be needed if we are to explain persistent individual differences in perceptual learning, and this must be a problem that relates to the stage-by-stage appearance of perceptual skills in "normal" development.

ACKNOWLEDGMENTS. In reviewing recent developments in neurobiology, the author received most helpful comments and suggestions from Michael Gaze, François Vital-Durand, and Giorgio Innocenti. The author accepts full responsibility for interpretations he has made of advice and information so generously given.

8. References

Adinolfi, A. M., 1971, The ultrastructure of synaptic junctions in developing cerebral cortex, *Anat. Rec.* **169**:226.

Adrian, E. D., 1946, *The Physical Basis of Perception,* Oxford University Press.

Alegria, J., and Noirot, E., 1978, Neonate orientation behaviour towards the human voice, *Early Hum. Dev.* **1**:291–312.

Allen, R. D., and Haberey, M., 1973, The behavior of amoebae, in: *Behaviour of Micro-Organisms* (A. Pérez-Miravete, ed.), pp. 157–167, Plenum Press, London and New York.

Allman, J. Baker, J., Newsome, W. and Petersen, S., 1981, Visual topography and function: Cortical visual areas in the owl monkey, in: *Multiple Visual Areas* (C. Woolsey, ed.), pp. 171–186, Humana Press, Clifton, New Jersey.

Alstermark, B., Eide, E., Gorska, T., Lundberg, A., and Pettersson, L.-G., 1984, Visually guided switching of forelimb target reaching in cats, *Acta Physiol. Scand.* **120**:151–153.

Altman, J., 1967, Postnatal growth and differentiation of the mammalian brain, with implications for a morphological theory of memory, in: *The Neurosciences: A Study Program* (G. Quarton, T. Melnechuk, and F. O. Schmitt, eds.), pp. 723–743, Rockefeller University Press, New York.

Angevine, J. B., Jr., 1970, Critical cellular events in the shaping of neural centers, in: *The Neurosciences: A Second Study Program* (F. O. Schmitt and T. Melnechuk, eds.), pp. 62–72, Rockefeller University Press, New York.

Anker, R. L., and Cragg, B. G., 1975, Development of the extrinsic connections of the visual cortex of the cat, *J. Comp. Neurol.* **154**:29–42.

Annett, M. A., 1964, A model of the inheritance of handedness and cerebral dominance, *Nature (London)* **204**:59–60.

Anokhin, P. K., 1964, Systemogenesis as a general regulator of brain development, *Prog. Brain Res.* **9**:54–86.

Arey, L. B., 1965, *Developmental Anatomy,* 7th ed., W. B. Saunders, Philadelphia.

Aslin, R. N., 1981, Development of smooth persuit in human infants, in: *Eye Movements; Cognition and Visual Perception* (D. F. Fisher, R. A. Monty, and J. W. Senders, eds.), Lawrence Erlbaum Associates, Hillsdale, New Jersey.

Aslin, R. N., and Jackson, R. W., 1979, Accommodative-convergence in young infants: Development of a synergistic sensory-motor system, *Can. J. Psychol.* **33**:222–231.

Åström, K. E., 1967, On the early development of the isocortex in fetal sheep, *Prog. Brain Res.* **26**:1–59.

Atkinson, J., and Braddick, O., 1979, New techniques for assessing vision, *Child Care Health Dev.* **5**:389–398.

Atkinson, J., Braddick, O., and Braddick, F., 1974, Acuity

and contrast sensitivity of infant vision, *Nature (London)* **247**:403–404.

Attardi, D. G., and Sperry, R. W., 1963, Preferential selection of central pathways by regenerating optic fibers, *Exp. Neurol.* **7**:46–64.

Axelrod, S., 1959, *The Effects of Early Blindness: Performance of Blind and Sighted Children on Tactile and Auditory Tasks* (Research Series, No. 7), American Foundation for the Blind, New York.

Baker, R. E., 1972, Biochemical specification versus specific regrowth in the innervation of skin grafts in anurans, *Nature New Biol.* **230**:235–237.

Baker, R. S., 1985, Horseradish peroxidase tracing of dorsal root ganglion afferents within fetal mouse spinal cord explants chronically exposed to tetrodotoxin, *Brain Res.* **334**:357–360.

Baker, R. S., Corner, M. A., and Habets, A. M. M. C., 1984, Effects of chronic suppression of bioelectrical activity on the development of sensory ganglion evoked responses in spinal cord explants, *J. Neurosci.* **4**:1187–1192.

Balázs, R., Lewis, P. D., and Patel, A. J., 1975, Effects of metabolic factors on brain development, in: *Growth and Development of the Brain* (M. A. B. Brazier, ed.), pp. 83–115, Raven Press, New York.

Ball, W., and Tronick, E., 1971, Infant responses to impending collision: Optical and real, *Science* **171**:818–820.

Banks, M. S., 1980, The development of visual accommodation during early infancy, *Child Dev.* **51**:646–666.

Banks, M. S., Aslin, R. N., and Letson, R. D., 1975, Critical period for the development of human binocular vision, *Science* **190**:675–677.

Barlow, H. B., 1975, Visual experience and cortical development, *Nature (London)* **258**:199–204.

Barlow, H. B., and Pettigrew, J. D., 1971, Lack of specificity of neurons in the visual cortex of young kittens, *J. Physiol. (London)* **218**:8P–100P.

Barlow, H. B., Blakemore, C., and Pettigrew, J. D., 1967, The neural mechanism of binocular depth discrimination, *J. Physiol. (London)* **193**:327–342.

Barondes, S. H., 1970, Brain glycomacromolecules in interneuronal recognition, in: *The Neurosciences: A Second Study Program* (F. O. Schmitt, G. C. Quarton, and T. Melnechuk, eds.), pp. 744–760, Rockefeller University Press, New York.

Bartelmez, G. W., 1922, The origin of the otic and optic primordia in man, *J. Comp. Neurol.* **34**:201–232.

Bartelmez, G. W., and Blount, M. P., 1954, The formation of neural crest from the primary optic vesicle in man, *Carnegie Inst. Wash. Publ. 603, Contrib. Embryol.* **35**:55–71.

Bartelmez, G. W., and Dekaban, A. S., 1962, The early development of the human brain, *Carnegie Inst. Wash. Publ. 621, Contrib. Embryol.* **37**:15–32.

Barth, L. G., 1953, *Embryology,* Dryden Press, New York.

Bartlett, R. C., 1932, *Remembering,* Cambridge University Press, Cambridge.

Basser, L. S., 1962, Hemiplegia of early onset and the faculty of speech with special reference to the effect of hemispherectomy, *Brain* **85**:427–460.

Bauer, J., and Held, R., 1975, Comparison of visual guided reaching in normal and deprived infant monkeys, *J. Exp. Psychol. Anim. Behav. Proc.* **1**:298–308.

Baxter, B. L., 1966, Effect of visual deprivation during postnatal maturation on the electroencephalogram of the cat, *Exp. Neurol.* **14**:224–237.

Beazley, L. D., 1975, Development of intertectal neuronal connections in *Xenopus.* The effects of contralateral transposition of the eye and of eye removal, *Exp. Brain Res.* **23**:505–518.

Beazley, L. D., Keating, M. J., and Gaze, R. M., 1972, The appearance, during development, of responses in the optic tectum following visual stimulation of the ipsilateral eye in *Xenopus laevis, Visual Res.* **12**:407–410.

Beintema, D. J., 1968, *A Neurlogical Study of Newborn Infants.* Spastics International–Heinemann, London.

Bellugi, U., Poizner, H., and Klima, E. S., 1983, Brain organization for language: Clues from sign aphasia, *Hum. Neurobiol.* **2**:155–170.

Benowitz, L. I., and Karten, H. J., 1976, Organization of the tectofugal visual pathway in the pigeon: A retrograde transport study, *J. Comp. Neurol.* **167**:503–520.

Bergström, R. M., 1969, Electrical parameters of the brain during ontogeny, in: *Brain and Early Behavior: Development in the Fetus and Infant* (R. J. Robinson, ed.), pp. 15–41, Academic Press, New York.

Berman, N., and Hunt, R. K., 1975, Visual projections to the optic tecta in *Xenopus* after partial extirpation of the embryonic eye, *J. Comp. Neurol.* **162**:23–42.

Bernstein, N., 1967, *The Coordination and Regulation of Movements,* Pergamon, Oxford.

Berry, M., 1974, Development of neocortex of the rat, in: *Studies in the Development of Behavior and the Nervous System* (G. Gottlieb, ed.). Vol. 2, pp. 19–115, Academic Press, New York.

Berry, M., and Rogers, A. W., 1966, Histogenesis of mammalian neocortex, in: *Evolution of the Forebrain* (R. Hassler and H. Stephan, eds.), pp. 197–205, Thieme, Stuttgart.

Beverley, K. I., and Regan, D., 1980, Visual sensitivity to the shape and size of a moving object: Implications for models of object perception, *Perception* **9**:151–160.

Bjorklund, A., and Stenevi, U., 1979, Regeneration of monoaminergic and cholinergic neurons in the mammalian central nervous system, *Physiol. Rev.* **59**:62–100.

Blakemore, C., 1974. Development of functional connexions in the mammalian visual system, *Br. Med. Bull.* **30**:152–157.

Blakemore, C. B., and Cooper, G. F., 1970, Development of the brain depends on the visual environment, *Nature (London)* **228**:477–478.

Blakemore, C., and Van Sluyters, R. C., 1974, Reversal of the physiological effects of monocular deprivation in kittens. Further evidence for a sensitive period, *J. Physiol. (London)* **237**:195–216.

Blakemore, C., and Vital-Durand, F., 1981, Postnatal development of the monkey's visual system, in: *The Fetus and Independent Life* (Ciba Foundation Symposium 86), pp. 152–171, Pitman, London.

Blakemore, C., and Vital-Durand, F., 1982, Development of contrast sensitivity by neurones in monkey cortex, *J. Physiol. (London)* **334**:18–19P.

Blakemore, C. B., Garey, L. J., and Vital-Durand, F., 1981, Orientation preferences in the monkey's visual cortex, *J. Physiol. (London)* **319**:78P.

Blakemore, C., Vital-Durand, F., and Garey, L. J., 1981, Recovery from monocular deprivation in the monkey. I. Reversal of physiological effects in the visual cortex, *Proc. R. Soc. Lond. B* **213**:399–423.

Boothe, R. G., Williams, R. A., Kiorpes, L., and Teller, D. Y., 1980, Development of contrast sensitivity in infant *Macaca nemestrina* monkeys, *Science* **208**:1290–1292.

Bornstein, M. H., 1975, Qualities of colour vision in infancy, *J. Exp. Child Psychol.* **19**:401–419.

Böving, B. G., 1965, Anatomy of reproduction, in: *Obstetrics,* 13th ed. (J. P. Greenhill, ed.), pp. 8–23, W. B. Saunders, Philadelphia.

Bower, T. G. R., 1974, *Development in Infancy,* W. H. Freeman, San Francisco.

Bower, T. G. R., 1978, Perceptual development: Object and space, in: *Handbook of Perception* (E. C. Carterette and M. P. Friedman, eds.), Vol. 8, pp. 83–103, Academic Press, New York.

Braddick, O., and Atkinson, J., 1979, Accommodation and acuity in the human infant, in: *Developmental Neurobiology of Vision* (R. D. Freedman, ed.), pp. 279–290, NATO Advanced Study Institutes Series, Plenum Press, New York.

Brazelton, T. B., Tronick, E., Adamson, L., Als, H., and Wise, S., 1975, Early mother–infant reciprocity, in: *Parent–Infant Interaction,* Ciba Foundation Symposium, No. 33, New Series (M. O'Connor, ed.), pp. 137–154, Elsevier–Excerpta Medica–North Holland, New York.

Brinkman, J., and Kuypers, H. G. J. M., 1973, Cerebral control of contralateral and ipsilateral arm, hand and finger movements in the split-brain rhesus monkey, *Brain* **96**:653–674.

Bronson, G., 1974, The postnatal growth of visual capacity, *Child Dev.* **45**:873–890.

Bronson, G. W., 1982, *The Scanning Patterns of Human Infants: Implications for Visual Learning,* Ablex, Norwood, New Jersey.

Bruner, J. S., 1975, Ontogenesis of speech acts, *J. Child Lang.* **2**:1–19.

Buisseret, P., and Gary-Bobo, E., 1979, Development of visual cortical orientation specificity after dark rearing: Role of extraocular proprioception, *Neurosci. Lett.* **13**:259–263.

Bunt, S. M., 1982, Retinotopic and temporal organization of the optic nerve and tracts in the adult goldfish, *J. Comp. Neurol.* **206**:209–226.

Busnel, M.-C., and Grenier-Deferre, C., 1983, And what of fetal audition, in: *The Behavior of Human Infants* (A. Oliverio and M. Zappella, eds.), pp. 93–126. Plenum Press, New York.

Carey, S., and Diamond, R., 1980, Maturational determination of the developmental course of face encoding, in: *Biological Studies of Mental Processes* (D. Caplan, ed.), pp. 60–93, M.I.T. Press, Cambridge, Massachusetts.

Carmichael, L., 1926, The development of behavior in vertebrates experimentally removed from the influence of external stimulation, *Psychol. Rev.* **33**:51–58.

Changeux, J.-P., 1979, Molecular interactions in adult and developing neuromuscular functions, in: *The Neurosciences: Fourth Study Program,* pp. 749–778, M.I.T. Press, Cambridge, Massachusetts.

Changeux, J.-P., and Danchin, A., 1976, Selective stabilization of developing synapses as a mechanism for the specification of neuronal networks, *Nature (London)* **264**:705–712.

Changeux, J.-P., Courrège, P., and Danchin, A., 1973, A theory of the epigenesis of neuronal networks by selective stabilization of synapses, *Proc. Natl. Acad. Sci. U.S.A.* **70**:2974–2978.

Chi, J. G., Dooling, E. C., and Gillies, F. H., 1977, Gyral development of the human brain, *Ann. Neurol.* **1**:86–93.

Chow, K. L., and Steward, D. L., 1972, Reversal of structural and functional effects of long-term visual deprivation in cats, *Exp. Neurol.* **34**:409–433.

Chung, S. H., and Cooke, J., 1975, Polarity of structure and of ordered nerve connections in the developing amphibian brain, *Nature (London)* **258**:126–132.

Chung, S. H., and Cooke, J., 1978, Observations on the formation of the brain and of nerve connections following embryonic manipulation of the amphibian neural tube, *Proc. R. Soc. Lond. B* **201**:335–373.

Chung, S. H., Keating, M. H., and Bliss, T. V. P., 1974, Functional synaptic relations during the development of the retino-tectal projection in amphibians, *Proc. R. Soc. Lond. B* **187**:449–459.

Clark, P. G. H., Ramachandran, V. S., and Whitteridge, D., 1979, The development of the binocular depth cells in the secondary visual cortex of the lamb, *Proc. R. Soc. Lond. B* **204**:455–465.

Coghill, G. E., 1924, Correlated anatomical and physiological studies of the growth of the nervous system of *Amphibia.* IV. Rates of proliferation and differentiation in the central nervous system of *Amblystoma, J. Comp. Neurol.* **37**:71–120.

Coghill, E. G., 1929, *Anatomy and the Problem of Behavior,* Cambridge University Press, London.

Conel, J. LeRoy, 1939–1963, *The Postnatal Development of the Human Cerebral Cortex,* Vols. I–VI, Harvard University Press, Cambridge, Massachusetts.

Constantine-Paton, M., 1983, Position and proximity in the development of maps and stripes, *Trends Neurosci.* **6(1)**:32–36.

Constantine-Paton, M., and Law, M. I., 1978, Eye-specific termination bands in tecta of three-eyed frogs, *Science* **202**:639–641.

Cook, J. E., 1982, Errant optic axons in the normal goldfish retina reach retinotopic tectal sites, *Brain Res.* **250**:154.

Cook, J., and Gaze, R. M., 1983, The positional coding system in the early eye rudiment of *Xenopus laevis* and its modification after grafting operations, *J. Emb. Exp. Morphol.* **77**:53–71.

Cook, J. E., and Horder, T. J., 1974, Interactions between optic fibers in their regeneration to specific sites in the goldfish tectum, *J. Physiol. (London)* **241**:89–90.

Corballis, M. C., and Beale, I. L., 1976, *The Psychology of Left and Right,* Lawrence Erlbaum Associates, Hillsdale, New Jersey.

Cowan, W. M., Martin, A. H., and Wenger, E., 1968, Mitotic patterns in the optic tectum of the chick during normal development and after early removal of the optic vesicle, *J. Exp. Zool.* **169**:71–92.

Cragg, B. G., 1972, The development of synapses in the cat visual cortex, *Invest. Ophthalmol.* **11**:377–385.

Cragg, B. G., 1975, The development of synapses in the visual system of the cat, *J. Comp. Neurol.* **160**:147–166.

Craik, K. J. W., 1943, *The Nature of Explanation,* Cambridge University Press, Cambridge.

Crossland, W. J., Cowan, W. M., Rogers, L. A., and Kelly, J. P., 1974, The specification of the retinotectal projection in the chick, *J. Comp. Neurol.* **155**:127–164.

Crossland, W. J., Cowan, W. M., and Rogers, L. A., 1975, Studies on the development of the chick optic tectum. IV. An autoradiographic study of the development of retinotectal connections, *Brain Res.* **91**:1–23.

Cynader, M., and Regan, D., 1982, Neurons in cat visual cortex tuned to the direction of motion in depth: Effect of positional disparity, *Vision Res.* **22**:967–982.

Daniel, P. M., and Whitteridge, D., 1961, The representation of the visual field on the cerebral cortex in monkeys, *J. Physiol. (London)* **159**:203–221.

Daniels, J. D., and Pettigrew, J. D., 1976, Development of neuronal responses in the visual system of cats, in: *Studies on the Development of Behavior and the Nervous System* (G. Gottlieb, ed.), Vol. 3, pp. 195–232, Academic Press, New York.

De Casper, A. J., and Fifer, W. P., 1980, Of human bonding: Newborns prefer their mothers' voices, *Science* **208**:1175–1176.

De Long, G. R., and Coulombre, A. J., 1965, Development of the retinotectal projection in the chick embryo, *Exp. Neurol.* **13**:351–363.

De Long, G. R., and Sidman, R. L., 1962, Effects of eye removal at birth on histogenesis of the mouse superior colliculus: An autoradiographic analysis with tritiated thymidine, *J. Comp. Neurol.* **118**:205–224.

Dennis, M., 1980, Language acquisition in a single hemisphere: Semantic organization, in: *Biological Studies of Mental Processes* (D. Caplan, ed.), pp. 159–185, M.I.T. Press, Cambridge, Massachusetts.

Dennis, M., 1980, Language in a congenitally acallosal brain, *Brain Lang.* **12**:33–53.

Dennis, M., 1983, The developmentally dyslexic brain and the written language skills of children with one hemisphere, in: *Neuropsychology of Language, Reading and Spelling* (U. Kirk, ed.), pp. 185–208, Academic Press, New York.

Diamond, I. T., and Hall, W. C., 1969, Evolution of the neo-cortex, *Science* **164**:251–262.

Dichgans, J., and Brandt, Th., 1974, The psychophysics of visually induced perception of self-motion and tilt, in: *The Neurosciences: Third Study Program* (F. O. Schmitt and F. G. Worden, eds.), pp. 123–129, M.I.T. Press, Cambridge, Massachusetts.

Dixon, J. S., and Cronly-Dillon, J. R., 1972, The fine structure of the developing retina in *Xenopus laevis, J. Embryol. Exp. Morphol.* **28**:659–666.

Dobbing, J., 1981, The later development of the brain and its vulnerability, in: *Scientific Foundations of Paediatrics,* 2nd ed. (J. A. Davis and J. Dobbing, eds.), pp. 744–759, Heinemann Medical Books, London.

Dobbing, J., and Smart, J. L., 1974, Vulnerability of developing brain and behavior, *Br. Med. Bull.* **30**:164–168.

Drager, U. C., 1975, Receptive fields of single cells and topography in mouse visual cortex, *J. Comp. Neurol.* **160**:269–290.

Drager, U. C., and Hubel, D. H., 1975, Physiology of visual cells in mouse superior colliculus and correlation with somatosensory and auditory input, *Nature (London)* **253**:203–204.

Drever, J., 1955, Early learning and the perception of space, *Am. J. Psychol.* **68**:605–614.

Dreyfus-Brisac, C., 1967, Ontogénèse du sommeil chez le prématuré humain: Étude polygraphique, in: *Regional Development of the Brain in Early Life* (A. Minkowski, ed.), pp. 437–457, Blackwell, Oxford.

Duffy, F. H., Snodgrass, S. R., Burchfiel, J. L., and Conway, J. L., 1976, Bicuculline reversal of deprivation amblyopia in the cat, *Nature (London)* **260**:256–257.

Dunnett, S. B., Björklund, A., and Stenevi, U., 1983, Dopamine-rich transplants in experimental parkinsonism, *Trends Neurosci.* **6**:266–275.

Eimas, P., Sigueland, E., Jusczyr, P., and Vigorito, J., 1971, Speech perception in infants, *Science* **171**:303–306.

Eisenberg, R. B., 1975, *Auditory Competence in Early Life: The Roots of Communicative Behavior,* University Park Press, Baltimore.

Evans, E. F., 1974, Neural processes for the detection of acoustic patterns and for sound localization, in: *The Neurosciences: Third Study Program* (F. O. Schmitt and G. Quarton, eds.), pp. 131–145, M.I.T. Press, Cambridge, Massachusetts.

Falkner, F., 1966, *Human Development,* W. B. Saunders, Philadelphia.

Fantz, R. L., 1963, Pattern vision in newborn infants, *Science* **140**:296–297.

Fantz, R. L., 1965, Ontogeny of perception, in: *Behavior of Non-human Primates: Modern Research Trends* (A. M. Schrier, H. F. Harlow, and F. Stollnitz, eds.), Vol. II, pp. 365–403, Academic Press, New York.

Fawcett, J. W., 1981, How axons grow down the *Xenopus* optic nerve, *J. Embryol. Exp. Morphol.* **65**:219–233.

Fawcett, J. W., and Gaze, R. M., 1982, The retino-tectal fibre pathways from normal and compound eyes in *Xenopus, J. Embryol. Exp. Morphol.* **72**:19–37.

Fawcett, J. W., and Willshaw, D. J., 1982, Compound eyes project stripes on the optic tectum in *Xenopus, Nature (London)* **296**:350–352.

Field, T. M., Woodson, R., Greenberg, R., and Cohen, D., 1982, Discrimination and imitation of facial expressions by neonates, *Science* **218**:179–181.

Finlay, B. L., Schneps, S. E., and Schneider, G. E., 1979, Orderly compression of the retinotectal projection following partial tectal ablation in newborn hamster, *Nature (London)* **280**:153–155.

Finlay, B. L., Schneps, S. E., Wilson, K. G., and Schneider, G. E., 1978, Topography of visual and somatosensory projections to the superior colliculus of the golden hamster, *Brain Res.* **142**:223–235.

Finlay, B. L., Wilson, K. G., and Schneider, G. E., 1979, Anomalous ipsilateral retinotectal projections in Syrian hamsters with early lesions: Topography and functional capacity, *J. Comp. Neurol.* **183**:721–740.

Flandrin, J. M., and Jeannerod, M., 1975, Superior colliculus: Environmental influences on the development of directional responses in the kitten, *Brain Res.* **89**:348–352.

Flechsig, P., 1901, Developmental (myelogenetic) localization of the cerebral cortex in the human subject, *Lancet* **2**:1027–1029.

Fodor, J. A., Garrett, M. F., and Brill, S. L., 1975, Pi Ka pu: The perception of speech sounds by prelinguistic infants, *Percept. Psychophysiol.* **18**:74–78.

Fraiberg, S., 1974, Blind infants and their mothers: An examination of the sign system, in: *The Effect of the Infant on Its Caregiver* (M. Lewis and L. A. Rosenblum, eds.), pp. 215–232, Wiley, New York.

Fraiberg, S., 1976, Development of human attachment in infants blind from birth, *Merrill-Palmer Q.* **21**:315–334.

Fraser, S. E., and Hunt, R. K., 1980, Retinotectal specificity: Models and experiments in search of a mapping function, *Annu. Rev. Neurosci.* **3**:319–352.

Freedman, D. G., 1964, Smiling in blind infants and the issue of innate vs. acquired, *J. Child Psychol. Psychiatry* **5**:171–184.

Frégnac, Y., Trotter, Y., Bienen Stock, E., Buisseret, P., Gary-Bobo, E., and Imbert, M., 1981, Effect of neonatal unilateral enucleation on the development of orientation selectivity in the primary visual cortex of normally and dark-reared kittens, *Exp. Brain Res.* **42**:453–466.

French, V., Bryant, P. J., and Bryant, S. V., 1976, Pattern regulation in epimorphic fields, *Science* **193**:969–981.

Friedman, S. L., and Sigman, M. (eds.), 1981, *Preterm Birth and Psychological Development: Psychological Needs of Infants Born Preterm,* Academic Press, New York.

Frost, D. O., and Schneider, G. E., 1979, Plasticity of retinofugal projections after partial lesions of the retina in newborn Syrian hamster, *J. Comp. Neurol.* **185**:517–568.

Galaburda, A. M., 1983, Definition of the anatomical phenotype, in: *Genetic Aspects of Speech and Language Disorders* (C. L. Ludlow and J. A. Cooper, eds.), pp. 71–84, Academic Press, New York.

Galaburda, A. M., Le May, M., Kemper, T. L., and Geschwind, N., 1978, Right–left asymmetries in the brain, *Science* **199**:852–856.

Ganz, L., 1975, Orientation in visual space by neonates and its modification by visual deprivation, in: *The Developmental Neuropsychology of Sensory Deprivation* (A. H. Riesen, ed.), pp. 169–210, Academic Press, New York.

Ganz, L., 1978, Sensory deprivation and visual discrimination, in: *Handbook of Sensory Physiology,* (H. Teuber, ed.), Vol. VIII, Springer-Verlag, Berlin.

Ganz, L., and Fitch, M., 1968, The effect of visual deprivation on perceptual behavior, *Exp. Neurol.* **22**:638–660.

Ganz, L., Fitch, M., and Satterberg, J. A., 1968, The selective effect of visual deprivation on receptive field shape determined neurophysiologically, *Exp. Neurol.* **22**:614–637.

Ganz, L., Hirsch, H. V. B., and Bliss-Tieman, S., 1972, The nature of perceptual deficits in visually deprived cats, *Brain Res.* **44**:547–568.

Gaze, R. M., 1970, *Formation of Nerve Connections,* Academic Press, New York.

Gaze, R. M., 1974, Neuronal specificity, *Br. Med. Bull.* **30**:116–121.

Gaze, R. M., and Fawcett, J. W., 1983, Pathways of *Xenopus* optic fibres regenerating from normal and compound eyes under various conditions, *J. Embryol. Exp. Morphol.* **73**:17–38.

Gaze, R. M., and Grant, P., 1978, The diencephalic course of regenerating retinotectal fibres in *Xenopus* tadpoles, *J. Embryol. Exp. Morphol.* **44**:201–216.

Gaze, R. M., and Hope, R. A., 1976, The formation of continuously ordered mappings, *Prog. Brain Res.* **45**:327–357.

Gaze, R. M., and Jacobson, M., 1962, The projection of the binocular visual field on the optic tecta of the frog, *Q. J. Exp. Physiol.* **47**:273–280.

Gaze, R. M., and Keating, M. J., 1972, The visual system and "neuronal specificity," *Nature (London)* **237**:375–379.

Gaze, R. M., and Sharma, S. C., 1970, Axial differences in the reinnervation of the goldfish optic tectum by regenerating optic nerve fibers, *Exp. Brain Res.* **10**:171–181.

Gaze, R. M., Jacobson, M., and Szekely, G., 1963, The retinotectal projection in *Xenopus* with compound eyes, *J. Physiol. (London)* **165**:484–499.

Gaze, R. M., Keating, M. J., Szekely, G., and Beazley, L., 1970, Binocular interaction in the formation of specific intertectal neuronal connections, *Proc. R. Soc. Lond B* **175**:107–147.

Gaze, R. M., Keating, M. J., and Chung, S. H., 1974, The evolution of the retinotectal map during development in *Xenopus, Proc. R. Soc. Lond. B* **185**:301–330.

Gaze, R. M., Feldman, J. D., Cooke, J., and Chung, S.-H., 1979*a,* The orientation of the visuotectal map in *Xenopus:* Developmental aspects, *J. Emb. Exp. Morphol.* **53**:39–66.

Gaze, R. M., Keating, M. J., Ostberg, A., and Chung, S.-H., 1979*b,* The relationship between retinal and tectal growth in larvae *Xenopus:* Implications for the development of the retino-tectal projection, *J. Embryol. Exp. Morphol.* **53**:103–143.

Gazzaniga, M. S., and Sperry, R. W., 1967, Language after section of the cerebral commissures, *Brain* **90**:131–148.

Geschwind, N., 1974, The anatomical basis of hemispheric differentiation, in: *Hemisphere Function in the Human Brain* (S. J. Dimond and J. G. Beaumont, eds.), Elek Science, London.

Geschwind, N., 1982, Autoimmunity in left-handers, *Science* **215**:141–146.

Geschwind, N., and Behan, P., 1982, Left-handedness: Association with immune disease, migraine and developmental learning disorder, *Proc. Natl. Acad. Sci. U.S.A.* **79**:5097–5100.

Geschwind, N., and Levitsky, W., 1968, Human brain, left–right asymmetries in temporal speech regions, *Science* **161**:186–187.

Gesell, A. L., Thompson, H., and Amatruda, C. S., 1934, *Infant Behavior: Its Genesis and Growth,* McGraw-Hill, New York.

Gibson, E. J., 1969, *Principles of Perceptual Learning and Development,* Appleton-Century-Crofts, New York.

Gibson, E. J., and Spelke, E. S., 1983, The development of perception, in: *Cognitive Development* (J. H. Flavell and E. M. Markman, series eds.), Vol. 3: *Handbook of Child Psychology* (P. H. Mussen, ed.), Wiley, New York.

Gibson, J. J., 1952, The relation between visual and postural determinants of the phenomenal vertical, *Psychol. Rev.* **59**:370–375.

Gibson, J. J., 1954, The visual perception of object motion and subjective movement, *Psychol. Rev.* **61**:304–314.

Gibson, J. J., 1966, *The Senses Considered as Perceptual Systems,* Houghton Mifflin, Boston.

Gibson, J. J., 1979, *An Ecological Approach to Visual Perception,* Houghton Mifflin, Boston.

Glickstein, M., 1984, Brain pathways in the visual guidance of movement and the behavioural functions of the cerebellum, in: *Brain Circuits and Function of the Mind: Essays in Honor of Roger W. Sperry* (C. Trevarthen, ed.), Cambridge University Press, New York.

Glickstein, M., and Gibson, A. R., 1976, Visual cells in the pons of the brain, *Sci. Am.* **234**:90–98.

Glucksmann, A., 1940, The development and differentiation of the tadpole eye, *Br. J. Ophthalmol.* **24**:153–178.

Glucksmann, A., 1965. Cell death in normal development, *Arch. Biol.* **76**:419–437.

Goldman, P. S., 1971, Functional development of the prefrontal cortex in early life and the problem of neuronal plasticity, *Exp. Neurol.* **32**:366–387.

Goldman, P. S., 1972, Developmental determinants of cortical plasticity, *Acta Neurobiol. Exp.* **32**:495–511.

Goldman, P. S., 1976, An alternative to developmental plasticity: Heterology of CNS structures in infants and adults, in: *Plasticity and Recovery of Function in the Central Nervous System* (D. G. Stein, J. J. Rosen, and N. Butters, eds.), pp. 149–174, Academic Press, New York.

Goldman, P. S., and Nauta, W. J. H., 1977, Columnar distribution of cortico-cortical fibers in the frontal association, limbic and motor cortex of the developing rhesus monkey, *Brain Res.* **122**:393–413.

Goldman, P. S., Rosvold, H. E., and Mishkin, M., 1970, Evidence for behavioral impairment following prefrontal lobectomy in the infant monkey, *J. Comp. Physiol. Psychol.* **70**:454–463.

Gottlieb, G., 1971*a, Development of Species Identification in Birds: An Inquiry into the Prenatal Determinants of Perception,* University of Chicago Press, Chicago.

Gottlieb, G., 1971*b,* Ontogenesis of sensory function in birds and mammals, in: *Biopsychology of Development* (E. Tobach, L. R. Aronson, and E. Shaw, eds.), pp. 67–128, Academic Press, New York.

Gottlieb, G., 1973, Introduction to behavioral embryology, in: *Studies on the Development of Behavior and the Nervous System,* Vol. I: *Behavioral Embryology* (G. Gottlieb, ed.), pp. 3–45, Academic Press, New York.

Grafstein, B., and Murray, M., 1969, Transport of protein in goldfish optic nerve during regeneration, *Exp. Neurol.* **25**:494–508.

Greene, P. H., 1972, Problems of organization of motor systems: in: *Progress in Theoretical Biology* (R. Rosen and F. M. Snell, eds.), Vol. II, pp. 303–338, Academic Press, New York.

Gregory, R. L., 1970, *The Intelligent Eye,* Weidenfeld, London.

Gregory, R. L., and Wallis, J. C., 1963, Recovery from early blindness: A case study, *Exp. Psychol. Soc. Monog., No. 2,* Heffer, Cambridge.

Gross, C. G., and Mishkin, M., 1977. The neural basis of stimulus equivalence across retinal translation, in: *Lateralization in the Nervous System* (S. R. Harnand, ed.), pp. 109–121, Academic Press, New York.

Guillery, R. W., 1972, Binocular competition in the control of geniculate cell growth, *J. Comp. Neurol.* **144**:117–130.

Gustafson, T., and Wolpert, L., 1967, Cellular movement

and contact in sea urchin morphogenesis, *Biol. Rev. Cambridge Philos. Soc.* **42**:442–498.

Guth, L., 1975, History of central nervous system regeneration research, *Exp. Neurol.* **48**(3, part 2):3–15.

Gwiazda, J., Brill, S., Mohindra, I., and Held, R., 1980, Preferential looking acuity in infants from two to fifty-eight weeks of age, *J. Opt. Physiol. Optics* **57**:428–432.

Haith, M. M., 1979, Visual competence in early infancy, in: *Handbook of Sensory Physiology* (R. Held, R. Leibowitz, and H. L. Teuber, eds.), Vol. VIII, pp. 311–356, Springer-Verlag, Berlin.

Haith, M. M., 1980, *Rules that Babies Look By,* Lawrence Erlbaum Associates, Hillsdale, New Jersey.

Haith, M. H., and Campos, J. J., 1977, Human infancy, *Annu. Rev. Psychol.* **28**:251–293.

Halliday, M. A. K., 1975, *Learning How to Mean,* Arnold, London.

Hamburger, V., 1948, The mitotic patterns in the spinal cord of the chick embryo and their relation to histogenetic processes, *J. Comp. Neurol.* **88**:221–284.

Hamburger, V., 1963, Some aspects of the embryology of behavior, *Q. Rev. Biol.* **38**:342–365.

Hamburger, V., 1969, Origins of integrated behavior, in: *The Emergence of Order in Developing Systems* (M. Locke, ed.), pp. 251–271, Academic Press, New York.

Hamburger, V., 1973, Anatomical and physiological basis of embryonic motility in birds and mammals, in: *Studies on the Development of Behavior and the Nervous System,* Vol. 1: *Behavioral Embryology* (G. Gottlieb, ed.), pp. 51–76, Academic Press, New York.

Hamburger, V., 1981, Historical landmarks in neurogenesis, *Trends Neurosci.* **4**:151–155.

Hamilton, W. J., Boyd, J. D., and Mossman, H. W., 1962, *Human Embryology,* 3rd ed., Heffer, Cambridge.

Harris, W. A., 1982, The transplantation of eyes to genetically eyeless salamanders: Visual projections and somatosensory interactions, *J. Neurosci.* **2**:339–353.

Harrison, R. G., 1907. Observations on the living developing nerve fiber, *Anat. Rec.* **1**:116–118.

Harrison, R. G., 1910, The outgrowth of the nerve fiber as a mode of protoplasmic movement, *J. Exp. Zool.* **9**:787–848.

Harrison, R. G., 1935, On the origin and development of the nervous system studied by the methods of experimental embryology, *Proc. R. Soc. Lond. B* **118**:155–96.

Harrison, R. G., 1945, Relations of symmetry in the developing embryo, *Trans. Conn. Acad. Arts Sci.* **36**:277–330.

Haynes, H., White, B. L., and Held, R., 1965, Visual accommodation in human infants, *Science* **148**:528–530.

Heacock, A. M., and Agranoff, B. W., 1982, Protein synthesis and transport in the regenerating goldfish *Carassius auratus* visual system, *Neurochem. Res.* **7**:771–788.

Hebb, D. O., 1949, *The Organization of Behavior,* Wiley, New York.

Hécaen, H., 1976, Acquired aphasia in children and the ontogenesis of hemispheric functional specification, *Brain Lang.* **3**:114–134.

Hécaen, H., and Albert, M. L., 1978, *Human Neuropsychology,* Wiley, New York.

Hein, A., 1970, Recovering spatial motor coordination after visual cortex lesions, in: *Perception and Its Disorders* (D. Hamburg, ed.), Research Publications of the Association for Research in Nervous Mental Disease, Vol. 48, pp. 163–185, Williams & Wilkins, Baltimore.

Hein, A., and Diamond, R. M., 1971, Independence of cat's scotopic and photopic systems in acquiring control of visually guided behavior, *J. Comp. Physiol. Psychol.* **76**:31–38.

Hein, A., and Held, R., 1967, Dissociation of the visual placing response into elicited and guided components, *Science* **158**:390–392.

Hein, A., Gower, E. C., and Diamond, R. M., 1970*a*, Exposure requirements for developing the triggered component of the visual-placing response, *J. Comp. Physiol. Psychol.* **73**:188–192.

Hein, A., Held, R., and Gower, E. C., 1970*b*, Development and segmentation of visually controlled movement by selective exposure during rearing, *J. Comp. Physiol. Psychol.* **73**:181–187.

Hein, A., Vital-Durand, F., Salinger, W., and Diamond, R., 1979, Eye movements initiate visual-motor development in the cat, *Science* **204**:1321–1322.

Held, R., and Bauer, J. A., 1967, Visual guided reaching in infant monkeys after restricted rearing, *Science* **155**:718–720.

Held, R., and Bauer, J. A., 1974, Development of sensorially guided reaching in infant monkeys, *Brain Res.* **71**:265–271.

Held, R., and Freedman, S. J., 1963, Plasticity in human sensorimotor control, *Science* **142**:455–462.

Held, R., and Hein, A., 1963, Movement-produced stimulation in the development of visually guided behavior, *J. Comp. Physiol. Psychol.* **56**:872–876.

Held, R., Birch, E., and Gwiazda, J., 1980, Stereoacuity of human infants, *Proc. Natl. Acad. Sci. U.S.A.* **77**:5572–5574.

Held, R., Shimojo, S., and Gwiazda, J., 1984, Gender differences in the early development of human visual resolution. Proceedings of the ARVO Meeting, April–May, 1984 (abstract No. 90), *Invest. Ophthalm. Vis. Sci.* **25**:220.

Herrick, C. J., 1948, *The Brain of the Tiger Salamander,* University of Chicago Press, Chicago.

Hess, E. H., 1956, Space perception in the chick, *Sci. Am.* **195**:71–80.

Hewitt, W., 1962, The development of the corpus callosum, *J. Anat.* **96**:355–358.

Hibbard, E., 1986, Retino-tectal connections made through ectopic optic nerves, in: *Brain Circuits and Functions of the Mind: Essays in Honor of Roger W. Sperry* (C. Trevarthen, ed.), Cambridge University Press, New York.

Hibbard, E., and Omberg, R. L., 1976, Restoration of vision in genetically eyeless axolotls, *Exp. Neurol.* **50**:113–123.

Hickey, 1981, The developing visual system, *Trends Neurosci.* **4**:41–44.

Hirsch, H. V. B., and Spinelli, D. N., 1970, Visual experience modifies distribution of horizontally and vertically oriented receptive fields in cats, *Science* **168**:869–871.

Hofsten, C., von, 1980, Predictive reaching for moving objects by human infants, *J. Exp. Child Psychol.* **30**:369–382.

Hofsten, C. von, 1982, Eye–hand coordination in the newborn, *Dev. Psychol.* **18**:450–467.

Hofsten, C., von, 1983, Catching skills in infancy, *J. Exp. Psychol. Hum. Percep. Perf.* **9**:75–85.

Hofsten, C., von, 1984, Developmental changes in the organization of prereaching movements, *Dev. Psychol.* **20**: 378–388.

Hofsten, C., von, and Fazel-Zandy, S., 1984, Development of visually guided hand orientation in reaching, *J. Exp. Child Psychol.* **38**:208–219.

Hohman, A., and Creutzfeldt, O. D., 1975, Squint and development of binocularity in humans, *Nature (London)* **254**:613–614.

Hollyfield, J. G., 1971, Differential growth of the neural retina in *Xenopus laevis* larvae, *Dev. Biol.* **24**:264–286.

Holst, E., von, 1954, Relations between the central nervous system and the peripheral organs, *Br. J. Anim. Behav.* **3**:89–94.

Holtfreter, J., 1939, Gewebeaffinitat, ein Mittel der embryonalen Formbildung, *Arch. Exp. Zellforsch.* **23**:169–209.

Hooker, D., 1952, *The Prenatal Origin of Behavior,* University of Kansas Press, Lawrence, Kansas.

Hope, R. A., Hammond, B. J., and Gaze, R. M., 1976, The arrow model: Retinotectal specificity and map formation in the goldfish visual system, *Proc. R. Soc. Lond. Ser. B.* **194**:447–466.

Hubel, D. H., and Wiesel, T. N., 1962, Receptive fields, binocular interaction and functional architecture in the cat's visual cortex, *J. Physiol. (London)* **160**:106–154.

Hubel, D. H., and Wiesel, T. N., 1963, Receptive fields of cells in striate cortex of very young, visually inexperienced kittens, *J. Neurophysiol.* **26**:994–1002.

Hubel, D. H., and Wiesel, T. N., 1965, Binocular interaction in the striate cortex of kittens reared with artificial squint, *J. Neurophysiol.* **28**:1041–1059.

Hubel, D. H., and Wiesel, T. N., 1968, Receptive fields and functional architecture of monkey striate cortex, *J. Physiol. (London)* **195**:215–243.

Hubel, D. H., and Wiesel, T. N., 1970, The period of susceptibility to the physiological effects of unilateral eye closure in kittens, *J. Physiol. (London)* **206**:419–436.

Hubel, D. H., and Wiesel, T. N., 1974*a*, Sequence, regularity and geometry of orientation columns in the monkey striate cortex, *J. Comp. Neurol.* **158**:267–294.

Hubel, D. H., and Wiesel, T. N., 1974*b*, Uniformity of monkey striate cortex: A parallel relationship between field size scatter and magnification factor, *J. Comp. Neurol.* **158**:295–306.

Hubel, D. H., Wiesel, T. N., and Le Vay, S., 1977, Plasticity of ocular dominance columns in monkey striate cortex, *Philos. Trans. Roy. Soc. London, Ser. B* **278**:131–163.

Hubel, D. H., Wiesel, T. N., and Stryker, M. P., 1978, Anatomical demonstration of orientation columns in macque monkey, *J. Comp. Neurol.* **177**:361–380.

Hughes, A. F. W., 1968, *Aspects of Neural Ontogeny,* Academic Press, New York.

Humphrey, A. L., Skeen, L. C., and Norton, T. T., 1980, Topographic organization of the striate cortex of the tree shrew *(Tupaia glis).* II. Deoxyglucose mapping, *J. Comp. Neurol.* **192**:549–566.

Humphrey, N., 1974, Vision in monkey without striate cortex: A case study, *Perception* **3**:241–255.

Humphrey, T., 1964, Some correlations between the appearance of human fetal reflexes and the development of the nervous system, *Prog. Brain Res.* **4**:93–135.

Humphrey, T., 1969, Postnatal repetition of human prenatal activity sequences with some suggestions of their neuroanatomical basis, in: *Brain and Early Behavior: Development in the Fetus and Infant* (R. J. Robinson, ed.), pp. 43–84, Academic Press, New York.

Hunt, R. K., 1975*a*, The cell cycle, cell lineage, and neuronal specificity, in: *The Cell Cycle and Cell Differentiation* (H. Holtzer and J. Reinart, eds.), pp. 43–62, Springer-Verlag, Berlin.

Hunt, R. K., 1975*b*, Developmental programming for retinotectal patterns, in: *Ciba Foundation Symposium on Cell Patterning, New Series* **29**:131–159.

Hunt, R. K., and Berman, N. J., 1975, Patterning of neuronal locus specificities in the retinal ganglion cells after partial extirpation of the embryonic eye, *J. Comp. Neurol.* **162**:43–70.

Hunt, R. K., and Cowan, W. M., 1986, The chemoaffinity hypothesis: An appreciation of Roger Sperry's contributions to developmental biology, in: *Brain Circuits and Functions of the Mind: Essays in Honor of Roger W. Sperry* (C. Trevarthen, ed.), Cambridge University Press, New York.

Huttenlocker, P. R., 1967, Development of cortical neuronal activity in the neonatal cat, *Exp. Neurol.* **17**:247–262.

Imbert, M., 1979, Maturation of visual cortex with and without visual experience, in: *Developmental Neurobiology of Vision* (R. D. Freeman, ed.), pp. 43–49, Plenum Press, New York.

Ingle, D., 1973, Two visual systems in the frog, *Science* **181**:1053–1055.

Ingle, D., 1976*a*, Behavioral correlates of central vision in anurans, in: *Frog Neurobiology* (R. Llinas and W. Precht, eds.), pp. 435–451, Springer-Verlag, Berlin.

Ingle, D., 1976*b*, Spatial vision in anurans, in: *The Am-*

phibian Visual System (K. Fite, ed.), pp. 119–140, Academic Press, New York.

Ingle, D., 1978, Visual behavior development in non-mammalian vertebrates, in: *Handbook of Sensory Physiology,* (M. Jacobson, ed.), Vol IX (a) *Development of Sensory Systems,* pp. 115–134, Springer-Verlag, Berlin.

Ingle, D., and Dudek, A., 1977, Aberrant retino-tectal projections in the frog, *Exp. Neurol.* **55**:567–582.

Ingle, D., and Sprague, J. M. (eds.), 1975, *Neurosciences Research Program Bulletin,* Vol. 13: *Sensorimotor Function of the Midbrain Tectum,* pp. 169–288, Neurosciences Research Program, Cambridge, Massachusetts.

Ingoglia, N. A., and Sharma, S. C., 1978, The effect of inhibition of axonal RNA transport on the restoration of retinotectal projections in regenerating optic nerves of goldfish, *Brain Res.* **156**:141–145.

Innocenti, G. M., 1981, Growth and reshaping of axons in the establishment of visual callosal connections, *Science* **212**:824–827.

Innocenti, G. M., 1983, Exuberant callosal projections between the developing hemispheres, in: *Advances in Neurotraumatology* (R. Villani, I. Papo, M. Giovanelli, S. M. Gaini, and G. Tomei, eds.), pp. 5–10, Excerpta Medica, Amsterdam.

Innocenti, G. M., and Clarke, S., 1983, Multiple sets of visual cortical neurones projecting transitorily through the corpus callosum, *Neurosci. Lett.* **41**:27–32.

Innocenti, G. M., and Frost, D. O., 1979, Effects of visual experience on the maturation of the efferent system to the corpus callosum, *Nature (London)* **280**:231–234.

Jacobson, C.-O., 1959, The localization of the presumptive cerebral regions in the neural plate of the axolotl larva, *J. Embryol. Exp. Morphol.* **7**:1–21.

Jacobson, M., 1968*a,* Development of neuronal specificity in retinal ganglion cells of *Xenopus, Dev. Biol.* **17**:202–218.

Jacobson, M., 1968*b,* Cessation of DNA synthesis in retinal ganglion cells correlated with the time of specification of their central connections, *Dev. Biol.* **17**:219–232.

Jacobson, M., 1970, *Developmental Neurobiology,* Holt, New York.

Jacobson, M., 1974, Premature specification of the retina in embryonic *Xenopus* eyes treated with ionophore X537A, *Science* **191**:288–289.

Jacobson, M., and Baker, R. E., 1969, Development of neuronal connections with skin grafts in frogs: Behavioural and electrophysiological studies, *J. Comp. Neurol.* **137**:121–142.

Jacobson, M., and Hirsch, H. V. B., 1973, Development and maintenance of connectivity in the visual system of the frog, I. The effect of eye rotation and visual deprivation, *Brain Res.* **49**:47–65.

Jacobson, M., and Hunt, R. K., 1973, The origins of nerve-cell specificity, *Sci. Am.* **228(2)**:26–35.

Jacobson, M., and Levine, R. L., 1975, Stability of implanted duplicate tectal positional markers serving as target for optic axons in adult frogs, *Brain Res.* **92**:468–471.

Jeannerod, M., 1982, A two-step model for visuo-motor development, in: *Regressions in Mental Development: Basic Phenomena* and *Theories* (T. G. Bever, ed.), pp. 299–310, Lawrence Erlbaum Associates, Hillsdale, New Jersey.

Jeeves, M. A., 1979, Some limits to interhemispheric integration in cases of callosal agenesis and its partial commissurotomy, in: *Structure and Function of the Cerebral Commissures* (I. Steele Russell, M. W. van Hof, and G. Berlucchi, eds.), pp. 449–474, Macmillan, New York.

Johansson, G., Hofsten, C. von, and Jansson, G., 1980, Event perception, *Annu. Rev. Psychol.* **31**:27–63.

Johnston, B. T., Schrameck, J. E., and Mark, R. F., 1975, Reinnervation of axolotl limbs. II. Sensory nerves, *Proc. R. Soc. Lond. B* **190**:59–75.

Jones, E. G., 1975, Varieties and distribution of non-pyramidal cells in the somatosensory cortex of the squirrel monkey, *J. Comp. Neurol.* **160**:205–268.

Kalil, R. E., and Schneider, G. E., 1975, Abnormal synaptic connections of the optic tract in the thalamus after midbrain lesions in newborn hamsters, *Brain Res.* **100**:690–698.

Kalnins, I. V., and Bruner, J. S., 1973, The coordination of visual observation and instrumental behavior in early infancy, *Perception* **2**:307–314.

Katz, M. J., and Lasek, R. J., 1979, Substrate pathways which guide growing axons in Xenopus embryos, *J. Comp. Neurol.* **183**:817–832.

Keating, M. J., 1976, The formation of visual neuronal connections: an appraisal of the present status of the theory of "neuronal specificity," in: *Neural and Behavioral Specificity, Studies on the Development of Behavior and the Nervous System* (G. Gottlieb, ed.), Vol. 3, pp. 59–110, Academic Press, New York.

Keating, M. J., and Gaze, R. M., 1970, Rigidity and plasticity in the amphibian visual system, *Brain Behav. Evol.* **3**:102–120.

Kennedy, H., Martin, K. A. C., and Whitteridge, D., 1983, Receptive field characteristics of neurons in newborn and adult sheep, *Neurosci.* **10**:295–300.

Kirk, U., 1983, *The Neuropsychology of Language, Reading and Spelling,* Academic Press, New York.

Klosovskii, B. N., 1963, *The Development of the Brain and Its Disturbance by Harmful Factors,* Pergamon, Oxford.

Koffka, K., 1931, *The Growth of the Mind,* Harcourt, New York.

Kollros, J. J., 1968, Endocrine influences in neural development, in: *Growth of the Nervous System* (G. E. W. Wolstenholme and M. O'Connor, eds.), pp. 179–192, Churchill, London.

Koppel, H., and Innocenti, G. M., 1983, Is there a genuine exuberancy of callosal projections in development? A quantitative electron microscopic study in the cat, *Neurosci. Lett.* **41**:33–40.

Kraschen, S., 1973, Lateralization of language learning and the critical period. Some new evidence, *Lang. Learning,* **23:**63–74.

Kraschen, S., 1975, The critical period for language acquisition and its possible bases, in: *Developmental Psycholinguistics and Communication Disorders* (D. R. Aaronson and R. W. Rieber, eds.), Vol. 263, pp. 211–224, Annals of the New York Academy of Science, New York.

Kuypers, H. G. J. M., 1973, The anatomical organisation of the descending pathways and their contributions of motor control, especially in primates, in: *New Developments in E.M.G. and Clinical Neurophysiology* (T. E. Desmedt, ed.), Vol. 3, pp. 38–68, Karger, Basel.

Landsdell, H., 1962, Laterality of verbal intelligence in the brain, *Science* **135:**922–923.

Landsdell, H., 1969, Verbal and non-verbal factors in right hemisphere speech, *J. Comp. Physiol. Psychol.* **69:**734–738.

Langworthy, O. R., 1928, The behavior of pouch young opossums correlated with myelinization of tracts, *J. Comp. Neurol.* **46:**201–240.

Larroche, J.-C., 1966, The development of the central nervous system during intrauterine life, in: *Human Development* (F. Falkner, ed.), pp. 257–276, W. B. Saunders, Philadelphia.

Larroche, J.-C., 1967, Maturation morphologique du système nerveux central: Ses rapports avec le développement pondéral du foetus et son age gestationnel, in: *Regional Development of the Brain in Early Life* (A. Minkowski, ed.), pp. 247–256, Blackwell, Oxford.

Lashley, K. S., 1951, The problem of serial order in behaviour, in: *Cerebral Mechanisms in Behaviour* (L. A. Jeffress, ed.), pp. 112–136, Wiley, New York.

Law, M. I., and Constantine-Paton, M., 1980, Right and left eye bands in frogs with unilateral tectal ablations, *Proc. Natl. Acad. Sci. U.S.A.* **77:**2314–2318.

Law, M. I., and Constantine-Paton, M., 1981, Anatomy and physiology of experimentally produced striped tecta, *J. Neurosci.* **1:**741–759.

Lawrence, D. G., and Kuypers, H. G. J. M., 1965, Pyramidal and non-pyramidal pathways in monkeys: Anatomical and functional correlation, *Science* **148:**973–975.

Lecours, A. R., 1975, Myelogenetic correlates of the development of speech and language, in: *Foundations of Language Development: A Multidisciplinary Approach* (E. H. Lenneberg and E. Lenneberg, eds.), pp. 75–94, University Publishers, New York.

Lecours, A. R., 1982, Correlates of Developmental Behavior in Brain Maturation, in: *Regressions in Mental Development: Basic Phenomena and Theories* (T. G. Bever, ed.), pp. 267–298, Lawrence Erlbaum Associates, Hillsdale, New Jersey.

Lee, D. N., 1978, The functions of vision, in: *Modes of Perceiving and Processing Information* (H. L. Pick and E. Saltzman, eds.), pp. 195–170, Lawrence Erlbaum Associates, Hillsdale, New Jersey.

Lee, D. N., and Lishman, J. R., 1975, Visual proprioceptive control of stance, *J. Hum. Movement Studies* **1:**87–95.

Lee, D. N., and Young, D. S., 1984, Visual timing of interceptive action, in: *Brain Mechanisms and Spatial Vision* (D. Ingle, M. Jeannerod, and D. M. Lee, eds.), Martinus Nijhoff, The Hague.

Leehey, S. C., Moskowitz-Cook, A., Brill, S., and Held, R., 1975, Orientational anisotropy in infant vision, *Science* **190:**900–902.

LeMay, M., 1976, Morphological cerebral asymmetries of modern man, fossil man, and nonhuman primate, *Ann. N.Y. Acad. Sci.* **280:**349–366.

Lenneberg, E. H., 1967, *Biological Foundations of Language,* Wiley, New York.

Lettvin, J. Y., Maturana, H. R., McCulloch, W. S., and Pitts, W. H., 1959, What the frog's eye tells the frog's brain, *Proc. I.R.E.* **47:**1940–1951.

LeVay, S., 1973, Synaptic patterns in the visual cortex of the cat and monkey: Electron microscopy of Golgi preparations, *J. Comp. Neurol.* **150:**53–86.

LeVay, S., Hubel, D. H., and Wiesel, T. N., 1975, The pattern of ocular dominance columns in macaque visual cortex revealed by a reduced silver stain, *J. Comp. Neurol.* **159:**559–576.

LeVay, S., Wiesel, T. N., and Hubel, D. H., 1980, The development of ocular dominance columns in normal and visually deprived monkeys, *J. Comp. Neurol.* **191:**1–52.

Leventhal, A. G., 1983, Relationship between preferred orientation and receptive field position of neurones in cat striate cortex, *J. Comp. Neurol.* **220:**476–483.

LeVere, T. E., 1975, Neural stability, sparing and behavioral recovery following brain damage, *Psychol. Rev.* **82:**344–358.

Levi-Montalcini, R., 1966, The nerve growth factor: Its mode of action on sensory and sympathetic nerve cells, *Harvey Lect.* **60:**217–259.

Levi-Montalcini, R., 1986, Ontogenesis of neuronal nets: Chemospecificity theory, 1963–1983, in: *Brain Circuits and Functions of the Mind; Essays in Honour of Roger W. Sperry* (C. Trevarthen, ed.), Cambridge University Press, New York.

Levy, J., 1969, Possible basis for the evolution of lateral specialization of the human brain, *Nature (London)* **224:**614–615.

Levy, J., 1976, A review of evidence for a genetic component in the determination of handedness, *Behav. Genet.* **6:**429–453.

Levy, J., and Nagylaki, T., 1972, A model for the genetics of handedness, *Genetics* **72:**117–128.

Levy, J., and Trevarthen, C., 1977, Perceptual, semantic and phonetic aspects of elementary language processes in split-brain patients, *Brain* **100:**105–118.

Levy, J., Trevarthen, C., and Sperry, R. W., 1972, Perception of bilateral chimeric figures following hemispheric deconnection, *Brain* **95:**61–78.

Lewis, M., 1969, Infants' responses to facial stimuli during the first year of life, *Dev. Psychol.* **1**:75–86.

Lewis, W. H., 1906, On the origin and differentiation of the optic vesicle in amphibian embryo, *Am. J. Anat.* **16**:141–45.

Lewis, W. H., 1907, Transplantation of the lips of the blastopore in *Rana palustris, Am. J. Anat.* **7**:139–141.

Liley, A. W., 1972, Disorders of amniotic fluid, in: *Pathophysiology of Gestational Disorders,* Vol. 2: *Fetal-Placental Maternal Disorders* (N. S. Assali, ed.), pp. 157–206, Academic Press, New York.

Lipsitt, L. P., 1969, Learning capacities of the human infant, in: *Brain and Early Behaviour: Development in the Fetus and Infant* (R. J. Robinson, ed.), pp. 227–245, Academic Press, New York.

Loeser, J. D., and Alvord, J. R., 1968, Agenesis of the corpus callosum, *Brain* **91**:553–570.

Loewenstein, W. R., 1968, Communication through cell junctions. Implications in growth control and differentiation, *Dev. Biol. (Suppl.)* **2**:151–183.

Luttenberg, J., 1966, Contributions to the foetal ontogenesis of the corpus callosum in man. III. Myelination in the corpus callosum, *Fol. Morphol.* **14**:192.

Magnin, M., Courin, J. H., and Flandrin, J. M., 1983, Possible visual pathways to the cat vestibular nuclei involving the nucleus prepositus hypoglossi, *Exp. Brain Res.* **51**:298–303.

Malpeli, J. G., and Baker, H., 1975, The representation of the visual field in the lateral geniculate nucleus of *Macaca mulatta, J. Comp. Neurol.* **161**:569–594.

Mann, I., 1964, *The Development of the Human Eye,* Grune & Stratton, New York.

Maratos, O., 1982, Trends in the Development of Imitation in Early Infancy, in: *Regressions in Mental Development: Basic Phenomena and Theories* (T.G., Bever, ed.) pp. 81–101, Lawrence Erlbaum Associates, Hillsdale, New Jersey.

Marin-Padilla, M., 1970, Prenatal and early post-natal ontogenesis of the cerebral cortex (neocortex) of the cat *(Felix domestica):* A Golgi study. I. The primordial neocortical organization, *Z. Anat. Entwicklungsgesch.* **134**:117–145.

Mark, R. F., 1974, *Memory and Nerve-Cell Connections,* Oxford University Press, London.

Mark, R. F., 1975, Topograhy and topology in functional recovery of regenerated sensory and motor systems, in: *Ciba Symposium on Cell Patterning, No. 29 (New Series),* pp. 289–313, American Elsevier, New York.

Mark, R. F., and Feldman, J., 1972, Binocular interaction in the development of optokinetic reflexes in tadpoles of *Xenopus laevis, Invest. Ophthalmol.* **11**:402–410.

Marr, D., 1982, *Vision,* W. H. Freeman, San Francisco.

Marty, R., and Scherrer, J., 1964, Critères de maturation des systèmes afférents corticaux, *Prog. Brain Res.* **4**:222–234.

Maturana, H. R., Lettvin, J. Y., McCulloch, W. S., and Pitts, W. H., 1959, Evidence the cut optic nerve fibers in a frog regenerate to their proper places in the tectum, *Science* **130**:1709–1710.

McGinnis, J. M., 1930, Eye movements and optic nystagmus in early infancy, *Genet. Psychol. Monogr.* **8**:321–430.

McGraw, M. B., 1943, *The Neuromuscular Maturation of the Human Newborn,* Columbia University Press, New York.

McKenzie, B. F., and Day, R. H., 1972, Object distance as a determinant of visual fixation in early infancy, *Science* **178**:1108–1110.

MacMahon, D., 1974, Chemical messengers in development: A hypothesis, *Science* **185**:1012–1021.

McMahon, D., and West, C., 1976, Transduction of positional information during development, in: *Cell Surface Interactions in Embryogenesis* (G. Poste and G. Nicolson, eds.), pp. 449–493, Elsevier North-Holland, New York.

Mehler, J., and Bertoncini, J., 1979, Infants' perception of speech and other acoustic stimuli, in: *Psycholinguistics,* Series II (J. Morton and J. C. Marshall, eds.), pp. 67–105, Elek Scientific Books, London.

Meltzoff, A. N., and Moore, M. H., 1977, Imitation of facial and manual gestures by human neonates, *Science* **198**:75–78.

Mesulam, M.-M., 1983, The functional anatomy and hemispheric specialization for directed attention, Trends Neurosci. **6**:384–387.

Metzger, W., 1974, Conscious perception and action, in: *Handbook of Perception* (E. C. Carterette and M. P. Friedman, eds.), Vol. 1, pp. 109–122, Academic Press, New York.

Meyer, R. L., 1975, Tests for field regulation in the retinotectal system of goldfish, in: *Developmental Biology, Pattern Formation, Gene Regulation* (D. McMahon and C. F. Fox, eds.), pp. 257–275, W. A. Benjamin, Menlo Park, California.

Meyer, R. L., 1979, "Extra" optic fibers exclude normal fibers from tectal regions in goldfish, *J. Comp. Neurol.* **183**:883–902.

Meyer, R. L., 1980, Mapping the normal and regenerating retinotectal projection of goldfish with autoradiographic methods, *J. Comp. Neurol.* **189**:273–289.

Meyer, R. L., 1983, Tetrodotoxin inhibits the formation of refined retinotopography in goldfish, *Dev. Brain Res.* **6**:293–298.

Meyer, R. L., 1986, The case for chemoaffinity in the retino-tectal system: Recent studies, in: *Brain Circuits and Functions of the Mind: Essays in Honour of Roger W. Sperry* (C. Trevarthen, ed.), Cambridge University Press, New York.

Meyer, R. L., and Sperry, R. W., 1973, Test for neuroplasticity in the anuran retinotectal system, *Exp. Neurol.* **40**:525–539.

Meyer, R. L., and Sperry, R. W., 1976, Retinotectal specificity: Chemospecificity theory, in: *Neural and Behavioral Specificity, Studies on the Development of Behav-*

ior and the Nervous System (G. Gottlieb, ed.), Vol. 3, pp. 111–149, Academic Press, New York.

Michotte, A., 1963, *The Perception of Causality* (translated by T. R. Miles and E. Miles), Methuen, London.

Milner, B., 1974, Interhemispheric differences and psychological processes, *Br. Med. Bull.* **27**:272–277.

Miner, N., 1956, Integumental specification of sensory fibers in the development of cutaneous local sign, *J. Comp. Neurol.* **105**:161–170.

Mishkin, M., Ungerleider, L. G., and Macko, K. A., 1983, Object vision and spatial vision: two cortical pathways, *Trends Neurosci.* **6**:414–417.

Mitchell, D. E., and Timney, B., 1984, Postnatal development of function in the mammalian visual system, in: *Handbook of Physiology—The Nervous System* (I. Darian-Smith, ed.), Vol. III, pp. 507–555, American Physiological Society, Bethesda, Maryland.

Molfese, D., 1983, Neural mechanisms underlying the processing of speech information in infants and adults: Suggestions of differences in development and structure from electrophysiological research, in: *Neuropsychology of Language, Reading and Spelling* (U. Kirk, ed.), pp. 109–128, Academic Press, New York.

Molfese, D. L., and Molfese, V. J., 1979, Hemisphere and stimulus differences as reflected in the cortical responses of newborn infants to speech stimuli, *Dev. Psychol.* **15**:505–511.

Molfese, D. L., Freeman, R. B., and Palermo, D. S., 1975, The ontogeny of brain lateralization for speech and nonspeech stimuli, *Brain Lang.* **2**:356–368.

Molliver, M., Kostovic, I., and Van der Loos, H., 1973, The development of synapses in cerebral cortex of the human fetus, *Brain Res.* **50**:403–407.

Morest, D. K., 1970, A study of neurogenesis in the forebrain of opossum pouch young, *Z. Anat. Entwicklungsgesch.* **130**:265–305.

Morgan, M. J., 1976, Embryology and the inheritance of asymmetry, in: *Lateralization of the Nervous System* (S. R. Harnand, R. W. Doty, L. Goldstein, J. Jaynes, and G. Krautheimer, eds.), pp. 173–194, Academic Press, New York.

Morgan, T. H., 1901, *Regeneration,* Macmillan, New York.

Mountcastle, V. B., 1957, Modality and topographic properties of single neurones of cats' somatic sensory cortex, *J. Neurophysiol.* **20**:408–434.

Mountcastle, V. B., 1974, Neural mechanisms in somesthesia, in: *Medical Physiology,* 13th ed. (V. B. Mountcastle, ed.), Vol. 1, pp. 307–347, C. V. Mosby, St. Louis.

Mountcastle, V. B., Lynch, J. C., Georgopolous, A., Sakata, H., and Acuna, C., 1975, Posterior parietal association cortex of the rhesus monkey: Command functions for operations within extrapersonal space, *J. Neurophysiol.* **38**:871–908.

Mundy-Castle, A., and Anglin, J., 1973, Looking strategies in infants, in: *The Competent Infant: Research and Commentary* (L. Stone, H. Smith, and L. Murphy, eds.), pp. 713–718, Basic Books, New York.

Muntz, W. R. A. 1962, Effectiveness of different colours of light in releasing positive phototactic behaviour of frogs, and a possible function of the retinal projection to the diencephalon, *J. Neurophysiol.* **25**:712–720.

Murray, M., and Forman, D. S., 1971, Fine structural changes in goldfish retinal ganglion cells during axonal regeneration, *Brain Res.* **32**:278–298.

Murray, M., and Grafstein, B., 1969, Changes in the morphology and amino acid incorporation of regenerating goldfish optic neurons, *Exp. Neurol.* **23**:544–560.

Nebes, R. D., 1974, Dominance of the minor hemisphere in commissurotomized man for the perception of part-whole relationships, in: *Hemispheric Disconnection and Cerebral Function* (M. Kinsbourne and W. L. Smith, eds.), pp. 155–164, Charles C Thomas, Springfield, Illinois.

Needham, J., 1942, *Biochemistry and Morphogenesis,* Cambridge University Press, London.

Neisser, U., 1967, *Cognitive Psychology,* Appleton-Century-Crofts, New York.

Neisser, U., 1976, *Cognition and Reality,* W. H. Freeman, San Francisco.

Neville, H. J., 1980, Event-related potentials in neuropsychological studies of language, *Brain Lang.* **11**:300–318.

Neville, H. J., and Bellugi, U., 1978, Patterns of cerebral specialization in congenitally deaf adults: A preliminary report, in: *Understanding Language Through Sign Language Research* (P. Siple, ed.), pp. 239–257, Academic Press, New York.

Nieuwkoop, P. D., and Faber, J., 1956, *Normal Table of Xenopus laevis (Daudin),* North-Holland, Amsterdam.

Oppenheim, R. W., 1973, Prehatching and hatching behavior: Comparative and physiological considerations, in: *Studies in the Development of Behavior and the Nervous System,* Vol. 1: *Behavioral Embryology* (G. Gottlieb, ed.), pp. 164–236, Academic Press, New York.

Oppenheim, R. W., 1974, The ontogeny of behavior in the chick embryo, in *Advances in the Study of Behavior* (D. H. Lehrman, J. S. Rosenblatt, R. A. Hinde, and E. Shaw, eds.), Vol. 5, pp. 133–172, Academic Press, New York.

Oppenheim, R. W. 1982, The neuroembryological study of behavior, *Curr. Top. Dev. Biol.* **17**(3):257–309.

Oppenheim, R. W., and Nunez, R., 1982, Electrical stimulation of hindlimb increases neuronal cell death in chick embryo, *Nature (London)* **295**:57–59.

Oppenheim, R. W., and Reitzel, J., 1975, Ontogeny of behavioral sensitivity to strychnine in the chick embryo: Evidence for the early onset of CNS inhibition, *Brain Behav. Evol.* **11**:130–159.

Orban, G. A., Kennedy, H., Maes, H., and Amblard, B., 1978, Cats reared in stroboscopic illumination: Velocity characteristics of Area 18 neurons, *Arch. Ital. Biol.* **116**:413–419.

Ordy, J. M., Massopust, L. C., and Wolin, L. R., 1962, Postnatal development of the retina, ERG, and acuity in the rhesus monkey, *Exp. Neurol.* **5**:364–382.

Papousek, H., 1961, Conditioned head rotation reflexes

in the first six months of life, *Acta Pediatr.* **50**:565–576.

Papousek, H., 1969, Individual variability in learned responses in human infants, in: *Brain and Early Behaviour: Development in Fetus and Infant* (R. J. Robinson, ed.), pp. 251–266, Academic Press, New York.

Pasik, T., and Pasik, P., 1971, The visual world of monkeys deprived of striate cortex: Effective stimulus parameters and the importance of the accessory optic system, *Vision Res. (Suppl.)* **3**:419–435.

Peiper, A., 1963, *Cerebral Function in Infancy and Childhood,* Consultants Bureau, New York.

Perenin, M. T., and Jeannerod, M., 1975, Residual vision in cortically blind hemifields, *Neuropsychologia* **13**:1–7.

Pettigrew, J. D., 1972, The importance of early visual experience for neurons of the developing geniculostriate system, *Invest. Ophthalmol.* **11**:386–394.

Pettigrew, J. D., and Freeman, R. D., 1973, Visual experience without lines: Effect on developing cortical neurons, *Science* **182**:599–601.

Piaget, J., 1953, *The Origins of Intelligence in Children,* Routledge and Kegan Paul, London (original French edition, 1936).

Piaget, J., 1970, Piaget's theory, in: *Carmichael's Manual of Child Psychology* (P. H. Mussen, ed.), pp. 703–732, Wiley, New York.

Pirchio, M., Spinelli, D., Fiorentini, A., and Maffei, L., 1978, Infant contrast sensitivity evaluated by evoked potentials, *Brain Res.* **141**:179–184.

Polyak, S. L., 1941, *The Vertebrate Visual System,* The University of Chicago Press, Chicago.

Pomaranz, B., 1972, Metamorphosis of frog vision: Changes in ganglion cell physiology and anatomy, *Exp. Neurol.* **34**:187–199.

Pöppel, E., Held, R., and Frost, D., 1973, Residual function after brain wounds involving the central visual pathways in man, *Nature (London)* **243**:295–296.

Prechtl, H. F. R., 1984 (ed.), *Continuity of Neural Functions from Prenatal to Postnatal Life,* Clinics in Developmental Medicine No. 94, Blackwell Scientific Publications, Oxford.

Prestige, M. C., 1970, Differentiation, degeneration, and the role of the periphery: Quantitative considerations, in: *The Neurosciences: Second Study Program* (F. O. Schmitt, ed.), pp. 73–82, Rockefeller University Press, New York.

Prestige, M. C., and Willshaw, D. J., 1975, On the role of competition in the formation of patterned neural connections, *Proc. R. Soc. London, Ser. B* **190**:77–98.

Probst, F. P., 1973, Congenital defects of the corpus callosum, *Acta Radiol. Suppl.* 331, Acta Radiologica, Stockholm.

Purves, D., 1983, Modulation of neuronal competition by postsynaptic geometry in antonomic ganglia, *Trends Neurosci.* **6**:10–16.

Rakic, P., 1972, Mode of cell migration to the superficial layers of fetal monkey neocortex, *J. Comp. Neurol.* **145**:61–84.

Rakic, P., 1974, Neurons in rhesus monkey visual cortex: Systematic relation between time of origin and eventual disposition, *Science* **183**:425–427.

Rakic, P., 1976, Prenatal genesis of connections subserving ocular dominance in the rhesus monkey, *Nature (London)* **261**:467–471.

Rakic, P., 1977, Prenatal development of the visual system in the rhesus monkey, *Philos. Trans. R. Soc. Lond.* **278**:245–260.

Rakic, P., 1979, Genesis of visual connections in the rhesus monkey, in: *Developmental Neurobiology of Vision* (R. D. Freeman, ed.), pp. 249–260. Plenum Press, New York.

Rakic, P., and Riley, K. P., 1983, Overproduction and elimination of retinal axons in the fetal rhesus monkey, *Science* **219**:1441–1444.

Rakic, P., and Yakovlev, P. I., 1968, Development of the corpus callosum and cavum septi in man, *J. Comp. Neurol.* **132**:45–72.

Ramón y Cajal, S., 1909–1911, *Histologie du système nerveux de l'homme et des vertébrés* (L. Azoulay, transl.), 2 vols., A. Maloîne, Paris (reprinted: Consejo Superior de Investigaciones Cientificas, Madrid, 1952 and 1955).

Ramón y Cajal, S., 1929, *Etude sur la neurogenèse de quelques vertébrés,* Madrid (reprinted: *Studies on Vertebrate Neurogenesis,* (L. Guth, translator), Charles C Thomas, Springfield, Illinois.

Rauschecker, J. P., and Singer, W., 1979, Changes in the circuitry of the kitten: Visual cortex are gated by postsynaptic activity, *Nature (London)* **280**:58–60.

Rauschecker, J. P., and Singer, W., 1981, The effects of early visual experience on the cat's visual cortex and their possible explanation by synapses, *J. Physiol. (London)* **310**:215–239.

Regan, D., and Beverley, K. I., 1982, How do we avoid confounding the direction we are looking and the direction we are moving? *Science* **215**:194–196.

Regan, D., and Cynader, M., 1982, Neurons in cat visual cortex tuned to the direction of motion in depth: Effect of stimulus speed, *Invest. Ophthalmol. Visual Sci.* **22**:535–550.

Reichardt, W. E., and Poggio, T., 1981 (eds.), *Theoretical Approaches in Neurobiology,* M.I.T. Press, Cambridge, Massachusetts.

Riesen, A. H., and Aarons, L., 1959, Visual movement and intensity discrimination in cats after early deprivation of pattern vision, *J. Comp. Physiol. Psychol.* **52**:142–149.

Roach, F. C., 1945, Differentiation of the central nervous system after axial reversals of the medullary plate of *Amblystoma, J. Exp. Zool.* **99**:53–77.

Robinson, J. S., and Fish, S. E., 1974, A cat's form experienced but visuo-motor deprived eye lacks focal vision, *Dev. Psychobiol.* **7**:331–342.

Rodieck, R. W., 1973, *The Vertebrate Retina: Principles of Structure and Function,* W. H. Freeman, San Francisco.

Roth, R. L., 1974, Retinotopic organization of goldfish optic nerve and tract, *Anat. Rec.* **178**:453.

Rudel, R., and Denkla, M. B., 1974, Relation of forward and backward digit repetition to neurological impairment in children with learning disabilities, *Neuropsychologia* **12**:109–118.

Rudel, R., Teuber, H. L., and Twitchell, T. E., 1974, Levels of impairment of sensorimotor early damage, *Neuropsychologia* **12**:95–108.

Saint-Anne Dargassies, S., 1966, Neurological maturation of the premature infant of 28–41 weeks gestational age, in: *Human Development* (F. Falkner, ed.), pp. 306–325, W. B. Saunders, Philadelphia.

Sarnat, H. B., and Netsky, M. G., 1974, *Evolution of the Nervous System,* Tavistock, Oxford.

Saul, R., and Sperry, R. W., 1968, Absence of commissurotomy symptoms with agenesis of the corpus callosum, *Neurology (New York)* **18**:307.

Saunders, J. W., Jr., and Fallon, J. F., 1966, Cell death in morphogenesis, in: *Major Problems in Developmental Biology* (M. Locke, ed.), pp. 289–314, Academic Press, New York.

Saxén, L., 1972, Interactive mechanisms in morphogenesis, in: *Tissue Interactions in Carcinogenesis* (D. Tarin, ed.), pp. 49–80, Academic Press, London.

Saxén, L., and Toivonen, S., 1962, *Primary Embryonic Induction,* Academic Press, New York.

Scalia, R., and Arango, V., 1983, The anti-retinotopic organization of the frog's optic nerve, *Brain Res.* **266**:121.

Scalia, F., and Fite, K., 1974, A retinotopic analysis of the central connections of the optic nerve in the frog, *J. Comp. Neurol.* **158**:455–478.

Scarf, B., and Jacobson, M., 1974, Development of binocularly driven single units in frogs raised with asymmetrical visual stimulation, *Exp. Neurol.* **42**:669–686.

Schaffer, H. R. (ed.), 1977, *Studies in Mother–Infant Interaction: The Loch Lomond Symposium,* Academic Press, London.

Scheibel, M. E., Lindsay, R. D., Tomiyasu, U., and Scheibel, A. B., 1975, Dendritic changes in aging human cortex, *Anat. Rec.* **181**:471.

Schmidt, J. T., 1978, Retinal fibers alter tectal positional markers during the expansion of the retinal projection in goldfish, *J. Comp. Neurol.* **177**:279–300.

Schneider, G. E., 1967, Contrasting visuomotor functions of tectum and cortex in the golden hamster, *Psychol. Forsch.* **31**:52–62.

Schneider, G. E., 1969, Two visual systems: Brain mechanisms for localization and discrimination are dissociated by tectal and cortical lesions, *Science* **163**:895–902.

Schneider, G. E., 1973, Early lesions of superior colliculus: Factors affecting the formation of abnormal retinal projections, *Brain Behav. Evol.* **8**:73–109.

Schneider, G. E., 1976, Growth of abnormal neural connections following focal brain lesions: Constraining factors and functional effects, in: *Neurosurgical Treatment in Psychiatry* (W. H. Sweet, S. Obrador, and J. G. Martin-Rodrigues, eds.), pp. 5–26, University Park Press, Baltimore.

Schneider, G. E., 1981, Early lesions and abnormal neuronal connections: Developmental rules can lead axons astray, with functional consequences, *Trends Neurosci.* **4**:187–192.

Schneider, G. E., and Jhaveri, S. R., 1974, Neuroanatomical correlates of spared or altered function after brain lesions in the newborn hamster, in: *Plasticity and Recovery of Function in the Central Nervous System* (D. G. Stein, J. J. Rosen, and N. Butters, eds.), pp. 65–109, Academic Press, New York.

Schneider, G. E., Jhaveri, S., Edwards, M. A., and So, K.-F., 1985, Regeneration, re-routing and redistribution of axons after early lesions: Changes with age and functional impact, in: *Recent Achievements in Restorative Neurology,* Vol. 1, *Upper Motor Neuron Functions and Dysfunctions* (J. E. Contra and M. R. Dimitrijevic, eds.), Karger, Basel.

Scholes, J. H., 1979, Nerve fiber topography in the retinal projection to the tectum, *Nature (London)* **278**:620–624.

Schwenk, G. C., and Hibbard, E., 1977, An autoradiographic study of optic fiber projections from eye grafts in eyeless mutant exolotls, *Exp. Neurol.* **55**:498–503.

Scott, M. Y., 1975, Functional capacity of compressed retinotectal projection in goldfish, *Anat. Rec.* **181**:474.

Scott, S. A., 1975, Persistence of foreign innervation on reinnervated goldfish extraocular muscles, *Science* **189**:644–646.

Sequeland, E. R., and DeLucia, C. A., 1969, Visual reinforcement of non-nutritive sucking in human infants, *Science* **165**:1144–1146.

Sharma, S. C., 1972, Reformation of retinotectal projections after various tectal ablations in adult goldfish, *Exp. Neurol.* **34**:171–182.

Sharma, S. C., 1972, Retinotectal connexions of a heterotopic eye, *Nature New Biol.* **238**:286–287.

Sharma, S. C., and Gaze, R. M., 1971, The retinotopic organization of visual responses from tectal reimplants in adult goldfish, *Arch. Ital. Biol.* **109**:357–366.

Sharma, S. C., and Tung, Y. L., 1979, Interactions between nasal and temporal hemiretinal fibers in adult goldfish tectum, *Neuroscience* **4**:113–119.

Shatz, C., 1977, Abnormal interhemispheric connections in the visual system of Boston Siamese cats: A physiological study, *J. Comp. Neurol.* **171**:229–246.

Sherman, S. M., 1972, Development of interocular alignment in cats, *Brain Res.* **37**:187–198.

Sherman, S. M., Guillery, R. W., Kaas, J. H., and Sanderson, K. J., 1974, Behavioral, electrophysiological and morphological studies of binocular competition in the development of the geniculo-cortical pathways of cats, *J. Comp. Neurol.* **158**:1–18.

Shkol'nik-Yarros, E. G., 1971, *Neurones and Interneuronal Connections of the Central Visual System* (B. Haigh, translation), Plenum Press, New York.

Shlaer, R., 1971, Shift in binocular disparity causes compensating change in the cortical structure of kittens, *Science* **173**:638–641.

Sidman, R. L., 1961, Histogenesis of mouse retina studied with thymidine-^{3}H, in: *Structure of the Eye* (G. K. Smelser, ed.), pp. 487–505, Academic Press, New York.

Sidman, R. L., and Rakic, P., 1973, Neuronal migration, with special reference to developing human brain: A review, *Brain Res.* **62**:1–35.

Slater, A. M., and Findlay, J. M., 1975, Binocular fixation in the newborn baby, *J. Exp. Child Psychol.* **20**:248–273.

Smith, A., 1966, Speech and other functions after left (dominant) hemispherectomy, *J. Neurol. Neurosurg. Psychiatry* **29**:467–471.

Spalding, D. A., 1873, Instinct with original observations on young animals, *Macmillans Magazine* **27**:282–293 (reprinted in *Br. J. Anim. Behav.* **2**:2–11).

Spemann, H., 1938, *Embryonic Development and Induction,* Yale University Press, New Haven.

Spemann, H., and Mangold, H., 1924, Über Induktion von Embronalanlagen durch Implantation artfremder Organisatoren, *Roux Arch.* **100**:599–638.

Sperry, R. W., 1943, Visuomotor coordination in the newt *(Triturus viridescens)* after regeneration of the optic nerves, *J. Comp. Neurol.* **79**:33–55.

Sperry, R. W., 1944, Optic nerve regeneration with return of vision in anurans, *J. Neurophysiol.* **7**:57–69.

Sperry, R. W., 1945, Restoration of vision after crossing of optic nerves and after contralateral transposition of the eye, *J. Neurophysiol.* **8**:15–28.

Sperry, R. W., 1951*a,* Mechanisms of neural maturation, in: *Handbook of Experimental Psychology* (S. S. Stevens, ed.), pp. 236–280, Wiley, New York.

Sperry, R. W., 1951*b,* Regulative factors in orderly growth of neural circuits, *Growth Symp.* **10**:63–87.

Sperry, R. W., 1952, Neurology and the mind-brain problem, *Am. Sci.* **40**:291–312.

Sperry, R. W., 1963, Chemoaffinity in the orderly growth of nerve fiber patterns and connections, *Proc. Natl. Acad. Sci. U.S.A.* **50**:703–710.

Sperry, R. W., 1965, Embryogenesis of behavioral nerve nets, in: *Organogenesis* (R. L. De Haan and H. Ursprung, eds.), pp. 161–186, Holt, New York.

Sperry, R. W., 1970, Perception in the absence of the neocortical commissures, *Res. Publ. Assoc. Res. Nerv. Ment. Dis.* **48**:123–138.

Sperry, R. W., 1974, Lateral specialization in the surgically separated hemispheres, in: *The Neurosciences: Third Study Program* (F. O. Schmitt and F. G. Worden, eds.), pp. 5–20, M.I.T. Press, Cambridge, Massachusetts.

Sperry, R. W., and Hibbard, E., 1968, Regulative factors in the orderly growth of retino-tectal connections, in: *Growth of the Nervous System* (G. E. W. Wolstenholme and M. O'Connor, eds.), pp. 41–52, Churchill, London.

Sperry, R. W., Gazzaniga, M. S., and Bogen, J. E., 1969, Interhemispheric relationships: The neocortical commissures; syndromes of hemisphere deconnection, in: *Handbook of Clinical Neurology* (P. J. Vinken and G. W. Bruyn, eds.), Vol. 4, pp. 273–290, North-Holland, Amsterdam.

Sprague, J. M., Levy, J., DiBerardino, A., and Berlucchi, G., 1977, Visual cortical areas mediating form discrimination in the cat, *J. Comp. Neurol.* **172**:441–488.

Sprague, J. M., Hughes, H. C., and Berlucchi, G., 1981, Cortical mechanisms in pattern and form perception, in: *Brain Mechanisms and Perceptual Awareness* (O. Pompeiano and C. Ajmore Marsan, eds.), pp. 107–132, Raven Press, New York.

Stanfield, B. B., 1984, Postnatal reorganization of cortical projections: The role of collateral elimination, *Trends Neurosci.* **7**:37–41.

Stein, B. E., Labos, E., and Kruger, L., 1973, Sequence of changes in properties of neurones of superior colliculus of the kitten during maturation, *J. Neurophysiol.* **36**:667–679.

Stern, D. N., 1985, *The Interpersonal World of the Infant: A View from Psychoanalysis and Developmental Psychology,* Basic Books, New York.

Stone, L. S., 1944, Functional polarization in retinal development and its reestablishment in regenerated retinae of rotated eyes, *Proc. Soc. Exp. Biol. Med.* **57**:13–14.

Stone, L. S., 1959, Experiments testing the capacity of iris to regenerate neural retina in eyes of adult newts, *J. Exp. Zool.* **142**:285–308.

Stone, L. S., 1960, Polarization of the retina and development of vision, *J. Exp. Zool.* **145**:85–93.

Straznicky, K., 1978, The acquisition of tectal positional specification in *Xenopus, Neurosci. Lett.* **9**:177–184.

Straznicky, K., and Gaze, R. M., 1971, The growth of the retina in *Xenopus laevis:* An autoradiographic study, *J. Embryol. Exp. Morphol.* **26**:67–79.

Straznicky, K., and Gaze, R. M., 1972, The development of the tectum in *Xenopus laevis:* An autoradiographic study, *J. Embryol. Exp. Morphol.* **28**:87–115.

Straznicky, K., and Gaze, R. M., 1982, The innervation of a virgin tectum by a double-temporal or a double-nasal eye in *Xenopus, J. Embryol. Exp. Morphol.* **68**:9–21.

Straznicky, K., and Tay, D., 1977, Retinal growth in normal double dorsal and double ventral eyes in *Xenopus, J. Embryol. Exp. Morphol.* **40**:175–185.

Stryker, M. P., 1981, *Soc. Neurosci. Abstr.* **7**:842.

Studdert-Kennedy, M., 1983, On learning to speak, *Hum. Neurobiol.* **2**:191–196.

Sturmer, C., 1981, Modified retinotectal projection in goldfish: A consequence of the position of retinal lesions, in: *Lesion-induced Neuronal Plasticity in Sensorimotor Systems* (H. Flohr and W. Precht, eds.), pp. 369–376, Springer-Verlag, Berlin.

Swindale, N. V., 1982, The development of columnar systems in the mammalian visual cortex. The role of innate and environmental factors, *Trends Neurosci.* **5**:235–241.

Székely, G., 1954, Untersuchung der Entwicklung optischer Reflex mechanismem an Amphibien larven, *Acta Physiol. Acad. Sci. Hung.* **6**(Suppl. 18).

Székely, G., 1957, Regulationstendenzen in der Ausbildung der "Funktionellen Spezifität" der Retinoanlage bei *Triturus vulgaris, Arch. Entwichlungs. Org.* **150**:48–60.

Szentagothai, M. J., and Arbib, M. A., 1975, The module concept in cerebral cortex architecture, *Brain Res.* **95**:475–496.

Talbot, S. A., and Marshall, W. H., 1941, Physiological studies on neural mechanisms of visual localization and discrimination, *Am. J. Ophthalmol.* **24**:1255–1263.

Taub, E., 1968, Prism compensation as a learning phenomenon: A phylogenetic comparison, in: *The Neuropsychology of Spatially Oriented Behavior* (S. F. Freedman, ed.), pp. 173–192, Dorsey Press, Homewood, Illinois.

Tees, R. C., 1976, Perceptual development in mammals, in: *Neural and Behavioral Specificity: Studies on the Development of Behavior and the Nervous System,* Vol. 3 (G. Gottlieb, ed.), pp. 281–326, Academic Press, New York.

Teller, D. Y., 1981, The development of visual acuity in human and monkey infants, *Trends Neurosci.* **4**:21–24.

Teszner, Tzavaras, A., Gruner, J., and Hécaen, H., 1972, L'asymétrie droite-gauche du planum temporale: àpropos de l'étude anatomique de 100 cerveaux, *Rev. Neurol.* **126**:444–449.

Teuber, H.-L., 1960, Perception, in: *Handbook of Physiology,* Section 1: *Neurophysiology* (J. Field, H. W. Magoun, and V. E. Hall, eds.), Vol. III, pp. 1595–1668, American Physiological Society, Washington, D.C.

Toivonen, S., and Saxén, L., 1968, Morphogenetic interaction of presumptive neural and mesodermal cells mixed in different ratios, *Science* **159**:539–540.

Trehub, S. E., 1973, Infants' sensitivity to vowel and tonal contrasts, *Dev. Psychol.* **9**:91–96.

Trevarthen, C., 1968*a,* Vision in fish: The origins of the visual frame for action in vertebrates, in: *The Central Nervous System and Fish Behavior* (D. Ingle, ed.), pp. 61–94, University of Chicago Press, Chicago.

Trevarthen, C., 1968*b,* Two mechanisms of vision in primates, *Psych. Forsch.* **31**:299–337.

Trevarthen, C., 1972, Brain bisymmetry and the role of the corpus callosum in behavior and conscious experience, in: *Cerebral Interhemispheric Relations* (J. Cernacek and F. Podovinsky, eds.), pp. 319–333, Slovak Academy of Sciences, Bratislava.

Trevarthen, C., 1974*a,* L'action dans l'espace et la perception de l'espace; mécanismes cérébraux de base, in: *De l'espace corporel à l'espace ecologique* (F. Bresson, ed.), pp. 65–80, Presses Universitaires de France, Paris.

Trevarthen, C., 1974*b,* Cerebral embryology and the split brain, in: *Hemispheric Disconnection and Cerebral Function* (M. Kinsbourne and W. L. Smith, eds.), pp. 208–236, Charles C Thomas, Springfield, Illinois.

Trevarthen, C., 1974*c,* The psychobiology of speech development, in: *Language and Brain: Developmental Aspects* (E. Lenneberg, ed.), Neurosciences Research Program Bulletin, **12**:570–585.

Trevarthen, C., 1977, Descriptive analyses of infant communication behavior, in: *Studies in Mother–Infant Interaction: The Loch Lomond Symposium* (H. R. Schaffer, ed.), pp. 227–270, Academic Press, London.

Trevarthen, C., 1978*a,* Modes of perceiving and modes of acting, in: *Modes of Perceiving and Processing Information* (H. Pick and E. Saltzman, eds.), pp. 99–136, U.S. Social Science Research Council.

Trevarthen, C., 1978*b,* Manipulative strategies of baboons and the origins of cerebral asymmetry, in: *The Asymmetrical Function of the Brain* (M. Kinsbourne, ed.), pp. 329–391, Cambridge University Press, Cambridge.

Trevarthen, C., 1979, Communication and cooperation in early infancy. A description of primary intersubjectivity, in: *Before Speech, The Beginnings of Human Communication* (M. Bullowa, ed.), pp. 321–346, Cambridge University Press, Cambridge.

Trevarthen, C., 1979, The tasks of consciousness, how could the brain do them?, in: *Brain and Mind,* (G. E. W., Wolstenholme, and M. O'Connor, eds.) CIBA Foundation Symposium 69 (New Series), pp. 187–217, Excerpta Medica, Amsterdam.

Trevarthen, C., 1980, Neurological development and the growth of psychological functions, in: *Developmental Psychology and Society* (J. Sants, ed.), pp. 46–95, Macmillan, London.

Trevarthen, C., 1982, Basic patterns of psychogenetic change in infancy, in: *Regressions in Mental Development: Basic Phenomena and Theories* (T. G. Bever, ed.), pp. 7–46, Lawrence Erlbaum Associates, Hillsdale, New Jersey.

Trevarthen, C., 1983*a,* Cerebral mechanisms for language: Prenatal and postnatal development, in: *Neuropsychology of Language, Reading and Spelling* (U. Kirk, ed.), pp. 45–80, Academic Press, New York.

Trevarthen, C., 1983*b,* Interpersonal abilities of infants as generators for transmission of language and culture, in: *The Behaviour of Human Infants* (A. Oliverio and M. Zapella, eds.), pp. 145–176, Plenum Press, New York.

Trevarthen, C., 1984*a,* How control of movements develops, in: *Human Motor Actions: Bernstein Reassessed* (H. T. A. Whiting, ed.), pp. 223–261, Elsevier–North-Holland, Amsterdam.

Trevarthen, C., 1984*b,* Hemispheric specialization, in: *Handbook of Physiology,* Section I: *The Nervous System,* Vol. III: *Sensory Processes* (I. Darian-Smith, ed.), pp. 1129–1190, American Physiological Society, Bethesda, Maryland.

Trevarthen, C., 1984*c,* Biodynamic structures, cognitive correlates of motive sets and development of motives in infants, in: *Cognition and Motor Processes* (W. Prinz and A. F. Saunders, eds.), pp. 327–350, Springer-Verlag, Berlin.

Trevarthen, C., 1984*d,* Emotions in infancy: Regulators of contacts and relationships with persons, in: *Ap-*

proaches to Emotion (K. Scherer and P. Edman, eds.), pp. 129–157, Lawrence Erlbaum Associates, Hillsdale, New Jersey.

Trevarthen, C., Murray, L., and Hubley, P., 1981, Psychology of infants, in: *Scientific Foundations of Clinical Pediatrics* (J. Davis and J. Dobbing), pp. 217–271, Heinemann Medical Books, London.

Trevarthen, C., and Sperry, R. W., 1973, Perceptual unity of the ambient visual field in human commissurotomy patients, *Brain* **96**:547–570.

Twitchell, T. E., 1965, The automatic grasping responses of infants, *Neuropsychologia* **3**:247–259.

Udin, S. B., 1977, Rearrangement of the retinotectal projection in *Rana pipiens* after unilateral caudal half-tectum ablation, *J. Comp. Neurol.* **173**:561–582.

Udin, S. B., 1983, Abnormal visual input leads to development of abnormal axon trajectories in frogs, *Nature (London)* **301**:336–338.

Ullman, S., 1979, *The Interpretation of Visual Motion,* M.I.T. Press, Cambridge, Massachusetts.

Uzgiris, I. C., 1974, Patterns of vocal and gestural imitation in infants, in: *The Competent Infant* (L. J. Stone, H. T. Smith, and L. B. Murphy, eds.), pp. 599–604, Tavistock, London.

Valverde, F., 1971, Rate and extent of recovery from dark rearing in the visual cortex of the mouse, *Brain Res.* **33**:1–11.

Van der Loos, H., 1977, Structural changes in the cerebral cortex upon modification of the periphery: Bands in somatosensory cortex, *Philos. Trans. R. Soc. (London) B.* **278**:373–376.

Van der Loos, H., 1979, The development of topological equivalences in the brain, in: *Neural Growth and Differentiation* (E. Meisami and M. A. B. Brazier, eds.), pp. 331–336, Raven Press, New York.

Van der Loos, H., and Dörfl, J., 1978, Does the skin tell the somatosensory cortex how to construct a map of the periphery?, *Neurosci. Lett.* **7**:23–30.

Van Essen, D. C., and Maunsell, J. H. R., 1983, Hierarchical organization and functional streams in the visual cortex, *Trends Neurosci.* **6**:370–375.

Van Sluyters, R. C., and Blakemore, C., 1973, Experimental creation of unusual neuronal properties in visual cortex of kitten, *Nature (London)* **246**:506–508.

Vinken, P. J., and Bruyn, R. W. (eds.), 1969, *Handbook of Clinical Neurology:* Vol. 3: *Disorders of Higher Nervous Activity;* Vol. 4: *Disorders of Speech Perception and Symbolic Behavior,* North-Holland, Amsterdam.

Vital-Durand, F., and Jeannerod, M., 1974, Maturation of the optkinetic response: Genetic and environmental factors, *Brain Res.* **71**:249–257.

Von Békésy, G., 1967, *Sensory Inhibition,* Princeton University Press, Princeton, New Jersey.

Von Senden, M., 1960, *Space and Sight: The Perception of Space and Shape in the Congenitally Blind Before and After Operation* (translated by P. Heath), Methuen, London.

Waber, D. P., 1976, Sex differences in cognition: A function of maturation rate? *Science* **192**:572–573.

Wada, J. A., Clarke, R., and Hamm, A., 1975, Cerebral hemispheric asymmetry in humans: Cortical speech zones in 100 adult and 100 infant brains, *Arch. Neurol.* **32**:239–246.

Waddington, C. H., 1966, *Principles of Development and Differentiation,* Macmillan, New York.

Walk, R. D., and Bond, E. K., 1971, The development of visually guided reaching in monkeys reared without sight of the hands, *Psychon. Sci.* **23**:115–116.

Walk, R. D., and Gibson, E. J., 1961, A comparative and analytic study of visual depth perception, *Psychol. Monogr.* **75**: No. 519.

Webster, W. R., and Aitkin, L. M., 1975, Central auditory processing, in: *Handbook of Psychobiology* (M. S. Gazzaniga and C. Blakemore, eds.), pp. 325–364, Academic Press, New York.

Weiskrantz, L., Warrington, E. K., Saunders, M. P., and Marshall, J., 1974, Visual capacity in the hemianopic field following a restricted occipital ablation, *Brain* **97**:709–728.

Weiss, P., 1939, *Principles of Development,* Holt, Rinehart and Winston, New York.

Werner, H., 1961, *Comparative Psychology of Mental Development* (revised ed.), Science Editions, New York.

White, M. J., 1969, Laterality differences in perception: A review, *Psychol. Bull.* **72**:387–405.

White, B. L., Castle, P., and Held, R., 1964, Observations on the development of visually-directed reaching, *Child Dev.* **35**:349–364.

Whitelaw, V. A., and Cowan, J. D., 1981, Specificity and plasticity of retinal connections: A computational model, *J. Neurosci.* **1**:1369–1387.

Whiting, H. T. A. (ed.), 1984, *Human Motor Actions—Bernstein Reassessed,* Elsevier–North-Holland, Amsterdam.

Whitteridge, D., 1973, Visual projections to the cortex, in: *Handbook of Sensory Physiology* (R. Jung, ed.), Vol. VII/3/B, pp. 247–268, Springer-Verlag, Berlin.

Wickelgren, B., 1972, Some effects of visual deprivation on the cat superior colliculus, *Invest. Ophthalmol.* **11**:460–467.

Wickelgren, B., and Sterling, P., 1969, Influence of visual cortex on receptive fields in the superior colliculus of the cat, *J. Neurophysiol.* **32**:16–23.

Wiesel, T. N., 1982, Postnatal development of the visual cortex and the influence of environment, *Nature (London)* **299**:583–591.

Wiesel, T. N., and Hubel, D. H., 1963, Single-cell responses in striate cortex of kittens deprived of vision in one eye, *J. Neurophysiol.* **26**:1003–1017.

Wiesel, T. N., and Hubel, D. H., 1965, Extent of recovery from the effects of visual deprivation in kittens, *J. Neurophysiol.* **28**:1060–1072.

Wiesel, T. N., and Hubel, D. H., 1974, Ordered arrangement of orientation columns in monkeys lacking visual experience, *J. Comp. Neurol.* **158**:307–318.

Willshaw, D. J., and von der Malsburg, C., 1976, How patterned neural connections can be set up by self-organization, *Proc. R. Soc. Lond. Ser. B.* **194**:431–445.

Wilson, P. D., and Riesen, A. H., 1966, Visual development in rhesus monkeys neonatally deprived of patterned light, *J. Comp. Physiol. Psychol.* **61**:87–95.

Windle, W. F., 1970, Development of neural elements in human embryos of four to seven weeks gestation, *Exp. Neurol. (Suppl.)* **5**:44–83.

Witelson, S. F., and Pallie, W., 1973, Left hemisphere specialization for language in the newborn: Neuroanatomical evidence of asymmetry, *Brain* **96**:641–646.

Wolff, P. H., 1966, The causes, controls and organization of behavior in the neonate, *Psychological Issues, Monograph Series,* Vol. 5, No. 1, Monogr. 17, International Universities Press, New York.

Wolff, P. H., 1968, Stereotypic behavior and development, *Can. Psychol.* **9**:474–484.

Wolff, P. H., 1969, The natural history of crying and other vocalizations in early infancy, in: *Determinants of Infant Behaviour* (B. M. Foss, ed.), Vol. IV, pp. 81–110, Methuen, London.

Wolpert, L., 1971, Positional information and pattern formation, *Curr. Top. Dev. Biol.* **6**:183–224.

Woods, B. T., 1980, Restricted effects of right hemisphere lesions after age one: Wechsler Test data, *Neuropsychologia* **18**:65–70.

Woods, B. T., and Teuber, H.-L., 1973, Early onset of complementary specialization of cerebral hemispheres in man, *Trans. Am. Neurol. Assoc.* **98**:113–115.

Woods, B. T., and Teuber, H.-L., 1978, Changing patterns of childhood aphasia, *Ann. Neurol.* **3**:273–280.

Yakovlev, P. I., 1962, Morphological criteria of growth and maturation of the nervous system in man, in: *Mental Retardation, Research Publications of the Association for Research on Nervous and Mental Diseases,* Vol. 39, pp. 3–46, Association for Research on Nervous and Mental Diseases, New York.

Yakovlev, P. I., and Lecours, A. R., 1967, The myelogenetic cycles of regional maturation of the brain, in: *Regional Development of the Brain in Early Life* (A. Minkowski, ed.), pp. 3–70, Blackwell, Oxford.

Yarbus, A. L., 1967, *Eye Movements and Vision,* Plenum Press, New York.

Yoon, M. G., 1971, Reorganization of retinotectal projection following surgical operations on the optic tectum in goldfish, *Exp. Neurol.* **33**:395–411.

Yoon, M. G., 1972, Reversibility of the reorganization of retinotectal projection in goldfish, *Exp. Neurol.* **35**:565–577.

Yoon, M. G., 1973, Retention of the original topographic polarity by the 180° rotated tectum reimplant in young adult goldfish, *J. Physiol. (London)* **233**:575–588.

Yoon, M. G., 1975, Readjustment of retinotectal projection following reimplantation of a rotated or inverted tectal tissue in adult goldfish, *J. Physiol. (London)* **252**:137–158.

Yoon, M. G., 1986, Neural reconnection between the eye and the brain in goldfish, in: *Brain Circuits and Functions of the Mind: Essays in Honor of Roger W. Sperry* (C. Trevarthen, ed.), Cambridge University Press, New York.

Zaidel, E., 1978, Auditory language comprehension in the right hemisphere following cerebral commissurotomy and hemispherectomy: A comparison with child language and aphasia, in: *Language Acquisition and Language Breakdown: Parallels and Divergences* (A. Caramazza and E. B. Zurif, eds.), pp. 229–275, Johns Hopkins University Press, Baltimore.

Zaporozhets, A. V., 1965, The development of perception in the preschool child, in: *European Research in Child Development* (P. H. Mussen, ed.), *Monogr. Soc. Res. Child Devel.* **30**(Ser. No. 100):82–101.

Zeki, S. M., 1973, Colour coding in Rhesus monkey prestriate cortex, *Brain Res.* **53**:422–427.

Zeki, S. M., 1974, The mosaic organization of the visual cortex in the monkey, in: *Essays on the Nervous System—Festschrift for J. Z. Young* (R. Bellairs and E. G. Grey, eds.), pp. 327–343, Oxford University Press, London.

14

The Differentiated Maturation of the Cerebral Cortex

T. RABINOWICZ

1. Introduction

A useful approach to our understanding of the development of the human cerebral cortex is to obtain quantitative data (e.g., neuronal densities, cortical depths) at given ages and on selected areas, thereby gaining a long-range representation of cortical development.

This work was first done by Conel. According to Tanner (1961), "His work has provided us with an extraordinarily valuable picture of the development of the cerebral cortex from birth to two years. Before birth, our information is scanty and qualitative, and after two years it is practically nonexistent." This was written more than 20 years ago. Meantime, Conel (1963, 1967) published his atlases of the cerebral cortex of children aged 4 and 6 years; my colleagues and I completed his last, unfinished, atlas of the child of 8 years (J.L. Conel and T. Rabinowicz, unpublished manuscript). We extended Conel's work both into the fetal period, at 8, 7, 6, and 5 months (22 weeks) and through adolescence at 10, 12, 14, 18, 20, and 22 years. Together with the data available from von Economo and Koskinas (1925) on the adult cerebral cortex, it is now possible to get at least an idea of the maturation of the cerebral cortex from the premature infant of 5 months gestation to adulthood on almost all the cortical areas.

Although we used the same criteria for maturation used by Conel (Rabinowicz, 1964, 1967*a,b*) in the present chapter we use only the following criteria: (1) The total width of the cortex; (2) The number of neurons per unit of volume; and (3) The histological structure on cresyl violet-stained sections. Some information has also been taken from our work with Leuba and Heumann on the quantitative cytoarchitectonic development of the cerebral cortex of mice.

T. RABINOWICZ • Division of Neuropathology, University Medical Center, Geneva 1205, Switzerland

2. Material and Methods

Almost all cases are selected from children without any brain lesion, with a complete autopsy and a medical record showing a normal development as well as a normal school record. More clinical data are given in each of Conel's atlases. At the following ages, only one or two cases were studied: Premature infants of the fifth and sixth month gestational age and cases aged 10–22 years. From the premature infants of the eighth gestational month up to the child of 8 years, counts were performed on four to nine cases, mostly on the left hemisphere, sometimes on both. Moreover, some cases were studied with nonquantitative methods; i.e., Cajal and Golgi–Cox impregnations and myelin stainings. For each age we have thus studied up to nine cases, and sometimes more. Each measurement (of cortical thickness or of cell dimension) was repeated 30 times or on 30 cells of each type in each area. Numbering to establish cell densities was repeated 30 times in each layer of each area in all the cases studied by Conel. In our cases, and in the premature infants, cell counts are repeated ten times. The results of each case are presented as simple arithmetic means of the 30 or 10 counts done in each area. Great care is taken to cut out blocks in exactly the same orientation (which depends on the location and size of each area) for all 50 areas studied. The entire histological procedure is carefully standardized. Further technical details are given in Conel (1939) and in Rabinowicz (1967*a*).

The techniques were exactly the same for experimental material (mice), but here, more precise stereological methods could be used: an Abercrombie connection was made, and nuclei were counted instead of whole cells on cresyl violet-stained sections of 25-μm thickness after paraffin embedding. Statistical methods for the correction of cell densities for each area on a minimum of 10 animals were also used. At each age the development of the cell densities of each layer was studied as well as that of the whole cortical thickness.

3. Frontal Lobe

3.1. Precentral Gyrus, FA_γ

The precentral gyrus was analyzed at different levels, together with the postcentral gyrus.

3.1.1. Thickness of the Cerebral Cortex

The evolution of the cortical thickness of the left hemisphere at the level commanding the muscles of the trunk is shown in Figure 1. In very young infants (premature up to full term) this block also comprises the areas commanding the movements of the arm and hand. The area of the hand was studied separately in subjects aged 1 month up to 22 years.

At first glance, the curve of the development of the area commanding the muscles of the trunk shows an increase that is not progressive. In this area, the increase of the whole depth of the cortex is rapid until 6 months postnatal, with the exception of the period of a month before the eighth fetal month. From 6 months to about 2 years, this area does not show an increase in thickness, reappearing only around 4 years. A sharper period of growth resumes between 6 and 10 years. From 10 years on, cortical depth decreases, the minimum being at 14 years. A sharp increase reappears between 18 and 20 years. We do not have any definitive data between 20 years and the 35-year value given by von Economo and Koskinas (1925) in their atlas, which refers to the age range 25–40 years.

If one compares this development with that of

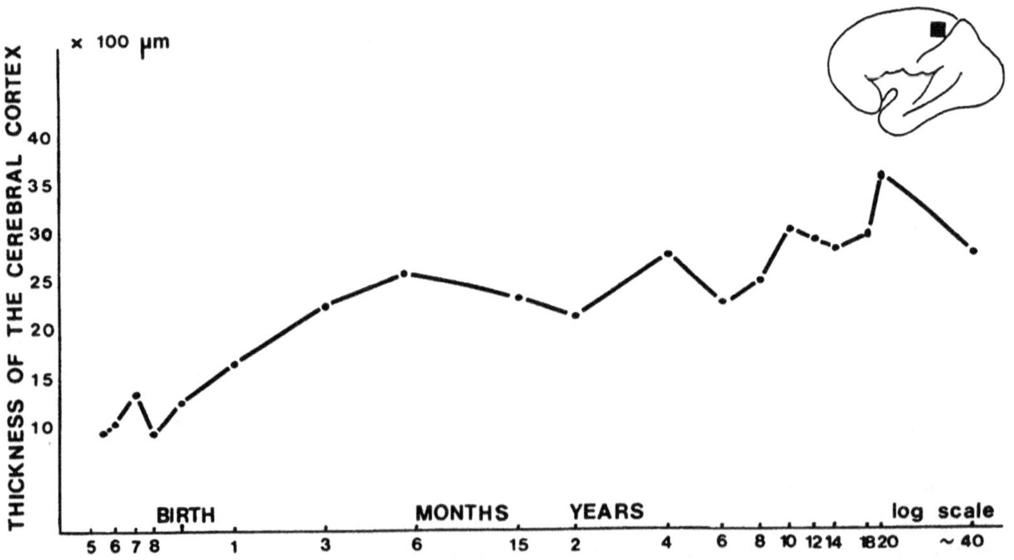

Fig. 1. Evolution through time of the depth of the human cerebral cortex in the area FA_γ of the left frontal lobe, measured in microns (average of 10–30 measurements done on the wall of the gyrus, in the middle of the area) on cresyl violet-stained, paraffin-embedded, 25-μm-thick sections. Similar techniques were employed in making the measurements, which served as the basis for the other figures in this chapter that portray the development of cerebral cortical thickness and neuronal density through time on the crown of the gyrus. Time is expressed on a logarithmic scale.

the other areas, one notices first that this is one of the thickest areas in the human cerebral cortex and also that this is one of the earliest areas to mature in the isocortical regions. On the curve shown in Figure 1, between 6 months and 4 years the increase in thickness is a slow one, as if most of the activities that have to be commanded by this area have been attained initially at 6 months. Other, more rapidly growing periods of development are shown here after 2 years and again after 6 years, with an almost adult value attained for the first time at around 10 years of age.

It is interesting to note that four periods of cortical-depth decreases are evidenced: at 1 month before birth (i.e., in the premature of 8 gestational months), at 2 years postnatal, and at 4 years and 14 years. We know from our research with mice (Rabinowicz *et al.*, 1977) that one can roughly consider periods of decrease of cortical depth to correspond to periods of remodeling of the dendritic pattern of neurones (see also Section 8).

3.1.2. Neuronal Density

The neuronal density (see Figure 2) is discussed for the precentral gyrus, FA_γ only for the second and fifth layers, which have been taken as examples for two different types of layers. In the second layer, the decrease in cell density is extraordinarily rapid between the premature infant of 6 months gestation and the full-term infant. At 6 months of fetal life, the cell density is more than 5000 per unit volume; 3000 at the seventh month, 2000 at the eighth month, and less than 400 at birth. This considerable decrease in cell density is a general phenomenon during the development of the cerebral cortex and is greater in the second layer than in almost any other layer of the other isocortical areas. After birth, the development of cell density goes at a much slower pace, with density reaching almost that of an adult at around 2 years.

The decrease in cell density can be seen not only in the cortex, but also in the various nuclei of the brain stem, the basal ganglia, and the cerebellum (Hamburger, 1975). The reduction in the number of neurons per unit volume gives good quantitative information on how rapidly maturation is progressing (Rabinowicz, 1967*b*, 1974, 1976). The fifth layer of the same area shows a much slower decrease in neuronal density from 6 postmenstrual months to birth. In the 6-month-old premature infant, the cell density in the fifth layer is about 800 cells, and this decreases to around 200 pyramidal cells at full-term birth. As we know from Yakovlev (1962), the fifth layer is a much earlier-maturing layer (in the precentral gyrus) than all the other lay-

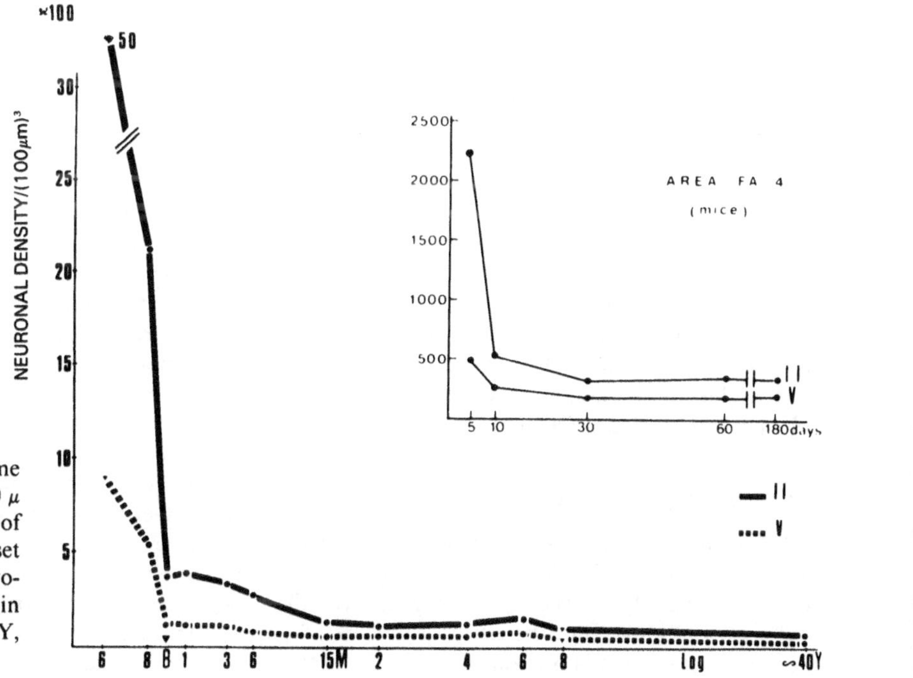

Fig. 2. Evolution through time of the neuronal density per 100 μm^3 of layers II (—) and V (---) of area FA_γ of the frontal lobe. Inset shows comparison with the evolution of the same area (FA 4) in mice. B, birth; M, months; Y, years.

ers, and some future Betz cells are recognizable as early as the fourth gestational month in the premature infant. At that time, these cells are located mostly in the fifth layer, but some of them can be seen in the sixth, and even in the fourth layer. In the fifth layer, an almost adult level of cell density is shown at around 6 months postnatally. Thus, an important difference exists between the second and the fifth layers in the speed of maturation in the same cortical area. The second layer is mature only around the 15th month to the 24th month after birth. The inset graph in Figure 2 shows that the phenomenon is exactly the same in the corresponding area of the cerebral cortex in mice (Leuba *et al.*, 1977).

Later during development, the graph of the neuronal density (Figure 2) shows a period, around 6 years, in which the density seems to be higher in both the second and fifth layers. In mice one sees that around 50 days there is also a slight increase in the cell density (Leuba *et al.*, 1977). It is most probable that there is no real increase in the number of neurons, as we know that the multiplication of neuroblasts ceases around the sixth prenatal month. As we have shown, using the data obtained on mice (Heumann *et al.*, 1977) and humans (Rabinowicz, 1976), this increase in density is most probably partly attributable to the decrease in size of pyramidal cells. In children around 6 years of age, the pyramidal cells of Betz are not as long from base to apex of the cell body as they are before that age, nor are they as long as they will become later on. Although this was seen for Betz cells (Rabinowicz, 1976), we assume that some of the other pyramidal cells follow more or less the same pattern (Rabinowicz *et al.*, 1977).

A comparison between the graphs of neuronal density and those of cortical thickness shows that there is a rather good correlation between the thickness of the cortex and the values of cell densities found at the same point for a given age. Still, the association of relatively higher cell density and lower cortical thickness at 6 years is most probably attributable to the fact that the same amount of cells packed together in a smaller volume.

3.1.3. Histological Development

The picture of the cerebral cortex of FA_γ at the level of the arm and hand is shown in Figure 3 at 8 months prenatally, in the full-term newborn, the 15-month-old child, and the child of 6 years. The increase in thickness is obvious between 8 months prenatally and the full-term newborn. Great differences exist, however, between these two stages of development. In the 8-month premature infant, the cell density is much higher especially in the second layer, the columns between cells are narrower, Betz cells are smaller, and the space around each cell is less. One also sees that between 8 months and term, the first and second layers do not increase in thickness, the third layer increases slightly, and most of the increase is in the fifth and sixth layers. Between term and 15 months of age, the differences are considerable. The whole thickness is not now present in the picture, and an increase in thickness occurs in all six layers, but more so in the pyramidal layers (third, fifth, and sixth layers) than in the granular layers (second and fourth). Between 15 months and 6 years, the differences in thickness are no longer so important, but the space around each cell becomes much greater, and the size of the Betz cells becomes quite different—broader and shorter at 6 years than at 15 months.

Cresyl violet-stained tissues show that Nissl bodies are already present at birth in Betz cells, although 1 month earlier there is only some powdery cresyl violet-positive dust in the whole cytoplasm of these Betz cells (Rabinowicz, 1964). At 6 years, Nissl bodies are well shown, the biggest ones being at the periphery of the cell and the smaller ones around the nucleus. The development of the size of Betz cells has been followed by Rabinowicz (1976).

3.2. Frontal Polar Area FE

3.2.1. Thickness of the Cerebral Cortex

In this area there is a rather rapid increase in thickness from the eighth prenatal month to the first month postnatally (see Figure 4). From the age of 1 month, this area shows a rather steady increase in thickness up to 4 years, followed by lower values from 8 to 18 years. The lower values for 10 to 18 years were obtained in one case each, and thus need confirmation. Between 8 and 20 years, this area shows, even on preliminary data, a rather great variability. This is noteworthy because, at the adult level, von Economo and Koskinas (1925) also found great variability in thickness among their cases. Generally, this area shows a progressive increase in thickness up to 4 years and a period between 8 and 18 years during which there is a sharp decrease. On the other hand, differences between adults were so large that von Economo and Koskinas (1925) did not give a single mean value, but

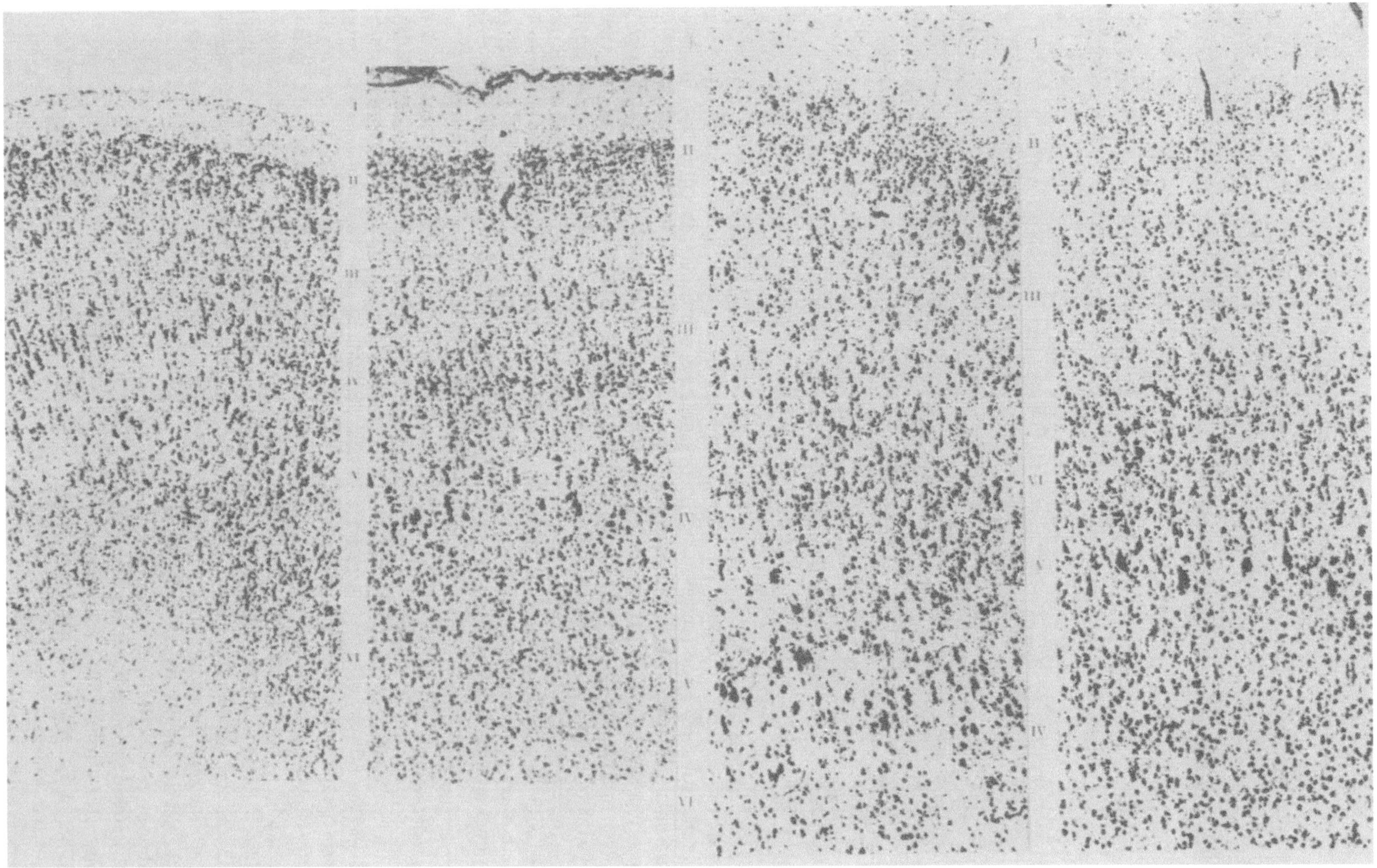

Fig. 3. Histological evolution of area FA_{γ}, hand. Cresyl violet stain. x50. From left to right: premature infant of 8 months, (Rabinowicz, 1964) full-term newborn, (Conel), child of 15 months (Conel, 1955), and child of 6 years (Conel, 1967).

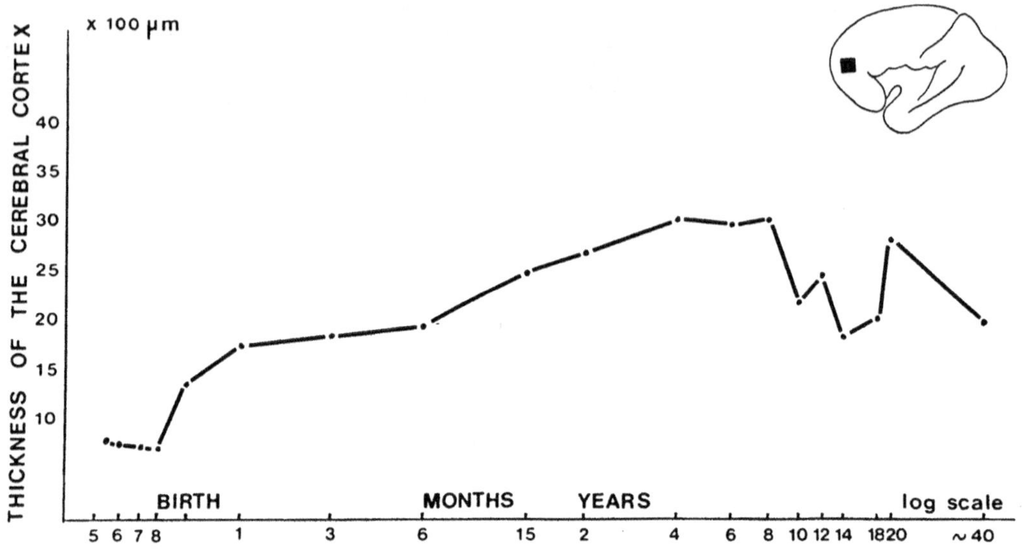

Fig. 4. Development through time of the thickness of the cerebral cortex in the frontal polar area (FE) of the left frontal lobe in human.

subdivided their cases into two groups, one showing and the other not showing an increase.

Since the frontal pole is thought to be concerned with various elaborate mental functions, the presence of a very irregular curve between 8 and 20 years may be related to great individual variability, which seems to persist even in adults.

Compared with the precentral motor area, FE appears to be thinner, sometimes much thinner. The reasons are unclear at present.

3.2.2. Neuronal Density

At 6 months prenatally, the cell densities of both the second and the fifth layers of this area are almost identical to those of the precentral motor area (see Figure 5). However, the decrease in cell density occurs less rapidly in both layers. The cell density is around 2800 in the second layer in the eighth prenatal month and around 800 at the same time in the fifth layer. Between the eighth prenatal month and birth, the density rapidly decreases. In the second layer, however, the cell density is higher, and thus at birth the frontal pole is more immature than the premotor area—around 700 cells, compared with about 400 in FA_γ trunk. On the other hand, the fifth layer shows almost the same density (200) seen in the precentral area, yet another example of differences in maturation between layers and areas.

For the second layer, the decrease is a more rapid one in FE than in FA_γ, but reaches, also around 2 years, the adult level; the fifth layer reaches this point 1 year earlier. The same phenomenon of slightly higher cell density appears at 6 years, as in the precentral gyrus, but we should be cautious about the data at the 22nd year because they have been obtained from only one case. Generally speaking, area FE shows a much slower maturation of its second layer compared with the precentral gyrus, while the fifth layer shows about the same speed of maturation in both areas.

As a whole, this area shows a progressive growth until 4 years. After 10 years, individual variability appears, resulting later in two groups of individuals: one without any further increase in thickness (and perhaps function?), and the other with an increase (von Economo and Koskinas, 1925). It is intriguing also that during the entire period between 4 years and adulthood, the period of schooling, there appears to be little further growth. Clearly, we need more information from additional cases.

3.3. Area FCB_m

3.3.1. Thickness of the Cerebral Cortex

Area FCB_m corresponds to the motor area of speech and seems to show three periods of growth in terms of total thickness of the cortex (see Figure 6). The first period is a rapidly increasing one, from

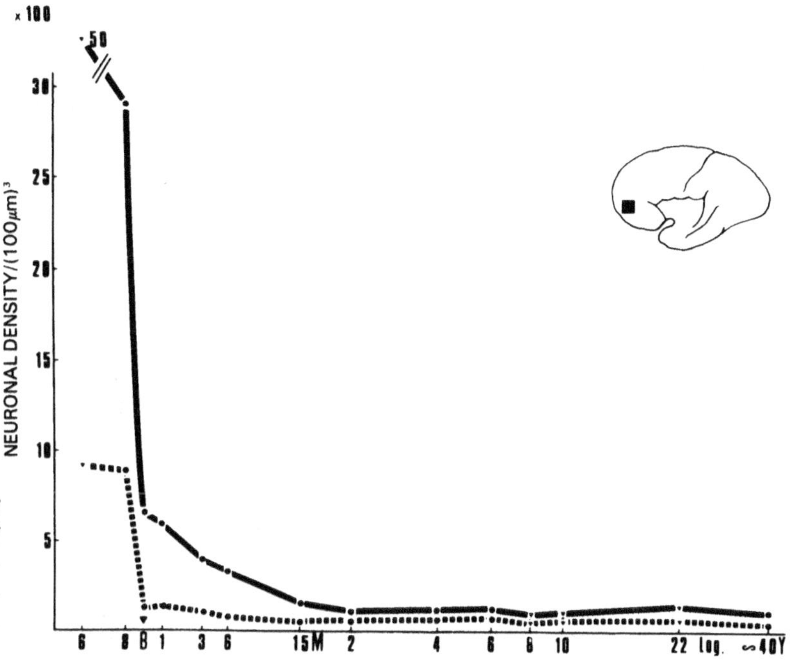

Fig. 5. Evolution through time of the neuronal density per 100 μm^3 of layers II (—) and V (---) of the frontal polar area (FE) of the frontal lobe in humans. B, birth; M, months; Y, years.

the fifth prenatal month to 1 month postnatally, with a decrease in the eighth prenatal month. The second period shows a slowly increasing thickness from 1 month to 2 years. The third period shows a rather rapid increase between the second and sixth years, and one can consider adult thickness to be attained at around 6 years. Again, we found a rather long period of decrease in cortical depth in children aged 6–18 years.

The evolution of this area corresponds fairly well to the development of motor speech abilities as described in children. We will see later that the posterior and inferior speech centers have a rather different development.

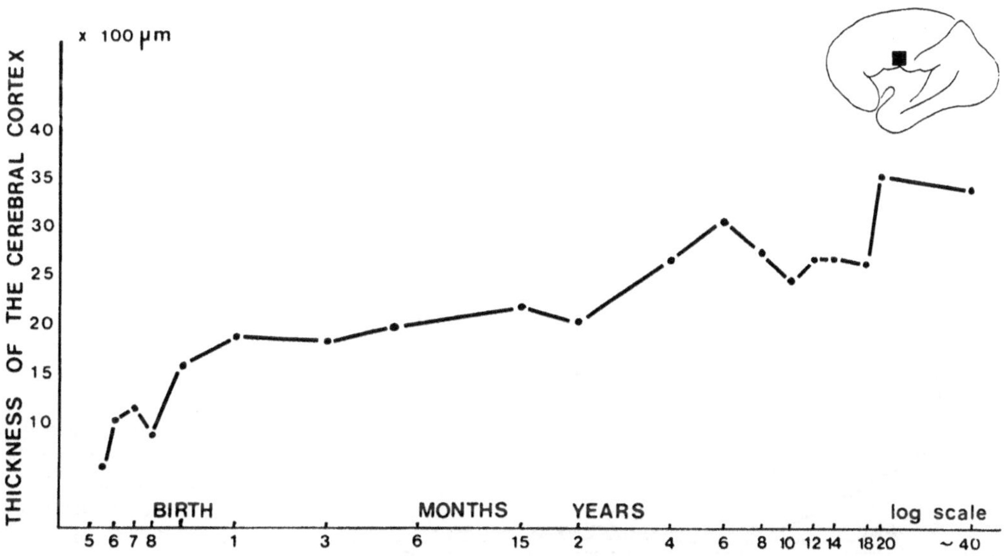

Fig. 6. Development through time of the thickness of the cerebral cortex in area FCB_m of the left lobe in humans.

3.3.2. Neuronal Density

If we compare the cell densities of the second layer in the eighth prenatal month among areas FA_γ, FCB_m, and FE, it becomes apparent that the precentral gyrus has at that time the lowest number of cells in the second layer (2200), while the less-well-developed area FE shows 2800 cells (see Figure 7). Area FCB_m is between, with 2500 cells per unit volume. These differences correspond well to the fact that the precentral gyrus is the earliest developed area of the frontal lobe, while FE is one of the latest. At birth, the neuronal density of the FCB_m area is also between that of the precentral motor and the frontal polar areas.

It is remarkable that some sort of topographical predeterminism is quantitatively noticeable at least 3 months before birth in the human frontal cortex. The decrease in cell density of area FCB_m shows that at around 15 months it is almost at the density level of the adult in the second layer, while this level is reached at 6 months for the fifth layer. For the latter, the decrease in cell density also shows a velocity that is between FA_γ and FE. Here we can see the same phenomenon at 6 years with a slight apparent increase in cell density both for the second and the fifth layer—already noted in the discussion on the FA_γ trunk area.

3.3.3. Histological Development

The motor center for speech is shown in Figure 8, in the eighth prenatal month, the full-term newborn, the child of 2 years, and the 6-year-old child. The histological picture corresponds quite well to our graphical depiction both of the evolution of the thickness of the cortex and of the neuronal density. The considerable difference in thickness of the cortex between the eighth prenatal month and the full-term newborn is clearly demonstrated, as well as the differences in cell densities. The difference is even more striking between the full-term newborn and the child of 2 years—not only is the thickness of the cortex at 2 years almost double that at birth, but also the density is much lower, meaning that the dendrites developed at a considerable rate during those first 2 years after birth. This is well demonstrated with the Golgi–Cox impregnations shown by Conel (1939, 1941, 1947, 1951, 1955, 1959). On closer scrutiny, there are some quite important, if less visible, differences between the cortex at 2 years and at 6 years. In the latter age group, the increase in space around the cells is a clearly visible indication that dendrites were still growing during that time. Even the first layer shows some noticeable increase in thickness, as does the second layer.

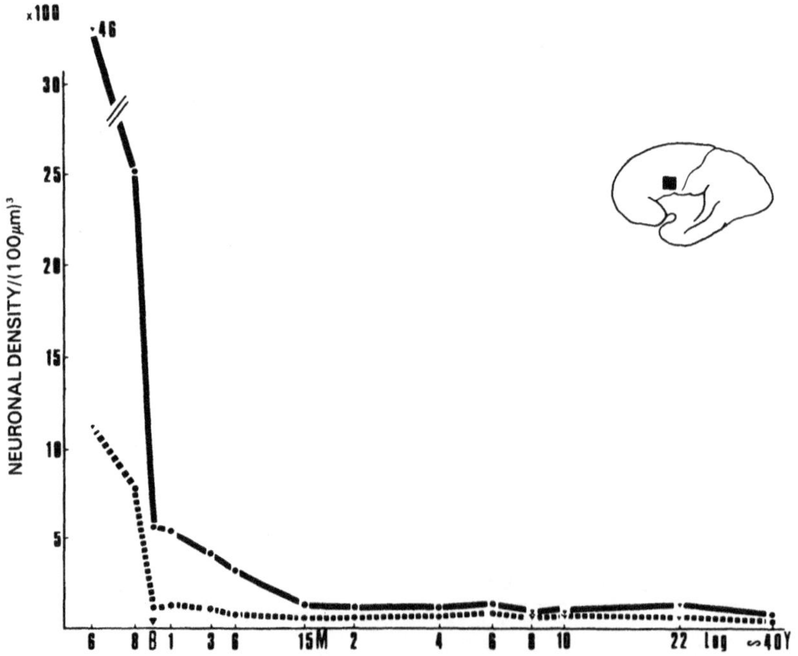

Fig. 7. Evolution through time of the neuronal density per 100 μm^3 of layers II (—) and V (---) of area FCB_m of the frontal lobe in humans. B, birth; M, months; Y, years.

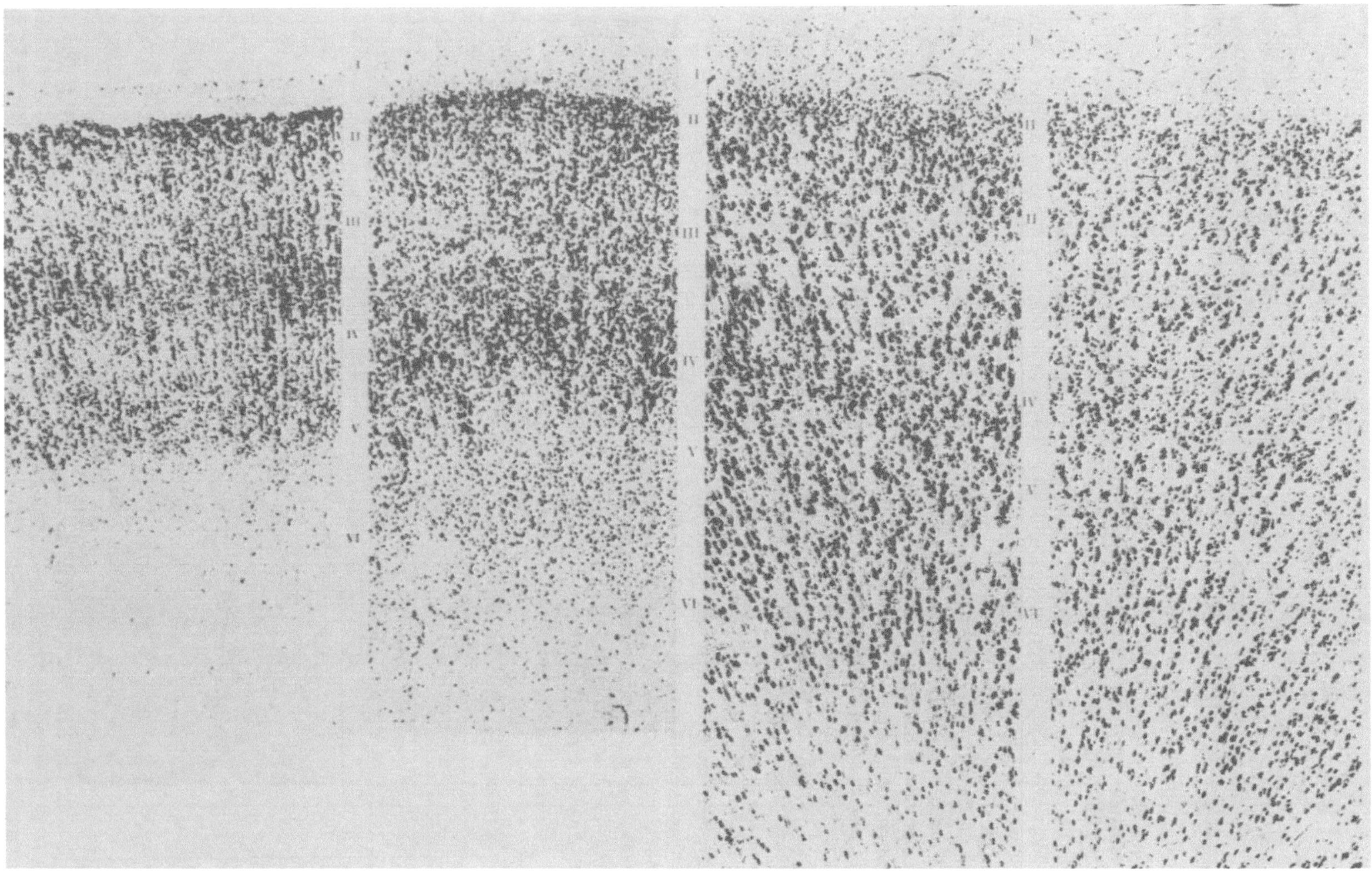

Fig. 8. Histological evolution of area FCB_m (Broca). Cresyl violet stain. x50. From left to right: premature infant of 8 months (Rabinowicz, 1963), full-term newborn (Conel, 1939), child of 24 months (Conel, 1959), and child of 6 years (Conel, 1967).

4. Parietal Lobe

4.1. Area PC, Trunk

4.1.1. Thickness of the Left Cerebral Cortex

The general pattern of growth of this area is relatively simple up to 15 months (See Figure 9). Increase in thickness is rapid from the fifth prenatal month to the full-term newborn, except around the eighth prenatal month. From that time, the increase in thickness is slow, but continuous, until 15 months. Between 15 months and adulthood, thickness varies greatly, with increases between 2 and 4 years, 6 and 8 years, 10 and 12 years, and 14 and 20 years. Compared with the growth of the motor area at the same level, the differences appear rather striking during adolescence—the motor area has a more rapid growth and is thicker. Its growth lasts longer between the eighth prenatal month and the sixth postnatal month, as compared with the post-central gryrus, where the period of rapid growth stops at birth.

If area PC is related to the complex development of motricity, there must be differences between the motor abilities, which are developing much more rapidly until the age of 6 months, and the sensory abilities, which are developing much more slowly from birth. Also noticeable is the rapid growth of this area during the prenatal months. Is there perhaps a tendency for the PC area not to develop concomitantly with motricity during the first 2 years postnatal? A concordance appears only after the age of 2 years with alternating periods of growth and decrease of development. Noticeable is the peak at 12 years, which has to be viewed with caution, since only one case was quantified at this age.

4.1.2. Neuronal Density

The postcentral area at the level of the trunk shows, as in other areas, a great difference between the rate of maturation of the second layer and that of the fifth—the former being, as usual, slower (see Figure 10). The second layer reaches an almost adult neuronal density only at 15 months. As in other areas, the cell density decreases very rapidly before birth. However, around 6 years, there is an apparent increase in cell density. Compared with the motor area at the same level, the differences are striking only for the second layer, while the fifth shows almost the same patterns of development.

4.2. Area PF (Posterior Speech Center, Wernicke)

4.2.1. Thickness of the Left Cerebral Cortex

In this area, the cerebral cortex shows a slight decrease in thickness from the sixth prenatal month to the eighth prenatal month (see Figure 11). From

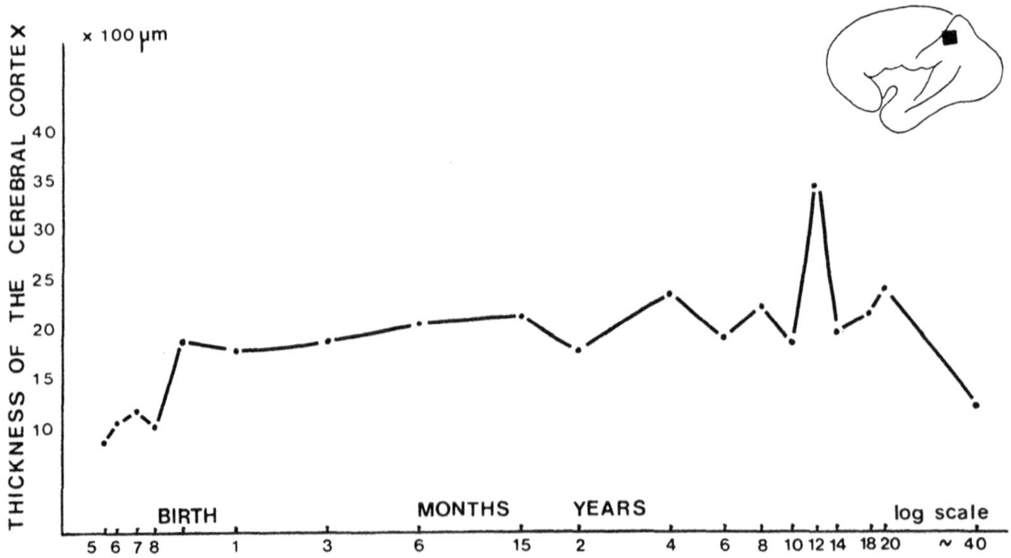

Fig. 9. Development through time of the thickness of the cerebral cortex in area PC of the left parietal lobe in humans.

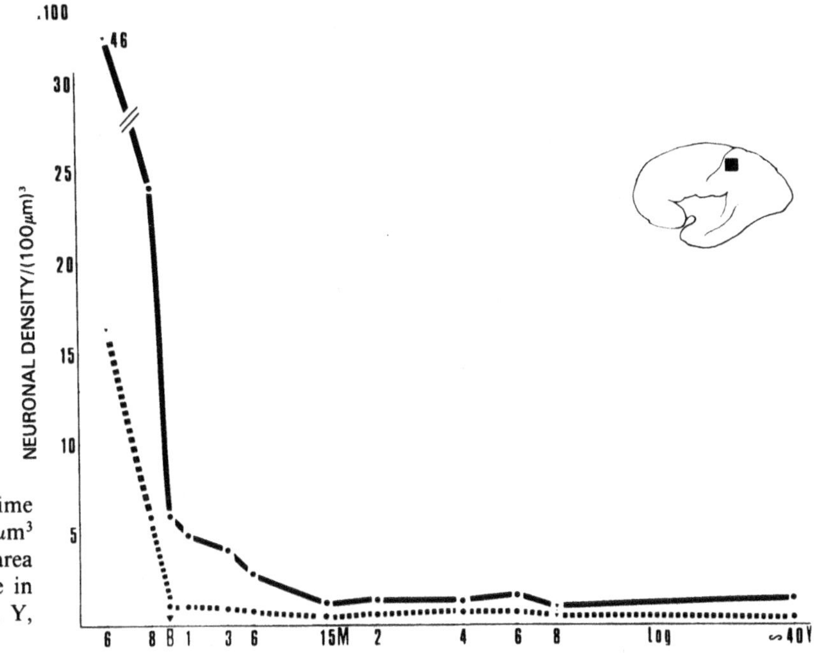

Fig. 10. Evolution through time of the neuronal density per 100 μm^3 of layers II (—) and V (---) of area PC (trunk) of the parietal lobe in humans. B, birth; M, months; Y, years.

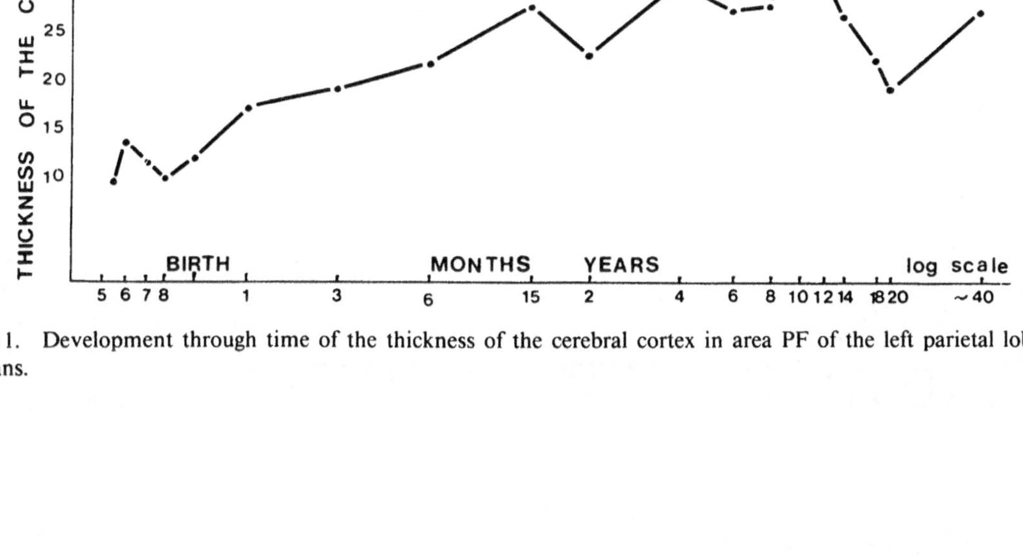

Fig. 11. Development through time of the thickness of the cerebral cortex in area PF of the left parietal lobe in humans.

birth until 15 months, the increase in thickness is rather steady, and from that point, the average thickness does not increase very much until 10 years. At adult levels, however, the total thickness is not greater, while between 15 months and 10 years the variability is notable with a clear and continuous decrease during adolescence.

This curve may reflect the fact that our material is limited and that variability between individuals is certainly high from one age to another, explaining the irregularities of the curve between 15 months and 12 years. As will be described in Section 5.1, another speech center, the temporal inferior area (TA), shows the same phenomenon even more markedly.

A striking fact in the evolution of the posterior speech center during adolescence is its uninterrupted and considerable decrease in depth between 10 and 20 years. Found in a succession of four cases (all boys), this suggests that there is a trend of decreasing dendritic pattern of this area during that period. Does this correspond to some speech habits during adolescence?

4.2.2. Neuronal Density

Area PF behaves rather similarly to area FCB_m (Broca) for the second and the fifth layers from birth to 15 months (see Figure 12). Area PF shows a much higher cell density in the second layer in the premature, showing this layer to be less developed prenatally. At the 15th month, cell density corresponds to an almost adult level, but the great variability between individual cases is not revealed by the averaging of results used. As in the other areas, the period around 6 years corresponds to a relative increase in cell density.

4.2.3. Histological Development

The area is shown in Figure 13 at the eighth prenatal month, the full-term newborn, and at 4 years. Each micrograph shows almost no increase in thickness between the eighth prenatal month and the full-term newborn. The latter still shows a high density of cells, but the space between columns the mean intercolumnar space, is clearly increasing, a feature that is quite well developed at 4 years of age. As seen in other areas, the cell density of the second layer is rapidly decreasing, while the thickness of the first, second, and third layers is increasing, as is the thickness of the fifth and even more so the sixth layers. Some large pyramidal cells are well developed at 4 years, while at birth they are hardly seen at all.

Not only is this area a rather variable one depending on the individual, but also the differences in degrees of maturation between layers are consid-

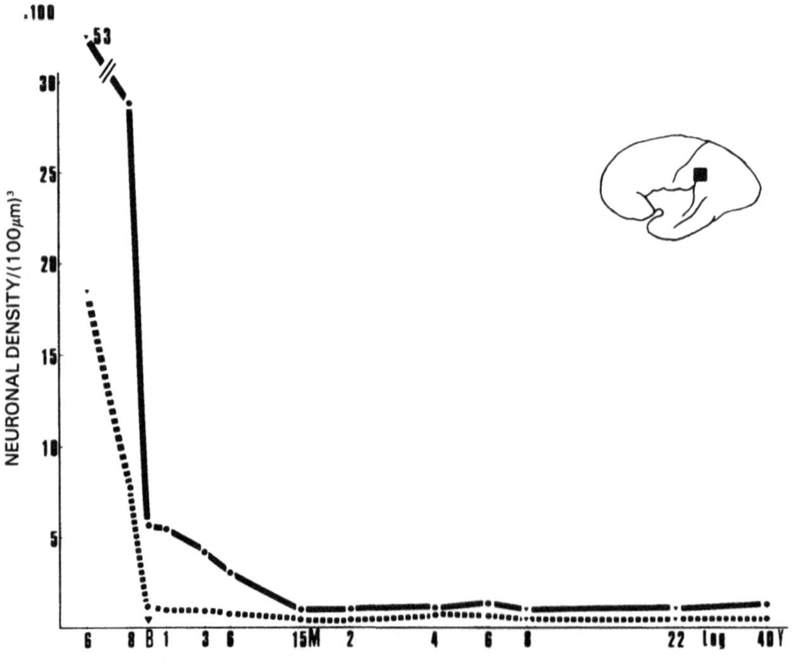

Fig. 12. Evolution through time of the neuronal density per 100 μm^3 of layers II (—) and V (---) of area PF of the parietal lobe in humans. B, birth; M, months; Y, years.

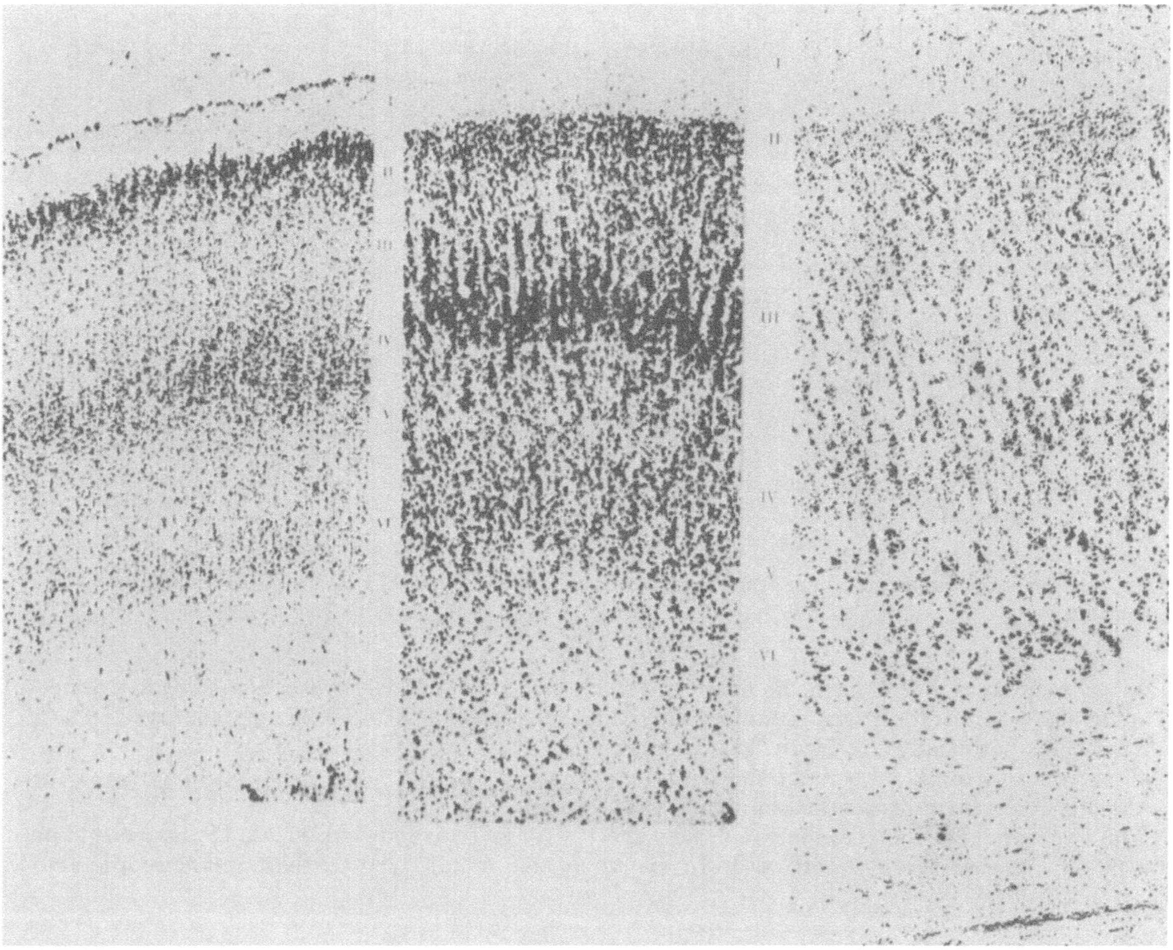

Fig. 13. Histological evolution of area PF. Cresyl violet stain. x50. From left to right: premature of 8 months (Rabinowicz, 1964), full-term newborn (Conel, 1939), child 4 years (Conel, 1963).

erable. Along with a differential growth in thickness of layers, there is also an important increase in the distance between cells, indicating an active dendritic development.

5. Temporal Lobe

5.1. Area TA: Anterior Lower Speech Center

5.1.1. Cortical Thickness

This area shows a rapid increase in thickness between the fifth and sixth prenatal months and a sharp decrease during the seventh and eighth fetal months (see Figure 14). After birth, the increase is a slow one up to 2 years. Between 8 and 20 years, the curve presents some variability at the different ages, which shows that individual variability can be noticeable during adolescence.

The posterior, as well as the anterior, mid-temporal speech cortex (TA) both show much variability and inhomogeneity in the cases used for the quantitative study. This is also reflected in the histology, in which one can see that the cortex of the 15-month-old child shown in Figure 16 is thicker than that of the 6-year-old.

5.1.2. Neuronal Density

The cell density in the second layer is relatively lower at the eighth prenatal month than in many

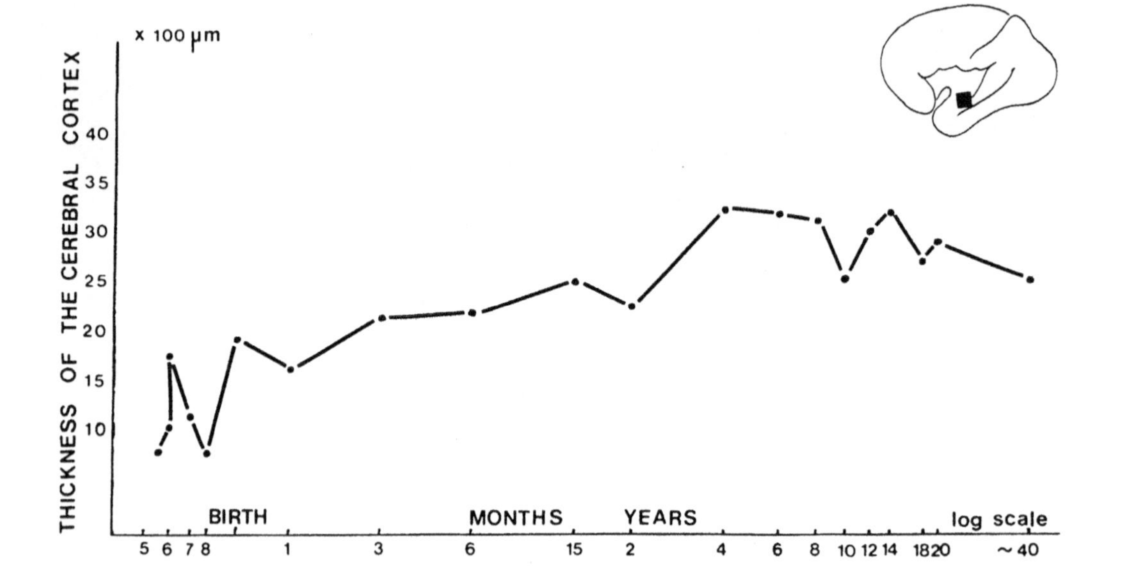

Fig. 14. Development through time of the thickness of the cerebral cortex in area TA of the left temporal lobe in humans.

other brain areas, but at birth the density is still quite high and then decreases rather rapidly until 15 months postnatally (see Figure 15). The usual differences in maturation between the second and the fifth layer are rather marked in area TA. The fifth layer is not far from its definitive density after 6 months. As elsewhere at 6 years, we find a relative increase in cell density corresponding to a decrease in the thickness of the cortex at that age.

5.1.3. Histological Development

Area TA is shown in Figure 16 at the eighth prenatal month, in the full-term newborn, at 15

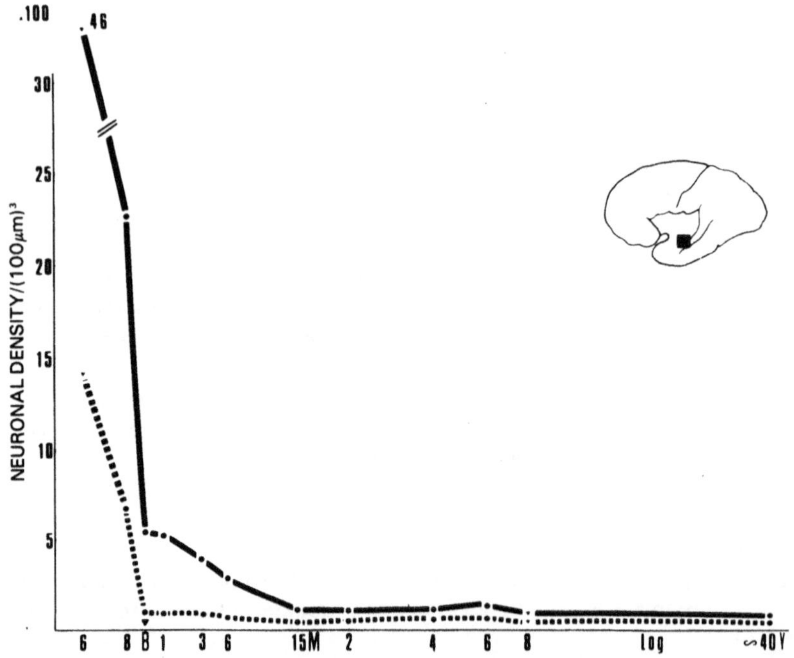

Fig. 15. Evolution through time of the neuronal density per 100 μm^3 of layers II (—) and V (---) of area TA of the temporal lobe in humans. B, birth; M, months; Y, years.

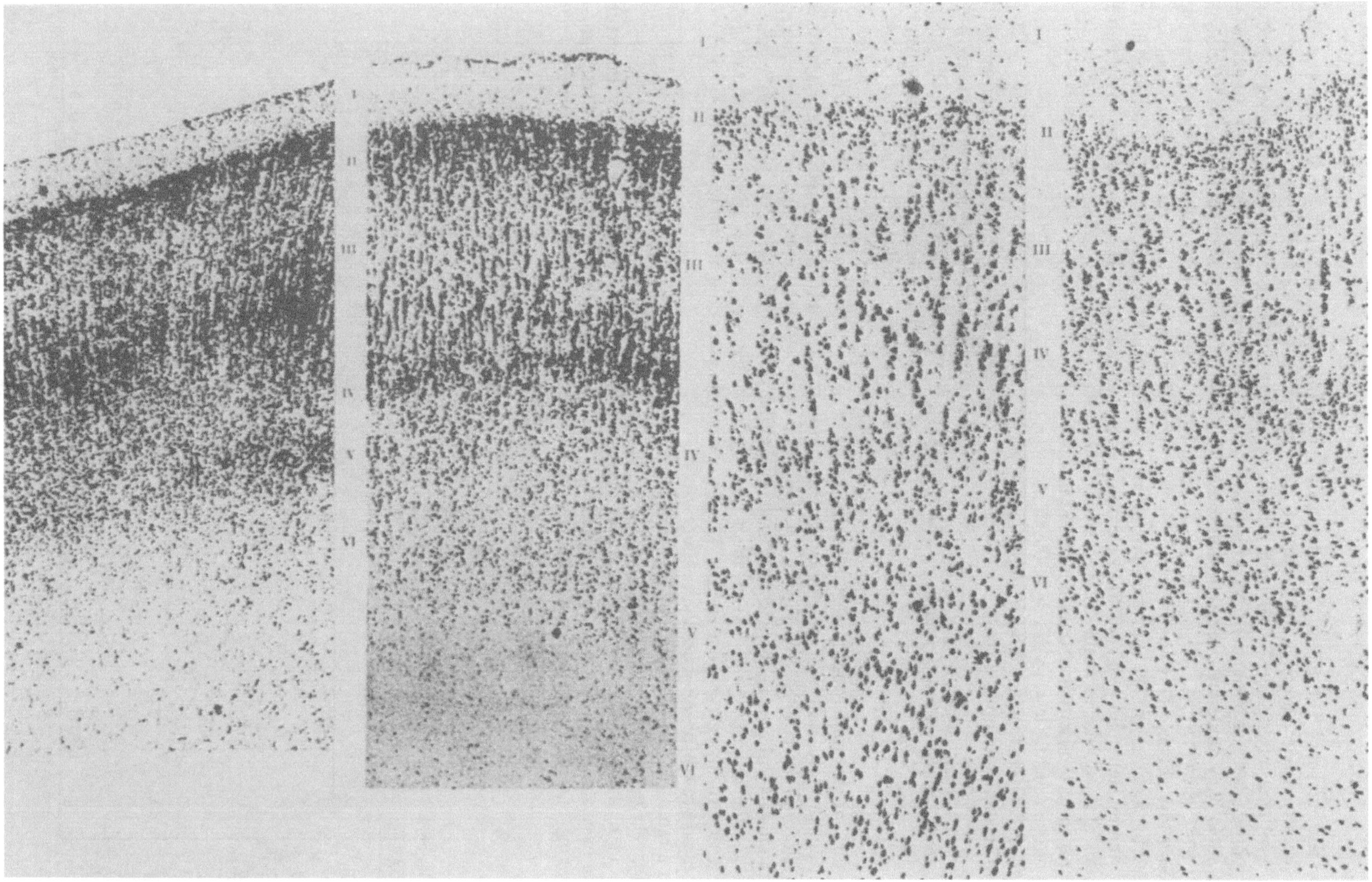

Fig. 16. Histological evolution of area TA. Cresyl violet stain. x50. From left to right: premature of 8 months (Rabinowicz, 1964), full-term newborn (Conel, 1939), child of 15 months (Conel, 1955), and child of 6 years (Conel, 1967).

months, and at 6 years. The decrease in cell density appears clearly if one compares the cortex of the newborn with the cortex at the 8th prenatal month. This important decrease in cell density corresponds only to a relatively moderate increase in thickness of the cortex, showing that, most probably, there has been a loss of cells. This has been effectively described by Hamburger (1975) in the brain stem of mice and by Heumann *et al.* (1978) in the cerebral cortex of mice. Our micrographs show that at 15 months the cell density is about the same as at 6 years or even slightly less. In this area, the third, fifth, and sixth layers increase most in depth in the full-term newborn and up to 6 years.

5.2. Area TC: Auditory Projection Area

5.2.1. Thickness of Area TC

This area shows a rather rapid increase in thickness during the last month of pregnancy and a slower one until 15 months (see Figure 17). Between 15 months and 2 years, there is a decrease in thickness, while between 2 and 8 years there is a second period of increase, which ends at 8 years after a rapid growth phase. The adult values are reached around 12 years followed by a period of rapid decrease between 14 and 18 years, with a return to adult values at 20 years. The number of cases we have studied is low; nevertheless, generally speaking, one can say that this is a rather steadily growing area in which the adult thickness is attained between 12 and 20 years.

Interestingly, there is a turning point of musical abilities between 4 and 8 years of age (see curves of TA and TC), which might correspond to the highest depth of the auditory speech center TA at these ages. It is also worth mentioning that variability of cortical depth is great during adolescence for the anterior motor (FCB_m), the inferior (TA), and the auditory (TC) speech areas.

5.2.2. Neuronal Density

This area shows a considerable decrease in cell density for the second layer between the sixth prenatal month and birth (see Figure 18). The same is true for the fifth layer. For the latter, however, the cell density for adult values is found around 6 months, if not even earlier, while for the second layer it is found between 15 months and 8 years. Thus, a quantitative analysis of the cytoarchitectonics shows that there must be some inequalities in the functional abilities of this area because of great differences in the speed of maturation between the second and fifth layers.

5.2.3. Histological Development

Area TC is illustrated in Figure 19 with pictures taken at the eighth prenatal month, at term, and at 15 months and 6 years. There is a striking differ-

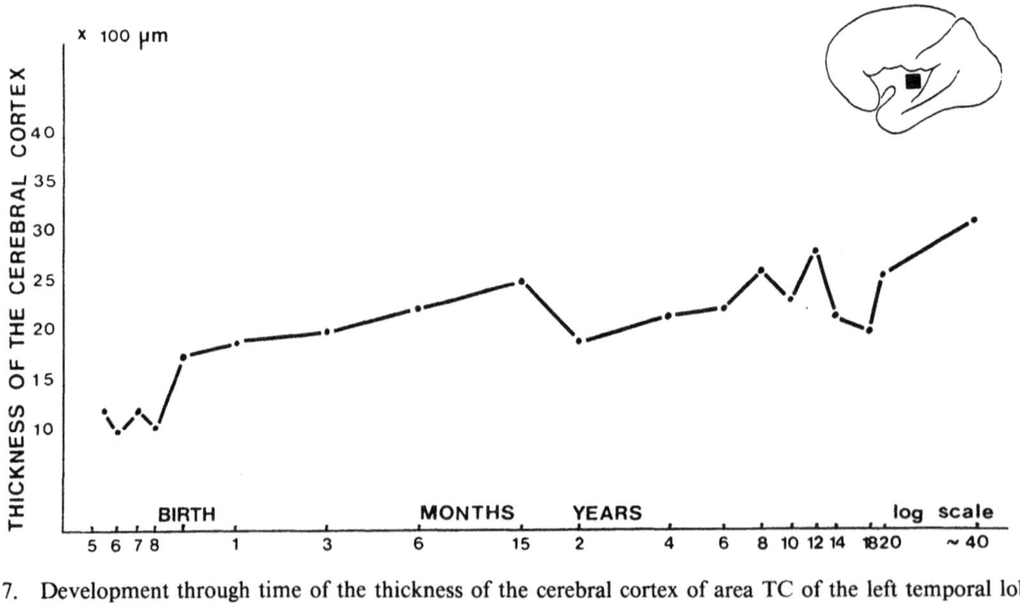

Fig. 17. Development through time of the thickness of the cerebral cortex of area TC of the left temporal lobe in humans.

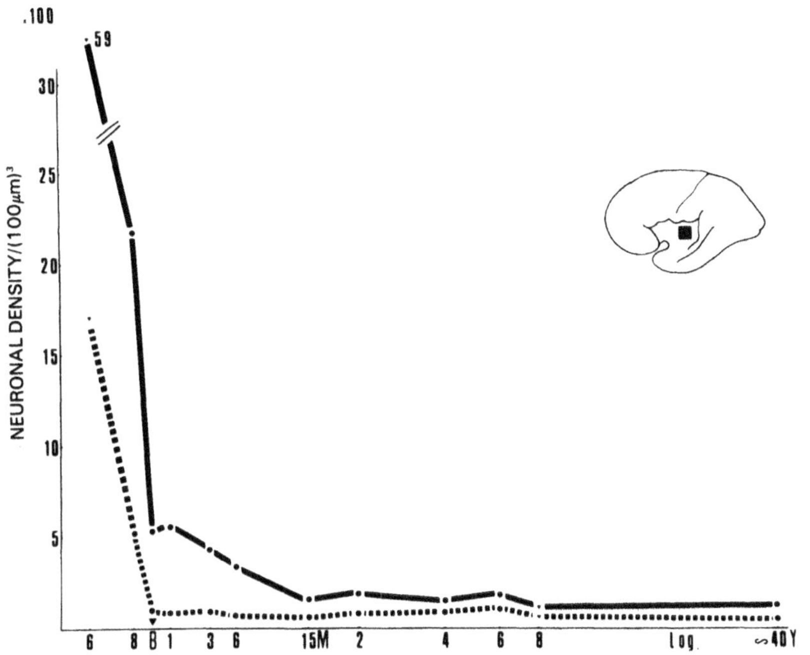

Fig. 18. Evolution through time of the neuronal density per 100 μ m³ of layers II (—) and V (---) of area TC of the temporal lobe in humans. B, birth; M, months; Y, years.

ence between the eighth prenatal month and the full-term newborn. The third layer of the full-term newborn has considerably increased, not only by the diminution of its cell density, but also by the development of the distance between cells. Once more, the child of 15 months has, in this series of pictures, a much thicker cortex than the child of 6 years.

6. Occipital Lobe

6.1. Area OC, Calcarine, Visual Projection Area

6.1.1. Thickness of the Cortex

The thickness of the visual primary projection area shows a sharp decrease between the sixth and eighth prenatal months and a rapid increase between the eighth prenatal month and the full term newborn (see Figure 20). From 1 to 3 months there does not seem to be an increase, but an increase occurs between 3 and 6 months. Remarkably, the thickness attained at 6 months is quite close to that found by von Economo and Koskinas (1925) in the adult, except for other increases at 8 and 14 years. One may consider that this area is most probably functioning very early in childhood. This possibility is emphasized if one studies the Golgi–Cox impregnations made in the premature infant (Rabinowicz, 1964) and in children from birth to 8 years (Conel, 1939, 1941, 1947, 1951, 1955, 1959, 1963, 1967). Another feature of this area is the small variability, which is expressed by an almost straight line from 6 months to adulthood except during adolescence. This phenomenon is illustrated in Figure 22, where there are almost no differences in cell densities and overall structure between the cortex of a child of 6 months and that of the child of 6 years.

6.1.2. Neuronal Density

Despite its early development, area OC shows a difference between the maturation of the second layer and the maturation of the fifth (see Figure 21). For the second layer, the decrease in cell density is considerable, going from 7700 cells per unit volume in the sixth prenatal month to about 600 cells at birth—probably the most rapid decrease seen in the cerebral cortex. From birth to the age of about 15 months, the decrease is slower, and probably the maturation of the second layer is almost completed at 15 months. Meanwhile, the fifth layer attains adult values as early as 3 months. Here also we have a rapid decrease in cell density for the fifth layer, and attention should be drawn to the fact that this layer is mature earlier than any other area we

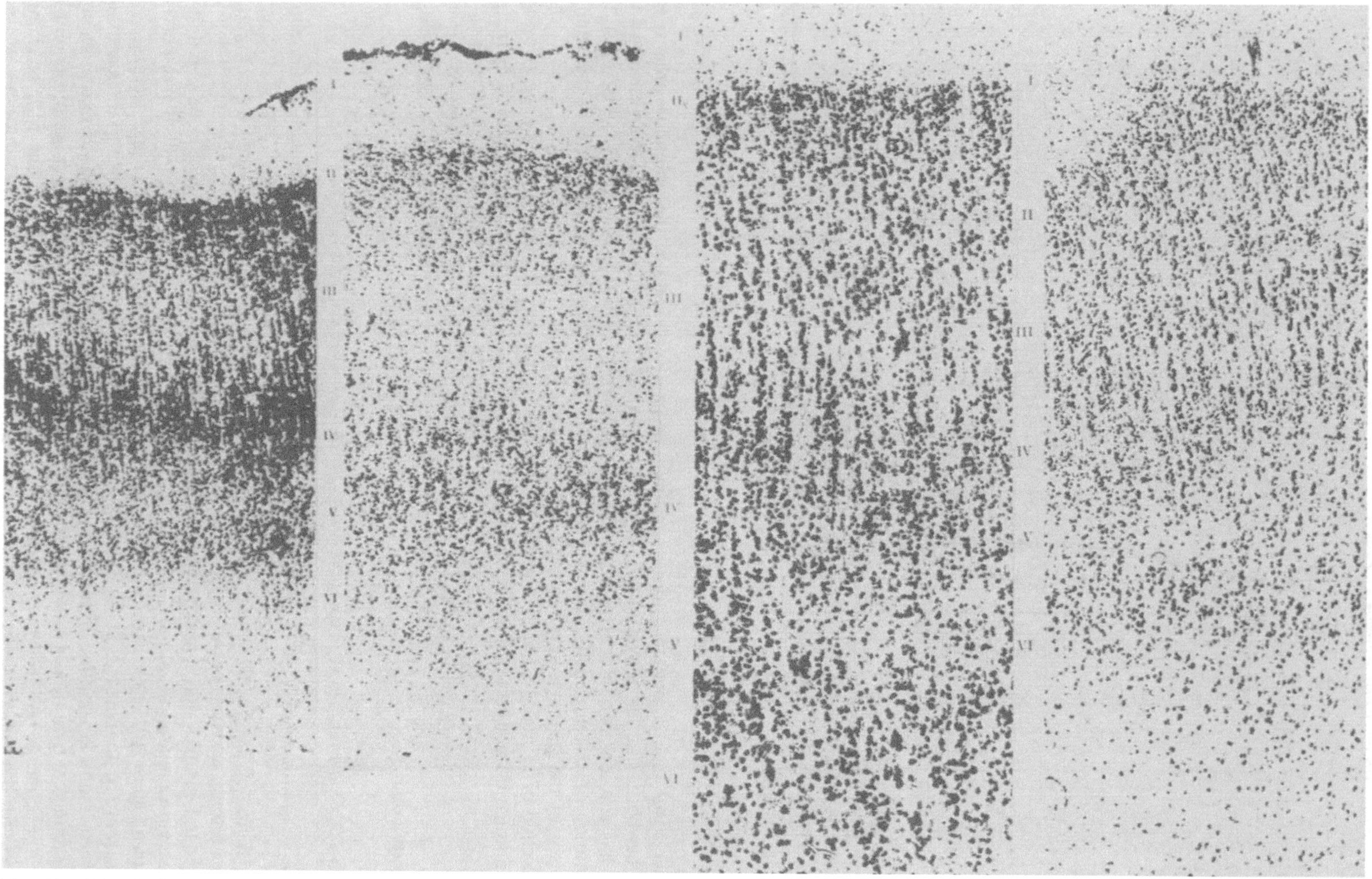

Fig. 19. Histological evolution of area TC. Cresyl violet, x50. From left to right: premature of 8 months (Rabinowicz, 1964), full-term newborn (Conel, 1939), child of 15 months (Conel, 1955), child of 6 years (Conel, 1967).

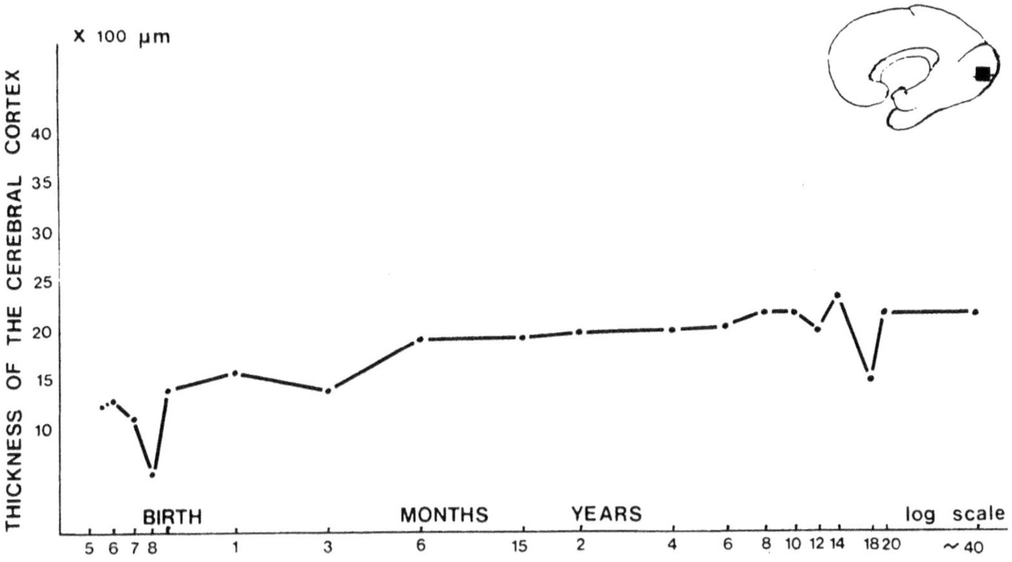

Fig. 20. Development through time of the thickness of the cerebral cortex in area OC of the left occipital lobe in humans.

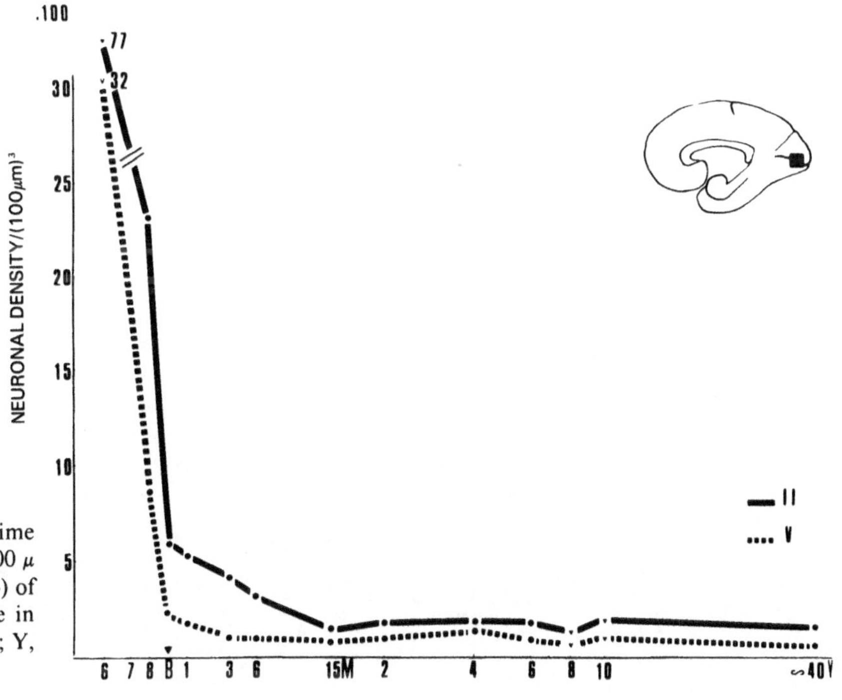

Fig. 21. Evolution through time of the neuronal density per 100 μm^3 of layers II (—) and V (---) of area OC of the occipital lobe in humans. B, birth; M, months; Y, years.

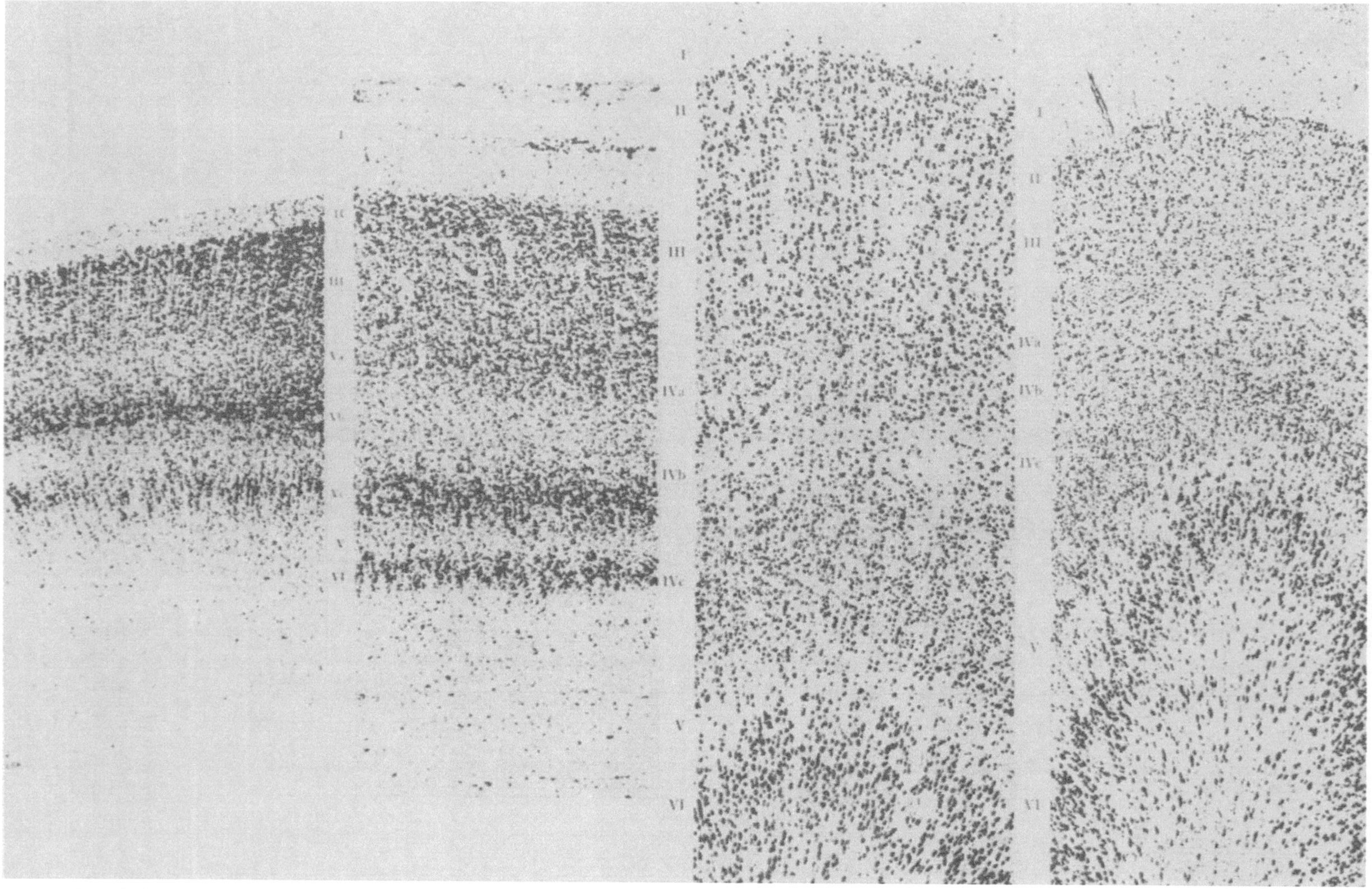

Fig. 22. Histological evolution of area OC. Cresyl violet stain. x50. From left to right: premature of 8 months (Rabinowicz, 1964), full-term newborn (Conel, 1939), child of 15 months(Conel, 1955), child of 6 years (Conel, 1967).

have studied to date. Golgi–Cox impregnations still demonstrate a clear difference in the degree of maturation of the pyramidal cells of the fourth B layer, which are much slower in developing than those of the sixth or the third layers (Rabinowicz, 1964).

6.1.3. Histological Development

Area OC shows an important increase in thickness during the last prenatal month, when all the layers are easily recognizable (see Figure 22). It is also interesting to note that our study of the sixth, seventh, and eighth prenatal months shows that the limits between OC and OB (Limes OB_γ) appear only at the seventh prenatal month, while at that same time, the different layers of OC are easily recognizable. This micrograph also shows a considerable decrease in cell densities, as well as an increase in size of the pyramidal cells between the full-term newborn and the child of 6 months. Comparison of the latter with the child of 6 years shows that the increase in cellular size is slight and not enough to alter the total cortical thickness significantly.

6.2. Area OA, Visual Association Area

6.2.1. Cortical Thickness

This area presents a remarkable development as compared with the development of OC (see Figure 23). OA shows an almost steady increase in thickness up to 15 months, at which time the thickness is greater than in the adult. From 15 months, there is a slow decrease, with a minimum around 4 years, followed by a sharp increase, peaking for the second time at 14 years. Compared with OC, it is obvious that this area reaches first an approximately adult level more than 9 months later than the primary visual cortex. This is not surprising, as we know that area OA is related to visual understanding. Noticeable also is the fact that four periods of thinner cortex are evidenced: at 8 months fetal; at 6 months postnatal; and at 4–6; and 18 years. A peak of cortical thickness is present at 15 months, at 14 years, and at 20 years. All in all, area OA is thicker than area OC and seems to decrease in thickness with age.

6.2.2. Neuronal Density

As shown in Figure 24, the differences in maturation between layers 2 and 5 are quite striking as well, but layer 5 seems to be slightly later, reaching the adult level later than layer 5 of area OC, while for layer 2 the development is very similar after birth to that seen in layer 2 of OC. The cell density, however, is slightly lower in the 6 prenatal months for both the second and the fifth layers in area OA.

7. Limbic Lobe

7.1. Anterior Limbic Area, LA

7.1.1. Thickness

The anterior limbic area (LA) shows a more irregular growth in that it is slow in the last gestational trimester and then very rapid during the last month before birth at full term (see Figure 25). From then on, there is almost no increase in thickness until 6 months, followed by a slow, progressive growth until 4 years, with a clear decrease at 6 years, reaching adult thickness around 8 years. This area shows a very rapid growth spurt 1 month before birth, with a maximum at birth followed by a slow increase up to 4 years. As in many other areas, a decrease is seen at 6, 12, and possibly around 20 years.

7.1.2. Neuronal Density

The anterior limbic area is remarkable for its very early maturation in terms of cell density (see Figure 26). The density of the second layer is around 3600 in the sixth prenatal month, decreasing rapidly to less than 400 at birth, thus showing a relatively rapid prenatal maturation. The second layer is also irregular in maturation between 3 and 6 months and reaches adult values at 15 months as elsewhere. For the fifth layer, the cell density of the premature infant is also very low, and the adult level is attained late, at 15 months, thus showing little difference between the maturation of the second and of the fifth layer, contrary to what is seen in many other areas.

7.2. Posterior Limbic Area, LC

7.2.1. Cortical Thickness

This area shows a more accentuated pattern of development than the anterior limbic area, with a rapid growth spurt before full-term birth, followed by a decrease in thickness at the first month and a

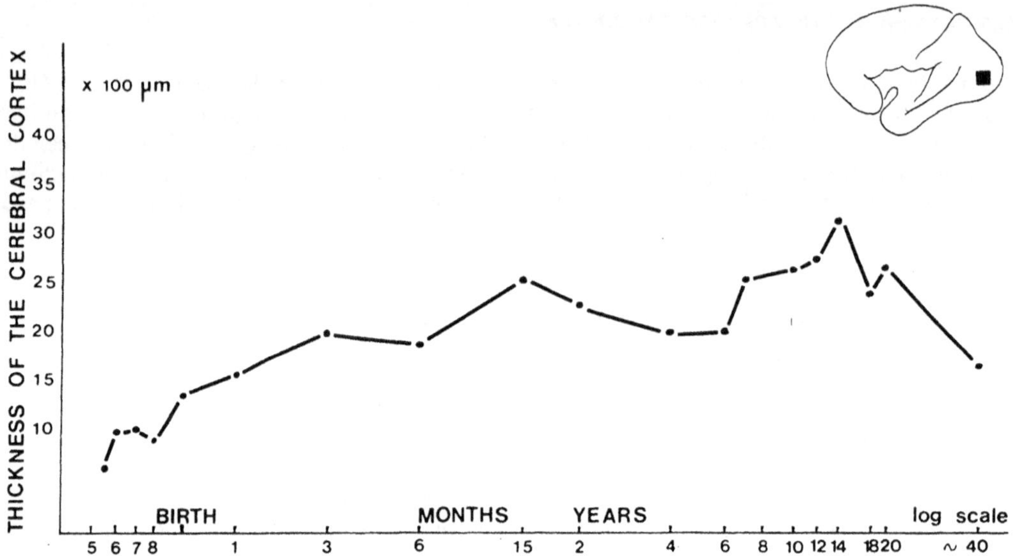

Fig. 23. Development through time of the thickness of the cerebral cortex in area OA of the left occipital lobe in humans.

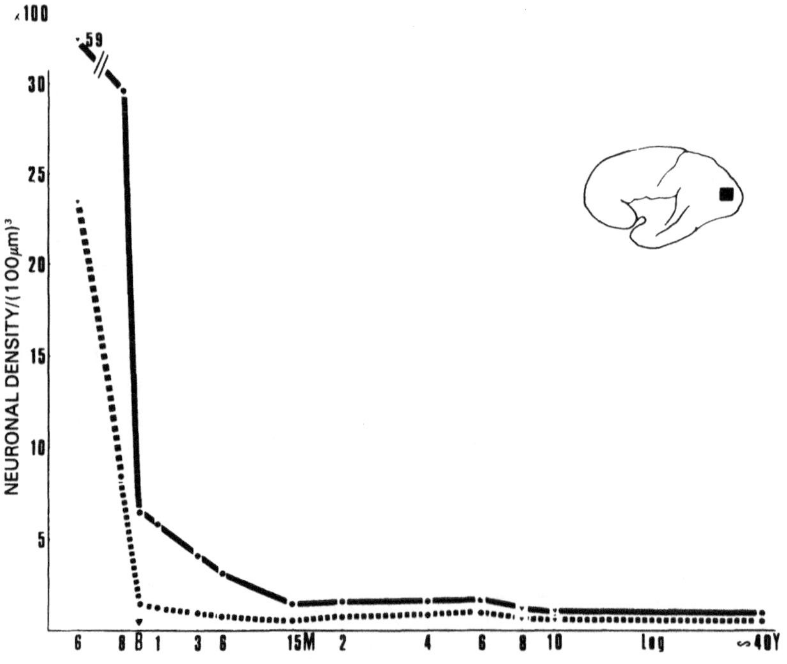

Fig. 24. Evolution through time of the neuronal density per 100 μm^3 of layers II (—) and V (---) of area OA of the occipital lobe in humans. B, birth; M, months; Y, years.

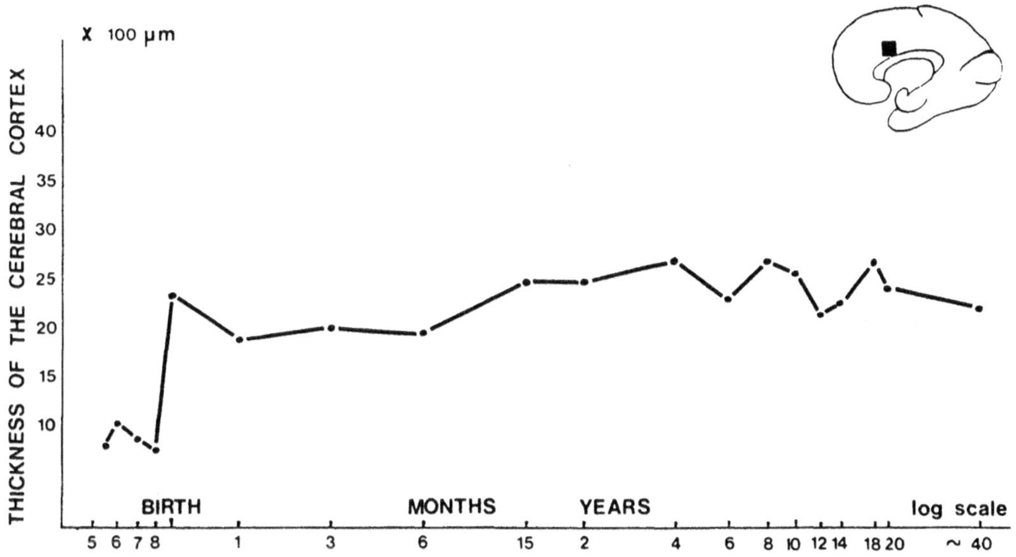

Fig. 25. Development through time of the thickness of the cerebral cortex in area LA of the left limbic lobe in humans.

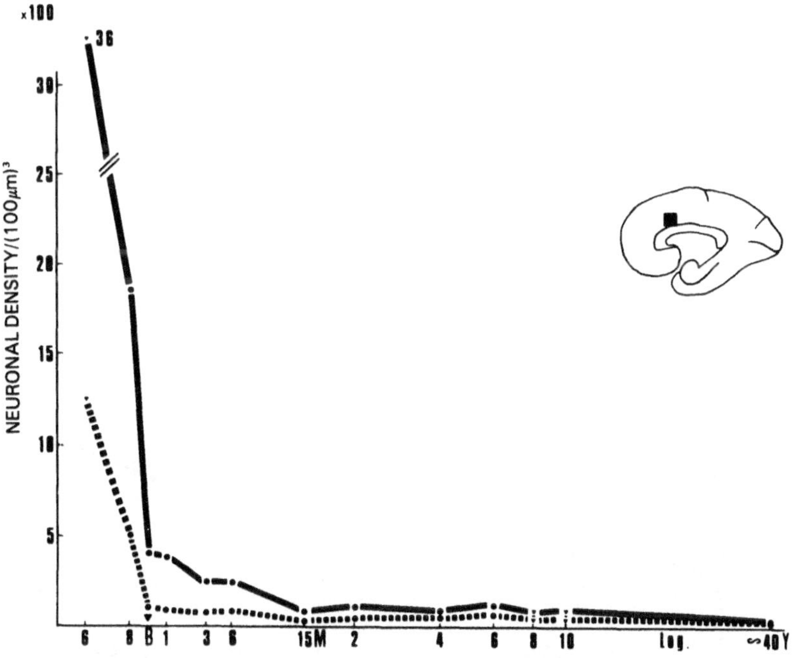

Fig. 26. Evolution through time of the neuronal density per 100 μm³ of layers II (—) and V (---) of area LA of the limbic lobe. B, birth; M, months; Y, years.

rather rapid increase up to 15 months (see Figure 27). At 15 months, the adult thickness level is attained for the first time but decreases once more at 2 years. Then the increase resumes up to 8 years, at which time it reaches again adult level. There is some similarity in the growth of both areas LA and LC from the sixth prenatal month to 14 years. The difference is mainly that, overall, LA has a more accelerated growth. But in both LA and LC, the ages of 1 month postnatal, 2, 6, 10, and 12 years correspond to lower points of the curve. It is only during adolescence, i.e., between 14 and 18 years that some differences in the pattern of growth can be seen between LA and LC.

7.2.2. Neuronal Density

Here the differences from the anterior limbic area LA are considerable (see Figure 28). The cell density at the sixth prenatal month is much greater for the second layer in LC than in LA, so maturation is slower. At birth, the cell density is still greater in this layer in LC than in LA, and the decrease is more rapid in LC than in LA. However, 15 months is also the point at which a lower density is attained. For the fifth layer, the difference is not so marked, except that the cell density is greater in the eighth prenatal month in area LC than in LA, and from birth the evolution of the cell density of the fifth layer is almost exactly the same for LC as for LA.

If, as is generally accepted, the limbic system is associated with emotional activities and mechanisms of memory, the data (both cortical thickness and cell density) provided by our study would indicate that some activity could already exist as early as 15 months. Both the curves of cortical thickness and of neuronal densities show that more adult patterns are attained between 8 and 10 years. Contrary to LA, area LC goes through a period of reduced cortical depth at ages 2–6 years.

8. General Considerations

Although the study of the development of the cortical thickness is still not completed, some general rules have already emerged. First, each hemisphere and each cerebral lobe has its own rate of development, but besides this, each area in each lobe also has its own developmental rate. Moreover, each area shows differences in the rate of development of each of its layers.

The development of cortical thickness shows important differences, depending on both the location of the area and most probably individual factors and sex as well. A period of rapid brain growth exists in humans and probably reaches its peak before the sixth prenatal month, if not earlier (Dobbing, 1974).

Up to 6–8 years, some areas are quite homogeneous in their development with no very great dif-

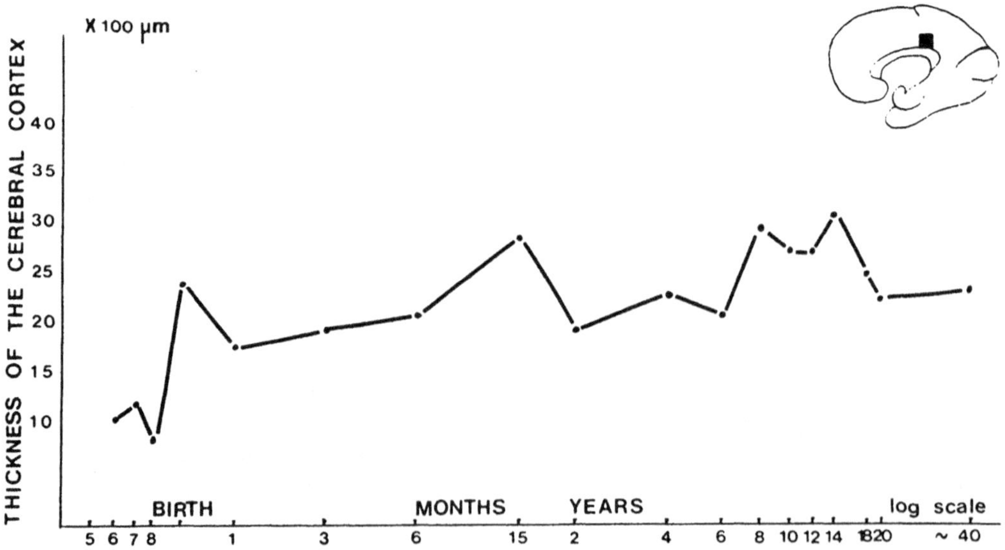

Fig. 27. Development through time of the thickness of the cerebral cortex in area LC of the left limbic lobe in humans.

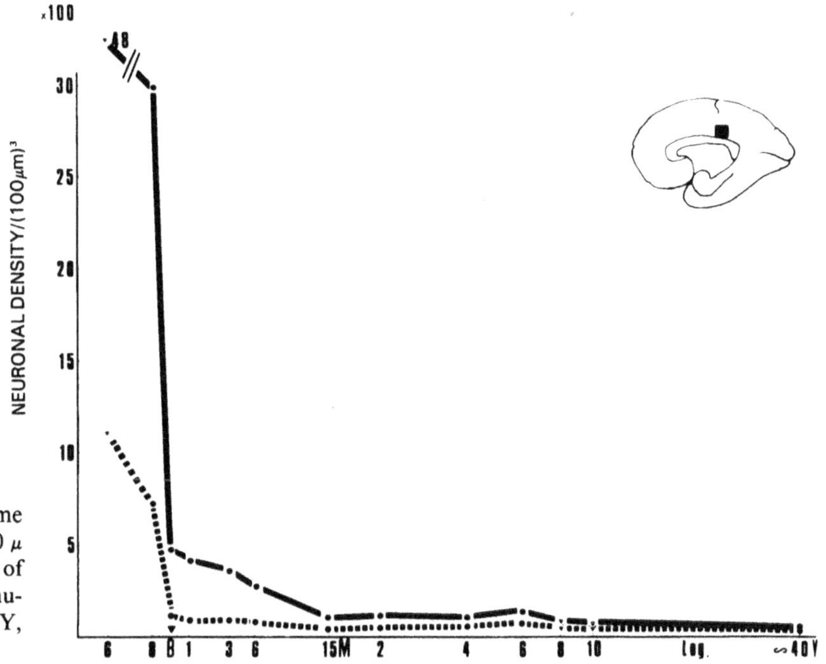

Fig. 28. Evolution through time of the neuronal density per 100 μ m³ of layers II (—) and V (---) of area LC of the limbic lobe in humans. B, birth; M, months; Y, years.

ferences between one individual and another of the same age. This is the case in the primary visual area and in the motor center for speech. By contrast, other areas show great differences between individuals of the same age, giving a very irregular curve of development with significant peaks and troughs. The posterior and anterior–inferior speech centers are in this category.

It is also noticeable that around 6 years of age, most areas show a decrease in the total thickness and an increase in the number of nerve cells in the different layers. This occurs also in mice around the age of 60 days. The fact that the Betz cells exhibit changes in dimensions during their development and, in particular, show at 6 years a decrease in length and an increase in width (Rabinowicz, 1976) might well correspond, at least partly, to that decrease of the total cortical thickness, provided we assume that many of the pyramidal cells of the cerebral cortex are going through the same development of their length and width. Moreover, the increased cell density at 6 years suggests a decrease in the dendritic pattern among the cells (Rabinowicz *et al.*, 1977) at that age.

Indeed, there seems to be some relationship between the functional development of the human cerebral cortex and the evolution of neuronal density, or cortical thickness, and of thickness of the layers.

In almost all areas, cell density decreases very rapidly until birth. From birth to between 3 and 6 months, the decrease is slower, ceasing altogether at about 15 months.

In a given area, differences exist in maturation between the layers. One has to conclude that a very important period in cortical maturation is that between 15 and 24 months, a period at which almost all the layers reach, for the first time, a similar state of maturation. Another important period seems to occur between 6 and 8 years, when a remodeling of the cortex (cortical thickness, number of neurons and dendrites) takes place (Rabinowicz *et al.*, 1977).

Finally, more recent data on cortical depths (Rabinowicz, unpublished data) suggest that adolescence harbors two periods of modification of growth spurts (possible cortical remodeling)—between 10 and 12 years and around 18 years.

ACKNOWLEDGMENTS. This work was supported by U.S. Public Health Service grants M-156-C1-C6, HD00326-07-08, and M 151 (from late Dr. J. L. Conel), by Swiss National Science Foundation grants 364,171 and 4.34-074, and by grants from the Swiss Foundation for Research on Mental Retardation. The author is indebted to Mrs. J. Mc D.-C. Petetot for histoquantitative and technical assistance, to Mr. B. Maurer and B. Gmunder for photography, and to Mrs. R. Clemençon for typing the manuscript.

9. References

Conel, J. L., 1939, *The Postnatal Development of the Human Cerebral Cortex,* Vol. I: *The Cortex of the Newborn,* Harvard University Press, Cambridge, Massachusetts.

Conel, J. L., 1941, *The Postnatal Development of the Human Cerebral Cortex,* Vol. II: *The Cortex of the One-Month Infant,* Harvard University Press, Cambridge, Massachusetts.

Conel, J. L., 1947, *The Postnatal Development of the Human Cerebral Cortex,* Vol. III: *The Cortex of the Three-Month Infant,* Harvard University Press, Cambridge, Massachusetts.

Conel, J. L., 1951, *The Postnatal Development of the Human Cerebral Cortex,* Vol. IV: *The Cortex of the Six-Month Infant,* Harvard University Press, Cambridge, Massachusetts.

Conel, J. L., 1955, *The Postnatal Development of the Human Cerebral Cortex,* Vol. V: *The Cortex of the Fifteen-Month Infant,* Harvard University Press, Cambridge, Massachusetts.

Conel, J. L., 1959, *The Postnatal Development of the Human Cerebral Cortex,* Vol. VI: *The Cortex of the Twenty-Four-Month Infant,* Harvard University Press, Cambridge, Massachusetts.

Conel, J. L., 1963, *The Postnatal Development of the Human Cerebral Cortex,* Vol. VII: *The Cortex of the Four-Year-Child,* Harvard University Press, Cambridge, Massachusetts.

Conel, J. L., and Rabinowicz, T., 1978, *The Postnatal Development of the Human Cerebral Cortex,* Vol. IX: *The Cortex of the Eight-Year Child* (unpublished manuscript).

Dobbing, J., 1974, The later growth of the brain and its vulnerability, *Pediatrics* **53**:2–6.

Hamburger, V., 1975, Cell death in the development of the lateral motor-column of the chick embryo, *J. Comp. Neurol.* **160**:535–546.

Heumann, D., Leuba, G., and Rabinowicz, T., 1977, Postnatal development of the mouse cerebral neocortex. II. Quantitative cytoarchitectonics of visual and auditory areas, *J. Hirnforsch.* **18**:483–500.

Heumann, D., Leuba, G., and Rabinowicz, T., 1978, Postnatal development of the mouse cerebral neocortex. IV. Evolution of the total cortical volume of the population of neurons and glial cells, *J. Hirnforsch.* **19**:411–416.

Leuba, G., Heumann, D., and Rabinowicz, T., 1977, Postnatal development of the mouse cerebral neocortex. I. Quantitative cytoarchitectonics of some motor and sensory areas, *J. Hirnforsch.* **18**:461–481.

Rabinowicz, T., 1964, The cerebral cortex of the premature infant of the 8th month, *Prog. Brain Res.* **4**:39–86.

Rabinowicz, T., 1967*a*, Techniques for the establishment of an atlas of the cerebral cortex of the premature of 8 months in: *Regional Development of the Brain in Early Life* (A. Minkowski, ed.), pp. 71–89, Blackwell, Oxford.

Rabinowicz, T., 1967*b*, Quantitative appraisal of the cerebral cortex of the premature of 8 months, in: *Regional Development of the Brain in Early Life* (A. Minkowski, ed.), pp. 91–118, Blackwell, Oxford.

Rabinowicz, T., 1974, Some aspects of the maturation of the human cerebral cortex, *Mod. Probl. Paediatr.* **13**:44–56.

Rabinowicz, T., 1976, Morphological features of the developing brain, in: *Brain Dysfunction in Infantile Febrile Convulsions* (M. A. B. Brazier and F. Coceani, eds.), pp. 1–23, II IBRO Symposium, Raven Press, New York.

Rabinowicz, T., Leuba, G., and Heumann, D., 1977, Morphologic maturation of the brain: A quantitative study, in: *Brain Fetal and Infant* (S. R. Berenberg, ed.), pp. 28–53, Martinus Nijhoff Medical Division, The Hague.

Tanner, J. M., 1961, Education and Physical Growth, University of London Press, London.

von Economo, C., and Koskinas, G. N., 1925, *Die Cytoarchitektonik der Hirnrinde des Erwachsenen Menschen,* Springer-Verlag, Vienna.

Yakovlev, P. I., 1962, Morphological criteria of growth and maturation of the nervous system in man, *Ment. Retard.* **39**:3–46.

15

Neuroanatomical Plasticity

Its Role in Organizing and Reorganizing the Central Nervous System

CHRISTINE GALL, GWEN IVY, and GARY LYNCH

1. Introduction

"Plasticity" is a necessary concept for any theory of brain function. The most complex products of the brain, the ongoing cognitive behaviors of the individual, are remarkable for their flexibility and their capacity for reorganization in the face of changing circumstances. Since behavior is characterized by its adaptability, it follows that the neural machinery that creates it must possess analogous features. But what, in neurobiological terms, is the property of the brain that gives it this plasticity? Suggested answers to this question have come from all the branches of the neurosciences. The idea most commonly advanced is that changes in the effectiveness of synaptic transmission are responsible for phenomena such as learning and memory. Physiological research, much of it quite recent, has shown that lasting changes can be created in monosynaptic systems by very brief trains of repetitive stimulation (Bliss and Lomo, 1973; Dunwiddie and Lynch, 1978). Neurochemical studies have indicated that some of the subcellular systems related to the transmitter and its actions are modifiable (Baudry and Lynch, 1980; Lynch *et al.*, 1982), and this certainly provides a means through which modification in the operation of neural circuits could be achieved. This chapter deals with still another mechanism by which the brain might gain its flexibility, specifically, that it is capable of modifying its very structure. This idea, which is quite old, has become the subject of intense interest in recent years, as newer methods (and the increased use of some more traditional procedures) have allowed anatomists to develop a clearer picture of the fine structure of neural tissue. Studies in the last decade indicate that the microanatomy of the neuron, as well as its dendritic ramifications and axonal arborization can be greatly modified and that under some circumstances the brain is capable of generating entirely new circuitry.

In the following pages, we shall review selected portions of this literature, discussing first anatomical plasticity in neonates and then turning to the adult brain. This division is dictated by the long-held suspicion of psychologists that the storage of experience somehow changes with development, as well as by the common (although by no means universal) observation that immature brains recover much more completely following damage than do those of adults. These two types of findings have suggested to many workers that the young brain is more plastic than that of the adult, and it is well that we bear this in mind while considering the anatomical data. We shall also separate the literature in terms of the type of manipulation used to uncover capabilities of neural tissue for change—in only a few instances have attempts been made to demonstrate that these capabilities are actually ex-

CHRISTINE GALL • Department of Anatomy, California College of Medicine, University of California, Irvine, California 92717. **GWEN IVY and GARY LYNCH** • Center for the Neurobiology of Learning and Memory, University of California, Irvine, California 92717.

pressed during the normal operation of the brain or as part of any repair process.

2. Effects of Experience and Environment on the "Molar" Composition of the Brain

Before discussing the plasticity of the constituents of the nervous system, it is appropriate that we briefly summarize the evidence that the physical size of brain areas is mutable. Several studies have demonstrated that the size of particular brain regions can be influenced by the degree of environmental complexity experienced by the animal. Generally, these experiments involve raising the subjects, most frequently rodents, either in an enriched environment with exposure to a variety of complex or novel stimuli or in a relatively impoverished environment such as isolated housing in a laboratory cage. Differential housing of this sort, for a period of several weeks, results in consistent differences in the gross dimensions of several brain areas that persist as long as differential housing is maintained (Rosenzweig *et al.,* 1972; Bennett, 1976). Although most studies compared the effect of different housing conditions on young rats (beginning the conditioning period at the time of weaning) similar effects can be obtained with differential housing of adult rats (Uylings *et al.,* 1978). The brains of the environmentally enriched animals have been found to be larger; one of the more frequent indexes of this difference is a thickening of the cerebral cortex, particularly in the occipital region (Diamond, 1967). These effects do not appear to be attributable to differential visual experience alone for they have been replicated with enucleated animals (Krech *et al.,* 1963). Conversely, pure sensory deprivation causes gross dimensional changes in the area related to the blocked modality. Visual deprivation by eyelid suture at birth has been shown to cause subnormal volume in both the lateral geniculate nucleus and striate cortex in comparison with similarly housed sighted littermates (Gyllensten *et al.,* 1964; Fifkova, 1970). The origins of these changes are obscure but seem to involve an increase in the volume of the cells which constitute the affected region. Which type of cellular element is primarily affected and the interrelationships among blood vessels, glia, and neurons that produce the final total volumetric increase or decrease are yet to be analyzed.

3. Morphological Plasticity of the Developing Brain

3.1. Altered CNS Development following Damage to Sensory Organs

A variety of structural changes may be effected in the mature nervous system by surgical intervention during its development. Sensory systems are especially amenable to such manipulation because of their accessibility and the topographical order of their connections. In particular, studies in the mammalian visual system have unmasked a number of the dynamic processes that occur during the construction of a mature nervous system and have shown which of these are pliable. In the fetal monkey the axons of retinal ganglion cells from the two eyes initially intermingle in the lateral geniculate nucleus of the thalamus. During the second half of gestation, these afferents gradually organize into six alternating laminae in which the neurons of one lamina receive input from only one eye (Rakic, 1976, 1977). Similarly, in layer IV of the visual cortex, the geniculocortical terminals initially overlap and later segregate into ocular dominance columns. Removal of one eye before this segregation takes place in the CNS prevents it from occurring at all. The lateral geniculate, then, does not develop cellular lamination, and the afferents from the intact eye remain dispersed throughout the nucleus, appearing to form synapses with all its neurons. Following suit, the geniculocortical axons fail to retract into ocular dominance columns (Rakic, 1981). If the enucleation is performed slightly later in development, nuclear lamination and afferent segregation in the thalamus, as well as ocular dominance column formation in neocortex, are permanently arrested at the stage they had achieved when the damage was inflicted.

Monocular enucleation demonstrates other principles operating in the developing nervous system as well. Throughout the period of axonal segregation in the geniculate, each optic nerve loses more than a million axons (Rakic and Riley, 1983*a,b*). Thus, the retina initially sends out many more axons than it will maintain and, after an apparent competition for terminal space in the geniculate, loses a portion of these. This axonal loss can be prevented in one optic nerve by removal of the opposite eye, thus giving competitive advantage to the remaining eye. The results of similar experiments in the neonatal rat suggest that the optic axons are lost as a result of the naturally occurring death of retinal ganglion cells, rather than to an elimination of axon collaterals from a fixed ganglion cell population. Monocular enucleation during the sensitive period in development drastically reduces the amount of cell death in the remaining retina (Sengelaub and Finlay, 1980; Jeffrey and Perry, 1982). Interestingly, the retinal cell death in the rat is paralleled by cell death in a main retinal target, the superior colliculus (Cunningham *et al.,* 1982). The cell death in both structures is temporally coincident and spatially proportional, occurring simultaneously to a greater extent in the nasal ganglion cell layer and its target cells in the superior colliculus than in other retinal or collicular areas. These data suggest a possible reciprocal control of cell survival between these two brain structures. It is not yet known whether monocular enucleation permits increased survival of tectal neurons. However, as discussed in Section 3.3, the matching of neuronal pool size to target tissue capacity *via* cell death is a common feature of the developing nervous system, and thus it is not unlikely that this form of efferent plasticity operates at many levels of the visual system.

Experiments in the rat somatosensory system illustrate some interesting facets of the developmental pliability of sensory systems. The sensory organs on a rat's face, whiskers, are arranged in five rows each containing five to seven whiskers. This unique pattern is faithfully replicated in afferent topography at all levels of the neuroaxis—brain stem, thalamus, and somatosensory cortex. If one row of whiskers is electrocauterized during a sensitive period shortly after birth, the representation of that row in each of the brain centers develops aberrantly into an amorphous cigar-shaped density rather than into the normal discrete row containing individual dense clusters of afferents (Belford and Killackey, 1979; Killackey and Belford, 1979). Physiological recordings in neocortex have shown that each normal, anatomically discrete cluster of afferents is also functionally discrete, as it is related to a single whisker on the face of the rat (Welker, 1976). By contrast, the amorphous area created by damage to a row of whiskers can be activated by stimulation of whiskers in adjacent rows, and often by whiskers several rows away (Killackey *et al.,* 1978). This functional mixing of normally discrete information might be accomplished by competition among the central processes of undamaged trigeminal ganglion cells for functionally denervated terminal space in the brain stem. In any case, while the damaged peripheral processes of the trigeminal ganglion cells may later regrow and innervate the same physical location on the face of the rat, the aberrant anatomical and functional organization that has developed in the CNS is not repaired.

Finally, there are both spatial and temporal limits on the modifiability of somatosensory patterns. For example, damage of one whisker follicle in each row (thus forming a line across the five rows) does not yield a duplicate pattern of afferents in the CNS; it thus appears that, at least at the cortical level, the axon terminals related to whiskers within a row, but not across rows, are capable of intermixing to form an anatomical unit (Killackey and Belford, 1980). The forces which dictate such spatial limitations on the pliability of afferent patterns are not yet understood but may involve differential adhesiveness among growing, fasciculating axon populations. In addition, there is a temporal limit on the pliability of the somatosensory system, as in the visual system, such that whisker follicle damage after the sensitive period can no longer modify central afferent patterns. Although the afferent patterns at the three levels develop sequentially, from caudal to rostral along the neuroaxis, they have a common sensitive period: peripheral damage inflicted after the development of a normal brain stem pattern but before the development of the neocortical pattern will not be reflected in the cortex. Rather, the afferent pattern that develops at any CNS level is a duplicate of the pattern that occurs at the immediately lower level (Belford and Killackey, 1979).

The studies detailed in this section indicate that sensory systems develop sequentially, from receptor to cortex, and that damage to sensory organs during a sensitive period is destined to be translated throughout the neuraxis into anatomically and functionally aberrant neural patterns. During normal development, the establishment of neural patterns is no less dynamic, involving a battle among axons for terminal space within a structure and the subsequent death of those neurons that fail to fight effectively. The epigenetic regulatory mechanisms at work may serve to ensure that sensory experience at the periphery will be channeled to each CNS level with the maintenance of a high degree of informational integrity. Local circuits might then process the information according to the unique design of that CNS station.

3.2. Growth of Aberrant Fiber Tracts and Expanded Terminal Fields after Early Brain Damage

Phenomena that have been accepted as demonstrations of axonal plasticity in immature animals generally involve the development of aberrant circuitry as a consequence of lesions placed within a still-developing system. As in many of the dendritic studies (Section 3.4), plasticity in these situations may not involve a change from one morphology to another but, rather, the production of an anatomical arrangement which differs from that which normally unfolds during development. Although demonstrations of this sort do not inform us of the expressed role of neuronal plasticity within the intact organism, they provide information on the capacity of axons to deviate from what might have been thought to be a genetically programmed termination pattern. Plasticity of this sort has been demonstrated in axonal systems following removal of their target structure, removal of neighboring axonal projections, and transection of the subject fibers directly.

The effect of target structure removal on the development of an ingrowing population of axons has been studied in several systems, including the retinal efferents to the superior colliculus following unilateral ablation of the colliculus (Schneider, 1970). Axons of the contralateral retina that would normally innervate this structure were found to terminate more heavily in the ventral lateral geniculate, another target of these fibers, which lies proximal to the lesion. In addition, fibers were found to extend their normal terminal field to include the medial portion of the intact superior colliculus, an area not normally receiving any input from the ipsilateral retina. Therefore, in this and similar studies on the olfactory system (Devor and Schneider, 1975), it appears that some axons will continue growing, if their normal targets are removed, until a termination zone is found. It has been theorized that the increased density of innervation observed in structures proximal to the lesion is a compensatory development following the principle of "conservation of the total axonal arborization." The essence of this idea is that an axon is genetically programmed to produce a certain number of branches and terminals and, if it cannot achieve this number at its normal target sites, it will develop them in some other region. As had been considered in Section 3.1, it may prove that a neuron must innervate some minimum amount of target area to survive.

It is of interest that the development of aberrant connections between retinal axons and the superior colliculus of the ipsilateral side alters the termination of the normal retinal afferents to that structure. Retinal fibers normally terminate over the entire superficial area of the contralateral superior colliculus. In the reorganized system, where ipsilateral retinal fibers occupy the most medial areas of the intact superior colliculus, contralateral retinal fibers do not include the medial regions in their terminal fields. The afferent fields seem to exclude each other, possibly by competition for available synaptic sites.

Ramon y Cajal (1959) was the first to report the effects of axonal transection in the immature CNS. He found that undercutting areas of neocortex just above the white matter, involving transection of the efferent axons beyond their first point of collateral branching, caused a proliferation of collaterals proximal to the lesion that deflected back into the undercut region. Similar observations were made by Purpura and Hausepian (1961), who suggested, as a result of electrophysiological analysis of such undercut cortical regions, that these collateral branches form functional synapses within the local cortical area, thereby generating aberrant functional circuits. These observations, together with those from target-removal studies, indicate that in several instances transection of immature axons may trigger increased collateralization of their surviving branches.

Abnormal axonal growth also takes place in intact fiber systems following the removal of adjacent input systems. These cases are particularly informative because they do not involve direct injury to the fibers of interest; damage is restricted to the neighbors of these axons. Partial deafferentation of target regions has been shown to increase the extent of innervation by the remaining intact afferents in several systems (Lund and Lund, 1971; Lynch *et al.*, 1972), in some cases involving the spread of the remaining afferents into regions they would not normally innvervate (Lynch *et al.*, 1973*b;* Zimmer, 1973, 1974). More dramatic demonstrations of the plasticity of developing axons are provided by those cases in which early lesions result in not only expanded terminal fields but entirely new fiber tracts. Following complete or partial unilateral retinal lesions, fibers of the intact retina have been found to grow into the deafferented region of the ipsilateral superior colliculus, a structure that they do not normally innervate (Lund *et al.*, 1973). Following unilateral hemispherectomy, which removes the normal crossed corticospinal tract to the contralateral spinal cord, Hicks and D'Amato (1970) demonstrated the development of a small uncrossed corticospinal tract from the intact hemisphere (D'Amato and Hicks, 1978).

One of the best studied examples of an afferent's hyperdevelopment following removal of a neighboring axonal system involves the commissural and associational projections to the dentate gyrus of the rat hippocampus (Figures 1 and 2). These two afferents, which arise as collaterals of the polymorph neurons of the dentate gyrus hilus, share the innervation of the proximal one-fourth of the dentate gyrus granule cell dendritic tree. The commissural and associational terminal field lies adjacent to, but does not overlap, the field of dense entorhinal cortical innervation which occupies the distal three-fourths of the granule cell dendritic tree. As such, the commissural/associational (C/A) and entorhinal afferents define distinct laminae within the dentate gyrus molecular layer. Any deviation in this orderly laminar arrangement is easily detected either by stains for the impregnation of healthy axons or by routine tract-tracing techniques, such as amino acid autoradiography or horseradish peroxidase (HRP) histochemistry.

The C/A and entorhinal afferent fibers grow into the dentate gyrus at about the same time during the first week of postnatal life and very rapidly set up the laminated pattern seen in the adult (see Figure 2) (Loy *et al.*, 1977; Fricke *et al.*, 1977). This is being accomplished while new granule cells are still being formed and before any of their dendrites have achieved even one-half their adult length. If the ipsilateral entorhinal cortex is removed in a rat 14 days of age or younger, the C/A fibers hyperdevelop and expand their innervation to include all of the granule cell dendritic tree (Figure 3). Subsequent lesion studies, allowing quantification of the density of commissural synapses in both the normal commissural terminal field and the expansion territory, have demonstrated that the massive growth of the commissural axons into "new" innervation fields does not occur at the expense of commissural innervation in their normal target zone. Rather, the commissural axonal system maintains its normal innervation density within both the proximal molecular layer in which it normally terminates and the distal molecular layer it occupies following the cortical lesion (Gall *et al.*, 1979*a*). Therefore, the lesion-induced axonal growth in this system represents additional growth rather than a redistribution of a fixed population of terminals.

Although full C/A expansion into what was normally the entorhinal cortical afferent terminal field occurs with entorhinal cortex removal in rats aged 14 days or younger, there are differences in the pattern of this abnormal growth depending on the age of the animal at lesion placement. If the lesion is placed in a 1-day-old rat, prior to the growth of entorhinal afferents into the dentate gyrus molecular layer, the C/A afferents spread evenly throughout the ipsilateral molecular layer (Staubli *et al.*, 1984). By contrast, if the lesion is placed in rats at 7 days of age, after entorhinal afferent ingrowth, the C/A axons form two plexuses within the molecular layer: an inner molecular layer plexus (as in the normal rat) and a secondary, entirely abnormal, plexus within the middle molecular layer (Figure 4) (Gall and Lynch, 1981). Small collaterals emerge from each plexus to innervate the surrounding molecular layer. The secondary C/A plexus in the mid-dendritic zone of the molecular layer forms within the same subfield normally occupied by axonal bundles of the entorhinal afferent fibers. This coincident location, and the observation that an aberrant secondary C/A plexus does not form with lesion placement prior to entorhinal innervation of the dentate, indicate that the C/A axons that grow into the distal dendritic field after deafferentation at 7 days of age are directed in their growth by the position of degenerating entorhinal afferent axons.

With entorhinal lesion placement in rats at 14 days of age, a third fiberarchitectural rearrangement is observed. Even though the C/A axons reinnervate the full depth of the molecular layer following this lesion, the C/A axons continue to course primarily within an inner molecular layer fiber plexus. There is, however, a modest marginal expansion of this plexus in the form of short collateral branches that invade the edge of the deafferented field as well as the occurance of a few long, straight, large-caliber axons that course for long distances within the depth of the entorhinal deafferented zone (Gall and Lynch, 1981). When axonal growth induced by deafferentation in the adult is discussed it will become clear the former marginal plexus expansion seems to represent the growth response of the more mature C/A axons at this age.

From the data thus far presented, the growth re-

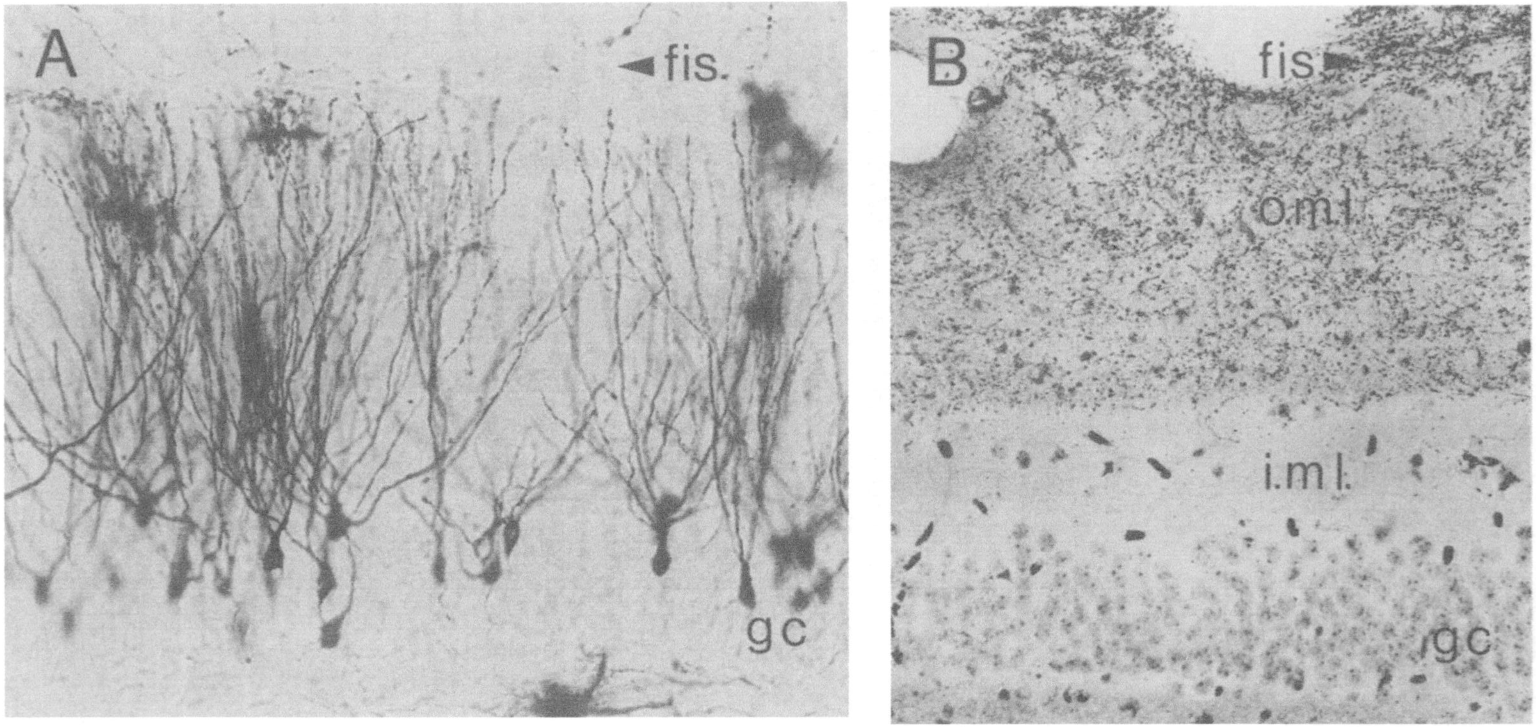

Fig. 1. Photomicrographs illustrating basic anatomical features of the dentate gyrus. (A) An aspect of the rat dentate gyrus shown impregnated with the Golgi technique. The dentate gyrus granule cells (gc) can be seen to form a compact neuronal cell layer and to extend dendritic processes between this cell layer and the hippocampal fissure (fis.). (B) Laminar termination of the entorhinal cortical afferents to this dendritic field. In this tissue, taken from a rat sacrificed 4 days after an ipsilateral entorhinal cortex lesion, the degenerating entorhinal afferent axons and terminals have been impregnated with silver by the Fink-Heimer technique. The coarse, black, granular entorhinal terminal degeneration can be seen to fill the distal molecular layer, whereas the inner molecular layer (i.m.l.) is devoid of terminal degeneration.

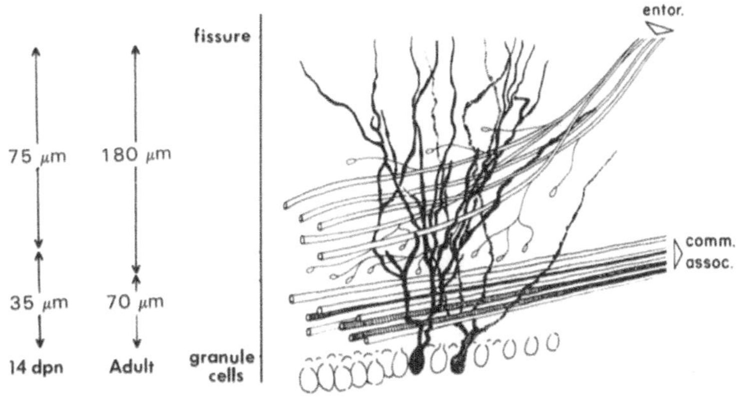

Fig. 2. Schematic illustration of the distribution of the principal afferents to the rat dentate gyrus molecular layer. The numbers on the left indicate the depth of each exclusive afferent field in the 14-day-old rat and the adult.

sponse of C/A axons to entorhinal deafferentation in the neonate has demonstrated that: (1) The C/A axons can deviate from their normal distribution pattern to functionally innervate "foreign" dendritic territory; and (2) The C/A system has a much greater growth capacity than is expressed in normal development. Furthermore, the differences seen in the pattern of aberrant growth with entorhinal afferent deprivation (1-day lesions) as compared to entorhinal deafferentation (7-day lesions) suggest degenerating axon fragments may "direct" the course of lesion-induced axonal growth.

Additional insight into the nature of the growth response in this system has been provided by analyses of its temporal parameters. As reviewed in Section 4.2, one of the intriguing aspects of lesion-induced axonal growth in the adult brain is the consistent observation of a lag of several days between the insult and the onset of the sprouting response. Just the opposite is true with young lesion placement; i.e., aberrant axonal growth begins very quickly and reinnervates the deafferented field extraordinarily rapidly. With entorhinal cortex ablation at 14 days postnatal, C/A axon invasion of the

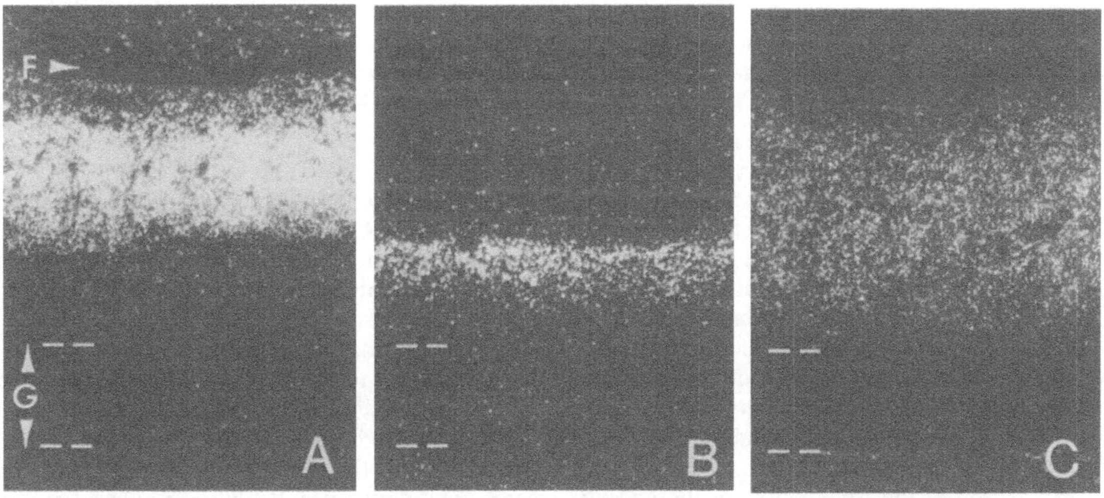

Fig. 3. Dark-field photomicrographs of equivalent sections of the rat dentate gyrus, extending from below the granule cell layer (G) to beyond the hippocampal fissure (F), illustrating the distribution of autoradiographically labeled afferents within this field. The white grains indicate the position of the afferent system labeled with [^{3}H]leucine in each case. In the normal dentate gyrus molecular layer, the entorhinal afferents (A) and commissural afferents (B) occupy adjacent nonoverlapping lamina. (C) Removal of the entorhinal cortical afferents in the 7-day-old rat results in an expanded commissural afferent terminal field.

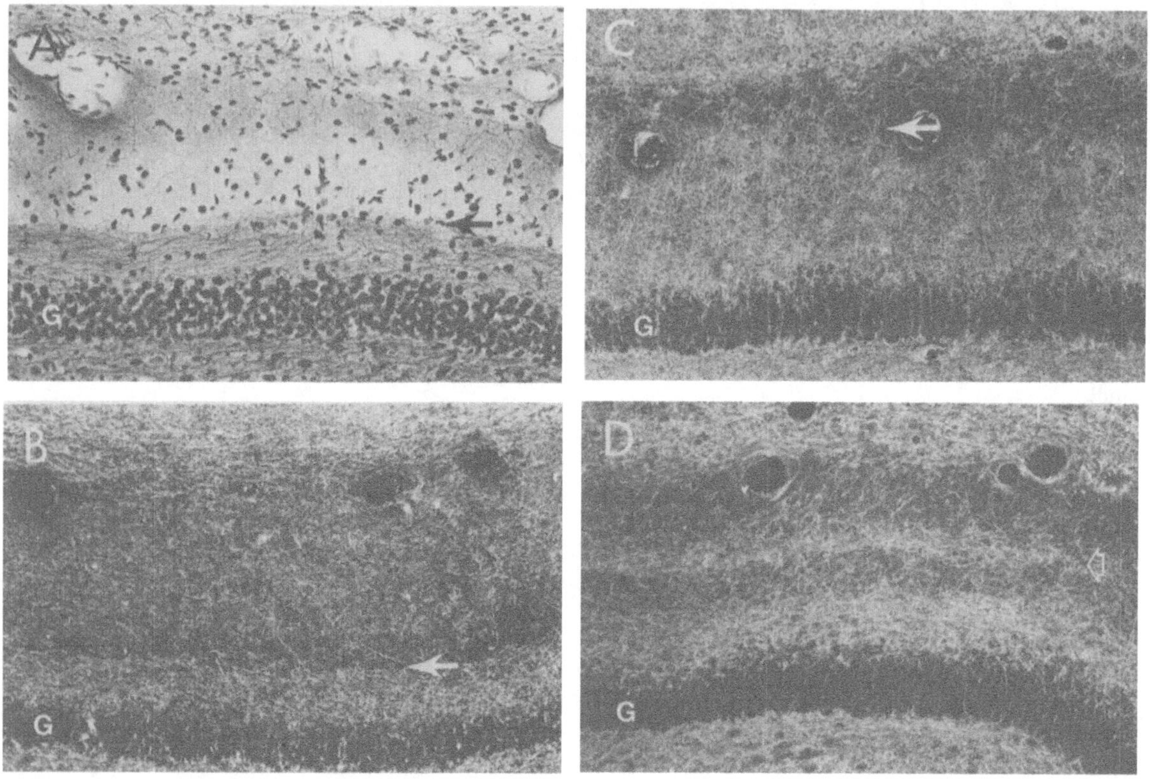

Fig. 4. Photomicrographs of equivalent sections of the rat dentate gyrus molecular layer processed with the Holmes technique for the metal impregnation of intact axons (as in Figure 8). (A, B) Normal axonal staining pattern as seen in light-field (A) and dark-field (B) microscopy. In each case an arrow indicates the upper (distal) limit of the inner molecular layer commissural/associational (C/A) fiber plexus that courses parallel to the granule cell layer (G). (C) With ipsilateral entorhinal cortex removal at 1 day of age, this C/A fiber plexus fills the entire molecular layer (upper limit indicated by arrow). (D) With entorhinal cortex lesion placement at 7 days of age, the C/A axons form two plexuses, including an entirely aberrant axonal plexus within what was the deafferented field (upper arrow).

deafferented field becomes evident between 10 and 18 hr postlesion and penetrates the full depth of the field by postlesion day 2. The rate at which the sprouting axons reinnervate this field exceeds that followed by the normal innervation process at this age; from 24 to 48 hr postlesion synapses were added at a rate of 4 per day/100 μm^2, whereas the normal middle molecular layer gains innervation at a rate of 1.3 per day/100 μm^2 during the same age interval (Gall *et al.*, 1980).

Examination of the time course of the fiber-architectural changes induced by deafferentation at 14 days postnatal reinforces the impression that two distinct forms of lesion-induced axonal growth are present at this age. By 4 days postlesion, the marginal expansion of the inner molecular layer fiber plexus is not yet in evidence, although a few long, straight C/A collaterals penetrate the full deafferented zone. Like axonal growth induced by lesion placement in the adult brain, the modest plexus expansion occurs between days 4 and 8 postlesion (Gall and Lynch, 1981).

With entorhinal cortex removal at 3 weeks of age a much more modest C/A growth response is observed. Autoradiographic data indicate that a few C/A axons reach some distance into the deafferented field but the modest marginal C/A field expansion predominates. With lesions placed at later ages, the C/A growth response is reduced to the shallow invasion of the deafferented field seen throughout adulthood.

This reduced proliferative response might be attributable to any of several variables. Quite conceivably, the C/A axons have lost some of their capacity for growth. This lost capacity does not seem to reflect genetic programming. There is ample evidence for the capacity for growth beyond the norm from the sprouting studies reviewed above. In looking for clues to this developmental decrement in the lesion-induced growth response, one should appreciate that the regulatory factors are most probably in evidence by 14 days postnatal. Although maximal field reinnervation was observed at this age, a good proportion of the sprouting axons exhibited a limited and delayed growth response such as that seen in the adult. Given that both forms of sprouting (robust and limited) occur within a given piece of tissue at the same age indicates that the limited response of the few is not likely to be caused by elements in the neuropil such as glia or degenerating debris that would presumably be encountered by all the axons but, rather, that the limitation reflects a property of the growing axon itself. In this regard, it seems quite possible that the accumulation of synaptic contacts by an axon may enforce certain maturational changes in that axon that reduce the probability, if not the possibility, of growth. One potential maturational change that might influence growth potential is cytoskeletal consolidation. Cytoskeletal elements are associated with the zones of membrane specialization that define the synaptic contact. The cytoskeletal link is thought to stabilize the membranes in these areas. As the number and density of synaptic contacts by an axon build, so does the cytoskeletal system—stabilizing the membrane and crosslinking cytoskeletal elements within the cytoplasm. It seems quite likely that this structural stability is achieved at the expense of the growth potential of the axon. It is also conceivable that nonneural elements are involved in the reduction of lesion-induced axonal growth with age. Glial cells are added to the dentate gyrus in increasing numbers during the first week of life and at some point shortly thereafter begin to take a very active role in the response of a brain region to the removal of a primary afferent. Understanding of the mechanisms responsible for the restrained lesion response of the C/A system at 3 weeks compared with their reaction at two weeks would surely provide important clues about the development of growth-regulatory influences of the brain.

Having shown that under several experimental circumstances (removal of targets or neighboring inputs) radically aberrant circuitry will form in the brain, it becomes relevant to ask whether such circuits really operate. It appears that the answer is yes. In his work with the golden hamster, Schneider (1970) used a simple behavioral test and found that his animals acted as though their new retinal projections were functional. Similar studies by Hicks and D'Amato indicate the aberrant corticospinal circuits are also operative. More direct evidence has been obtained using neurophysiological methods in the studies of C/A afferent hyperdevelopment in the hippocampus. In the normal rat, electrical stimulation of the commissural projection produces monosynaptic field potentials restricted to the inner dendritic tree of the granule cells. Following entorhinal lesions at 10 days of age, these potentials can be readily recorded from even the most distal portions of the dendrite (Lynch *et al.*, 1973*a*). This finding strongly suggests that the axonal projections that expand into the outer dendritic zones form competent synaptic connections.

3.3. Efferent and Afferent Plasticity in the Neocortex

The experiments discussed above illustrate mainly the malleability of axon populations. Peripheral or CNS damage during development can stabilize a normally transient terminal pattern or cause the growth of aberrant fiber tacts and the sprouting of axon collaterals. On the other hand, a form of efferent plasticity was illustrated in the retina. Retinal ganglion cells that apparently fail to compete effectively for terminal space within the tectum die unless the competing retina is removed. This type of efferent plasticity is common in the developing nervous system; its properties are best illustrated by motor neurons of the ventral horn of the spinal cord (reviewed in Jacobson, 1978). Briefly, motor neurons are initially overproduced and later die in large numbers unless additional target tissue (muscle) is artificially supplied. Conversely, artificially decreasing the amount of target tissue causes an increase in the amount of cell death. Indeed, the variable size of motorneuron populations in the ventral horn is proportional to the number and size of the muscles they innervate (Landmesser, 1980). Such initial overproduction of neurons and later matching of the neuronal pool size to target tissue capacity *via* cell death is a basic feature of the developing vertebrate nervous system.

By contrast, the neocortex demonstrates what may be a unique form of efferent plasticity, one that is not mediated by cell death. In neonatal rodents and cats, the neurons connecting the two hemispheres *via* the corpus callosum are distributed in uninterrupted horizontal bands that span the tangential extent of the neocortex. However, by several weeks after birth in these animals the callosal projection neurons have gaps in their distribution: they are now confined to column-like vertical arrays in certain portions of the neocortex (Innocenti *et al.*, 1977; Ivy *et al.*, 1979). This phenomenon is demonstrated in Figure 5, which shows the pattern of callosal projection neurons in the visual cortex of mouse at postnatal day 3, and in the adult. The neurons have been retrogradely labeled with HRP which was injected into the contralateral neocortex. It is obvious from these data that certain neocortical areas lose their projections across the corpus callosum during development. The question as to how these areas accomplish this has been partially answered by experiments utilizing the retrograde transport of fluorescent dyes. The dye fast blue injected into the neocortex of a neonatal rat or cat is quickly transported back to callosal projection neurons in the contralateral hemisphere and can remain in the perikarya for weeks, acting as a fate marker for these neurons. Such studies have shown that the neonatally marked callosal projection neurons located in areas that do not send callosal projections in the mature animal are still present; they have neither died nor migrated to an adjacent callosal zone (Innocenti, 1981; Ivy and Killackey, 1982; O'Leary *et al.*, 1981). Rather, as shown in a series of experiments using two different dyes, many neurons in the rat parietal cortex initially send axons both across the corpus callosum and to the ipsilateral motor cortex. As development proceeds, the neurons lose their contralateral but maintain their ipsilateral processes (Ivy and Killackey, 1982). A similar phenomenon has been demonstrated with regard to neurons that send axons subcortically. In the neonatal rat, the large pyramidal cells that project to the spinal cord are distributed throughout the neocortex in a single lamina (Bates and Killackey, 1984; Leong, 1983; Stanfield *et al.*, 1982). Many of these neurons initially send axons to both the spinal cord and the tectum but, as development proceeds, the neurons lose either the tectal or the spinal projection, depending on the area of neocortex in which they reside (Ivy and Killackey, 1981*b*). Since each cortical

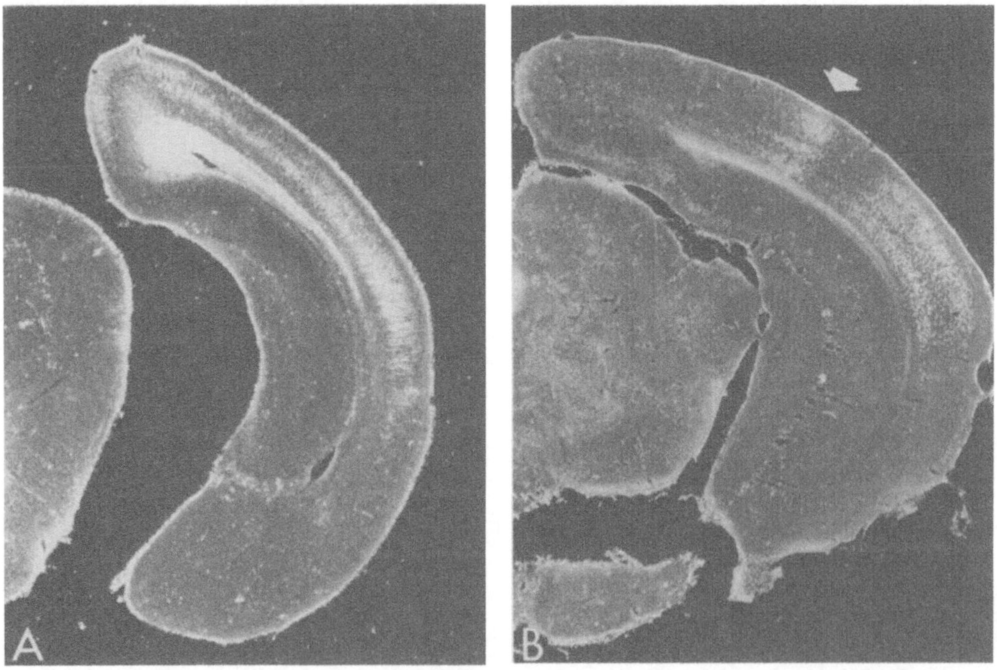

Fig. 5. Photomicrographs illustrating the pattern of callosal projection neurons in the mouse visual cortex (A) at postnatal day 3 and (B) in the adult. The neurons appear white under dark-field illumination, as they contain horseradish peroxidase (HRP), which was retrogradely transported from the injection site in the contralateral hemisphere. Note the continuous distribution of the brightly labeled cells in (A) and the more discrete distribution in (B). The vertical array of labeled cells in (B) denotes the area 17/18 border (white arrow).

lamina contains callosal or subcortical projection neurons, one may conclude that neocortical neurons, in general, appear to distribute more axon collaterals initially than they will maintain.

The basis on which a cortical neuron "chooses" one target area over another is not known, but several studies in the visual callosal system offer insight into the factors that may be at work. Both the number and the regional distribution of callosal efferent neurons in visual cortex can be modified by altering visual input. For example, bilateral enucleation or suturing of the eyelids in kittens causes a reduction in the number of callosal projection neurons in the visual cortex (Innocenti and Frost, 1980). Also, several manipulations such as monocular deprivation or enucleation or the induction of convergent or divergent strabismus *via* section of the lateral or medial rectus muscles lead to an abnormally wide distribution of callosal projection neurons in visual cortex (Innocenti and Frost, 1979, 1980).

Taken together, the above data indicate that the developmental loss of callosal axon collaterals is malleable and that it can be prevented by changing the amount or quality of sensory input. This efferent plasticity is a function of the selective maintenance of a normally transient developmental pattern of projection neurons. Interestingly, the terminal ends of these same callosal projection neurons appear to be governed by different rules. The patterns of callosal afferents in a mature cat or rodent mimics that of the projection neurons. Furthermore, the distribution of the callosal afferents is modified by monocular enucleation and induction of strabismus in a manner that, again, mimics that of the cell bodies (Innocenti and Frost, 1979, 1980; Lund *et al.,* 1978; Lund and Mitchell, 1979; Rhoades and Delacroce, 1980). For example, removal of one eye leads to the establishment of a wider than normal matched distribution of callosal perikarya and afferents in the visual cortex contralateral to that eye. The cortex receiving input from the remaining eye contains a normal distribution of callosal perikarya and axon terminals. Thus, in these animals, a normal discrete callosal area in one hemisphere can form an abnormally wide terminal pattern contralaterally. Conversely, a population of neurons that is arrested in its more widely distributed developmental state can still terminate in an apparently normal discrete manner on the opposite side of the brain.

In contrast to callosal efferent plasticity, however, afferent plasticity does not result from the gradual restriction of an earlier pattern. Instead, during development the callosal axons innervate only those portions of the neocortex destined to become callosal projection zones in the adult (Ivy *et al.,* 1979). The afferent innervation of the cortical target area is temporally coincident with the winnowing of the distribution of the callosal perikarya. This same principle apparently operates in the spinal projection system as well. During the period in which some cortical neurons are losing their collaterals to the spinal cord, the future functional motor neurons are beginning to invade their target zones in the spinal gray matter (Donatelle, 1977; Ivy and Killackey, 1981*b*). There is evidence that the transient corticospinal projections, such as those arising from neurons in the visual cortex, never enter the gray matter (Distel and Hollander, 1980).

It thus appears that neocortical neurons may send axons to several brain regions during development but that only one or a few of these collaterals prove capable of entering the target tissue and forming synapses. As we have seen, changing the amount or quality of visual experience can cause callosal axons to enter cortical areas that would not normally be considered target tissue and can stabilize young callosal projection neurons located in cortical areas that do not normally transmit information through the callosum.

While the qualities characterizing an "appropriate" target area remain an enigma, several aspects of target invasion by spinal and callosal axons may provide clues as to their nature. With regard to the corticospinal system, all neurons in layer Vb, but not in other cortical laminae, initially grow caudally through the nervous system, presumably because of their special chemical affinity for a substrate pathway and/or the availability of such a pathway at that time. However, the axons do not appear to grow out simultaneously. Rather, rostral portions of the neocortex generally develop earlier than those located more caudally (Jensen and Altman, 1982). This rostrocaudal gradient dovetails with the gradient of invasion of target areas of the subcortical projection neurons (Distel and Hollander, 1980). On the basis of these data, Bates and Killackey (1984) hypothesized that temporal matching of axon outgrowth and target availability may be the key to the appropriateness of a target area for a given population of layer Vb neurons.

With regard to the target areas of the callosal projections, there is evidence that the axons initially cross the midline on a glial bridge and then fasciculate upon each other in an orderly fashion (Silver *et al.*, 1982), thus allowing the maintenance of topographic ordering of the axons. While this finding may help account for the fact that callosal connections in mammals are largely homotopic, it cannot account for the fact that certain cortical areas—those that receive innervation from specific sensory nuclei of the thalamus—are not invaded by callosal terminals. For two reasons, it appears that the specific thalamic input dictates the regions of cortex that are to become callosal. First, at least in the rat somatosensory cortex, the thalamic input may arrive in the cortical white matter before the callosal input and certainly becomes established in layer IV before the callosal axons ascend to superficial cortical layers (Killackey and Akers, 1980; Killackey and Belford, 1979; Wise and Jones, 1976); they may thus be outcompeting the callosal afferents on a temporal basis (however, see Wise and Jones, 1978). Second, the thalamic input to the visual cortex is altered in genetically mutant Siamese cats (reviewed by Shatz, 1979) such that geniculocortical axons purveying information from similar, binocular portions of the visual field terminate in mirror-image portions of visual cortex in the two hemispheres. However, these projections are abnormally expanded on both sides of the normal discrete termination site at the area 17/18 border. Callosal connections generally serve to connect bits of related information that enter from opposite sides of the body; thus in the visual cortex, callosal connections link cortical regions "seeing" similar portions of the visual field. As would be expected, callosal connections in Siamese cats develop throughout the expanded cortical areas. These results are similar to those found in genetically normal cats after the experimental induction of strabismus and, interestingly, Siamese cats are known to be strabismic (Innocenti and Frost, 1979). Taken together, the data suggest that the tangential arrangement of callosal connections varies as a function of thalamic input.

The experiments detailed above have immediate relevance for brain development in children afflicted with strabismus or with any form of ocular occlusion. Furthermore, they indicate that patterns of neocortical connectivity are particularly susceptible to modification by events that occur at the periphery and that these events include aberrant experience that is not caused by receptor damage.

It should be noted that the above data do not rule out cell death as an event in the development of the neocortex. Indeed, there is now good evidence that cell death does take place in the upper layers of the rat neocortex postnatally. On the basis of their cell counts in five cortical areas in the hamster, Finlay and Slattery (1983) propose that cortical neurons are initially overproduced and are assigned to laminae as they arrive in the cortex, forming first the deep and then the superficial layers (Angevine and Sidman, 1961). The neurons are then contacted by afferents in this same sequence, with considerable normal variability among the cortical areas as to the amount of afferents arriving. Any mismatch between the numbers of neurons and the available afferents (or efferent targets) would then be corrected by cell death in the superficial layers. This hypothesis provides a mechanism for the normal differential laminar organization among cortical fields as well as a potential substrate for experimental manipulation.

Finally, it should be noted that the particular form of efferent plasticity demonstrated by neocortical neurons is not exhibited by cells that project through the other major pathway linking the hemispheres, the anterior commissure. If the corpus callosum is transected in a mouse, and HRP is injected into one hemisphere, it will be transported across the midline by axons in the anterior commissure. The patterns of labeled cells resulting from such an experiment performed in a mouse on its day of birth and in another adult mouse are shown in Figure 6a and 6b, respectively). The tangential distribution of labeled neurons does not appear to be wider in a neonatal than in an adult mouse (Ivy and Lynch, 1982). Also, unless additional events have transpired prenatally, it seems that while many neocortical neurons project to different brain regions, those accessed *via* the anterior commissure are not among them.

On the basis of data presented here, as well as in Section 3.1, two major conclusions can be drawn. First, the bulk of evidence indicates that the neocortex exhibits an additional, perhaps novel, evolutionary mechanism for determining connectivity patterns, that of the deployment of long axon collaterals that may later be eliminated. Second, it is clear that sensory experience during development creates effects that cascade up the nervous system, causing alterations of afferent patterns at several levels. Such afferent anomalies at the cortical level may be responsible for the changes observed in the distribution of callosal projection neurons and could conceivably influence cortical lamination as well. In any case, the ability of the neocortex to modify its connectivity patterns in response to sen-

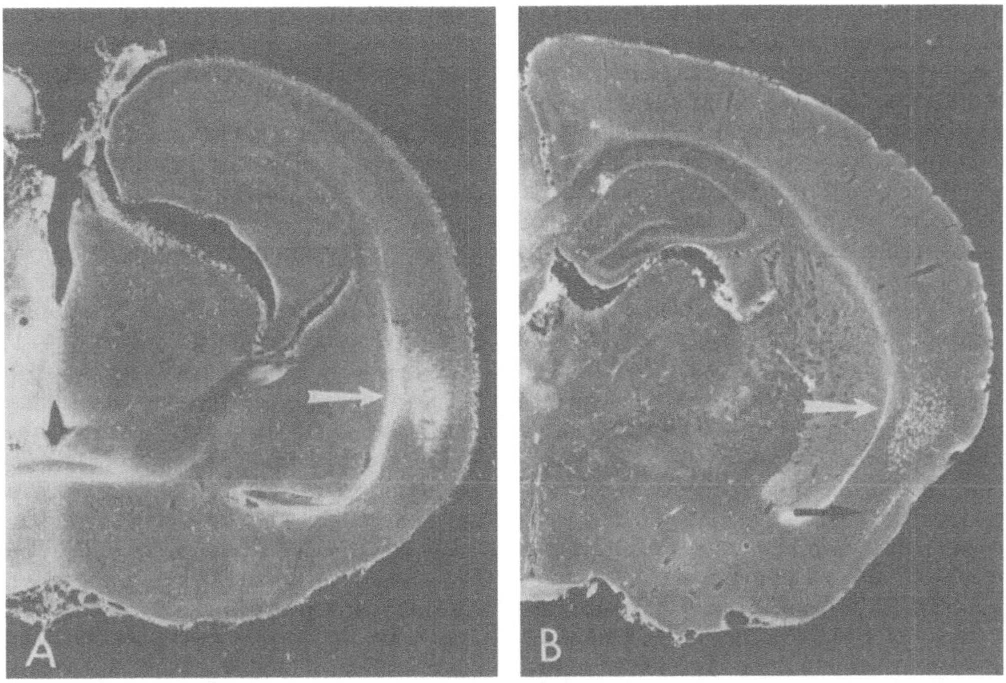

Fig. 6. Dark-field photomicrographs showing the pattern of neurons that send axons through the anterior commissure of the mouse (A) at postnatal day 1 and (B) in the adult. Horseradish peroxidase (HRP) was injected throughout the contralateral hemisphere immediately following transection of the corpus callosum. Note the similar distributions of brightly labeled cells in inferotemporal cortex (white arrows). The additional band of labeled neurons in the pyriform cortex of the adult (black arrow) develops on postnatal days 3–4.

sory experience is a property that must certainly be expressed during the normal development of the brain.

3.4. *Experimental Manipulation of Size, Shape, and Orientation of Dendritic Trees*

Fundamental to studies on the plastic properties of immature dendrites is the question of the extent to which dendritic morphology is fixed by the cell's genetic program. The alternate perspective might ask to what extent extrinsic influences determine the morphology of developing dendrites; i.e., the extent to which their morphology is plastic. Investigation of this latter question has involved the use of a wide variety of experimental circumstances ranging from severe afferent deprivation to the more indirect influences of differential experience. In general, Golgi studies of immature dendrites indicate a loss or nondevelopment of spines following removal of their normal supply of axonal input (Globus and Scheibel, 1967; Globus, 1975). Such experiments suggest that the presence of a presynaptic component is essential to the development and/or maintenance of spines on immature dendrites. Although this seems to be the rule in studies involving forebrain structures, an exception has been demonstrated in the cerebellum. The cerebellar Purkinje cell is densely innervated by the parallel fibers (axons of the cerebellar granule cell) that terminate on spines of the more distal dendritic branches. The granule cells germinate postnatally from an external germinal layer that covers the immature cerebellum. This arrangement leaves the granule cell population vulnerable to selective destruction by treatments such as X-irradiation and viral infection, treatments that leave the Purkinje cells intact. Under these circumstances, the Purkinje cells never receive their normal parallel fiber input. As a result, their dendritic growth is retarded in length and tertiary branching, but spines develop nonetheless (Altman and Anderson, 1972; Hirano *et al.*, 1972). Therefore, in this case, spines develop and persist without presynaptic contact. With further maturation, the Purkinje cell dendrites of the agranular cerebellum assume a reverse-normal orientation. Rather than directing themselves toward the cerebellar surface, they curve inward, toward the now dominant afferent, the climbing fibers (Herndon and Oster-Granet, 1975). Although the presence of the parallel fibers seems to be essential to the full expression of normal dendritic morphology, the dendritic spines, unlike those of several forebrain systems, are not entirely dependent on direct interaction with a presynaptic element.

Neonatal X-irradiation has also been used to demonstrate that under certain conditions the postsynaptic element can hyperdevelop. The granule cells of the hippocampal formation germinate to a large extent postnatally. Therefore, like the cerebellar granule cells, the hippocampal granule cells can be removed by neonatal X-ray treatment. Normally the granule cell layer defines the limit of the basal dendritic field of hippocampal CA3c pyramidal cells. With X-irradiation removal of the granule cell layer, the basal dendrites of the pyramidal cells grow all the way to the surface of the hippocampus, virtually doubling their length (Figure 7) (Gall and Lynch, 1980). Electron microscopic studies have further demonstrated that the somata of the granule cells that survive this X-irradiation become hyperinnervated (Lee and Lynch, 1982). Thus, this paradigm indicates that CNS neurons do not fully exhaust either their dendritic growth capacity or maximal density of postsynaptic elements during normal development. Both measures can be modified dramatically under experimental conditions.

Studies of sensory deprivation suggest that the activity of the presynaptic component, the afferent of the dendrite under study, can influence dendritic morphology. In an analysis of the development of pyramidal cells in the visual cortex, Valverde (1971; Valverde and Ruiz-Marcos, 1969) described the growth of three classes of dendritic spines. One group followed a normal pattern of development despite visual deprivation throughout the maturation period. The second class of spines appeared when light was first supplied some days after eye opening. The third group of spines developed normally only if light was supplied at the time of eye opening, thus demonstrating dependence upon some sort of physiological activity during a critical period for the expression of the spine morphology.

These and several other studies indicate that the development of CNS dendrites is dependent not merely on the presence of afferent contact but is in a morphological sense responsive to the activity of those afferents. One might therefore question whether heightened or specialized afferent activity might also shape the morphology of the target dendrite. This idea is indirectly supported by a study of Ryugo *et al.* (1975) in which it was found that enucleation or whisker removal in newborn rats, producing sensory deprivation of either visual cortex or specific areas of somatosensory cortex, caused an increase in spine density on pyramidal cells of the adjacent auditory cortex. It was suggested that hyperdevelopment of the auditory system represented anatomical adjustment to a heightened dependence on audition possibly involving increased physiological activity within this system.

The question of the influence of afferent activity in molding the morphology of immature dendrites has also been addressed in much more subtle circumstances, approximating more closely the experiential range normally encountered by the animal. The influence of environmental complexity on dendritic morphology was first studied by Holloway (1966), who found that stellate neurons in layer II visual cortex of animals reared in enriched environments display a higher degree of dendritic branching than do those of animals reared in isolation. Subsequently other investigators replicated and extended these results for several classes of cortical neurons (Greenough *et al.*, 1973; Globus *et al.*, 1973). Similar results were obtained in Golgi studies of animals subjected to stressful stimulation during early postnatal life (Schapiro and Vulkovich, 1970). These morphological differences have only been found in those areas in which differential environment has been shown to influence cortical thickness. The most consistent differences are reported for occipital cortex, while in frontolateral cortex, neither thickness effects nor dendritic changes were observed (Greenough *et al.*, 1973). Although these observations indicate a much larger dendritic surface area for the affected cortical neurons, Globus (1975) and Greenough (1976) failed to find increased dendritic field volume for cells in which increased dendritic branching is observed, suggesting that glial rather than dendritic growth is responsible for the cortical thickness effect.

4. Anatomical Plasticity in the Adult Brain

4.1. Changes in the Detailed Anatomy of Dendrites following Various Treatments

Deafferentation of mature dendrites has been shown to initiate plastic changes similar to those

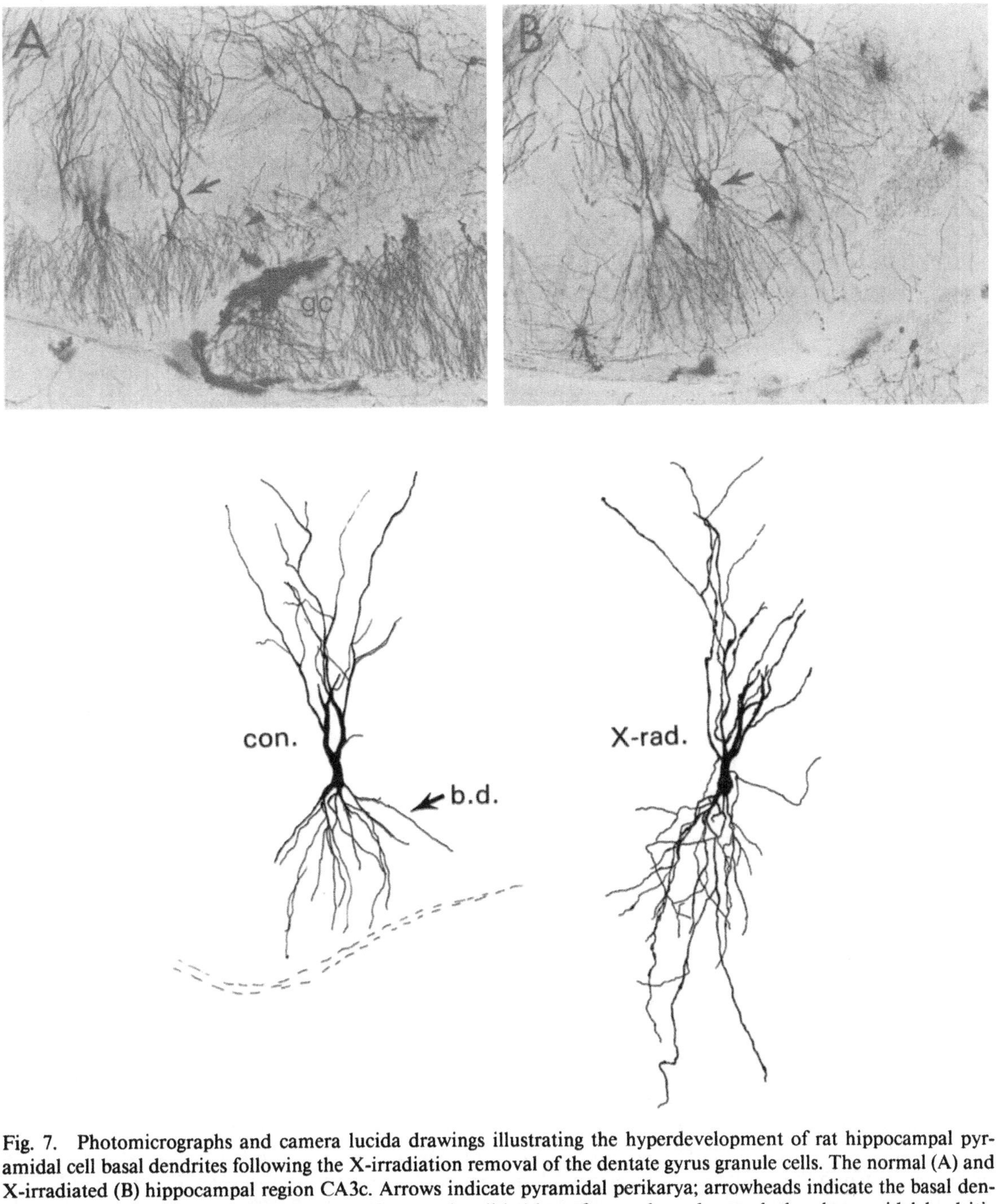

Fig. 7. Photomicrographs and camera lucida drawings illustrating the hyperdevelopment of rat hippocampal pyramidal cell basal dendrites following the X-irradiation removal of the dentate gyrus granule cells. The normal (A) and X-irradiated (B) hippocampal region CA3c. Arrows indicate pyramidal perikarya; arrowheads indicate the basal dendritic arborization. In the normal structure, the granule cell (gc) layer forms a boundary to the basal pyramidal dendritic field. In the X-irradiated hippocampus, this boundary is removed and the pyramidal cell basal dendrites can be seen to grow well beyond normal lengths. Camera lucida drawings of individual neurons from the normal (con.) and X-irradiated (X-rad.) hippocampus illustrate this effect on the basal dendrites (b.d.) as well. The dashed line below the control cell indicates the position of the granule cell layer in this case.

seen in the developing animal. Removal of afferents to the spines of cells throughout the neuraxis results in the apparent loss of those spines, at least as examined in Golgi-impregnated material (Kemp and Powell, 1971; White and Westrum, 1964). Both light microscopic and electron microscopic studies have shown that adult spinal motor neurons exhibit a loss of dendritic field and the formation of varicosities in the primary and secondary dendrites following severe afferent deprivation (Bernstein and Bernstein, 1973). Partial deafferentation of the granule cells of the hippocampal dentate gyrus, by lesions of the entorhinal cortex, results in a severe decrease in the spine density of the affected dendritic zone (Parnavelas *et al.*, 1974). These hippocampal dendritic regions become repopulated with spines over time following the lesion, an event that coincides with reinnervation of the dendrites by healthy presynaptic elements (Matthews *et al.*, 1976; Lee *et al.*, 1977; Caceres and Steward, 1983; Steward and Vinsant, 1983).

The Purkinje cells of the cerebellum normally develop two morphologically distinct types of spines. The large central dendrites, innervated by the climbing fibers, exhibit short broad spines, whereas the distal tertiary branches are densely covered by longer, thinner spines known to be contacted by the parallel fibers. Sotelo *et al.* (1975) studied the effects of climbing-fiber removal in the adult animal by selective chemical lesions of the inferior olive. Following such lesions, the primary and secondary branches have been shown to be reinnervated by parallel fibers. Concurrently, the climbing-fiber-type spine of the primary and secondary dendrites is lost in favor of the development of the longer, thinner, parallel-fiber-type spine that ultimately attains a density in these regions exceeding that seen prior to the lesion.

Therefore, in these studies of dendritic morphology following deafferentation, mature dendrites have demonstrated the capacity to lose and develop dendritic spines in coordination with the loss and redevelopment of presynaptic contact. In the case of the cerebellum, the postsynaptic specialization reflects the change of the type of presynaptic element. Unlike direct deafferentation, sensory deprivation instituted in the adult has not been found to influence the morphology of neocortical dendrites (Rothblat and Schwartz, 1979; Chang and Greenough, 1982).

As in the developing system, the morphology of mature CNS dendrites has been shown to be responsive to the physiological activity of direct afferents. Electrical stimulation of the neocortex of the adult cat has been reported to cause an increase in dendritic branching, dendritic field, and spine density of the pyramidal cells of the contralateral homotopic region receiving innervation from the stimulated area (Ruttledge *et al.*, 1974; Ruttledge, 1976). The interpretation of these observations, made on Golgi material, is open to dispute. One possibility is that hyperactivity of existing axodendritic synapses led to the induction of spines; alternatively, it is conceivable that the effect is caused by an increase in the number of presynaptic elements, leading to an increase in the number of spines. With the latter interpretation, the increase in spine density represents a secondary result of stimulation, with the primary effect being on the stimulated axon population. It is also possible that the stimulation altered the staining properties of the postsynaptic neurons such that they became more clearly defined by the Golgi method. If this were to be the case, dendritic elements not detected in the normal situation might have become apparent following stimulation. A possibility of this sort, while probably remote, serves as a cautionary note in interpreting the many experiments cited above that used the Golgi stain to reconstruct dendritic architecture after experimental manipulations.

Electron microscopic studies obviate some of the reservations that must be observed when interpreting Golgi studies of dendritic morphology. Using this technique, subtle changes in spine morphology and dendritic innervation have been observed in an instance of physiologically induced changes in synaptic efficacy. As first described by Bliss and Lomo (1973), and subsequently replicated by a number of investigators in several mammalian forebrain systems, brief trains of repetitive high-frequency stimulation can induce a change in the size of the extracellular evoked potential recorded in the target region of the stimulated fibers. That is, a stimulus pulse of a given size will evoke a larger potential in the target field after high-frequency stimulation. This long-lasting increase in synaptic efficacy, referred to as long-term potentiation (LTP), is considered by some to reflect a capacity for synaptic modulation that might be involved in learning and memory processes (Lynch, 1983). The stability of the LTP effect has led to speculation that changes in structure underlie the change in physiological processing. Studies have been conducted in an effort to examine this possibility.

Two separate laboratories (Lee *et al.*, 1980, 1981; Chang and Greenough, 1984) have conducted comprehensive electron-microscopic analyses of the terminal fields of low-frequency stimulated (nonpotentiated) and high-frequency stimulated (physiologically verified to be potentiated) afferents to region CA1 of the rat hippocampus. In this paradigm, the stimulation activated at most 30% of the synapses in the area studied. In one laboratory, the study was conducted on both perfusion-fixed tissue following *in vivo* stimulation and on immersion-fixed tissue following *in vitro* stimulation (Lee *et al.*, 1980, 1981). In all cases, two very subtle morphological differences between the nonpotentiated and potentiated tissue were detected. In the potentiated tissue, there was a higher incidence of synapses on dendritic shafts (as opposed to dendritic spines) and a lower variability in several morphological indexes of spine shape (spine neck width, spine area within a section, length of postsynaptic density). The latter observation indicates that the spines had changed their shape such that random measurements through them were less variable. This could be accomplished by a shift from an irregular to a more round shape. Changes in spine shape could alter the coupling between the spine and the dendrite and result in changes in the level of excitation in the target cell following spine–synapse activity.

As mentioned earlier, gross volumetric influences of differential housing have been observed in adult animals. Dendritic responsiveness to this treatment instituted during maturity has been demonstrated as well. Analyses of Golgi-impregnated neurons from the occipital cortices of adult animals reared in either enriched or isolated environments have detected an increase in layer IV stellate cell and layer III pyramidal cell pyramidal cell dendritic branching and terminal branch length (Green *et al.*, 1983; Juraska *et al.*, 1980). These effects are observed when differential housing conditions are maintained for only 45 days and begun as late as 15 months of age (Green *et al.*, 1983). Clearly, the morphology of neurons in the occipital cortex remain responsive to the complexity of the environment in which they live well into middle age. What is even more extraordinary are data indicating that the morphology of cortical dendrites can be influenced by specific training situations. This has been demonstrated in a number of laboratories. One of the more dramatic demonstrations was obtained by Ruttledge (1976) in a classic conditioning paradigm. In this experiment, stimulation to the suprasylvian gyrus was followed after a brief interval by a foot shock, creating a classic conditioning situation. The stimulated cortex and the contralateral efferent gyrus were both analyzed in Golgi preparations. Although the stimulation alone, in the presence of nonassociated foot shock, caused dendritic development in the target cortex, a significant increase in the number of spines on vertical and oblique branches of the contralateral pyramidal cells was noted only in trained animals. These results indicate that some aspect of the training experience caused dendritic development in addition to the adjustments produced by repeated monosynaptic activation.

Using a very different training situation, Greenough (1976) and colleagues (Chang and Greenough, 1982) provided data leading to similar conclusions. In these studies, adult rats were trained on Hebb Williams maze problems. In some cases, rats were trained with fixed opaque contact lenses for constant unilateral visual occlusion; in other cases, the contact lenses were alternated between eyes on subsequent days. Rapid Golgi analysis of neocortical occipital cortex demonstrated an increase in dendritic branching of distal apical dendrites of layer IV and V pyramidal neurons in the trained cortex (contralateral to the sighted eye) as compared to the untrained cortex in cases with fixed occluders. In the animals with alternating occlusion some smaller bilateral effects were observed. This study, and that of Ruttledge, suggest that some aspect of learning, rather than the training experience, can initiate elaboration of the dendrites of adult cortical neurons.

4.2. *Induction of Axonal Sprouting, Terminal Proliferation, and New Circuitry by Discrete Lesions*

It will be recalled that a substantial body of evidence indicates that removal of one afferent to an immature brain region will cause that region's remaining afferents to begin growing new branches and connections. There is also reason to believe that comparable phenomena occur in the adult brain. Several studies have shown that a given fiber projection will increase the density of its terminal field in partially deafferented brain areas. Goodman and Horel (1966) found elimination of the cortical projections to the lateral geniculate nucleus to result in an apparent proliferation of the terminals of the optic tract in the geniculate. Subsequent studies using several different anatomical methods have reported comparable effects in the colliculus (Stenevi *et al.*, 1972), septum (Moore *et al.*, 1971), and hippocampus (Lynch *et al.*, 1972; Steward *et al.*, 1973). Electron microscopic experiments have provided evidence that the number of synaptic connections generated by one afferent can increase dramatically following removal of its neighbors; this effect has been obtained in the brain stem (Westrum and Black, 1971), colliculus (Lund and Lund, 1971), septum (Raisman, 1969; Raisman and Field, 1973), and hippocampus (Matthews *et al.*, 1976; McWilliams and Lynch, 1979). Problems inherent in this type of experiment indicate the need for caution in interpreting the results. Deafferentation causes considerable shrinkage in the target area (Raisman and Field, 1973; Lynch *et al.*, 1975); this fact alone will produce an apparent increase in terminal and synaptic density. That is, the residual population of intact afferents will be compressed into a smaller space, resulting in an apparent increase in density but an increase that does not require any growth response. Light microscopic experiments are further plagued by a problem discussed in regard to dendritic changes after experimental manipulation: the residual elements may be rendered more visible with the histological stain. Deafferentation could produce local changes that alter the biochemistry of intact terminals in such a way as to increase the probability that they will be detected by the particular tracing method being used. It is also conceivable that the residual terminals might increase in size but not number; this would tend to increase their likelihood of being counted as an ending. The electron microscopic experiments typically count synapses, not boutons—if the number of synapses per bouton were to increase in the deafferented area, this could be misinterpreted as an increase in bouton number.

Lesion-induced axonal growth is much more easily detected and interpreted if it involves growth of a given axonal population into an area it does not normally innervate. Growth of this sort has been demonstrated to occur in a number of areas including the deafferented dentate gyrus of the rat. It will be recalled that the C/A axons expand their field of termination to include the entire granule cell dendritic tree after its more distal regions have been deafferented by a lesion of the entorhinal cortex in the neonate. After the same lesion in the adult, these two fiber systems expand outward, but not nearly to the degree that they do in the immature brain (Lynch *et al.*, 1973*c*, 1976). Studies using both light microscopic and electron microscopic techniques have demonstrated that this growth begins on the fifth or sixth day postlesion (Lynch *et al.*, 1977; Lee *et al.*, 1977; Steward and Vinsant, 1983) (Figures 8 and 9) and is followed 2–3 days later by physiological changes suggesting that functional synapses have been formed (West *et al.*, 1975).

These data demonstrate that intact axons in the mature brain retain the capacity to grow new branches, although it is evident that for unknown reasons this process is considerably reduced from that seen in younger animals. However, lesion-induced axonal growth within the deafferented hippocampus is fundamentally different between the young and mature rat in two respects. With deafferentation of the young hippocampus, reactive axonal growth is virtually immediate and as extensive as the field of deafferentation. With deafferentation in the adult, growth is delayed and limited, leaving a persistent paucity of innervation throughout most of the deafferented field. So stated, with an emphasis on striking differences, one is inclined to suppose that two categories of response exist: immature and mature. However, closer analysis of the temporal parameters of growth within each phase suggests a gradual change in growth potential throughout the life of the animal. Section 3.2 described the time course for axonal distribution (fiberarchitectural) changes within the dentate deafferented in the 14-day-old rat. At that age, one component of the growth response exhibits the spatiotemporal parameters of adult sprouting, despite maximal reinnervation (Gall and Lynch, 1981). Thus, the robust and limited forms of axon sprouting do not occupy exclusive phases in early development; both forms are present during the transitional period.

Analysis of the rate of innervation of the deafferented dentate gyrus inner molecular layer indicates that gradual changes in lesion-induced axonal growth occur during adulthood as well (Table I). It will be remembered that the inner molecular layer is innervated by axons from both the contralateral and ipsilateral hippocampus (the commissural and associational systems, respectively). The normal innervation density in this field is quite constant from 1 to 18 months of age and declines only slightly (8%) by 22 months, which is considered aged in the rat. If the commissural input to this zone is removed in the young adult rat, the associational axons sprout to reinnervate the field such that near-normal synaptic bouton densities are recovered by postlesion day 8 in 35-day-old rats and

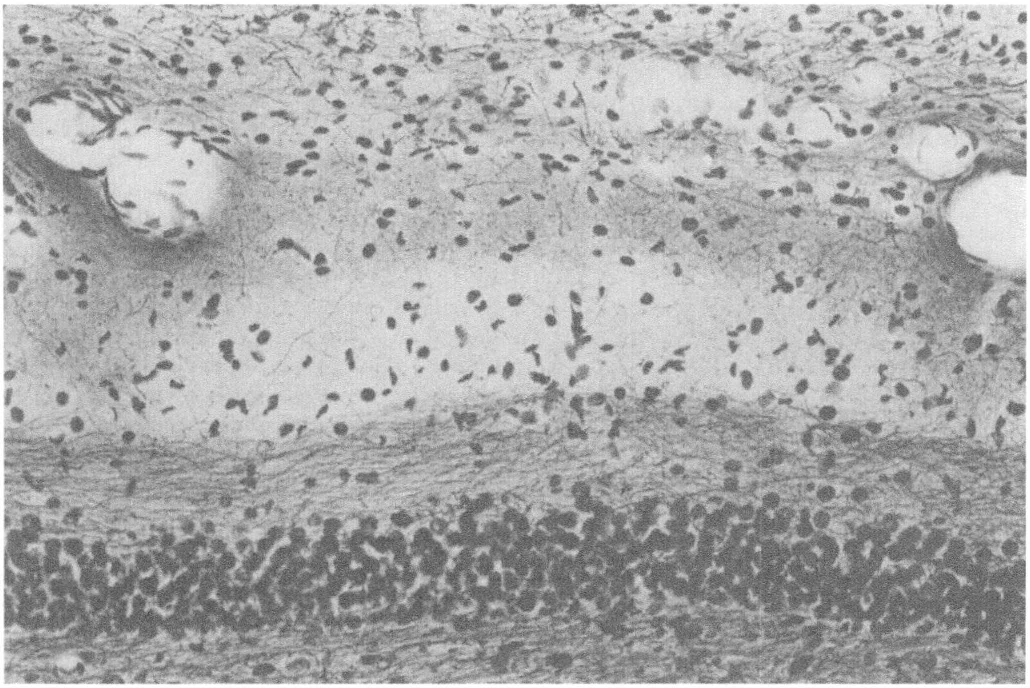

Fig. 8. Photomicrograph of a portion of the dentate gyrus molecular layer impregnated with the Holmes stain for normal axons. The plexus of commissural/associational axons can be seen coursing in a discrete lamina close to the granule cells at the bottom of the frame. The outer 75% of the molecular layer is virtually axon free in this rat, which received an ipsilateral entorhinal cortex lesion 2 days before sacrifice.

by postlesion day 15 in 60-day-old rats. By contrast, if the commissural input is removed in 90- or 180-day-old rats, very little reinnervation has occurred by 8 days postlesion, and less than 50% of the lost synaptic density has returned by postlesion day 15 (McWilliams and Lynch, 1983). Clearly, there is a drop in what might be called the vigor of the sprouting response from young adulthood to middle age. From 6 to 12 months of age, the rate of associational sprouting is fairly constant, but another decline in growth capacity is seen from late-middle to old age. If the commissural input is removed in a 12-month-old rat, 39% of the lost synapses are recovered, presumably by the sprouting of associational axons, by 15 days postlesion. However, with similar deafferentation in an 18- to 24-month-old rat, synaptic recovery is barely detectable by postlesion day 15 (Table I) (McWilliams and Lynch, 1984).

Studies on lesion-induced axon sprouting in the rat dentate gyrus have demonstrated that CNS

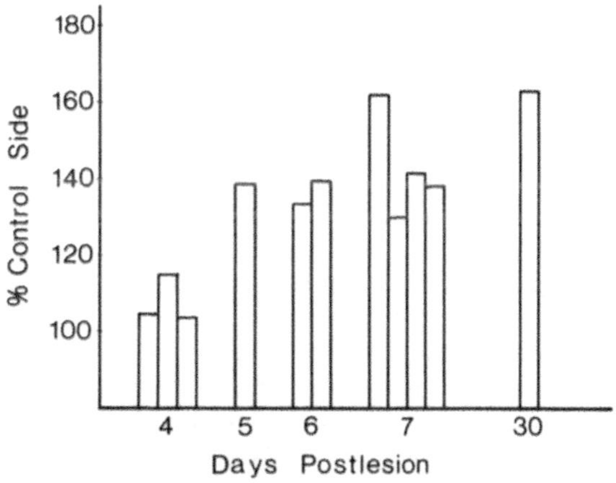

Fig. 9. Bar graph of depth measurements of the commissural/associational fiber plexus in rat dentate gyrus molecular layer ipsilateral to an entorhinal cortex lesion. Measurements are expressed as a percentage of the plexus width within the contralateral nondeafferented dentate gyrus.

Table I. Number (Mean ± Standard Deviation) of Synaptic Boutons in the Inner Molecular Layer of the Dentate Gyrus following a Contralateral Hippocampal (Commissural) Lesion at Different Postnatal Ages (per 100 μm^2)[a]

Age	Control	2 days postlesion	15 days postlesion
35 days	24.9 ± 0.3	16.5 ± 0.1	24.4 ± 0.5
2 months	25.1 ± 0.5	16.4 ± 0.3	22.5 ± 1.3
3 months	25.2 ± 0.6	16.1 ± 0.3	20.4 ± 1.1
6 months	25.2 ± 0.8	16.2 ± 0.3	19.7 ± 0.7
12 months	25.0 ± 1.2	—	19.6 ± 0.8
18 months	24.9 ± 1.4	—	16.5 ± 0.9
22 months	23.2 ± 0.3	—	16.3 ± 0.4
24 months	22.8 ± 0.5	—	15.8 ± 0.4

[a]Based on data from McWilliams and Lynch (1983, 1984).

axons are capable of much more extensive axonal growth than is expressed during normal development and can functionally innervate dendritic fields outside their normal area of termination. However, axonal growth capacity does appear to diminish over the life of the animal such that the speed and extent of axon sprouting and reinnervation would be expected to be radically different following brain damage in youth, mid-adulthood, or old age.

4.3. *Proliferation, Migration, and Hypertrophy of Glial Cells at Deafferented Brain Sites*

The capacity of the nonneural elements of the brain to undergo plastic changes through the entire life of the animal is rarely discussed under the heading of brain plasticity. Nonetheless, the glial cells comprise a major portion of the brain's cell population and, if one must be neurocentristic, they undoubtedly constitute a critical influence in the neuroenvironment. In almost all circumstances in which neural plasticity has been observed, glial cells have been found to undergo concurrent morphological transformations. The participation of glial cells may therefore be integral to a larger process of which neuroplasticity is also a part. Any attempt to evaluate this idea requires data on the temporal correlation of glial and neuronal events. This section reviews an attempt of this sort that used the rat hippocampus as a model system.

Starting at about 24 hr after removal of the entorhinal cortex, the astrocytes in the deafferented dentate gyrus begin to hypertrophy, an effect that reaches its climax 48–72 hr later (i.e., 3–4 days postlesion). At this point, the astrocytes have greatly enlarged cell bodies, longer thicker processes, and, strangely enough, processes that appear to be aligned with each other (Figure 10). Electron microscopic analyses (Matthews *et al.*, 1976; Lee *et al.*, 1977) have shown that the astrocytes are actively engaged in phagocytosis of the degenerating terminals during the period of hypertrophy. Between 96 and 120 hr postlesion, the astroglia begin to atrophy, an effect that continues for several days thereafter. At no time was any increase in the total number of astrocytes in the dentate gyrus detected, although their distribution in the molecular layer was changed. That is, the number of astroglia in the inner molecular layer declined, while the population in the outer (deafferented) zone increased. The time courses and magnitude of the two effects were nearly identical, leading to the hypothesis that the astrocytes were migrating from the inner to the outer molecular layer (see Rose *et al.*, 1976, for further discussion).

The microglia of the deafferented dentate gyrus display a very different type of behavior. It should be understood that these ubiquitous cells have been a source of controversy since their discovery by del Rio Hortega (1932), in some measure because of their similarity to oligodendrocytes. The studies reviewed in this section used autoradiographic methods, and it is not possible to be completely certain that the oligodendrocytes did not make some contribution to the results; in fact, it is likely they did.

The microglia are scattered randomly throughout the normal hippocampus and, as shown in autoradiographic experiments using isotope-labeled thymidine (which marks cells that are synthesizing DNA and hence dividing), some subgroup of their population is dividing in the naive animal. At 20 hr postlesion, but not before, the number of dividing microglial cells increases dramatically, an effect that continues for some 40–60 hr. These mitotic cells are found scattered throughout the hippocampus, even in areas well removed from degeneration or deafferentation. At some point in the period of cell division or shortly afterward, the microglia begin to migrate toward the deafferented zones. This is evident from studies in which the rats were allowed either 6 hr, 4 days, or 8 days survival after a brief pulse of [^{3}H]thymidine. In the former group, the labeled cells are randomly located, while in the latter longer-survival group, they are concentrated in the zones of degeneration. La-

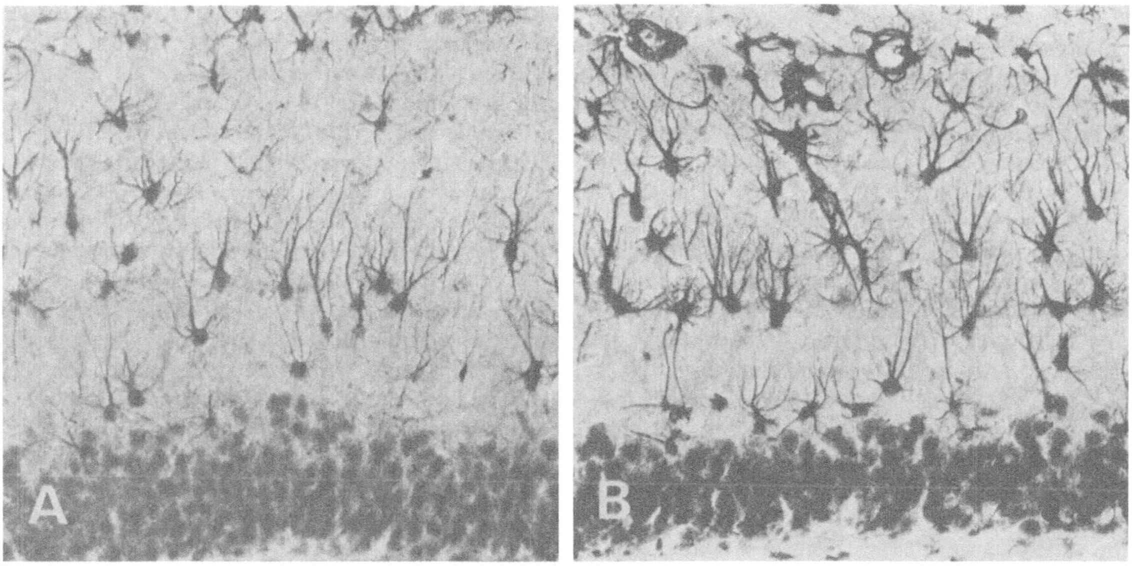

Fig. 10. Photomicrographs of a portion of the adult rat dentate gyrus molecular layer stained with Cajal's gold sublimate method for astrocytes. Pictured are equivalent sections from a naive animal (A) and an animal sacrificed 4 days after lesion of the ipsilateral entorhinal cortex (B). The granule cell layer can be seen at the bottom of both sections; the fissure, representing the outer limit of the molecular layer, lies near the top.

beled cells are also found accumulated at sites that might logically serve as barriers to migration; e.g., the densely packed row of granule cells (Gall *et al.*, 1979*b*).

Both microglia and astroglia undergo dramatic responses to deafferentation, and significant aspects of their responses (e.g., hypertrophy, proliferation) are concluded prior to the onset of any sprouting response (Figures 5 and 11). This leads naturally to the hypothesis that the earlier glial events facilitate the axonal growth, either by removing obstacles that normally retard this growth or by actively stimulating it (Lynch, 1976). This hypothesis is rendered more plausible by the several *in vitro* studies that have suggested that glial cells release nerve-growth-promoting substances (e.g., see review in Varon and Saier, 1975). Finally, it is worth noting that the functional significance of the rather extraordinary behavior of the microglia remains a complete mystery. The hypertrophy of the astrocyte is undoubtedly associated with their phagocytic role, but the consequences of the proliferation and migration of the microglia are entirely unknown.

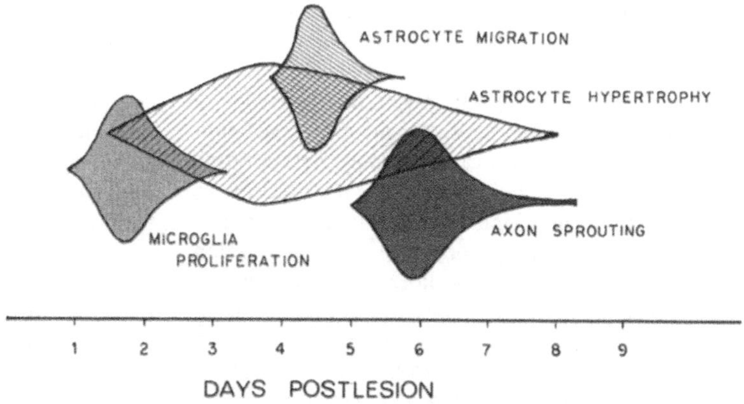

Fig. 11. Graphic illustration of the temporal position of various cellular events occurring in the dentate gyrus of the adult rat following partial deafferentation by lesion of the ipsilateral entorhinal cortex. The height of each area indicates the magnitude of the dynamic process involved relative to its own maxima.

5. Summary. The Utility of Neuroanatomical Plasticity across the Stages of Life

We have discussed the adaptability of the CNS from about the time neurons stop dividing to the onset of senescence. Early in this period, when cells in the telencephalon are only beginning to form connections, modifications involving virtually every aspect of growth seem possible. These include the selective elimination of neurons and their processes as well as alterations in the direction and magnitude of the growth of both axons and dendrites. This plasticity strongly suggests that the nervous system that ultimately emerges from postmitotic ontogeny is not simply the product of genetic programming, but instead reflects to a considerable measure the outcome of interactions between developing elements. As growth continues the variety as well as the extent of potential alterations become progressively limited. Axons, for example, gradually lose the capacity to grow lengthy projections following removal of competitors while still retaining their ability to rapidly form new synapses.

Despite the concentrated effort that has gone into defining these age-related changes, little is known of the mechanisms that produce them. It is clear that the genetic limits on growth for axonal and dendritic extension or for synapse formation are not achieved during normal development. Possibly the interactions so critical in regulating the direction of growth also serve to stabilize the system. Thus as contacts are added, cellular processes might be initiated that inhibit further additions and, as these accumulate, the cells become increasingly constrained. In any event, the stabilization process is not total, and some forms of plasticity persist into adulthood and are not lost until old age. Understanding the cellular factors responsible for growth suppression and stabilization represents one of the great challenges facing developmental neurobiology.

In early development, plasticity provides a means through which sensory and motor specializations in the periphery can generate corresponding modifications of brain circuitries with little or no genetic alteration. Consider the facial whiskers of rodents. Evolutionary changes in the organization of these critical sensory specializations must be accompanied by different types of modifications in the entire somatosensory system from brain stem through cerebral cortex. Rather than accomplish this through multiple changes in genetic instructions, ontogeny appears to arrive at an appropriate format by permitting each stage of the somatosensory chain, beginning with the whiskers themselves, to exert a powerful influence on the subsequent station in the system. Many of the experiments described in the present review suggest that this theme is repeated throughout the nervous system. If so, many of the cross-species differences found in the brain would reflect evolution not of the brain itself but of the inputs it processes and the targets it operates on.

If there is a role in normal brain functioning for the neuroanatomical plasticity that persists into adulthood, it must be quite unlike that described immediately above. An attractive and often discussed idea is that the capacity for structural change plays a role in the encoding of experience.

Contrary to this hypothesis is the fact that memories are formed quickly in response to trivial signals, while studies of plasticity in the adult typically use severe treatments and then find evidence of growth only after several days. But recent work strongly suggests that certain types of physiological stimulation can produce new synapses in a matter of minutes. It remains to be determined whether these physiologically induced effects involve the same cellular events that occur after a lesion in the mature nervous system. However, if this proves to be so, the plasticity that remains after development is concluded might well provide a substrate for long-term memory storage.

To summarize, we are suggesting that plasticity in development serves to link brain to body, while in adulthood it couples brain to the environment. In some respects, the stabilization process may be critical to the emergence of long-term memory, as memory is a form of plasticity that remains after development. That is, any contributions of synaptic growth and modification to memory can be maintained only if they occur against a background of relative stability—otherwise, they could be rendered insignificant or eliminated by changes occurring as part of ongoing growth. From this perspective, development sets the stage for long-term memory by first producing an unchanging neuron with very limited domains in which lasting structural modifications can be induced.

ACKNOWLEDGMENTS. This work was supported in part by National Science Foundation grant BNS 82-00319 and by an Alfred P. Sloan Fellowship to C. G., by National Institute of Mental Health grant NS 18950 to G. I., and by National Science Foundation grant BNS 81-11994 to G. L.

6. References

Altman, J., and Anderson, W. J., 1972, Experimental reorganization of the cerebellar cortex. I. Morphological effects of elimination of all microneurons with prolonged x-radiation started at birth, *J. Comp. Neurol.* **146**:355–406.

Angevine, J. B., and Sidman, R. L., 1961, Autoradiographic study of cell migration during histogenesis of cerebral cortex in the mouse, *Nature (London)* **192**:766–768.

Bates, C., and Killackey, H. P., 1984, The emergence of a discretely distributed pattern of corticospinal projection neurons, *Dev. Brain Res.* **13**:265–273.

Baudry, M., and Lynch, G., 1980, Regulation of hippocampal glutamate receptors: Evidence for the involvement of a calcium-activated protease, *Proc. Natl. Acad. Sci. U.S.A.* **11**:2298–2302.

Belford, G. R., and Killackey, H. P., 1979, The development of vibrissae representation in subcortical trigeminal centers of the neonatal rat, *J. Comp. Neurol.* **188**:63–74.

Bennett, E. L., 1976, Cerebral effects of differential experience and training, in: *Neural Mechanisms of Learning and Memory* (M. R. Rosenzweig and E. L. Bennett, eds.), pp. 279–287, MIT Press, Cambridge, Massachusetts.

Bernstein, M., and Bernstein, J. 1973, Regeneration of axons and synaptic complex formation rostral to the site of hemisection in the spinal cord of the monkey, *Int. J. Neurosci.* **5**:15–26.

Bliss, T., and Lomo, T., 1973, Long-lasting potentiation of synaptic transmission in the dentate area of the anaesthetized rabbit following stimulation of the perforant path, *J. Physiol. (London)* **232**:331–356.

Caceres, A., and Steward, O., 1983, Dendritic reorganization in the denervated dentate gyrus of the rat following entorhinal cortical lesions: A golgi and electron microscopic analysis, *J. Comp. Neurol.* **214**:387–403.

Chang, F.-L. F., and Greenough, W. T., 1982, Lateralized effects of monocular training on dendritic branching in adult split-brain rats, *Brain Res.* **232**:283–292.

Chang, F.-L. F., and Greenough, W. T., 1984, Transient and enduring morphological correlates of synaptic activity and efficacy change in the rat hippocampal slice, *Brain Res.* **309**:35–46.

Cunningham, T. J., Mohler, I. M., and Giordano, D. L., 1982, Naturally occurring neuron death in the ganglion cell layer of the neonatal rat: Morphology and evidence for regional correspondence with neuron death in superior colliculus, *Dev. Brain Res.* **2**:203–215.

D'Amato, C. J., and Hicks, S. P., 1978, Normal development and post-traumatic plasticity of corticospinal neurons in rats, *Exp. Neurol.* **60**:557–569.

Devor, M., and Schneider, G. E., 1975, Neuroanatomical plasticity: The principle of conservation of total axonal arborization, in: *Aspects of Neural Plasticity/Plasticité Nerveuse* (F. Vital-Durand and M. Jeannerod, eds.), Vol. 43, pp. 191–200, INSERM, Paris.

del Rio Hortega, P., 1932, Microglia, in: *Cytology and Cellular Pathology of the Nervous System* (W. Penfield, ed.), Vol. 2, pp. 483–534, Hoeber, New York.

Diamond, M. C., 1967, Extensive cortical depth measurements and neuron size increases in the cortex of environmentally enriched rats, *J. Comp. Neurol.* **131**:357–364.

Distel, H., and Hollander, H., 1980, Autoradiographic tracing of developing subcortical projections of the occipital region in fetal rabbits, *J. Comp. Neurol.* **192**:505–518.

Donatelle, J. M., 1977, Growth of the corticospinal tract and the development of placing reactions in the postnatal rat, *J. Comp. Neurol.* **175**:207–232.

Dunwiddie, T. V., and Lynch, G. S., 1978, Long term potentiation and depression of synaptic responses in the rat hippocampus: Localization and frequency dependency, *J. Physiol. (London)* **276**:353–367.

Fifkova, E., 1970, The effect of unilateral deprivation on visual centers in rats, *J. Comp. Neurol.* **140**:431–438.

Finlay, B. L., and Slattery, M., 1983, Local differences in the amount of early cell death predict adult local specializations, *Science* **219**:1349–1351.

Fricke, R., and Cowan, W. M., 1977, An autoradiographic study of the development of the entorhinal and hippocampal afferents to the dentate gyrus of the rat, *J. Comp. Neurol.* **173**:231–250.

Gall, C. M., and Lynch, G., 1980, The regulation of fiber growth and synaptogenesis in the developing hippocampus, in: *Current Topics in Developmental Biology,* Vol. 15: *Neural Development,* Part I (R. K. Hunt, ed.), pp. 159–180, Academic Press, New York.

Gall, C. M., and Lynch, G., 1981, Fiber architecture of the dentate gyrus following removal of the entorhinal cortex in rats of different ages: Evidence that two forms of axon sprouting occur after lesions in the immature rat, *Neuroscience* **6**:903–910.

Gall, C. M., McWilliams, R., and Lynch, G., 1979*a*, The effect of collateral sprouting on the density of innervation of normal target sites: Implications for theories on the regulation of the size of developing synaptic domains, *Brain Res.* **178**:37–47.

Gall, C. M., Rose, G., and Lynch, G., 1979*b*, Proliferative and migratory activities of glial cells in the partially deafferented hippocampus, *J. Comp. Neurol.* **183**:539–550.

Gall, C. M., McWilliams, R., and Lynch, G., 1980, Accelerated rates of synaptogenesis by "sprouting" afferents in the immature hippocampal formation, *J. Comp. Neurol.* **193**:1047–1061.

Globus, A., 1975, Brain morphology as a function of presynaptic morphology and activity, in: *The Developmental Neuropsychology of Sensory Deprivation* (A. Riesen, ed.), pp. 9–91, Academic Press, New York.

Globus, A., and Scheibel, A. B., 1967, Synaptic loci on parietal cortical neurons: Terminations of corpus callosum fibers, *Science* **156**:1127–1129.

Globus, A., Rosenzweig, M. R., Bennett, E. L., and Diamond, M. C., 1973, Effects of differential experience on dendritic spine counts in rat cerebral cortex, *J. Comp. Physiol. Psychol.* **82**:175–181.

Goodman, D. C., and Horel, J. A., 1966, Sprouting of optic tract projections in the brain stem of the rat, *J. Comp. Neurol.* **127**:71–88.

Green, E. J., Greenough, W. T., and Schlumpf, B. E., 1983, Effects of complex or isolated environments on cortical dendrites of middle-aged rats, *Brain Res.* **264**:233–240.

Greenough, W. T., 1976, Enduring brain effects of differential experience and training, in: *Neural Mechanisms of Learning and Memory* (M. R. Rosenzweig and E. L. Bennett, eds.), pp. 255–278, MIT Press, Cambridge, Massachusetts.

Greenough, W. T., and Volkman, F. R., 1973, Pattern of dendritic branching in occipital cortex of rats reared in complex environments, *Exp. Neurol.* **40**:491–504.

Greenough, W. T., Volkman, F. R., and Juraska, J. M., 1973, Effects of rearing complexity on dendritic branching in frontolateral and temporal cortex of the rat, *Exp. Neurol.* **41**:371–378.

Gyllensten, L., Malmfors, T., and Narrlin, M. L., 1964, Effect of visual deprivation on the optic centers of growing and adult mice, *J. Comp. Neurol.* **122**:79–90.

Herndon, R., and Oster-Granet, M., 1975, Effect of granule cell destruction on development and maintenance of the Purkinje cell dendrite, in: *Physiology and Pathology of Dendrites* (G. W. Kreutzberg, ed.), pp. 361–379, Raven Press, New York.

Hicks, S., and D'Amato, C., 1970, Motor-sensory and visual behavior after hemispherectomy in newborn and mature rats, *Exp. Neurol.* **29**:416–438.

Hirano, A., Dembitzer, H. M., and Jones, M., 1972, An electron microscopic study of cycasin induced cerebellar alterations, *J. Neuropathol. Exp. Neurol.* **31**:113–125.

Holloway, R. L., Jr., 1966, Dendritic branching: Some preliminary results of training and complexity in rat visual cortex, *Brain Res.* **2**:393–396.

Innocenti, G. M., 1981, Growth and reshaping of axons in the establishment of visual callosal connections, *Science* **212**:824–827.

Innocenti, G. M., and Frost, D. O., 1979, Effects of visual experience on the maturation of the efferent system to the corpus callosum, *Nature (London)* **280**:231–234.

Innocenti, G. M., and Frost, D. O., 1980, The postnatal development of visual callosal connections in the absence of visual experience or the eyes, *Exp. Brain Res.* **39**:365–375.

Innocenti, G. M., Fiore, L., and Caminiti, R., 1977, Exuberant projection into the corpus callosum from the visual cortex of newborn cats, *Neurosci. Lett.* **4**:237–242.

Ivy, G. O., and Killackey, H. P., 1981*a*, The ontogeny of the distribution of callosal projection neurons in the rat parietal cortex, *J. Comp. Neurol.* **195**:367–389.

Ivy, G. O., and Killackey, H. P., 1981*b*, Corticospinal and corticotectal neurons: Mechanisms of ontogenetic changes in distribution, *Soc. Neurosci. Abs. 7:178.*

Ivy, G. O., and Killackey, H. P., 1982, Ontogenetic changes in the projections of neocortical neurons, *J. Neurosci.* **2**:735–743.

Ivy, G. O., and Lynch, G. S., 1982, Postnatal development of the corpus callosum and anterior commissure in the mouse, *Soc. Neurosci. Abs.* **8**:300.

Ivy, G. O., Akers, R. M., and Killackey, H. P., 1979, Differential distribution of callosal projection neurons in the neonatal and adult rat, *Brain Res.* **173**:532–537.

Jacobson, M., 1978, *Developmental Neurobiology,* Plenum Press, New York.

Jeffery, G., and Perry, V. H., 1982, Evidence for ganglion cell death during development of the ipsilateral retinal projection in the rat, *Dev. Brain Res.* **2**:176–180.

Jensen, K. F., and Altman, J., 1982, The contribution of late-generated neurons to the callosal projection in the rat: A study with prenatal x-irradiation, *J. Comp. Neurol.* **209**:113–122.

Juraska, J., Greenough, W. T., Elliott, C., Mack, K. J., and Berkowitz, R., 1980, Plasticity in adult rat visual cortex: An examination of several cell populations after differential rearing, *Behav. Neural Biol.* **29**:157–167.

Kemp, J. M., and Powell, T. P. S., 1971, The termination of fibers from the cerebral cortex and thalamus upon dendritic spines in the caudate nucleus: A study with the Golgi method, *Philos. Trans. R. Soc. London Ser. B.* **262**:429–439.

Killackey, H. P., and Akers, R. M., 1980, Patterns of corticocortical fiber development in the neonatal rat, *Soc. Neurosci. Abs.* **6**:638.

Killackey, H. P., and Belford, G. R., 1979, The formation of afferent patterns in the somatosensory cortex of the neonatal rat, *J. Comp. Neurol.* **183**:285–304.

Killackey, H. P., and Belford, G. R., 1980, Central correlates of peripheral pattern alterations in the trigeminal system of the rat, *Brain Res.* **183**:205–210.

Killackey, H. P., Ivy, G. O., and Cunningham, T. J., 1978, Anomalous organization of SMI somatotopic map consequent to vibrissae removal in the newborn rat, *Brain Res.* **155**:136–140.

Krech, P., Rosenzweig, M., and Bennett, E., 1963, Effects of complex environment and blindness on rat brain, *Arch. Neurol.* **8**:403–412.

Landmesser, L. T., 1980, The generation of neuromuscular specificity, *Annu. Rev. Neurosci.* **3**:279–302.

Lee, K., and Lynch, G., 1982, Axo-somatic synapses in the normal and x-irradiated dentate gyrus: Factors affecting the density of afferent innervation, *Brain Res.* **249**:51–56.

Lee, K., Stanford, E., Cotman, C., and Lynch, G., 1977, Ultrastructural evidence for bouton sprouting in the dentate gyrus of adult rats, *Exp. Brain Res.* **29**:475–485.

Lee, K., Schottler, F., Oliver, M., and Lynch, G., 1980, Brief bursts of high frequency stimulation produce two types of structural change in rat hippocampus, *J. Neurophysiol.* **44**:247–258.

Lee, K. S., Oliver, M., Schottler, F., and Lynch, G., 1981, Electron microscopic studies of brain slices: The effects of high frequency stimulation on dendritic ultrastructure, in: *Electrical Activity in Isolated CNS Preparations*

(G. Kerkut, ed.), pp. 189–212, Academic Press, London.

Leong, S. K., 1983, Localizing the corticospinal neurons in neonatal, developing and mature albino rat, *Brain Res.* **265:**1–9.

LeVay, S., Wiesel, T., and Hubel, D., 1980, The development of ocular dominance columns in normal and visually deprived monkeys, *J. Comp. Neurol.* **191:**1–51.

Loy, R., Lynch, G., and Cotman, C., 1977, Development of afferent lamination in the fascia dentata of the rat, *Brain Res.* **121:**229–243.

Lund, R. D., and Lund, J. S., 1971, Synaptic adjustment after deafferentation of the superior colliculus of the rat, *Science* **171:**804–807.

Lund, R. D., and Mitchell, D. E., 1979, Asymmetry in the visual callosal connections of strabismic cats, *Brain Res.* **167:**176–179.

Lund, R. D., Cunningham, T. S., and Lund, J. S., 1973, Modified optic projections after unilateral eye removal in young rats, *Brain Behav. Evol.* **8:**51–72.

Lund, R. D., Mitchell, D. E., and Henry, G. H., 1978, Squint-induced modification of callosal connections in cats, *Brain Res.* **144:**169–172.

Lynch, G., 1976, Neuronal and glial responses to the destruction of input: The "deafferentation syndrome," in: *Cerebrovascular Diseases* (P. Scheinberg, ed.), pp. 209–227, Raven Press, New York.

Lynch, G., 1983, The cell biology of neuronal plasticity: Implications for mental retardation, in: *Curative Aspects of Mental Retardiation: Biomedical and Behavioral Advances* (F. Menolascino, R. Neman, and J. Stark, eds.), pp. 99–109, Paul H. Brooks Publishing Co., Baltimore.

Lynch, G., Matthews, D. A., Mosko, S., Parks, T., and Cotman, C. W., 1972, Induced acetylcholinesterase-rich layer in rat dentate gyrus following entorhinal lesions, *Brain Res.* **42:**311–318.

Lynch, G., Deadwyler, S., and Cotman, C. W., 1973*a*, Postlesion axonal growth produces permanent functional connections, *Science* **180:**1364–1366.

Lynch, G., Mosko, S., Parks, T., and Cotman, C., 1973*b*, Relocation and hyperdevelopment of the dentate gyrus commissural system after entorhinal lesions in immature rats, *Brain Res.* **50:**174–178.

Lynch, G., Stanfield, B., and Cotman, C. W., 1973*c*, Developmental differences in postlesion axonal growth in the hippocampus, *Brain Res.* **59:**155–168.

Lynch, G., Rose, G., Gall, C., and Cotman, C. W., 1975, The response of the dentate gyrus to partial deafferentation, in: *The Golgi Centennial Symposium* (M. Santini, ed.), pp. 305–317, Raven Press, New York.

Lynch, G., Gall, C., Rose, G., and Cotman, C. W., 1976, Changes in the distribution of the dentate gyrus associational system after unilateral and bilateral entorhinal lesions in adult rats, *Brain Res.* **110:**57–71.

Lynch, G., Gall, C., and Cotman, C., 1977, Temporal parameters of axon "sprouting" in the brain of the adult rat, *Exp. Neurol.* **54:**179–183.

Lynch, G., Halpain, S., and Baudry, M., 1982, Effects of high-frequency stimulation on glutamate receptor binding studied with an *in vitro* hippocampal slice preparation, *Brain Res.* **244:**101–111.

Matthews, D. A., Cotman, C., and Lynch, G., 1976, An electron microscopic study of lesion induced synaptogenesis in the dentate gyrus of the adult rat. II. Reappearance of morphologically normal synaptic contacts, *Brain Res.* **115:**23–41.

McWilliams, J. R., and Lynch, G., 1979, Terminal proliferation in the partially deafferented dentate gyrus: Time course for the appearance and removal of degeneration and the replacement of lost terminals, *J. Comp. Neurol.* **187:**191–198.

McWilliams, J. R., and Lynch, G., 1983, Rate of synaptic replacement in denervated rat hippocampus declines precipitously in the juvenile period to adulthood, *Science* **221:**572–574.

McWilliams, J. R., and Lynch, G., 1984, Synaptic density and axonal sprouting in rat hippocampus: Stability in adulthood and decline in late adulthood, *Brain Res.* **294:**152–156.

Moore, R. Y., Bjorklund, A., and Stenevi, U., 1971, Plastic changes in the adrenergic innervation of the rat septal area in response to denervation, *Brain Res.* **33:**13–35.

O'Leary, D. D. M., Stanfield, B. B., and Cowan, W. M., 1981, Evidence that the early postnatal restriction of the cells of origin of the callosal projection is due to the elimination of axonal collaterals rather than to the death of neurons, *Dev. Brain Res.* **1:**607–617.

Parnavelas, J., Lynch, G., Brecha, N., Cotman, C., and Globus, A., 1974, Spine loss and regrowth in the hippocampus following deafferentation, *Nature (London)* **248:**71–73.

Purpura, D. P., and Hausepian, E. M., 1961, Morphological and physiological properties of chronically isolated immature neocortex, *Exp. Neurol.* **4:**377–401.

Raisman, G., 1969, Neuronal plasticity in the septal nuclei of the adult rat, *Brain Res.* **24:**25–48.

Raisman, G., and Field, P., 1973, A quantitative investigation of the development of collateral regeneration after partial deafferentation of the septal nuclei, *Brain Res.* **50:**241–264.

Rakic, P., 1976, Prenatal genesis of connections subserving occular dominance in the rhesus monkey, *Nature (London)* **261:**467–471.

Rakic, P., 1977, Prenatal development of the visual system in rhesus monkey, *Phil. Trans. R. Soc. Lond. Ser. B.* **278:**245–260.

Rakic, P., 1981, Development of visual centers in the primate brain depends on binocular competition before birth, *Science* **214:**928–929.

Rakic, P., and Riley, K. P., 1983*a*, Overproduction and elimination of retinal axons in the fetal rhesus monkey, *Science* **219:**1441–1444.

Rakic, P., and Riley, K. P., 1983*b*, Regulation of axon number in primate optic nerve by prenatal binocular competition, *Nature (London)* **305:**135–137.

Ramon y Cajal, S., 1959, *Degeneration and Regeneration*

of the Nervous System (R. M. May, translation), reprinted by Hafner, New York.

Rhoades, R. W., and Dellacroce, D. D., 1980, Neonatal enucleation induces an asymmetric pattern of visual callosal connections in hamsters, *Brain Res.* **202**:189–195.

Rose, G., Lynch, G., and Cotman, C., 1976, Hypertrophy and redistribution of astrocytes in the deafferented dentate gyrus, *Brain Res. Bull.* **1**:87–93.

Rosenzweig, M. R., Bennett, E. L., and Diamond, M. C., 1972, Chemical and anatomical plasticity of brain: Replications and extensions, in: *Macromolecules and Behavior,* 2nd ed. (J. Gaito, ed.), pp. 205–277, Appleton-Century-Crofts, New York.

Rothblat, L., and Schwartz, M. L., 1979, The effect of monocular deprivation on dendritic spines in visual cortex of young and adult albino rats: Evidence for a sensitive period, *Brain Res.* **161**:156–161.

Ruttledge, L. T., 1976, Synaptogenesis: Effects of synaptic use, in: *Neural Mechanisms of Learning and Memory* (M. R. Rosenzweig and E. L. Bennett, eds.), pp. 329–339, MIT Press, Cambridge, Massachusetts.

Ruttledge, L. T., Wright, C., and Duncan, J., 1974, Morphological changes in pyramidal cells of mammalian neocortex associated with increased use, *Exp. Neurol.* **44**:209–228.

Ryugo, D., Ryugo, R., Globus, A., and Killacky, H., 1975, Increased spine density in auditory cortex following visual or somatic deafferentation, *Brain Res.* **90**:143–146.

Schapiro, S., and Vulkovich, K. R., 1970, Early experience effects upon cortical dendrites. A proposed model for development, *Science* **167**: 292–294.

Schneider, G. E., 1970, Mechanisms of functional recovery following lesions of visual cortex or superior colliculus in neonate and adult hamsters, *Brain. Behav. Evol.* **3**:295–323.

Sengelaub, D. R., and Finlay, B. L., 1980, Retinal ganglion cell death during normal development in the Syrian Hamster, *Soc. Neurosci. Abs.* **6**:290.

Shatz, C. J., 1979, Abnormal connections in the visual system of Siamese cats, *Soc. Neurosci. Symp.* **4**:121–141.

Silver, J., Lorenz, S. E., Wahlsten, D., and Coughlin, J., 1982, Axonal guidance during development of the great cerebral commissures: Descriptive and experimental studies, in vivo, on the role of preformed glial pathways, *J. Comp. Neurol.* **210**:10–29.

Sotelo, C., Hillman, D., Zamora, A., and Llinas, R., 1975, Climbing fiber deafferentation: Its action on Purkinje cell dendritic spines, *Brain Res.* **98**:574–581.

Stanfield, B. B., O'Leary, D. D. M., and Fricks, C., 1982, Selective collateral elimination in early postnatal development restricts cortical distribution of rat pyramidal tract neurons, *Nature (London)* **298**:371–373.

Staubli, V., Gall, G., and Lynch, G., 1984, The distribution of the commissural-associational afferents of the dentate gyrus after perforant path lesions in one-day-old rats, *Brain Res.* **292**:156–159.

Stenevi, U., Bjorklund, A., and Moore, R. Y., 1972, Growth of intact central adrenergic axons in the denervated lateral geniculate body, *Exp. Neurol.* **35**:290–299.

Steward, O., Cotman, C. W., and Lynch, G., 1973, Reestablishment of electrophysiologically functional entorhinal cortical input to the dentate gyrus deafferented by ipsilateral entorhinal lesions: Innervation by the contralateral entorhinal cortex, *Exp. Brain Res.* **18**:396–414.

Steward, O., and Vinsant, S., 1983, Terminal proliferatioı and reactive synaptogenesis in the rat dentate gyru after entorhinal lesions. An electron microscopic stud} *J. Comp. Neurol.* **214**:370–386.

Uylings, H. B. M., Kuypers, K., Diamond, M. C., an Veltman, W. A. M., 1978, Effects of differential env ronments on plasticity of dendrites of cortical pyram dal neurons in adult rats, *Exp. Neurol.* **62**:658–677.

Valverde, F., 1971, Rate and extent of recovery from dark rearing in the visual cortex of the mouse, *Brain Res.* **33**:1–11.

Valverde, F., and Ruiz-Marcos, A., 1969, Dendritic spines in the visual cortex of the mouse. Introduction to a mathematical model, *Exp. Brain Res.* **8**:269–283.

Varon, C., and Saier, M., 1975, Culture techniques and glial-neuronal interrelationships in vitro, *Exp. Neurol.* **48**:135–162.

Welker, C., 1976, Receptive fields of barrels in the somatosensory neocortex of the rat, *J. Comp. Neurol.* **166**:173–190.

West, J., Deadwyler, S., Cotman, C., and Lynch, G., 1975, Time-dependent changes in commissural field potentials in the dentate gyrus following lesions of the entorhinal cortex in adult rats, *Brain Res.* **97**:215–233.

Westrum, L., and Black, R., 1971, Fine structural aspects of the synaptic organization of the spinal trigeminal nucleus (pars interpolaris) of the cat, *Brain Res.* **25**:265–288.

White, L. E., Jr., and Westrum, L. E., 1964, Dendritic spine changes in prepyriform cortex following olfactory bulb lesions—Rat, Golgi method, *Anat. Rec.* **148**:410–411.

Wise, S. P., and Jones, E. G., 1976, Organization and postnatal development of the commissural projection in the rat somatic sensory cortex, *J. Comp. Neurol.* **168**:313–343.

Wise, S. P., and Jones, E. G., 1978, Developmental studies of thalamocortical and commissural connections in the rat somatic sensory cortex, *J. Comp. Neurol.* **178**:187–208.

Zimmer, J., 1973, Extended commissural and ipsilateral projections in postnatally dentorhinated hippocampus and fascia dentata demonstrated in rats by silver impregnation. *Brain Res.* **64**:293–311.

Zimmer, J., 1974, Proximity as a factor in the regulation of aberrant growth in postnatally deafferented fascia dentata, *Brain Res.* **72**:137–142.

16

Sexual Differentiation of the Brain

PAMELA C. B. MACKINNON and BEN GREENSTEIN

1. Introduction

It is abundantly clear that men and women are different—different in their genetic composition, their structure, and their behavior. Whether some of these differences can be attributed to differentiation of CNS mechanisms, brought about as a result of changes in the hormonal environment at a critical stage of brain development, is an important and intriguing question. But before considering this issue, it is as well to recall all the fundamental bases of the differences between sexes.

The genetic basis of sex determination and differentiation in mammals has been well documented; experiments have shown that the development of a testis from the indifferent gonad depends on the presence of a gene or set of genes on the Y chromosome (Welshons and Russell, 1959). It has recently been suggested that the product of such a gene, or set of genes, may be a protein with antigenic properties (the histocompatibility antigen), which is present on the surface of the germ cells (Wachtel *et al.*, 1975; Bennett *et al.*, 1975).

The process of testicular differentiation is characterized by a rapid proliferation of medullary cords of cells that attract and surround the germ cells at the same time as a regression of cortical tissue is taking place. In the absence of the Y chromosome, a prospective ovary is formed by a much slower proliferation of cortical elements and a regression of medullary tissue (Witschi, 1962). Once the developing testis begins to function, a local diffusion of androgens takes place that is responsible for differentiating the male (or wolffian) reproductive tract (Siiteri and Wilson, 1974), while a further nonandrogenic testicular factor, possibly protein in nature (Josso, 1972, 1973), causes the presumptive female or mullerian tract to regress. By contrast, the development of the female reproductive tract does not require a hormonal stimulus as experiments with antiandrogens have shown (Elgers, 1966; Neumann *et al.*, 1966; see also Neumann *et al.*, 1970); thus, in the absence of either gonad, a female type of reproductive tract will develop (Jost, 1970; Price, 1970; Jost *et al.*, 1973).

At a later stage of embryonic growth, when the circulatory system is becoming established, testicular hormones are carried to the region of the urogenital sinus, where masculinization of the external genitalia takes place. Again, in the absence of testes or of ovarian tissue, the ultimate appearance will be female in type.

Of particular interest from the point of view of the present chapter is the considerable body of evidence, much of it related to rodents, which shows that the presence of testicular hormones at an early stage of neural development also leads to sexual differentiation of those mechanisms that are responsible for the control of both gonadotropin output and sexual behavior (Harris, 1970; Breedlove *et al.*, 1982). The early presence of androgens ensures an almost constant or tonic output of gonadotropins in adult life and predisposes the animal toward certain male patterns of behavior. On the other hand, the lack of such an early hormonal stimulus results in a superimposed cyclic or phasic output of gona-

PAMELA C. B. MACKINNON • Department of Human Anatomy, University of Oxford, Oxford OX1 3QX England. **BEN GREENSTEIN** • Department of Pharmacology, St. Thomas's Hospital Medical School, University of London, London SE1 7EH, England.

dotropins and in certain types of behavior that are characteristically female. Stated in a somewhat different way, the presence of a testis is required to prevent certain functions and activities that are typically female, while in mammals the presence of ovaries does not appear to be essential for the normal development of the reproductive tract, external genitalia, and brain of the female (Figure 1). However, it is worth noting that a role for adrenal progesterone in the organization of the neonatal female brain has been suggested (see Shapiro *et al.*, 1976*a,b*).

A fascinating aspect of sexual evolution and development is that despite the normal dependence of testicular formation and secretion on the presence of a Y chromosome, the masculinizing action of testosterone is, according to Ohno (1971), mediated by a single product of a gene on the X chromosome. Such a product (or products) may be represented by a protein that forms part of both the cytosol and nuclear androgen receptors and on which the manifestations of the male phenotype depend (Fang *et al.*, 1969; Mueller *et al.*, 1972; King and Mainwaring, 1974; Kato, 1975).

In general, the body of work in this field suggests that the overall pattern of sexual development is similar in both the primate and the nonprimate. Nevertheless, the consequences of fetal sexual differentiation of the brains and spinal cord of lower species such as rodents are more dramatic in terms of subsequent sexual function in adult life. There is, however, evidence for sexual differentiation of the primate brain, including that of humans, and there are intriguing implications for behavioral patterns, both sexually and socially.

2. Evidence for Sexual Differentiation of the Brain in Nonprimates

2.1. Gonadal Steroids and the Brain

2.1.1. Effects of Neonatal Steroids

The most important contribution to the early literature on sexual differentiation of the brain was provided by Pfeiffer (1936). Principally he showed that when genetic male and female rats castrated at birth were allowed to grow up and were then transplanted with an ovary, corpora lutea (indicative of cyclic gonadotropin secretion and ovulation) would be formed. On the other hand, when male rats were castrated at birth in which testicular tissue was transplanted to the neck upon reaching maturity, ovarian grafts failed to show true corpora lutea. Similarly, many females that received a testicular transplant at birth lost the capacity to form corpora lutea at adulthood and, moreover, the vaginal epithelium of these animals was in an almost constant state of vaginal cornification. From this work Pfeiffer concluded that rats of either sex were born with undifferentiated pituitary tissue capable of secret-

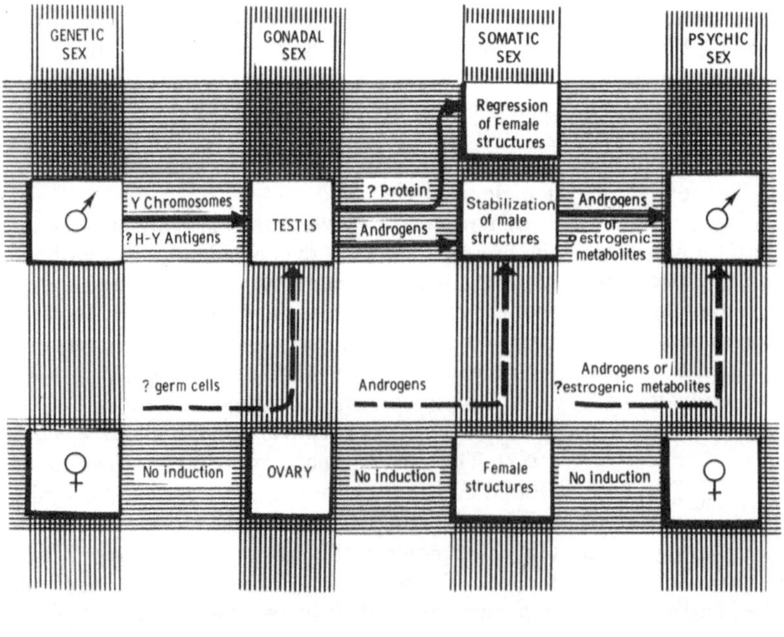

Fig. 1. Schematic representation of the different stages of sexual differentiation. (After Dr. N. Neumann.)

ing gonadotropins in a cyclic fashion; however, if, at an early stage of development, the pituitary gland was exposed to testicular secretion, the tissue became differentiated and unable to support cyclic ovarian and reproductive function. At the time that this concept was proposed, the importance of the hypothalamohypophysial portal system had not been appreciated, still less the possibility of pituitary hormone-releasing factors.

Almost two decades later, Harris and Jacobsohn (1952) in a classic experiment based on earlier work by Greep (1936), showed that male pituitary tissue when transplanted to the sella turcica of hypophysectomized females, gained a specific blood supply from the median eminence and was thereafter capable of sustaining normal female reproductive function. With confirmation of this work in mice (Martinez and Bittner, 1956) and the accumulation of further evidence derived from pituitary transplantation or stalk section on the one hand and the effects of brain stimulation or lesioning on the other, the importance of the CNS in the control of gonadotropin output and reproductive activity was substantiated (Harris, 1955).

The tide of scientific thought then turned to the consideration of the brain rather than the anterior pituitary as a possible site of sexual differentiation. Subsequent work was greatly facilitated when Barraclough and Leathem (1954), using mice, and Barraclough (1961), using rats, showed that a single injection of testosterone propionate (TP) into neonates sufficed to "masculinize" the female and cause postpubertal anovulation and constant vaginal estrus. This simple technique permitted a definition of the presumptive period of neural plasticity during which sexual differentiation takes place, i.e., the maximal effects of a single TP (1.25 mg) injection were found to extend from the late stages of fetal life (H. E. Swanson and van der Werff ten Bosch, 1964, 1965) to the fifth postnatal day, while the effects became progressively less effective between 5 and 10 days of age, and after about 10 days of age only very heavy doses caused sterility (Brown-Grant, 1974*a*). However, if much smaller doses of TP (e.g., 0.125 mg) were given between 0 and 5 days of age, the onset of sterility in the adult female rat was greatly delayed and was preceded by a period of apparently normal cyclicity and reproductive activity (H. E. Swanson and van der Werff ten Bosch, 1965). This effect, which does not occur if a testis is transplanted to a normal female neonate, has been termed the *delayed anestrous syndrome* and is thought to be caused by an incomplete or partial differentiation of neurons by a subthreshold dose of androgen (Gorski, 1968).

The results of studies relating to the period of sexual differentiation of neural tissue in male rats were commensurate with those in neonatally androgenized females. Yazaki (1960) castrated groups of male rats of different ages from birth to 40 days of age and subsequently grafted ovarian and vaginal tissue beneath the abdominal skin. Daily vaginal smears and ovarian histology showed the occurrence of 4-day ovarian cycles in most of the animals that had been castrated before the third day of life, whereas in animals castrated later in life vaginal cornification and polyfollicular ovaries persisted.

The effect of testosterone on neonatal females in causing postpubertal sterility can be elicited with equal facility by injecting instead its immediate precursor, androstenedione (Stern, 1969; Luttge and Whalen, 1970*a;* Edwards, 1971). With respect to the breakdown products of testosterone, however, the ring A reduced metabolite, dihydrotestosterone (DHT), appears incapable of causing masculinization of the brain (Luttge and Whalen, 1970*a;* Brown-Grant *et al.*, 1971), while the ring A aromatized products; i.e., estrogens (Figure 2) are far more potent than testosterone. A single injection of estradiol benzoate (EB) will cause anovulation and constant vaginal estrus in adulthood even when a dose equivalent to only 1/10 that required for testosterone to have the same effect is given (Gorski, 1968; Gorski and Barraclough, 1963). A persuasive argument in favor of the concept that estrogen, rather than testosterone, is the ultimate effective agent in masculinizing the brain is the observation that pretreatment of neonatal female rats with antiestrogens blocks the effects of testosterone (MacDonald and Doughty, 1972). However, in one study, hypothalamic implants of an estrogen antagonist in the neonatal brain failed to block neonatal androgen sterilization (Hayashi, 1976); other evidence that runs counter to the estrogen hypothesis has been discussed by Brown-Grant (1972). Masculine differentiation of a sex center in the lumbar cord of the rat, however, is clearly dependent on testosterone rather than on estrogen (Breedlove *et al.*, 1982).

Apart from the masculinizing effects on adult female rats of a single small dose of EB in neonatal life, repeated small doses of EB during the first few days of life (Arai, 1971), or a single large dose on the fourth postnatal day (Brown-Grant, 1974*a*) can cause a much more severe impairment of gonado-

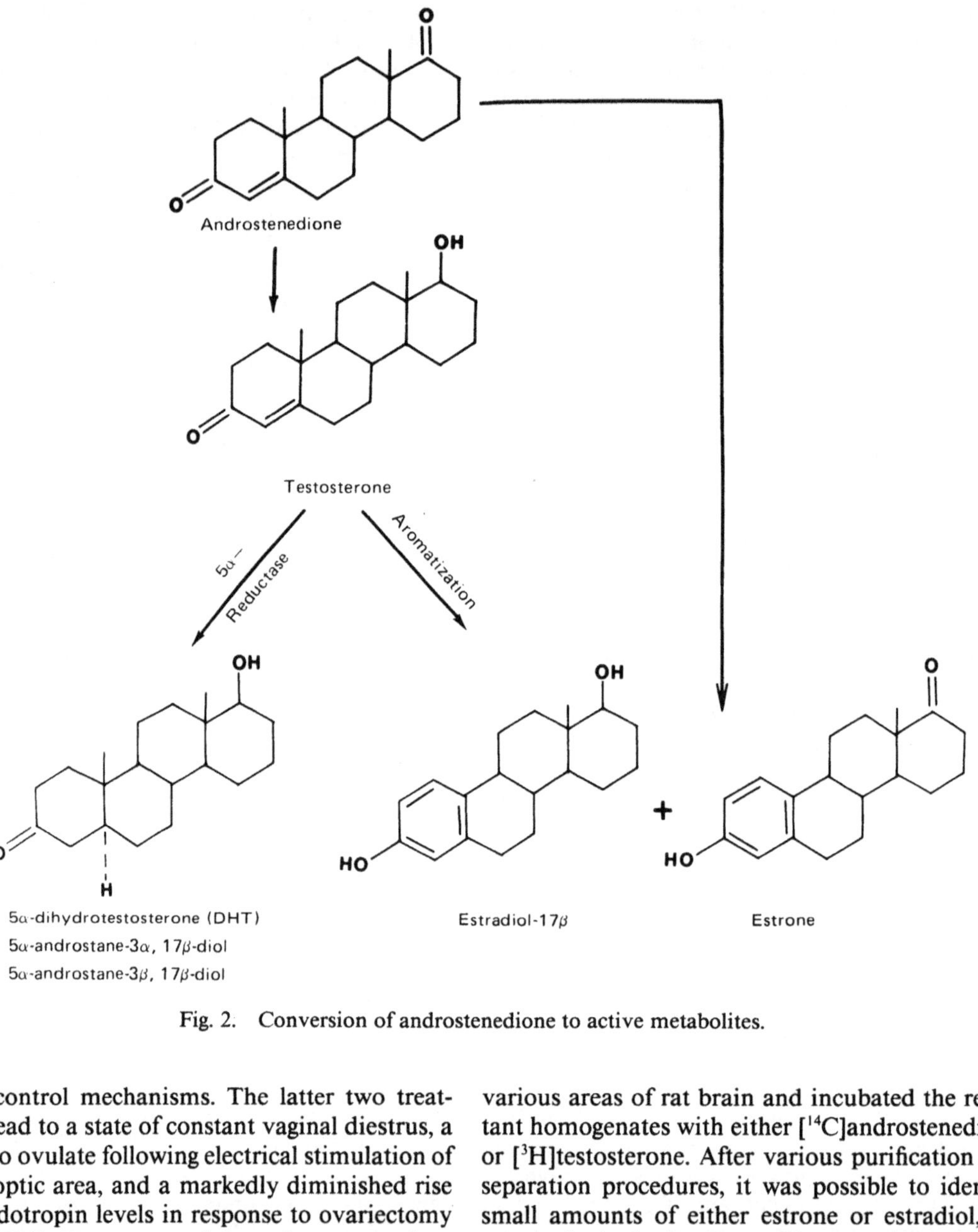

Fig. 2. Conversion of androstenedione to active metabolites.

tropin control mechanisms. The latter two treatments lead to a state of constant vaginal diestrus, a failure to ovulate following electrical stimulation of the preoptic area, and a markedly diminished rise in gonadotropin levels in response to ovariectomy (Arai, 1971). Since none of these features is characteristic of the female that has been masculinized either by TP or by a single small dose of EB, the possibility remains that experimental high levels of estrogen may differentiate (or perhaps damage) neural or extraneural sites additional to those at which testosterone or small doses of estrogen may normally be expected to act.

Strong support for a possible role of estrogen in "masculinizing" the brain has been provided by Naftolin *et al.* (1975). These workers homogenized various areas of rat brain and incubated the resultant homogenates with either [^{14}C]androstenedione or [^{3}H]testosterone. After various purification and separation procedures, it was possible to identify small amounts of either estrone or estradiol, respectively, in the incubates. These results suggested that aromatization of the androgens had taken place, and subsequent work substantiated this view. The presence of aromatizing systems in the hypothalamus and the limbic system (i.e., the hippocampus and amygdala) has been demonstrated not only in the guinea pig and rabbit but in the rhesus monkey and the human fetus as well. The data obtained from *in vitro* preparations was later supported by *in vivo* studies (Knapstein *et al.*, 1968; Weisz and Gibbs, 1974). Of particular interest were those in which an isolated cranial vault containing the brain and pituitary gland of immature male rhesus monkeys were perfused. In this preparation, both free and conjugated [^{3}H]estrone and free [^{3}H]estradiol were identified in the venous effluates after injection of [^{3}H]androstenedione into the perfusate; moreover, dissection and monitoring of the tissues showed that these labeled steroids were present only in the hypothalamus and the limbic system (Flores *et al.*, 1973).

2.1.2. Plasma Androgen and Estrogen Levels in the Fetal Rat

A fundamental requirement for the postulation of androgen and/or estrogen action in the fetal animal is the demonstration of their presence in the circulation or in the target cell. Systematic radioimmunoassay measurements of circulating testosterone have been made in rats. Despite conflicting reports of absolute values, it appears that both male and female fetuses possess similar circulating concentrations of testosterone during the neonatal "critical period" for sexual differentiation of the brain (Resko *et al.*, 1968; Döhler and Wuttke, 1975). Although the last mentioned investigators described a surge of testosterone on day 18 of gestation in the male only, circulating levels of testosterone were found to be relatively low (0.5 ng/ml) in comparison with those measured in young adult male rats (2 ng/ml). Nevertheless, the fetal brain is capable of concentrating the steroid or its metabolites (see Section 2.2.1), and concentrations of testosterone in the target cell may well be within the adult range.

If indeed estrogen is the effective hormone that masculinizes the brain, then it is curious that infant rats do not succeed in sterilizing themselves, since total plasma estrogen levels are very high in the infant female (e.g., see Ojeda *et al.*, 1975). However, this state of affairs may in part be accounted for by the presence of a protein; i.e., α-fetoprotein (AFP), which has a high affinity for estrogen but does not bind testosterone.

AFP is present in the plasma of neonatal rats and may help sequester the hormone (Raynaud *et al.*, 1971; Nunez *et al.*, 1971; Puig-Duran *et al.*, 1978). An additional factor may be the presence of intracellular AFP, which has been localized in presumed selective neurons of the hypothalamus and amygdala of the developing rat brain by means of the unlabeled antibody peroxidase–antiperoxidase technique (Benno and Williams, 1978). A further feature in developing rats is the high plasma follicle-stimulating hormone (FSH) values (see, e.g., MacKinnon *et al.*, 1976), which decline at about the same time that AFP levels are diminishing. Although there is no rigorous evidence to suggest that estrogen binding to AFP accounts entirely for the failure of negative feedback inhibition, nevertheless it is reasonable to assume that as the estimated concentrations of the biologically active estrogen are low between 0 and 23 days of age (Puig-Duran *et al.*, 1978), while its affinity for AFP is very strong, the resultant activity of the hormone is too weak either to inhibit FSH levels or to masculinize the brain. The possible role of AFP in this respect is shown in Figure 3.

Since testosterone has been detected in the circulation of the fetal female rat, it seems simplistic to assume a scenario in which perinatal sexual differentiation of the rat brain is caused by any one steroid hormone. Furthermore, there are grounds for believing that active "feminization" of the female brain may occur as well, especially with regard to sexual behavior.

2.2. Neural Sites of Steroid Action

2.2.1. The Mechanism of Steroid Action

Increasing knowledge of steroid action in various tissues has provided a powerful adjunct to the study of steroid hormone action in the brain (see King and Mainwaring, 1974; Gorski and Gannon, 1976). The current view is that steroids travel in the

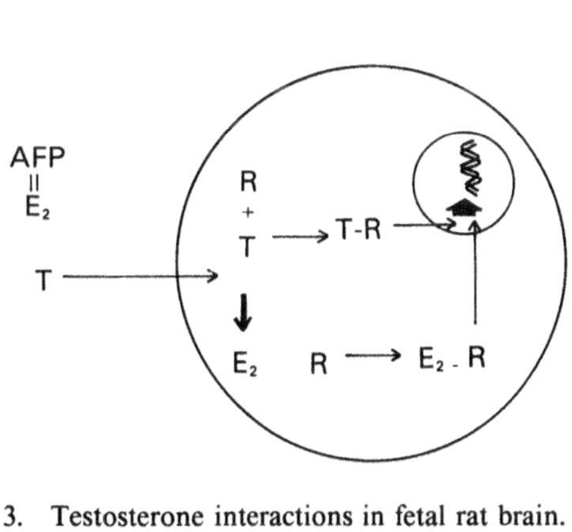

Fig. 3. Testosterone interactions in fetal rat brain. Circulating estradiol (E_2) but not testosterone (T) is bound to plasma α-fetoprotein (AFP). T diffuses into the cell and combines with its receptor. T is also metabolized to E_2, which binds with its receptor. Both complexes enter the cell nucleus to modify neural growth.

circulation largely bound to plasma proteins. On arrival at a target tissue, the steroid dissociates rapidly from the plasma protein and diffuses into the cell, where it combines with intracellular cytoplasmic receptor proteins. The steroid–receptor complex enters the nucleus and interacts with the nuclear chromatin to alter transcription and subsequent protein synthesis (Figure 4). These receptor systems concentrate the steroid, and radioactive steroids of high specific activity can be used to locate the tissue distribution of the receptors.

2.2.2. *Androgen Receptors in Neonatal Rat Brain*

To identify the possible site or sites at which androgens may act on brain mechanisms, Sar and Stumpf (1972) investigated the localization of high specific activity [^{3}H]testosterone in different areas of the brain using the technique of dry-mount autoradiography. These workers showed that in both immature and adult male rats, radioactivity concentrated and was retained in cell nuclei of certain hypothalamic nuclei, in addition to the cortical and medial nuclei of the amygdala and the subiculum of the hippocampus (Figure 5). Although the chemical nature of the radioactivity in the brain was not determined in this study, the administration of an antiandrogen, which reduced the retention and concentration of the label in the tissue, supported the possibility of its still being attached to testosterone, rather than a product of its metabolism. The presence of cytosolic and nuclear receptors for dihydrotestosterone (DHT), a 5α-reduced metabolite

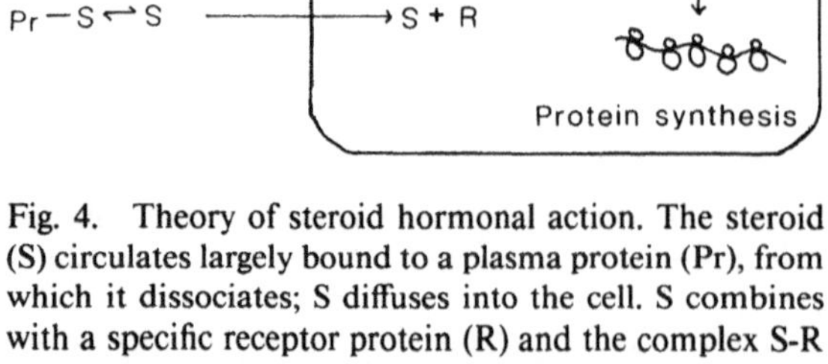

Fig. 4. Theory of steroid hormonal action. The steroid (S) circulates largely bound to a plasma protein (Pr), from which it dissociates; S diffuses into the cell. S combines with a specific receptor protein (R) and the complex S-R interacts with the genome to modify transcription and subsequent protein synthesis.

of testosterone, have been found in the hypothalamus and amygdala of both neonatal males and females (Kato *et al.*, 1974; Barley *et al.*, 1975). But DHT is relatively inert as an agent for brain masculinization, and more significant was the demonstration of receptors for [^{3}H]testosterone in the perinatal rat brain (Lieberburg *et al.*, 1980).

In an attempt to localize the intracellular effects of neonatally administered androgens, Clayton *et al.* (1970) studied the incorporation of [^{3}H]uridine into RNA in the brain of 2-day-old female rats that had been injected 4 hr before death with a low (nonmasculinizing) dose of testosterone. In autoradiographs of brain sections, the uptake of uridine in most parts of the brain was found to be reduced in comparison with uninjected control females; in the preoptic and amygdaloid areas, however, the reduction of uridine uptake was significantly less than that observed in the rest of the brain. The interpretation of these data was that the administered androgens had affected RNA mechanisms at those sites which, in the male, would normally be expected to differentiate. The possibility that mechanisms of DNA, RNA, and probably also protein synthesis are involved at an early stage in the process of sexual differentiation was supported by Salaman and Birkett (1974), who were able to protect neonatal female rats from the effects of TP by the additional injection of α-amanitin, an inhibitor of nucleoplasmic or messenger–precursor RNA synthesis; hydroxyurea, an inhibitor of DNA synthesis with no effect on overall RNA synthesis, was also able to provide a high degree of protection. On the other hand, inhibition of protein synthesis was less effective, while inhibition of ribosomal RNA synthesis alone was apparently completely ineffective.

2.2.3. *Estrogen Receptors in Neonatal Rat Brain*

Dry-mount autoradiographs of brain sections obtained from adult female rats previously injected with [^{3}H]estradiol-17β have shown that the isotope concentrates in the cell nucleus of several different hypothalamic nuclei (Stumpf, 1968), in the medial preoptic nucleus, and in nuclei of the corticomedial complex of the amygdala (Stumpf and Sar, 1971). Similar accumulations of [^{3}H]estradiol-concentrating neurons have been seen in the mouse, tree shrew, and squirrel monkey (Stumpf *et al.*, 1974). A study of [^{3}H]estradiol uptake in the ovariectomized rhesus monkey has also shown concentrations of the steroid in the medial preoptic-anterior

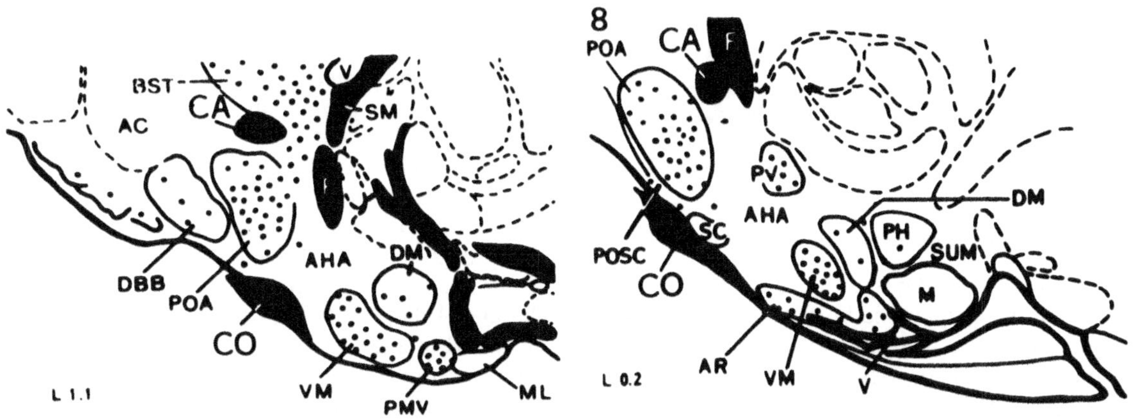

Fig. 5. Schematic drawings prepared after serial section autoradiographs from immature intact and mature castrated male rats 1 hr after the injection of 0.5–1.0 μg of [1,2-^{3}H]testosterone, specific activity 50 Ci mM. Sagittal sections according to the atlas of DeGroot at the level of L 1.1 at the level of 0.2. The black dots represent areas of concentration of neurons radiolabeled after [^{3}H]testosterone administration. AC, nucleus (n.) accumbens septi; AHA, area anterior hypothalami; AR, n. arcuatus; BST, n. interstitialis striae terminals (bed nucleus); CA, commissura anterior; CO, chiasma opticum; DBB, gyrus diagonalis (diagonal band of Broca); DM, n. dorsomedialis hypothalami; F, fornix; M, n. mammillaris medialis; ML, n. mammillaris lateralis; PH, n. posterior hypothalami; PMV, n. premammillaris ventralis; POA, area preoptica; POSC, n. preopticus suphrachiasmaticus; PV, n. paraventricularis hypothalami; SC, n. suprachiasmaticus; SM, stria medullaris thalami; SUM, area supramammilaris; V, ventricle; VM, n. ventromedialis hypothalami. (From Sar and Stumpf, 1972.)

hypothalamic areas, the ventromedial n., arcuate n., medial n. of amygdala, bed n. of the stria terminalis, and both basophil and acidophil cells of the anterior pituitary (Pfaff *et al.,* 1976).

In 2-day-old female rats, a similar pattern of uptake was observed following injection of labeled estrogen, while competition experiments showed that nuclear concentration of radioactivity was inhibited by either "cold" estradiol or testosterone (except in the medial preoptic area where it was only partly reduced) but not by DHT (Sheridan *et al.,* 1974). This characteristic distribution overlaps to some extent with that observed in autoradiographs obtained from [^{3}H]testosterone-injected male rats, particularly with respect to the medial preoptic area (Sar and Stumpf, 1972), but this might be explained by the androgen having at least in part metabolized to estradiol. Nevertheless, the possibility that certain cells may concentrate both hormones, and that separate binding sites for both androgens and estrogens might exist, cannot be ignored. Also worth mentioning is the observation that uptake of labeled estradiol in the hypothalamus, preoptic area, and cortex (relative to that in the plasma) in gonadectomized male and female rats that had been injected with the tritiated hormone and killed 2 or 24 hr later was the same in both sexes (Whalen and Luttge, 1970). Likewise, there were no sex differences in the regional distribution of estrogen metabolites in the brain of [^{3}H]estradiol-17β (Luttge and Whalen 1970*b*).

The presence of estrogen receptors in the cytosol fractions of brain homogenates is well documented (Kato, 1971; Kato *et al.,* 1974; see also Zigmond, 1975). Westley and Salaman (1976, 1977) demonstrated the presence of high-affinity nuclear estrogen receptors in the ventral portion of the hypothalamus and the amygdala of the neonatal rat. In the intact neonatal male or the androgen-injected neonatal female, the numbers of available (unoccupied) nuclear estrogen receptor sites in the hypothalamoamygdaloid region of the brain were significantly fewer than in 4-day-old females or 4-day-old males that had been castrated 24 hr earlier. According to Westley and Salaman (1977), these findings can be attributed to the occupation of nuclear binding sites by endogenous estrogen arising as a result of the aromatization of androgens in the hypothalamoamygdaloid area.

The ontogeny of the estrogen receptors in the fetal and neonatal rat brain has been studied (MacLusky *et al.,* 1979; White *et al.,* 1979), and these were first detected at day 21 of gestation. Significantly perhaps, nuclear estrogen receptors could be detected only in the male brain (MacLusky *et al.*, 1979).

2.3. Possible Sites of Sexual Differentiation: Mechanisms Controlling Gonadotrophin Output

2.3.1. Preoptic–Anterior Hypothalamic Area

The technique devised by Halasz and Pupp (1965), in which a curved knife held in a stereotactic instrument enabled the hypothalamus to be disconnected entirely or in part from other areas of the brain, promoted the concept that the integrity of the preoptic–anterior hypothalamic area is essential for the phasic output of gonadotropins. Much further experimentation has supported this work (see Schwartz and McCormack, 1972), and recent methods using immunohistochemical techniques have shown that GnRh-synthesizing cell bodies are localized in and around the preoptic–anterior hypothalamic area (Silverman *et al.*, 1979).

Neural mechanisms outside the mediobasal–hypothalmus–pituitary unit also appear to be concerned with gonadotropin output and the elicitation of a steroid-stimulated preovulatory-type LH surge. However, although there is evidence to suggest that the brain stem may have a part to play in these mechanisms, as a site of sexual differentiation, the preoptic area appears to be the stronger candidate. Indeed, a number of workers have provided evidence of metabolic changes (Moguilevski *et al.*, 1969; Moguilevski and Rubinstein, 1967) and of changes in nuclear and nucleolar size of nerve cells in the anterior hypothalamic–preoptic area of normal male and female rats and in experimentally treated animals (Docke and Kolocyek, 1966; Pfaff, 1966; Arai and Kusama, 1968; Dorner and Staudt, 1968). The cell nuclear size of neuron cells in this area appears to be less in male rats than in normal females and in males castrated on the first day of life. However, the animals were investigated at only one time of day, and possible changes in nuclear size due to sexual differences in circadian periodicities (see ter Haar *et al.*, 1974) were unaccounted for.

Compelling evidence for the sexual differentiation of the preoptic area has been provided by Gorski *et al.* (1978), who reported a marked sexual dimorphism in the volume of cells in this brain region (Figure 6a,b). The overall volume of this cerebral nucleus is larger in brains of adult males than in those of females, and this difference can be eliminated by neonatal castration of the male; on the other hand, injection of the neonatal female with testosterone propionate increased its size (Jacobson *et al.*, 1981). It was also established that the sexually differentiated nucleus is still forming at day 18 gestation when the testis is actively secreting testosterone (Jacobson and Gorski, 1981).

The effects of steroids on developing preoptic neurons have been demonstrated *in vitro* (Toran-Allerand, 1980). Both testosterone and its metabolite estradiol accelerated and increased the degree of neurite proliferation in explants from neonatal mouse hypothalamus and preoptic area responsive cells containing estrogen receptors (Toran-Allerand *et al.*, 1978). However, the functional role of the area has yet to be clearly elucidated.

Of considerable interest is the observation that electrical or chemical stimulation of the preoptic area will cause ovulation not only in female rats but also in neonatally androgenized females that have been kept under conditions of normal or continuous lighting (Terasawa *et al.*, 1969; Everett *et al.*, 1970). A similar effect can be obtained in adult male rats bearing ovarian grafts (Quinn, 1966). This combination of results suggests that the mechanisms lying between the preoptic area and the anterior pituitary gland that are concerned with goadotrophin output are similar in both the male and female, and it has been argued that an afferent connection to the preoptic area rather than the preoptic neurons themselves might be sexually differentiated (Everett, 1969).

By cutting the projection pathway (stria terminalis) from the amygdala to the preoptic area (POA) and allowing orthograde degeneration to occur, Raisman and Field (1971, 1973) demonstrated that strial fibers establish synaptic contacts in the neuropil of both the dorsal part of the POA and the region surrounding the ventromedial nucleus. Furthermore, most nondegenerated terminals (which were not of amygdaloid origin) were found to contact either dendritic shafts or spines in these areas. Of particular interest was the observation that in females a significantly greater number of fibers made synaptic contacts with dendritic spines in the POA than in males. No such differences were found in the ventromedial area (Table I).

The significance of this potentially important finding is not yet clear, since lesions of the sexually differentiated part of the preoptic area, although causing an acute blockade of ovulation and a high

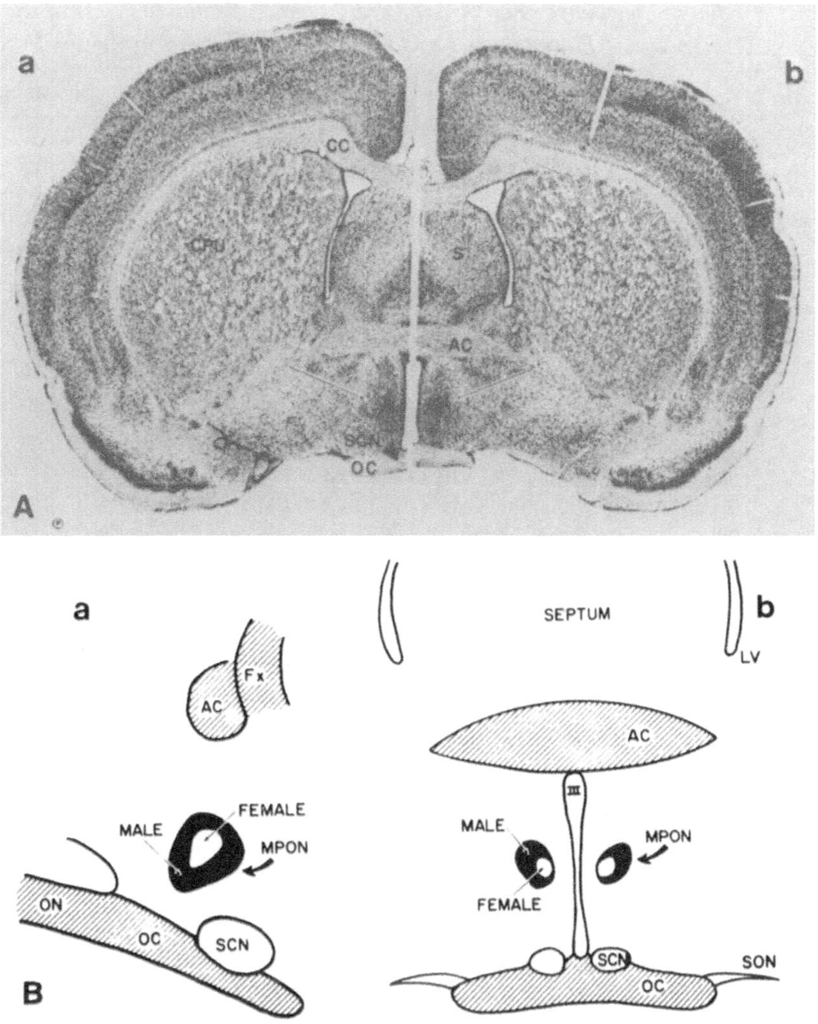

Fig. 6. (A) Coronal sections through the brain of the adult female (a) and male (b) rat sacrificed 2 weeks after gonadectomy; thionine. The arrows indicate that portion of the medial preoptic nucleus that exhibits a marked sexual dimorphism. Both at the same magnification. The absence of the suprachiasmatic nucleus (SCN) in b is an artifact of the plane of tissue sectioning. Abbreviations: AC, anterior commissure; CC, corpus callosum; CPU, caudate and putamen; OC, optic chiasma; S, septum. (From Gorsky *et al.*, 1978.) (B) Localization of the sexually dimorphic components of the medial preoptic nucleus (MPON) in the sagittal (a) and coronal (b) planes. The nucleus of the female rat is drawn completely within the volume of the nucleus of the male. Fx, fornix; LV, lateral ventricle; ON, optic nerve; SON, supraoptic nucleus; III, third ventricle; other abbreviations as in (A). Drawn from slides magnified × 37. (From Gorsky *et al.*, 1978.)

incidence of both immediate and delayed pseudopregnancies, are nevertheless commensurate with ultimate cyclicity and reproductive function (Brown-Grant and Raisman, 1972).

Investigations of a quite different nature have been reported in which single unit recordings were made from neurons in the preoptic and anterior hypothalamic areas of rats of different sexual status following electrical stimulation of the mediobasal hypothalamus and the amygdala (Dyer *et al.*, 1976). The results showed that in males or in masculinized females antidromically identified cells of the POA were more likely to receive a synaptic connection from the amygdala than were the same cells in females and neonatally castrated males. From these and other findings, Dyer and associates suggested that exposure to androgens in neonatal life might facilitate the formation of both excitatory and inhibitory synaptic contacts in the POA hypothalamic area.

Although sexual dimorphism of the POA has been shown by both neuroanatomical and neurophysiological studies, the two sets of data do not appear to be directly compatible. Dyer *et al.* (1976) found a sexually differentiated neural input from the amygdala, whereas the sexual dimorphism reported by Raisman and Field (1971) involved fibers of nonamygdaloid origin. But, as Dyer points out, "it is possible that a fixed number of amygdaloid afferents could be more effective in altering

Table I. Incidences Per Grid Square (Mean ± Standard Error) of the Four Types of Synapse in the Preoptic Area and Ventromedial Nucleus in the Six Groups of Animals[a]

	Endocrine status[b]					
	Cyclic[c]			Noncyclic[d]		
Cyclic	F(16)	F16(7)	MO(9)	M(11)	M7(7)	F4(14)
Preoptic area						
Nonstrial						
Shaft	50.0 ± 2.0	53.1 ± 3.0	54.3 ± 3.0	55.2 ± 1.4	52.6 ± 1.3	48.4 ± 2.4
Spine	5.3 ± 0.3	5.4 ± 0.5	5.0 ± 0.3	3.3 ± 0.2	3.9 ± 0.4	3.5 ± 0.2
Shaft	1.2 ± 0.1	1.0 ± 0.1	1.2 ± 0.1	1.2 ± 0.1	0.7 ± 0.1	0.9 ± 0.1
Spine	1.6 ± 0.2	1.4 ± 0.3	1.8 ± 0.2	1.7 ± 0.1	1.3 ± 0.1	1.6 ± 0.2
Ventromedial nucleus						
Nonstrial						
Shaft	46.3	44.4 ± 1.6	43.6 ± 1.6	42.4 ± 1.5	42.3 ± 1.4	42.8 ± 1.8
Spine	12.9	12.6 ± 0.8	13.2 ± 0.9	12.0 ± 0.7	13.9 ± 0.9	13.1 ± 0.9
Strial						
Shaft	2.3	1.6 ± 0.2	1.9 ± 0.3	1.9 ± 0.3	1.8 ± 0.2	1.9 ± 0.2
Spine	6.2	6.2 ± 0.6	6.6 ± 0.5	7.0 ± 0.6	6.7 ± 0.4	6.5 ± 0.5

[a]From Raisman and Field (1973).
[b]Number of rats is given in parentheses. Statistical comparisons between the mean incidence of spine synapses of nonstrial fibers in the different groups of cyclic and noncyclic animals were, in almost all cases, significant.
[c]This group consists of F (normal females), F16 (females treated with androgen on day 16), and MO (males castrated within 12 hr of birth).
[d]This group consists of M (normal males), M7 (males castrated on day 7), and F4 (females androgenized on day 4).

the firing pattern of a cell if that cell receives less spine synapses from another, nonamygdaloid, pathway."

Sexual differentiation of steroid receptors has also been reported by Rainbow *et al.* (1982), who found fewer estrogen and progestin receptors in the periventricular region of the POA in the male than in the female (see also Section 2.5.3). But whether this reflects reduced sensitivity of the male POA to these steroids remains to be seen.

2.3.2. Amygdala

With respect to sexual dimorphism of amygdaloid neurons, there is certain further, although less direct, support for the involvement of the amygdala in sexual differentiation. Staudt and Dorner (1976) reported that neuronal cell nuclear size in the medial and central parts of the amygdala was significantly larger in adult male rats castrated within 24 hr of birth than in adult intact males (with presumably normal levels of circulating androgens). The nuclear size of rats castrated at 1 day of age approximated that of the female; furthermore, the administration of neonatal androgens prevented the difference.

A sexual dimorphism of synaptic organization in the medial amygdaloid nucleus (AMN) has also been described by Nishizuka and Arai (1981). Masculinization of neonatal female rats led to increased formation of dendritic shaft synapses. In a later study, Mizukami *et al.*, (1983) compared the nuclear volume of the medial and lateral nuclei of the amygdala in male and female rats. The volume was larger in the male and in masculinized females. The presumptive role of the amygdala in the modulation of the hypothamohypophyseal axis, however, is unclear.

Electrical stimulation of the corticomedial complex of the amygdala has been shown to increase plasma LH levels and to provoke ovulation in constant-estrus females exposed to constant illumination (Velasco and Taleisnik, 1969). Furthermore, it has been reported that stimulation in the same area can cause a rise in plasma LH levels in estrogen-primed ovariectomized animals (Kawakami and Terasawa, 1973). It is therefore worth noting that LH responses to electrical stimulation were unobtainable in either males transplanted with ovarian tissue (Arai, 1971) or masculinized females that had been ovariectomized and primed with estrogen as adults (Kawakami and Terasawa, 1973).

2.3.3. Suprachiasmatic Nucleus

Remarkably scant attention has been paid to the role that the suprachiasmatic nucleus (Sch.n.), which lies within the preoptic–anterior hypothalamus area, may play in the control of the preovulatory surge in cyclic rats, more especially since electrical stimulation of the nucleus apparently causes as great an increase in the concentration of luteinizing hormone releasing factor (LHRF) in pituitary portal blood as does stimulation of the medial preoptic area (Chiappa *et al.*, 1977).

Ever since Everett and Sawyer (1950) first demonstrated a critical period of the proestrous day during which prevention of a neural event by a central nervous depressant, sodium pentobarbital (Nembutal), led to a 24-hr delay in ovulation, the importance of a daily neural signal for the maintenance of regular estrous cycles has been recognized (see Schwartz and McCormack, 1972). It is notable that lesions of the Sch.n., which receives fibers directly from the retina (Moore, 1974), prevent circadian rhythms of both neural and hormonal origin (Moore and Eichler, 1972; Raisman and Brown-Grant, 1977; Coen and MacKinnon, 1980). Furthermore, the integrity of the Sch.n. appears to be essential for the estrogen-stimulated preovulatory LH surge (Coen and MacKinnon, 1976, 1977). It is therefore tempting to think that this nucleus or mechanisms controlling its function might in some way be sexually differentiated. There is evidence from anatomical studies to support this speculation. Guldner (1982) reported that the Sch.n. of male rats contains more axospinal synapses, more postsynaptic density, and more asymmetrical synapses, while Le Blond *et al.* (1982) reported a greater degree of synaptic volume. A vasopressinergic projection from the Sch.n. also appears to be sexually dimorphic (de Vries *et al.*, 1981).

2.3.4. The Arcuate Nucleus

The arcuate nucleus (AN) (see Figure 5) is of particular interest, since in the primate it appears to be the site of neuronal cell bodies that synthesize GnRH, and the integrity of this region is critical for the normal functioning of the anterior pituitary gland. Furthermore, unlike the situation in the rodent, the POA can be removed from the primate brain without interruption of normal sexual cyclicity or behavior (see Knobil, 1980). In the rat, the AN appears to be sexually differentiated, since in females there are more spine synapses and fewer axosomatic synapses than in males. Moreover, these sex differences can be eliminated by neonatal castration of the male or by treatment of the neonatal female with testosterone propionate (Matsumoto and Arai, 1980, 1981).

2.3.5. Other Areas of the Rat Central Nervous System

Sexual differentiation of the rat brain now appears to be more generalized than was previously thought. Rethelyi (1977) measured a larger neurovascular contact surface along the median eminence and proximal pituitary stalk of adult female rats, and Akmaev and Fidelina ((1978) found sex differences in the activity of the tanycyte ependymal lining of the lateral walls and floor of the third ventrical close to the arcuate nucleus in neonatal rats.

The hippocampus, too, has not escaped attention and appears to undergo sexual dimorphism on one side only, the right hippocampus in male rats being thicker than the left and *vice versa* in the female (Diamond *et al.*, 1982). This unilateral difference appears to be independent of the neonatal hormonal milieu.

A discrete and sexually differentiated nucleus to which considerable attention has recently been paid is the spinal nucleus of the bulbocavernosus (SNB), which is situated in the ventral region of the lumbar spinal cord. It innervates two perineal muscles attached to the penis that have a sexual function. In the female, the cell bodies of this nucleus are both fewer in number and smaller (Breedlove and Arnold, 1980). In adult males, the size of the neurons depends on the presence of testosterone, which concentrates in the cell bodies; the perineal muscles are also sensitive to androgens. Exposure of the female to testosterone, but not to estrogens, shortly after birth masculinizes the nucleus–perineal muscle complex. Interestingly, the critical periods of neonatal development over which cell number and cell size are influenced by the presence of androgens are different; thus, the regulation of these two processes may be independent (Breedlove and Arnold, 1983).

2.3.6. Anterior Pituitary

The possibility of sexual dimorphism of the anterior pituitary gland has also been examined. Pituitaries of neonatally androgenized rats are capable of releasing an ovulatory quota of LH in response both to exogenous GnRH (Borvendeg *et al.*, 1972) and to stimulation of the medial preoptic area, which causes a release of endogenous releasing hormone (Gorski and Barraclough, 1963; Terasawa *et al.*, 1969; Jamieson and Fink, 1976). However, the pituitary response to GnRH is dependent on its hormonal environment (Debeljuk *et al.*, 1972; Aiyer and Fink, 1974). A study of its responsiveness after injection of GnRH in ovariectomized, androgen-sterilized rats that had been stimulated with estrogen alone, or with progesterone after estrogen priming, showed a significant reduction in the increment of LH compared with that found in either similarly treated castrated males or nonsterilized females (Barraclough and Turgeon, 1974). These results were supported by those of Fink and Henderson (1977), although not by Mennin *et al.* (1974) using different experimental protocols.

The combined results of these studies suggest that there may be differences between the pituitary responses to GnRH in rats of different sexual status under different steroid regimens, but whether these are caused by direct effects on neural mechanisms concerned with endogenous GnRH release and/or on pituitary tissue is unclear.

In an investigation of cytoplasmic estrogen receptors in the hypothalamus and anterior pituitary of adult and immature rats of both sexes and of neonatally androgen-treated females, no differences in receptor levels were found among any of the groups belonging to the same strain (Cidlowski and Muldoon, 1976). By contrast, when receptor kinetics were observed between the sexes the rate of replenishment of the receptor was apparently much faster in the androgenized female than in the untreated female, while the level 15 hr after estrogen injection was appreciably greater.

In the hypothalamus, receptor depletion was far greater in the untreated than in the androgenized animals. During the course of the same investigation, Cidlowski and Muldoon (1976) could not find evidence of a defective synthesis of estrogen receptors. These data therefore fail to support an earlier hypothesis that had suggested that nuclear receptor activity and the ability to synthesize new estrogen receptor are impaired by androgenization (Vertes *et al.*, 1973).

On the other hand, the progesterone receptor in the pituitary gland may be sexually differentiated, since progesterone was found to increase estrogen uptake in female but not in male lactotrophs (see Dluzen and Ramirez, 1982; Gunnet and Freeman, 1983). There is evidence that the pituitary gland is responsible for the maintenance of sex differences in hepatic steroid metabolism (Colby *et al.*, 1973; Denef, 1974; Gustafsson and Stenberg, 1974; Lax *et al.*, 1974) and that this difference occurs at about 30 days of age (Stenberg, 1976). If pituitary tissue obtained from female rats aged 28 days and older is placed beneath the kidney capsule of hypophysectomized adult males, it causes a change in hepatic steroid metabolism to that of female type; if donor tissue is taken from females below 28 days of age, no such change is seen. The transitional period during which a "feminizing factor" begins to be secreted by anterior pituitary tissue appears to take place between 28 and 35 days of age. Evidence based on results of electrothermic lesions of the hypothalamus has indicated that the release of a feminizing factor from the pituitary is controlled by a release-inhibiting factor from the hypothalamus of male but not of female rats (Gustafsson *et al.*, 1976). The combined results suggest that, with respect to hepatic steroid metabolism, the hypothalamic pituitary axis in Sprague-Dawley rats matures between 28 and 35 days of age.

It may also be noteworthy that in both rats and monkeys, characterization studies of pituitary LH and FSH have shown that after gonadectomy the properties of these hormones are modified (Bogdanove *et al.*, 1974; Peckham and Knobil, 1976*a*). Furthermore, the molecular size of LH and FSH from male pituitaries is apparently smaller and their disappearance from the circulation faster than that of the corresponding hormones in female monkeys (Peckham and Knobil, 1976*b*). By contrast, the molecular size of FSH in the female rat is apparently smaller than that in the male, while the clearance of endogenous FSH in the androgen-treated orchidectomized rat takes longer than that in the androgen-deprived orchidectomized rat (Bogdanove *et al.*, 1974).

2.4. Gonadotropin Response to a Steroid Stimulus

In those spontaneously ovulating animals from which experimental evidence has been forthcoming, it appears that the central nervous system of the female differs from that of the male in that the control of gonadotropin output, which is responsible for ovarian function and ovulation, is of a cyclic nature. In contradistinction, the secretory pattern of the male is relatively constant. The preovulatory LH surge that occurs in the evening of proestrus is characteristic of the cycling female rat (Brown-Grant *et al.*, 1970) and can be stimulated in the ovariectomized adult female by either two spaced injections of EB (Caligaris *et al.*, 1971) or by a single injection of progesterone after estrogen priming (Brown-Grant, 1974*b*). A simpler experimental model using either the adult ovariectomized female or the intact immature female (Figure 7) is that in which only single dose of EB is given at noon on the first experimental day (day 1). In response to the steroid, LH concentrations are initially inhibited (negative feedback) until about 54 hr later, in the evening of day 2, when a surge of LH occurs (so-called positive feedback) (Table II), which is repeated on successive evenings over the next few days (Burnet and MacKinnon, 1975).

Although the LH response to estrogen, or to progesterone after estrogen priming, can be readily elicited in castrated females and in males that have been castrated at birth, no such effect can be obtained in either adult castrated male or neonatally androgenized females castrated as adults (Brown-Grant, 1972). In the immature intact female in which circulating free estrogen levels are apparently negligible (Puig-Duran *et al.*, 1978), the response seems to develop at about 14 days of age, since it cannot be provoked before that time either by progesterone in the estrogen-primed animal or by synthetic estrogens, for which plasma receptor proteins have little affinity (Puig-Duran and MacKinnon, 1978). These results suggest that although mechanisms underlying the preovulatory LH surge may be sexually differentiated at an early age, they become overtly so between 12 and 15 days of age.

With respect to serum prolactin concentrations, Neill (1972) observed an increased evening output of the hormone 23 hr after a second daily noon injection of estrogen in ovariectomized adult females but not in adult castrate males. It is of particular note that the LH response to an estrogen stimulus that is central to the question of sexual differentiation in spontaneous ovulators has not been elicited

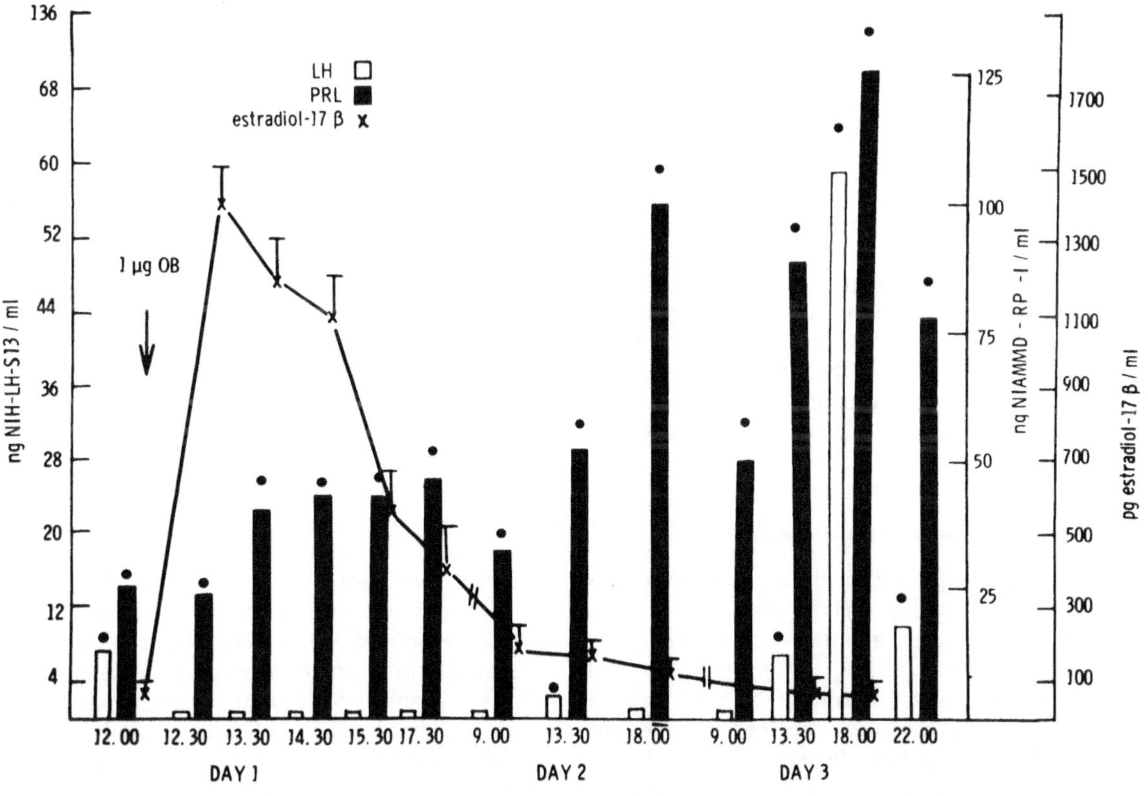

Fig. 7. Serum luteinizing hormone (LH), prolactin, (PRL), and 17β-estradiol concentrations (mean ±SEM) in 21-day-old female rats injected with 1 µg estradiol benzoate at 1200 hr on day 1 and killed at variable intervals of time thereafter; 4–8 animals per group. (From Puig-Duran and MacKinnon, 1978.)

Table II. Serum LH and Prolactin (ng/ml) in 21-Day-Old Male, Female, Neonatally Androgenized Female, and 1-Day-Castrate Male Rats Following a Single Injection of 1 μg Estradiol Benzoate or Oil at Noon (1200 hr) on Day 1[a,b]

		Basal values		Day 2 (1200 hr) 24 hr postinjection		Day 3 (1800 hr) 54 hr postinjection	
Group	Treatment[c]	LH	PRL	LH	PRL	LH	PRL
♀	Oil	6.0 ± 4.4	16.5 ± 5.1(6)	3.6 ± 2.9	18.4 ± 6.0(6)	5.7 ± 3.8	17.1 ± 9.4(8)
	EB			0.3 ± 0.05	60.5 ± 13.3(6)	59.5 ± 9.1	116.3 ± 11.7(8)
♂	Oil	2.5 ± 0.9	25.1 ± 4.9(6)	1.9 ± 0.8	28.3 ± 5.6(5)	0.9 ± 0.8	30.1 ± 4.2(5)
	EB			0.3 ± 0.02	25.1 ± 2.1(8)	0.4 ± 0.03	30.1 ± 2.6(8)
♀ A	Oil	1.6 ± 0.1	42.5 ± 3.6(6)	1.4 ± 0.6	40.2 ± 4.2(5)	0.1 ± 0.9	38.9 ± 4.7(5)
	EB			0.9 ± 0.02	69.8 ± 7.7(8)	0.4 ± 0.06	115.6 ± 2.4(8)
1-day ♂	Oil	24.4 ± 3.3	26.2 ± 2.7(6)	19.7 ± 6.7	30.2 ± 4.1(5)	20.3 ± 5.9	29.5 ± 11.2(5)
	EB			2.3 ± 0.7	59.6 ± 3.6(8)	28.8 ± 1.4	125.8 ± 12.0(8)

[a] Values are mean ±SEM. Numbers of animals in parentheses.
[b] From Puig-Duran and MacKinnon.
[c] EB, estradiol benzoate; LH, luteinizing hormone; PRL, prolactin.

in voles (Milligan, 1978) or apparently in any other reflex ovulator.

2.5. Sexual Behavior: Sexual Differentiation

A further manifestation of the sexually differentiated brain is that of sexual behavior. However, sexual differentiation is not as well defined in terms of this parameter as it is with respect to patterns of gonadotropin output. The reason for this difference is that changes in behavior, far from being clear cut, tend to occur as a result of alterations in the thresholds of response to certain hormonal stimuli. It is also necessary to distinguish masculinizing and defeminizing components of the sexual differentiation of sexual behavior in the rat, particularly with reference to the organizational effects of the sex steroids and of their reversible effects on adult sexual behavior.

2.5.1. Female-Type Behavior in the Rat

2.5.1a. Females. Although the characteristics of mammalian female behavior may best be considered in terms of sexual attractiveness, proceptivity, and receptivity (Beach, 1976), the behavioral feature to which most attention has been paid in the rat is that of lordosis. The lordosis quotient (LQ), defined as the ratio between the number of lordoses to the number of mounts in response to an experienced male, can be elicited with maximal ease around the time of ovulation. After ovariectomy, the response is depleted, but it can be rapidly restored by estrogen either alone or in combination with progesterone (Whalen and Edwards, 1967).

The effects of prenatal androgens or of estrogen on subsequent adult behavior indicate that lordosis is markedly depressed by such treatment (see, e.g., Hendricks and Welton, 1976), but the results of a further extensive and careful study have suggested otherwise (Brown-Grant, 1975; see also Mullins and Levine, 1968) in that adult females treated with varying doses of TP or EB neonatally showed a marked and progressive capacity for lordosis on exposure to concentrated periods of mounting activity. The failure of perinatal androgen treatment to disrupt female receptive behavior in rats has also been reported in ferrets (Baum, 1976) and in rhesus monkeys (Goy, 1970) and is supported by the results of a study in women with the adrenogenital syndrome (Money and Ehrhardt, 1972). An interesting suggestion put forward to account for the progressive increase of the LQ in neonatally androgenized animals during mating tests was the possibility of a mating-induced release of GnRH into the CNS (Brown-Grant, 1975). This hypothesis was based on previous observations of facilitated LQs in ovariectomized estrogen-primed rats injected systemically with GnRh (Moss and McCann, 1973; Pfaff, 1973) and is supported by recent work that has shown that lordosis is facilitated by hypophysectomy (Crawley *et al.*, 1976), which increases the hypothalamic content of LHRH.

The effect of estrogen on adult neonatally androgenized female rats was observed to increase the LQ still further; however, if the steroid was followed by an additional dose of progesterone a few hours before testing, then the LQ was significantly depressed, as it is in normal male rats (Brown-Grant, 1975). Sodersten (1976) examined the effects of progesterone on the LQ in estrogen-primed animals and in general confirmed Brown-Grant's findings.

2.5.1b. Males. The normal adult male rat is fully capable of exhibiting lordosis when mounted by vigorous male partners. This behavior is greatly enhanced if the animals are treated with estrogen and particularly if the animals have been previously castrated (Davidson, 1969; Davidson and Levine, 1969; Brown-Grant, 1975).

In adult genetic male rats treated with antiandrogens both before and after birth, elicitation of lordosis was reported to be achieved as frequently in response to EB, followed by an injection of progesterone, as it was in normal females (Nadler, 1969). This effect has also been observed in adult males castrated within a few hours of birth (Harris and Levine, 1965; Grady *et al.,* 1965). In adult 1-day castrate males primed with estrogen, injected with progesterone, and tested 4–6 hr later, the lordotic response was enhanced, although according to Gerall *et al.* (1967) it did not compare in quantity and quality with that observed in similarly treated ovariectomized females.

There is far less information about sexual differentiation of proceptivity, apart from an interesting attempt to assess sexual preferences (see Baum, 1979). In these experiments, variously treated animals were placed in the vicinity of restrained, sexually active males or females and preferences assessed. Neonatally castrated male rats displayed no clear preference. But if they were pretreated shortly before the test with EB and progesterone, they oriented toward the stimulus male rats. Females treated neonatally with testosterone propionate showed less proceptivity toward males than did their normal counterparts.

2.5.2 Male-Type Behavior in the Rat

Male patterns of sexual behavior are analyzed in terms of the latency and frequency of mounting, thrusting, intromission, and ejaculation when the animal is placed with a receptive female. But some of these male parameters are as difficult to assess as female behavior patterns, since mounting is quite commonly observed in normal females of various species. A further complication in the appraisal of male sexual behavior patterns is the fact that intromission and ejaculation are dependent on sensory feedback information derived from the penis. Ossification of the penile bone and its sensitization are both heavily dependent on exposure of the organ to androgens in neonatal life (Beach *et al.,* 1969), and the degree of development of the phallus is of considerable importance in the interpretation of data derived from behavioral studies, more particularly in animals exposed to manipulations of the hormonal environment in neonatal life.

2.5.2a. Males. Most normally developing postpubertal males will attempt to mount other animals of the same species, and a high frequency of mounting can be obtained if large doses of testosterone (or estrogen) are administered beforehand (Davidson, 1969). If mating tests are undertaken after the testes have been removed in adulthood, mating will eventually cease, although it can be restored with the appropriate dose of testosterone. Adult male rats castrated at birth will exhibit mounting behavior in response to estrous females, an effect that can be increased by injections of either testosterone or EB (Gerall *et al.,* 1967; Beach and Holz, 1946; Beach, 1971), but they are less likely than are testosterone-treated adult castrates to achieve intromission and ejaculation (see Whalen, 1968; Davidson and Levine, 1972). However, a high ejaculation score can be achieved if neonatally castrated males are injected with either testosterone or DHT in addition to RU 2858 (a synthetic estrogen) during the first few days of life (Booth, 1977); DHT or RU 2858 alone do not have this effect. These results give full support to the concept that neonatal estrogen is of vital importance in the neural organization of male behavioral patterns provided that DHT is also present to complete differentiation of the male external genitalia.

2.5.2b. Females. Mounting of estrous females by other females is commonplace, and in every mammalian species that has been studied there would seem to be a prenatal or perinatal period during which exposure to androgens enhanced some aspect of male-type behavior, but here again criticism of the data with regard to phallic development and sensory feedback obtains. Nevertheless, studies in the female ferret have been useful, since differentiation of the phallus in response to perinatal androgen treatment occurs in this species at an earlier time of development than that at which the potential for masculine responses is heightened (Baum, 1976). In postnatally androgenized female ferrets lacking phallic sheaths, estrogen administration in adulthood significantly stimulated male-type behavioral patterns in the absence of any clitoral stimulation (Baum, 1976). Thus, in this species at least, there is an effect of perinatal androgens on adult male-type behavior that is divorced from any effects on the penis.

Recently the concept that female behavior is basic in type and that its development is not dependent on differentiation of neural mechanisms by steroid hormones has been brought to question. Apparently, the androgen-insensitive genetically male rat (testicular feminization) fails to display either masculine or feminine sexual behavior when primed with the appropriate sex hormones. The implication of these results is that "female differentiation" of the brain with respect to mechanisms controlling behavior may require active imprinting by perinatal hormones (Shapiro *et al.*, 1976*a*).

Evidence for a sexual dimorphism of sexual behavior in the rat is supported by work derived from three other species of rodent, namely, the guinea pig (Phoenix *et al.*, 1959), golden hamster (Swanson and Crossley, 1971), and the mouse (Edwards and Burge, 1971). Moreover, there are other forms of behavioral patterns which show sexual dimorphism, such as open-field behavior in the rat (Gray *et al.*, 1969) and hamster (Swanson, 1967; Carter *et al.*, 1972), aggression in the mouse (Bronson and Desjardins, 1969; Edwards, 1971), saccharin preference in drinking water in the rat (Valenstein *et al.*, 1967; Wade and Zucker, 1969), urination with hind-leg raising in the dog (Beach, 1975), urination with squatting in the ewe (Short, 1974; Clarke *et al.*, 1976), and courtship song in the canary and zebra-finch (Gurney and Konishi, 1980; Gurney, 1981).

2.6. Possible Sites of Sexual Differentiation of Mechanisms Controlling Sexual Behavior

The existence of a site or sites in the rat brain that respond to androgen by permitting a manifestation of male sexual behavior was originally demonstrated by stereotactic implantation of the steroid in the preoptic–anterior hypothalamic area (Davidson, 1966; Lisk, 1967) and by the elimination of mating behavior by lesions of the medial preoptic nucleus (Larsson and Heimer, 1964). Whether this result was a direct effect of androgen or a product of its metabolism has been questioned, although systemic administration of the 5α-reduced metabolite DHT is as ineffective in provoking male behavior (Davidson and Bloch, 1972) as it is in causing subsequent masculinization when injected into the female neonate (Luttge and Whalen, 1970*b*; Brown-Grant *et al.*, 1971). By contrast, DHT–estrogen combinations are highly effective (Feder *et al.*, 1974). That aromatization of androgen to estrogen may be essential for eliciting changes in behavior is further supported by the results of studies in which antiestrogens were found to block androgen-induced sex behavior in both rats (Whalen *et al.*, 1972; Sodersten and Larsson, 1974) and rabbits (Beyer and Vidal, 1971).

Whether estrogen is the effective agent at the intracellular level that triggers the complex pattern of male sexual behavior, it is unlikely that only the preoptic area (POA) is concerned. Early work showed that male rats, following destruction of the neocortex, fail to respond to, and will not mate with, receptive females (Beach, 1940). Moreover, parasagittal cuts that sever the connections between the POA and the medial forebrain bundle markedly inhibit male sexual behavior (Paxinos and Bindra, 1973). Yet further studies employing lesions or electrical stimulation have implicated the amygdala and, perhaps importantly, the midbrain (for a review see Gorski, 1974).

With respect to female behavior in the rat, early work showed that although the cortex is not essential for mating, almost complete removal of the neocortex interferes with the integration of normal lordotic behavior (Beach, 1944), and this has been supported by more recent studies by Merari *et al.* (1975). It is generally accepted that the POA, which contains estrogen receptors, is important in the integrity of elicitation or lordosis; although there have been contradictory reports, it appears that lesions of the POA abolish this behavior, while estrogen implants restore it (see Gorski, 1974).

Other neural areas are implicated in the control of female receptive behavior as well. Yanasi and Gorski (1976) implanted estrogen in the preoptic-diagonal band area of ovariectomized rats followed by progesterone (3 days later in either the midbrain or the preoptic area). The results of the study demonstrated that progesterone in either area facilitates the estrogen-primed lordotic response, while control cholesterol implants displayed no such effect. The results of further experiments using implants of antiestrogen in the midbrain suggested that although exposure of either the POA or the midbrain to estrogen facilitates the response to systemically injected progesterone, prior exposure of the midbrain to estrogen may be required for the action of locally implanted progesterone. Gorski (1976) recently reviewed the possible sites of steroid facilitation of lordosis in the rat and proposed a model for regulation of lordosis behavior. Since the midbrain appears to be an important site of progesterone action, it may also prove to be a site of sexual differentiation.

2.6.1. Masculinization and Defeminization

The wealth of information that has accumulated about sex hormone influences on neural primordia, coupled with the clear differences that exist between certain behavioral responses point to a distinction between anatomical sites of masculinization and defeminization.

Since female rats show mounting behavior, they are to some extent "masculinized," presumably because of the presence of circulating androgens *in utero*. Interestingly, female rats that were adjacent to males *in utero* showed significantly more mounting behavior as adults as well as an increased urogenital distance (Clemens *et al.*, 1978).

Although the processes of defeminization and masculinization are probably independently regulated in the rat, both occur during sexual maturation, although not necessarily at the same time.

The defeminization of sexual behavior in male rats is established during the perinatal period and can be established in females by neonatal treatment with TP. The net result is a decreased ability to display lordosis as adults and a decreased sensitivity to administered estradiol in this respect. The steroid metabolites responsible for defeminization are unknown; more than one may be involved, since estradiol and DHT act synergistically in neonatally castrated male rats to suppress female sexual behavior (Booth, 1979).

A neural site important in the control of female sexual behavior in the rat appears to be the medial basal hypothalamus (MBH) (Christensen and Gorski, 1978; Davis and Barfield, 1979; Nordeen and Yahr, 1982). This area contains estradiol and DHT receptors, although fewer estradiol receptors were measured in the MBH of adult males (Nordeen and Yahr, 1983). This may form part of the basis of reduced sensitivity to estradiol in terms of lordosis responses in males.

The masculinizing effects of testosterone are undoubtedly complex, in view of the many components of masculine behavior, ranging from mounting to thrusting to intromission, quite apart from the permanent disruption of the neural mechanisms required for ovulation. Beach *et al.* (1969) argued that normal penile development could explain many of the sexual differences in these forms of behavior, but it seems more likely that testosterone and/or metabolites may organize neural structures responsible for mating and ejaculatory behavior. For example, genetic females exposed prenatally to testosterone exhibited the complete pattern of male sexual behavior including an ultrasonic postcoital vocalization (Sachs *et al.*, 1973). Also, normal male sexual behavior could be invoked in male rats despite suppression of penile development (Hart, 1972). A further important site of perinatal masculinization of the brain appears to be localized in or near the POA (Christensen and Gorski, 1978; Nordeen and Yahr, 1982). This brain area is markedly sexually dimorphic in rats (see Section 2.3.1), and dimorphism at this site is dependent on perinatal exposure to testosterone.

The possibility that a hormonally directed neural differentiation may occur in the female brain at a critical period of development, similar in a sense to that observed in the male, has recently been raised. The question was highlighted by the observation that sexual behavior patterns of testicular-feminized rats that had been gonadectomized and subsequently injected with steroids were neither male nor female in type as compared with those of similarly treated littermate males and nonlittermate females (Shapiro *et al.*, 1976*b*). Since both serum and adrenal progesterone concentrations are significantly higher in 3-day old female rats than in males of the same age, it has been suggested that the gonadotropin–adrenal axis may be required for adequate feminization of the brain.

3. Sexual Differentiation of the Avian Brain

Avian systems are particularly well suited to the study of sexual differentiation of the brain, since in some species well-defined neural pathways and nuclei have been identified and correlated with stereotyped patterns of sexual behavior. The courtship songs of the male canary and zebrafinch (females sing simpler songs than males or in the case of the zebrafinch they do not sing at all) have been extensively studied (see Arnold and Gorski, 1984). Both birds have an interconnected series of brain regions that project to and control the muscles of the vocal organ, the syrinx (Nottebohm *et al.*, 1976). The nuclei are larger in the male than in the female brain (Nottebohm and Arnold, 1976), and lesions of some markedly reduce song (Nottebohm *et al.*, 1976). The sex differences concern both the number of cells and their size and are dependent on the organizational effects of steroids during neonatal life. In the zebrafinch, sexually dimorphic nuclei are differentiated around the time of hatching, a process that depends on the presence of testosterone and its metabolites, which appear to have separate effects (Gurney and Konishi, 1980). In adult life, testosterone has an activational function in that castration leads to a loss of courtship song that can be restored by the administration of testosterone or DHT. In adult female canaries (Nottebohm, 1980) and white crowned sparrows (Konishi, 1965), testosterone increases the size of the sexually dimorphic nuclei and stimulates song. These observations indicate that a degree of brain plasticity in adult male birds may exist that is virtually unknown in mammals, the increased nuclear size apparently attributable to growth of dendritic trees and not to increased cell number (De Voogt and Nottebohm, 1981).

There is less information about sexual differentiation of other functions governed by the avian brain. However, release of LH is sexually differentiated in chickens (Sharp, 1975), while the Japanese quail shows evidence of sexual dimorphism of the photoperiodic response system governing the release of LH (Urbanski and Follett, 1982).

4. Evidence for Sexual Differentiation of the Brain in Primates

4.1. Gonadotrophin Response to a Steroid Stimulus

Using the rhesus monkey, which has a similar pattern of hormonal output during its menstrual cycle as that observed in the human, Knobil (1974) and associates have attempted to delineate those areas of the brain that are essential to the control of phasic gonadotropin output. "Hypothalamic islands," which ensure complete disconnection of the mediobasal hypothalamus and its attached pituitary complex from the preoptic-Sch.n. area and all other extrahypothalamic structures were made in female monkeys (Krey *et al.*, 1973) with the use of a Halasz knife (Halasz and Pupp, 1965). Radioimmunoassay measurements of plasma LH and FSH levels in blood samples obtained before and after the operation showed that basal levels were unaffected by the cerebral trauma and that in some animals normal menstrual cycles and ovulation continued unabated. Furthermore, in ovariectomized animals in which pulsatile release of LH are characteristically found (Dierschke *et al.*, 1970), a similar operative procedure was entirely consistent with the retention of this 'tonic' rhythm. With respect to the simulated preovulatory LH surge, the administration of EB in the early part of the follicular phase, in dosages designed to imitate the rise of estrogen at midcycle, occasioned an initial inhibition of LH levels followed closely by a facilitatory release. Not surprisingly therefore, the presence of GnRH-Synthesizing neurons in the monkey have been located not in the POA, as in the rat, but in the AN. Lesions of the AN lead to a fall in gonadotropin levels while their restoration depends on a pulsatile infusion of GnRH. The neural and humeral control of the amplitude and frequency of these pulses, however, is not known (Hess *et al.*, 1977; Knobil, 1980).

It is clear that spontaneous LH surges in the monkey are not dependent, as in the rodent, on mechanisms associated with the preoptic area and the Sch.n. and its connections. A question then arises: If these areas, which are important in the rat in terms of the LH surge, are sexually differentiated, is their apparent unimportance in the primate associated with a lack of sexual differentiation in the primate brain and therefore an ability to elicit an LH surge in both sexes? Knobil and associates have gone some way to answering this question by showing that the integrity of the hypothalamus close to the median eminence, which apparently controls pulsatile release of LH, is all that is required apart from the preovulatory increase in secretion of estrogen (see Knobil, 1980).

Karsch *et al.* (1973) castrated groups of adult male and female rhesus monkeys and suppressed the resulting high serum LH levels with crystalline estradiol-17β contained in Silastic capsules placed beneath the skin. A subsequent subcutaneous injection of EB led to unambiguous discharges of LH 24 hr later, which were similar in all respects in the two sexes (Figure 8). These results were supported by similar findings in humans. Stearns *et al.* (1973) demonstrated a stimulatory effect on LH output of progestin in estrogen-primed castrate adult men, while other workers (e.g., Yen *et al.*, 1972) observed a similar rise in gonadotropins after administration of progestin to estrogen-pretreated castrate and postmenopausal women. Collectively, these results suggest that mechanisms controlling the LH surge in primates may not be sexually differentiated; on the other hand, more subtle stimuli may be required to expose such a potential difference.

In line with the above data are the results of studies on the offspring of rhesus monkeys treated with androgen repeatedly from day 24 of pregnancy to just prior to birth (average gestation 168 days) (Goy, 1970; Goy and Resko, 1972). Although the onset of puberty in these animals was delayed, their subsequent menstrual cycles were normal as regards hormonal output and were no more irregular than those observed in control animals.

Although there are no substantial indications that mechanisms subserving positive feedback responses of LH to estrogen are sexually differentiated in primates, it is possible that mechanisms underlying negative feedback responses of LH to estrogen may be. Evidence has shown that in groups of gonadectomized male, female, and "androgenized" rhesus monkeys in which Silastic capsules containing estradiol had been implanted beneath the skin one week earlier LH levels in the female were significantly lower than in the other two groups (Steiner *et al.*, 1976). Negative feedback mechanisms in the female would therefore appear to be more sensitive to the effects of long-term feedback inhibition of estrogen than are those in males or androgenized females.

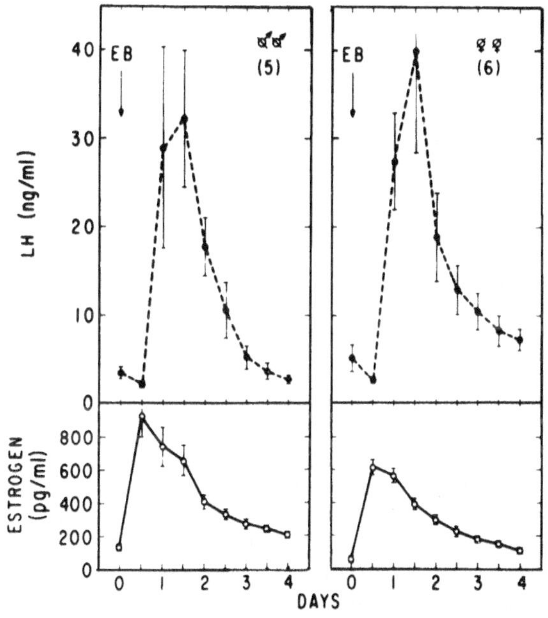

Fig. 8. Induction of luteinizing hormone (LH) surges in gonadectomized male (left) and female (right) rhesus monkeys by the administration of estradiol benzoate (EB). Before EB was injected, plasma LH concentrations were suppressed by the subcutaneous implantation of Silastic capsules containing crystalline 17β-estradiol. Plasma concentrations of estrogen effected by these treatments are shown in the lower panels. The mean $\pm$SE is shown; the number of observations is in parentheses. (From Karsch *et al.*, 1973.)

4.2. Sexual Behavior and Sexual Differentiation

4.2.1. Rhesus Monkeys

In an attempt to alter the hormonal environment of the fetus, van Wagenen and Hamilton (1943) and later Wells and van Wagenen (1954) administered heavy doses of testosterone to pregnant rhesus monkeys. When the offspring were born, the genetic females were clearly hermaphroditic with morphological changes of the external genitalia. Using a similar technique, Goy (1970) and other workers (Goy and Resko, 1972; Phoenix, 1974) likewise produced a small number of hermaphroditic rhesus monkeys reared from 3 months of age to 4 years of age with other hormone-treated or -untreated monkeys of similar age. At frequent intervals during development, various behavioral patterns were carefully assessed; the outcome was that rough-and-tumble play, threats, play initiation, and chasing play were observed significantly more often in males and in hermaphrodites than in females. On the other hand, such forms of behavior as grooming, huddling, or fear-grimacing did not show signs of being sexually differentiated. Castration of males at birth or at 90 days of age did not affect the frequencies with which these behavioral patterns were exhibited when a comparison was made with scores obtained from intact male peers. Likewise, the behavioral patterns of ovariectomized monkeys were similar to those observed in intact control females.

Mounting behavior too was studied, and again this behavior was more frequently observed in hermaphrodites than in normal females. There was also a qualitative difference in this behavior; when normal male rhesus monkeys mount a female, they clasp the partner's shanks with their hind legs while holding her back with the forepaws. This characteristic shank-clasp position (Figure 9) was more often displayed by the small number of hermaphrodites that had been reared with mothers and their peer group than by the normal control females (Goy and Goldfoot, 1973).

Tests of male-type behavior in castrated adult

Fig. 9. Illustration of shank-clasp position of adult male monkey during coitus. Note position of hindpaws on hindlimbs of female.

hermaphrodites showed that the hermaphrodites were significantly more aggressive and more likely to sit next to an EB-stimulated female than were control females; this difference was not altered by administration of TP (Eaton *et al.*, 1973).

The period when sexual differentiation of sexual behavior occurs is likely to coincide with that of masculinization of the genitalia is between days 45 and 100 after conception, a time when testosterone levels in umbilical artery plasma are approximately 10 times higher in male than in female rhesus monkeys (Resko *et al.*, 1973). It is presumably during this prenatal period that development of the neuronal systems that favor the eventual adoption of the shank-clasp position occurs. There appears to be a sliding scale in tendency with respect to this behavioral reflex, since female hermaphrodites were intermediate to control male and female monkeys (Goy, 1978). Interestingly, there was no correlation between the extent of masculinization of the external genitalia and the degree of sexual behavior displayed, which points to a separate action on the periphery and the brain.

A pertinent question is whether structural differences exist between the male and female monkey brain that might underlie differentiation of sexual behavior. If there are structural differences, they may be subtle. For example, Ayoub *et al.* (1983) could find no differences in gross measurements of the preoptic area, although they did report some fine structural differences, in that neurons of males had more dendritic bifurcations and a higher frequency of spines. Bubenik and Brown (1973) measured the nuclear diameters of neurons in the medial nucleus of the amygdala of squirrel monkeys and found them to be smaller in females.

The combined evidence suggests that prenatal exposure to androgens or to the products of aromatization (see Naftolin *et al.*, 1975) modifies social interaction patterns independently of the hormonal environment during the late stages of development. Circulating androgen levels in the fetal monkey are therefore of considerable interest, and radioimmunoassay measurements of umbilical artery samples have shown that testosterone levels in the fetal male rhesus are higher on average than in the female from day 59 to just before birth (Resko *et al.*, 1973).

4.2.2. Humans

There are two available sources of human data from which an attempt can be made to ascertain whether certain components of human sexuality (see Whalen, 1966) are determined by prenatal or postnatal hormonal influences on central neural development or whether they depend entirely on the current hormonal status. The first is a variety of prenatal abnormalities associated with pseudohermaphroditism, and the second is the condition of transsexualism, which is not associated with any physical or genetic anomaly (Money and Ehrhardt, 1972; Bancroft, 1977).

Of the prenatal states of hormonal dysfunction, the most common is the adrenogenital syndrome, characterized by a genetic deficiency of the hydroxylating enzyme at the 21-position of the steroid molecule. This deficiency leads to a failure of cortisol production, with a subsequent increase in adrenocorticotrophin (ACTH) output and, in turn, an excessive production of androgens. Should the dysfunction be sufficiently severe, the external genitalia of the female fetus become masculinized *in utero* and the child might be assigned at birth to the genetically incorrect sex. Some of these children have therefore been reared as boys and others as girls, with the consequent development of a self-image or gender identity that is relevant to the sex of assignment. Of particular interest are those in whom the endocrine state and the external genitalia are of male type, in spite of which they have been reared as girls. Two such groups of children have been studied (Ehrhardt *et al.*, 1968*b*). In one group, the diagnosis was made at birth and cortisol treatment was instituted immediately. In the other group, recognition of the condition was delayed, and treatment was started in adolescence or in adulthood (Ehrhardt *et al.*, 1968*a*). Behavioral patterns in these two groups of children were studied, and the results were reminiscent of those observed in masculinized infant monkeys in that they exhibited more tomboy behavior than did a matched control group. These observations are further supported by data obtained from a group of children who had inadvertently been exposed to excess progestins during intrauterine development because their mothers had been treated with these drugs in the early weeks of pregnancy in order to prevent abortion (Ehrhardt and Money, 1967). Again, an analysis of play patterns indicated distinct tomboyish qualities.

In the late-treatment group of children with adrenogenital dysfunction, there appeared to be differences in regard to sexual preference as pertaining to characteristics of a desirable sexual partner or erotic stimulus. A greater degree of bisexual fantasy and bisexual experience was shown by this group than by subjects in the early-treatment group (Ehrhardt *et al.*, 1968*b*). In another study, Lev-Ran (1974) reported that in women born with the adrenogenital syndrome in whom treatment designed to lower androgen levels was started (i.e., women similar to the late-treatment group), sexual dreams virtually ceased.

A naturally occurring experimental model in genetic males that is of considerable interest is the state of testicular feminization in which there is a deficiency of androgenic binding proteins, hence a failure of hormonal expression. The phenotype of these testes-carrying subjects is typically female, and most of these subjects are reared from birth as girls and thus develop a female gender identity. In adulthood, their behavior is very close to the female stereotype, for among other parameters they are said to make good mothers to their adopted children. On the criteria of sexual preference, they also appear to be strongly feminine, with little evidence of homosexuality or of bisexual imagery (Money and Ehrhardt, 1972).

Equivalent perhaps to the girls born to progestin-treated mothers are males born to diabetic mothers who, in order to reduce the high fetal mortality rate associated with the disorder, had been treated with high doses of estrogen and low doses of progestins during pregnancy. It is notable that these boys were found to be less aggressive and less athletic than were a control group of children born to untreated mothers (Yalom *et al.*, 1973).

From these observations of genetic males and females exposed to a variety of naturally occurring or drug-induced endocrine abnormalities, it appears that gender identity is of paramount importance in the development of human sexuality; moreover, it is strongly dependent on parental attitudes and social learning. Yet there are, in addition, indications that prenatal influences may also play a part in human sexuality in that exposure to androgens during development appears to be conducive to the formation of a masculine gender identity. Of relevance to such a consideration is a report of a group of genetic males with a 5α-reductase deficiency (Imperato-McGinley *et al.*, 1974). Such an enzyme deficiency prevents the metabolism of testosterone to DHT and thus leads to a failure of the external genitalia to masculinize. As a consequence of the appearance of female-type genitalia at birth, these 5α-reductase-deficient children were reared as girls. But, unexpectedly, hormonal changes occurred at puberty that led to a partial masculinization of the genitalia and subsequent alterations toward a masculine gender identity; in addition, interest increased with respect to the opposite genetic sex.

The second source of human data from which evidence might be gleaned in support of a prenatal or postnatal hormonal influence on gender identity is that of transsexualism. This condition is one in which the individual believes himself to be, or has a very strong desire to be, the gender opposite that of his genetic and anatomical sex. The characteristic behavioral pattern is often evolved from fetishistic transvestism in early childhood and becomes more marked at puberty (Bancroft, 1972). There is, however, little evidence to link the condition with hormonal changes during development either in the genetic male (Migeon *et al.*, 1969) or in the genetic female (Jones and Saminy, 1973; Fulmer, 1973). But in the knowledge that these subjects are susceptible to taking exogenous hormones, a carefully controlled hormonal investigation would be of considerable interest and importance.

In respect to a further aspect of human sexuality, namely, sexual preference, several groups of clinical research workers have sought to account for homosexuality on the basis of prenatal or postnatal hormonal imbalance. So far these attempts have proved negative, with one possible exception, in which the estrogen-stimulated LH response was investigated in groups of heterosexual, homosexual, and bisexual men (Dorner *et al.*, 1975). Of 21 homosexuals, 13 appeared to show a facilitatory LH response to a single injection of estrogen in comparison with 2 of the 20 heterosexuals and none of the bisexuals. The facilitatory LH response of all the male groups was said to be less marked than that observed in normal women, but this was accounted for by androgen levels that are higher in the male than the female and that may alter the pituitary response to estrogen. In the light of previous results that have shown that a facilitatory LH response can be elicited by injection of gonadal steroids in castrated male and female monkeys or castrated men and women, this report requires confirmation.

A structural or biochemical basis of human sexual differentiation remains to be established, and reports are fragmentary. De Lacoste-Utiamsing and Holloway (1982) presented evidence for sexual dimorphism of the splenium of the corpus collosum, that of the female being larger and more bulbous. Women appear to be more susceptible than men to sexual sensations associated with temporal lobe epilepsy (Remillary *et al.*, 1983), and Gur *et al.* (1982) found evidence for more fast-perfusing brain tissue, i.e., gray matter, in women than in men.

5. Summary and Conclusions

There is considerable evidence for sexual differentiation of the brain in the rat, some of which is supported by data obtained in other nonprimate species, including cow (Jost *et al.*, 1963), sheep (Short, 1974; Karsch and Foster, 1975), dog (Beach, 1970, 1975), guinea pig (Brown-Grant *et al.*, 1971), hamster (Swanson, 1967; Alleva *et al.*, 1969), gerbil (Turner, 1976), mouse (Barraclough and Leathem, 1954), and bird (see Arnold and Gorski, 1984).

It appears that testosterone, perhaps by virtue of its conversion to estrogen and DHT, can at an early stage of development organize and differentiate basal female-type mechanisms into those of male type. The presence of testes in male rats or the administration of testosterone to neonatal females prevents the possibility of a preovulatory LH surge and therefore of regular estrous cycles after puberty. On the other hand, if the male is castrated within 24 hr of birth, it retains the potential to exhibit a phasic output of LH.

Maintenance of phasic gonadotrophin output in the rat depends not only on the presence of neurons that synthesize GnRH but on neural mechanisms that appear to signal the time of day and control its release. Although the number of reported sites of sexual differentiation within the CNS is increasing, those pertaining to phasic control of gonadotrophin output are still unclear. However, stimulation of the preoptic area in male rats leads to an LH surge and ovulation if the animals have been implanted with ovarian tissue; it may therefore be expected that an important site of differentiation may lie outside the preoptic-median eminence-anterior pituitary unit.

In the monkey, menstrual cyclicity does not appear to depend on the time of day. Menstrual cycles continue unabated in this species even if the mediobasal hypothalamus (which apparently contains GnRH-synthesizing neurons) is surgically separated from the rest of the brain, provided that an adult pattern of pulsatile GnRH release is present. Furthermore, the LH response to an estrogen (or estrogen/progesterone) stimulus, which is so clearly sexually differentiated in rodents and certain other nonprimates, can be elicited with ease in both castrated male and female monkeys or in castrated humans. Commensurate with these findings are those in which normal menstrual cycles or LH surges were observed in female monkeys exposed to increased androgen levels during intrauterine development.

With respect to sexual behavior of the rat, castration of the male on the day of birth or administration of testosterone or estrogen to neonatal females tends to change thresholds of sexual behavioral patterns toward those of the opposite sex. Sexually differentiated neural pathways responsible for sexual behavior are still unclear, with the exception of the sexually differentiated nucleus in the lumbar spinal cord. In the rhesus monkey, there are apparently significant differences in social interaction patterns and mounting displays between males and females treated with androgens *in utero* and untreated females. These patterns of behavior are considered to have a counterpart in the tomboy activities and attitudes displayed by female children who have, as a result of hormonal dysfunction or inadvertent drug treatment, also been exposed to the effects of androgen in uterine life.

In summary, evidence for the organizational effects of sex steroids on the developing brain in nonprimates is plentiful. So too are the activational effects of these steroids on the differentiated neural pathways in later life. It is also becoming apparent that different hormones exert specific and differing effects in this process; moreover, the critical periods over which their influences extend are temporarily dissimilar. Evidence for sexual differentiation of the primate brain, however, is not as strong. But since such fundamental differences in neural organisation have been found in lower mammals, it might be expected that further evidence for sexual differentiation of the CNS of primates will in time accrue.

6. References

Aiyer, M. S., and Fink, G., 1974, The role of sex steroid hormones in modulating the responsiveness of the anterior pituitary gland to luteinizing hormone releasing factor in the female rat, *J. Endocrinol.* **62**:553–572.

Aiyer, M. S., Chiappa, S. A., and Fink, G., 1974, A priming effect of luteinizing hormone releasing factor on the anterior pituitary gland in the female rat, *J. Endocrinol.* **62**:573–588.

Akmaev, I. G., and Fidelina, O. V., 1978, The role of tanycytes in the mechanisms of sexual differentiation of the brain, *Neurosci. Behav. Physiol.* **9**:23–235.

Alleva, F. R., Alleva, J. J., and Umberger, E. J., 1969, Effect of a single prepubertal injection of testosterone propionate on later reproductive functions of the female golden hamster, *Endocrinology* **85**:312–318.

Arai, Y., 1971, Some aspects of the mechanisms involved in steroid induced sterility, in: *Steroid Hormones and Brain Function* (C. H. Sawyer and R. A. Gorski, eds.) pp. 185–191, University of California Press, Los Angeles.

Arai, Y., and Kusama, T., 1968, Effect of neonatal treatment with estrone on hypothalamic neurons and regulation of gonadotrophin secretion, *Neuroendocrinology* **3**:107–114.

Arnold, A. P., and Gorski, R. A., 1984, Gonadal steroid induction of structural sex differences in the central nervous system, *Annu. Rev. Neurosci.* **7**:413–442.

Ayoub, D. M., Greenbough, W. T., and Juraska, J. M., 1983, Sex differences in dendritic structure in the preoptic area of the juvenile macaque monkey brain, *Science* **219**:197–198.

Bancroft, J. H. J., 1972, The relationship between gender identity and sexual behaviour: Some clinical aspects, in: *Gender Differences; Their Ontogeny and Significance* (C. Ounsted and D. C. Taylor, eds.), pp. 57–72, Churchill Livingstone, Edinburgh.

Bancroft, J. H. J., 1977, Biological determinants of sexual behavior, in: *Hormones and Sexual Behavior in the Human* (J. Hutchinson, ed.), pp. 493–519, Wiley, New York.

Barley, J., Ginsburg, M., Greenstein, B. D., MacLusky, N. J., and Thomas, P. J., 1975, An androgen receptor in rat brain and pituitary, *Brain Res.* **100**:373–393.

Barraclough, C. A., 1961, Production of anovulatory, sterile rats by single injections of testosterone propionate, *Endocrinology* **68**:62–67.

Barraclough, C. A., and Leathem, J. H., 1954, Infertility induced in mice by a single injection of testosterone propionate, *Proc. Soc. Exp. Biol. Med.* **85**:673–674.

Barraclough, C. A., and Turgeon, J. L., 1974, Further studies of the hypothalamo-hypophyseal-gonadal axis of the androgen-sterilized rat, International Symposium on Sexual Endocrinology of the Perinatal Period, *INSERM* **32**:339–356.

Baum, M. J., 1976, Effects of testosterone propionate administered perinatally on sexual behavior of female ferrets, *J. Comp. Physiol. Psychol.* **90**:399–410.

Baum, M. J., 1979, Differentiation of coital behavior in mammals: A comparative analysis, *Neurosci. Biobehav. Rev.* **3**:265–284.

Beach, F. A., 1940, Effects of cortical lesions upon the copulatory behavior of male rats, *J. Comp. Physiol. Psychol.* **29**:193–244.

Beach, F. A., 1944, Effects of injury to the cerebral cortex upon sexuality-receptive behavior in the female rat, *Psychosomat. Med.* **6**:40–45.

Beach, F. A., 1970, Hormonal effects of socio-sexual behavior in dogs, in: *Mammalian Reproduction* (H. Gibian and E. J. Plotz, eds.), pp. 437–466, Springer-Verlag, New York.

Beach, F. A., 1971, Hormonal factors controlling the differentiation development and display of copulatory behaviour in the ramstergig and related species, in: *Biopsychology of Development* (E. Tobach, L. R. Aronson, and E. Shaw, eds.), pp. 247–296. Academic Press, New York.

Beach, F. A., 1975, Hormonal modification of sexually dimorphic behaviour, *Psychoneuroendocrinology* **1**:3–23.

Beach, F. A., 1976, Sexual attractivity, proceptivity and receptivity in female mammals, *Horm. Behav.* **7**:105–138.

Beach, F. A., and Holz, A. M., 1946, Mating behavior in male rats castrated at various ages and injected with androgen, *J. Exp. Zool.* **101**:91–142.

Beach, F. A., Noble, R. G., and Arndorf, R. K., 1969, Effects of perinatal androgen treatment on responses of male rats to gonadal hormones in adulthood, *J. Comp. Physiol. Psychol.* **68**:490:497.

Bennett, D., Boyse, E. A., Lyon, M. F., Mathieson, B. J., Scheid, M., and Yanagisawa, K., 1975, Expression of H-Y (male) antigen in phenotypically female Tfm/Y mice, *Nature (London)* **257**:236–238.

Benno, R. H., and Williams, T. H., 1978, Evidence for intracellular localization of alpha-fetoprotein in the developing rat brain, *Brain Res.* **142**:182–186.

Beyer, C., and Vidal, N., 1971, Inhibitory action of MER-25 on androgen-induced oestrous behaviour in the ovariectomised rabbit, *J. Endocrinol.* **51**:401–402.

Bogdanove, E. M., Campbell, G. T., Blair, E. D., Mula, M. E., Miller, A. E., and Grossman, G. M., 1974, Gonad–pituitary feedback involves qualitative change: Androgens alter the type of FSH secreted by the rat pituitary, *Endocrinology* **95**:219–228.

Booth, J. E., 1977, Sexual behaviour of neonatally castrated rats injected during infancy with oestrogen and dihydrotestosterone, *J. Endocrinol.* **72**:135–141.

Booth, J. E., 1979, Evidence for the involvement of dihydrotestosterone in sexual differentiation of the brain in neonatally castrated rats, *J. Endocrinol.* **80**:43P–44P.

Borvendeg, J., Hermann, H., and Bajusz, S., 1972, Ovulation induced by synthetic luteinizing hormone releasing factor in androgen-sterilized female rats, *J. Endocrinol.* **55**:207–208.

Breedlove, S. M., and Arnold, A. P., 1980, Hormone accumulation in a sexually dimorphic motor nucleus of the rat spinal cord, *Science* **210**:564–566.

Breedlove, S. M., and Arnold, A. P., 1983, Hormonal control of a developing neuromuscular system. II. Sensitive periods for the androgen induced masculinization of the rat spinal nucleus of the bulbocavernosus, *J. Neurosci.* **3**:424–432.

Breedlove, S. M., Jacobson, C. D., Gorski, R. A., and Arnold, A. P., 1982, Masculinization of the female rat spinal cord following a single neonatal injection of testosterone propionate but not estradiol benzoate, *Brain Res.* **237**:173–181.

Bronson, F. H., and Desjardins, C., 1969, Aggressive behavior and seminar vesicle function in mice: Differential sensitivity to androgen given neonatally, *Endocrinology* **85**:971–974.

Brown-Grant, K., 1972, Recent studies on the sexual differentiation of the brain, in: *Foetal and Neonatal Physiology* (K. S. Comline, K. W. Cross, G. S. Dawes and P. W. Nathanielsz, eds.), pp. 527–545, Cambridge University Press, Cambridge.

Brown-Grant, K., 1972*b*, Reproductive function in the rat following selective destruction of afferent fibres to the hypothalamus from the limbic system, *Brain Res.* **46**:23–42.

Brown-Grant, K., 1974*a*, On "critical periods" during the post-natal development of the rat, in: *Les colloques de l'institut National de la Santé et de la Recherche Medicale, INSERM* **32**:357–376.

Brown-Grant, K. 1974*b*, Steroid hormone administration and gonadotrophin secretion in the gonadectomised rat, *J. Endocrinol.* **62**:319–332.

Brown-Grant, K., 1975, A re-examination of the lordosis response in female rats given high doses of testosterone propionate or estradiol benzoate in the neonatal period, *Horm. Behav.* **6**:351–378.

Brown-Grant, K., Exley, D., and Naftolfin, F., 1970, Peripheral plasma oestradiol and luteinizing hormone concentrations during the oestrous cycle of the rat, *J. Endocrinol.* **48**:295–296.

Brown-Grant, K., Munck, A., Naftolin, F., and Sherwood, M. R., 1971, The effects of the administration of testosterone propionate alone or with phenobarbitone and of testosterone metabolites to neonatal female rats, *Horm. Behav.* **2**:173–182.

Bubenik, G. A., and Brown, G. M., 1973, Morphologic sex differences in primate brain areas involved in regulation of reproductive activity, *Experimentia* **29**:619–621.

Burnet, F. R. B., and MacKinnon, P. C. B., 1975, Restoration by oestradiol benzoate of a neural and hormonal rhythm in the ovariectomized rat, *J. Endocrinol.* **64**:27–35.

Caligaris, L., Astrada, J. J., and Taleisnik, S., 1971, Release of luteinizing hormone induced by estrogen injection into ovariectomised rat, *Endocrinology* **88**:810–815.

Carter, C. S., Clemens, L. G., and Hockemeyer, D. J., 1972, Neonatal androgen and adult sexual behavior in the golden hamster, *Physiol. Behav.* **9**:89–95.

Chiappa, S. A., Fink, G., and Sherwood, N. M., 1977, Immunoreactive luteinizing hormone releasing factor (LRF) in pituitary stalk plasma from female rats: Effects of stimulating diencephalon, hippocampus and amygdala, *J. Physiol. (London)* **267**:625–640.

Christensen, L. W., and Gorski, R. A., 1978, Independent masculinization of neuroendocrine systems by intracerebral implants of testosterone or estradiol in the neonatal female rat, *Brain Res.* **146**:325–340.

Cidlowski, J. A., and Muldoon, T. G., 1976, Sex related differences in the regulation of cytoplasmic estrogen receptor levels in responsive tissues of the rat, *Endocrinology* **98**:833–841.

Clark, J. M., Campbell, P. S., and Peck, E. J., Jr., 1972, Receptor–oestrogen complex in the nuclear fraction of the pituitary and hypothalamus of male and female immature rats, *Neuroendocrinology* **10**:218–228.

Clarke, J., Scaramuzzi, R. J., and Short, R. V., 1976, Effects of testosterone implants in pregnant ewes on their female offspring, *J. Embryol. Exp. Morphol.* **36**:87–99.

Clayton, R. B., Kogura, J., and Kraemer, H. C., 1970, Sexual differentiation of the brain: Effects of testosterone on brain RNA metabolism in newborn female rats, *Nature (London)* **226**:810–812.

Clemens, L. G., Gladue, B. A., and Goniglio, L. P., 1978, Prenatal endogenous androgenic influences on masculine sexual behavior and genital morphology in male and female rats, *Horm. Behav.* **10**:40–53.

Coen, C., and MacKinnon, P. C. B., 1980, Lesions of the suprachiasmatic nuclei in the serotonin-dependent phasic release of luteinizing hormone in the rat: Effects on the drinking rythmicity on the consequences of preoptic area stimulation, *J. Endocrinol.* **84**:231–236.

Colby, H. D., Gaskin, J. H., and Kitay, J. I., 1973, Effect of anterior pituitary hormones in hepatic corticosterone metabolism in rats, *Steroids* **24**:679–686.

Crawley, W. R., Rodriguez-Sierra, J. P., and Komisaruk, B. R., 1976, Hypophysectomy facilitates sexual behavior on female rats, *Neuroendocrinology* **20**:328–338.

Davidson, J. M. 1966, Activation of the male rat's sexual behavior by intracerebral implantation of androgen, *Endocrinology* **79**:783–794.
Davidson, J. M., 1969, Effects of oestrogen on the sexual behavior of male rats, *Endocrinology* **84**:1365–1372.
Davidson, J. M., and Bloch, G. J., 1972, Neuroendocrine aspects of male reproduction, *Biol. Reprod. (Suppl. 1)* **1**:67–92.
Davidson, J. M., and Levine, S., 1969, Progesterone and heterotypical sexual behavior of male rats, *J. Endocrinol.* **44**:129–130.
Davidson, J. M., and Levine, S., 1972, Endocrine regulation of behavior, *Annu. Rev. Phys.* **34**:375–408.
Davis, P. G., McEwen, B. S., and Pfaff, D. W., 1979, Localized behavioral effects of tritiated estradiol implants in the ventromedial hypothalamus of female rats, *Endocrinology* **104**:898–903.
Debeljuk, L., Arimura, A., and Schally, A. V., 1972, Effect of testosterone and estradiol on the LH and FSH release induced by LH-releasing hormone (LH-RH) in intact male rats, *Endocrinology* **90**:1578–1581.
De Lacoste-Utiamsing, C., and Holloway, R., 1982, Sexual dimorphism in the human corpus callosum, *Science* **216**:1431–1432.
De Voogd, T. J. and Nottebohm, F., 1981, Sex difference in dendritic morphology of a song control nucleus in the canary. *J. Comp. Neurol.* **196**: 309.
Denef, C., 1974, Effect of hypophysectomy and pituitary implants at puberty on the sexual differentiation of testosterone metabolism in rat liver, *Endocrinology* **94**:1577–1582.
De Vries, G. J., Buijs, R. M., and Swaab, D. F., 1981, Ontogeny of the vasopressinergic neurons of the suprachiasmatic nucleus and their extrahypothalamic projections in the rat brain—Presence of a sex difference in the lateral septum, *Brain Res.* **218**:67–78.
Diamond, M. C., Murphy, G. M., Akiyama, K., and Johnson, R. E., 1982, Morphologic hippocampal asymmetry in male and female rats, *Exp. Neurol.* **76**:553–565.
Dierschke, D. J., Bhattacharya, A. N., Atkinson, L. E., and Knobil, E., 1970, Circhoral oscillations of plasma LH levels in the ovariectomised rhesus monkey, *Endocrinology* **87**:850–853.
Dluzen, D. E., and Ramirez, V. D., 1982, A functional dimorphism in the response of the hypothalamic pituitary axis of prepubertal rats to steroid treatment, *Biol. Reprod.* **27**:456–461.
Docke, F., and Kolocyek, G., 1966, Einfluss einer postnatalen Androgen-behandlung auf den Nucleus Hypothalamicus anterior, *Endokrinologie* **50**:225–230.
Dohler, K. D., and Wuttke, W., 1975, Changes with age in levels of serum gonadotropins, prolactin, and gonadal steroids in prepubertal male and female rats, *Endocrinology* **97**:898–907.
Dorner, G., and Staudt, J., 1968, Structural changes in the preoptic anterior hypothalamic area of the male rat, following neonatal castration and androgen substitution, *Neuroendocrinology* **3**:136–140.
Dorner, G., Rohde, W., Stahl, F., Krell, L., and Masius, W. G., 1975, A neuroendocrine predisposition for homosexuality in men, *Arch. Sex. Behav.* **4**:1–8.
Dyer, R. G., MacLeod, N. K., and Ellendorff, F., 1976, Electrophysiological evidence for sexual dimorphism and synaptic convergence in the preoptic and anterior hypothalamic areas of the rat, *Proc. R. Soc. London* **193**:142–440.
Eaton, G. G., Goy, R. W., and Phoenix, C. H., 1973, Effects of testosterone treatment in adulthood on sexual behaviour of female pseudohermaphrodite rhesus monkeys. *Nature New Biol.* **242**:119–120.
Edwards, D. A., 1971, Neonatal administration of androstenedione, testosterone or testosterone propionate. Effects on ovulation, sexual receptivity and aggressive behavior in female mice, *Physiol. Behav.* **6**:223–228.
Edwards, D. A., and Burge, K. G., 1971, Early androgen treatment and male and female sexual behavior in mice, *Horm. Behav.* **2**:249–58.
Ehrhardt, A. A., and Money, J., 1967, Progestin-induced hermaphroditism I.Q. and psychosexual identity in a study of ten girls, *J. Sex Res.* **3**:83–100.
Ehrhardt, A. A., Evers, K., and Money, J., 1968*a*, Infulence of androgen and some aspects of sexually dimorphic behavior in women with late-treated adrenogenital syndrome, *Johns Hopkins Med. J.* **123**:115–112.
Ehrhardt, A. A., Epstein, R., and Money, J., 1968*b*, Fetal androgens and female gender identity in the early treated adrenogenital syndrome, *Johns Hopkins Med. J.* **122**:160–167.
Elger, W., 1966, Die Rolle der fetalen Androgene in der Sexual differenzierung des Kaninchens und ihre Abgrenzung gegen andere hormonale und somatische Faktoren durch Anwendung eines starken Antiandrogens, *Arch. Anta. Microsc. Morphol. Exp.* **55**:657–743.
Everett, J. W., 1969, Neuroendocrine aspects of mammalian reproduction, *Annu. Rev. Physiol.* **31**:383–416.
Everett, J. W., and Sawyer, C. H., 1950, A 24-hour periodicity in the LH-release apparatus of female rats disclosed by barbiturate sedation, *Endocrinology* **47**:198–218.
Everett, J. W., Holsinger, J. W., Zeilmaker, G. H., Redmond, W. C., and Quinn, D. L., 1970, Strain differences for preoptic stimulation of ovulation in cyclic, spontaneously persistent-oestrus, and androgen-sterilized rats, *Neuroendocrinology* **6**:98–108.
Fang, S., Anderson, K. M., and Liao, S., 1969, Receptor proteins for androgens. On the role of specific proteins in selective retention of 17β-hydroxy-5α-androstan-3-one by rat ventral prostate in vivo and in vitro, *J. Biol. Chem.* **244**:6584–6595.
Feder, H. H., Naftolin, F., and Ryan, K. J., 1974, Male and female sexual responses in male rats given estradiol benzoate and 5α-androstan-17β-ol-3-one propionate, *Endocrinology* **94**:136–141.

Fink, G., and Henderson, S. R., 1977, Steroids and pituitary responsiveness in female, androgenized female and male rats, *J. Endocrinol.* **73:**157–164.

Flores, F., Naftolin, F., Ryan, K. J., and White, R. J., 1973, Estrogen formation by the isolated perfused rhesus monkey brain, *Science* **180:**1074–1075.

Fulmer, G. P., 1973, Testosterone levels and female-to-male transsexualism, *Arch. Sex. Behav.* **2:**399–400.

Gerall, A. A., Hendricks, S. E., Johnson, L. L., and Bounds, T. W., 1967, Effects of early castration in male rats on adult sexual behavior, *J. Comp. Physiol. Psychol.* **64:**206–212.

Gorski, R. A., 1968, Influence of age on the response to paranatal administration of a low dose of androgen, *Endocrinology* **82:**1001–1004.

Gorski, R. A., 1976, The possible neural sites of hormonal facilitation of sexual behavior in the female rat, *Psychoneuroendocrinology* **1:**371–387.

Gorski, R. A., and Barraclough, C. A., 1963, Effects of low dosages of androgen on the differentiation of hypothalamic regulatory control of ovulation in the rat, *Endocrinology* **73:**210–216.

Gorski, J., and Gannon, F., 1976, Current models of steroid hormone action: A critique, *Annu. Rev. Physiol.* **38:**425–450.

Gorski, R. A., Gordon, J., Shryne, J. E., and Southam, A. M., 1978, Evidence for a morphological sex difference within the medial preoptic area of the rat brain. *Brain Res.* **148:**333–346.

Goy, R. W., 1970, Experimental control of psychosexuality, *Philos. Trans. R. Soc. London Ser. B* **259:**149–162.

Goy, R. W., 1978, Development of play and mounting behavior in female rhesus monkeys virilized prenatally with esters of testosterone or dihydrotestosterone, in: *Recent Advances in Primatology* (D. Chivers and J. Herbert, eds.), Vol. I, pp. 449–462, Academic Press, New York.

Goy R. W., and Goldfoot, D. A., 1973, Hormonal influences on sexually dimorphic behavior, in: *Handbook of Physiology, Endocrinology* (R. O. Greep and E. B. Astwood, eds.), Section 7, Vol. 2, Part 1, pp. 169–186, Williams & Wilkins, Baltimore.

Goy, R. W., and Resko, J., 1972, Gonadal hormones and behaviour of normal and pseudohermaphrodite non human, female primates, *Rec. Prog. Horm. Res.* **28:**707–733.

Grady, K. L., Phoenix, C. H., and Young, W. C., 1965, Role of the developing rat testis in differentiation of the neural tissues mediating mating behavior, *J. Comp. Physiol. Psychol.* **59:**179–182.

Gray, J. A., Lean, J., and Keynes, A., 1969, Infant androgen treatment and adult openfield behavior: Direct effects and effects of injections to siblings, *Physiol. Behav.* **4:**177–181.

Greep, R. O., 1936, Functional pituitary grafts in rats, *Proc. Soc. Exp. Biol. N.Y.* **34:**754–755.

Guldner, F. H., 1982, Sexual dimorphisms of the axo-spine synapses and postsynaptic density material in the suprachiasmatic nucleus of the rat, *Neurosci. Lett.* **28:**145–150.

Gunnet, J. W., and Freeman, M. E., 1983, The mating-induced release of prolactin: A unique neuroendocrine response, *Endocrinol. Rev.* **4:**44–61.

Gur, R. C., Gur, R. E., Obrist, W. D., Hungerbuhler, J. P., Younkin, D., Rosen, A. D., Skolnick, B. E., and Reivich, M., 1982, Sex and handedness differences in cerebral blood flow during rest and cognitive activity, *Science* **217:**659–661.

Gurney, M. E., 1981, Hormonal control of cell form and number in the zebra finch song system, *J. Neurosci* **1:**658–673.

Gurney, M. E., and Konishi, M., 1980, Hormone induced sexual differentiation of brain and behavior in zebra finches, *Science* **208:**1380–1382.

Gustafsson, J. A., and Stenberg, A., 1974, Masculinization of rat liver enzyme activities following hypophysectomy, *Endocrinology* **95:**891–896.

Gustafsson, J. -A., Ingelman-Sundberg, M., Stenberg, A., and Jokfelt, T., 1976, Feminization of hepatic steroid metabolism in male rats following electrothermic lesion of the hypothalamus, *Endocrinology* **98:**922–926.

Halasz, B., and Pupp, L., 1965, Hormone secretion of the anterior pituitary gland after physical interruption of all nervous pathways to the hypophysiotropic area, *Endocrinology* **77:**553–562.

Harris, G. W., 1955, *Neural Control of the Pituitary Gland,* Edward Arnold, London.

Harris, G. W., 1970, Hormonal differentiation of the developing central nervous system with respect to patterns of endocrine function, *Philos. Trans. R. Soc. London Ser. B* **259:**165–177.

Harris, G. W., and Jacobsohn, D., 1952, Functional grafts of the anterior pituitary gland, *Proc. R. Soc. London Ser. B* **139:**263–276.

Harris, G. W., and Levine, S., 1965, Sexual differentiation of the brain and its experimental control, *J. Physiol.* **181:**379–400.

Hart, B. L., 1972, Manipulation of neonatal androgen. Effects on sexual responses and penile development in male rats, *Physiol. Behav.* **8:**841–845.

Hayashi, Y., 1976, Failure of intrahypothalamic implants of an estrogen antagonist, ethamoxytriphetol (MER-25), to block neonatal androgen-sterilization, *Proc. Soc. Exp. Biol. Med.* **152:**389–392.

Hendricks, S. E., and Welton, M., 1976, Effects of estrogen given during various periods of prepubertal life on the sexual behaviour of rats, *Physiol. Psychol.* **4(1):**105–110.

Hess, D. L., Wilkins, R. H., Moossy, J., Chang, J. L., Plant, T. M., McCormack, J. T., Nakai, Y., and Knobil, E., 1977, Estrogen-induced gonadtropin surges in decerebrated female rhesus monkeys with medial basal hypothalamic (MBH) peninsulae, *Endocrinology* **101:**1264–1271.

Imperato-McGinley, J., Gerrero, L., Gautier, T., and Peterson, R. E., 1974, Steroid 5α-reductase deficiency in man: An inherited form of male pseudohermaphroditism, *Science* **186**:1213–1215.

Jacobson, C. D., and Gorski, R. A., 1981, Neurogenesis of the sexually dimorphic nucleus of the preoptic area in the rat, *J. Comp. Neurol.* **196**:519–529.

Jacobson, C. D., Csernus, V. J., Shryne, J. E., and Gorski, R. A., 1981, The influence of gonadectomy, androgen exposure, or a gonadal graft in the neonatal rat on the volume of the sexually dimorphic nucleus of the preoptic area, *J. Neurosci.* **1**:1142–1147.

Jamieson, M., and Fink, G., 1976, Immunoreactive luteinizing hormone releasing factor in rat pituitary stalk blood: Effects of electrical stimulation of the medial preoptic area, *J. Endocrinol.* **68**:71–87.

Jones, J. R., and Saminy, J., 1973, Plasma testosterone levels and female transsexualism, *Arch. Sex. Behav.* **2**:251–256.

Josso, N., 1972, Permeability of membranes to the Mullerian inhibiting substance synthetized by the human fetal testis in vitro: A clue to its biochemical nature, *J. Clin. Endocrinol. Metab.* **34**:265–270.

Josso, N., 1973, In vitro synthesis of Mullerian inhibiting hormone by seminiferous tubules isolated from the calf fetal testis, *Endocrinology* **93**:829–834.

Jost, A., 1970, Hormonal factors in the sex differentiation of the mammalian foetus, *Philios. Trans. R. Soc. London Ser. B* **259**:119–130.

Jost, A., Chodkiewicz, H., and Mauleon, P., 1963, Intersexualité du foetus de veau produite par des androgènes. Comparaison entre l'hormone foetale responsable du free-martinisme et l'hormone testiculaire adulte, *C.R. Seanc. Acad. Sci. Paris* **256**:274–276.

Jost, A., Vigier, B., Prepin, J., and Perchellet, J. P., 1973, Studies on sex differentiation in mammals, *Rec. Prog. Horm. Res.* **29**:1–35.

Karsch, F. J., and Foster, D. L., 1975, Sexual differentiation of the mechanism controlling the preovulatory discharge of luteinising hormone in sheep, *Endocrinology* **97**:373–379.

Karsch, F. J., Dierschke, D. J., and Knobil, E., 1973, Sexual differentiation of pituitary function: Apparent difference between primates and rodents, *Science* **179**:848.

Kato, J., 1971, Estrogen receptors in the hypothalamus and hypophysis in relation to reproduction, in: *Proceedings of the Third International Congress on Hormonal Steroids* (V. H. T. James and L. Martini, eds.), pp. 764–773, Excerpta Medica, Amsterdam.

Kato, J., 1975, The role of hypothalamic and hypophyseal 5α-dehydrotestosterone, estradiol and progesterone receptors in the mechanism of feedback action, *J. Steroid Biochem.* **6**:979–987.

Kato, J., Atsumi, Y., and Inaba, M., 1974, Estradiol receptors in female rat hypothalamus in the developmental stages and during pubescence, *Endocrinology* **94**:309–317.

Kawakami, M., and Terasawa, E., 1973, Further studies on sexual differentiation of the brain: Response to electrical stimulation in gonadectomized and estrogen primed rats, *Endocrinol. Jpn.* **20**:595–607.

King, R. J. B., and Mainwaring, W. I. P., 1974, *Steroid-Cell Interactions*, 245 pp. Butterworths, London.

Knapstein, P., David, A., Wu, C. H., Archer, D. F., Flickinger, G. L., and Touchstone, J. C., 1968, Metabolism of free and sulfoconjugated Dhea in brain tissue in vivo and in vitro, *Steroids* **11**:885–896.

Knobil, E., 1974, On the control of gonadtrophin secretion in the rhesus monkey, *Rec. Prog. Horm. Res.* **30**:1–36.

Knobil, E. C., 1980, The neuroendocrine control of the menstrual cycle, *Rec. Prog. Horm. Res.* **36**:53–88.

Konishi, M., 1965, The role of auditory feedback in the control of vocalization in the white crowned sparrow, *Z. Tierpsychol.* **22**:770–783.

Krey, L. C., Butler, W. R., Weiss, G., Weick, R. F., Dierschke, D. J., and Knobil, E., 1973, Influences of endogenous and exogenous gonadal steroids on the actions of synthetic LRF in the rhesus monkey in: *Hypothalamic Hypophysiotrophic Hormones* (C. Gaul and E. Rosenberg, eds.), pp. 39–47, Int. Congr. Ser. No. 263, Excerpta Medica, Amsterdam.

Larsson, K., and Heimer, L., 1964, Mating behaviour of male rats after lesions in the preoptic area, *Nature (London)* **202**:413–414.

Lax, E. R., Hoff, H. G., Ghraf, R., Schroder, E., and Schreifers, H., 1974, The role of the hypophysis in the regulation of sex differences in the activities of enzymes involved in hepatic steroid hormone metabolism, *Hoppe-Seylers Z. Physiol. Chem.* **355**:1325–1331.

Le Blond, C. B., Morris, S., Karakiulakis, G., Powell, R., and Thomas, P. J., 1982, Development of sexual dimorphism in the suprachiasmatic nucleus of the rat, *J. Endocrinol.* **95**:137–145.

Lev-Ran, A., 1974, Sexuality and educational level of women with the late-treated adreno-genital syndrome, *Arch. Sex. Behav.* **3**:27–32.

Lieberburg, I., MacLusky, N. J., and McEwen, B. S., 1980, Androgen receptors in the perinatal rat brain, *Brain Res.* **196**:125–138.

Lisk, R. D., 1967, Neural localization for androgen activation of copulatory behavior in the male rat, *Endocrinology* **80**:754–761.

Luttge, W. G., and Whalen, R. E., 1970*a*, Dihydrotestosterone, androstenedione, testosterone; comparative effectiveness in masculinizing and defeminizing reproductive systems in male and female rats, *Horm. Behav.* **1**:265–281.

Luttge, W. G., and Whalen, R. E., 1970*b*, Regional localization of estrogenic metabolites in the brain of male and female rats, *Steroids* **15**:605–612.

MacDonald, P. G., Doughty, C., 1972, Inhibition of androgen-sterilization in the female rat by administration of an anti-oestrogen, *J. Endocrinol.* **55**:455–456.

MacKinnon, P. C. B., Mattock, J. M., and ter Haar, M. B., 1976, Serum gonadotrophin levels during development in male, female and androgenized female rats and the effect of general disturbance on high luteinizing hormone levels, *J. Endocrinol.* **70**:361–371.

MacLusky, N. J., Lieberburg, I., and McEwen, B. S., 1979, The development of estrogen receptor systems in the rat brain: Perinatal development, *Brain Res.* **178**:129–142.

Martinez, C., and Bittner, J. J., 1956, A non-hypophysial sex difference in oestrous behaviour of mice bearing pituitary grafts, *Proc. Soc. Exp. Biol. Med.* **91**:506–509.

Matsumoto, A., and Arai, Y., 1980, Sexual dimorphism in "wiring pattern" in the hypothalamic arcuate nucleus and its modification by neonatal hormonal environment, *Brain Res.* **190**:238–242.

Matsumoto, A., and Arai, Y., 1981, Effect of androgen on sexual differentiation of synaptic organization in the hypothalamic arcuate nucleus: An ontogenetic study, *Neuroendocrinology* **33**:166–169.

Mennin, S. P., Kubo, K., and Gorski, R. A., 1974, Pituitary responsiveness to luteinizing hormone-releasing factor in normal and androgenized female rats, *Endocrinology* **95**:412–416.

Merari, A., Frenk, C., Hirwig, M., and Ginton, A., 1975, Female sexual behavior in the male rat: Facilitation by cortical application of potassium chloride, *Horm. Behav.* **6**:159–164.

Migeon, C. J., Rivarola, M. A., and Forest, M. G., 1969, Studies of androgens in male transsexual subjects; effects of oestrogen therapy, in: *Transsexualism and Sex Re-assignment* (R. Green and J. Money, eds.), pp. 203–211, Johns Hopkins University Press, Baltimore.

Milligan, S. R., 1978, The feedback of exogenous steroids on LH release and ovulation in the intact female vole *(Microtus agrestis), J. Reprod. Fertil.* **54**:309–311.

Mizukami, S., Nishizuka, M., and Arai, Y., 1983, Sexual difference in nuclear volume and its ontogeny in the rat amygdala, *Exp. Neurol.* **79**:569–575.

Moguilevski, J. A., and Rubinstein, L., 1967, Glycolytic and oxidative metabolism of hypothalamic areas in prepubertal androgenized rats, *Neuroendocrinology* **2**:213–221.

Moguilevski, J. A., Libertun, C., Schiaffini, O., and Scacchi, P., 1969, Metabolic evidence of the sexual differentiation of the hypothalamus, *Neuroendocrinology* **4**:264–269.

Money, J., and Ehrhardt, A. A., 1972, *Man and Women: Boy and Girl. The Differentiation and Dimorphism of Gender Identity from Conception to Maturity,* Johns Hopkins University Press, Baltimore.

Moore, R. Y., 1974, Visual pathways and the central neural control of diurnal rhythms, in: *The Neurosciences. Third Study Program* (F. O. Schmitt and F. G. Worden, eds.), pp. 537–542, MIT Press, Cambridge, Massachusetts.

Moore, R. Y., and Eichler, V. B., 1972, Loss of a circadian adrenal corticosterone rhythm following suprachiasmatic lesions in the rat, *Brain Res.* **42**:201–206.

Moss, R. L., and McCann, S. M., 1973, Induction of mating behavior in rats by luteinizing hormone-releasing factor, *Science* **181**:177–179.

Mueller, G. C., Vanderhaar, B., Kim, U. H., and Le Mahieu, M., 1972, Estrogen Action: An in road to cell biology, *Rec. Prog. Horm. Res.* **281**:1–45.

Mullins, R. F., and Levine, S., 1968, Hormonal determinants during infancy of adult sexual behaviour in the female rat, *Physiol. Behav.* **3**:333–343.

Nadler, R. D., 1969, Differentiation of the capacity for male sexual behavior in the rat, *Horm. Behav.* **1**:53–63.

Naftolin, F., Ryan, K. H., Davies, I. J., Reddy, V. V., Flores, F., Petro, Z., Kuhn, M., White, R. J., Takaoka, Y., and Wolin, L., 1975, The formation of oestrogens by central neuroendocrine tissues, *Rec. Prog. Horm. Res.* **31**:295–315.

Neill, J. D., 1972, Sexual differences in the hypothalamic regulation of prolactin secretion, *Endocrinology* **90**:1154–1159.

Neumann, F., Elger, W., and Kramer, M., 1966, The development of a vagina in male rats by inhibiting the androgen receptors with an anti-androgen during the critical phase of organogenesis, *Endocrinology* **78**:628–632.

Neumann, F., von Berswordt-Wallrabe, R., Elger, W., Steinbeck, M., Hahn, J. D., and Kramer, M., 1970, Aspects of androgen-dependent events studied by antiandrogens, *Rec. Prog. Horm. Res.* **26**:337–405.

Nishizuka, M., and Arai, Y., 1981, Organizational action of estrogen on synaptic pattern in the amygdala: Implications for sexual differentiation of the brain, *Brain Res.* **213**:422–426.

Nordeen, E. J., and Yahr, P., 1982, Hemispheric asymmetries in the behavioral and hormonal effects of sexually differentiating mammalian brain, *Science* **218**:391–394.

Nordeen, E. J., and Yahr, P., 1983, A regional analysis of estrogen binding to hypothalamic cell nuclei in relation to masculinization and defeminization, *J. Neurosci.* **3**:933–941.

Nottebohm, F., 1980, Testosterone triggers growth of brain vocal control nuclei in adult female canaries, *Brain Res.* **189**:429–436.

Nottebohm, F., and Arnold, A. P., 1976, Sexual dimorphism in vocal control areas of the song bird brain, *Science* **194**:211–213.

Nottebohm, F., Stokes, T. M., and Leonard, C. M., 1976, Central control of song in the canary *(Serenus canarius), J. Comp. Neurol.* **165**:467–486.

Nunez, E., Engelmann, F., Benessayag, C., Savu, L., Crepy, O., and Jayle, M. -F., 1971, Identification et purification préliminaire de la foeto-proteine liant les oestrogenes dans le serum des rats nouveaunes, *C. R. Acad. Sci. D* **273**:831–834.

Ohno, S., 1971, Simplicity of mammalian regulatory systems inferred by single gene determination of sex phenotypes, *Nature (London)* **234**:134–137.

Ojeda, S. R., Kalra, P. S., and McCann, S. M., 1975, Further studies on the estrogen negative feedback on gonadotropin release in the female rat, *Neuroendocrinology* **18**:242–255.

Paxinos, G., and Bindra, D., 1973, Hypothalamic and midbrain neural pathways involved in eating, drinking, irritability, aggression and copulation in rats, *J. Comp. Physiol. Psychol.* **82**:1–14.

Peckham, W. D., and Knobil, E., 1976*a*, Qualitative changes in the pituitary gonadotropins of the male rhesus monkey following castration, *Endocrinology* **98**:1061–1064.

Peckham, W. D., and Knobil, E., 1976*b*, The effects of ovariectomy, estrogen replacement and neuraminidase treatment on the properties of the adenohypophysial glycoprotein hormone of the rhesus monkey, *Endocrinology* **98**:1054–1060.

Pfaff, D. W., 1966, Morphological changes in the brains of adult male rats after neonatal castration, *J. Endocrinol.* **36**:415–416.

Pfaff, D. W., 1973, Luteinising hormone-releasing factor potentiating lordosis behavior in hypophysectomized ovariectomized female rats, *Science* **182**:1148–1149.

Pfaff, D. W., Gerlach, J. L., McEwen, B. S., Ferin, M., Carmel. P., and Zimmerman, E. A., 1976, Autoradiographic localization of hormone and concentrating cells in the brain of the female rhesus monkey, *J. Comp. Neurol.* **170**:279–291.

Pfeiffer, C. A., 1936, Sexual differences of the hypophyses and their determination by the gonads, *Am. J. Anat.* **58**:195–225.

Phoenix, C. H., 1974, Prenatal testosterone in the non human primate and its consequences for behavior, in: *Sex Differences in Behavior* (R. C. Friedman, R. M. Richart, and R. L. Vande Wiele, eds.), pp. 19–32, Wiley, New York.

Phoenix, C. H., Goy, R. W., Gerall, A. A., and Young, W. C., 1959, Organizing action of prenatally administered testosterone propionate on the tissue mediating mating behavior in the female guinea pig, *Endocrinology* **65**:369–382.

Price, D., 1970, In vitro studies on differentiation of the reproductive tract, *Philos. Trans. R. Soc. London Ser. B* **259**:133–139.

Puig-Duran, E., and MacKinnon, P. C. B., 1978, The ontogeny of luteinizing and prolactin hormone responses to oestrogen stimuli and to a progesterone stimulus after priming with either a natural or synthetic oestrogen in immature female rats, *J. Endocrinol.* **76**:311–320.

Puig-Duran, E., Greenstein, B. D., and MacKinnon, P. C. B., 1978, Serum oestrogen-binding components in developing female rats and their effects on the unbound oestadiol-17β fraction, serum FSH concentrations and uterine uptake of ^{3}H-oestradiol-17β, *J. Reprod. Fertil.* **56**:707–714.

Quinn, D. L., 1966, Luteinising hormone release following preoptic stimulation in the male rat, *Nature (London)* **209**:891–892.

Rainbow, T. C., Parsons, B., and McEwen, B. S., 1982, Sex differences in rat brain oestrogen and progestin receptors, *Nature (London)* **300**:648–649.

Raisman, G., and Brown-Grant, K., 1977, The suprachiasmatic syndrome. Endocrine and behavioral abnormalities following lesions of the suprachiasmatic nuclei in the female rat, *Proc. R. Soc. London (Biol.)* **198**:297–314.

Raisman, G., and Field, P. M., 1971, Sexual dimorphism in the preoptic area of the rat, *Science,* **173**:731–733.

Raisman, G., and Field, P. M., 1973, Sexual dimorphism in the neuropil of the preoptic area of the rat and its dependence on neonatal androgen, *Brain Res.* **54**:1–29.

Raynaud, T. P., Mercier-Bodard, C., and Baulieu, E. E., 1971, Rat estradiol binding plasma protein (EBP), *Steroids* **18**:767–788.

Remillard, G. M., Andermann, F., Testa, G. F., Gloor, P., Aube, M., Martin, J. B., Feindel, W., Guberman, A., and Simpson, C., 1983, Sexual ictal manifestations predominate in women with temporal lobe epilepsy: A finding suggesting sexual dimorphism in the human brain. *Neurology (New York)* **33**:323–330.

Resko, J. A., Feder, H. H., and Goy, R. W. 1968, Androgen concentrations in plasma and testis of developing rats, *J. Endocrinol.* **40**:485–491.

Resko, J. A., Malley, A., Begley, D. E., and Hess, D. L., 1973, Radioimmunoassay of testosterone during fetal development of the rhesus monkey, *Endocrinology* **93**:156–161.

Rethelyi, M., 1977, Regional and sexual differences in the size of the neuro-vascular contact surface of the rat median eminence and pituitary stalk, *Neuroendocrinology* **28**:82–91.

Sachs, B. D., Pollack, E. I., Krieger, M. S., and Barfield, R. J., 1973, Sexual behavior: Normal male patterning in androgenized female rats, *Science* **181**:770–771.

Salaman, D. F., and Birkett, S., 1974, Androgen-induced sexual differentiation of the brain is blocked by inhibitors of DNA and RNA synthesis, *Nature (London)* **247**:109–112.

Sar, M. and Stumpf, W. E., 1972, Cellular localization of androgens in the brain and pituitary after the injection of tritiated testosterone, *Experimentia* **28**:1364–1366.

Sawyer, C. H., 1957, Activation and blockade of the release of pituitary gonadotropin as influenced by the reticular formation, in: *Reticular Formation of the Brain* (H. H. Jasper, L. D. Proctor, R. S. Knighton, W. C. Noshay, and R. T. Costello, eds.), pp. 221–223, J. and A. Churchill, London.

Schwartz, N. B., and McCormack, C. E., 1972, Reproduction: Gonadal function and its regulation, *Annu. Rev. Physiol.* **34**:425–472.

Shapiro, B. H., Goldman, A. S., Steinbeck, H. F., and Neumann, F., 1976*a*, Is feminine differentiation of the brain hormonally determined?, *Experientia* **32**:650–651.

Shapiro, B. H., Goldman, A. S., Bongiovanni, A. M., and

Marino, J. M., 1976*b*, Neonatal progesterone and feminine sexual development, *Nature (London)* **264**:795–796.

Sharp, P. J., 1975, A comparison of variations in plasma luteinizing hormone concentrations in male and female domestic chickens *(Gallus domesticus)* from hatch to sexual maturity. *J. Endocrinol.* **67**:211–223.

Sheridan, P. J., Sar, M., and Stumpf, W. E., 1974, Autoradiographic localization of ^{3}H-estradiol or its metabolites in the central nervous system of the developing rat, *Endocrinology* **94**:1386–1390.

Short, R. V., 1974, Sexual differentiation of the brain of sheep, in: *Les Colloques de l'Institut National de la Santé et de la Recherche Medicale, INSERM* **32**:121–142.

Siiteri, P. K., and Wilson, J. D., 1974, Testosterone formation and metabolism during male sexual differentiation in the human embryo, *J. Clin. Endocrinol. Metab.* **38**:113–129.

Silverman, A. J., Krey, L. C., and Zimmerman, E. A., 1979, A comparative study of the luteinizing hormone releasing hormone (LHRH) neuronal networks in mammals, *Biol. Reprod.* **20**:98–110.

Sodersten, P., 1976, Lordosis behaviour in male, female and androgenized female rats, *J. Endocrinol.* **70**:409–420.

Sodersten, P., and Larsson, K., 1974, Lordosis behavior in castrated male rats treated with estradiol benzoate or testosterone propionate in combination with an estrogen antagonist, MER-23, and in intact male rats, *Horm. Behav.* **5**:13–18.

Staudt, J., and Dorner, G., 1976, Structural changes in the medial and central amygdala of the male rat, following neonatal castration and androgen treatment, *Endocrinologie* **67**:296–300.

Stearns, E. L., Winter, J. S. D., and Faiman, C., 1973, Positive feedback effect of progestin upon serum gonadotropins in estrogen primed castrate men, *J. Clin. Endocrinol.* **37**:635–648.

Steiner, R. A., Clifton, D. K., Spies, H. G., and Resko, J. A., 1976, Sexual differentiation and feedback control of luteinizing hormone secretion in the rhesus monkey, *Biol. Reprod.* **15**:206–212.

Stenberg, A., 1976, Developmental, diurnal and oestrous cycle dependent changes in the activity of liver enzymes, *J. Endocrinol.* **68**:265–272.

Stern, J. J., 1969, Neonatal castration, androstenedione and the mating behavior of the male rat, *J. Comp. Physiol. Psychol.* **69**:608–612.

Stumpf, W. E., 1968, Estradiol concentrating neurons, Topography in the hypothalamus by dry-mount autoradiography, *Science* **162**:1001–1003.

Stumpf, W. E., and Sar, M., 1971, Estradiol concentrating neurons in the amygdala, *Proc. Soc. Exp. Biol. Med.* **136**:102–114.

Stumpf, W. E., Sar, M., and Kepper, D. A., 1974, Anatomical distribution in the central nervous system of mouse, rat, tree shrew and squirrel monkey, *Adv. Biosci.* **15**:77–84.

Swanson, H. E., and van der Werf ten Bosch, J. J., 1964, The "early androgen" syndrome: Differences in response to pre-natal and post-natal administration of various doses of testosterone propionate in female and male rats, *Acta Endocrinol.* (Copenhagen) **47**:37–50.

Swanson, H. E., and van der Werff ten Bosch, J. J., 1965, The "early androgen" syndrome; effects of prenatal testosterone propionate. *Acta Endocrinol. (Copenhagen)* **50**:379–390.

Swanson, H. H., 1967, Alteration of sex typical behavior of hamsters in open field and emergence tests by neonatal administration of androgen or oestrogen, *Anim. Behav.* **15**:209–216.

Swanson, H. H., and Crossley, D. A., 1971, Sexual behaviour in the golden hamster and its modification by neonatal administration of testosterone propionate, in: *Proceedings of the International Conference on Hormones in Development* (M. Hamburgh and E. J. W. Barrington, eds.), pp. 677–687, Appleton-Century-Crofts, New York.

Terasawa, E., Kawakami, M., and Sawyer, C. H., 1969, Induction of ovulation by electrochemical stimulation in androgenized and spontaneously constant oestrous rats, *Proc. Soc. Exp. Biol. N.Y.* **132**:497–501.

ter Haar, M. B., MacKinnon, P. C. B., and Bulmer, H. G., 1974, Sexual differentiation in the phase of the circadian rhythm of ^{35}S-methionine incorporation into cerebral proteins, and of serum gonadotrophin levels, *J. Endocrinol.* **62**:257–265.

Toran-Allerand, C. D., 1983, Sex steroids and the development of the newborn mouse hypothalamus and preoptic area *in vitro.* II. Morphological correlates and hormonal specificity. *Brain Res.* **189**:413–427.

Toran-Allerand, C. D., Gerlach, J. L., and McEwen, B. S., 1978, Autoradiographic localization of ^{3}H-estradiol in relation to steroid responsiveness in cultures of the hypothalamus/preoptic area, *Neurosci. Abs.* **4**:129.

Turner, J. W., 1976, Influence of neonatal androgen on the display of territorial marking behavior in the gerbil, *Physiol. Behav.* **15**:265–276.

Urbanski, J. F., and Follett, B. K., 1982, Sexual differentiation of the photoperiodic response in Japanese quail, *J. Endocrinol.* **92**:279–282.

Valenstein, E. S., Kakolewski, J. W., and Cox, V. C., 1967, Sex differences in taste. Preference for glucose and saccharine solutions, *Science* **146**:942–943.

van Wagenen, G., and Hamilton, J. B., 1943, Experimental production of pseudohermaphroditism, in: *Essays in Biology* (T. Cowles, ed.), pp. 581–607, University of California Press, Berkeley.

Velasco, M. E., and Taleisnik, S., 1969, Release of gonadotropins induced by amygdaloid stimulation in the rat, *Endocrinology* **84**:132–139.

Vertes, M., Barnea, A., Lindner, H. R., and King, R. J. B., 1973, Studies on androgen and estrogen uptake by rat hypothalamus, in: *Receptors for Reproductive Hormones* (B. W. O'Malley and A. R. Means, eds.), pp. 137–173, Plenum Press, New York.

Wachtel, S. S., Ohno, S., Koo, G. C., and Boyse, E. A., 1975, Possible role for H-Y antigen in the primary determination of sex, *Nature (London)* **257**:235–236.

Wade, G. H., and Zucker, I., 1969, Taste preferences of female rats modification by neonatal hormones, food deprivation and prior experience, *Physiol. Behav.* **4**:935–943.

Weisz, J., and Gibbs, C., 1974, Metabolites of testosterone in the brain of the newborn female rat after an injection of tritiated testosterone, *Neuroendocrinology* **14**:72–86.

Weisz, J., and Ward, I. L., 1980, Plasma testosterone and progesterone titers of pregnant rats, their male and female fetuses and neonatal offspring, *Endocrinology* **106**:306–316.

Wells, L. J., and van Wagenen, H. G., 1954, Androgen induced female pseudohermaphroditism in the monkey *(Macaca mulatta)* anatomy of the reproductive organs, *Contrib. Embryol.* **35**:93–106.

Welshons, W. J., and Russell, L. B., 1959, the Y-chromosome as the bearer of male determining factors in the mouse, *Proc. Natl. Acad. Sci. U.S.A.* **45**:560–566.

Westley, B. R., and Salaman, D. F., 1976, Role of oestrogen receptor in androgen-induced sexual differentiation of the brain, *Nature (London)* **262**:407–408.

Westley, B. R., and Salaman, D. F., 1977, Nuclear binding of the oestrogen receptor of neonatal rat brain after injection of oestrogens and androgens: Localization and sex differences, *Brain Res.* **119**:375–388.

Whalen, R. E., 1966, Sexual motivation, *Psychol. Rev.* **73**:151–163.

Whalen, R. E., 1968, Differentiation of the neural mechanisms which control gonadotropin secretion and sexual behavior, in: *Perspectives in Reproduction and Sexual Behavior* (M. Daiamond, ed.), pp. 303–340., Indiana University Press, Bloomington.

Whalen, R. E., and Edwards, D. A., 1967, Hormonal determinants of the development of masculine and feminine behavior in male and female rats, *Anat. Rec.* **157**:173–180.

Whalen, R. E., and Luttge, W. G., 1970, Long-term retention of tritiated estradiol in brain and peripheral tissues of male and female rats, *Neuroendocrinology* **6**:255–263.

Whalen, R. E., Battle, C., and Luttge, W. G., 1972, Testosterone androstenedione and dihydrotestosterone: Effects on mating behavior of male rats, *Behav. Biol.* **2**:117–125.

White, T. O., Hall, C., and Lim, L., 1979, Developmental changes in the content of oestrogen receptors in the hypothalamus of the female rat, *Biochem. J.* **184**:465–468.

Witschi, E., 1962, Embryology of the ovary, in: *The Ovary* (H. G. Grady and D. E. Smith, eds.), pp. 1–10, International Academic Pathology Monograph No. 3, Williams & Wilkins, Baltimore.

Yalom, I., Green, R., and Fisk, H., 1973, Prenatal exposure to female hormones—Effect on psychosexual development in boys, *Arch. Gen. Psychiatry* **28**:554–561.

Yanasi, M., and Gorski, R. A., 1976, Sites of oestrogen and progesterone facilitation of lordosis behavior in the spayed rat, *Biol. Reprod.* **15**:536–543.

Yazaki, I., 1960, Further studies on endocrine activity of subcutaneous ovarian grafts in male rats by daily examination of smears from vaginal graft, *Annot. Zool. Jpn.* **33**:217–225.

Yen, S. S. C., Tsai, C. C., Vandenberg, G., and Rebar, R., 1972, Gonadotropin dynamics in patients with gonadal dysgenesis: A model for the study of gonadotropin regulation, *J. Clin. Endocrinol. Metab.* **35**:897–904.

Zigmond, R. E., 1975, Binding, metabolism and action of steroid hormones in the central nervous system, in: *Handbook of Psychopharmacology* (L. L. Iversen, S. D. Iversen, and S. H. Snyder, eds.), Vol. 5, pp. 239–328, Plenum Press, New York.

17

Patterns of Early Neurological Development

INGEBORG BRANDT

1. Introduction

1.1. Relevance of Early Neurological Development

There is now greater understanding of the hitherto secret life of the fetus. Many recent advances have been made, such as continuous real-time ultrasound observations of early fetal movements (i.e., from 7 postmenustrual weeks) and behavior (de Vries *et al.*, 1982; Prechtl, 1983*b*). Since fetal motor behavior represents central nervous system (CNS) output, a limited neurological examination of the young fetus is now possible (see Section 3). Careful longitudinal examinations of normal fetal behavior and its variability by real-time ultrasound during the first half of pregnancy (de Vries *et al.*, 1982) serve as a basis for this assessment. Up to 20 weeks, ultrasound examination is facilitated by the fact that the entire fetus can be visualized within the field of one ultrasound transducer.

The fetal intrauterine movements are similar to movements after birth if the effects of the amniotic fluid are recognized. The fetus moves like an astronaut. de Vries *et al.* (1982) classified 16 distinct movement patterns that can be seen by 14 weeks.

The frequency and quality of early fetal movements are used as prognostic signs by obstetricians. Analyzing 4313 pregnancies before 20 postmenstrual weeks, Schmidt *et al.* (1981) found that more abortions occurred in a group with infrequent fetal movements. Prenatal brain pathology can also be studied by ultrasound and then related to fetal motor behavior.

As most cases of severe mental retardation (Chamberlain, 1975; Hagberg, 1979) and many cases of cerebral palsy (Michaelis and Hege, 1982) are of prenatal origin, analysis of fetal behavior by sequential ultrasound, especially in high-risk pregnancies, should be of diagnostic aid. Ideally, an abnormality should be detected at the time of its occurrence in the hope of identifying its cause and of preventing it in future. Recently, "biophysical profiles" have been developed as a method for antenatal evaluation of fetal well-being. Manning *et al.* (1981) combined observations of fetal movement, tone, reactivity, and breathing with qualitative amniotic fluid volume for scoring during the last trimester of pregnancy. The extrauterine behavior of preterm infants during the last weeks before term can now be compared with the intrauterine neurological development of the fetus at the corresponding time.

We know that many primary reflexes and reactions have intrauterine functions (Langreder, 1949) and that they remain active for some time after birth. Of the five senses, the sense of touch seems to be of greatest importance for the fetus. By this it can avoid obstacles and is therefore prevented from becoming entangled with the umbilical cord. After intrauterine death, secondary umbilical cord entanglements are frequently observed (Langreder, 1949).

INGEBORG BRANDT • University Children's Hospital, University of Bonn, 5300 Bonn 1, Federal Republic of Germany.

Active fetal movements and responses develop earlier and resemble postnatal behavior much more than previously realized. The CNS is not just switched on at birth; the neonate has been well prepared for extrauterine life.

After birth the pattern of early neurological development is used for maturational assessment in preterm infants, since there are phases of some reactions and reflexes that appear at clear-cut times before term. Repeated neurological examination in the perinatal period permits the achievement of successive neuromuscular "milestones" to be followed. In contrast, external characteristics, used in the assessment of gestational age (Farr *et al.*, 1966*a,b*), need to be scored within a time limit from 12 to 36 hr postnatal age because the "results apply only to babies examined at this age."

Knowledge of the postmenstrual age at birth is a prerequisite for proper evaluation of an infant and for differentiating between small-for-gestational-age (SGA) infants and appropriate-for-gestational-age (AGA) infants. Infants of the same weight but of different gestational age behave quite differently as regards risks in the newborn period (Battaglia and Lubchenco, 1967), neurological development (Robinson, 1966), and prognosis (Brandt, 1975; Schröder, 1977; Hagberg, 1979; Ounsted *et al.*, 1983).

Also, for later developmental assessment of preterm infants, at least in the first and second year of life, the age must be corrected for prematurity; i.e., by subtracting the time of prematurity from chronological age giving corrected age. Therefore, knowledge of postmenstrual age at birth is essential (Gesell and Amatruda, 1967; Parmelee and Schulte, 1970). According to the results of the Bonn longitudinal study (Brandt 1983; Schröder, 1977), the developmental and intelligence quotients (Griffiths, 1964) and Stanford-Binet (Terman and Merrill, 1965) differ significantly from 3 to 54 months if chronological and corrected age are compared.

Today ultrasound is increasingly used for head scanning through the open fontanelle in young infants, particularly neonates, and recently *in utero,* in order to evaluate anatomy and pathology of the brain (Babcock and Han, 1981; Couture and Cadier, 1983). Consequently, knowledge of normal early neurological development and its variability in preterm and full-term neonates is gaining increasing attention and interest.

Nuclear magnetic resonance (NMR) tomography is another noninvasive technique for imaging of the CNS. It has the unique advantage of differentiating gray from white matter. Even in preterm infants, a more precise morphological diagnosis is now possible. Since we know that very small preterm infants frequently suffer from intracranial hemorrhage (Pape and Wigglesworth, 1979) a continuing refinement of current examination techniques and development of new ones, which take into account the rapid development of the CNS at this early age, are badly needed.

In preterm infants of very early postmenstrual age, pathological intracranial findings, e.g. intraventricular hemorrhage (IVH) may occur without associated neurological abnormality (Brand, 1983). This may be because the corresponding brain structures are immature and also because the particular developmental characteristics of the preterm infant are not well tested in a neurological examination designed for the full-term neonate. This points to the need for more detailed knowledge of neurological development during the last 13 postmenstrual weeks.

There is an increasing interest in the functional equivalents of pathological findings in the brain; e.g., IVH or periventricular leucomalacia in preterm neonates. Dubowitz and Dubowitz (1981) correlated clinical signs including limb tone, popliteal angle, motility, and visual orientation with the presence of IVH in very preterm infants. However limb tone, for example, was only indicative of IVH in babies older than 31 weeks.

Nowadays, with better neonatal care and early and high-energy feeding, preterm infants are discharged from hospital earlier than previously, and often before their expected date of delivery. In many cases their postmenstrual age is only 36–37 weeks. As muscle tone increases until the expected date of delivery (see Section 4.3), there is the risk that this phenomenon could be misinterpreted as the beginning of spasticity. Unnecessary measures, such as physiotherapy, may be initiated and additional anxiety produced. On the other hand, preterm infants with underlying pathology and developing hypotonia (i.e., when the tone does not increase physiologically) may be considered normal, and again the diagnosis is missed.

Here the knowledge of normal neurological development in the last weeks before term is important not only for the neonatologist but also for all health professionals.

A further reason for improving neurological examination methods is the increased number of at-risk preterm infants who survive today without an overall increase of handicap (Hagberg *et al.*, 1982).

Preterm infants are an easily identifiable group in whom an early diagnosis of handicap is especially desirable with the goal of early intervention.

This chapter deals only with neurological examinations that are clearly defined and easy to perform. The more complex techniques for neuromaturational assessment, such as motor nerve conduction velocity (Schulte, 1968; Schulte *et al.*, 1968*a*, 1969) and electromyographic evaluation of the reactions and reflexes (Schulte *et al.*, 1968*b*; Schulte and Schwenzel, 1965), which require special equipment, will not be discussed.

1.2. Definitions

1.2.1. Postmenstrual Age, Conceptional Age, Synonyms

Postmenstrual, fetal, or gestational age is the time from the first day of the mother's last menstrual period until birth. In the preterm infant, postmenstrual age should be used at all examinations up to, and often after, 40 postmenstrual weeks, irrespective of birth. The definitions of postmenstrual age and postconceptional age (time from conception to birth) should be clearly distinguished. The confusing use of "conceptional age" for the time from the first day of the mother's last menstrual period until birth, plus the postnatal period of life (e.g., by Saint-Anne Dargassies, 1966; Graziani *et al.*, 1968; Schulte, 1974; Prechtl and Beintema, 1976) is to be avoided.

André-Thomas and Saint-Anne Dargassies (1952) use the term conceptional age, but they clearly define it as time from conception. They distinguish exactly the different "ages" of a preterm infant: (1) intrauterine life *(vie utérine)*; (2) life outside the uterus *(vie aérienne)*; and (3) total age *(vie totale)*. An infant, born at 6 months intrauterine life, and examined at 3 months outside the uterus, is considered as 9 months old (total age) and may be compared with a full-term infant. As long as these clear definitions are observed, conceptional age is considered by André-Thomas to be of great importance.

If, in the present literature, the term conceptional age is used as postmenstrual age plus postnatal age combined, this confusion adds to the complication of newborn neurology. This chapter shows that there are considerable differences between infants of, for example, 34 postmenstrual weeks and 36 postmenstrual weeks; i.e., the difference between postmenstrual age and conceptional age of 14 days, provided that conception is calculated as occurring 14 days after the first day of the last menstrual period.

For differentiation between a newborn infant of 35 postmenstrual weeks and an infant of the same age born at 28 postmenstrual weeks and with an age of 7 weeks postnatally, it is suggested that two ages be given: the postmenstrual age at birth and the postnatal age.

1.2.2. Embryo and Fetus

There is no general agreement on the anatomical or physiological developmental stage that marks the end of the embryonic stage. Nesbitt (1966) considers the period until the end of the second month from conception the stage of the embryo, the time during which the major differentiation of organs takes place. Thereafter, during the rest of intrauterine life, the growing organism is referred to as a fetus.

1.2.3. Perinatal Period

The perinatal period is defined as the time from 28 postmenstrual weeks to the 7th day after term (Psychyrembel and Dudenhausen, 1972).

1.2.4. Appropriate-for-Gestational-Age and Small-for-Gestational-Age Infants

Preterm infants are defined as those born with a postmenstrual age of less than 38 weeks (i.e., 37 weeks + 6 days or less), irrespective of birth weight (McKeown and Gibson, 1951; Battaglia and Lubchenco, 1967; Brandt, 1978).

To quantitate normal fetal growth and to diagnose fetal growth retardation, intrauterine growth standards, based on postnatal measurements of infants born at different postmenstrual ages, are necessary. A further prerequisite is the knowledge of postmenstrual age. The differentiation between AGA and SGA preterm infants is necessary, since infants of the same weight behave neurologically quite differently if they are not of the same postmenstrual age. Neurological development—with few exceptions—is a function of age and not of weight.

Preterm infants with normal intrauterine development and with birth weights between the tenth and ninetieth percentiles of Lubchenco *et al.* (1963) are defined here as appropriate for gestational age. Small-for-gestational-age infants are defined as those with a birth weight below the tenth percentile of Lubchenco *et al.* (1963) or the tenth percentile of Hosemann (1949).

1.2.5. Early Neurological Development

"Early" may be defined differently, depending on the starting point; i.e., conception, preterm birth, or birth at expected date of delivery. Three periods of early development may be differentiated: (1) intrauterine; (2) extrauterine after preterm birth from 28 postmenstrual weeks to the seventh day after term—the perinatal period; and (3) from birth at term to the seventh day—the newborn period. This chapter emphasizes the perinatal period.

2. Survey of Methods

2.1. Assessment of Gestational Age before Birth: Maternal History and Ultrasound Examinations

Knowledge of the gestational age of the fetus is important for assessment of the intrauterine development. Primarily this is determined by the maternal history; i.e., last menstrual period. A basal temperature curve may be used to show the date of conception.

Recently more precise determination of fetal age has become possible by ultrasound measurements, which offer today the best opportunities for serial studies of fetal development. The earliest assessment is made by measurement of the diameter of the gestational sac (Hackelöer and Hansmann, 1976). Measurements of fetal crown–rump length (CRL) in the first trimester of pregnancy, introduced by Robinson (1973; Robinson and Fleming, 1975), prove to be the most reliable method for gestational age assessment. Between 11 and 18 postmenstrual weeks there is a period of very rapid growth velocity of CRL that amounts to 10–12 mm/week with a peak of 12.6 mm in the fourteenth postmenstrual week (Hansmann *et al.*, 1979). This rapid growth velocity enables an accurate estimation of gestational age; one single measurement yields an accuracy of ± 4.7 days and three independent measurements even of ± 2.7 days with a reliability of 95% (Robinson and Fleming, 1975).

These results are more precise than assessment of the expected date of delivery from the last menstrual period of the mother after the rule of Naegele; i.e., to subtract 3 months from the first day of the mother's last menstrual period, and to add 7 days.

In the second trimester ultrasound measurements of the biparietal diameter—before it exceeds 78 mm, corresponding to the twenty-ninth week of gestation—enable the obstetrician to estimate fetal age with confidence limits ± 10 days in 96% of cases (Hansmann, 1974).

Determination of fetal age in the third trimester by a biparietal diameter greater than 90 mm is clinically no longer reliable (range of two standard deviations = ± 24 days) (Hansmann, 1976).

2.2. Observation of Intrauterine Fetal Movements by Ultrasonography

Intrauterine fetal movements can be observed directly using real-time scanning, permitting neurophysiological examinations in the normal physiological environment (Reinold, 1971, 1976; de Vries *et al.*, 1982), in contrast to earlier studies on the fetus, made outside the uterus after premature termination of pregnancy (Humphrey, 1964). This method is considered a useful diagnostic tool for the early detection of fetal distress and abnormalities.

2.3. Tests Other Than Neurological Signs Used as Maturation Criteria after Birth

External superficial criteria have been introduced by von Harnack and Oster (1958) and by Farr *et al.* (1966*a,b*) for maturational assessment in the first 12–36 hr. The results apply only to infants examined at this age. The examination of joint mobility is also considered helpful by some workers (Dubowitz *et al.*, 1970; Dubowitz and Dubowitz, 1977; Saint-Anne Dargassies, 1966, 1974). Extensibility of the joints in the full-term newborn infant may be demonstrated; e.g., by the dorsiflection angle of the foot, and the volarflection angle of the hand. This is related to the maturational age reached *in utero.* With increasing gestational age, near the expected date of delivery, joint mobility is increased due to synergistic influences of placental hormones. This hormonal factor is absent in preterm infants.

2.4. Neurological Examination of Reflexes and Reactions in the Perinatal Period

The great contribution of the Paris School (André-Thomas and Saint-Anne Dargassies, 1952; Saint-Anne Dargassies, 1955, 1966, 1974) is to have demonstrated age-dependent changes in the neurologic behavior of preterm infants before their expected date of delivery. Periodic testing of the preterm infant offers the opportunity to observe extrauterine neurological maturation, which normally occurs *in utero* in the third trimester of pregnancy, since many responses develop similarly under both conditions. The close relationship of gestational age with many developmental patterns allows maturational assessment based on clinical-neurological examination.

The neurological examination elaborated for full-term newborn infants by Prechtl and Beintema (1964) is well defined and quantifiable. Moreover, Beintema (1968) demonstrated the inter- and intra-individual variability of these neurological responses in normal newborn infants in the first 9 postnatal days.

In many studies of neurological development of preterm infants (Robinson, 1966; Graziani *et al.*, 1968; Meitinger *et al.*, 1969; Finnström, 1971; Michaelis *et al.*, 1975), the examination methods are based on those of Prechtl and Beintema (1964). The design of a neurologic examination must take into account the age-specific properties of the developing nervous system, and this postulate of Prechtl (1970) is especially valid for preterm infants. Furthermore, the neurological tests must be centered on sufficiently complex functions of the nervous system and not on artificially fragmented responses (Prechtl, 1970).

2.5. Consideration of Infant State

Behavioral states are now recognized even in the fetus (Nijhuis *et al.*, 1982). Following the strategy used by Prechtl and Beintema (1964) for the newborn, four categories of behavioral states have been designated. These four behavioural states can be categorized as follows:

State 1F: Quiescence, which can be regularly interrupted by brief gross body movements, mostly startles; eye movements absent; heart rate stable, with a small oscillation bandwidth, isolated accelerations occur; these are strictly related to movements; this heart rate pattern is called FHRPA.

State 2F: Frequent and periodic gross body movements—mainly stretches and retroflexions—and movements of the extremities; eye movements continually present (REMs and SEMs); heart rate (called FHRPB) with a wider oscillation bandwidth than FHRPA; and frequent accelerations during movements.

State 3F: Gross body movements absent; eye movements continually present; heart rate (called FHRPC) stable, but with a wider oscillation bandwidth than FHRPA and no accelerations.

State 4F: Vigorous, continual activity including many trunk rotations. eye movements continually present (when observable); heart rate (called FHRPD) unstable, with large and long-lasting accelerations; frequently fused into a sustained tachycardia (Nijhuis *et al.*, 1982).

In the normal fetus these states begin to appear between 36 and 38 weeks gestation and remain until delivery.

Prechtl (1970) emphasized that it is necessary to standardize the examination technique for an infant and that particular functions should only be assessed in particular states because of the wide but systematic fluctuations in the readiness of particular nervous system functions to act. Since the behavioral state of an infant is often influenced by the examination procedure, strict adherence to a rigid sequence of tests is recommended (Prechtl, 1970; Prechtl and Beintema, 1964).

The concept of state is a prerequisite for a quantitative neurological assessment (Prechtl, 1974). The criteria and definitions of states vary between investigators. Prechtl and Beintema (1964) distinguish five different behavioral states in newborn infants:

State 1: Eyes closed, regular respiration, no movements;
State 2: Eyes closed, irregular respiration, no gross movements;
State 3: Eyes open, no gross movements;
State 4: Eyes open, gross movements, no crying;
State 5: Eyes open or closed, crying.

Beintema (1968) defined the term "state of an infant" according to Ashby (1956): "By a state of a system is meant any well-defined condition or property that can be recognized if it occurs again." In *Neurological Study of Newborn Infants,* Beintema (1968) used the states described by him and Prechtl (1964). In addition, he used the following scores for alterability of a state by handling:

0: Almost impossible to alter the state of the infant at any time in the examination;
1: Usually difficult to obtain the optimal state at any time;
2: The examiner sometimes succeeds in altering the state;
3: Fairly easy to obtain the optimal state;
4: Very easy to obtain the optimal state.

The relationship of 14 reflexes and reactions in the newborn to behavioral state was studied by Lenard *et al.* (1968). The reflexes were elicited in three states only: state 1, regular sleep; state 2, irregular sleep; and state 3, quiet wakefulness.

According to the relationship with state, three groups of reflexes are distinguished:

1. Reflexes equally strong during regular sleep and wakefulness, but weak or absent during irregular sleep. To this group belong, among others, the knee jerk and the Moro reflex.
2. Responses mostly absent during regular sleep, weak during irregular sleep, and strongest during wakefulness; all are exteroceptive reflexes. To this group belong the palmar grasp and the glabella reflex.
3. Reflexes that do not show any alterations but that are easily obtained in all three states. To this group belongs, among others, the Babinski reflex.

Robinson (1966) used "state of arousal" for each response following the system of Prechtl and Beintema (1964) for preterm infants. If there existed a "required state" for a response, the babies were only tested by him in this state. Also, Finnström (1971) applied the states, as defined by Prechtl and Beintema (1964), for the preterm infants in his study of maturity in newborn infants. Graziani *et al.* (1968) did not consider different groups of states, but described the general state during testing, e.g., "awake state: gross body movements struggling and stretching, but in general not crying, interfered with reliable scores." Dubowitz and Dubowitz (1981) used the six gradings of state defined by Brazelton (1973) in the neurological assessment of newborn infants.

Michaelis, in his studies of SGA infants (Michaelis, 1970; Michaelis *et al.*, 1970), used the states defined by Prechtl and Beintema (1964).

Saint-Anne Dargassies (1966, 1974) described, instead of states, the quality of vigilance of an infant; i.e., 0 = no vigilance during the examination; 1 = good and lasting; 2 = good but of short duration; 3 = bad but lasting; and 4 = bad and disappearing *par éclipse.*

Since there exists no adequate classification of states for preterm infants, those described by Prechtl (1974) and Prechtl and Beintema, (1964) for full-term infants are used in our Bonn study. The state is recorded as that demonstrated by the infant for most of the duration of examination; i.e., more than 50%. Since with every neurological examination anthropometric measurements are also made, sometimes it is possible to alter the state to the optimal one for the corresponding neurological response. Throughout every examination 6 to 8 photographs are made, so that the states can be checked in evaluating the results.

2.6. Drawbacks of Early Neurological Examination

The neurological examination of a preterm infant during the first 2 days of life is of little value for the determination of gestational age, because factors greatly influencing the general condition of the infant will prejudice the results. This is a serious shortcoming, since the early determination of gestational age—especially in the SGA infant—is important because of the need for special care immediately after birth. Clinical external characteristics and nerve conduction velocity assessment are more valuable in these conditions.

If the examination is performed after the third day, short-acting factors directly related to birth are usually eliminated (Escardó and de Coriat, 1960). Furthermore, in infants with a normal Apgar (1953) score at birth, the exmination results from the third to fifth postnatal day correspond to the true gestational age.

In infants in a poor general state of health; e.g., with intracranial hemorrhage, hyperbilirubinemia, general hypotonia, marked edema, infections, or electrolyte disturbances, only unreliable results are to be expected from the neurological examination. Hypotonia may lead one erroneously to consider the infant as younger than it is. As soon as the general state improves, the infant exhibits the reactions and reflexes corresponding to postmenstrual age. This catchup should not be misinterpreted as "accelerated" neurological development.

Some neurological responses are impaired in infants born in the breech position (Prechtl and Knol, 1958). Depressant drugs, such as barbiturates or intravenous narcotics, given the mother immediately before or during birth, accentuate and prolong the disorganizing effects that seem to occur as a normal result of the birth process (Brazelton, 1961). The extent of relative CNS disorganization seems positively correlated with the type, amount, and timing of the medication given to the mother, whereas inhalant anesthesia has a more transient effect than premedication, for example, with a barbiturate.

According to Schulte (1970*a*), on the basis of neurological criteria, in about 70% of normal newborn infants the gestational age may be assessed with an accuracy of 2 weeks; for 97% of infants this is possible to 3–4 weeks. In abnormal infants the responses are abnormal too, and an age assessment becomes impossible because the abnormal responses may imitate immature behavioral patterns. This pertains for motor development and also for EEG. Even nerve conduction velocity assessment does not allow a very precise age determination. Schulte (1970*a*) further stated that neurological criteria are reliable only in completely normal infants but, in abnormal infants, for whom knowledge of postmenstrual age is of clinical significance, they are useless. Abnormal results from an examination, then, may either signify that the infant is younger than reported by the mother or that the infant is abnormal.

The critical objections of Schulte (1970*a*) hold when neurological development phases are used for the determination of gestational age in ill infants. In cases in which the age of the newborn infant is known with some certainty, the neurological behavior is a valuable diagnostic tool and permits a prognosis (Schulte, 1973). Moreover, with increasing use of ultrasound measurements in early pregnancy, especially in high-risk cases, there will be an increasing number of newborn infants with a precise age assessment, and neurological development will regain its significance for longitudinal observations in the perinatal period.

Comparison of some of the neurological responses described in the literature is made difficult by differences in examination methods, in methods of reporting results, and in definition of abnormal responses.

2.7. Longitudinal Neurological Examinations of AGA Preterm Infants from the Bonn Study

Out of the Bonn longitudinal study (Brandt, 1975, 1976, 1978), a group of 29 AGA preterm infants (16 girls and 13 boys), born between 28 and 32 postmenstrual weeks (mean 30 weeks) were studied. The selection was made according to the following criteria:

1. The menstrual cycles of the mother were normal. No oral contraceptives were used in the last 6 months before conception.
2. Data on the last menstrual period of the mother were reliable, and regular obstetrical examinations during pregnancy were done. In some cases the date of conception was known because of measurement of the basal temperature; in others ultrasound follow-up examinations in early pregnancy were performed.
3. The mother had a normal prenatal course and no severe complications of labor and delivery.
4. The recognition of fetal movements by the mother was within the normal range; i.e., for primipara between 18 and 22 postmenstrual weeks and for multipara between 16 and 20 weeks (Hansmann, personal communication, 1977).
5. The neonatal course was uncomplicated.
6. All the infants had a birth weight (mean 1350 g) between the tenth and nintieth percentiles of Lubchenco *et al.* (1963) at their gestational age.
7. The long-term outcome of the infants was normal; i.e., their neuromotor and psychological development, followed longitudinally at least until the age of 6 years, was normal as compared with full-term control infants (Schröder, 1977).

These preterm infants represent a relatively homogeneous group with respect to postmenstrual age, birth weight, and risk factors. The optimality criteria from R. Michaelis (personal communication, 1975) were used—an extension of those introduced by Prechtl (1968), giving 50 instead of 42 items. The fiftieth percentile for the Bonn preterm infants was found to be 41.4 (8.6 nonoptimal criteria), similar to that of the preterm infants in a study by Michaelis *et al.* (1979). It must be noted that two items for the preterm infant can never be optimal: postmenstrual age and birth weight. Excluding these two criteria from the total of 50 gives a reduction in optimality of 6.6 out of 48; i.e., 13.8%, which corresponds to Prechtl's (1968) middle-risk group of full-term infants.

After birth, the external maturational criteria

were assessed according to von Harnack and Oster (1958) and Farr *et al.* (1966*a,b*).

The first neurological examination was carried out on the fourth to tenth postnatal day, 1½–2 hr after the last feed if possible. Until discharged, the infants were examined once a week, or every 2–3 weeks. All examinations were made by the author, a total of 153, performed on the 29 infants until the expected date of delivery. As in the study by Robinson (1966), the responses were tested as described by Prechtl and Beintema (1964) for the full-term infant, with slight modifications according to the developmental particularities of preterm infants of low postmenstrual age. An attempt was made to test the babies in the state required for a response to be elicited.

The state of the infants during examination was recorded as defined by Prechtl and Beintema (1964). Although judgment of states is considered problematic in preterm infants of low postmenstrual age, they were described during each examination. The sequence of the examination was the same for all infants.

Out of the 55 reflexes and reactions examined in the Bonn study, only 20 will be reported and discussed. Measurements of head circumference, supine length, and body weight were made at the time of each neurological examination.

The results presented here are based on further experiences from about 600 additional longitudinal observations on 35 AGA, 51 SGA preterm infants, and 85 full-term control infants in the perinatal period.

3. *Intrauterine Development—Fetal Movements*

Embryo–fetal motor activity starts during postmenstrual week 7 (Reinold, 1976; Vecchietti and Borruto, 1982). The first and just discernible movements occurring at 7–8.5 weeks are small shifts in the contours of the embryo that last ½ to 2 sec (de Vries *et al.*, 1982) (see also Section 1.1). This early initiation of motor activity is in accordance with the classic findings of Humphrey (1964), who reported a contralateral flexion of the neck elicited by perioral tactile stimulation. The CRL of the 8-week-old fetus is only 1.1 cm (Hansmann *et al.*, 1979). In the brain at 6–8 weeks, some of the neuroepithelial cells (ventricular cells) of the ventricular zone, which are the progenitors of all neurons and macroglial cells of the adult CNS (Boulder Committee, 1970), are undergoing differentiating mitosis and become immature nerve cells. They migrate farther outward and, at 8–10 weeks, as a first wave forming a new accumulation of cells, which is the beginning of the neocortical plate (stage I, the primary cortical plate according to Sidman and Rakic, 1973). Larroche (1981) succeeded in identifying a few axodentritic synapses in the neocortex of fetuses as early as 9 postmenstrual weeks.

In fetuses of 14 mm CRL corresponding to 8–9 postmenstrual weeks (Robinson and Fleming, 1975; Hansmann *et al.*, 1979), Okado *et al.* (1979) demonstrated the appearence of a small number of axodendritic synapses in the motor neuropil of the cervical cord. Between 9 and 10 postmenstrual weeks, there is a substantial increase in the numbers of axodendritic synapses from 1.4 synapses per 200 μm^2 to 10 synapses per 200 μm^2 in the lateral motor column (Okado, 1980). Precise interpretation of Okado's data is made difficult because of his use of Streeter's (1920) fetal growth standards for CRL which up to 13 postmenstrual weeks give too great a length for given fetal age (Brandt, 1984).

At 8 and 9 weeks, general movements can be observed, when the whole body is moved without recognizable patterning of body parts (de Vries *et al.*, 1982).

Since the fetal body floats in the amniotic fluid—its specific weight being only slightly above that of the fluid—very little effort is needed for the fetus to move. At 10 to 12 weeks, the general movements become forceful. They are of large amplitude, frequently lead to changes in the fetal position, and are on average, 10 per hour (de Vries *et al.*, 1982). Between 8 and 12 postmenstrual weeks, the fetus has quadrupled its CRL, which is now 4.32 cm (Hansmann *et al.*, 1979). In the brain at 10–11 weeks, the cortical plate increases in thickness, becomes more compact, and is clearly demarcated from the fiber-rich part of the intermediate zone (stage II, according to Sidman and Rakic, 1973).

After 12 weeks, general fetal movements are more variable in speed and amplitude and may last about 1–4 min but wax and wane during this period; an important characteristic is their gracefulness (de Vries *et al.*, 1982). Sixteen different movement patterns are described by de Vries *et al.* (1982), all of which can also be observed after birth. From 11 to 13 fetal weeks, the cortical plate in the brain, having increased in thickness, becomes subdivided into an inner zone, occupied mainly by cells with relatively large somewhat widely spaced nuclei and an outer zone of cells with densely packed oval nuclei elongated in the axis perpendicular to the cortical surface. This is the stage of bilaminate cortex (stage III, according to Sidman and Rakic, 1973). At 12 postmenstrual weeks, Larroche *et al.* (1981) and Larroche and Houcine (1982) demonstrated that there are already numerous synapses of the axodendritic type above and below the cortical plate.

At 13–15 postmenstrual weeks, the number of positional changes of the fetus shows a peak: 10–20 changes per hour, including rotations around the sagittal or transverse axis with somersaults. However, the supine position exceeds all other positions during this time (de Vries *et al.*, 1982). de Vries *et al.* (1982) described a complete change in position around the transverse axis with a backward somersault, achieved by a general movement with alternating leg movements resembling neonatal stepping. Also, rotations around the longitudinal axis occur by leg movements with hip rotation or by rotation of the head, followed by trunk rotation. Complex phenomena such as sucking and swallowing are also seen in 14-week-old fetuses. From 13 to 15 weeks, there is a period of secondary condensation in the brain. This second wave involves more cells than previously. A key question, with implications for the organization of normal cortex as well as for the pathogenesis of developmental disease, concerns the mode of cell migration. How do cells find their way from proliferative centers to their final locations, which are becoming increasingly more distant as the fetal brain enlarges? Is the mechanism of migration the same during the initial stages of cortex formation, when the pathways are only about 100 μm long, as in the later stages, when the radial trajectories may reach more than 5000 μm?

By 15 weeks, the duration of movements increases; they last about 5 min with pauses up to 14 min (de Vries *et al.*, 1982).

At 15 weeks, the CRL of the fetus has increased to 10.11 cm (Hansmann *et al.*, 1979).

After 15 weeks, the rate of changes in the fetal position decreases to about 10 per hour on average. This is not the result of a decrease of motor activity but might be attributable to the spatial conditions in the uterus for the rapidly growing fetus (de Vries *et al.*, 1982).

Reinold (1976), a pioneer in this field, introduced a classification of spontaneous fetal movements for clinical use: (1) strong and brisk movements, forceful initial motor impulse, movement involving entire body, and change in location and posture; and (2) slow and sluggish movements, no initial motor impulse, movements confined to fetal parts, and no change in location and posture. Reinold (1976) distinguished further between "spontaneous movements" and "passive fetal movements" elicitable in the absence of spontaneous movements; i.e., in the resting fetus. However, de Vries *et al.* (1982) pointed out the difficulty in distinguishing between spontaneous fetal movements and those that are elicited, because of the very high basic rate of spontaneous activity. According to Reinold (1976), who studied fetal behavior by the "motor provocation test," the absence of any response of the fetus appears to be a sign of being at risk. Similarly, Schmidt *et al.* (1981) identified fetal pathology in fetuses presenting with slow movements when not reacting appropriately to provocation by pushing the uterus *(Stoßprovokation).* Haller *et al.* (1973) showed that in ultrasound examinations, between the ninth and twenty-first weeks of gestation the frequency of movements increases up to the twenty-first week in normal pregnancies. It is difficult to describe fetal movements quantitatively (Henner *et al.*, 1975). A new method of quantifying and analyzing fetal movements was described by Henner *et al.* (1975) in which two-dimensional fetal movements, seen on a TV screen, are transferred to a geometrical diagram that is then quantified.

At about 20 postmenstrual weeks, fetal movements become more vigorous and are felt by the mother as quickening (H. Hoffbauer, personal communication, 1977). From 24 weeks, the amount of somersaults and loopings decreases, and the fetus, in turning, exhibits marked motility of the extremities (H. Hoffbauer, personal communication, 1977).

Timor-Tritsch *et al.* (1976) studied fetal movements from 26 postmenstrual weeks to term by continuous recording with a tocodynamometer. They distinguished four types of movement. The first type is a rolling movement. The second type is simple movement; i.e., short and easily palpable, possibly originating from an extremity. The third type is high-frequency movement and the fourth respiratory movement.

In the third trimester, fetal movements become more and more limited with decreasing space in the uterus. The extent is dependent on the amount of amniotic fluid; the findings are variable and are not appropriate diagnostic criteria (M. Hansmann, personal communication, 1977).

Patrick *et al.* (1982) made continuous measurements of fetal body movements with an ultrasonic real-time scanner for periods of 24 hr in 31 healthy pregnant women and reported that the number of active gross fetal body movements (rolling and stretching) was similar for the gestational age groups 30–31, 34–35, and 38–39 weeks—on average, 31 per hour and of a mean duration of 11 sec. Only 1% of intervals of inactivity were greater than 45 min, and the longest period of absence of gross fetal body movements was 75 min.

4. Extrauterine Development in the Perinatal Period after Premature Birth

4.1. Posture in Supine Position

4.1.1. Spontaneous Body Posture

In the same way as occurs in the intrauterine fetus, the extrauterine preterm infant shows spontaneous movements and changes of position. No age-typical body posture occurs before tone of the flexor muscles has increased to such a degree that the extremities remain in a more or less flexed position. Manning *et al.* (1981) even tried to judge fetal tone in the last trimester of pregnancy, by the speed of return to flexion of an extended limb. Meitinger *et al.* (1969) studied 26 preterm infants between 30 and 38 weeks gestation who showed no typical age-dependent posture. In some infants they observed the asymmetrical tonic neck reflex or a paradoxical pattern of the reflex; i.e., flexion of the arm toward which the face is turned.

A similar age independency was shown by Prechtl *et al.* (1975) in a pilot study of motor behavior and posture of 12 preterm infants born between the twenty-eighth and thirty-sixth postmenstrual weeks. Before 35 postmenstrual weeks there was hardly any particular posture prevailing within the 120 min of each observation. After the postmenstrual age of 37 weeks, they found no infant in an arms-extended posture. Others (Saint-Anne Dargassies, 1955, 1966, 1974; Koenigsberger, 1966; Amiel-Tison, 1968; Graziani *et al.*, 1968; Dubowitz *et al.*, 1970) reported typical age-related changes in posture in preterm infants: At 28 postmenstrual weeks, the completely hypotonic infant lies with arms and legs extended and with the head in a markedly lateral position. The less mature the infant, the more pronounced the lateral position of the head. This weakness of the neck muscles leads to the dolichocephalus in preterm infants (Brandt, 1976).

At 30 postmenstrual weeks, there is a beginning of flexion of the thigh at the hip; this flexion becomes stronger at 32 weeks (Amiel-Tison, 1968). Saint-Anne Dargassies (1974) described a slight flexion of the legs of the infant at 32 postmenstrual weeks. According to Koenigsberger (1966), there is "total extension" at 28–32 weeks.

At 34–35 postmenstrual weeks, the froglike (batrachian) attitude is prevalent with flexion of the lower limbs and an abduction or outward rotation of the upper thighs. This flexion contrasts with the extension of the upper limbs. At 36–37 postmenstrual weeks, the attitude of flexion of the four extremities occurs. This posture is also prevalent at the expected date of delivery.

At term in 74% of the 150 full-term infants—all born in the vertex position—in the study by Saint-Anne Dargassies (1974), the four extremities were flexed; in 22% only the arms were flexed, with the legs extended; 1% showed a froglike position; and 1% showed an extension of the upper, and flexion of the lower, extremities.

4.1.2. Posture after Passive Extension of the Extremities: Imposed Posture

Although a typical age-dependent posture was rejected by Prechtl *et al.* (1975), there is a close age relationship with imposed posture in the preterm infant; i.e., posture after a slow passive extension of the extremities or after putting the head in the midline of the trunk. The posture that the baby assumes after this manipulation for at least 2–3 min—before being changed by spontaneous movements—can be recorded. The typical changes in imposed posture with increasing postmenstrual age are due to increasing resistance against gradual stretching of the extremities, caused by tonic myotatic reflexes (Schulte, 1974). Imposed posture is closely related to the recoil of the lower and upper extremities. These responses are also ascribed at the tonic myotatic reflex activity of the infant (Schulte, 1974).

Imposed posture of the lower and upper extremities in the Bonn study was evaluated longitudinally in 29 AGA preterm infants of 32 weeks and under, until the expected date of delivery. An attempt was made to get the infant in Prechtl states 3, 4, or 5 for this examination. We excluded evaluation of the posture of the lower extremities for the first 2 postnatal weeks of the 6 infants born in breech position because the responses of the lower extremities are different from those of infants born in the vertex position. Out of 19 newborn infants born in complete breech position, Prechtl and Knol (1958) showed that in 18 cases there was a resting posture with the lower limbs in an extended, or nearly extended, position.

Photographs at each observation are correlated with the records made (Figure 1):

1. *Imposed posture, lower extremities* (results shown in Table I): These findings contrast with reports in the literature where at 35–36 weeks a froglike attitude prevails, and at the expected date of delivery there is flexion of the legs. No clear-cut changes of the imposed posture of the lower extremities in relation to gestational age were observed. The frequency of flexed position from 35 to 38 weeks of 70% was not significantly different from 50% at 39–40 weeks. Posture of the lower extremities is of little value for gestational age assessment.

Table I. Imposed Posture, Lower Extremities—Results of the Bonn Study

	Postmenstrual age in weeks			
Response	28–33	34	35–38	39–40
Flexion	5	6	31	18
Semiflexion	28	8	12	18
Extension	4	0	1	0
No. of observations	37	14	44	36
No. of infants	23	14	29	29

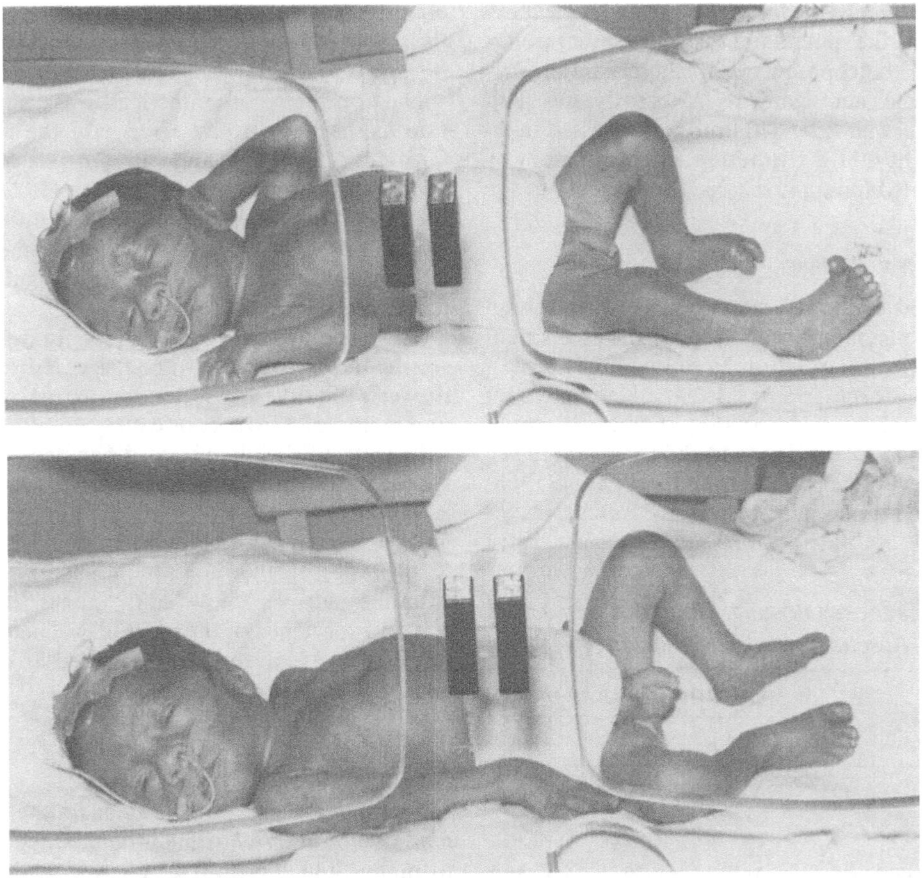

Fig. 1. (Top) Spontaneous posture with the arms flexed; the left leg is flexed due to the asymmetric tonic neck reflex. (Bottom) Imposed posture, i.e., after passive extension of the extremities, the arms remain in the extended position, and the legs return to flexion. Preterm infant born at 32 postmenstrual weeks, 4 days after birth.

2. *Imposed posture, upper extremities* (results shown in Table II): "Imposed posture" of the upper extremities may be used to differentiate infants of 33 weeks and under, and 37 weeks and over. At 34–36 weeks the results are intermediate, and no conclusions as to gestational age may be drawn. Obviously, the results of slowly extending the arms by gentle manipulation are different from those following a quick extension and release to elicit recoil.

4.2. Recoil

The quick recoil of the extremities into the flexed position of the normal full-term infant after extension depends on the tonic myotatic reflex activity. The motor neuron activity demonstrated electromyographically by Schulte (1974) during a quick recoil maneuver of the forearms "provides clear evidence that the phenomenon is a spinal reflex rather than the consequence of the elastic properties of muscles and ligaments." Recently, the quality of recoil of an extended limb was assessed in fetuses in the third trimester of pregnancy by ultrasound (Manning *et al.*, 1981).

4.2.1. Lower Extremities

The recoil of the lower extremities after a brief passive extension has been used for the assessment of neurological maturation by Saint-Anne Dargassies (1955, 1966, 1970, 1974), Koenigsberger (1966), Graziani *et al.* (1968), Dubowitz *et al.* (1970), and Dubowitz and Dubowitz (1977). Prechtl and Beintema (1964) and Beintema (1968) do not mention this response. Koenigsberger (1966) observed a "slight" recoil of the lower extremities at 32 postmenstrual weeks and a good recoil from 34 weeks. Saint-Anne Dargassies (1966) reported that for the gestational age of 35 weeks, extension of the lower extremities is difficult, and an infant returns to its customary flexed position like a spring. Graziani *et al.* (1968), on 16 AGA preterm infants and 15 SGA infants, showed the midpoint of age range of rapid recoil of lower extremities to be 34 postmenstrual weeks, whereas a slow response was observed from 30 weeks with a midpoint age of 31 weeks. There was no difference between AGA and SGA infants. Dubowitz *et al.* (1970), and Dubowitz and Dubowitz (1977) used "leg recoil" in their scoring system for neurological criteria but did not give the age range involved.

Saint-Anne Dargassies (1974) reported an excellent return into flexion of the lower extremities at 35 postmenstrual weeks. In her quantitative study on the neurological behavior of full-term infants, the response is not mentioned (Saint-Anne Dargassies, 1974).

Data on the recoil of the legs in the supine position in the Bonn study agree with the findings of Graziani *et al.* (1968) and with the data in the table of Michaelis, published by Schulte (1974). Infants born in the breech position were excluded. For the reaction, four scores are given: slow, medium, or quick recoil, and response absent (Figure 2). The results are shown in Table III.

This response can seemingly be used for the assessment of neuromaturational age; it distinguishes between infants of 33 postmenstrual weeks or under, and of 34 weeks or over.

In spite of the significant differences of this response in the different gestational age groups, a disadvantage of the test is that it is not valid for infants delivered in the breech position (Prechtl and Knol, 1958); furthermore, there has to be a differentiation between "slow" and "quick," implying a certain experience of the examiner. Therefore, this response can only be recommended as of limited value in assessment of gestational age.

Table II. Imposed Posture, Upper Extremities—Results of the Bonn Study[a,b]

Response	Postmenstrual age in weeks 28–33	34–36	37–40
Flexion	0	13	53
Semiflexion	1	8	2
Extension	40	18	0
No. of observations	41	39	55
No. of infants	29	29	29

[a]Nie *et al.* (1975).
[b]Difference between the three age groups, highly significant, $p < 0.001$, Least significant difference procedure.

4.2.2. Upper Extremities

Recoil of the forearms depends on the tonic myotatic reflex activity of the infant, which is similar to many other "primitive reflexes" and reactions (Schulte, 1974). In a wakeful full-term infant, a brief extension of the forearms at the elbow is followed by a quick recoil into flexion, which is due to biceps muscle activity, as has been demonstrated

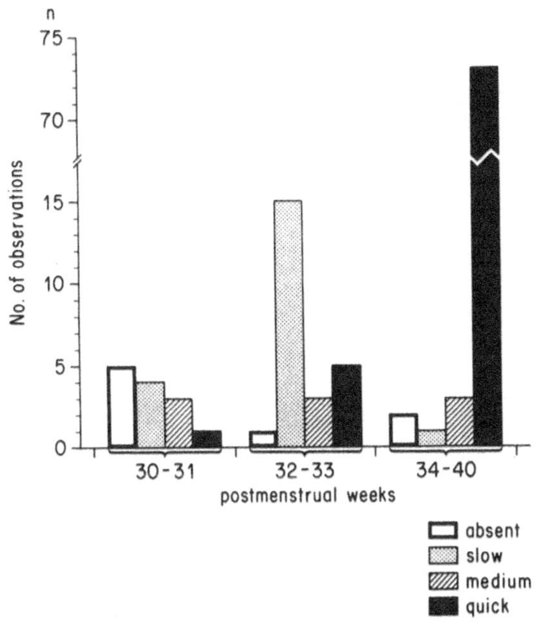

Fig. 2. Recoil of legs. From 30 to 31 weeks, all kinds of responses are seen. From 32 to 33 weeks, a slow recoil prevails in most of the observations. From 34 to 40 weeks, a quick recoil is seen in most cases (73 out of 79).

electromyographically by Schulte (1974). Recoil of the forearms at the elbow is listed among the recognized neuromaturational criteria by Koenigsberger (1966), Amiel-Tison (1968), Graziani *et al.* (1968), Dubowitz *et al.* (1970), Finnström (1971), and Saint-Anne Dargassies (1955, 1970, 1974). Koenigsberger (1966) reported a negative response at 34 postmenstrual weeks, a "slow recoil" of the upper extremities at 37 weeks, and a "good recoil" at 41 weeks. According to Graziani *et al.* (1968) the "midpoint of age range" for slow recoil of the upper extremities is at 31 postmenstrual weeks, and of rapid recoil at 35 weeks.

Table III. Recoil of Legs—Results of the Bonn Study

	Postmenstrual age in weeks		
Response	30–31	32–33	34–40
Quick	1	5	73
Medium	3	3	3
Slow	4	15	1
Absent	5	1	2
No. of observations	13	24	79

In the table on "neurological evaluation of the maturity" of Amiel-Tison (1968) at 34 weeks, a "flexion of forearms begins to appear, but very weak." A "strong return to flexion" is reported at 36 postmenstrual weeks: "the flexion tone is inhibited if forearm is maintained for 30 seconds in extension." From 38 to 40 weeks, the forearms return very promptly to flexion after being extended for 30 sec.

Saint-Anne Dargassies (1970, 1974) regarded a quick recoil of the upper extremities as among the six most reliable neuromaturational signs. She found that at 37 postmenstrual weeks recoil of the upper extremities could occur but was inhibited by a slow and prolonged passive extension. At 41 weeks, there is a quick recoil of the upper extremities; a strong resistance against extension is exhibited.

Finnström (1971) studied preterm and full-term infants divided into gestational age groups of 2-week intervals, six of which had a gestational age of 32 weeks and below. The recoil of the upper extremities was among the 6 responses out of 26 neurologic tests with a good correlation with postmenstrual age. There were only small differences between AGA and SGA infants. In all full-term infants (n = 49) between 1 and 9 days in study by Beintema (1968), recoil of the forearms at the elbow—mostly a marked, quick recoil in both arms—was present. Out of the 150 full-term infants in the study of Saint-Anne Dargassies (1974) 136 showed a positive response, and in 14 infants the reaction was missing on both sides.

In the Bonn study, the recoil of the forearms at the elbow in the supine position was tested according to Prechtl and Beintema (1964); both forearms were simultaneously and briefly passively extended at the elbows and then released (Figure 3). The required state was 3, 4, or 5. The results for this reaction were similar to those reported by Amiel-Tison (1968) and by Saint-Anne Dargassies (1970, 1974). The data agreed also with the figures in the developmental table of Michaelis cited by Schulte (1974).

Because of increase in muscle tone (i.e., resistance against passive movements) in a caudocephalic direction, one would expect that recoil of the upper extremities after a brief passive extension would occur at a later age than that of the lower extremities. Reactions are classified in terms of recoil: quick, medium, slow, or absent. The results are shown in Table IV and Figure 4.

The differences among the three gestational age

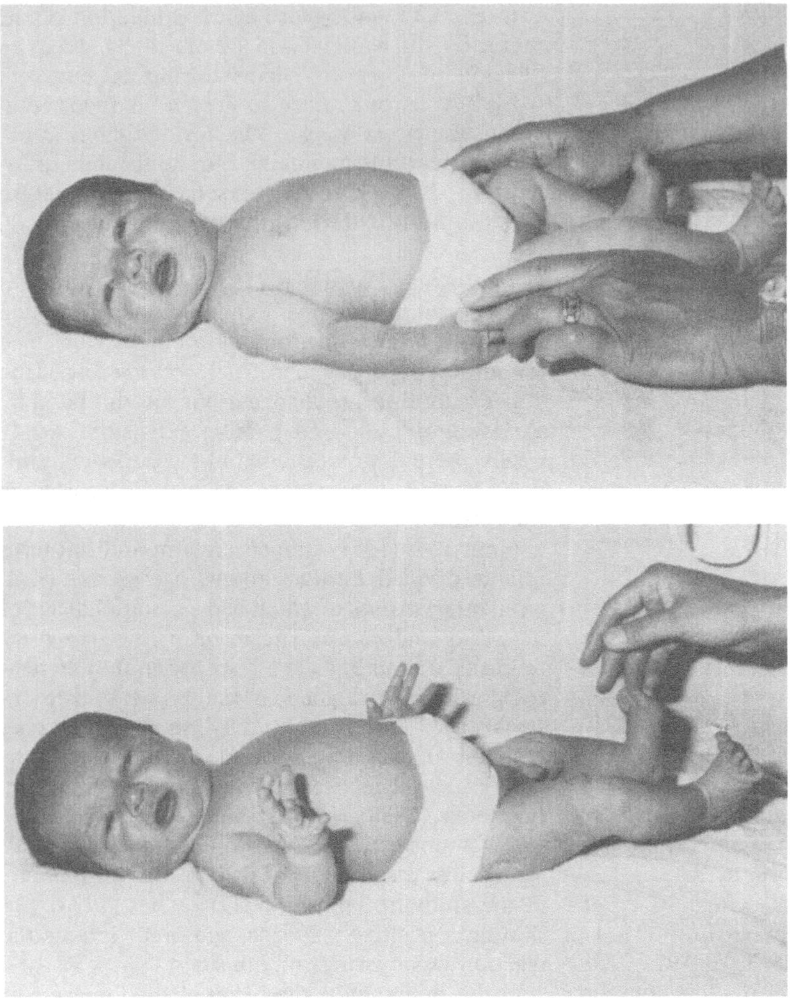

Fig. 3. Recoil of the forearms at the elbow. Full-term infant at the fourth postnatal day.

Table IV. Recoil of the Forearms at the Elbow—Results of the Bonn Study[a]

Response	Postmenstrual age in weeks		
	28–33	34–36	37–40
Quick recoil	0	4	27
Medium recoil	0	2	8
Slow recoil	0	15	6
No recoil	41	18	0
No. of observations	41	39	41

[a]Nie *et al.* (1975).
[b]Difference between the three age groups highly significant, $p < 0.001$; least significant difference procedure of the subprogram one way.

groups, 28–33 weeks, 34–36 weeks, and 37–40 weeks, are highly significant, $p < 0.001$ (Figure 4, Table IV). Recoil of the forearms has a relatively clear-cut time of appearance and is a useful indicator of gestational age in infants in states 3–5. The test allows one to distinguish between infants of 33 weeks and under, where the reaction is absent completely, and infants of 37 weeks and over, where the reaction is quick, or medium, in 85% of infants.

4.3. Resistance against Passive Movements—Tone in the Limbs

The term "resistance against passive movements" is used to describe muscle tone. In their manual for *The Neurological Examination of the*

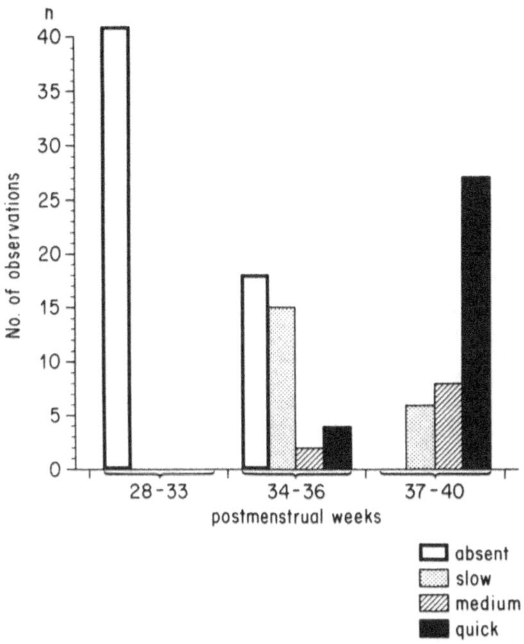

Fig. 4. Recoil of forearms at the elbow. This reaction distinguishes between infants at 36 weeks and under, where the response is absent or slow (85%), and of 37 weeks and over, where a quick (66%) or medium (19%) recoil is seen in most cases.

Full-Term Newborn Infant, Prechtl and Beintema (1964) avoided the words "tone" and "tonus" because so much confusion has arisen about them. Beintema (1968) gave scores separately for the resistance against passive movements of shoulders, elbows, wrists, hips, knees, and ankles. A scoring system for "general tone" has been elaborated by Brazelton (1973). Muscle tone comprises both cellular and contractile components (Schulte, 1974), and posture of the infant reflects tone to a large extent (Saint-Anne Dargassies, 1966, 1974; Brazelton, 1973).

In a study of muscle tone in preterm infants compared with full-term infants at the gestational ages of 38–41 weeks, Michaelis *et al.* (1975) showed significant differences between the groups. Passive tone, examined at shoulder, elbow, trunk, hip, and knee, was significantly less in preterm infants than in full-term control infants. Also, the active tone, evaluated by the flexion angle of elbows and knees after traction, was decreased. The duration of extrauterine life had no influence on muscle tone between 38 and 41 postmenstrual weeks. Forslund and Bjerre (1983) also found no differences in the resistance to passive movement between low-risk preterm infants at term and full-term infants.

Saint-Anne Dargassies (1955, 1966, 1970, 1974), Koenigsberger (1966), Amiel-Tison (1968), and Finnström (1971) used increasing muscle tone in a caudocephalic direction as maturational criteria. André-Thomas *et al.* (1960; André-Thomas and Saint-Anne Dargassies, 1952) and Saint-Anne Dargassies (1955) elaborated some responses for evaluating muscle tone.

Dubowitz and Dubowitz (1981), who assessed limb tone in early preterm infants (up to 31 weeks) with intraventricular hemorrhage (IVH), found no statistical significantly difference in muscle tone from infants without IVH of the same postmenstrual age. Their explanation is that tone is normally low in these infants and that further diminution may be difficult to recognize.

4.3.1. Lower Extremities

The amount of resistance against passive movements in the lower extremities is usually tested by the popliteal angle, the heel-to-ear maneuver, and the abduction angle of the hips.

The *popliteal angle* decreases between 28 and 32 postmenstrual weeks from 180° to 150°, between 33 and 34 weeks to 120°, between 35 and 37 weeks to 90°, and remains so until the expected date of delivery (Saint-Anne Dargassies, 1966, 1974; Koenigsberger, 1966). Saint-Anne Dargassies (1974) counted this response among the "particularly unreliable signs" of maturity. The figures of Amiel-Tison (1968) showed an earlier closing of the popliteal angle than reported by Saint-Anne Dargassies: at 32 weeks, 110°; at 34 weeks, 100°. Dubowitz *et al.* (1970; Dubowitz and Dubowitz, 1977) included the changes of the popliteal angle in their scoring system, but absolute values were not given. In her study of full-term infants, Saint-Anne Dargassies (1974) reported a considerable variability: only 63% have a popliteal angle of 90° or less, whereas in 37%, the angle is between 100° and 180°.

Dubowitz and Dubowitz (1981) reported an abnormally tight popliteal angle in a highly significant proportion of infants with IVH below 36 postmenstrual weeks, compared with infants without IVH.

The maturational changes of the popliteal angle (i.e., decrease from 150° to 90°) are not valid for the infant born in the breech position. As an extended spontaneous position of the knees is often found (Prechtl and Knol, 1958), this may persist for some weeks after birth.

According to the Bonn results, the popliteal angle decreases in the last weeks before term; but the relationship of the angle to postmenstrual age is not clear-cut before 37 postmenstrual weeks (Figure 5). In evaluation, we excluded the 6 infants born in breech position in the first 2–3 weeks. The results are shown in Table V.

The popliteal angle distinguishes between infants of 33 weeks and under and 37 weeks and over. But even if this response shows clear-cut changes from 37 postmenstrual weeks in this homogeneous group of preterm infants, it may behave more variably in infants of the same postmenstrual age but of different postnatal age. Moreover, the interval of 4 weeks for a clear-cut change is insufficient for maturational assessment. The differences among the three gestational age groups, 28–33 weeks, 34–36 weeks, and 37–40 weeks, are significant, $p < 0.001$, by the least significant difference procedure of the subprogram one way (Nie *et al.*, 1975).

The heel-to-ear maneuver has been considered useful for gestational age assessment by Koenigsberger (1966), Amiel-Tison (1968), Dubowitz *et al.* (1970), and Saint-Anne Dargassies (1966, 1974). At 28–32 postmenstrual weeks there is no resistance against drawing the infant's foot as near to the ing postmenstrual age, this maneuver becomes more and more difficult and is impossible in the normal infant at term. Among the 150 full-term infants in the study of Saint-Anne Dargassies (1974), strong resistance, a normal response, was observed in 40%. Forty-three percent of the infants showed only slight resistance. Since this response is closely related to the popliteal angle, and its relation to

Table V. Popliteal Angle—Results of the Bonn Study[a,b]

Response (degrees)	Postmenstrual age in weeks		
	28–33	34–36	37–40
Less than 90	0	0	1
90–110	1	12	51
120–140	10	20	2
150 and greater	24	7	1
No. of observations	35	39	55

[a]Nie *et al.* (1975).
[b]Difference between the three age groups, highly significant, $p < 0.001$, Least significant difference procedure.

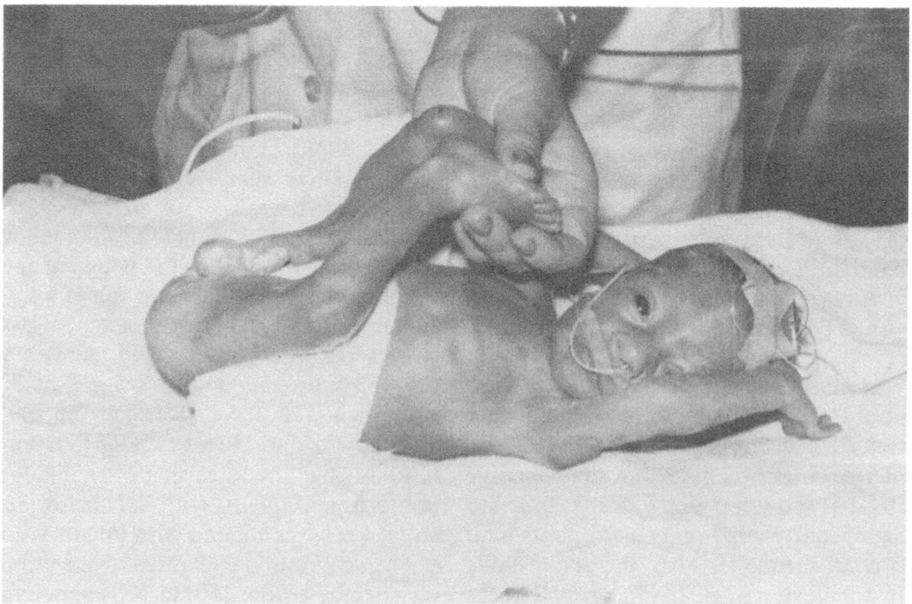

Fig. 5. Popliteal angle 130° to 140° at 32 postmenstrual weeks in a preterm infant of 4 days. The arms—in a reflex movement—are also moved upwards.

postmenstrual age is not very clear-cut, it cannot be recommended as a reliable maturational criterion.

4.3.2. Upper Extremities

Resistance against passive movements in the upper extremities and in the shoulder girdle is usually tested with an approximation of the "hand-to-the-opposite-shoulder," the well-known *scarf maneuver (mouvement du foulard)*, as described by André-Thomas *et al.* (1960), Saint-Anne Dargassies (1966, 1974), Amiel-Tison (1968), and Dubowitz *et al.* (1970; Dubowitz and Dubowitz, 1977). One observes in this response how far the elbow will move across when one draws the hands to the opposite shoulder, and then attempts to put them around the neck (like a scarf). Up to a postmenstrual age of 32 weeks, no resistance is found against the scarf maneuver (Amiel-Tison, 1968). According to Koenigsberger (1966), the maneuver is possible until 33 weeks without resistance and minimal resistance against approximation of the hands to the opposite shoulder occurs at 34 weeks; Graziani *et al.* (1968) report a "midpoint of age range" of 33 weeks for slight to moderate resistance. From 36 to 38 weeks, a moderate to marked resistance is demonstrated (Graziani *et al.*, 1968), with the elbows just passing the midline (Amiel-Tison, 1968). At term, the scarf maneuver is limited (Amiel-Tison, 1968), and the elbows will not reach the midline. Because of the changes with increasing postmenstrual age, this response is recommended for assessment of maturity by Koenigsberger (1966), Amiel-Tison (1968), Graziani *et al.* (1968), Dubowitz *et al.* (1970), Capurro *et al.* (1978), Ballard *et al.* (1979), and Dubowitz and Dubowitz (1981). Contrary to her earlier publications, Saint-Anne Dargassies counts the scarf sign (1974) among the six particularly unreliable criteria of maturity.

As an example and representative for the longitudinal development of the preterm infants in the Bonn study, the results of the scarf maneuver are plotted in Figure 6. Each of the 29 different symbols represents an individual with its repeated observations. The attitude of the elbows is scored; that is, how far do the elbows cross the midline: 1 = no resistance, with elbows widely crossing midline and nearly reaching opposite axillary line; 2 = elbows between midline and opposite axillary line (slight resistance); 3 = elbows reaching midline (moderate resistance); and 4 = elbows only reaching axillary line (strong resistance).

Figure 6 shows that the resistance in the upper extremities increases with increasing postmenstrual age until term. There is a highly significant correlation of r = 0.72 between the score and the post-

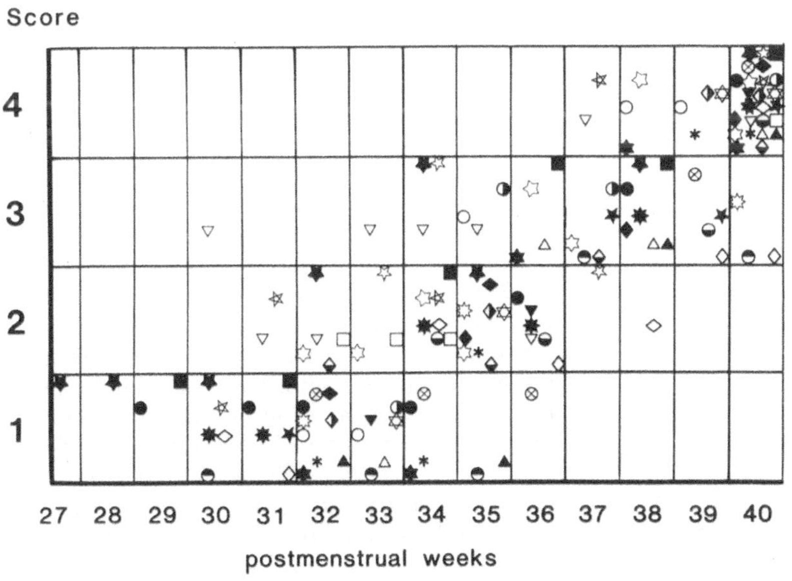

Fig. 6. Scarf maneuver, longitudinal results. Each of the 29 different symbols represents an individual with its repeated observations, altogether 135. Scoring of resistance: 1 = none; 2 = slight; 3 = moderate; 4 = strong.

menstrual age, $p < 0.001$; (Kendall's tau, Nie *et al.*, 1975). It is possible to differentiate four age groups: 28–33 weeks, with mostly no resistance (73%); 34–36 weeks, when a slight resistance prevails (56%); 37–39 weeks, with a moderate resistance in most of the infants (59%); and 40 weeks and over, when most infants (89%) exhibit a strong resistance (Table VI). The differences among the four gestational age groups are significant, $p < 0.001$. The Bonn results agree with the reports of Koenigsberger (1966), Amiel-Tison (1968), and Graziani *et al.* (1968). The scarf sign may be used for the assessment of gestational age provided that the attitude of the elbows is recorded and the infant is in states 3–5. The scoring presupposes that the examiner has some experience. The scarf maneuver permits one to differentiate among infants of 33 weeks and under (mostly no resistance), 34–39 weeks (mostly slight or moderate resistance), and 40 weeks and over (mostly strong resistance).

4.4. Spontaneous Motor Activity

Unlike the fetus *in utero*, the preterm infant is exposed to the force of gravity, and this influences motility. The degree of spontaneous movements (i.e., speed, intensity, and amount) is used as a maturational criterion by Koenigsberger (1966) and Finnström (1971); Saint-Anne Dargassies (1955, 1966, 1974) counts it among the six most reliable signs.

Precht *et al.* (1975, 1979) conducted a longitudinal investigation of the quality and quantity of the spontaneous motor activity of low-risk preterm infants and found a significant decrease after 35 postmenstrual weeks. Postnatal age had no influence on the amount of motility; i.e., there was a high degree of independence of motility from the duration of extrauterine experience, indicating that maturation of inhibitory mechanisms is highly independent of environmental conditions. A similar decline in general movements is observed *in utero* by ultrasound examinations (H. Hoffbauer, personal communication, 1977).

Table VI. Scarf Maneuver—Results of the Bonn Study[a,b]

Response	Postmenstrual age in weeks			
	28–33	34–36	37–39	40
Strong resistance	0	0	9	25
Moderate resistance	2	10	16	3
Slight resistance	9	22	2	0
No resistance	30	7	0	0
No. of observations	41	39	27	28

[a]Nie *et al.* (1975).
[b]Difference between the four age groups highly significant, $p < 0.001$, least significant difference procedure.

Saint-Anne Dargassies (1955, 1966) described as characteristic of the stage of development at 28 postmenstrual weeks "repetitive movements of the upper limbs that have a catatonic appearance," which are later incorporated into a generalized movement and further bursts of movements. The movements are more frequent in the lower extremities, where alternating pedaling movements are often seen. At 32 weeks, the spontaneous motility has become generalized and is "dominated by vigorous movements of the whole body," especially the trunk with lateral incurvations. Typical is the raising of an extended leg which is held up by the infant at right angles to the pelvis, and the pelvis may be raised. At 35 weeks, spontaneous motility has increased and continues. The preterm infant is able to change his position completely in the incubator and can raise the pelvis on heels and occiput. In the following weeks until the expected date of delivery, motility of the extremities and the trunk becomes more limited in time and space, caused by increasing muscle tone, whereas good lateral rotation of the head can be observed. At term, motility is more market in a preterm infant than in a recently born full-term infant (Saint-Anne Dargassies, 1966, 1974).

In the Bonn study we observed no easily quantifiable age-dependent changes in spontaneous motor activity during the examinations evaluated to date. Only the amount of spontaneous head rotation showed clear-cut changes, as detailed in Table VII.

Precise scoring of spontaneous motor activity necessitates continuous monitoring. A reliable and objective method for this activity has been introduced by Bratteby and Andersson (1976) with two instruments (Animex I and II). All movements observed are classified according to amplitude, velocity, and also in relation to the size of the moving object (i.e., hand, arm). The advantage of the instruments is that they provide an integrated record of all movements without any attachment of cables or electrodes to the infant. To date, no longitudinal observations on preterm infants have been reported.

Table VII. Spontaneous Head Rotations—Results of the Bonn Study

Response	Postmenstrual age in weeks		
	28–33	34–39	40
Absent	30	3	0
Moderate	11	58	8
Good but not against resistance	0	3	13
Against slight resistance	0	2	7
No. of observations	41	66	28

4.5. Reflexes and Reactions Dependent on Postmenstrual Age

The designation "reflex" is not quite appropriate for these specific reactions or automatims, which have been known for a long time. The pathway of the reflex arc is not known in detail for most of the reflexes studied (Robinson, 1966).

To this group belong primary reflexes and reactions with a relatively clear-cut time of appearance in the perinatal period, which is the same in AGA and SGA infants. Thus, they can be used as an indicator of postmenstrual age. These reactions and reflexes allow one to follow neurological maturation in preterm infants until 40 weeks, the expected date of delivery, when comparison with full-term infants may be made. Each examination improves the value of the preceding one. The value of the different reactions for maturity assessment is rated differently by different investigators.

4.5.1. Moro Reflex

The Moro response (Moro, 1918; McGraw, 1937; Saint-Anne Dargassies, 1955; Parmelee, 1964; Schulte *et al.*, 1968*b*) is one of the most frequently examined. There is no general agreement about the different components of the reflex. The classification of Prechtl and Beintema (1964, 1976) with its four phases for the pattern of arm movements is most used: (1) abduction at the shoulder; (2) extension at the elbow; (3) adduction of the arms at the shoulder; and (4) flexion at the elbow. Phases (3) and (4) are described as bowing by McGraw (1937). In phase (2) an observation of the hands is useful, the fingers may be all fanned, fanned with flexion of the distal phalanges of the thumb and index finger, all semiflexed, or all flexed. A cry as a further component of the reflex has been described by McGraw (1937) and Saint-Anne Dargassies (1955, 1966).

The Moro response is usually elicited by a rapid but gentle head drop of a few centimeters with the baby suspended horizontally and the head in the midline (Prechtl and Beintema, 1964). In order to observe all phases of the test, it should be carried out at least three times. Another method of elicitation is pulling the infant up by the wrists so that the shoulders and head are lifted a few centimeters off the table or bed; then the hands are released, so that the head falls back to the examination table. There is no agreement about the main afferent pathway of the Moro reflex. According to André-Thomas and Saint-Anne Dargassies (1952) and Saint-Anne Dargassies (1954), it is a response to proprioceptive stimulation from the neck by a sudden change of the position of the head in relationship to the trunk. Peiper (1964) considered the Moro response a labyrinthine reaction. Further details were given by Parmelee (1964) in a critical evaluation.

There are characteristic development changes of the Moro response with increasing postmenstrual age. At 28 postmenstrual weeks, only extension of the arms and spreading of the fingers occur (Saint-Anne Dargassies, 1966, 1974). Amiel-Tison (1968) describes the Moro response at 28–30 weeks as "weak, obtained just once and not elicited every time."

At 32 weeks, Saint-Anne Dargassies (1966) and Amiel-Tison (1968) reported that the Moro reflex "has become complete" whereby abduction and extension of the arms with spreading of the fingers and a cry are described; the phases of adduction and flection are not mentioned. From 32 to 34 postmenstrual weeks, Finnström (1971), using the scoring system of Saint-Anne Dargassies (1966), reported that in 100% of the infants in his study an extension of forearms at the elbow with extension of the fingers and an abduction of the arms is observed and in no case adduction. Schulte *et al.* (1968*b*) showed electromyographically that at 32 postmenstrual weeks the Moro response consists of the extension and abduction of the upper extremities.

From 34 to 36 weeks, Finnström (1971) reported an adduction (embrace) in 56% of the infants tested. Graziani *et al.* (1968) observed a partial embrace at a mean age of 35 weeks.

From 36 to 38 weeks, 80% of the infants in Finnström's study showed an adduction, and 98–100% in the weeks thereafter. Graziani *et al.* (1968) reported a midpoint of age range for "complete embrace" at 39 weeks. Amiel-Tison (1968) did not mention the adduction phase. Saint-Anne Dargassies (1970, 1974) reported "complete embrace" at 41 weeks but did not consider this phase in her quantitative neurological study on full-term infants.

Robinson (1966) reported the adduction phase of the Moro response as "more constantly present after the period 32–36 weeks than before" but considered the time of appearance too unpredictable to be an accurate measure of postmenstrual age.

Babson and McKinnon (1967) observed the embrace component of the Moro reflex at 35 ± 1 weeks of postmenstrual age in 82% of infants tested.

In summary, it may be said that at 28 weeks there is only an extension of the arms and fingers; from 32 to 35 postmenstrual weeks, the Moro response consists of an extension and abduction of the arms with extension of the fingers and occasional crying. From 36 to 40 weeks, the phases of arm adduction and flexion are observed more and more distinctly (complete embrace) with increasing postmenstrual age, whereas the phases of extension and abduction become less prominent. These results agree with the descriptions in the table on the development sequence of various reflexes (Michaelis, cited by Schulte, 1974).

The age-dependent changes of the Moro reaction of AGA preterm and full-term infants and of SGA full-term infants have been evaluated by Schulte *et al.* (1968*b*) in an electromyographic study. The duration and amount of biceps and triceps activity in at least four Moro responses were tested. For calculation of the overall reaction pattern, a triceps/biceps Moro response activity quotient is used. With increasing postmenstrual age, duration and amount of biceps activity increased significantly, whereas triceps activity showed no significant alteration. The triceps/biceps activity quotients were <1 in full-term newborn infants and >1 in preterm newborn infants of a mean postmenstrual age of 36.2 weeks—a highly significant difference ($p < 0.001$)—but the variance in each group is very high. The SGA full-term infants with a mean weight of 2118 g exhibited a mean triceps/biceps activity quotient intermediate between full-term infants of the same gestational age and preterm infants of the same weight. In spite of the significant increase in biceps activity, the Moro response is considered unsuitable for an exact estimation of postmenstrual age because of "considerable variability in all parameters" in infants of the same postmenstrual age. Schulte *et al.* (1968*b*) suggested that this might be due, in part, to the fact that even minimal pathological influences in the newborn period affect the Moro response more than age differences. Schulte's electromyographic findings have confirmed the results of behavioral studies of preterm infants.

According to Lenard *et al.* (1968), the Moro reflex is equally strong during regular sleep and wakefulness but is weak or absent during irregular sleep. The infants of the Bonn study were tested only when awake.

Fredrickson and Brown (1980), who compared AGA and SGA full-term neonates, found a significant difference in the Moro flexion and adduction phase; i.e., SGA babies exhibited less flexion and adduction than AGA. In the extension and abduction phase there was no difference between the two groups.

The Bonn results agree with the developmental trend of the Moro response as reported in the literature. The adduction phase of the Moro appears at a relatively clear-cut time; one reason might be that the preterm infants who were included represented a relatively homogeneous group with respect to gestational age at birth; i.e., 32 weeks and under, and of middle risk (see Chapter 2). In evaluation of the results, each phase was scored separately. The results are shown in Table VIII and in Figure 7.

At 28 weeks only extension of the arms and fingers is observed. From 29 to 35 postmenstrual weeks the Moro response consists of extension of the arms with abduction, and with spreading and extension of the fingers (68 examinations).

At 36 weeks, the response was intermediate: in 4 of 11 cases an adduction was observed and in 7 cases extension of the arms with abduction, and extension and spreading of the fingers only (Figure 7). From 37 to 40 weeks, adduction of the arms with flexion was observed in 39 out of 48 examinations; however, at 37 weeks, the flexion component is not complete in all cases (Figure 7; Table VIII).

Table VIII. Moro Response; Adduction of the Arms and Flexion at the Elbow—Results of the Bonn Study[a,b]

Response	Postmenstrual age in weeks 28–35	36	37–40
No adduction and flexion	66	7	9
Adduction and flexion	2	4	39
No. of observations	68	11	48

[a]Nie *et al.* (1975).
[b]Difference between the three age groups highly significant, $p < 0.001$; least significant difference procedure.

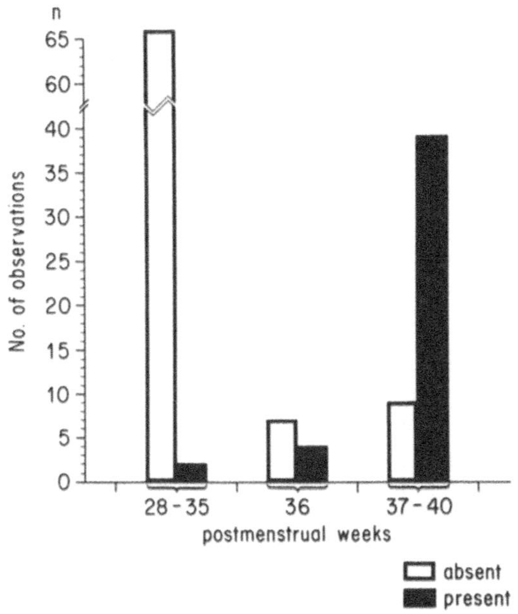

Fig. 7. Moro response, adduction phase. The response distinguishes between infants of 35 postmenstrual weeks or under and 37 weeks or over.

The position of the fingers—scored separately—yields the following results: up to 34 postmenstrual weeks extension and fanning was observed; from 35 to 38 weeks in 20% of infants a flexion of the thumb and index finger occurred; and from 39 to 40 weeks, this occurred in 50% (as described by McGraw, 1937, and Saint-Anne Dargassies, 1974).

Crying during the Moro response has not been recorded up to 32 postmenstrual weeks; at 33 weeks, 50% of the infants cried; and from 34 weeks to term, a cry was registered in 80–90% of examinations. Saint-Anne Dargassies (1974) reported crying in 87% of normal newborn full-term infants.

The timing of the adduction phase of the Moro response allows one to distinguish between infants of 28–35 postmenstrual weeks and of 37 weeks and over, the difference between these two gestational age groups being highly significant, $p < 0.01$ (Table VIII). The adduction/flexion phase of the Moro reflex may also be recommended as a maturational criterion according to the results of Babson *et al.* (1967), Graziani *et al.* (1968), Finnström (1971), and Michaelis (cited by Schulte, 1974). This is contrary to the findings of Robinson (1966), Schulte *et al.* (1968*b*), and Saint-Anne Dargassies (1974), who count the Moro reflex among the six particularly unreliable maturational signs.

4.5.2. Neck-Righting Reflex

With this reflex, Schaltenbrand (1925) introduced into the neurological examination of human infants the results of the animal experiments conducted by Magnus and de Kleijn (1912) and Magnus (1924) concerning neck and righting reflexes. The neck-righting reflex is elicited with the infant in a supine position and the head being rotated by 90° to the left or right side. This lateral head movement is followed by a relfex rotation of the spine; i.e., trunk and legs, in the same direction (Figure 8). In the full-term newborn infant, the marked neck-righting reflex leads to a "vehement roll over of the body in line with head rotation" (Schaltenbrand, 1925). Nowadays, even in the fetus of about 12 postmenstrual weeks, a rotation around the longitudinal axis is observed resulting from a rotation of the head, followed by trunk rotation (de Vries *et al.*, 1982).

Robinson (1966) first published data on the maturational changes of the neck-righting reflex in AGA and SGA preterm infants. According to his

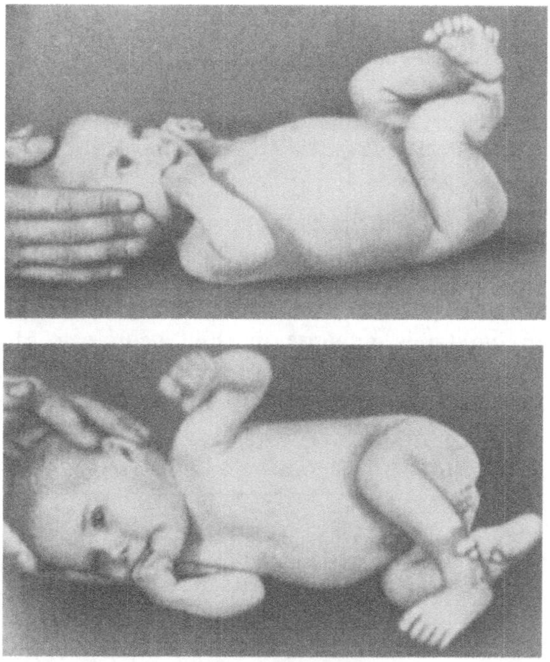

Fig. 8. Neck-righting reflex. (Top) Supine position. (Bottom) After turning the head of the baby to the right side, the whole body is turned to the right. (Original photograph from Schaltenbrand, 1925.)

results, this reflex usually appears between 34 and 37 postmenstrual weeks in both AGA and SGA infants. Between the postmenstrual age periods less than 37 weeks and more than 37 weeks, there is a significant difference ($p < 0.01$).

Graziani *et al.* (1968) scored the movement of the opposite shoulder, and of the trunk and pelvis, after head rotation toward the left and then to the right. At 37 postmenstrual weeks a "sustained shoulder elevation of >1.5 in., trunk and pelvis may turn" response is described. The midpoint of age range for "sustained shoulder, trunk and pelvis response" is 39 weeks.

Dubowitz *et al.* (1970; Dubowitz and Dubowitz, 1977) consider the neck-righting reflex "inconsistent because of the difficulty in defining a positive response."

The Bonn results, which are similar to those of Robinson (1966), are shown in Table IX and Figure 9.

The difference between the gestational age groups 34–35 weeks and 37–40 weeks is highly significant, $p < 0.001$ (Table IX). The neck-righting reflex allows one to differentiate between infants of 35 postmenstrual weeks or under and 37 weeks or over. It is therefore recommended as being useful for maturational assessment.

The neck-righting reflex requires state 3, 4, or 5, and it is not elicitable in states of hypotonia of the neck muscles.

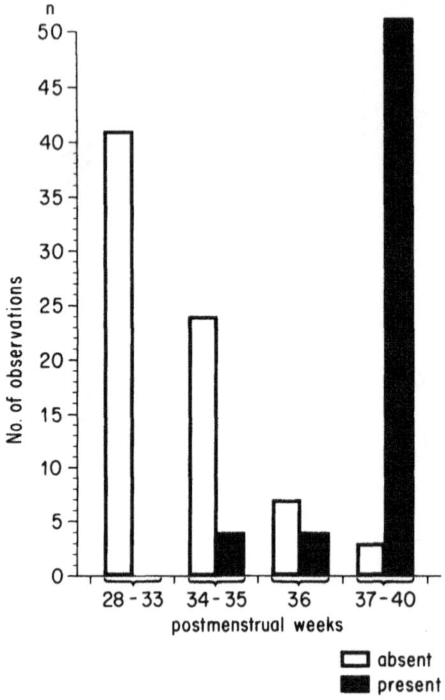

Fig. 9. Neck-righting reflex. The response distinguishes between infants of 35 postmenstrual weeks or under and 37 weeks or over.

4.5.3. Traction Response—Head Control in Sitting Position

Head control in the sitting position is listed among the labyrinthine righting reactions. On the basis of animal experiments of Magnus (1924) and Rademaker (1926), Schaltenbrand (1925) first introduced this response in child neurology. His method of elicitation was slightly different from that used now: The infants were blindfolded in order to eliminate optical righting reactions, and held by the examiner at the pelvis in an upright position. Of the 28 full-term infants tested in the first week by Schaltenbrand (1925), in 12 cases a moderate head control was observed, 13 infants were not able to lift the head up, and in 3 infants the response was questionable.

Head control in the sitting position represents the second part of the traction response as described by Prechtl and Beintema (1964). In the first part the infant is slowly pulled up by the wrists to a sitting position from being supine with the head

Table IX. Neck-Righting Reflex—Results of the Bonn Study[a,b]

Response	Postmenstrual age in weeks			
	28–33	34–35	36	37–40
No following of trunk and legs	41	24	7	3
Trunk and legs follow	0	4	4	51
No. of observations	41	28	11	54

[a]Nie *et al.* (1975).
[b]Difference between 34–35 and 37–40 weeks highly significant, $p < 0.001$; least significant difference procedure.

in the midline. The resistance to extension of the arms at the elbow is observed.

For testing the head control in the sitting position in small preterm infants before the expected date of delivery, the method described by Saint-Anne Dargassies (1954) and Beintema (1968) is more practical and is used in the Bonn study: The infant, after being brought to the sitting position, is supported at the shoulders and trunk with both hands of the examiner (Figure 10). From shortly before term and thereafter, head control in the sitting position is recorded with the infant held at the wrists, as described by Prechtl and Beintema (1964).

The recording of the response is as recommended by Prechtl and Beintema (1964). The four scores are: (1) head hangs passively down; (2) head is raised once or twice momentarily, or head is lifted up and drops backward without remaining in the upright position; (3) head is raised and remains in the upright position for at least 3 sec, oscillations may occur; and (4) head is maintained in the upright position for more than 5 sec with no, or only little, oscillation (Figure 11).

In preterm infants head control in the sitting position develops with increasing postmenstrual age. Results from other studies are hardly comparable because methods of elicitation of the reflex and scoring were different.

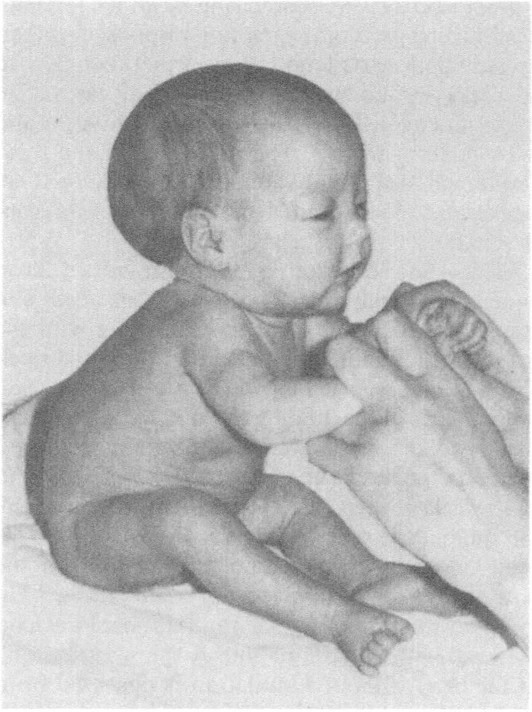

Fig. 11. Traction response, good head control in sitting position. Appropriate-for-gestational age (AGA) preterm infant at expected date of delivery, born at 32 postmenstrual weeks with a birth weight of 1370 g.

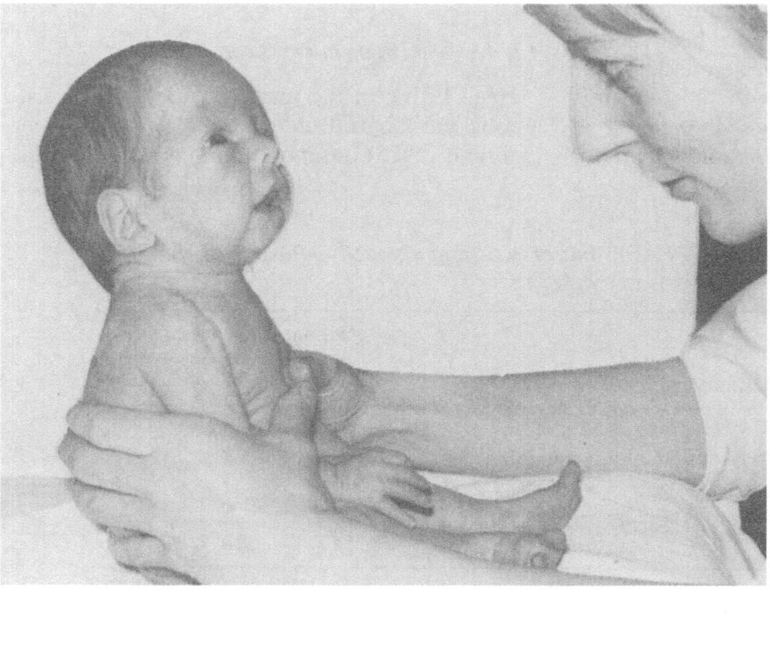

Fig. 10. Traction response, head control in sitting position. The infant, after being brought to the sitting position, is supported at the trunk. Appropriate-for-gestational age (AGA) preterm boy at 39 postmenstrual weeks with a postnatal age of 8 weeks.

Robinson (1966) scored the arm flexion and head-raising parts of the traction response together, because both parts tend to appear between 33 and 36 weeks gestation and are present in all cases of 36 weeks and over. The results are similar in AGA and SGA preterm infants. The difference in the three gestational age groups under 33 weeks, 33–36 weeks, and 36 weeks and over are highly significant ($p < 0.001$).

Saint-Anne Dargassies (1966) described head control "fleetingly" at 35 postmenstrual weeks and stated that the head can be kept in line with the trunk in the sitting position at 41 weeks. Amiel-Tison (1968) reported a good righting of the head, but inability of the infant to hold it, at 36 postmenstrual weeks. At 38 weeks, the infant begins to maintain the head's position for a few seconds, and at 40 weeks it keeps the head in line with the trunk for more than a few seconds. Finnström (1971) used "strong" or "moderate" stimulation for the observation of "head extension when sitting." Up to 34 postmenstrual weeks no infant could extend the head without stimulation. A "slow extension" of the head without stimulation is observed from 34.1 to 36 weeks in 33% and from 38.1 to 40 weeks in 48% of the infants. Babson and McKinnon (1967), who scored the "maintenance of head nearly parallel to trunk" during traction, reported that "neck flexion on traction developed at 36±1 weeks of postmenstrual age in 68 percent" of the infants.

In their comparative neurological study of 97 preterm infants, 32 of whom were of a gestational age of 32 weeks and below, and 97 full-term infants at term, Howard *et al.* (1976) reported head control in the sitting position (posterior neck muscles) to be among the neurologic items with "significantly greater incidence of weak responses" in the preterm than in the full-term infants. In this study "anterior and posterior neck muscle control" of the infant, brought to a sitting position and supported at the shoulders, was scored. A similar observation was reported by Saint-Anne Dargassies (1966) that in preterm infants who have reached the expected date of delivery, the "straightening responses will be less firm and lasting."

The full-term infants in the study of Beintema (1968) have lower scores for head control in the sitting position at 1 and 2 postnatal days than thereafter. From three days in about 50% of infants, the head remains in the upright position for some seconds.

Good head control in the sitting position was reported by Saint-Anne Dargassies (1974) in 53% of full-term infants and a medium reaction in 25%.

In the Bonn study we observed clear-cut changes of the head-raising part of the traction response in relation to postmenstrual age. The results are shown in Table X and Figure 12. The required state of the infants was 3, 4, or 5. The arm flexion part of the response behaves less predictably and was omitted from this report.

The difference between the gestational age periods 28–34 weeks, 35 weeks, and 36–40 weeks is significant ($p < 0.05$). Age group 36–40 weeks differs highly significantly from the other two groups, ($p < 0.001$). Because of these maturational changes, head control in the sitting position is considered useful for the assessment of postmenstrual age, in agreement with Robinson (1966) and Amiel-Tison (1968). The response differentiates between infants of 34 weeks or under and of 36–40 weeks.

4.5.4. Head Lifting in the Prone Position

Head lifting in the prone position is also listed among the labyrinthine righting reactions. Schaltenbrand (1925) observed momentary head lifting

Table X. Traction Response—Head Control in Sitting Position—Results of the Bonn Study[a,b]

Response	Postmenstrual age in weeks		
	28–34	35	36–40
Head hangs passively down	47	6	2
Head raising once or twice without remaining upright	7	7	19
Head upright at least 3 sec	0	1	16
Head upright more than 5 sec	0	0	20
No. of observations	54	14	57

[a]Nie *et al.* (1975).
[b]Difference between 28–34 and 36–40 weeks highly significant, $p < 0.001$; least significant difference procedure.

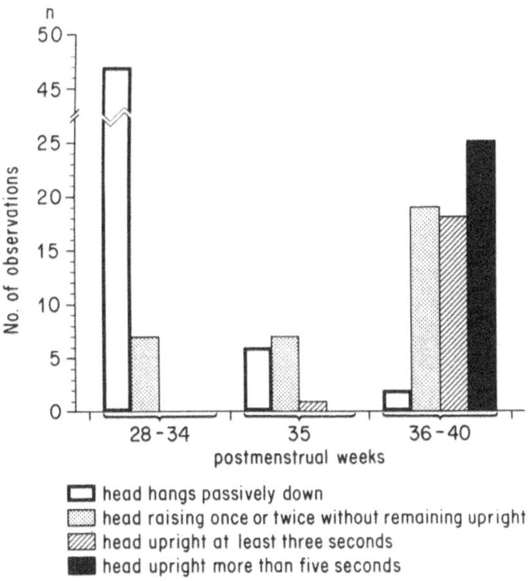

Fig. 12. Traction response, head control in sitting position. The response distinguishes between infants of 34 weeks or under and 36 weeks or over.

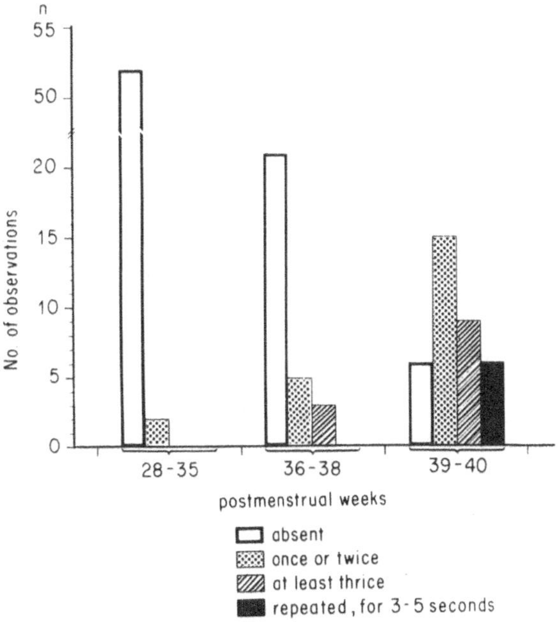

Fig. 13. Head lifting in the prone position. The response distinguishes between infants of 38 postmenstrual weeks or under and 39 weeks or over.

in 13 out of 30 full-term infants in the first postnatal week; in 13 cases the response was absent; and in four, doubtful.

Neck extension in the prone position occurred at 37±1 weeks in 84% of the preterm infants in the study of Babson and McKinnon (1967), who scored "raising of head off the bed." According to the table on "developmental sequence of various reflexes" of Michaelis (cited by Schulte, 1974), head raising is present from 38 postmenstrual weeks; from 35 to 37 weeks the response is "not yet present in all cases."

Beintema (1968) observed significantly lower scores for head lifting in the prone position of full-term infants in the first 5 days than at 9 days. From day 1-5, 69-25% of infants were unable to lift the head in the prone position, but by day 6-9, 12.5-14% were unable to do so. From 6 days, about 50% of infants are able to lift the head for at least 2 sec.

In the Bonn study, head lifting in the prone position was scored according to Prechtl and Beintema (1964) and Beintema (1968), with little modification for the preterm infant (see Table XI and Figure 13). The infants were in states 3, 4, or 5.

For this reaction, photographs of the infants, made at each examination, serve as controls in sta-

Table XI. Head Lifting in the Prone Position—Results of the Bonn Study[a,b]

Response	Postmenstrual age in weeks		
	28–35	36–38	39–40
No head lifting	52	21	6
A short lift of the head once or twice	2	5	15
Momentary head lifting at least three times	0	3	9
Repeated head lifting for 3–5 sec or more	0	0	6
No. of observations	54	29	36
No. of infants	29	29	29

[a]Nie *et al.* (1975).
[b]Difference between the three age groups significant, $p < 0.05$; and between 39–40 weeks and the other two groups highly significant, $p < 0.001$ (least significant difference procedure).

tistical evaluation of the results. Figure 14 shows head lifting in the prone position in a preterm girl at the expected date of delivery.

Up to 35 postmenstrual weeks, no head lifting was observed in 96% of examinations (52 of 54), and from 36 to 38 weeks in 72% (21 of 29). From 39 to 40 weeks, no head lifting was observed in 17% of examinations (6 of 36); in 41% the head was lifted shortly once or twice; in 25% at least three times; and in 17% repeated head lifting for 3–5 sec was seen (see Table XI and Figures 13 and 14). The difference among the three gestational age periods (28–35, 36–38, and 39–40 weeks) is significant ($p < 0.05$); group 3 (39–40 weeks) shows a highly significant difference from the other two groups ($p < 0.001$). A differentiation is possible between infants of 38 postmenstrual weeks and under, and of 39 weeks or over, using this response.

In the prone position, a further lateral rotation of the head, when placed in mid-position, can be seen. This response may be considered as a protective movement in order to free the nose and mouth for unhindered respiration.

From 31 to 33 postmenstrual weeks, such lateral rotation of the head in the prone position was observed in about 60% of the Bonn cases (24 examinations) and from 34 to 40 weeks, it was present in 80–100% of 72 examinations.

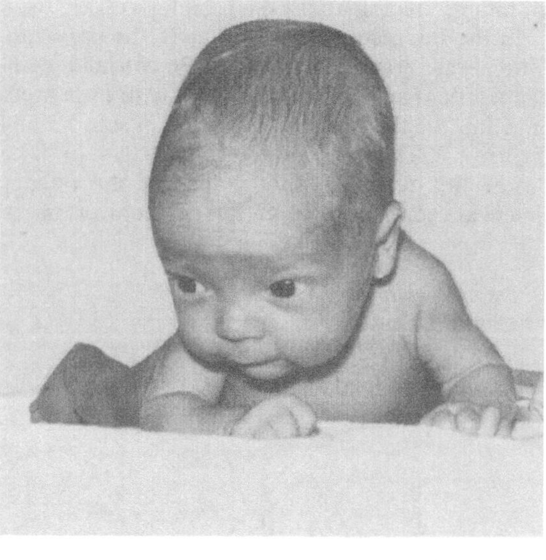

Fig. 14. Head lifting in the prone position for 3–5 sec. Appropriate-for-gestational age (AGA) preterm girl at expected date of delivery, born at 32 postmenstrual weeks.

4.5.5. *Ventral Suspension—Head Lifting*

Ventral suspension was used as a maturational criterion by Dubowitz *et al.* (1970) in their scoring system. The pattern in relation to postmenstrual age was not given. In the quantitative study by Beintema (1968), most full-term infants from day 1 to 9 were able to lift their heads to the horizontal plane for at least several seconds when held in prone suspension. However, maintaining the head in the horizontal plane for at least 5 sec was never attained by most infants. Ventral suspension is also used in the estimation of gestational age (Lubchenco, 1976; L. O. Lubchenco, personal communication, 1977). At term she observes a straight back and the head held in line with the body. The post-term infant holds the head above body level.

In the Bonn study, the relationship of the head to the trunk was scored; i.e., whether the head lifted or not. The following scores have been used: (1) no head lifting; (2) head lifting to the horizontal plane sustained for 1– 3 sec; (3) head lifting to the horizontal plane sustained for more than 3 sec; and (4) sustained head lifting with retroflexion of the head above the horizontal plane for 1–3 sec (never attained). The results are shown in Table XII. The head lifting in ventral suspension may be used as a reliable maturational criterion. The differences between the gestational age groups 28–37 weeks and 39–40 weeks are highly significant ($p < 0.001$). This reaction allows one to distinguish between infants of 37 weeks and under and of 39–40 weeks. At 38 weeks the results are intermediate.

4.5.6. *Righting Reactions—Lower Extermities, Trunk, Head, and Arms*

These reactions demonstrate a righting of: (1) the legs; (2) the trunk; and (3) the head; (4) in the Bonn

Table XII. Ventral Suspension—Head Lifting—Results of the Bonn Study[a,b]

Response	Postmenstrual age in weeks 28–37	38	39–40
No head lifting	46	7	9
Head lifting for 1–3 sec	4	5	14
Head lifting for 3 sec	2	2	13
No. of observations	52	14	36

[a]Nie *et al.* (1975).
[b]Difference between 28–37 and 39–40 weeks highly significant, $p < 0.001$; least significant difference procedure.

study, a righting of the arms was also scored. A full-term newborn infant is able to right himself when his feet are placed on a table and to maintain an upright posture; therefore, this response has also been called "primary standing" by André-Thomas *et al.* (1960). This "standing response" is produced by a tonic myotatic stretch reflex (Schulte, 1970*b*).

The righting reactions were introduced by André-Thomas and Saint-Anne Dargassies (1952), André-Thomas *et al.* (1960), and Saint-Anne Dargassies (1954, 1955, 1966) to newborn neurology. On the basis of maturational changes of these reactions with increasing postmenstrual age they have been used for age assessment (Brett, 1965; Koenigsberger, 1966; Amiel-Tison, 1968; Saint-Anne Dargassies, 1966, 1970).

For an infant of 28 postmenstrual weeks, Saint-Anne Dargassies (1966) stated that "the active tone is already well enough developed to permit extension of the lower limbs (although weak and supporting little weight) when the infant is held upright." "The beginning of extension of the lower leg on the thigh upon stimulation of the soles in a lying position" at 30 weeks is reported by Amiel-Tison (1968).

At 32 weeks, according to Saint-Anne Dargassies (1966), "straightening of the lower limbs when the infant is held in the upright position happens more quickly and extension of the legs on the pelvis is gradually firmer." Straightening of the trunk appears for the first time. Amiel-Tison (1968) reported "good support" of body weight with the baby in the standing position, but only very briefly, at 32 weeks. She did not describe trunk righting.

At 35 weeks, "straightening of the lower limbs in the upright position becomes stronger and begins to include the muscles of the trunk" (Saint-Anne Dargassies, 1966). At 34 weeks, Amiel-Tison (1968) observed excellent righting reaction of the leg and a transitory righting of the trunk; at 36 weeks there is good righting of the trunk with the infant held in vertical suspension.

According to Koenigsberger (1966), who scored "trunk elevation," the reaction is "slight" at 34 weeks, and from 37 weeks there is good trunk elevation on the hips.

At 37 weeks, Saint-Anne Dargassies (1966) reported that "all straightening responses are . . . excellent. Straightening of the lower limbs leads to straightening of the trunk" which leads to "similar response of the whole body axis." Amiel-Tison (1968) observed at 38 weeks "good righting of trunk with the infant held in a walking position." At 40 weeks, she reported a "straightening of the head and trunk together," as did Saint-Anne Dargassies (1970, 1974), who observed a solid righting reaction or holding of the head in alignment with the trunk. In her quantitative study of full-term newborn infants, a righting reaction of the trunk and head was observed in 68% of 150 full-term infants; in 18% the reaction is incomplete; in 13% absent; and in 1% "abnormal."

The righting reactions were mostly tested in the Bonn study with the infant in the standing position as described by Amiel-Tison (1968). Then, after exerting a short light pressure on the soles of the feet, we observed an extension of the legs, a plantar support of body weight, and a righting of the trunk, head, and arms. Near the expected date of delivery, the righting reaction was also tested by having the infant on a firm surface, then placed squatting on the haunches with the feet flat against the surface (André-Thomas *et al.,* 1960; André-Thomas and Saint-Anne Dargassies, 1952.) Intermittent pressure of the feet against the resting plane resulted in the righting reactions. Figure 15 demonstrates the righting reaction in an AGA preterm infant at the expected date of delivery.

In evaluation of the Bonn results, in most of the cases the type of reaction and the scoring could be checked with the help of photographs made at each examination in a standardized position. A righting of the lower extremities, trunk, head, and arms is scored separately on the examination forms:

1. From 30 postmenstrual weeks, a righting reaction of the legs on the thighs upon stimulation of soles was observed in all the infants (105 observations). With the infant lying in an incubator, at least an extension of the lower extremities could be tested. Near term, a righting of the lower extremities has become less pronounced in some preterm infants, possibly because of increasing flexor tone. Sometimes there is a marked resistance against full extension of the legs in full-term newborn infants.
2. The righting reaction of the trunk in the infant held in a standing position was seen from 34 postmenstrual weeks (73 examinations). Up to 33 weeks, the reaction was negative in most of the cases (20 out of 22 examinations). This is in accordance with the results of the Paris school (Saint-Anne Dargassies, 1966; Amiel-Tison, 1968).
3. The righting reaction of the head in the infant

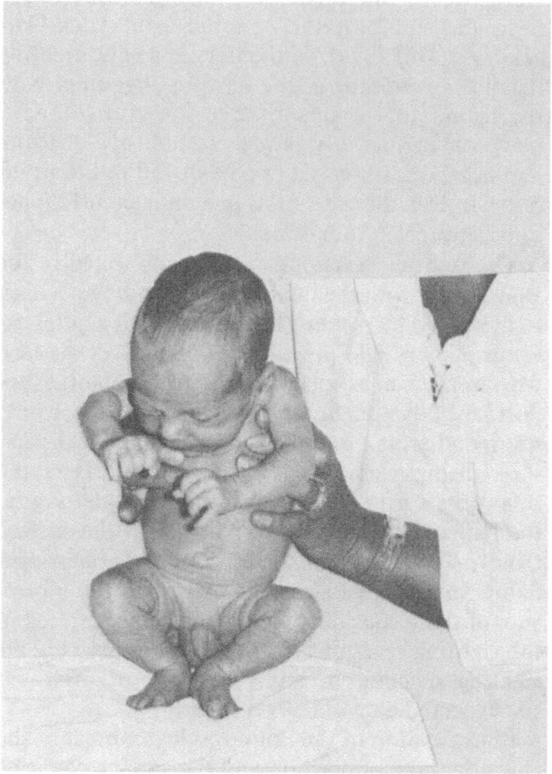

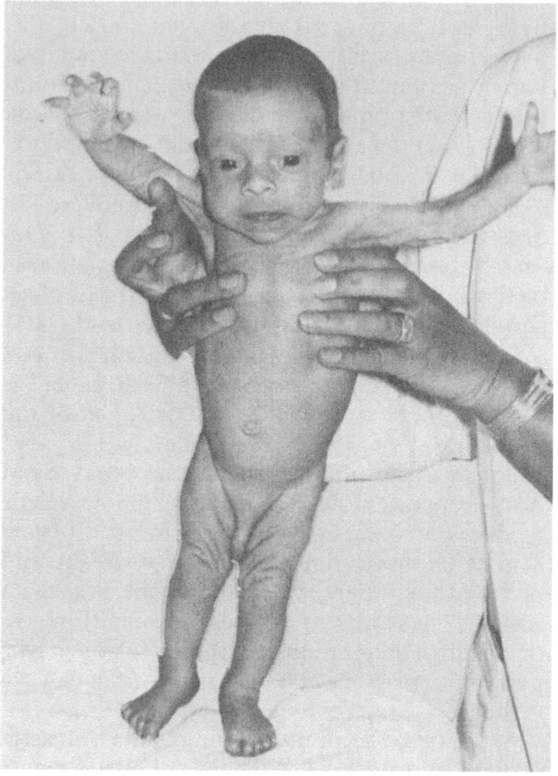

Fig. 15. (Left) Eliciting of the righting reaction according to André-Thomas and Saint-Anne Dargassies (1952). (Right) Righting of lower extremities, trunk, head, arms. Appropriate-for-gestational age (AGA) preterm infant at expected date of delivery, born at 32 postmenstrual weeks.

held in a standing position was scored as absent if there was no true head righting and present if there was head righting for a few seconds (Table XIII). Up to 35 weeks, head righting is absent in most of the cases (38 out of 43 examinations). At 36 weeks, the response is intermediate, a head righting being observed in half the cases. From 37 to 40 weeks, there is a head righting in most of the infants (in 50 out of 53 examinations, Figure 15). There is a highly significant correlation of 0.67 between the righting reaction of the head and the postmenstrual age, ($p < 0.001$ Kendall's tau, Nie *et al.*, 1975). The righting reaction of the head allows one to differentiate between infants of 35 weeks or under and 37 weeks or over.

4. The righting reaction of the arms was scored as present when they are lifted above the horizontal (Figure 15; Table XIV); at 35 weeks and under, it was present in 7% of examinations (3 of 42); and at 36 weeks and over in 93% (54 of 58). It is therefore a good maturational criterion. There is a highly significant correlation of 0.6 between the righting reaction of the arms and the postmenstrual age ($p < 0.001$, Kendall's tau, Nie *et al.*, 1975).

Table XIII. Righting Reaction of the Head—Results of the Bonn Study

	Postmenstrual age in weeks	
Response	28–35	37–40
Absent	38	3
Present	5	50
No. of observations	43	53

In spite of the clear-cut changes demonstrated in relationship to postmenstrual age, the righting re-

Table XIV. Righting Reaction of the Arms—Results of the Bonn Study

Response	Postmenstrual age in weeks	
	28–35	36–40
Absent	39	4
Present	3	54
No. of observations	42	58

actions can be recommended for maturational assessment only with reservation, since they are too easily influenced by any impairment of the infant's general state. Elicitation needs a healthy infant who may be examined outside an incubator in an upright position. All states that reduce muscle tone have to be rejected; states 4 or 5 are required.

4.5.7. Head Turning to Diffuse Light—Phototropism

This reaction was described as long ago as 1859 by Kussmaul (1859) and was observed by him in an infant of "seven months gestation." Also, Stirnimann (1940) observed a tropism to soft light in newborn infants who often turn their heads to the window. With a bright light—as coming from the snow-covered Alps—the infants turn their heads away as in a protective reaction. Peiper (1964) reported that "subdued light is generally so attractive for infants that, already at an early age, not only the eyes but also the head is turned toward it."

According to Saint-Anne Dargessies (1954) and Robinson (1966), in order to elicit the response, the infant, in a state of quiet alertness, is held with the back against the upper part of the body of the examiner in a comfortable position with the head supported. Then by slow turning, the infant is brought near to a diffuse light (e.g., a window) so that diffuse light falls on the lateral side of his face. In a positive reaction the infant slowly turns the eyes and head toward the light. While turning slowly away, with the baby from the window to the other side, the infant maintains the direction of gaze by turning his head until the light of the window disappears. If the light then comes from the other side, the baby orientates again to it. The turning toward light of the eyes, and of the eyes and the head, is scored separately on both sides. It is difficult to test this reflex in an incubator, so it cannot be evaluated in very immature infants.

Robinson (1966) tested the head turning to light reflex longitudinally, and reported that in four babies born at less than 32 weeks gestation, in whom the response was initially absent, the response first appears at 32, 33, 34, and 36 weeks. He stated that "the response usually appears between 32 and 36 weeks" in AGA and probably also in SGA preterm infants. According to his data: under 32 weeks the reflex is absent in all observations; from 32 to 36 weeks the reflex is present in most (11 AGA and 8 SGA preterm infants); and at 36 weeks and over, head turning to light is present in nearly all observations (10 AGA and 15 SGA).

Saint-Anne Dargassies (1966, 1974) described turning of the eyes and then of the head toward a soft light source (phototropism) to be present first at 37 postmenstrual weeks as a new acquisition. The pattern of this response in the earlier weeks is not given.

Studying full-term infants, she (1974) observed in 90 of 150 infants an "active and complete" head turning to light; in 30 infants the reaction was incomplete (i.e., turning only of the eyes to light or only to one side); it was absent in 30. She stated that phototropism also exists in blind infants.

Dubowitz and Dubowitz (1977) considered it impossible to obtain any consistent response to diffuse light, even in full-term infants.

In the Bonn study, the head-turning-to-light reaction elicited according to Saint-Anne Dargassies (1954) (Figure 16). The response is scored for each side separately. Turning of the eyes (1) and turning of the eyes and the head (2) are scored. The reflex has rarely been tested in very young infants because they could not be moved from the incubator without risk. The observed response was bilaterally symmetric in all 29 infants of the study group. The results are shown in Table XV.

There are significant differences among the three gestational age groups 28–33, 34, and 35–40 weeks ($p < 0.01$), (see Table XV, Figure 17), group 3 (35–40 weeks) differs highly significantly from the other two groups, ($p < 0.001$). This reaction allows one to differentiate between infants of 33 weeks or under and 35 weeks or over. We therefore, agree with Robinson (1966) that phototropism may be counted among the reflexes of value in assessing gestational age.

In some other infants of the Bonn study, a unilateral response was observed (as reported by Saint-Anne Dargassies, 1974); i.e., the head-turning response appeared first on one side and 1 or 2 weeks later also on the other.

Some preterm infants of more than 34 weeks ges-

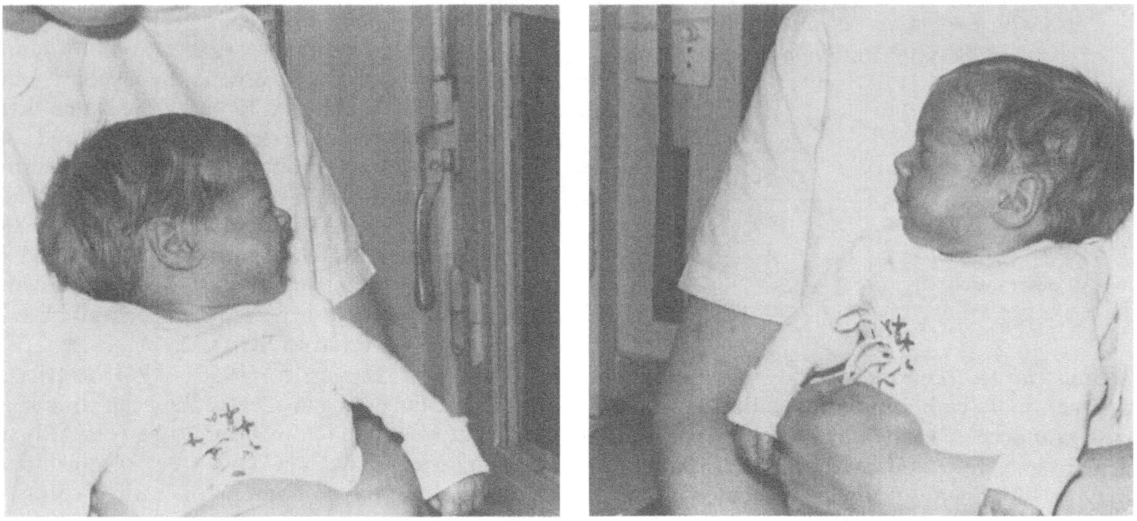

Fig. 16. Head turning to light (positive phototropism) in a preterm boy at expected date of delivery, born at 31 postmenstrual weeks. The response is present to both sides.

tation who are still in the incubator turn their heads constantly to the window, under the influence of phototropism. This may induce the asymmetrical tonic neck reflex (see Section 4.7.) with a marked flexion of the leg on the side toward which the occiput is turned. If this position is maintained for several days, there exists a risk of hip dysplasia in the constantly flexed and adducted leg.

The head turning to light belongs to the primitive reactions. In normal infants,it becomes negative at the corrected age of 2–4 months; however, in cases of brain damage it may persist longer. Perhaps the appearance of this primitive reaction is more related to the developing ability of spontaneous head rotation (present from 34 weeks in most of the Bonn infants) than to the phototropism itself.

Table XV. Head Turning to Light (Phototropism)—Results of the Bonn Study[a,b]

Response	Postmenstrual age in weeks		
	28–33	34	35–40
Absent	28	6	7
Turning of eyes	0	4	11
Turning of eyes and head	0	2	57
No. of observations	28	12	75
No. of infants	28	12	29

[a]Nie *et al.* (1975).
[b]Difference between the three age groups significant, $p < 0.01$; and between 35–40 weeks and the other two groups highly significant, $p < 0.001$ (least significant difference procedure).

4.6. Reflexes and Reactions Almost Constantly Present in the Perinatal Period Independent of Postmenstrual Age

Some of these reactions are also called "fetal reflexes" (M. Minkowski, 1921, 1924; Humphrey, 1964) because they are elicitable in immature fetuses and become negative within the first 3 months. Langreder (1949) discussed an intrauterine function of such "reflexes;" e.g., active adaptation to attitude, stabilization of attitude, and active participation during birth. These "reflexes" are already present in infants from 28 postmenstrual weeks and do not essentially change until the expected date of delivery. Most of these movement patterns can be observed spontaneously in fetuses as early as 13 weeks (de Vries *et al.,* 1982) (see Section 3).

To this group belongs the well-known lateral curving of the trunk, or *Galant's* response, which is elicited with the infant in the prone position by softly scratching the paravertebral region on the right or left side. This results in a curving of the trunk to the stimulated side. The response was elicited by Galant (1917) in all but one of 164 infants from birth to 6 months; with increasing age, the response became feeble. According to Paine (1960) the lateral trunk curving is seen as early as 26 postmenstrual weeks and remains positive until 2

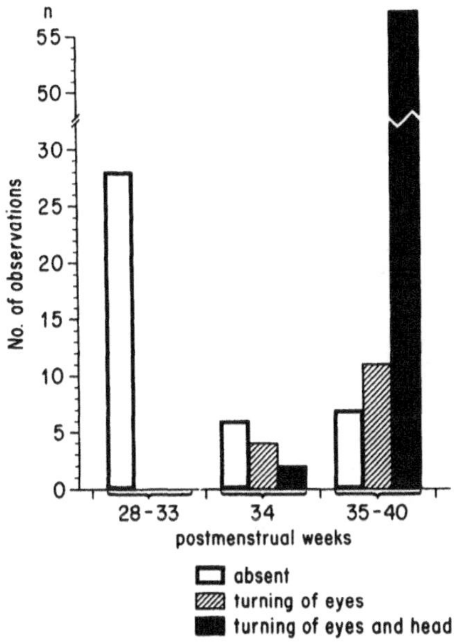

Fig. 17. Head turning to light (phototropism). The response distinguishes between infants of 33 weeks or under and 35 weeks or over.

months. Robinson (1966) observed a Galant response in 87% of preterm infants, including babies of 26 weeks gestation; therefore, it is of little value in estimating fetal age.

A fetal precursor of the Babinski (1896) reflex has been elicited by M. Minkowski (1921) in a fetus at the fifth month by stimulation of the sole of the foot, resulting in flexion of the toes. The Babinski reflex, with extension of the big toe and fanning of the other toes, is present in all infants in the early postnatal days (Galant, 1917).

The methods used to elicit the plantar response and the descriptions of resulting movements vary greatly. In the Bonn study, the Babinski reflex was elicited according to Wolff (1930) by passing the smooth handle of a reflex hammer softly over the lateral part of the soles of the feet. The toe responses to the plantar stimulus were scored as: (1) Dorsal extension of all toes with or without fanning; (2) Plantar flexion of the toes; and (3) Isolated slow hyperextension of the big toe. Dorsal extension of all toes with or without fanning was the most frequent response from 28 to 40 postmenstrual weeks. It was observed in 125 of 135 examinations. Eight infants showed plantar flexion of the toes, and in two cases an isolated slow hyperextension of the big toe was seen.

According to Lenard *et al.* (1968) the Babinski reflex of the infant may be regarded as part of a withdrawal reaction to a nociceptive stimulus; it is easily obtainable during regular sleep, irregular sleep, and quiet wakefulness.

The knee jerk belongs to the reflexes "which are equally strong during regular sleep and wakefulness, but are weak or absent during irregular sleep" (Lenard *et al.,* 1968), and which may be elicited even in very immature infants. According to Schulte (1974), this tendon reflex is usually seen in newborn infants beyond 30 weeks gestation. He has elicited the patellar reflex also in a preterm infant of 25 postmenstrual weeks. Robinson (1966) reported that the knee jerks correlate loosely with gestational age and therefore cannot be recommended for age assessment. Beintema (1968) never found an absent response in newborn infants. However, Finnström (1971) reported a correlation coefficient of 0.51 for the patellar reflex and gestational age. The response is scored absent in 50% of the infants below 32 weeks, and in 10% of the gestational age group of 32.1–36 weeks. In the Bonn study, knee jerks could be elicited in all infants from 28 postmenstrual weeks.

We will discuss in greater detail a few examples out of the numerous reactions and reflexes which are almost constantly present from birth also in preterm infants. They are selected because their value for maturational assessment is disputed in the literature.

4.6.1. Glabella Tap Reflex

This reflex is elicited by tapping sharply on the glabella (Prechtl and Beintema, 1964). The response is a tight closure of the eyes of short duration. The glabella reflex is mostly absent during regular sleep, weak during irregular sleep, and strongest during wakefulness (Lenard *et al.,* 1968).

Robinson (1966) scored a "blink of the eyelids in response to a tap by a finger on the glabella." A grimace of the whole face, sometimes seen in very immature infants, is not considered a positive response. In his study of AGA and SGA preterm infants, the glabella reflex was absent under 32 gestational weeks in 17 of 21 infants. At 32–34 weeks the result was intermediate, in half of the cases the reflex being present. Serial examinations on five preterm infants in whom the reflex was initially absent showed that it first appeared between 32 and 34 postmenstrual weeks. From 34 weeks, the reflex was present in most of the observations, with no difference between AGA and SGA infants. According to Joppich and Schulte (1968), the glabella reflex appears at 32–34 weeks gestational age. It probably consists of two components: the short blink; a monosynaptically mediated proprioceptive reflex of the musculus orbicularis oculi (cranial nerve VII); and the following tonic contraction, maintained by polysynaptic activation of the same motor neurons (Schulte, 1970*b*).

Koenigsberger (1966), who used the expressions supraciliary tap or MacCarthy reflex, elicited it on each side by tapping the supraciliary region above the eyebrow, which produces a homolateral and sometimes bilateral blink. He suggested that this is "probably the same reflex arc as the corneal reflex (afferent cranial nerve V, efferent VII)." Koenigsberger found the reflex inconstant at 28 postmenstrual weeks and present from 32 weeks.

Graziani *et al.* (1968) also elicited the reflex—which they called "supraorbital blink response"—on both sides by tapping the supraorbital area lightly when the infant's eyes are open. The midpoint of age range for unilateral blink, or asymmetrical responses, is 30 weeks.

Dubowitz *et al.* (1970) saw the glabellar tap consistently present in all infants over 30 weeks gestation and thus regarded it as of little value for assessment of age. Neither Beintema (1968), Peiper (1964), nor Saint-Anne Dargassies (1974) mentioned this reaction.

In the Bonn study, the glabella tap reflex was elicited as described by Prechtl and Beintema (1964). Before 31 postmenstrual weeks, the reflex was not tested regularly in all infants. From 29 to 30 weeks, a positive response was observed in 5 of 6 infants. From 31 to 32 weeks, the reflex was seen in 18 of 19 cases. From 33 to 40 weeks, there was a positive response in all examinations ($n = 95$).

A similar blink of the eyelids as a response to a bright light (flashlight for the photographs) was observed in the Bonn study in all preterm infants from 28 to 40 postmenstrual weeks (153 examinations). This blink reflex in its response is very similar to the glabella tap reflex. It is constantly present from birth, even at an early age. Even during sleep, the eyes are closed more tightly and it is sometimes accompanied by grimacing in infants of very low gestational age. This reflex is considered a defense reaction by Peiper (1964).

4.6.2. Crossed Extensor Reflex

The crossed extensor reflex is counted among the "crossed reflexes;" i.e., where the stimulus of the stimulated extremity overlaps the other one, and there elicits an effect. In a fetus of 4 months Minkowski (1924) reported distinct signs of antagonistic innervation in the lower extremities, where pinching of one leg produces a flexion of it, and an extension of the crossed one. Peiper (1964) mentioned only a crossed flexion reflex. The reflex is weak or absent after breech delivery with extended legs (Prechtl and Beintema, 1964).

To elicit the response, one leg is extended and kept in this position by pressing the knee down. After stimulating the lateral part of the sole of the foot of the passively extended leg, the unstimulated (free) leg is first flexed, then extended and adducted slowly to the stimulated one with fanning and dorsiflexion of the toes.

Saint-Anne Dargassies (1954, 1966, 1974) attached great importance to the crossed extensor reflex. She distinguished three phases: flexion, extension, and then adduction of the free leg, which vigorously approaches the stimulated one. When the stimulation is continued, the free leg is placed on the stimulated one with fanning of the toes. This phase of adduction is considered the most important phase of the reflex. She demonstrated its gradual appearance in the course of the last postmenstrual month, and it is not complete before near term (41 weeks). This phase is characteristic of this fetal age. A prerequisite for eliciting the reflex is that the infant does not cry, as the reaction is inhibited by the cry, by pain, and by hypertonia (Saint-Anne Dargassies, 1974).

The crossed extensor reflex was counted by Saint-Anne Dargassies (1974) among the six maturational signs, which are reliable in a high percentage of cases. At 28 postmenstrual weeks, the real reflex does not exist and instead only a gesture of defense is observed. At 32 weeks, the unstimulated free leg is flexed and then abducted. At 35 weeks, the reflex is said to be excellent in its two first phases, flexion and extension, but without adduction. The abduction, observed at 32 weeks, has become less. At 37 weeks, the crossed extensor reflex remains imperfect in its adduction movement. At 41 weeks, the adduction phase of the reflex is perfect.

In her quantitative study with 150 full-term infants, examined between the fifth and seventh day, Saint-Anne Dargassies (1974) reported a perfect adduction in 90% of cases (135), and the reaction was absent in only 2 infants. In 7% of infants, the crossed extensor reflex is incomplete and weak.

Robinson (1966) stated the crossed extensor reflex proved an unreliable index of maturity. He suggested "that the reflex is more commonly present after 32 weeks gestation than before." In longitudinal studies the extension of the opposite leg did not appear at a particular postmenstrual age and remain so, but varied in its presence or absence.

Graziani *et al.* (1968) scored the reflex differently. They distinguished: (1) flexion of contralateral extremity; (2) flexion, then delayed extension; and (3) flexion, then rapid extension and occasional adduction. The midpoints of age range for these three reflex phases are 30 weeks for (1), 34 weeks for (2), and 39 for (3).

Beintema (1968) scored the extension phase of the crossed extensor reflex as recommended by Prechtl and Beintema (1964) in their study of full-term newborn infants. On the first and second day, the reflex is absent or doubtful in 12–25% of observations, and thereafter it is present in 95–100%.

In the Bonn study, the crossed extensor reflex was elicited as described by Prechtl and Beintema (1964) and the following phases were scored: (1) flexion; (2) extension; (3) abduction; and (4) adduction. The flexion phase (1) was present in most of the observations from 29 postmenstrual weeks (89 of 92 examinations). The extension phase (2) was present in most of the cases from 29 postmenstrual weeks (90 of 92 examaminations). From 31 to 36 weeks, the abduction phase (3) was observed in about one-half the examinations (21 of 45), and from 37 to 40 weeks in most cases (38 of 42). Up to 35 weeks, the adduction phase (4) was absent in most of the examinations (38 of 41). From 36 to 37 weeks, the results were intermediate, the response being positive in half of the cases (7 of 13), and from 38 to 40 weeks, this fourth phase was present in most of the cases (35 of 38 examinations).

This reflex is of limited value because the infants object to the elicitation and cry and resist. For many preterm infants a fixation of the leg and stimulation of the sole of the foot represents a nociceptive stimulus because of frequent extractions of blood mostly from the heel. Moreover, in infants with a low postmenstrual age (i.e., 31 weeks and under) who are upset by this umpleasant stimulation, the degree of excitement is easy to recognize; a monitor showing a remarkable rise of heart rate and respiration irregularities. In these infants, the reaction cannot be considered as normal for the reflex, but as a reaction to pain. Therefore, in recent years the reflex has no longer been performed and it cannot be recommended for age assessment.

4.6.3. Palmar Grasp

A palmar grasp has been shown to occur already in early fetal life. Humphrey (1964) reported a quick partial closure of all fingers on stimulation of the palm of the hand, first elicited at 10–10.5 postmenstrual weeks. At 13–14 weeks, finger closure is complete but momentary. At 27 postmenstrual weeks, Humphrey (1964) described a "maintained grasp with ability to support almost the entire body weight momentarily."

The grasp reflex of the hands is among the responses that are constantly present in preterm infants from 28 postmenstrual weeks (Robinson, 1966). It is mostly absent during regular sleep, weak during irregular sleep, and strongest during wakefulness (Lenard *et al.*, 1968). With increasing postmenstrual age, the grasp reflex of the hands becomes firmer and the reaction extends to the shoulder and neck muscles; permitting the infant to be lifted completely, whereby the head is also raised. No clear-cut changes that permit age assessment are observed.

Saint-Anne Dargassies (1966, 1970) described developmental stages of palmar grasp: at 28 postmenstrual weeks an isolated flexion of the fingers; at 32 weeks an involvement of the muscles of the forearms; at 37 weeks a steady and firm grasping; and at 41 weeks a firm grasp reflex by which the infant may be lifted. More recently Saint-Anne Dargassies (1974) counted the palmar grasp among the six particularly unreliable neurological signs.

In Beintema's (1968) quantitative study of full-term infants, an absent response for palmar grasp was never found, in accordance with most previous reports, but the intensity of the palmar grasp increased with postnatal age. Saint-Anne Dargassies (1974) reports an absent palmar grasp in 2 of 150 full-term infants in her study.

Palmar grasp is observed in all preterm infants of the Bonn study from 28 postmenstrual weeks. A scoring of the firmness of grasp did not yield additional information for maturational assessment.

4.6.4. Rooting Response

The rooting reflex is counted by Minkowski, who observed a rooting response already in fetuses of 2 and 3 months gestational age, among the earliest and most constant fetal reflexes (cited by Prechtl, 1958).

According to Humphrey (1964), the earliest response to an exteroceptive stimulation was observed in a human fetus of 7.5 weeks after stroking the perioral region with a hair. The stimulation caused a contralateral flexion in the neck region; i.e., away from the stimulus. At 9.5 postmenstrual weeks, she describes mouth opening by lowering of the mandible after stimulation of the edge of the lower lip; and at 10 weeks after stimulation of the lower lip area and over the mandible, a ventral flexion, (i.e. toward the stimulus). At 11–11.5 weeks, a perioral stimulation is followed by an ipsilateral rotation of face.

Prechtl (1958) gave a comprehensive review of this response. In preterm infants of 28 postmenstrual weeks, he observed weak side-to-side movements of the head with or without perioral stimulation. The "physiological conditions of sleep" (acting as a brake) "and hunger" (facilitating the reaction) "are the main factors influencing" the rooting response.

According to the method of Prechtl and Beintema (1964), the rooting response is elicited by stimulation of the perioral region by fingertip at the angles of the mouth, the upper lip and the lower lip. Stimulation at the edges elicits a head turning to the stimulated side. Stimulation of the upper lip results in opening of the mouth and retroflexion of the head. After stimulation of the lower lip, the mouth is opened and the infant tries to grasp the stimulating finger with the lips and to flex the neck. In preterm infants of low postmenstrual age (i.e., 31 weeks and under) who cannot keep the head in the midline, the head has to be supported a little.

Amiel-Tison (1968) considered the rooting reflex dependent on postmenstrual age. From 28 to 30 weeks, the observed response is slow and imperfect with a long latency period. At 32 weeks, rooting is "complete and more rapid." From 34 to 40 weeks, rooting is "brisk, complete, and durable."

Beintema's (1968) quantitative study of full-term newborn infants found that on the first day 10% of infants had an absent response and 12% showed a lip movement only. On the first 6 days, the rooting response was significantly weaker than at 9 days, where 88% of the infants showed "full turn towards the stimulated side, sustained for at least three seconds while the finger maintained contact with the angle of the mouth, and lip-grasping."

Saint-Anne Dargassies (1974) differentiated three phases of the rooting response—lateral rotation of the head, extension, and flexion—and counted this response among the six maturational criteria, which are reliable in a high percentage of cases. At 28 postmenstrual weeks, where support of the head is indispensable for eliciting the reflex, she reported lateral rotation of the head and limited extension, and a stimulation at the midpoint of the lower lip produces no response (i.e., no flexion). Yawning is still found as a primary response to perioral excitation (Saint-Anne Dargassies, 1966). Yawning instead of rooting as a response to stimulation of the perioral region has also been described in full-term infants (Peiper, 1964). In the Bonn study, a rooting response was observed in all examinations (n = 153) from 28 postmenstrual weeks.

Besides postmenstrual age, other factors might influence a rooting response being weaker in infants of a postmenstrual age of 30 weeks and under. Since there is a high correlation for rooting with the level of alertness (Prechtl and Beintema, 1964), a weaker response is to be expected in states 1 (eyes closed, regular respiration, no movements) and 2 (eyes closed, irregular respiration, rarely gross movements) prevailing in these babies. Another factor may be that the infants are not hungry because they are fed every hour and often receive intravenous glucose in addition.

Since marked intraindividual variations in the rooting response have been attributed to factors like hunger and state (Beintema, 1968), low postmenstrual age alone cannot be identified with certainty as being responsible for a slow and imperfect response. Therefore, the rooting response as the earliest response to an exteroceptive stimulation in fetal life, cannot be recommended for maturational assessment of the preterm infant. The reported changes in this reaction during the perinatal period cannot be ascribed to the rooting relfex itself.

4.6.5. Stepping Movement

The stepping movements are elicited, according to Prechtl and Beintema (1964), by holding the infant upright with the soles of the feet touching the surface of a table; the infant is then moved forward to accompany the stepping. The response requires state 4 (eyes open, gross movements, no crying) or state 5 (eyes open or closed, crying). Stepping movements are "absent in infants born after breech delivery, who either extend or flex the legs" (Prechtl and Beintema, 1964).

Peiper (1964) described the newborn infant making stepping movements in every body position, even with the head downward. Automatic stepping has been observed in decerebrated animals that walk until they collapse (Peiper, 1964). This automatic function of the CNS is inhibited with increasing maturity of the brain.

According to Langreder (1958), stepping movements of the fetus can already be demonstrated *in utero* by impulses on the abdominal wall and by experiences with women in labor. It is suggested that they, like other newborn reflexes such as spontaneous crawling, may serve as an active intrauterine "reflex participation (or aid) in birth" of the fetus.

Milani-Comparetti (1981) considered *automatic walking* a residual sign of the fetal locomotion, modified by gravity. Prechtl (1983*a*), using real-time ultrasound, demonstrated rhythmic flexion and extension movements of the legs in 15-week-old fetuses; unequivocally the stepping movements of the newborn infant, which are considered a specific intrauterine adaptation, and remain after birth for a few weeks.

The stepping movements are recommended for maturational assessment by Brett (1965), Koenigsberger (1966), Amiel-Tison (1968), and Graziani *et al.* (1968). Finnström (1971) reported a low correlation coefficient of 0.30 with gestational age, and Saint-Anne Dargassies (1974) counts the stepping movement among the six particularly unreliable maturational signs.

Stirnimann (1938, 1940) studied stepping movements in 75 newborn infants in the first 24 hr after birth and at 9–14 days. On the first day, only 16% of the infants make stepping movements. Some infants lift only one leg or both without a tendency to move forward; 40% of the infants show no reaction.

At the age of 9–14 days, 35% of infants "stepped" adequately, 18% stepped with crossed legs, and some infants made only one step. Altogether stepping movements were seen in 58% of infants, contrary to 16% within the first 24 hr. According to Stirnimann (1938), neither the type of delivery nor the maturational state is of great significance. Only dyspnea seems to inhibit the reaction.

Beintema (1968) observed stepping movements in 66.5–91.5% of full-term infants 4–9 days after birth. On the first day, stepping movements were seen in only 36.5%, these consisting mostly of one or two steps.

In her study of 150 full-term infants, examined at the fifth to seventh postnatal day, Saint-Anne Dargassies (1974) reported regular stepping movements in 58%, incomplete or suggestive in 15%, abnormal or with crossing over of the feet in 9%, and absent in 18%.

In the Bonn study, following the examination of the righting reactions, the walking reflex is tested by leaning the infant a little forward. By alternating flexion and extension movements of the legs, infants bring themselves forward when the trunk is supported. The full-term infant "walks" with the knees and hips slightly flexed and more on the heels. The response is scored present when at least two steps are made; furthermore, the position of the legs is scored. In order to avoid unnecessary disturbance, the infants are rarely tested before 32 postmenstrual weeks. From 32 to 40 weeks, stepping movements were present in most observations (65 of 69). No preterm infant exhibited "tip-toeing." In 65 observations they walked on the sole of the foot (Figure 18), which is different from the full-term newborn who starts stepping movements on the heel (Figure 18). Only at 36–37 weeks do a few infants (n = 4) walk more on the anterior part of the foot than on the sole. Contrary to reports of others (e.g., Illingworth, 1967; Saint-Anne Dargassies, 1974) the Bonn preterm infants did not walk on tip-toe. The greater dorsiflexion angle of the foot of 20–40° (on average) in preterm infants compared with 0° in normal full-term infants, is no hindrance to normal walking movements; since the ankle dorsiflexion in normal infants during the early years is also at least 30°.

4.6.6. Auropalpebral Reflex

The auropalpebral reflex is elicited by an auditory stimulus and consists of a contraction of the orbicularis oculi muscles; the child quickly and distinctly shuts the eyelids if the eyes are open or screws them up if they are shut. It is considered an objective test of hearing (Fröding, 1960; Northern and Downs, 1978). Using high-resolution ultrasound imaging, Birnholz and Benacerraf (1983), reported a positive auropalpebral reflex after vibroacoustic stimulation to be consistently present in fetuses after 28 postmenstrual weeks.

Fröding (1960) found a negative auropalpebral reflex in only 3.3% (n = 66) of newborns in the first postnatal half-hour and after 48 hr in only 1.6% (n = 32). Among these remaining infants with a negative reflex were 22 (68.7%) premature infants. Except for one infant with defective hearing, all children subsequently showed a positive reflex after 3 days to 8 weeks. Saint-Anne Dargassies (1974) observed this reflex only rarely, i.e., in 11 of 150 full-term newborn infants studied, and considered it not specific for hearing. In the Bonn study, out of 75 repeated examinations in 29 preterm infants, the auropalpebral reflex, elicited by hand clapping, was negative in 6 infants on 10 occasions. There was no correlation with age. All infants showed at least one positive reaction before term, except for one in whom the reflex became positive at two months (corrected age). At follow-up, the reaction to acoustical stimuli, using the Griffiths test (Griffiths, 1964), was normal in all infants.

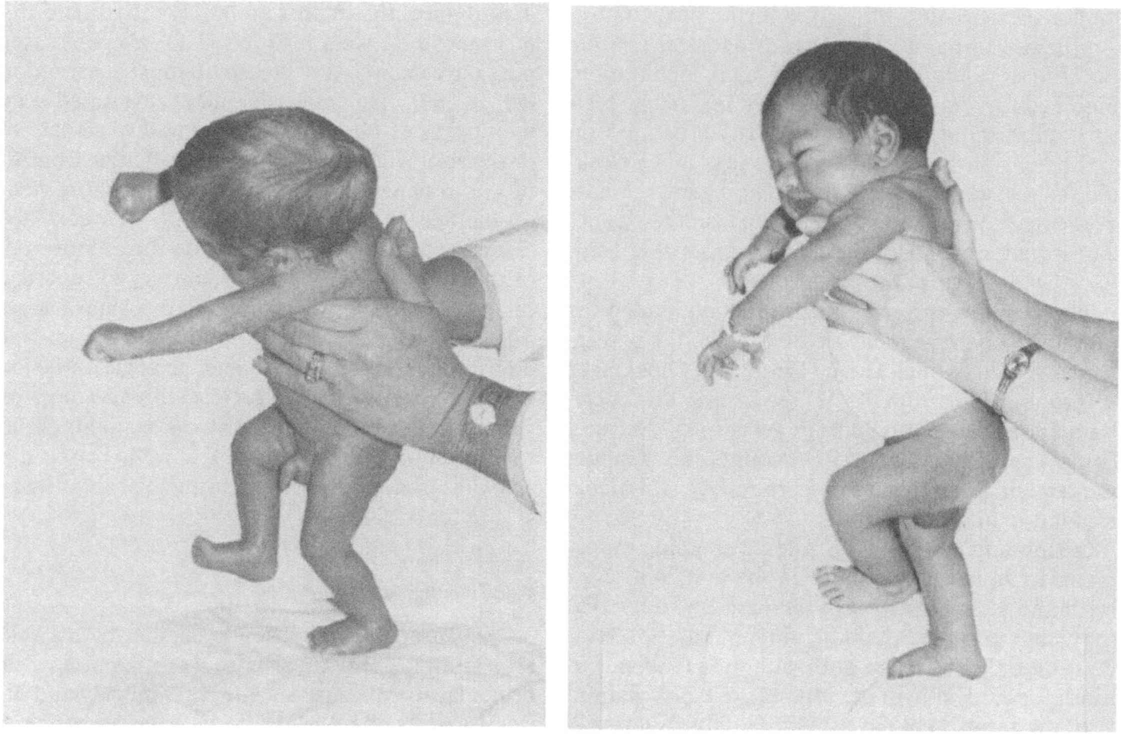

Fig. 18. Stepping movements on the sole of the foot. (Left) Appropriate-for-gestational age (AGA) preterm infant at expected date of delivery, born at 31 postmenstrual weeks, birth weight 1100 g. (Right) Full-term infant at the postnatal age of 7 days.

4.7. Asymmetrical Tonic Neck Reflex

This reflex belongs to those reactions that may be spontaneously present in the perinatal period or elicitable but are not demonstrable in all infants. By turning the head to one side, the asymmetrical tonic neck reflex may be elicited, consisting of an extension of the extremities on the side toward which the face is turned and of a flexion on the contralateral side; i.e., of the occiput. The flexion pattern may be present only in the arm, only in the leg, or in both. While turning the head to one side, an elicitation of the neck-righting reflex pattern has to be avoided; i.e., the trunk has to be kept in a supine position. The reflex pattern is similar to a fencing stance.

Magnus and de Kleijn (1912), who examined the influence of head position on tone and posture of the extremities in decerebrate cats and dogs, showed that after head turning to one side, the extremities toward which the face is turned have an increased extensor tone, whereas on the contralateral side (of the occiput) extensor tone is decreased. This neck reflex persists as long as the head is kept in its specific position. M. Minkowski (1921) observed tonic neck reflexes in fetuses of low postmenstrual age. Schaltenbrand, who introduced the results of animal experiments of Magnus and de Kleijn (1912) and Magnus (1924) in infant neurology, reported that in young infants "asymmetric neck reflexes are often to be seen"; 4 of 21 newborns in the first week show an asymmetric tonic neck reflex, and in 5 the pattern is doubtful.

Gesell (1938) found the asymmetrical tonic neck response "an ubiquitous, indeed, a dominating characteristic of normal infancy in the first three months of life," in contrast to the findings of Magnus (1924), who was unable to elicit the reflex in normal infants from the newborn period to 16 postnatal weeks and considered it a pathological phenomenon. Gesell documented his findings in a series of photographs of normal infants. He induced the asymmetrical tonic neck reflex indirectly by slowly moving a stimulus across the field of vision, thereby inducing head rotation. In longitudinal observations during the first 3 months, he found that at least one-third of the infants consistently assumed a right tonic neck reflex, one-third the left, and the remainder either a right or a left reflex in an ambivalent manner (Gesell, 1938). The asymmetrical tonic neck posture is considered by Gesell (1952) to be "a reflex attitude so basic that it is already present in the fetal infant, born two months prematurely." Robinson (1966) reported only a weak correlation with gestational age in his study of preterm infants.

Saint-Anne Dargassies (1974) distinguished two kinds of reflex patterns: (1) spontaneous, constant, quick, inexhaustible; and (2) slow, inconstant, incomplete, needing a long time of latency. The first is considered a highly pathological sign, whereas the second may be observed in normal newborn babies. Contrary to Gessell (1938), André-Thomas (cited by Saint-Anne Dargassies, 1974) reported the reflex to be inconstant in normal infants.

In full-term infants, Saint-Anne Dargassies (1974) rarely observed the asymmetrical tonic neck reflex. The first reaction pattern, described as slow or asymmetrical, was absent in 118 of 150 infants. Eighteen infants showed the reflex only in the legs, nine only in the arms, and five in both. The second pattern of reaction, called rapid, was absent in 135 of 150 infants, 12 infants showed the reflex in the upper and lower extremities, and three only in the upper ones.

In the Bonn study, the asymmetrical tonic neck reflex was examined to the right and to the left side and was evaluated for each side separately. It was scored present when a positive response was observed, either in the arms, the legs, or both. On examination form, arms and legs are scored separately. Furthermore, a differentiation is made between "spontaneously assumed" and "reflex can be elicited." "Spontaneously assumed" means that either the infant is lying in the pattern of the reflex at the beginning of examination or assumes this position spontaneously during the observation without being moved. The two photographs in Figure 19 show an AGA preterm infant of 32 weeks, assuming the reflex posture spontaneously. "Reflex can be elicited" is scored when the reflex is present after isolated turning of the head to one side. It is also scored if the reflex pattern is surmounted easily, or with difficulty, and if it is a prevailing posture during the examination. For evaluation of the results, the photographs, taken at each examination, again serve as a control. The reflex needs a few seconds for elicitation.

We found that the asymmetrical tonic neck reflex showed no clear-cut changes in relation to postmenstrual age, but there was a trend toward more positive reactions with increasing age, the correlation coefficient being only 0.22 (Kendall's tau). The results are shown in Table XVI. The reflex pattern was scored "easy to surmount" in most of the infants at all examinations (69 of 73).

The response was examined to the right and to the left, and a comparison yielded no difference in the group of 29 AGA preterm infants, followed longitudinally; i.e., it was always bilateral, except in one infant at 39 weeks (unilateral to the right).

The asymmetrical tonic neck reflex is too unpre-

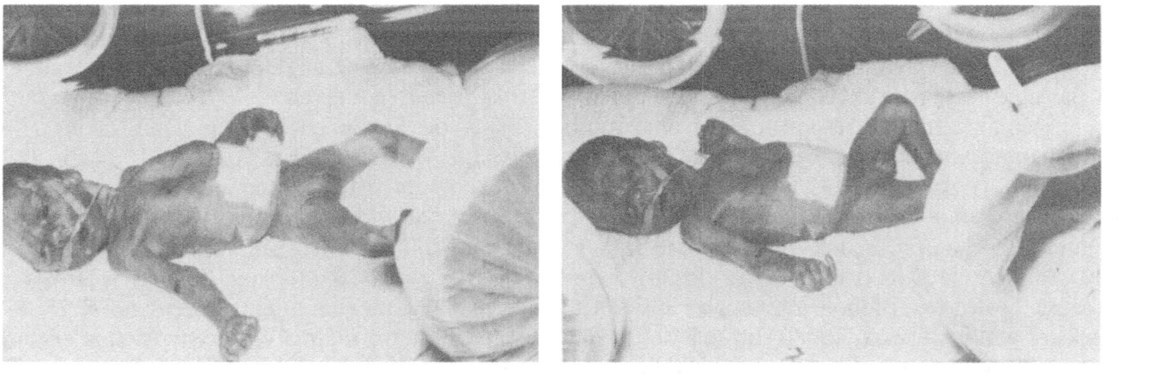

Fig. 19. Appropriate-for-gestational age (AGA) preterm infant of 7 days born at 31 postmenstrual weeks. The asymmetrical tonic neck reflex is assumed spontaneously. The face arm and face leg are in extension, the contralateral extremities in flexion. Between the two photographs a time of 10–20 sec has elapsed.

Table XVI. Asymmetrical Tonic Neck Reflex—Results of the Bonn Study

Response	Postmenstrual age in weeks		
	28–32	33–36	37–40
Absent	21	21	17
Spontaneously assumed	3	13	18
Elicitable	5	14	20
No. of observations	29	48	55

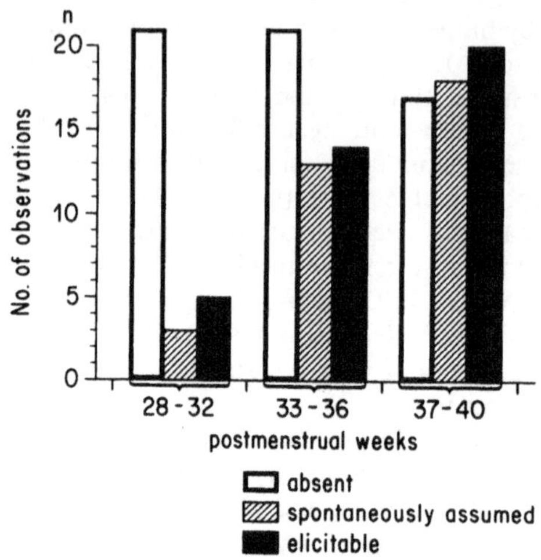

Fig. 20. Asymmetrical tonic neck reflex. This reaction is inconsistent and has no clear-cut time of appearance; no gestational age can be excluded with certainty by a negative response.

dictable and cannot be recommended for maturational assessment, which agrees with Robinson (1966). As demonstrated in Figure 20, this reaction has no clear-cut time of appearance and, by a negative response, no gestational age can be excluded with certainty.

In most of the infants (two-thirds) with a positive reaction, the reflex is consistent (i.e., the response remains positive at every examination) in one-third, the reflex is inconsistent; i.e., present in one examination, absent in the next, and present or absent in the following. In most of the infants with a negative reaction, the finding is consistent in the perinatal period. Later assessment showed the infants of each group to be developing normally.

4.8. Nonneurological Tests as Maturational Criteria

Joint mobility in the hand and foot—the degrees of flexion of the ankle and wrist—are used as maturational criteria but they are not neurologically based since they are not dependent upon the development of resistance to passive movements or of muscle tone. These angles are related more to the general flexibility of the joints, which are relatively stiff early in gestation and become relaxed closer to term, allowing the infant to mold himself or herself to the uterine space (Lubchenco, 1976). Possibly the placental hormones responsible for relaxation of maternal joints late in pregnancy also influence fetal joints (Lubchenco, 1976; L. O. Lubchenco, personal communication, 1977).

Purely mechanical factors—as supposed in the past (Saint-Anne Dargassies, 1974)—do not seem to cause the decrease of these joint angles in the last weeks of gestation, from 20–40° up to 0° at term. These tests of joint mobility are valuable for age assessment only within the first postnatal days since their intrauterine development is different from the extrauterine development in the corresponding time. They are not applicable for longitudinal observations before term; at the expected date of delivery, there is a difference between preterm and full-term infants.

4.8.1. Volarflection Angle of the Hand—Square Window

The angle between the hypothenar eminence and the ventral aspect of the forearm or volarflexion angle of the hand, called "square window" by Dubowitz *et al.* (1970; Dubowitz and Dubowitz, 1977) is tested by flexing the hand on the forearm between the thumb and index finger of the examiner. In the scoring system for the neurologic criteria of Dubowitz *et al.* (1970) scores 0–4 are given for angles of 90–0°. The angles in relation to postmenstrual age are not given.

The volarflexion angle of the hand is considered by Saint-Anne Dargassies (1974) to be particularly unreliable for maturational assessment because of its great individual variability. At term she observes oscillations between 0° and 45°. At 28 postmenstrual weeks, Saint-Anne Dargassies (1974) reported a volarflexion angle of the hand of 25–30°. In the full-term infants of her study, this angle is zero in 40% of cases, and 10–30° in 50%, i.e., in 90% less than 30°. In 9% of the infants she observed an angle between 30° and 45° and in 1% an angle between 46 and 70°.

4.8.2. Dorsiflexion Angle of the Foot

For this reaction, the foot is dorsiflexed onto the anterior aspect of the leg. The dorsiflexion angle between the dorsum of the foot and the anterior aspect of the leg (ankle dorsiflexion) is measured according to Dubowitz *et al.* (1970; Dubowitz and Dubowitz, 1977). These workers used the ankle dorsiflexion as a maturational criterion; for a decrease from 90 to 0°, scores 0–4 are given. There are no figures in relationship to postmenstrual age.

The dorsiflexion angle of the foot is counted by Saint-Anne Dargassies (1974) among the six maturational criteria reliable in a high percentage of cases. In her study of 150 full-term infants, she ovserved in 82% an angle of 0°; 18% exhibited an angle of 10–20°. The dorsiflexion angle of the foot is considered characteristic for a full-term newborn infant at term. According to Saint-Anne Dargassies (1974), the dorsiflexion angle of the foot is greater the more premature the infant is at birth. At 28 premenstrual weeks, she observed a dorsiflexion angle of the foot of 35–40°, which remains so at 32 and 37 weeks. At 41 weeks, she reports an angle of 40–60° in preterm infants contrary to 0° in full-term infants, and she attributes this difference to mechanical reasons, i.e., compression in the uterus.

5. Comparison between SGA and AGA Infants

Robinson (1966), comparing AGA and SGA preterm infants, showed a lack of influence of intrauterine growth retardation on neurological maturation, confirming the results of Saint-Anne Dargassies (1955, 1974).

Michaelis *et al.* (1970), however, reported differences in motor behavior of SGA infants, born at term, and AGA full-term infants as a weak adduction and flexion phase of the Moro reflex, a marked asymmetrical tonic neck reflex of the legs, and poor or absent head lifting in the prone position.

Also Schulte *et al.* (1971), who compared 21 SGA infants of toxemic mothers with a mean gestational age of 39 weeks, examined between the third and eighth postnatal day, with 21 control infants, demonstrated differences between the groups. In the SGA "skeletal muscle tone and general excitability were significantly reduced and the motor behavior was usually not consistent with the infant's conceptional age."

Finnström (1971) showed SGA infants to have a mean maturity score of 10 days gestational age less than AGA infants; however, it cannot be said whether this observed reduction in neurological score applies only to certain types of SGA infants. Leijon *et al.* (1980) found in their neurological examination lower muscle tone and fewer optimal responses in SGA than in AGA infants. By contrast, Dubowitz and Dubowitz (1981) observed increased tone, especially in the extensors of the neck and trunk in SGA infants.

Saint-Anne Dargassies (1974) stated that nutrition and fetal growth do not play a role in the characteristic quality of muscle tone (righting reactions, dorsiflexion angle of the foot) and do not alter the neurological maturational criteria.

Ounsted *et al.* (1978) observed no differences between the neurological scores for SGA and AGA full-term infants but stated that their findings cannot necessarily be extrapolated to earlier periods of gestational age.

Since SGA preterm infants represent a heterogeneous group in respect to pathogenesis and prognosis (Brandt, 1975, 1977), *it is also to be expected that their neurological behavior will be diverse.* This may explain the different findings cited in the literature.

According to the Bonn results, the neurological maturation of SGA preterm infants with normal development at later follow-up is similar to that of the AGA group. SGA infants born at term are not part of the study. Figure 21 shows head lifting in the prone position in an SGA preterm boy, born at 34 postmenstrual weeks with a birth weight of 850 g, and at the expected date of delivery with weight of 1620 g. He belongs to a group of SGA infants with catch-up growth of head circumference after early and high-energy feeding (Brandt, 1975; Weber *et al.*, 1976).

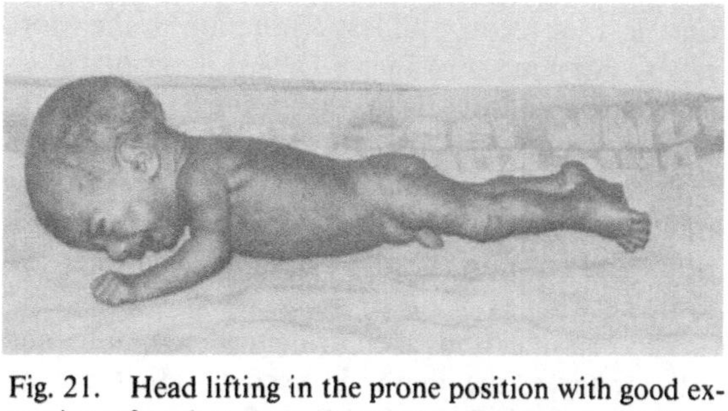

Fig. 21. Head lifting in the prone position with good extension of neck and trunk muscles. Small-for-gestational age (SGA) preterm boy at expected date of delivery, born at 34 postmenstrual weeks with a birth weight of 850 g.

6. Value for Determination of Postmenstrual Age: A Critical Evaluation

Which reflexes and reactions show clear-cut changes in their response in relationship to increasing postmenstrual age and may serve as indices of maturity?

To begin with the prerequisites: (1) Extrauterine development in preterm infants until the expected date of delivery has to be similar to the intrauterine development in full-term infants, so that the test will be completely comparable over the whole gestational range; (2) The developmental course is not affected by premature environmental influences; (3) The reaction is clearly defined and easily elicitable; and (4) The component of the response considered the mature phase has also to be constantly present in normal full-term infants, at least from the third postnatal day. If a reaction varies greatly in a group of normal full-term newborn infants, it cannot be recommended for developmental assessment of preterm infants.

A. Minkowski (1966) stated that an infant born prematurely at 28 weeks will, at the postnatal age of 6 weeks, have the same neurological development as an infant born at 34 weeks.

The results of Robinson (1966) virtually support the concept of Saint-Anne Dargassies (1955, 1974) that neurological maturation in the third trimester of pregnancy is neither accelerated nor retarded by premature extrauterine environmental influences. Using the Brazelton scales, Paludetto *et al.* (1982) found a perfect correspondence for motor performance between preterm infants at term and full-term neonates. Palmer *et al.* (1982) compared preterm infants born at 27–35 postmenstrual weeks when they reached their expected date of delivery with full-term infants at day 5, and found the only significant differences between the groups were that the preterm infants showed a more extended posture, less leg recoil, and less adduction in the Moro reflex. Forslund and Bjerre (1983) described a selected group of low-risk preterm infants at term as having responses similar to those of full-term controls except for a more pronounced head lag during traction, a better head control in the sitting position, and a more easily elicitable asymmetrical tonic neck reflex.

The classification of neurological reactions and reflex phases, the methods used to elicit the response, and the descriptions of resulting movements vary greatly in the literature. This makes an evaluation of the results of different studies difficult. Moreover, often missing are the age range of appearance of reactions and information about the constancy of the response and about its interindividual and intraindividual variability.

Figure 22 presents a critical evaluation of the neurological signs, found to be useful by earlier observers, based on our own experience from some 100 examinations in the perinatal period. An attempt is made to quantify the signs, which have been proved reliable by statistical evaluation thus far. Of 153 examinations of 29 AGA preterm infants born at 32 postmenstrual weeks and below with a normal development until school age (Figure 22).

Table XVIIA reviews different reflexes and reactions in the perinatal period considered *not useful* for determination of gestational age according to the Bonn results. Most of these reactions belong to "fetal reflexes." A plus (+) signifies a reported good correlation, and a minus (−) signifies a weaker or lower correlation with gestational age. "No statement" means that the reflex or reaction considered is not referred to in the cited reports.

In agreement with the Bonn results, most of the reactions are shown to be useful only by a few of the investigators cited, with the exception of the crossed extensor reflex, which is considered valuable in five of eight reports, and the rooting response in half the studies. The stepping movements, for instance, that are considered useful for age assessment by three of eight investigators in Table XVIIA cannot be recommended for determination of gestational age because they, may be absent in normal full-term infants and can be observed even in young preterm infants and fetuses (see Section 4.6.5).

Table XVIIB reviews different reflexes and reactions in the perinatal period that are considered *useful* for determination of gestational age according to the Bonn results compared with the view of different investigators. The key for the signs is the same as for Table XVIIA. Recoil of the upper extremities is considered useful by six of eight investigators. The traction response—head control in sitting position—is evaluated positively in seven of eight reports. The adduction phase of the Moro reflex is evaluated contradictorily: four investigators report a close correlation with gestational age, two consider this reaction useless, and two give no statement.

Head lifting in ventral suspension in Table XVIIB is considered only by Dubowitz *et al.* (1970; Dubowitz and Dubowitz, 1977). This reaction has

Table XVII. Synopsis of Reflexes and Reactions in the Perinatal Period for Determination of Gestational Age—View of Different Investigators[a]

	Koenigsberger (1966)	Robinson (1966)	Babson and McKinnon (1967)	Amiel-Tison (1968)	Graziani *et al.* (1968)	Dubowitz and Dubowitz (1977)	Finnström (1971)	Saint-Anne Dargassies (1974)
A. Considered not useful (according to the Bonn results)								
Degree of spontaneous movements	+	n.st.	n.st.	n.st.	n.st.	n.st.	+	+
Heel-to-ear maneuver	+	n.st.	n.st.	+	n.st.	+	−	n.st.
Popliteal angle	+	n.st.	n.st.	+	n.st.	+	n.st.	−
Moro relfex abduction, extension	+	−	n.st.	+	+	−	+	−
Glabella tap reflex	+	+	n.st.	n.st.	+	−	−	n.st.
Crossed extensor reflex	+	−	n.st.	+	+	−	+	+
Knee jerk	n.st.	−	n.st.	n.st.	n.st.	−	+	n.st.
Palmar grasp	+	−	n.st.	+	+	−	−	−
Rooting response	+	n.st.	n.st.	+	n.st.	−	+	+
Stepping movements	+	n.st.	n.st.	+	+	−	−	−
B. Considered useful (according to the Bonn results)								
Posture[b]	+	n.st.	n.st.	+	−	+	−	+
Recoil lower extremities	+	n.st.	n.st.	n.st.	+	+	n.st.	n.st.
Recoil upper extremities	+	n.st.	n.st.	+	+	+	+	+
Scarf maneuver	+	n.st.	n.st.	+	+	+	n.st.	−
Moro reflex adduction, flexion	+	−	+	n.st.	+	−	+	n.st.
Neck-righting reflex	−	+	n.st.	n.st.	+	−	n.st.	n.st.
Traction response,[c] head control	+	+	+	+	n.st.	+	+	+
Head lifting in prone position	n.st.	n.st.	+	n.st.	n.st.	n.st.	n.st.	n.st.
Righting reactions legs, trunk, head	+	n.st.	n.st.	+	n.st.	n.st.	−	+
Ventral suspension, head lifting	n.st.	n.st.	n.st.	n.st.	n.st.	+	n.st.	n.st.
Head turning to light, positive phototropism	n.st.	+	n.st.	n.st.	n.st.	−	−	+

[a](+), useful for determination of gestational age; (−), not useful for determination of gestational age; n.st., no statement
[b]Imposed posture in the Bonn study.
[c]Head control during traction or when sitting.

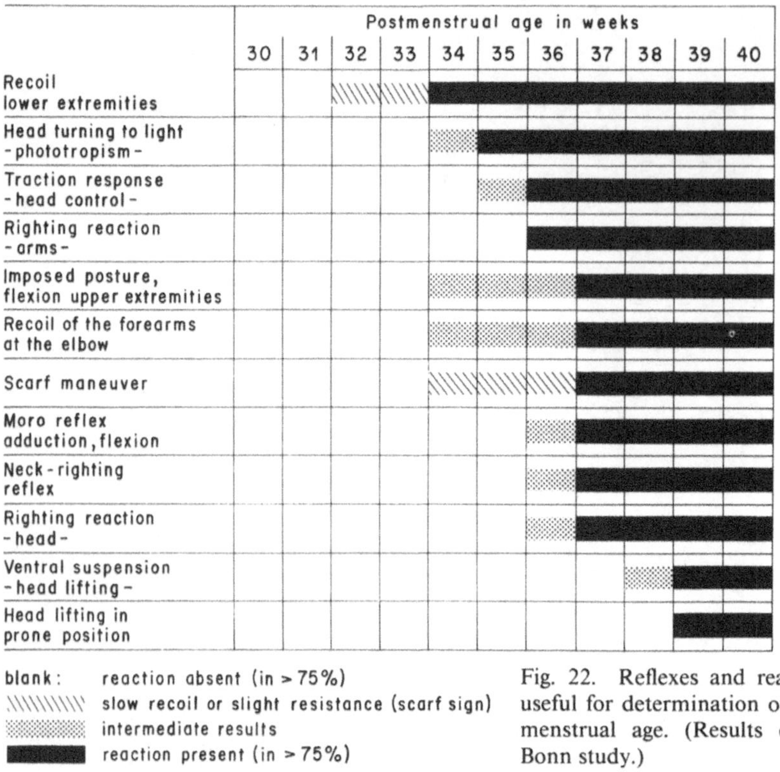

Fig. 22. Reflexes and reactions useful for determination of postmenstrual age. (Results of the Bonn study.)

proved to have a clear-cut appearance according to the Bonn results and also is considered useful for maturational assessment by L. O. Lubchenco (personal communication, 1977).

Figure 22 summarizes the reflexes and reactions considered useful for determination of gestational age according to the Bonn results. Reactions that may be present at birth and from 28 postmenstrual weeks, the so-called fetal reflexes (e.g., palmar grasp, rooting response, crossed extensor reflex, and stepping movements), are omitted, because they are not predictive of maturational development. If a reaction has been graded for scoring, the positive results are combined (e.g., for head lifting in prone position, the scores for all types of head lifting are combined).

The order of the items listed in Figure 22 corresponds to the order of their appearance. The age placement of a reaction is made according to its presence in at least 75% of infants. If different responses are present in approximately equal parts, the results are called intermediate:

1. *Recoil lower extremities:* Distinguishes infants of 33 postmenstrual weeks or under with absent or slow recoil from those of 34 weeks or over with quick recoil. The results agree well with the findings of Koenigsberger (1966), Saint-Anne Dargassies (1966), Graziani *et al.* (1968), and Michaelis (cited by Schulte, 1974).
2. *Head turning to light, phototropism:* Distinguishes between infants of 33 weeks or under and 35 weeks or over. The Bonn results agree with the findings of Robinson (1966).
3. *Traction response, head control in sitting position:* Distinguishes between infants of 34 postmenstrual weeks or under whose head is hanging passively down and infants of 36 weeks or over who raise the head at least once or keep it upright for at least 3 sec. At 35 weeks, the results are intermediate, with the head hanging passively down in about one-half of cases and being raised at least once in the remaining half. This response agrees with the findings of Robinson (1966), Babson and McKinnon (1967), and Finnström (1971).
4. *Righting reaction, arms:* Distinguishes between 35 weeks and under and 36 weeks and

over. Only the Bonn study has scored this to-date obviating comparison with other studies.

5. *Imposed posture, flexion upper extremities:* Differentiates infants of 33 weeks and under where the arms remain extended, and 37 weeks and over where the arms are flexed. At 34–36 weeks, the results are intermediate; i.e., the arms are either flexed, semiflexed, or remain extended. There are no published results from other studies.
6. *Recoil of the forearms at the elbow:* Distinguishes between infants of 33 weeks or under with an absent recoil and of 37 weeks and over with a quick or medium recoil. From 34–36 postmenstrual weeks, the results are intermediate; i.e., in most infants recoil is either absent or slow. The results are similar to the studies of Koenigsberger (1966), Finnström (1971), and Saint-Anne Dargassies (1974); according to Amiel-Tison (1968), the response is present at 36 weeks and Graziani *et al.* (1968) report a rapid recoil at 35 weeks.
7. *Scarf maneuver:* Differentiates infants of 33 postmenstrual weeks and under with no resistance (elbows widely crossing midline and nearly reaching opposite axillary line); 34–36 postmenstrual weeks with slight resistance (elbows between midline and opposite axillary line); 37–40 postmenstrual weeks with moderate to strong resistance (elbows reaching midline or only axillary line). This maneuver shows similar responses in relationship to age in the studies of Koenigsberger (1966) and Graziani *et al.* (1968).
8. *Moro reflex, adduction phase:* Distinguishes between infants of 35 postmenstrual weeks and under who display no adduction and of 37 weeks or over where an adduction can be observed. At 36 weeks, the results are intermediate, an adduction is present in about half of cases and is absent in the remaining. The appearance of the adduction phase of this reflex is similar to that reported by Finnström (1971); in the Babson study (1967), it is present at 35 ± 1 weeks, where a "partial embrace" is reported by Graziani *et al.* (1968).
9. *Neck righting reflex:* Distinguishes between infants of 35 postmenstrual weeks or under without following of trunk and legs after lateral head rotation by 90° and infants of 37 weeks and over with following of trunk and legs; at 36 weeks, the results are intermediate, after a lateral head rotation, following of trunk and legs is observed in about half of cases and is missing in the remaining. The timing of this reflex is the same as in the London study of Robinson (1966). Graziani *et al.* (1968) located the response at 39 weeks.
10. *Righting reaction, head:* Distinguishes between 35 weeks or under and 37 weeks or over. At 36 weeks, the results are intermediate. This reaction occurs at a similar time in the study of Amiel-Tison (1968).
11. *Ventral suspension, head lifting:* Distinguishes between infants of 37 postmenstrual weeks or under who cannot lift their head and infants of 39 weeks and over who lift their head for at least 1–3 sec. At 38 weeks, the results are intermediate, about half of cases display no head lifting whereas the remainder lift their head for at least 1 sec. This reaction is not considered in relationship to postmenstrual age in other studies.
12. *Head lifting in prone position:* Distinguishes between infants of 38 postmenstrual weeks or under who display no head lifting and infants of 39 weeks or over who lift their head at least once or twice. This is observed earlier in the Babson study (37 ± 1 weeks) and also according to the table of Michaelis (cited by Schulte, 1974).

As demonstrated in Figure 22, most of the clear-cut age-dependent changes in the phases of the cited reflexes and reactions occur only in the last postmenstrual weeks. This could suggest that some of them serve as preparation and facilitation for birth at the expected date of delivery.

On the basis of their study, Prechtl *et al.* (1975) concluded that "there is no indication that preterm infants benefit from the prolonged exposure to extrauterine life." "Environmental conditions, provided they are not adverse, seem to have very little influence on the development of various neural functions" before 40 postmenstrual weeks.

7. Prognostic Value

In their foreword to the book by André-Thomas *et al.*, 1960; Polani and MacKeith stated that the lack of correlation between subsequent development and states considered abnormal in the early days of life is at times leveled as a criticism against the value of neurological examination of the neonate. "We are, it appears, still at the stage of finding out what is normal and what the normal variation is." These remarks made in 1960 still seem valid. The prognostic value is further obscured by the plasticity and adaptability of the developing brain—as often stressed by Prechtl (1973).

Prechtl (1965) demanded three criteria for a neurological examination: (1) It must be reliable: i.e., it should give the same results when carried out by different examiners; (2) It must be valid: i.e., it must be sensitive enough not only to detect gross disturbances, but minor abnormalities as well; and (3) It should predict the risk of permanent damage. Criteria (2) and (3) cannot be fulfilled completely in the newborn period because most of the reflexes and reactions studied belong to the "primitive" ones, some of which are observed even in decerebrate animals. In any case, it may be assumed that the influence of the cerebrum on the reactions in the neonatal period is small. Inversely, a pathological "primitive" reaction may signify that damage to the inferior brain centers may have occurred and to the cerebrum as well.

Long-term prognostic value can be evaluated only by a follow-up study over many years. Prechtl (1965) reported a high correlation between the neonatal findings of hyperexcitability syndrome, apathy syndrome, and hemisyndrome, and the results of the follow-up examination at 2–4 years. In 102 neonates without abnormal neurological signs in the newborn period, 14 were found to be abnormal and 88 normal at the follow-up. Of 150 neonates with abnormal neurological signs, 110 were still abnormal and 40 normal between 2 and 4 years. Prechtl (1965) showed "that neurological abnormalities diagnosed by a rigorously standardized and reliable examination at birth are highly prognostic signs."

According to Prechtl (1968), a strongly positive correlation exists between nonoptimal obstetrical conditions and neurological abnormality as measured by his standardized procedure (Prechtl and Beintema, 1964; Prechtl, 1977). Using this technique, these workers consider neonatal abnormalities predictive of continuing abnormality and future behavioral disturbance.

On the basis of a detailed neurological study of 192 mature babies by Donovan *et al.* (1962) as part of a group of 2100 babies in the Collaborative Project of the NIH, it was concluded that the conventional neurological signs and postural automatisms in the newborn infant seem of only very limited predictive value with regard to abnormalities at one year. Among the few responses that may be more ominous than others are: persistent overall depression of the Moro reflex, universal muscular hypotonia, persistently shrill cry, and asymmetric automatisms.

Illingworth (1967) also counted a persistently absent Moro reflex as one of the neurological responses suggesting a poor prognosis. Furthermore, the absence of oral reflexes, blink reflexes and the absence of the Galant response are considered ominous.

Prechtl (1972) stated that people often want to make an early assessment of neural dysfunction and to identify neurological conditions, such as spasticity or mental deficiency, at a time when these signs or conditions are not yet present as such. Therefore, it is necessary to know which signs indicate the risk of the development of these handicapping conditions and which have high prognostic value.

Schulte (1973) recommended the use of the term "pathological" only rarely in the newborn period. All pathological findings are abnormal, but these may be seen in normal healthy infants as well. Prognostic value has nothing in common with the definition "abnormal" or "pathological." Schulte (1973) stated that the more precise (accurate) the examination in the newborn period, the safer a prognosis on later development will be. The prognosis on the basis of neurological examination of preterm infants is often considered too favorably, since the chronic encephalopathies do not manifest themselves, or only by small deviations, in motor behavior (Schulte, 1973).

Brown *et al.* (1974) found a high correlation between the pattern of muscle tone and outcome in newborn infants (mostly full-term) with perinatal asphyxia. Infants with persisting hypotonia had the poorest prognosis; 43% died and only 16% were normal. However, the much lower general tone in very preterm infants makes the use of hypotonia as a neurological and prognostic sign difficult. Even in those preterm infants with brain pathology, as demonstrated by routine ultrasound, it may still be difficult to find corresponding neurological signs.

DeReuck *et al.* (1972) related neuropathological findings in newborn preterm and full-term infants with periventricular leucomalacia to neurological signs and observed hypotonia in the newborn period in most cases. Dubowitz *et al.* (1981) also found decreased tone in a considerable proportion of preterm infants with intraventricular hemorrhage (IVH). This was more marked in infants between 32 and 35 weeks, probably because their initial tone is better and the change more striking than in those 31 weeks and under.

According to Saint-Anne Dargassies (1977), a "symptomatic gap" after the newborn period may occur before definite signs of damage appear later. This causes much difficulty in establishing an early prognosis. The particularly severe sequelae (i.e., the "bedridden vegetative children") do not manifest such a gap. She reported an extensive symptomatic gap in cases with cerebral palsy; 80% of infants first appeared to be normal or were improving, and 20% had nonspecific symptoms of sequelae.

Biermann-van Eendenburg *et al.* (1981) evaluated the prognostic value of the neonatal neurological examination in relationship to later outcome. Of 79 infants who were neurologically abnormal in the neonatal period, 13 were abnormal at 18 months, 5 of them severely, pointing to the very high rate of false positives in the neonatal period. On the other hand, among the 80 controls (normal neonates), 2 had mild neurological abnormality later, emphasizing that the absence of signs of neonatal neurological morbidity does not guarantee later uneventful development.

The value of neurological signs for maturational assessment and their prognostic value are increased by repeated examinations. In infants in whom the age is known from reliable dates of the mother, from ultrasound examinations in early pregnancy, or from age assessment at birth using the external maturational criteria, assessment of the neurological behavior (appropriate for age or not) can lead to conclusions about the infant's well-being.

Before the infant shows neurological responses corresponding to postmenstrual age, the timing, extent, and duration of abnormal neurological signs in the perinatal period enables conclusions about the residual neurological deficit to be drawn. This is an important contribution neonatal neurology can make.

According to the fundamental research of Prechtl (1977) and Beintema (1968) in full-term infants, a better standardization of the neurological reactions and their phases in preterm infants, together with a quantification of the results—which is more than simple recording of the presence or absence of a response—in relationship to postmenstrual age, is a major task for the future.

In conclusion, knowledge of early neurological development and examination methods for preterm infants until their expected date of delivery serves three main purposes:

First, it permits reasonably accurate assessment of gestational age to be made.

Second, new noninvasive techniques for imaging of the brain throw light on the normal and pathological development processes of the central nervous system. Appropriate and more refined neurological examination methods are therefore required.

Third, earlier discharge of preterm infants from hospital implies increasing contact with physicians and health professionals. These, as well as neonatologists also need information about developmental patterns and peculiarities.

ACKNOWLEDGMENTS. This research was carried out with grants from the Bundesminister für Jugend, Familie und Gesundheit, from the Minister für Wissenschaft und Forschung des Landes Nordrhein-Westfalen, from the Fritz Thyssen Stiftung, and from the Milupa A.G., International Scientific Department, which are gratefully acknowledged. For support of this study at the Universitäts-Frauenklinik Bonn (Director, Professor Dr. Ernst Jürgen Plotz) and at the Universitäts-Kinderklinik Bonn (Director, Professor Dr. Heinz Hungerland until 1974, and Professor Dr. Walther Burmeister from 1975) my thanks are due to my colleagues and assistants. I particularly wish to thank all the mothers and fathers in the study for their faithful cooperation. My thanks are due to Dr. Phil. Elisabeth Sticker of the Psychological Institute of the Bonn University for help in writing the computer programs. The computations were performed on the IBM/370-168 of the Regionales Hochschul-Rechenzentrum der Universität Bonn.

8. References

Amiel-Tison, C., 1968, Neurological evaluation of the maturity of newborn infants, *Arch. Dis. Child.* **43**:89–93.

André-Thomas, and Saint-Anne Dargassies, S., 1952, *Etudes Neurologiques sur le Nouveau-Né et le Jeune Nourrisson,* Masson, Paris.

André-Thomas, Chesni, Y., and Saint-Anne Dargassies, S., 1960, The neurological examination of the infant, *Little Club Clinics Develop. Med.,* No. 1.

Apgar, V., 1953, A proposal for a new method of evaluation of the newborn infant, *Curr. Res. Anesth. Analg.* **32**:260–267.

Babcock, D. S., and Han, B. K., 1981, *Cranial Ultrasonography of Infants,* Williams & Wilkins, Baltimore.

Babinski, M. J., 1896, Sur le réflexe cutané plantaire dans certaines affections organiques du systemè nerveux central, *C. R. Soc. Biol.* **3**:207–208.

Babson, S. G., and McKinnon, C. M., 1967, Determination of gestational age in premature infants, *J. Lancet* **87**:174–177.

Ballard, J. L., Kazmaier Novak, K., and Driver, M., 1979, A simplified score for assessment of fetal maturation of newly born infants, *J. Pediatr.* **95**:769–774.

Battaglia, F., and Lubchenco, L. O., 1967, A practical classification of newborn infants by weight and gestational age, *J. Pediatr.* **71**:159–163.

Beintema, D. J., 1968, A neurological study of newborn infants, *Clin. Dev. Med.* **28**:1–178.

Biermann-van Eendenburg, M. E. C., Jurgens-van der Zee, A. D., Olinga, A. A., Huisjes, H. H., and Touwen, B. C. L., 1981, Predictive value of neonatal neurological examination: A follow-up study at 18 months, *Dev. Med. Child Neurol.* **23**:296–305.

Birnholz, J. C., and Benacerraf, B. R., 1983, The development of human fetal hearing, *Science* **222**:516–518.

Boulder Committee, 1970, Embryonic vertebrate central nervous system: Revised terminology, *Anat. Rec.* **166**:257–261.

Brand, M., 1983, Fetal and neonatal intracranial hemorrhages—Incidence, prognosis, prevention, presented at the *Eleventh Deutscher Kongress für Perinatale Medizin, Berlin, 1983.*

Brandt, I., 1975, Postnatale Entwicklung von Früh-Mangelgeborenen, *Gynaekologe* **8**:219–233.

Brandt, I., 1976, Dynamics of head circumference growth before and after term, in: *The Biology of Human Fetal Growth* (D. F. Roberts and A. M. Thomson, eds.), pp. 109–136. Taylor & Francis, London.

Brandt, I., 1978, Growth dynamics of low birth weight infants with emphasis on the perinatal period, in: *Human Growth, A Comprehensive Treatise* (F. Falkner and J. M. Tanner, eds.), Vol. 2, pp. 557–617, Plenum Press, New York.

Brandt, I., 1983, *Griffiths-Entwicklungsskalen (GES) zur Beurteilung der Entwicklung in den ersten beiden Lebensjahren,* Beltz Verlag, Weinheim, pp. 34–36, 45–50.

Brandt, I., 1984, Human growth up to six years—Distance and velocity, in: *Genetic and Environmental Factors During the Growth Period* (S. Susanne, ed.), pp. 101–122, Plenum Press, New York.

Brandt, I., and Schröder, R., 1976, Postnatales Aufholwachstum des Kopfumfanges nach intrauteriner Mangelernährung, *Monatsschr. Kinderheilkd.* **124**:475–477.

Bratteby, L. E., and Anderson, L., 1976, Continuous monitoring of motor activity in the newborn, in: *Fifth European Congress of Perinatal Medicine,* pp. 141–149, Uppsala.

Brazelton, T. B., 1961, Psychophysiologic reactions in the neonate. II. Effect of maternal medication in the neonate and his behavior, *J. Pediatr.* **58**:513–518.

Brazelton, T. B., 1973, Neonatal behavioral assessment scale, *Clin. Dev. Med.* **50**:1–66.

Brett, E. M., 1965, The estimation of foetal maturity by the neurological examination of the neonate, in: *Gestational Age, Size and Maturity* (M. Dawkins and B. MacGregor, eds.), *Clin. Dev. Med.,* **19**:105–115.

Brown, J. K., Purvis, R. J., Forfar, J. O., and Cockburn, F., 1974, Neurological aspects of perinatal asphyxia, *Dev. Med. Child Neurol.* **16**:567–580.

Capurro, H., Konichezky, S., Fonseca, D., and Caldeyro-Barcia, R., 1978, A simplified method for diagnosis of gestational age in the newborn infant, *J. Pediatr.* **93**:120–122.

Chamberlin, H. R., 1975, Mental retardation, in: *Pediatric Neurology,* (T. W. Farmer, ed.), 2nd ed., pp. 90–117, Harper & Row, Hagerstown, Maryland.

Cooke, R. W. I., 1983, Intraventricular haemorrhage, *Lancet* **1**:878.

Couture, A., and Cadier, L., 1983, *Echographie Cérébrale par Voie Transfontanellaire,* Vigot, Paris.

Dawes, G. S., 1976, The potential of fetal breathing measurements in clinical research, in: *Proceedings of Fifth European Congress of Perinatal Medicine,* (G. Reoth and L. E. Bratteby, eds.) pp. 175–176, Uppsala.

DeReuck, J., and Richardson, E. P. 1972, Pathogenesis and evolution of periventricular leukomalacia in infancy, *Arch. Neurol.* **27**:229–236.

Donovan, D. E., Coues, P., and Paine, R. S., 1962, The prognostic implications of neurological abnormalities in the neonatal period. *Neurology (New York)* **12**:910–914.

Dubowitz, L. M. S., and Dubowitz, V., 1977, *Gestational Age of the Newborn,* Addison-Wesley, London.

Dubowitz, L. M. S., Dubowitz, V., and Goldberg, C., 1970, Clinical assessment of gestational age in the newborn infant, *J. Pediatr.* **77**:1–10.

Dubowtiz, L. M. S., and Dubowitz, V., 1981, The Neurological Assessment of the Preterm and Full-term Newborn Infant, *Clin. Dev. Med.,* **79**:1–103.

Dubowitz, L. M. S., Dubowitz, V., Palmer, P., and Verghote, M., 1980, A new approach to the neurological assessment of the preterm and full-term newborn infant, *Brain Dev.* **2**:3–14.

Dubowitz, L. M. S., Levene, M. I., Morante, A., Palmer, P., and Dubowitz, V., 1981, Neurologic signs in neonatal intraventricular hemorrhage: A correlation with real-time ultrasound, *J. Pediatr.* **99**:127–133.

Escardó, F., and de Coriat, L. F., 1960, Development of postural and tonic patterns in the newborn infant, *Pediatr. Clin. North Am.* **7**:511–525.

Farr, V., Mitchell, R. G., Neligan, G. A., and Parkin, J. M., 1966*a*, The definition of some external characteristics used in the assessment of gestational age in the newborn infant, *Dev. Med. Child Neurol.* **8**:507–511.

Farr, V., Kerridge, D. F., and Mitchell, R. G., 1966*b*, The value of some external characteristics in the assessment of gestational age at birth, *Dev. Med. Child Neurol.* **8**:657–660.

Ferrari, F., Grosoli, M. V., Fontana, G., and Cavazzuti,

G. B., 1983, Neurobehavioural comparison of low-risk preterm and fullterm infants at term conceptional age, *Dev. Med. Child Neurol.* **25**:450–458.

Finnström, O., 1971, Studies on maturity in newborn infants, III, Neurological examination, *Neuropaediatrie* **3**:72–96.

Forslund, M., and Bjerre, I., 1983, Neurological assessment of preterm infants at term conceptional age in comparison with normal full-term infants, *Early Hum. Dev.* **8**:195–208.

Fredrickson, W. T., and Brown, J. V., 1980, Gripping and Moro responses: Differences between small-for-gestational age and normal weight term newborn, *Early Hum. Dev.* **4**:69–77.

Fröding, C.-A., 1960, Acoustic investigation of newborn infants, *Acta Oto-Laryngol.* **52**:31–40.

Galant, S., 1917, *Der Rückgratreflex, (Ein neuer Reflex im Säuglingsalter), Mit besonderer Berücksichtigung der anderen Reflexvorgänge bei den Säuglingen,* Diss, Basel.

Gesell, A., 1938, The tonic neck reflex in the human infant, *J. Pediatr,* **13**:445–464.

Gesell, A., 1952, *Infant Development, The Embryology of Early Human Behavior,* Harper & Row, New York.

Gesell, A., and Amatruda, C. S., 1967, *Developmental Diagnosis, Normal and Abnormal Child Development,* 2nd ed., Harper & Row, New York.

Graziani, L. J., Weitzman, E. D., and Velasco, M. S. A., 1968, Neurologic maturation and auditory evoked responses in low birth weight infants, *Pediatrics* **41**:483–494.

Griffiths, R., 1964, *The Abilities of Babies, A Study in Mental Measurement,* University of London Press, London.

Hackelöer, B.-J., and Hansmann, M., 1976, Ultraschalldiagnostik in der Frühschwangerschaft, *Gynaekologe* **9**:108–122.

Hagberg, B., 1979, Epidemiological and preventive aspects of cerebral palsy and severe mental retardation in Sweden, *Eur. J. Pediatr.* **130**:71–78.

Hagberg, B., Hagberg, G., and Olow, I., 1982, Gains and hazards of intensive neonatal care: An analysis from Swedish cerebral palsy epidemiology, *Dev. Med. Child Neurol.* **24**:13–19.

Haller, U., Rüttgers, H., Wille, F., Heinrich, D., Müller, P., and Kubli, F., 1973, Aktive Kindsbewegungen im schnellen Ultraschall-B-Bild, *Gynaekol. Rdsch.* **13**(Suppl. 1):118–119.

Hansmann, M., 1974, Kritische Bewertung der Leistungsfähigkeit der Ultraschalldiagnostik in der Geburtshilfe heute, *Gynaekologe* **7**:26–35.

Hansmann, M., 1976, Ultraschall-Biometrie im II. und III. Trimester der Schwangerschaft, *Gynaekologe* **9**:133–155.

Hansmann, M., Schuhmacher, H., Foebus, J., and Voigt, U., 1979, Ultraschallbiometrie der fetalen Scheitelsteißlänge in der ersten Schwangerschaftshälfte, *Geburtshilfe Frauenheilkd.* **39**:656–666.

von Harnack, G.-A., and Oster, H., 1958, Quantitative Reifebestimmung von Frühgeborenen, *Monatsschr. Kinderheilkd.* **106**:324–328.

Henner, H., Haller, U., Wolf-Zimper, O,. Lorenz, W. J., Bader, R., Müller, B., and Kubli, F., 1975, *Quantification of Fetal Movement in Normal and Pathologic Pregnancy,* Excerpta Medica International Congress Series, No. 363, pp. 316–319, Excerpta Medica. Amsterdam.

Hosemann, H., 1949, Schwangerschaftsdauer und Neugeborenengewicht, *Arch. Gynaekol.* **176**:109–123.

Howard, J., Parmelee, A. H., Kopp, C. B., and Littman, B., 1976, A neurologic comparison of pre-term and full-term infants at term conceptional age, *J. Pediatr.* **88**:995–1002.

Humphrey, T., 1964, Some correlations between the appearance of human fetal reflexes and the development of the nervous system, *Prog. Brain Res.* **4**:93–135.

Illingworth, R. S., 1967, *The Development of the Infant and Young Child, Normal and Abnormal,* 3rd ed., E. & S. Livingstone, Edinburgh.

Joppich, G., and Schulte, F. J., 1968, *Neurologie des Neugeborenen,* Springer-Verlag, Heidelberg.

Koenigsberger, R. M., 1966, Judgement of fetal age. I. Neurologic evaluation, *Pediatr. Clin. North Am.* **13**:823–833.

Kussmaul, A., 1859, *Untersuchungen über das Seelenleben des neugeborenen Menschen,* Verlag Franz Pietzcker, Tübingen.

Langreder, W., 1958, Die geburtshilflich bedeutsame Motorik des Neugeborenen, *Kinderärztl. Prax.* **26**:559–560.

Langreder, W., 1958, Die geburtshilflich bedeutsame Motorik des Neugeborenen, *Kinderärztl. Prax.* **26**:559–560.

Larroche, J.-C., 1981, The marginal layer in the neocortex of a 7-week-old human embryo. A light and electron microscopic study, *Anat. Embryol.* **162**:301–312.

Larroche, J.-C., and Houcine, O., 1982, Le néo-cortex chez l'embryon et le foetus humain. Apport du microscope électronique et du Golgi, *Reprod. Nutr. Dev.* **22**:163–170.

Larroche, J.-C., Privat, A., and Jardin, L., 1981, Some fine structures of the human fetal brain, in: *Physiological and Biochemical Basis for Perinatal Medicine* (M. Monset-Couchard and A. Minkowski, eds.), pp. 350–358, S. Karger, Basel.

Latis, G. O., Simionato, L., and Ferraris, G., 1981, Clinical assessment of gestational age in the newborn infant. Comparison of two methods, *Early Hum. Dev.* **5**:29–37.

Leijon, I., Finnström, O., Nilsson, B., and Rydén, G., 1980, Neurology and behaviour of growth-retarded neonates. Relation to biochemical placental function tests in late pregnancy, *Early Hum. Dev.* **4**:257–270.

Lenard, H.-G., Bernuth, H. von, and Prechtl, H. F. R., 1968, Reflexes and their relationship to behavioral state in the newborn, *Acta Paediatr. Scand.* **57**:177–185.

Lubchenco, L. O., 1976, *The High Risk Infant,* W. B. Saunders, Philadephia.

Lubchenco, L. O., Hansman, C., Dressler, M., and Boyd,

E., 1963, Intrauterine growth as estimated from live born birth-weight data at 24 to 42 weeks of gestation, *Pediatrics* **32**:793–800.

Magnus, R., 1924, *Körperstellung,* Springer-Verlag, Berlin.

Magnus, R., and de Kleijn, A., 1912, Die Abhängigkeit des Tonus der Extremitätenmuskulatur von der Kopfstellung, *Pfluegers Arch. Ges. Physiol.* **145**:455–458.

Manning, F. A., Baskett, T. F., Morrison, I., and Lange, I., 1981, Fetal biophysical profile scoring: A prospective study in 1,184 high-risk patients, *Am. J. Obstet. Gynecol.* **140**:289–294.

McGraw, M. B., 1937, The Moro reflex, *Am. J. Dis. Child.* **54**:240–251.

McKeown, T., and Gibson, J. R., 1951, Observations on all births (23,970) in Birmingham 1947. IV. Premature birth, *Br. Med. J.* **2**:513–517.

Meitinger, C., Vlasch, V., and Weinmann, H. M., 1969, Neurologische Untersuchungen bei Frühgeborenen, *Münch. Med. Wochenschr.* **20**:1158–1168.

Michaelis, R., 1970, Risikofaktoren, neonatale Adaption und motorische Automatismen bei untergewichtigen Neugeborenen, *Fortschr. Med.* **88**:525–528.

Michaelis, R., and Hege, U., 1982, Welche spastischen zerebralen Paresen könnten pränatal bedingt sein?, *Paediatr. Fortbildungskurse Prax.* **53**:63–69.

Michaelis, R., Schulte, R. J., and Nolte, R., 1970, Motor behavior of small for gestational age newborn infants, *J. Pediatr.* **76**:208–213.

Michaelis, R., Nolte, K., and Sonntag, I., 1975, Der Muskeltonus Frühgeborener im Vergleich zum Muskeltonus reifer Neugeborener, presented at the Congress of South German pediatricians in Ulm.

Michaelis, R., Dopfer, R., Gerbig, W., Dopfer-Feller, P., and Rohr, M., 1979, I. Die Erfassung obstetrischer und postnataler Risikofaktoren durch eine Liste optimaler Bedingungen, *Monatsschr. Kinderheilkd.* **127**:149–155.

Milani-Comparetti, A., 1981, The neurophysiologic and clinical implications of studies on fetal motor behavior, *Semin. Perinatol.* **5**:183–189.

Minkowski, A., 1966, Development of the nervous system in early life. I. Introduction, in: *Human Development* (F. Falkner, ed.), pp. 254–257, W. B. Saunders, Philadephia.

Minkowski, M., 1921, Sur les mouvements, les réflexes et les réactions musculaires du foetus humain de 2 à 5 mois et leurs rélations avec le système nerveux foetal, *Rev. Neurol.* **37**:1105–1118; 1235–1250.

Minkowski, M., 1924. Zum gegenwärtigen Stand der Lehre von den Reflexen in entwicklungsgeschichtlicher und anatomisch-physiologischer Beziehung, *Schweiz. Arch. Neurol. Psychiatr.* **15**:239–259.

Moro, E., 1918, Das erste Trimenon, *Muench. Med. Wochenschr.* **65**:1147–1150.

Nesbitt, R. E. L., 1966, Perinatal development, in: *Human Development* (F. Falkner, ed.), pp. 123–149, W. B. Saunders Company, Philadephia.

Nie, N. H., and Hull, C. H., 1983, *SPSS 8, Statistik-Programm-System für die Sozialwissenschaften,* Eine Beschreibung der Programmversionen 8 und 9 (P. Beutel, H. Küffner, and W. Schubö, eds.), 4.Aufl., Fischer, Stuttgart.

Nie, N. H., Hull, C. H., Jenkins, J. G., Steinberger, K., and Bent, D. H., 1975, *SPSS, Statistical Package for the Social Sciences,* 2nd ed., McGraw-Hill, New York.

Nijhuis, J. G., Prechtl, H. F. R., Martin, C. B., Jr., and Bots, R. S. G. M., 1982, Are there behavioural states in the human fetus?, *Early Hum. Dev.* **6**:177–195.

Northern, J. L., and Downs, M. P., 1978, *Hearing in Children,* 2nd ed. Williams & Wilkins, Baltimore, Maryland.

Okado, N., 1980, Development of the human cervical spinal cord with reference to synapse formation in the motor nucleus, *J. Comp. Neurol.* **191**:495–513.

Okado, N., Kakimi, S., and Kojima, T., 1979, Synaptogenesis in the cervical cord of the human embryo: Sequence of synapse formation in a spinal reflex pathway, *J. Comp. Neurol.* **184**:491–518.

Ounsted, M. K., Chalmers, C. A., and Yudkin, P. L., 1978, Clinical assessment of gestational age at birth: The effects of sex, birthweight, and weight for length for gestation, *Early Hum. Dev.* **2**:73–80.

Paine, R. S., 1960, Neurologic examination of infants and children, *Pediatr. Clin. North Am.* **7**:471–510.

Palmer, P. G., Dubowitz, L. M. S., Verghote, M., and Dubowitz, V., 1982, Neurological and neurobehavioural differences between preterm infants at term and full-term newborn infants, *Neuropediatrics* **13**:183–189.

Paludetto, R., Mansi, G., Rinaldi, P., De Luca, T., Corchia, C., De Curtis, M., and Andolfi, M., 1982, Behaviour of preterm newborns reaching term without any serious disorder, *Early Hum. Dev.* **6**:357–363.

Pape, K. E., and Wigglesworth, J. S., 1979, Haemorrhage, Ischaemia and the Perinatal Brain, *Clin. Dev. Med.,* **69–70**:1–196.

Papile, L.-A., Burstein, J., Burstein, R., and Koffler, H., 1978, Incidence and evolution of subependymal and intraventricular hemorrhage: A study of infants with birth weights less than 1,500 gm, *J. Pediatr.* **92**:529–534.

Parmelee, A. H., 1964, A critical evaluation of the Moro reflex, *Pediatrics* **33**:773–788.

Parmelee, A. H., and Schulte, F. J., 1970, Developmental testing of pre-term and small-for-date infants, *Pediatrics* **45**:21–28.

Patrick, J., Campbell, K., Carmichael, L., Natale, R., and Richardson, B., 1982, Patterns of gross fetal body movements over 24-hour observation intervals during the last 10 weeks of pregnancy, *Am. J. Obstet. Gynecol.* **142**:363–371.

Peiper, A., 1964, *Die Eigenart der kindlichen Hirntätigkeit,* 3rd ed., Thieme, Leipzig.

Polani, P. E., and MacKeith, D. M., 1960, Foreword, in: *The Neurological Examination of the Infant* (André-Thomas, Y. Chesni, and S. Saint-Anne Dargassies, eds.), pp. 2–8, *Little Club Clinics in Developmental Medicine, No. 1.*

Prechtl, H. F. R., 1958, The directed head turning response and allied movements of the human baby, *Behaviour* **13**:212–242.

Prechtl, H. F. R., 1965, Prognostic value of neurological signs in the newborn infant, *Proc. R. Soc. Med.* **58**:1–4.

Prechtl, H. F. R., 1968, Neurological findings in newborn infants after pre- and paranatal complications, in: *Aspects of Praematurity and Dysmaturity* (J. H. P. Jonxis, H. K. A. Visser, and J. A. Troelstra, eds.), pp. 303–321, Nutricia Symposium, Groningen, 1967, H. E. Stenfert N. V., Kroese Leiden.

Prechtl, H. F. R., 1970, Hazards of oversimplification, *Dev. Med. Child Neurol.* **12**:522–524.

Prechtl, H. F. R., 1972, Strategy and validity of early detection of neurological dysfunction, in: *Mental Retardation, Prenatal Diagnosis and Infant Assessment* (C. P. Douglas and K. S. Holt, eds.), pp. 41–46, Butterworths, London.

Prechtl, H. F. R., 1973, *Das leicht hirngeschädigte Kind. Theoretische Überlegungen zu einem praktischen Problem,* pp. 282–305, Van Loghum Slaterus Deventer.

Prechtl, H. F. R., 1974, The behavioural states of the newborn infant (A review), *Brain Res.* **76**:185–212.

Prechtl, H. F. R., 1977, The Neurological Examination of the Full-Term Newborn Infant, 2nd ed., *Clin. Dev. Med.,* **63**:1–68.

Prechtl, H. F. R., 1983*a*, Entwicklungsneurologie vor und nach der Geburt, presented at the *Ninth Jahrestagung der Gesellschaft für Neuropädiatrie, Tübingen.*

Prechtl, H. F. R., 1983*b*, Towards a prenatal neurology, in: *Dilemmas in Gestosis. Controversies in Etiology, Pathology, Therapy. International Symposium, Vienna 1982* (H. Janisch and E. Reinold, eds.), pp. 214–217, Thieme, Stuttgart.

Prechtl, H. F. R., and Beintema, D. J., 1964, *The Neurological Examination of the Full-Term Newborn Infant,* Little Club Clinics in Developmental Medicine, No. 12. Heinemann Medical, London.

Prechtl, H. F. R., and Beintema, D. J., 1976, Die neurologische Untersuchung des reifen Neugeborenen, 74 pp., 2nd ed., Thieme, Stuttgart.

Prechtl, H. F. R., and Knol, A. R., 1958, Der Einfluss der Beckenendlage auf die Fusssohlenreflexe beim neugeborenen Kind, *Arch. Psychiatr. Nervenkr.* **196**:542–553.

Prechtl, H. F. R., Fargel, J. W., Weinmann, H. M., and Bakker, H. H., 1975, Development of motor function and body posture in pre-term infants, in: *Aspects of Neural Plasticity/Plasticité nerveuse* (F. Vital-Durand and M. Jeannerod, eds.), Vol. 43, pp. 55–66, INSERM, Paris.

Prechtl, H. F. R., Fargel, J. W., Weinmann, H. M., and Bakker, H. H., 1979, Postures, motility and respiration of low-risk pre-term infants, *Dev. Med. Child Neurol.* **21**:3–27.

Pschyrembel, W., and Dudenhausen, J. W., 1972, *Grundriss der Perinatalmedizin,* de Gruyter, Berlin.

Rademaker, G. G. J., 1926, Die Bedeutung der Roten Kerne und des übrigen Mittelhirns für Muskeltonus, Körperstellung und Labyrinthreflexe, *Monogr. Gesamtgeb. Neurol. Psychiatr.* **44**:344 pp.

Reinold, E., 1971, Beobachtung fetaler Aktivität in der ersten Hälfte der Gravidität mit dem Ultraschall, *Paediatr. Paedol.* **6**:274–279.

Reinold, E., 1976, Ultrasonics in early pregnancy. Diagnostic scanning and fetal motor activity, *Contrib. Gynecol. Obstet.* **1**:1–148.

Robinson, H. P., 1973, Sonar measurement of fetal crown-rump length as means of assessing maturity in first trimester of pregnancy, *Br. Med. J.* **4**:28–31.

Robinson, H. P., and Fleming, J. E. E., 1975, A critical evaluation of sonar "crown–rump length" measurements, *Br. J. Obstet. Gynaecol.* **82**:702–710.

Robinson, R. J., 1966, Assessment of gestational age by neurological examination, *Arch. Dis. Child.* **41**:437–447.

Saint-Anne Dargassies, S., 1954, Méthode d'examen neurologique du nouveau-né, *Etud. Néo-Natales* **3**:101–123.

Saint-Anne Dargassies, S., 1955, La maturation neurologique des prématurés, *Etud. Néo-Natales* **4**:71–116.

Saint-Anne Dargassies, S., 1966, Neurological maturation of the premature infant of 28 to 41 weeks' gestational age, in: *Human Development* (F. Falkner, ed.), pp. 306–325, W. B. Saunders, Philadelphia.

Saint-Anne Dargassies, S., 1970, Détermination neurologique de l'âge foetal néonatal (tableaux—Schéma évolutif—grille d'examen maturatif), *J. Paris. Pediatr.* **1**:310–326.

Saint-Anne Dargassies, S., 1974, *Le développement neurologique du nouveau-né à terme et prématuré,* Masson, Paris. (English translation 1977), *Neurological Development in the Full-term and Premature Neonate,* Elsevier North-Holland, Amsterdam.

Saint-Anne Dargassies, S., 1977, Long-term neurological follow-up study of 286 truly premature infants. I. Neurological sequelae, *Dev. Med. Child Neurol.* **19**:462–478.

Schaltenbrand, G., 1925, Normale Bewegungs- und Lagereaktionen bei Kindern, *Dtsch. Z. Nervenheilkd.* **87**:23–59.

Schmidt, W., Garoff, L., Heberling, D., Zaloumis, M., Cseh, I., Haller, U., and Kubli, F., 1981, Überwachung der fetalen Bewegungsaktivität mit Real-Time-Ultraschall und deren Bedeutung für den Schwangerschaftsverlauf, *Geburtshilfe Frauenheilkd.* **41**:601–606.

Schröder, R., 1977, Longitudinalstudie über Wachstum und Entwicklung von früh- und reifgeborenen Kindern von der Geburt bis zum 6. Lebensjahr. Ergebnisse der bisherigen Auswertung der psychologischen Untersuchungsbefunde, *Forschungsbericht des Landes Nordrhein-Westfalen,* Westdeutscher Verlag GmbH, Opladen.

Schulte, F. J., 1968, Gestation. Wachstum und Hirnentwicklung, in: *Fortschritte der Pädologie* (F. Linneweh, ed.), Vol. II, pp. 46–64, Springer-Verlag, Heidelberg.

Schulte, F. J., 1970*a*, Sinn und Unsinn der Altersschätzung von Neugeborenen aufgrund neurologischer Kriterien, *Perinat. Med.* **2**:261–262.

Schulte, F. J., 1970*b*, Neonatal brain mechanisms and the

development of motor behavior, in: *Physiology of the Perinatal Period* (U. Stave, ed.), Vol. 2, pp. 797–841, Appleton-Century-Crofts, New York.

Schulte, F. J., 1971, Das motorische Verhalten von Früh- und Neugeborenen, in: *Handbuch der Kinderheilkunde* (H. Opitz and F. Schmid, eds.), I/1, pp.108–124. Springer Verlag, Berlin.

Schulte, F. J., 1973, Besonderheiten bei der neurologischen Untersuchung und Beurteilung von Neugeborenen und Säuglingen, in: *Neuropädiatrie* (A. Matthes and R. Kruse, eds.), pp. 169–191, Thieme, Stuttgart.

Schulte, F. J., 1974, The neurological development of the neonate, in: *Scientific Foundations of Paediatrics* (J. A. Davis and J. Dobbing, eds.), pp. 587–615, Heinemann Medical, London.

Schulte, F. J., and Schwenzel, W., 1965, Motor control and muscle tone in the newborn period. Electromyographic studies, *Biol. Neonate* **8**:198–215.

Schulte, F. J., Michaelis, R., Linke, I., and Nolte, R., 1968*a*, Motor nerve conduction velocity in term, preterm and small-for-dates newborn infants, *Pediatrics* **42**:17–26.

Schulte, F. J., Linke, I., and Nolte, R., 1968*b*. Electromyographic evaluation of the Moro reflex in preterm, term, and small-for-dates newborn infants, *Dev. Psychobiol.* **1**:41–47.

Schulte, F. J., Albert, G., and Michaelis, R., 1969, Gestationsalter und Nervenleitgeschwindigkeit bei normalen und abnormen Neugeborenen, *Dtsch. Med. Wochenschr.* **94**:599–601.

Schulte, F. J., Schrempf, G., and Hinze, G., 1971, Maternal toxemia, fetal malnutrition, and motor behavior of the newborn, *Pediatrics* **48**:871–882.

Sidman, R. L, and Rakic, P., 1973, Neuronal migration, with special reference to developing human brain: A review, *Brain Res.* **62**:1–35.

Siegel, S., 1956, *Nonparametric Statistics for the Behavioral Sciences,* International Student Edition, McGraw-Hill, Tokyo.

Stirnimann, F., 1938, Das Kriech- und Schreit-phaenomen der Neugeborenen, *Schweiz, Med. Wochenschr.* **19**:1374–1476.

Stirnimann, F., 1940, *Psychologie des neugeborenen Kindes,* Kindler Verlag GmbH, Munich.

Streeter, G. L., 1920, Weight, sitting height, head size, foot length, and menstrual age of the human embryo, *Contrib. Embryol. Carneg. Inst.* **11**:143–170.

Terman, M. L., and Merrill, M. A., 1965, *Stanford-Binet Intelligenz-Test, Deutsche Bearbeitung von H. R. Lückert,* Verlag für Psychologie C. J. Hogrefe, Göttingen.

Timor-Tritsch, I., Zador, I., Hertz, R. H., and Rosen, M. G., 1976, Classification of human fetal movement, *Am. J. Obstet. Gynecol.* **126**:70–77.

Touwen, B. C. L., Huisjes, H. J., Jurgens-v.d.Zee, A.D., Bierman-van Eendenburg, M. E. C., Smrkovsky, M., and Olinga, A. A., 1980, Obstetrical condition and neonatal neurological morbidity. An analysis with the help of the optimality concept, *Early Hum. Dev.* **4**:207–228.

Vecchietti, G., and Borruto, F., 1982, Fetal behavior, in: *Fetal Ultrasonography: The Secret Prenatal Life* (F. Borruto, M. Hansmann, and J. W. Wladimiroff, eds.), pp. 25–37, Wiley, New York.

de Vries, H. J. I. P., Visser, G. H. A., and Prechtl, H. F. R., 1982, The emergence of fetal behaviour. I. Qualitative aspects, *Early Hum. Dev.* **7**:301–322.

Weber, H. P., Kowalewski, S., Gilje, A., Mollering, M., Schnaufer, I., and Fink, H., 1976, Unterschiedliche Calorienzufuhr bei 75 "low birth weights": Einfluss auf Gewichtszunahme, Serumeiweiss, Blutzucker und Serumbilirubin, *Eur. J. Pediatr.* **122**:207–216.

Wolff, L. V., 1930, The response to plantar stimulation in infancy, *Am. J. Dis. Child.* **39**:1176–1185.

18

Development of Newborn Behavior

T. BERRY BRAZELTON

1. Neonatal Behavior

As a newborn lies undressed and uncovered immediately after delivery, his color begins to change, with mottled uneven acrocyanosis of his trunk and extremities as the autonomic nervous system attempts to control the loss of body heat. The infant begins to shiver, and then to cry and flail his limbs in jerky, thrusting movements in an effort to maintain his own body temperature. In spite of the enormous demands of a new uncontrolled environment, if one speaks gently and insistently into one ear, his movements become smoother, and slower, and the infant gradually quiets completely. His color improves as his face softens and brightens; his eyes widen and, with an alert look, both eyes turn smoothly to the side from which the voice is coming. His head follows with a sudden smooth turn toward the voice, and he searches for the face of the speaker. His face brightens as he fixes on the eyes of the examiner and listens intently for several minutes. If the examiner moves his face slowly to the baby's midline and then across to the other side, the newborn will track the examiner's face, head, and eyes, smoothly turning in an 180° arc.

This behavior is neither unusual nor unique but usually goes unnoticed or unsolicited in a busy delivery room. This complex integration of visual, auditory, and motor behavior to respond to human stimuli is managed by the neonate despite the enormous physiological demands of being undressed and unrestrained in a cold, overstimulating nursery. To ignore the importance of this capacity, as one evaluates the neurological and physiological integrity of the neonate, is to fail to grasp the powerful effect of the neonate's control of autonomic and physiological systems from birth. As physicians, we must become more aware of the interactions between physiological and psychological mechanisms, as they represent integrity in the neonate and predict differences in developmental outcome.

The neonate has been thought of as subcortical in his behavior (Flechsig, 1920; Peiper, 1963). The examinations commonly used during the neonatal period for assessing and predicting the integrity of the CNS rely heavily on the newborn's capacity to respond at the brainstem level. The Apgar scores (Apgar, 1960), the pediatric evaluation, and the commonly used neurological evaluations score the physiological, autonomic, and midbrain responses. The classic pediatric neurological examination evaluates the newborn's capacity to respond to painful or intrusive stimuli and, as such, the resultant reflex behavior mediated by the brainstem or midbrain.

But the neonate has marvelous capacities for alerting and attending to positive or interesting stimuli and for enlisting the cardiorespiratory and autonomic systems in a total response to an interaction with his environment. If we are to assess the neonate's integrity at birth and predict his future potential for development, we must enlist his behavioral reactions to all kinds of stimuli—positive as well as negative. We must understand his reactions within a range of expectable behavior. There is now much literature available on neonatal behavior of importance to those interested in neonatal assessment.

The infant's behavior is built up from a genetic

T. BERRY BRAZELTON • Child Development Unit, Children's Hospital Medical Center, Harvard Medical School, Boston, Massachusetts 02115.

base that has evolved through adaptation to the environments in which our species evolved. Much of the reflexive behavior seen in neonates can be related to similar behavior in other mammals. The Moro reflex, for instance, is still adaptive for some monkeys; the placing reflex is a residuum of the adaptation to the tree-living primate's need to grab for a branch as it brushes the plantar aspect of the hand or foot as it falls. This reflex behavior unfolds with maturation of the CNS in the uterus, and during the *in utero* period "external" events influence these developments. For example, maternal nutrition, infection, drugs, and neuroendocrine factors all affect the development of fetal behavior. We have every reason to believe that the emotional state of the mother may play an important role in shaping the fetus's behavioral reactions. Hence, the neonate's behavior is phenotypic, and his individuality at birth is the result of an interaction among his genotype, maturational processes, and the intrauterine environment.

The neonate is in active transaction with his environment, shaping his caretakers as much as they shape him. Since the newborn is able to select input from the environment by the mechanisms of habituation and orientation, he is able to show behavioral preferences among these external stimuli. However, while self-regulation is gradually achieved, the neonate has important behaviors that provide information to the caretaker so that he or she will provide the appropriate nuturing. The newborn is a sophisticated elicitor.

2. Intrauterine Psychological Capabilities

Sontag and Richardo (1938) first suggested that conditioning of fetal behavioral responses represents learning *in utero* that contributes to the striking individual differences seen in the neonate's behavior. When the heart rate of the fetus was monitored by a cardiotachometer placed on the mother's abdomen, the fetal heart responded to three kinds of stimuli: (1) auditory; (2) cigarette smoking; and (3) emotional shocks administered to the mother. As these stimuli were repeated, the fetal heart rate response became diminished and its latency prolonged. The infant was learning not to respond to such stimuli. In another study, Ando and Hattori (1970) demonstrated that infants near an airfield in Okinawa who were exposed to high noise levels during the first 4 months of gestation were significantly less reactive to loud sounds after birth. Thus, one can expect that all kinds of information and stimulation—those directly received by the fetus or those received *via* the neurological and chemical responses of the mother—shape his behavior. But the effects may not always be the same. For example, the fetus may respond to anxiety in the mother by becoming more active and reactive; others respond instead by learning to cope with the stress induced by her anxiety by becoming quieter, having learned to shut down their own responses to the mother's signals. At birth, the infant may be intensely driving and overreactive. On the other hand the infant may be able to handle stimulation by becoming quiet. The infant demonstrates a mixture of both mechanisms.

3. Intrauterine Influences

The neonate has already been shaped by his experience in the uterus. Gottlieb (1971) reported a sequence in the onset of function of the sensory systems—tactile, vestibular, auditory, and visual. Approximately 2 months before birth, changes in fetal heart rate can be detected in response to auditory stimuli (Sontag and Richardo, 1938). Visually evoked cortical potentials are recorded on infants 28 weeks of age. Prenatal experience is likely to be important in the establishment and maintenance of neurobehavioral structures in the fetus. The first evaluation in the delivery room can offer us an irreplaceable set of observations that tell us about this experience.

We must be alert to signs of chronic stress that have interfered with developing organs at critical periods in their development. A physical resemblance to prematurity out of proportion to the infant's gestational age may point to a chronically stressed circulation to the fetus with resulting underdevelopment. Those organs (in the placental conduit system) that are situated near the uterine os, hence not well vascularized, are stunted in body size at birth (Widdowson, 1971). All organs (including the brain) are correspondingly small; their cellular content (as reflected by DNA count) is decreased and remains so throughout life. Rosso and Winick (1973) and Zamenhof *et al.* (1968) pointed to the significant decreases in cell number (as reflected by DNA) and cellular content (as measured by RNA) in the brain of neonatal animals whose mothers have had an inadequate protein intake during pregnancy. As much as 40% of the potential DNA content of the brain and other vital organs may never by made in the developing human fetus if the mother's protein intake is below a critical level during pregnancy.

Behavior in the neonate may be a major step toward understanding some of these effects. In a recent study of the offspring of malnourished women in Guatemala, we were able to clinically evaluate the offspring of the chronically depleted women by the immature and dysmature appearances of their neonates at full term. Their behavior was also affected by their state of undernutrition. The infants were quieter and more difficult to rouse, alert, and bring to responsive states for interaction. It was easy to predict that they would elicit less response from their already depleted stressed mothers. As a result, we could predict with 80% certainty in the neonatal period which infants would suffer from marasmus in the latter half of the first year or from kwashiorkor in the second year of life. We believed we were seeing neonatal behavior that reflected the effects of chronic intrauterine depletion, with reduced number of cells and cellular contents (Brazelton *et al.,* 1975).

Other maternal conditions affect behavioral outcome of the fetus as well. Drugs administered to the pregnant mother may have powerful lasting effects to which we are now becoming attuned. Tranquilizers, such as reserpine, meprobamate, and chlorpromazine, given to pregnant animals adversely affect birthweight and response to stress in the neonatal period, as well as learning tasks in later infancy. The time at which these drugs are administered in pregnancy has an important influence—controls < mid < late < early. The controls were not affected; infants whose mothers received drugs in mid-pregnancy were less affected than were infants whose mothers received drugs in late pregnancy. Those who suffered most, however, were infants whose mothers received doses in early pregnancy. We do not know the effect of commonly used tranquilizers on human offspring, but the possibility of risk should be in our minds as we assess the neonates of mothers who have been chronically exposed (Hoffeld *et al.,* 1968).

Narcotic addition carries with it a high incidence of congenital anomalies in the offspring as well as rather frightening withdrawal symptoms during the neonatal period (Strauss *et al.,* 1975). For several weeks after delivery, addicted babies demonstrate major difficulties, with extreme irritability and an inability to be comforted except by constant feeding, pacifiers, and swaddling. They overreact to all environmental stimuli with violent, agitated motor activity. Their lability of behavior and inability to quiet or control themselves make them very difficult to care for. Their hyperexcitability and hyperkinesis coupled with poor response to being comforted increase their vulnerability when they are returned to their already overstressed mothers. The incidence of neglect and abuse is expectably high in this group. An evaluation of the neonate's behavior can predict his contribution to the at-risk interaction.

When stimuli are presented repeatedly, the responses of the fetal heart decrease rapidly, and the latency to response becomes significantly prolonged. The infant seems to be "learning" to decrease his responses to these stimuli. This decrement in response learned *in utero* persists into the neonatal period.

We have, however, wondered whether this learning about stress is not costly to the fetus. Animal work by Keeley (1962) indicated that severe behavioral and physiological stress to cats and mice during pregnancy (conditioned anxiety, crowding, epinephrine injection) produced permanent changes in the behavior of their offspring. The newborn's open-field activity and maze learning were severely impaired.

4. Premature Behavior

Prenatal behavioral organization has been visualized both *in utero* and after emergency cesarean sections by Gesell (1945), Hooker (1969), and Windle (1950). As expected, the sequence of responses is highly correlated with the degree of maturation of the CNS. At birth, the degree of immaturity of a premature infant can be assessed by applying a continuum of neurological and behavioral responses. The Dubowitzes (1970) documented a scoring system in which the external signs of immaturity such as edema, skin texture, skin color, skin opacity, lanugo, plantar creases, nipple formation, breast size, ear form, ear firmness, and genitalia appearance are given a score, which is added to a score for the infant's motor behavior—such as his spontaneous posture, the flexion angles producible in his arms and feet, the recoil after extension, the degree of resistance to flexion of the arms in a "scarf" maneuver and of the legs in a heel-to-ear maneuver, head lag on being pulled to sit, and truncal tone in ventral suspension. The combination of these two scores assessed in the first 5 days produces a number that accurately coincides with the infant's gestational age. Other, more refined tests of neurological maturity, such as nerve-conduction velocity (Schulte *et al.*, 1968), sleep states as recorded by EEG (Parmelee *et al.*, 1968), and latency measures of visual evoked responses by EEG (Hrbek and Mares, 1964), are more difficult to document in premature babies but may be even more accurate in determining gestational age. Since the premature neonate is often at the mercy of the demands of recovery from stress, these behavioral signs are often difficult to elicit, and more objective measures become necessary in an effort to determine his age accurately.

There has been little documentation, so far, of sensorimotor behavior in the premature infant. Thirty-four to thirty-six-week infants do fix upon and track a visual object. They can quiet, attend to, and turn their heads toward a human voice or a soft rattle, although these responses are often more obligatory than they are in full-term infants. The revised Brazelton Neonatal Assessment Scale (Brazelton, 1973) documents this alerting behavior. Although sensory responses are more difficult to elicit, due in large part to the excessive physiological demands of immaturity, we have found that when premature infants are alert and responsive, their motor organization and physiological responses stabilize for the period, as if the infant were indeed capable of overcoming systemic interference in order to attend to sensory stimuli.

Early examination of the neonate should include looking for signs of dysmaturity such as dried or peeling skin. A small-for-dates baby is more at risk than a well-nourished one. Evidences of extracellular depletion and speed of recovery may reflect the duration and depth of his deprivation *in utero*. Short-term signs such as skin texture and subcutaneous fat depletion may not be as significant as are the long-term effects; e.g., decreased linear bony growth, decreased head circumference, or minor congenital development defects, that may point to chronic intrauterine depletion that affected the fetus at critical periods of development.

An estimate of the stage of maturity using such physical signs as size of breast nodule, palm and sole creases, scalp and body hair distribution, ear lobe, testes, and scrotum (Lubchenco *et al.*, 1963), and behavioral signs that measure reflex behavior on a fetal continuum (Dubowitz), can be measured against careful dating from the mother's last menstrual period. When these do not coincide, the significance of relative immaturity becomes of predictive importance, suggesting long-term intrauterine deprivation that has affected cellular development in the fetus. Although we have not given enough credence in the past to mothers' histories, we are finding them more reliable as we begin to understand the characteristics of small-for-dates and large-for-dates infants. A behavioral assessment of the maturation of reflex behavior, as outlined by Dubowitz, can be coupled with an evaluation of the relationship of gestational age to height, weight, and head circumference on normative intrauterine growth charts developed by Lubchenco *et al.* (1963).

5. Assessment of Behavioral States

The newborn infant has been thought of as neurologically insufficient and subcortical in behavior—a blank slate to be written on by a new world of buzzing confusion. None of these descriptions fits the kind of predictable, directed responses one sees in a neonate when he is in a social interaction with a nurturing adult or as he responds to an attractive auditory or visual stimulus. For when positive rather than intrusive stimuli are utilized, the neonate from birth has amazing capacities for alerting and attention, for suppressing interfering reflex responses in order to attend, and, with very predictable behavior, for responding to and interacting with his environment. But this predictability requires a knowledge of his ongoing state of consciousness; when the state is accounted for, most of his reactions are predictable—both to negative and to positive stimuli, from internal as well as external sources. State becomes a matrix for understanding his reactions. It qualifies stimulation as appropriate or inappropriate to his ongoing organization. Thus, for almost any maturational level, the behavior produced by appropriate stimuli in appropriate states will demonstrate the complexity of an intact and adaptable CNS. Perhaps the concept that the neonate is functioning at a subcortical level has worked too long and has blinded observers to this complexity and to its implications for assessment and prediction of neurological integrity.

The classic pediatric neurological assessment of the neonate is based on his responses to painful or intrusive stimuli, and, as such, the resulting reflex behavior is, indeed, mediated by the midbrain. A standard neurological evaluation simply evaluates the integrity of his brainstem as it copes with these stimuli. But such an examination neglects the available organized behavior that the infant can demonstrate as he suppresses reflexive behavior in order to attend to more "interesting" stimuli, such as the human face or voice, a soft rattle, or a light caress.

6. Sleep States

Sleep states have been recognized and defined by Wagner (1937). She described some of the behaviors seen in deep sleep (i.e., jerky startles and relative unresponsiveness to external stimuli) as well as more regulated, smooth movements accompanied by responsiveness to stimuli in lighter sleep. Wolff (1959) added his observations of regular, deep respirations with sudden spontaneous motor patterns such as sobs, mouthing and sucking, and erections, which occurred in deep sleep at fairly regular intervals in an otherwise inactive baby. He observed that babies were more responsive to stimuli in light sleep. Aserinsky and Kleitman (1955) described cycles of quiet, regular sleep followed by active periods of body movements and rapid eye movements (REMs) under closed lids. Light sleep has therefore come to be designated REM sleep. More recently, electrophysiologists have differentiated quiet and active sleep into four stages as designated by electroencephalographic (EEG) activity (Anders and Weinstein, 1972). By behavioral observational techniques, two main sleep states—quiet and active—must suffice in the full-term neonate. In prematures and abnormal infants, and to some extent in term infants, a third, indeterminate sleep state is observable (Anders and Weinstein, 1972). At term, active sleep (REM) occupies 45–50% of total sleep time, indeterminate sleep 10%, and quiet sleep 35–45%. The predominance of active sleep has led to the hypothesis that REM sleep mechanisms stimulate the growth of the neural systems by cyclic excitation, and it is in REM sleep that much of the differentiation of neuronal structures and neurophysiological discharge patterns occurs (Anders and Roffwarg, 1973). Quiet sleep seems to serve the purpose of inhibiting CNS activity and is truly an habituated state of rest.

The length of sleep cycles (REM active and quiet sleep) changes with age. At term sleep cycles occur in a periodicity of 45–50 min, but immature babies have even shorter, less well-defined cycles. Newborn infants have as much active REM in the first half of the deep sleep period as in the second half. Initial, brief, individualized sleep-and-wake patterns coalesce as the environment presses the neonate to develop diurnal patterns of daytime wakefulness and night sleep. Appropriate feeding patterns, diet, absence of excessive anxiety, sufficient nurturing stimulation, and a fussing period prior to a long sleep have all been implicated as reinforcing to CNS maturation, necessary for the development of diurnal cycling of sleep and wakefulness. Sleep polygrams, consisting of both EEG and activity monitoring, are sensitive indicators of neurological maturation and integrity in the neonatal period. Steinschneider (1973) analyzed the occurrence of apnea episodes during sleep as part of a study of sudden infant death syndrome (SIDS). He found that these episodes are more likely to occur during REM sleep and has suggested that prolonged apnea, a concomitant of sleep, is part of a final physiological pathway culminating in SIDS.

In our behavioral assessment of neonates, we utilized the two states of deep, regular and active REM sleep described by Wolff (1959) and Prechtl (1963) as well as four states of wakefulness. These states can be reliably determined by observation and without any instrumentation.

State 1. Deep sleep with regular breathing; eyes closed; spontaneous activity confined to startles and jerky movements at quite regular intervals. Responses to external stimuli are partially inhibited, and any response is likely to be delayed. No eye movements. State changes are less likely after stimuli or startles than in other states.

State 2. Active REM sleep with irregular breathing; sucking movements; eyes closed but rapid eye movements underneath the closed lids; low activity level; irregular, smooth, organized movements of extremities and trunk. Startles or startle equivalents as response to external stimuli often with change of state.

Alert states have been separated into four states for our behavioral assessment:

State 3. Drowsy, a transitional state with semidozing; eyes may be open or closed, eyelids often fluttering; activity level variable, with interspersed mild startles, possibly fussy vocalizations, and slow, smoothly monitored movements of extremities at periodic intervals; reactive to sensory stimuli but with some delay; state change frequently after stimulation.

State 4. Quiet alert with bright look; focused attention on sources of auditory or visual stimuli; motor activity suppressed in order to attend to stimuli. Impinging stimuli break through with a delayed response.

State 5. Active awake with eyes open; considerable motor activity; thrusting movements of extremities; occasional startles set off by activity and with fussing. Reactive to external stimulation with an increase in startles or motor activity; discrete reactions difficult to distinguish because of general high-activity level.

State 6. Intense crying with jerky motor movements; difficult to break through with stimulation.

The waking states are easily influenced by fatigue, hunger, or other organic needs and may last for variable amounts of time. However, in the neonatal period, they are at the mercy of the sleep cycles and surround them in a fairly regular fashion. Waking states are infrequently observed in noisy, overlit neonatal nurseries, but in a rooming-in situation or at home they become a large part of each cycle, and the neonate lies in his crib looking around for as much as 20–30 min at a time. Appropriate stimulation can bring him to a responsive state 4. Rocking, gentle jiggling, crooning, stroking, setting off vestibular responses by bringing him upright or by rotating him all serve to open his eyes. Then his interest in visual and auditory stimuli helps him maintain a quiet alert state. In the alert state, his respirations are regular (50–60 per min); his cardiac rate, too, is regular and fairly slow (around 100–120 per min); his eyes are wide, shiny, and capable of conjugate movements to scan and to follow, with head turning to appropriate objects; his limbs, trunk, and face are relaxed and inactive; his skin color is uniform.

Alert inactive states occur in the first 30–60 min after delivery but then are likely to decrease in duration and occurrence over the next 48 hr as the infant recovers. They return after the first 2 days and constitute as much as 8–16% of total observation time in the first month (Wolff, 1959).

Kleitman (1963) speaks of wakefulness of "necessity" and wakefulness of "choice." Wakefulness of necessity is brought about by stimuli such as hunger, cold, bowel movements, or external stimuli that disturb sleep cycles. Sleep recurs as soon as the response to the disturbing stimuli is completed. Wakefulness of choice is related to neocortical activity and is a late acquisition that coincides with the emergence of voluntary motor actions and a mature capacity to achieve and maintain full consciousness. Absence of hunger and fatigue, bowel and bladder activity, as well as of gross motor activity are necessary to this state. After a few weeks, this state can accompany gross motor activity as long as the infant is not too active. Pursuit movements of his extremities accompany visual fixation and following as early as 2–3 weeks (Brazelton *et al.*, 1966). The occurrence and duration of these quiet inactive states may be highly correlated with intactness and mature organization of the neonate's CNS.

7. Crying

Crying serves many purposes in the neonate, not the least of which is to shut out painful or disturbing stimuli. Hunger and pain are also responded to with crying, which brings the caretaker. Fussy crying occurs periodically throughout the day, usually in cyclic fashion. It seems to act as a discharge of energy and as an organizer of the states that ensue. After a period of such fussy crying, the neonate may be more alert, or he may sleep more deeply.

As a behavior of organizing his day and reducing disturbance within the CNS, crying seems to be of real importance in the neonatal period. Most parents can distinguish cries of pain, hunger, and fussiness by 2–3 weeks, and learn quickly to respond appropriately (Wolff, 1968). So the cry is of ethological significance, with the function of eliciting appropriate caretaking for the infant.

The neonate's cry has been analyzed by Lind *et al.* (1966), Wolff (1969), and others for diagnostic purposes. It is a complex, serially organized acoustical pattern that is directly regulated by the CNS. Wolff (1969) found three distinct acoustical patterns of cries in normal infants, signifying hunger, anger, and pain. Lind *et al.* (1966) found four patterns: birth, hunger, pain, and pleasure. The fundamental frequency of the normal infant's cries is between 250 and 600 Hz, predominantly between 350 and 400 Hz. Lind *et al.* (1966) found that experienced researchers can discriminate pain cries in normal infants from the cries of infants who had asphyxia or CNS afflictions. The discriminating feature appears to be an increase in the pitch of the infant's cry with a fundamental frequency around 650–800 Hz in brain-damaged infants (Wolff, 1969; Wasz-Hockert *et al.*, 1968). Moreover, brain-damaged infants require more stimuli to elicit a cry, show a larger latency to their cry, and have a less sustained and more arrhythmical cry.

8. Sensory Capacities

8.1. Visual

The newborn infant is equipped at birth with a marvelous capacity for processing complex visual information and for demonstrating visual motor movements to track an object in space. Even more important to his survival is the fact that he can defend himself from visual stimuli that might otherwise force him to responses that would make excessive demands on his immature physiological system. When a bright light is flashed into a neonate's eyes, not only do his pupils constrict but he blinks, his eyelids and whole face contract, he withdraws his head by arching his whole body, often setting off a complete startle as he withdraws, his heart rate and respirations increase, and an evoked response is registered on his visual occipital EEG. Repeated stimulation of this nature induces diminishing responses because of his capacity to shut down on his responses. For example, in a series of 20 bright-light stimuli presented at 1-min intervals, we found that the infant rapidly "habituated" out the behavioral responses; by the tenth stimulus not only had his observable motor responses decreased but his cardiac and respiratory responses had begun to decrease markedly as well. The latency to evoked responses as measured by EEG tracings was increasing and, by the fifteenth stimulus, the EEG reflected the induction of a quiet, unresponsive behavioral state accompanied by trace alternans and spindles (Brazelton, 1962). The newborn's capacity to shut out repetitious, disturbing visual stimuli protects him from having to respond to visual stimulation and at the same time frees him to save his energy to meet physiological demands.

This capacity on the part of the neonate to suppress his responses to visual stimuli has been designated as a kind of neurological habituation and has been found to be present in neonates with intact CNS. In immature infants, this capacity is decreased, although it is present. It is affected by medication such as barbiturates given to mothers as premedication at the time of delivery. The infant finally falls deeply asleep, with tightened, flexed extremities, little movement except jerky startles, no eyeblinks, deep regular respirations, and rapid regular heart rate. This state of habituation seems to signal a defensive state against the assaults of the environment and, on cessation of the stimulation, the infant almost immediately goes back to his initial state or even to a more alert state. Often one can see neonates in noisy, brightly lit neonatal nurseries in this "defensive" sleep state.

It is important to realize that any newborn baby in a brightly lit, noisy nursery is likely to be in such an habituated state. In such a state, he is unlikely to be as responsive to loud noises or a flashlight if they are used as a test stimulus. Consequently, to test the newborn infant's vision or hearing it is necessary to move him to a darker, quieter room, where his complex visual responses can be captured more easily and reliably. Just as he is equipped with the capacity to shut out certain stimuli, he also demonstrates the capacity to become alert and to turn his eyes and head to follow and fix on a stimulus that appeals to him. Fantz (1965) first pointed out neonatal preference for certain kinds of complex visual stimuli. For instance, he found that sharply contrasting colors, larger squares, and medium brightly lit objects were appealing to the neonate and brought him to a prolonged alert state of fixation. He and others found the neonate to prefer an ovoid object, particularly one having eyes and a mouth. The kind of attention and the length of fixation were found to be markedly reduced in infants whose mothers had been medicated before delivery.

More recently, Goren *et al.* (1975) showed that, immediately after delivery, a human neonate would not only fix on a drawing that resembled a human face but would follow it for an 180° arc, with eyes and head turning to follow. A scrambled face did not demand the same kind of attention, nor did the infant follow the distorted face. We found the capacity of neonates to fix on and follow a red ball to be a good predictive sign of neurological integrity; the absence of this response could not be thought of as a serious predictor, however, because it was so dependent on the infant being in an alert state. Visual fixing and following during alert periods may be a more sensitive predictor of neurological and visual integrity than are many of the neurological tests that we now rely on (Tronick and Brazelton, 1975).

Visual acuity of the newborn still is difficult to determine. Gorman *et al.* (1957) used the neonate's opticokinetic responses to a moving drum lined with stripes and found that 93 of 100 infants responded preferentially to stripes subtending a visual angle of only 33.5 min of arc. We found that prematures were less reliable but could also fix on and follow the same lined drum (Brazelton *et al.*, 1966). Dayton *et al.* (1964) found at least 20/150 vision in newborn infants by this same technique.

It does not seem, however, that newborn infants are able to accommodate well, but rather that they have a fixed focal length of about 19 cm (Haynes, *et al.*, 1965). In order to capture visual interest, an examiner must present a bright object at this distance. Acute brightness discrimination has been reported for infants in the first month (Hershenson, 1964). Peeples and Teller (1975) demonstrated that newborn infants have dichromatic vision. Normal adults have trichromatic vision, which enhances all wavelengths of the visible spectrum. Dichromates cannot discriminate all wavelengths from white, but they can and do have color preferences, such as reds and blues but not yellows or greens.

Newborn infants are unable to accommodate the range of stimuli that they will respond to later on. Targets at distances of 75 cm or less are preferred until 3 months of age (Braddick *et al.*, 1979), and adult accommodation is reached by 6 months of age (Banks, 1980). The newborn infant's inability to accommodate, however, does not rule out his gross ability to discriminate. It does, though, severely limit his spatial sensitivity and his capacity to register a good image at a distance beyond 75 cm (infants have a greater depth of focus than do adults, because of their smaller eye size and greater relative depth of the eyeball). As a result, contrasts in visual form and color become the preferred visual targets. Both behavioral patterns and evoked cortical responses can be used to determine preferential fixation on contrasting visual stimuli. Preferential looking at paired stimuli, as well as a measurement of oculomotor fixation and movements by electrodes that register oculomotor movements, become the paradigms for testing visual behavior and preference. Newborn infants prefer the corners of a checkerboard (Salapatek and Kessen, 1966), on which contrasts are high, or the edges of the human face, as well as the brightness of the eyes and the mouth. They are processing visual information from birth.

Extremes of brightness and noise in the environment have been found to interfere with the neonate's capacity to respond. In a noisy, overlighted nursery, the neonate tends to shut down his capacity to attend, but in a semidarkened room, a normal neonate in an alert state can be brought to respond to the human face as well as to a red or shiny object. The following description indicates the behaviors the infant is capable of displaying.

The neonate is held or propped up at a 30° angle. Vestibular responses enhance eye opening and tend to bring him to a more alert state (Korner and Thoman, 1970). His eyes begin to scan the environment with a dull look, wide pupils, and saccadic lateral movements of both eyes. As a bright red ball is brought into his line of vision and is moved slowly up and down to attract his attention, the infant's pupils contract slightly. As the ball is moved slowly from side to side, his face begins to brighten, his eyes widen, his limbs are stilled, and he stares fixedly at the object, beginning to track the ball slowly from side to side. He maintains the stilled posture in order to attend to the ball. His eyes first track in small arcs that take him off the target but, as he becomes more invested, the eye movements become smoother. As his eyes move laterally, his head begins to turn, and he begins to move his head from side to side in order to facilitate the tracking of the object. He is able to follow it for as much as 120°, to the right and left, and will even make eye and head movements to follow it 30° up or down. Meanwhile, interfering body movement or startles seem to be actively suppressed. The neonate can maintain this intense visual involvement for several minutes before he startles, becomes upset or dull, and loses the alert state necessary to this kind of visual behavior.

The involvement of the neonate's state behavior as well as the coordination of visual motor and head turning seem to point to the entrainment of rather complex nervous system pathways. It is difficult not to believe that his cortex is maintaining this alert state of consciousness and controlling any interfering motor behavior. The visual and motor cortex certainly are involved as the following is engaged in by the head and eyes. In a recent long-term study, Sigman *et al.* (1973) found that visual behavior may be one of the best predictors of intactness of the CNS in the neonate. These investigators found that summary scores on the neonate's neurological examination were significantly related to the length of first fixation on a black-and-white checkerboard as a visual stimulus. This capacity to stare at a complex object was, of course, related to an alert state. All these capacities—to alert, to maintain an alert state, to fix on and attend to a visual display—were correlated with the infant's state of maturity, the optimal condition of his CNS (i.e., no evidences of hypoxia, illness, or other impinging risk factors); on follow-up examination, they predict future CNS integrity. Thus, visual performance in the neonatal period becomes a promising area for assessing and predicting future outcome.

Sigman and coworkers (1973) also found that

prolonged obligatory attention to a visual display might point to a lack of inhibitory mechanisms and might be a sign of less-than-optimal CNS functioning. We have also found this phenomenon in an anencephalic infant whom we assessed with the lined drum of Gorman *et al.* (1957). Although anencephalic, this baby was able to fix and follow the moving lines but with a fixed stare that did not change in quality. After several stimulus presentations, his fixations on the moving lines appeared to be part of an obligatory response that he was unable to modulate, and he was finally able to shut out the stimuli only by crying (Brazelton *et al.,* 1966). His inability to habituate or to modulate his responses to repeated monotonous stimuli seemed to be dependent on inadequate CNS controls. Thus, an optimal response to visual stimulation in a neonate can be described as: (1) an initial alerting; (2) attention that increases but is followed by; (3) a gradual decrease in interest, and (4) a final turning away from a monotonous presentation. This behavior can be elicited in the neonatal nursery during examination of a baby's heart and lungs.

Adamson (1976) demonstrated the importance of vision to the neonate by covering an alert baby's eyes with both an opaque and a clear plastic shield. The baby first swipes at the opaque shield and vigorously attempts to remove it, building up to frantic activity to do so, but quieting suddenly when it is removed. When the clear shield is substituted, he calms to look interestedly through it.

Bower (1965) pointed out that visual attention to an object brings about abortive bits of reaching behaviors, even parts of real swipes directed toward the object. These swipe behaviors occur during the first few weeks, when a baby is propped up and alert and an attractive object is brought into his "reach space" (about 9–12 in. in front of him in the midline).

Ball and Tronick (1971) demonstrated that neonates will actively defend themselves from looming visual targets with a startle, a withdrawal, or part of a defensive reaction as an object approaches too rapidly. This defense can include such organized behaviors as head turning and directed arm movements toward the object as if to fend it off.

Several observers (Hershenson, 1964; Parmelee, 1962) demonstrated that neonates prefer visual patterns that are moving and somewhat complex to stationary ones. If the object can be moved slowly parallel to the natural, lateral movements of the neonate's eyes, it is more likely to capture his interest. Furthermore, the duration and degree of his attention may be correlated with both a middle range of complexity and the similarity of the target to an ovoid shape, with structures similar to the human face.

Salapatek and Kessen (1966) found the highest points of concentration of fixation in the neonatal period to be on the contrasting edges of an object. For the neonate, the eyes or sides of the head seem to be the most compelling features of the face. Thus, the neonate seems to be highly programmed for learning visually about human faces from birth. The visual stimuli that are the most appealing seem to be the features of the human face, and the shiny eyes, as with any other moving object, seem to be adapted to capture his attention very early. If his physiological systems are not overwhelmed with demands of too much information or for too prolonged a period of attention, he will attend for long alert periods.

A pediatrician can elicit these complex behaviors during routine neonatal examination. I always pick up a neonate at a 30° angle, head up. As I hold him, I swing him gently from side to side in order to produce eye opening; then, as I talk to him, I attempt to get him to fix his gaze on my face and to follow me as I move slowly from side to side within his range of vision. If he alerts to me and follows my face, I consider him a healthy baby. If he does not alert or interact visually with me, I reserve judgment.

To prove the significance of visual interaction, Klaus and Kennel (1970) demonstrated the importance of en face behavior in mothers of at-risk infants. These mothers attempt early to elicit eye-to-eye contact with their premature infants, even while they are in their isolettes, in order to reassure themselves that the baby is intact and in order to feel that the baby belongs to them despite his separation and difficulties. So, visual behavior in the neonate has powerful, adaptive significance as well as importance in predicting CNS intactness.

8.2 Auditory

The neonate's auditory responses are also specific and well organized. Bredberg (1968) reported that the organ of Corti is fully differentiated and that the cochlear duct is innervated by the sixth month of fetal development. Experience for the baby is amassing *in utero,* and his auditory preferences may be pretty well established by the neonatal period. In 1888, Preyer had doubts about the neonate's capacity to hear, but few observers have questioned it since. All too commonly, assessments are not sensitive to the complexity of newborn behavior. For example, the loud clackers used in the Collaborative Project of the National Institute of Neurological Disease and Blindness, used for early detection of CNS defects, were ineffective in loud, noisy nurseries. A large percentage of neonates tested with such a routine were unresponsive, as they appeared to have shut out or to have habituated to the ambient auditory stimuli. Another approach under these conditions would have been to use a soft rattle. This stimulus would have been more appropriate to break through the habituated states of these neonates.

The mechanism by which the neonate would have responded is termed dishabituation. In an habituated, nonresponding state, any change in stimulus properties is likely to cause a return of alertness and responsiveness. With an interesting auditory stimulus, such as a rattle, we see an infant move from a sleeping state to an alert state. His breathing becomes irregular, his face brightens, his eyes open, and, when he is completely alert, his eyes and head will turn toward the sound. In the case of a well-organized neonate, head turning will be followed by a searching look on his face—a scanning of his eyes to find a source for the auditory stimulus. In order to find out whether the neonate can respond in this way, a full test of hearing should include several stimuli-animate as well as inanimate—with careful attention to the neonate's ongoing state of consciousness, so that it will break into his state. For example, one should use a rattle in light sleep, a voice in awake states, and a hand clap in deep sleep. Respirations and eyeblinks can be monitored for change as he responds, as well as more obvious behavioral startles to auditory stimuli.

Eisenberg (1965) determined the differential responses to different ranges of sound available to the infant. In the range of human speech (500–900 Hz), the neonate will inhibit behavior; he often will demonstrate cardiac deceleration as evidence of his attention, and alerting and head turning are likely to occur as he turns to the source of sound. Outside this humanoid range, there is a less complex behavioral response. With a high-pitched and too-loud sound, the infant will startle initially and turn his head away from the sound, his heart rate and breathing will accelerate, and his skin will redden. He will attempt to shut out the repetitious sound by habituation and, if that proves unsuccessful, he will start to cry in order to control the disturbing behavioral effects of such auditory input. The strikingly narrow range of stimuli for positive, attending responses can be demonstrated by linking devices for recording sucking to the auditory input (Lipsitt, 1967). Within this narrow "humanoid" range, sucking will cease as an initial response to the stimulus, to be followed by a burst–pause pattern of sucking, as if the infant were pausing off and on to receive more of the interesting auditory input.

Lower and higher frequencies have different functional properties. Signals above 4000 Hz are more effective in producing a response—even in crying or sleep states—but they are likely to produce distress. Lower frequencies (35–40 db) are effective inhibitors of distress, especially as continuous "white noise" (Lipton *et al.*, 1966). White noise at these levels will most often induce a sleep state after a period, even in a crying neonate (Wolff, 1966). The importance of rise time of the sound on the neonate's behavioral responses was demonstrated by Kearsley (1973). Sounds with prolonged onset times and low frequencies produced eye opening and cardiac deceleration, followed by an attentive look, whereas sounds with rapid onset and high-frequency produced eye closing, cardiac acceleration, and increased head movements and aversion.

In assessing neonatal responses to sound, one must be aware of and account for what Lacey has called the "law of initial values" (Lacey, 1967). He pointed out that a response to an external stimulus, such as an auditory one, must be seen as pressing the neonate's autonomic response level toward a homeostatic resting level. Using heart rate responses as his measure, Lacey demonstrated that when the cardiac rate was high at the stimulus onset, an appropriate response brought the heart rate down; when the cardiac rate was low, the response brought the rate up. This same paradigm can be seen in all behavioral systems; i.e., when the baby is active or crying, an appropriate or interesting auditory stimulus is likely to produce quieting in the baby as he attends to it. When he is quiet or asleep, he will startle initially and his heart rate will accelerate; he will become more active as he attends to the stimulus. Thus, an appropriate response becomes one that leads him behaviorally or autonomically in the direction of his normal, alert, resting level and homeostasis. In a sinusodial fashion, the neonate can be seen to build up from being down and will go down when he becomes too wide awake as a response to an "appropriate" stimulus. This enables the neonate to overcome his ambient "state" to attend to and interact with an auditory stimulus.

In sum, as one defines the neonate's behavioral response to an appropriate sound, one sees a series of regular steps. As the sound is localized (Lipsitt, 1967), cardiac rate first increases (Drillien, 1972) and may be accompanied by a mild startle. If the auditory stimulus is attractive to the infant, his face will brighten, his heart rate will decelerate, his breathing will slow, and he will alert and search with his eyes until the source of the sound is localized to the en face midline of the baby. This train of behavior, which occurs as a response to an attractive auditory stimulus (e.g., a rattle or the human voice), becomes a measure of the neonate's capacity to organize his central and autonomic nervous systems.

Bridger (1961) found that the heart rate acceleration as an initial response to an auditory stimulus ceased after several trials, and the baby essentially habituated behaviorally and autonomically to this repetitive stimulus. But a change in frequency or a tonal change brought about an immediate increase in motor activity as well as in heart rate change, representing dishabituation. These changes in discrimination tests could be as minor as the change from 200 to 250 Hz. Eisenberg *et al.* (1966, 1974) found that the frequency of the heart rate as a response increased as the tone increased, then decreased accordingly when it went down. In other words, cardiac response can be used to study the infant's mental organization for repetitive, novel, and contingent responses. A paradigm of habituation registers his response to repetition of auditory, visual, and tactile stimuli. Not only does the change in heart rate decrease as the same stimulus is presented repeatedly to a newborn infant, but he begins to lose interest and actively shuts out the repeated stimulus by going to sleep or by crying. He seems to have appreciated and "learned" about similarity. If the stimulus is varied slightly, he begins to become interested again; he rouses or quiets, begins to brighten, and his heart rate changes return to the "new" stimulus. One can document his capacity to detect differences in duration, intensity, and rise time by monitoring his heart rate at such a time.

The sensitivity of such methods for detecting intactness and differences in neonatal equipment seems to be represented by a study of cardiac responses to habituation of auditory signals in malnourished neonates in Guatemala. Lester (1975) found that well-nourished neonates exhibited a substantial cardiac response to a 90-db pure tone, followed by rapid habituation to the initial onset of the tone and dishabituation or return of responses to two changes in tonal frequency. By contrast, infants undernourished *in utero* showed an attenuation or, in more severe cases, a complete absence of the cardiac response to the initial stimulus and no evidence of dishabituation to changes in tonal frequency. Since these babies were deemed neurologically and pediatrically intact on other tests, this finding became the only predictor during the neonatal period of attentional deficits attributable to an insult such as intrauterine deprivation. It might be seen to predict the later deficits in learning to which this malnourished group was subject. Such paradigms of sensory responsiveness may become increasingly sensitive measures of "soft" CNS dysfunction in the future and show real promise for differentiating minor CNS difficulties during the neonatal period. Eye movements demonstrate how beautifully a neonate can localize the spatial position of a sound source (Wertheimer, 1961). Initially, the eyes turn as the head turns to localize the source. A neonate can fix on and follow a sound to each side of the head for 90°, as well as vertically up and down for 30°.

Cairns and Butterfield (1975) documented the differences in the neonate's responses to human versus nonhuman sounds by using a sucking pardigm to detect subtle information-processing differences (discussed in Section 9). These investigators believe that monitoring for a burst–pause pattern in sucking in response to changes in auditory stimuli can differentiate between CNS impairments of receptive processing and peripheral impairment found in the nerve deafness of rubella, congenital malformations, hyperbilirubinemia, and other disorders.

8.3. Olfactory

Engen *et al.* (1963) demonstrated observably differentiated responses to odors in the neonate, from which one can conclude that the infant is equipped with a highly developed sense of smell, ready to pick up the odors that will help him adapt to his new world. For example, he behaves as though he is offended by acetic acid, asafetida, and alcohol during the neonatal period but is attracted to sweet odors such as milk and sugary solutions. More recently, MacFarlane (1975) showed that 7-day-old neonates can reliably distinguish their own mother's breast pads from those of other lactating mothers, although this ability was not present at 2 days.

These neonates turn their heads toward their own mother's breast pads with 80% reliability, after controls for laterality are imposed. In other words, their sense of smell incorporated the differential odor of their own mother by the fifth day, as a learned response. This refined capacity, present at birth, takes surprisingly little information from the environment to set up a reliable pattern for adaptive differentiation. We have seen that breast-feeding infants at 3 weeks may refuse to accept a formula from their mothers. This refusal seems to be related to the infant's ability to choose the available breast by odor. Yet, fathers are successful in giving a bottle to these same babies. Thus, it seems that the neonate does have the capacity to make choices, and this is used to enhance the attachment process to the mother.

8.4. Taste

The newborn has fine differential responses to taste. Pratt *et al.* (1930) observed differentiated sucking responses to sugar and decreased sucking to other tastes. More recently, Johnson and Salisbury (1975) reported that a newborn's taste preferences are expressed in an even more complex fashion. When an infant is fed different fluids through a monitored nipple, his sucking pattern is recorded. Saline solution causes such resistance that the baby is likely to aspirate. With a cow's milk formula, he will suck in a rather continuous fashion, pausing at irregular intervals. If breast milk then is fed to him in this same system, he will register his recognition of the change in taste after a short latency, then suck in bursts, with frequent pauses at regular intervals. The pauses seem to be directly related to the taste of breast milk. The burst–pause pattern seems to indicate a change to a kind of program for other kinds of stimuli (such as social communication) as these are added to the feeding situation during the pauses. Lipsitt (1967) showed that the criterion for response to changes in sweetness of liquids is only 0.02 ml (two sucks). The rapidity with which the neonate registers his recognition of changes in taste is also registered by the depth of sucking (increasing for sweeter fluids) and the effectiveness of sucking. For milk, he sucks more effectively and less rapidly. He also increases his burst–pause pattern for breast milk as if he were waiting for social stimuli in the pauses (Kaye and Brazelton, 1971). With sugar water, the pauses decrease, and the sucks increase in number with increasing sweetness (Lipsitt, 1967). Experience increases the complexity of these responses.

8.5. Tactile

The sensitivity of the infant to handling and to touch is quite apparent. A mother's first response to an upset baby is to contain him, to shut down on his disturbing motor activity by touching or holding him. By contrast, fathers are more likely to tap in a playful, rhythmic fashion or to use tactile methods to excite the infant to interact Dixon *et al.* 1981. Touch becomes a message system between the caregiver and the infant, both for calming him down and for building him up in order to attend to cues. We found that a patting motion of three times per minute is soothing, whereas five to six times per minute becomes an alerting stimulus (Brazelton *et al.,* 1975). As with auditory stimuli, the law of initial values seems to be of primary importance. When a baby is quiet, a rapid, intrusive tactile stimulus serves to bring him to an alert state. When he is upset, a slow, modulated tactile stimulus seems to reduce his activity. Tactile stimulation around the mouth elicits rooting and sucking, as well as gastric motility of the upper GI tract. Pressure on the palms of one of the hands elicits head turning and mouth opening to the ipsilateral side (Babkin reflex). Stroking on one side of the mouth also stimulates the ipsilateral hand to flex and be brought up to the mouth. Thus, the hand-to-mouth cycle is established before birth, and the stage is set for oral–tactile feedback and exploration. The hand-to-mouth organization serves as a major motor control system as well.

Swaddling is used in many cultures to replace the important constraints offered first by the uterus and then by mothers and caregivers. As a restraining influence on the overreactions of hyperractive neonates, the supportive control offered by a steady hand on a baby's abdomen or by holding his arms so that he cannot startle reproduces the swaddling effects of holding or wrapping a neonate. This added control of disturbing motor responses allows the neonate to attend and interact with his environment.

We use a tactile maneuver to contain any newborn after we have undressed him. So disturbing is it to him to be undressed and at the mercy of temperature changes, and the freedom of unrestrained movement, that we have found it wise to hold one arm tightly in flexion up against the newborn infant's body. This keeps him from flying off into random, excited startling behavior that upsets him, makes him cry, and sets off more cyclical Moro startling. This maneuver can be taught to mothers for use at bath times.

If an infant is unable to use soothing tactile stimuli to help him adapt his state behavior, the examiner must consider a diagnosis of CNS irritability. A baby with CNS irritation from a bleed or from infection demonstrates constantly increasing irritability with stimuli, especially tactile. If the infant is constantly upset by all tactile stimulation, this response should become a signal to the examiner to investigate further for evidence of CNS difficulties.

8.6. Vestibular Stimulation

Korner and Thoman (1970) demonstrated the importance of vestibular stimulation in both soothing and in alerting the newborn. These workers found that using a rocking waterbed to soothe a disorganized premature infant not only reduced his overactivity and his sleep apnea but also promoted the more rapid development of state controls. We found a semiupright (30° angle) optimal for producing prolonged alertness in newborns. Korner and Thoman (1970) demonstrated that bringing the newborn to the upright position has an effect on bringing him to an alert state.

9. Sucking

The awake, hungry newborn exhibits rapid searching movements in response to tactile stimulation in the region around the mouth, and even as far out on the face as the cheek and sides of the jaw and head (Peiper, 1963; Blauvelt and McKenna, 1961). This is called the rooting reflex, which is present in the premature infant even before sucking itself is effective (Prechtl, 1963). Peiper described three sets of oral pads in the cheeks and mouth that help maintain and establish negative pressure. Sucking is facilitated by the thorax in inspiration (Jensen, 1932) and by fixing the jaw to maintain it between respirations. A second mechanism, "expressing," is made by the tongue as it moves up against the hard palate and from the front to the back of the mouth. Swallowing and respiration must be coordinated, and the depth and rate of respiration are handled differentially in nutritive and nonnutritive sucking. Peiper (1963) argued for a hierarchical control of swallowing, sucking, and breathing, in which swallowing controls sucking and sucking controls breathing. Absence of coordination among these three systems in a neonate is indicative of discoordination within the CNS, and it may be seen in brain-damaged infants and in the very immature infant.

Gryboski (1965) described a technique of monitoring the three components of sucking with three transducers: one at the front of the tongue, another at the base of the tongue, and a third in the upper esophagus. The timing among these three components is a measure of the immaturity of the CNS of a premature infant, as well as a measure of the disruption of the central processes controlling these mechanisms. Since one can set off and feel the three components by inserting a finger into the newborn infant's mouth when he is sucking, an examiner can determine for himself the presence and coordination of the three components. This, then, is an easily available and valuable measure of the infant's stage of relative maturity and CNS functioning.

The infant sucks in a more or less regular pattern of bursts and pauses. During nonnutritive sucking, his rate varies around two sucks per second (Wolff, 1968). Bursts seem to occur in packages of 5–24 sucks per burst (Kaye, 1967). The pause between bursts has been considered a rest-and-recovery period as well as a period during which cognitive information is being processed by the neonate (Halverson, 1944; Bruner, 1969). Kaye and Brazelton (1971) found that the pauses have ethological significance, since they are used by mothers as signals to stimulate the infant to return to sucking. However, the mother's jiggling actually prolongs the pause, as the infant responds to the information given to him by his mother.

Factors that affect sucking, such as age, hunger, fatigue, rate of milk flow, and state of arousal, affect but do not change the individual burst–pause pattern that seems to be a relatively stable inborn pattern (Kron *et al.*, 1963).

Because of this stability produced through central controk, sucking is used to measure sensory discrimination, conditioning, and learning (Kessen *et al.*, 1970), as well as attention and orienting (Haith *et al.*, 1969; Kaye, 1967; Sameroff, 1968).

Finger sucking is common during the neonatal period, and there is evidence that the insertion of parts of the hand in the mouth occurs commonly in the uterus. The importance of sucking as a regulatory response can be seen in a newborn as he begins to build up from a quiet state to crying. His own attempts to achieve hand-to-mouth contact in order to keep his activity under control are fulfilled when he is able to insert a finger into his mouth, suck on it, and quiet himself. The sense of satisfaction and gratification at having achieved this self-regulation are apparent. His face softens and alerts as he begins to concentrate on maintaining this kind of self-regulation. A pacifier can achieve this same kind of quieting in an upset baby but may not serve the self-regulatory feedback system as richly as the baby's own maneuver.

10. Neonatal Behavioral Assessment

In order to record and evaluate some of the integrated processes evidenced in certain kinds of neonatal behavior, we developed a behavioral examination that tests and documents the infant's organized responses to various environmental events, as well as his use of state behavior (states of consciousness) (Brazelton, 1973).

Since the neonate's reactions to all stimuli are dependent on his ongoing "state," any interpretation must be made with this in mind. His use of state to maintain control of his reactions to environmental and internal stimuli is an important mechanism and reflects his potential for organization. State no longer need be treated as an error variable or only as a modifier of discrete reactions but serves to set a dynamic pattern to allow for the full behavioral repertoire of the infant. Specifically, our examination tracks state changes over the course of the examination—its liability and direction. The variability of state points to the infant's capacities for self-organization. His ability to quiet himself as well as his need for stimulation also measure this adequacy. The 26 behavioral responses to environmental stimuli include the kind of interpersonal stimuli that mothers use in their handling of the infant as they attempt to help him adapt to his new environment. The examination includes a graded series of procedures—talking, hand on belly, restraint, holding, and rocking—designed to soothe and alert the infant. His responsiveness to animate stimuli (e.g., voice and face) and to inanimate stimuli (e.g., rattle, ball, red bell, white light, temperature changes) is assessed. Estimates of vigor and attentional excitement are measured and motor activity and tone as well as autonomic responsiveness as he changes state are assessed. In addition to the 26 behavioral items assessed on a 9-point scale, 20 reflexes are assessed with a 3-point scale (Table I). These items are included to elicit gross neurological abnormalities. As part of the behavioral examination, they also offer methods for systematic handling and evaluation of the infants.

The 26 behavioral items assess the neonate's capacity to: (1) Organize his states of consciousness; (2) Habituate reactions to disturbing events; (3) Attend to and process simple and complex environmental events; (4) Control motor tone and activity while attending to these events; and (5) Perform integrated motor acts. These reflect the range of behavioral capacities of the normal neonate. They seem to demand control and CNS organization, which are dependent on either the cortex or higher brain centers of the neonate. Moreover, the accomplishment of these behaviors by the neonate requires that the infant successfully manage the impinging physiological demands of this early adjustment period. As such, they reflect the neonate's capacity to attend to his environment despite the major disorganization and demands of this recovery period. A condensed list of the behavioral

Table I. Elicited Responses[a]

Response	O	L	M	H	A
Plantar grasp		1	2	3	
Hand grasp		1	2	3	
Ankle clonus		1	2	3	
Babinski reflex		1	2	3	
Standing		1	2	3	
Automatic walking		1	2	3	
Placing		1	2	3	
Incurvation		1	2	3	
Crawling		1	2	3	
Glabella		1	2	3	
Tonic deviation of head and eyes		1	2	3	
Nystagmus		1	2	3	
Tonic neck reflex		1	2	3	
Moro		1	2	3	
Rooting (intensity)		1	2	3	
Sucking (intensity)		1	2	3	
Passive movement					
Arms					
Right		1	2	3	
Left		1	2	3	
Legs					
Right		1	2	3	
Left		1	2	3	

[a]A, asymmetry; O, response not elicited (omitted). H, high; L, low; M, medium.

Table II. Neonatal Behavioral Assessment[a]

1.	Response decrement to repeated visual stimuli
2.	Response decrement to rattle
3.	Response decrement to bell
4.	Response decrement to pinprick
5.	Orienting response to inanimate visual stimuli
6.	Orienting response to inanimate auditory stimuli
7.	Orienting response to animate visual—examiner's face
8.	Orienting response to animate auditory—examiner's voice
9.	Orienting responses to animate visual and auditory stimuli
10.	Quality and duration of alert periods
11.	General muscle tone in resting and in response to being handled (passive and active)
12.	Motor maturity
13.	Traction responses as he is pulled to sit
14.	Cuddliness—responses to being cuddled by the examiner
15.	Defensive movements—reactions to a cloth over child's face
16.	Consolability with intervention by examiner
17.	Peak of excitement and capacity to control self
18.	Rapidity of buildup to crying state
19.	Irritability during examination
20.	General assessment of kind and degree of activity
21.	Tremulousness
22.	Degree of startling
23.	Lability of skin color (measuring autonomic lability)
24.	Lability of states during entire exam
25.	Self-quieting activity—attempts to console self and control state
26.	Hand-to-mouth activity

[a]See text for discussion of the 26-item list.

items is indicated in Table II. These 26 items can be grouped together into four behavioral dimensions of newborn organizations:

1. Interactive capacities; i.e., the newborn infant's capacity to attend to and process simple and complex environmental events.
2. Motoric capacities, which assess the infant's ability to maintain adequate tone, to control motor behavior, and to perform integrated motor activities (e.g., ability to bring hands to mouth, insert thumb or finger, and maintain it there long enough to establish a good suck.)
3. Organizational capacities with respect to state control, which detects how well the infant maintains a calm, alert state despite increased stimulation.
4. Organizational capacities with respect to physiological responses to stress (How much is the infant at the mercy of the physiological demands of his immaturity and the recovery from labor and delivery at this time? How well is he able to inhibit startles, tremors, and interfering movement as he becomes aroused or attends to social and inanimate stimuli? How much exhaustion enters into the picture of state modulation? How much is he at the mercy of the environment? How vulnerable is he to continued stimulation?)

We believe that the behavioral items elicit more important evidences of cortical control and responsiveness, even during the neonatal period. The neonate's capacity to manage and overcome the physiological demands of this adjustment period in order to attend, differentiate, and habituate to the complex stimuli of an examiner's maneuvers may be an important predictor of his future CNS organization. Certainly, the curve of recovery of these responses over the first neonatal week must be of greater significance than the midbrain responses detectable in routine neurological examinations. Repeated behavioral examinations on any 2 or 3 days in the first 10 days after delivery might be expected to be sensitive predictors of future CNS function.

In this examination are couched behavioral tests of such important CNS mechanisms as: (1) Habituation or the neonate's capacity to shut out disturbing or overwhelming stimuli; (2) Choices in attention to various objects or human stimuli (a neonate shows clear preferences for female versus male voices and for human versus nonhuman visual stimuli); and (3) Control of his state in order to attend to information from his environment (the effort to complete a hand-to-mouth cycle in order to attend to objects and people around him). These are all evidenced in the neonate and even in the premature infant and seem to be more predictive of CNS intactness than are reflex responses.

11. Timing of Administration

The most important indication of the infant's integrity may be his capacity to integrate responsive behavior during the first days of life. The curve of recovery of overall organization may be the most important measure of prediction. Does the baby become increasingly more mature and organized in these behaviors over time, or does he remain the same? Does the infant become even more disorganized as time goes on? In order to be able to assess this recovery, at least two examinations are needed: on day 3 or 4, when the immediate stresses of delivery and some of the medication effects have begun to wear off, and again on day 9 or 10, when the baby has been at home and has adjusted to his home environment. The day immediately after discharge from the hospital should be avoided as an examination day, since the change from hospital to home environment tends to make for more instability in behavior. These two points on a recovery curve may well predict CNS and psychophysiological integrity, as well as the capacity to cope with future stress.

12. Repeated Examinations

When the scale was conceptualized it was not intended to provide a one-shot assessment of the neonate. Rather, it was designed to reflect the neonate's capacity to recover from the stress of events such as labor and delivery and the influences of the new environment. Since I see the neonate as an organism who is recovering from a stressful series of events, I think the best test of his capacity to cope in the future may be reflected by the pattern of recovery as responses move from less than optimal just after delivery, to rapid improvement over the next few days. Thus, a "recovery curve" as demonstrated in improving behavioral responses, becomes a test of physiological stress as well as a test of the infant's capacity to organize, in spite of it, and to interact with his environment. I believe that repeated examinations are necessary to evaluate the neonate's organizational adjustments and coping strategies in response to the demands of a new environment. Thus, the maximum predictive validity of the scale lies in the pattern of recovery reflected in repeated assessments over the first few weeks of life rather than in any one assessment. By plotting each neonate's performance over time one can calculate individual recovery curves that may in turn be used to predict the coping capacities of the neonate.

This recovery-curve approach implies that it would be foolish to expect more than moderate test–retest stability of the 26 scale items. Although some consistency in performance can be expected from day to day, it may be more appropriate to think of fluctuations in performance as indicative of how the neonate organizes around changing environmental demands. Moreover, those demands are likely to vary from infant to infant. There are probably individualized styles of recovery and ways of coping with equal demands.

The neurological reflex items would be least likely to be influenced by stress of recovery or by extrauterine events. Consequently, they would be expected to be the least stable in day-to-day test–retest scoring. In time, after data on many, many babies have been evaluated, perhaps all of us who are using the scale can set up *a priori* curves of recovery on all items against which individual babies can be measured. In this way, by predicting several expectable curves of recovery, we can see on which items the individual deviates and thereby understand the ingredients of each infant's unique recovery curve.

13. Evidence for Learning in the Neonate

There are two questions that might shed light on the potential for recovery and development of neonates. If these complex responses are present, does the neonate learn from experiencing them? If he can learn from the experience of controlling input, from his choices, from his efforts at state control, does his attempt to integrate his CNS in order to respond become an organizing influence?

Studies of the effects of stimulation in the neonatal period on the recovery of premature and other high-risk neonates have been limited thus far to groups of neonates who are in our present deprivation system, but there is increasing evidence that sensory input appropriate to the state of physiological recovery of that neonate may have a powerful positive effect on his weight gain, his sensory integrity, and even on his outcome.

Early stimulation must be individualized to each subject. We have evidence that leads us to believe that each premature or recovering neonate must be observed for the possibility of sensory overloading. A premature infant will respond to a soft rattle with head turning away from the rattle (and other evidences of shutting it out), whereas a normal neonate will turn toward the rattle and search for it. This crude example of the finely defined thresholds for appropriate sensory stimuli, as opposed to those that must be "coped with" or shut out, must be taken as seriously as to whether or not stimulation is offered.

14. Behavior of the Infant in Context: The Infant–Adult Communication System

Observational studies of mother–infant interactions clearly demonstrate that a complex communication system exists during the first months of the infant's life (Bowlby, 1969; Richards and Bernal, 1972; Brazelton *et al.*, 1975; Richards, 1971; Goldberg, 1971, unpublished results). Most of the existing studies have analyzed the interactions to show the one-to-one contingencies between maternal and infant behavioral displays. These displays were at least several seconds and even minutes long and were typically observed while the pair was engaged in functional or task-oriented situations (Lytton, 1971). Impressive reciprocity exists between the infant and the mother under normal conditions. Examples of mothers responding differentially are their latency of response, varying according to the type of infant cry (Wolff, 1969) or access to the breast (Blauvelt and McKenna, 1961), and the infant's cessation of sucking producing a jiggling of the infant by the mother (Kaye and Brazelton, 1971). Furthermore, although the relative frequencies of specific behaviors vary widely, there is nonetheless a similar relationship between infant and maternal behaviors across many cultures and social classes (Rebelsky, 1971; Korner and Thoman, 1970; Freedman and Freedman, 1969; Brazelton *et al.*, 1966).

There is considerable variation in the expressive qualities of individual normal infants. Moreover, infants who are small for gestational age (Als *et al.*, 1975), malnourished (Brazelton *et al.*, 1976), premature (H. Als, E. Tronick, L. Adamson, and T. B. Brazelton, 1976, unpublished results), drug addicted (Strauss *et al.*, 1975), or who have received high levels of drugs administered to the mother during delivery (Standley *et al.*, 1974; Tronick and Scanlon, 1976; Scanlon *et al.*, 1974) have markedly modified behaviors and appearances. There are also variations in infants of different cultural groups (Brazelton *et al.*, 1976, 1977; Freedman and Freedman, 1969).

These variations in infant behavioral characteristics have been shown to have significant effects on caretakers. In normal hospital settings, Als (1975) found that mothers fit their reactions to their newborn infant's visual or auditory responsiveness and to their feeding patterns. Similarly, cuddly infants produced more cuddling by adults (Schaffer and Emerson, 1964). State of arousal, activity levels, and irritability markedly effect parental responses (Lewis, 1972; Brazelton *et al.*, 1975; Korner and Thoman, 1972; Escalona, 1968; Als and Lewis, 1975).

Bowlby (1969), Bell (1971), Goldberg (1971, unpublished results), and others have theoretical positions that describe how the behavioral signals emitted by the infant produce effects on the caretaker. For the most part, they still view the infant as a passive emitter of signals. Bowlby does not describe the infant as making goal-corrected actions until he is more mobile at 8 months of age. Before that time, although the infant has potent signaling capacities, he does not modify them signficantly to serve his own purposes. Bowlby's view of interactive behavior can be compared with Lorenz's view of the infant's physical appearance. The infant's "Kewpie" doll physiognomy and size elicit and inhibit particular responses from adults. The infant has this appearance but has no control over it. Similarly, for Bowlby, the infant emits signals, both physiognomic and behavioral, but does not modify them. These signals affect the adult, but the infant is not aware of their effects, nor does he intend them (Tronick *et al.*, 1977).

15. Communication during Neonatal Assessment

One of the most remarkable performances that one can observe in neonates is seen as the infants change from a quiet state to a state in which they could become distressed if they were unable to control themselves. As they begin to rouse, they make real efforts to turn their heads to one side, then perform a cycle of hand-to-mouth movements (Babkin, 1960). When they are able to bring their fist up to their mouth and to hold it there, and even to suck on it, they quiet down, their agitated motor activity subsides, and, as they relax, they begin to alert, looking around for auditory and visual stimuli. This active attempt to control disturbing motor activity and to maintain an alert state using the ability to bring the hand to the mouth seems to be a process designed to permit infants to attend to their environment. The observer who watches a neonate achieve this becomes struck with how "programmed" infants are for interaction with their environment.

When they are alerted, infants respond with periods of active fixation on an attractive visual stimulus such as a bright, shiny, red ball. They will quiet, maintain a quiet inactive state in order to follow the ball through complete 180° arcs of movement, and turn their heads as well as follow it with their eyes. If, then, they are presented with a human face, infants will act "hungry" as they follow the face laterally and vertically. In this process of registering a preference for human stimuli, it is impossible for an adult interactant not to become "hooked" to the infant. The infant's ability to communicate this preference by facial and eye "softening" and his increasing attention is reflected in prolonged suppression of motor activity in the rest of his body, as well as in increased state control. The examiner or adult interactant becomes aware of his own involvement with the infant, as he too maintains an intense period of eye-to-eye and face-to-face communication.

In the same way, an infant can register auditory preferences. An infant can react to the sound of a bell or a loud rattle by turning away from it with startling, jerky movements that either propel him into a crying state or into an inactive state resembling sleep. In such a state, he shuts out stimuli, holds his extremities and body tight, his eyes tightly closed, his face masked, and his respirations deep, jagged, and regular (Brazelton, 1961). Thus, sleep can be a protective state. In the same way, he can use crying as another way of regulating his environment as he attempts to control the input of stimulation from those around him. If in either of these shutting-out states, crying or sleep, he is offered the sound of a soft rattle or a soothing human voice, he is likely to quiet from agitation or rouse from light sleep and become alert, gradually turning toward the attractive sound. If the stimulus is nonhuman, he will search for it while maintaining an alert facial expression and become quietly inactive. When the stimulus is the human voice, the neonate not only searches for the observer's face but, upon finding it, his face and eyes become wide, soft, and eager, and he may even crane his neck, lifting the chin gently toward the source of the voice. As he does so, the infant's body tension gradually increases, but he is quietly inactive. A nurturing adult feels impelled to respond to these signals by cuddling the baby.

As a well-organized, alert neonate is held in a cuddled position, he molds into the adult's body, turning gently toward the chest. He may even grab hold of the adult's clothing with his free hand, and his legs may mold around the side of the adult's body. This molding response cannot help but become a reinforcing signal to the adult for more active cuddling, for looking down to engage the infant in face-to-face contact for rocking or singing.

When the infant is held upright at the adult's shoulder, he first lifts his head to look around. As he does so, he actively holds on more tightly with all four extremities. After a period of alert scanning of the environment triggered by the vestibular stimulation of the upright position (Korner and Thoman, 1972), he is most likely to tire and to put his head against the adult's shoulder, nestled in the crook of the adult's neck. Mothers tell me that as the soft fuzz of the infant's head makes contact with the skin of the crook of the neck, a tightening sensation is felt in their breasts, followed by a "let-down" reflex of milk. No adult is likely to resist the feeling of a soft head resting on the shoulder.

Another powerful set of communicative signals occurs as infants build up to crying. If an adult keeps on talking at one side of the infant's head, the baby will probably stop crying, quiet down, and gradually turn toward the voice. The adult can use the infant's capacity to alert to voices by changing his own vocal behavior. If an adult's voice softens, the infant will maintain his focused searching and scanning. He will remain quietly alert and may "smile" or may soften his face into the precursor of a "smile face." As adult and infant continue to communicate, the adult can bring the infant up to a more active state by gradually increasing the pitch and timbre of the voices; on the other hand, the adult can cause the infant to "overload" and return to a crying state by changing the tempo to a staccato rhythm or may help the infant maintain the quiet, alert state by speaking softly in a slow, rhythmic fashion.

The infant adjusts and changes his actions in relationship to the ongoing actions of his partner, actively regulating these evolved behavioral adaptations to regulate the adult–infant communication system. Als (1975) demonstrated this during the first 3 days of life of newborn infants when with their mothers. Infants as young as 3 weeks of age showed significantly different attentional cycles and behavior while interacting with objects as contrasted with people (Brazelton, *et al.,* 1975). When interacting with people during face-to-face exchanges, there are smooth, rhythmical accelerations and decelerations of behavioral responses as the infants greet, then interact with, a caretaker. During these interchanges, the hands are used gesturally and are linked rhythmically to the infant's vocalizations. In contrast, with objects the infant evidences prolonged attentional cycles broken by abrupt turning away. His body is still but his movements are jerky, and the movements of his extremities come in bursts accompanied by swipes at the object. With the object the infant's goal seems to be to reach for it, while with people the goal is to reciprocate in an affective interchange.

Joint regulation of these interactive systems in order to produce optimal performance in the neonate led us to see the powerful signaling systems, the regulatory systems of state control, and the nonverbal communicative systems available during the first few weeks in the human infant. Brazelton and associates (Yogman *et al.,* 1976) demonstrated that by 4 weeks, reliable, reproducible patterns of rhythmic behavioral responses are already reserved for the mother and father. These responses are differentiated for each parent and are significantly different from the patterns or nonpatterns that already signal the baby's responses to a stranger, as reported by the same investigators (Dixon *et al.,* 1981). If these patterns are shaped so early, they must be predetermined *in utero,* to some extent, and shaped in the first few days and weeks by important messages and cues offered by each parent and quickly reorganized by the infant. These allow for: (1) Early patterning toward individual differences in both partners; (2) The importance of individual cues on the part of important adults and individual responses on the part of the infant; (3) Rapid recognition of the important parameters of context, the parental envelope; and (4) Physiological and psychological control.

This latter goal of the infant is evidenced in studies in which the normal behavior of the adult in relationship to the infant is distorted. For example, if the mother remains still-faced and immobile in front of an infant, the infant attempts to get the interaction back on track (Tronick *et al.,* 1977*a,* 1978*a,b*). Initially the infant orients to the adult and smiles, but when the adult fails to respond, the infant sobers. His facial expression is serious, and he becomes still. He stares at the adult and smiles again, but briefly. Then he looks away. He repeatedly looks toward and away from the adult, smiling briefly in conjunction with the look toward, and sobering with the look away. Eventually he slumps in his seat with his chin tucked, his head and eyes oriented away from the adult, looking hopeless and helpless in the face of an unresponsive adult. This pattern of behavior occurs reliably within 2–3 min of a distorted interaction.

16. Evaluation of Mother–Infant Attachment Behavior

Bowlby (1969) stressed the importance of observing the earliest interactions between mother and infant as predictive of the kind of attachment a mother may form for the infant. He suggested that there is a kind of "imprinting" of responses from her that may be triggered by the neonate's behavior. Goldberg (1971) pointed to the triggerlike value in setting off mothering activities—the newborn infant's small size, helpless appearance, and distress cries. Klaus and Kennell (1970) described the kinds of initial contacts that mothers make with their newborn infants as well as the distortions in this behavior when the mother is depressed by abnormalities in the baby; e.g., prematurity and illness in the neonatal period. Eye-to-eye contact, touching, handling, and nursing behavior on the part of the mother may be assessed and judged for predicting her ability to relate to the new baby. Changes in these behaviors over time are stressed as indicators of recovery or nonrecovery of maternal capacity to attach to the baby by mothers who have been depressed and unable to function optimally by having produced an infant at risk.

17. Summary

The importance of an immediate assessment of the neonate—his stage of maturity or dysmaturity and his responsiveness to his new environment in order to estimate his intrauterine conditions—is stressed as an irreplaceable predictor of the infant's future potential. A pediatric examination, a neurological evaluation, and a behavioral assessment of his capacity to respond to stimulation are of vital importance in assessing the infant at risk. Apgar scores simply reflect his capacity to respond to the immediate stresses of labor and delivery. Repeated assessment of his neurological and behavioral status furnishes a curve of recovery or nonrecovery. This curve may be the best single predictor of future CNS function. Finally, observations of mother–infant interaction, with the neonate's behavioral responses in mind, may permit the examiner to predict the kind of attachment that will help or hinder the infant at risk as he attempts to adjust to the problems of his environment.

ACKNOWLEDGMENTS. This work was funded by the Robert Wood Johnson Foundation, the National Institute of Mental Health, and the Amy S. Cohen Foundation.

18. References

Adamson, L., 1976, *Infants Response to Visual and Tactile Occlusions,* Ph.D. dissertation, University of California, Berkeley.

Als, H., 1975, *The Newborn and His Mother: An Ethological Study of Their Interaction,* Ph.D. dissertation, University of Pennsylvania, Philadelphia.

Als, H., and Lewis, M., 1975, The contribution of the infant to the interaction with his mother, presented at SRCD meetings, Denver.

Als, H., Tronick, E., Adamson, L., and Brazelton, T. B., 1976, The behavior of the fullterm yet underweight infant, *Dev. Med. Child Neurol.* **18**:590–602.

Anders, T. F., and Roffwarg, H., 1973, The effects of selective interruption and total sleep deprivation in the human newborn, *Dev. Psychobiol.* **6**:77–89.

Anders, T. F., and Weinstein, P., 1972, Sleep and its disorders in infants and children: A review, *Pediatrics* **50**:312.

Ando, Y., and Haltori, H., 1970, Effects of intense noise during fetal life upon postnatal adaptability, *J. Acoust. Soc. Am.* **47(4)**:1128.

Apgar, V. A., 1960, A proposal for a new method of evaluation of the newborn infant, *Curr. Res. Anesth.* **32**:260.

Aserinsky, E. and Kleitman, N. A., 1955, A motility cycle in sleeping infants as manifested by ocular and gross motor activity, *J. Appl. Physiol.* **8**:11.

Babkin, P. S., 1960, The establishment of reflex activity in early postnatal Life, in: *Central Nervous System and Behavior,* pp. 24–31, Third Conference, Princeton, New Jersey, U.S. Department of Health Education and Welfare (OTS 62-43772).

Ball, W., and Tronick, E., 1971, Infant response to impending collision: Optical and real, *Science* **171**:818–820.

Banks, M. S., 1980, The development of visual accommodation during early infancy. *Child Dev.* **51**:646–666.

Bell, R. Q., 1971, Stimulus control of parent or caretaker behavior by offspring, *Dev. Psychol.* **4**:63.

Blauvelt, H., and McKenna, J., 1961, Mother neonate interaction: Capacity of the human newborn for orientation, in: *Determinants of Infant Behavior* (B. M. Foss, ed.).

Bowlby, J., 1969, *Attachment and Loss, Vol. I: Attachment,* Basic Books, New York.

Braddick, O., Atkinson, J., French, J., and Howland, H. C., 1979, A photorefractive study of infant accommodation. *Vision Res.* **19**:1319–1330.

Brazelton, T. B., 1962, Observations of the neonate, *J. Am. Acad. Child Psychiatry* **1**:38.

Brazelton, T. B., 1984, Neonatal behavioral assessment scale, *Clin. Dev. Med.,* **88**:1–24.

Brazelton, T. B., Scholl, M. L., and Robey, J. S., 1966, Visual responses in the newborn, *Pediatrics* **37**:284.

Brazelton, T. B., Tronick, E., Adamson, L., Als, H., and Wise, S., 1975, Early mother–infant reciprocity, in: *Parent–Infant Interaction,* CIBA Foundation Symposium, p. 33, Elsevier, Oxford.

Brazelton, T. B., Koslowski, B., and Tronick, E., 1976, Neonatal behavior among urban Zambians and Americans, *J. Acad. Child Psychiatry* **15**:97–105.

Brazelton, T. B., Tronick, E., Lechtig, A., and Lasky, R., 1977, The behavior of nutritionally deprived Guatemalan infants, *Dev. Med. Child Neurol.* **19**:364–372.

Bredberg, G., 1968, Cellular patterns and nerve supply of the human organ of Corti, *Acta Oto-Laryngol. (Suppl.)* Whole No. 236.

Bridger, W. H., 1961, Sensory habituation and discrimination in the human neonate, *Am. J. Psychiatry* **117**:991.

Bruner, J. S., 1969, Eye, hand and mind, in *Studies in Cognitive Development* (D. Elkind and J. G. Flavell, eds.), pp. 223–243, Oxford University Press, New York.

Cairns, G. F., and Butterfield, E. C., 1975, Assessing infant's auditory functioning, *Exceptional Infant* B. Z. Friedlander, G. M. Sterritt and G. E. Kirk, eds.), Vol. 3, p. 84, Brunner/Mazel, New York.

Dayton, G. O., Jr., Jones, M. H., Air, P., Rawson, R. A., Steele, B., and Rose, M., 1964, Developmental study of coordinated eye movements in the human infant, *Arch. Ophthalmol.* **71**:865–869.

Dixon, S., Yogman, M. W., Tronick, E., Als, H., Adamson, L., and Brazelton, T. B., 1981, Early social interaction of parents and strangers, *J. Am. Acad. Child Psychiatry,* **20**:32–52.

Drillien, C. M., 1972, Aetiology and outcome in low birthweight infants, *Dev. Med. Child Neurol.* **14**:563.

Dubowitz, L., Dubowitz, V., 1970, Clinical assessment of gestational age in the newborn infant, *J. Pediatr.* **77**:1.

Eisenberg, R. B., 1965, Auditory behavior in the human neonate: Methodologic problems, *J. Audit. Res.* **5**:159.

Eisenberg, R. B., Coursin, D. B., and Rupp, N. R., 1966, Habituation to an acoustic pattern as an index of differences among human neonates, *J. Audit. Res.* **6**:239.

Eisenberg, R. B., Marmarou, A., and Giovachino, P., 1974, EEG changes to a synthetic speech sound, *J. Audit. Res.* **14**:29.

Engen, T., Lipsitt, L. P., and Kaye, H., 1963, Olfactory responses and adaptation in the human neonate, *J. Comp. Physiol. Psychol.* **56**:73.

Escalona, S. K., 1968, *Roots of Individuality,* Aldine, Chicago.

Flechsig, P. E., 1920, *Anatomie des Menschlichen Gehirns und Ruckenmarks auf Myelogenetischer Grundlage,* Thieme, Leipsig.

Fantz, R. L., 1965, Visual perception from birth as shown by pattern selectivity, *Ann. N. Y. Acad. Sci.* **118**:793.

Freedman, D. G., and Freedman, N., 1969, Behavioral differences between Chinese American and American newborns, *Nature (London)* **244**:227.

Gesell, A., 1945, *The Embryology of Behavior,* Harper, New York,

Goren, C., Sarty, M., and Wu, P., 1975, Visual following and pattern discrimination of face-like stimuli by new born infants, *Pediatrics* **56**:544.

Gorman, J. J., Cogan, D. G., and Gellis, S. S., 1957, An apparatus for grading the visual acuity of infants on the basis of opticokinetic nystagmus, *Pediatrics* **19**:1088.

Gottlieb, G., 1971, Ontogenesis of sensory function, in: *The Biopsychology of Development* (E. Tobach, L. R. Aronson, and E. Shaw, eds.), Academic Press, New York.

Gryboski, J. D., 1965, The swallowing mechanism of the neonate: Esophageal and gastric motility, *Pediatrics* **35**:445.

Haith, M. M., Kessen, W., and Collins, D., 1969, Response of the human infant to level of complexity of intermittent visual movement, *J. Exp. Child Psychol.* **7**:52.

Halverson, H. M., 1944, Mechanisms of early feeding, *J. Genet. Psychol.* **64**:185.

Haynes, H., White, B. L., and Held, R., 1965, Visual accommodation in human infants, *Science,* **148**:528.

Hershenson, M., 1964, Visual discrimination in the human newborn, *J. Comp. Physiol. Psychol.* **58**:270.

Hoffeld, B. R., McNew, J., and Webster, R. L., 1968, Effect of tranquilizing drugs during pregnancy on the activity of the offspring, *Nature (London)* **218**:357.

Hooker, D., 1969, *The Prenatal Origin of Behavior,* 3rd ed., Hafner, New York.

Hrbek, A., and Mares, P., 1964, Cortical evoked responses to visual stimulation in full term and premature infants, *EEG Clin. Neurophyiol.* **16**:575.

Jensen, K., 1932, Differential reactions to taste and temperature stimuli in newborn infants, *Genet. Psychol. Monogr.* **12**:361.

Johnson, P., and Salisbury, D. M., 1975, Breathing and sucking during feeding in the newborn, in: *Parent–Infant Interaction,* CIBA Foundation Symposium, p. 119, Elsevier, Oxford.

Kaye, H., 1967, Infant sucking and its Modification, in: *Advances in Child Development and Behavior* (L. P. Lipsitt and C. C. Spiker, eds.), Vol. III, pp. 1–52, Academic Press, New York.

Kaye, K., and Brazelton, T. B., 1971, The ethological significance of the burst–pause pattern in infant sucking, presented at the Society for Research in Child Development, Minneapolis.

Kearsley, R. B., 1973, Newborn's response to auditory stimulation: A demonstration of orienting and defensive behavior, *Child Dev..* **44**:582.

Keeley, K., 1962, Prenatal influences on the behavior of offspring of crowded mice, *Science* **135**:44.

Kessen, W., Haith, M. M., and Salapatek, P. H., 1970, Human infancy: A bibliography and guide, in: *Carmichael's Manual of Child Psychology,* (P. Mussen, ed.), Vol. 1, p. 287, Wiley, New York.

Klaus, M. H., and Kennell, J. H., 1970, Mothers separated from their newborn infants, *Pediatr. Clin. North Am.* **17**:1015.

Kleitman, N., 1963, *Sleep and Wakefulness,* 2nd ed., Chicago University Press, Chicago.

Korner, A. F., and Thoman, E. B., 1970, Visual alertness in neonates as evoked by maternal care, *J. Exp. Child Physiol.* **10**:67.

Korner, A. F., and Thoman, E. B., 1972, Relative efficacy of contact and vestibular-proprioceptive stimulation in soothing neonates, *Child Dev.* **43**:443.

Kron, R. E., Stein, M., and Goddard, K. E., 1963, A method of measuring sucking behavior of newborn infants, *Psychosom. Med.* **25**:181.

Lacey, J. I., 1967, Somatic Response Patterning and Stress, in: *Psychological Stress,* (M. H. Appley and R. Thumbell, eds.), pp. 14–42, Appleton-Century-Crofts, New York.

Lester, B. M., 1975, Cardiac habituation of the orienting response to an auditory signal in infants of varying nutritional states, *Devel. Psychol.* **11**:432.

Lewis, M., 1972, State as an infant–environment interaction: An analysis of mother–infant behavior as a function of sex, *Merrill-Palmer Q.* **18**:95.

Lind, J., Wasz-Hockert, O., Vuorenkoski, F., Partanen, T., Theorell, K., and Valanne, E., 1966, Vocal responses to painful stimuli in newborn and young infants, *Ann. Paediatr. Fenn.* **12**:55.

Lipsitt, L. P., 1967, Learning in the human infant, in: *Early Behavior: Comparative and Behavioral Approaches* (H. W. Stevenson, H. L. Rheingold, and E. Hess, eds.), pp. 225–247, Wiley, New York.

Lipton, E. L., Steinschneider, A., and Richmond, J., 1966, Auditory sensitivity in the infant: Effect of intensity on cardiac and motor responsivity, *Child Dev.* **37**:233.

Lubchenco, L. O., Hansman, C., Dressler, M., and Boyd, E., 1963, Intrauterine growth estimated from liveborn, birthweight data at 24 to 42 weeks of gestation, *Pediatrics* **32**:793.

Lytton, H., 1971, Observational studies of parent–child interaction: A metholodological review, *Child Dev.* **42**:3.

MacFarlane, A., 1975, Olfaction in the development of social preferences in the human neonate, in: *Parent–Infant Interaction,* CIBA Foundation Symposium, p. 103, Elsevier, Oxford.

Parmelee, A. H., Jr., 1962, European neurological studies of the newborn, *Child Dev.* **33:**169.

Parmelee, A. H., Jr., Schulte, F. J., Akiyama, Y., Wenner, W. H., Schultz, M. A., and Stern, E., 1968, Maturation of EEG activity during sleep in premature infants, *EEG Clin. Neurophysiol.* **24:**319.

Peeples, D. R., and Teller, D. Y., 1975, Color vision and brightness discrimination in two-month old human infants, *Science* **189:**1102–1103.

Peiper, A., 1963, *Cerebral Function in Infancy and Childhood,* Consultants Bureau, New York.

Pratt, K. C., Nelson, A. K., and Sum, K. H., 1930, The behavior of the newborn infant, Ohio State Univ. Stud. Contrib. Psychol., Vol. 10.

Prechtl, H. F. R., 1963, The mother–child interaction in babies with minimal brain damage, in: *Determinants of Infant Behavior* (B. M. Foss, ed.), Vol. II, pp. 53–59, Methuen, London.

Preyer, W., 1888, *The Mind of the Child,* New York, Appleton.

Rebelsky, F. G., 1971, Infancy in two cultures, in: *Child Development and Behavior* (F. G. Rebelsky and E. Dorman, eds.), Knopf, New York.

Richards, M. P. M., 1971, Social interactions in the first two weeks of human life, *Psychiatr. Neurol. Neurochir.* **14:**35.

Richards, M. P. M., and Bernal, J. R., 1972, An observational study of mother–infant interaction, in: Ethological Studies of Child Behavior (N. J. Blurton-Jones, ed.), pp. 175–197, Cambridge University Press, London.

Rosso, P., and Winick, M., 1973, Relation of nutrition to physical and mental development, *Pediatr. Ann.* **12:**33.

Salapatek, P. H., and Kessen, W., 1966, Visual scanning of triangles by the human newborn, *J. Exp. Child Psychol.* **3:**155.

Sameroff, A. J., 1968, The components of sucking in the human newborn, *J. Exp. Child Psychol.* **6:**607.

Scanlon, J. W., Brown, W. V., Weiss, J. B., and Alper, N. H., 1974, Neurobehavioral responses of newborns after maternal epidural anesthesia, *Anesthesiology* **40:**121.

Schaffer, H. R., and Emerson, P. E., 1964, Patterns of response to physical contact in early development. *J. Child Psychol. Psychiatry* **5:**1.

Schulte, F., Michaelis, R., Linke, I., and Nolte, R., 1968, Motor nerve conduction velocity in term, preterm and small for dates newborn infants, *Dev. Psychobiol.* **1:**41.

Sigman, M., Kopp, C. B., Parmelee, A. H., and Jeffrey, W. E., 1973, Visual attention and neurological organization in neonates, *Child Dev.* **44:**461.

Sontag, L. W., and Richardo, T. W., 1938, Studies in fetal behavior: Fetal heart rate as a behavioral indicator, *Monogr. Soc. Res. Child Dev.* **3**(4):1–38.

Standley, K., Soule, A. B., Copans, S. A., and Duchowny, M. S., 1974, Local–regional anesthesia during childbirth: Effects on newborn behavior, *Science* **186:**634.

Steinschneider, A., 1973, Prolonged apnea and the sudden infant death syndrome: Clinical and laboratory observations, *Pediatrics* **50:**646.

Strauss, M. E., Lessen-Firestone, J., Starr, R. H., and Ostrea, E. M., 1975, Behavior of narcotics addicted neonates, Presented at Society for Research in Child Development, Denver.

Tronick, E., and Brazelton, T. B., 1975, Clinical uses of the Brazelton neonatal scale, in: *Exceptional Infant* (B. Z. Friedlander, G. M. Sterritt and G. E. Kirk, eds.), Vol. 3, p. 137, Brunner/Mazel, New York.

Tronick, E., Als, H., and Brazelton, T.B., 1977, Mutuality in mother–infant interaction, *J. Commun.* **27(2):**74.

Tronick, E. Als, H., and Adamson, L., 1978*a,* Structure of early face-to-face communicative interactions, in: *Before Speech: The Beginnings of Communication* (M. Bullowa, ed.), pp. 349–373, Cambridge University Press, London.

Tronick, E., Als, H., Adamson, L., Wise, S., and Brazelton, T. B., 1978*b,* The infant's response to entrapment between contradictory messages in face-to-face interaction, *J. Am. Acad. Child Psychiatry* **17:**1–13.

Wagner, I. F., 1937, The establishment of a criterion of depth of sleep in the newborn infant, *J. Genet. Psychol.* **51:**17.

Wasz-Hockert, O., Lind, J., Vuorenkoski, V., Partanen, T., and Valanne, E., 1968, The infant cry, *Clin. Dev. Med.* **29:**68–73.

Wertheimer, M., 1961, Psychomotor coordination of auditory and visual space at birth, *Science* **134:**1962.

Windle, W., 1950, Reflexes of mammalian embryos and fetuses, in: *Genetic Neurology* (P. Weiss, ed.), p. 214, University of Chicago Press.

Wolff, P., 1966, The causes, controls, and organization of behavior in neonates, *Psychol. Monogr.* **5:**17.

Wolff, P. H., 1959, Observations on newborn infants, *Psychosomat. Med.* **21:**110.

Wolff, P. H., 1968, The serial organization of sucking in the young infant, *Pediatrics* **42:**943.

Wolff, P. H. 1969, The natural history of crying and other vocalizations in early infancy, in: *Determinants of Infant Behavior* (B. M. Foss, ed.), Vol. 4, p. 81, Methuen, London.

Yogman, M., and Dixon, S., 1976, Father–infant interaction, presented at American Pediatric Society—SPR meetings, St. Louis, April 1976.

Zamenhof, S., Van Maihens, E., and Margolis, F. L., 1968, DNA (cell number) and protein in neonatal brain, *Science* **160:**322.

Index

GPSR Compliance

The European Union's (EU) General Product Safety Regulation (GPSR) is a set of rules that requires consumer products to be safe and our obligations to ensure this.

If you have any concerns about our products, you can contact us on ProductSafety@springernature.com

In case Publisher is established outside the EU, the EU authorized representative is:

Springer Nature Customer Service Center GmbH
Europaplatz 3
69115 Heidelberg, Germany

Batch number: 10762209

Printed by Printforce, the Netherlands